A Color Guide to the Petrography of Sandstones, Siltstones, Shales and Associated Rocks

Memoir 109

By

Dana S. Ulmer-Scholle, Peter A. Scholle,
Juergen Schieber and Robert J. Raine

Published by

The American Association of Petroleum Geologists

ISBN: 978-0-89181-389-7

AAPG Editor: Michael L. Sweet
AAPG Geoscience Director: James B. Blankenship

COVER: Primary and secondary porosity within a volcanic arenite from Pliocene–Pleistocene Gila Group from southern New Mexico.

This publication is available from:

The AAPG Bookstore
P.O. Box 979
Tulsa, OK U.S.A. 74101-0979
Phone: 1-918-584-2555 or 1-800-364-AAPG (U.S.A. only)
E-mail: bookstore@aapg.org
www.aapg.org

Canadian Society of Petroleum Geologists
600, 640 – 8th Avenue S.W.
Calgary, Alberta T2P 1G7
Canada
Phone: 1-403-264-5610
Fax: 1-403-264-5898
E-mail: reception@cspg.org
www.cspg.org

Geological Society Publishing House
Unit 7, Brassmill Enterprise Centre
Brassmill Lane, Bath BA13JN
United Kingdom
Phone: +44-1225-445046
Fax: +44-1225-442836
E-mail: sales@geolsoc.org.uk
www.geolsoc.org.uk

Affiliated East-West Press Private Ltd.
G-1/16 Ansari Road, Daryaganj
New Delhi 110-002
India
Phone: +91-11-23279113
Fax: +91-11-23260538
E-mail: affiliate@vsnl.com

The American Association of Petroleum Geologists

The American Association of Petroleum Geologists Books Refereeing Procedures

The Association makes every effort to ensure that the scientific and production quality of its books matches that of its journals. Since 1937, all book proposals have been refereed by specialist reviewers as well as by the Association's Publications Committee. If the referees identify weaknesses in the proposal, these must be addressed before the proposal is accepted.

Once the book is accepted, the Association Book Editors ensure that the volume editors follow strict guidelines on refereeing and quality control. We insist that individual papers can only be accepted after satisfactory review by two independent referees. The questions on the review forms are similar to those for the AAPG Bulletin. The referees' forms and comments must be available to the Association's Book Editors on request.

Although many of the books result from meetings, the editors are expected to commission papers that were not presented at the meeting to ensure that the book provides a balanced coverage of the subject. Being accepted for presentation at the meeting does not guarantee inclusion in the book.

More information about submitting a proposal and producing a book for The American Association of Petroleum Geologists can be found on its web site: www.aapg.org.

Table of Contents

About the Authors

Dana S. Ulmer-Scholle
Scholle Petrographic, LLC, P. O. Box 386, Socorro, NM 87825 U. S. A.
(e-mail: ulmerscholle@gmail.com)

Dana developed an early love for geology while growing up on the classic Upper Ordovician outcrops around Cincinnati, Ohio. She received a B.S. degree in 1981 from the University of Cincinnati. Dana completed a M.S. degree (1983) at Southern Methodist University working on the Mississippian Arroyo Peñasco Group of New Mexico. After a stint working for ARCO Exploration Co., she returned to SMU for a Ph.D. (1992). Her dissertation research concentrated on evaporite-related diagenesis in upper Paleozoic carbonate rocks from New Mexico, Wyoming and Greenland.

Dana has worked, or consulted, for a number of companies including Amoco Oil and Gas Co., ARCO Exploration, ARCO International, Mobil Research, and Maersk Oil and Gas. Currently, she is the co-owner of Scholle Petrographic, LLC, a petrographic consulting company. Dana is also an Associate Research Professor at the New Mexico Institute of Mining and Technology. She teaches carbonate-related courses including petrography and depositional/diagenetic models. Her research interests include CO_2 sequestration in carbonate and siliciclastic systems, carbonate sedimentology and diagenesis, petrography, low-temperature isotope and trace element geochemistry, fluid inclusion analysis and fluid flow histories in sedimentary rocks. While at New Mexico Tech, she has also been involved in environmental investigations that include heavy-metals bioremediation and fate-and-transport of heavy minerals in the environment. She is the author, coauthor or editor on numerous papers, reports, books and CD-ROMs with AAPG Memoir 77 receiving the Robert H. Dott, Sr., Memorial Award (2005).

Peter A. Scholle
Scholle Petrographic, LLC, P. O. Box 386, Socorro, NM 87825 U. S. A.
(e-mail: scholle1@gmail.com)

Peter received his B.S. in geology from Yale University in 1965. After a year at the University of Munich in Germany, and another year at the University of Texas at Austin, he received a Ph.D. in geology from Princeton University in 1970. His dissertation work, on deep-water carbonate turbidites in the Italian Apennines, was supervised by Al Fischer.

Peter's professional employment included state and federal government, the petroleum industry, and academia. He worked for five years for various oil companies (Cities Service, Gulf and Chevron) and consulted for other oil companies for many years. Nine years were spent with the U. S. Geological Survey in Reston (VA) and Denver (CO), including three years as chief of the Oil and Gas Branch. He taught at the University of Texas at Dallas for three years and was Albritton Professor of Geology at Southern Methodist University in Dallas from 1985 to 1999. From 1999-2011, he was the New Mexico State Geologist and director of the New Mexico Bureau of Geology and Mineral Resources. He is now the co-owner of Scholle Petrographic, LLC.

Peter has devoted much of his time to carbonate research and writing. His major interests are in deepwater carbonates (especially chalks) as well as the diagenesis and petroleum potential of Permian rocks in many areas of the world. He has worked in nearly 30 countries and has written, coauthored, or edited nine books, about 200 papers and abstracts, 23 CD-ROMs, and a number of other computer or audio-visual products. Peter has been a member of AAPG and SEPM since 1976-77. He was an AAPG Distinguished Lecturer (1975-76) and received the AAPG President's award twice, the Sproule Memorial Award, the AAPG Certificate of Merit and the AGI Ian Campbell Medal for Superlative Service to the Geosciences (2013). He served as president and special publications editor of SEPM and is an honorary member of that society. He was also president of AGI and AASG (the Association of American State Geologists).

Juergen Schieber
Indiana University, Department of Geological Sciences, 1001 E. 10th St., Bloomington, Indiana, 47405, U.S.A. (e-mail: jschiebe@indiana.edu)

Juergen received a B.S. degree (1978) in geology from the University of Tübingen, Germany and a Ph.D. (1985) in geology from the University of Oregon where he worked on the relationship between basin evolution and the genesis of stratiform sulfide deposits of mid-Proterozoic sediments of central Montana. He is a Professor of Geology at Indiana University (2002-present) and previously worked at the University of Texas-Arlington (1996-2002).

Juergen is a specialist on shales. He has published extensively (100 papers, 20 guidebook chapters, two books, 215 conference abstracts) and has been an invited lecturer at universities in the US, Canada, Europe and Asia, at research organizations, industry short courses and symposia. His research interests include: basin analysis and sedimentology, sedimentology of shales, the genesis of black shales and sediment-hosted mineral deposits, evolution of the Belt Basin and the Devonian basins of the eastern US, geochemistry of sediments and planetary geology and sedimentary geology of Mars. Juergen is a member of the science team that currently explores the geology of Gale Crater on Mars with NASA's Curiosity rover.

Juergen's research is characterized by a holistic approach to shales; integrating field studies (facies, stratigraphy) with lab studies (thin sections, electron microscopy and geochemistry) in order to understand the various factors that are involved in the formation of shales. A key area of focus for his work is the experimental study of shale sedimentology via flume studies and related work. Funding for his research is provided by government agencies (NSF, DOE, NASA), foundations (Petroleum Research Fund), and industry via the Indiana University Shale Research Consortium (ExxonMobil, Anadarko, Marathon, Shell, Chevron, ConocoPhillips, Wintershall, Whiting, Statoil) and separate research agreements (Schlumberger/TerraTek; Pioneer Natural Resources). He consults on matters pertaining to shale sedimentology, shale fabric and pore structure, and also teaches short courses on shale sedimentology and facies analysis, as well as microscope-based petrography.

Robert J. Raine
Geological Survey of Northern Ireland, Belfast, Ireland, United Kingdom BT9 5BF (email: robertr@bgs.ac.uk)

Rob graduated in 2003 with an M.S. degree in geology from the University of Birmingham and remained there to study for a Ph.D. in geology. He studied the Cambrian and Ordovician carbonate sedimentology and stratigraphy of Laurentian passive margin sediments in northwest Scotland. During this time, he was also the assistant curator for the Lapworth Museum of Geology and took time out to work as a museum conservator.

In 2008, Rob joined Ichron Limited as a sedimentologist and petrographer and conducted studies on marine, lacustrine and terrestrial sediments from India, Sri Lanka, Algeria and the UK and Norwegian sectors of the North Sea. He also led a number of field trips for oil companies to Portugal and Yorkshire and core logging courses in India.

Rob recently joined the Geological Survey of Northern Ireland as a petroleum and energy geologist to provide scientific support to the Department of Enterprise, Trade and Investment on petroleum licenses and to conduct research to promote the petroleum and geothermal exploration potential of Northern Ireland. His main research interests are peritidal carbonates and evaporites in mixed carbonate siliciclastic settings, peritidal and lacustrine microbialites, reservoir quality and diagenesis of fluvial and eolian sediments, Permian and Triassic rocks and geoarcheology.

INTRODUCTION

This book covers the microscopic study of sandstones, mudstones and associated lithologies but excludes carbonate rocks which were covered in AAPG Memoir 77 (Scholle and Ulmer-Scholle, 2003). Sandstone petrography, and sedimentary petrography in general, is considered by many to be a science in decline. As a consequence, it is being taught at fewer universities, or at least commonly is subsumed into broader petrology or sedimentology classes, where it receives less time and less focus.

In reality, however, the petrography of clastic terrigenous rocks remains a vital area of study, contributing to both academic and applied economic discoveries. It is a science in transition, however. In the mid-20th century, the research focus was largely on the use of textural properties to define mechanisms of transport, environments of deposition and the use of grain identification and rock classification to shed light on source terranes, paleoclimates and paleotectonic settings. Since that time, the research forefront has shifted to studies of diagenesis in sandstones and to an increased understanding of the finer-grained terrigenous rocks—siltstones, mudstones and shales (see, for example, the insightful historical review by Steel and Milliken, 2013). That transition has been driven largely by the fact that hydrocarbon exploration has shifted to deeper targets, more stratigraphic traps and unconventional, finer-grained reservoir rocks.

The shift in focus for petrography has led to amazing discoveries regarding the character and extent of diagenetic processes as well as to the development of predictive models of subsurface alteration and porosity retention, loss or creation. At the same time, the increased recognition of the significance of subsurface alteration has forced a reevaluation of long-established principles of sandstone provenance based on petrographically-determined sediment composition. Composition is now much more clearly understood to be not just a product of its source terrane as modified by the impacts of surficial weathering and the rigors of transport and deposition. Diagenesis has taken a prominent place as at least an equal partner in controlling the ultimate composition of sandstones through the burial-associated processes of dissolution and replacement (Table 1).

The studies of fine-grained terrigenous deposits (shales and siltstones), driven in part by the revolution in oil and gas recovery from "unconventional" reservoirs, has also led to remarkable new insights and discoveries, many of them made using light and electron microscopy. When looking at sandstones and carbonate rocks, outcrops can be counted on to provide a wide array of interesting observations. Features that inform about transport processes, sedimentary environments, and depositional history are readily observed and by the time we move on from those outcrops, we have arrived at some preliminary rock assessments. Fine-grained rocks, on the other hand, typically are not very impressive in outcrop, because they weather easily and turn into a crumbly mass. When examined in thin section and SEM, however, shales truly reveal their character.

More than 100 years ago, H. C. Sorby (1908) observed that "examination in a natural condition is enough to show that the structure of clays differs enormously, and indicates formation under very different conditions; but there is always some doubt as to their true structure, when not made into thin sections". On this score nothing has changed.

In a well-prepared thin section of a shale, one can easily observe a variety of layers and multiple rock types in spatial context and with high detail on contact relationships. With these observations, one can then proceed to derive critical insights about depositional processes and environments from a few square centimeters of cross-section. A well-chosen thin section of a shale unit has the potential to deliver as much sedimentologic insight as one might be able to derive from an entire outcrop of sandstones and carbonates. When working with shales, thin sections and SEM views are indispensable for developing an understanding of depositional processes, environments of deposition and diagenetic history.

The focus on fine-grained rocks, on diagenetic concerns and on subsurface studies, has proceeded hand-in-hand with remarkable scientific and instrumental advances in analytical chemistry. That, in turn, has made petrography less of a "stand-alone" science and more of a "team sport". Various flavors of electron microscopy have extended the reach of petrographic examination into the submicron range. More sophisticated and automated x-ray diffraction, microprobe, ion probe, cathodoluminescence and other analytical techniques, along with laser ablation and

Table 1: Major stages in the life of sands and sandstones (factors that influence ultimate rock composition and make provenance determinations progressively more difficult)

Birth: **Source area conditions**

Meet the parents—the initial influence is the source-rock composition and pre-exposure (*in-vitro*) alteration. Selective weathering, breakage and erosion, along with tectonic, topographic and climatic influences, further shape initial sediment composition.

Leaving home: **Transport**

Fracturing, abrasion and geochemical alteration, along with sorting and mixing, selectively alter sediments. Possible interim deposition or exposure events (again, with strong tectonic/topographic and climatic controls) can produce further changes.

Settling down: **Deposition**

Continued weathering and abrasion occurs in terrestrial settings; localized sea-floor diagenesis occurs in marine settings; potential addition of locally-derived intraclasts, carbonate skeletal or oolitic components along with formation of associated chemical/biochemical deposits (limestones, cherts, evaporites, greensands or ironstones).

A first job: **Early burial (gen. < 500 m [1,600 ft] burial and temperatures <40°C)**

Considerable dewatering and compaction; potential pedogenic, meteoric, brine, or mixed-water diagenetic alterations (partial to complete dissolution of unstable minerals; selective cementation, and/or replacement, especially by calcite, dolomite or siderite, with kaolinite, chlorite and, in some settings, zeolites). Biogenic (microbial) alteration of organic matter in some cases.

Growing maturity: **Intermediate burial (0.5-2.5 km [1,600-8,200 ft]; temperatures 40 to 100°C)**

Lesser compaction but accelerated diagenetic alterations (increased cementation by quartz and illitic clays; partial dissolution and/or replacement of unstable calcic and potassium feldspars, smectitic clays and unstable to moderately stable heavy minerals). Thermochemical alteration of opal to opal-CT. Onset of thermochemical hydrocarbon generation.

Creeping senility: **Deep burial (typically 2.5-10 km [8,200-33,000 ft] with temperatures of 100 to 250°C)**

Extensive dissolution of remaining unstable feldspars and heavy minerals; near-complete transformation of smectite to illite (and neoformation of pure illite); kaolinite transformation to dickite or illite; overall formation of a stable and fully lithified quartz, albite, illite assemblage or, in some cases, conversion to a "diagenetic quartzarenite". Completion of conversion of opal-CT sediments to quartz (chert or chalcedony). Peak to end of thermochemical hydrocarbon generation and thermochemical sulfate reduction (with potential for calcite dissolution and saddle dolomite formation).

Off to the crematorium (or back to the surface for reincarnation): **Burial to >10 km (33,000 ft) and temperatures >250°C**

Onset of low-grade metamorphism with chlorite and mica formation. Or uplift and perhaps ultimate exposure and erosion.

other microsampling methods, allow acquisition of geochemical data at a microscopic scale.

Yet as undeniably useful and sophisticated as these techniques are, petrography (at light and electron imaging levels) remains the underlying science that informs these analytical studies—in its simplest form, you need to know exactly what you are analyzing before the analysis can make much sense. For example, is the material detrital or authigenic? This has become especially important in an era when improvements in microscale imaging and analysis allow elemental and isotopic studies of small portions of single grains; for example, the detrital core and authigenic overgrowths on even silt-sized or finer material.

Overall, petrography benefits enormously from the wide range of analytical data available today, ranging from routine bulk x-ray data to detailed spectral chemical analysis, to more sophisticated (and expensive) studies such as radiometric or fission-track dating of clays and stable, detrital heavy minerals including zircon, tourmaline and apatite. Such analyses can assign well-defined ages to specific events in a paragenetic sequence and provide quantitative information about the age and composition of source rocks.

Petrography also still crucially and fundamentally informs much applied day-to-day work in the petroleum industry. To give a specific example of the value of the "teaming" of petrography and geochemistry to applied sedimentology, take a hypothetical x-ray or spectral geochemical result for a fine-grained rock sample: 48% quartz, 7% plagioclase feldspar, 3% orthoclase feldspar, 3% calcite, 39% combined clays. So what is all that quartz and from where does it come? Is it mainly detrital quartz grains, detrital quartz-rich rock fragments, altered planktic organisms (radiolarians or diatoms), microcrystalline quartz (chert), quartz overgrowth cement or yet something else? You will never know what that quartz means unless you do petrography, and in terms of rock fracability and hydrocarbon production it very much matters what form the quartz takes. And what about the clays? Are they detrital matrix in a poorly-sorted sandstone, are they squashed sedimentary rock fragments or are they diagenetic precipitates? Such knowledge makes a difference in exploration models, and is impossible to acquire without petrography.

About This Book

Sedimentary petrography is an art as much as it is a science. At its core, you need to be able to differentiate grains, matrix, cements and pores and each of those components are covered in one or more chapters that are illustrated with numerous photomicrographs. Quartz, feldspar and rock fragments dominate most sandstones and siltstones, and the identification of the many varieties of these grains occupies a major part of the book. There also are many accessory minerals that should be identified in any useful petrographic study, so the most important and common of those minerals are covered as well.

Grain identifications provide an understanding of sediment provenance. Beyond that, one needs to understand the textural parameters of sedimentary deposits (size, sorting, rounding, and others) and to distinguish detrital matrix from pseudomatrix and precipitated clays. With those topics covered, one will have the information needed to understand how clastic terrigenous deposits are classified (and why there have been so many different approaches to their classification, especially in the "era of nomenclatural searching" [the 1950s-1970s]). Finally, the complexities of diagenesis, once poorly understood or largely confined to studies of near surface alteration, are now far better understood although much still remains to be learned. So chapters on compaction, cementation, dissolution and replacement will help one to understand diagenetic products and the important distinctions between primary and secondary porosity.

There are a variety of useful resources in this book, beyond the numerous photographs. General utility tables showing standard grain-size and sorting terminology, along with visual comparators for sorting, shape and rounding of grains, are found in Chapter 6. A comparison chart for visual grain percentage estimation is provided at the end of this introduction. A glossary is located near the back of the book, and it supplies definitions for many of the technical terms used throughout the volume. Considerable effort also went into creating a thorough index that should allow you to quickly find the information that matches your interest. Finally, a Michel-Lévy foldout color chart at the end of this introduction provides accurate birefringence colors for most minerals found in sedimentary rocks.

The overarching purpose of this book is not to be a textbook in sedimentology, but to be a pictorial guide to microscopic studies — that is, to provide you with annotated illustrations of the grains, matrix, cements, pores, fabrics and textures of clastic terrigenous rocks so that you can learn to recognize these features yourself. We have tried to provide good and clear, but not exceptional, examples in our pictures so that what you see in the book will be most like what you will see in the rocks you are studying. However, this book is not a treatise on optical microscopy or mineralogy, and those who have no experience in either of those subjects may want to look at some of the references given at the end of this introduction. It is not absolutely necessary to have formal knowledge about either subject, but some background certainly helps.

It is important to note that this book is directed primarily towards recognition and identification of the components of clastic terrigenous sediments and only secondarily toward interpretation. In that regard, it parallels its companion volume on carbonate rocks. The features that allow you to identify source terranes for eroded material or to distinguish primary (detrital) from secondary (diagenetic) materials are illustrated and discussed. However, full interpretation of all the grains and fabrics shown is simply beyond the scope of this book, especially because many such interpretations remain quite controversial. So a list of relevant citations is provided at the end of each section. They have been chosen to contain a mix of papers: topical overviews, specific scientific studies, and studies that have applied the kinds of data covered in that section to broader geological problems. The overview papers, in particular, are worth reading because they provide many additional references that will take you further into the subject matter.

Likewise, there is a section at the end of the book that illustrates some of the analytical techniques, microscopic, geochemical and other, that complement and commonly are combined with petrography. That section simply gives a hint as to the rich panoply of tools that exist — whole books have been written on many of these techniques, so we clearly cannot do justice to those subjects. Therefore, we have provided references for your use in filling in the gigantic gaps we have left. Photographs using some of the more directly petrographic techniques, such as cathodoluminescence or fluorescence microscopy, are shown throughout the book.

A word about the photography in this book — unlike Memoir 28, the predecessor to this volume, the photographs in this book were almost entirely taken in digital formats (even in the relatively few photos that show the same field of view as in Memoir 28 were generally rephotographed in digital format). Digital photography has revolutionized petrographic documentation, allowing one to photograph features at much lower light levels. The digital format also allows later manipulation of images in software such as Adobe Photoshop. In the interest of full disclosure, it should be said that almost all images in this book have had minor color correction, contrast correction or cropping to bring the photographs into line with what was actually seen under the microscope. Plane and cross-polarized light photos have been digitally combined into side-by-side pairs, a process that would be far more difficult with film. In some cases, irrelevant and distracting blemishes such as gas bubbles in the thin sections, were removed, but only when doing so did not change any of the critical features in the photograph.

Explanation of Picture Captions

Each photograph in this book has a caption in standard format. The top lines in bold text give the stratigraphic unit and geologic age followed by a generalized location (typically the region, province or state and, if not from the U.S.A., the country of origin). The main part of the caption describes the image and is followed, at the bottom of the caption box, by lighting, special techniques, staining, impregnation and scale information.

The following caption abbreviations and scale bar codes are used: Stratigraphic units are abbreviated with Gp. for Group, Fm. for Formation and Mbr. for Member, and stratigraphic ages are abbreviated with Lo. for Lower, Mid. for Middle and Up. for Upper.

This symbol † at the end of the stratigraphic information designates an image that was kindly contributed by Ichron Limited.

Standard microscopic illumination

PPL – Transmitted, plane-polarized light

XPL – Transmitted, cross-polarized light

PXPL – Transmitted, partially cross-polarized light

RL – Incident reflected light (using a reflected light microscope)

ORL – Oblique reflected light (using an external light source adjacent to a transmitted light microscope)

GP – Gypsum (or quartz red I) plate inserted with cross-polarized light

PPL & RL – Combined transmitted and reflected light

PPL | XPL – Split image with plane- and cross-polarized light images

Other imaging techniques

MAC – Macroscopic photograph (not using a microscope)

CL – Imaging using standard cathodoluminescence excitation

CCL – Cathodoluminescence image acquired with Gatan ChromaCL instrumentation

FL365 – Epi-fluorescent excitation centered at 365 nm wavelength

FL470 – Epi-fluorescent excitation centered at 470 nm wavelength

SEM – Scanning electron microscopy

Col SEM – Hand-colorized SEM

SE(SEM) – Secondary electron (on SEM)

BSE(SEM) – Back-scattered electron imaging (on SEM)

BSE(MP) – Back-scattered electron imaging (on microprobe)

SE-CCI(SEM) – Secondary electron / charged secondary electron imaging (on SEM)

CCI(SEM) – Charged secondary electron imaging (on SEM)

Most of the cathodoluminescence photomicrographs for this volume were taken with a Gatan Chroma CL system on a FEI QuantaFEG 400 SEM.

Staining and impregnation

AS – Alizarin red-S staining of carbonate (calcite stains pink or red)

AFeS – Alizarin red-S plus potassium ferricyanide stain for ferroan carbonates

PFS – Sample stained for plagioclase feldspars

KFS – Sample stained for potassium feldspars

BDI – Blue (or blue-green) dyed impregnation material in pores

RDI – Red (or pink) dyed impregnation material in pores

Image sizes

All images have the same size scale bar in the lower right-hand corner (a scale bar that is 1/10th the horizontal length of the full image) and the metric length for the scale in that picture is given at the bottom of each caption.

Acknowledgments

Figure 6.1: Reprinted from Computers & Geosciences, vol. 27/4, Dougal A. Jerram, Visual comparators for degree of grain-size sorting in two and three-dimensions, p. 485-492, Copyright (2001), with permission from Elsevier.

Lower right corner of Figure 6.4: Reprinted from Continental Shelf Research, v. 20, Flemming, B. W., 2000, A revised textural classification of gravel-free muddy sediments on the basis of ternary diagrams, p. 1125-1137, Copyright (2000), with permission from Elsevier.

Figure 8.1: Used with permission of Dr. Tom Anderson, Energy & Geoscience Institute, University of Utah.

Ichron Limited, United Kingdom allowed the authors to use the photomicrographs marked with the symbol (†) in the photo headers.

Cairn India Limited provided permission to use the photomicrographs from India and Sri Lanka.

SM Energy Company allowed the authors to use photomicrographs from various consulting projects.

We wish to thank The US Department of Energy (DOE), National Energy Technology Laboratory (NETL) and the Southwest Partnership for the use of thin sections from projects DE-FE0001812 and DE-FC26-05NT42591.

Drs. Robert G. Loucks of Texas Bureau of Economic Geology and Steven M. Cather, New Mexico Bureau of Geology and Mineral Resources performed the heroic task of reviewing the entire book. We hope they recover soon and are deeply in their debt for all their suggestions that greatly improved this book.

Drs. David W. Houseknecht of the US Geological Survey, Kenneth P. Helmold of the Alaskan Department of Natural Resources, Oil and Gas Division, Peter S. Mozley of New Mexico Institute of Mining and Technology, Joann E. Welton of RQ Consultant, and Virgil W. Lueth and Nelia W. Dunbar of the New Mexico Bureau of Geology and Mineral Resources reviewed many individual chapters (in some cases more than half the book). They caught an incredible number of errors and their expertise and suggestions for improvements were invaluable!

Additionally, Peter S. Mozley and Steven M. Cather allowed the authors access to their thin section collections. Virgil W. Lueth provided photographs of mineral specimens from the mineral collection at the New Mexico Bureau of Geology and Mineral Resources, and helped with the reflected light microscopy.

Albert V. Carozzi, Steve Cather, Julie Dumoulin, Nelia W. Dunbar, Shirley P. Dutton, Cortland F. Eble and the Kentucky Geological Survey, Sara Gallagher, Kenneth P. Helmold, Glenn Izett, Noel P. James, Stanley A. Kling, E. D. Pittman, Peir K. Pufahl, B. Charlotte Schreiber, R. F. Sipple, Joann E. Welton, E. L. "Jerry" Winterer, and Jeremy Young kindly provided images (in some cases, ones reproduced from the earlier Memoir 28 version of this book).

Maersk Oil and Gas for permission to use SEM and other images.

FEI, Incorporated analyzed one of our thin section and provided the data and images utilized in Chapter 24. InGrain and Gatan, Incorporated, also, supplied images and text for that chapter.

Zeiss kindly provided the digital file for the Michel-Lévy foldout color chart.

Supplementary Teaching Materials

A DVD at the back of the book contains Microsoft Powerpoint files for each of the chapters that contains all of the photomicrographs (generally without the arrows or mineral labels shown in the book). The notes section for each slide contains the full image captions from the book. These files, however, do not contain any of the introductory text, diagrams or bibliographic citations.

Comparison chart for visual percentage estimation (after Terry and Chilingar, 1955).

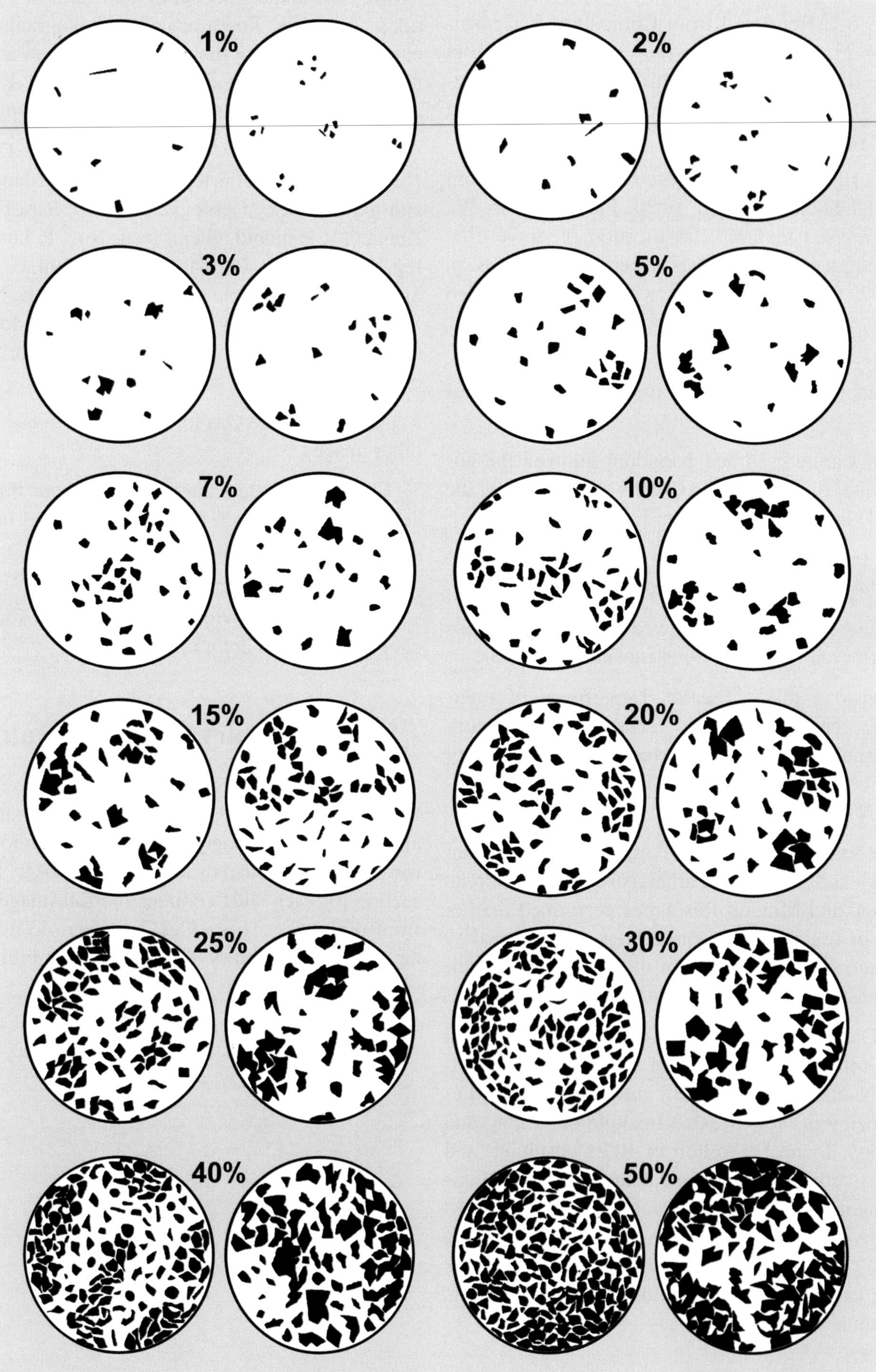

Introductory Petrographic References

General petrography and optical mineralogy:

Deer, W. A., R. A. Howie, and J. Zussman, 1992, An Introduction to the Rock-forming Minerals (2nd Edition): London, Longman Group Ltd., 696 p.

Hurlbut, C. S., and W. E. Sharp, 1998, Dana's Minerals and How to Study Them (After Edward Salisbury Dana), 4th Edition: New York, John Wiley & Sons, 336 p.

Larsen, E. S., and H. Berman, 1934, The microscopic determination of the nonopaque minerals (2nd Edition): USGS Bulletin 848, 209 p.

MacKenzie, W. S., C. H. Donaldson, and C. Guilford, 1982, Atlas of Igneous Rocks and their Textures: New York, John Wiley & Sons, 148 p.

MacKenzie, W. S., and C. Guilford, 1980, Atlas of Rock-forming Minerals in Thin Section: New York, John Wiley & Sons, 98 p.

Nesse, W. D., 2003, Introduction to Optical Mineralogy (3rd Edition): New York, Oxford University Press, 348 p.

Perkins, D., and K. R. Henke, 2003, Minerals in Thin Section (2nd Edition): Upper Saddle River, NJ, Prentice Hall, 176 p.

Pichler, H., and C. Schmitt-Riegraf, 1997, Rock-forming Minerals in Thin Section: New York, Chapman and Hall, 220 p.

Raith, M. M., P. Raase, and J. Reinhardt, 2011, Guide to thin section microscopy: 107 p. Free online download at: <http://www.minsocam.org/msa/content/OpenAccess_publications/Thin_Section_Microscophy.pdf>

Rost, F. D. W., and R. J. Oldfield, 2000, Photography with a Microscope: New York, Cambridge University Press, 288 p.

Saggerson, E. P., 1975, Identification Tables for Minerals in Thin Sections: New York, Longman, 378 p.

Sorby, H. C., 1908, On the application of quantitative methods to the study of the structure and history of rocks: Quarterly Journal of the Geological Society, v. 64, p. 171-233, doi: 10.1144/GSL.JGS.1908.064.01-04.12.

Stoiber, R. E., and S. A. Morse, 1994, Crystal Identification with the Polarizing Microscope: London, Chapman & Hall, 358 p.

Tröger, W. E., H. U. Bambauer, F. Taborszky, and H. D. Trochim, 1979, Optical Determination of Rock-Forming Minerals. Part 1: Determinative Tables: Stuttgart, E. Schweizerbart'sche Verlagsbuchhandlung, 188 p.

Vernon, R. H., 2004, A Practical Guide to Rock Microstructure: Cambridge, UK, Cambridge University Press, 606 p.

Sedimentology and sedimentary petrography:

Arribas, J., S. Critelli, and M. J. Johnsson, eds., 2007, Sedimentary Provenance and Petrogenesis: Perspectives from Petrography and Geochemistry: Boulder, CO, GSA Special Paper 420, 396 p.

Blatt, H., 1982, Sedimentary Petrology: San Francisco, CA, W. H. Freeman & Co., 564 p.

Blatt, H., G. V. Middleton, and R. C. Murray, 1980, Origin of Sedimentary Rocks: Englewood Cliffs, NJ, Prentice Hall Inc., 782 p.

Blatt, H., R. Tracy, and B. Owens, 2006, Petrology: Igneous, Sedimentary, and Metamorphic (3rd Edition): New York, NY, W. H. Freeman, 530 p.

Boggs, S., Jr., 2011, Principles of Sedimentology and Stratigraphy (5th Edition): Upper Saddle River, NJ, Prentice Hall, 600 p.

Boggs, S., Jr., 2009, Petrology of Sedimentary Rocks (2nd Edition): New York, NY, Cambridge University Press, 610 p.

Carozzi, A. V., 1993, Sedimentary Petrography: Englewood Cliffs, NJ, Prentice-Hall, 330 p.

Carver, R. E., 1971, Procedures in Sedimentary Petrology: New York, Wiley-Interscience, 672 p.

Folk, R. L., 1951, A comparison chart for visual percentage estimation: Journal of Sedimentary Research, v. 21, p. 32-33.

Folk, R. L., 1980, Petrology of Sedimentary Rocks: Austin, TX, Hemphill's Book Store, 184 p. [free download at: https://www.lib.utexas.edu/geo/folkready/entirefolkpdf.pdf]

Füchtbauer, H., and G. Müller, 1970, Sedimente und Sedimentgesteine: Stuttgart, E. Schweizerbart'sche Verlagsbuchhandlung, 726 p.

Greensmith, J. T., 1988, Textbook of Petrology, Vol 2: Petrology of the Sedimentary Rocks (7th Edition): London, Unwin Hyman, 262 p.

Larsen, G., and G. V. Chilingar, eds., 1979, Diagenesis in Sediments and Sedimentary Rocks: New York, Elsevier Scientific Publishing, 579 p.

Larsen, G., and G. V. Chilingar, eds., 1983, Diagenesis in Sediments and Sedimentary Rocks, 2: New York, Elsevier Scientific Publishing Co., 572 p.

Lewis, D. W., and D. McConchie, 1994, Practical Sedimentology [2nd Edition]: New York, Chapman & Hall, 213 p.

Milliken, K. L., S.-J. Choh, and E. F. McBride, Sandstone Petrology:: A Tutorial Petrographic Image Atlas (2nd Edition): Tulsa, OK, AAPG/Datapages Discovery Series, No. 10, PC-based CD.

Milner, H. B., 1962a, Sedimentary Petrography. Volume I, Methods in Sedimentary Petrography (4th Edition): London, George Allen & Unwin, 643 p.

Milner, H. B., 1962b, Sedimentary Petrography. Volume II, Principles and Applications (4th Edition): London, George Allen & Unwin, 715 p.

Pettijohn, F. J., 1957, Sedimentary Rocks (2nd Edition): New York, Harper Brothers, 718 p.

Pettijohn, F. J., P. E. Potter, and R. Siever, 1987, Sand and Sandstones (2nd Edition): New York, Springer-Verlag, 553 p.

Scholle, P. A., 1979, A Color Illustrated Guide to Constituents, Textures, Cements, and Porosities of Sandstones and Associated Rocks: Tulsa, OK, AAPG Memoir 28, 201 p.

Scholle, P. A., and D. S. Ulmer-Scholle, 2003, A Color Guide to the Petrography of Carbonate Rocks: Grains, textures, porosity, diagenesis: Tulsa, OK, AAPG Memoir 77, 474 p.

Steel, R. J., and K. L. Milliken, 2013, Major advances in siliciclastic sedimentary geology, 1960–2012., *in* M. E. Bickford, ed., The Web of Geological Sciences: Advances, Impacts, and Interactions: Boulder, CO., GSA Special Paper 500, p. 121-167, doi: 10.1130/2013.2500(04).

Terry, R. D., and G. V. Chilingar, 1955, Summary of "Concerning some additional aids in studying sedimentary formations," by M. S. Shvetsov: Journal of Sedimentary Research, v. 25, p. 229-234.

Tucker, M. E., 2001, Sedimentary Petrology: An Introduction to the Origin of Sedimentary Rocks (3rd Edition): Oxford, Wiley-Blackwell, 272 p.

Thanks to the following......

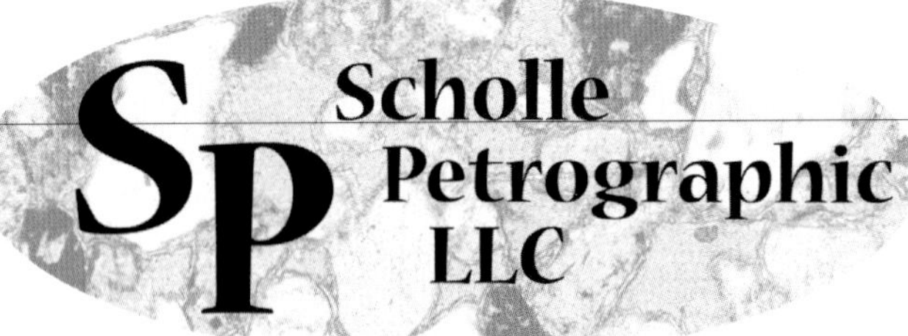

INGRAIN

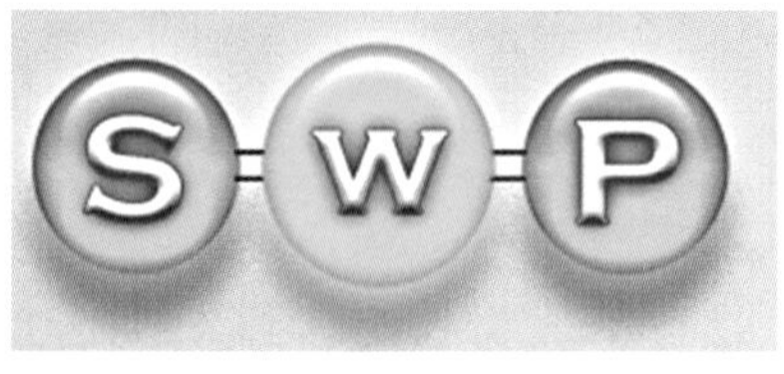

Chapter 1 Title Page: Top: Embayed volcanic quartz from Tertiary Monzonite intrusive from Clark Co., Nevada (PPL | XPL). Bottom: Quartz crystals with adularia (orthoclase) from Dona Ana Co., New Mexico. Photograph courtesy of the New Mexico Bureau of Geology and Mineral Resources.

GRAINS:
QUARTZ AND SILICA

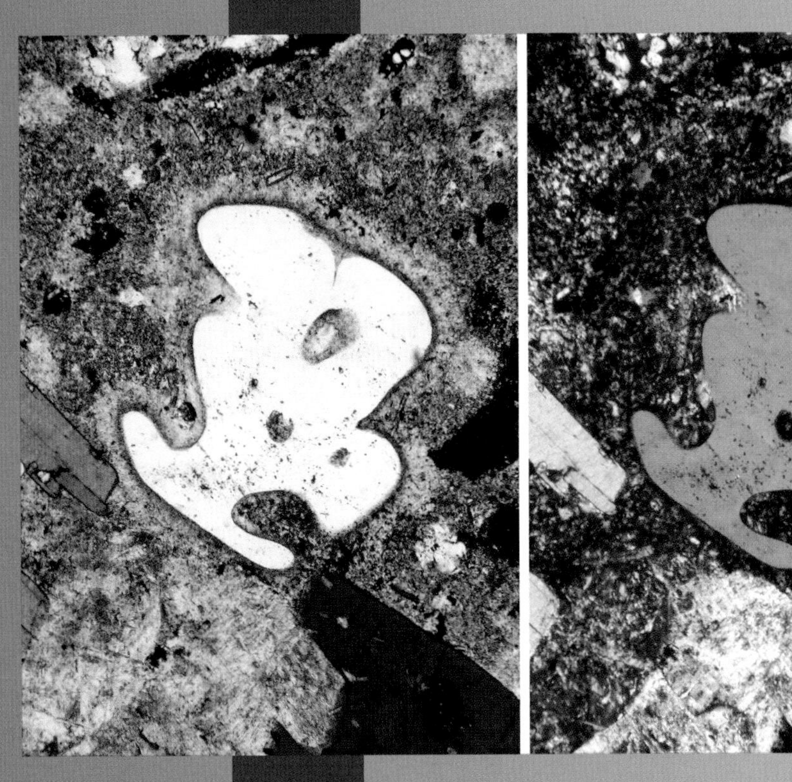

CHAPTER 1

QUARTZ AND SILICA

Quartz (SiO_2) is the most abundant mineral in terrigenous sedimentary rocks and is exceedingly durable (surviving multiple generations of weathering and deposition). Quartz and silica occur in many varieties—true quartz in the form of megaquartz, chert, microquartz, or chalcedony and various other forms of silica, mainly opal (opal-A and opal-CT [cristobalite]).

Major characteristics:

Color: Typically clear, but cloudy and colored varieties occur where crystals are especially rich in water or mineral inclusions. In thin section, these colors are rarely seen, although red and pale brown varieties related to included hematite and limonite are sometimes found.

Common crystal habit: Hexagonal, bipyramidal.

Pleochroism: None.

Cleavage/fracture: No cleavage, but fractures and healed fractures can be common.

Relief and optic sign: Low; uniaxial (+).

Birefringence: First-order (in a 30-μm section) typically first-order white and gray to pale straw yellow.

Other: No twinning; may contain trains of water, gas or mineral inclusions including ones, such as rutile and tourmaline needles, that may provide clues to grain provenance. Optical behavior varies based on deformation history; grains can have undulatory (undulose or sweeping) or nonundulatory (nonundulose or straight) extinction. Quartz grains commonly have syntaxial quartz overgrowths that form diagenetically.

Minerals that may have similar appearance and distinguishing differences:

Feldspars: Commonly show some cleavage, more alteration (vacuolization, sericitization, partial to complete dissolution), presence of characteristic twinning patterns and are biaxial. Can be differentiated by staining.

Apatite: Normally lower birefringence, has a higher relief and is uniaxial (-).

Gypsum and barite: Both have birefringence similar to quartz, but barite has slightly higher relief and gypsum has lower relief; both gypsum and barite have strong cleavage.

Zeolites: Numerous zeolite minerals can be confused with fibrous varieties of quartz. Most of the common zeolite minerals are biaxial (+/) and most have somewhat lower birefringence than quartz.

Provenance indicators:

Quartz can occur as single crystals or polycrystalline aggregates that may provide clues to the provenance of the grains, but quartz is common to most rock types other than basalts (Table 1.1). Semicomposite and polycrystalline quartz can be found in metamorphic and plutonic rocks as well as hydrothermal vein deposits and fractures. For metamorphic quartz, the size of the crystals may represent increasing metamorphic grade; larger crystals form under higher temperatures and pressures (Fig. 1.1).

Grain size can make provenance determination more difficult. With decreasing grain size, the ability to see undulatory quartz or polycrystalline/composite grains becomes more difficult. Since crystal sizes within polycrystalline grains may be large, grains formed from their breakdown may not exhibit polycrystallinity or undulatory extinction.

Chert and opaline silica can be found in either sedimentary or volcanic rocks — in the former as direct precipitates and in the latter as recrystallized glass.

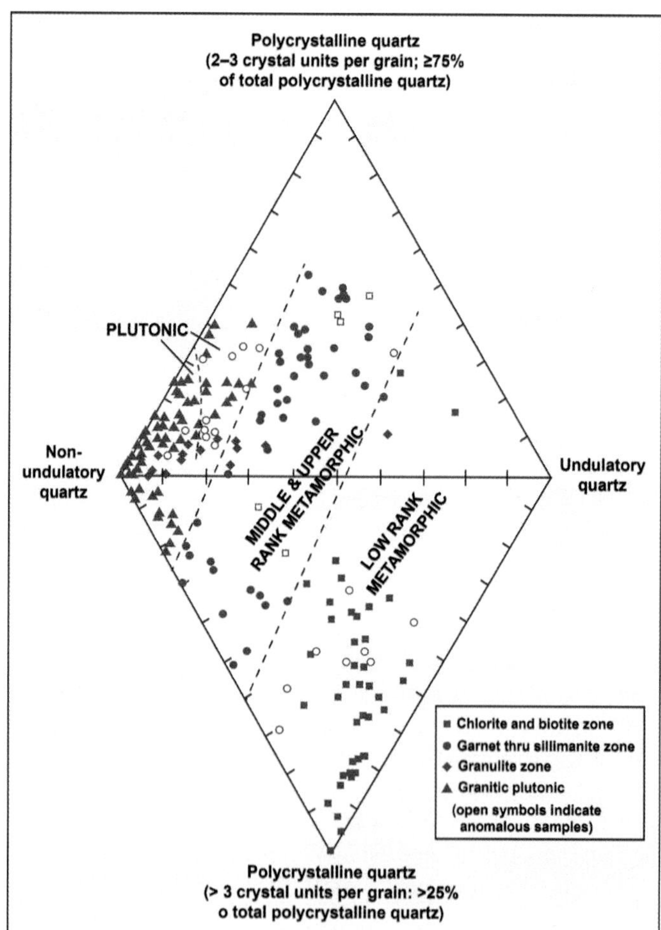

Figure 1.1: Plot of modern beach sands and their source terranes shows how the amount of nonundulatory, undulatory and polycrystalline quartz grains and the number of crystals in polycrystalline grains can help determine quartz provenance (from Basu et al., 1975).

Common inclusions and provenance –

Amphiboles (actinolite and others): metamorphic, intermediate to mafic igneous rocks and skarn deposits.

Apatite: felsic to intermediate volcanic rocks, skarns, carbonatites, calcium-rich metamorphic rocks, pegmatites and hydrothermal vein deposits.

Chlorite: fracture fills in low to medium grade metamorphic rocks, skarns and pegmatites.

Iron oxides (hematite, goethite and others): very common, and they form in a variety of environments ranging from anoxic (magnetite) to highly oxidizing (hematite, goethite) and low (goethite), medium (hematite) to high temperature (magnetite) conditions.

Sillimanite: metamorphic rocks.

Titanium minerals (such as rutile, titanite, anatase and others): metamorphic rocks, pegmatites and hydrothermal vein deposits.

Tourmaline: pegmatites.

Zircon: common in igneous and metamorphic rocks.

Hydrocarbons, bitumen and methane: trapped during growth of the quartz crystals from fluids containing organics. These can sometimes be used to reconstruct basin temperature and fluid histories.

Table 1.1: Overview of quartz types, characteristics and provenance (adapted from Krynine, 1946 and Folk, 1980).

Quartz Types / Source	Extinction Characteristics	Other Features
Common or plutonic (granitic or many others)	Unit (straight) to slightly undulatory	Some vacuoles and/or microlites; subequant to xenomorphic grains
Vein or pegmatitic	Unit or semi-composite to undulatory, possible shearing	Abundant vacuoles, possible vermicular chlorite; comb structure common; typically found in large clasts
Volcanic	Unit (straight)	Euhedral bipyramidal shapes, round corners, and straight sides with embayments; virtually no inclusions; posible negative crystals
Schistose metamorphic	Unit to slightly undulatory	Elongate composite grains with straight borders; common mica inclusions
Stretched metamorphic	Strongly undulatory; subunit borders crenulate, granulate or smooth	Elongate, lenticular-shaped crystal sub-units; some microlites and vacuoles
Recrystallized metamorphic	Unit to slightly undulatory	Straight boundaries between equant interlocking grains in a mosaic; some microlites or vacuoles

Mid. Ordovician Lander Sandstone, Johnson Co., Wyoming

Well-rounded, medium-sized quartz grains in a very fine grained, bimodal quartzarenite (interpreted to be an eolian deposit). In cross-polarized light, most of the grains exhibit unit (nonundulatory) extinction—i.e., the entire grain goes to extinction at the same time. The range of birefringence colors (gray to yellow) in the cross-polarized light photomicrograph is slightly "high" for a standard thickness (30 μm) thin section; thus, this section is slightly thicker than standard.

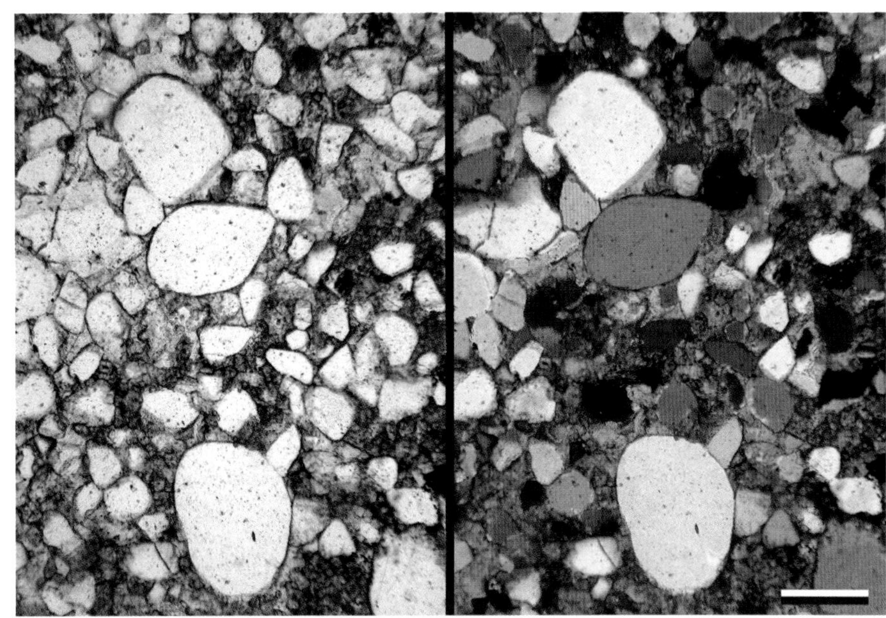

PPL | XPL, Scale bar = 0.20 mm

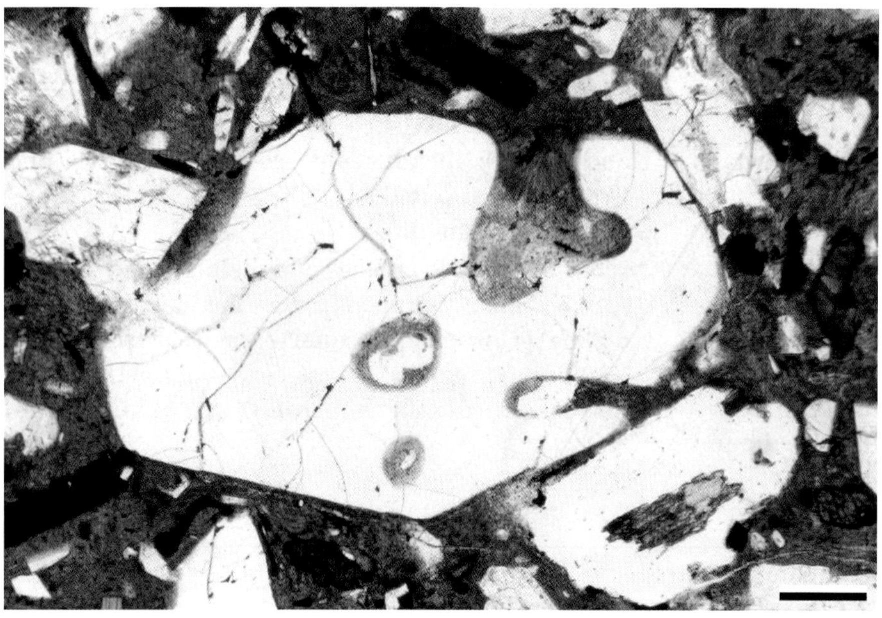

Paleogene Needles Range Fm., Millard Co., Nevada

A volcanic quartz grain with euhedral, bipyramidal outline. Euhedral shape, embayments, unit extinction and scarcity of inclusions are all indicative of an extrusive igneous (volcanic) provenance, but none, by itself, is conclusive evidence. This crystal started out as β-quartz, a hexagonal, high-temperature polymorph (573–870°C) of quartz. β-quartz is not stable at surface temperatures, and it inverts to α-quartz upon cooling.

PPL, Scale bar = 0.36 mm

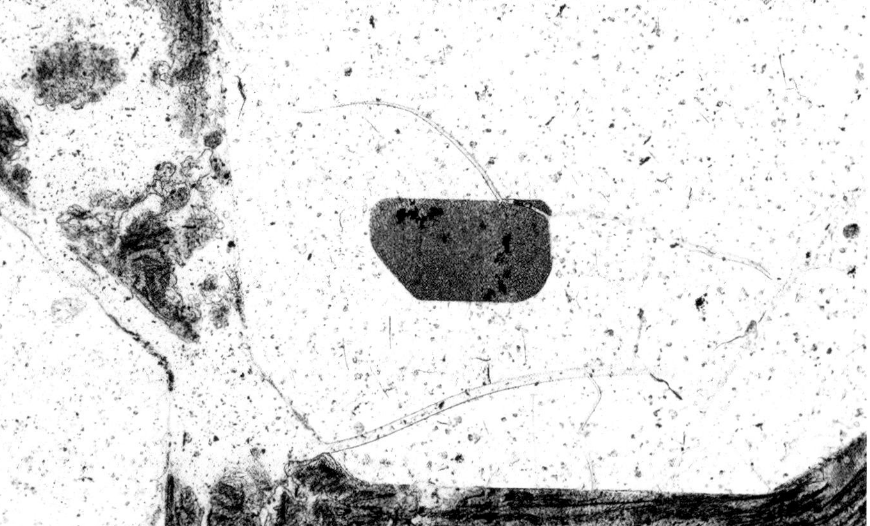

Neogene Yellowstone Tuff, Park Co., Wyoming

A higher magnification view of a volcanic quartz crystal. This grain has unit extinction, a euhedral outline and a large "negative crystal" or vacuole. The vacuole (or hole) has the same crystallographic orientation as the complete quartz grain, hence the term "negative crystal". This feature is common but not ubiquitous in volcanic quartz crystals.

PPL, KFS, Scale bar = 0.14 mm

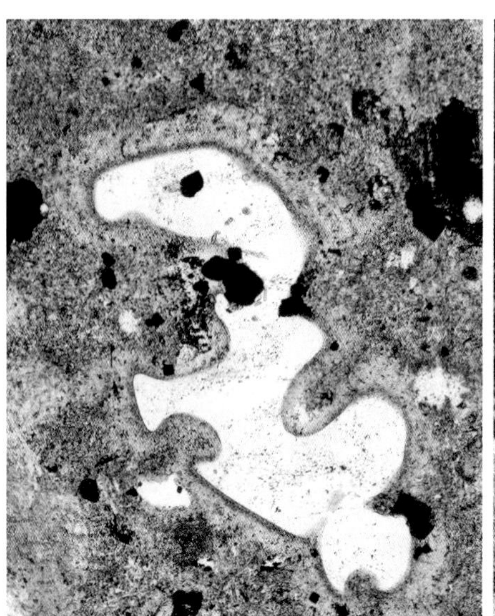

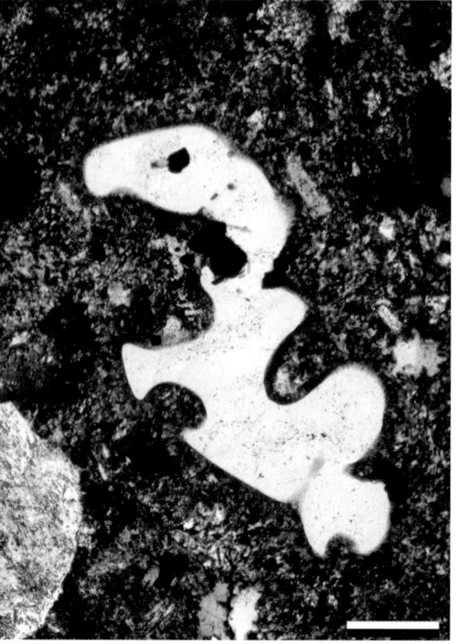

Cenozoic monzonite intrusive, Clark Co., Nevada

A deeply (monstrously) embayed volcanic quartz crystal within a monzonite porphyry. Embayed quartz crystals are common in acidic volcanics and have been ascribed to quartz disequilibrium with the melt causing the crystals to be reabsorbed (dissolution). Embayments normally have rounded edges. Like most volcanic quartz, this grain has unit extinction and few inclusions. Even though this is a source rock, it contains oddly shaped and rounded grains that can be liberated during weathering.

PPL | XPL, Scale bar = 0.26 mm

Paleocene, Rajasthan, India

An embayed quartz crystal (pyroclast) within a crystal tuff. The embayments are not crystallographically controlled, suggesting that they result from magmatic corrosion. The corrosion occurs during changes in melt composition. Some embayments may form as partial inclusions, but it is not interpreted to be the case here. The quartz crystal is surrounded by devitrified volcanic glass and stretched pumice. The matrix has taken a stain from the sodium cobaltinitrite, suggesting that the alteration product of the glass includes K-feldspar.

PPL, KFS, Scale bar = 0.49 mm

Paleogene Vieja Gp., Presidio Co., Texas

A single crystal, unit extinction, detrital quartz grain ("common quartz" or "normal igneous quartz" of Krynine, 1946). The grain goes to complete extinction under crossed polarizers with less than one degree of stage rotation. Such quartz is supplied by many types of source rocks and may be selectively concentrated during weathering and transportation. This grain may have been derived from a volcanic source as indicated by the embayments.

XPL, Scale bar = 0.26 mm

Up. Pennsylvanian – Lo. Permian Granite Wash, southern Oklahoma

An example of a thin section that was cut thicker than normal (i.e., > 30 μm). Instead of the gray to pale yellow birefringence expected for quartz, many of the quartz grains are deep yellow. Conversely, if the thin section is too thin, then the highest quartz birefringence colors might only be light gray. Thus, one has to "adjust one's expectations" to accurately identify these grains in thin section. Many of the other grains in this sample are microcline feldspars, recognizable because of their twinning (example indicated by white arrow).

XPL, Scale bar = 0.22 mm

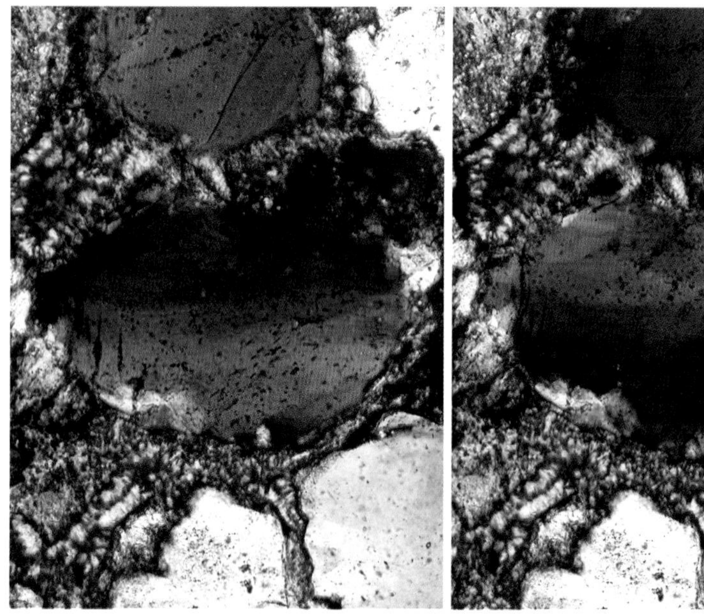

Lo. Cretaceous Patula Arkose, Coahuila, Mexico

This strained quartz grain has strongly undulatory extinction. The difference in stage rotation between the two photomicrographs is 8°, with the extinction zone sweeping from top to bottom. The straining here was produced *in situ* during burial and compaction causing crystal lattice dislocations. However, grains like these are common in source terranes that have undergone substantial deformation or metamorphism (Basu, 1985) and thus can arrive as prestrained detrital grains. The distinction is made by seeing if most or just a few grains show straining and if there is or is not a consistent pattern or fabric of strain features.

XPL | XPL, Scale bar = 0.10 mm

Silurian Tuscarora Sandstone, Lebanon Co., Pennsylvania

Undulatory extinction in a composite quartz grain. This quartz grain consists of a number of separate quartz crystals. Such grains can be derived from a range of source terranes, including plutonic, metamorphic and hydrothermal.

XPL, Scale bar = 0.14 mm

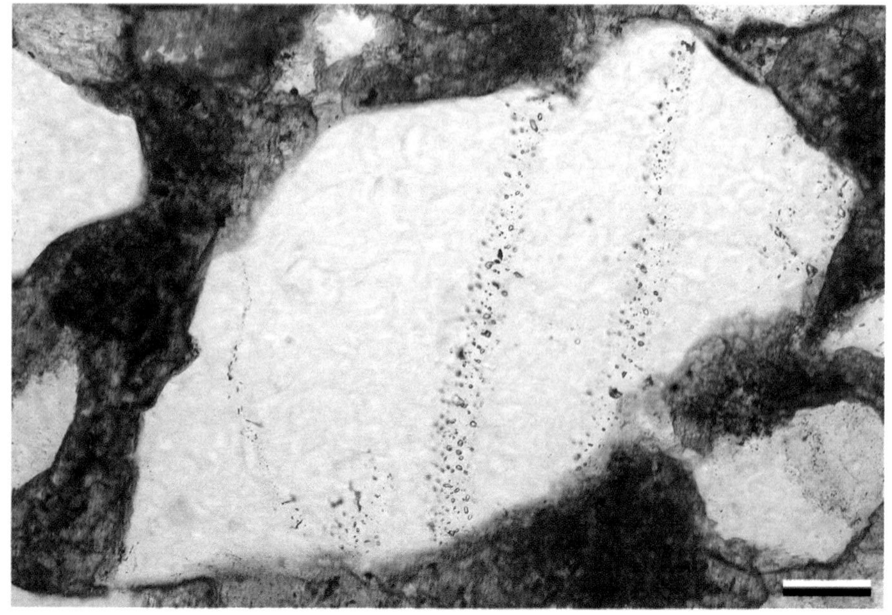

Jurassic Tilje Fm., Norwegian sector, North Sea †

Trails of fluid inclusions (vacuoles) within an otherwise inclusion-free quartz grain. Although aligned, these inclusion trains do not represent Boehm lamellae. The crosscutting of the trails suggests that the vacuoles are secondary and occurred from fluid flow along fractures that were subsequently healed. The quartz grain is surrounded by siderite cement.

PPL, KFS, BDI, Scale bar = 0.07 mm

Lo. Cretaceous Travis Peak Fm., Eastland Co., Texas

A quartz grain with exceptionally abundant vacuoles (probably liquid-filled inclusions) and semicomposite extinction (Krynine's "hydrothermal" quartz). Such vacuole-rich quartz grains are most commonly derived from hydrothermal-vein sources. The quartz overgrowth (cement) around the upper part of the grain has far fewer inclusions.

XPL, Scale bar = 0.26 mm

Lo. Paleozoic andalucite schist, Carroll Co., New Hampshire

Multiple quartz crystals in a schist that have abundant needle-like mineral inclusions. These inclusions are probably sillimanite or tourmaline on the basis of their moderate relief and birefringence. When sillimanite occurs in fibrous masses as in this sample, it is called fibrolite.

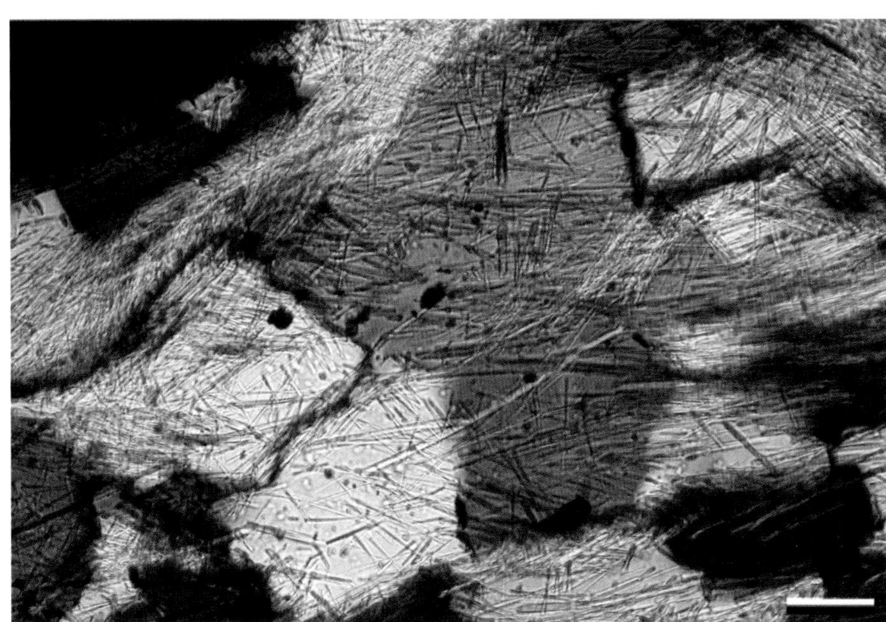

XPL, Scale bar = 0.05 mm

Lo. Paleozoic andalucite schist, Carroll Co., New Hampshire

Quartz crystals with abundant mineral inclusions. The largest inclusions are biotite crystals (e.g., the brown crystal indicated by the red arrow). The smaller needle-fiber inclusions are probably tourmaline on the basis of their greenish crystal color, shape and birefringence.

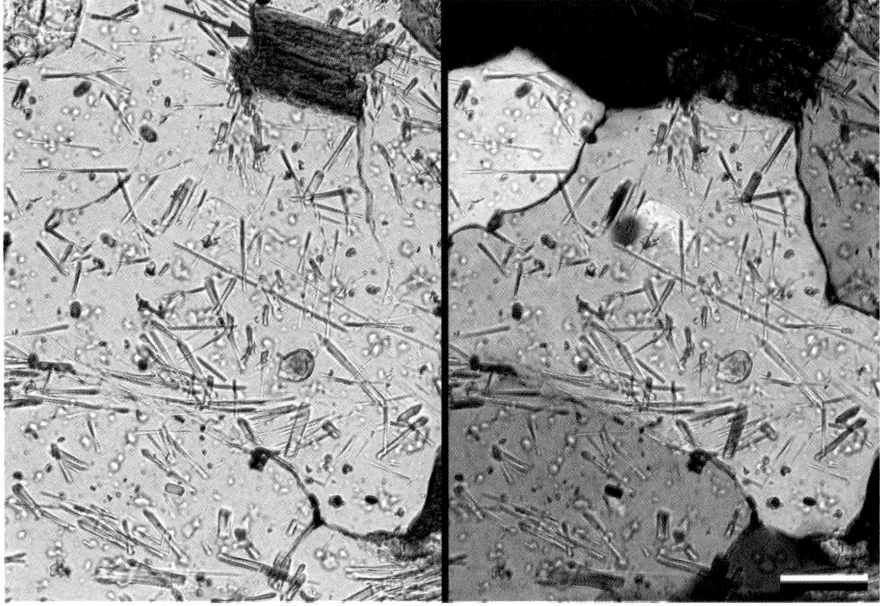

PPL | XPL, Scale bar = 0.05 mm

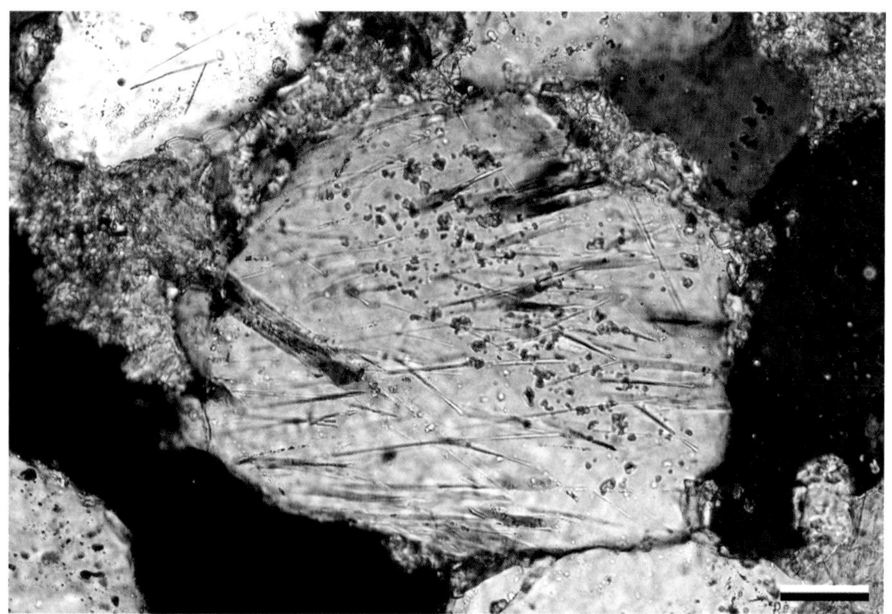

Mid. Permian (Guadalupian) Brushy Canyon Fm., Culberson Co., Texas

A quartz grain with numerous mineral inclusions. The needle-shaped mineral inclusions probably are rutile and/or tourmaline, but there also are fluid inclusions which make the grain appear speckled or dusty. These quartz grains are surrounded by carbonate cement.

XPL, Scale bar = 0.05 mm

Mid. Jurassic Ness Fm., Norwegian sector, North Sea †

Acicular mineral inclusions within a quartz grain. The crystals are likely to be rutile but are too fine crystalline to determine their optical properties. The crystals show no preferred orientation.

PPL, BDI, Scale bar = 0.12 mm

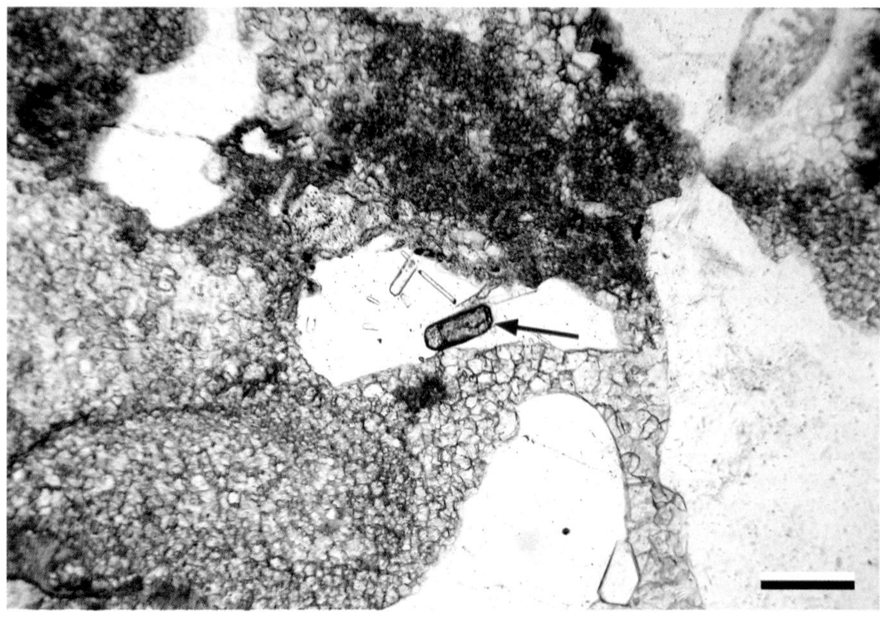

Pennsylvanian Alamitos Fm., San Miguel Co., New Mexico

A quartz grain containing both zircon and apatite inclusions. The zircon inclusion (indicated by red arrow) is the large greenish crystal with very high relief. The apatite inclusions are the colorless, moderate relief crystals (they have more relief than the surrounding quartz grain). This grain may be from a pegmatite or hydrothermal vein deposit. To see examples of zircon's very high and apatite's very low birefringence under crossed polarizers, see Chapter 4.

PPL, Scale bar = 0.16 mm

Up. Miocene (Tortonian), Tagus Basin, Setúbal Peninsula, Portugal †

A quartz grain displaying mineral inclusions of biotite. The host grain is nonundulatory and inclusion free, suggesting a possible igneous origin. The biotite crystals are in optical alignment with each other. They show a single, well-defined cleavage, which distinguishes them from tourmaline.

PPL | XPL, BDI, Scale bar = 0.04 mm

Silurian Tuscarora Sandstone, Lebanon Co., Pennsylvania

Vermicular chlorite inclusions in a detrital quartz grain. The chlorite booklets are pale green, have anomalous "ultrablue" colors under polarized light and appear to be worm-like masses. These inclusions are typical of hydrothermal quartz.

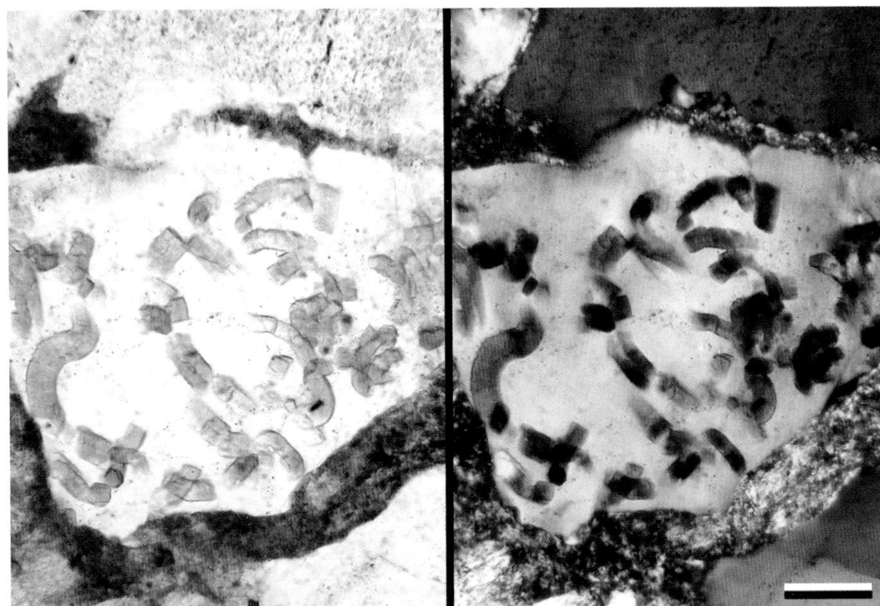

PPL | XPL, Scale bar = 0.14 mm

Silurian Tuscarora Sandstone, Lebanon Co., Pennsylvania

Vermicular chlorite inclusions in a detrital quartz grain. Vermicular chlorites can form during hydrothermal alteration or within pegmatites, but they also can form during metamorphism. These photomicrographs were taken with the substage condenser raised, thereby highlighting the individual chlorite plates that make up the vermicules.

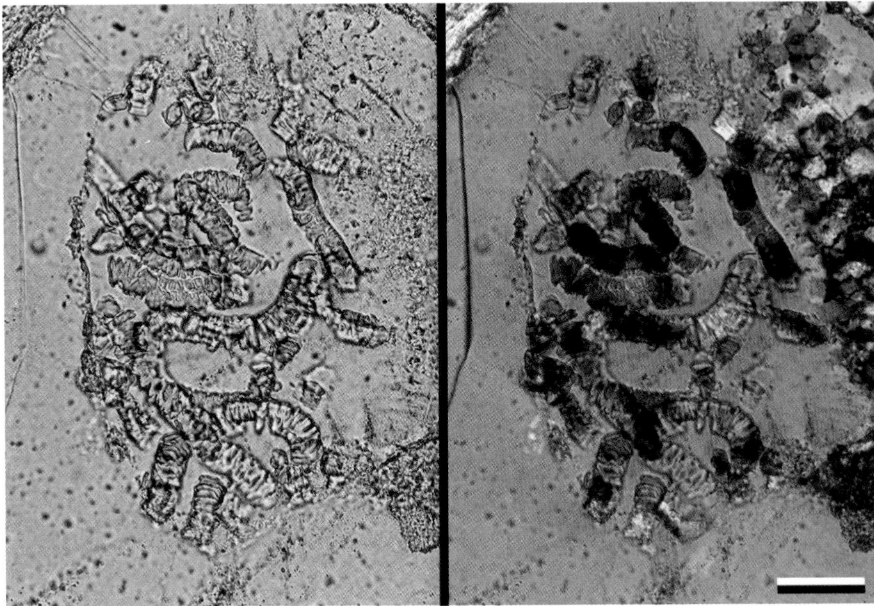

PPL | XPL, Scale bar = 0.05 mm

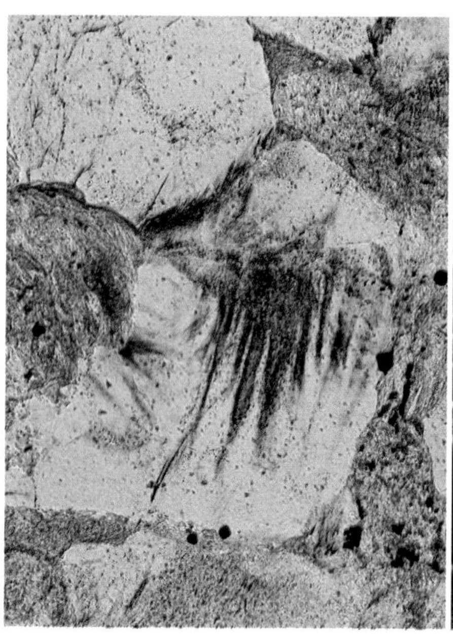
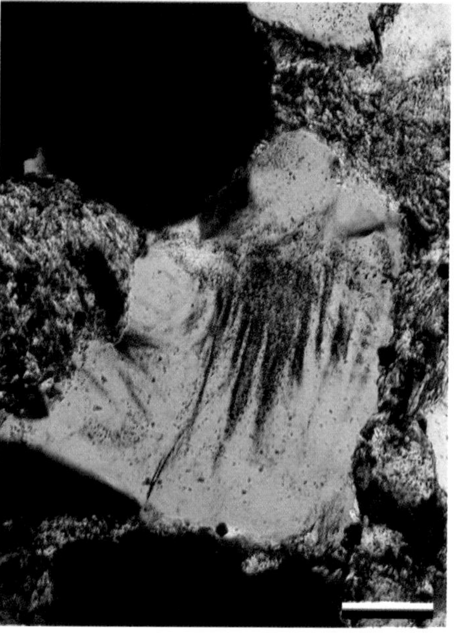

Silurian Tuscarora Sandstone, Lebanon Co., Pennsylvania

A detrital quartz grain with subparallel lines of submicron- to micron-sized bubbles (vacuoles). These features, a product of intense strain deformation, are termed Boehm lamellae. When identifying these structures in thin section, it is important to distinguish between reworked detrital strained grains and those that were strained *in situ*. Also, one must be careful not to confuse them with twinned feldspars.

PPL | XPL, Scale bar = 0.10 mm

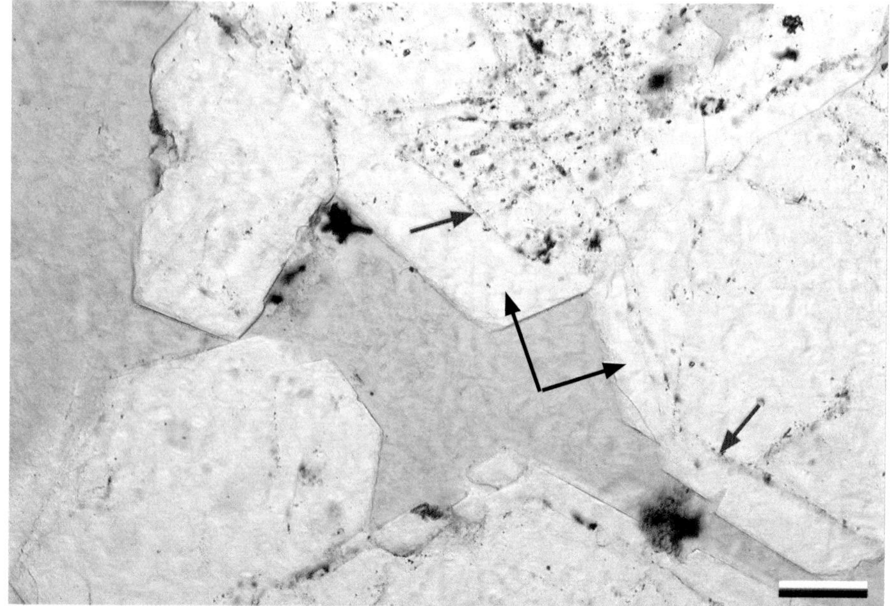

Jurassic Fulmar Fm., United Kingdom sector, North Sea †

Quartz grains that display well-formed, syntaxial quartz overgrowths extending into pore space (examples shown by black arrows). Some of the grains display a marked grain boundary, whilst others do not. The distinction between the grain and the overgrowth relies on the presence of thin clay or iron oxide coatings or a line of inclusions along the grain-overgrowth contact, such as in this case (red arrows).

PPL, BDI, Scale bar = 0.07 mm

Up. Devonian Cairn Fm., Jasper area, Alberta, Canada

Euhedral quartz overgrowths on very fine sand grains in a dolomitized reefal limestone. Such replacement crystals can be a major source of quartz grains with reworked (second-cycle or inherited) overgrowths after they are released from their carbonate host rock during weathering. Transport and abrasion during reworking yields rounded overgrowths that are the main diagnostic feature of second-cycle quartz (quartz grains from a sedimentary source).

XPL, Scale bar = 0.26 mm

Jurassic Tilje Fm., Norwegian sector, North Sea †

A quartz grain with an inherited quartz overgrowth. The overgrowth is distinguished from the grain by a line of vacuoles (indicated by black arrow). The irregular (abraded?) margin of the overgrowth is an indicator of reworking but is not conclusive. The lack of overgrowths on the vast majority of the other detrital quartz grains in this sample, however, supports the conclusion that these overgrowths were formed in a precursor deposit. The main cement in this sample is early-formed siderite.

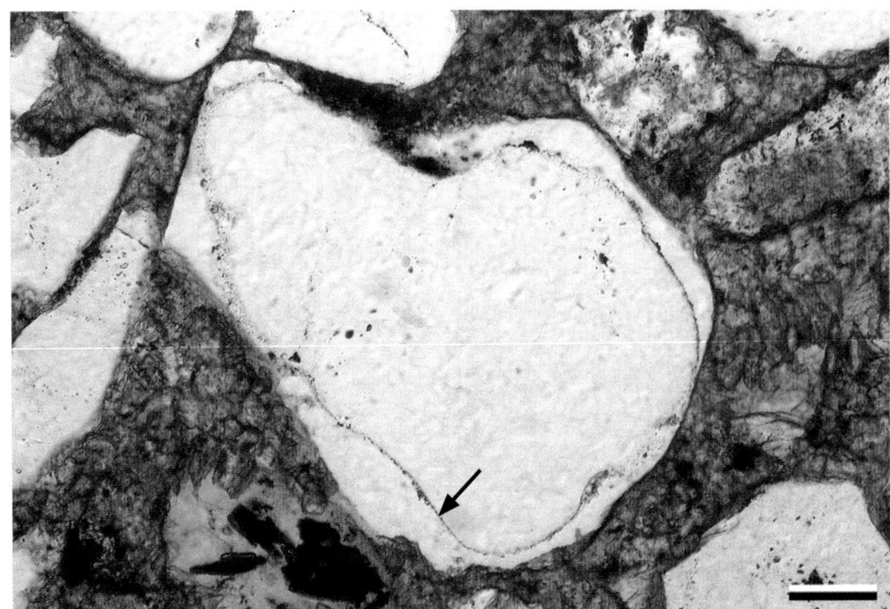

PPL, KFS, BDI, Scale bar = 0.07 mm

Paleogene Vieja Gp., Presidio Co., Texas

A detrital grain with authigenic quartz overgrowths that have been rounded during a second cycle of erosion followed by deposition. Recycled or second-cycle quartz grains can be difficult to recognize if quartz overgrowths are not present. The original grain is largely inclusion free whereas the surrounding, rounded overgrowth is rich in inclusions (carbonate, clays and iron oxides).

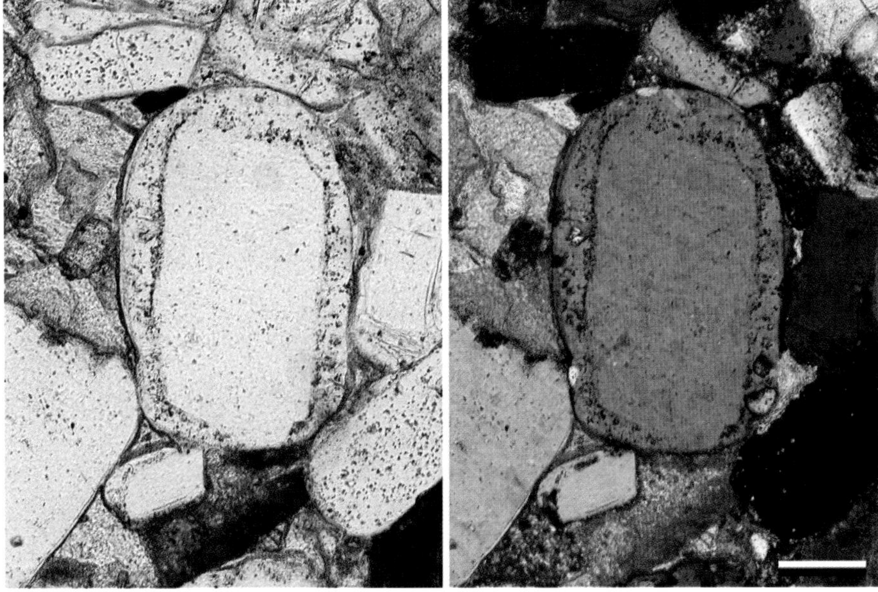

PPL | XPL, Scale bar = 0.05 mm

Up. Triassic – Lo. Jurassic New Haven Arkose, New Haven Co., Connecticut

This semicomposite quartz grain exhibits undulatory extinction. Grains like this are composed of numerous quartz crystals that have closely aligned optic c-axes. Such quartz grains are common in metamorphic, plutonic and hydrothermal environments.

XPL | XPL, Scale bar = 0.64 mm

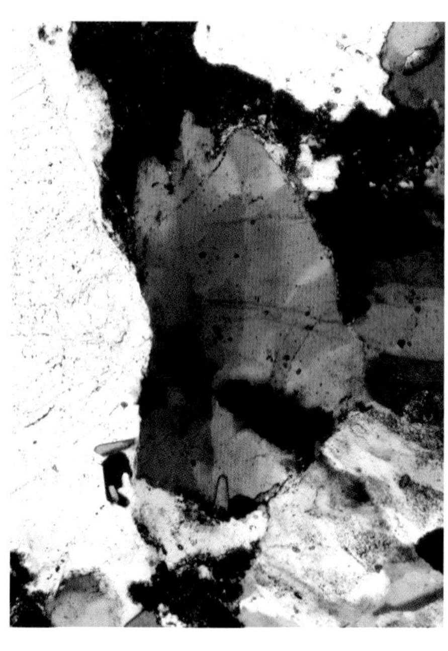

Up. Triassic – Lo. Jurassic New Haven Arkose, New Haven Co., Connecticut

A semicomposite quartz grain, with slightly undulatory extinction, in a lithic arenite. Note the changes between the two images that show sweeping extinction with less than five degrees of stage rotation.

XPL | XPL, Scale bar = 0.17 mm

Up. Triassic – Lo. Jurassic New Haven Arkose, New Haven Co., Connecticut

A polycrystalline quartz grain with strong elongation or stretching of the quartz crystals. This may indicate that the origin of this grain is from a metamorphic source terrane. This rock would be classified as a lithic sandstone on the basis of the abundance of rock fragments (lithic clasts). Polycrystalline quartz grains, however, are grouped differently in various classifications—some assign them to lithic fragments and others to quartz grains. Such differences reflect, in part, whether one is more interested in accurate description of composition or the determination of provenance.

PPL | XPL, Scale bar = 0.64 mm

Mid.–Up. Ordovician Martinsburg Fm., Berks Co., Pennsylvania

Polycrystalline (also known as composite) quartz grains in a poorly sorted sandstone. The large size of the individual crystals in the grains may indicate that the origin of the grains is from an igneous or high-grade metamorphic source terrane.

XPL, Scale bar = 0.51 mm

Paleocene Tang Fm., Møre Basin, Norwegian Sea †

A polycrystalline quartz grain that neither displays sutured internal crystal/boundaries nor truly polygonal boundaries. This would support an origin from weakly metamorphosed sandstone or quartzite.

XPL, BDI, Scale bar = 0.12 mm

Triassic Sherwood Sandstone Gp., Co. Antrim, Northern Ireland, U.K. †

A polycrystalline quartz grain of metamorphic origin. The grain has sutured internal boundaries between composite crystals and a preferential alignment of the constituent crystals. In addition, crystal boundaries are diffuse.

XPL, AFeS, KFS, BDI, Scale bar = 0.07 mm

Miocene Topanga Fm., Ventura Co., California

A rounded polycrystalline quartz grain in a lithic arenite. The elongation of the quartz crystals in the grain indicates that this grain probably formed in a metamorphic terrane. High-density lattice dislocations in quartz produce polycrystallinity and are indicative of metamorphic environments (Basu, 1985).

XPL, Scale bar = 0.26 mm

Jurassic Morrison Fm., Moffat Co., Colorado

A polycrystalline quartz grain (center) surrounded by chert clasts. Based on the strong elongation of the quartz crystals, this was most likely sourced from a metamorphic terrane. Surrounding the stretched polycrystalline quartz grains are several large chert clasts that have a dark gray birefringence and speckled appearance related to the microcrystalline character of chert. These indicate that sedimentary rocks were also present in the source area. Numerous silt-sized monocrystalline quartz grains also are present in the interstices between the larger grains.

XPL, RDI, Scale bar = 0.26 mm

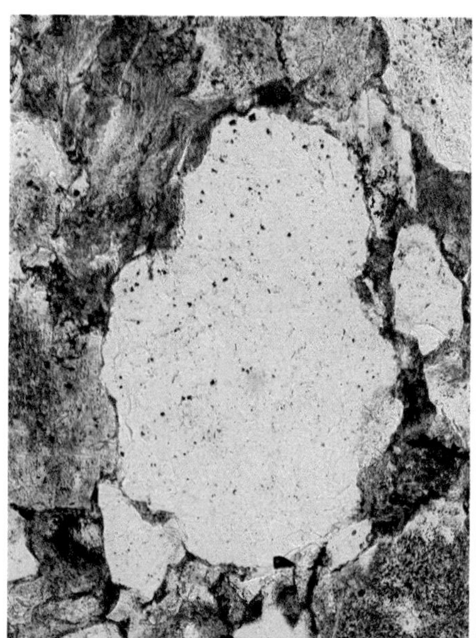

Miocene Topanga Fm., Ventura Co., California

A polycrystalline quartz grain in a feldspathic sandstone. As in the previous photomicrograph, there is strong elongation to the quartz crystals in the grain suggesting a high-grade metamorphic origin for the grain. This sample also shows, especially in plane-polarized light, how feldspar can be sometimes be distinguished from quartz simply based on its tendency to be more "dusty".

PPL | XPL, Scale bar = 0.12 mm

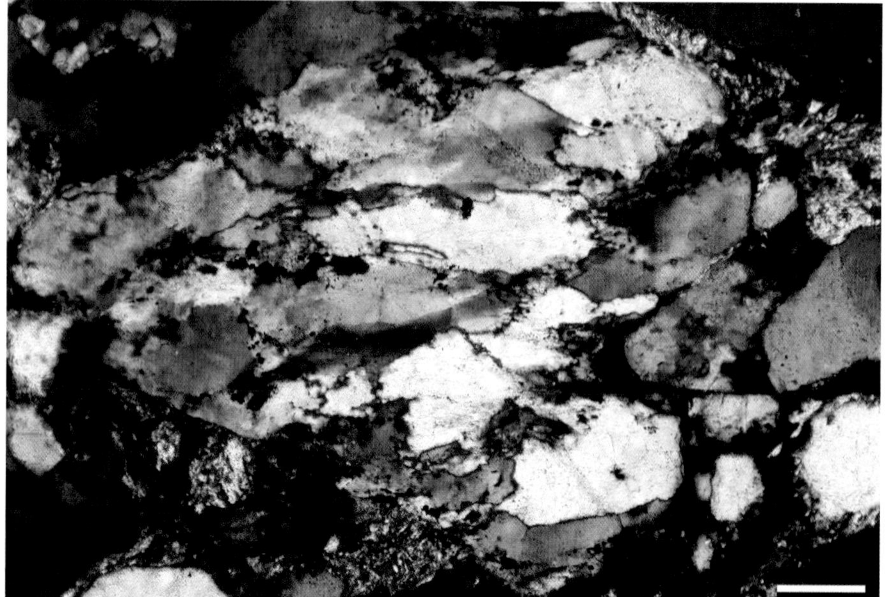

Up. Silurian Bloomsburg Fm., Warren Co., New Jersey

A polycrystalline quartz grain with strong elongation and sutured crystal boundaries. The probable origin of this grain is in a high-grade metamorphic terrane.

XPL, Scale bar = 0.10 mm

Mid. Permian Wegener Halvø Fm., Karstryggen area, East Greenland

Polycrystalline or composite quartz in a feldspathic arenite. The origin of this grain is problematic; it may be metamorphic, but plutonic or hydrothermal sources are equally likely. The inability to assign a unambiguous source terrane is common if one only has quartz indicators. The calcite cements in this sample are stained and therefore appear to be red even in cross-polarized light.

XPL, AFeS, Scale bar = 0.26 mm

Up. Permian Schuchert Dal Fm., Jameson Land, East Greenland

A well-rounded polycrystalline quartz granule. The blue-dyed fracture porosity surrounding the grain may have been produced by pressure release during uplift and unloading of overburden.

PPL | XPL, AFeS, BDI, Scale bar = 0.51 mm

Precambrian Neihart Quartzite, Little Belt Mountains, Cascade Co., Montana

A live-color cathodoluminescent SEM photomicrograph of a quartz arenite. Blue grains have the highest formation temperatures, suggesting derivation from plutonic or high-grade metamorphic sources (e.g., Augustsson and Reker, 2012). Reddish grains are likely from lower-grade metamorphic source rocks (greenschist). Reddish blotches and healed fractures in the grains may reflect lower temperature, hydrothermal overprint. Low-temperature quartz cements (deep blue) have a late (hotter) overprint (reddish) along grain boundaries and residual pore spaces in overgrowths.

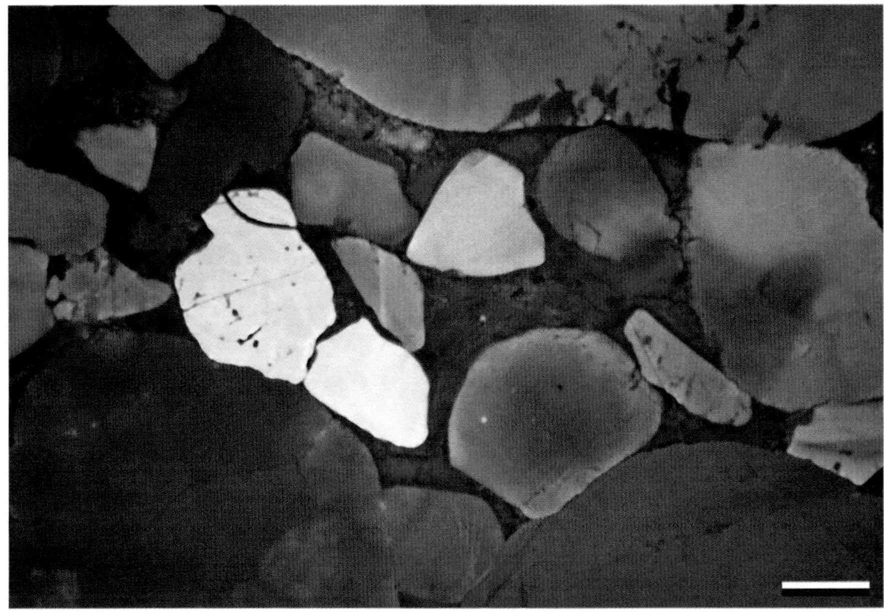

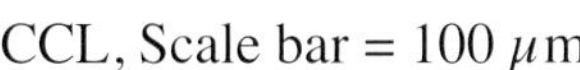

CCL, Scale bar = 100 μm

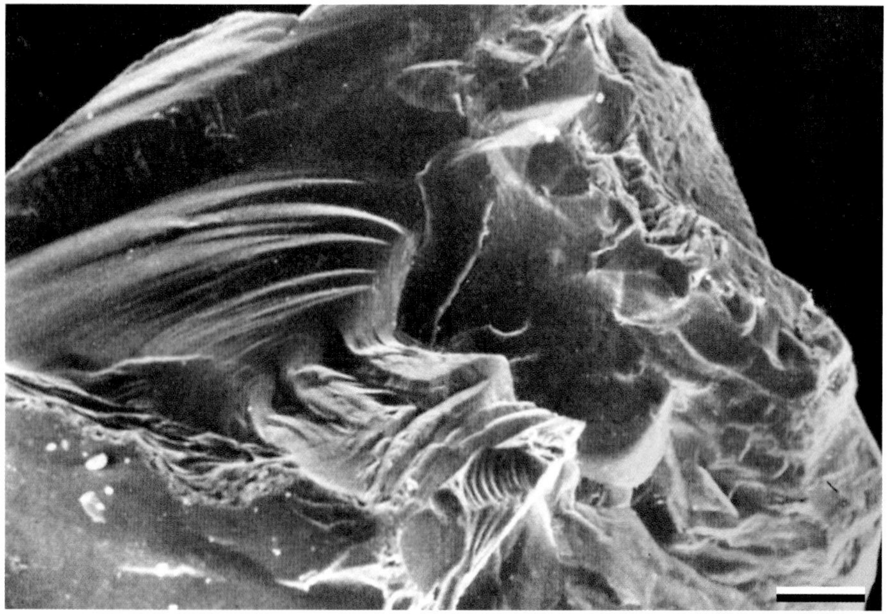

Holocene grus on Mesozoic granite, California

Quartz grains are commonly studied using scanning electron microscopy, either to obtain more three-dimensional images of grain shape or to examine surface features. Quartz grains are normally recognizable in the scanning electron microscope on the basis of their conchoidal fracture and lack of cleavage. Photo by David E. Hoyt.

SEM, Scale bar = 50 μm

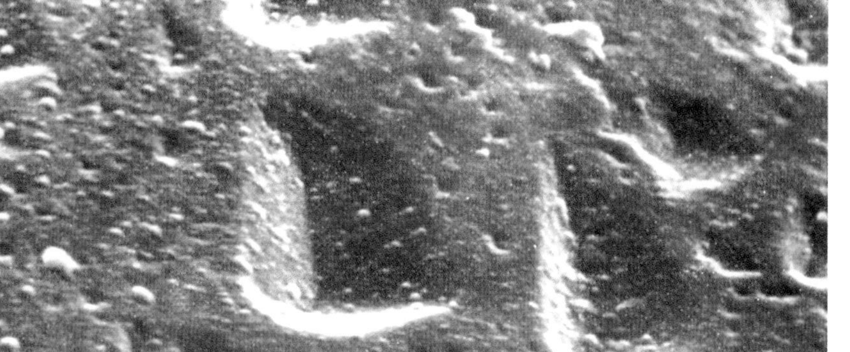

Quaternary grus on Precambrian Packsaddle Schist, Texas

Quartz grains have a wide variety of surface textures when examined in SEM. V-shaped indentations are seen in this example and are probably formed by chemical dissolution. Various structures have been interpreted as being indicative of particular source rocks, transport processes, depositional environments or diagenetic settings (see Krinsley and Doornkamp, 2011). Photo by David E. Hoyt.

SEM, Scale bar = 0.07 μm

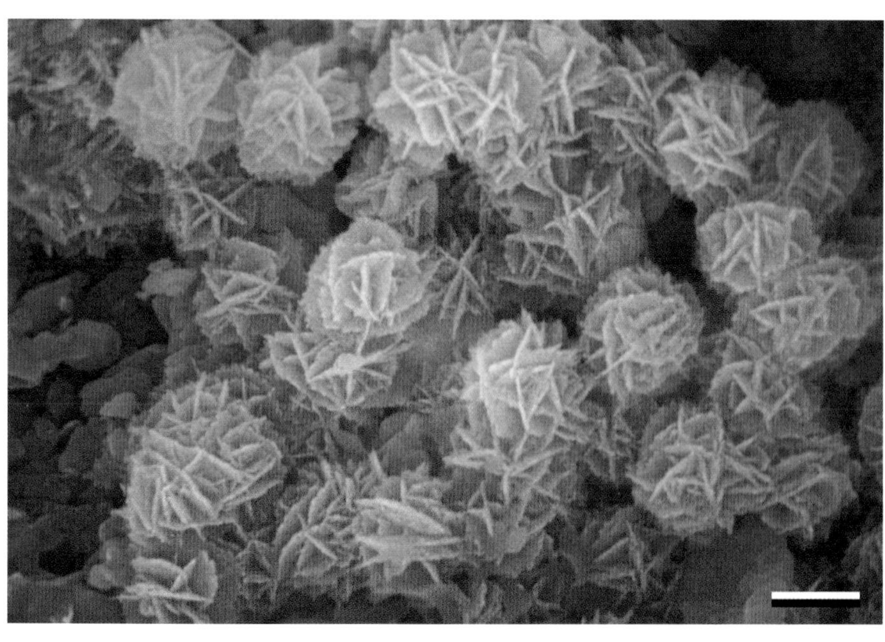

Up. Cretaceous Craie Grise, Limburg, The Netherlands

One of the polymorphs of quartz is cristobalite (opal-CT). In this photomicrograph, the cristobalite occurs as spherulites that are commonly termed lepispheres. The alteration of biogenic opal to quartz generally goes through the intermediate stage of cristobalite. The transitions between these three phases are largely temperature controlled, although time and associated lithologies also play a large role. In general, the presence of clay minerals retards the transformation; $CaCO_3$ accelerates it (e.g., Isaacs, 1982). The opal-CT pictured here forms part of larger replacive nodules in a sandy limestone.

SEM, Scale bar = 2.4 μm

Lo. Permian Park City Fm., middle Franson Mbr., Daggett Co., Utah

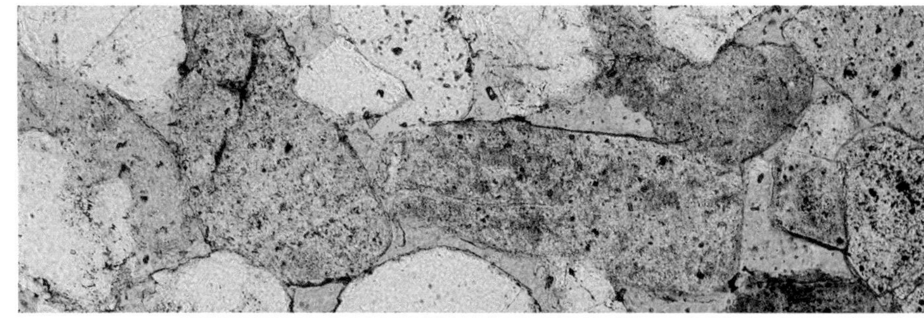

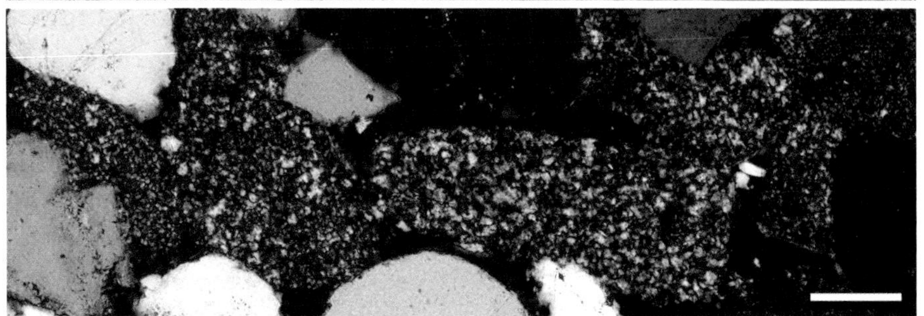

Detrital chert clasts in a quartz arenite. Chert is microcrystalline quartz, and it typically contains abundant inclusions. These are mostly water-filled, but inclusions also can be organics, iron oxides, iron sulfides, carbonates and other insolubles that make the grains look cloudy and brownish in plane light. In cross-polarized light, chert appears to be finely speckled due to its microcrystallinity.

PPL / XPL, BDI, Scale bar = 0.10 mm

Mid. Jurassic Curtis Fm., Utah

Fragments of detrital chert (left) and detrital chalcedony (right). Chalcedony is a microscopically fibrous form of quartz that is common as void fillings or replacement fabrics in sedimentary rocks. Here, the chalcedony is surrounded by carbonate (dolomite) cement crystals. As with cherts, a wide variety of chalcedonic textures exist and chalcedony occurs in length-fast and length-slow forms (those can be distinguished using a gypsum or other sensitive tint plate).

XPL, Scale bar = 0.03 mm

Oligocene – Miocene Horse Spring Fm., Clark Co., Nevada

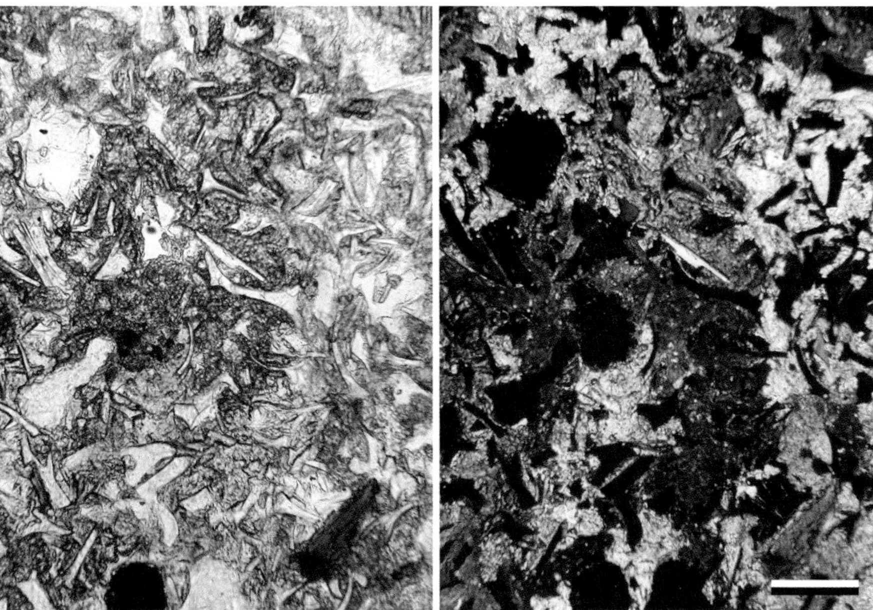

Shards of volcanic glass in a calcite-cemented tuffaceous sandstone. The glass shards (volcanic glass) are isotropic (black under cross-polarized light). They are commonly dissolved or are replaced by quartz or other minerals during diagenesis, in part because, as hydrophilic particles, they are especially susceptible to alteration. Because shards are isotropic, it can be difficult to distinguish dissolved shards (shard-shaped pores) from intact ones.

PPL | XPL, Scale bar = 0.15 mm

Cited References and Additional Information Sources

Adams, S. F., 1920, A microscopic study of vein quartz: Economic Geology, v. 15, p. 623-644, doi: 10.2113/gsecongeo.15.8.623.

Akhaven, A. C., 2013, The Quartz Page: http://www.quartzpage.de/about.html

Arribas, J., R. Marfil, and J. A. de la Peña, 1985, Provenance of Triassic feldspathic sandstones in the Iberian Range (Spain): Significance of quartz types: Journal of Sedimentary Research, v. 55, p. 864-868, doi: 10.1306/212F8823-2B24-11D7-8648000102C1865D.

Augustsson, C., and A. Reker, 2012, Cathodoluminescence spectra of quartz as provenance indicators revisited: Journal of Sedimentary Research, v. 82, p. 559-570, doi: 10.2110/jsr.2012.51.

Basu, A., 1985, Reading provenance from detrital quartz, *in* G. G. Zuffa, ed., Provenance of Arenites (NATO Science Series C): Dordrecht, Netherlands, D. Reidel Publishing, p. 231-248, doi: 10.1007/978-94-017-2809-6_11.

Basu, A., S. W. Young, L. J. Suttner, W. C. James, and G. H. Mack, 1975, Re-evaluation of the use of undulatory extinction and polycrystallinity in detrital quartz for provenance interpretation: Journal of Sedimentary Research, v. 45, p. 873-882, doi: 10.1306/212F6E6F-2B24-11D7-8648000102C1865D.

Bernet, M., and K. Bassett, 2005, Provenance analysis by single-quartz-grain SEM-CL/optical microscopy: Journal of Sedimentary Research, v. 75, p. 492-500, doi: 10.2110/jsr.2005.038.

Bernet, M., D. Kapoutsos, and K. Bassett, 2007, Diagenesis and provenance of Silurian quartz arenites in south-eastern New York State: Sedimentary Geology, v. 201, p. 43-55, doi: 10.1016/j.sedgeo.2007.04.006.

Blatt, H., 1967, Original characteristics of clastic quartz grains: Journal of Sedimentary Research, v. 37, p. 401-424 , doi: 10.1306/74D716E0-2B21-11D7-8648000102C1865D.

Blatt, H., and J. M. Christie, 1963, Undulatory extinction in quartz of igneous and metamorphic rocks and its significance in provenance studies of sedimentary rocks: Journal of Sedimentary Research, v. 33, p. 559-579, doi: 10.1306/74D70EBB-2B21-11D7-8648000102C1865D.

Boggs, S., Jr., Y. I. Kwon, G. G. Goles, B. G. Rusk, D. Krinsley, and A. Seyedolali, 2002, Is quartz cathodoluminescence color a reliable provenance tool? A quantitative examination: Journal of Sedimentary Research, v. 72, p. 408-415, doi: 10.1306/102501720408.

Bokman, J. W., 1952, Clastic quartz particles as indices of provenance: Journal of Sedimentary Research, v. 22, p. 17-24, doi: 10.1306/D426949F-2B26-11D7-8648000102C1865D.

Chandler, F. W., 1988, Quartz arenites: Review and interpretation: Sedimentary Geology, v. 58, p. 105-126, doi: 10.1016/0037-0738(88)90065-6.

Conolly, J. R., 1965, The occurrence of polycrystallinity and undulatory extinction in quartz in sandstones: Journal of Sedimentary Research, v. 35, p. 116-135, doi: 10.1306/74D71208-2B21-11D7-8648000102C1865D.

Culver, S. J., P. A. Bull, S. Campbell, R. A. Shakesby, and W. B. Whalley, 1983, Environmental discrimination based on quartz grain surface textures: A statistical investigation: Sedimentology, v. 30, p. 129-136, doi: 10.1111/j.1365-3091.1983.tb00655.x.

Donaldson, C. H., and C. M. B. Henderson, 1988, A new interpretation of round embayments in quartz crystals: Mineralogical Magazine, v. 52, p. 27-33, doi: 10.1180/minmag.1988.052.364.02.

Götte, T., T. Pettke, K. Ramseyer, M. Koch-Müller, and J. Mullis, 2011, Cathodoluminescence properties and trace element signature of hydrothermal quartz: A fingerprint of growth dynamics: American Mineralogist, v. 96, p. 802-813, doi: 10.2138/am.2011.3639.

Götte, T., and K. Ramseyer, 2012, Trace element characteristics, luminescence properties and real structure of quartz, *in* J. Götze, and R. Möckel, eds., Quartz: Deposits, Mineralogy and Analytics: New York, Springer, p. 265-285, doi: 10.1007/978-3-642-22161-3_12.

Götte, T., and D. K. Richter, 2006, Cathodoluminescence characterization of quartz particles in mature arenites: Sedimentology, v. 53, p. 1347-1359, doi: 10.1111/j.1365-3091.2006.00818.x.

Götze, J., and R. Möckel, eds., 2012, Quartz: Deposits, Mineralogy and Analytics: New York, Springer, 360 p., doi: 10.1007/978-3-642-22161-3.

Harrell, J., and H. Blatt, 1978, Polycrystallinity: Effect on the durability of detrital quartz: Journal of Sedimentary Research, v. 48, p. 25-30, doi: 10.1306/212F73E7-2B24-11D7-8648000102C1865D.

Hesse, R., 1988, Diagenesis #13. Origin of chert: Diagenesis of biogenic siliceous sediments: Geoscience Canada, v. 15, p. 171-192.

Isaacs, C. M., 1982, Influence of rock composition on kinetics of silica phase changes in the Monterey Formation, Santa Barbara area, California: Geology, v. 10, p. 304-308, doi: 10.1130/0091-7613(1982)10<304:IORCOK>2.0.CO;2.

Keller, W. D., and R. F. Littlefield, 1950, Inclusions in the quartz of igneous and metamorphic rocks: Journal of Sedimentary Research, v. 20, p. 74-84, doi: 10.1306/D42693A0-2B26-11D7-8648000102C1865D.

Krinsley, D. H., and J. C. Doornkamp, 2011, Atlas of Quartz Sand Surface Textures: Cambridge, UK, Cambridge University Press, 102 p.

Krinsley, D. H., and P. Trusty, 1985, Environmental interpretation of quartz grain surface textures, *in* G. G. Zuffa, ed., Provenance of Arenites (NATO Science Series C): Dordrecht, Netherlands, D. Reidel Publishing, p. 213-229, doi: 10.1007/978-94-017-2809-6_10.

Krynine, P. D., 1946, Microscopic morphology of quartz types: Annals of 2nd Pan-American Congress of Mining and Geological Engineers, p. 35-49.

Mahaney, W. C., 2002, Altas of Sand Grain Surface Textures and Applications: New York, Oxford University Press, 256 p.

Matter, A., and K. Ramseyer, 1985, Cathodoluminescence microscopy as a tool for provenance studies of sandstones, *in* G. G. Zuffa, ed., Provenance of Arenites (NATO Science Series C): Dordrecht, Netherlands, D. Reidel Publishing, p. 191-211, doi: 10.1007/978-94-017-2809-6_9.

McBride, E. F., 2012, Petrology of the Eureka Quartzite (Middle and Late Ordovician), Utah and Nevada, USA: Rocky Mountain Geology, v. 47, p. 81-111, doi: 10.2113/gsrocky.47.2.81.

Moss, A. J., 1966, Origin, shaping and significance of quartz sand grains: Journal of the Geological Society of Australia, v. 13, p. 97-136, doi: 10.1080/00167616608728607.

Sawatzky, C. C., and G. Pe-Piper, 2013, Detrital quartz sources in the Scotian Basin, eastern Canada, using hot-cathode cathodoluminescence: Availability of coarse-grained sand for reservoirs: AAPG Bulletin, v. 97, p. 1503-1520, doi: 10.1306/03251312185.

Seyedolali, A., D. H. Krinsley, S. Boggs, Jr., P. F. O'Hara, H. Dypvik, and G. G. Goles, 1997, Provenance interpretation of quartz by scanning electron microscope–cathodoluminescence fabric analysis: Geology, v. 25, p. 787-790, doi: 10.1130/0091-7613(1997)025<0787:PIOQBS>2.3.CO;2.

Sippel, R. F., 1968, Sandstone petrology, evidence from luminescence petrography: Journal of Sedimentary Research, v. 38, p. 530-554, doi: 10.1306/74D719DD-2B21-11D7-8648000102C1865D.

Suttner, L. J., A. Basu, and G. H. Mack, 1981, Climate and the origin of quartz arenites: Journal of Sedimentary Research, v. 51, p. 1235-1246, doi: 10.1306/212F7E73-2B24-11D7-8648000102C1865D.

Tortosa, A., M. Palomares, and J. Arribas, 1991, Quartz grain types in Holocene deposits from the Spanish Central System: Some problems in provenance analysis, *in* A. C. Morton, S. P. Todd, and P. D. W. Haughton, eds., Developments in Sedimentary Provenance Studies: London, GSL Special Publication 57, p. 47-54, doi: 10.1144/GSL.SP.1991.057.01.05.

Walderhaug, O., and J. Rykkje, 2000, Some examples of the effect of crystallographic orientation on the cathodoluminescence colors of quartz: Journal of Sedimentary Research, v. 70, p. 545-548, doi: 10.1306/2DC40926-0E47-11D7-8643000102C1865D.

Young, S. W., 1976, Petrographic textures of detrital polycrystalline quartz as an aid to interpreting crystalline source rocks: Journal of Sedimentary Research, v. 46, p. 595-603, doi: 10.1306/212F6FFA-2B24-11D7-8648000102C1865D.

Facing Page: Top – Thin section view of Carlsbad-twinned orthoclase feldspar, Mid. Ordovician Newtown Gneiss, New Haven Co., Connecticut (XPL). Bottom – Twinned orthoclase feldspar crystal, Otero Co., New Mexico. Photo from Mineral Museum, New Mexico Bureau of Geology and Mineral Resources.

GRAINS:
FELDSPARS

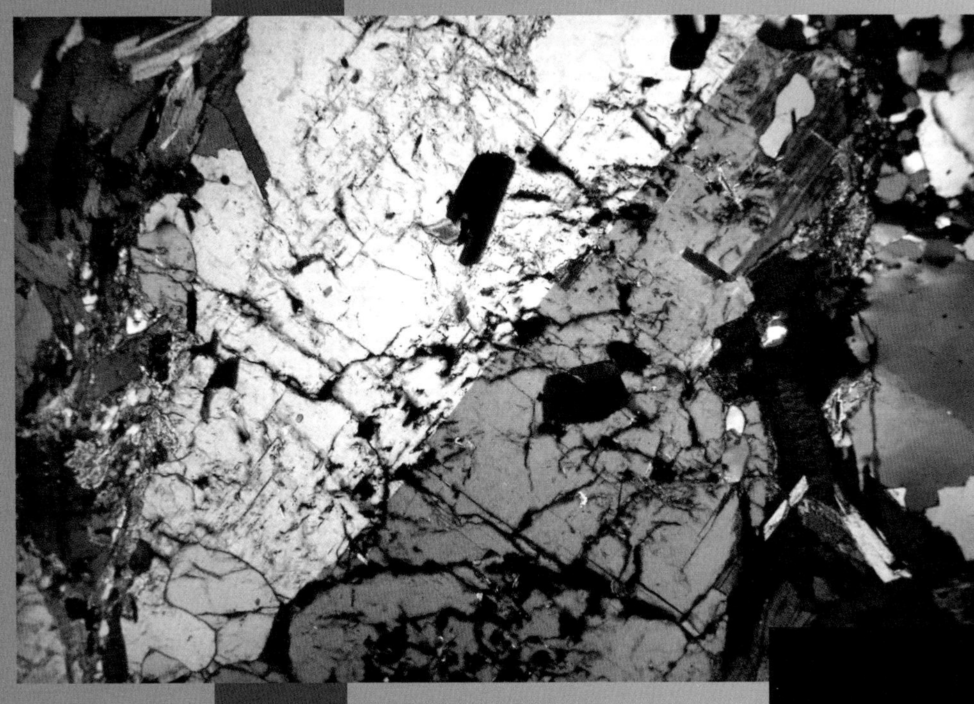

CHAPTER 2

FELDSPARS

Feldspars ($XAl_{(1-2)}Si_{(2-3)}O_8$) are the most common rock-forming minerals in the Earth's crust, and they occur in many varieties (Fig. 2.1) — ranging from sodium- and calcium-rich (plagioclase) to potassium-rich (K-feldspar or alkali feldspar). K-feldspars may also contain significant amounts of sodium in their crystal lattices. Feldspars are far less resistant than quartz to chemical and physical destruction and thus are altered or removed by weathering, transport and diagenesis (e.g., Land et al., 1987), yielding secondary pores or alteration products (illite, white mica/sericite, albite or kaolinite). Even so, they are the second most abundant grains in sandstones, and identifying their mineralogy is crucial for accurate sandstone classification and provenance studies.

Major characteristics:

Color: Typically clear to cloudy in thin section, sometimes with pinkish color.

Common crystal habit: Tabular, euhedral to anhedral crystals are common.

Pleochroism: None.

Cleavage/fracture: K-feldspars have two main cleavage planes at roughly 90° to each other. Plagioclase cleavages are 93–94°.

Relief and optic sign: K-feldspars are biaxial (-) and have low (-) relief. Plagioclases are biaxial, and the optic sign and 2V varies with composition. Plagioclase has moderate relief.

Extinction angles: Varies systematically with composition.

Birefringence: First-order gray and white in a 30-μm section, except for calcic compositions which may go up to first-order yellow.

Feldspar minerals:

Plagioclase – Plagioclase minerals are all triclinic. Albite twinning occurs only in triclinic feldspars (especially plagioclase), and Carlsbad twinning is also fairly common (Fig. 2.2). Both are parallel to the crystallographic axis. Pericline twins are less common and are normal to albite twins. In igneous rocks, plagioclase phenocrysts commonly display zoning (oscillatory, continuous or discontinuous) and are twinned (although untwinned volcanic plagioclase is quite common); metamorphic plagioclase tends to lack zoning and twinning.

Orthoclase – is a monoclinic alkali (K-rich) feldspar. It is often overlooked or misidentified in thin section. It has low relief, commonly is untwinned and cleavages may be hard to see; therefore, it is easily confused with quartz. If twinning is present, it is normally Carlsbad twinning. Orthoclase sometimes contains vacuoles.

Microcline – is also a potassium-rich feldspar, but it belongs to the triclinic crystal system. It is usually easy to identify because it has two directions of polysynthetic twins (a combination of albite and pericline twinning) that commonly is called tartan, grid, cross-hatched or microcline twinning.

Sanidine – is also a monoclinic alkali feldspar similar to orthoclase. Unlike orthoclase, it has a smaller 2V and is inclusion free. It is common in high-temperature acidic volcanic rocks.

Anorthoclase – is a triclinic Na-rich alkali feldspar that grades into microcline. It, too, can have grid twinning, but it can be distinguished from microcline by its smaller 2V. It is common in Na-rich volcanic rocks.

Minerals that may have similar appearance and distinguishing differences:

Quartz: Has similar birefringence, but no twinning or cleavage, and it is uniaxial (+), whereas all feldspars are biaxial.

Gypsum and barite: Both have birefringence similar to feldspars; barite has slightly higher relief and gypsum has lower relief; both gypsum and barite have strong cleavage.

Staining is a common method used to identify feldspars and to distinguish them from quartz. Plagioclase stains pink in a potassium rhodizonate solution, and K-feldspars stains yellow in a sodium cobaltinitrite solution. For provenance studies, where detailed compositional data are needed, the K-feldspars should not be stained, since staining alters the refractive index by partly dissolving the feldspars.

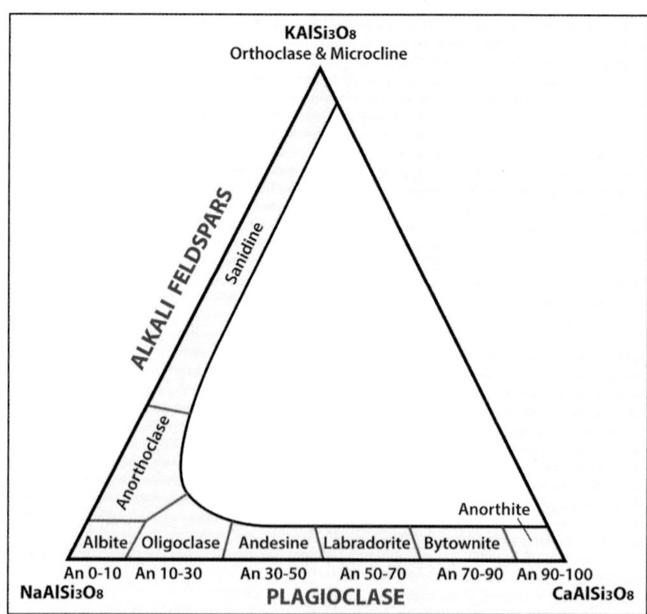

Fig. 2.1: Ternary feldspar compositional diagram.

Because twinning typically is still visible despite the stain, microcline can be recognized in a modal count and thus can yield some provenance information. In SEM, CL and on ion-milled surfaces, many of the features shown in this chapter can be recognized in silt-size feldspars (i.e., mineral inclusions, twinning, alteration, corrosion, overgrowths).

Provenance indicators:

Almost all detrital feldspars are igneous or metamorphic in origin, with orthoclase and microcline being the most common. Na-rich plagioclase, the next most common feldspar, is usually from volcanic rocks. Sanidine, from high-temperature felsic volcanic rocks, and Ca-rich plagioclase, from mafic to intermediate igneous rocks, are relatively uncommon.

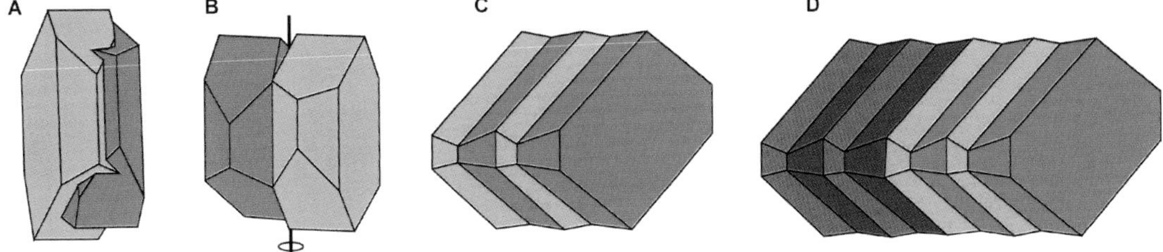

Fig. 2.2: Feldspar twinning: A) Interpenetrating Carlsbad twins, B) Carlsbad contact twins, C) Polysynthetic albite twins, and D) Carlsbad and albite twins (modified from Deer et al., 2001).

Mid. Eocene – Lo. Oligocene Spears Gp., Socorro Co., New Mexico

The grain on the right is a euhedral, complexly zoned plagioclase feldspar in a volcaniclastic sandstone. This type of zoning is referred to as "oscillatory" and represents abrupt (not continuous) changes in plagioclase composition during crystallization. Zoning is very common in plagioclase feldspars, and this type of oscillatory zoning normally occurs in crystals of volcanic origin. In addition to compositional zoning, the other plagioclase grains in this sample exhibit both Carlsbad and albite twinning.

XPL, Scale bar = 0.14 mm

Mid. Permian Wegener Halvø Fm., Karstryggen area, East Greenland

This plagioclase feldspar (derived from a volcanic source) shows zonation that is marked by vacuoles and melt inclusions. No compositional changes are visible in this phenocryst, but changes in the crystal growth caused melt inclusions to be formed.

PPL | XPL, Scale bar = 0.33 mm

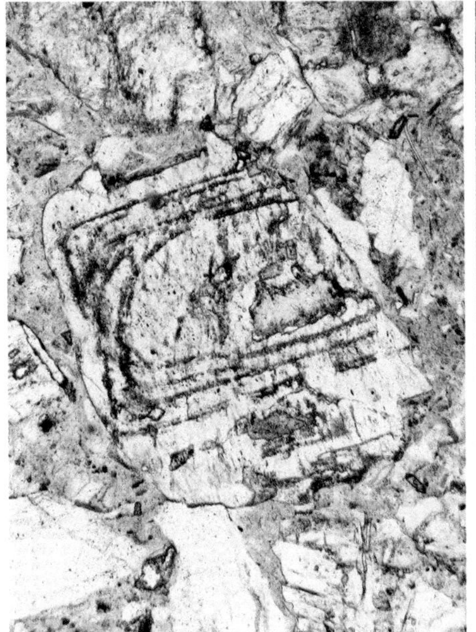

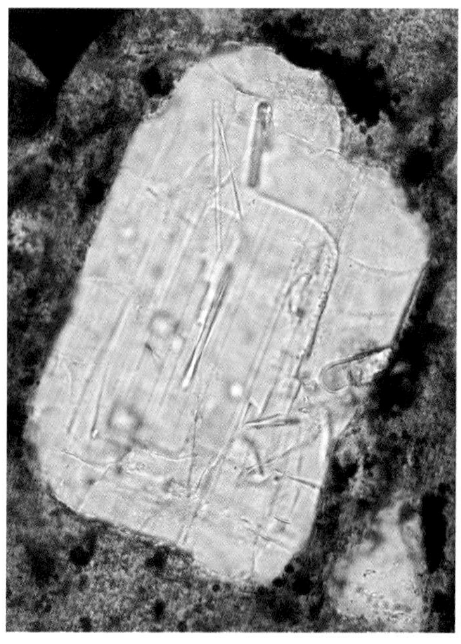

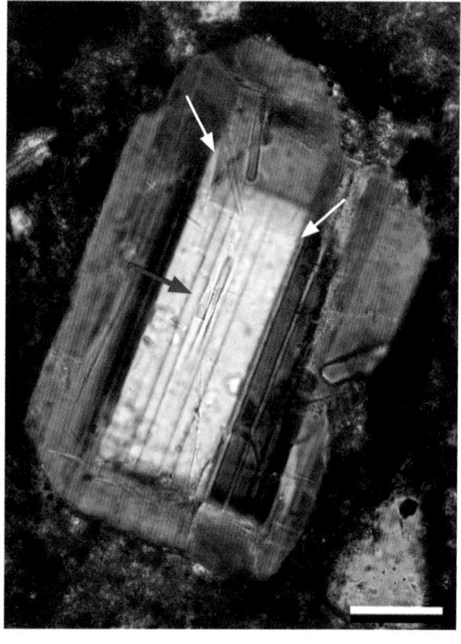

Cretaceous Mowry Fm., Moffat Co., Colorado

A zoned plagioclase crystal in a litharenite. This grain of plagioclase exhibits both zoning and twinning. The zoning is discontinuous—there is a distinct break in the composition of the core relative to the rim. The long, thin, linear twins are polysynthetic albite twins (red arrow shows one of several) and are overlain by broader Carlsbad twins (white arrows).

PPL | XPL, Scale bar = 0.06 mm

Lo. Miocene Harmony Hills Tuff, Lincoln Co., Nevada

A compositionally zoned plagioclase feldspar that contains mineral inclusions of quartz and compositionally different plagioclase. The large grain is a complexly zoned phenocryst with oscillatory and discontinuous zoning. Notice how the zoning and crystal growth were affected by the feldspar inclusion (white arrow). The inclusion produced an hourglass-shaped crystal that was ultimately encased by later crystal growth. Mineral inclusions are not common in plagioclase crystals, but they can occur.

XPL, KFS, Scale bar = 0.51 mm

Mid. Eocene – Lo. Oligocene Spears Gp., Catron Co., New Mexico

Plagioclase crystals in a volcanic rock fragment from a coarse conglomerate. The two large, intersecting crystals are contact (or penetration) twins. All of the feldspar crystals are euhedral and contain oscillatory zoning, and many of the feldspars in this clast also contain albite and Carlsbad twins.

XPL, Scale bar = 0.26 mm

Pliocene Verde Fm., Yavapai Co., Arizona

A volcanic plagioclase (center) in a carbonate-rich matrix. The grain shows excellent albite and Carlsbad twinning. The polysynthetic albite twins are the long, thin, light and dark planes within the crystal. The Carlsbad (also known as simple) twin lies between the light area and the mostly dark area of the crystal (red arrow). There may be pericline twinning at right angles to the albite twinning (in the bright side of the crystal). Zoning is also present—a combination of continuous (where crystal composition changes gradually) and discontinuous (abrupt) changes.

XPL, Scale bar = 0.10 mm

Neogene (Miocene?) Horse Spring Fm., Clark Co., Nevada

Plagioclase crystals in a volcaniclastic sandstone. The plagioclase crystals exhibit euhedral crystal outlines and are twinned. The larger crystals have both excellent albite and Carlsbad twinning. Some faint zoning is also present.

XPL, Scale bar = 0.38 mm

Paleocene Tang Fm., Møre Basin, Norwegian Sea †

A plagioclase grain that shows an embayed margin. In this case, it is not likely to be related to corrosion by changes in magma composition, but more likely resulted from dissolution of a mineral intergrowth. The original inclusions and/or intergrowths were, in part, formed of euhedral crystals, but the original mineral that constituted them is unknown.

XPL, KFS, BDI, Scale bar = 0.07 mm

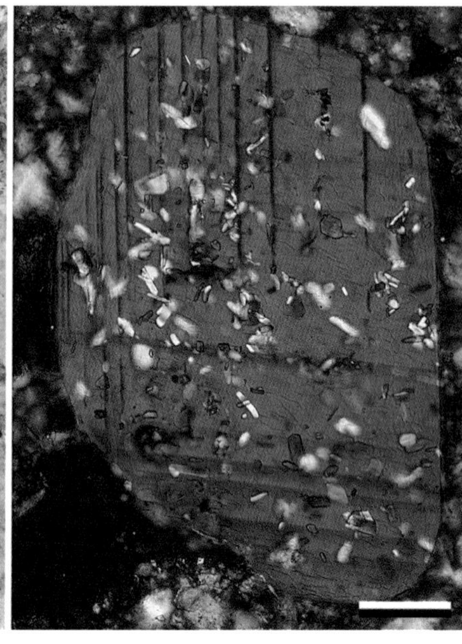

Mid. Jurassic Fulmar Fm., North Sea, United Kingdom †

A plagioclase feldspar that displays two directions of twinning (albite and pericline). Without staining of K-feldspar grains within the sample, this may be superficially confused with microcline. The twins seen in this grain are growth twins rather than deformation twins as they terminate abruptly and generally maintain their width.

PPL | XPL, AS, BDI, Scale bar = 0.07 mm

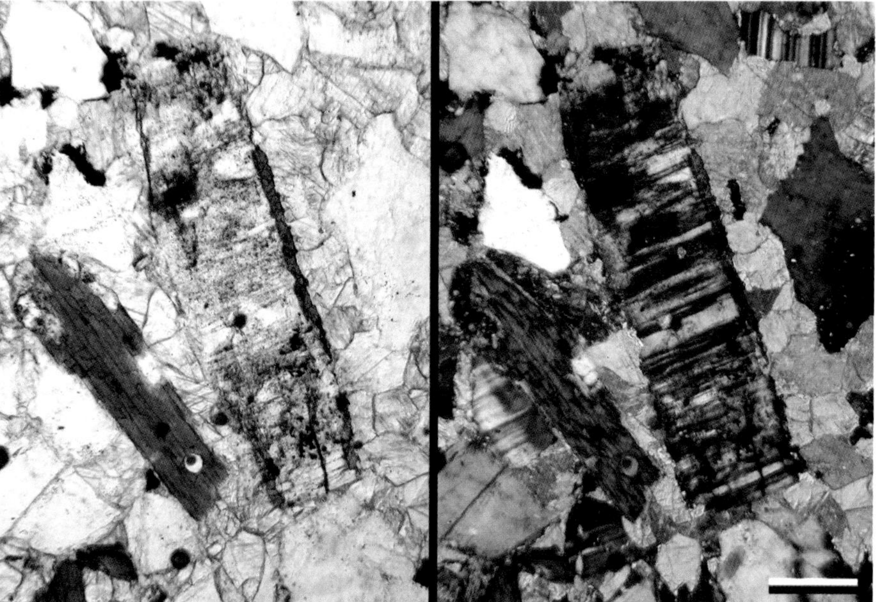

Lo. Miocene Arikaree Fm., Platte Co., Wyoming

An altered plagioclase crystal in a volcaniclastic sandstone. Feldspars as a group are far more easily altered by chemical and mechanical processes than quartz, so it is common to see them degraded or even partially to completely dissolved in thin section. Thus, alteration commonly is a key factor in the recognition of feldspars. In this pair of photomicrographs, the central grain exhibits more alteration than the other nearby plagioclase grains. This may be a function of compositional variations. The nearby dull green grain is a pyroxene crystal.

PPL | XPL, Scale bar = 0.20 mm

Cenozoic monzonite intrusive, Clark Co., Nevada

These plagioclase feldspar crystals are still in their source rock (a monzonite) but the crystals exhibit extensive *in-situ* alteration to sericite (a term for fine-grained white mica). Sericite is one of the major minerals that replaces feldspars, but calcite replacement also frequently occurs. These crystals have both albite and Carlsbad twins, and despite the alteration, strong zoning is still visible.

XPL, Scale bar = 0.26 mm

Paleocene sandstone, Sri Lanka

A plagioclase grain displaying extensive patches of alteration to albite (red arrow points to one such area). The albite shows a lighter gray birefringence. Albitization is a common subsurface replacement process in sandstones (e.g., Land et al., 1987; see also Chapter 12 on feldspar cementation as well Chapters 19 and 20 on dissolution and replacement and the citations at the ends of those chapters).

XPL, KFS, BDI, Scale bar = 0.12 mm

Lo. Permian Cutler Fm., Mesa Co., Colorado

A highly altered plagioclase grain in a lithic-rich, arkosic sandstone. This plagioclase fragment has undergone alteration that includes vacuolization and illitization (the small but brightly birefringent crystals). Despite the alteration, the albite twins are still clearly visible in cross-polarized light. Highly altered grains like this are easily distinguishable from quartz grains, but it is important to try to determine if the grains have undergone alteration *in situ* or were already altered prior to deposition.

PPL | XPL, Scale bar = 0.16 mm

Mid. Ordovician Newtown Gneiss, New Haven Co., Connecticut

A complex plagioclase grain with a core and outer rim of different compositions. Albite twinning extends throughout the grain, but the core shows considerable alteration (mainly vacuolization and illitization) whereas the rim is largely unaltered. This crystal clearly illustrates that alteration can take place in source rocks and can be highly compositionally dependent. In a sandstone, the clear outer rim of the feldspar might well be interpreted as an authigenic overgrowth, but because this crystal is still in its metamorphic source, the rim clearly is a primary feature.

XPL, PFS, Scale bar = 0.09 mm

Lo. Cretaceous Patula Arkose, Coahuila, Mexico

Highly altered, twinned plagioclase feldspars in an arkosic (feldspathic) sandstone. The grains have been extensively altered by vacuolization and sericitization. The alteration partially obliterates the twinning. The small feldspar grain in the center and others in the thin section are largely unaffected by alteration. This probably is related to compositional variations, but it could indicate that the alteration occurred in the source terrane and affected only some of the grains.

XPL, Scale bar = 0.26 mm

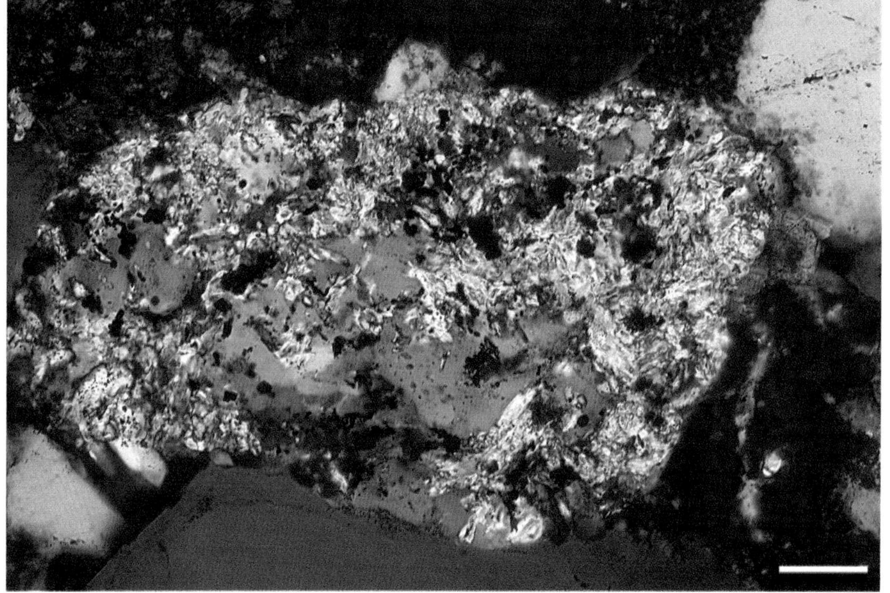

Up. Cretaceous (Campanian) Nise Fm., Møre Basin, Norwegian Sea †

A grain of ambiguous origin that may be a lithic grain or a plagioclase containing mineral inclusions of quartz and K-feldspar. The plagioclase has been altered to white mica 'sericite', but in this case it is relatively coarser than normal, suggesting that the feldspar may have been hydrothermally altered prior to deposition as a grain.

XPL, KFS, BDI, Scale bar = 0.07 mm

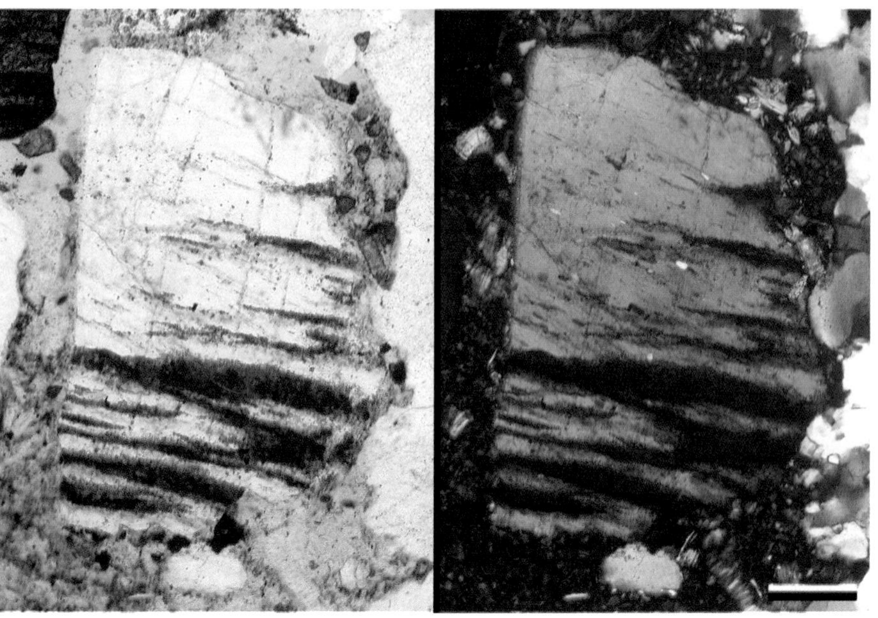

Mid. Jurassic Ness Fm., Norwegian sector, North Sea †

An antiperthite that consists of an albite grain with K-feldspar exsolution lamellae. The irregular lamellar intergrowths formed in the crystal as a consequence of reduction of temperatures below the solvus during crystallizaton, causing the solid solution series to break down.

PPL | XPL, KFS, BDI, Scale bar = 0.12 mm

Up. Pennsylvanian Bursum Fm., Socorro Co., New Mexico

Nearby erosion of Precambrian granites brought to the surface by block faulting produced this grain. The central grain appears to be composed of two different feldspars, but it is entirely albite with albite twinning clearly visible in the colorless part of the grain. This sample shows how difficult feldspar identification can be. The fact that the entire grain is albite is based on quantitative geochemical analysis (provided by FEI) and is supported by detailed petrographic observations. The alteration may reflect variations in the grain chemistry or albitization of a precursor feldspar.

PPL | XPL, Scale bar = 0.26 mm

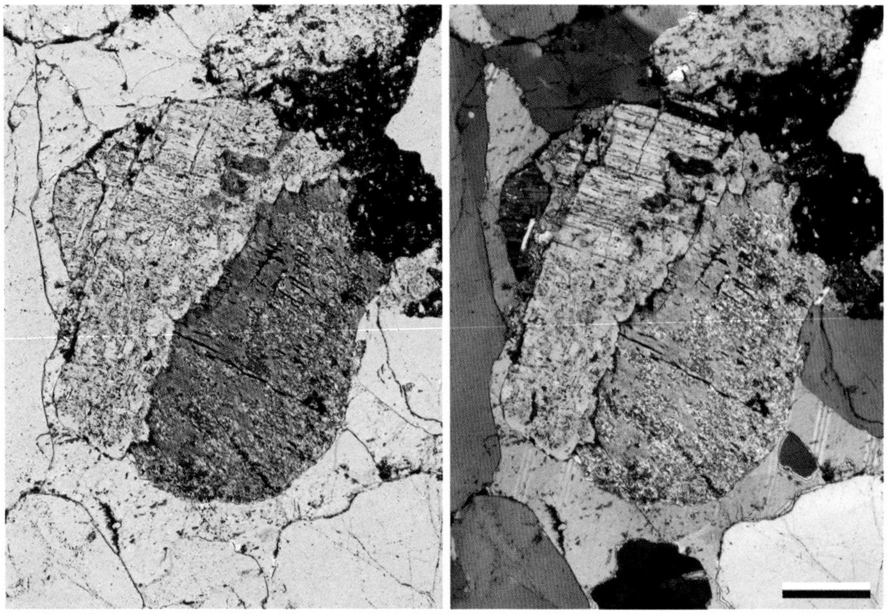

Up. Pennsylvanian Bursum Fm., Socorro Co., New Mexico

Another example from the Bursum Formation of an albite grain. This grain has well-developed cleavage and contains abundant mineral inclusions of quartz and precursor albites. In addition, it has a dusty brown appearance due to abundant vacuoles. The lack of twinning might suggest orthoclase, but standard petrographic techniques (optic sign and 2V) should be utilized to double check initial conclusions, and microprobe or other quantitative analytical techniques (or inexpensive staining) can be further employed to characterize the sample.

PPL | XPL, Scale bar = 0.34 mm

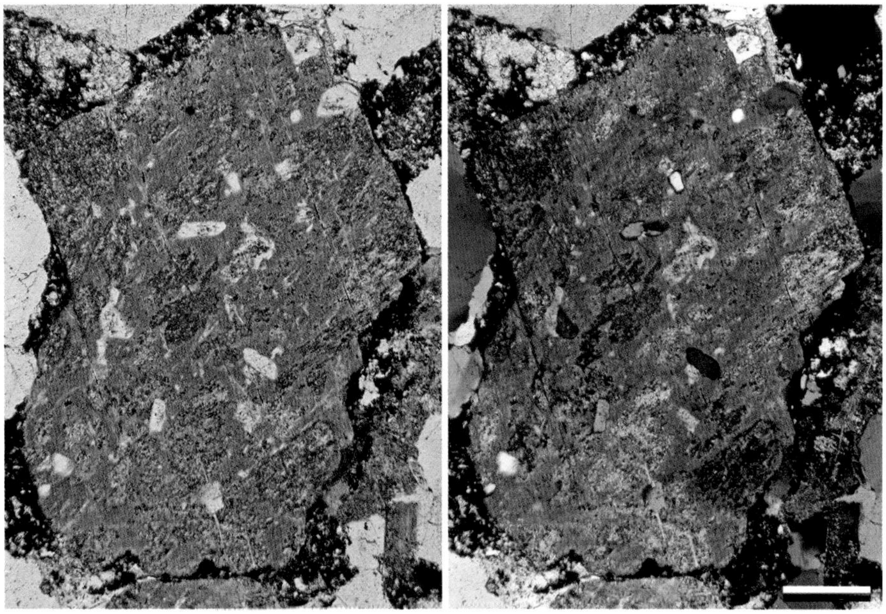

Up. Pennsylvanian Bursum Fm., Socorro Co., New Mexico

This albite grain contains exsolution lamellae. Such lamellae form during cooling of a grain below the stability field of the grain's composition. As this occurs, the mineral exsolves to form two minerals. This fabric is also called perthite. This grain is currently all albite, but originally the grain probably started out as K-spar (brown) and albite (white; one example indicated by yellow arrow). During the later conversion of K-spar to albite, vugs or pores formed that were later filled in by calcite spar cements (the brightly birefringent mineral in cross-polarized light).

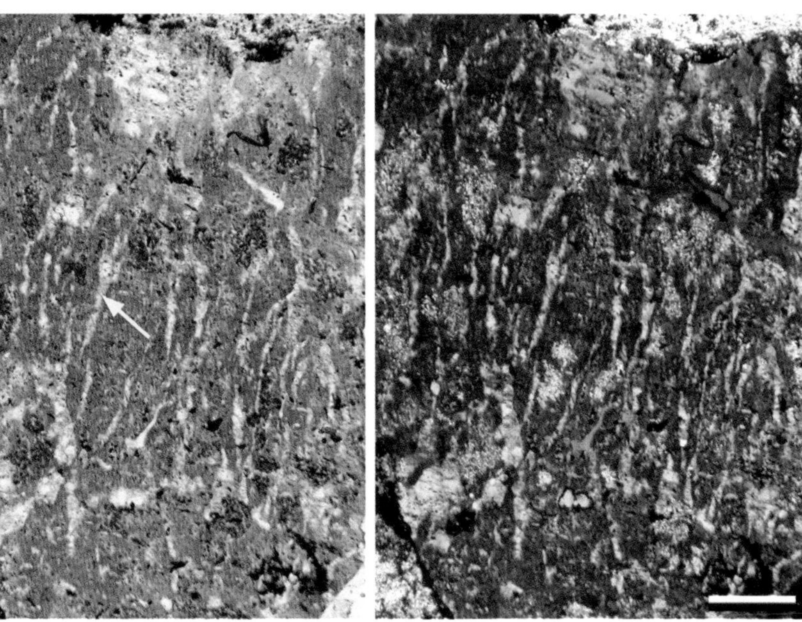

PPL | XPL, Scale bar = 0.26 mm

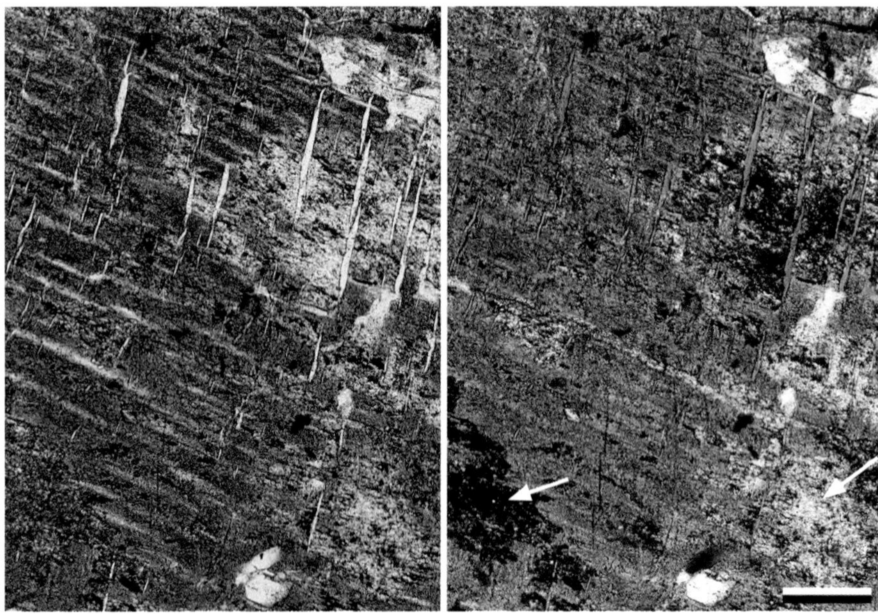

Up. Pennsylvanian Bursum Fm., Socorro Co., New Mexico

A closeup view of lamellar fabric (perthite) exhibiting two directions of exsolution. As in the previous example, the original K-feldspar grain was albitized during mesogenetic diagenesis, and parts of the grain were infilled or replaced by calcite (arrows point to calcitic areas that appear either dark or bright in cross-polarized light depending on extinction position). Calcitization is a common diagenetic alteration product in feldspars and is discussed more fully in Chapter 20. The red coloration in this and the two previous feldspars probably comes from minute hematite inclusions (Putris et al., 2007).

PPL | XPL, Scale bar = 0.10 mm

Up. Triassic – Lo. Jurassic Huizachal Fm., Tamaulipas, Mexico

A perthitic plagioclase feldspar grain with exsolution lamellae of albite (dark gray first-order birefringence) and anorthite (yellow to orange first-order birefringence; example shown by red arrow). This sandstone has been cemented by illitic clays (example shown by white arrow) that have birefringence colors similar to those of the anorthite and by calcite that has very high birefringence (as shown by the yellow arrow).

XPL, Scale bar = 0.05 mm

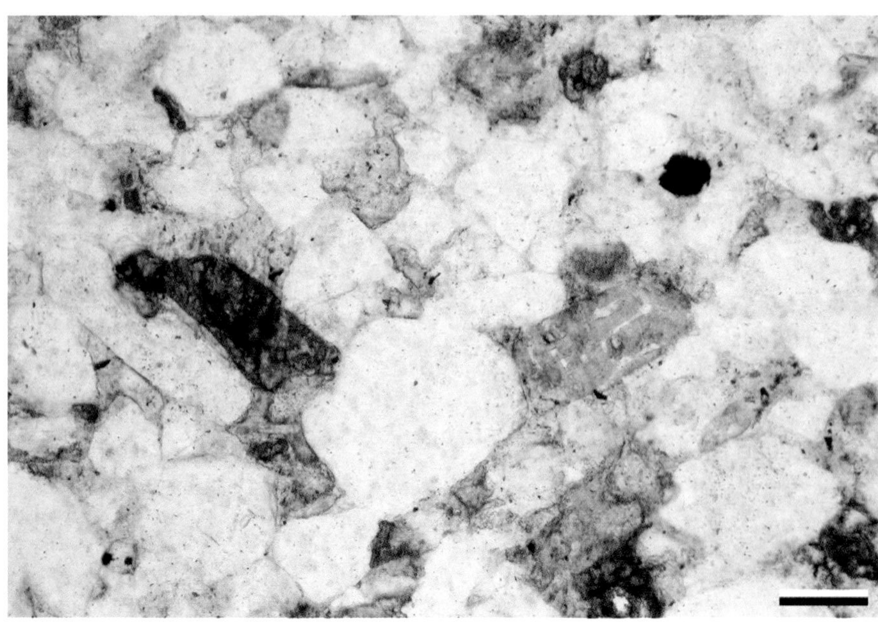

Mid. Jurassic Entrada Fm., Moffat Co., Colorado

Staining can help in the identification of feldspars in thin section as well as in hand specimen (Houghton, 1980). This thin section has been stained for both plagioclase and potassium feldspars. Plagioclase stains pink, and the more calcium in the plagioclase crystal, the deeper pink the stain. The potassium feldspars turn a greenish to brownish yellow as seen in this sample. Albite, a feldspar with intermediate composition, may not pick up a stain or be very lightly stained. The clear white (unstained) grains in this sample are quartz.

PPL, KFS, PFS, BDI, Scale bar = 0.10 mm

Modern sediment, Llano Uplift, Texas

This cathodoluminescent image shows multiple feldspar grains from creek sediments derived from the Town Mountain Granite of the Llano Uplift. Most of the feldspar grains are albite and show considerable variation in color and texture, probably related to their position within the granitic pluton. The K-spar/ microcline grain has considerable brighter CL than the albite, a characteristic distinction between the two feldspar types.

CCL, Scale bar = 24.9 μm

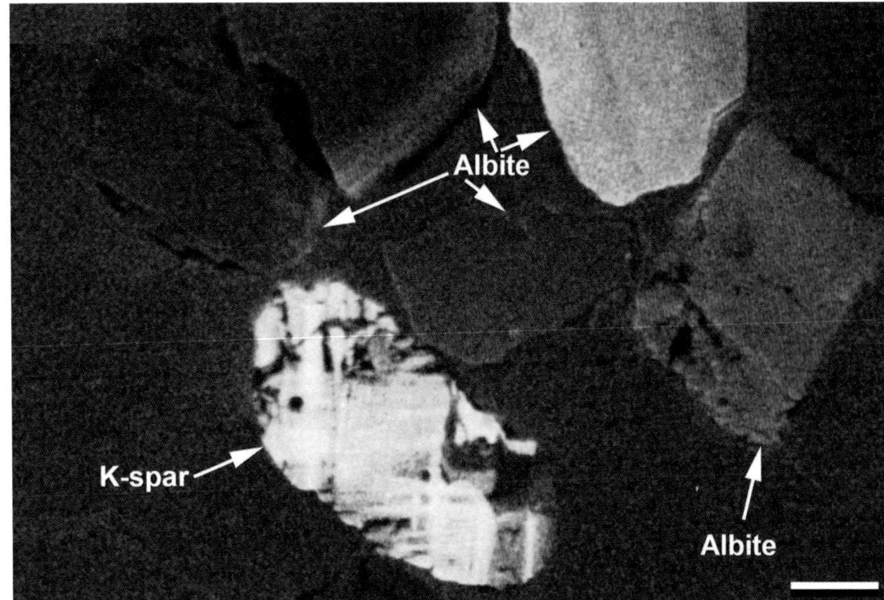

Eocene Cub Mountain Fm., Lincoln Co., New Mexico

Feldspars are inherently less stable than quartz during burial. In this photomicrograph, a plagioclase grain was preferentially dissolved along the crystal's twin planes. The dissolution has created secondary intragranular porosity. Diagenetic alteration or partial to complete dissolution of feldspars is so common that it can substantially change the composition of sandstones. For proper evaluation of original sediment composition, it is important to recognize altered feldspars, even ones far more completely altered than this one (as in the next image).

PPL, BDI, Scale bar = 0.10 mm

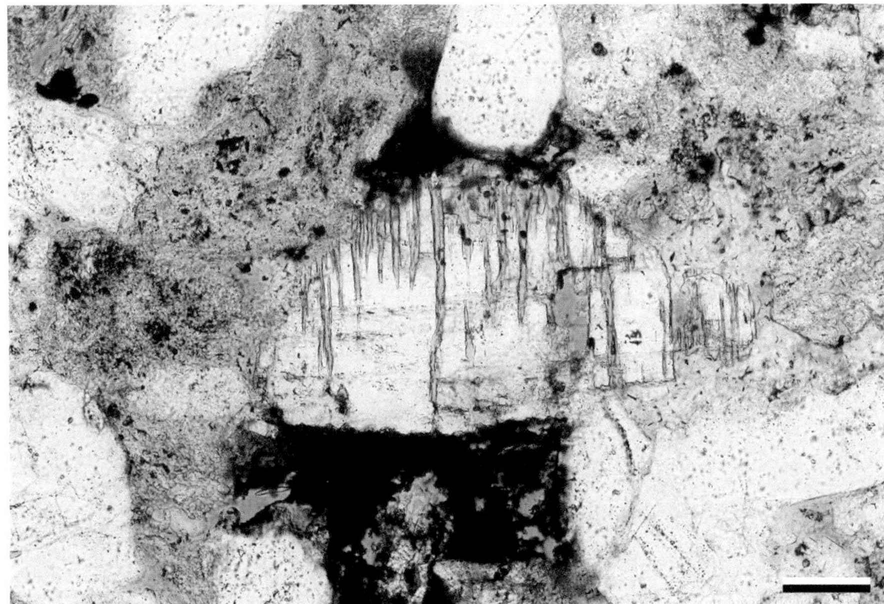

Mid. Jurassic Entrada Fm., Moffat Co., Colorado

In this sample, plagioclase grains were almost completely dissolved. The grain in the center of this view contains remnants of plagioclase along twinning planes and crystallographic axes, indicating the influence both can play on the selective dissolution of feldspars during diagenesis. Secondary porosity created by feldspar dissolution is an important component in numerous sandstone reservoirs in the Gulf Coast, the North Sea and other petroleum provinces.

PPL, PFS, KFS, BDI, Scale bar = 0.05 mm

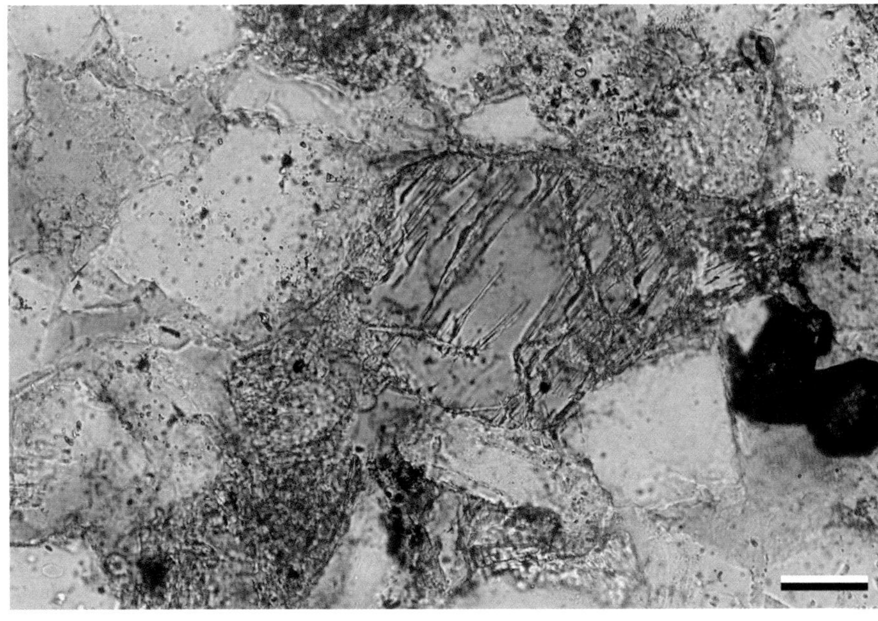

Up. Jurassic Spekk Fm., Møre Basin, Norwegian Sea †

A plagioclase grain in which there are no twins visible. The recognition as a feldspar in the absence of twins is through the cleavage and shape of the grain. The sample was stained for K-feldspars and so the grain is known to be plagioclase, without examining other optical properties. The rock is largely cemented by ankerite (the blue-stained cements).

XPL, KFS, AFeS, Scale bar = 0.07 mm

Triassic Snadd Fm., Barents Sea, offshore Norway †

A microcline grain that only took a weak stain and so displays twinning quite clearly. In this example the twins taper, rather than terminating abruptly and are therefore deformation twins. The grain is also partly perthitic and shows exsolved lamellae of albite, which are more irregular than the twins.

XPL, KFS, BDI, Scale bar = 0.07 mm

Mid. Ordovician Newtown Gneiss, New Haven Co., Connecticut

Microcline feldspars with typical microcline, grid or tartan twinning. Although such twinning is characteristic of most alkali feldspars, it is most commonly shown by microcline and is a combination of albite and pericline twins. Some small inclusions of plagioclase also are present.

XPL, PFS, Scale bar = 0.16 mm

Lo. Oligocene Nile Gp., Kongahu Breccia, Westland, New Zealand

Detrital microcline with both tartan and Carlsbad twinning. Microcline is a potassium feldspar that normally occurs in anhedral crystal masses in plutonic and gneissic source rocks — they rarely exhibit crystal terminations. This large microcline grain has plagioclase crystal inclusions that may be a calcic plagioclase based on the higher birefringence of those inclusions.

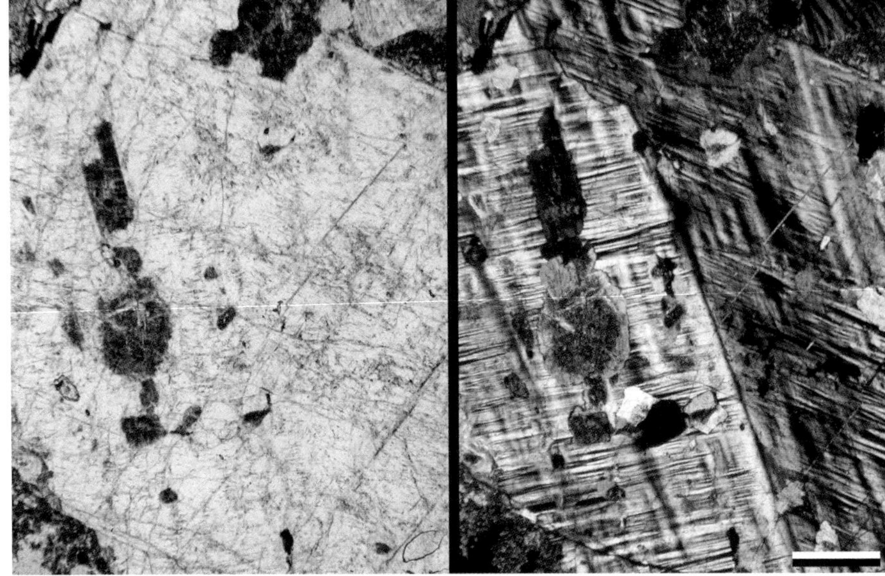

PPL | XPL, Scale bar = 0.64 mm

Lo. Miocene Arikaree Fm., Platte Co., Wyoming

The microcline grain in this photomicrograph has microperthite intergrowths. The microcline or grid twinning is interrupted by a more brightly birefringent (orange) mineral. It may be anorthite because calcium-rich plagioclase has higher birefringence.

XPL, Scale bar = 0.16 mm

Cambrian Hitchock Lake Schist, New Haven Co., Connecticut

The yellowish grain (stained) in center is a microcline feldspar with spindle twinning—the irregular microcline twins can often be used to distinguish microcline. The brown, elongate grain directly to the right of the microcline is biotite.

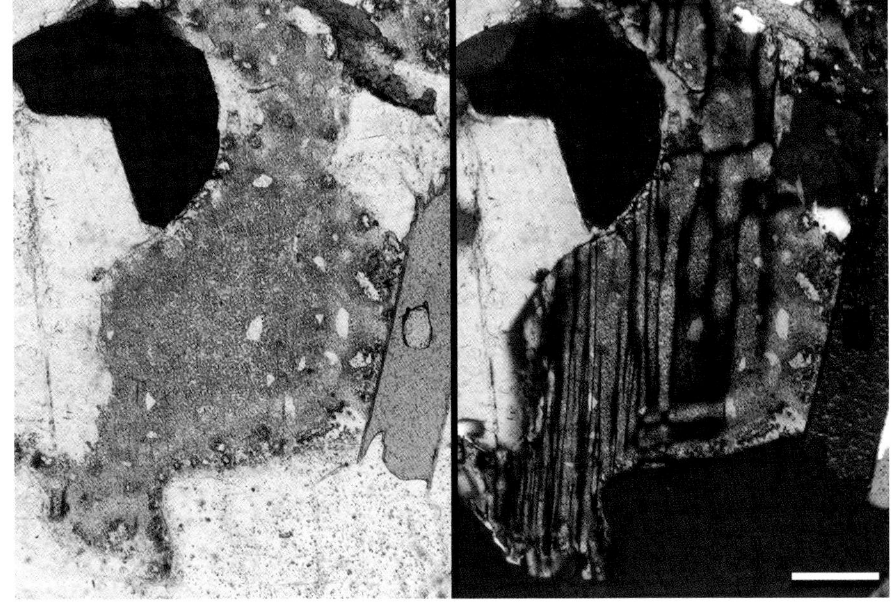

PPL | XPL, KFS, Scale bar = 0.26 mm

Up. Pennsylvanian Alamitos Fm., San Miguel Co., New Mexico

A microcline grain with plagioclase inclusions. The cores of the plagioclase crystals are more altered than the rims, indicating that they are compositionally zoned. The twinning in the microcline is not well developed, possibly related to how the grain was cut or more likely it may be an intergrowth between potassium- and sodium-rich feldspars (mesoperthite).

XPL, Scale bar = 0.26 mm

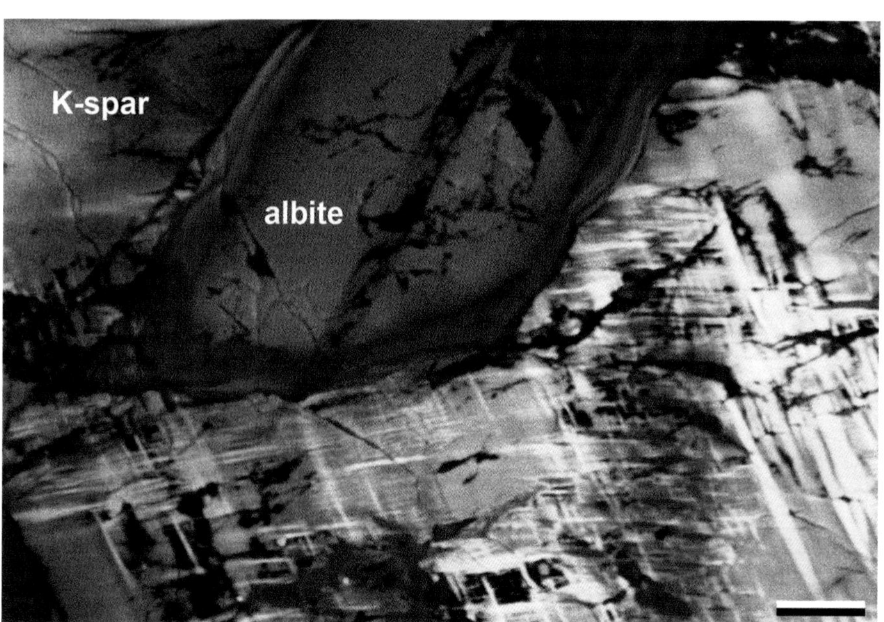

Precambrian Town Mountain Granite, Llano Uplift, Texas

The cathodoluminescence of feldspars can be highly informative with regard to provenance and formational processes. In this image, the K-feldspars (blue tones) have characteristic cross-hatch twinning and contain albite inclusions (purple tones). The reddish blotches and rims of the albite crystal as well as the black blotches and veinlets in the K-feldspar reflect hydrothermal alteration as the magma cooled.

CCL, Scale bar = 4.7 μm

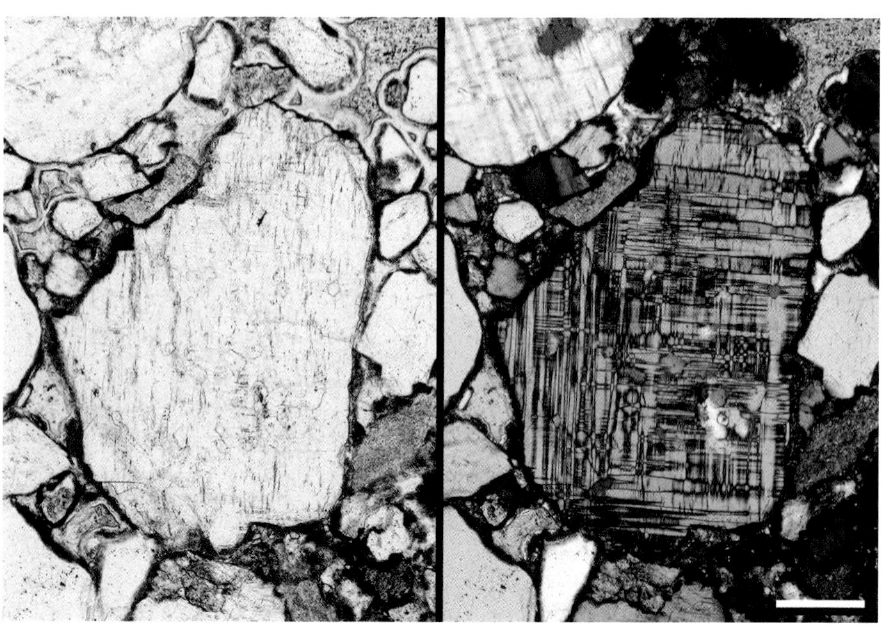

Mid. Eocene Baca Fm., Socorro Co., New Mexico

Anorthoclase contains more sodium than microcline or orthoclase and commonly has two directions of polysynthetic twinning (albite and pericline), forming a distinctive tartan or grid pattern. Although similar to microcline twinning, the twins in anorthoclase have especially thin lamellae.

PPL | XPL, Scale bar = 0.26 mm

Devonian Old Red Sandstone Gp., Dumbarton, Scotland, U.K.

Orthoclase has very low relief and two cleavages. It is biaxial negative with a 2V commonly between 35° to 85°. It lacks the distinctive grid twinning of microcline or the polysynthetic albite twinning of plagioclase, but orthoclase may have Carlsbad twins, as in this example.

XPL, Scale bar = 0.08 mm

Lo. Miocene Arikaree Fm., Platte Co., Wyoming

Orthoclase's low relief can make cleavage recognition difficult, but not impossible. Orthoclase can be easily confused with quartz, but determining if it is biaxial (feldspars) or uniaxial (quartz) is a definitive way to ascertain its identity. Where the grains are in contact with a mounting-medium-filled pore, relative relief can be used for identification (quartz having slightly higher relief than orthoclase). However, staining generally is the most reliable and cost-effective way to distinguish orthoclase from quartz.

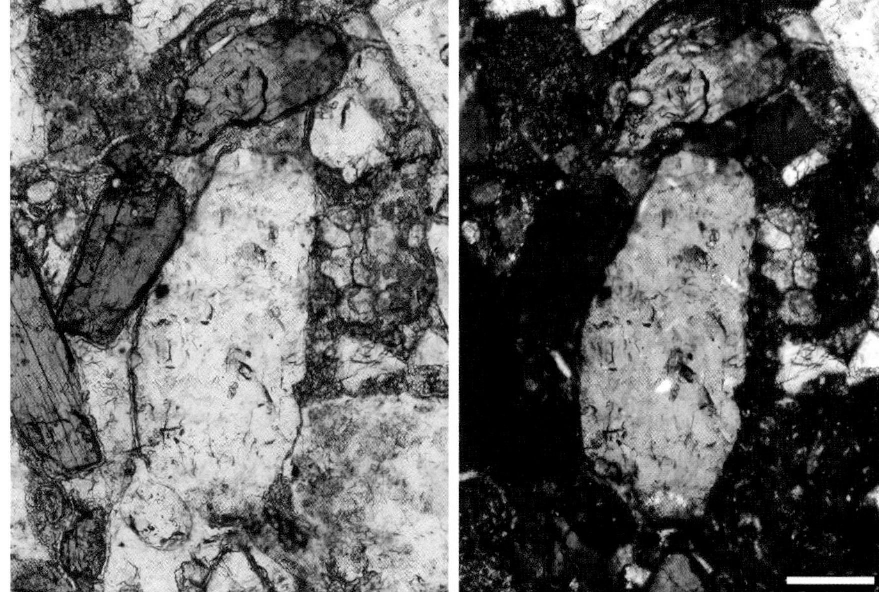

PPL | XPL, Scale bar = 0.26 mm

Up. Cretaceous Mesaverde Gp., McKinley Co., New Mexico

This orthoclase grain was probably intergrown with another feldspar—a feature called perthite. The partial dissolution of the intergrowths (albite?) during diagenesis formed secondary intragranular porosity. Because of the perthite structure within the grain, orthoclase is easy to distinguish from quartz (example at top center).

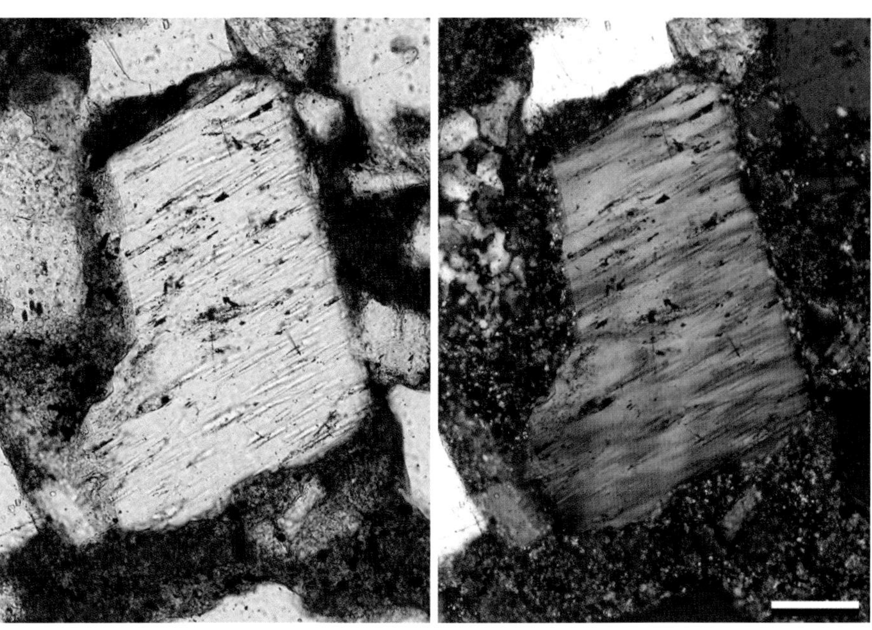

PPL | XPL, BDI, Scale bar = 0.08 mm

Up. Cretaceous Prince Creek Fm., North Slope, Alaska

Orthoclase grains in a chert- and quartz-rich sandstone. The orthoclase grains are cloudy because they contain abundant vacuoles related to alteration. The orthoclase feldspars are surrounded by euhedral K-feldspar overgrowths. These overgrowths appear clearer because they contain fewer inclusions.

PPL | XPL, BDI, Scale bar = 0.04 mm

Mid. Permian Wegener Halvø Fm., Karstryggen area, East Greenland

An orthoclase grain with numerous quartz inclusions within it (yellow arrow points to one of many such inclusions). Unlike quartz, feldspars are structurally less competent and, during compaction, they can break along cleavage planes. In this grain, the quartz inclusions show no sign of breakage; whereas, the orthoclase feldspar is shattered.

XPL, Scale bar = 0.46 mm

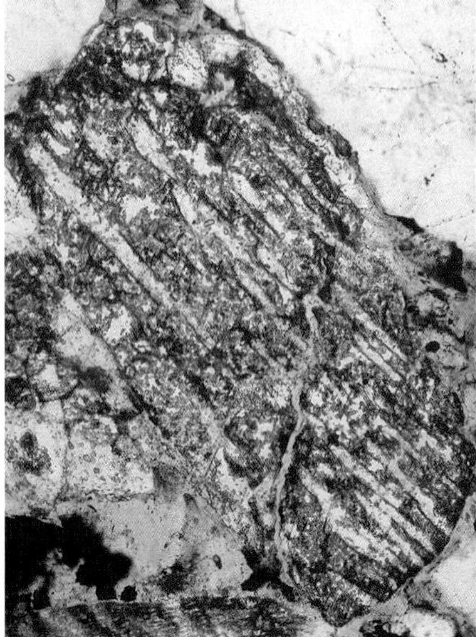

Mid. Jurassic Ness Fm., Norwegian sector, North Sea †

A grain of perthite containing elongate lamellae and blebs of exsolved albite within K-feldspar. The albite lamellae were not stained by the sodium cobaltinitrite.

PPL | XPL, KFS, BDI, Scale bar = 0.12 mm

Modern sediment, Llano Uplift, Texas

Small creeks that carry sediment from the Town Mountain Granite complex contain abundant weathered feldspathic debris. This pair of BSE and CL images shows a fine-sand-size K-feldspar grain that is clearly identifiable as a microcline and has albite exsolution lamellae. Even silt-size grains will retain this textural imprint, and thus be traceable to their source.

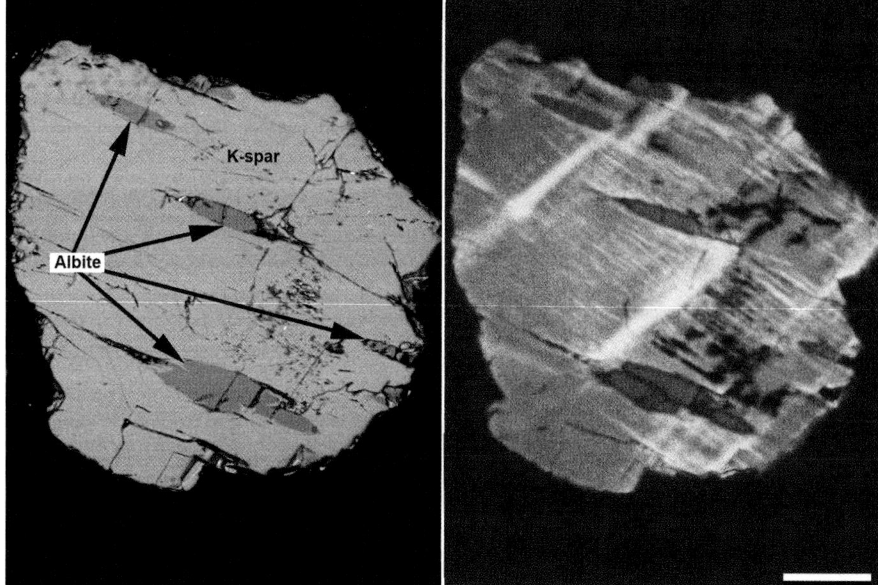

BSE(SEM) | CCL, Scale bar = 14.2 μm

Up. Cretaceous (Coniacian) Blålange Fm., Møre Basin, Norwegian Sea †

A perthite grain that has had albite lamellae preferentially dissolved. This dissolution has created secondary intragranular pore space. The shape of the lamellae appears to have been tapering and wavy, similar to 'flame perthite'. The dissolution appears to have postdated the main period of compaction, because the partly dissolved grain is unaffected at grain-to-grain contacts.

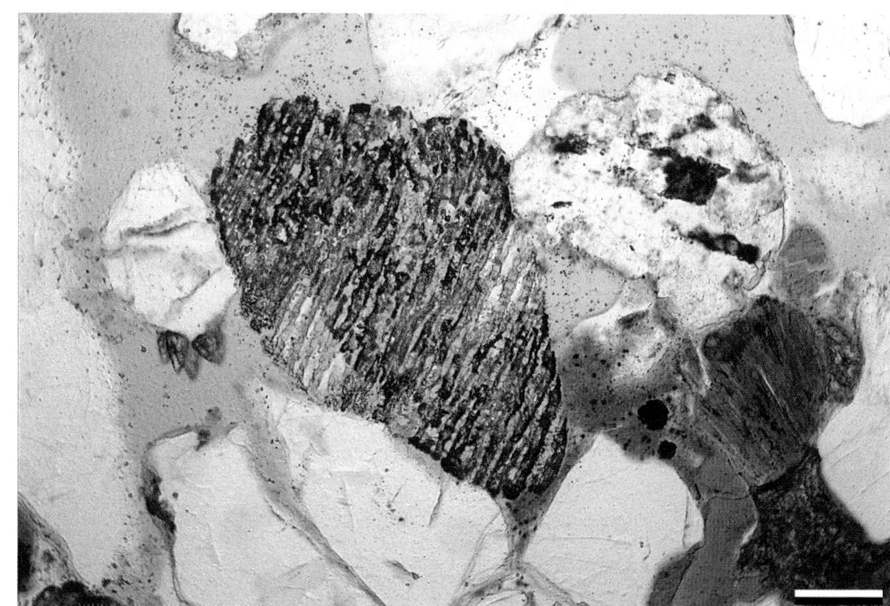

PPL, KFS, BDI, Scale bar = 0.12 mm

Cenozoic ignimbrite, Nye Co., Nevada

A sanidine crystal in a weakly welded volcanic tuff. The slide is stained for alkali feldspars, so the feldspar phenocrysts and some of the groundmass picks up the yellow stain. Sanidine is a high-temperature mineral commonly found in volcanic rocks that are rich in silica and contain variable amounts of sodium and potassium (rhyolites, rhyodacites, phonolites and trachytes); in addition, it is sometimes, but not commonly, found in contact metamorphic rocks. Carlsbad twins are common in sanidine, as shown here.

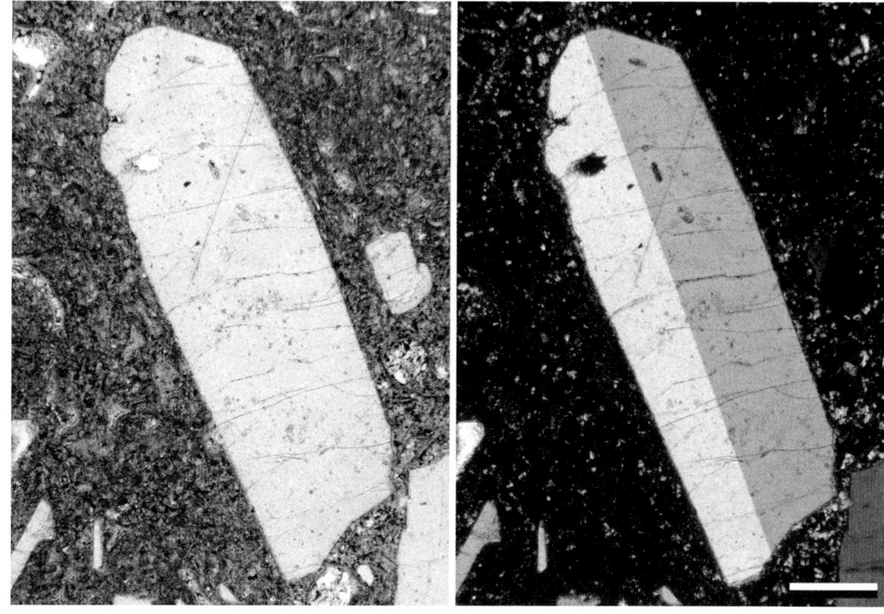

PPL | XPL, KFS, Scale bar = 0.51 mm

Cenozoic ignimbrite, Nye Co., Nevada

Sanidine can easily be confused with quartz, but it is biaxial (-) and has two directions of cleavage. It also can be distinguished from other alkali feldspars by its small 2V. The sanidine crystal in the upper right side of the photomicrograph is euhedral and zoned. Within the sanidine, there is a twinned crystal that has been partially resorbed prior to sanidine crystallization. This crystal is likely to be either anorthoclase (part of the solid solution series between albite and sanidine) or albite. Anorthoclase has grid twinning like microcline and occurs in high temperature sodium-rich melts.

PPL | XPL, KFS, Scale bar = 0.10 mm

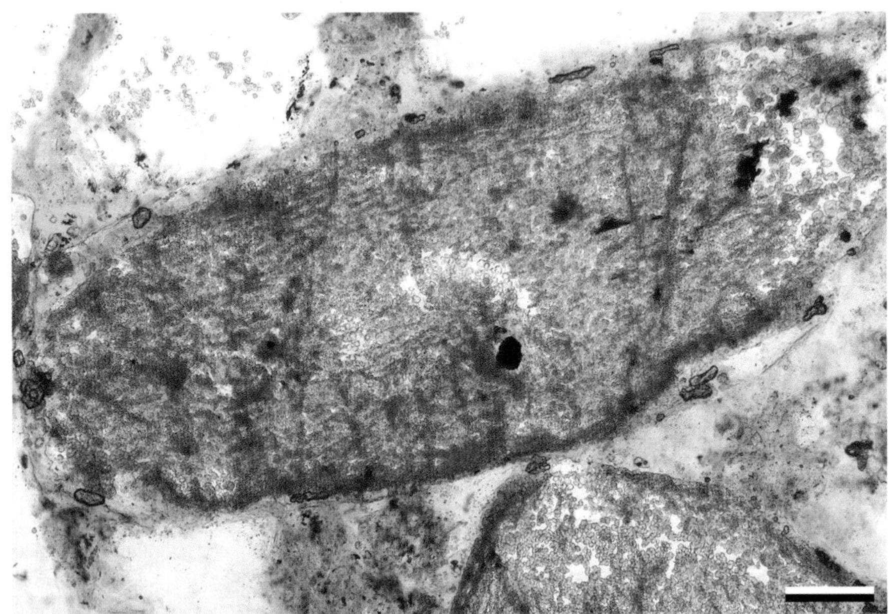

Mid. Jurassic Fulmar Fm., United Kingdom sector, North Sea †

A grain of K-feldspar (microcline or orthoclase based on twinning) that has taken a strong stain. The stain intensity may represent composition as the grain displays a thin overgrowth, which is a paler yellow. The overgrowth has also been subjected to partial dissolution.

PPL, KFS, AFeS, BDI, Scale bar = 0.07 mm

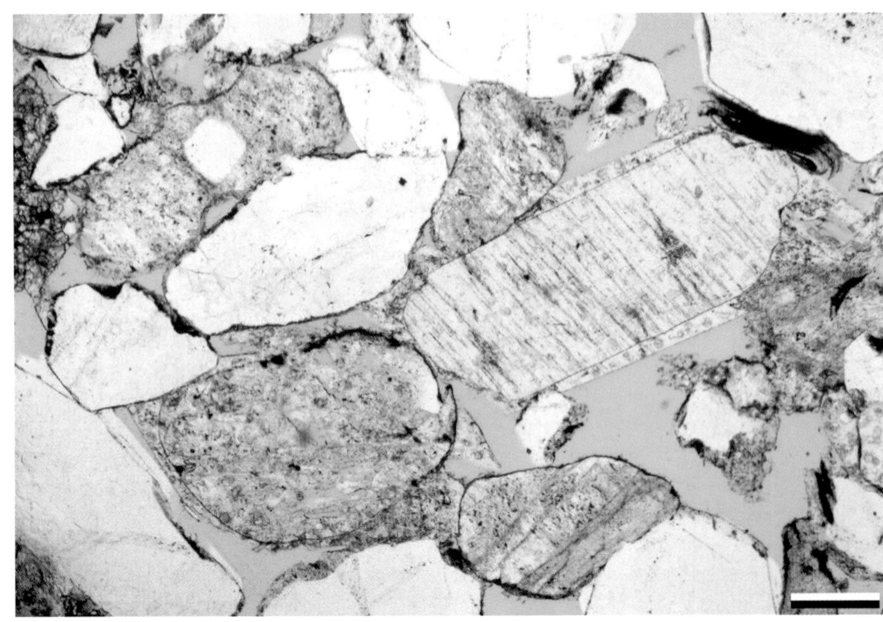

Triassic Sherwood Sandstone Gp., Co. Antrim, Northern Ireland, U.K. †

A view of K-feldspar grains (stained yellow) with well-developed diagenetic overgrowths. The larger grain exhibits a fine-scale perthite texture, which results from the presence of multiple exsolution lamellae of albite.

PPL, KFS, BDI, Scale bar = 0.17 mm

Lo.–Mid. Eocene Cub Mountain Fm., Lincoln Co., New Mexico

The dissolution of orthoclase grains was controlled by cleavage and fracture planes within the grains. Partial to complete dissolution of these grains has produced secondary intragranular porosity. The rounded pore (yellow arrow) is a grain mold produced by the complete dissolution of a feldspar or rock fragment. The sample was cemented by clays, quartz overgrowths, iron oxides and calcite.

PPL, KFS, BDI, Scale bar = 0.10 mm

Lo. Pennsylvanian Morrow B Sandstone, Ochiltree Co., Texas

This was likely a perthite grain of orthoclase and albite lamellae. During diagenesis, most of the orthoclase was dissolved leaving thin bands of inclusion-rich albite and creating secondary intragranular porosity. Later, the albite stringers were overgrown by inclusion-poor albite cements.

PPL, RDI, Scale bar = 0.51 mm

Mid. Jurassic Entrada Fm., Moffat Co., Colorado

A feldspar-stained arkosic sandstone. The central feldspar grain has a plagioclase core that is stained pink surrounded by gold colored (stained) orthoclase. Due to the pale pink nature of the stain, the core may be albite. Staining can help quickly tell the difference between plagioclase, K-feldspars and quartz. For large projects, staining can increase productivity and cost-effectiveness as well as reliability of the information.

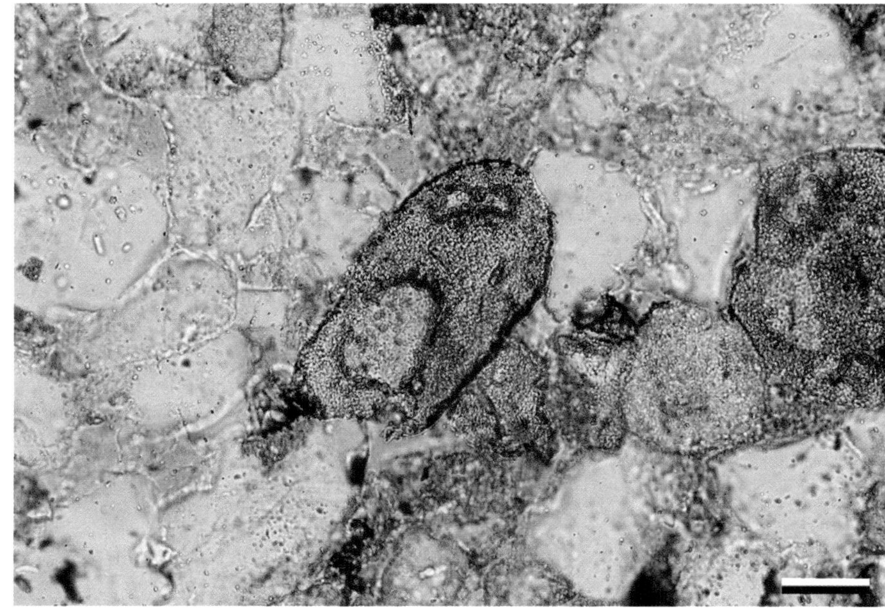

PPL, PFS, KFS, BDI, Scale bar = 0.05 mm

Cited References and Additional Information Sources

Barth, T. F. W., 1969, Feldspars: New York, Wiley-Interscience, 259 p.

Blatt, H., 1985, Provenance studies and mudrocks: Journal of Sedimentary Research, v. 55, p. 69-75, doi: 10.1306/212F8611-2B24-11D7-8648000102C1865D.

Brown, W. L., ed., 1984, Feldspars and Feldspathoids: Structures, Properties and Occurrences: Dordrecht, Netherlands, D. Reidel Publishing, NATO ASI Series C 137, 541 p.

Deer, W. A., R. A. Howie, and J. Zussman, 2001, Rock-Forming Minerals, Volume 4A, Framework Silicates: Feldspars: London, GSL, 972 p.

Dickinson, W. R., 1985, Interpreting provenance relations from detrital modes of sandstones, *in* G. G. Zuffa, ed., Provenance of Arenites (NATO Science Series C): Dordrecht, Netherlands, D. Reidel Publishing, p. 333-361, doi: 10.1007/978-94-017-2809-6_15.

Emmons, R. C., R. M. Gates, S. E. Clabaugh, R. M. Crump, K. B. Ketner, V. Mann, C. D. Reynolds, C. Bradley, and E. J. Lyons, eds., 1953, Selected Petrogenic Relationships of Plagioclase: New York, GSA Memoir 52, 142 p., doi: 10.1130/MEM52-pv.

Götze, J., M. R. Krbetschek, D. Habermann, and D. Wolf, 2000, High-resolution cathodoluminescence studies of feldspar minerals, *in* M. Pagel, V. Barbin, P. Blanc, and D. Ohnenstetter, eds., Cathodoluminescence in Geosciences: Berlin, Springer, p. 245-270, doi: 10.1007/978-3-662-04086-7_10.

Helmold, K. P., 1985, Provenance of feldspathic sandstones—the effect of diagenesis on provenance interpretations: A review, *in* G. G. Zuffa, ed., Provenance of Arenites (NATO Science Series C): Dordrecht, Netherlands, D. Reidel Publishing, p. 139-164, doi: 10.1007/978-94-017-2809-6_7.

Hersey, J. B., 1959, Optical properties of potassic feldspars: GSA Bulletin, v. 70, p. 511-538, doi: 10.1130/0016-7606(1959)70[511:OPOPF]2.0.CO;2.

Houghton, H. F., 1980, Refined technique for staining plagioclase and alkali feldspars in thin section: Journal of Sedimentary Research, v. 50, p. 629-631.

James, W. C., G. H. Mack, and L. J. Suttner, 1981, Relative alteration of microcline and sodic plagioclase in semi-arid and humid climates: Journal of Sedimentary Research, v. 51, p. 151-164, doi: 10.1306/212F7C34-2B24-11D7-8648000102C1865D.

Krynine, P. D., 1942, Provenance versus mineral stability as a controlling factor in the composition of sediments: GSA Bulletin, v. 53, p. 1850-1851.

Land, L. S., K. L. Milliken, and E. F. McBride, 1987, Diagenetic evolution of Cenozoic sandstones, Gulf of Mexico sedimentary basin: Sedimentary Geology, v. 50, p. 195-225, doi: 10.1016/0037-0738(87)90033-9.

Laves, F., 1950, The lattice and twinning of microcline and other potash feldspars: Journal of Geology, v. 58, p. 548-571, doi: 10.1086/625762.

Laves, F., 1952, Phase relations of the alkali feldspars: I. Introductory remarks: Journal of Geology, v. 60, p. 436-450, doi: 10.1086/625996.

Le Pera, E., J. Arribas, S. Critelli, and A. Tortosa, 2001, The effects of source rocks and chemical weathering on the petrogenesis of siliciclastic sand from the Neto River (Calabria, Italy): Implications for provenance studies: Sedimentology, v. 48, p. 357-378, doi: 10.1046/j.1365-3091.2001.00368.x.

Lee, M. R., and I. Parsons, 1998, Microtextural controls of diagenetic alteration of detrital alkali feldspars: A case study of the Shap conglomerate (Lower Carboniferous), northwest England: Journal of Sedimentary Research, v. 68, p. 198-211, doi: 10.2110/jsr.68.198.

Milliken, K. L., 1992, Chemical behavior of detrital feldspars in mudrocks versus sandstones, Frio Formation (Oligocene), South Texas: Journal of Sedimentary Research, v. 62, p. 790-801, doi: 10.1306/D42679DD-2B26-11D7-8648000102C1865D.

Morad, S., 1988, Albitized microcline grains of post-depositional and probable detrital origins in Brøttum Formation sandstones (Upper Proterozoic), Sparagmite Region of southern Norway: Geological Magazine, v. 125, p. 229-239, doi: 10.1017/S0016756800010177.

Nesbitt, H. W., C. M. Fedo, and G. M. Young, 1997, Quartz and feldspar stability, steady and non-steady-state weathering, and petrogenesis of siliciclastic sands and muds: Journal of Geology, v. 105, p. 173-192, doi: 10.1086/515908.

Odom, I. E., T. W. Doe, and R. H. Dott, 1976, Nature of feldspar-grain size relations in some quartz-rich sandstones: Journal of Sedimentary Research, v. 46, p. 862-870, doi: 10.1306/212F7077-2B24-11D7-8648000102C1865D.

Parsons, I., 2010, Feldspars defined and described: A pair of posters published by the Mineralogical Society. Sources and supporting information: Mineralogical Magazine, v. 74, p. 529-551, doi: 10.1180/minmag.2010.074.3.529.

Parsons, I., and M. R. Lee, 2009, Mutual replacement reactions in alkali feldspars I: Microtextures and mechanisms: Contributions to Mineralogy and Petrology, v. 157, p. 641-661, doi: 10.1007/s00410-008-0355-4.

Parsons, I., P. Thompson, M. R. Lee, and N. Cayzer, 2005, Alkali feldspar microtextures as provenance indicators in siliciclastic rocks and their role in feldspar dissolution during transport and diagenesis: Journal of Sedimentary Research, v. 75, p. 921-942, doi: 10.2110/jsr.2005.071.

Pittman, E. D, 1970, Plagioclase feldspar as an indicator of provenance in sedimentary rocks: Journal of Sedimentary Research, v. 40, p. 591-598, doi: 10.1306/74D71FDC-2B21-11D7-8648000102C1865D.

Plymate, T. G., and L. J. Suttner, 1983, Evaluation of optical and X-ray techniques for detecting source-rock-controlled variation in detrital potassium feldspars: Journal of Sedimentary Research, v. 53, p. 509-519, doi: 10.1306/212F821A-2B24-11D7-8648000102C1865D.

Putnis, A., R. Hinrichs, C. V. Putnis, U. Golla-Schindler, and L. G. Collins, 2007, Hematite in porous red-clouded feldspars: Evidence of large-scale crustal fluid–rock interaction: Lithos, v. 95, p. 10-18, doi: 10.1016/j.lithos.2006.07.004.

Ribbe, P. H., 1995, The crystal structures of the aluminum-silicate feldspars, *in* I. Parsons, ed., Feldspars and Their Reactions (NATO ASI Series, Series C, v. 421): Dordrecht, Netherlands, Kluwer Academic Publ., p. 1-50.

Smith, J. V., and W. L. Brown, 1988, Feldspar Minerals, Volume 1. Crystal Structure, Physical, Chemical and Microtextural Properties (2nd Edition): New York, Springer-Verlag, 828 p.

Stewart, D. B., and P. H. Ribbe, 1975, Optical properties of feldspars, *in* P. H. Ribbe, ed., Feldspar Mineralogy (Reviews in Mineralogy 2): Washington, DC, Mineralogical Society of America, p. 121-140.

Suttner, L. J., and P. K. Dutta, 1986, Alluvial sandstone composition and paleoclimate: I. Framework mineralogy: Journal of Sedimentary Research, v. 56, p. 329-345, doi: 10.1306/212F8909-2B24-11D7-8648000102C1865D.

Tobi, A. C., 1962, Characteristic patterns of plagioclase twinning: Norsk Geologisk Tidsskrift, v. 42, p. 264-271.

Todd, T. W., 1968, Paleoclimatology and the relative stability of feldspar minerals under atmospheric conditions: Journal of Sedimentary Research, v. 38, p. 832-844, doi: 10.1306/74D71A87-2B21-11D7-8648000102C1865D.

Trevena, A. S., and W. P. Nash, 1979, Chemistry and provenance of detrital plagioclase: Geology, v. 7, p. 475-478, doi: 10.1130/0091-7613(1979)7<475:CAPODP>2.0.CO;2.

Trevena, A. S., and W. P. Nash, 1981, An electron microprobe study of detrital feldspar: Journal of Sedimentary Research, v. 51, p. 137-149, doi: 10.1306/212F7C2F-2B24-11D7-8648000102C1865D.

Tyrrell, S., A. K. Souders, P. D. W. Haughton, J. S. Daly, and P. M. Shannon, 2010, Sedimentology, sandstone provenance and palaeodrainage on the eastern Rockall Basin margin: Evidence from the Pb isotopic composition of detrital K-feldspar, *in* B. Vining, and S. C. Pickering, eds., Global Petroleum Systems in Space and Time (Petroleum Geology Conference series 7): London, GSL, p. 937-952, doi: 10.1144/0070937.

van de Kamp, P. C., 2010, Arkose, subarkose, quartz sand, and associated muds derived from felsic plutonic rocks in glacial to tropical humid climates: Journal of Sedimentary Research, v. 80, p. 895-918, doi: 10.2110/jsr.2010.081.

Van Der Plas, L., 1966, The Identification of Detrital Feldspars: New York, Elsevier Publ. Co., 305 p.

Vance, J. A., 1961, Polysynthetic twinning in plagioclase: American Mineralogist, v. 46, p. 1097-1119.

Wright, T. L., and D. B. Stewart, 1968, X-ray and optical study of alkali feldspar: I. Determination of composition and structural state from refined unit-cell parameters and 2V: American Mineralogist, v. 53, p. 38-87.

Facing Page: Lithic clasts from the shore of Jackson Lake, near Moran, Teton Co., Wyoming.

GRAINS:

ROCK FRAGMENTS (LITHIC FRAGMENTS)

CHAPTER 3

SEDIMENTARY (SRFs)

METAMORPHIC (MRFs)

IGNEOUS (IRFs)

ROCK FRAGMENTS

Rock fragments (also called lithic fragments or composite grains) can be derived from a wide variety of lithotypes and commonly have source-specific textures and compositions that can be recognized in thin section. Because of their multicrystalline/granular nature, rock fragments tend to be more common in the coarser grain-size modes of clastic terrigenous rocks (although, under the right circumstances, they can even be seen in mudrocks). Given the composite character of lithic fragments, many petrographers use the Gazzi-Dickinson method of point counting to record the constituent crystals within the fragments, rather than counting the fragments as such (Ingersoll et al., 1984). Rock fragments should be very common in sediments, and they are in many deposits, but because of their multi-crystalline or multi-granular nature, many succumb to the effects of weathering, abrasion or later mechanical or chemical diagenesis. But because the surviving rock fragments yield some of the most direct evidence of contributions from igneous, metamorphic or sedimentary terranes, it is especially important that such grains be accurately identified.

SEDIMENTARY ROCK FRAGMENTS

Sedimentary rock fragments (SRFs) consist of eroded and transported clasts of biochemical (cherts, carbonates and coals), chemical (evaporites and oolitic limestones) or siliciclastic rocks (sandstone, siltstone and claystone). The type of fragments can affect the mechanical stability of the resulting rock (especially mudstone clasts that can disaggregate to form pseudomatrix); therefore, differentiating between ductile and rigid fragments is important to the compactional history as well as to the potential to preserve primary porosity.

Major characteristics:

Composition:

Biochemical – formed by organisms or from the remains of organisms (including trapping and binding by microbes). Carbonate clasts are composed of calcite or dolomite primarily produced, directly or indirectly, by biological or biochemical processes. Most such SRFs are composed of fine-grained carbonate (mudstones) or are skeletal or pelletal limestones or dolomites. Siliceous organisms (diatoms, radiolarians and sponges) are a major source of silica for cherts. Soft organic matter, including plant remains, are common in organic-rich shales, coals and peat deposits.

Chemical – deposits which form by direct precipitation from water. This includes a variety of calcareous deposits including tufas, soil nodules, oolitic limestones, cave deposits and travertines. Siliceous materials, including silcretes and inorganic cherts (along with associated megaquartz and chalcedony), and a variety of evaporite and halide minerals and iron oxides also can form SRFs.

Siliciclastic – can be a variety of mineral particles, but most sedimentary rock fragments are dominated by quartz, feldspars, heavy minerals, micas and clays. Fine-grained sedimentary rock fragments sometimes contain a high proportion of mica, clay and organic material, making them brown in plane light.

Other:

Diagenesis can alter the composition of grains. Bright birefringence in argillaceous grains, for example, may be an indication of an illitic composition (either inherited from the source rock or locally induced).

Other similar grains:

Some clayey grains may originate as fecal pellets and therefore are not rock fragments. Likewise, grains of cemented sediment (typically calcareous, but potentially also phosphatic) that formed penecontemporaneously within the depositional setting of the host sediment (as opposed to having been eroded and transported from older rocks outside the depositional area) are commonly termed carbonate intraclasts or phosphatic nodules and are not counted by many petrographers as true SRFs in rock classification. Carbonate clasts of older material reworked from outside the depositional setting are termed "extraclasts" (as opposed to the internally-derived "intraclasts") and are viewed by all workers as true SRFs.

Cemented and compacted sandstones grade into metamorphic quartzite. Chert is dominantly sedimentary in origin, but it also can result from diagenetic replacement of volcanic grains (pumice in particular) or may form from hydrothermal fluids. Texture within the clasts sometimes can give clues to their origin. The compaction of fragments of altered igneous rock and micas, or of shales and claystones may lead to the production of pseudomatrix; more rigid grains forcing softer grains into available pore spaces. The presence or absence of quartz silt can help to distinguish the squashed detrital grains from true matrix.

Lo. Permian Wolfcamp Shale, Reeves Co., Texas

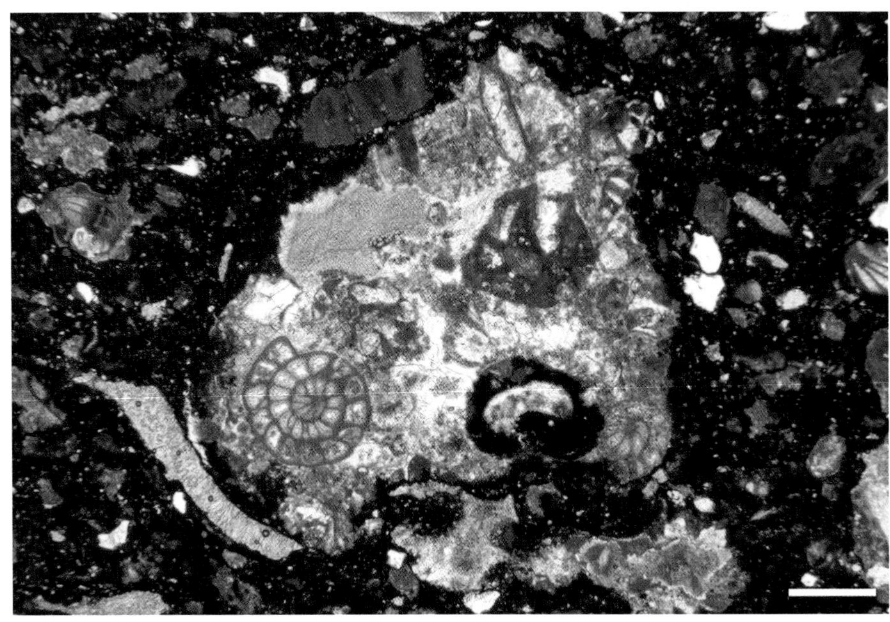

A fossiliferous carbonate rock fragment in an organic-rich shaly matrix. Fragments (extraclasts) of a carbonate shelf margin were probably slumped or eroded into the adjacent basin. Based on the angularity of this grain, it was not transported a long distance from its source in light of the fact that carbonate rock fragments round or break down rapidly during transport. However, there is an important exception to that rule—gravity flows (debris flows, mud slides and the like) where grains are cushioned by a finer-grained matrix during transport.

PPL, Scale bar = 0.51 mm

Paleogene Vieja Gp., Presidio Co., Texas

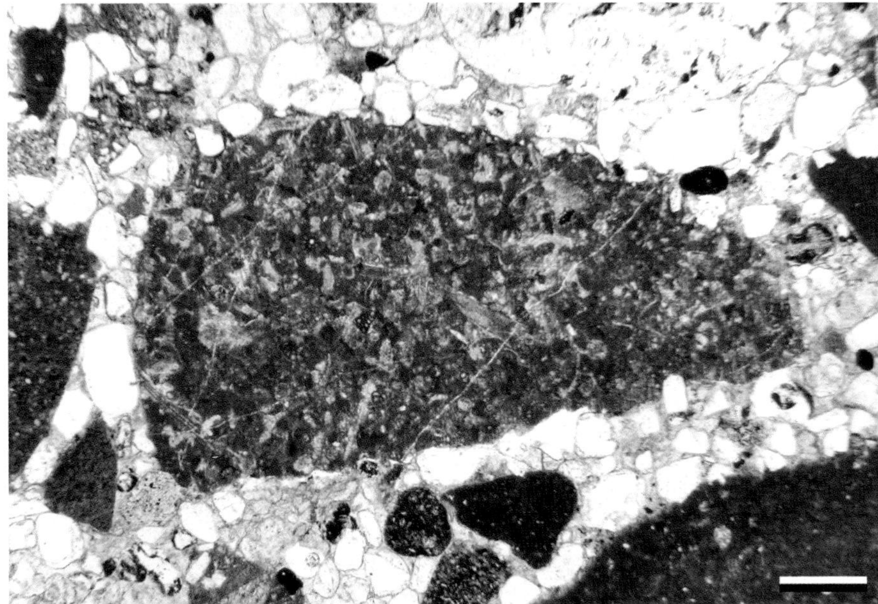

A variety of carbonate clasts in a quartz arenite. The carbonate clasts are all substantially rounded indicating that they were transported over enough distance to abrade their edges. The clasts have a variety of compositions ranging from mudstones to packstones, indicating that the grains were supplied by the erosion of multiple units or multiple facies. Since carbonate SRF's weather and abrade easily, there are a limited number of environments in which they occur in abundance. Those include many arid-region settings and especially alluvial fan deposits.

PPL, Scale bar = 0.47 mm

Paleogene Vieja Gp., Presidio Co., Texas

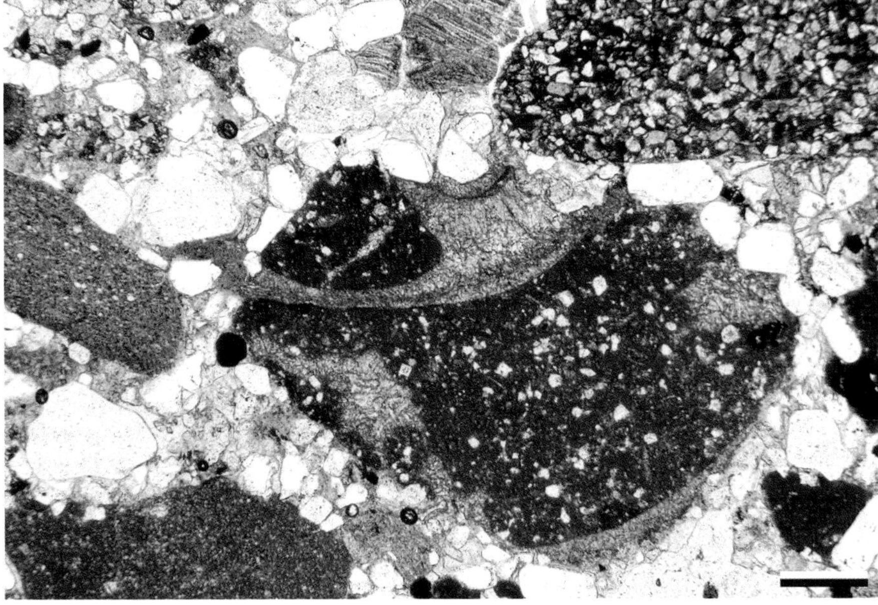

Diverse, rounded sedimentary lithoclasts in a fluvial or alluvial fan sandstone. The SRFs include a dolomitic limestone clast with a large truncated gastropod shell, carbonate mudstones (micrites), and terrigenous siltstones and claystones. Again, the diversity of lithoclasts suggests that multiple units or facies were being eroded to source the clasts, but transport distance was probably minimal. If these were intraclasts, most classifications would not count them as SRFs; but because they most likely are extraclasts, they generally are counted as sedimentary rock fragments.

PPL, Scale bar = 0.51 mm

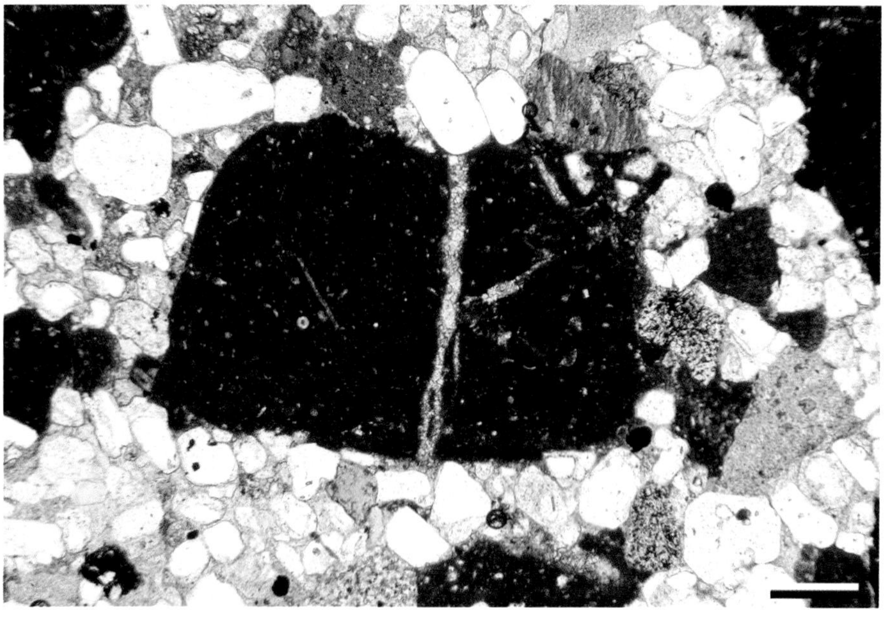

Paleogene Vieja Gp., Presidio Co., Texas

A variety of carbonate lithoclasts in a quartz-rich sandstone. The dark wackestone fragment shows that these rocks were well lithified prior to weathering and transport. The clast contains a calcite-filled fracture that was clearly formed prior to reworking. These grains are all extraclasts, eroded from older, but not very distant, limestone source rocks and deposited in an alluvial to fluvial environment.

PPL, Scale bar = 0.51 mm

Up. Eocene Oberaudorf Beds, Tirol, Austria

Spicular chert and carbonate rock fragments make up this rock. These carbonate clasts are all extraclasts; they have a variety of different carbonate fabrics and textures and were fully lithified prior to erosion and transport. Some clasts are recrystallized and one has a healed fracture. All of these features indicate an extra-basinal origin for the carbonate rock fragments (the definition of an extraclast). A rock, such as this, that is dominated by extraclasts is termed a "calclithite".

PPL | XPL, Scale bar = 0.26 mm

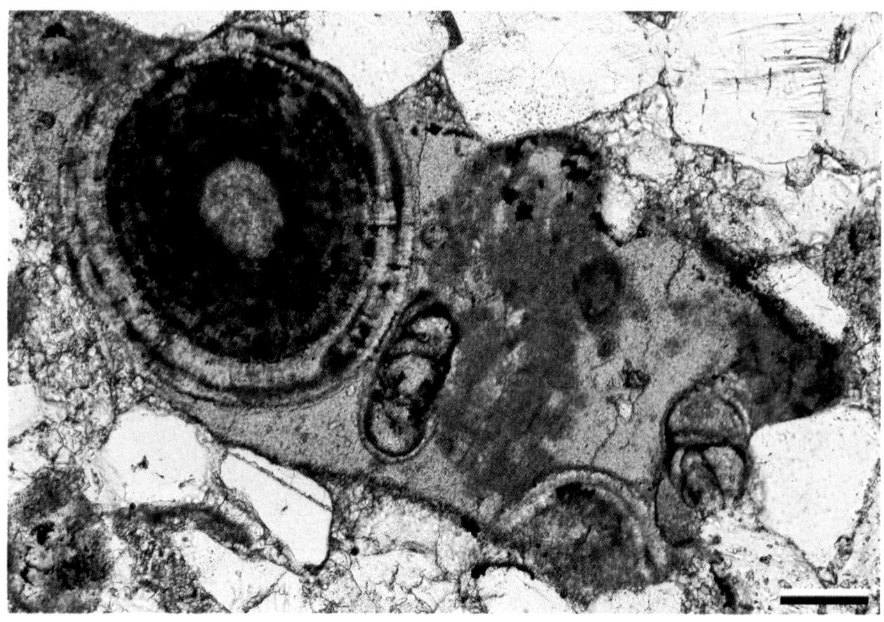

Mid. Jurassic Curtis Fm., Moffat Co., Colorado

A carbonate extraclast in a quartz sandstone. This SRF is a partially glauconitized ooid and foraminiferal grainstone. The rock was lithified and partially replaced prior to being ripped up and reworked into the sediment. The clast probably was transported a short distance into a deeper-water environment with an abundance of clastic material. Note the abundant microfractures filled with pink-stained epoxy.

PPL, RDI, Scale bar = 0.10 mm

Mid. Jurassic Curtis Fm., Moffat Co., Colorado

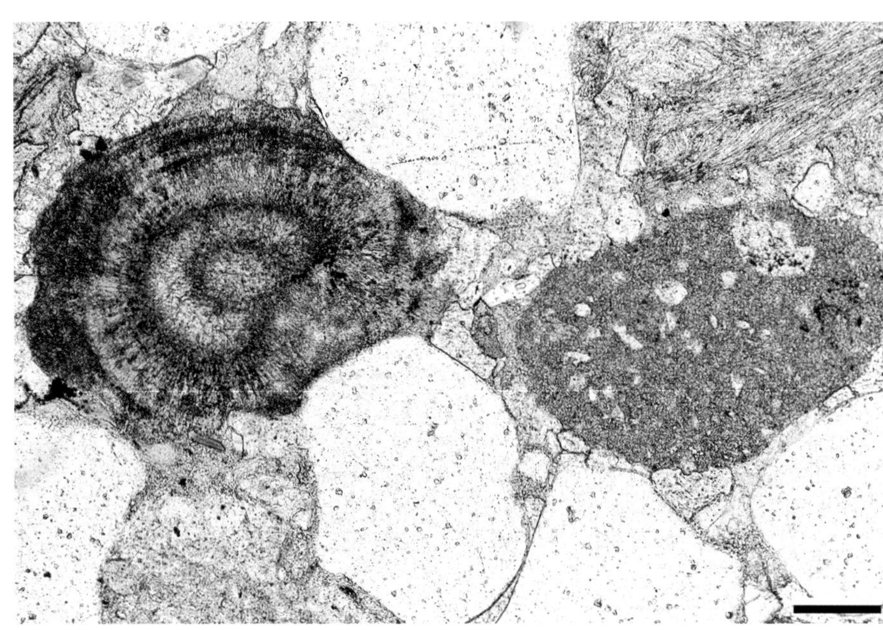

This sample shows a calcitic ooid, a shell fragment and a micritic limestone clast set in a quartz sandstone. It commonly is difficult to differentiate intraclasts (contemporaneous limestone fragments formed within the depositional setting) from extraclasts (older carbonate clasts transported from outside the depositional basin). Therefore, it is important to integrate both broader petrographic and field data to determine the origin of the clasts. These grains probably are all extraclasts based both upon their diverse character and the presence of clear-cut extraclasts in other examples from this rock (see previous photo).

PPL, RDI, Scale bar = 0.10 mm

Up. Cambrian Sillery Gp., Gaspé Peninsula, Quebec, Canada

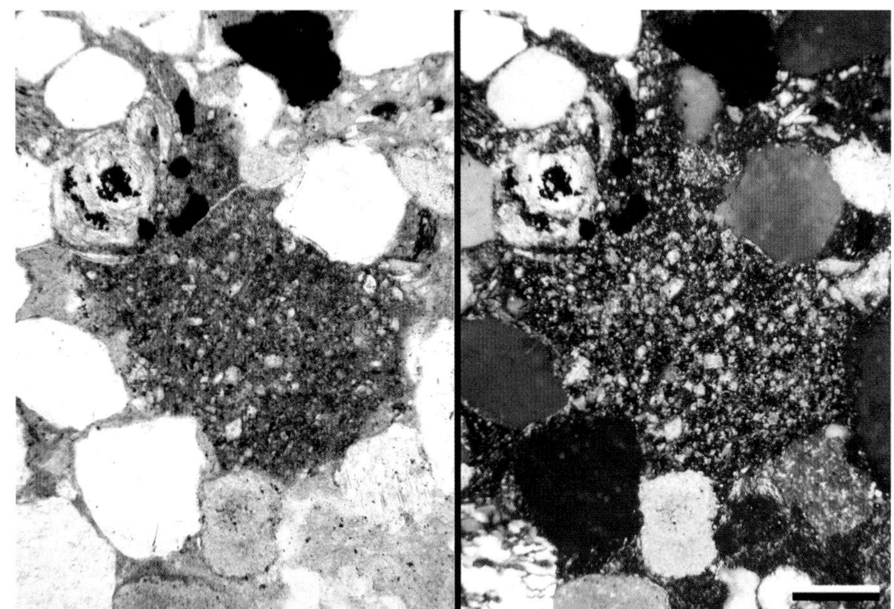

A fine-grained, marly, carbonate rock fragment. This clast has been partially dolomitized with several larger dolomite rhombs clearly visible. During compaction, this grain behaved in a ductile fashion, deforming plastically into surrounding pore space.

PPL | XPL, Scale bar = 0.16 mm

Triassic Skagerrak Fm., United Kingdom sector, North Sea

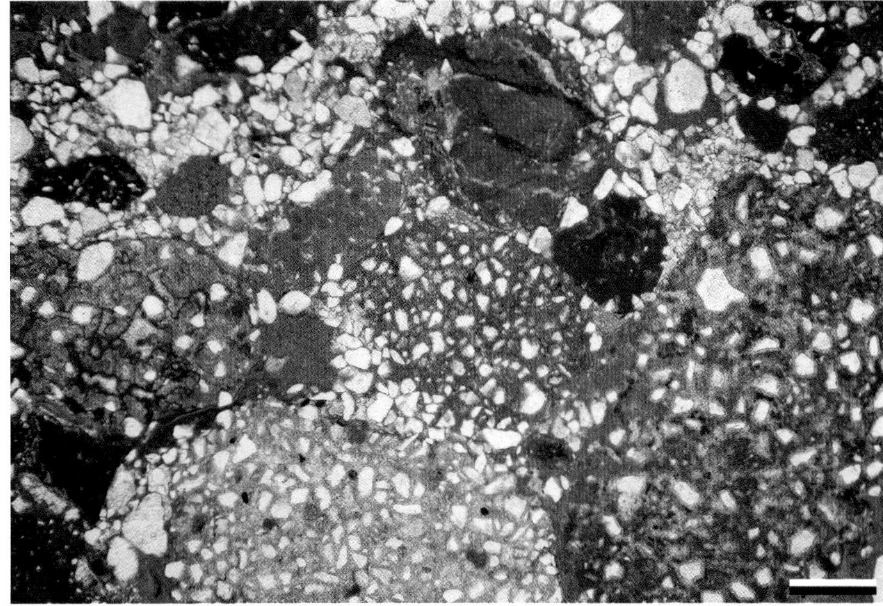

Numerous large sedimentary rock fragments within a conglomerate. This intraformational conglomerate contains intraclasts that range from carbonate mudstones to very sandy mudstones. These clasts were derived from caliche horizons within the formation and were reworked into this considerably younger unit. In the absence of such stratigraphic information, these compositionally relatively similar extraclasts could be mistaken for intraclasts.

PPL, BDI, Scale bar = 0.70 mm

Triassic Dockum Gp., New Mexico

A compacted calclithite composed almost entirely of carbonate rock fragments (extraclasts). These grains are less varied in composition and texture than those in the previous examples. This sediment is derived from a weathered and reworked caliche (a calcium carbonate-rich soil crust), but one of substantially older age than the rock into which it was redeposited. Thus, these grains can legitimately be called extraclasts. The red-brown color was caused by infiltration of iron-oxides into the matrix and some of the clasts.

PPL, Scale bar = 0.38 mm

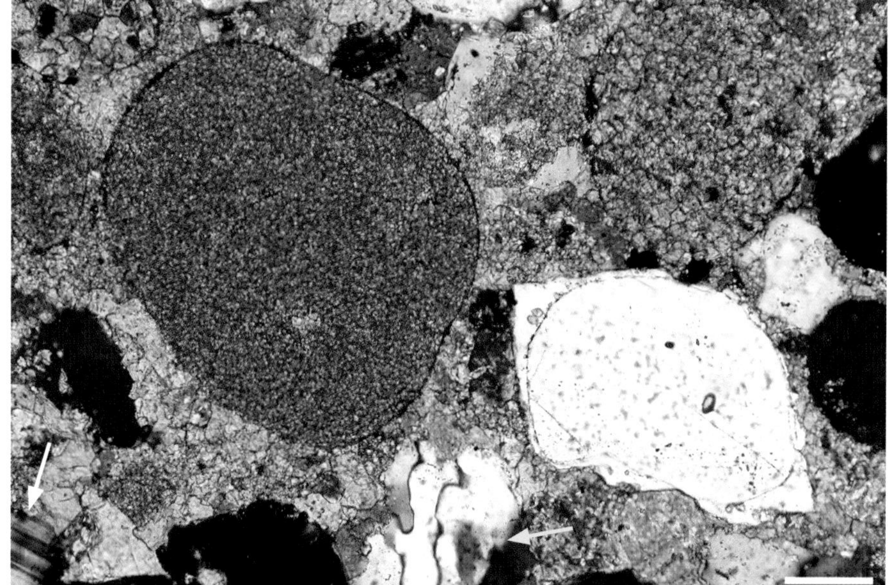

Mid. Permian Brushy Canyon Fm., Culberson Co., Texas

Several well-rounded detrital dolomite rock fragments are visible in this quartzose sandstone. Dolomite is slightly harder than calcite and survives transport and compaction better. In this sample, the dolomite clasts have iron oxide stain on their exterior surface. The quartz grain in the lower right has well-developed bipyramidal overgrowths that are in optical continuity with the detrital core of the grain. Other grains include polycrystalline quartz (yellow arrow) and plagioclase (white arrow). The rock is cemented by both authigenic quartz and dolomite.

XPL, Scale bar = 0.10 mm

Mid. Oligocene Molasse, Ton Mergel Beds, Bavaria, Germany

The iron-stained dolomite rhombs in this photomicrograph are of probable detrital origin. The iron staining, the slightly rounded corners and their size-equivalence with the other detrital components of this rock are all evidence of a probable detrital origin for the individual dolomite crystals and multicrystalline dolomite clasts. The detrital dolomites are partly enclosed by authigenic carbonate cements.

XPL, Scale bar = 0.04 mm

Eocene Sanders Canyon Fm., Lincoln Co., New Mexico

A detrital dolomite rhomb in a feldspathic sandstone. The cloudy dolomite rhomb is considered to be detrital based on the exceptionally well rounded corners on this rhomb (and occurrence of other similarly rounded rhombs in the same sample); the detrital grain here was overgrown by less inclusion-rich dolomite cement that filled porosity in the surrounding area.

XPL, Scale bar = 0.14 mm

Mid. Permian Huledal Fm., Canning Land, East Greenland

A detrital sandstone clast within a lithic sandstone. Such sandstone clasts are good indicators of a sedimentary source, but sandstone fragments are relatively scarce in the rock record, especially considering the high percentage of the Earth's surface covered by terrigenous clastic rocks. Most sandstone and siltstone fragments (other than quartzite) tend to break down into their component grains after even moderate transport distances. This example shows subangular to angular quartz fully cemented by carbonate and clay.

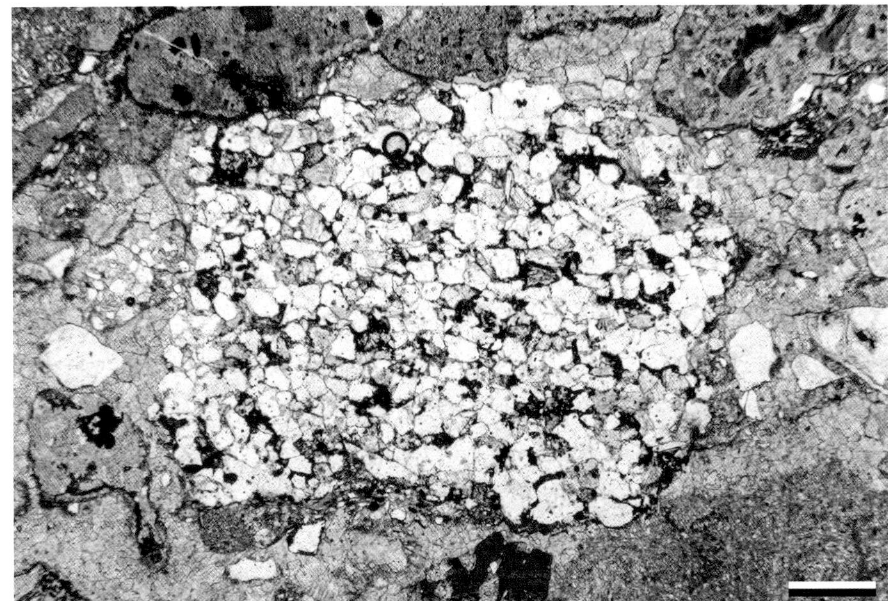

PPL, BDI, Scale bar = 0.51 mm

Carboniferous (Kulm) Teuschnitzer Conglomerate, Frankenwald, Bavaria, Germany

A large sedimentary rock fragment in a matrix-rich sandstone (a possible debris-flow deposit). This fragment is a very fine grained sandstone clast with clay cements. Like all sandstone clasts, this rock probably did not travel very far prior to deposition and/or what transport it underwent may have been as a largely cushioned, dense gravity flow. Many of the grains in the matrix are similar to the grains within the clast.

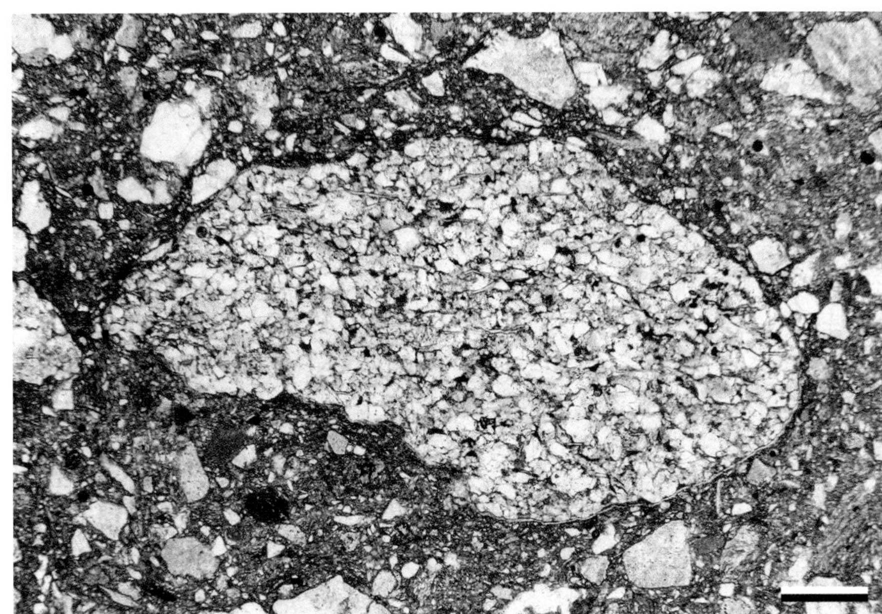

PPL, Scale bar = 0.51 mm

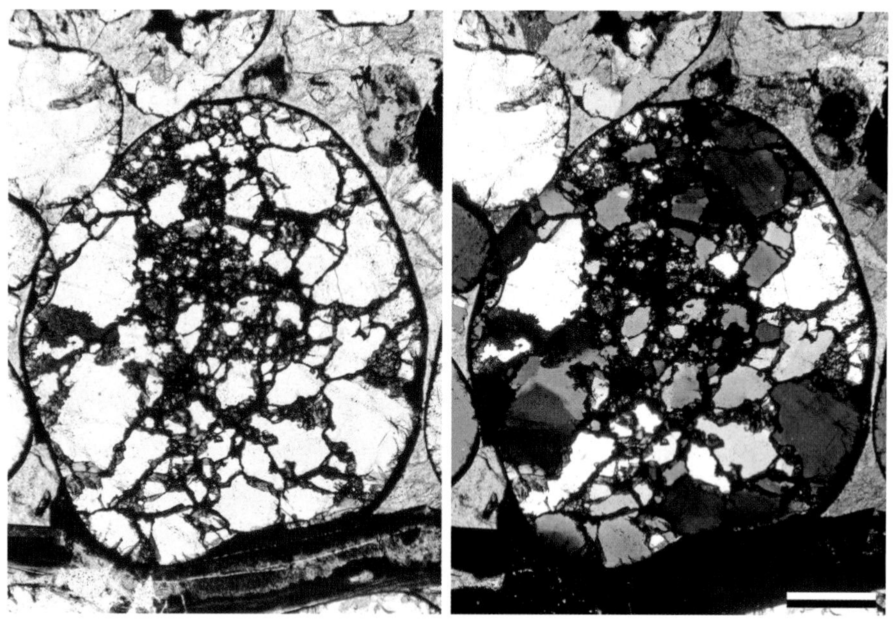

Mid. Cambrian Riley Fm., Hickory Sandstone Mbr., Llano Co., Texas

A supermature sandstone containing a well-rounded sedimentary rock fragment. The quartz grains on the edge of the SRF are broken and worn smooth, and the internal grains are cemented by iron oxides. Because the grain is exceptionally well rounded, it may well have survived more than one cycle of erosion and deposition. The grain has embayed a chitonophosphatic brachiopod shell and is encased in calcite cement. This rock directly overlies a major unconformity and although it sits directly on Precambrian basement, it probably was derived from erosion of earlier sandstones that sat on the unconformity.

PPL | XPL, BDI, Scale bar = 0.26 mm

Silurian Clinton Ironstone, Giles Co., Virginia

A detrital siltstone rock fragment (center) and numerous extremely well-rounded (second-cycle) quartz grains are cemented together by hematite. The presence of very well-rounded quartz, especially when accompanied by SRFs, helps to identify sedimentary source terranes. The angular quartz silt grains in the SRF are embedded in an exceptionally light-colored, phosphatic matrix.

PPL, Scale bar = 0.26 mm

Paleogene Vieja Gp., Presidio Co., Texas

This rock contains a variety of lithic fragments ranging from a large siltstone clast in the center to smaller volcanic rock (black arrow), siltstone and limestone (orange arrow) fragments in the surrounding sediment (dark-colored and light-colored respectively in the plane-polarized image). All the lithic fragments are rounded, indicating at least moderate-distance transport, although most of these lithologies are fairly easily abraded.

PPL | XPL, Scale bar = 0.64 mm

Mid. Permian Huledal Fm., Canning Land, East Greenland

Rounded, laminated siltstone and sandstone rock fragments within a lithic arenite. This calcite-cemented conglomerate contains both sedimentary and igneous rock fragments reflecting a complex source terrane. This rock is interpreted as having been deposited in a fluvial to marginal marine environment that was fairly proximal to the sources of the clasts. Note the circumgranular fracturing (marked by blue-dyed epoxy) that is probably an artifact of sample handling or thin-section preparation.

PPL, BDI, Scale bar = 0.26 mm

Silurian Tuscarora Sandstone, Lebanon Co., Pennsylvania

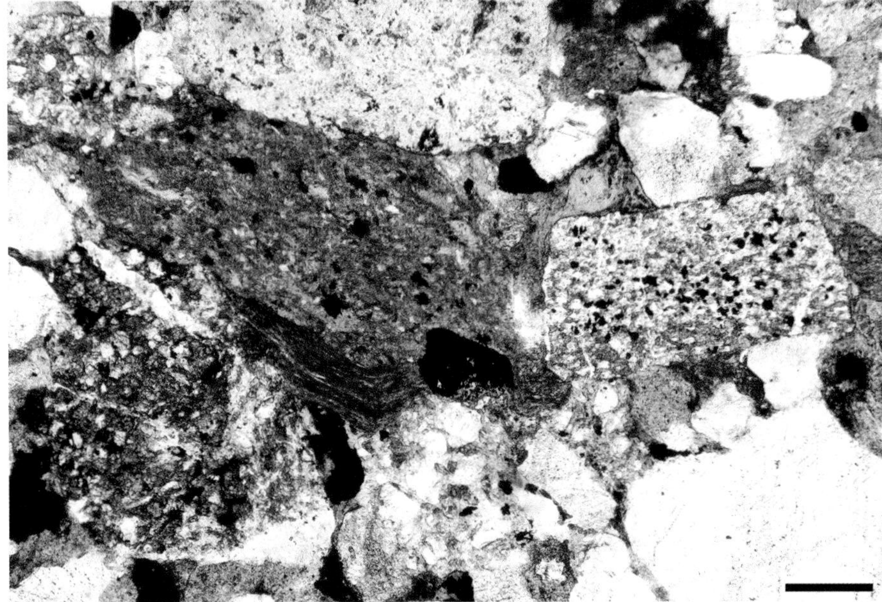

A lithic sandstone with a variety of sedimentary rock fragments that range from shale to siltstone and sandstone. The shale fragments underwent ductile deformation during compaction, in some places getting squeezed into adjacent pore spaces. The black crystals in this view are pyrite, formed both as replacement and as intergranular cement.

PPL, Scale bar = 0.26 mm

Mid. Permian Huledal Fm., Canning Land, East Greenland

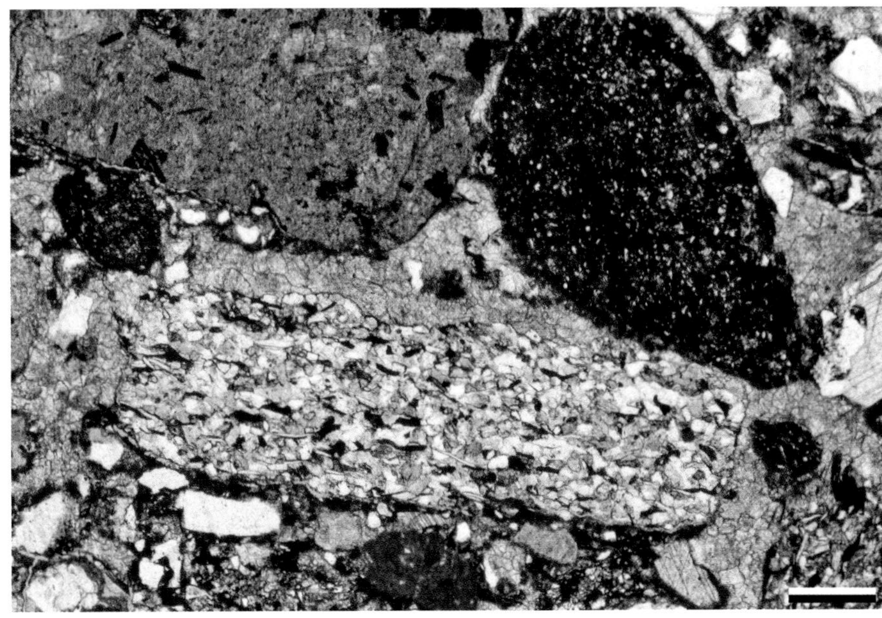

A variety of rock fragments make up this conglomeratic (granule-grade) rock. Some of the lithic fragments are igneous (mostly volcanic); however, most are sedimentary rock fragments. Visible sedimentary clasts include argillaceous sandstone, silty shale and limestone. The shale clasts are extremely dark, probably related to a high organic-carbon content.

PPL, BDI, Scale bar = 0.45 mm

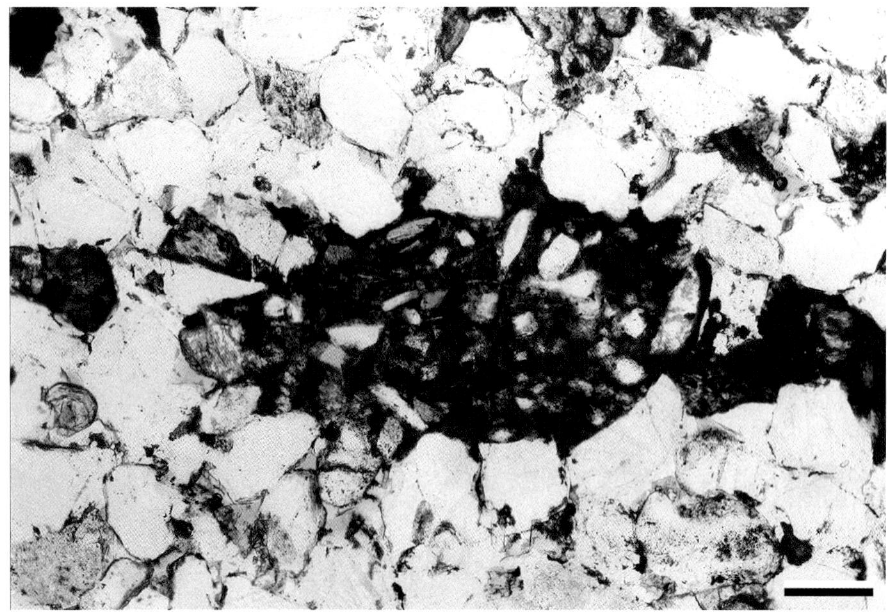

Mid. Jurassic Fulmar Fm., United Kingdom sector, North Sea †

A shallow marine sandstone with a silty claystone clast containing grains of K-feldspar, quartz and glauconite. The SRF was partly compacted between rigid grains. The size of this rock fragment is larger than surrounding detrital quartz. The clay matrix is dark in color, due, at least in part, to its organic-carbon content. It is sometimes the case that feldspars preferentially survive diagenesis within clay-rich areas, such as this clast, because the clays minimize contact with reactive pore fluids.

PPL, KFS, BDI, Scale bar = 0.12 mm

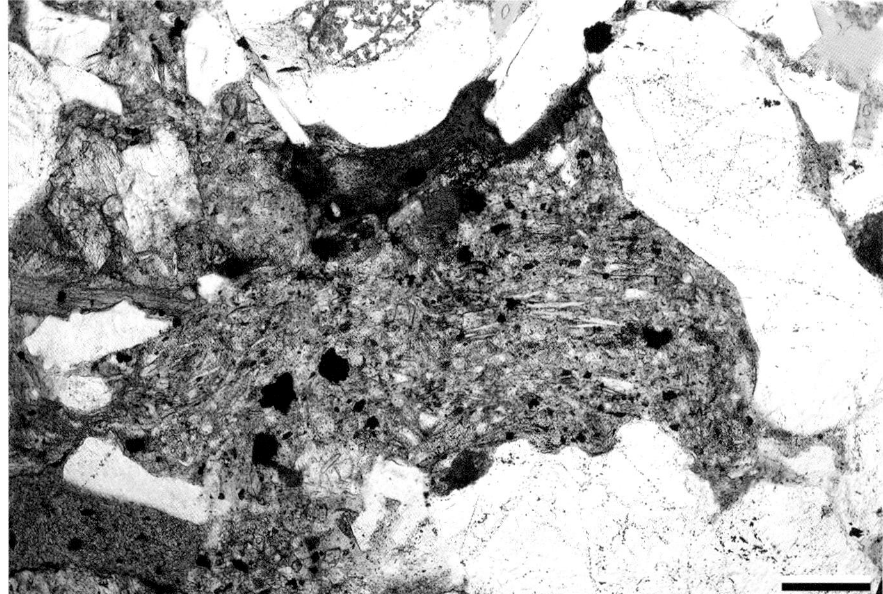

Mid. Jurassic Fulmar Fm., United Kingdom sector, North Sea †

A sandstone containing a large siltstone clast (center), surrounded by other ductile siltstone and claystone grains. The aligned micas in the grain may have resulted from hydrodynamic processes or compaction and are not in optical alignment as determined under cross-polarized light. The grain also contains quartz and K-feldspar as well as authigenic pyrite. Identification of the true number of lithic grains in a rock is important for proper sandstone classification—several slightly deformed to extensively degraded SRFs are visible in this field of view.

PPL, KFS, BDI, Scale bar = 0.12 mm

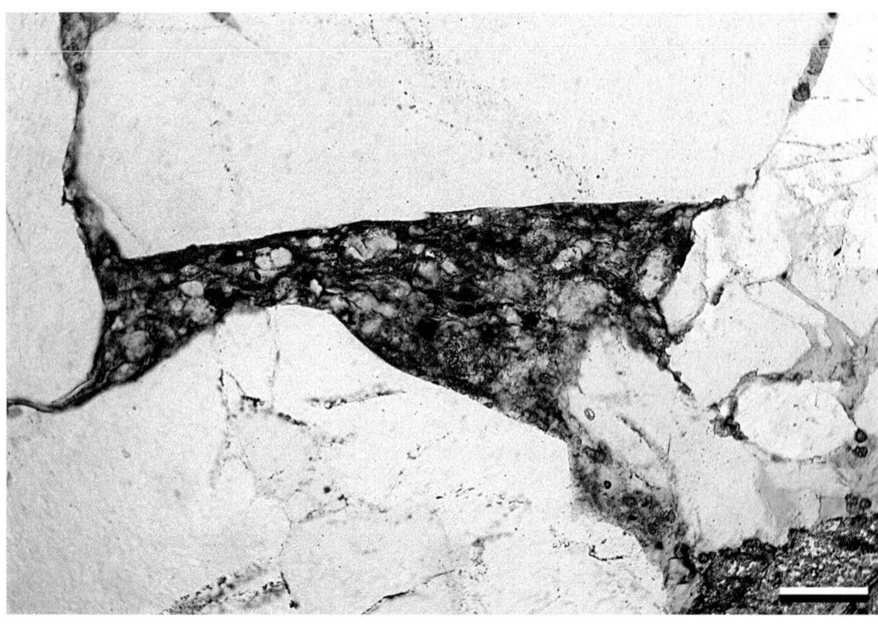

Mid. Jurassic Ness Fm., Norwegian sector, North Sea †

A siltstone grain that has been heavily deformed and squeezed between grains, to the point where it could be classed as pseudomatrix or could be mistaken for detrital clay and silt. The "clean" pores in the rest of this sample confirm that this was once a detrital lithic fragment.

PPL, KFS, BDI, Scale bar = 0.12 mm

Mid. Jurassic Fulmar Fm., United Kingdom sector, North Sea †

A shale clast within a litharenite. Many of the other grains are degraded and compacted. The darker streaks are an artifact of resin impregnation along shrinkage fractures in the grain. The claystone grain contains numerous, small pyrite framboids (the tiny dark dots best seen in the light-colored layer).

PPL, AFeS, BDI, Scale bar = 0.49 mm

Triassic Chinle Fm., eastern Arizona

Shale sedimentary rock fragments can be difficult to distinguish from low-rank metamorphic fragments, like slate. SRFs tend to be somewhat softer and more malleable than metamorphic clasts, so they more commonly deform in a ductile manner, essentially "flowing" into surrounding pore space. As very soft grains, they also are likely to be embayed by adjacent harder grains. In addition to the deformational properties, the relatively poor internal clay crystal orientation also supports a sedimentary origin for these clasts.

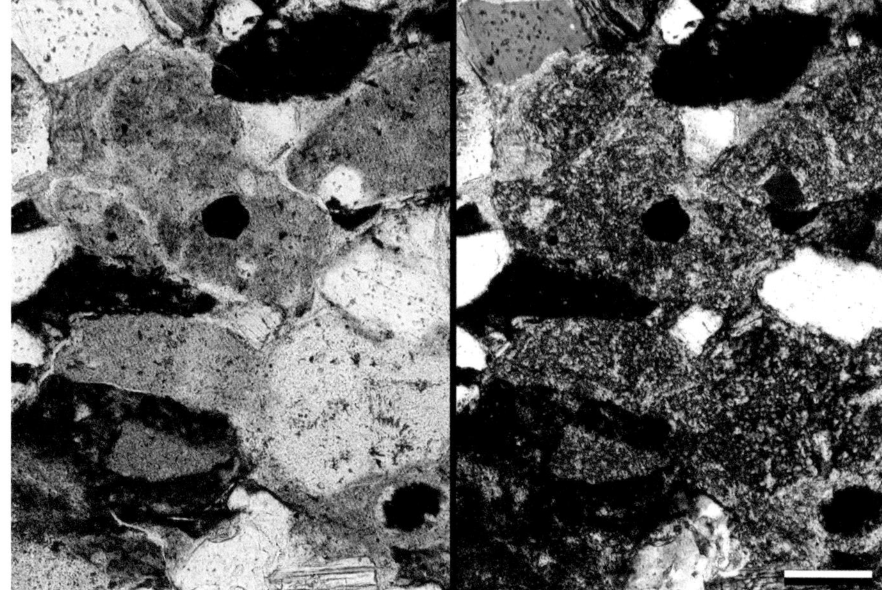

PPL | XPL, Scale bar = 0.10 mm

Triassic Chinle Fm., eastern Arizona

As in the previous photomicrograph, this shale fragment was plastically deformed and embayed by surrounding more rigid grains, mostly chert clasts, during burial and compaction. In the cross-polarized light view, the clay minerals show relatively little consistency of orientation.

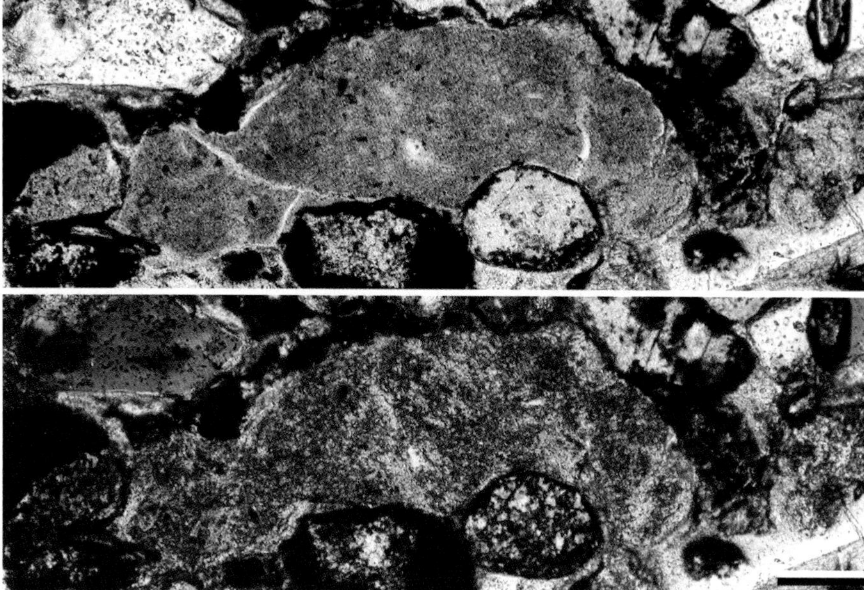

PPL / XPL, Scale bar = 0.10 mm

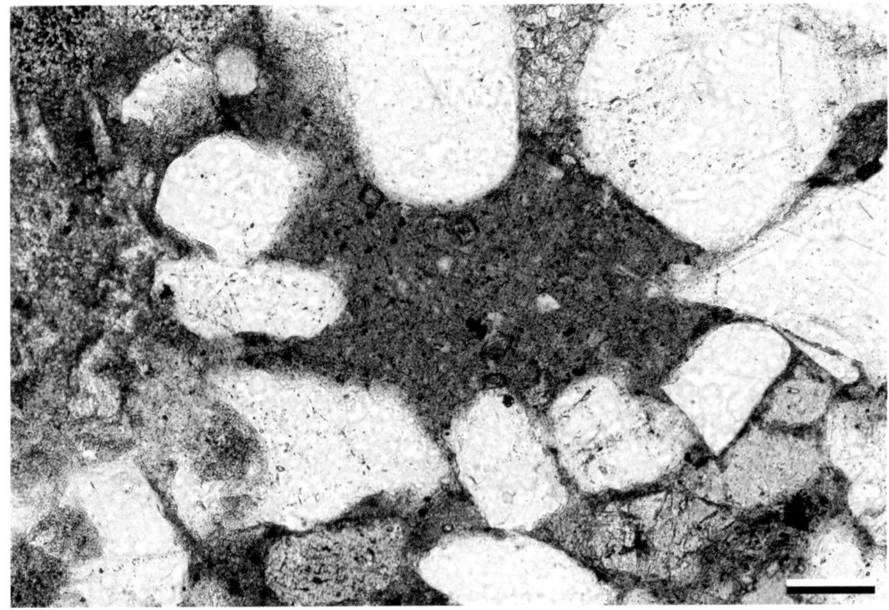

Up. Cambrian Sillery Gp., Gaspé Peninsula, Quebec, Canada

In addition to being easily confused with metamorphic rock fragments, semiconsolidated shale fragments also can be confused with rock matrix (finer-grained interstitial material originally deposited along with the coarser grains). In this example, a slightly silty and dolomitic shale fragment has been so deformed between adjacent harder grains that the original boundaries of the clast are no longer visible, producing pseudomatrix. One of the keys to recognition is to look at the rest of the slide to see if the material between all the grains looks similar (matrix) or varied and more isolated (pseudomatrix).

PPL, Scale bar = 0.09 mm

Up. Cambrian Sillery Gp., Gaspé Peninsula, Quebec, Canada

A variety of squashed shale fragments yielding what appears to be a matrix-rich rock. Compaction caused the shale grains to deform plastically and "flow" into surrounding pores to form pseudomatrix. In this example, one can still see some original clast outlines as well as abrupt differences in the character of the matrix from one area to another—both are keys to recognizing pseudomatrix.

PPL, Scale bar = 0.09 mm

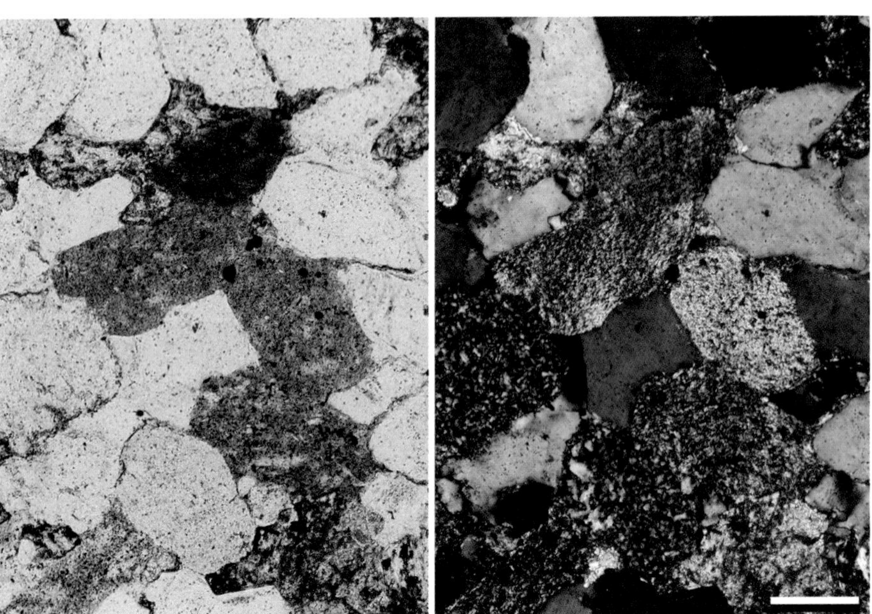

Permian basinal sandstone, Presidio Co., Texas

These sedimentary rock fragments are relatively hard shale (possibly slate) fragments. The clay minerals in the clasts exhibit considerable preferred orientation. In plane-polarized light, the center of the image appears to be one large, albeit oddly-shaped, grain or a patch of deformed matrix. However, in cross-polarized light, three distinct grains are visible based on the different orientations of the strongly aligned clay minerals in each clast.

PPL | XPL, Scale bar = 0.26 mm

Permian basinal sandstone, Presidio Co., Texas

Several squashed rock fragments have flowed together to form pseudomatrix. The color variations within the pseudomatrix probably represent different, now largely merged shale clasts. In the absence of those variations, it would be difficult to distinguish this pseudomatrix from true primary matrix.

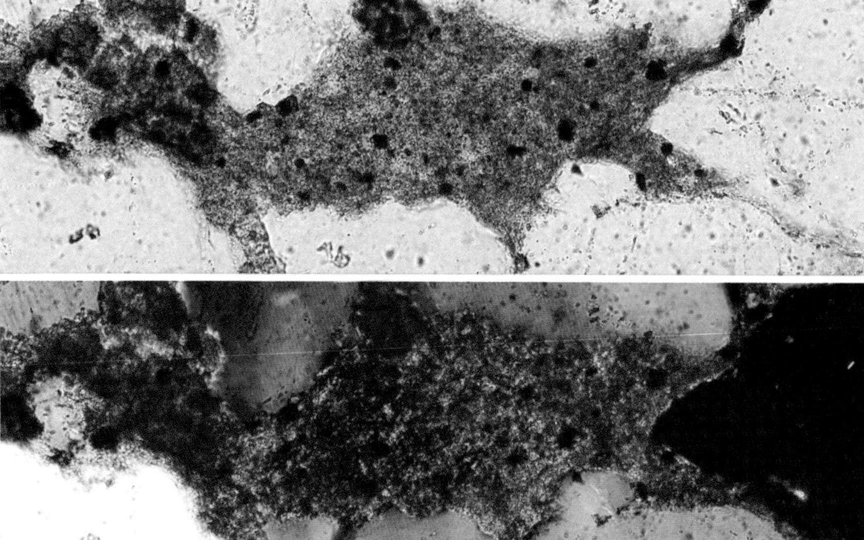

PPL / XPL, Scale bar = 0.05 mm

Lo. Cambrian Unicoi Fm., Shenandoah Valley, Virginia

Multiple deformed shale or slate sedimentary rock fragments are deeply embayed by adjacent, more rigid grains causing the clay minerals to deform plastically into open pore spaces. The strong clay mineral orientation in these shale clasts may be secondary (that is, produced *in situ* during the deformation of this rock and not inherited from the source of the clasts).

PPL | XPL, Scale bar = 0.10 mm

Mid. Jurassic Entrada Fm., Emery Co., Utah

A well-rounded chert SRF in a quartz arenite. The plane-polarized light view of the chert shows a typical brownish color and mottled appearance that is most likely due to the abundance of water-filled inclusions common in cherts and crystal boundary effects in the microcrystalline fabric of chert grains. In polarized light, the microcrystalline to fibrous fabric (representing a mix of chert and chalcedony) is clearly visible. The dark material in part of this photomicrograph is hematite cement.

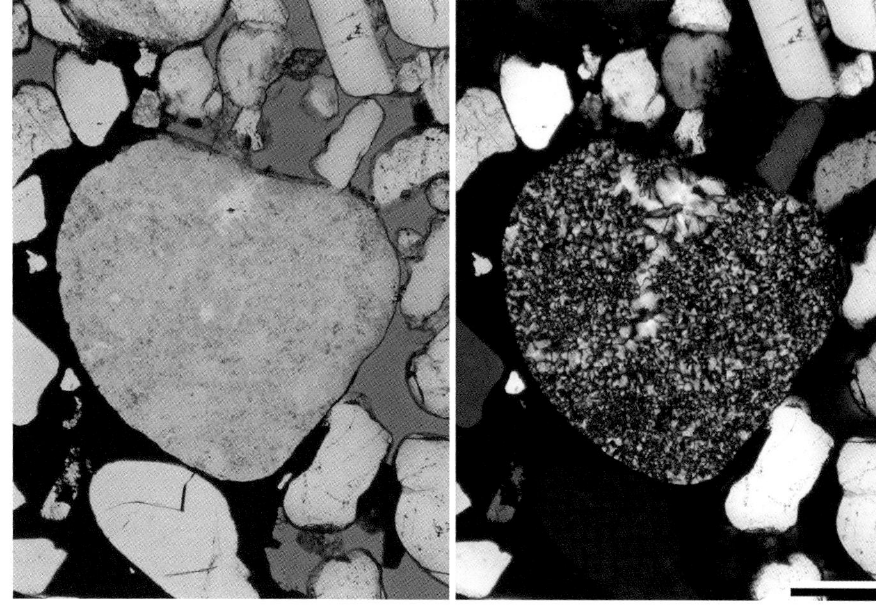

PPL | XPL, RDI, Scale bar = 0.17 mm

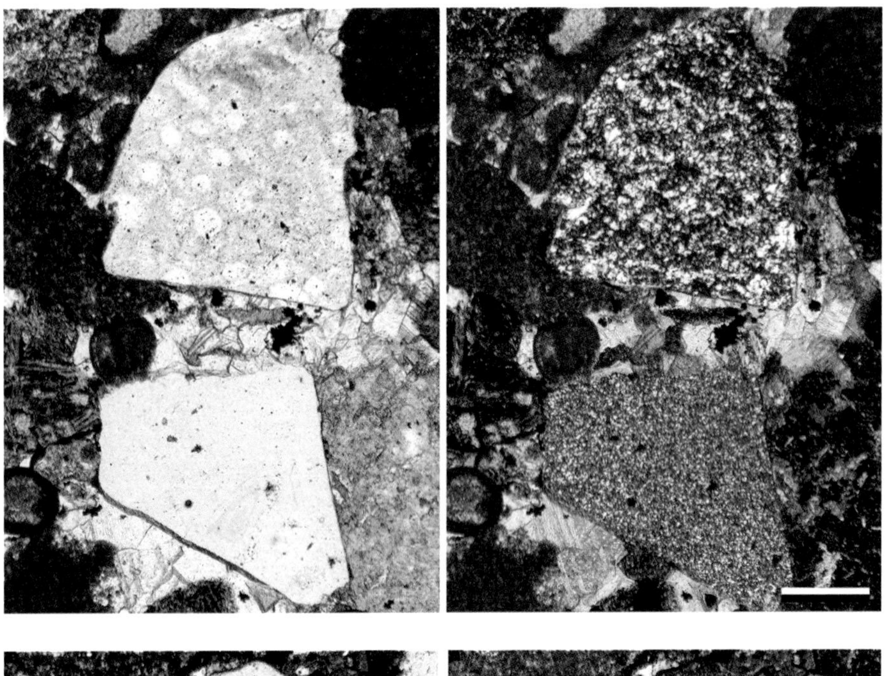

Mid. Permian (former Ufimian) Solikamskaya Horizon, near Berezniki, Perm region, Russia

Large and slightly rounded chert clasts in a calclithite. The lower chert grain is uniformly microcrystalline quartz, but the upper grain has more variability in crystal size. The circular ghosts in the upper chert clast are likely after sponge spicules. Cherts generally are the most durable of all sedimentary rock fragments; they survive both substantial transport and considerable weathering. Thus, they are the most common rock fragment indicators of sedimentary provenance and in a mature sandstone they may be the only indicators of a sedimentary source component.

PPL | XPL, Scale bar = 0.33 mm

Lo. Permian (Wolfcampian) Hueco Fm., Powwow Conglomerate, Hudspeth Co., Texas

This lithic arenite contains limestone and chert SRFs as well as fragmented bioclasts. Some of the chert clasts contain molds of siliceous sponge spicules and possible bivalve fragments (black arrow). The conglomerate has small, nonferroan carbonate clasts (red); ferroan calcite cement (blue) and euhedral dolomite cement (unstained) have occluded all porosity. Although limestone clasts fare poorly during transport and chert clasts are more durable, they commonly occur together where transport has been short because they typically co-occur in the source terranes.

PPL | XPL, AFeS, Scale bar = 0.33 mm

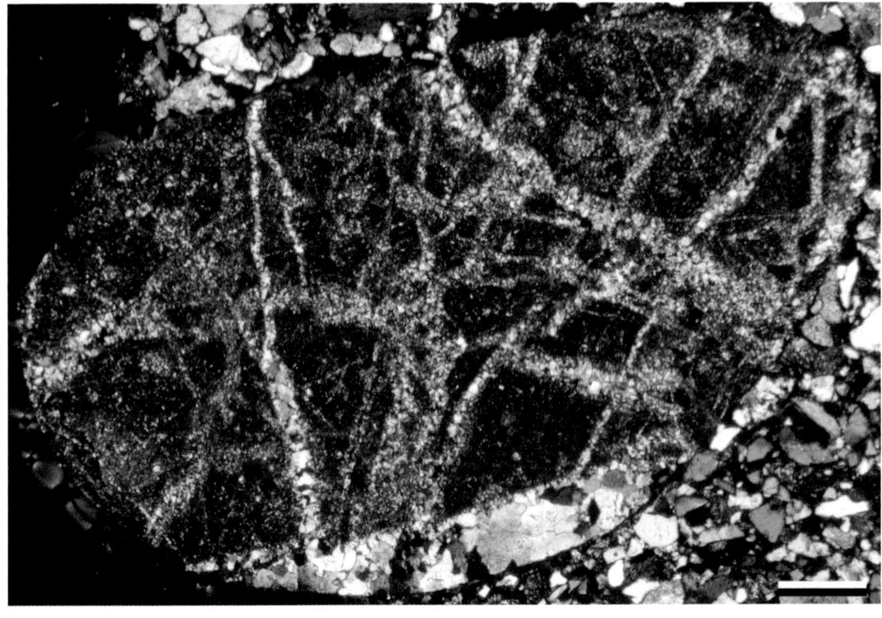

Oligocene – Miocene tuffaceous sandstone, Jackson Co., Oregon

A large, well-rounded chert clast that is far larger than any of the other grains in this rock. The chert is highly fractured (fractures inherited from the source rock and not produced *in situ* in this deposit). Fracturing is very common in cherts because they are among the hardest and most brittle of all sedimentary rocks. Thus, fractures are one characteristic that helps to distinguish chert grains from more ductile fine-grained shale clasts composed of low-birefringence clays (e.g., kaolinite). It is, alas, of little help in distinguishing chert SRFs from finely crystalline basaltic or silicic VRFs that also can be extensively fractured.

XPL, Scale bar = 0.51 mm

Up. Cretaceous Prince Creek Fm., North Slope, Alaska

An SRF-rich, lithic and feldspathic arenite with abundant chert and shale clasts. The variety of chert clasts is difficult to recognize in plane-polarized light because they range from dark or pale brown to virtually clear white grains (depending on the number of fluid-filled or organic inclusions in the chert) — but see the cross-polarized light photo below for better identification. Importantly, clay cements rim the pores and may have helped to prevent subsequent cementation in this rock.

PPL, BDI, Scale bar = 0.14 mm

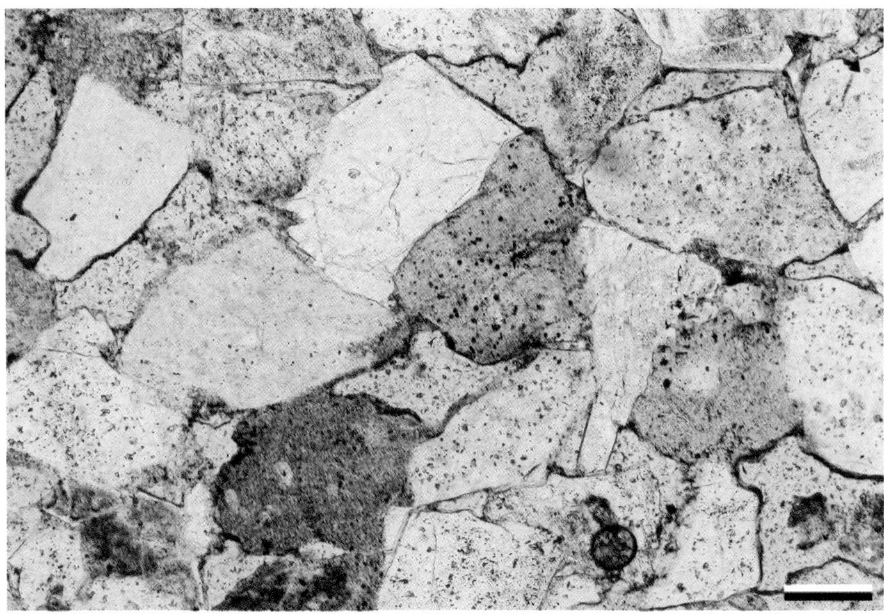

Up. Cretaceous Prince Creek Fm., North Slope, Alaska

Same field of view as previous photomicrograph, but under cross-polarized light. This image shows the characteristic speckled extinction of the tiny quartz crystals that compose chert. It also shows the variety of crystal sizes possible in chert grains and those variations greatly affect how chert appears under cross-polarized light. Comparison with the image above shows a correlation between grain color and birefringence patterns since both reflect crystal structure and size. The clay-cement rims noted above are illite/smectite and show bright birefringence.

XPL, BDI, Scale bar = 0.14 mm

Up. Eocene Oberaudorf Beds, Tirol, Austria

The chert clast in this sample contains numerous siliceous sponge spicules—such spicules are a common source of opaline silica for chert formation and a major contributor to "deeper-water" marine cherts. Dissolution of the original opal in the sponge spicules was probably the source of the fibrous silica in the spicules and the chert that replaced the original carbonate matrix. Visible in some of the spicules is a central canal that characterizes originally siliceous sponge spicules; also apparent are the long- and short-axis cuts through these cigar-shaped fossil remains.

PPL | XPL, Scale bar = 0.51 mm

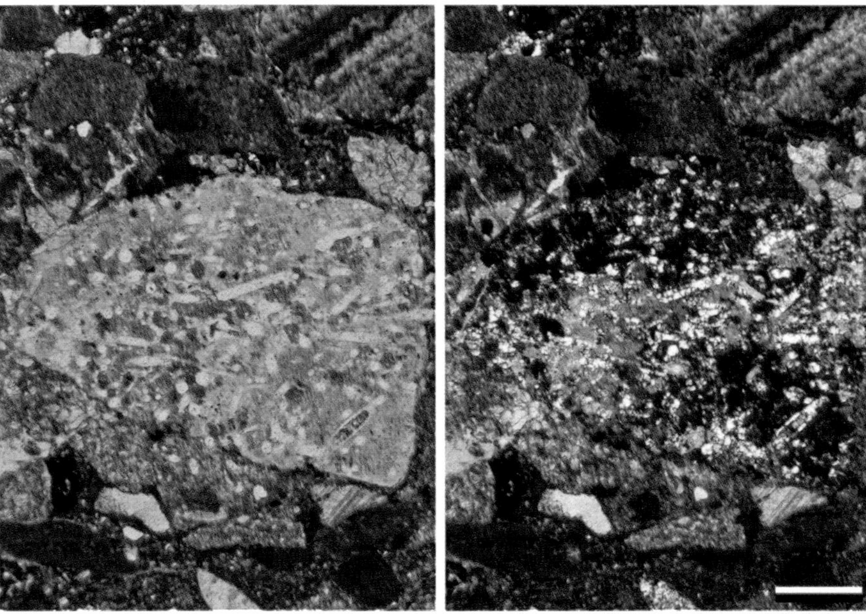

Cretaceous Dakota Gp., Moffat Co., Colorado

A chalcedony SRF in a quartz arenite. Chalcedony was formerly considered to consist entirely of fibrous microcrystalline quartz. Now it is thought to be an intergrowth of triclinic cryptocrystalline quartz and another quartz polymorph, monoclinic moganite. The shape and crystal fabric of this grain indicate that it was a pore- or vug-filling cement, probably in a limestone. Some dissolution of the encasing limestone (note carbonate remnants at yellow arrow) most likely occurred after deposition of the entire clast; otherwise, this grain probably would not have survived transport intact.

PPL / XPL, RDI, Scale bar = 0.16 mm

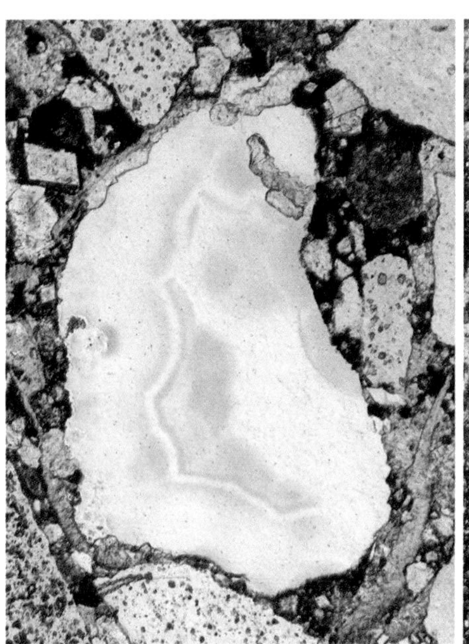

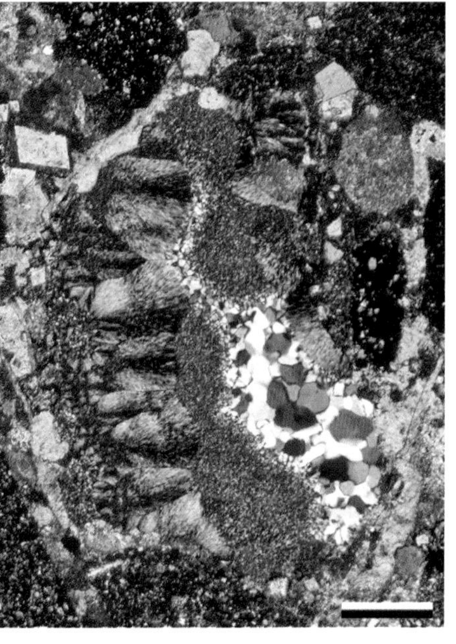

Lo. Permian (Wolfcampian) Hueco Fm., Powwow Conglomerate, Hudspeth Co., Texas

This is a chalcedony SRF from a cherty calclithite. Carbonate extraclasts and chert fragments are the dominant grains in this rock that formed a basal lag deposit on an unconformity atop eroded Pennsylvanian strata. This SRF shows an eroded vug-fill clast. Botryoids of fibrous chalcedony and moganite line the vug (note the pale brown lumpy surfaces visible in plane-polarized light) and the last stage of filling is megaquartz. The fibrous and speckled areas are individual chalcedony and moganite fibers that were cut parallel or perpendicular to fiber long axes.

PPL | XPL, Scale bar = 0.26 mm

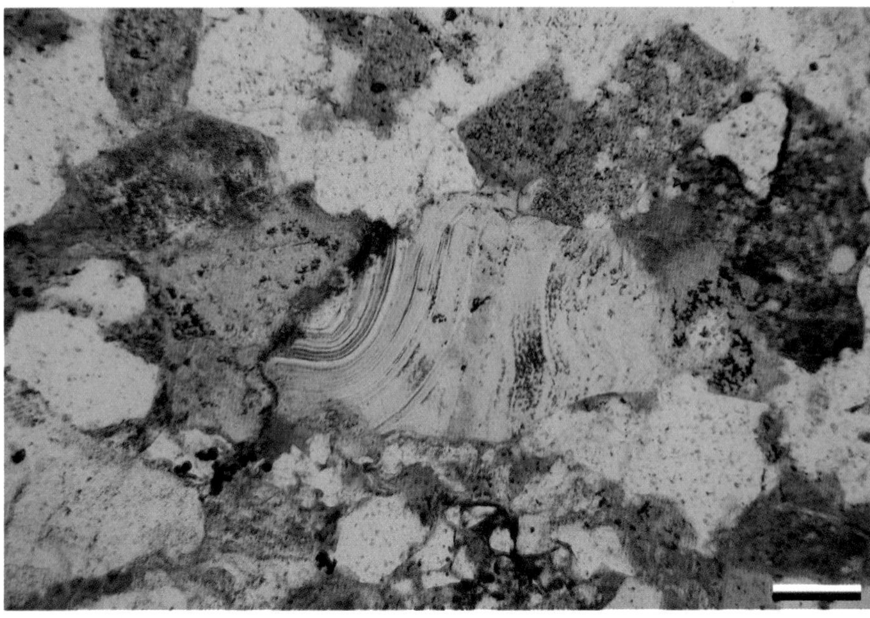

Up. Cretaceous Mesaverde Gp., Piceance Basin, Colorado

A sedimentary rock fragment consisting mainly of chalcedony and moganite. The amount of moganite can vary greatly within the growth zones of chalcedony. This fragment shows numerous growth bands that had subtle compositional variations (possibly zones with more moganite) making some layers more soluble than others. Partial leaching of these layers has produced secondary porosity within the grain (and filled with blue epoxy giving the grain an atypical color). Overall, chalcedony is nearly twice as soluble as quartz under most conditions.

PPL, BDI, Scale bar = 0.16 mm

Metamorphic Rock Fragments

Metamorphic rock fragments (MRFs) cover a wide spectrum of grain types derived from high- to low-grade metamorphic terranes. Classification of argillaceous/ micaceous metamorphic rock fragments is based on the metamorphic rank, which is a function of texture and sheet silicate mineralogy (Table 3.1). During modal analysis, MRFs can be separated into metasedimentary or meta-igneous or foliated and nonfoliated types. Common grain types recognized as MRFs include schist, gneiss and quartzite clasts. Metachert is sometimes recognizable, but most chert fragments are classed as SRFs.

Major characteristics:

Composition: Dominated by quartz and sheet silicates (muscovite, biotite and chlorite); gneiss fragments also contain feldspar. Some exotic constituents are diagnostic of a metamorphic origin, such as staurolite, kyanite, sillimanite, zoisite, glaucophane and andalusite.

Texture: A preferred orientation of crystals suggests either flow banding (volcanic or intrusive) or growth of crystals under pressure (metamorphic). Metamorphic grains commonly display some foliation (e.g., slaty cleavage) or schistosity. Quartz crystals in MRFs may display sutured contacts if highly strained. MRF grains may contain high aspect-ratio crystals or may be composed of polygonal equant crystals, depending whether they have been annealed.

Other: Metamorphic rock fragments grade into sedimentary and igneous rock fragments, and their differentiation can sometimes be a fine art. The crystal sizes for gneisses make them difficult to identify in finer-grain clast sizes. Most of the grains in this chapter are sand sized, but silt-sized MRFs can still provide important provenance information.

Table 3.1: Characteristics of argillaceous/micaceous metamorphic rock fragments based on their metamorphic rank (modified from Garzanti and Vezzoli, 2003).

Metamorphic Rank	Texture	Phyllosilicates
None	unoriented	clay minerals
Very low	rough cleavage	illite, chlorite
Low	strong cleavage	sericite
Medium	schistosity	tiny micas
High	crystals <62.5 μm	muscovite
Very high	crystals >62.5 μm	biotite

Lo. Paleozoic andalucite schist, Carroll Co., New Hampshire

Foliated and crenulated micas in a schist. This is an example of the type of schistose texture commonly seen in high-rank metamorphic rock fragments in which elongate quartz grains are separated by thin mica plates. Detrital fragments of such rock types are normally not very durable and thus rarely survive extensive transport; however, when such fragments are found, they are excellent indicators of a metamorphic source. In this example, the crenulation effectively represents a secondary cleavage, resulting from a second deformation event in the source terrane.

PPL | XPL, Scale bar = 0.51 mm

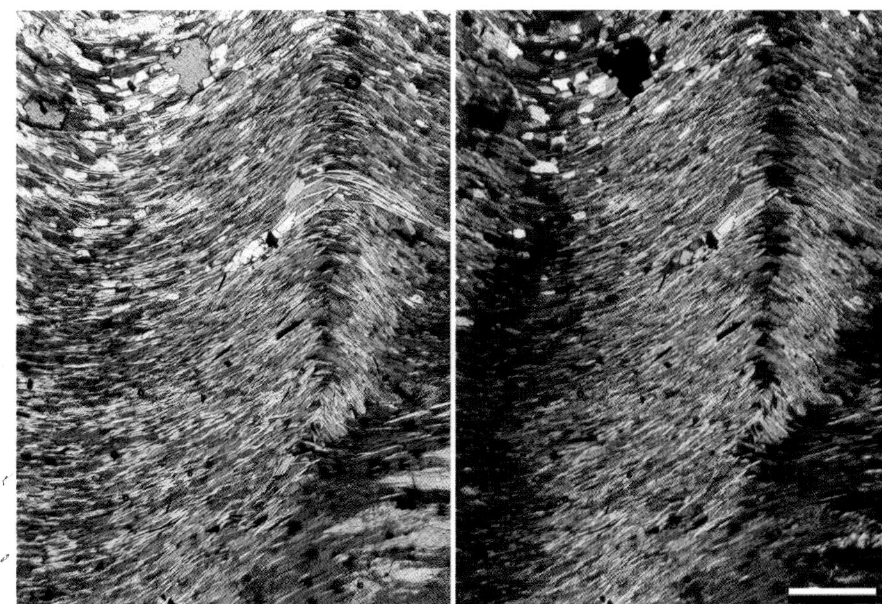

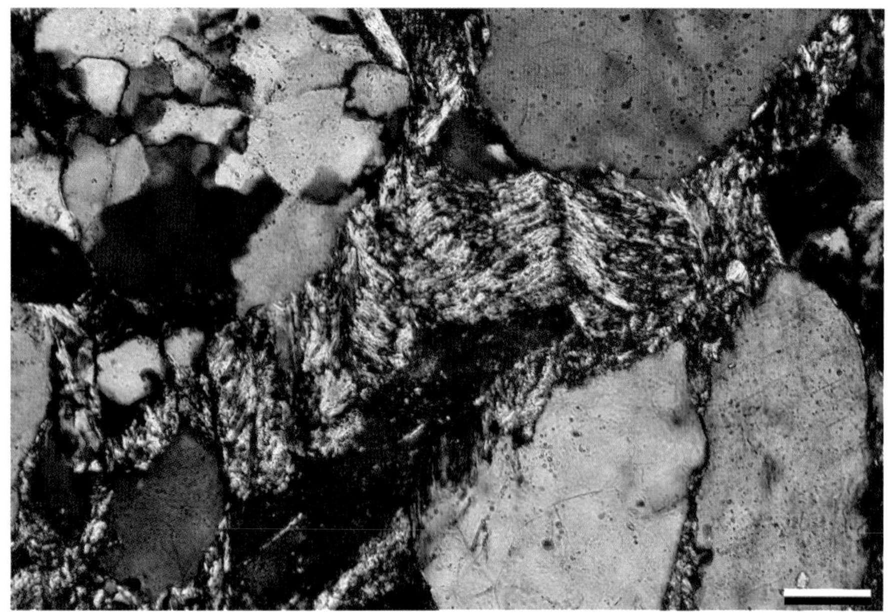

Up. Silurian Bloomsburg Fm., Warren Co., New Jersey

A crenulated metamorphic rock fragment reworked into a sandstone. The well-foliated micas in this grain are biotite and chlorite. This grain is from a high-grade metamorphic source rock.

XPL, Scale bar = 0.04 mm

Up. Triassic – Lo. Jurassic Portland Fm., Middlesex Co., Connecticut

This schistose metamorphic rock fragment contains scattered, elongate quartz crystals in a matrix of foliated micas. The micas and quartz have a preferred orientation normal to the direction of compression. Schist fragments tend not to survive extensive transport; therefore, this clast was probably deposited in relatively close proximity to its source.

PPL, Scale bar = 0.26 mm

Triassic Sherwood Sandstone Gp., Co. Antrim, Northern Ireland, U.K.

A detrital schist fragment (MRF). This grain consists almost entirely of muscovite crystals that are all cut parallel to their foliation, indicating that this metamorphic rock fragment was cut almost precisely parallel to schistosity along a mica-rich layer in the rock fragment.

PPL / XPL, Scale bar = 0.51 mm

Jurassic Ula Fm., Norwegian sector, North Sea †

An amphibole in a quartz schist rock fragment. The amphibole is likely hornblende and is recognized by its green color, pleochroism and second-order birefringence. The typical cleavages for amphiboles are not visible because it is an "inconvenient" orientation. The rock is also foliated.

PPL | XPL, KFS, BDI, Scale bar = 0.07 mm

Up. Triassic – Lo. Jurassic Portland Fm., Middlesex Co., Connecticut

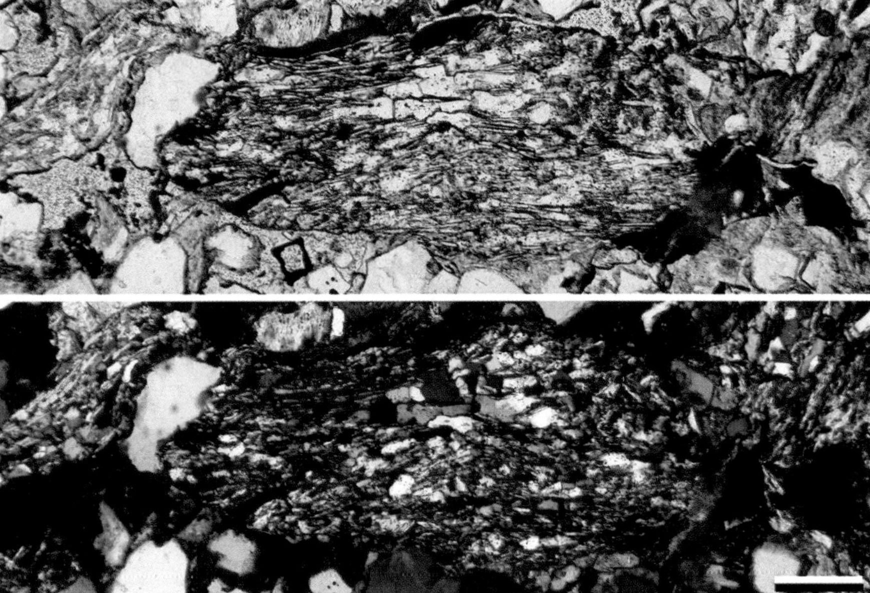

A foliated, detrital schist MRF fragment composed of quartz, muscovite and biotite. The quartz crystals are elongate, with micas interspersed between the crystals. The proportion of micas relative to quartz in such fragments affects the clast's preservation potential—the more micas the grain contains, the shorter the distance over which the grain is likely to survive weathering and abrasion. Notice also how most of the muscovite crystals show the same blue birefringence (an indication of their uniformity of size and orientation).

PPL / XPL, Scale bar = 0.26 mm

Up. Triassic – Lo. Jurassic Portland Fm., Middlesex Co., Connecticut

A metamorphic rock fragment consisting of foliated quartz-rich schist. Foliation is visible due to the presence of highly birefringent micas interspersed with the elongate quartz crystals that are predominant in this fragment.

XPL, Scale bar = 0.26 mm

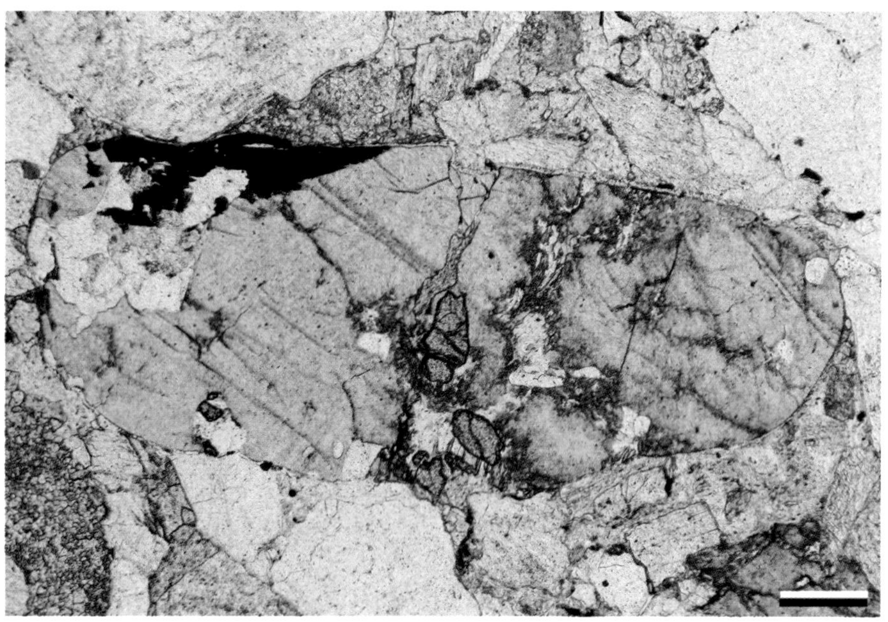

Mid. Permian Wegener Halvø Fm., Karstryggen area, East Greenland

A metamorphic rock fragment made up of biotite mica, quartz and garnets. The garnets (gray crystals) have very high relief and are isotropic in cross-polarized light. This is a fragment that is somewhere between a schist and a gneiss from a medium- to high-grade metamorphic source terrane. Fragments like this break down during weathering and transport and release grains of garnets and quartz.

PPL, BDI, Scale bar = 0.26 mm

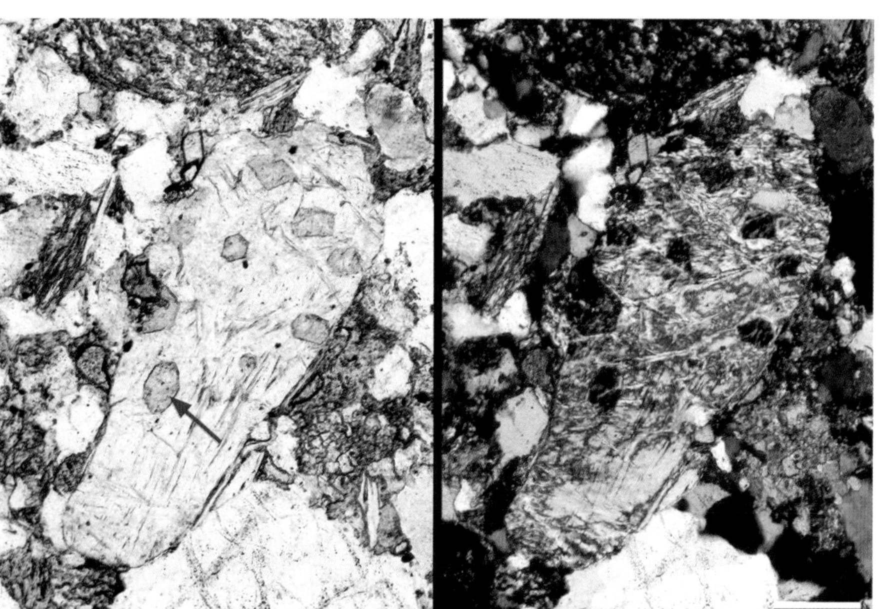

Up. Cretaceous, Rieselberger Flysch, Bavaria, Germany

A schist fragment from a medium- to high-grade metamorphic terrane. This grain is composed primarily of muscovite micas (white with high birefringence) and chlorite (the pale green mineral; example at arrow). The chlorite clusters are probably retrograde replacements of higher-grade garnets and/or hornblendes. Mica-rich schist grains rarely survive significant transport unless carried in suspension with muddy sediments. Note that the chlorite crystals have anomalous colors in cross-polarized light. This rock fragment was cut parallel to the main muscovite foliation.

PPL | XPL, Scale bar = 0.26 mm

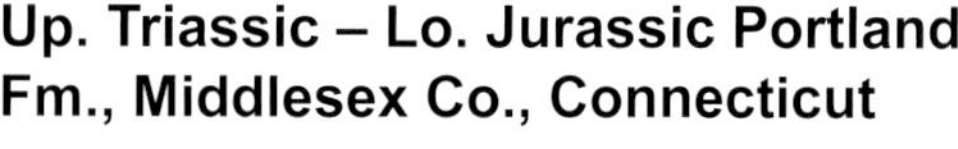

Up. Triassic – Lo. Jurassic Portland Fm., Middlesex Co., Connecticut

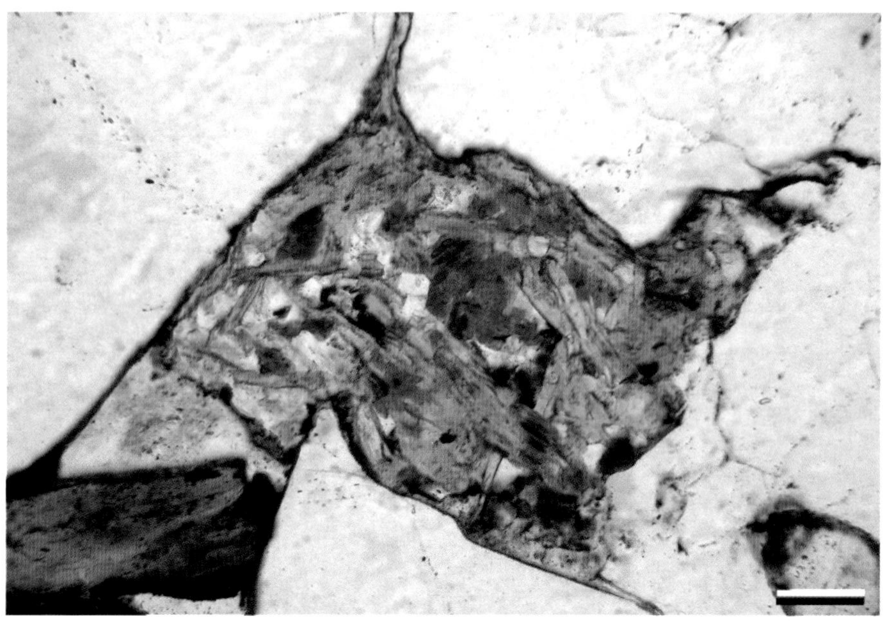

A chlorite, biotite mica and quartz schist fragment. The disruption of the original foliation is related to deformation during burial. Because this grain is structurally weaker than surrounding quartz grains, compaction has caused the rock fragment to deform in a ductile manner and fill the primary intergranular porosity.

PPL, Scale bar = 0.08 mm

Up. Triassic – Lo. Jurassic Huizachal Fm., Nuevo Leon, Mexico

The chloritic rock fragments (indicated by red arrows) in this sandstone were derived from a probable low-grade metamorphic source that probably was relatively close to the site of deposition. Internally, the grains still preserve foliation and because they are almost completely chlorite, they have anomalous ultrablue birefringence. The thin, more brightly birefringent minerals surrounding some of the grains are carbonate and illite cements.

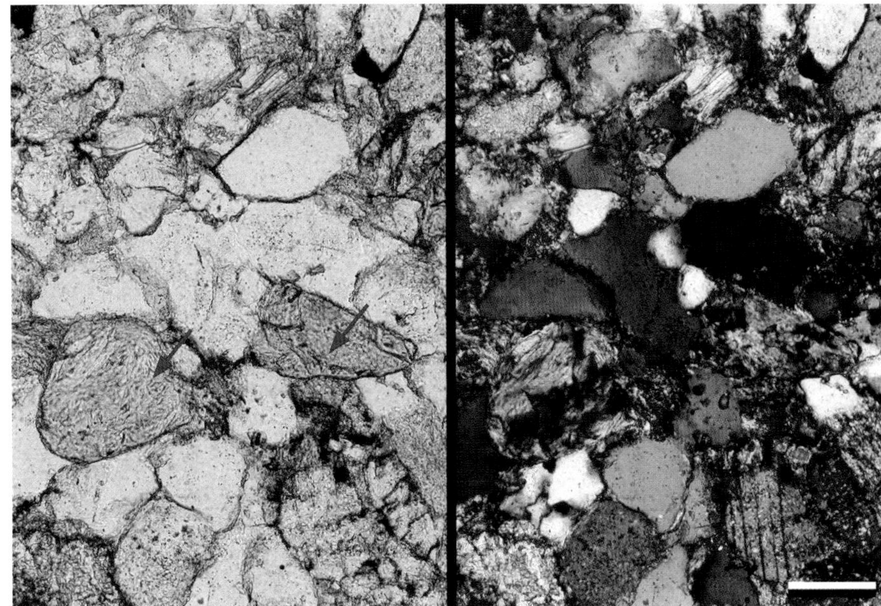

PPL | XPL, Scale bar = 0.10 mm

Up. Triassic – Lo. Jurassic Portland Fm., Middlesex Co., Connecticut

Several quartz-rich schist and schistose quartzite metamorphic rock fragments. This is a medium- to high-grade MRF. Most of the micas in these grains are muscovite. During more extensive transport, such grains would probably break down into silt-size fragments or their individual constituent quartz crystals.

XPL, Scale bar = 0.26 mm

Up. Jurassic Eldfisk Fm., Norwegian sector, North Sea †

This schist fragment has acted in a ductile manner during compaction. Superficially the grain resembles a siltstone, but the alignment of micas and sutured contacts between constituent quartz grains identifies it as a MRF.

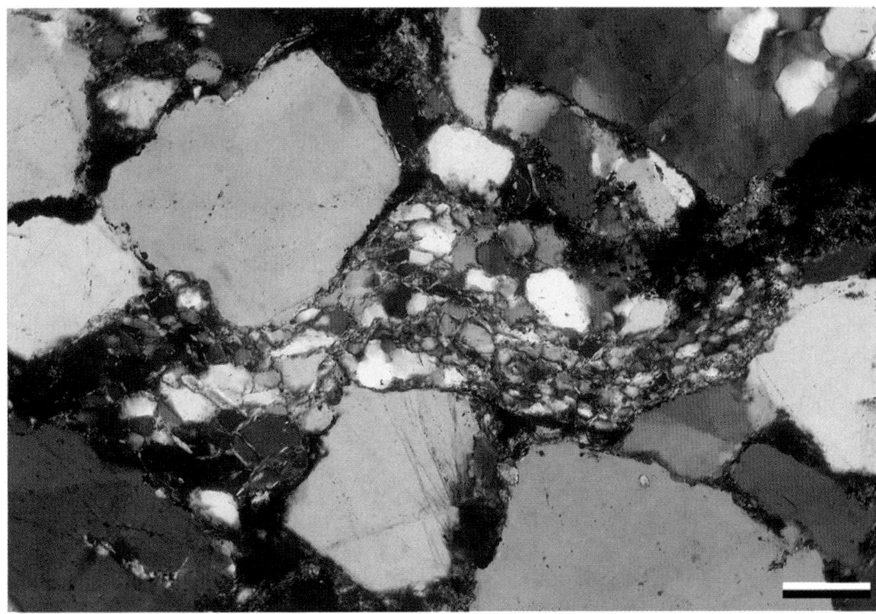

XPL, Scale bar = 0.12 mm

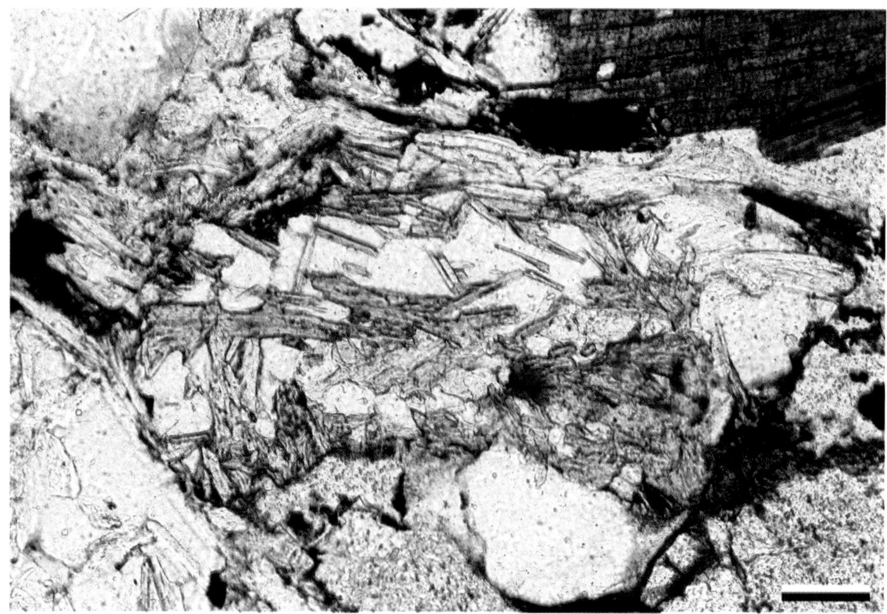

Up. Triassic – Lo. Jurassic Portland Fm., Middlesex Co., Connecticut

A large, quartz-rich, chlorite schist fragment is present at the center of this image. It is surrounded by other grains probably also derived from metamorphic sources, including a partially chloritized biotite in the upper right. The schistose grains have deformed in either ductile or rigid fashions depending on their competence and rigidity.

PPL, Scale bar = 0.10 mm

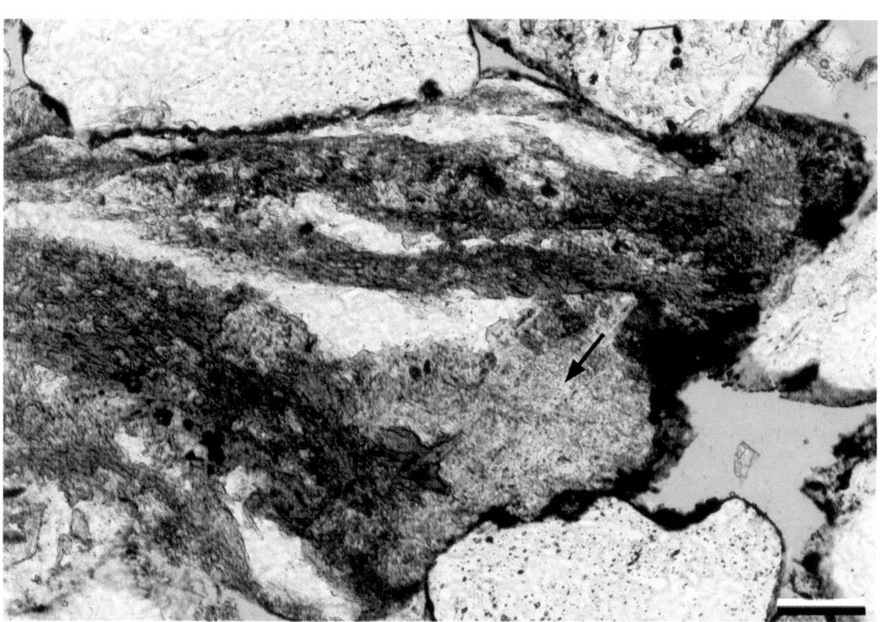

Triassic Sherwood Sandstone Gp., Co. Antrim, Northern Ireland, U.K. †

A metamorphic rock fragment (schist) with well-defined foliation. The rock consists of layers of K-feldspar (stained yellow) and biotite sandwiched between quartz rich layers. The larger area of K-feldspar contains a plagioclase (black arrow) in the center that may be a porphyroblast.

PPL, KFS, BDI, Scale bar = 0.17 mm

Up. Permian Schuchert Dal Fm., Jameson Land, East Greenland

A polycrystalline quartz grain that has a metamorphic (quartzite) origin as indicated by the elongation and undulatory extinction of the individual subcrystals as well as their sutured boundaries. Quartzites are normally very stable rocks, but inclusions within individual rock fragments can decrease resistance to chemical weathering and mechanical abrasion.

PPL | XPL, BDI, Scale bar = 0.26mm

Triassic Sherwood Sandstone Gp., Co. Antrim, Northern Ireland, U.K.

A fragment of metamorphic quartzite or polycrystalline quartz. This is probably a intermediate- to high-grade metamorphic rock fragment based on the equidimensional, crenulated quartz crystals; scattered biotite flakes and dark-green chlorite crystals also are present. Foliation is visible in both the micas and the alignment of the quartz crystals.

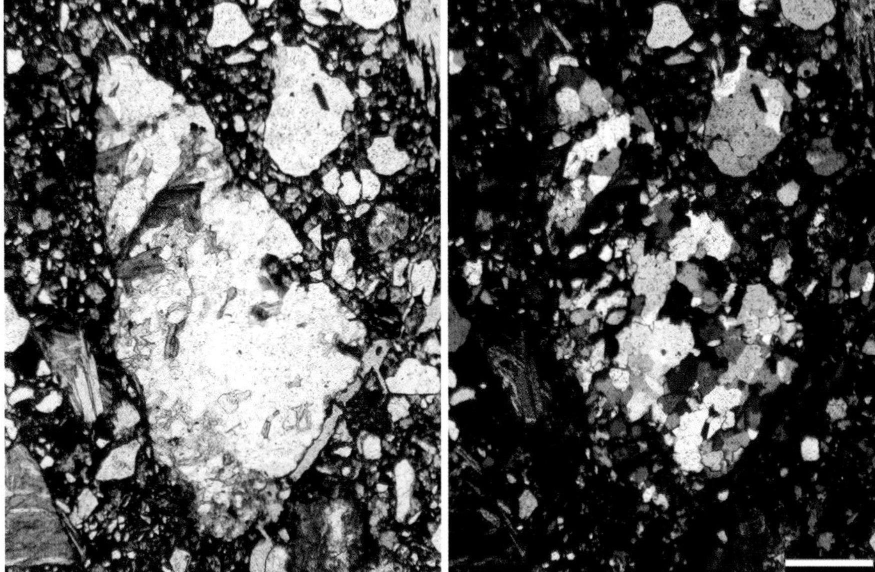

PPL | XPL, Scale bar = 0.51 mm

Triassic Sherwood Sandstone Gp., Co. Antrim, Northern Ireland, U.K.

A fragment of metamorphic schistose quartzite. This is identifiable as a medium-grade metamorphic rock based on the predominantly equant shapes of the included quartz crystals. There is foliation present, defined by large green chlorite crystals and muscovite flakes.

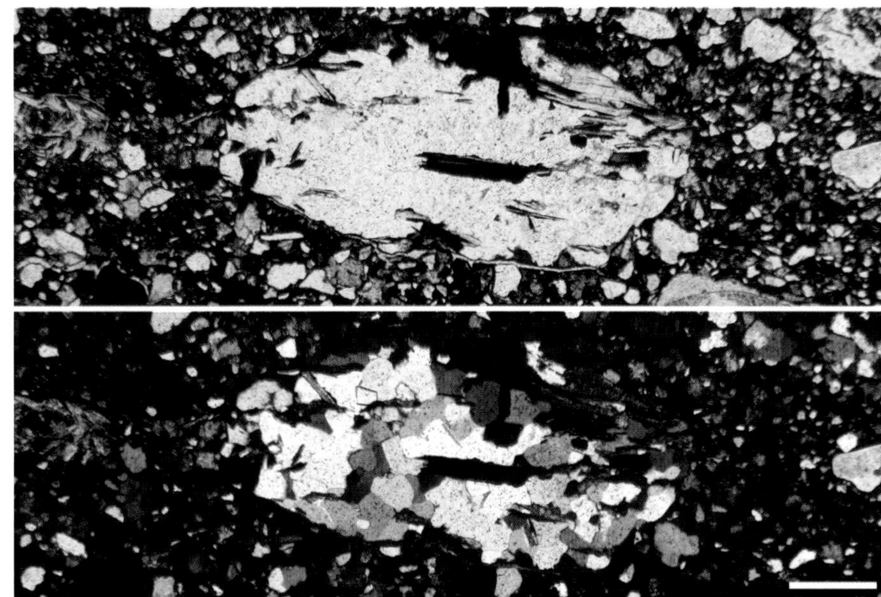

PPL / XPL, Scale bar = 0.51 mm

Mid. Permian Wegener Halvø Fm., Karstryggen area, East Greenland

A large grain of schistose quartzite. This metamorphic rock fragment contains numerous, elongate, crenulate quartz crystals that are sutured together and interspersed with scattered flakes of mica. Because tightly welded quartz dominates this grain, it is less likely to break down during weathering and transport than the more micaceous or schistose MRFs.

PPL / XPL, Scale bar = 0.51 mm

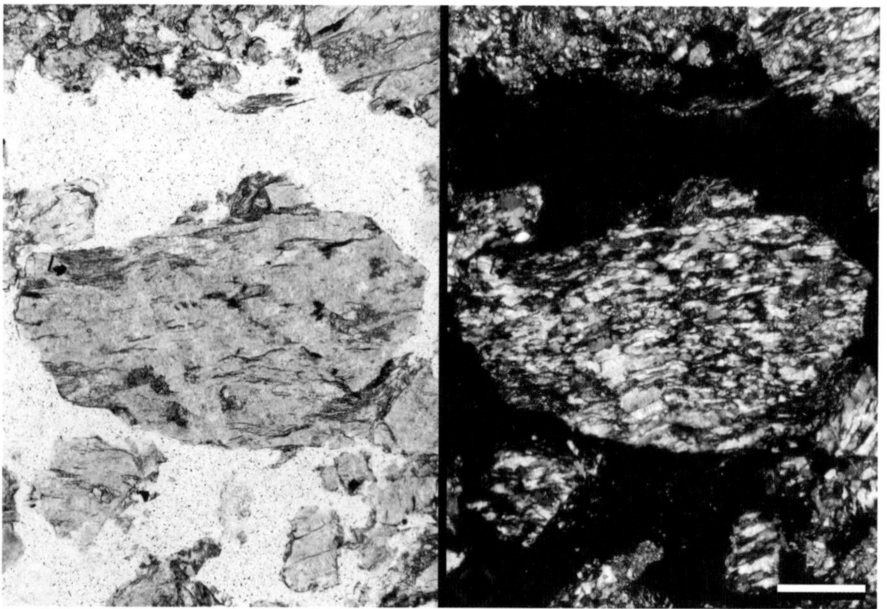

Oligocene Tongriano Conglomerate, Liguria, Italy

These quartzite grains are composed of finely-crystalline elongate quartz crystals that are welded together. Quartzite fragments get preferentially preserved in the rock record, because they tend to lack the micas that weaken grains. Although these grains may break into smaller fragments, they are normally still recognizable.

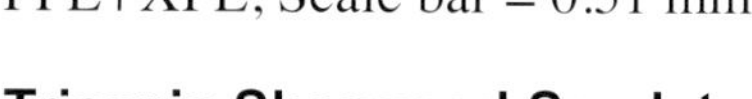

PPL | XPL, Scale bar = 0.51 mm

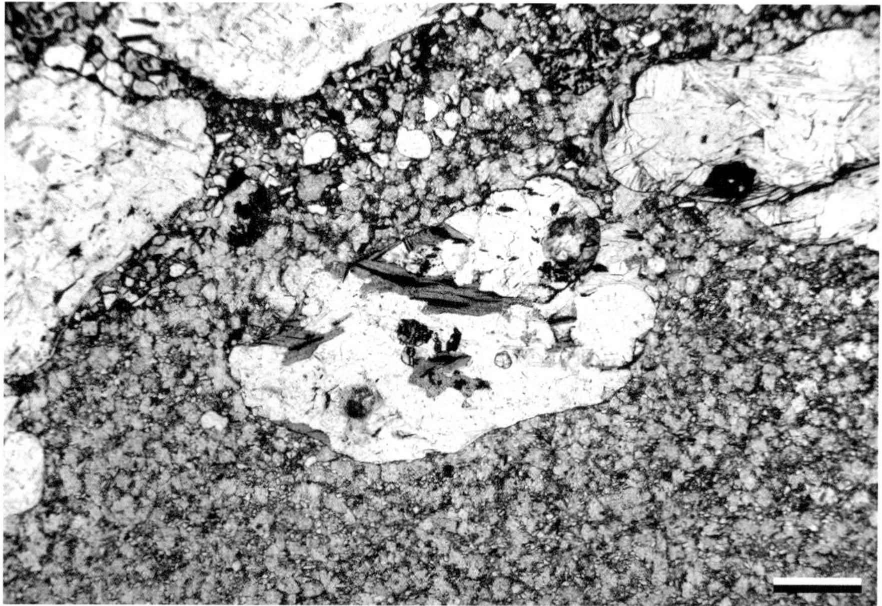

Triassic Sherwood Sandstone Gp., Co. Antrim, Northern Ireland, U.K.

Metamorphic rock fragments are derived from a continuum of source rock types — especially between commonly co-located schist and gneiss. The small size of many MRFs make it especially difficult to definitively identify clasts from such settings. In this example, the clast is a schistose quartzite or gneissic rock fragment—the individual constituent quartz crystals are coarsely crystalline as in a gneiss, but the fragment is dominated by quartz and contains no feldspar and thus is more likely derived from a metaquartzite.

PPL, Scale bar = 0.38 mm

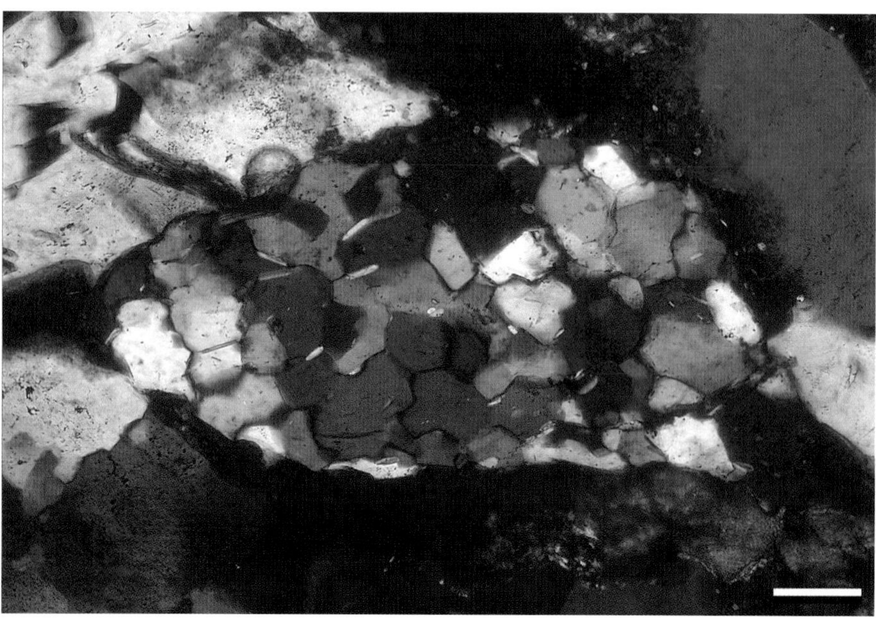

Triassic Snadd Fm., Barents Sea, offshore Norway †

A quartz-rich metamorphic rock fragment consisting almost entirely of quartz with interspersed small muscovite crystals. The quartz displays some sutured contacts, but is approaching a granoblastic polygonal aggregate texture typical of high-grade metamorphism.

XPL, BDI, Scale bar = 0.07 mm

Mid. Jurassic Fensfjord Fm., Norwegian sector, North Sea †

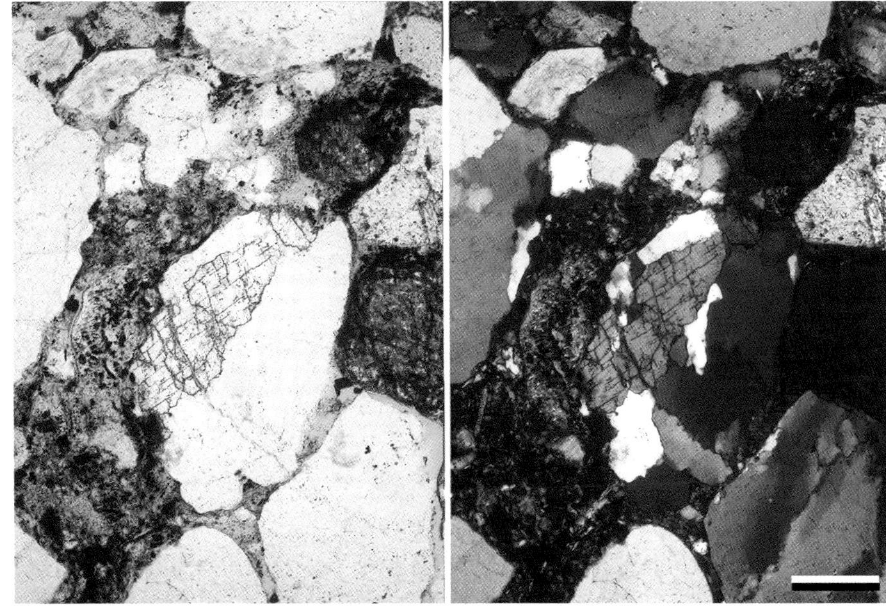

A grain (center left) comprising several crystals, but composed of sutured quartz and untwinned plagioclase feldspar. This may be from a metamorphosed igneous rock or from a gneiss. The coarser crystal size of gneisses, sometimes limit their identification, especially in finer- grained sandstones.

PPL | XPL, KFS, BDI, Scale bar = 0.07 mm

Mid. Jurassic Fulmar Fm., United Kingdom sector, North Sea †

A granitic-looking grain composed of K-feldspar (microcline), quartz and plagioclase. The large number of small quartz crystals suggests that the grain may be metamorphic (?gneiss), although it could have an igneous plutonic origin.

XPL, KFS, BDI, Scale bar = 0.12 mm

Up. Triassic – Lo. Jurassic Portland Fm., Middlesex Co., Connecticut

The metamorphic rock fragments in this photomicrograph are dominated by hornblende gneisses. The gneiss clasts are composed of bands of hornblende, quartz and feldspar. The pleochroic, blue-green to olive hornblendes display the typical cleavage (124°/56°) for amphiboles. Gneisses form as a result of medium- to high-grade regional metamorphism. In addition to this clast, there is a biotite mica that was replaced by chlorite (red arrow). This lithic-rich sandstone is cemented by calcite (especially well seen in the right-hand quarter of the image).

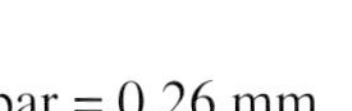

PPL / XPL, Scale bar = 0.26 mm

Mid.? Permian Huledal Fm., Jameson Land, East Greenland

This gneissic-looking rock fragment is composed mostly of microcline crystals. While it is possible that this grain is an plutonic igneous rock fragment, most of the other clasts within this lithic arenite are either volcanic or metamorphic rock fragments. This sample has been partially altered (contact metamorphosed) by a nearby Cenozoic dike.

PPL | XPL, BDI, Scale bar = 0.51 mm

Up. Permian Schuchert Dal Fm., Jameson Land, East Greenland

A metamorphic rock fragment composed of quartz, feldspar and chlorite in a lithic arenite. Within this rock fragment, the chlorites are aligned and the quartz crystals exhibit undulatory and crenulated crystal boundaries. These fabrics all indicate a metamorphic origin for the grain. Feldspar (the cloudy grains in plane-polarized light) has been largely altered to sericite.

PPL | XPL, AFeS, BDI, Scale bar = 0.51 mm

Up. Permian Schuchert Dal Fm., Jameson Land, East Greenland

A gneiss fragment composed of feldspar and polycrystalline quartz crystals. The quartz crystals are strongly undulose, elongated and have crenulated crystal boundaries. Feldspars (the cloudy, brownish grains) have been sericitized and are partially leached. In addition to MRFs, there are numerous carbonate bioclasts in this sample that act as substrates for the zoned, slightly ferroan to very ferroan calcite cements (stained purple).

PPL, AFeS, BDI, Scale bar = 0.51 mm

Igneous Rock Fragments

Rock fragments derived from an igneous parent rock include volcanic and plutonic clasts. They can be further subdivided for modal analysis into compositional types (basalt, rhyolite, granite, gabbro, etc.) and ductile or rigid variants. Volcanic grains are more likely than most other fragments to be degraded and to act in a ductile manner. Because igneous rock fragments can be composed of a number of minerals with a range of sizes, determinations of their mineralogic composition, crystallinity and texture are key to identifying them.

Plutonic Rock Fragments:

Major characteristics:

Crystal size – Plutonic grains must contain two or more crystals to qualify as a rock fragment, and they are composed of phaneritic crystals (large enough to be seen with the naked eye). Crystals tend to be roughly the same size and commonly are anhedral to subhedral. As with other rock fragments, they are more common in coarser-grained modes.

Composition – Dominated by clasts with granitic composition (quartz, feldspar, iron-titanium oxides and mica or hornblende).

Textures – Include micrographic and graphic granites, granophyres, myrmekites and symplectites, perthites and antiperthites.

Other – Coarsely crystalline meta-igneous rocks such as gneisses can resemble granite at the scale of individual grains, and it may be difficult to distinguish between them.

Volcanic Rock Fragments:

Major characteristics:

Crystal size – Groundmass typically is glassy (non-crystalline) to aphanitic (very finely crystalline). Porphyritic fabrics are common (large crystals in a much finer groundmass). Phenocrysts in porphyries tend to be subhedral to euhedral.

Composition – May contain feldspar, quartz, hornblende, olivine, pyroxene, iron-titanium oxides and glass.

Texture – Spherulitic or perlitic fabrics or radiating clusters of crystals are common. Vesicles may be filled (amygdales) or empty, and glass shards can display curved faces. Lava clasts may show lath-like crystals, which are locally aligned (trachytic texture). Melt inclusions and resorption features may be seen in some crystals.

Other – Volcanic grains commonly alter to clays (smectite, illite, chlorite, etc.), and their identification may become difficult. Unaltered volcanic glass is isotropic, but it may alter either to quartz or to K-feldspar. Staining (for K-feldspar) may help to identify the presence of what was once volcanic glass within lava, ignimbrite or pumice grains.

Up. Pennsylvanian Alamitos Fm., San Miguel Co., New Mexico

A granule-sized granitic rock fragment. The grain is composed of quartz and microcline. These grains were shed from a granite-cored Ancestral Rockies uplift into a nearby basin. Note how the crystals are roughly equicrystalline and anhedral in the fragment. Large grains like this, that haven't been broken down into their constituent crystals during transport, can help in source area determination.

PPL | XPL, AFeS, Scale bar = 0.64 mm

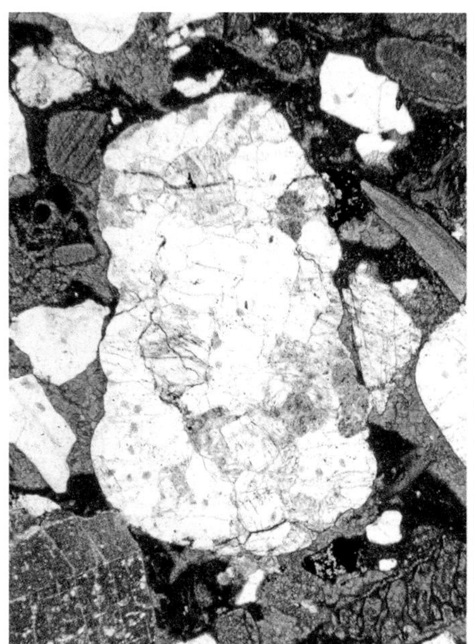

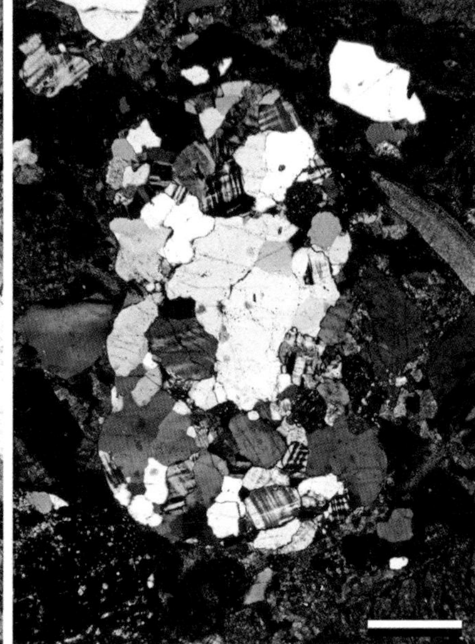

Mid. Permian Huledal Fm., Karstryggen area, East Greenland

A plutonic rock fragment from a sandstone that has been somewhat altered by contact metamorphism along a dike wall. The feldspars, in particular, have been substantially vacuolized and sericitized during alteration (cloudy brownish areas). The vacuolization emphasizes the need to distinguish predepositional from postdepositional alteration. This commonly is done by observing whether all the potentially susceptible grains are altered or only a few.

PPL | XPL, AFeS, Scale bar = 0.64 mm

Lo. Miocene Arikaree Fm., Platte Co., Wyoming

A granitic rock fragment composed of crystals of plagioclase and quartz. The plagioclase crystals have albite twinning and are partially altered to sericite. During weathering, grains like this can potentially decompose into their component crystals.

XPL, Scale bar = 0.16 mm

Up. Pennsylvanian Bursum Fm., Socorro Co., New Mexico

Plutonic rock fragments composed of quartz (as at red arrow) and altered plagioclase (as at yellow arrow). The plagioclase crystals are vacuolized and partially replaced by sericite. Many of the grains surrounding this clast are similar in size to the crystals in the grain, emphasizing the fact that rock fragments are found predominantly in the coarser-grained fractions.

XPL, Scale bar = 0.38 mm

Precambrian Town Mountain Granite, Llano Co., Texas

An example of a quartz and feldspar intergrowth with highly embayed quartz from a granitic source rock. Weathering of the rock, especially with feldspar dissolution, can create quartz clasts that look similar to volcanic quartz, except that these have more inclusions and may have undulatory extinction.

XPL, Scale bar = 0.38 mm

Paleocene Tang Fm., Møre Basin, Norwegian Sea †

A grain composed of plagioclase feldspar (cloudy white) and chlorite (dull green). The grain may either be derived from a metagranite or metagabbro, or a high-grade metamorphic gneiss which has undergone retrograde metamorphism. The chlorite contains areas of brown biotite, which the chlorite may have replaced, but there is evidence from crystal shapes that the original mineral was hornblende.

PPL, KFS, BDI, Scale bar = 0.12 mm

Mid. Jurassic Fulmar Fm., United Kingdom sector, North Sea †

The grain in the center of the image is entirely composed of feldspar that is dominantly microcline, but contains areas of albite. The microcline and the albite are both single crystals so this could be classed as a perthite grain. However, because this is likely to be an intergrowth rather than exsolution lamellae, it is better classed as an igneous rock fragment.

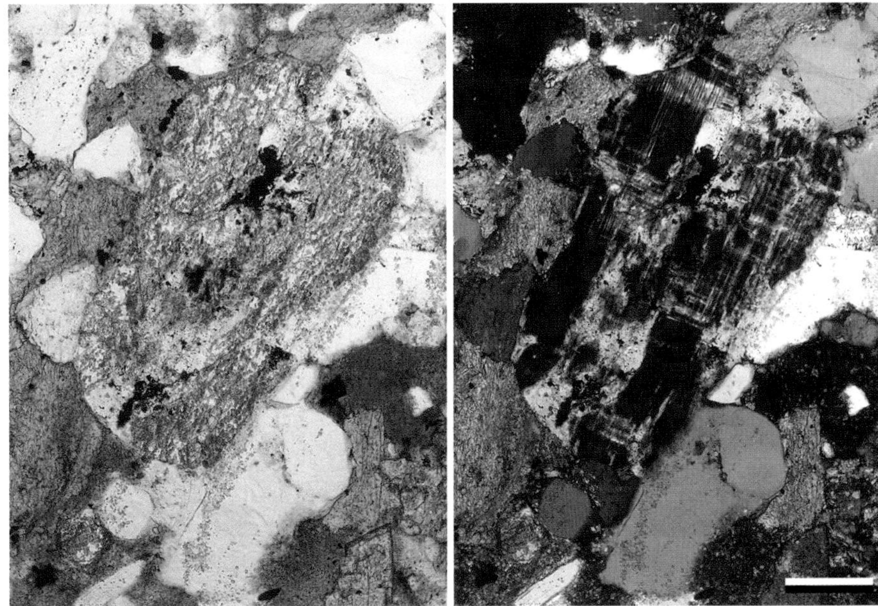

PPL | XPL, KFS, AFeS, BDI, Scale bar = 0.12 mm

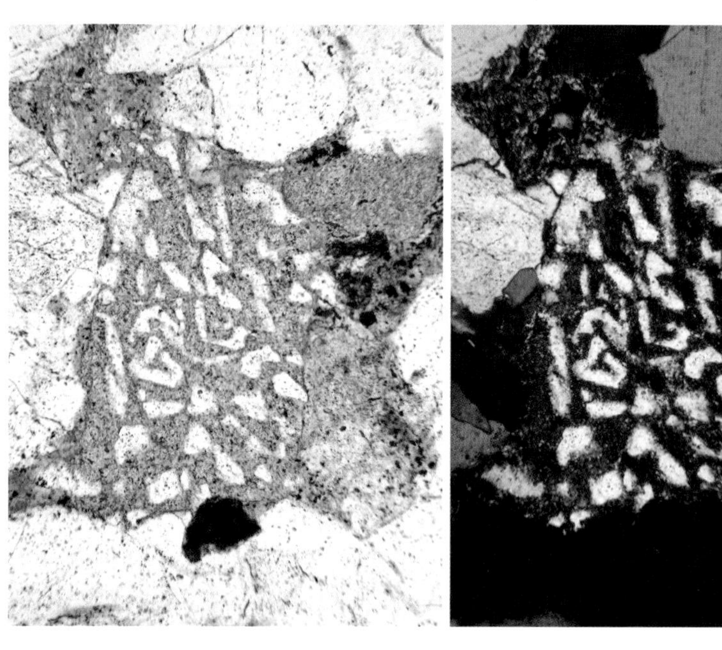

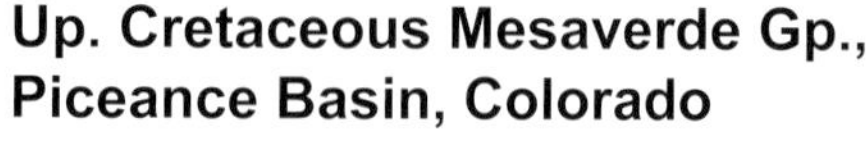

Up. Cretaceous Mesaverde Gp., Piceance Basin, Colorado

Micrographic and graphic granite textures are intergrowths of two minerals that have the appearance of runic writing. In this sample, quartz crystals (the inclusion-poor crystals) are intergrown with an alkali feldspar (inclusion-rich). If the fabric is only visible in thin section, then it is called micrographic; if it is visible to the naked eye, then it is termed graphic granite.

PPL | XPL, BDI, Scale bar = 0.20 mm

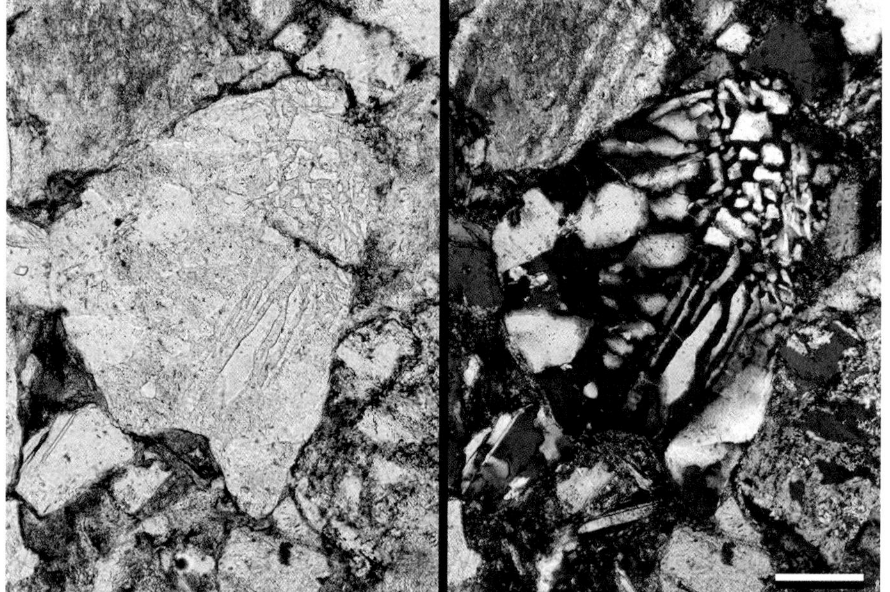

Miocene Topanga Fm., Ventura Co., California

This granite clast contains granophyric fabric, another type of micrographic intergrowth. Granophyric texture is an intergrowth in which the intergrown crystals appear to be radiating outward. Quartz intergrowths are in optical continuity with each other because they represent a single crystal. Micrographic and graphic granite is thought to form during the last stages of crystallization of felsic magmas.

PPL | XPL, Scale bar = 0.10 mm

Up. Pennsylvanian – Lo. Permian Granite Wash, southern Oklahoma

Identification of feldspar/quartz intergrowths can be problematic, because the textures can overlap. In this clast, some of the intergrowth textures might be described as a myrmekitic (worm-like to globular) or granophyric (radiating). These intergrowths can provide information on the provenance of the grains. Granophyric textures are thought to form in cooler magma chambers containing volatiles/water or to be associated with bolides. Myrmekite is thought to be a product of metasomatism (alteration of felsic rocks by hot fluids) or to form during deformation.

XPL, Scale bar = 0.22 mm

Up. Pennsylvanian – Lo. Permian Granite Wash, southern Oklahoma

A granitic clast with a granophyric texture. Note the well-developed radiating fabrics of quartz crystals intergrown with alkali feldspars. The quartz is relatively inclusion-free compared to the feldspars. In this case, the grains have fabrics that are traceable back to a specific source rock, the Long Mountain Granite. The granite is thought to have formed by under-cooling of the felsic melt by 70–150°C (Morgan and London, 2012).

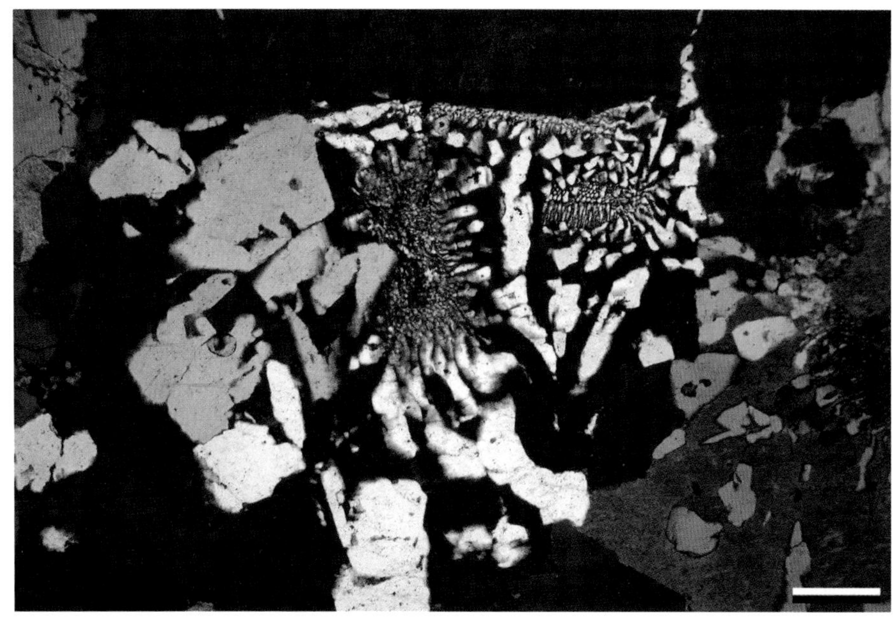

XPL, Scale bar = 0.22 mm

Eocene Thumbli Fm., Rajasthan, India

A grain showing micrographic intergrowth of K-feldspar and quartz. The feldspar is in optical continuity, but the quartz is not, suggesting that the feldspar is a single crystal and has intergrown with several quartz crystals. Porosity has been created in the grain through selective dissolution of parts of the K-feldspar, a process which could ultimately leave clusters of angular quartz "grains" in a secondary pore.

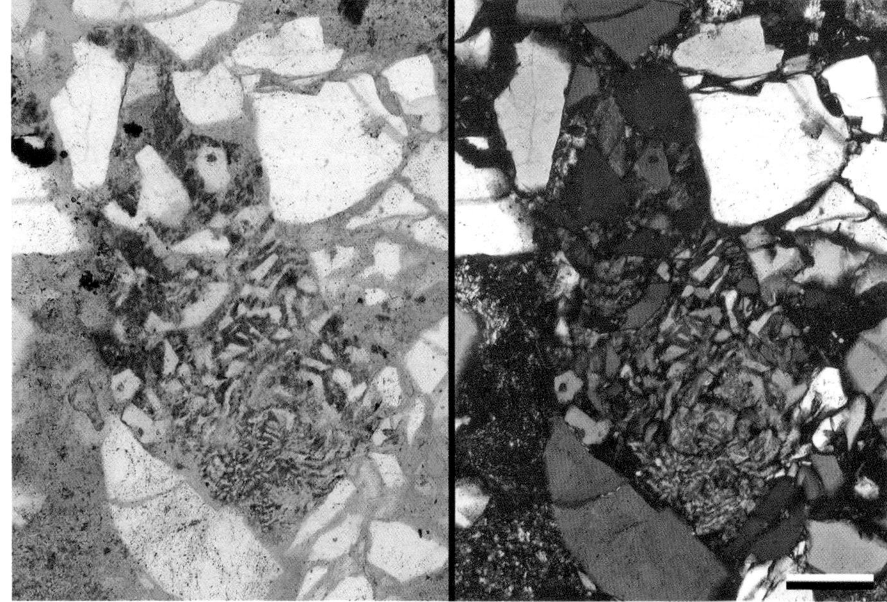

PPL | XPL, KFS, BDI, Scale bar = 0.06 mm

Mid.? Permian Huledal Fm., Canning Land, East Greenland

A very large volcanic rock fragment, probably a rhyolite porphyry fills the entire field of view in this image. This fragment contains euhedral, embayed volcanic quartz (example at yellow arrow) and altered euhedral plagioclase phenocrysts (example at red arrow) in an aphanitic matrix. Within the embayed quartz crystal near the center of each view, a melt inclusion (the dark rhombic feature) is visible under cross-polarized light. Chlorite (green crystals) occurs as a replacement of precursor amphibole.

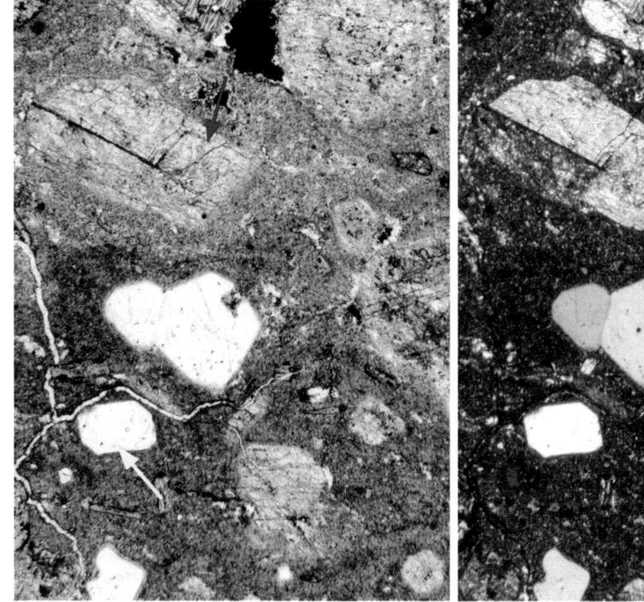

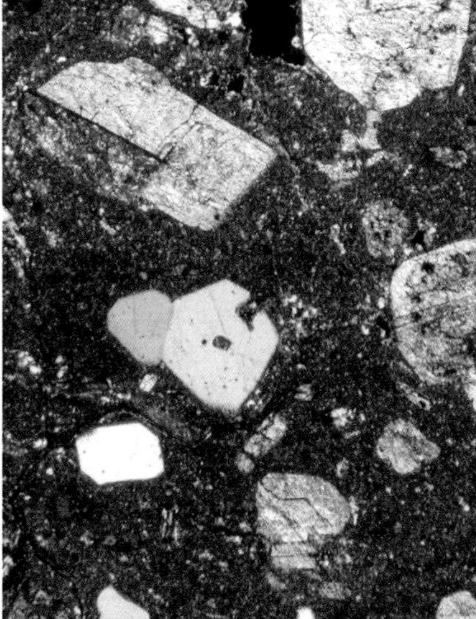

PPL | XPL, BDI, Scale bar = 0.51 mm

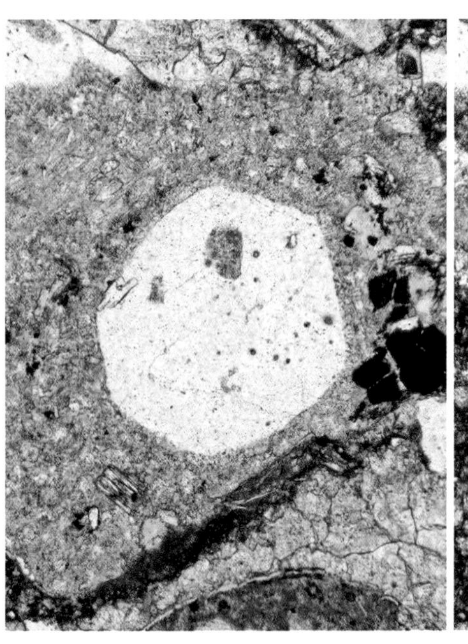

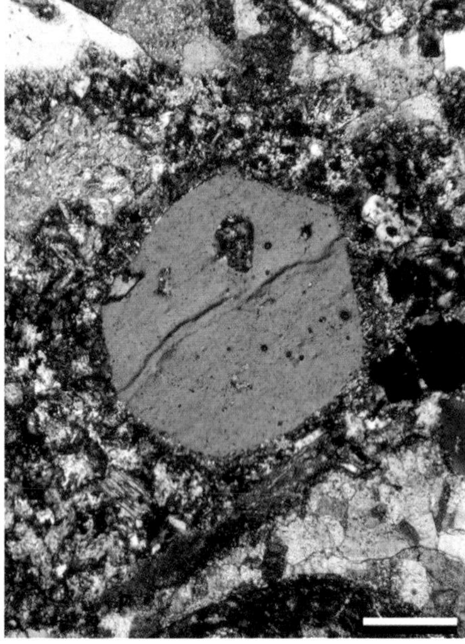

Mid.? Permian Huledal Fm., Canning Land, East Greenland

A large volcanic rock fragment, probably a rhyolite porphyry. This fragment contains a euhedral (hexagonal), embayed, volcanic quartz grain and altered phenocrysts in an aphanitic matrix. Chlorite (green crystals) and iron oxides occur as replacements of precursor amphiboles. The quartz crystal contains a melt inclusion; a clear indicator of this lithic fragment's volcanic origin.

PPL | XPL, BDI, Scale bar = 0.26 mm

Mid. Eocene – Lo. Oligocene Spears Gp., Socorro Co., New Mexico

A sandstone with abundant volcanic clasts of andesitic porphyry and loose plagioclase crystals derived from a volcanic source. The lithic fragments also contain crystals of clinopyroxene (brightly birefringent) and plagioclase within an aphanitic groundmass. These sediments were deposited as part of an Eocene caldera complex in central New Mexico.

XPL, Scale bar = 0.34 mm

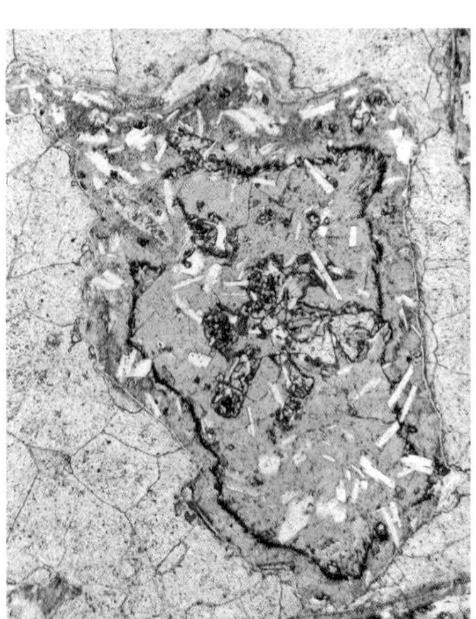

Lo. Oligocene Deborah Volcanic Fm., near Oamaru, Otago, New Zealand

These volcanic rock fragments formed through marine erosion of pillow basalts. The glass matrix is less stable than the entrained plagioclase and pyroxene (augite?) crystals, and alteration rims are common, here consisting of palagonite (bright yellow) and an outer discontinuous, fibrous zeolite mineral, mainly phillipsite. The palagonite is a heterogeneous alteration product of basaltic glasses, that may, in part, be smectitic. Micro-fibrous, spherulitic calcite cements fill the remaining porosity.

PPL | XPL, Scale bar = 0.64 mm

Lo. Oligocene Deborah Volcanic Fm., Otago, New Zealand

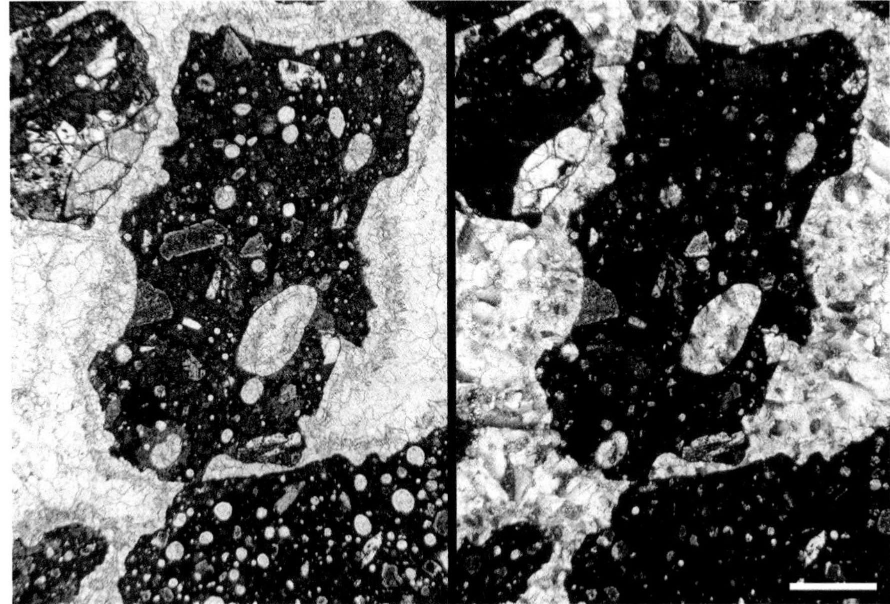

Amygdaloidal basaltic glass (tachylite) clasts containing slightly altered pyroxene (augite?) phenocrysts. Unlike the previous example, the glass does not have a well-developed alteration rim, but minor alteration (orange-brown) within the grains may be palagonite. The vesicles (holes) are mainly filled by isopachous, marine calcite cements with slightly undulatory extinction. The same type of calcite cement fills the primary intergranular porosity. Some of the vesicles are additionally filled with low birefringence to near-isotropic zeolite minerals.

PPL | XPL, AFeS, Scale bar = 0.64 mm

Mid. Eocene – Lo. Oligocene Spears Gp., Socorro Co., New Mexico

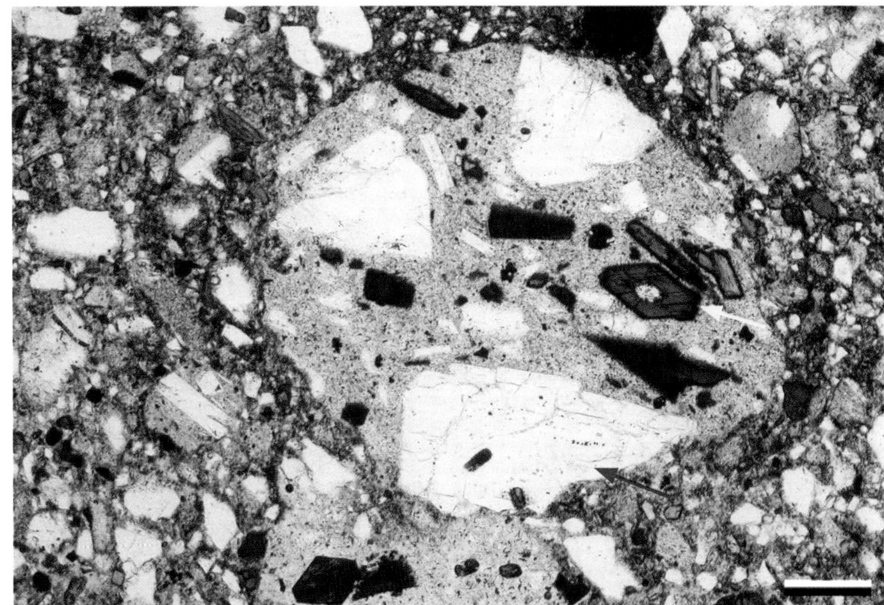

Reworked volcanic rock fragments in a lithic arenite. The volcanic clasts contain euhedral plagioclase (example at red arrow) and amphibole (example at yellow arrow) phenocrysts in an aphanitic groundmass. These grains are well rounded, indicating some transport prior to burial, but the feldspar exhibits little alteration, possibly related to inhibition of pore-water entry because of the extensive fine-grained matrix. Note also the opaque reaction rims on the amphibole grains (also visible at yellow arrow), caused by dehydration of ascending magma.

PPL, Scale bar = 0.34 mm

Mid. Eocene – Lo. Oligocene Spears Gp., Socorro Co., New Mexico

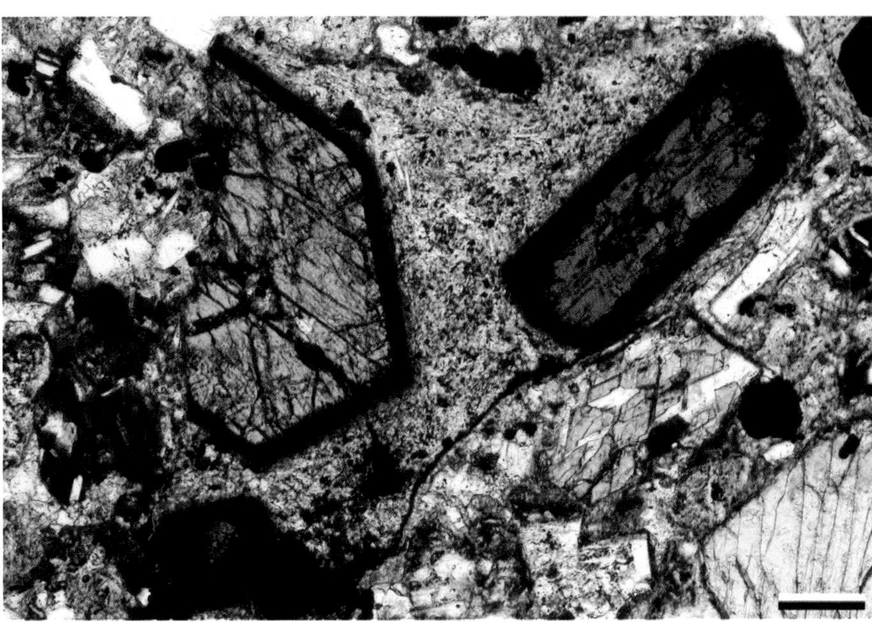

Plagioclase and euhedral amphibole phenocrysts in porphyritic volcanic rock fragments. As in the previous example, these amphibole grains have opaque reaction rims, caused by dehydration of ascending magma.

PPL, BDI, Scale bar = 0.14 mm

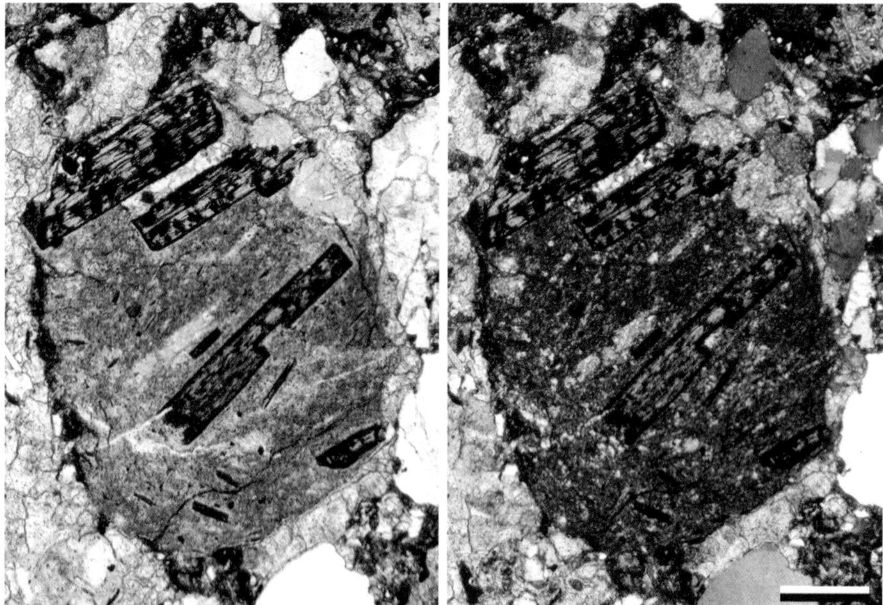

Mid.? Permian Huledal Fm., Canning Land, East Greenland

A volcanic rock fragment within the basal conglomerate between Permian sedimentary rocks and Devonian basement rocks. This andesitic porphyry consists of large phenocrysts floating in a finely crystalline to glassy matrix. The phenocrysts are mainly pleochroic biotite crystals that have been slightly altered (note the dark reaction rims around the crystals that were produced by hot, magmatic alteration).

PPL | XPL, BDI, Scale bar = 0.26 mm

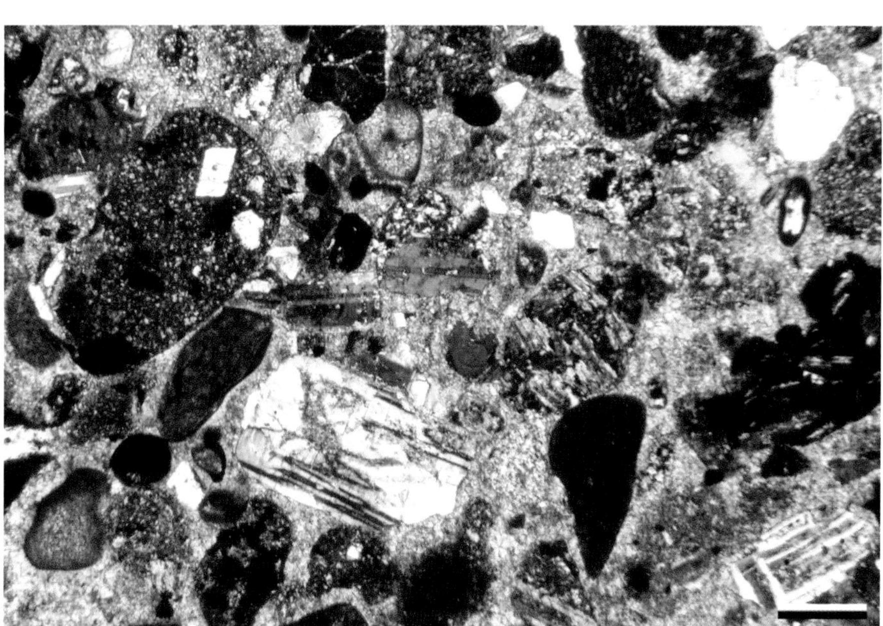

Up. Cretaceous Ildefonso Fm., Ponce-Coamo area, Puerto Rico

Carbonate-cemented volcanic rock fragments in a lithic arenite. Large euhedral phenocrysts of plagioclase are mixed with well-rounded aphanitic volcanic and carbonate rock fragments.

XPL, Scale bar = 0.51 mm

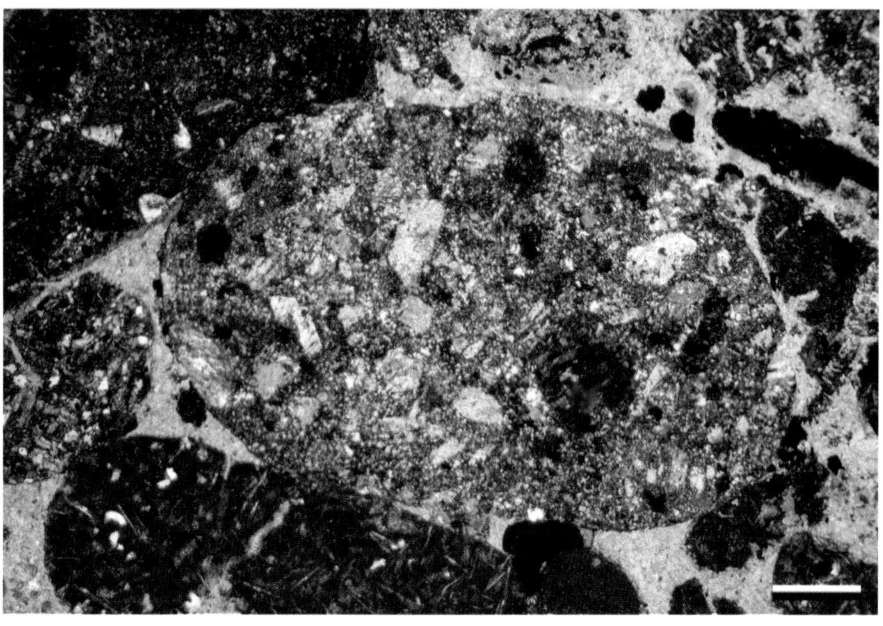

Up. Cretaceous Ildefonso Fm., Ponce-Coamo area, Puerto Rico

These well-rounded fragments of basic to intermediate volcanic rocks were derived from island arc volcanism in the Caribbean region. The large grain in the center contains laths of altered feldspar in a finely crystalline (aphanitic) matrix. The other grains contain laths of plagioclase but are more finely crystalline and must be carefully distinguished from detrital chert or clay clasts. The cement in this sample is calcite.

XPL, Scale bar = 0.26 mm

Mid. Eocene – Lo. Oligocene, Spears Gp., Catron Co., New Mexico

A rounded volcanic rock fragment in a volcanic arenite. The grain is an andesite porphyry containing scattered clinopyroxenes and larger feldspar phenocrysts. The groundmass is made up of a felted mass of finely crystalline plagioclase crystals and altered volcanic glass.

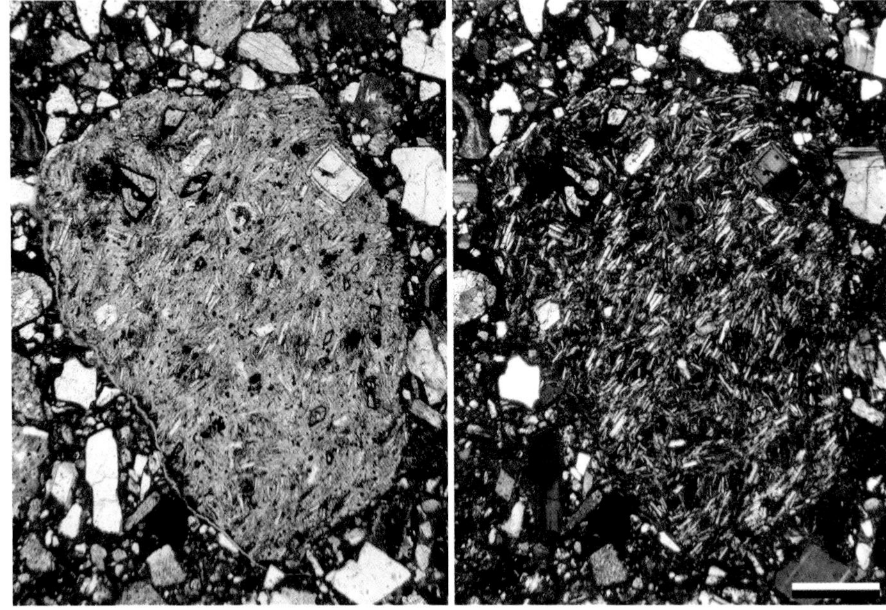

PPL | XPL, BDI, Scale bar = 0.26 mm

Mid.? Permian Huledal Fm., Jameson Land, East Greenland

A rounded porphyritic volcanic rock fragment in a calcite-cemented lithic arenite. The green phenocrysts in the aphanitic matrix are chlorite. Chlorite is a common replacement mineral within mafic rocks; here, it is probably replacing pyroxene phenocrysts. The green color, strong pleochroism and anomalous birefringence of chlorite generally make it easy to identify in thin section. In addition to the chlorite, iron oxides may also have formed during diagenetic alteration.

PPL, BDI, Scale bar = 0.51 mm

Up. Permian Schuchert Dal Fm., Jameson Land, East Greenland

Volcanic rock fragments (mostly devitrified glass) in a lithic arenite. The large grain on the left is a VRF containing clasts of devitrified pumice (lighter areas) with a few scattered phenocrysts. Pumice is highly vesicular and is formed during explosive volcanic eruptions. These fragments appear to have been engulfed by a later lava or ash flow.

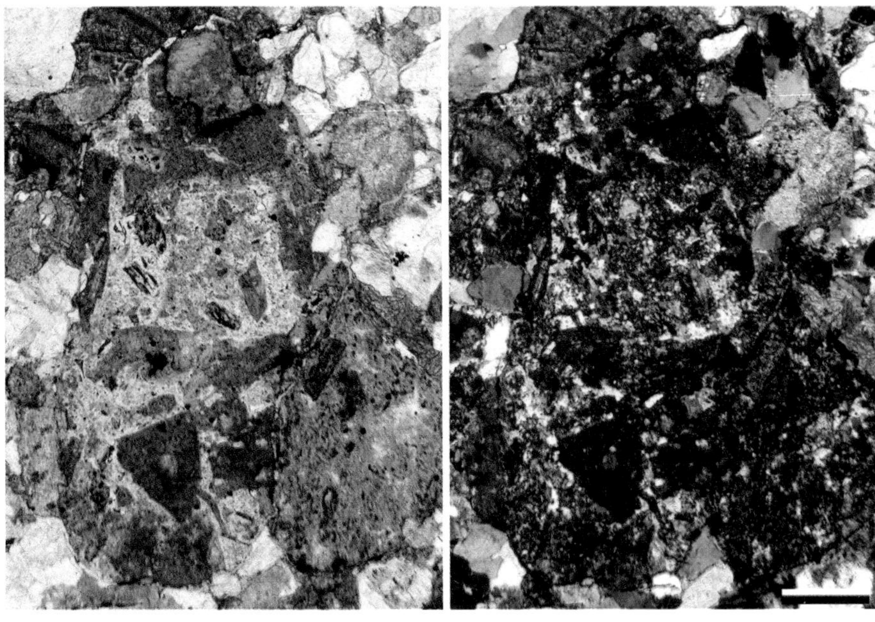

PPL | XPL, AFeS, BDI, Scale bar = 0.51 mm

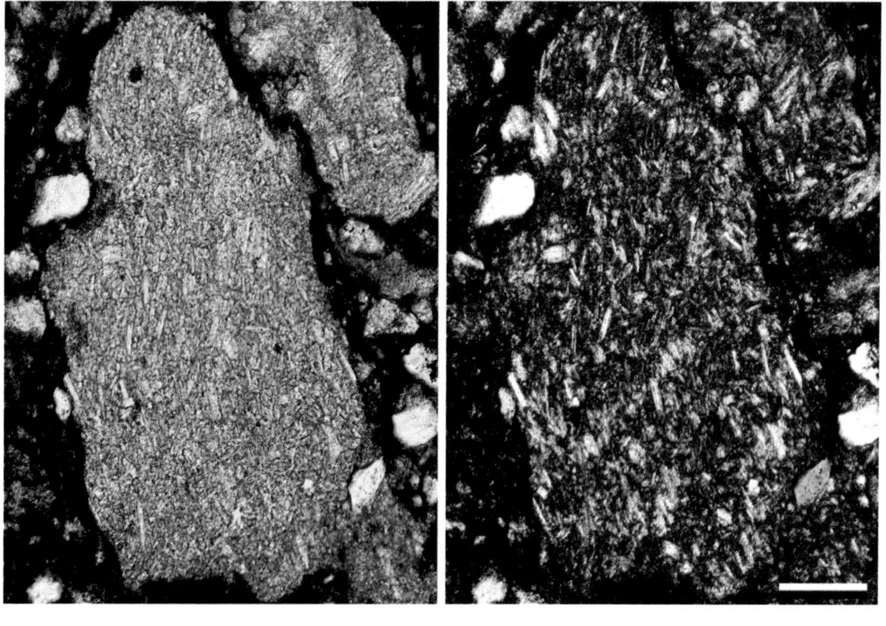

Up. Cretaceous Shumagin Fm., Shumagin Islands, Alaska

These trachyte fragments consists of a felted network of small, twinned plagioclase laths in an aphanitic matrix. These clasts were derived from a basaltic to andesitic island-arc terrane and, like virtually all VRFs, were susceptible to destruction during transport and weathering. Because these clasts are abundant, this unit may have been deposited close to its volcanic source. Clasts like this can, at first glance, be confused with chert or shale/claystone clasts, but the plagioclase laths are clear indicators of a volcanic origin.

PPL | XPL, Scale bar = 0.12 mm

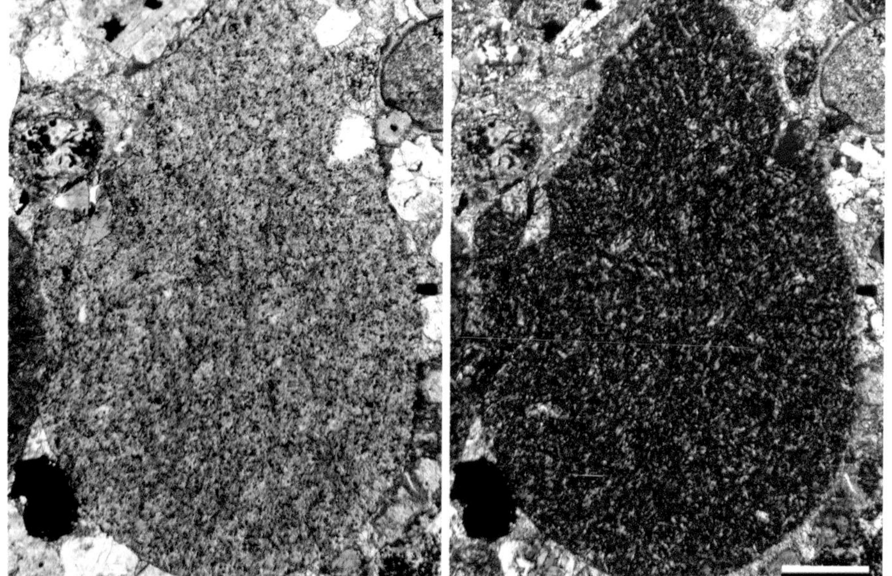

Up. Cretaceous Ildefonso Fm., Ponce-Coamo area, Puerto Rico

A fragment of a trachyte with an even more finely-crystalline texture, here displayed in clasts derived from an andesitic island-arc volcanic complex. Both the brownish color in plane-polarized light and the granular, black to gray fabric in cross-polarized light make these grains appear very similar to chert fragments at first glance. Examination at higher magnification is needed to resolve the small, but distinctive, lath-shaped plagioclase crystals that characterize these grains but would not be found in chert clasts.

PPL | XPL, Scale bar = 0.24 mm

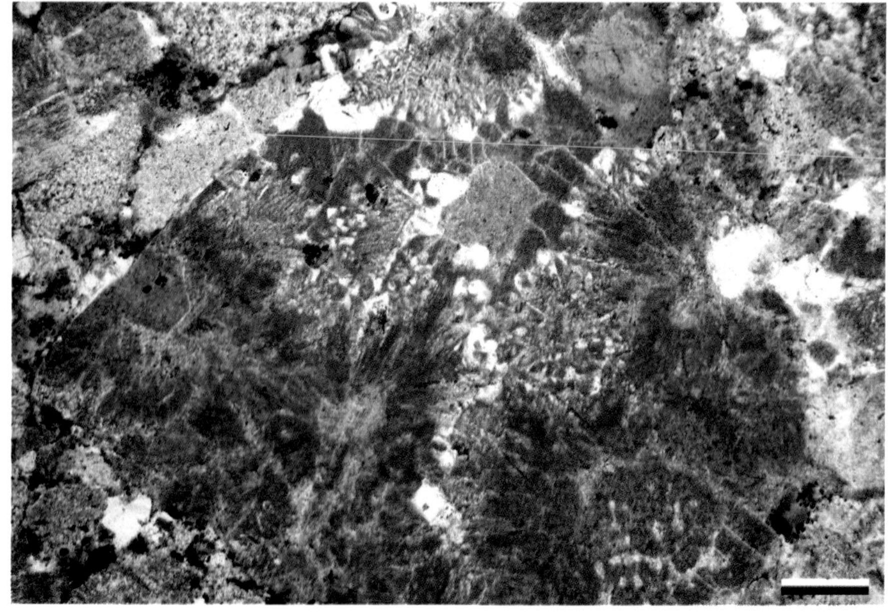

Up. Pennsylvanian – Lo. Permian Granite Wash, southern Oklahoma

A volcanic rock clast containing spherulitic fabrics. These features are common to felsic melts that cooled quickly. In this case, the spherulites appear to have nucleated on previously formed crystals.

PPL, BDI, Scale bar = 0.44 mm

Up. Pennsylvanian – Lo. Permian Granite Wash, southern Oklahoma

A volcanic rock fragment of devitrified glass. This grain contains colorless, needle-like crystals that may be recrystallized tridymite. Tridymite is one of the high-temperature polymorphs of quartz and commonly forms randomly-aligned networks of crystals in glass. During devitrification (changing of an amorphous material to a crystalline one), the tridymite crystals convert to quartz. Rock fragments like this are difficult to tell apart from sedimentary cherts, but needle-like plagioclase crystals and spherulitic fabrics are key to identifying them as volcanic rock fragments.

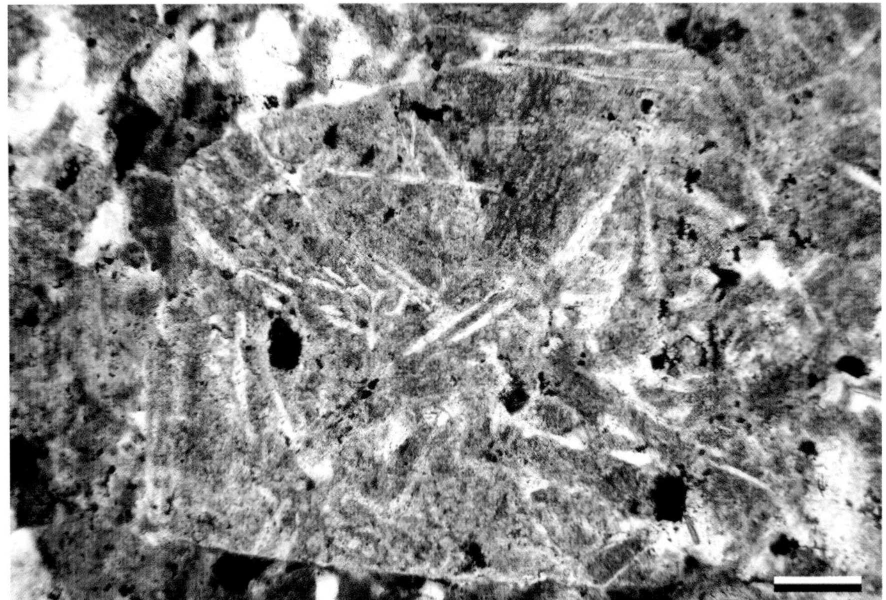

PPL, BDI, Scale bar = 0.09 mm

Triassic Snadd Fm., Barents Sea, offshore Norway †

A rock fragment of lava, which is partly degraded but has acted as a rigid grain during compaction. Areas of groundmass which contained glass are now K-feldspar (shown by the pale yellow stain) and many of the plagioclase phenocrysts are replaced by authigenic chlorite. The area of chlorite-rich clay to the left of the lava grain is likely to be a volcanic grain that has been extensively degraded and compacted.

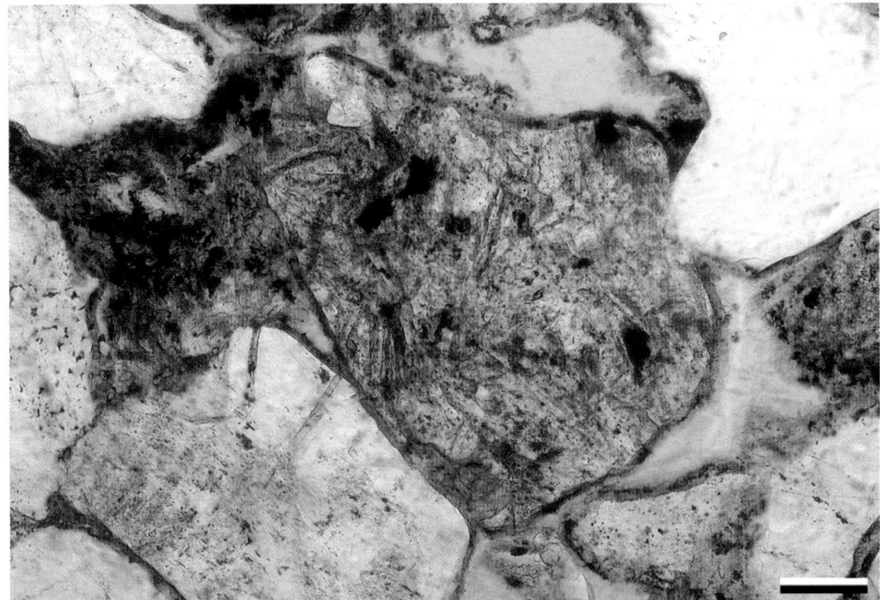

PPL, KFS, BDI, Scale bar = 0.07 mm

Paleocene, Rajasthan, India

A volcanic rock fragment with a possible dacitic composition. The grain is porphyritic and contains lath-shaped phenocrysts of plagioclase in a largely plagioclase groundmass. The plagioclase phenocrysts within this clast exhibit a slight preferred orientation because they were aligned by lava flowage.

PPL, Scale bar = 0.49 mm

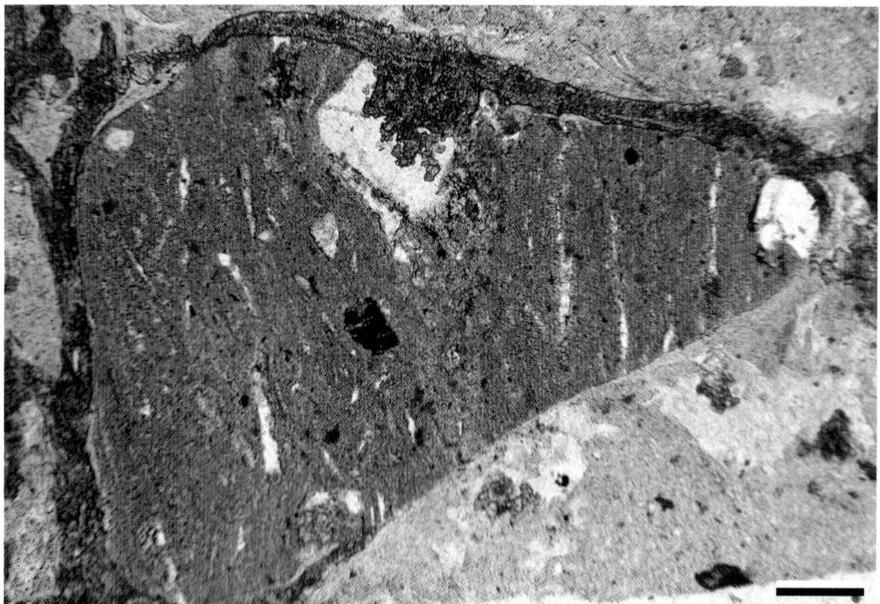

Paleocene, Rajasthan, India

A former glassy, welded tuff (ignimbrite) which is now devitrified to K-feldspar and microcrystalline quartz. The tuff contains euhedral quartz crystals and curved shards of glass in a welded glassy matrix. Some of the longer, aligned quartz areas may be stretched vesicles. The grain is surrounded by, and partly replaced by, ferroan calcite that has precipitated along fractures between the grain and the matrix.

PPL, AFeS, KFS, BDI, Scale bar = 0.40 mm

Paleocene, Rajasthan, India

A devitrified, welded tuff grain in a lithic sandstone composed of abundant volcanically-derived material. The grain shows deformation of glass shards and bubbles (vesicles). During welding, the glass shards and bubbles collapse and then become aligned and stretched. An axiolitic texture is also seen as welding probably occurred before devitrification began, and the devitrification extended across individual shard boundaries, obliterating shard structures.

PPL, KFS, Scale bar = 0.49 mm

Paleocene, Rajasthan, India

Spherulitic lava (probably rhyolite), with amygdales. The amygdales are filled by chalcedony and megaquartz cements. The spherulites themselves include single and compound forms. They are commonly composed of microlites of both feldspar and quartz, radiating from a nucleus. When elongate spherulites are observed, they are termed axiolites.

PPL, KFS, Scale bar = 0.49 mm

Pleistocene San Antonio Pumice, Socorro Co., New Mexico

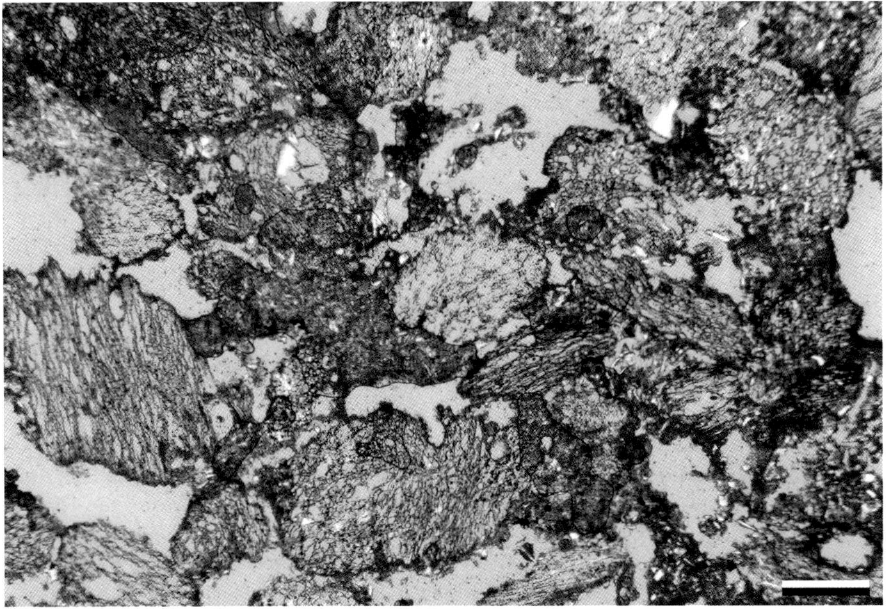

This rock is made up of remarkably delicate and porous pumice fragments derived from a 1.6 Ma Jemez volcanic field eruption. The fragments are thought to have been transported by low-viscosity debris flows more than 220 km (137 mi) from their source (Cather, 1988). In addition to pumice fragments, this rock contains some euhedral to subhedral volcanic quartz crystals and precipitated clay and carbonate cements. Pumice forms during explosive volcanic events where gases rapidly exsolve from magmas, forming and preserving enormous numbers of micro-vessicles as the magma cools.

PPL, BDI, Scale bar = 0.51 mm

Pleistocene San Antonio Pumice, Socorro Co., New Mexico

The center of the photomicrograph shows a highly vesiculated pumice clast from the same debris flow shown in the prior image. It is surrounded by a mixture of fragmented glass shards and volcanic quartz that have been cemented by microcrystalline carbonate cements and clays. Most pumice has a rhyolitic composition, and because of the very porous, vesicular nature of this volcanic glass, it is able to float on water. Normally, pumice is too fragile to survive long-distance transport with other rock fragments, but it clearly can survive when cushioned in debris flows.

PPL, BDI, Scale bar = 0.26 mm

Pleistocene San Antonio Pumice, Socorro Co., New Mexico

The fragmented and rounded pumice grains here are surrounded by glass shards and volcanic quartz grains. Unlike the previous examples, this sample is from a unit that may have undergone fluvial reworking. The shards in this sample also are the result of explosive eruptions, where vesicle walls were shattered into arcuate fragments. Depending on the size of the fragments, they can be carried by wind or water great distances from their source. Glass shards are hydrophyllic and are therefore unstable in near surface environments, ultimately becoming devitrified, replaced or dissolved.

PPL, BDI, Scale bar = 0.26 mm

Cited References and Additional Information Sources

Abbott, P. L., and G. L. Peterson, 1978, Effects of abrasion durability on conglomerate clast populations: Examples from Cretaceous and Eocene conglomerates of the San Diego area, California: Journal of Sedimentary Research, v. 48, p. 31-42, doi: 10.1306/212F73EC-2B24-11D7-8648000102C1865D.

Allen, J. R. L., 1962, Petrology, origin and deposition of the highest Lower Old Red Sandstone of Shropshire, England: Journal of Sedimentary Research, v. 32, p. 657-697, doi: 10.1306/74D70D49-2B21-11D7-8648000102C1865D.

Arribas, J., S. Critelli, E. Le Pera, and A. Tortosa, 2000, Composition of modern stream sand derived from a mixture of sedimentary and metamorphic source rocks (Henares River, Central Spain): Sedimentary Geology, v. 133, p. 27-48, doi: 10.1306/212F8823-2B24-11D7-8648000102C1865D.

Boggs, S., Jr., 1968, Experimental study of rock fragments: Journal of Sedimentary Research, v. 38, p. 1326-1339, doi: 10.1306/74D71B72-2B21-11D7-8648000102C1865D.

Burns, L. K., and F. G. Ethridge, 1979, Petrology and diagenetic effects of lithic sandstones: Paleocene and Eocene Umpqua Formation, southwest Oregon, *in* P. A. Scholle and P. R. Schluger, eds., Aspects of Diagenesis: Tulsa, OK, SEPM Special Publication 26, p. 307-317, doi: 10.2110/pec.79.26.0307.

Cameron, K. L., and H. Blatt, 1971, Durabilities of sand size schist and 'volcanic' rock fragments during fluvial transport, Elk Creek, Black Hills, South Dakota: Journal of Sedimentary Research, v. 41, p. 565-576, doi: 10.1306/74D722D4-2B21-11D7-8648000102C1865D.

Cather, S. M., 1988, Jemez-derived pumice near San Antonio, New Mexico: Depositional processes and implications: New Mexico Geology, v. 10, p. 65-66.

Critelli, S., and R. V. Ingersoll, 1995, Interpretation of neovolcanic versus palaeovolcanic sand grains: An example from Miocene deep-marine sandstone of the Topanga Group (Southern California): Sedimentology, v. 42, p. 783-804, doi: 10.1111/j.1365-3091.1995.tb00409.x.

Dickinson, W. R., 1970, Interpreting detrital modes of graywacke and arkose: Journal of Sedimentary Research, v. 40, p. 695-707, doi: 10.1306/74D72018-2B21-11D7-8648000102C1865D.

Dickinson, W. R., 1985, Interpreting provenance relations from detrital modes of sandstones, *in* G. G. Zuffa, ed., Provenance of Arenites (NATO Science Series C): Dordrecht, Netherlands, D. Reidel Publishing, p. 333-361, doi: 10.1007/978-94-017-2809-6_15.

Dickinson, W. R., K. P. Helmold, and J. A. Stein, 1979, Mesozoic lithic sandstones in central Oregon: Journal of Sedimentary Research, v. 49, p. 501-516, doi: 10.1306/212F777A-2B24-11D7-8648000102C1865D.

Fisher, R. V., 1966, Rocks composed of volcanic fragments and their classification: Earth-Science Reviews, v. 1, p. 287-298, doi: 10.1016/0012-8252(66)90010-9.

Fontana, D., 1991, Detrital carbonate grains as provenance indicators in the Upper Cretaceous Pietraforte Formation (northern Apennines): Sedimentology, v. 38, p. 1085-1096, doi: 10.1111/j.1365-3091.1991.tb00373.x.

Garzanti, E., and G. Vezzoli, 2003, A classification of metamorphic grains in sands based on their composition and grade: Journal of Sedimentary Research, v. 73, p. 830-837, doi: 10.1306/012203730830.

Graham, S. A., R. V. Ingersoll, and W. R. Dickinson, 1976, Common provenance for lithic grains in Carboniferous sandstones from Ouachita Mountains and Black Warrior Basin: Journal of Sedimentary Research, v. 46, p. 620-632, doi: 10.1306/212F7009-2B24-11D7-8648000102C1865D.

Ingersoll, R. V., T. F. Fullard, R. L. Ford, J. P. Grimm, J. D. Pickle, and S. W. Sares, 1984, The effect of grain size on detrital modes: A test of the Gazzi-Dickinson point-counting method: Journal of Sedimentary Research, v. 54, p. 103-116, doi: 10.1306/212F83B9-2B24-11D7-8648000102C1865D [see also: discussions and replies in: Journal of Sedimentary Research, v. 55, p. 616-621].

Lee, Y. I., and G. de V. Klein, 1986, Diagenesis of sandstones in the back-arc basins of the western Pacific Ocean: Sedimentology, v. 33, p. 651-675, doi: 10.1111/j.1365-3091.1986.tb01968.x.

MacKenzie, W. S., C. H. Donaldson, and C. Guilford, 1982, Atlas of Igneous Rocks and their Textures: New York, John Wiley Sons, 148 p.

Marsaglia, K. M., 1993, Basaltic island sand provenance, *in* M. J. Johnsson, and A. Basu, eds., Processes Controlling the Composition of Clastic Sediments: Boulder, CO, GSA Special Paper 284, p. 41-66, doi: 10.1130/SPE284-p41.

Marsaglia, K. M., and R. V. Ingersoll, 1992, Compositional trends in arc-related, deep-marine sand and sandstone: A reassessment of magmatic-arc provenance: GSA Bulletin, v. 104, p. 1637-1649, doi: 10.1130/0016-7606(1992)104<1637:CTIARD>2.3.CO;2.

Mathisen, M. E., and J. G. McPherson, 1991, Volcaniclastic deposits: Implications for hydrocarbon exploration, *in* R. V. Fisher, and G. A. Smith, eds., Sedimentation in Volcanic Settings: Tulsa, OK, SEPM Special Publication 45, p. 27-36, doi: 10.2110/pec.91.45.0027.

McBride, E. F., and M. D. Picard, 1987, Downstream changes in sand composition, roundness, and gravel size in a short-headed, high-gradient stream, northwestern Italy: Journal of Sedimentary Research, v. 57, p. 1018-1026, doi: 10.1306/212F8CD3-2B24-11D7-8648000102C1865D.

Morgan, G. B., VI, and D. London, 2012, Process of granophyre crystallization in the Long Mountain Granite, southern Oklahoma: GSA Bulletin, v. 124, p. 1251-1261, doi: 10.1130/B30569.1.

Pittman, E. D., and R. E. Larese, 1991, Compaction of lithic sands: Experimental results and applications: AAPG Bulletin, v. 75, p. 1279-1299.

Sabins, F. F., Jr., 1962, Grains of detrital, secondary, and primary dolomite from Cretaceous strata of the Western Interior: GSA Bulletin, v. 73, p. 1183-1196, doi: 10.1130/0016-7606(1962)73[1183:GODSAP]2.0.CO;2.

Schmid, R., 1981, Descriptive nomenclature and classification of pyroclastic deposits and fragments: Geologische Rundschau, v. 70, p. 41-43, doi: 10.1007/BF01822152.

Sneed, E. D., and R. L. Folk, 1958, Pebbles in the lower Colorado River, Texas a study in particle morphogenesis: Journal of Geology, v. 66, p. 114-150, doi: 10.1086/626490.

Walton, A. W., 1977, Petrology of volcanic sedimentary rocks, Vieja Group, southern Rim Rock Country, Trans-Pecos Texas: Journal of Sedimentary Research, v. 47, p. 137-157, doi: 10.1306/212F7117-2B24-11D7-8648000102C1865D.

Wolf, K. H., 1971, Textural and compositional transitional stages between various lithic grain types (with a comment on "Interpreting detrital modes of graywacke and arkose'): Journal of Sedimentary Research, v. 41, p. 328-332. (see also erratum in v. 41, p. 889).

Worden, R. H., M. J. Mayall, and I. J. Evans, 1997, Predicting reservoir quality during exploration: Lithic grains, porosity and permeability in Tertiary clastic rocks of the South China Sea basin, *in* A. J. Fraser, and S. J. Matthews, eds., Petroleum Geology of Southeast Asia: London, GSL Special Publication 126, p. 107-115, doi: 10.1144/GSL.SP.1997.126.01.08.

Young, S. W., 1976, Petrographic textures of detrital polycrystalline quartz as an aid to interpreting crystalline source rocks: Journal of Sedimentary Research, v. 46, p. 595-603, doi: 10.1306/212F6FFA-2B24-11D7-8648000102C1865D.

Zuffa, G. G., 1991, On the use of turbidite arenites in provenance studies: Critical remarks, *in* A. C. Morton, S. P. Todd, and P. D. W. Haughton, eds., Developments in Sedimentary Provenance Studies: London, GSL Special Publication 57, p. 23-29, doi: 10.1144/GSL.SP.1991.057.01.03.

Facing Page: Top – Garnet (pyrope) grains. Photograph from the Mineral Museum at the New Bureau of Geology and Mineral Resources. Bottom – A photomicrograph of pseudohexagonal biotite crystals in a Cenozoic intrusive from Clarke Co., Nevada (PPL).

GRAINS:

ACCESSORY MINERALS

CHAPTER 4

ACCESSORY MINERALS

Accessory minerals include all the many detrital minerals that are found in clastic terrigenous rocks that do not contribute directly to rock classification (thus, primarily minerals other than quartz and feldspar). Although thousands of minerals could potentially fall under that definition; practically, a limited number are found with any great frequency. Accessory minerals as a whole typically make up less than 1% (rarely more than 2%) of most terrigenous sedimentary rocks. Quartz arenites commonly have the fewest accessory minerals (as little as 0.05% in some cases); arkoses are somewhat richer in accessories, and lithic arenites generally have the highest levels. This results from the fact that most accessory minerals, like some feldspars and lithic fragments, lack the abrasion resistance or chemical stability to survive erosion, transport and diagenesis.

Accessory minerals can be examined in thin sections; alternatively, they can be concentrated by mechanical (shaker table) or flotation (heavy liquid) methods (see, for example, Munsterman and Kerstholt, 1996; Koroznikova et al., 2008) and can then be viewed with stereoscopic microscopes, SEM or other methods. Thin-section examination shows the grains in the context of rock fabric, but such minerals can be quite scarce in any single section. Disaggregation and concentration is much more effective for evaluating the full assemblage of such minerals in rock or sand samples and also allows identification by x-ray or geochemical methods.

Because accessory minerals are so commonly studied as separates, they generally are divided into light and heavy minerals with a boundary drawn by various workers at specific gravities between 2.85 and 3 (they also are commonly divided into opaque and nonopaque minerals). The most commonly encountered detrital light accessory minerals are micas (mainly muscovite and also biotite). Heavy minerals are vastly more numerous, and can be grouped into ultrastable, intermediate stability, and unstable categories. The ultrastable minerals are the ultimate survivors, even more stable than quartz under most conditions — thus, they are found in most clastic terrigenous rocks. The intermediate group has varied levels of survivability, but most such minerals can be degraded or removed under specific conditions; minerals in the unstable group survive only under very favorable conditions (minimal mechanical and chemical stresses).

There are so many detrital accessory minerals that occur in clastic terrigenous rocks that it is simply impossible to provide a brief yet usable summary of the mineralogical features and optical characteristics of all these minerals. We provide instead a chart of the relative stabilities of the most common accessory minerals (Table 4.1). In addition, characteristic mineral properties are described in the individual photo captions for each mineral illustrated. For additional information readers are encouraged to consult the references at the end of this chapter or the more general mineralogy texts listed in the bibliography in the introduction to this book.

Provenance:

Again, with the abundance of minerals involved, it is impossible in this text to detail the provenance implication of the entire spectrum of accessory minerals. Nonetheless, they are among the most important clues to the recognition of source terranes, despite the fact that many of the most diagnostic minerals are subject to alteration or removal through chemical and mechanical processes acting during the erosion, transport, deposition and burial stages in the life of these grains.

An overall summary of the diagnostic sources for some heavy minerals is shown in Table 4.2. Beyond that,

Table 4.1: The common accessory minerals and their relative stability (adapted from material summarized in Pettijohn et al., 1987).

Group	Stability level	Most common minerals
Light	Intermediate	Muscovite, biotite (less stable than muscovite)
Heavy	Ultrastable	Zircon, tourmaline, rutile, anatase (scarce)
	Intermediate	Apatite, epidote, garnet, kyanite, monazite, sillimanite, staurolite, titanite (sphene), zoisite
	Unstable	Hornblende, actinolite, augite, diopside, hypersthene, andalucite, pyroxene, olivine (very unstable)
Opaque	Intermediate	Hematite, ilmenite, magnetite

two approaches have been shown to be most useful in provenance studies (Morton and Hallsworth, 1999). The first uses the entire heavy mineral suite and looks at relative abundance ratios of hydraulically equivalent grains within the same relative stability suites (Table 4.1). The second, almost opposite yet complementary, approach is to examine in detail the color, habit and internal structural varieties within single mineral groups (tourmaline, zircon, etc.) and also to do single–crystal geochemical work, including geochronology, on those grains. A number of recent papers have tied specific compositional patterns in amphibole, apatite, clinopyroxene, garnet, tourmaline and zircon, or geochronologic determinations on such grains, to provenance delineation (see papers in the specific mineral sections of the bibliography for this chapter).

For ease of use, this chapter has been arranged following the mineral groupings in Table 4.1. Light micaceous minerals are followed by heavy, non-opaque minerals in order of their decreasing stability, and opaque minerals complete the chapter.

Table 4.2: Source terranes of common accessory minerals (adapted from Feo-Codecido, 1956 and Pettijohn et al., 1987). Names in red are the ultrastable minerals, and the names in green generally are classed as light minerals.

Source Terrane	Common Accessory Minerals
Acidic igneous rocks	Apatite, brookite, hornblende, monazite, rutile, titanite, tourmaline (pink varieties), zircon; muscovite, biotite
Pegmatites	Cassitierite, dumortierite, fluorite, garnet, monazite, topaz, tourmaline (blue varieties), wolframite, xenotime; muscovite
Basic igneous rocks	Augite, chromite, diopside, hypersthene, ilmenite, magnetite, olivine, chrome-spinel, Fe- and Mg-rich spinel; pyrope garnet in ultrabasics
Contact metamorphic rocks	Andalusite, chondrodite, corundum, garnet, phlogopite, staurolite, topaz, vesuvianite, wollastonite, zoisite
Regional metamorphic rocks	Andalusite, biotite, chloritoid, epidote, garnet, glaucophane, kyanite, muscovite, sillimanite, staurolite, titanite, zoisite-clinozoisite
Reworked sediments	Barite, iron ores, leucoxene, rutile, tourmaline (as rounded grains), zircon (as rounded grains)

Up. Pennsylvanian Bursum Fm., Socorro Co., New Mexico

A foliated grain of detrital muscovite. Muscovite is the most common detrital mica in sedimentary rocks, because it has the greatest chemical stability. Unlike other micas, it is colorless in plane-polarized light and shows bright blue to yellow second-order birefringence under cross-polarized light. Muscovite flakes commonly have a layered internal structure (one perfect cleavage) when cut across foliation and show a "speckled" appearance, similar to birch bark, in plane- and cross-polarized light.

PPL | XPL, BDI, Scale bar = 0.16 mm

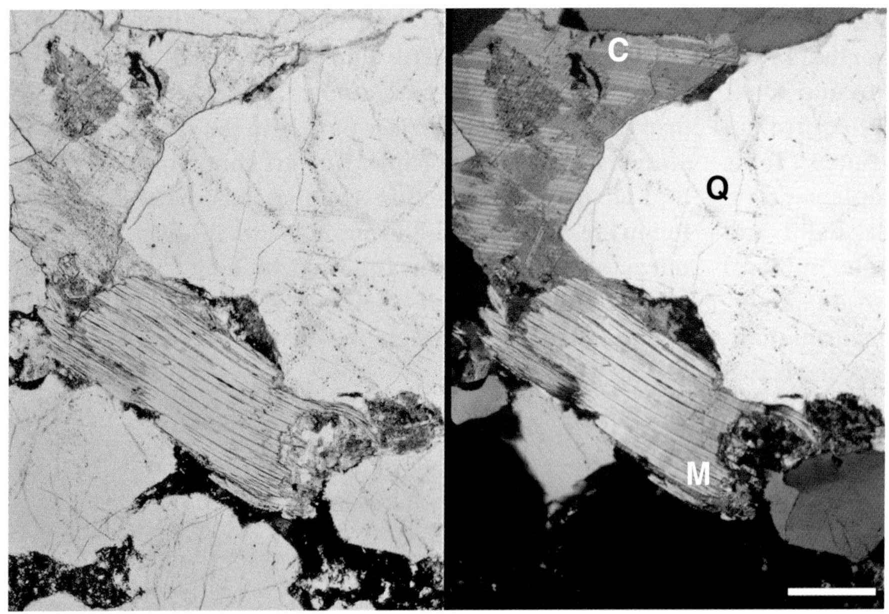

Up. Pennsylvanian Bursum Fm., Socorro Co., New Mexico

A large, detrital muscovite (M) surrounded by monocrystalline quartz grains (Q) and cemented by highly birefringent calcite (C) and opaque iron-oxides. Muscovite is biaxial negative, and it has low positive relief and one direction of perfect cleavage. Muscovite is a common rock-forming mineral in granites, schists, gneisses, pegmatites and is a replacement mineral in hydrothermal systems. Sericite, a normally finely crystalline alteration product of feldspars, is optically and chemically similar to muscovite.

PPL | XPL, BDI, Scale bar = 0.64 mm

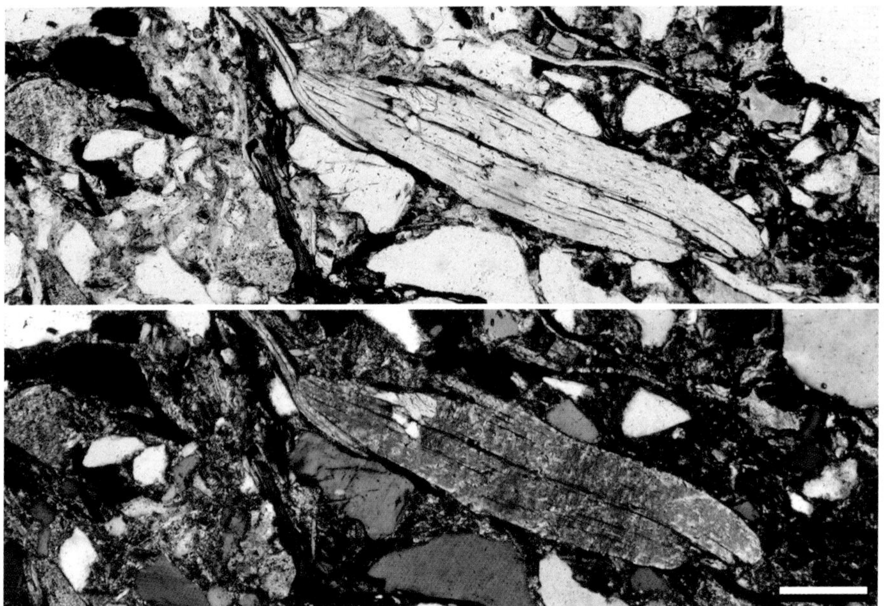

Up. Triassic New Haven Arkose, New Haven Co., Connecticut

Detrital muscovite (and smaller biotite) grains in a lithic sandstone. Muscovite is usually colorless in thin section, but some grains can be slightly colored (pale brown, red, yellow). The colored varieties also are sometimes weakly pleochroic. This grain has the "birds-eye" or "birch bark" speckling in cross-polarized light that is common to both muscovite and biotite micas. Muscovite is a very ductile mineral, and during burial and compaction, muscovites are commonly indented or deformed by stronger and more rigid grains.

PPL / XPL, Scale bar = 0.10 mm

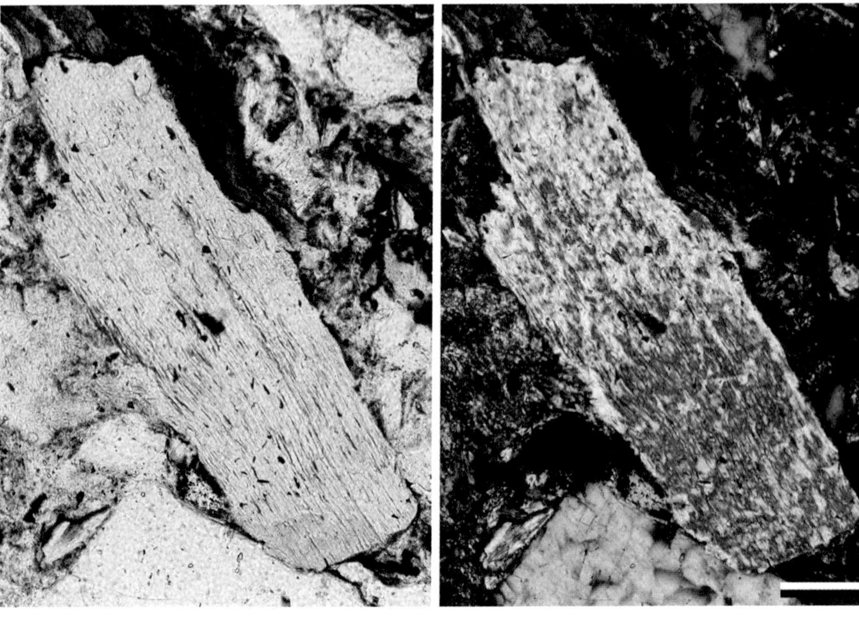

Up. Triassic New Haven Arkose, New Haven Co., Connecticut

A large detrital muscovite grain with slightly lower than normal birefringence related to the grain being cut too thin during thin-section preparation. Mica grains that are oriented parallel to the basal cleavage (which this one is not) also can have lower birefringence colors. In addition to muscovite, this rock also contains small, deformed green chlorite and biotite grains.

PPL | XPL, Scale bar = 0.10 mm

Mid. Jurassic Ness Fm., Norwegian sector, North Sea †

Sandstones rarely contain more than 2% mica. This example is particularly micaceous, with concentrated layers of both muscovite and biotite. The micas are largely unaltered but are compacted and deformed between rigid detrital grains. The opaque areas are aggregates of authigenic, cubic pyrite crystals. Porosity is present within the rest of the sample, but where the micas are abundant and matrix is absent, compaction is more pronounced, and macroporosity was nearly completely destroyed.

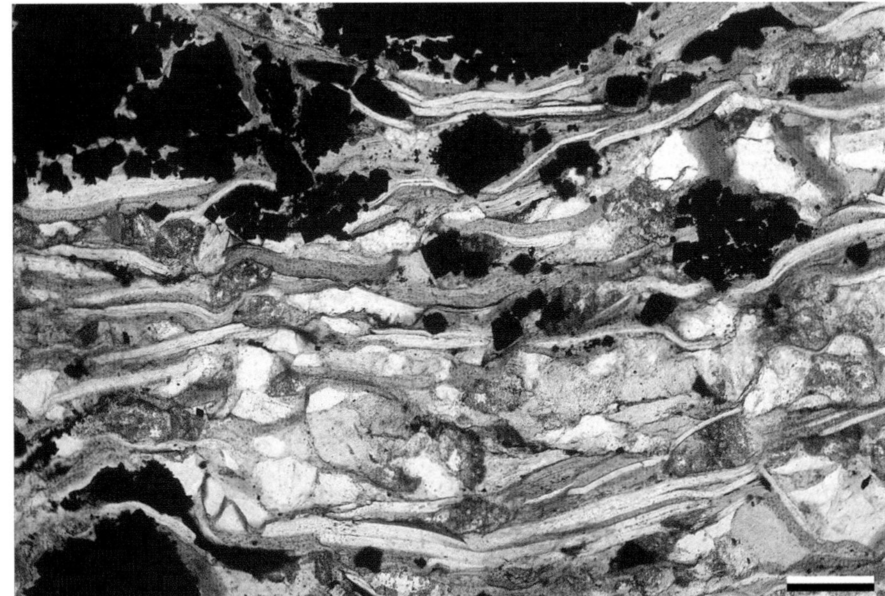

PPL, BDI, KFS, Scale bar = 0.12 mm

Lo. Permian Cutler Fm., Mesa Co., Colorado

Abundant detrital biotite grains in a sandstone. Like muscovite, biotite commonly has a speckled appearance in both plane- and cross-polarized light. Biotite is biaxial negative, but those two minerals are easily distinguished by color (muscovite is colorless or very weakly colored; biotite typically is strongly colored brown or green). In addition, biotite is strongly pleochroic in most thin-section orientations. Although muscovite has slightly higher birefringence, the color of the biotite masks the true interference colors and gives anomalous birefringence.

PPL / XPL, Scale bar = 0.26 mm

Oligocene? volcanic sandstone, Jackson Co., Oregon

Biotite is normally strongly pleochroic. Pleochroism is the changing of a mineral's color as the microscope stage is rotated under plane-polarized light. As light travels through a grain, it is absorbed variably by the crystal axes resulting in a change of color. In the first image, the biotite grain is a dark brown, but after turning the microscope stage 90°, the grain has turned a much lighter yellowish brown.

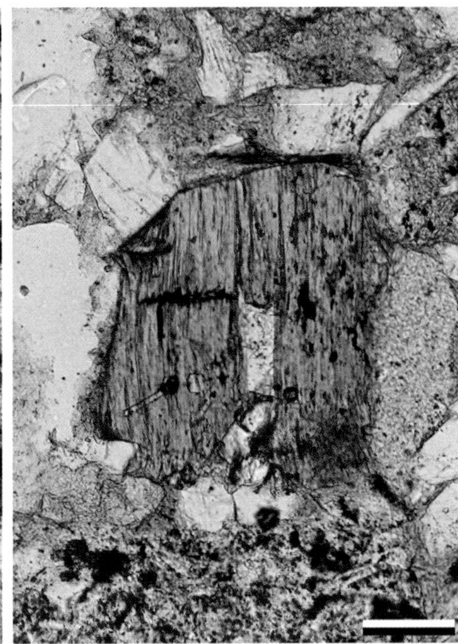

PPL | PPL, BDI, Scale bar = 0.08 mm

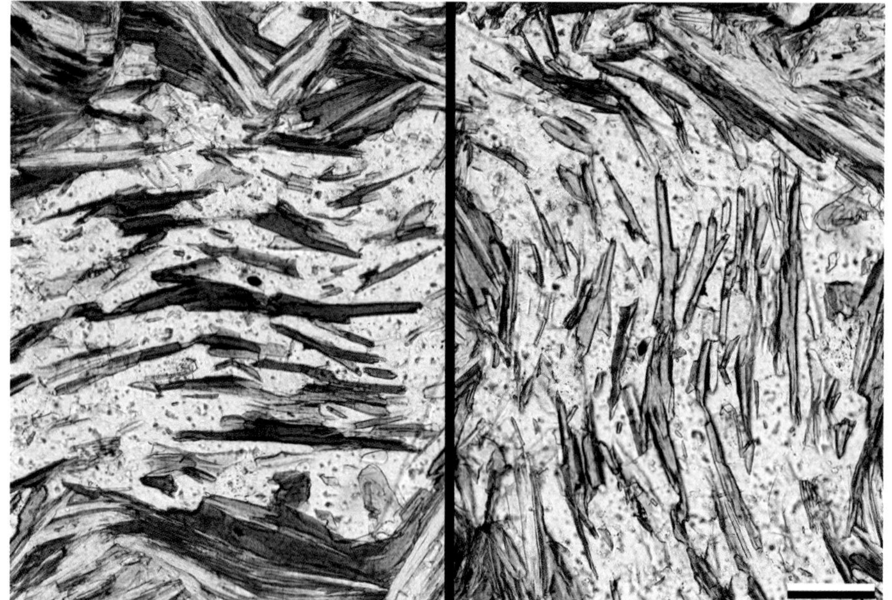

Ordovician? mica schist, New Haven Co., Connecticut

An example of strongly pleochroic biotite from a schistose metamorphic source rock. In this case, the biotite grains shift from a dark green color to a pale yellow/brown color through the 90° stage rotation. When rocks such as this erode, individual biotite grains can be liberated or biotites can be retained as inclusions in quartzose rock fragments. Biotite weathers readily, and if especially abundant in a sediment, one should suspect a relatively proximal volcanic or metamorphic source.

PPL | PPL, PFS, Scale bar = 0.10 mm

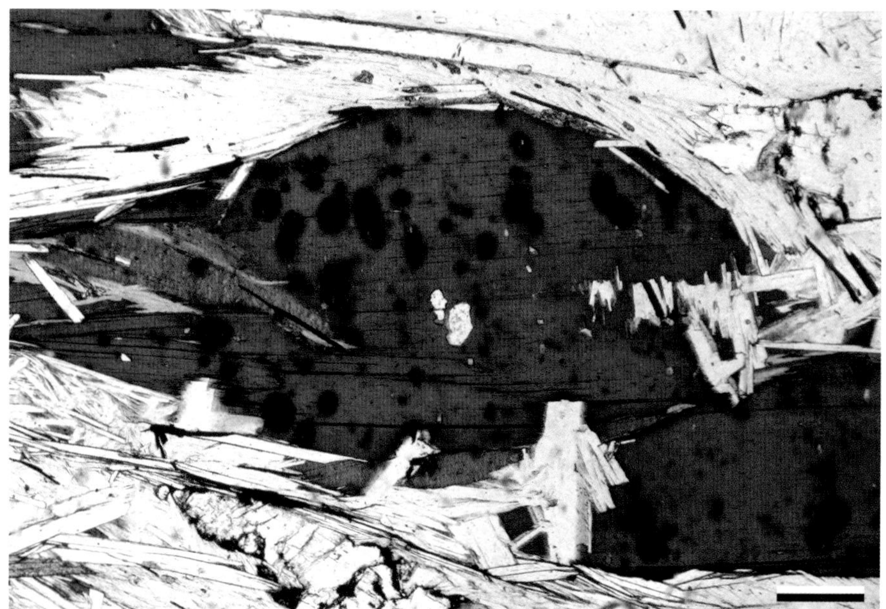

Ordovician? mica schist, New Haven Co., Connecticut

Pleochroic halos (also termed radiation halos) in biotite crystals. These black circular areas within the dark-brown biotite crystals represent zones of radiation damage that form around minute inclusions of zircon, apatite, monazite or other minerals that contain radioactive elements. These large biotite crystals are surrounded by colorless muscovite.

PPL, PFS, Scale bar = 0.26 mm

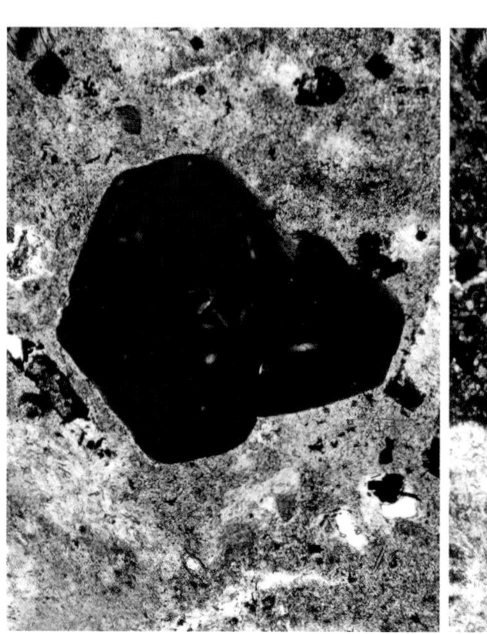

Cenozoic intrusive rock, Clarke Co., Nevada

Euhedral pseudohexagonal biotite crystals are found mainly in volcanic rocks (as in this shallow intrusive). Such euhedral crystals are rarely seen in sedimentary rocks (in part because, as platy minerals, most mica flakes such as these lie parallel to bedding and most thin sections are not cut parallel to bedding). When found in sedimentary rocks, however, pseudohexagonal biotite is a good indicator of a volcanic source. More generally, biotite can be derived from a wide variety of source rocks—schists, gneisses, contact metamorphic rocks, granites, rhyolites, basalts and hydrothermally altered deposits.

PPL | XPL, Scale bar = 0.26 mm

Up. Cretaceous (Maastrichtian) Springar Fm., Møre Basin, Norwegian Sea †

A mechanically deformed biotite grain that has been partly compacted, so that the end of the grain is kinked and splayed. Micas are prone to kinking and ductile deformation, and they commonly start to expand into available pore space with increasing temperature or changes in pore water chemistry.

PPL | XPL, KFS, BDI, Scale bar = 0.08 mm

Lo. Permian Cutler Fm., Mesa Co., Colorado

Chlorite (another phyllosilicate) forms on its own or as a replacement of biotite (as in this case). Chlorite also has to be distinguished from green biotites (crystals that have lost Fe). Chlorite group minerals have variable chemistry (e.g., Mg-rich clinochore or Fe-rich chamosite), and they can be biaxial negative or positive. Chlorite can be colorless, green or yellow, and pleochroism varies from weak to strong. Chlorite commonly has anomalous birefringence colors, in particular ultrablue (also called Berlin blue), as seen here, or brownish colors. Other chlorites have fairly characteristic first-order birefringence colors.

PPL | XPL, Scale bar = 0.08 mm

Up. Cretaceous Mesaverde Gp., Piceance Basin, Colorado

In this view, the large green chlorite grain is probably a replacement of detrital biotite. Like other phyllosilicate minerals, this grain has well-developed cleavage and occurs in booklets. This grain has anomalous birefringence ranging from blue to greenish brown. There also are some pale green authigenic chlorite cements between detrital quartz grains in this sample.

PPL | XPL, BDI, Scale bar = 0.08 mm

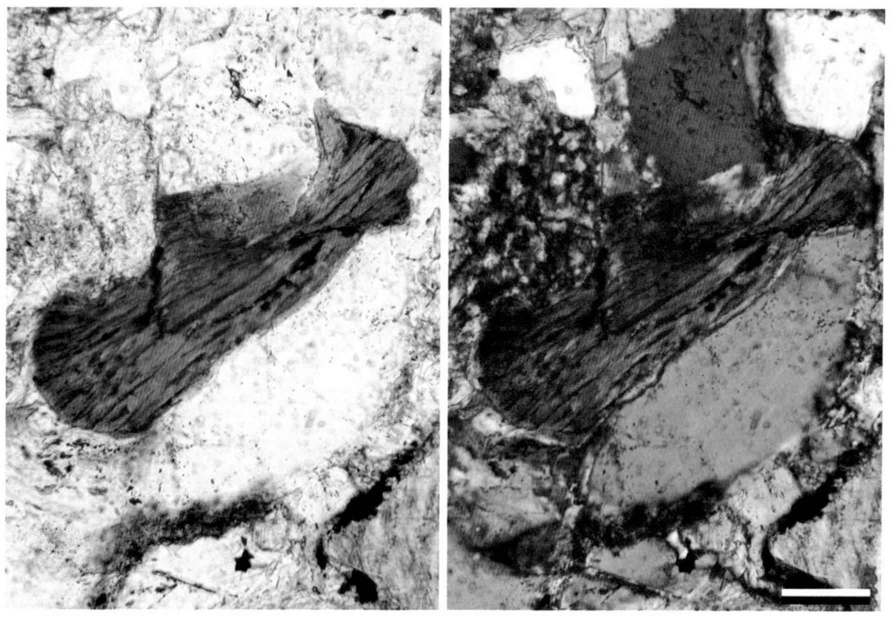

Miocene Upper Marine Molasse, Switzerland

Chlorite behaves in a ductile manner, much like muscovite and biotite. Here, a chloritic grain (probably an altered biotite) is compressed between adjacent grains of chert (above left) and quartz (below right).

PPL | XPL, BDI, Scale bar = 0.08 mm

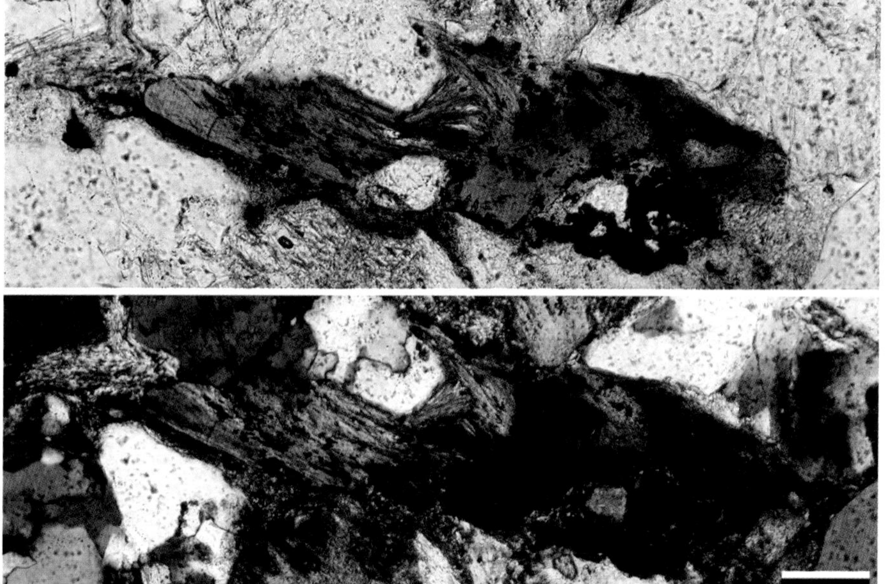

Carboniferous Dilsburger Beds, Saar Basin, Germany

This deformed biotite grain is more extensively replaced by chlorite (note the anomalous birefringence colors). This grain was deformed during burial causing it to expand into adjacent porous areas.

PPL / XPL, Scale bar = 0.10 mm

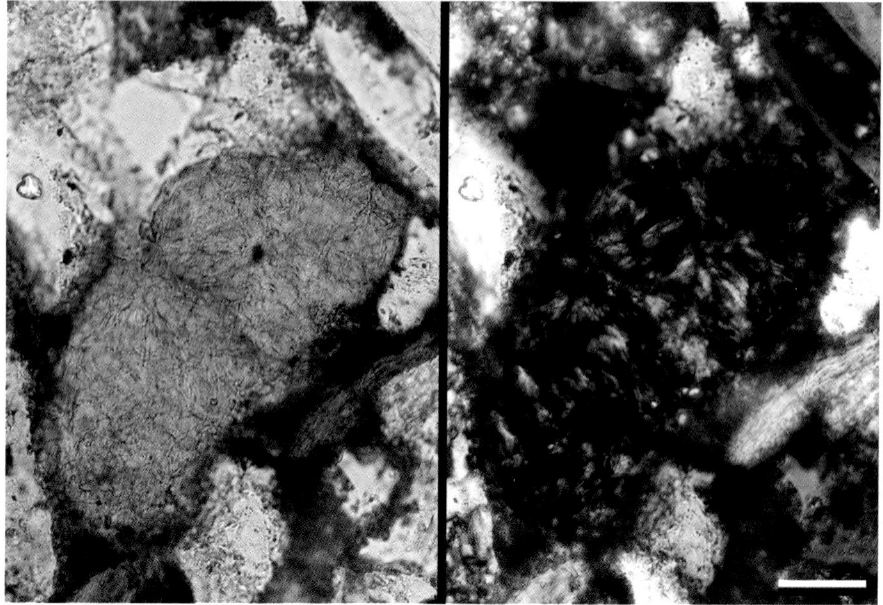

Oligocene? volcanic sandstone, Jackson Co., Oregon

This grain of chlorite is likely a replacement of another grain of uncertain origin. Chlorite is a common replacement in many ferromagnesian minerals. In this grain, the anomalous "ultrablue" birefringence is clearly visible.

PPL | XPL, Scale bar = 0.05 mm

Holocene beach sediment, Orange Co., California

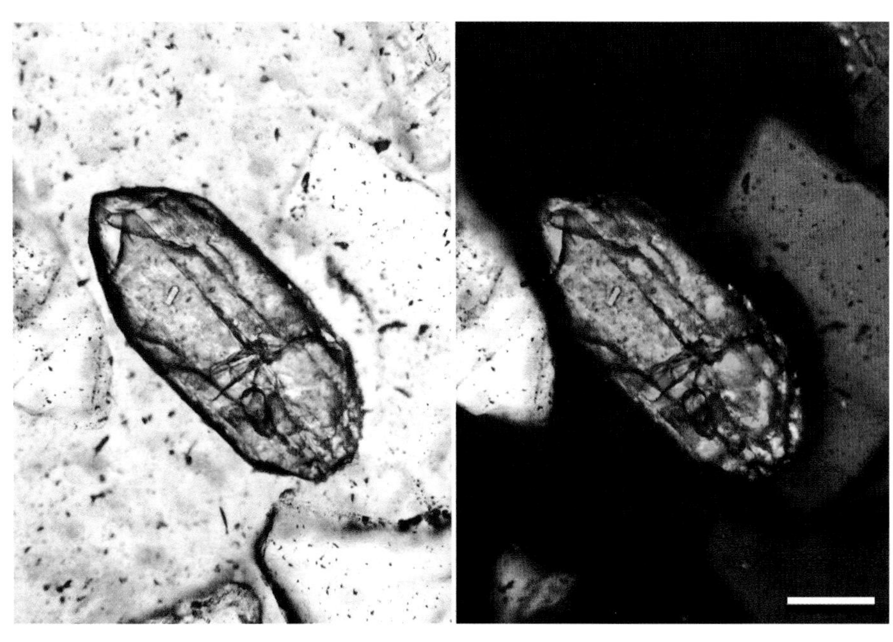

A very slightly rounded, detrital zircon crystal in a thick grain mount. Zircon grains are among the most durable of heavy minerals and can exhibit almost unaltered dipyramidal prismatic forms, as in this grain. Zircon is uniaxial positive and has extremely high relief and birefringence. Cleavage is imperfect to poor and zircon, like quartz, commonly displays conchoidal fracture (note the curved surfaces in this grain). Most zircons are colorless, but some are pale yellow, brown, pink or purple and may show slight pleochroism in grain mounts. Zircon can be confused with monazite, but monazite is biaxial positive.

PPL | XPL, Scale bar = 0.03 mm

Mid. Jurassic Entrada Fm., Moffat Co., Colorado

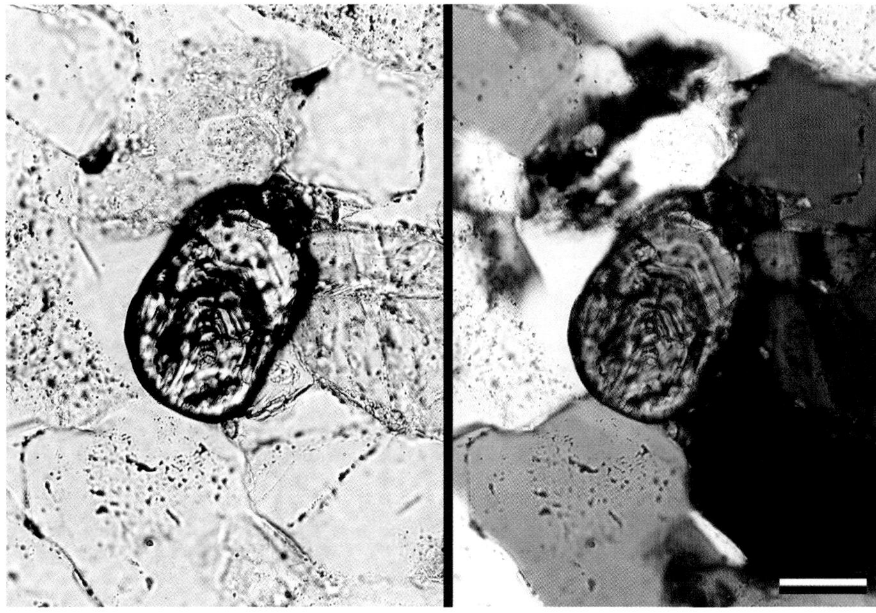

A cut through the c-axis of a rounded zircon crystal showing growth zoning within the crystal. Zircon can be initially sourced from igneous and metamorphic rocks, but because of its durability, it also can survive multiple cycles of reworking from sedimentary sources. The radiation emitted by uranium and thorium contained in zircons can cause pleochroic halos in associated minerals (as in a biotite crystal shown earlier in this chapter). The dark margins around the zircon are a result of the great relief difference between zircon and the adjacent quartz and feldspar.

PPL | XPL, RDI, Scale bar = 0.05 mm

Up. Cretaceous Mesaverde Gp., McKinley Co., New Mexico

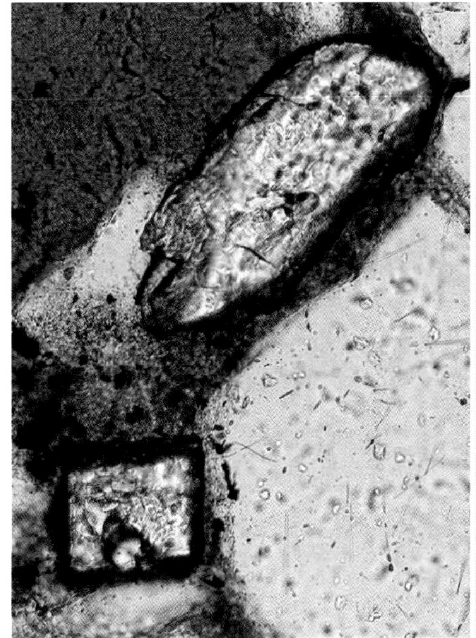

Two zircons in a glauconitic sandstone. Zircons, although rarely abundant, are among the most ubiquitous components of sedimentary rocks because of their great physical and chemical stability. Zircon is a tetragonal mineral, and the zircon in the lower left corner exhibits a euhedral tetragonal outline. The other zircon is slightly rounded but still shows a dipyramidal outline. In high-energy environments, including beaches and rivers, zircons can be concentrated in placer deposits rich in heavy minerals because zircons have a high specific gravity (4.6–4.7 g/cc as compared with 2.65 g/cc for quartz).

PPL | XPL, BDI, Scale bar = 0.05 mm

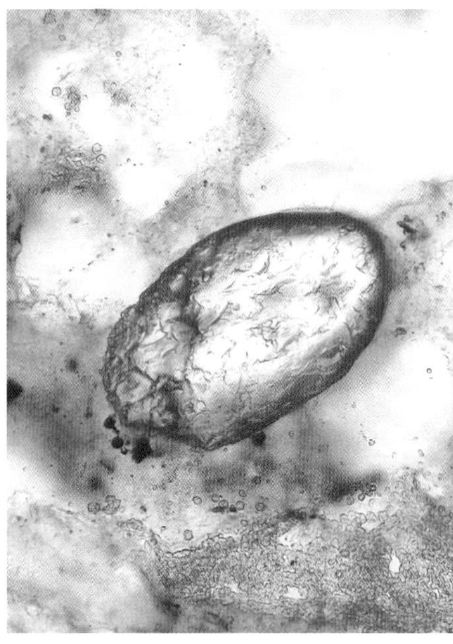

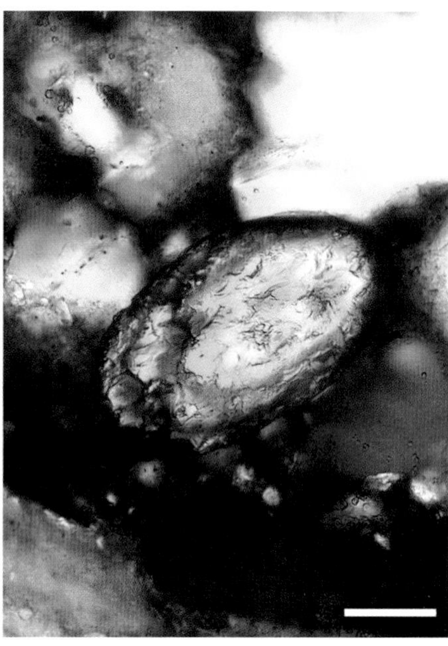

Mid. Jurassic Fulmar Fm., United Kingdom sector, North Sea †

This well-rounded zircon grain, most likely derived from a preexisting sandstone (second cycle), has numerous small depressions resulting from grain collisions. The high relief of zircon is evident by the black line around the grain and the surrounding quartz grains being out of focus when the zircon is in focus. The color bands (in cross-polarized light) are an indication of the interference order. This example is third order.

PPL | XPL, KFS, BDI, Scale bar = 0.03 mm

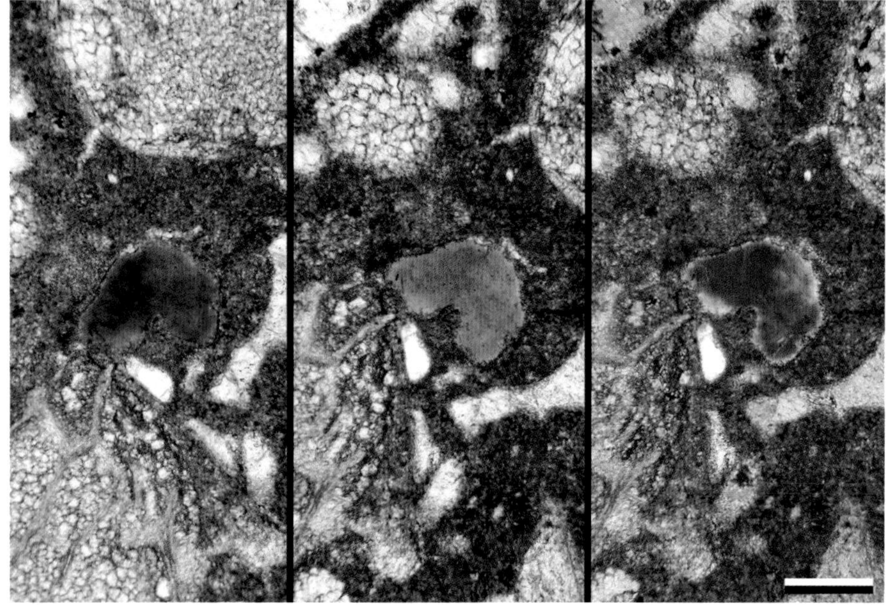

Pennsylvanian Alamitos Fm., San Miguel Co., New Mexico

Tourmalines are a group of ultrastable heavy minerals that are very common in sediments. In plane-polarized light, tourmaline displays a wide variety of colors from colorless, pink, yellow, green, blue, brown and shades in between. Tourmaline commonly is very visible in thin section because of its color. Colored crystals are strongly pleochroic, and only the very pale to colorless grains lack pleochroism. Note the color change as this tourmaline was rotated about 45° between the left and center images. The third image shows tourmaline's birefringence—high first-order to low second-order colors.

PPL | PPL | XPL, Scale bar = 0.08 mm

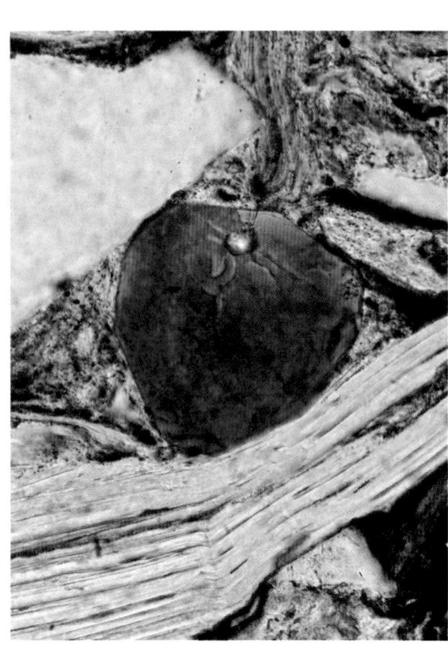

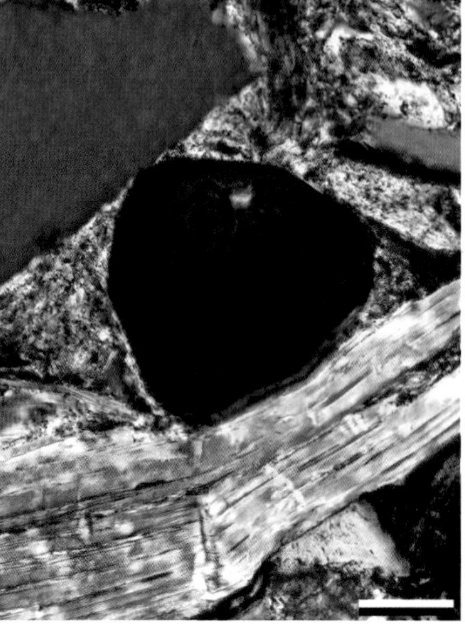

Up. Triassic New Haven Arkose, New Haven Co., Connecticut

This euhedral tourmaline grain is wedged between quartz and muscovite grains. Tourmaline is a uniaxial negative mineral, occurs as prismatic crystals, commonly with a beveled trigonal outline and moderately high relief. This grain is cut perpendicular to the c-axis, making it look isotropic. Tourmalines are common in heavy mineral placer deposits because of their chemical and mechanical stability. Tourmalines form in a variety of environments including pegmatites, felsic plutonic rocks, metamorphic rocks and skarns; additionally, grains can show authigenic overgrowths that form during burial diagenesis.

PPL | XPL, Scale bar = 0.05 mm

Mid. Jurassic Fulmar Fm., United Kingdom sector, North Sea †

A brown tourmaline grain (probably an Mg-rich variety called dravite), which displays a marked pleochroism, where it turns from dark brown to very pale brown through a 90° stage rotation. The grain displays moderately high relief and some minor inclusions. There is a hint of a poorly developed cleavage, but generally tourmaline can be distinguished from biotite or hornblende because it is unixial negative and has poor cleavage.

PPL I PPL, KFS, BDI, Scale bar = 0.07 mm

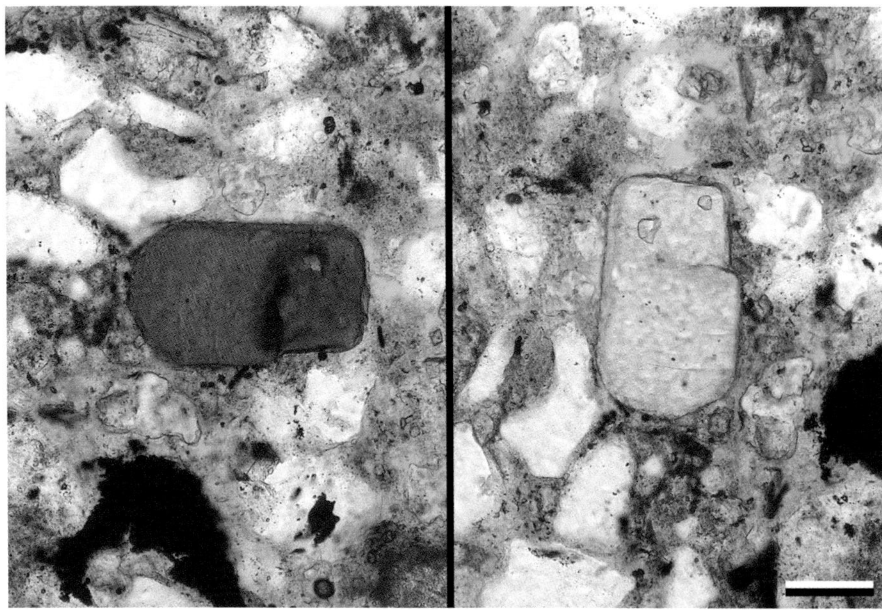

Jurassic (Oxfordian) Melke Fm., Møre Basin, Norwegian Sea †

This green tourmaline (probably common schorl) grain is a fractured crystal with inclusions of smaller euhedral opaques and quartz. The birefringence of the grain is lower than usual due to its orientation, and it superficially resembles some hornblende grains.

PPL I XPL, BDI, Scale bar = 0.07 mm

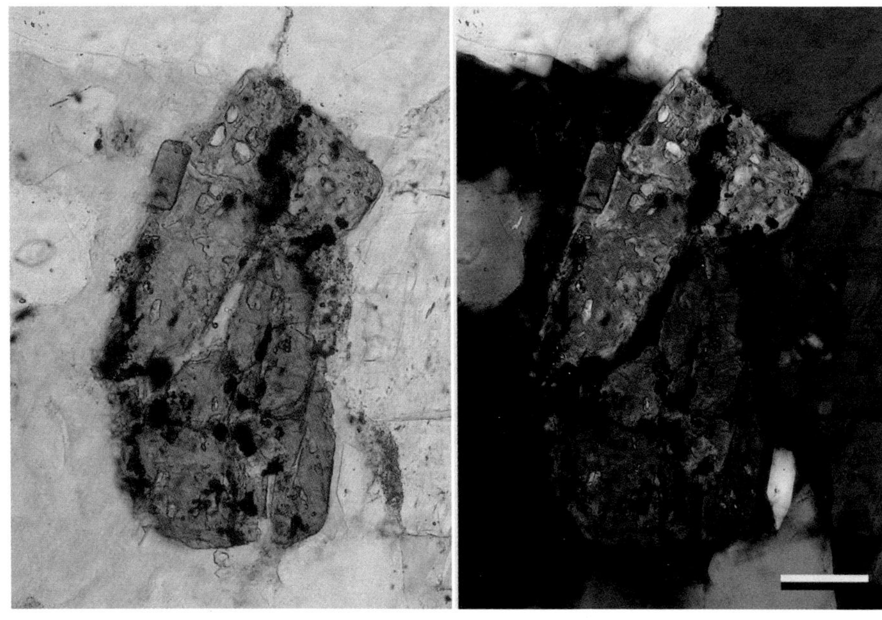

Up. Triassic – Lo. Jurassic Portland Fm., Middlesex Co., Connecticut

Rutile is another heavy mineral that is common, both as inclusions in other minerals and as individual detrital grains. Rutile is a uniaxial positive mineral that has very high relief and in plane light ranges in color from yellow to red-brown to black. It has extreme birefringence (commonly difficult to see because of mineral coloration), some pleochroism and two directions of cleavage. Rutile forms as prismatic crystals with striations or felted masses of acicular crystals. Rutile commonly has contact twins with two, six or eight intersecting crystals. In this example, most of the twins are two crystals.

PPL I XPL, Scale bar = 0.12 mm

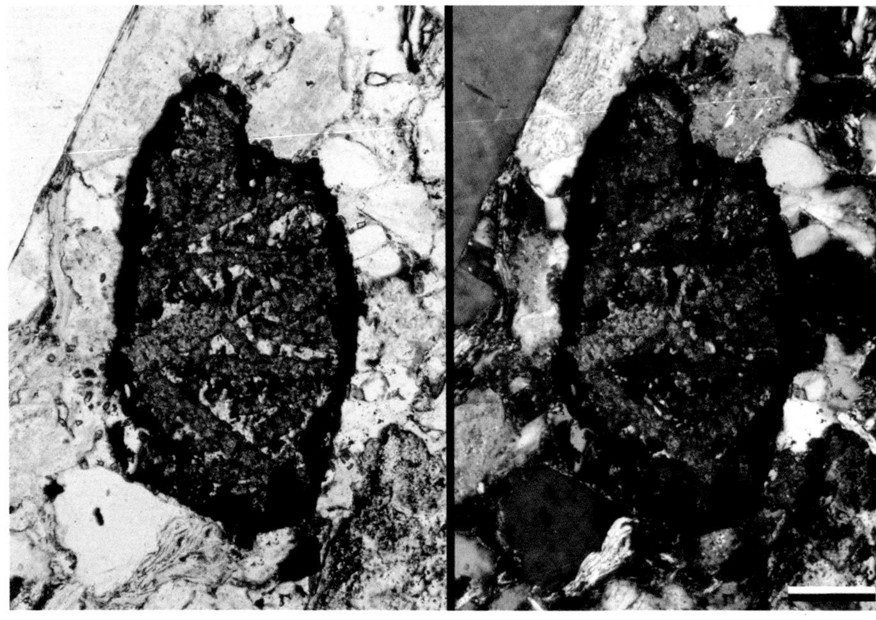

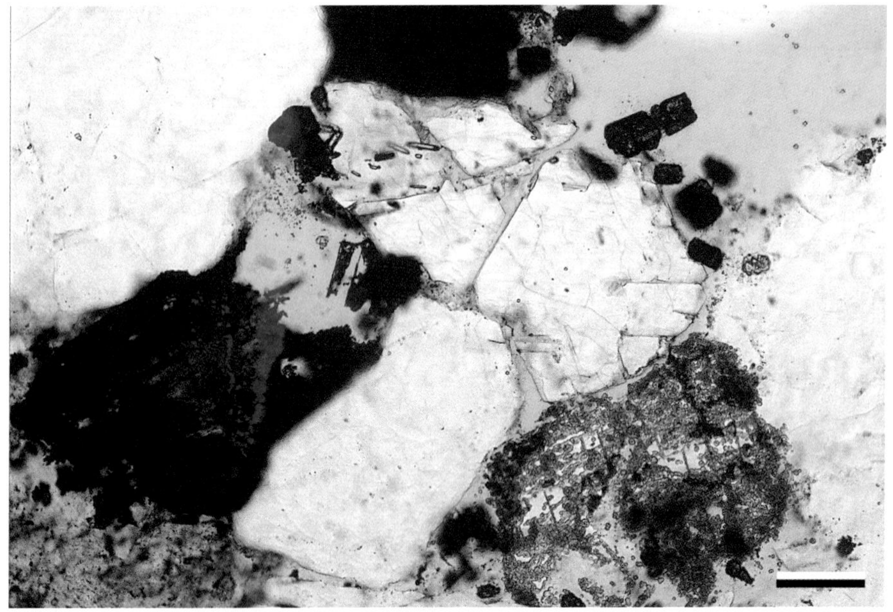

Mid. Jurassic Ness Fm., Norwegian sector, North Sea †

A partially dissolved rutile grain (far left). Rutile, although mechanically very stable, is prone to dissolution during burial diagenesis. A single cleavage is seen in this rutile, although the color is so deep, it is almost opaque. This rutile grain is associated with a number of small crystals of anatase (titanium oxide) that have precipitated in nearby pores (upper right). It is likely that the titanium for the anatase was sourced from the rutile. Anatase is a uniaxial negative mineral with high relief and birefringence, and ranges from colorless to brown, red or blue/green.

PPL, KFS, BDI, Scale bar = 0.07 mm

Jurassic Ile Fm., Norwegian sector, North Sea †

This yellow-brown rutile grain looks similar in both plane- and cross-polarized light, because the high birefringence is masked by the deep color of the mineral. Rutile is an extremely common detrital heavy mineral derived mainly from high-temperature acidic igneous rocks, crystalline metamorphic rocks, pegmatites or *in-situ* decomposition of ilmenite.

PPL | XPL, BDI, Scale bar = 0.03 mm

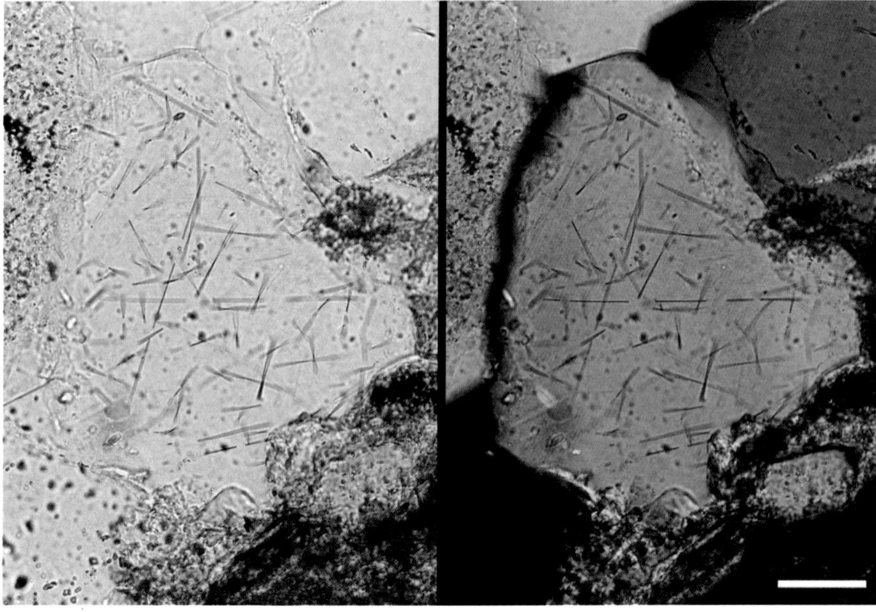

Mid. Jurassic Entrada Fm., Moffat Co., Colorado

Rutile commonly occurs as long acicular needle-shaped inclusions in other minerals, as in this example in a quartz grain. Such inclusions can be easily overlooked at lower magnifications in either plane- or cross-polarized light. Rutilated quartz grains can originate in a wide variety of source terranes, especially in hydrothermal, plutonic and metamorphic rocks, but they are uncommon in extrusive volcanics.

PPL | XPL, PFS, KFS, Scale bar = 0.05 mm

Cretaceous Dakota Gp., Moffat Co., Colorado

Long acicular needles of rutile here stand out under fluorescence. Although these rutile inclusions may appear to be contact twins, the crystals are not in the same plane with each other and therefore do not intersect.

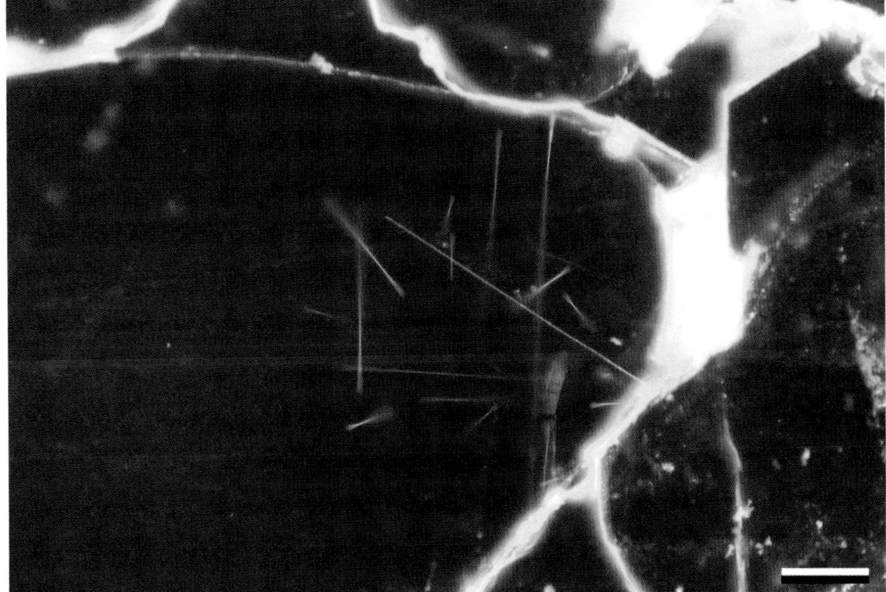

FL 470, RDI, Scale bar = 0.05 mm

Pennsylvanian Alamitos Fm., San Miguel Co., New Mexico

This detrital rutile is made up of prismatic crystals of rutile with contact twins. The color of these reddish brown crystals masks their very high birefringence. In this example, the rutile crystals may have been intergrown with another mineral or they may have replaced a detrital ilmenite grain. Later, that mineral was leached and the pores were subsequently infilled by slightly ferroan calcite cement.

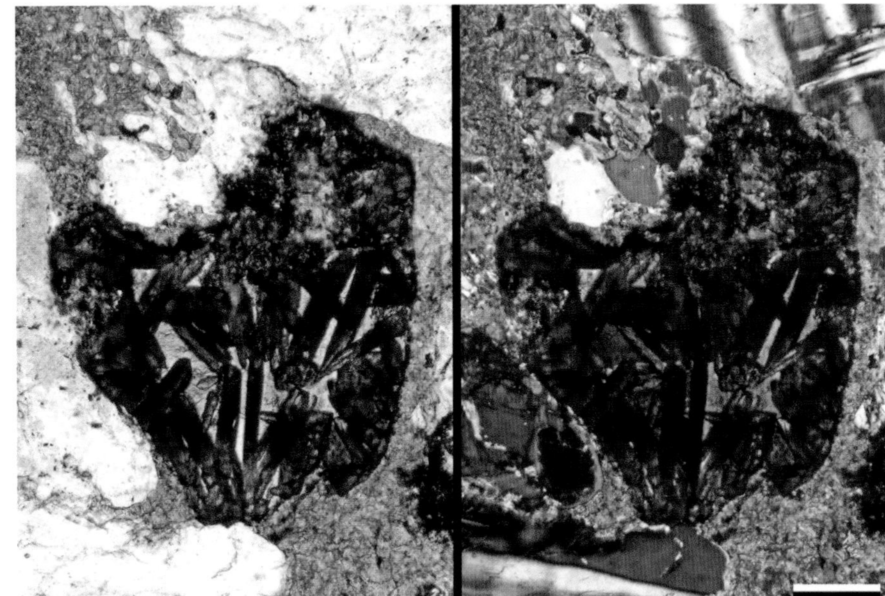

PPL, AFeS, Scale bar = 0.16 mm

Eocene Thumbli Fm., Rajasthan, India

This detrital rutile grain displays a very finely crystalline lattice of rutile contact twins under mixed back-scattered electron/scanning electron microscopic imaging. The rutile, which probably replaces ilmenite, is intergrown with iron oxides (hematite or magnetite).

BSE(SEM), Scale bar = 12.7 μm

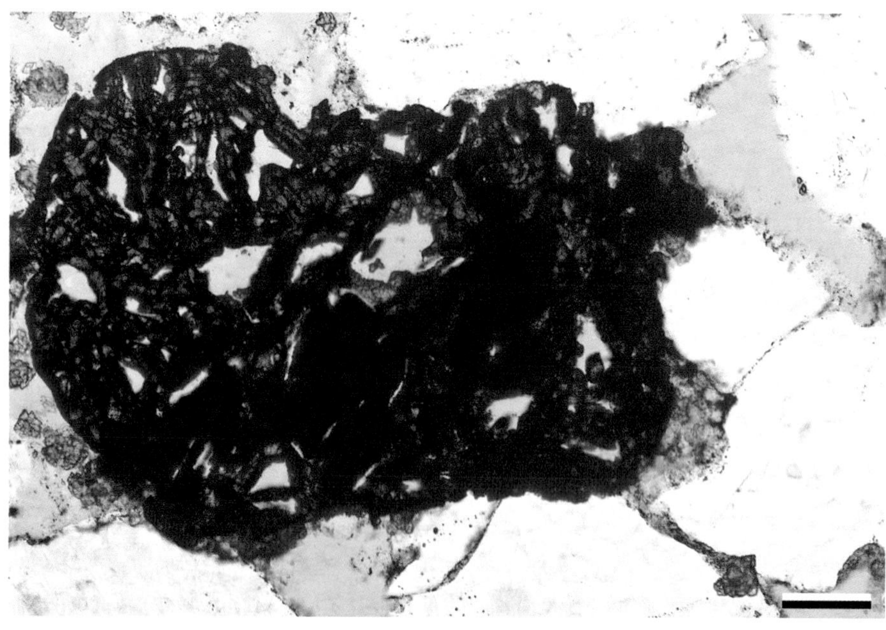

Mid. Jurassic Ness Fm., Norwegian sector, North Sea †

A partly degraded and dissolved rutile grain. It is not known what the intergrown mineral was, but it could have been another titanium-rich mineral, such as ilmenite.

PPL, KFS, BDI, Scale bar = 0.07 mm

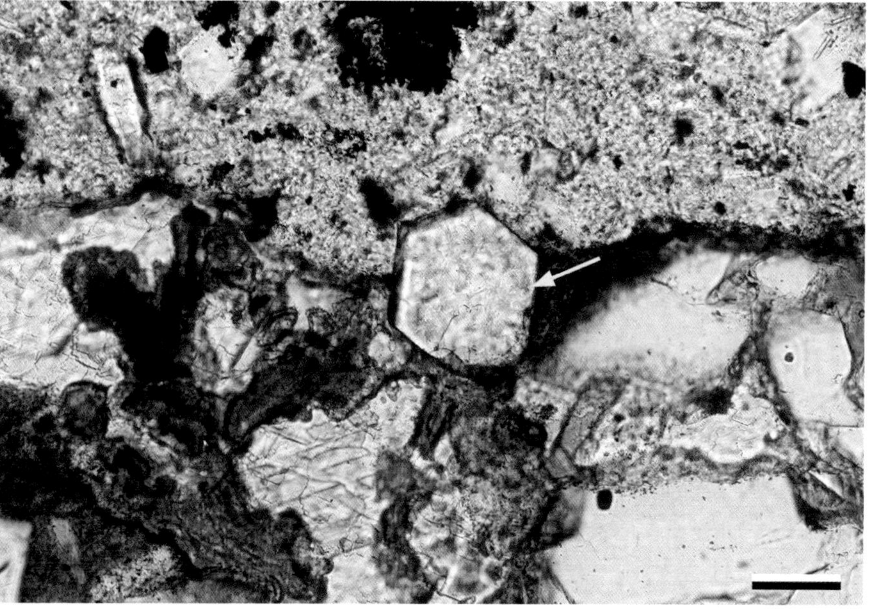

Mid. Eocene – Lo. Oligocene Lower Spears Gp., Socorro Co., New Mexico

A euhedral (hexagonal) apatite crystal in a volcaniclastic sandstone (yellow arrow). Apatite is actually a group of minerals that include, among others, calcium fluorapatite and hydroxylapatite. Apatite group minerals are the most common calcium phosphate accessory mineral. They are found in a wide variety of environments including almost every type of igneous rock, contact metamorphosed carbonate rocks, hydrothermal veins, sedimentary phosphorites, laterites and bone material. Apatite can look like colorless tourmaline, but apatite has lower birefringence.

PPL, BDI, Scale bar = 0.07 mm

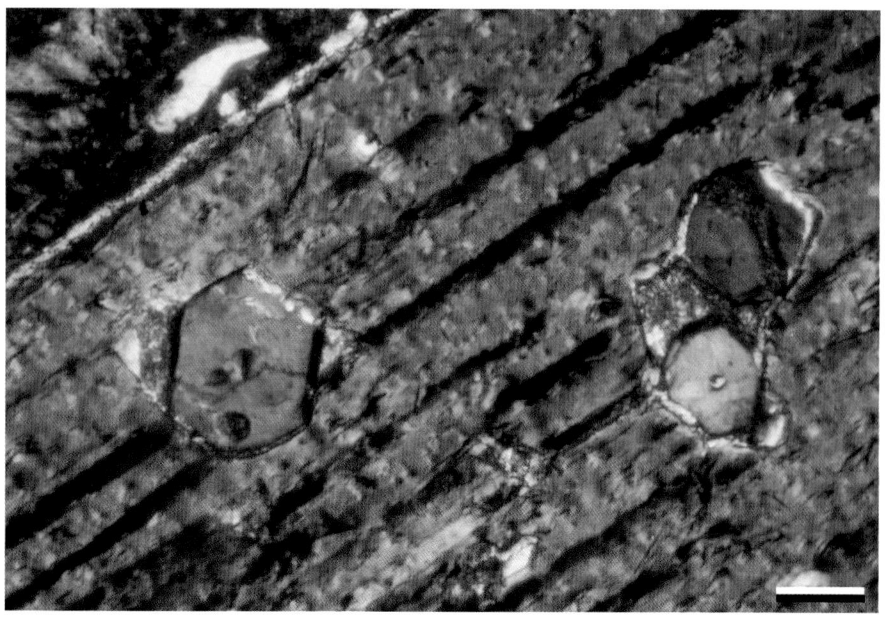

Oligocene Fish Canyon Tuff, Colorado

These apatite inclusions in biotite have hexagonal outlines, imperfect cleavage, moderate relief and very low birefringence. Apatite is uniaxial negative and commonly has weak to no pleochroism. Apatite commonly is found as detrital grains in sedimentary rocks (it generally is stable during burial although it can be significantly degraded during prolonged surficial weathering). Apatite is widely used in provenance and thermal history studies (e.g., Green et al., 1989; Morton and Yaxley, 2007).

XPL, Scale bar = 0.01 mm

Mid. Jurassic Fulmar Fm., United Kingdom sector, North Sea †

A well-rounded detrital apatite grain that displays higher relief than the surrounding quartz grains. Apatite shows a characteristic dull gray birefringence that is almost bluish. As a heavy mineral, apatite is moderately stable and is relatively common, even in mature sandstones.

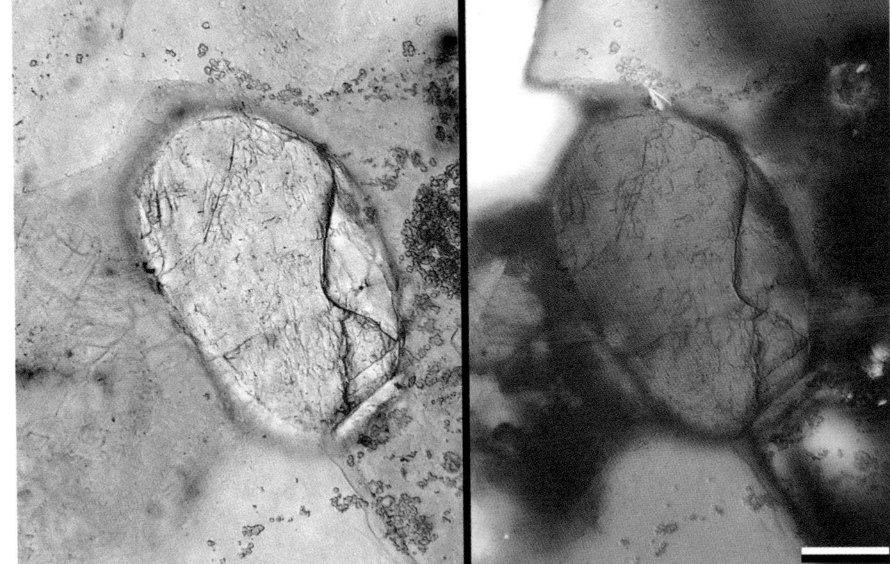

PPL | XPL, KFS, BDI, Scale bar = 0.04 mm

Pliocene sandstone, Santa Barbara Co., California

Apatite frequently occurs as inclusions in other minerals. Apatite, here, is seen as inclusions in a detrital feldspar grain within this bitumen-impregnated sandstone. The inclusions are clearly visible, because they have a slightly lower birefringence and a higher relief than the feldspar grain.

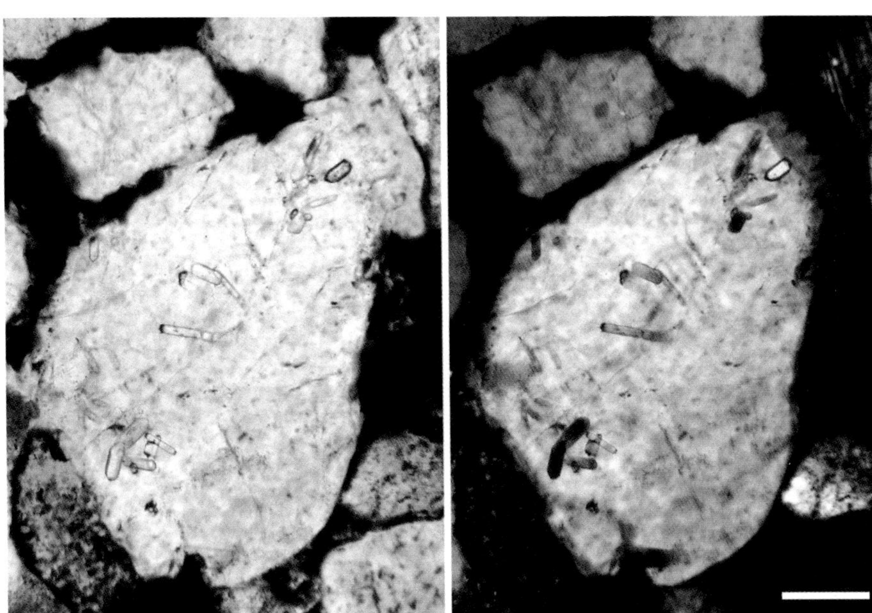

PPL | XPL, Scale bar = 0.08 mm

Precambrian (Proterozoic) Ortega Fm., New Mexico

This image shows epidote crystals from a quartz vein in a metaquartzite rock. Epidote has high relief, strong pleochroism, perfect basal cleavage and yellow to green color. It is a biaxial negative mineral with very high (third-order) birefringence. Epidote is derived mainly from altered igneous rocks, hydrothermal veins and crystalline metamorphic terranes.

PPL | XPL, Scale bar = 0.03 mm

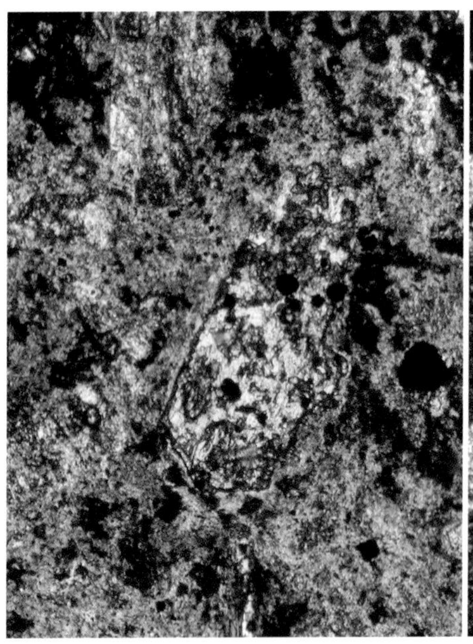
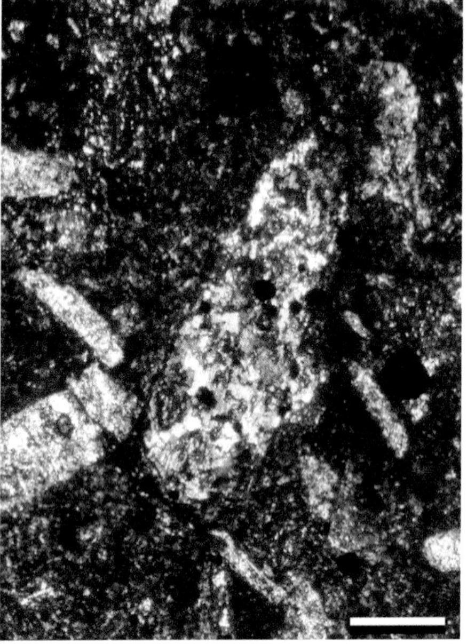

Eocene Lo. Spears Gp., Socorro Co., New Mexico

This volcaniclastic sandstone has undergone substantial hydrothermal (propylitic) alteration. Under such conditions, epidote commonly forms as an alteration product of hornblende, plagioclase and clinopyroxene and thus can be mistaken for detrital epidote grains (as with the epidote-replaced grain in the center of these images). The feldspars in this view were substantially altered to sericite (a fine-grained white mica). In thin section, epidote can be difficult to identify because of the granular character of replacements such as this, but the birefringence is the key factor in identification.

PPL | XPL, Scale bar = 0.17 mm

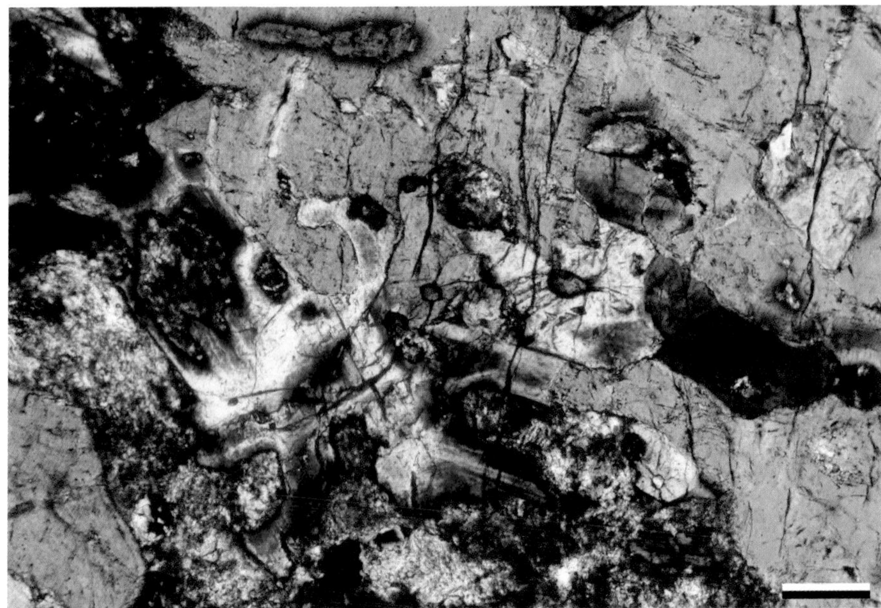

Paleozoic schist, Massachusetts

Zoisite, a member of the epidote group, is shown here in a source rock. It is an orthorhombic, biaxial positive mineral and occurs as prismatic crystals to massive clusters. In plane-ploarized light, zoisite is colorless, pink to pale blue and has high relief. Under cross-polarized light, it has anomalous birefringence colors (ultrablue) that are clearly shown in this example. Zoisite has perfect basal cleavage and conchoidal fracture. It is a mineral commonly found in medium-grade metamorphic schists and some blueschists.

XPL, Scale bar = 0.14 mm

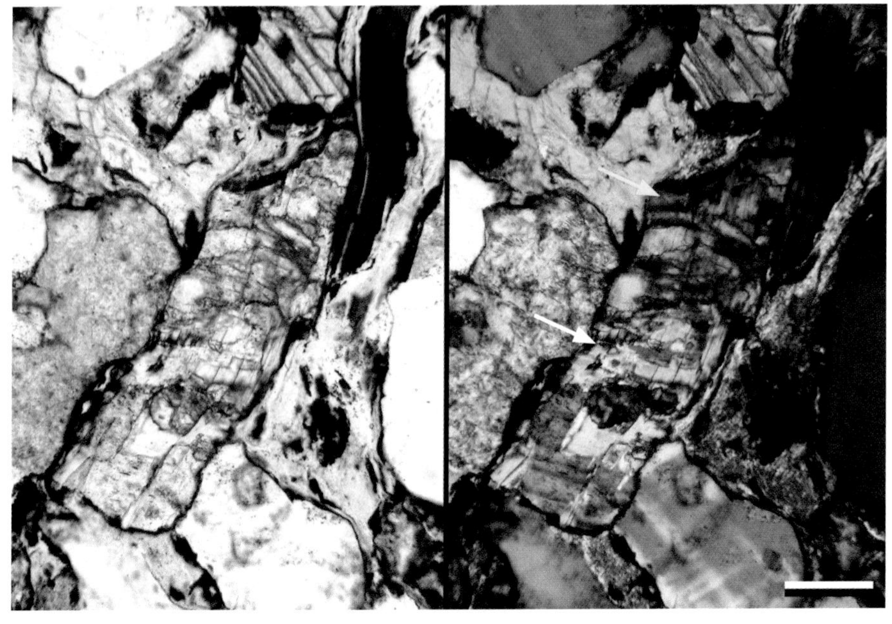

Lo. Permian Cutler Fm., Mesa Co., Colorado

This detrital metamorphic grain, comprised of zoisite (yellow arrow) and muscovite (white arrow), is in a compacted immature sandstone. Zoisite displays perfect basal cleavage and anomalous ultrablue birefringence. Zoisite is brittle and does not withstand significant mechanical transport. In this view, the clays behaved in a ductile manner during compaction, bending around more rigid grains; the zoisite, however, shows brittle deformation.

PPL | XPL, Scale bar = 0.10 mm

Holocene beach sediment, Orange Co., California

Clinozoisite is another member of the epidote group and is seen here in a grain mount. It is pale green to colorless and has anomalous pale ultrablue to high first-order birefringence. Birefringence is dependent on the iron content of the crystal, the more iron, the higher the birefringence. It has perfect basal cleavage, high relief, slightly oblique extinction (unlike zoisite) and is biaxial positive. Clinozoisite is relatively common in sediments, and is derived mainly from schists and metamorphosed basic volcanic rocks. It is frequently mistaken for zoisite.

PPL | XPL, Scale bar = 0.05 mm

Pennsylvanian Alamitos Fm., San Miguel Co., New Mexico

Garnet is a general name applied to large group of isometric minerals. Garnet has dodecahedral or trapezohedral habit, very high relief and conchoidal fracture. In plane light, garnet has a wide range of colors, from colorless to shades of red, yellow, green, brown and black. Under cross-polarized light, garnet is isotropic (i.e., has no birefringence), so it generally is easy to identify. The garnet grain at center of this image is euhedral and colorless with conchoidal fracture and mineral inclusions. Garnet has intermediate stability and commonly shows crystal faces even on detrital grains. The grain at far left is rutile (yellow arrow).

PPL, Scale bar = 0.08 mm

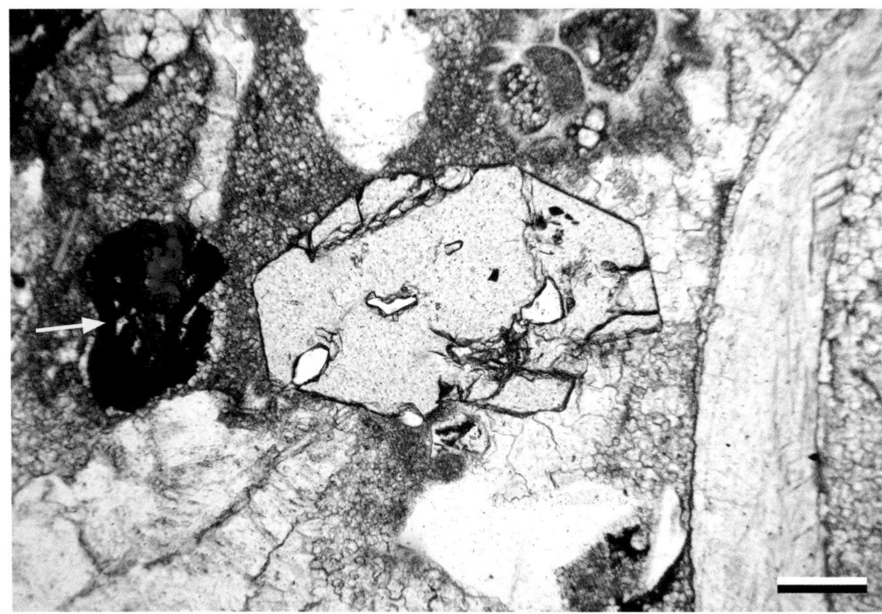

Up. Triassic – Lo. Jurassic Portland Fm., Middlesex Co., Connecticut

The reworked garnet porphyroblast (a crystal of metamorphic origin) in this image has some of its original crystal faces preserved, is isotropic (in cross-polarized light) and contains abundant mineral inclusions (mainly quartz). It is common for garnet to contain solid (quartz, mica or other minerals) and/or fluid inclusions. Garnets are found in a wide variety of regional and contact metamorphic rocks as well as igneous rocks. They also form in granitic pegmatites, and pyrope is a common constituent of ultrabasic igneous rocks.

PPL | XPL, Scale bar = 0.07 mm

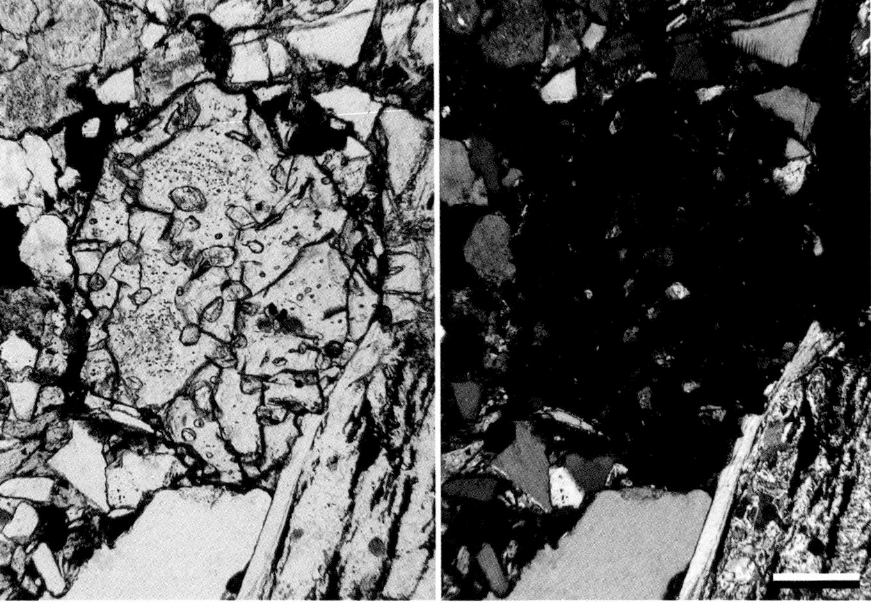

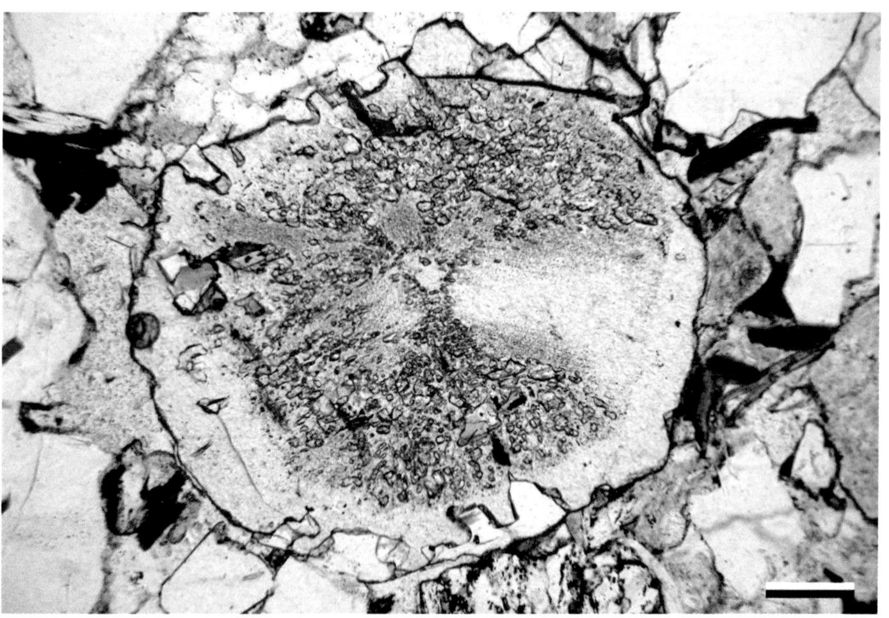

Up. Triassic – Lo. Jurassic Portland Fm., Middlesex Co., Connecticut

Garnet is frequently zoned with normal, oscillatory and sector zoning as the most common forms. This reworked porphyroblast has a core that has textural sector zoning followed by further growth with no evident zoning. Textural sector zoning is marked here by inclusion-rich (hourglass shaped) and inclusion-poor sectors.

PPL, Scale bar = 0.20 mm

Up. Triassic – Lo. Jurassic Portland Fm., Middlesex Co., Connecticut

A garnet grain with linear trends of inclusions that formed during crystal growth. Garnet has moderate grain stability, and porphyroblasts commonly are rounded or broken during chemical weathering and mechanical abrasion.

PPL, Scale bar = 0.26 mm

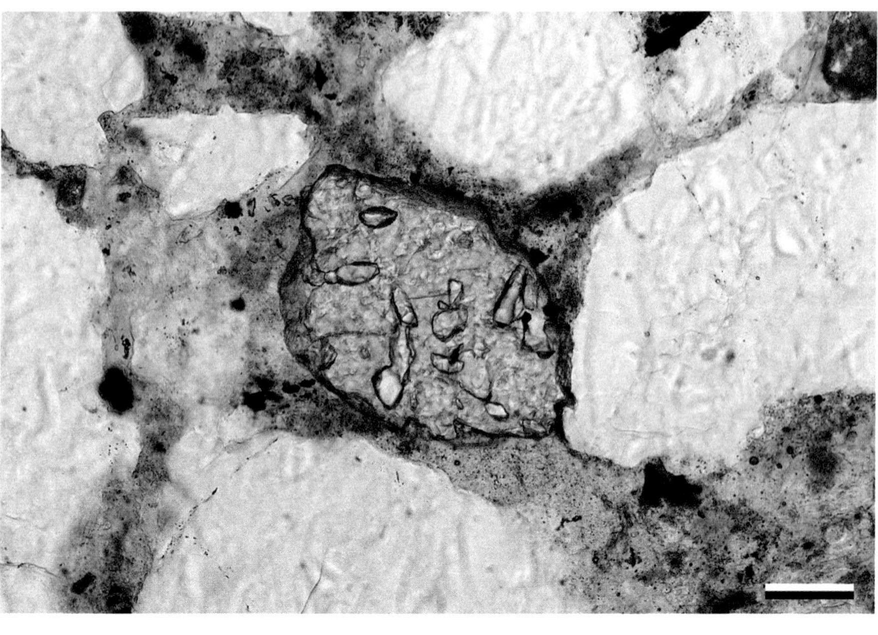

Jurassic (Oxfordian) Melke Fm., Møre Basin, Norwegian Sea †

A garnet grain with inclusions, which suggest that it may be a porphyroblast from a metamorphic rock. The grain shows a pale pinkish color and high relief in plane-polarized light. Not all garnets exhibit color light, and colors of most garnets tend to be quite pale at the thickness of a thin section. The ragged margins are something that typify many garnet grains and help identify them.

PPL, KFS, BDI, Scale bar = 0.07 mm

Holocene beach sediment, New Haven Co., Connecticut

A brownish pink garnet grain as viewed in a grain mount. Because of their relative stability, garnet is common in placer deposits such as this. Garnets of this color generally are almandine or grossular, and the color here is more intense than in previous photographs, because this is a complete 3-D grain from a grain mount it is far thicker than the 30-μm-thick grains seen in previous thin-section photomicrographs. Another advantage of grain mounts is that individual grains can be sampled for a variety of optical or geochemical analyses.

PPL, Scale bar = 0.08 mm

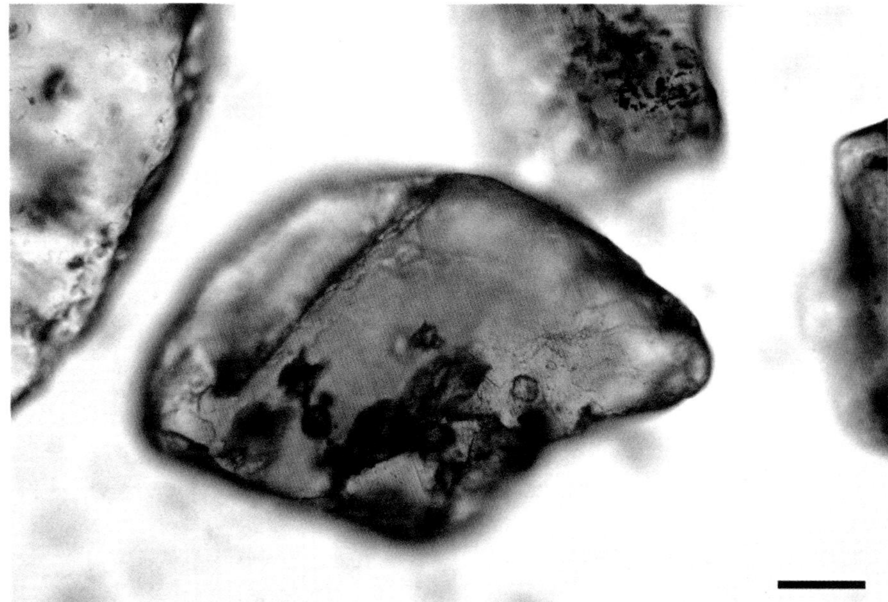

Ordovician? Zoar Gneiss, New Haven Co., Connecticut

A kyanite crystal surrounded by quartz crystals in a metamorphic rock. Kyanite has high relief and two cleavages (one perfect and one good, but uneven). It is normally colorless, but it also occurs in pale shades of blue, yellow and red. Kyanite is weakly pleochroic, biaxial negative and has up to first-order orange-red birefringence. Kyanite is only found in moderately high-grade metamorphic source areas and thus is a valuable provenance indicator. It has moderate chemical stability, but relatively low abrasion resistance.

PPL | XPL, Scale bar = 0.10 mm

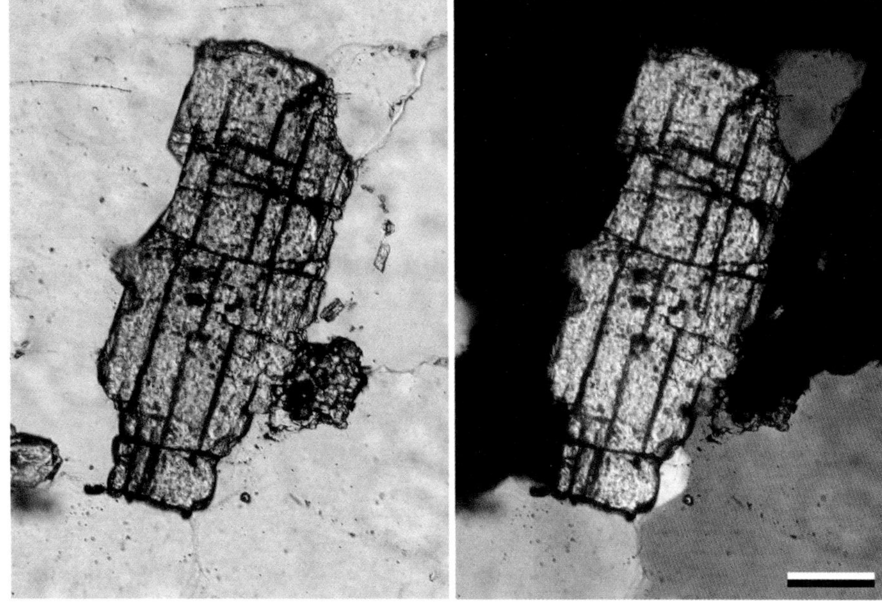

Mesozoic (Jurassic?) metaquartzite, Imperial Co., California

Kyanite typically forms long and bladed crystals in metamorphic rocks (in this case, a metaquartzite). These kyanite crystals also contain abundant quartz inclusions, that may be present here because the premetamorphic rock was a sandstone. Kyanite and pyroxene can be hard to distinguish from one another, but kyanite has higher relief, 85° cleavages, and lower birefringence.

PPL | XPL, Scale bar = 0.26 mm

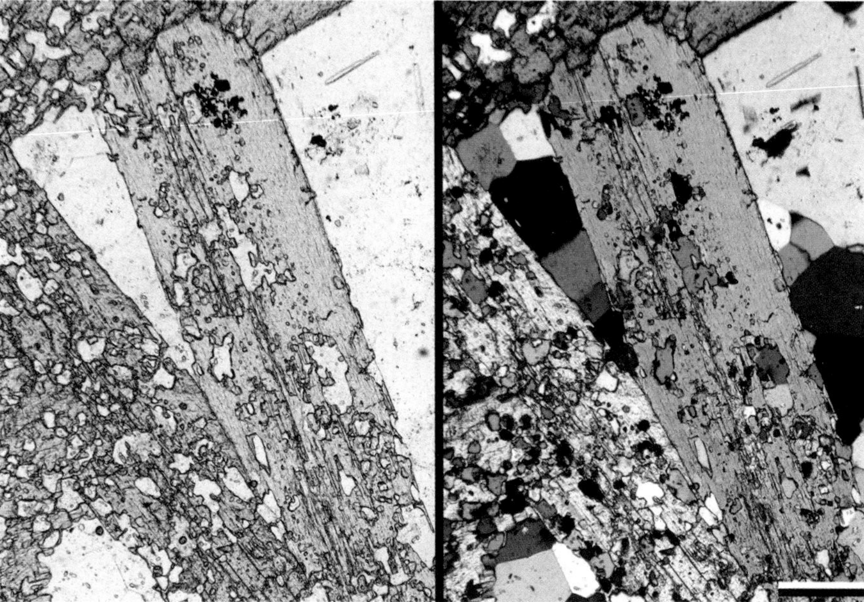

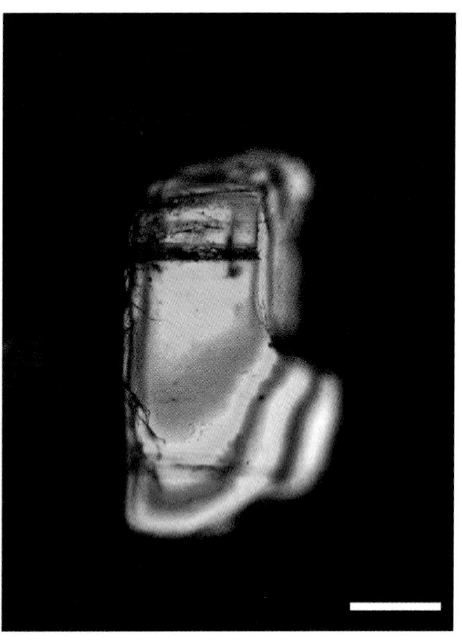

Holocene beach sediment, New Haven Co., Connecticut

This detrital kyanite crystal in a grain mount has high relief, excellent cleavage, bladed habit and very pale blue color. Because this is a grain mount, the crystal is much thicker than 30 μm, and therefore, the pleochroism of the thickest part of grain is much higher than in a standard thin section. The angularity and lack of alteration indicate that the grain is first-cycle and has probably undergone little chemical weathering and short transport (indeed, the site of deposition is just kilometers to tens of kilometers from kyanite-bearing source rocks).

PPL | XPL, Scale bar = 0.08 mm

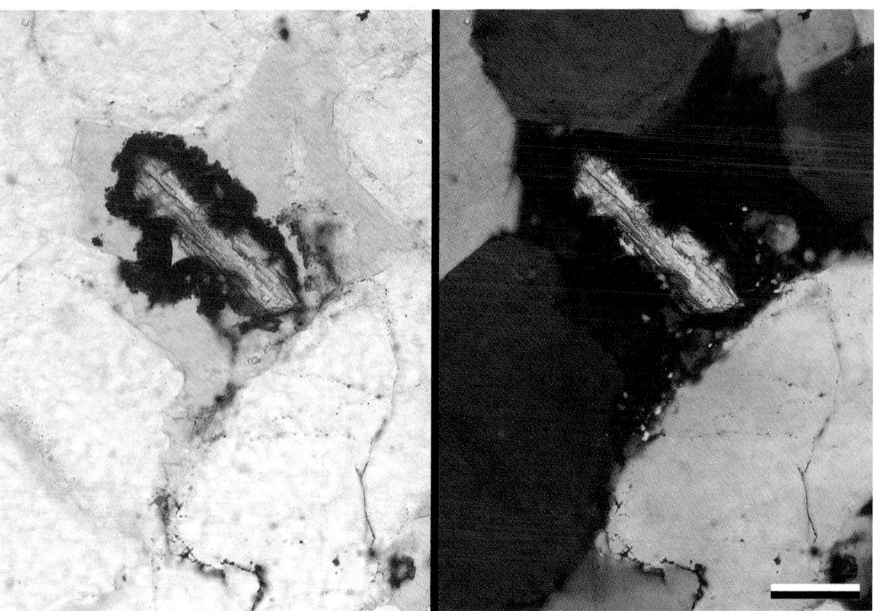

Mid. Jurassic Fulmar Fm., United Kingdom sector, North Sea †

A monazite grain surrounded by degraded oil. The emission of radiation from the monazite grain has degraded the bitumen. Both zircon and monazite are variably radioactive, and this feature is often seen with both minerals. Monazite has very similar relief and birefringence to zircon and titanite but can be distinguished by its inclined extinction angle and biaxial positive crystals. They can also be differentiated by the fact that titanite has deeper color and zircon has a straight extinction angle and is uniaxial.

PPL | XPL, BDI, Scale bar = 0.07 mm

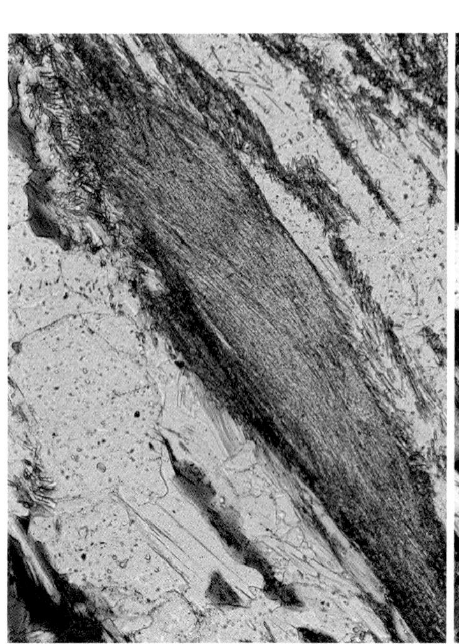
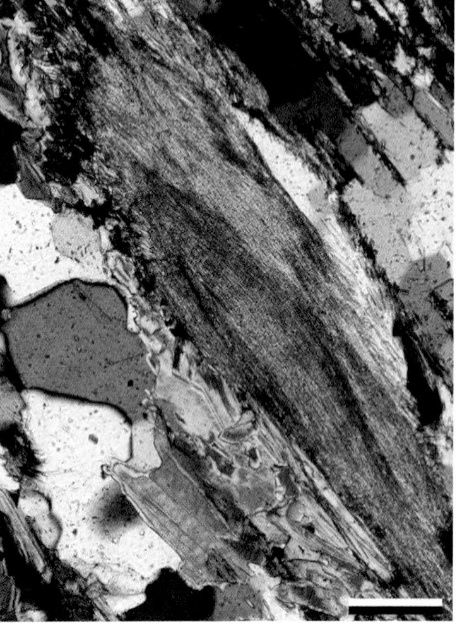

Ordovician-Silurian sillimanite schist, New Haven Co., Connecticut

Sillimanite generally is finely fibrous (called fibrolite) to acicular. It is light brown to colorless, has high relief and is slightly pleochroic. Under cross-polarized light, it is biaxial positive and has parallel extinction with up to low second-order colors (more commonly it has upper first-order colors where it is finely crystalline). Sillimanite is found mainly in high-grade aluminous schists and some contact metamorphic rocks. It has moderate chemical stability but relatively low abrasion resistance. In this sample, biotite, muscovite and quartz also are visible.

PPL | XPL, Scale bar = 0.10 mm

Ordovician? Zoar Gneiss, New Haven Co., Connecticut

A euhedral staurolite crystal (center) is surrounded by quartz (colorless) and muscovite (stained red in this section). Staurolite has brownish yellow color, moderate relief, moderate pleochroism, abundant inclusions and prismatic crystal habit with one direction of poorly developed cleavage and conchoidal fracture. Staurolite birefringence is first-order orange, and it is biaxial positive with positive elongation. Detrital specimens commonly show some surficial alteration to chlorite. It is an excellent indicator of a schistose metamorphic source. During transport, staurolite abrades quickly.

PPL | XPL, KFS, Scale bar = 0.51 mm

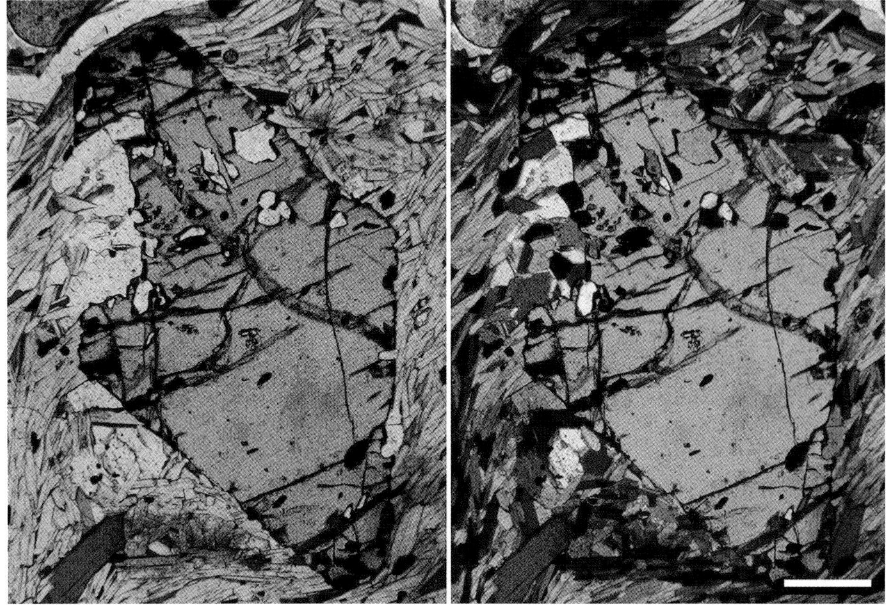

Precambrian (1.1 Ga) Fordham Gneiss, Westchester Co., New York

These wedge-shaped to prismatic titanite crystals (formerly known as sphene) are significant constituents of this gneissic/skarn source rock. Titanite is a common accessory heavy mineral in felsic to intermediate plutonic rocks, pegmatites, skarns, gneisses and schists. In sedimentary rocks, it occurs as detrital grains, cements and replacements. Titanite is a biaxial positive mineral with very high relief and birefringence. In thin section, it can be colorless, yellow, brown or reddish brown and is pleochroic. It has one direction of cleavage and one direction of parting; in addition, it frequently has contact twins.

PPL | XPL, Scale bar = 0.43 mm

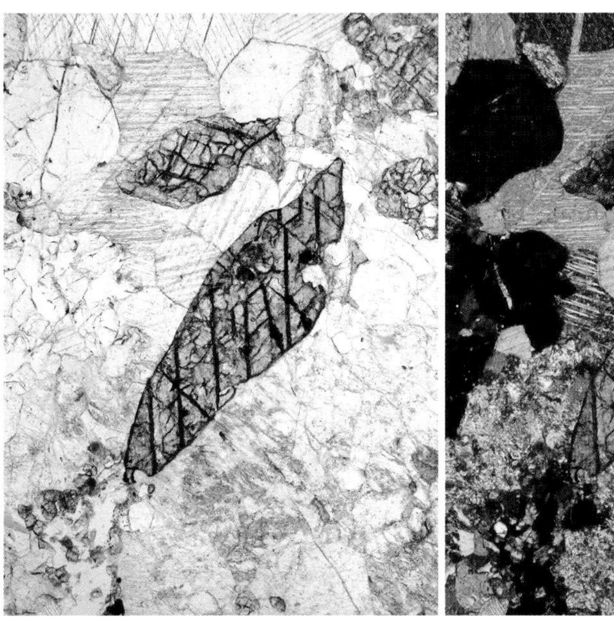

Mid. Jurassic Ness Fm., Norwegian sector, North Sea †

A detrital grain of titanite that displays very high relief. The brownish color is common, and the birefringence, although high, is sometimes masked by the color of the mineral. This grain displays possible cleavage with hints of a diamond shaped cross-hatched pattern.

PPL, KFS, BDI, Scale bar = 0.07 mm

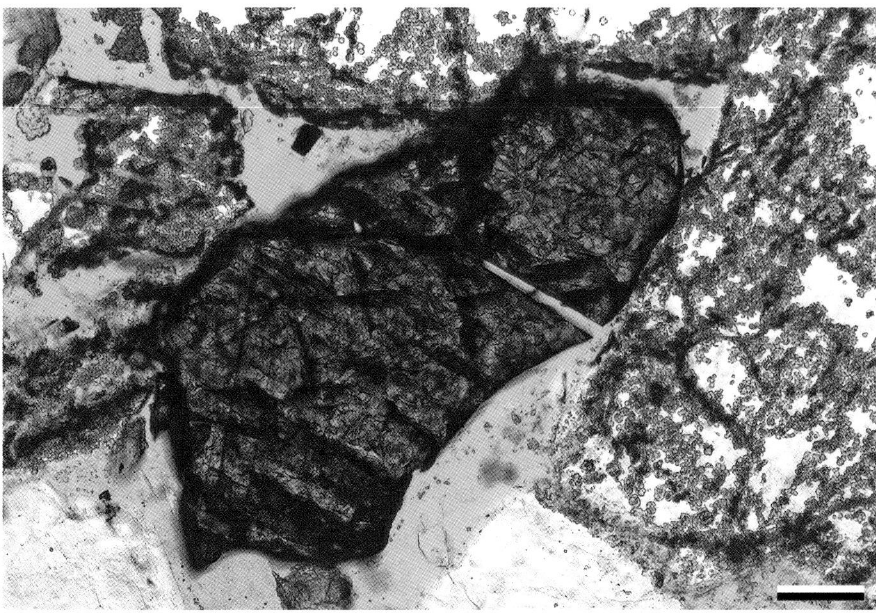

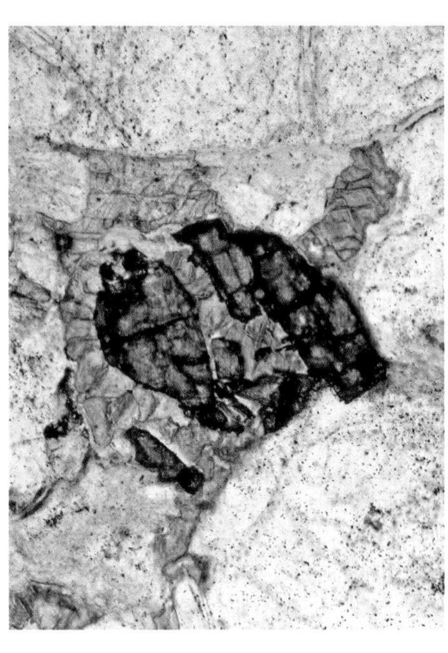
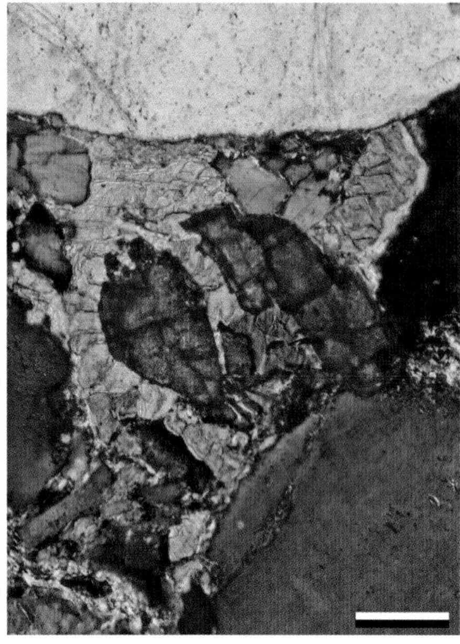

Mid. Permian Wegener Halvø Fm., Karstryggen area, East Greenland

A rounded titanite grain in a sandstone; the titanite was shattered along cleavage/parting planes during burial and compaction. Because of its strong color, it looks similar in plane- and cross-polarized light. The grain was later surrounded by nonferroan to ferroan calcite cements.

PPL | XPL, AFeS, BDI, Scale bar = 0.10 mm

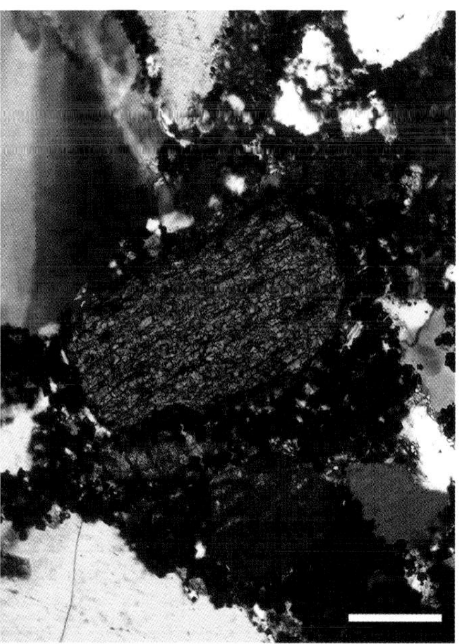

Jurassic Tilje Fm., Norwegian sector, North Sea †

The birefringence of titanite is very high, making it difficult to determine interference color order (especially because the expected high-order, pastel interference colors are easily mask). One distinguishing factor is that its appearance in cross-polarized light is very similar to that in plane-polarized light.

PPL | XPL, KFS, BDI, Scale bar = 0.12 mm

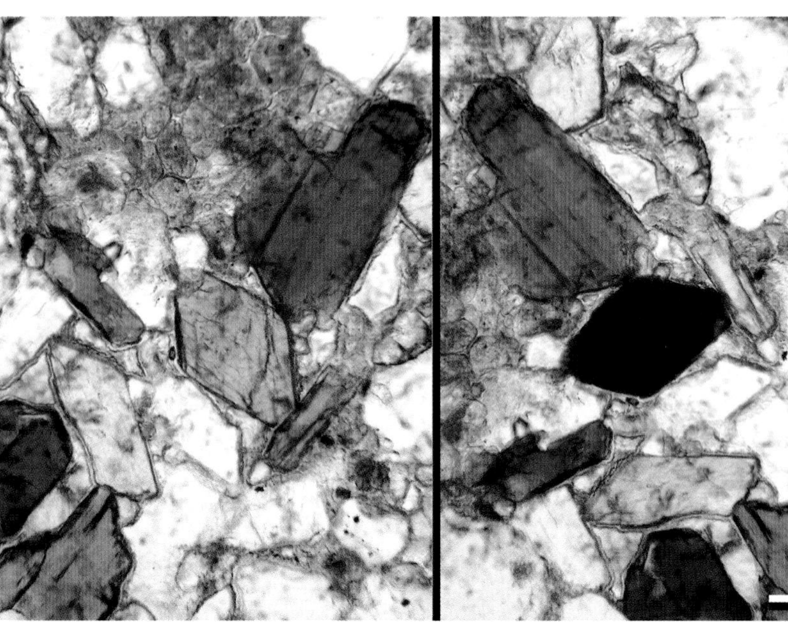

Lo. Miocene Arikaree Fm., Platte Co., Wyoming

Amphiboles are a large group of orthorhombic and monoclinic minerals that generally have two cleavages intersecting at 124° (as in the center yellow grain); amphiboles also have stronger pleochroism than pyroxenes. In this pair of photomicrographs, all the colored grains are hornblende (a common amphibole). The yellow grain is a typical basal section through a hornblende crystal. These crystals exhibit strong pleochroism (visible in these two images with approximately 90° of stage rotation). Amphiboles are a large, diverse group of minerals, and they occur in a wide range of igneous and metamorphic environments.

PPL | PPL, Scale bar = 0.09 mm

Oligocene Up. Spears Gp., Socorro Co., New Mexico

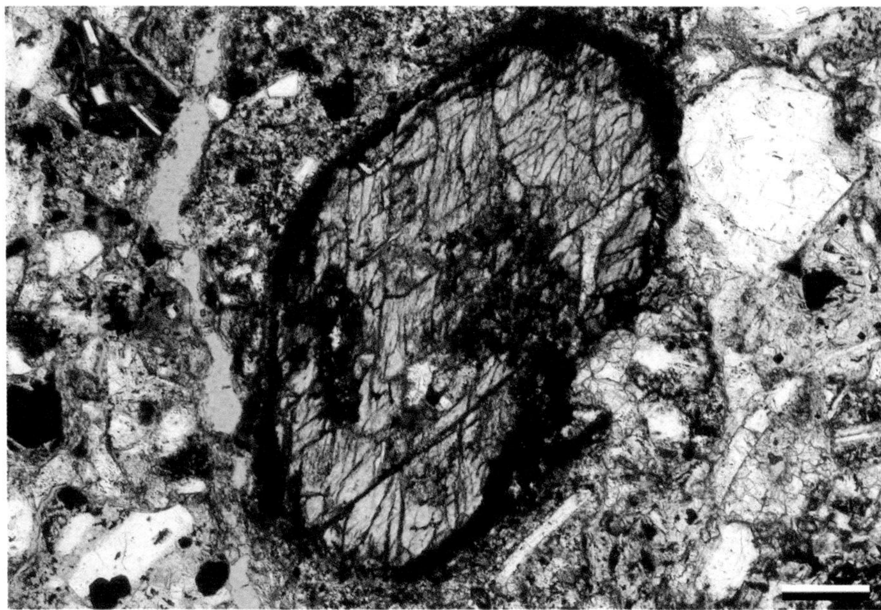

This hornblende grain may have started out as a euhedral crystal from a volcanic source, but it was extensively rounded during transport. The dark alteration rim was formed by hydrothermal alteration. The two characteristic cleavages of the basal section (124°/56°) are clearly displayed here. Hornblende can be biaxial negative or positive. Hornblende is the most common amphibole in acidic to intermediate igneous rocks, and it is common in moderate- to high-grade regional metamorphic rocks. Because of its chemical and mechanical instability, hornblende tends to occur only in relatively immature sediments.

PPL, BDI, Scale bar = 0.14 mm

Precambrian (1.1 Ga) Fordham Gneiss, Westchester Co., New York

The amphiboles in this section are hornblende. Once again, the high relief of hornblende is visible as is the strong pleochroism and the two cleavage directions at 124° and 56°. Horn blende is a common metamorphic mineral occurring in metamorphosed sedimentary and plutonic igneous rocks. Surficial alteration of hornblende to chlorite and epidote is common.

PPL | XPL, Scale bar = 0.26 mm

Holocene beach sediment, Orange Co., California

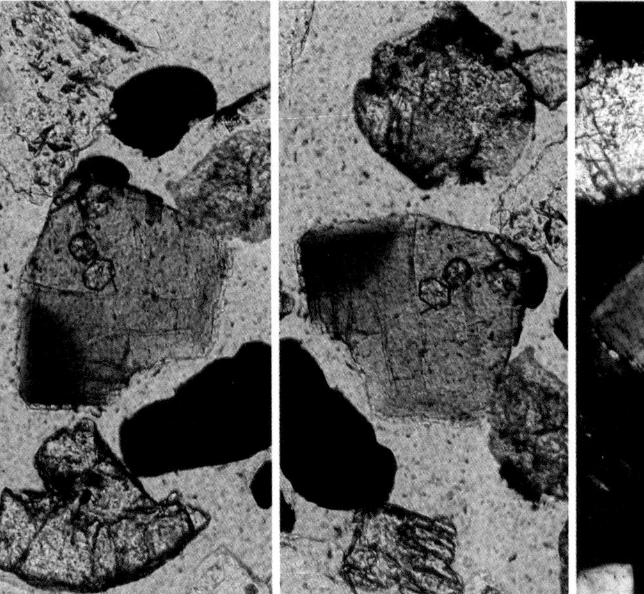

Glaucophane is a monoclinic amphibole with distinctive coloration, both in plane- and cross-polarized light. It is biaxial negative, has moderate to high relief, is strongly pleochroic in shades of blue, lavender and yellow and has high first-order birefringence or anomalous ultrablue color. The pleochroism is visible in the left and center images that have 90° relative rotation; the ultrablue birefringence is seen in the last image. Glaucophane is common in blueschist facies (low temperature, high pressure), greenschists, ecologites and as an alteration product of pyroxenes.

PPL | PPL | XPL, Scale bar = 0.10 mm

Oligocene Up. Spears Gp., Socorro Co., New Mexico

A zoned orthopyroxene crystal (center) in a volcaniclastic sandstone. Orthopyroxenes can be biaxial positive (enstatite and ferrosilite) or biaxial negative (bronzite, hypersthene, eulite). They have high relief, two cleavages at 90° and their birefringence ranges from gray to high first-order colors with increasing iron content. They also have parallel extinction, are pleochroic and are colorless to pale green, brown or pink in thin section. Orthopyroxenes are found in basic to ultrabasic igneous rocks and high-grade metamorphics. Most of the rest of the grains in this rock are plagioclase.

PPL | XPL, Scale bar = 0.17 mm

Oligocene Up. Spears Gp., Socorro Co., New Mexico

A volcaniclastic sandstone containing clinopyroxene (up to second-order blue to green; examples at white arrows) along with laths of plagioclase and an amphibole (red arrow). Clinopyroxenes are monoclinic, occur as tabular crystals, have two cleavages at 90°, and are most commonly biaxial positive. All clinopyroxenes have high relief, inclined extinction, can be non- to strongly pleochroic and are colorless to yellow, green or brown. They are sourced from intermediate to ultrabasic volcanic rocks, acidic igneous rocks and intermediate- to high-grade metamorphic rocks.

PPL | XPL, Scale bar = 0.26 mm

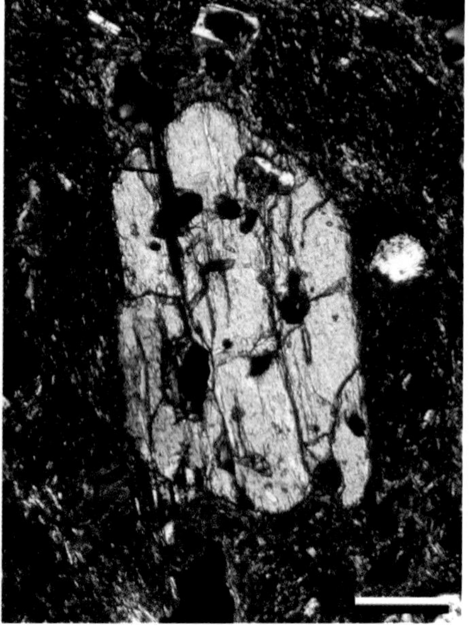

Mid. Eocene – Lo. Oligocene Spears Gp., Catron Co., New Mexico

This euhedral crystal of augite is very pale green in plane-polarized light and has conchoidal fracture. Augite, a clinopyroxene, can be pleochroic in shades of pale green to pale brown. It has high relief and two directions of cleavage (at 90°). Augite is biaxial positive, and under cross-polarized light, it has high first- to low second-order colors. Augite is common in basic and ultrabasic igneous rocks and high-grade metamorphic rocks. This phenocryst is set in an andesitic volcanic rock fragment.

PPL | XPL, Scale bar = 0.26 mm

Up. Triassic – Lo. Jurassic, Huizachal Fm., Tamaulipas, Mexico

An igneous rock fragment with an intergrowth of radiating augite crystals encased by plagioclase. This rock fragment is a plume-like symplectic intergrowth of the two minerals and such intergrowths commonly form as late crystallization products within magma chambers or within metamorphic reaction rims.

XPL, Scale bar = 0.05 mm

Lo. Miocene Arikaree Fm., Platte Co., Wyoming

Two members of the clinopyroxene group are strongly pleochroic, aegirine-augite and aegirine. In this example, there are three clinopyroxene grains at different optical orientations (similar to rotating the stage) exhibiting strong pleochroism. Based on their pleochroism, birefringence (first-order yellow), and characteristic 90° cleavages these grains are aegirine-augite (aegirine has a much higher birefringence with second- to third-order colors). Aegirine-augite is found in alkaline igneous rocks and pegmatites as well as some glaucophane schists.

PPL | XPL, Scale bar = 0.09 mm

Holocene beach sediment, Orange Co., California

This olivine grain in modern beach sediment is pale green and has up to third-order birefringence, high relief and conchoidal fracture (curved cracks). Olivine is colorless, pale yellow to pale green, is nonpleochroic to slightly pleochroic and has two poor cleavages. Fayalite forms eight-sided crystals, and forsterite usually occurs as granular masses. They are common in basic and ultrabasic igneous rocks and far less common in intermediate and acidic igneous rocks and metamorphosed carbonate and iron-rich deposits. It is very rare to find detrital olivine in sedimentary rocks following even moderate burial.

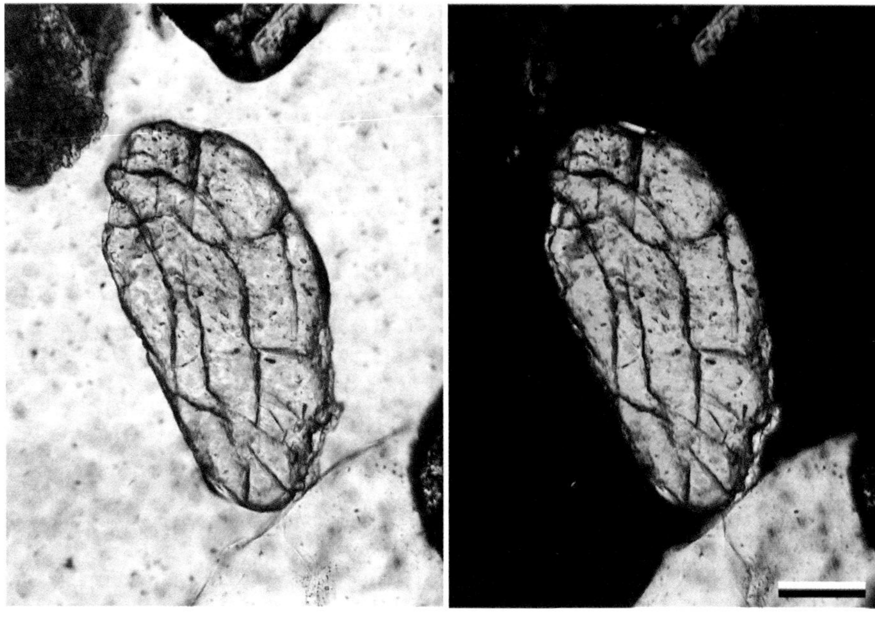

PPL | XPL, Scale bar = 0.05 mm

OPAQUES

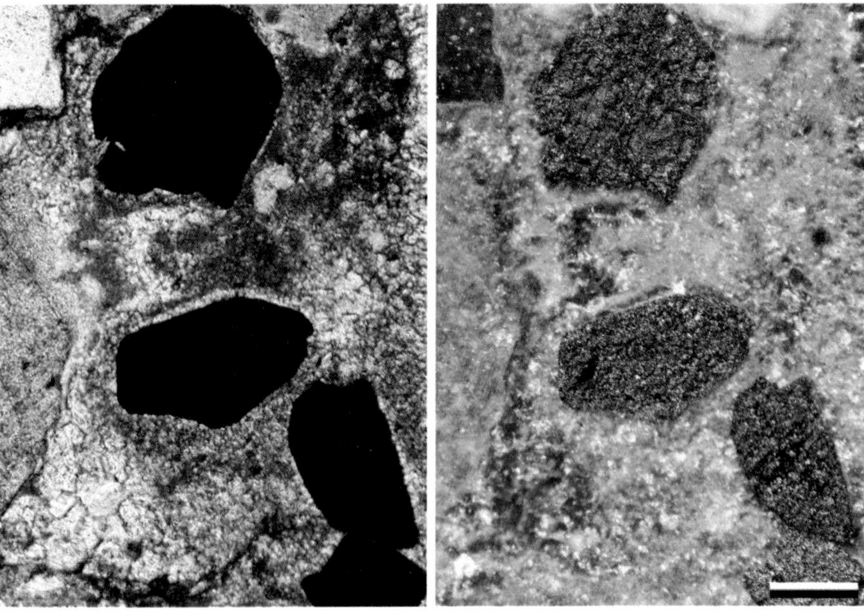

Lo. Permian Abo Fm., Otero Co., New Mexico

Opaque detrital heavy minerals can be abundant in sedimentary rocks, especially where concentrated in placers, and they are best examined in reflected light. The iron oxides seen here were illuminated with oblique reflected light (a simple LED light directed between the slide and the microscope objective). The grains here are magnetite (silvery) with some limonitic alteration (brownish yellow). Many opaque grains (especially pyrite) are of authigenic origin. Thus, thin section study can be useful in distinguishing detrital grains from authigenic precipitates.

PPL | ORL, Scale bar = 0.08 mm

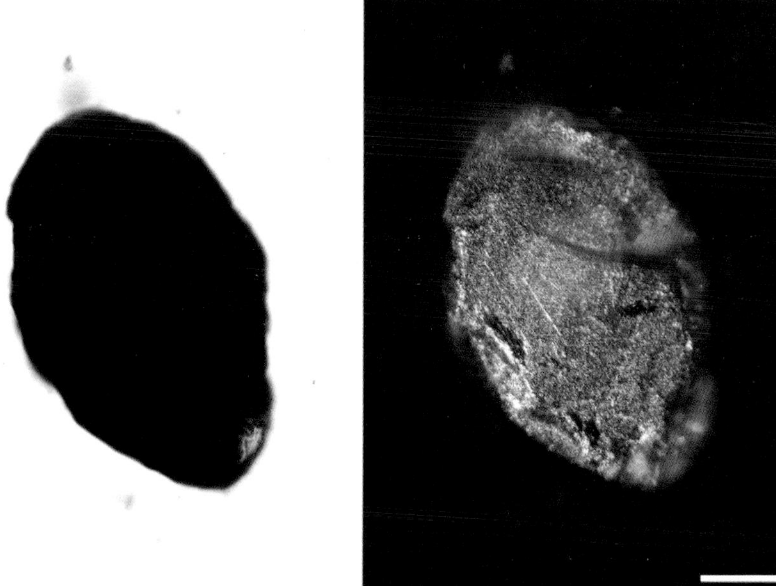

Holocene beach sediment, New Haven Co., Connecticut

A grain mount of detrital magnetite in transmitted and oblique reflected light. Identification of heavy minerals is facilitated, in many cases, by crushing the rock and concentrating the heavy minerals by heavy liquid flotation or by mechanical mineral-separation techniques. Separated grains can be examined in refractive index oils or in polished mounts. Magnetite shows a metallic, silver-black color when viewed in reflected light, although colors can vary with the color spectrum of the light source used.

PPL | ORL, Scale bar = 0.10 mm

Up. Triassic – Lo. Jurassic Huizachal Fm., Nuevo León, Mexico

The detrital opaque minerals visible here were concentrated into laminae that are typical of placer deposits. These grains are mostly magnetite, a mineral that is silver-black in reflected light. Placer deposits are made up of chemically stable heavy minerals that have been gravitationally concentrated during transport.

PPL, Scale bar = 0.26 mm

Cited References and Additional Information Sources

Accessory detrital minerals - general:

Briggs, L. I., 1965, Heavy mineral correlations and provenances: Journal of Sedimentary Research, v. 35, p. 939-955, doi: 10.1306/74D713B6-2B21-11D7-8648000102C1865D.

Cerveny, P. F., N. M. Johnson, R. A. K. Tahirkheli, and N. R. Bonis, 1989, Tectonic and geomorphic implications of Siwalik Group heavy minerals, Potwar Plateau, Pakistan, *in* L. L. Malinconico, Jr., and R. J. Lillie, eds., Tectonics of the Western Himalayas: Boulder, GSA Special Paper 232, p. 129-136, doi: 10.1130/SPE232-p129.

Feo-Codecido, G., 1956, Heavy-mineral techniques and their application to Venezuelan stratigraphy: AAPG Bulletin, v. 40, p. 984-1000.

Garzanti, E., and S. Andò, 2007, Heavy mineral concentration in modern sands: Implications for provenance interpretation, *in* M. A. Mange, and D. T. Wright, eds., Heavy Minerals in Use (Developments in Sedimentology 58): New York, Elsevier, p. 517-545, doi: 10.1016/S0070-4571(07)58020-9.

Hubert, J. F., 1962, A zircon-tourmaline-rutile maturity index and the interdependence of the composition of heavy mineral assemblages with the gross composition and texture of sandstones: Journal of Sedimentary Research, v. 32, p. 440-450, doi: 10.1306/74D70CE5-2B21-11D7-8648000102C1865D.

Komar, P. D., K. E. Clemens, Zhenlin Li, and Shyuer-ming Shih, 1989, The effects of selective sorting on factor analyses of heavy-mineral assemblages: Journal of Sedimentary Research, v. 59, p. 590-596, doi: 10.1306/212F8FF8-2B24-11D7-8648000102C1865D.

Koroznikova, L., C. Klutke, S. McKnight, and S. Hall, 2008, The use of low-toxic heavy suspensions in mineral sands evaluation and zircon fractionation: Journal of the South African Institute of Mining and Metallurgy, v. 108, p. 25-34.

Mange, M. A., and H. F. W. Maurer, 1992, Heavy Minerals in Colour: London, Chapman & Hall, 147 p., doi: 10.1007/978-94-011-2308-2.

Mange, M. A., and A. C. Morton, 2007a, Geochemistry of heavy minerals, *in* M. A. Mange, and D. T. Wright, eds., Heavy Minerals in Use (Developments in Sedimentology 58): Amsterdam, Elsevier, p. 345-391, doi: 10.1016/S0070-4571(07)58013-1.

Mange, M. A., and A. C. Morton, 2007b, High resolution heavy mineral analysis (HRHMA): A brief summary, *in* M. A. Mange, and D. T. Wright, eds., Heavy Minerals in Use (Developments in Sedimentology 58): Amsterdam, Elsevier, p. 433-436, doi: 10.1016/S0070-4571(07)58016-7.

Mange, M. A., and D. T. Wright, eds., 2007, Heavy Minerals in Use (Developments in Sedimentology 58): Amsterdam, Elsevier, 1283 p.

Mange-Rajetzky, M. A., 1995, Subdivision and correlation of monotonous sandstone sequences using high-resolution heavy mineral analysis, a case study: the Triassic of the Central Graben, *in* R. E. Dunay, and E. A. Hailwood, eds., Non-biostratigraphical Methods of Dating and Correlation: London, GSL Special Publication 89, p. 23-30, doi: 10.1144/GSL.SP.1995.089.01.03.

Milliken, K. L., 1988, Loss of provenance information through subsurface diagenesis in Plio-Pleistocene sandstones, northern Gulf of Mexico: Journal of Sedimentary Research, v. 58, p. 992-1002, doi: 10.1306/212F8EE0-2B24-11D7-8648000102C1865D.

Milliken, K. L., 2007, Provenance and diagenesis of heavy minerals, Cenozoic units of the northwestern Gulf of Mexico sedimentary basin, *in* M. A. Mange, and D. T. Wright, eds., Heavy Minerals in Use (Developments in Sedimentology 58): New York, Elsevier, p. 247-261, doi: 10.1016/S0070-4571(07)58008-8.

Milliken, K. L., and L. E. Mack, 1990, Subsurface dissolution of heavy minerals, Frio Formation sandstones of the ancestral Rio Grande Province, South Texas: Sedimentary Geology, v. 68, p. 187-199, doi: 10.1016/0037-0738(90)90111-6.

Molinaroli, E., and A. Basu, 1993, Toward quantitative provenance analysis: A brief review and case study, *in* M. J. Johnsson, and A. Basu, eds., Processes Controlling the Composition of Clastic Sediments: Boulder, CO, GSA Special Paper 284, p. 323-334, doi: 10.1130/SPE284-p323.

Morton, A. C., 1984, Stability of detrital heavy minerals in Tertiary sandstones from the North Sea Basin: Clay Minerals, v. 19, p. 287-308, doi: 10.1180/claymin.1984.019.3.04.

Morton, A. C., 1985, Heavy minerals in provenance studies, *in* G. G. Zuffa, ed., Provenance of Arenites (NATO Science Series C): Dordrecht, Netherlands, D. Reidel Publishing, p. 249-277, doi: 10.1007/978-94-017-2809-6_12.

Morton, A. C., and C. R. Hallsworth, 1999, Processes controlling the composition of heavy mineral assemblages in sandstones: Sedimentary Geology, v. 124, p. 3-29, doi: 10.1016/S0037-0738(98)00118-3.

Morton, A. C., and A. Hurst, 1995, Correlation of sandstones using heavy minerals: An example from the Statfjord Formation of the Snorre Field, northern North Sea, *in* R. E. Dunay, and E. A. Hailwood, eds., Non-biostratigraphical Methods of Dating and Correlation: London, GSL Special Publication 89, p. 3-22, doi: 10.1144/GSL.SP.1995.089.01.02.

Morton, A. C., and M. J. Johnsson, 1994, Identifying provenance-specific features of detrital heavy mineral assemblages in sandstones: Sedimentary Geology, v. 90, p. 241-256, doi: 10.1016/0037-0738(94)90041-8.

Morton, A. C., S. McFadyen, A. Hurst, J. Pyle, and P. Rose, 2014, Constraining the origin of reservoirs formed by sandstone intrusion: Insights from heavy mineral studies of the Eocene in the Forties area, UK Central North Sea: AAPG Bulletin, v. 98, p. 545-561, doi: 10.1306/06141312191.

Munsterman, D., and S. Kerstholt, 1996, Sodium polytungstate, a new non-toxic alternative to bromoform in heavy liquid separation: Review of Palaeobotany and Palynology, v. 91, p. 417-422, doi: 10.1016/0034-6667(95)00093-3.

Naeser, N. D., and T. H. McCulloh, eds., 1988, Thermal History of Sedimentary Basins: New York, Springer-Verlag, 500 p.

Pettijohn, F. J., P. E. Potter, and R. Siever, 1987, Sand and Sandstones (2nd Edition): New York, Springer-Verlag, 553 p.

Uddin, A., and N. Lundberg, 1998, Unroofing history of the eastern Himalaya and the Indo-Burmese ranges: Heavy-mineral study of Cenozoic sediments from the Bengal Basin, Bangladesh: Journal of Sedimentary Research, v. 68, p. 465-472, doi: 10.2110/jsr.68.465.

van Andel, T. H., 1959, Reflections on the interpretation of heavy mineral analyses: Journal of Sedimentary Research, v. 29, p. 153-163, doi: 10.1306/74D708B7-2B21-11D7-8648000102C1865D.

Vital, H., K. Stattegger, and C.-D. Garbe-Schoenberg, 1999, Composition and trace-element geochemistry of detrital clay and heavy-mineral suites of the lowermost Amazon River: A provenance study: Journal of Sedimentary Research, v. 69, p. 563-575, doi: 10.2110/jsr.69.563.

von Eynatten, H., and R. Gaupp, 1999, Provenance of Cretaceous synorogenic sandstones in the Eastern Alps: Constraints from framework petrography, heavy mineral analysis and mineral chemistry: Sedimentary Geology, v. 124, p. 81-111, doi: 10.1016/S0037-0738(98)00122-5.

Walderhaug, O., and K. W. Porten, 2007, Stability of detrital heavy minerals on the Norwegian continental shelf as a function of depth and temperature: Journal of Sedimentary Research, v. 77, p. 992-1002, doi: 10.2110/jsr.2007.091.

Weibel, R., and H. Friis, 2007, Alteration of opaque heavy minerals as a reflection of the geochemical conditions in depositional and diagenetic environments, *in* M. A. Mange, and D. T. Wright, eds., Heavy Minerals in Use (Developments in Sedimentology 58): Amsterdam, Elsevier, p. 277-303, doi: 10.1016/S0070-4571(07)58010-6.

Amphibole and pyroxene:

Cawood, P. A., 1983, Modal composition and detrital clinopyroxene geochemistry of lithic sandstones from the New England Fold Belt (east Australia): A Paleozoic forearc terrane: GSA Bulletin, v. 94, p. 1199-1214, doi: 10.1130/0016-7606(1983)94<1199:MCADCG>2.0.CO;2.

Krawinkel, H., S. Wozazek, J. Krawinkel, and W. Hellmann, 1999, Heavy-mineral analysis and clinopyroxene geochemistry applied to provenance analysis of lithic sandstones from the Azuero–Soná Complex (NW Panama): Sedimentary Geology, v. 124, p. 149-168, doi: 10.1016/S0037-0738(98)00125-0.

Schäfer, J., 1997, Electron microprobe study of detrital amphibole from Famennian synorogenic clastic sediments of the Saxothuringian belt (Erbendorf Paleozoic, NE-Bavaria, Germany): Consequences for provenance and geotectonic development: Geodinamica Acta, v. 10, p. 125-136.

Apatite:

Gleadow, A. J. W., D. X. Belton, B. P. Kohn, and R. W. Brown, 2002, Fission track dating of phosphate minerals and the thermochronology of apatite: Reviews in Mineralogy and Geochemistry, v. 48, p. 579-630, doi: 10.2138/rmg.2002.48.16.

Green, P. F., I. R. Duddy, A. J. W. Gleadow, and J. F. Lovering, 1989, Apatite fission-track analysis as a paleotemperature indicator for hydrocarbon exploration, *in* N. D. Naeser, and T. H. McCulloh, eds., Thermal History of Sedimentary Basins: Methods and Case Histories: New York, Springer-Verlag, p. 181-195, doi: 10.1007/978-1-4612-3492-0_11.

Morton, A., and G. Yaxley, 2007, Detrital apatite geochemistry and its application in provenance studies, *in* J. Arribas, S. Critelli, and M. J. Johnsson, eds., Sedimentary Provenance and Petrogenesis: Perspectives from Petrography and Geochemistry: Boulder, CO, GSA Special Paper 420, p. 319-344, doi: 10.1130/2006.2420(19).

Naeser, C. W., 1979, Thermal history of sedimentary basins: Fission track dating of subsurface rocks, *in* P. A. Scholle, and P. Schluger, eds., Aspects of Diagenesis: Tulsa, SEPM Special Publication 26, p. 109-112, doi: 10.2110/pec.79.26.0109.

Garnet:

Morton, A. C., 1985, A new approach to provenance studies: Electron microprobe analysis of detrital garnets from Middle Jurassic sandstones of the northern North Sea: Sedimentology, v. 32, p. 553-566, doi: 10.1111/j.1365-3091.1985.tb00470.x.

Morton, A. C., 1987, Influences of provenance and diagenesis on detrital garnet suites in the Paleocene Forties Sandstone, central North Sea: Journal of Sedimentary Research, v. 57, p. 1027-1032, doi: 10.1306/212F8CD8-2B24-11D7-8648000102C1865D.

Morton, A. C., G. Borg, P. L. Hansley, P. D. W. Haughton, D. H. Krinsley, and P. Trusty, 1989, The origin of faceted garnets in sandstones: Dissolution or overgrowth?: Sedimentology, v. 36, p. 927-942, doi: 10.1111/j.1365-3091.1989.tb01754.x.

Morton, A., C. Hallsworth, and B. Chalton, 2004, Garnet compositions in Scottish and Norwegian basement terrains: A framework for interpretation of North Sea sandstone provenance: Marine and Petroleum Geology, v. 21, p. 393-410, doi: 10.1016/j.marpetgeo.2004.01.001.

Takeuchi, M., 1994, Changes in garnet chemistry show a progressive denudation of the source areas for Permian-Jurassic sandstones, southern Kitakami Terrane, Japan: Sedimentary Geology, v. 93, p. 85-105, doi: 10.1016/0037-0738(94)90030-2.

Rutile and other Fe/Ti minerals:

Basu, A., and E. Molinaroli, 1989, Provenance characteristics of detrital opaque Fe-Ti oxide minerals: Journal of Sedimentary Research, v. 59, p. 922-934, doi: 10.1306/212F90B6-2B24-11D7-8648000102C1865D.

Jamieson, J. C., and B. Olinger, 1969, Pressure–temperature studies of anatase, brookite, rutile and TiO_2 (II): A discussion: American Mineralogist, v. 54, p. 1477-1481.

Morad, S., and A. A. Aldahan, 1986, Alteration of detrital Fe-Ti oxides in sedimentary rocks: GSA Bulletin, v. 97, p. 567-578, doi: 10.1130/0016-7606(1986)97<567:AODFOI>2.0.CO;2.

Morton, A., and S. Chenery, 2009, Detrital rutile geochemistry and thermometry as guides to provenance of Jurassic–Paleocene sandstones of the Norwegian Sea: Journal of Sedimentary Research, v. 79, p. 540-553, doi: 10.2110/jsr.2009.054.

Triebold, S., G. L. Luvizotto, R. Tolosana-Delgado, T. Zack, and H. von Eynatten, 2011, Discrimination of TiO_2 polymorphs in sedimentary and metamorphic rocks: Contributions to Mineralogy and Petrology, v. 161, p. 581-596, doi: 10.1007/s00410-010-0551-x.

Zack, T., H. Von Eynatten, and A. Kronz, 2004, Rutile geochemistry and its potential use in quantitative provenance studies: Sedimentary Geology, v. 171, p. 37-58, doi: 10.1016/j.sedgeo.2004.05.009.

Tourmaline:

Krynine, P. D., 1946, The tourmaline group in sediments: Journal of Geology, v. 54, p. 65-87.

Marschall, H. R., and Shao-Yong Jiang, 2011, Tourmaline isotopes: No element left behind: Elements, v. 7, p. 313-319, doi: 10.2113/gselements.7.5.313.

Preston, J., A. Hartley, M. Mange-Rajetzky, M. Hole, G. May, S. Buck, and L. Vaughan., 2002, The provenance of Triassic continental sandstones from the Beryl Field, northern North Sea: Mineralogical, geochemical, and sedimentological constraints: Journal of Sedimentary Research, v. 72, p. 18-29, doi: 10.1306/042201720018.

van Hinsberg, V. J., D. J. Henry, and H. R. Marschall, 2011, Tourmaline: An ideal indicator of its host environment: The Canadian Mineralogist, v. 49, p. 1-16, doi: 10.3749/canmin.49.1.1.

Zircon:

Bernet, M., and J. I. Garver, 2005, Fission-track analysis of detrital zircon: Reviews in Mineralogy and Geochemistry, v. 58, p. 205-237, doi: 10.2138/rmg.2005.58.8.

Corfu, F., J. M. Hanchar, P. W. O. Hoskin, and P. Kinny, 2003, Atlas of zircon textures, *in* J. M. Hanchar, and P. W. O. Hoskin, eds., Zircon (Reviews in Mineralogy and Geochemistry 53): Chantilly, VA, Mineralogical Society of America, p. 469-500, doi: 10.2113/0530469.

Dickinson, W. R., and G. E. Gehrels, 2008, Sediment delivery to the Cordilleran foreland basin: Insights from U-Pb ages of detrital zircons in Upper Jurassic and Cretaceous strata of the Colorado Plateau: American Journal of Science, v. 308, p. 1041-1082, doi: 10.2475/10.2008.01.

Fedo, C. M., K. N. Sircombe, and R. H. Rainbird, 2003, Detrital zircon analysis of the sedimentary record: Reviews in Mineralogy and Geochemistry, v. 53, p. 277-303, doi: 10.2113/0530277.

Hoskin, P. W. O., and T. R. Ireland, 2000, Rare earth element chemistry of zircon and its use as a provenance indicator: Geology, v. 28, p. 627-630, doi: 10.1130/0091-7613(2000)28<627:REECOZ>2.0.CO;2.

Košler, J., H. Fonneland, P. Sylvester, M. Tubrett, and R.-B. Pedersen, 2002, U–Pb dating of detrital zircons for sediment provenance studies—a comparison of laser ablation ICPMS and SIMS techniques: Chemical Geology, v. 182, p. 605-618, doi: 10.1016/S0009-2541(01)00341-2.

McAteer, C. A., J. S. Daly, M. J. Flowerdew, J. N. Connelly, T. B. Housh, and M. J. Whitehouse, 2010, Detrital zircon, detrital titanite and igneous clast U–Pb geochronology and basement–cover relationships of the Colonsay Group, SW Scotland: Laurentian provenance and correlation with the Neoproterozoic Dalradian Supergroup: Precambrian Research, v. 181, p. 21-42, doi: 10.1016/j.precamres.2010.05.013.

Naeser, N. D., P. K. Zeitler, C. W. Naeser, and P. F. Cerveny, 1987, Provenance studies by fission-track dating of zircon—etching and counting procedures: International Journal of Radiation Applications and Instrumentation, Part D, Nuclear Tracks and Radiation Measurements, v. 13, p. 121-126, doi: 10.1016/1359-0189(87)90022-7.

Owen, M. R., 1987, Hafnium content of detrital zircons, a new tool for provenance study: Journal of Sedimentary Research, v. 57, p. 824-830, doi: 10.1306/212F8C74-2B24-11D7-8648000102C1865D.

Facing Page – Top: A chamositic ooid from the Lo. Cretaceous Nahr Umr Fm., offshore Qatar (PPL). Bottom – Interbedded Up. Jurassic radiolarites and siliceous mudstones, Elba, Italy. Photograph from E. L. Winterer.

GRAINS:

Miscellaneous Detrital Grains & Associated Rock Types

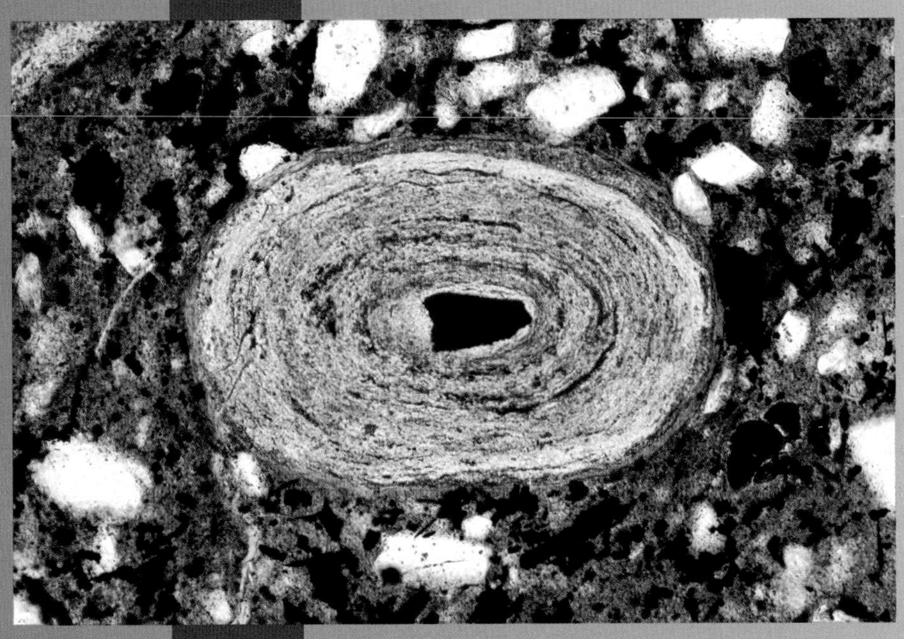

CHAPTER 5

INTRODUCTION TO MISCELLANEOUS GRAINS AND ROCKS

Many clastic terrigenous rocks contain variable, but in some cases substantial, amounts of primarily nondetrital constituents. These include biogenic/skeletal grains such as calcareous shells, siliceous tests, phosphatic vertebrate or invertebrate material and organic matter (from plant remains down to plankton and microbial filaments). Nonskeletal, but still biogenic grains, primarily fecal pellets, also can be abundant in some deposits, especially bioturbated ones. Other materials, such as phosphate, gypsum, green marine clays (glauconite, berthierine, chamosite) and ferrous oxides and hydroxides, are found in terrigenous deposits as minerals formed by direct precipitation, through alteration of other minerals or as detrital grains. Most of these grains, in some circumstances, can be sufficiently abundant to be the major constituents of rocks. Even where such grains are not the major rock constituent, however, it is important to recognize them and, if deemed important, one can add a descriptive adjective to any rock name (e.g., glauconitic quartz arenite or radiolarian-bearing arkosic siltstone).

Grain types by composition:

Calcium carbonate (limestones) – An enormous diversity of skeletal grains are found in clastic terrigenous rocks or associated limestones. These range from large benthic skeletal organisms to minute plankton (see Scholle and Ulmer-Scholle, 2003 for images). Such grains are composed of calcite, high-Mg calcite or aragonite in Pleistocene to Recent deposits, and generally are predominantly low-Mg calcite in older Phanerozoic rocks (with potential for either selective leaching or replacement by dolomite). Carbonate material also can be found as ooids, pellets or intraclasts/extraclasts. Carbonate micro- and nannoplankton (especially planktic foraminifers, calcispheres and coccolithophores) can be abundant, especially in Cretaceous and younger marine shales and marls. Carbonate grains have a wide variety of shapes and internal structures, but all show high birefringence. Carbonate rocks, limestones and dolomites, are not covered in this volume because they are fully detailed in AAPG Memoir 77.

Silica (bedded cherts) – Precambrian cherts may have formed by direct precipitation from seawater (e.g., Stefurak et al., 2014). The source of silica for Phanerozoic cherts, however, is mainly from planktic organisms (radiolarians, diatoms and silicoflagellates), although benthic, spicular siliceous sponges and volcanic ash also can be significant silica contributors. Radiolarians range from Middle Cambrian to Recent and generally occur as small (hundredths to tenths of a millimeter, but with some exceeding 1 mm) spherical, highly perforate skeletons with or without spines. They are fully marine and are very common in basinal shales. Diatoms are unicellular algae that range from at least the Early Jurassic to Recent (although they are rare before the Late Cretaceous). They consist of pairs of overlapping, highly perforate valves (similar to tiny Petri dishes but with more varied shapes). Marine forms are especially common in high-latitude and upwelling-zone deposits; nonmarine species are common in lacustrine deposits. All siliceous marine fossils originated as opaline silica (isotropic) but commonly alter to opal-CT and then to quartz (mainly chert or chalcedony, both of which show low birefringence). Rates of silica transformation are a function of temperature, pressure and clay content of the surrounding rock (Keller and Isaacs, 1985). Those transformations ultimately yield a brittle, cherty rock that typically is highly fractured even with minor deformation. Most bedded cherts are of intermediate to deep-marine origin (hundreds to thousands of meters water depth) although shallower marine deposits are not uncommon; some diatomaceous cherts are lacustrine as are cherts from sodic playa lakes (where they form as replacement of hydrous sodium silicates). Cherts also are commonly found as nodular to bedded replacements of limestones and, less commonly, as replacements of evaporite minerals. Detrital chert grains in sandstones typically are sourced from either bedded cherts or from limestone-replacement nodules, the latter mainly because of the enormous difference in solubility and mechanical durability between chert and limestone.

Phosphate (phosphorites) – Phosphatic grains can be of organic or inorganic origin; both are primarily composed of one of the varieties of apatite (especially carbonate-fluorapatite or hydroxylapatite). The organic particles include bones, teeth, and scales of vertebrate animals, the remains of some invertebrates (e.g., conodonts and lingulid brachiopods), and fecal pellets of vertebrate and invertebrate organisms; inorganic grains include ooids or coated grains and pellets/peloids of varied origin. Phosphatic deposits can be associated with upwelling nutrient-rich waters and with intervals of slow sedimentation (condensed sections) or hiatus surfaces (commonly marked by hardgrounds, with phosphatic crusts, nodules and granular lag deposits). Generally, they are found as brownish, very low birefringence to virtually isotropic grains, sometimes with patterned internal fabrics. Phosphorites also can be produced in shallow-marine to coastal environments with concentration of phosphate during subaerial exposure and weathering (as in the Miocene of Florida).

Organic materials (coals, peats, organic mudstones) – Can consist of relatively large particles, such as plant leaves and woody fragments as well as much smaller spores, pollen and organic-walled plankton; commonly also found as amorphous, disseminated organic matter along with microbial filaments and peloids. Organic particles are a significant component of many black shales and are the primary constituent of peat and coal. High organic-carbon contents can be found in marine as well as nonmarine settings; exceptionally high concentrations generally are associated with poorly-oxygenated (dysaerobic to anaerobic) settings. Organic matter generally appears light yellow, yellowish-brown or black (depending, in part, on thermal maturity) and is nonbirefringent; more detailed typing of organic matter is commonly done using fluorescence microscopy.

Evaporites – Formed through evaporation of seawater and other saline and subsaline fluids. This group of deposits includes gypsum ($CaSO_4 \cdot 2H_2O$), anhydrite ($CaSO_4$), halite (NaCl), barite ($BaSO_4$), celestine ($SrSO_4$) and many other less common minerals. All the above named minerals can be found as bedded deposits, but gypsum is the only evaporite mineral stable enough to be reworked and preserved as detrital grains in clastic terrigenous deposits. In such instances, gypsum typically is found as large crystal fragments or, less commonly, as ooids—both showing low first-order birefringence. Other evaporite minerals are found mainly as nondetrital bedded deposits, as cold-seep deposits (barite) or as displacive or replacive minerals and/or cements in sandstones and claystones. Formation of evaporites is common in coastal marine settings (salinas and sabkhas) as well as in nonmarine settings (permanent lakes and playa lakes), primarily in arid regions. Restricted influx of seawater into large basins during early phases of oceanic rifting (e.g., the Red Sea or early Atlantic) or other tectonic or eustatic events (e.g., the Miocene of the Mediterranean) can generate enormous evaporite accumulations. Because evaporite minerals generally are quite soluble and easily remobilized, it is important to distinguish between primary and secondary deposits.

Green marine clays (greenstones/glaucony facies) – These terms refer to deposits of a group of iron-rich clays variously referred to as glauconite (consisting of K-poor glauconitic smectites to K-rich glauconitic micas depending on degree of diagenetic maturity), berthierine (a ferrous and magnesian, serpentine-type clay) or chamosite (a ferrous and magnesian, chlorite-type clay)—the latter two are commonly also classed as part of the next rock group, ironstones. The most common of the green clays, glauconite, occurs as light- to dark-green particles with anomalous birefringence. These consist of ovoid, pelletal, lumpy or vermicular grains formed on the seafloor, as infills of chambers in bioclasts, or through alteration of precursor clays. Glauconite-rich deposits (termed glaucony facies or greenstones, see Odin and Matter, 1981) are thought to indicate deposition in fully marine waters, typically at 10–50 m (30–165 ft) water depths (but with some formation at, or transport to, greater depths). Formation conditions are inferred to have been warm to cool in settings with a low sedimentation rate that allowed winnowing and concentration of glauconite grains. Such conditions also are favorable for phosphorite and ironstone formation, and these deposits often occur in association with each other. Glauconite is also found as grains reworked from older deposits.

Ironstone/iron formations – Found in Precambrian to Recent strata, these deposits include Phanerozoic oolitic ironstones (which overlap with those described under green marine clays) as well as mainly Precambrian banded-iron formations (BIFs). Phanerozoic iron deposits that do not fall in the "green marine clays" category commonly contain iron oxides (goethite/limonite, hematite, or magnetite) or ferroan carbonates (primarily siderite) and are characterized by oolitic grains with grayish opaque appearance (or a reddish-brown ferruginous appearance where weathered). Environments of formation of Phanerozoic iron deposits are probably diverse and still much disputed, but typically center around shallow-marine, marginal-marine or lateritic settings with low sedimentation rates. Precambrian BIFs generally consist of millimeter to centimeter-thick bands of iron oxides (magnetite or hematite) repeatedly interspersed with layers of quartz, chert or shale. The environments and processes of BIF formation are too varied and controversial to summarize here.

Tuffaceous units – The distinction between eruptive volcanic and sedimentary processes is often difficult to make in ash flows, lahars, bentonites and other deposits in which atmospheric or aqueous transport was involved. Furthermore, such volcanic layers can be intimately interbedded with clastic terrigenous materials—thus, they are described here as an associated rock type. Most tuffaceous units are recognized by the altered remains of glass shards (with various degrees of welding and devitrification) and/or by the presence of brecciated volcanic rock fragments; rocks rich in such clasts are termed volcanic agglomerates. The presence of other clay-, silt- or sand-sized particles in tuffaceous units results from mixing with other sediment and helps to delineate an at least partially sedimentary origin of the deposit. Tuffaceous units also generally are rich in secondary zeolites as cements or grain replacements.

Photographs in this section are grouped into the above categories with marginal headings to note the start of each section.

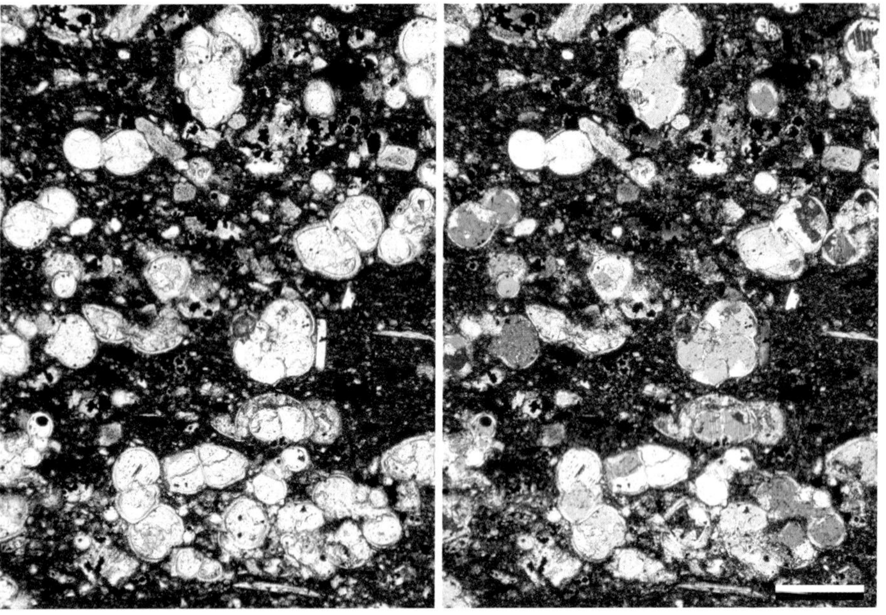

Up. Cretaceous Mancos Shale, San Juan Basin, New Mexico

Abundant planktic foraminifera in a calcareous shale. Planktic foraminifers evolved in the Jurassic and are very common in Cretaceous to Recent open marine deposits. Although there are a wide variety of test shapes in planktic foraminifera, many (including the ones shown here) have very simple, globular chambers with thin, light walls that allow the organisms to float in the water column. The living chambers in essentially all these grains have been filled with diagenetic calcite cement. The large single crystals of high-birefringence calcite that fill or nearly fill many chambers give the unusual patchy colors in this image.

PPL | XPL, Scale bar = 0.24 mm

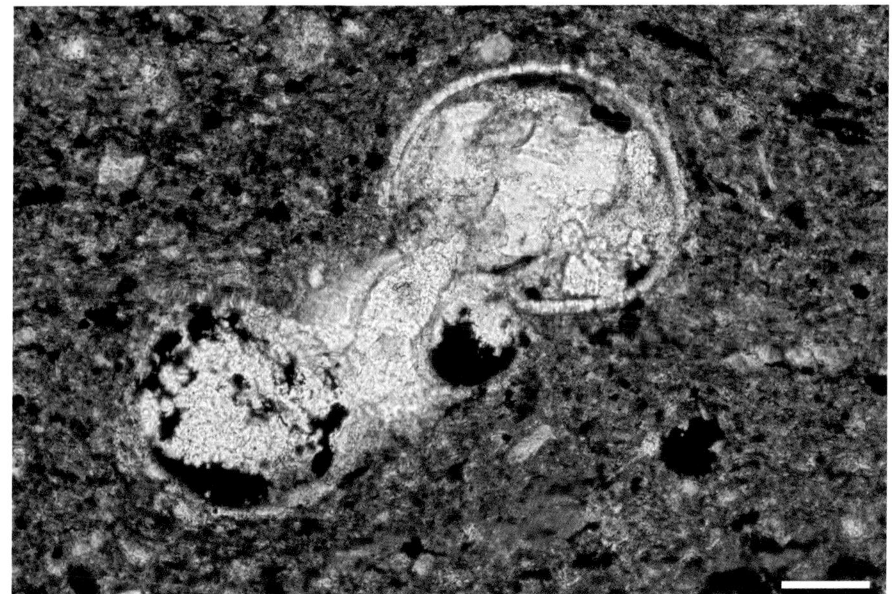

Cretaceous Mowry Fm., Moffat Co., Colorado

A closeup view of a single planktic foraminifer in a calcareous shale—the chambers here are filled by polycrystalline calcite cement, pyrite and minor kaolinite. This planispiral foraminifer has multiple chambers aligned on a plane and is thin-walled and globular (an adaptation for a floating lifestyle). In contrast, benthic foraminifera, which extend back to the Cambrian, typically have thicker calcareous, organic or even agglutinated walls and lived within sediment or as encrusters on other organisms.

PPL, Scale bar = 0.05 mm

Up. Cretaceous – Paleocene unknown unit, Sri Lanka

This is a nummulitid foraminifer which belongs to an informal group often called the "larger benthic foraminifers". They generally are, although not always, larger in size than planktic foraminifers and have internally complex tests. While small benthic foraminifers are identified from their external shape, identification of larger forms, when done in thin section, is based both on shape and internal test architecture. This type of foraminifers were most abundant in the Paleogene.

PPL, KFS, AFeS, BDI, Scale bar = 0.40 mm

Miocene (Tortonian), Tagus Basin, Setúbal Peninsula, Portugal †

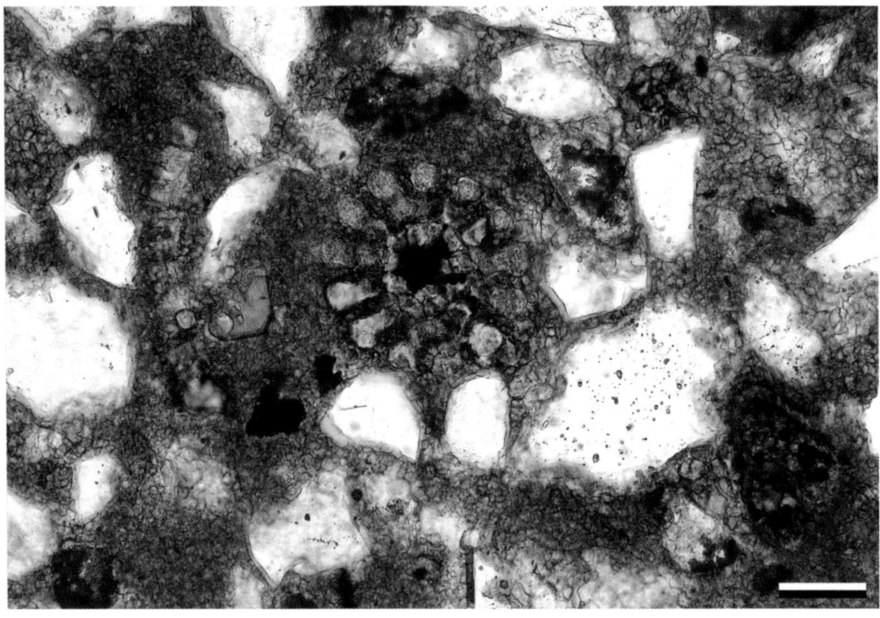

An echinoid spine (center) in transverse section. The spine is composed of nonferroan calcite (red) and later was cemented by ferroan calcite (dark blue). The spine shows the characteristic internal radial pattern and is hollow (diadematoid form), suggesting an origin from an irregular echinoid, as opposed to solid forms (cidaroid). The key to identifying all echinoderms (echinoids, crinoids and blastoids) is that each of their fossil segments, plates or spines is composed of a single crystal of calcite. Under cross-polarized light, the entire grain will go to extinction at the same time, even in a complex grain like this one.

PPL, KFS, AFeS, BDI, Scale bar = 0.07 mm

Up. Cretaceous Mancos Shale, San Juan Basin, New Mexico

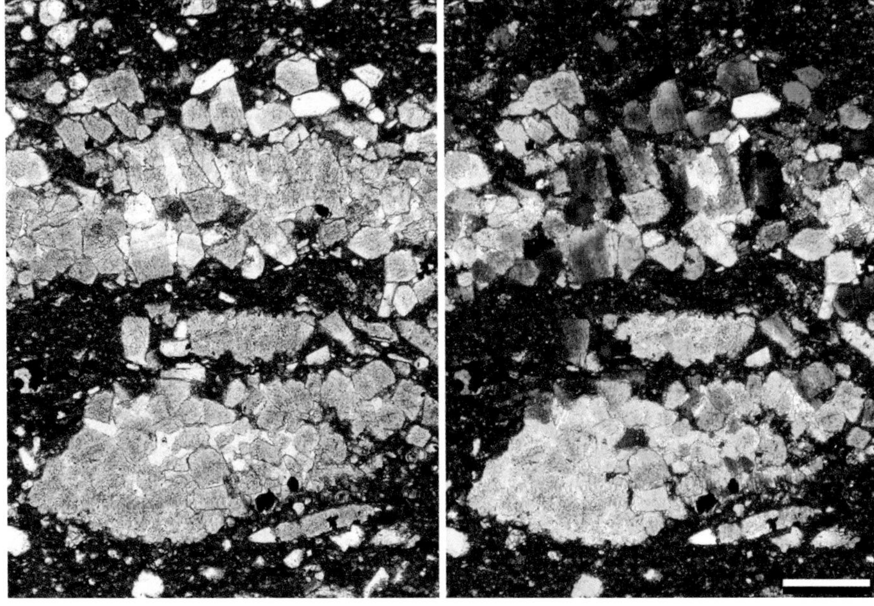

Essentially all the carbonate fragments in this view are individual prismatic crystals, or clusters of such prisms, that constituted the main part of the shell of a Cretaceous bivalve of the genus *Inoceramus*. These shells broke apart easily during transport and burial, and in thin section, they commonly occur as isolated polygonal blocks. The large (up to 2 m or 6 ft across) and relatively thin shells of this group of benthic bivalves made them especially well suited to live on soft and muddy seafloors. Thus, their remains are very common in Cretaceous chalks and shales in a wide variety of settings.

PPL | XPL, Scale bar = 0.20 mm

Mid. Ordovician Reedsville Fm., Mifflin Co., Pennsylvania

Trilobites are common bioclasts in Paleozoic rocks. Since these benthic arthropods molted, fossil fragments can be very common in sediments. Trilobite fragments are complexly curved, and the thoracic segments commonly have a "shepherd's crook" shape, like the fragment in the photomicrograph. Skeletal fragments are composed of minute prismatic crystals of calcite oriented perpendicular the carapace's surface causing the fragment to have a sweeping extinction. While they share sweeping extinction with ostracodes and some bivalves, their complexly-curved shape and size are diagnostic.

PPL, Scale bar = 0.26 mm

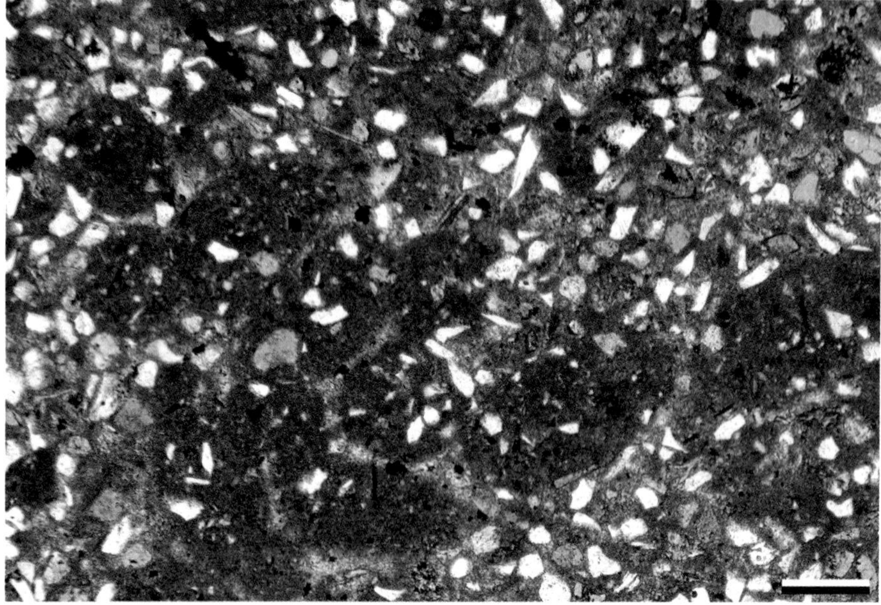

Up. Cretaceous Fox Hills Sandstone, Dewey Co., South Dakota

Rounded grains of carbonate micrite with random internal fabric are seen here in a quartz and glauconite-rich sediment. Based on their large and uniform size, ovoid shape and consistent composition, these are considered fecal (possibly crustacean) pellets rather than intraclasts. The inclusion of silt- to sand-sized terrigenous particles within the grains differentiates them from micritized ooids, and the randomness of distribution of the included particles rules out tests of arenaceous foraminifers.

PPL, Scale bar = 0.26 mm

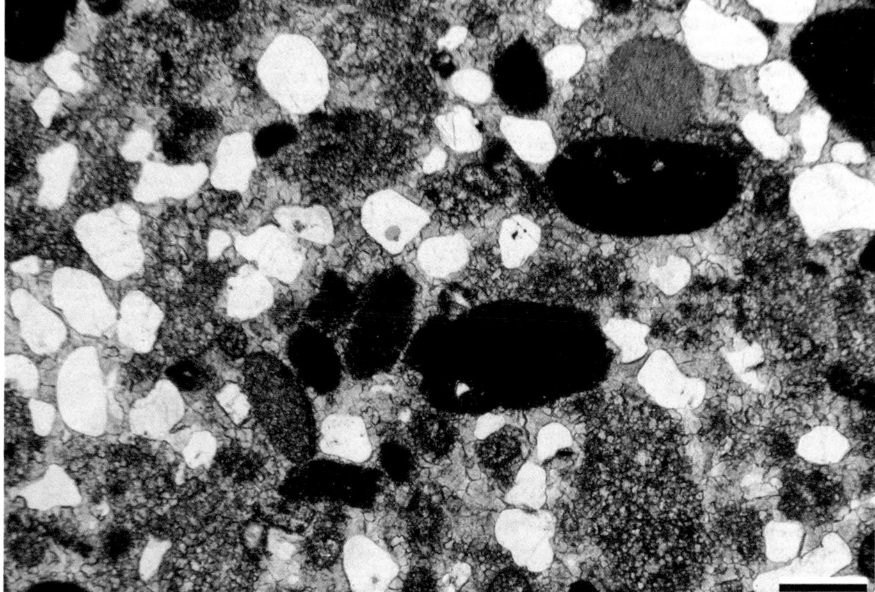

Up. Cambrian Gatesburg Fm., Pennsylvania

This sample contains a variety of grains that are probably pellets based on their uniform circular to ovoid shape. The size of the pellets vary, but that is not unusual in areas where there is a diverse population of benthic organisms. Because organisms like crustaceans and worms may burrow through multiple horizons, pellets also can exhibit some compositional variability. Pellets can sometimes be confused with oriented cross sections of micritic burrow fills, but burrows typically are more elongate, more consistently oriented and have less regular shape than the objects in this photo.

PPL, Scale bar = 0.26 mm

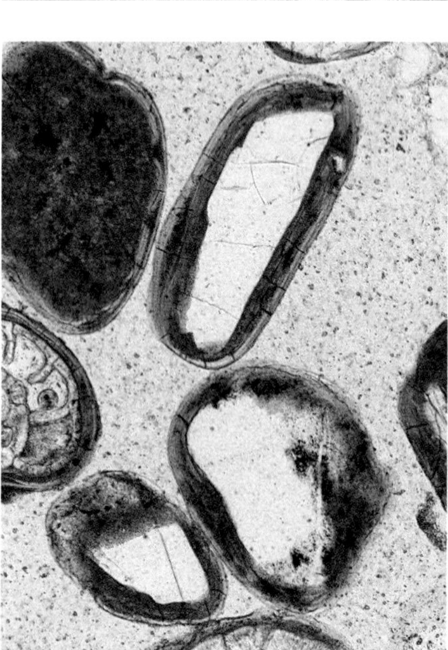

Recent sediment, Baffin Bay, Kleberg Co., Texas

Ooids (coated grains < 2mm in diameter) are found in both marginal marine and lacustrine sandstones. Here, quartz, bioclasts and carbonate pellets form the nuclei of these thinly coated ooids. Modern and ancient ooids were originally composed of aragonite and/or high-Mg calcite (and, in some time periods, low-Mg calcite). In environments with variable salinity and episodic high energy, like this one, ooids may form with incomplete cortical coatings, thin coatings and mixed mineralogy. Here, aragonite coatings are highly birefringent; calcite parts are granular or micritic.

PPL | XPL, Scale bar = 0.10 mm

Up. Cambrian Gatesburg Fm., central Pennsylvania

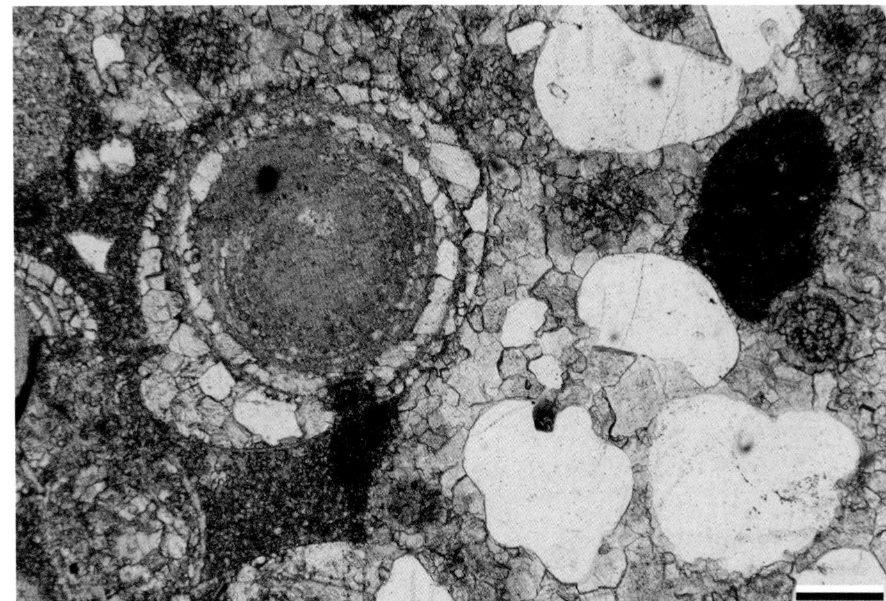

The original composition of the cortices of these ooids controlled their diagenesis in this sandy limestone. The core and some of the cortical layers preserve traces of the original ooid fabric; other parts of the coatings were dissolved and later filled with coarser crystalline carbonate. Both high-Mg calcite and aragonite are less stable than low-Mg calcite, and they behave differently during diagenesis. High-Mg calcite tends to be fabric preserving (altering to low-Mg calcite or replacement dolomite) whereas aragonite dissolves creating pore space. These ooids were mostly high-Mg calcite with a few aragonitic layers.

PPL, Scale bar = 0.26 mm

Recent sediment, North Atlantic Ocean

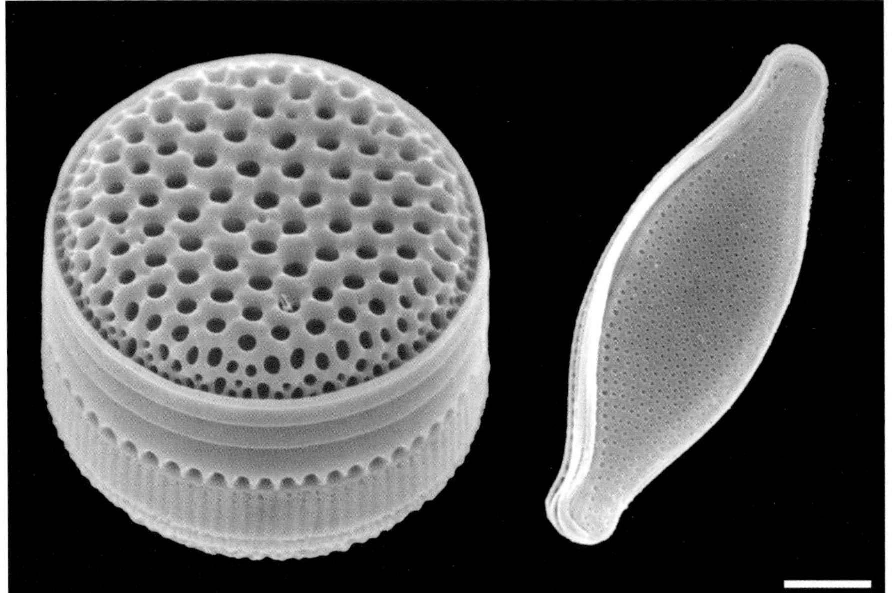

An SEM image of a typical centric diatom with radial symmetry (left) and a typical pennate diatom with bilateral symmetry (right). The diatom on the left shows upper and lower valves held together by a central girdle. The minute pores found in all diatoms explain why diatomaceous sediments are so widely used as filtration material (for swimming pools and many other applications). Diatoms are composed of opal-A that is subject to dissolution or conversion to more stable forms of silica (opal-CT and/or quartz). Photograph from Jeremy R. Young.

SEM, Scale bar = ~2.9 μm

Paleocene Barmer Hill Fm., Rajasthan, India

A mold of a diatom frustule in an opal-CT porcellanite. The frustule is a centric form, and the valve is undulatory or "sectored" so that alternate sectors are elevated or depressed. This one shows ten sectors. The presence of opal-CT lepispheres surrounding the mold suggests that they formed while the opal-A frustule was intact and that both forms of opal were present. The rock now contains quartz and opal-CT, and other than rare molds, there is little evidence of its diatomaceous origin.

SEM, Scale bar = 12 μm

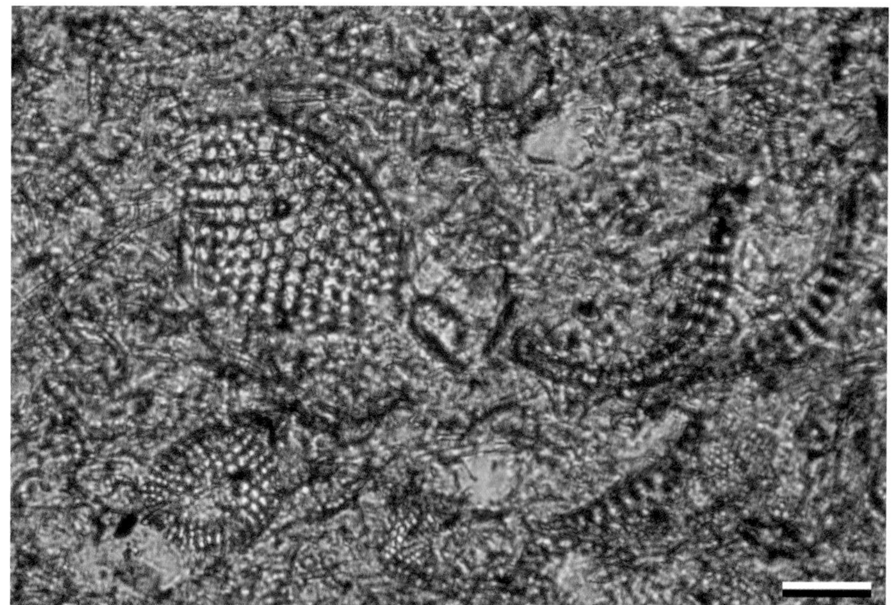

Eocene Oamaru Diatomite, Otago, New Zealand

Abundant, partially fragmented diatoms in a siliceous marine sediment. Diatoms are unicellular algae that have been extant since the Early Jurassic in marine to marginal marine environments and since the Paleocene for freshwater forms. Note the simple circular to ovoid shapes and the small, regularly arranged pores that give the grains the distinctive appearance of small sieves. As mentioned previously, their opaline silica frustules commonly dissolve or are replaced by opal-CT, quartz or calcite, especially during burial diagenesis.

PPL, Scale bar = 0.07 mm

Up. Miocene Virgin Valley Fm., Harney Co., Oregon

A variety of cuts through freshwater diatoms in a diatomite. Because their opaline tests were unstable, they have been replaced by a thin, brownish layer of chalcedony. Later, the interiors were filled with more extensive chalcedony.

XPL, Scale bar = 0.05 mm

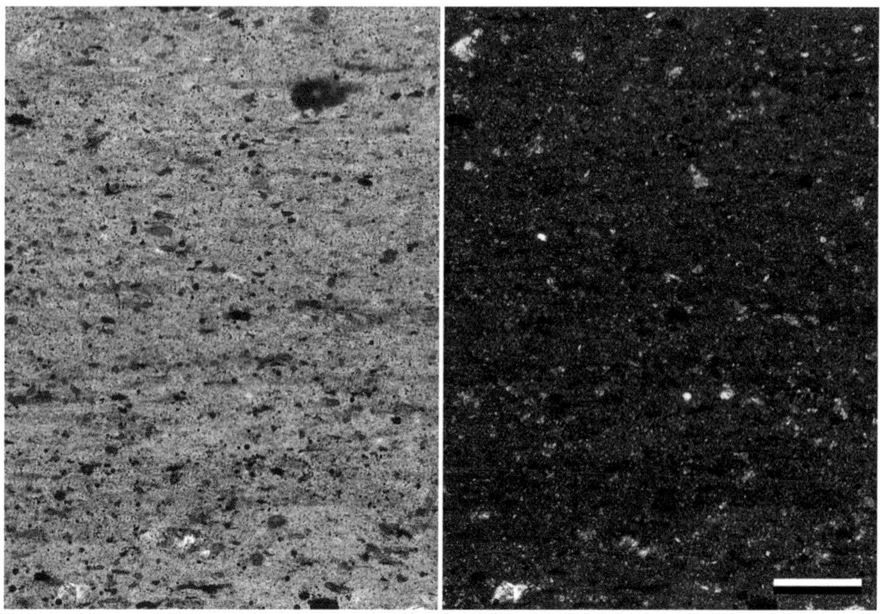

Paleocene Barmer Hill Fm., Rajasthan, India

A porcellanite (chert) composed of microcrystalline quartz (confirmed in XRD and SEM). This rock is a diatomite that has been recrystallized from opal to quartz. In plane-polarized light, the thin section shows some of the color of the blue resin, which has infiltrated the abundant microporosity. Organic fragments are associated with residual oil. In cross-polarized light, the birefringence of the microquartz is anomalous dark gray because of its small crystal size (much smaller than the thickness of a thin section). Small authigenic crystals of dolomite and siderite (higher birefringence) are visible.

PPL | XPL, BDI, Scale bar = 0.14 mm

Up. Oligocene ooze, Caroline Ridge, Pacific Ocean

An SEM image of opaline silica tests from a diverse assemblage of well-preserved, mainly nassellarian radiolarians. This sample is from a deep-sea siliceous ooze in 2,850 m (9,350 ft) water depth (present-day). Photograph from Stanley A. Kling.

SEM, Scale bar = ~37.5 μm

Up. Jurassic Franciscan Gp., Point Sal ophiolite, Santa Barbara Co., California

A longitudinal cross section through several radiolarian tests in a bedded chert. Radiolarians are marine plankton that have tests composed of opaline silica. This example has been diagenetically altered to microcrystalline quartz (chert) but still shows the coarse pore structure in the external skeleton. Radiolarians range from Middle Cambrian to Recent. They are extremely important in the formation of deep-water siliceous deposits, although they can be found in all marine environments.

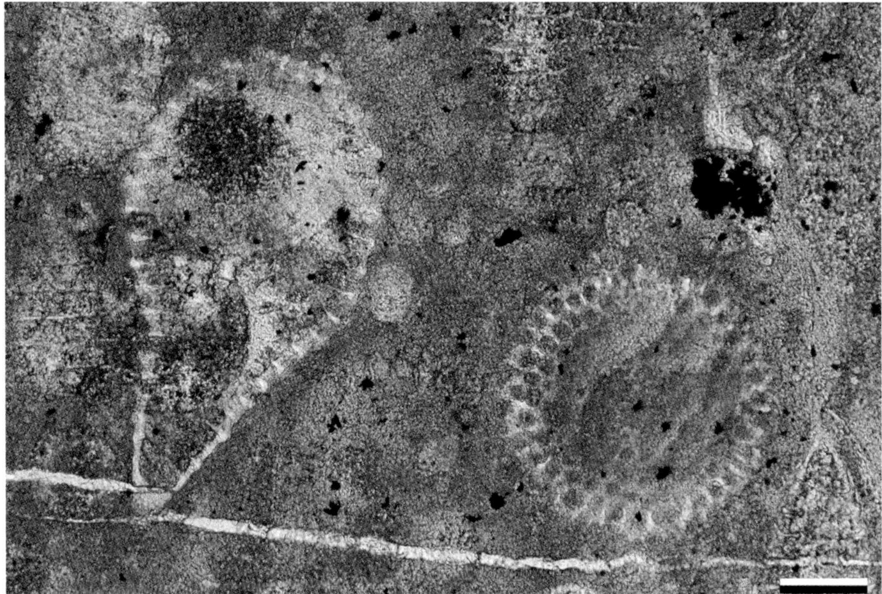

PPL, Scale bar = 0.26 mm

Cretaceous Mowry Fm., Moffat Co., Colorado

A cross section through a relatively well-preserved radiolarian test in a calcareous shale. This example has been replaced by chert, but still preserves the original perforate wall structure of the organism. The dark material in this view is organic debris (probably mainly plant remains) within the shale.

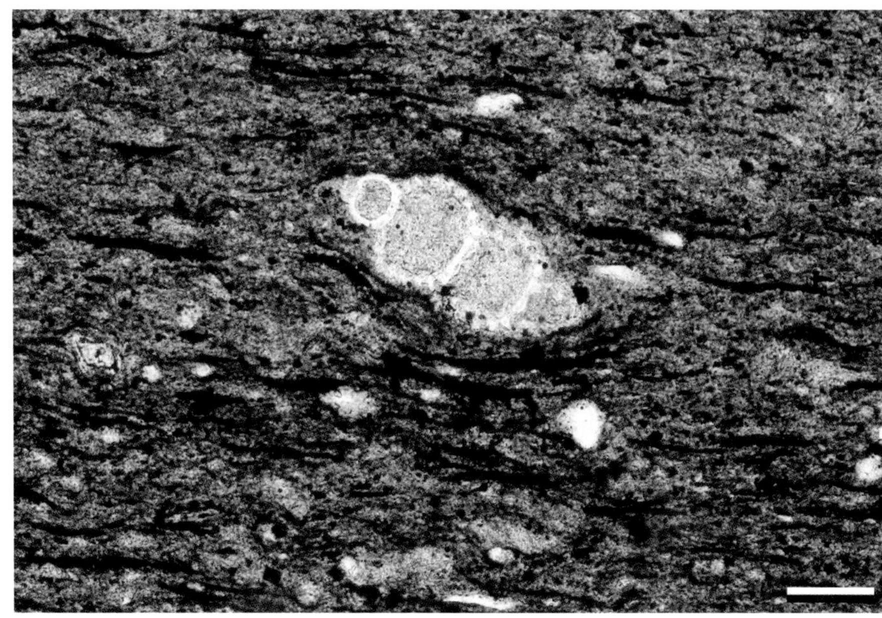

PPL, RDI, Scale bar = 0.08 mm

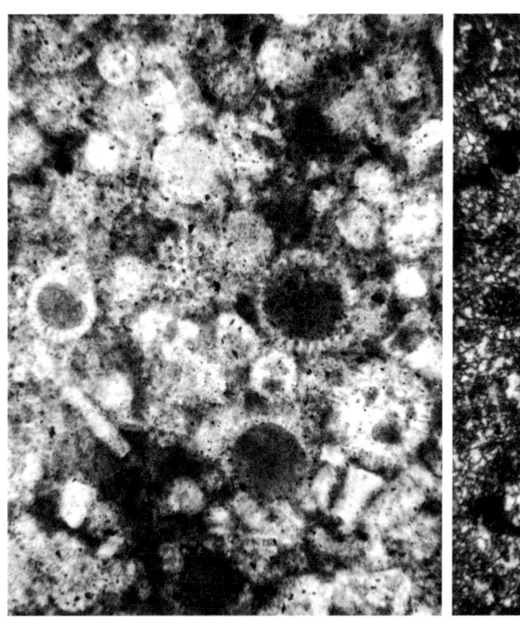

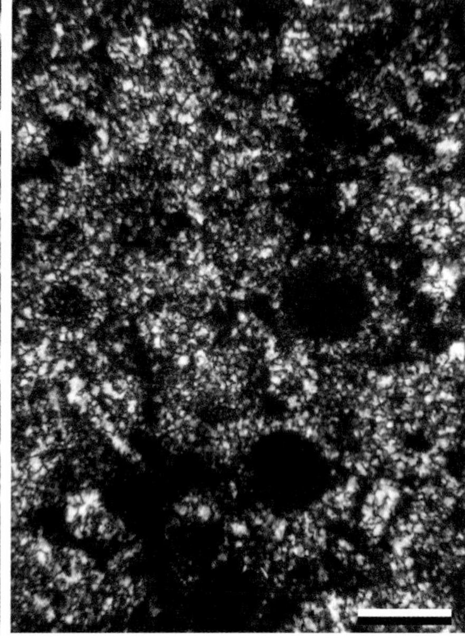

Up. Jurassic – Lo. Cretaceous radiolarite, Mahadah area, Oman

A moderately well-preserved assemblage of radiolarians (and subordinate sponge spicules) in an oceanic radiolarite. Spherical-shaped forms are readily visible, as are the coarse pores that typify radiolarian skeletons. In deep-water deposits like this sample, the slower sedimentation rates for terrigenous clastic sediment allow higher concentration of radiolaria to accumulate (as compared with most shelfal settings), thus leading to the formation of a biogenic ooze.

PPL | XPL, Scale bar = 0.16 mm

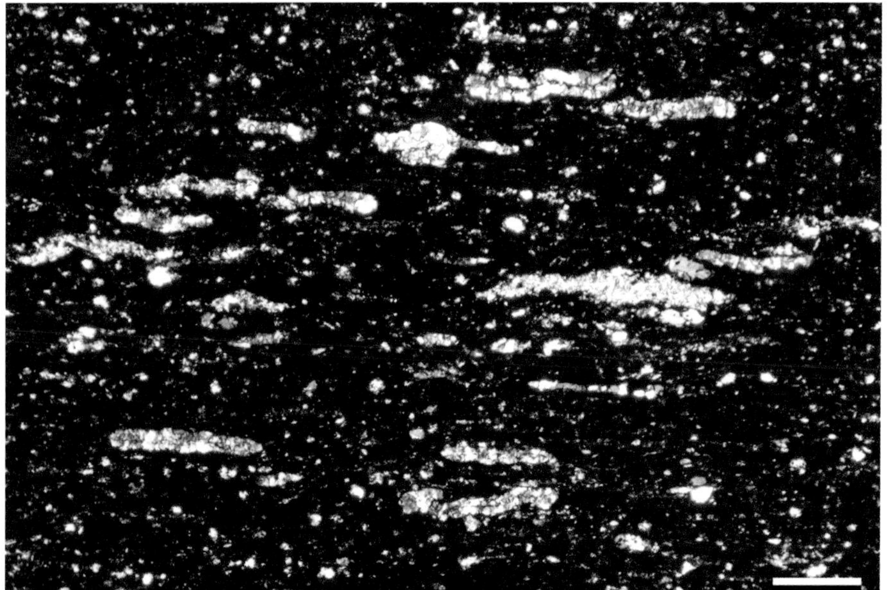

Up. Devonian – Lo. Mississippian Woodford Shale, Reeves Co., Texas

A view of numerous crushed arenaceous foraminifers within a silty shale. Some benthic foraminifers create their tests by gluing (agglutinating) grains together. The building materials may include terrigenous clastics, sponge spicules or other available grains. When crushed like the ones in this view, they can be confused with collapsed burrows that had arenaceous linings (although burrows typically show more diversity of shapes; see Milliken et al., 2007).

XPL, BDI, Scale bar = 0.10 mm

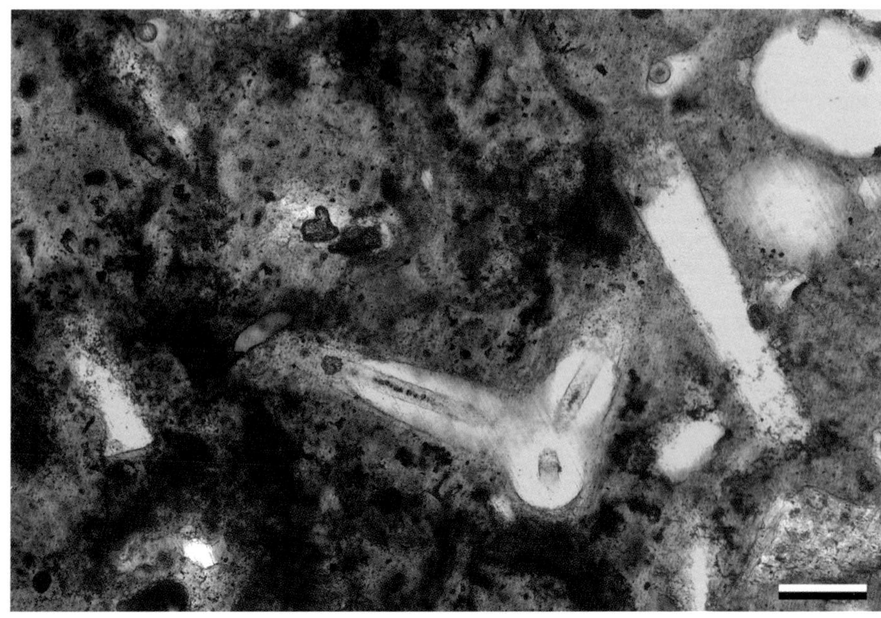

Jurassic Draupne Fm., Norwegian offshore continental shelf †

These former sponge spicules are seen in a phosphate-cemented unit. The thin section has cut across a spicule to show that it has a triaxon (or polyaxon) form. The central canal is phosphatized, but the opal that would have formed the sheath is missing. The dissolution of the original opaline silica has formed extensive secondary porosity within this rock.

PPL, BDI, Scale bar = 0.12 mm

Up. Jurassic – Lo. Cretaceous radiolarite, near Mahda, Oman

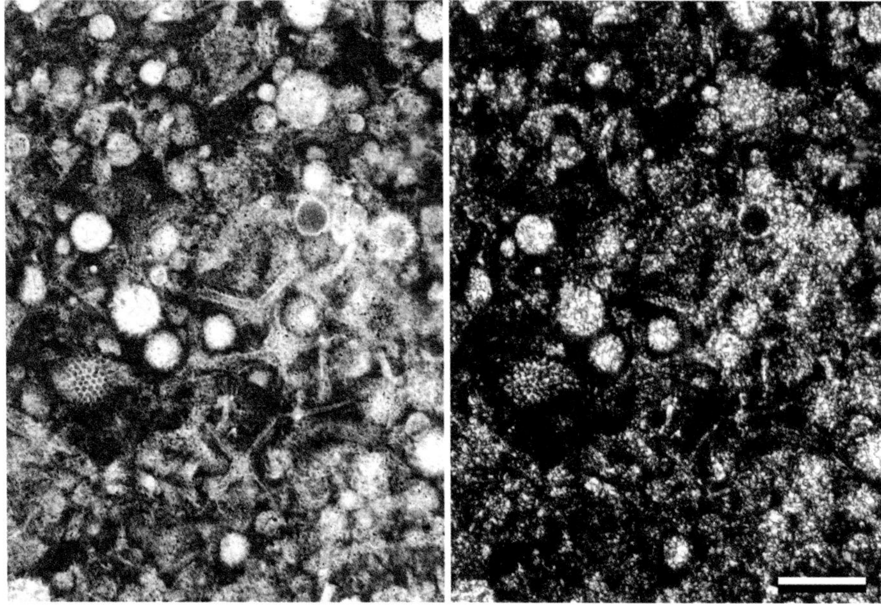

A radiolarian chert composed of polyaxon siliceous sponge spicules and radiolarians. Siliceous sponges contain networks of opaline spicules. The spicules have a distinctive central canal and monaxon or polyaxon shapes. Sponges usually disaggregate upon death, forming spicule-rich sediments. Within the sediments, the opaline grains convert during progressive burial to opal-CT and then one or more forms of quartz (typically chert or chalcedony).

PPL | XPL, Scale bar = 0.26 mm

Jurassic Morrison Fm., Moffat Co., Colorado

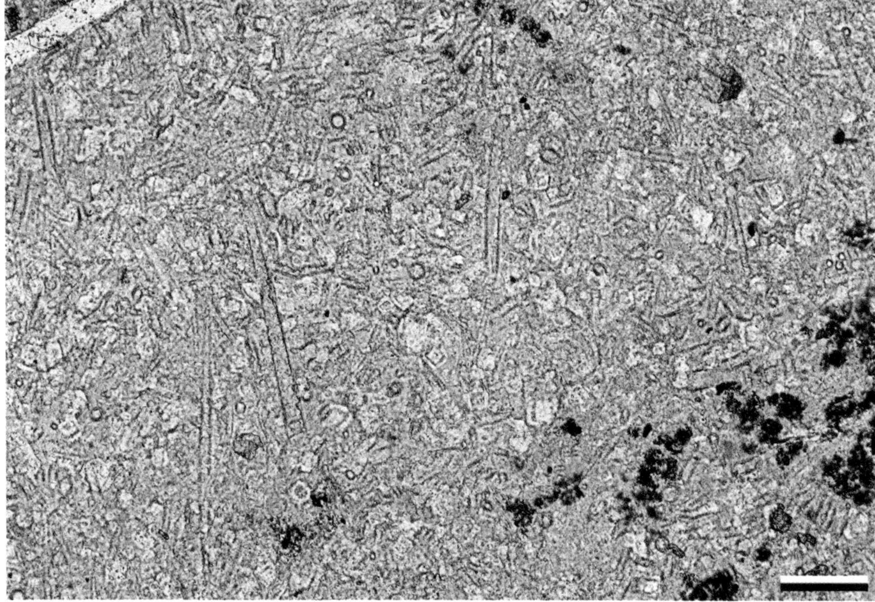

A sedimentary rock fragment composed of a dolomitized spiculitic chert. The grain contains numerous monaxon sponge spicules that have all been replaced by chert. The circular cross sections through spicules can be confused with calcispheres, but the sediment should also include elongate longitudinal sections (like the ones in this grain) if they are spicules.

PPL, Scale bar = 0.10 mm

Lo. Pennsylvanian Marble Falls Fm., Burnet Co., Texas

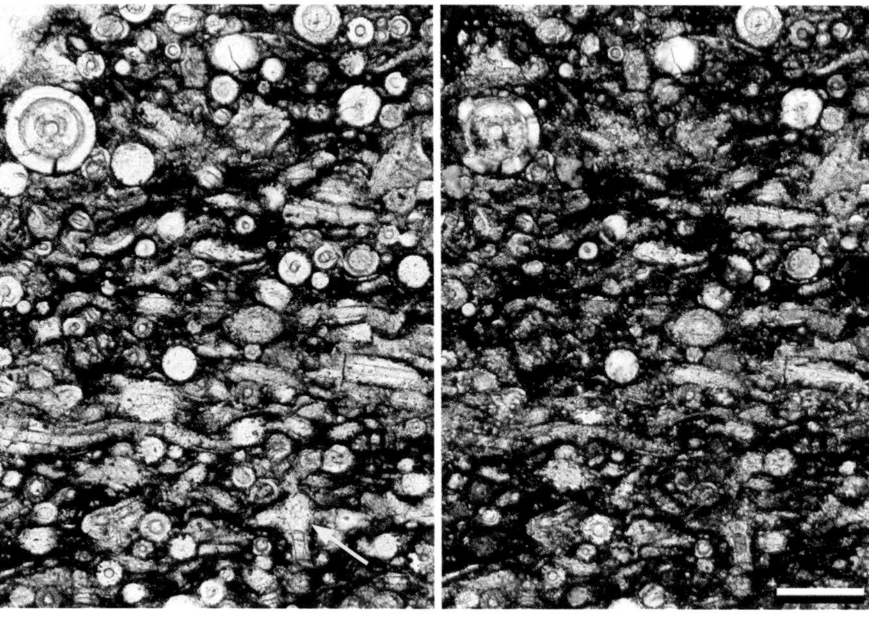

A shallow-shelf spiculitic shale interbedded with crinoidal limestones (not visible in this image). Smooth circular outlines with remnants of original central canals are visible in most of the spicules despite replacement of the original opaline silica by chert, microquartz and, in places, also by calcite. Some cuts show simple elongate sections (with central canals) indicating that these were mostly monaxon spicules, although a few multi-axon junctions are visible (example at arrow). The presence of central canals is a reliable indicator of an originally siliceous composition for the spicules.

PPL | XPL, Scale bar = 0.07 mm

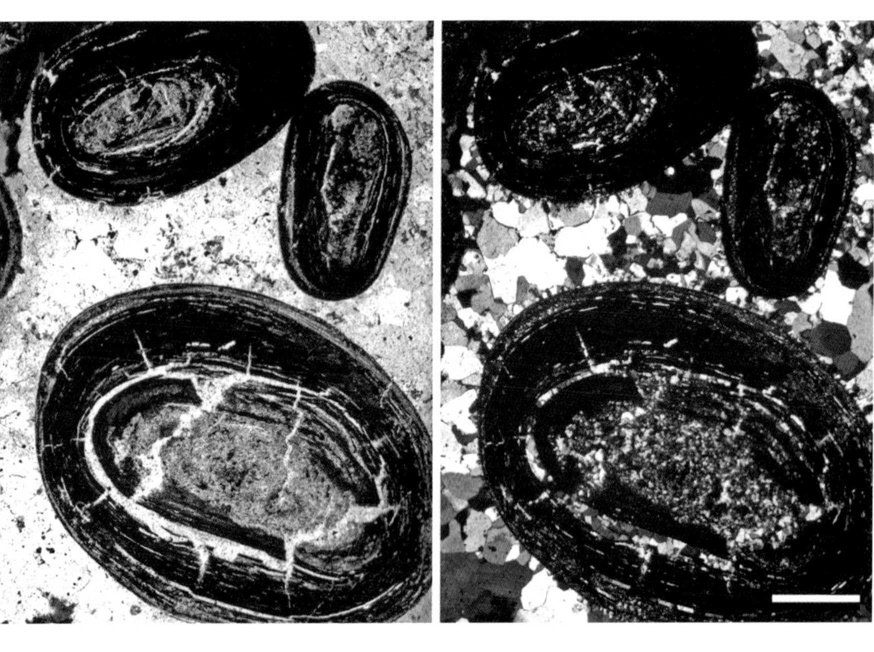

Precambrian (Paleoproterozoic) Gunflint Fm., Thunder Bay, Ontario, Canada

A coated-grain layer in a chert. The coated grains include ooids (<2 mm diameter) and oncoids (>2 mm diameter); both are irregular, algally- or microbially-coated grains. In this example, the coated grains have chert intraclasts as nuclei. The grainstone has been cemented first by a layer of drusy to chalcedonic quartz, followed by blocky quartz. Photomicrograph from Peir K. Pufahl.

PPL | XPL, Scale bar = 0.24 mm

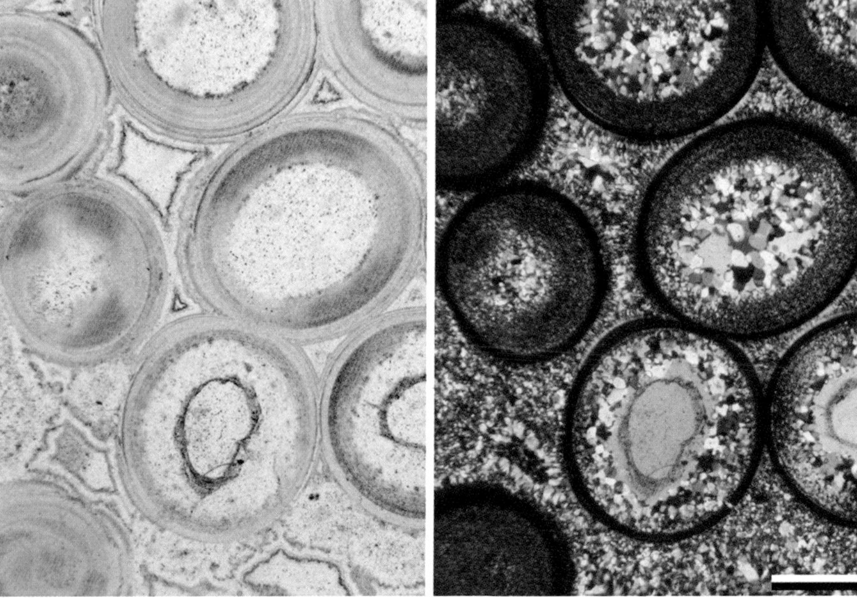

Up. Cambrian Beekmantown Gp., Mines Dolomite, Centre Co., Pennsylvania

Some of these silicified ooids have detrital quartz grains as their cores; these grains acted as nucleation sites for syntaxial overgrowths. Chert and megaquartz replaced the remainder of the probably originally calcitic ooids. Silica also appears to be cementing the ooids (as opposed to replacing a precursor cement), because there is a consistent silica cement stratigraphy (from chalcedony to megaquartz and finally, in a few places, to chert).

PPL | XPL, Scale bar = 0.51 mm

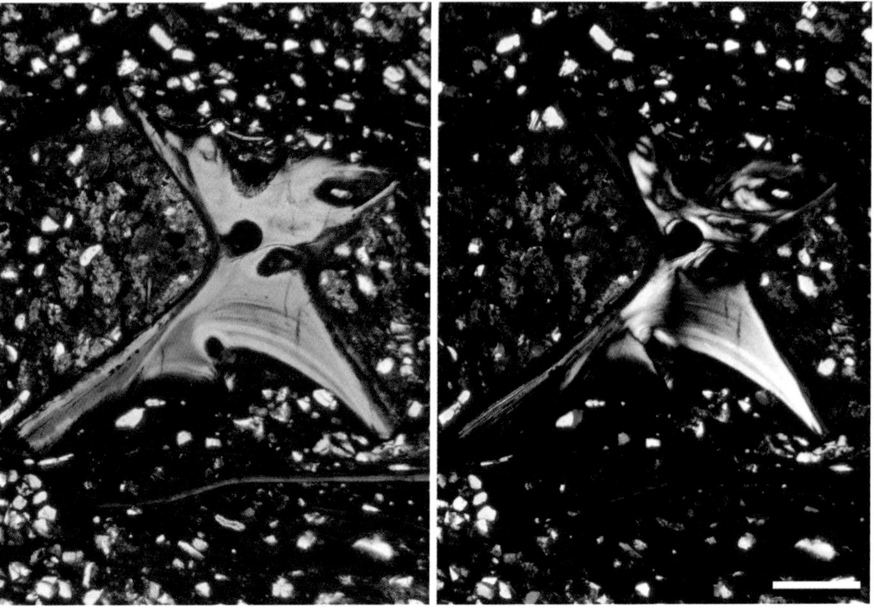

Cretaceous Mowry Fm., Moffat Co., Colorado

A phosphatic fish vertebra in a silty shale matrix. The bone is made of calcium fluorapatite, and it is normally non-birefringent (isotropic) to first-order white. Organics within the bone give it a reddish-brown coloration in plane-polarized and, to a lesser degree, in cross-polarized light. Based on the birefringence, this bone is francolite. Note the differential compaction of the shale around the vertebra. The surrounding shale generally is carbonate free, but in areas sheltered by the vertebra, the sediment contains carbonate, indicating cementation around the bone or pressure solution removal of carbonate from unsheltered areas.

PPL | XPL, RDI, Scale bar = 0.51 mm

Cretaceous Mowry Fm., Moffat Co., Colorado

A fish tooth and miscellaneous phosphatic debris displaying the intense coloration of organically-formed phosphate. The shape, the exterior wall structure (dentine and enamel) and pulpy cavity (dentine tubules) in the center are key to recognizing this as a fish tooth. Teeth and fish scales are the most common phosphatic debris found in most rocks—because they contain organic material, they are subject to thermal maturation and become progressively darker with exposure to higher temperatures.

PPL, RDI, Scale bar = 0.26 mm

Permian Phosphoria Fm., Retort Shale Mbr., Hot Springs Co., Wyoming

Abundant phosphatic debris (mainly shell and bone fragments) in a glauconitic quartz sandstone. In plane-polarized light, shell fragments show multiple structural layers and are composed of collophane. In cross-polarized light, the collophane fragments are all nonbirefringent to very slightly birefringent. This sample is from a world-class phosphate deposit.

PPL | XPL, Scale bar = 0.26 mm

Eocene Green River Fm., Lincoln Co., Wyoming

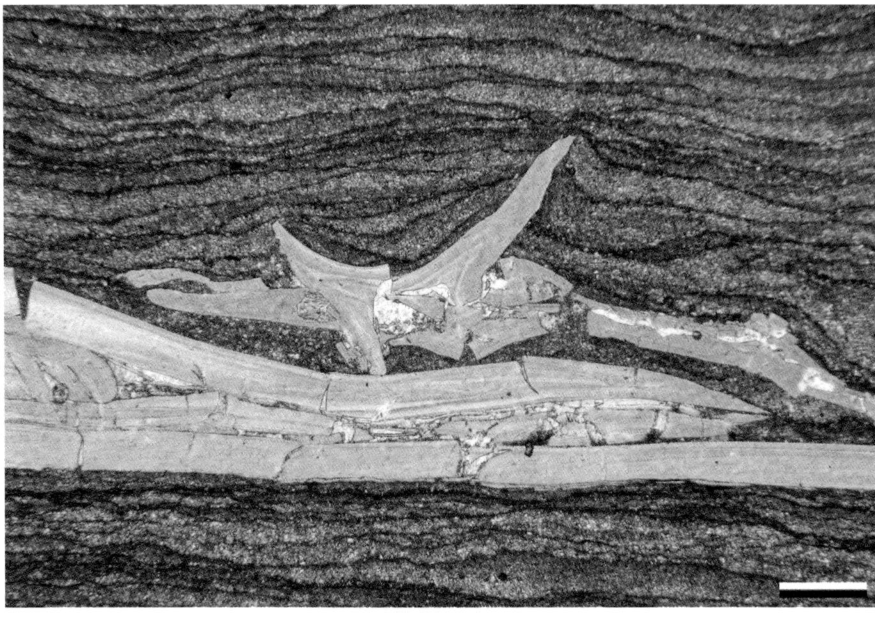

Partly crushed, porous fish bones (yellowish-brown) in a laminated calcareous shale. In cross-polarized light, the collophane is slightly birefringent with low gray colors. Note the slight compactional drape of the laminated soft sediment around the harder skeletal material. A cross-polarized light view of this photomicrograph is in Chapter 8 (page 199).

PPL, Scale bar = 0.51 mm

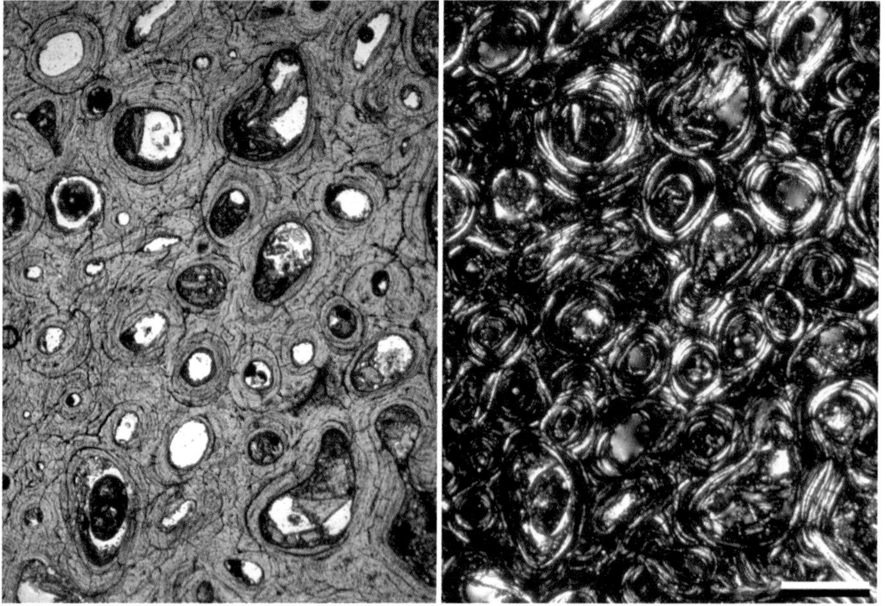

Miocene Yorktown Fm., New Hanover Co., North Carolina

This whale rib has large haversian canals, concentric structure in the surrounding phosphatic material and barely visible traces of caniculi. The large pores and concentrically laminated phosphatic structure of mammal bones are accentuated in the cross-polarized light view. Note the streaky, concentrically layered birefringence patterns common to many phosphatic bone fragments. Photomicrograph from Albert V. Carozzi.

PPL I XPL, Scale bar = 0.30 mm

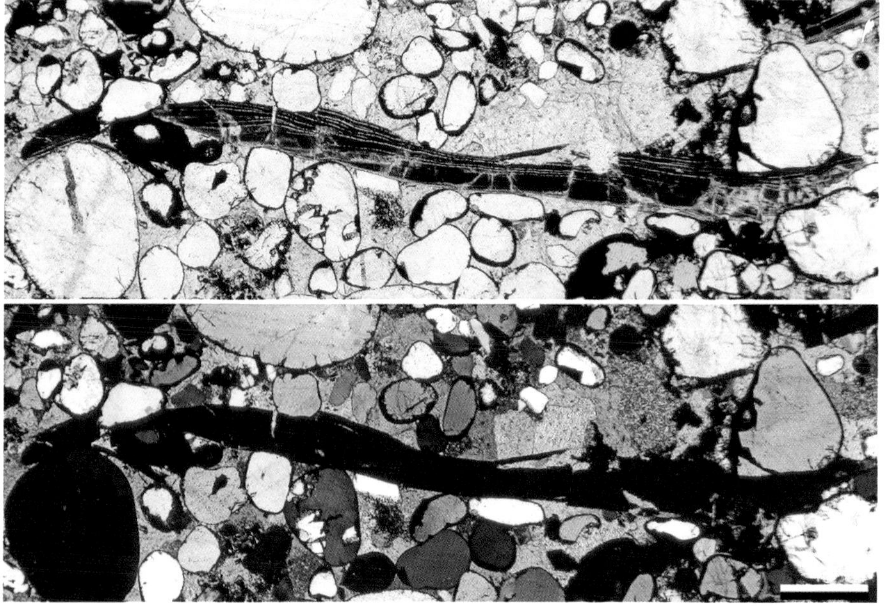

Mid. Cambrian Riley Fm., Lion Mountain Mbr., Llano Co., Texas

A view of a chitinophosphatic, inarticulate brachiopod shell. Characteristic features include the inclined laminar foliation within the shell, the brownish color of the chitinous and phosphatic material in plane-polarized light and the slight curvature of the shell. In cross-polarized light, the phosphatic shell has isotropic or near-isotropic behavior. Well-rounded quartz grains and minor glauconite pellets are cemented together by poikilotopic calcite in the surrounding rock.

PPL / XPL, BDI, Scale bar = 0.51 mm

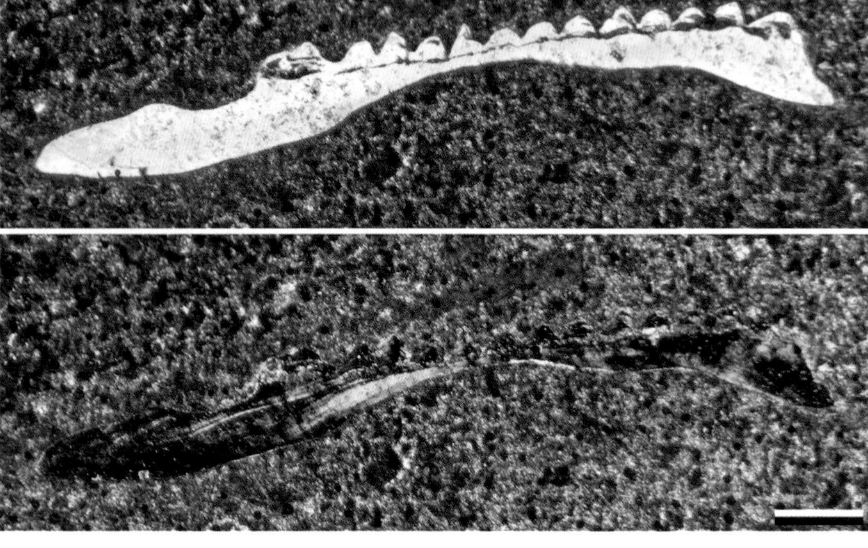

Permian Goose Egg Fm., Ervay Mbr., Big Horn Co., Wyoming

A conodont element in a calcareous mudstone. Conodont elements are made of carbonate fluorapatite and range from colorless to black in plane-polarized light (based on the unit's thermal maturity). These elements exhibit a platform shape with multiple denticles and are isotropic to dark gray in cross-polarized light. They are interpreted to be the earliest vertebrate remains—teeth of jawless fish. Conodont elements are extremely important for biostratigraphic work in Paleozoic sections. Over time (late Precambrian? to Triassic), they changed from simple cones to more complex, denticulate forms.

PPL / XPL, AFeS, Scale bar = 0.51 mm

Up. Cretaceous (Campanian) Alhisa Phosphorite Fm., Jordan

A reworked, granular phosphorite with abundant vertebrate bone fragments (including a prominent vertebra on the left) and reworked, authigenic francolite peloids (dark honey-brown grains). The honey-brown color is characteristic of francolite in thin section. These beds constitute the richest economic deposits in Jordan and adjacent countries in the South Tethyan Phosphogenic Province. Photomicrograph from Peir K. Pufahl.

PPL, Scale bar = 0.24 mm

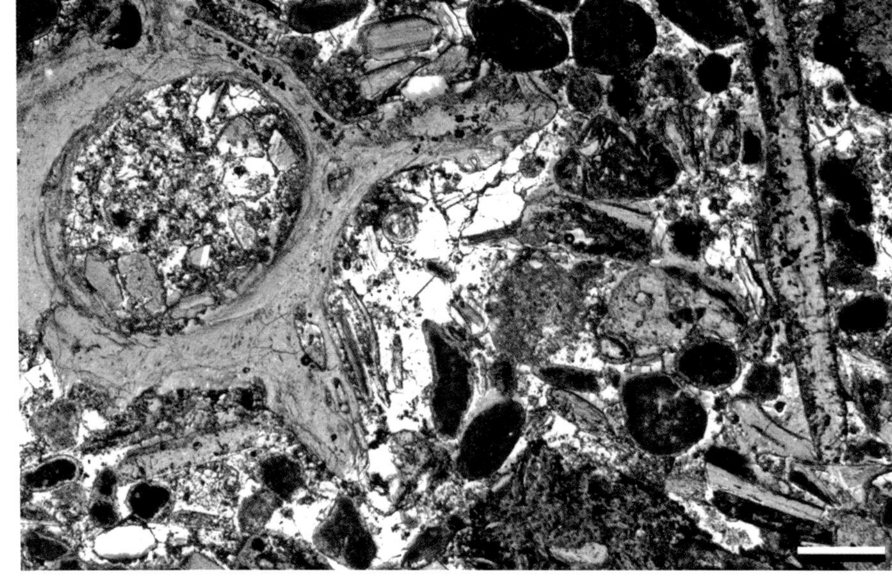

Up. Cretaceous (Campanian) Alhisa Phosphorite Fm., Jordan

A very granular phosphorite with abundant vertebrate bone fragments and reworked, authigenic francolite peloids (the dark honey-brown ovoid to spherical grains). Under cathodoluminescence, the francolite pellets are the purple-blue grains. The reddish-orange areas (in CL) are low magnesium calcite cements. Photomicrograph from Peir K. Pufahl.

PPL | CL, AS, Scale bar = 0.24 mm

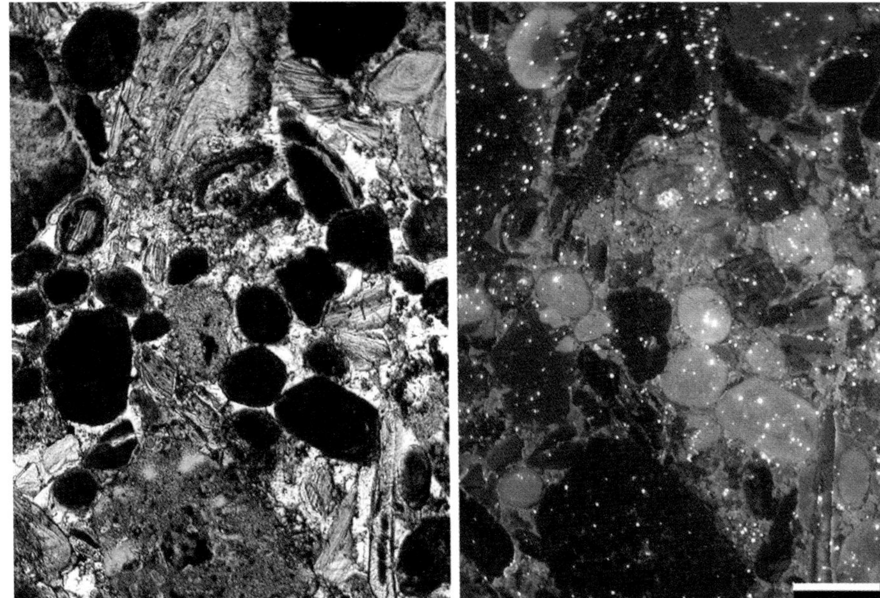

Lo.–Mid. Permian Phosphoria Fm., Caribou Co., Idaho

This sample is from one of North America's largest phosphate deposits and shows a phosphatic oolite in which phosphatic shells, scales and bone fragments (collophane) act as cores (nuclei) of many of the ooids. The skeletal debris is relatively clear (white to pale yellow); the precipitated phosphatic coatings of the ooids are yellowish brown in this example.

PPL | XPL, Scale bar = 0.16 mm

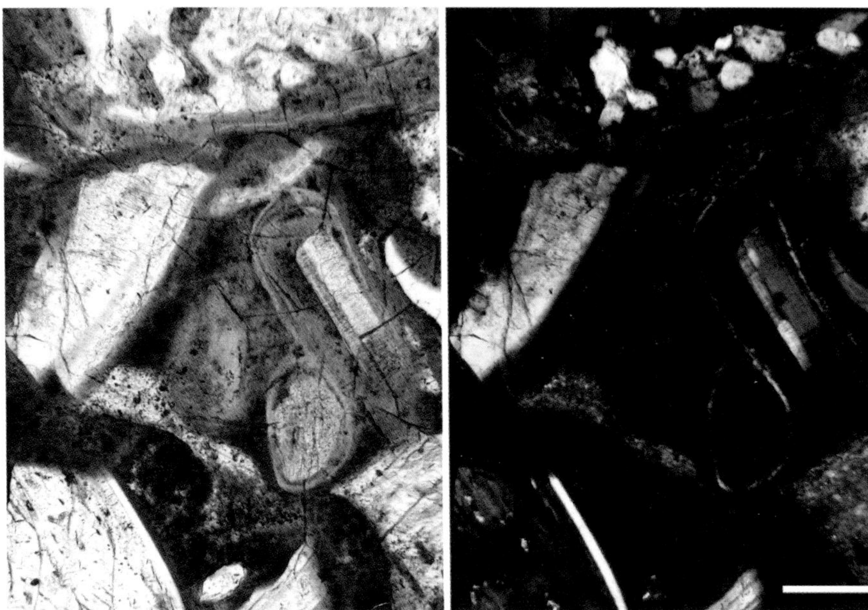

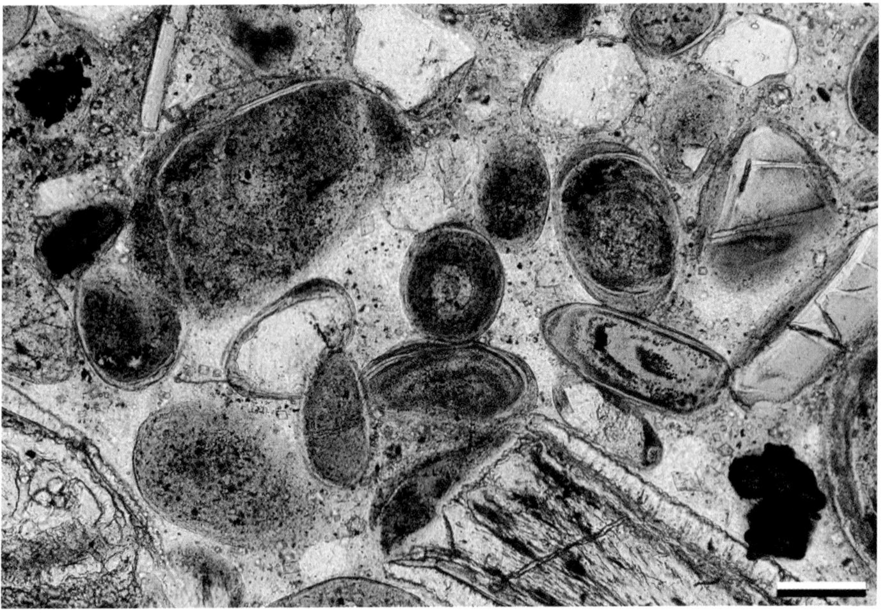

Permian Phosphoria Fm., Retort Shale Mbr., Hot Springs Co., Wyoming

Well-formed phosphatic ooids with nuclei composed of phosphatic pellets, shells or bone fragments. Sedimentation of this unit took place on a cool-water ramp during transgressive cycles across broad carbonate banks. The coated grains with primary phosphate cores and the relative absence of matrix can be indicators of strong wave or current action, implying a high-energy (although not necessarily a shallow-water) environment of phosphate formation. See the next example for an alternate interpretation of similar grains.

PPL, Scale bar = 0.10 mm

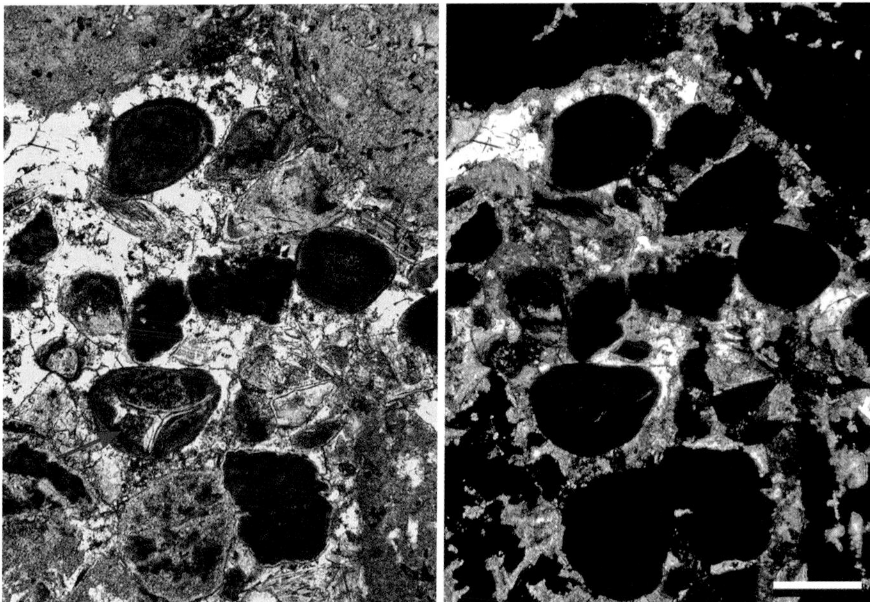

Up. Cretaceous (Campanian) Alhisa Phosphorite Fm., Jordan

A reworked phosphorite sand with coated, authigenic francolite peloids. The nucleus of the grain at the red arrow is a vertebrate bone fragment. These coated francolite grains are not ooids. They have been interpreted to precipitate in sediment just beneath the seafloor when pore waters become saturated with phosphate through the microbial degradation of accumulating organic matter (Pufahl and Grimm, 2003). The francolite grains are pseudoisotropic and thus dark in cross-polarized light. Photomicrograph from Peir K. Pufahl.

PPL | XPL, Scale bar = 0.24 mm

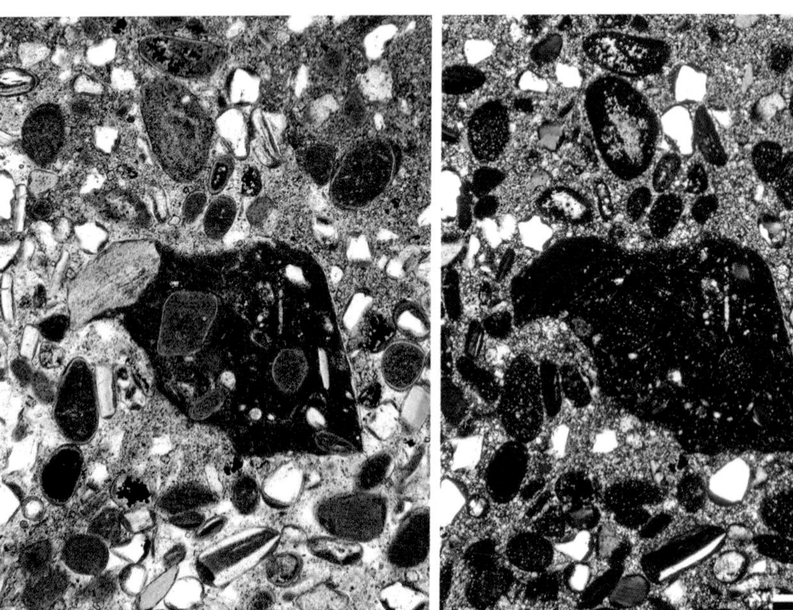

Permian Phosphoria Fm., Retort Shale Mbr., Hot Springs Co., Wyoming

A phosphorite with an outsized phosphatic intraclast (in the center of this view) that clearly had been phosphatized prior to being ripped up and reworked. The other grains in this view are phosphatic ooids and pellets, along with shell and bone fragments. Quartz grains locally act as cores for coated grains. The lack of matrix, the coated grains and the reworked intraclast all indicate a moderately high-energy environment in an area where nutrient-rich marine waters were upwelling onto the continental shelf.

PPL | XPL, AFeS, BDI, Scale bar = 0.33 mm

Pliocene Bone Valley Fm., Polk Co., Florida

Rounded phosphatic grains associated with a major pebble-phosphate deposit. In this example, phosphate accumulated (mainly in the form of vertebrate bones, teeth and scales) in a shallow, coastal shelf setting. The phosphate was then concentrated and reworked into granules and pebbles by later subaerial exposure and weathering. Note the clear color of the pebble phosphates in plane-polarized light and virtual isotropism under cross-polarized light. The grains are cemented by wavellite, an aluminous phosphate mineral.

PPL | XPL, Scale bar = 0.26 mm

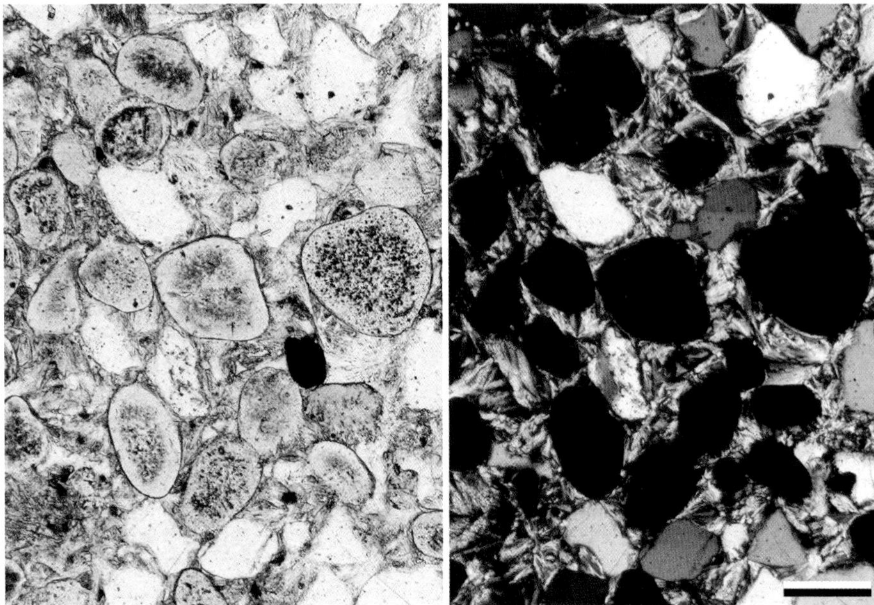

Oligocene – Miocene Otekaike Fm., Oamaru, Otago, New Zealand

A stromatolitic crust atop a marine hardground. The lumpy, digitate laminated crust is largely phosphatic, hence the brownish color in plane-polarized light. Although a biogenic (microbial) origin of such structures is likely, it is difficult to prove in the absence of preserved microbial remains.

PPL, AS, Scale bar = 4.0 mm

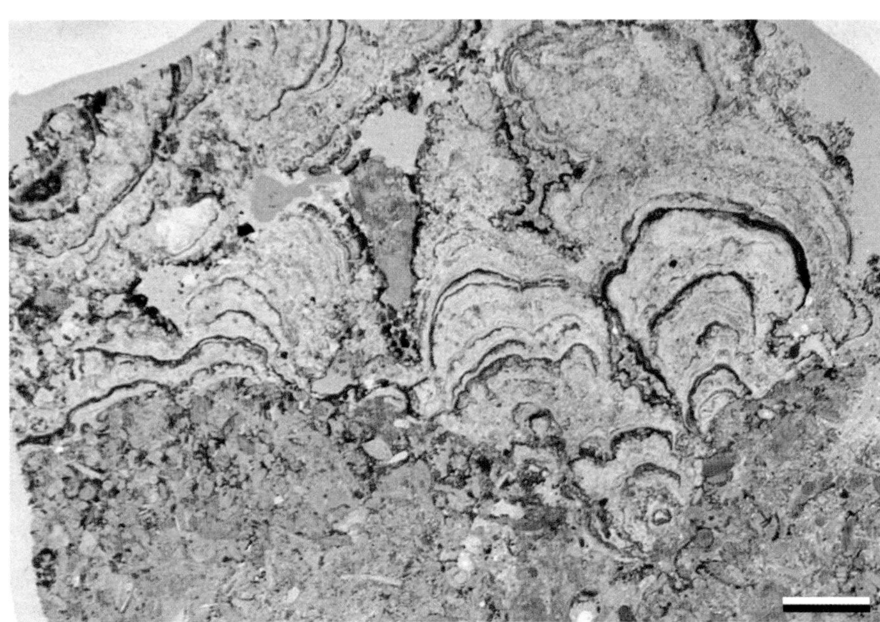

Oligocene Lower Globigerina Limestone Fm., Malta

These phosphate-impregnated carbonate nodules were formed by the concentration of phosphatic debris in *Thalassinoides* burrows. The burrows were then exposed, broken up and rounded by periodic, strong bottom currents (Rehfeld and Janssen, 1995), forming a rubbly hardground. The nodules exhibit varying degrees of phosphate staining, a reflection of the time the pebbles spent exposed on the seafloor. The clasts contain abundant globigerinid foraminifers that were concentrated within burrows.

PPL | XPL, Scale bar = 0.64 mm

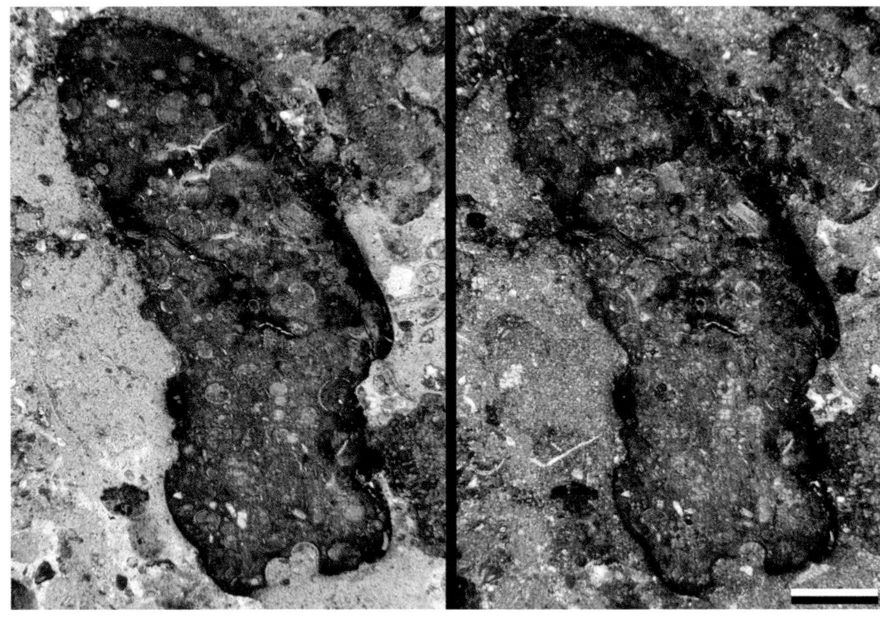

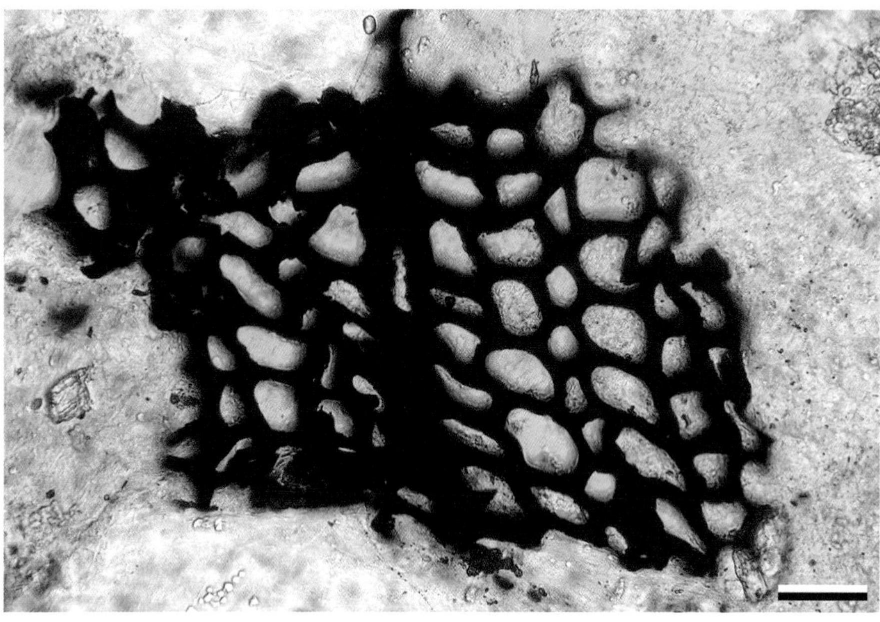

Mid. Jurassic Ness Fm., Norwegian sector, North Sea †

A fossil wood fragment displaying the cellular structure of tracheids. A ray can be seen in the nonvascular part, and this contains a ray tracheid suggesting that the tree from which this came may have been a gymnosperm. The wood fragment has abundant macroporosity "intraskeletal", which is shown by the infiltration of the blue resin, but the pores would not be well connected in three dimensions.

PPL, KFS, BDI, Scale bar = 0.03 mm

Cretaceous Mowry Fm., Moffat Co., Colorado

A fragment of wood in a sandy shale—it is identifiable by its diagnostic cellular structure. In plane-polarized light, wood fragments can range from amber brown to black, depending in part on the thermal maturity of the sample. Because it is normally opaque, the best way to view plant material is usually by reflected light (and that also helps to determine if it has been replaced by pyrite or other minerals). This fragment is probably from a gymnosperm (conifer), based on the regular arrangement of the tracheids (pores).

PPL, PFS, RDI, Scale bar = 0.10 mm

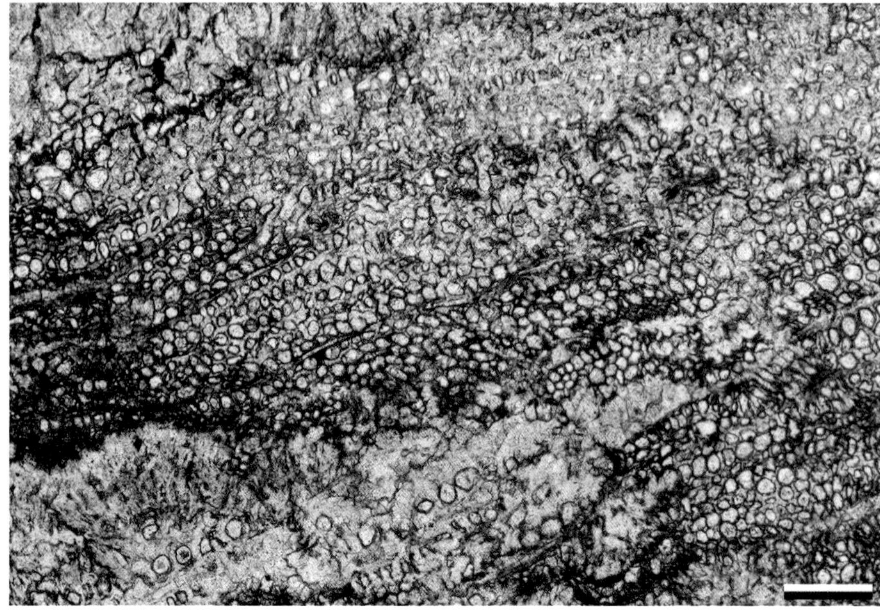

Mid.–Up. Permian (Kazanian–Tatarian) Ochor Sandstone, Perm region, Russia

Unlike most petrified wood, this piece of wood was replaced by both iron oxides and calcite, rather than silica. This piece is probably also from a gymnosperm, as in the previous examples, but unlike the other examples, iron minerals (probably pyrite or marcasite) initially replaced the tree's tracheid structures. Later, calcite filled and/or replaced the remnant pores and organic material. Finally, the early-formed iron minerals weathered to ferruginous oxides or hydroxides on outcrop.

PPL, Scale bar = 0.26 mm

Cretaceous Mowry Shale, Uintah Co., Utah

A fairly typical view of disseminated organic matter in a laminated shale. In this view, the isolated grains are plant fragments that have been flattened and slightly deformed during burial. The plant debris is reddish brown, a color reflective of the thermal maturity and/or the type of organic material.

PPL, Scale bar = 0.05 mm

Up. Devonian – Lo. Mississippian Woodford Shale, Lincoln Co., Oklahoma

Tasmanites are organic-walled cysts from a group of planktonic green algae (*Chlorophyta, Prasinophyceae*) that have existed since Precambrian time. These organic spheres typically are yellow to golden-brown in thin section and are commonly collapsed. *Tasmanites* are especially common in Paleozoic marine shales and in places are so abundant that they form coal deposits (termed "white coal" or "tasmanite").

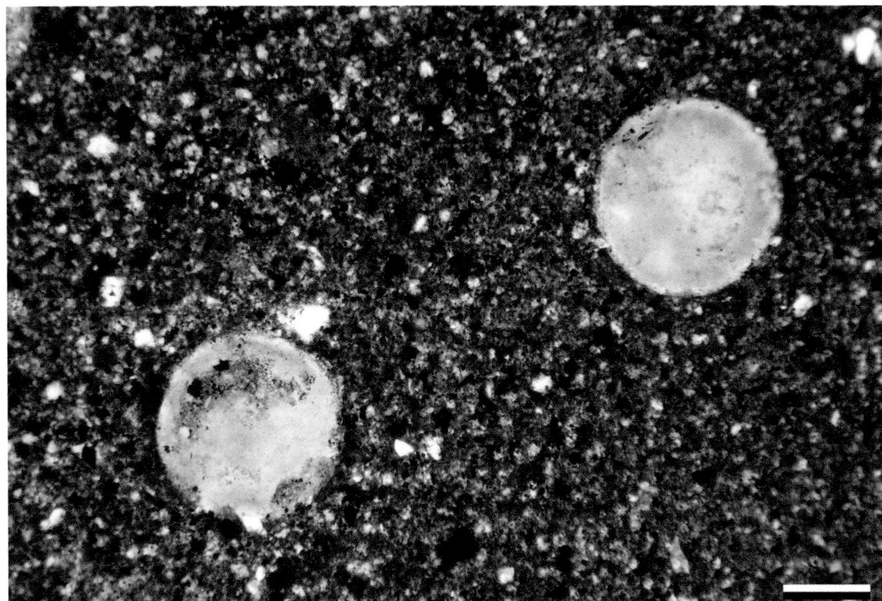

PPL, Scale bar = 0.09 mm

Up. Devonian – Lo. Mississippian Woodford Shale, Lincoln Co., Oklahoma

A close-up view of a *Tasmanites* cyst on bedding plane surface. It is not uncommon for these organic cysts to be replaced or filled by pyrite or silica. *Tasmanites* and their diagenesis are discussed further in Chapter 8.

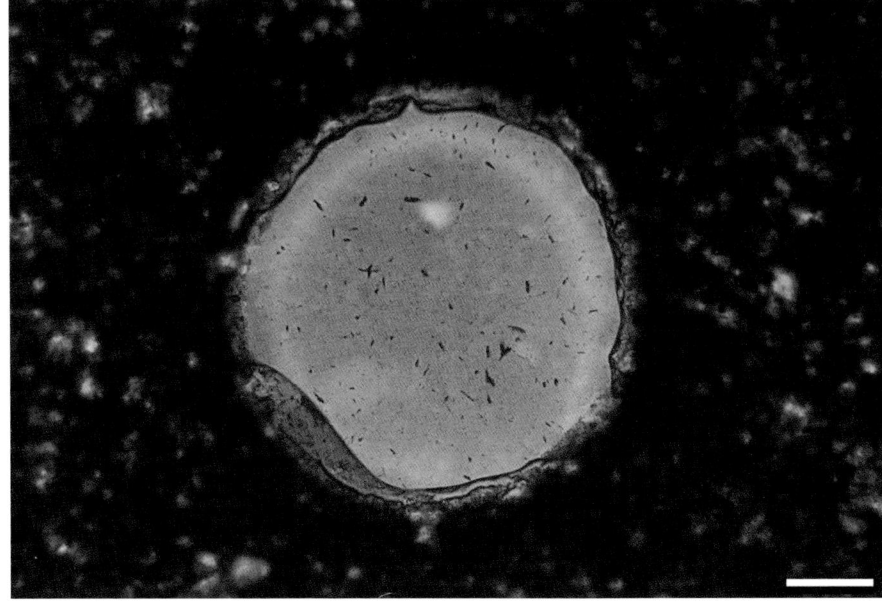

PPL, Scale bar = 0.05 mm

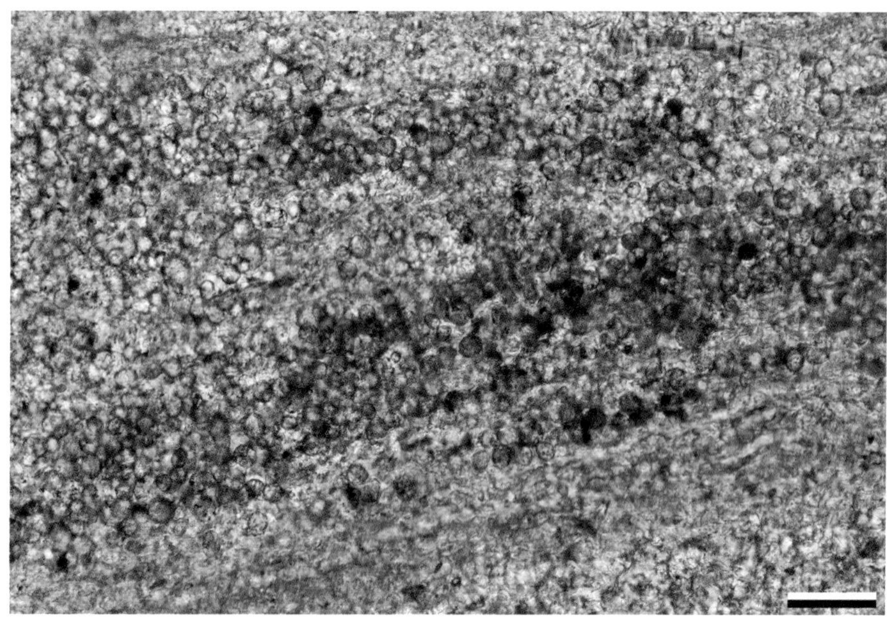

Up. Eocene diatomaceous shale, northern Bohemia, Czech Republic

A diatomite containing horizons of concentrated spores and pollen. In this view, there are hundreds of such spherical, organic-walled objects. Spores are the reproductive bodies of nonflowering plants, algae, fungi and other organisms and are common in nonmarine and marine settings. Both terrestrial and marine spores are resistant to chemical breakdown. Pollen is a male reproductive structure of flowering plants and some trees, with smooth to spinose grains that range from elongate spheres to tetrahedral forms. Spores and pollen are usually yellow to brown in thin section, depending on degree of thermal maturity.

PPL, Scale bar = 0.05 mm

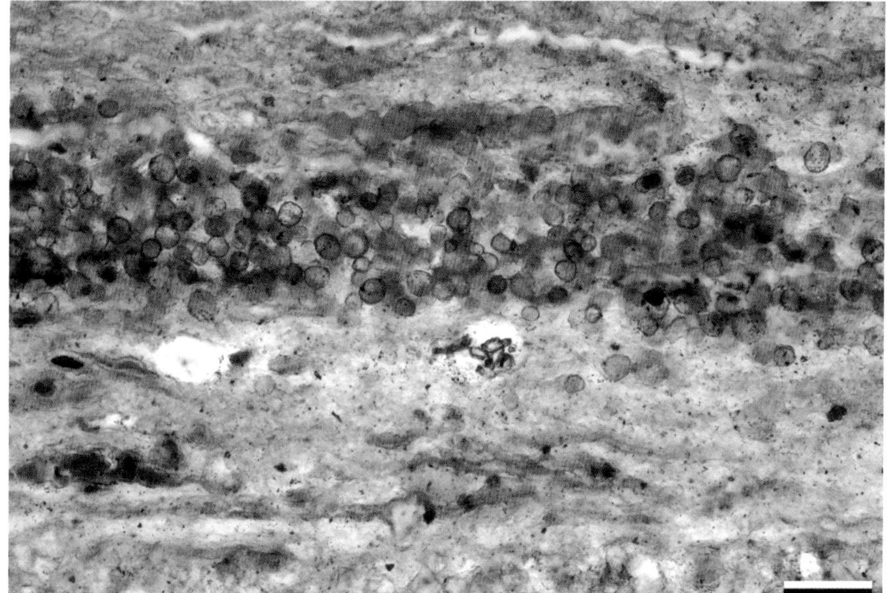

Up. Eocene diatomaceous shale, northern Bohemia, Czech Republic

A close-up of the pollen or spore cases. Note the uniform size and shape of the cysts as well as their golden color. Pollen and spores are difficult to tell apart in standard thin sections, but separates can be viewed under a microscope or with a SEM to make a definitive identification.

PPL, Scale bar = 0.04 mm

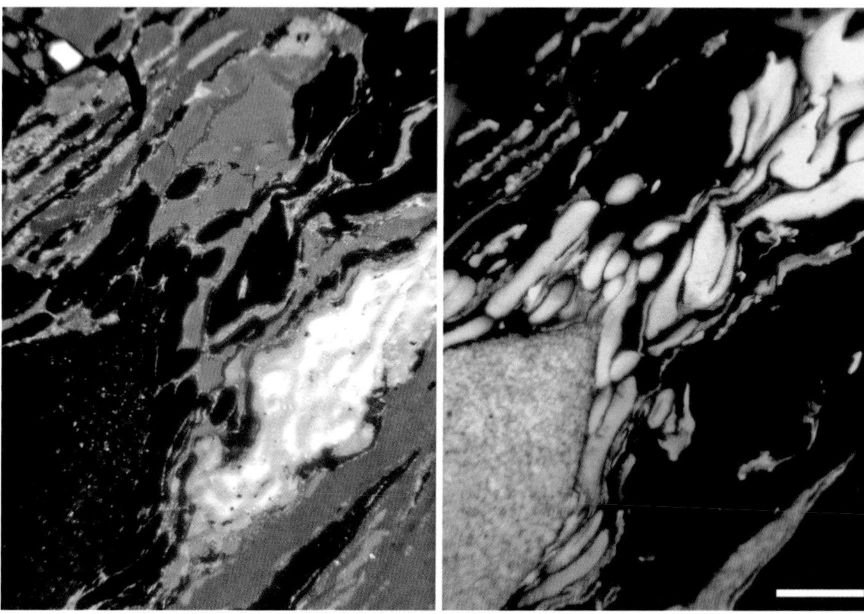

Mid. Pennsylvanian Four Corners Fm., Leatherwood Coal Bed, Perry Co., Kentucky

This coal has the rank of high volatile bituminous A. By utilizing reflected (incident) white light on a polished surface, the coal can be seen to be composed of vitrinite (gray areas), inertinite (white areas) and liptinite (dark brown to black areas). Under fluorescent light, vitrinite and inertinite do not fluoresce, and appear black, whereas, the liptinite fluoresces bright yellow. Photomicrograph from Cortland Eble.

RL | FL 470, Scale bar = 35 μm

Mid. Pennsylvanian Breathitt Gp., Hyden Fm., Fire Clay Coal bed, Perry Co., Kentucky

This coal has the rank of high volatile bituminous A (R_o maximum = 0.88%). Using reflected white light, coal macerals in this sample are inertinite (white material with cellular detail) and liptinite (dark brown to black material). Under fluorescent light, the inertinite does not fluoresce, although the internal cell wall's vitrinite fluoresces weakly. Liptinite fluoresces bright yellow. Photomicrograph from Cortland Eble.

RL | FL 470, Scale bar = 35 μm

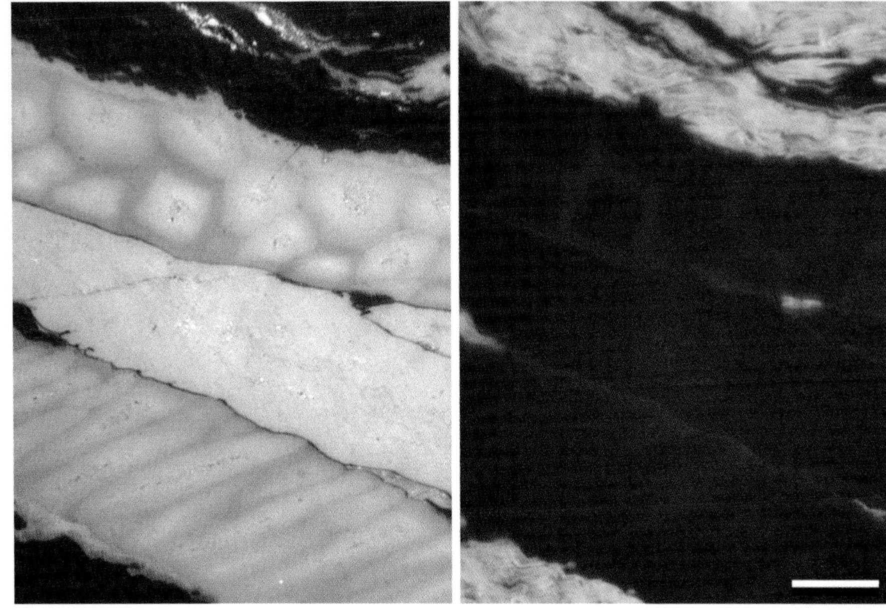

Up. Devonian (Famennian) Ohio Shale, Knox Co., Kentucky

In reflected light, the elongated dark bands are maceral bituminite, and the dark gray center mass is solid bitumen. The white, granular streaks are comprised of micrinite and are mainly a devolatilization by-product of bituminite thermal maturation. The elongated dark bands of bituminite fluoresce weakly, as does the solid bitumen. The two bright yellow fluorescing particles (at one o'clock) are spores. This sample is from the Turner Mine shale, which directly overlies the Springfield coal bed in the Illinois Basin. Photomicrograph from Cortland Eble.

RL | FL 470, Scale bar = 19.5 μm

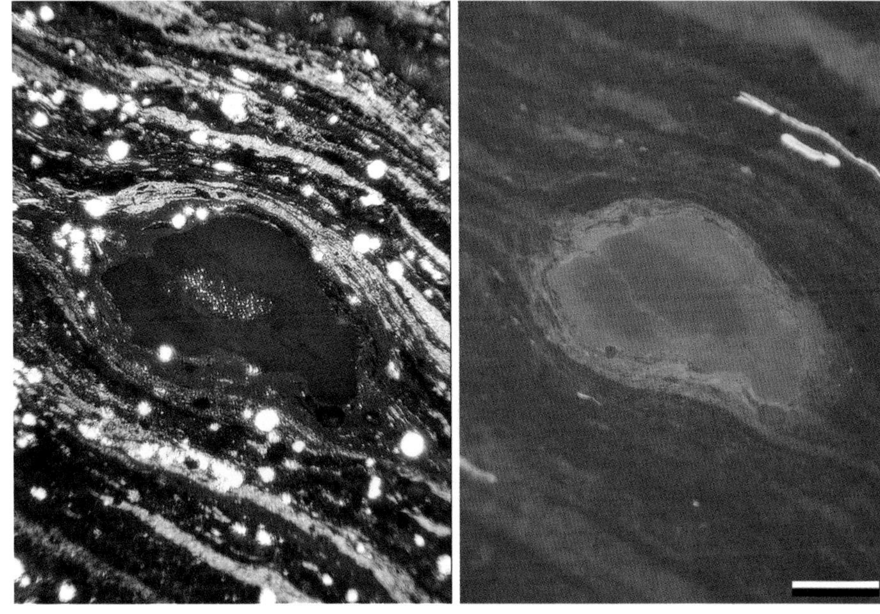

Up. Devonian (Famennian) Ohio Shale, Knox Co., Kentucky

Solid bitumen in this shale sample appears gray to dark gray in reflected light. Solid bitumen is a common component of oil/gas shales like this one. Under fluorescent light, bitumen fluoresces yellow. The fluorescence level of solid bitumen in oil/gas shale varies and is in part dependent on the thermal maturation of the strata. Photomicrograph from Cortland Eble.

RL / FL 470, Scale bar = 15 μm

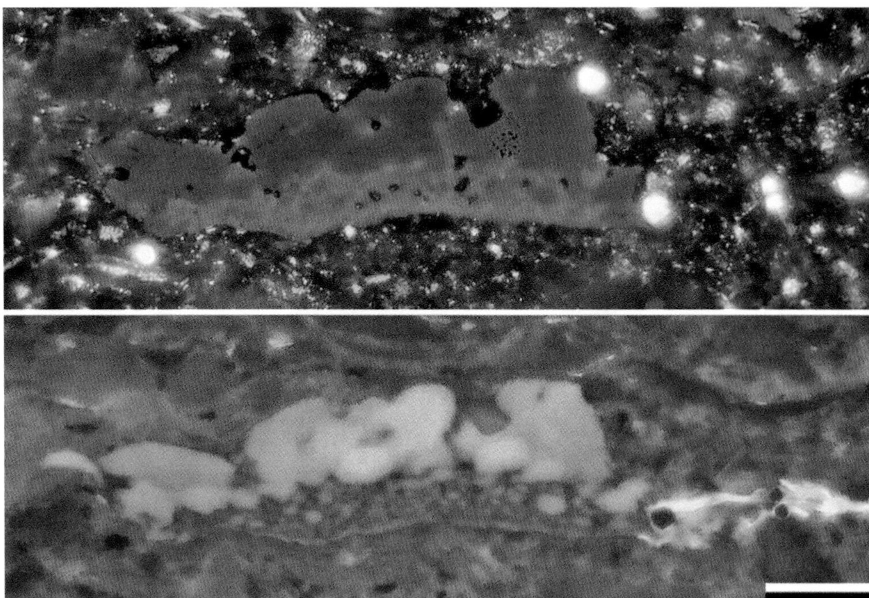

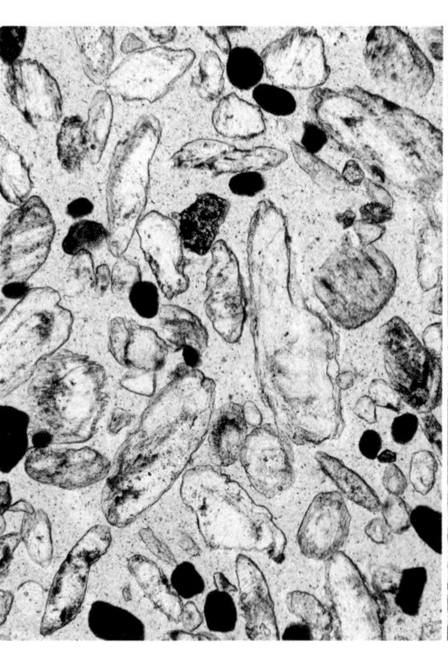

Holocene gypsum dunes, Tooele Co., Utah

Although quite soluble and soft, gypsum ($CaSO_4 \cdot 2H_2O$) can, under exceptional circumstances be reworked into detrital deposits. In this example from an arid setting (the Great Salt Lake region), discoidal gypsum crystals (clear with low birefringence) were transported by eolian processes from playa surfaces (deflationary salt flats) into small dunes. Although the grains show evidence of rounding despite less than a kilometer (< 0.6 mi) of transport, they have remained sufficiently unaltered to form fairly pure gypsarenite deposits. The darker grains are small carbonate peloids from associated lacustrine beds.

PPL | XPL, Scale bar = 0.24 mm

Up. Miocene (Messinian) Gesso Solfifera, southern Italy

Detrital gypsum crystals and carbonate matrix. Several of the gypsum crystals are twinned and have euhedral crystal outlines; all have characteristic low birefringence. The gypsum crystals are interpreted to have precipitated in shallow water prior to reworking into a deeper-water setting by turbidity currents (e.g., Schreiber et al., 1976; Schlager and Bolz, 1977). Photomicrograph from B. Charlotte Schreiber.

XPL, Scale bar = 0.03 mm

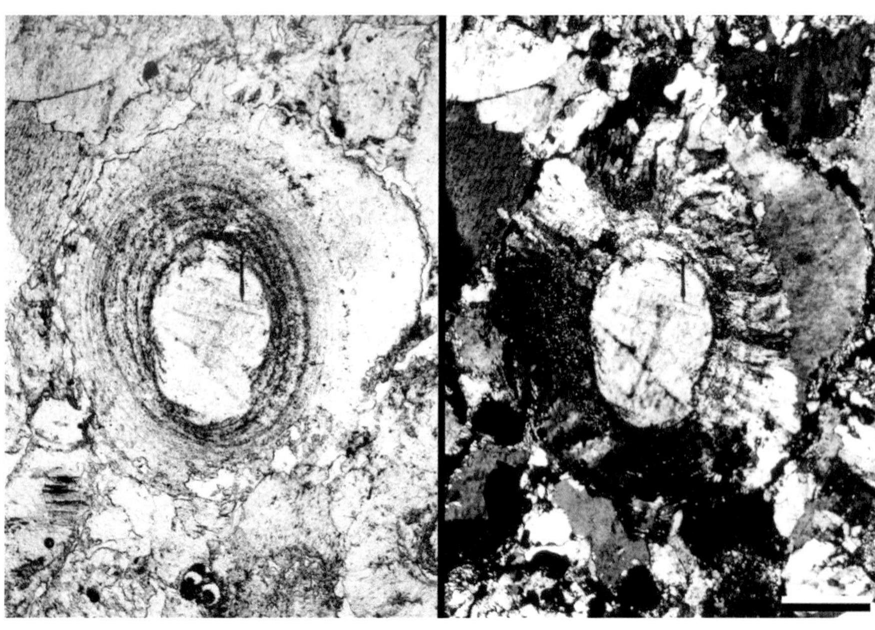

Up. Miocene (Messinian) Gesso Solfifera, southern Italy

On occasion, gypsum can form oolitic growths as seen here. The nucleus is a rounded gypsum crystal fragment which is coated with numerous concentric (and slightly eccentric) layers of gypsum marked by brown bands of inclusions. These ooids are incorporated in a groundmass of gypsum cement. These grains formed in agitated, shallow hypersaline water Photomicrograph from B. Charlotte Schreiber.

PPL | XPL, Scale bar = 0.03 mm

Up. Permian Castile Fm., Eddy Co., New Mexico

These interlaminated gypsum and calcite layers in varved evaporite deposits probably represent deposition in a deep-water basin behind a barrier that restricted seawater inflow. The light and dark layers in this unit have been interpreted to reflect annual climate cycles. The gypsum layers (light-colored in plane-polarized light) precipitated during dryer and more evaporative seasons, and the thinner, darker, organic-rich calcite layers precipitated during wetter seasons. The entire evaporite section in this basin contains approximately 260,000 varve couplets (Anderson et al., 1972).

PPL | XPL, Scale bar = 0.51 mm

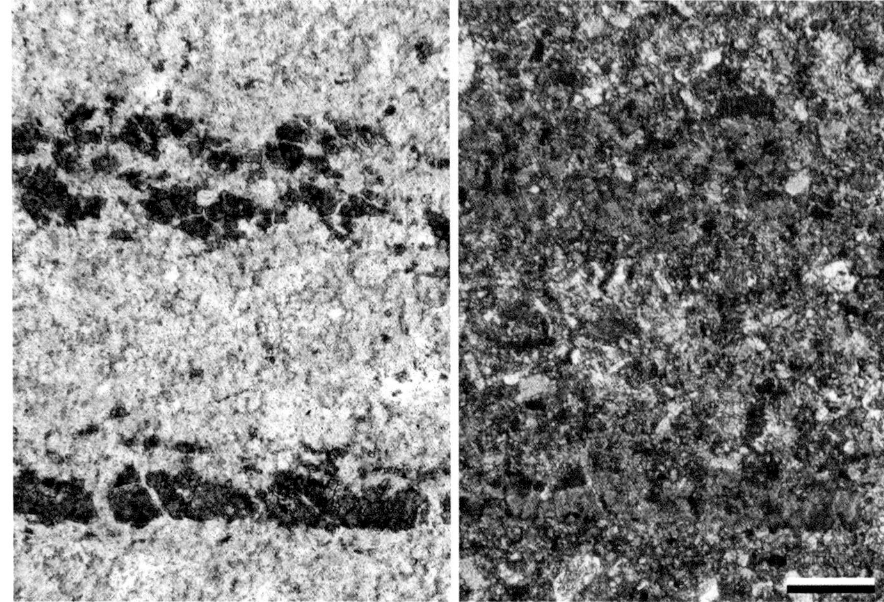

Mid. Permian Zechstein Z1, western Silesia, Poland

An example of bedded gypsum that contains abundant anhydrite inclusions. Bedded gypsum and anhydrite can both form at or near the earth's surface. When gypsum dehydrates during burial, it becomes anhydrite, and during uplift, anhydrite typically rehydrates to gypsum. It is not uncommon to find inclusions of the other mineral in what ever phase is currently present, as in this example with intermediate-birefringence inclusions of anhydrite in low-birefringence gypsum crystals.

XPL, Scale bar = 0.16 mm

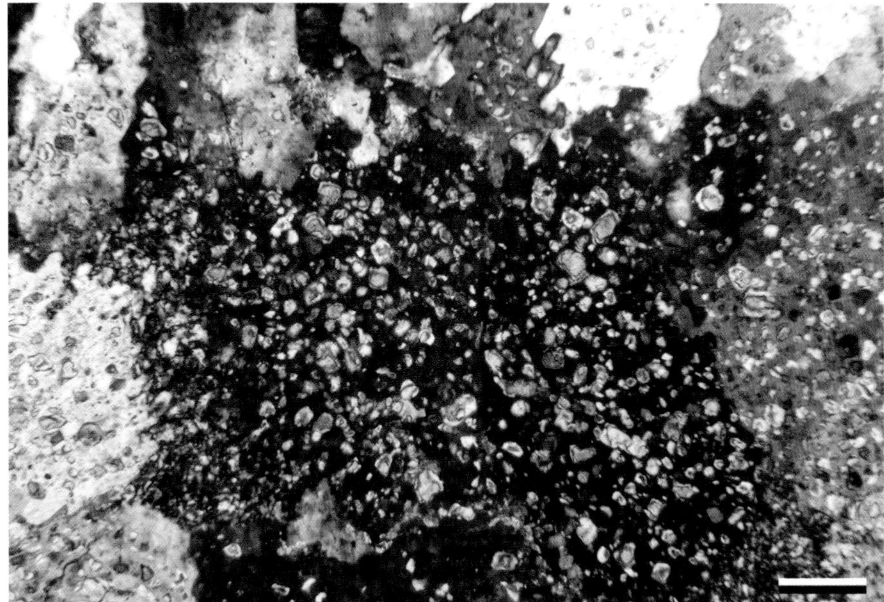

Devonian Keg River Fm., Zama North oil field, Alberta, Canada

In this sample, the growth of anhydrite ($CaSO_4$) within the sabkha sediments has displaced quartz silt laminae. Nodular or "chicken-wire" anhydrite morphologies develop as the nodules grow and coalesce, moving and compressing the silt into thin, inter-nodular zones. The felted micro-lath texture suggests primary (synsedimentary) displacive growth rather than replacement through burial-related dehydration of gypsum.

XPL, Scale bar = 0.16 mm

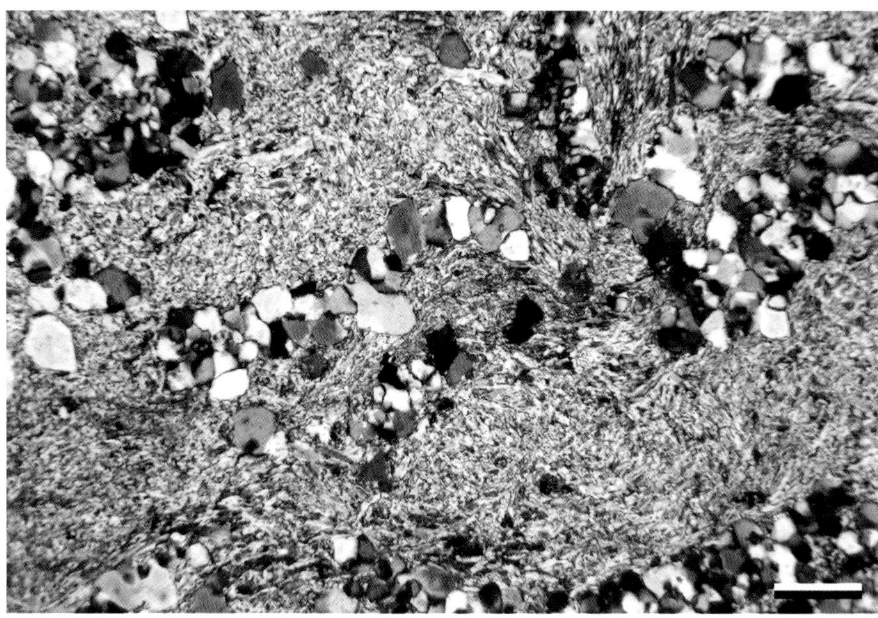

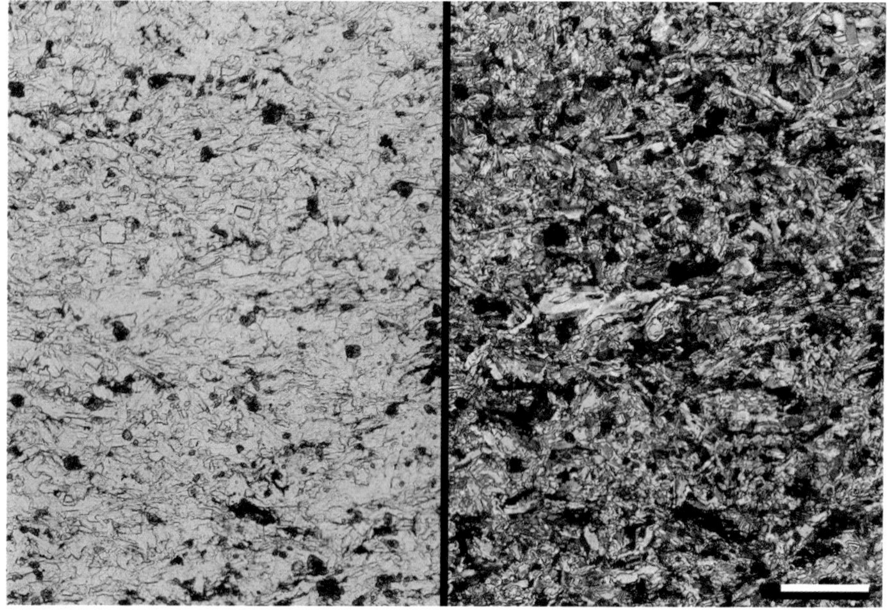

Up. Permian Zechstein A1, Września Co., Poland

Anhydrite nodules commonly consist of felted masses of tiny cleavage fragments of anhydrite. This texture typically is produced during synsedimentary displacive crystal growth within relatively soft host sediment, although it also can be produced during burial diagenesis of bedded sulfates (e.g., Warren, 2006). In plane-polarized light, anhydrite has low relief and is sometimes weakly pleochroic with shades of pale yellow, pink and violet. In cross-polarized light, anhydrite can show up to third-order birefringence colors.

PPL | XPL, Scale bar = 0.26 mm

Up. Permian Zechstein Ca1, Września Co., Poland

Intergrown with the brightly birefringent, felted anhydrite crystals are several crystals with low birefringence that look like gypsum. However, gypsum is an unlikely mineral at the 3,940 m (~12,950 ft) depth of this sample. In fact, those gray to black crystals are not gypsum, but are anhydrite. Mineral birefringence is dependent on crystal orientation and views down the c-axis always show the lowest birefringence. To be certain that this is anhydrite and not gypsum, one needs to determine the 2V of the crystal—both gypsum and anhydrite are biaxial positive, but anhydrite has a 2V = 40-44°, and gypsum has a 2V = 58°.

XPL, Scale bar = 0.10 mm

Up. Permian Zechstein A1, Września Co., Poland

This coarsely crystalline anhydrite appears to be intergrown with gypsum, but as in the previous example, these crystals are all anhydrite, and the ones with gray to black birefringence are ones that are oriented with the c-axis perpendicular to the slide surface. As in the previous photo, this section comes from a deep core sample taken from approximately 3,865 m (~12,680 ft) depth—a depth at which gypsum is an unlikely mineral phase.

XPL, Scale bar = 0.26 mm

Up. Permian Zechstein A1, Września Co., Poland

Evaporite minerals in clastic terrigenous or carbonate sediments commonly grow displacively or replacively and, of course, also occur as primary precipitates forming bedded evaporite units. In this subsurface example, anhydrite laths formed coarse, oriented crystal masses within a shear band in a bedded evaporite unit. Flowage of the evaporite unit in the subsurface probably added to the disruption of these precipitates.

XPL, Scale bar = 0.51 mm

Mid.? Permian (former Ufimian) Solikamskaya Horizon, near Berezniki, Perm region, Russia

Halite (NaCl) is a common evaporite mineral and this image illustrates its cubic, hopper-shaped crystal habit as well as the striking zonation of inclusion-rich and inclusion-poor zones. Halite is isotropic, and thus is black under cross-polarized light. Halite is normally colorless in plane-polarized light, but it can be brown due to inclusions. In hand specimens, halite can display shades of red, blue, orange, yellow, purple and black based on the mineral and fluid inclusions within the crystals. This sample, and subsequent Russian halide examples, are from subsurface mines or cores.

PPL, Scale bar = 0.51 mm

Mid.? Permian (former Ufimian) Solikamskaya Horizon, near Berezniki, Perm region, Russia

This image shows numerous hopper crystals of halite that clearly show their cubic habit. Abundant minute bubbles (mainly water/brine-filled inclusions) mark the outlines of the hoppers. The triangular outlines of some of the inclusion trains are due to the plane of the thin section cutting the corners of the cubic crystals.

PPL, Scale bar = 0.51 mm

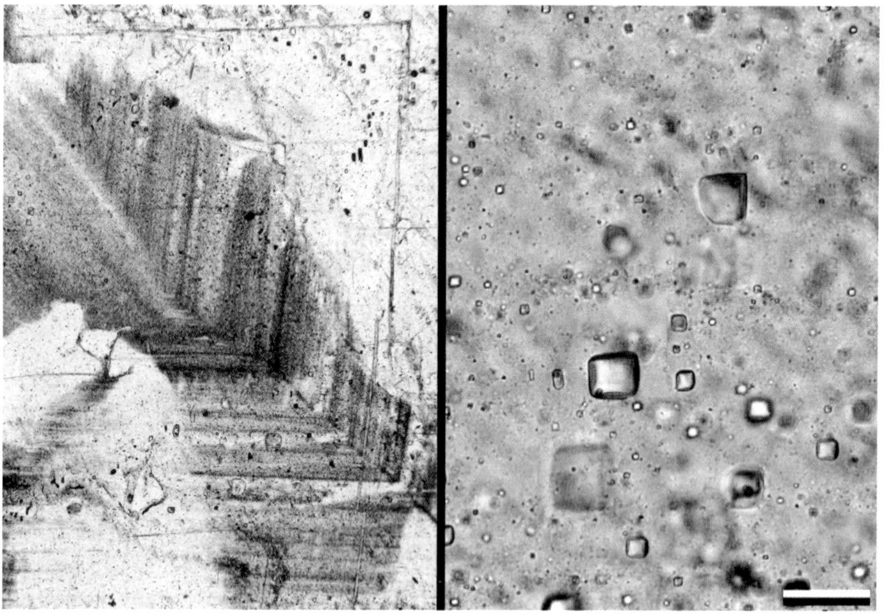

Mid.? Permian (former Ufimian) Solikamskaya Horizon, near Berezniki, Perm region, Russia

Inclusions can be abundant in most evaporite minerals. The view on the left is a zoned cubic halite crystal in which the zoning is due to a higher concentration of fluid inclusions within the darker bands. On the right is a closeup of some of the 1-phase (liquid only) and 2-phase (gas and liquid) fluid inclusions that make up the zoning. It is typical for these inclusions to mirror the cubic crystal structure of the host mineral (thus, they are called negative crystals). These negative crystals can be useful in identifying halite in thin sections that are not impregnated with dyed epoxy.

PPL | PPL, Scale bar = 0.64 | 0.03 mm

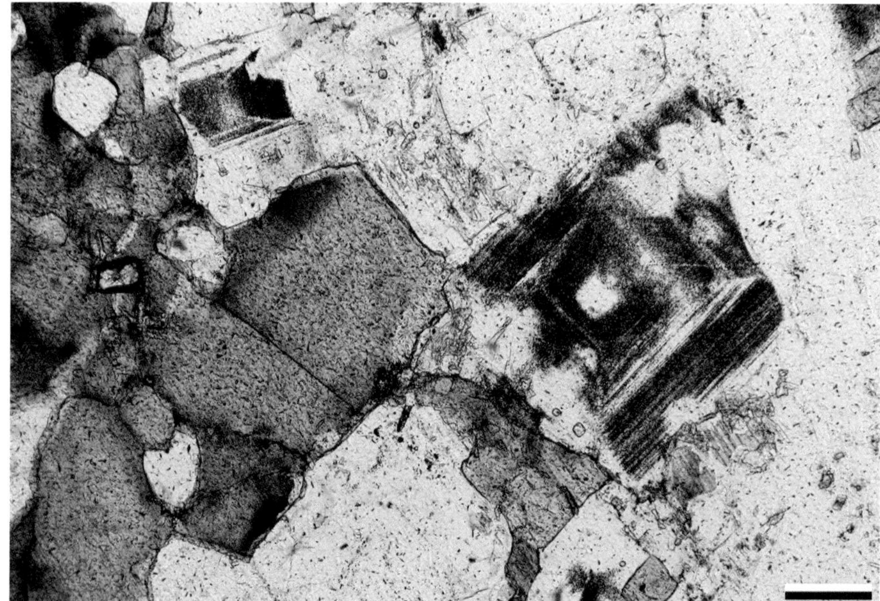

Mid.? Permian (former Ufimian) Solikamskaya Horizon, near Berezniki, Perm region, Russia

In this view, halite and sylvite are intergrown. Sylvite (KCl) is the reddish brown phase concentrated mainly on the left side of the photomicrograph, and it also occurs as solid inclusions within the colorless halite. Within the halite, cubic hopper crystals are clearly visible due to the abundant fluid inclusions delineating growth zones.

PPL, Scale bar = 0.26 mm

Up. Permian Salado Fm., Eddy Co., New Mexico

Sylvite has a very similar crystallographic structure to halite (both are isometric cubic with three perfect cleavages). Sylvite has a lower index of refraction than halite and frequently contains inclusions which result in color tinting of the crystal, especially the reddish-brown hues shown in this example from a subsurface mine. Sylvite also contains radioactive potassium, which can produce elevated gamma-ray log response in evaporite sections.

PPL, Scale bar = 0.03 mm

Mid.? Permian (former Ufimian) Solikamskaya Horizon, near Berezniki, Perm region, Russia

Sylvite has three well-developed cleavages and it is an isometric, cubic mineral, like halite. As in the previous sample, the color tinting of the crystal (reddish-brown hues) is due to inclusions (mainly hematite). No cross-polarized light views of these samples are shown, because sylvite, like halite, is isotropic and the image would be entirely black.

PPL, Scale bar = 0.26 mm

Mid.? Permian (former Ufimian) Solikamskaya Horizon, near Berezniki, Perm region, Russia

In this sample, the sylvite is more granular and rounded and contains inclusions that color the rock both gray and red. The area where the sylvite crystals abut each other contain fewer inclusions and may represent recrystallization (dissolution and reprecipitation) during burial and compaction.

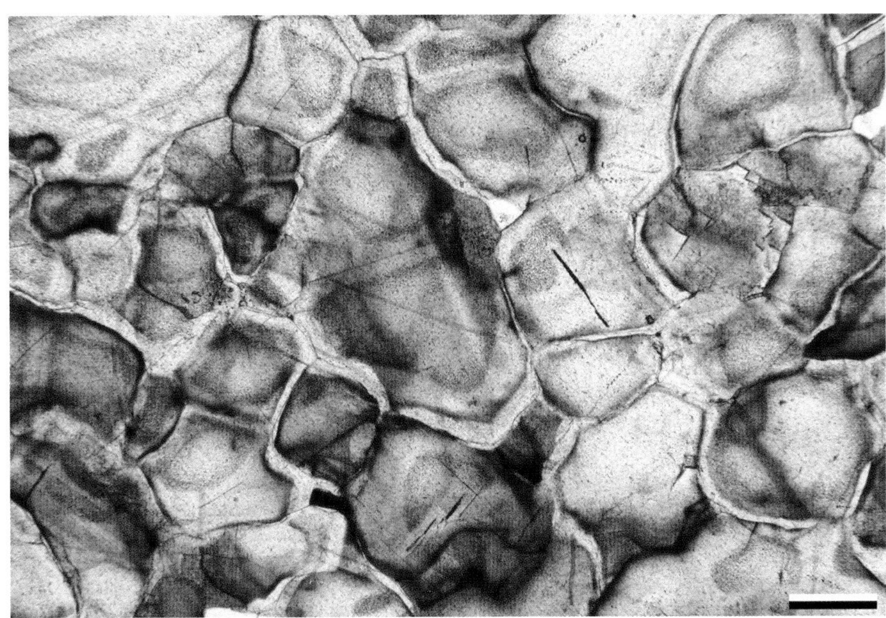

PPL, Scale bar = 0.51 mm

Mid. Cambrian Riley Fm., Lion Mountain Mbr., Llano Co., Texas

This sample shows glauconite "pellets" and quartz grains cemented by poikilotopic calcite cement that nucleated on trilobite fragments not visible in the field of view of this image (McBride, 1988). Normally, one might equate rounding in grains with abrasion and transport (note the well-rounded quartz grains here), but the rounded glauconite grains most likely formed through *in-situ* alteration and coating of fecal pellets. Such glauconite grains are commonly found in shallow-marine environments that have low-sedimentation rates and sufficient wave or current energy levels to winnow and concentrate the pellets.

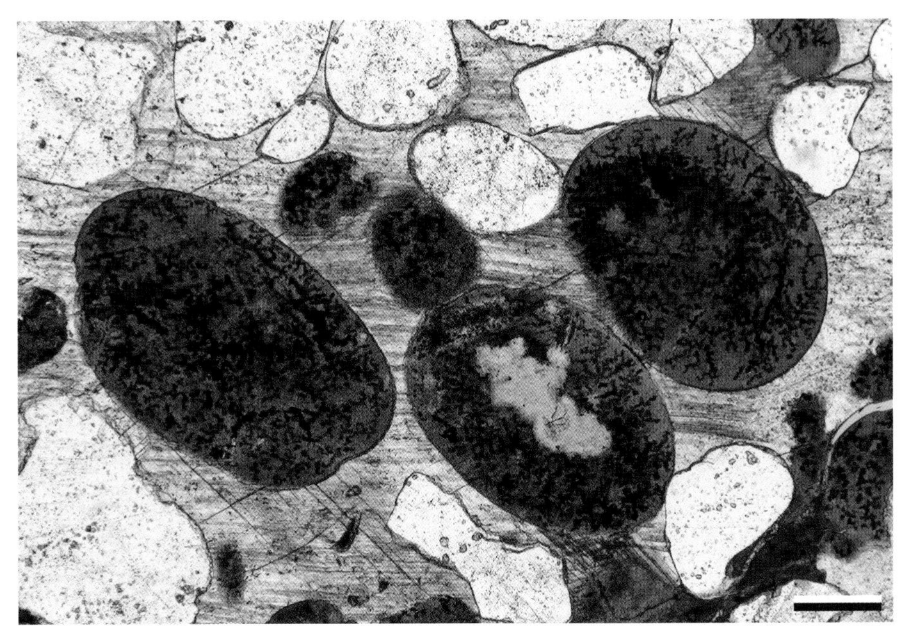

PPL, BDI, Scale bar = 0.21 mm

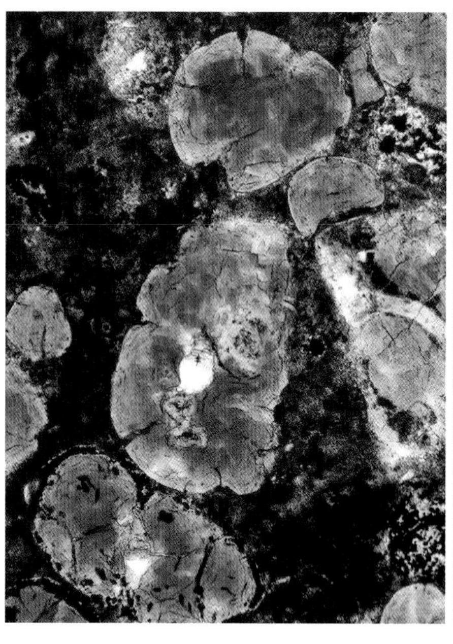
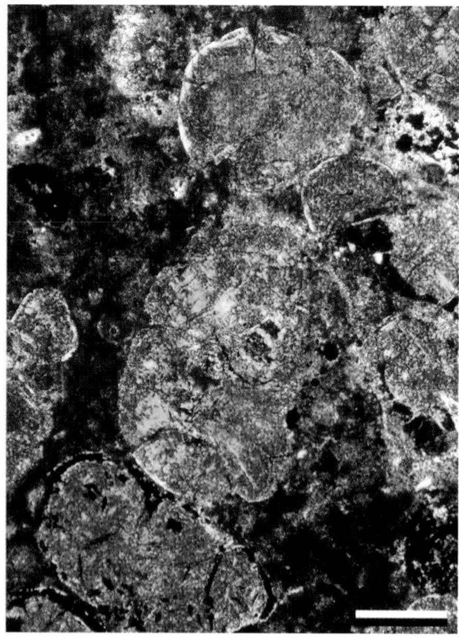

Up. Cretaceous Navesink Fm., Monmouth Co., New Jersey

Large pelletal glauconite grains in a marl. Characteristically, glauconite has a light green color in plane-polarized light and a speckled, or grainy, dark green appearance in cross-polarized light. Glauconite is an iron- and magnesium-rich, smectite-type mica, and the word, as commonly used, really refers to a family of related minerals. Glauconite grains can have varying degrees of mineral ordering, as well as a complex range of interlayered clay minerals (especially smectite).

PPL | XPL, Scale bar = 0.26 mm

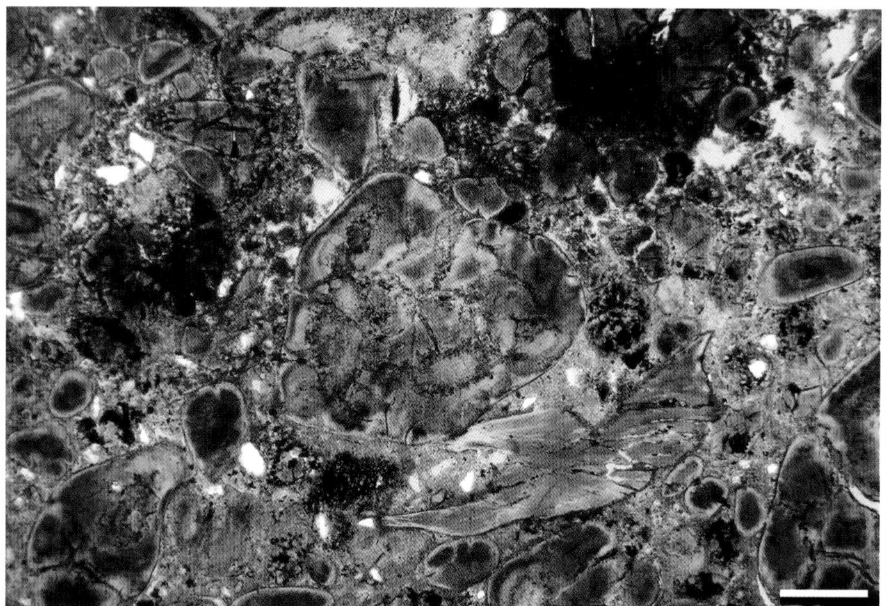

Up. Cretaceous Manesquan Fm., Monmouth Co., New Jersey

When a rock is glauconite rich, it is called a greensand (or glaucony facies). This is an altered greensand with minor quartz sand. Alteration of the glauconite grains is pervasive throughout the slide and is seen as cracking and brownish discoloration. The alteration involves loss of Fe and K and enrichment or concentration of Al (Sánchez-Navas et al., 2008).

PPL, BDI, Scale bar = 0.38 mm

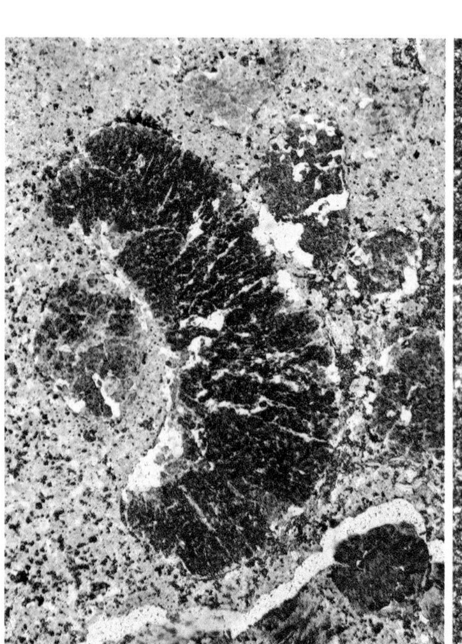
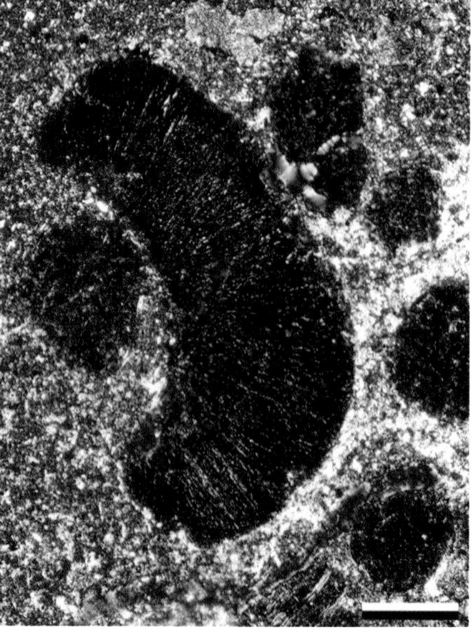

Up. Cretaceous (Maastrichtian) Tinton Fm., Monmouth Co., New Jersey

Brown vermicular glauconite. Brown glauconites generally are Al-rich and Fe- and Mg-poor. They can be found either as early stage products of seafloor alteration of precursor aluminous clays or they can represent the products of outcrop weathering of glauconites (Courbe, 1981), in which the formation process is essentially reversed. Vermicular forms give a better indication of the true birefringence of glauconite, because the finely crystalline nature of pelletal glauconites yields an anomalous green birefringence.

PPL | XPL, Scale bar = 0.26 mm

Oligocene – Miocene Otekaike Fm., Canterbury, New Zealand

An example of a greensand that has not undergone significant burial. In this relatively uncompacted sample, glauconite grains have remained undeformed and have point contacts. Their uniform size and shaped indicates that they are mainly pelletal in origin, although several grains near the bottom of the photomicrograph are platy and may represent altered clays or micas—possibly glauconite-replaced biotite.

PPL | XPL, AFeS, BDI, Scale bar = 0.26 mm

Mid. Cambrian Riley Fm., Lion Mountain Mbr., Llano Co., Texas

Glauconite grains are structurally weak and, with overburden loading or shear, the grains deform plastically. Deformation can occur even at shallow burial depths, but with greater burial, the grain contacts become increasingly fused, fitted and linear, as in this greensand. Grain boundaries are still clearly visible (because they trap thin, ferruginous early-formed cements), but the soft glauconite grains were flattened essentially into a single large mass. Later load release during uplift (or desiccation of the sample) allowed development of fractures (filled here with blue dyed epoxy).

PPL, BDI, Scale bar = 0.26 mm

Lo. Cretaceous (Albian) Nahr Umr Fm., offshore Qatar

Chamositic ooids in a glauconitic shallow-shelf sandstone. The term "chamositic" is used for grains, such as these ooids, with an uncertain composition of berthierine (an Fe-rich serpentine) and/or chamosite (an Fe-rich chlorite). Berthierine typically is the initial mineral in marine ooids, but layers can be selectively replaced by phosphate, siderite or chamosite during diagenesis. These ooids probably have altered to chamosite during burial. Chamositic ooids commonly form in inner shelf environments at times of slow sedimentation, moderate wave or current activity, and mildly reducing pore fluid conditions.

PPL | XPL, AFeS, BDI, Scale bar = 0.26 mm

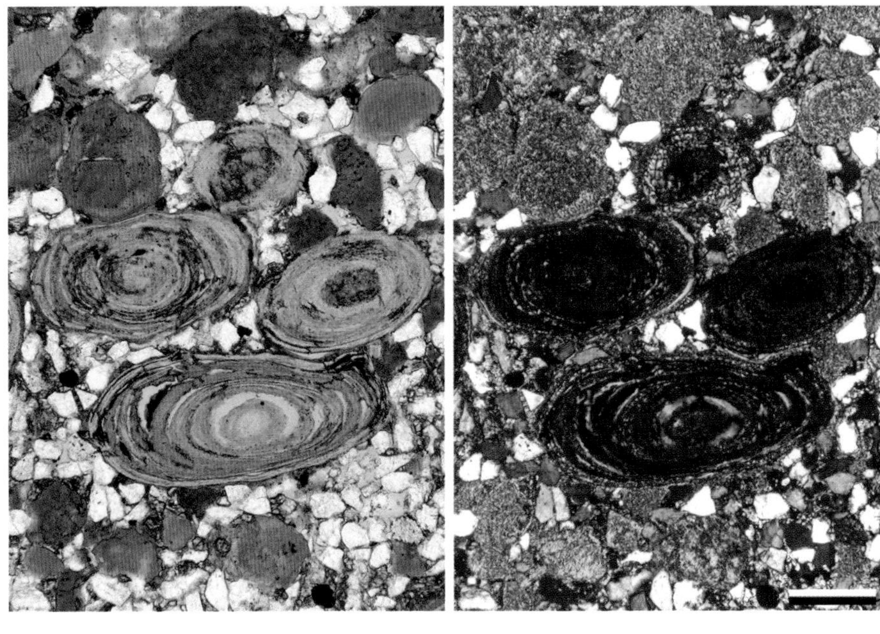

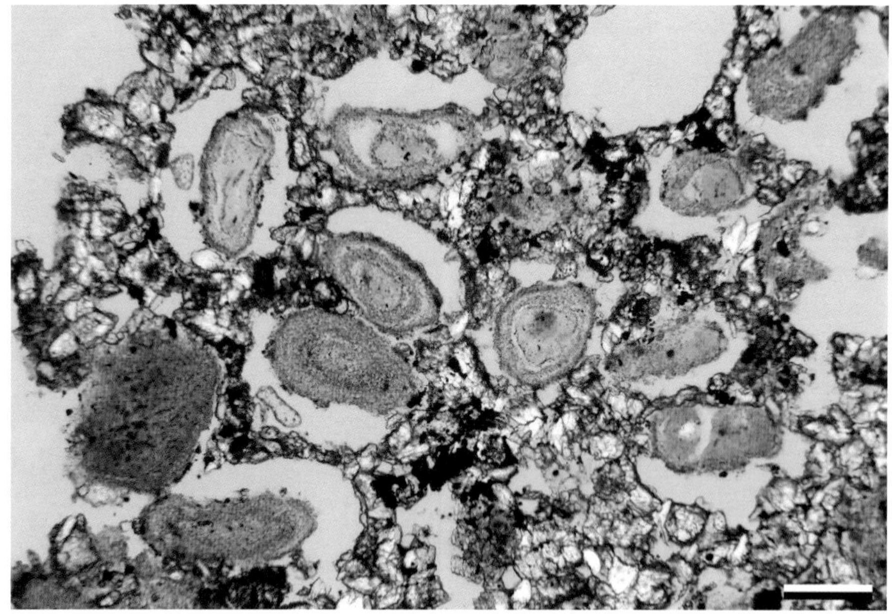

Lo. Cretaceous (Albian) Nahr Umr Fm., offshore Qatar

A siderite-cemented chamositic oolite (the siderite consists of the light-colored, lozenge-shaped crystals). Later alteration of the ooids created deformed grains and secondary porosity. "Chamosite" ooids can have complex mineralogies, consisting of chamosite, berthierine, phosphate, siderite and iron oxides. Because such mineral variations can occur between individual cortical layers, chamositic ooids commonly show selective leaching, allowing the ooids to collapse into the bottom of resultant pores.

PPL, BDI, Scale bar = 0.16 mm

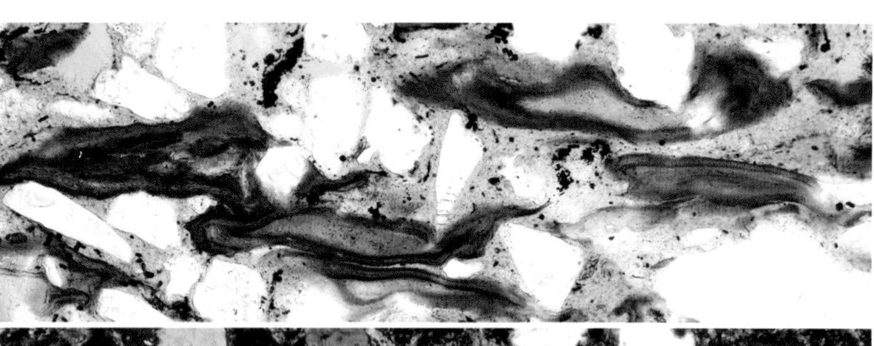

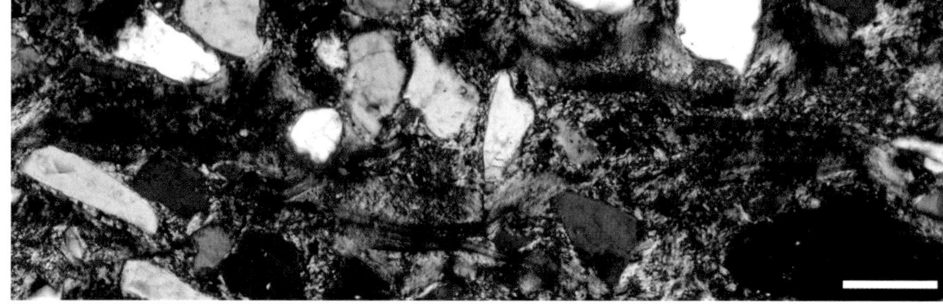

Lo. Cretaceous (Albian) Nahr Umr Fm., offshore Qatar

Squashed chamositic ooids. Because these ooids can be weakened by dissolution of interlayers, as noted above, and because they are relatively soft, they commonly occur as highly compacted, sheared or otherwise deformed grains. Such grains are termed "spastoliths" (Young, 1989). Here the surviving layers are probably no longer mainly composed of chamosite but instead contain a significant fraction of more highly birefringent (probably smectitic) clay minerals.

PPL / XPL, BDI, Scale bar = 0.1 mm

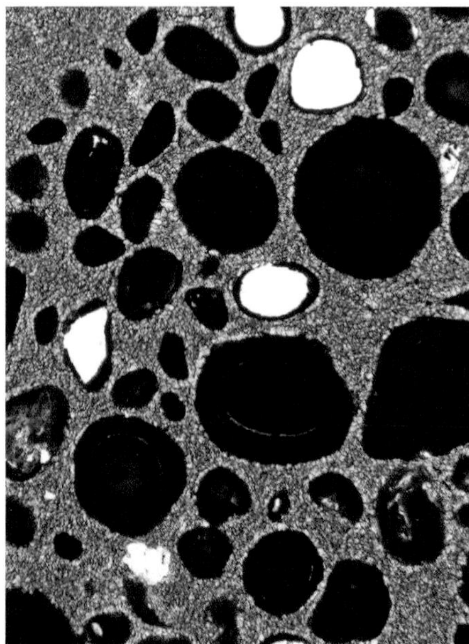

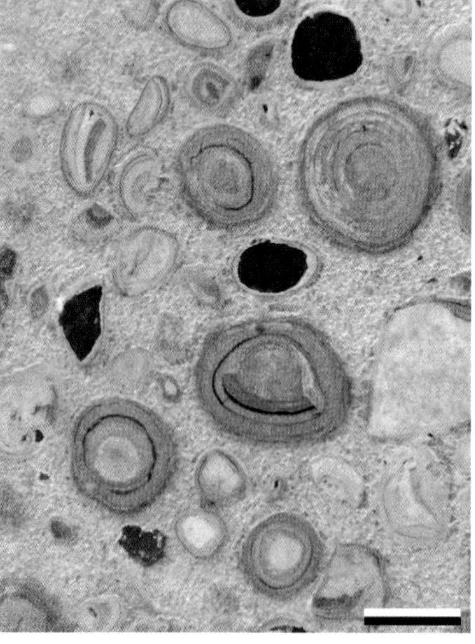

Lo. Cretaceous (Albian) Nahr Umr Fm., offshore Qatar

This example shows coexisting chamositic and iron ooids, or chamositic ooids that have been partially to completely replaced by iron oxides ("limonite"). The larger grains, although heavily stained in plane-polarized light, are chamosite. The iron staining obscures the original fabric of the ooids in plane-polarized light, but the cortices are clearly visible in reflected light. The ooids are "floating" in a dolomitized matrix (although they probably touch in 3-D). The white (PPL)/black (ORL) objects are grains dissolved or torn out during thin-section preparation, something one always must be aware of when doing petrography.

PPL | ORL, BDI, Scale bar = 0.26 mm

Mid. Jurassic (Dogger) Aalenian iron ore, near Besançon, France

Iron formations are found throughout the rock record. Here, ooids and ooid fragments composed of goethite (and minor "limonite") are visible along with goethite that infilled and partially replaced bioclasts. The goethite ooids in this sample are semiopaque, but in very thinly cut sections, or with the conoscopic condenser in place, the dark red to brown color characteristic of it is visible as well as the ooid coatings. Goethite ooids are normally indicative of oxidizing shallow-marine environments with slow sedimentation rates (like phosphate, chamosite and glauconite), but they also form under lateritic conditions.

PPL, Scale bar = 0.09 mm

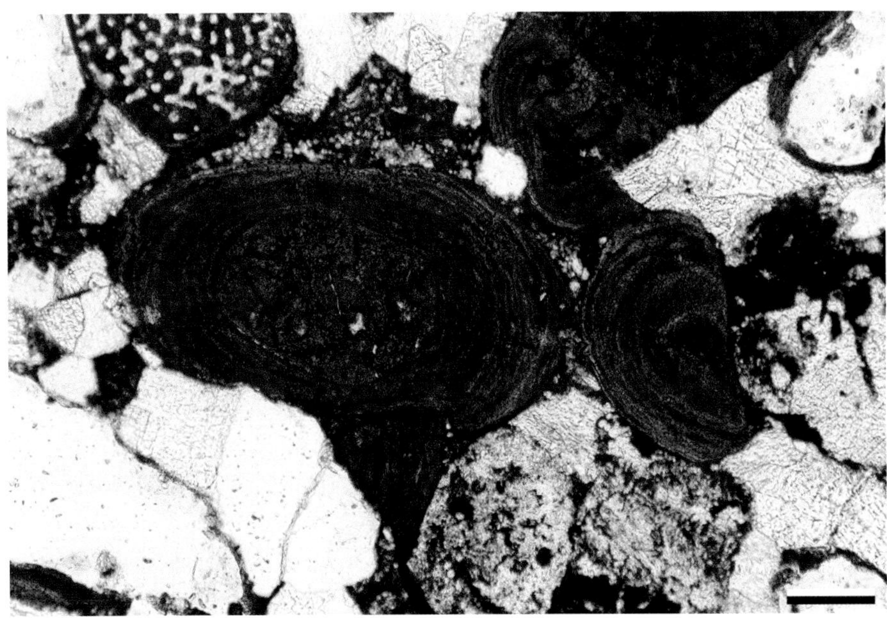

Jurassic Barrow Sandstone, North Slope, Alaska

Because iron minerals generally are opaque and nonbirefringent, often the best way to view them is using reflected light. In this sample, the dark rod-shaped grains in plane-polarized light are clearly seen to be ooids (coated grains) in oblique-reflected light. These "limonite" (iron oxyhydroxide) ooids are set in a red hematite-stained matrix.

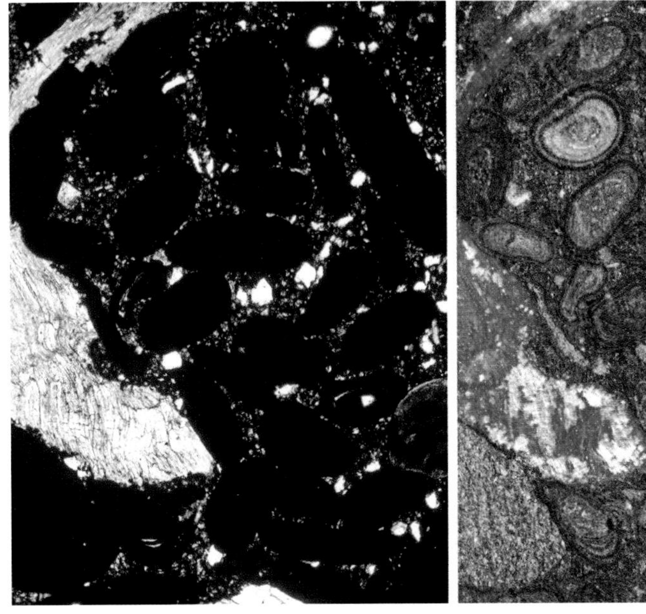

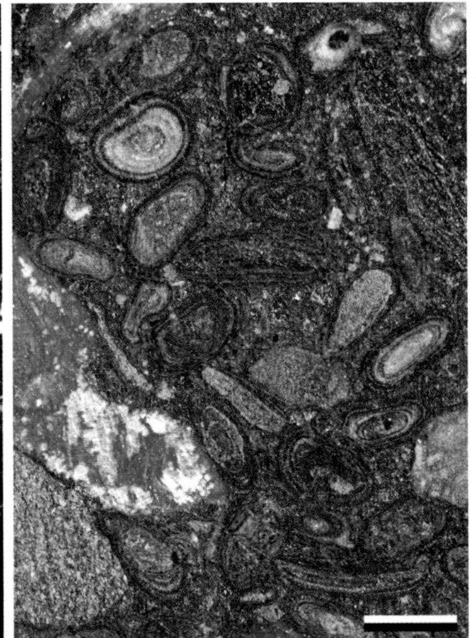

PPL | ORL, Scale bar = 0.26 mm

Mid. Jurassic Dogger Eisenoolith, Bavaria, Germany

"Limonite"/goethite ooids illuminated with the conoscopic condenser in place—this produces a very intense transmitted light beam that shows the brownish-yellow color of these slightly weathered iron oxyhydroxides (nearly opaque under normal illumination). These are inferred to have been originally carbonate ooids that were replaced by various iron minerals. Two major periods of oolitic ironstone formation are recognized: Early Ordovician to Devonian and Early Jurassic to mid-Cenozoic (Van Houten and Bhattacharyya, 1982). These were generally times of high sea-level stands with reduced terrigenous sediment influx.

PPL, Scale bar = 0.07 mm

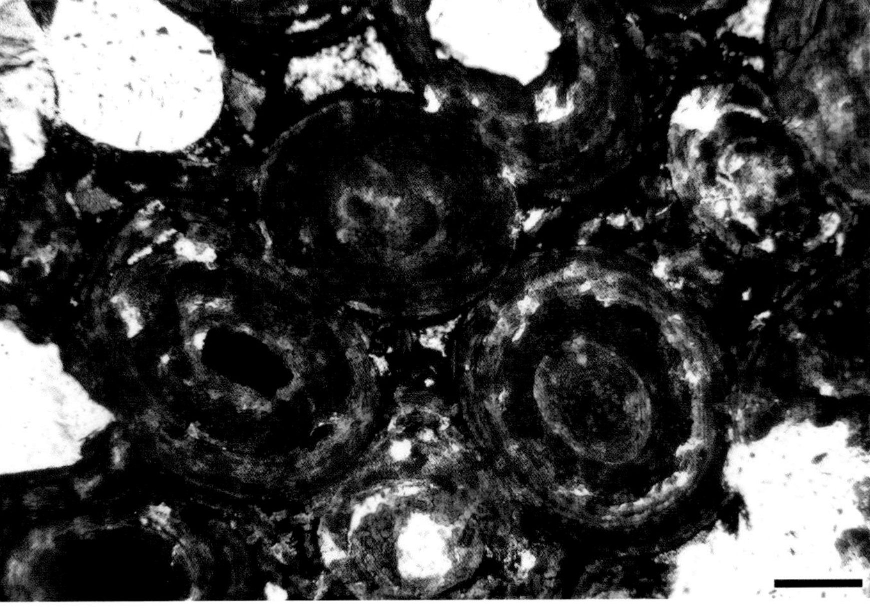

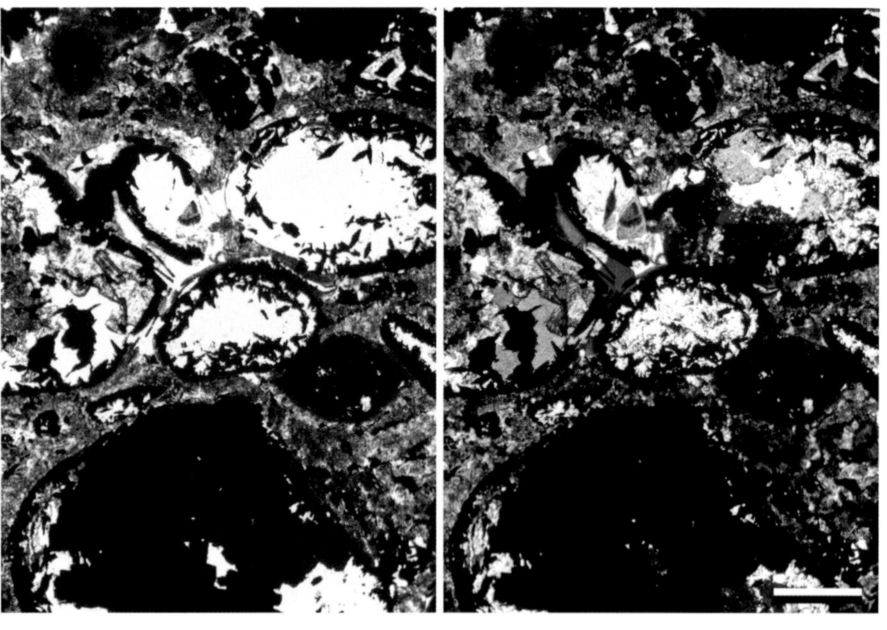

Paleoproterozoic Sokoman Fm., Schefferville, Quebec, Canada

In this chert and hematite grainstone, the chert grains have hematite rims, are recrystallized to blocky quartz and are variably replaced by Fe-rich carbonate. Hematitic grains also are recrystallized and have hematite rims. Iron-rich carbonate cements are the dull brown areas between the grains. Photomicrograph from Peir K. Pufahl.

PPL | XPL, Scale bar = 0.24 mm

Precambrian unknown unit, Canada

Precambrian banded iron formations generally consist of millimeter- to centimeter-thick bands of iron oxides (magnetite or hematite) repeatedly interspersed with layers of quartz, chert or shale as can be seen in this photomicrograph. The inferred environments and processes for banded iron formation are remarkably diverse (from deep-marine to nonmarine and most environments in between). Thus, the topic is far too complex to cover here and the reader is referred to references at the end of this chapter.

PPL, Scale bar = 0.38 mm

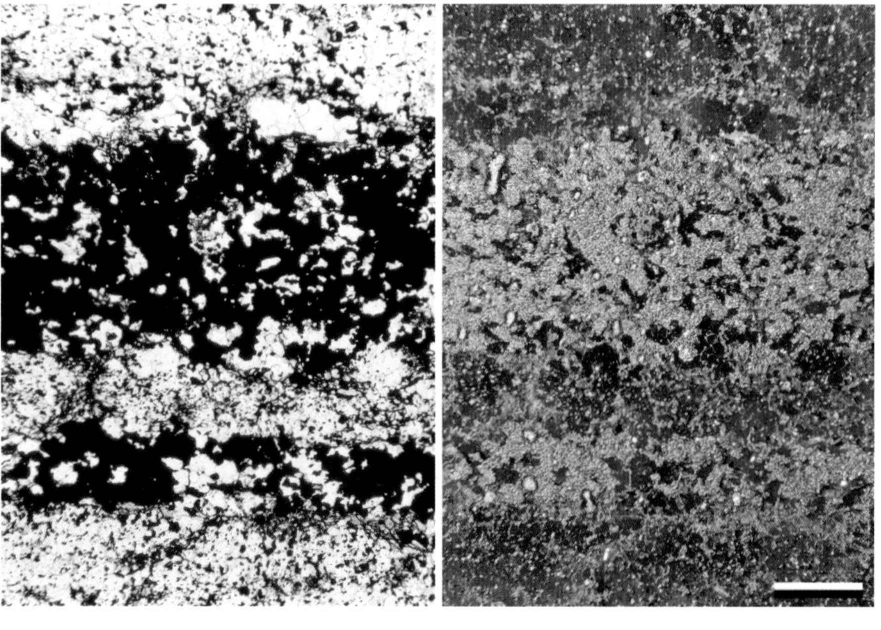

Precambrian (earliest Paleoproterozoic) main iron unit, Fremont Co., Wyoming

The predominant iron mineral in banded iron formations commonly is magnetite. In the oblique reflected light view, the silver gray reflectance of magnetite is visible as well as the oxidized iron staining within the sandy laminae. The yellow oxide staining is "limonite" (iron oxyhydroxides), and the red oxide staining is hematite.

PPL | ORL, Scale bar = 0.51 mm

Neogene Yellowstone Tuff, Park Co., Wyoming

Where tuffaceous/pyroclastic material is deposited while hot, as in ignimbrites, the glassy shards can fuse together to form a welded tuff. Such deposits commonly exhibit a "stratigraphy" of sorts, consisting of a thin, quenched zone at the base, overlain by a glassy core zone followed by progressively less welding towards the top of the unit (Smith, 1960). During welding, the heated fragments can behave plastically, causing the shards to compress and flow rheomorphically, distorting the layering, as in this example.

PPL, KFS, Scale bar = 0.34 mm

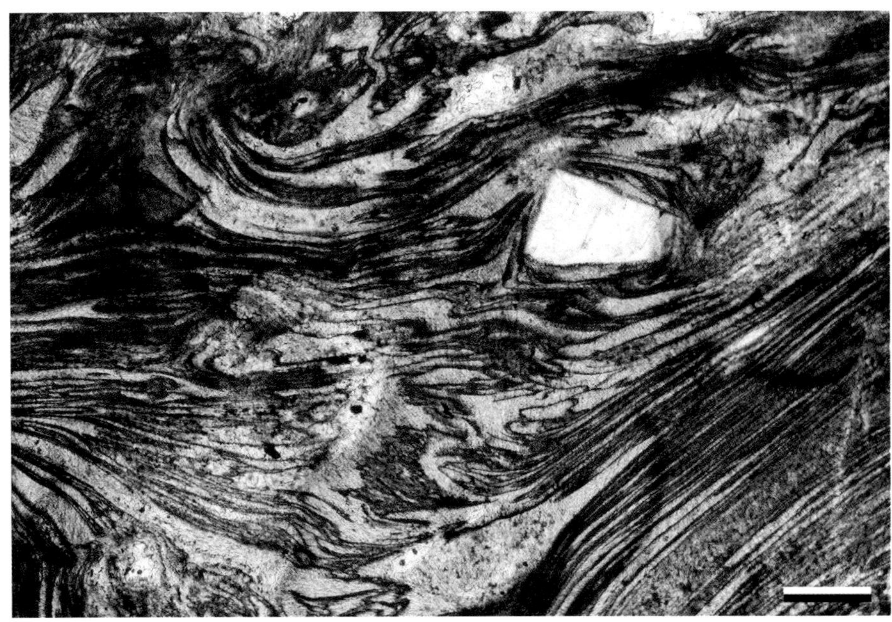

Neogene Yellowstone Tuff, Park Co., Wyoming

This flow-banded welded tuff contains embayed volcanic quartz crystals and numerous spherulitic aggregates. Note how the flow banding is compressed and contorted around the embayed volcanic quartz crystals. Within the welded tuff, there are numerous spherulitic aggregates that formed during devitrification of the glass, possibly while the material was still hot. These spherules are typically made of cristobalite and feldspar.

PPL, KFS, Scale bar = 0.38 mm

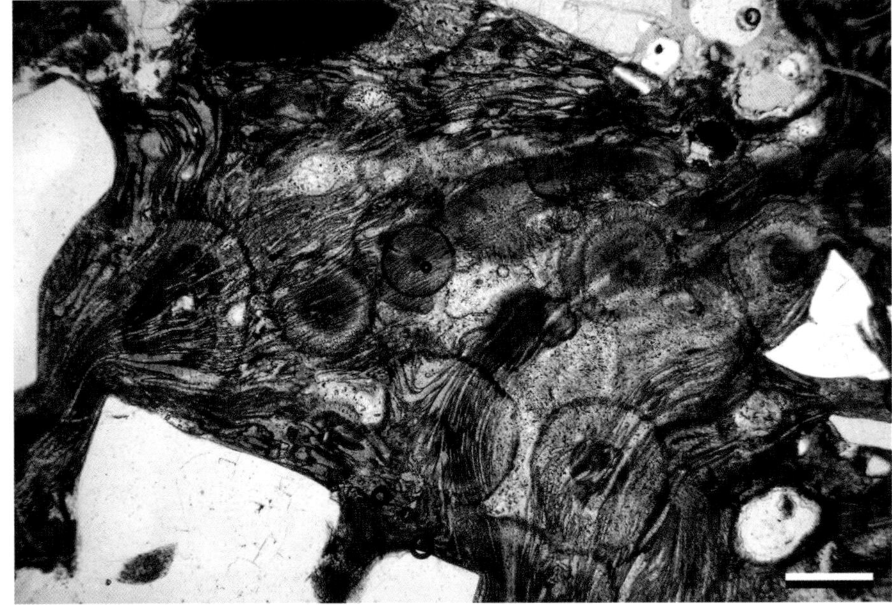

Cenozoic ignimbrite, Nye Co., Nevada

A tuff with partially welded glass shards showing flow fabric. This fabric is a result of deposition by a pyroclastic flow in which tuffaceous material was deposited while still hot; the tuff was then compressed by rapidly added overburden, causing the glassy shards to partially fuse together.

PPL / XPL, Scale bar = 0.34 mm

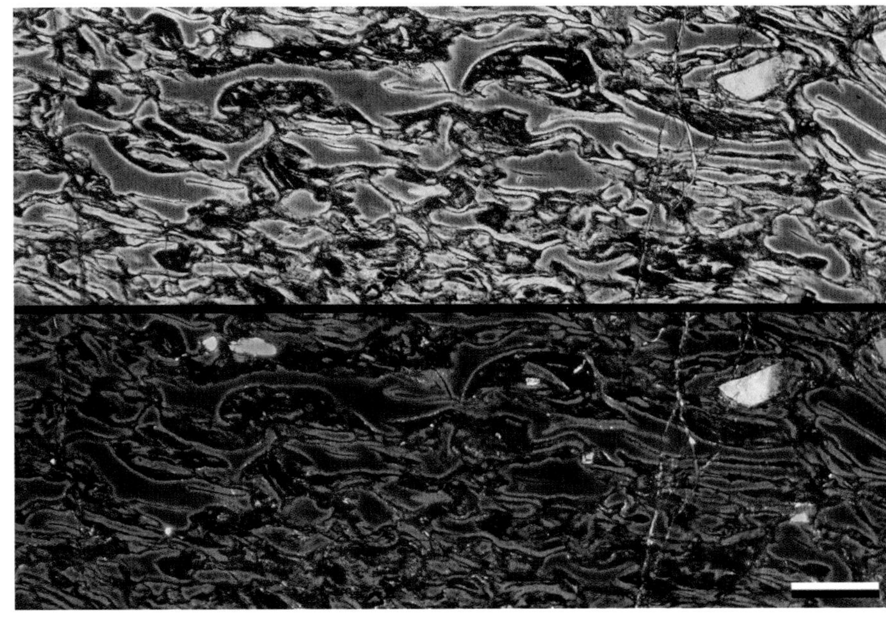

Precambrian Malani Igneous Suite, Rajasthan, India

Tuffs (rocks composed of ash-sized pyroclastic material) can look superficially like volcaniclastic sandstones, since there commonly is some sorting through ash fall and ash flow. The key distinguishing factors are the presence of compacted pumice clasts (although these also can occur through burial compaction), veining and genetically-related volcanic clasts (here the clasts are acidic glassy lava, monocrystalline angular quartz and pumice). This example would be classed as a weakly-welded lithic and vitric tuff.

PPL, KFS, AFeS, Scale bar = 0.49 mm

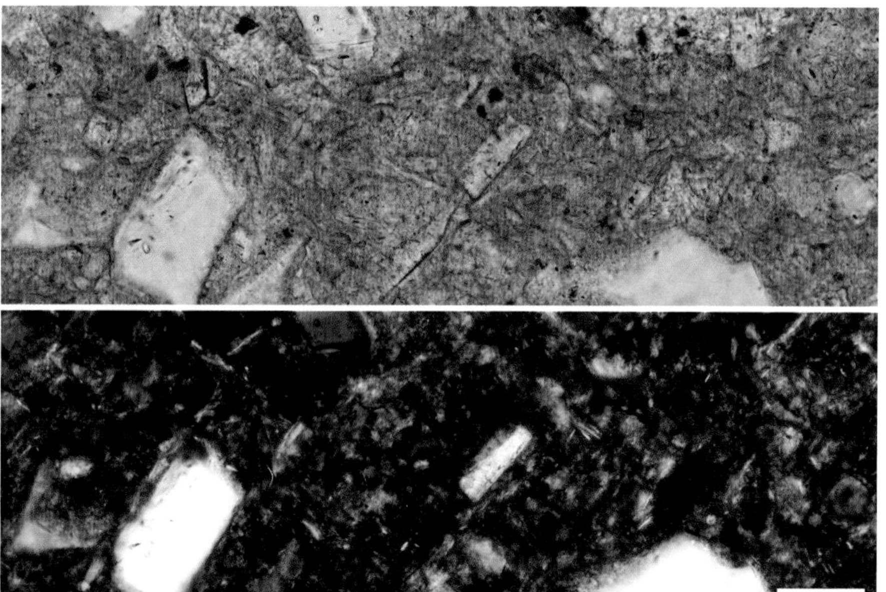

Mid.–Up. Eocene Wagon Bed Fm., Wind River Basin, Wyoming

Tuffs are commonly interbedded within clastic terrigenous sections. Here is an example of a bentonite/crystal-rich tuff with silt- to sand-size fragments of volcanic quartz and feldspars in a diagenetic chlorite matrix. In the plane-polarized light view, there are faint Y-shaped and curved forms present—ghosts of altered glass shards. These deposits were probably sourced from the Absaroka volcanic field.

PPL / XPL, Scale bar = 0.05 mm

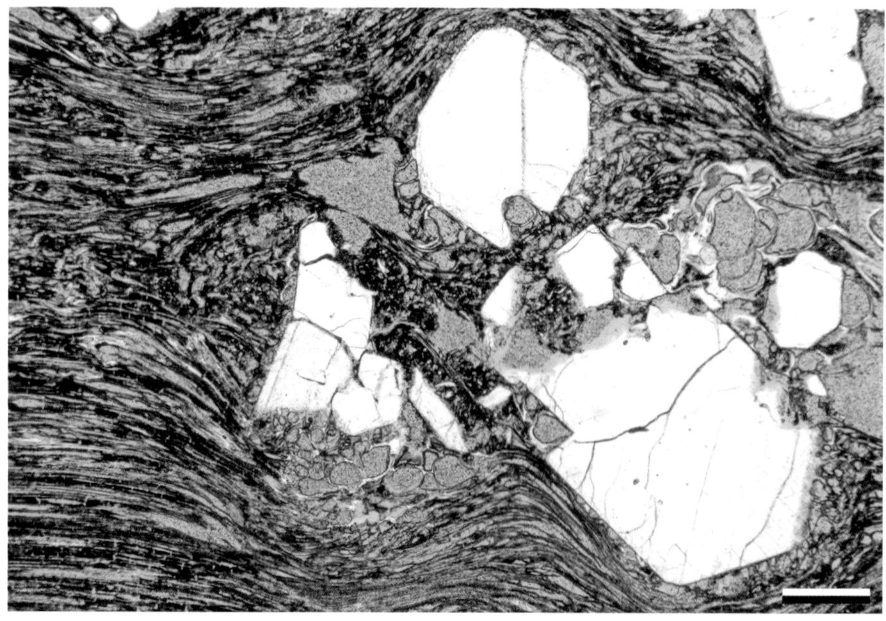

Pleistocene San Antonio Pumice, Socorro Co., New Mexico

Volcanic quartz phenocrysts within a debris-flow-transported pumice fragment associated with the 1.6 Ma Jemez (Valles Caldera) eruption (Cather, 1988). Euhedral terminations and embayments are typical for volcanic quartz as shown in this example. The vesicles within the pumice fragment have flattened and deformed around the quartz crystals. The pumice grains here survived long distance transport (over 200 km) only because they were cushioned within debris flows. Otherwise, fluvial or marine transport would quickly abrade pumice fragments, leaving only the quartz crystals plus some shards.

PPL, BDI, Scale bar = 0.51 mm

Oligocene – Miocene Horse Spring Fm., Clark Co., Nevada

Abundant, only slightly altered volcanic glass shards here are cemented by poikilotopic calcite. Normally, shards alter quickly, however, being encased in the calcite cements may have protected these shards from significant alteration. In addition to the shards, this sample also contains clasts (crystals) of biotite, plagioclase and hornblende. Glass shards can be difficult to see in thin section, because just like the thin section glass, they are isotropic. If the thin section is impregnated with colored epoxy, shard molds are easily distinguished from intact shards. Otherwise, inclusions and variations in coloration may help.

PPL | XPL, Scale bar = 0.26 mm

Oligocene – Miocene Horse Spring Fm., Clark Co., Nevada

Abundant, unaltered volcanic glass shards cemented by poikilotopic calcite. Because they are amorphous and noncrystalline, the shards are isotropic and appear black under cross-polarized light Volcanic glasses vary in composition, but typically are about 70% silica; they also contain iron, magnesium and other cations. Large biotite phenocrysts (example at yellow arrow) and faintly zoned plagioclase phenocrysts (example at red arrow) also are visible.

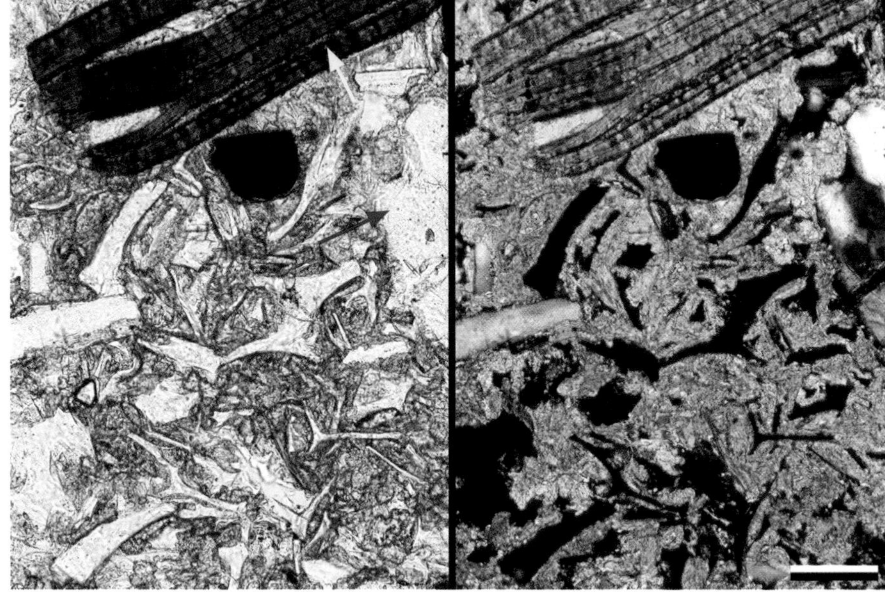

PPL | XPL, Scale bar = 0.10 mm

Pleistocene San Antonio Pumice, Socorro Co., New Mexico

Glass shards are fragments of chilled volcanic glass walls surrounding gas bubbles that form during explosive eruptions, commonly of felsic magmas. This is a shard-rich layer within a pumice unit. The shards formed as gases within the magma quickly expanded during eruptive degassing or by pumice abrasion. As the gas bubbles expand, they break into arcuate fragments or shards. Depending on the size of fragments, they can be carried by wind or water great distances from their source. Glass shards are hydrophyllic and are therefore unstable in near surface environments. During diagenesis, shards are devitrified, replaced or dissolved.

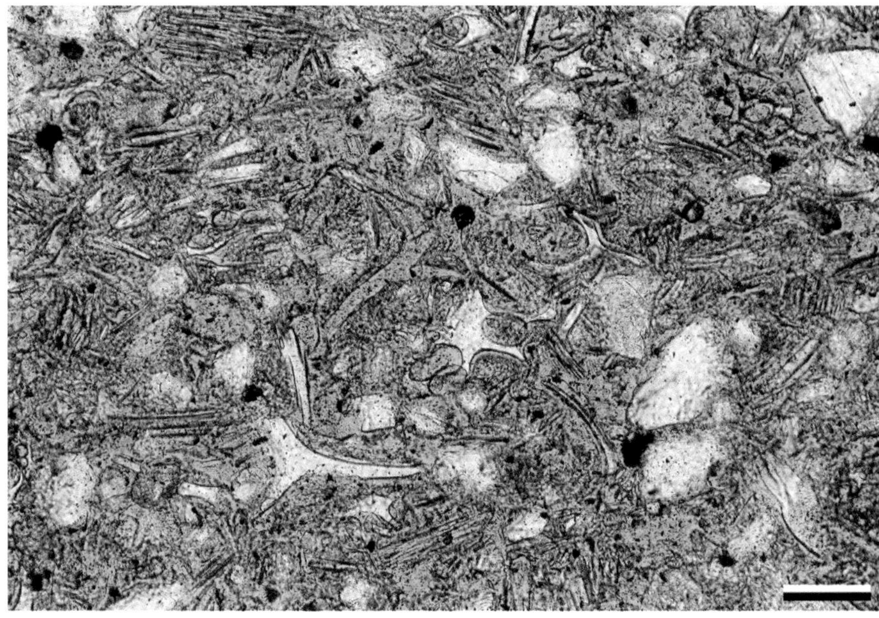

PPL, BDI, Scale bar = 0.10 mm

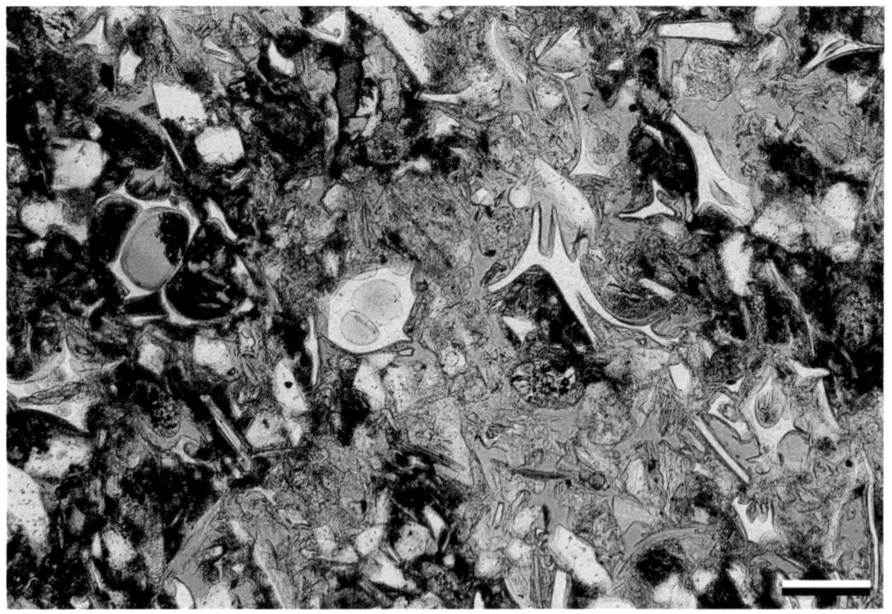

Pleistocene San Antonio Pumice, Socorro Co., New Mexico

The volcanic glass shards in this example have started to undergo alteration by hydrating and partially dissolving leaving Y-shaped and arcuate molds visible throughout this rock and recognizable only where they had thin clay or carbonate cement coatings. Because of their silica content, dissolution of shards is a source of silica during diagenesis.

PPL, BDI, Scale bar = 0.14 mm

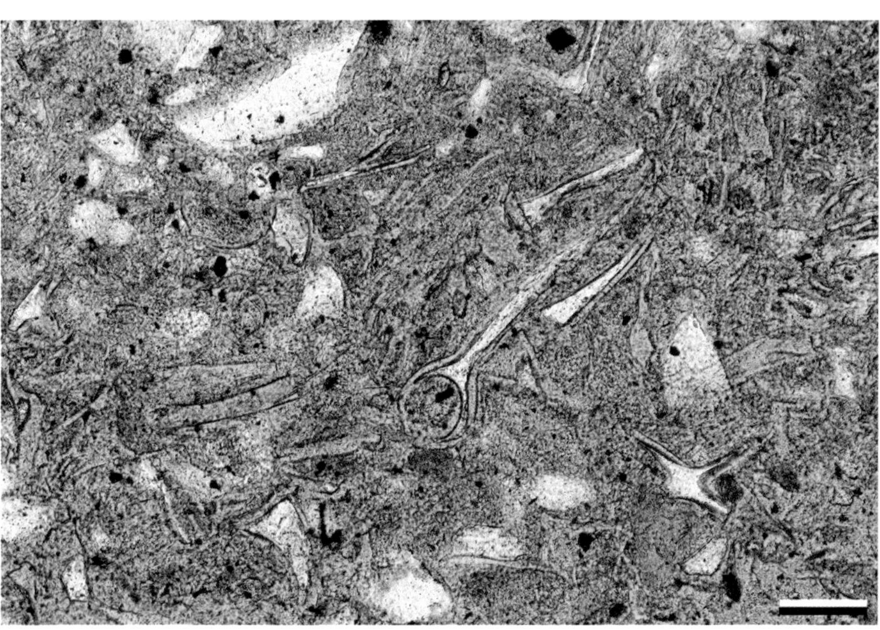

Pleistocene San Antonio Pumice, Socorro Co., New Mexico

These glass shards are in an intermediate stage of the devitrification process. Most of glass shards have either been dissolved (forming arcuate pores) or they have been hydrated. Because glass shards are unstable in near-surface environments, the longer these sediments remain unburied, the more alteration they will exhibit.

PPL, BDI, Scale bar = 0.10 mm

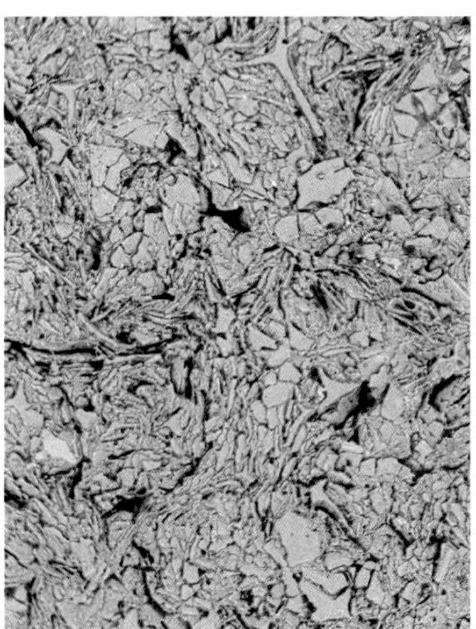

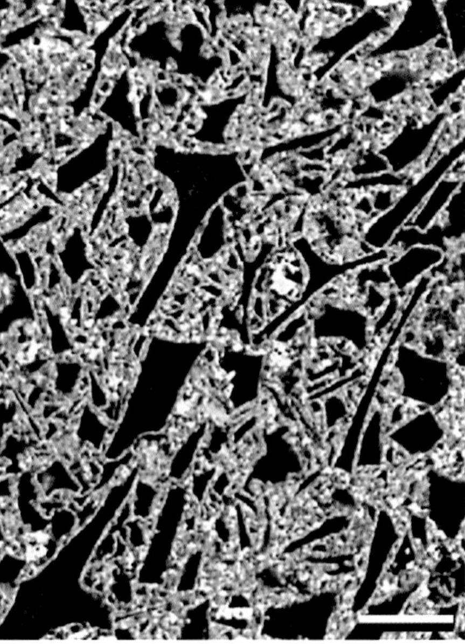

Oligocene – Miocene Horse Spring Fm., Lake Mead, Nevada *and* Eocene Huambos Fm., Andes, Peru

In thin section, ash fabrics can be difficult to see because of later alteration of the glass shards to zeolites and other minerals. Other techniques, like BSE, show details of compositional and textural variations within the sample not visible in thin section. In both examples, numerous glass shards, with their diagnostic Y-shape, are readily visible. In the Horse Spring Fm., glass shards are relatively unaltered but are compacted (see other examples); in the Peruvian sample, the glass shards have all been leached, creating distinctively-shaped pores with alteration rims. Images from Nelia Dunbar.

BSE(MP), Scale bars = 234 μm | 27.3 μm

Mid. Eocene – Lo. Oligocene Spears Gp., Socorro Co., New Mexico

This is an example in which the non-crystalline volcanic glass shards have undergone devitrification. Devitrification is the alteration of a non-crystalline (glassy) material to a crystalline mineral phase. Since these glasses are dominated by SiO_2, various forms of quartz (cristobalite, micro-quartz, chert or chalcedony) are the common replacement minerals. Despite the alteration, the shard shapes are well preserved and easy to identify here. The zoning within the large glass shards indicates multiple generations of quartz alteration or void filling.

PPL | XPL, Scale bar = 0.10 mm

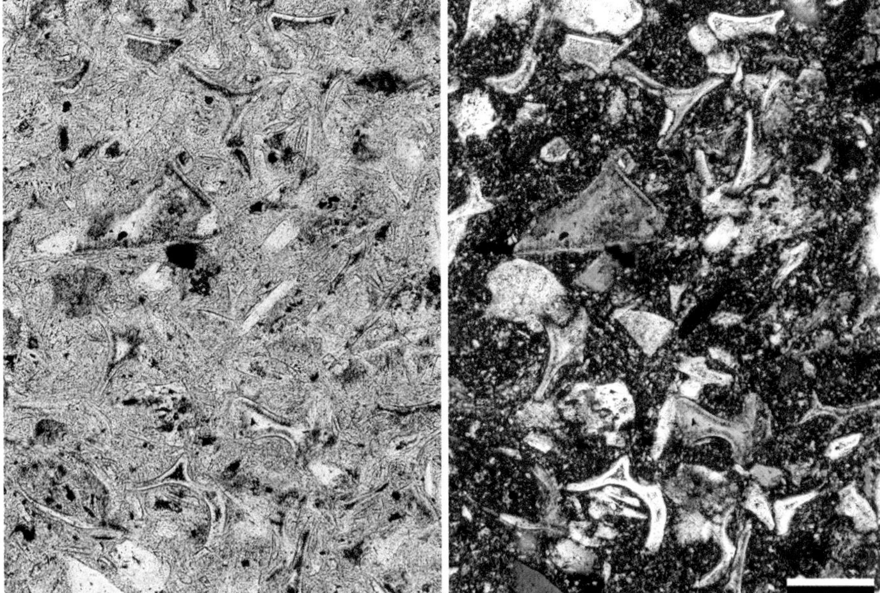

Mid. Eocene – Lo. Oligocene Spears Gp., Socorro Co., New Mexico

A close-up of a glass shard that has been devitrified. The shard filling is zoned, indicating that multiple generations of quartz cement have filled open voids rather than directly replacing the shards. In other words, the shard dissolved, producing an open pore that later was filled by silica cement (probably sourced from the dissolution of other shards).

PPL, Scale bar = 0.10 mm

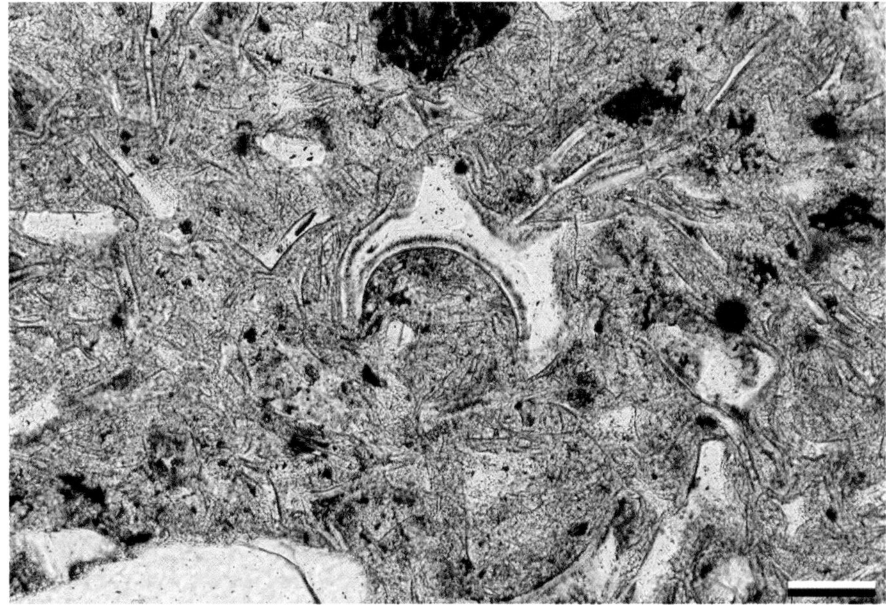

Mid. Eocene – Lo. Oligocene Up. Spears Gp., Rock House Canyon Tuff, Socorro Co., New Mexico

Another devitrified welded tuff containing glass shards. The radiating spherulites, formed after glass shards, are composed of fibrous quartz (chalcedony). The groundmass is mainly a combination of microcrystalline quartz and feldspar. The glass shards and bubbles show little evidence of compaction prior to welding, but the devitrification process has destroyed some of the shard structure.

PPL | XPL, Scale bar = 0.26 mm

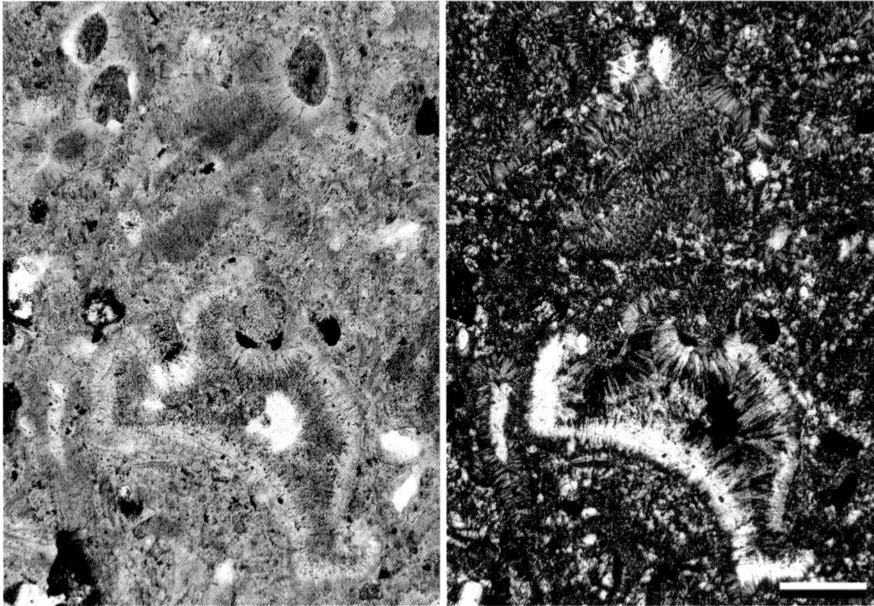

Cited References and Additional Information Sources

Limestones:

Flügel, E., 2004, Microfacies of Carbonate Rocks: Analysis, Interpretation and Application: New York, Springer, 976 p.

Scholle, P. A., and D. S. Ulmer-Scholle, 2003, A Color Guide to the Petrography of Carbonate Rocks: Grains, textures, porosity, diagenesis: Tulsa, OK, AAPG Memoir 77, 474 p.

Scoffin, T. P., 1987, An Introduction to Carbonate Sediments and Rocks: New York, Chapman & Hall, 274 p.

Tucker, M. E., 1991, Carbonate Petrology: An Introduction: Oxford, Blackwell Scientific Publications, 272 p.

Bedded cherts and siliceous shales:

Baumgartner, P. O., 2013, Mesozoic radiolarites–accumulation as a function of sea surface fertility on Tethyan margins and in ocean basins: Sedimentology, v. 60, p. 292-318, doi: 10.1111/sed.12022.

Behl, R. J., and R. E. Garrison, 1994, The origin of chert in the Monterey Formation of California (USA), *in* A. Iijima, A. M. Abed, and R. E. Garrison, eds., Siliceous, Phosphatic and Glauconitic Sediments of the Tertiary and Mesozoic: (Proceedings of the 29th International Geological Congress, Part C): Utrecht, Netherlands, VSP, p. 101-132.

De Wever, P., P. Dumitrica, J. P. Caulet, C. Nigrini, and M. Caridroit, 2002, Radiolarians in the Sedimentary Record: Amsterdam, CRC Press, 533 p.

Eugster, H. P., 1969, Inorganic bedded cherts from the Magadi area, Kenya: Contributions to Mineralogy and Petrology, v. 22, p. 1-31, doi: 10.1007/BF00388011.

Folk, R. L., and C. E. Weaver, 1952, A study of the texture and composition of chert: American Journal of Science, v. 250, p. 498-510, doi: 10.2475/ajs.250.7.498.

Hein, J. R., and J. T. Parrish, 1987, Distribution of siliceous deposits in space and time, *in* J. R. Hein, ed., Siliceous Sedimentary Rock-hosted Ores and Petroleum: New York, Van Nostrand Reinhold, p. 10-57.

Hesse, R., 1989, Silica diagenesis: Origin of inorganic and replacement cherts: Earth-Science Reviews, v. 26, p. 253-284, doi: 10.1016/0012-8252(89)90024-X.

Keller, M. A., and C. M. Isaacs, 1985, An evaluation of temperature scales for silica diagenesis in diatomaceous sequences including a new approach based on the Miocene Monterey Formation, California: Geo-Marine Letters, v. 5, p. 31-35, doi: 10.1007/BF02629794.

Maliva, R. G., A. H. Knoll, and R. Siever, 1989, Secular change in chert distribution: A reflection of evolving biological participation in the silica cycle: Palaios, v. 4, p. 519-532, doi: 10.2307/3514743.

Matheney, R. K., and L. P. Knauth, 1993, New isotopic temperature estimates for early silica diagenesis in bedded cherts: Geology, v. 21, p. 519-522, doi: 10.1130/0091-7613(1993)021<0519:NITEFE>2.3.CO;2.

McBride, E. F., and A. Thomson, 1970, The Caballos Novaculite, Marathon Region: Boulder, CO, GSA Special Paper 122, 129 p., doi: 10.1130/SPE122-p1.

Milliken, K. L., S.-J. Choh, P. Papazis, and J. Schieber, 2007, "Cherty" stringers in the Barnett Shale are agglutinated foraminifera: Sedimentary Geology, v. 198, p. 221-232, doi: 10.1016/j.sedgeo.2006.12.012.

Minoura, K., T. Susaki, and K. Horiuchi, 1996, Lithification of biogenic siliceous sediments: Evidence from Neogene diatomaceous sequences of northeast Japan: Sedimentary Geology, v. 107, p. 45-59, doi: 10.1016/S0037-0738(96)00017-6.

Murray, R. W., D. L. Jones, and M. R. Buchholtz ten Brink, 1992, Diagenetic formation of bedded chert: Evidence from chemistry of the chert–shale couplet: Geology, v. 20, p. 271-274, doi: 10.1130/0091-7613(1992)020<0271:DFOBCE>2.3.CO;2.

Schubel, K. A., and B. M. Simonson, 1990, Petrography and diagenesis of cherts from Lake Magadi, Kenya: Journal of Sedimentary Research, v. 60, p. 761-776, doi: 10.1306/212F9269-2B24-11D7-8648000102C1865D.

Smol, J. P., and E. F. Stoermer, 2010, The Diatoms: Applications for the Environmental and Earth Sciences (2nd Edition): New York, Cambridge University Press, 667 p.

Stefurak, E. J. T., D. R. Lowe, D. Zentner, and W. W. Fischer, 2014, Primary silica granules—A new mode of Paleoarchean sedimentation: Geology, v. 42, p. 283-286, doi: 10.1130/G35187.1.

Phosphatic grains and phosphorites:

Baird, G. C., 1978, Pebbly phosphorites in shale: A key to recognition of a widespread discontinuity in the Middle Devonian of New York: Journal of Sedimentary Research, v. 48, p. 545-555, doi: 10.1306/212F74CD-2B24-11D7-8648000102C1865D.

Baturin, G. N., 1982, Phosphorites on the Sea Floor. Origin, Composition, and Distribution: New York, Elsevier, 343 p.

Bitter, M. R., 1986, Origin of phosphatic nodules and cements in the Upper Cretaceous Frontier Formation, Natrona County, Wyoming: The Mountain Geologist, v. 23, p. 45-50.

Boskey, A. L., 2007, Mineralization of bones and teeth: Elements, v. 3, p. 385-391.

Carter, R. M., J. K. Lindqvist, and R. J. Norris, 1982, Oligocene unconformities and nodular phosphate—hardground horizons in western Southland and northern West Coast: Journal of the Royal Society of New Zealand, v. 12, p. 11-46, doi: 10.1080/03036758.1982.10427164.

Glenn, C. R., K. B. Follmi, S. R. Riggs, G. N. Baturin, K. A. Grimm, J. Trappe, A. M. Abed, C. Galliolivier, R. E. Garrison, and A. V. Ilyin, 1994, Phosphorus and phosphorites—sedimentology and environments of formation: Eclogae Geologicae Helvetiae, v. 87, p. 747-788.

Graber, K. K., and H. S. Chafetz, 1990, Petrography and origin of bedded barite and phosphate in the Devonian Slaven Chert of central Nevada: Journal of Sedimentary Research, v. 60, p. 897-911, doi: 10.1306/D4267640-2B26-11D7-8648000102C1865D.

Gulbrandsen, R. A., 1960, Petrology of the Meade Peak phosphatic shale member of the Phosphoria Formation at Coal Canyon, Wyoming, Washington, DC, USGS Bulletin 1111-C, p. 71-146.

Heckel, P. H., 1977, Origin of phosphatic black shale facies in Pennsylvanian cyclothems of mid-continent North America: AAPG Bulletin, v. 61, p. 1045-1068.

Herring, J. R., 1995, Permian phosphorites: A paradox of phosphogenesis, *in* P. A. Scholle, T. M. Peryt, and D. S. Ulmer-Scholle, eds., The Permian of Northern Pangea. Volume 2, Sedimentary Basins and Economic Resources: New York, Springer, p. 292-312, doi: 10.1007/978-3-642-78590-0_15.

Horton, A., H. C. Ivimey-Cook, R. K. Harrison, and B. R. Young, 1980, Phosphatic öoids in the Upper Lias (Lower Jurassic) of central England: Journal of the Geological Society, v. 137, p. 731-740, doi: 10.1144/gsjgs.137.6.0731.

Jarvis, I., W. C. Burnett, J. Nathan, F. S. M. Almbaydin, A. K. M. Attia, L. N. Castro, R. Flicoteau, M. E. Hilmy, V. Husein, A. A. Quitwanah, A. A. Serjani, and Y. Zanin, 1994, Phosphorite geochemistry: State-of-the-art and environmental concerns: Eclogae Geologicae Helvetiae, v. 87, p. 643-700.

Notholt, A. J. G., and I. Jarvis, eds., 1990, Phosphorite Research and Development: London, GSL Special Publication 52, 326 p.

Notholt, A. J. G., R. P. Sheldon, and D. F. D. Davidson, eds., 2005, Phosphate Deposits of the World: Volume 2, Phosphate rock resources: Cambridge, UK, Cambridge University Press, 600 p.

Pufahl, P. K., and K. A. Grimm, 2003, Coated phosphate grains: Proxy for physical, chemical, and ecological changes in seawater: Geology, v. 31, p. 801-804, doi: 10.1130/G19658.1.

Rehfeld, U., and A. W. Janssen, 1995, Development of phosphatized hardgrounds in the Miocene *Globigerina* Limestone of the Maltese Archipelago, including a description of *Gamopleura melitensis* sp. nov. (Gastropoda, Euthecosomata): Facies, v. 33, p. 91-106, doi: 10.1007/BF02537445.

Riggs, S. R., 1980, Intraclast and pellet phosphorite sedimentation in the Miocene of Florida: Journal of the Geological Society, v. 137, p. 741-748, doi: 10.1144/gsjgs.137.6.0741.

Ruffell, A. H., 1990, The mineralogy and petrography of the Sulphur Band phosphates (Aptian-Albian), at Folkestone, Kent: Proceedings of the Geologists' Association, v. 101, p. 79-84, doi: 10.1016/S0016-7878(08)80207-6.

Soudry, D., 1992, Primary bedded phosphorites in the Campanian Mishash Formation, Negev, southern Israel: Sedimentary Geology, v. 80, p. 77-88, doi: 10.1016/0037-0738(92)90033-N.

Soudry, D., and Y. Nathan, 1980, Phosphate peloids from the Negev phosphorites: Journal of the Geological Society, v. 137, p. 749-755, doi: 10.1144/gsjgs.137.6.0749.

Swett, K., and R. K. Crowder, 1982, Primary phosphatic oolites from the Lower Cambrian of Spitsbergen: Journal of Sedimentary Research, v. 52, p. 587-593, doi: 10.1306/212F7FA9-2B24-11D7-8648000102C1865D.

Organic-rich sediments and coals:

Cook, A. C., and N. R. Sherwood, 1991, Classification of oil shales, coals and other organic-rich rocks: Organic Geochemistry, v. 17, p. 211-222, doi: 10.1016/0146-6380(91)90079-Y.

Crelling, J. C., 1983, Current uses of fluorescence microscopy in coal petrology: Journal of Microscopy, v. 132, p. 251-266, doi: 10.1111/j.1365-2818.1983.tb04591.x.

Hutton, A. C., 1987, Petrographic classification of oil shales: International Journal of Coal Geology, v. 8, p. 203-231, http://dx.doi.org/10.1016/0166-5162(87)90032-2.

Hutton, A. C., 1995, Organic petrography of oil shales, *in* C. E. Snape, ed., Composition, Geochemistry and Conversion of Oil Shales (NATO ASI Series Volume 455): Dordrecht, Netherlands, Kluwer Academic Publishers, p. 17-33, doi: 10.1007/978-94-011-0317-6_2.

Scott, A. C., 2002, Coal petrology and the origin of coal macerals: A way ahead?: International Journal of Coal Geology, v. 50, p. 119-134, doi: 10.1016/S0166-5162(02)00116-7.

Taylor, G. H., and D. C. Glick, 1998, Organic Petrology: A new handbook incorporating some revised parts of Stach's Textbook of Coal Petrology: Berlin, Gebrüder Borntraeger, 704 p.

Evaporites:

Anderson, R. Y., W. E. Dean, Jr., D. W. Kirkland, and H. I. Snider, 1972, Permian Castile varved evaporite sequence, west Texas and New Mexico: GSA Bulletin, v. 83, p. 59-86, doi: 10.1130/0016-7606(1972)83[59:PCVESW]2.0.CO;2.

Barone, M., R. Dominici, and S. Lugli, 2007, Interpreting gypsarenites in the Rossano basin (Calabria, Italy): A contribution to the characterization of the Messinian salinity crisis in the Mediterranean, *in* J. Arribas, S. Critelli, and M. J. Johnsson, eds., Sedimentary Provenance and Petrogenesis: Perspectives from Petrography and Geochemistry: Boulder, CO, GSA Special Paper 420, p. 135-148, doi: 10.1130/2006.2420(09).

Borchert, T. H., and R. O. Muir, 1964, Salt Deposits: The Origin, Metamorphism, and Deformation of Evaporites: New York, D. van Nostrand, 338 p.

Briggs, L. I., Jr., 1960, Petrography of salt, *in* V. W. Kaufman, ed., Sodium Chloride: New York, Reinhold Publishing Co., p. 22-27.

Hardie, L. A., 1984, Evaporites: Marine or non-marine?: American Journal of Science, v. 284, p. 193-240, doi: 10.2475/ajs.284.3.193.

Hardie, L. A., T. K. Lowenstein, and R. J. Spencer, 1985, The problem in distinguishing between primary and secondary features in evaporites, *in* B. C. Schreiber, and H. L. Harner, eds., Sixth International Symposium on Salt, Vol. 1: Alexandria, VA, Salt Institute, p. 11-39.

Holliday, D. W., 1970, The petrology of secondary gypsum rocks: A review: Journal of Sedimentary Research, v. 40, p. 734-744, doi: 10.1306/74D7202C-2B21-11D7-8648000102C1865D.

Lowenstein, T. K., and R. J. Spencer, 1990, Syndepositional origin of potash evaporites: Petrographic and fluid inclusion evidence: American Journal of Science, v. 290, p. 1-42, doi: 10.2475/ajs.290.1.1.

Maiklem, W. R., D. G. Bebout, and R. P. Glaister, 1969, Classification of anhydrite—a practical approach: Bulletin of Canadian Petroleum Geology, v. 17, p. 194-233.

Schlager, W., and H. Bolz, 1977, Clastic accumulation of sulphate evaporites in deep water: Journal of Sedimentary Research, v. 47, p. 600-609, doi: 10.1306/212F71F3-2B24-11D7-8648000102C1865D.

Schreiber, B. C., G. M. Friedman, A. Decima, and E. Schreiber, 1976, Depositional environments of upper Miocene (Messinian) evaporite deposits of the Sicilian Basin: Sedimentology, v. 23, p. 729-760, doi: 10.1111/j.1365-3091.1976.tb00107.x.

Schreiber, B. C., ed., 1988, Evaporites and Hydrocarbons: New York, Columbia University Press, 475 p.

Schreiber, B. C., M. S. Roth, and M. L. Helman, 1982, Recognition of primary facies characteristics of evaporites and differentiation of the forms from diagenetic overprints, *in* C. R. Handford, R. G. Loucks, and G. R. Davies, eds., Depositional and Diagenetic Spectra of Evaporites - a core workshop (Calgary, 1982): Tulsa, OK, SEPM Core Workshop No. 3, p. 1-32, doi: 10.2110/cor.82.01.0001.

Spencer, R. J., 2000, Sulfate minerals in evaporite deposits: Reviews in Mineralogy and Geochemistry, v. 40, p. 173-192, doi: 10.2138/rmg.2000.40.3.

Warren, J. K., 2006, Interpreting evaporite texture, *in* J. K. Warren, Evaporites: Sediments, Resources and Hydrocarbons: Berlin, Springer, p. 1-57, doi: 10.1007/3-540-32344-9_1.

Greensands/glauconitic sands/glaucony facies:

Amireh, B. S., G. Jarrar, F. Henjes-Kunst, and W. Schneider, 1998, K-Ar dating, X-ray diffractometry, optical and scanning electron microscopy of glauconics from the early Cretaceous Kurnub Group of Jordan: Geological Journal, v. 33, p. 49-65, doi: 10.1002/(SICI)1099-1034(199801/03)33:1<49::AID-GJ759>3.0.CO;2-Y.

Amorosi, A., 1997, Detecting compositional, spatial, and temporal attributes of glaucony: A tool for provenance research: Sedimentary Geology, v. 109, p. 135-153, doi: 10.1016/S0037-0738(96)00042-5.

Amorosi, A., 2012, The occurrence of glaucony in the stratigraphic record: Distribution patterns and sequence-stratigraphic significance, *in* S. Morad, J. M. Ketzer, and L. F. De Ros, eds., Linking Diagenesis to Sequence Stratigraphy: West Sussex, UK, John Wiley & Sons, p. 37-53, doi: 10.1002/9781118485347.ch2.

Amouric, M., and C. Parron, 1985, Structure and growth mechanism of glauconite as seen by high resolution transmission electron microscopy: Clays and Clay Minerals, v. 33, p. 473-482, doi: 10.1346/CCMN.1985.0330601.

Chafetz, H. S., and A. Reid, 2000, Syndepositional shallow-water precipitation of glauconitic minerals: Sedimentary Geology, v. 136, p. 29-42, doi: 10.1016/S0037-0738(00)00082-8.

Courbe, C., B. Velde, and A. Meunier, 1981, Weathering of glauconites: Reversal of the glauconitization process in a soil profile in western France: Clay Minerals, v. 16, p. 231-243, doi: 10.1180/claymin.1981.016.3.02.

Huggett, J. M., and A. S. Gale, 1997, Petrology and palaeoenvironmental significance of glaucony in the Eocene succession at Whitecliff Bay, Hampshire Basin, UK: Journal of the Geological Society, v. 154, p. 897-912, doi: 10.1144/gsjgs.154.5.0897.

Ireland, B. J., C. D. Curtis, and J. A. Whiteman, 1983, Compositional variations within some glauconites and illites and implications for their stability and origins: Sedimentology, v. 30, p. 769-786, doi: 10.1111/j.1365-3091.1983.tb00710.x.

Kelly, J. C., and J. A. Webb, 1999, The genesis of glaucony in the Oligo–Miocene Torquay Group, southeastern Australia: Petrographic and geochemical evidence: Sedimentary Geology, v. 125, p. 99-114, doi: 10.1016/S0037-0738(98)00149-3.

McBride, E. F., 1988, Contrasting diagenetic histories of concretions and host rock, Lion Mountain Sandstone (Cambrian), Texas: GSA Bulletin, v. 100, p. 1803-1810, doi: 10.1130/0016-7606(1988)100<1803:CDHOCA>2.3.CO;2.

McConchie, D. M., and D. W. Lewis, 1980, Varieties of glauconite in late Cretaceous and early Tertiary rocks of the South Island of New Zealand, and new proposals for classification: New Zealand Journal of Geology and Geophysics, v. 23, p. 413-437, doi: 10.1080/00288306.1980.10424113.

Odin, G. S., ed., 1988, Green Marine Clays: Oolitic Ironstone Facies, Verdine Facies, Glaucony Facies and Celadonite-Bearing Rock Facies—A Comparative Study: New York, Elsevier Science Publ. Co., 445 p.

Odin, G. S., 1990, Clay mineral formation at the continent-ocean boundary: The verdine facies: Clay Minerals, v. 25, p. 477-483, doi: 10.1180/claymin.1990.025.4.06.

Odin, G. S., and R. Letolle, 1980, Glauconitization and phosphatization environments: A tentative comparison, *in* Y. K. Bentor, ed., Marine Phosphorites: Tulsa, OK, SEPM Special Publication 29, p. 227-237., doi: 10.2110/pec.80.29.0227.

Odin, G. S., and A. Matter, 1981, De glauconarium origine: Sedimentology, v. 28, p. 611-641, doi: 10.1111/j.1365-3091.1981.tb01925.x.

Odom, I. E., 1984, Glauconite and celadonite minerals, *in* S. W. Bailey, ed., Micas: Chantilly, VA, Reviews in Mineralogy and Geochemistry 13, p. 545-572.

Porrenga, D. H., 1967, Glauconite and chamosite as depth indicators in the marine environment: Marine Geology, v. 5, p. 495-501, doi: 10.1016/0025-3227(67)90056-4.

Rao, V. P., M. Lamboy, and P. A. Dupeuble, 1993, Verdine and other associated authigenic (glaucony, phosphate) facies from the surficial sediments of the southwestern continental margin of India: Marine Geology, v. 111, p. 133-158, doi: 10.1016/0025-3227(93)90193-Y.

Sánchez-Navas, A., A. Martín-Algarra, V. Eder, B. J. Reddy, F. Nieto, and Y. N. Zanin, 2008, Color, mineralogy and composition of Upper Jurassic West Siberian glauconite: Useful indicators of paleoenvironment: The Canadian Mineralogist, v. 46, p. 1249-1268, doi: 10.3749/canmin.46.5.1249.

Triplehorn, D. M., 1966, Morphology, internal structure, and origin of glauconite pellets: Sedimentology, v. 6, p. 247-266, doi: 10.1111/j.1365-3091.1966.tb01894.x.

Van Houten, F. B., and M. E. Purucker, 1984, Glauconitic peloids and chamositic ooids—favorable factors, constraints, and problems: Earth-Science Reviews, v. 20, p. 211-243, doi: 10.1016/0012-8252(84)90002-3.

Ferruginous grains, ironstones and iron formations:

Bekker, A., J. F. Slack, N. Planavsky, B. Krapež, A. Hofmann, K. O. Konhauser, and O. J. Rouxel, 2010, Iron formation: The sedimentary product of a complex interplay among mantle, tectonic, oceanic, and biospheric processes: Economic Geology, v. 105, p. 467-508, doi: 10.2113/gsecongeo.105.3.467.

Bhattacharyya, D. P., and P. K. Kakimoto, 1982, Origin of ferriferous ooids: An SEM study of ironstone ooids and bauxite pisoids: Journal of Sedimentary Research, v. 52, p. 849-857, doi: 10.1306/212F8071-2B24-11D7-8648000102C1865D.

Dimroth, E., and J.-J. Chauvel, 1973, Petrography of the Sokoman Iron Formation in part of the central Labrador Trough, Quebec, Canada: GSA Bulletin, v. 84, p. 111-134, doi: 10.1130/0016-7606(1973)84<111:POTSIF>2.0.CO;2.

Gross, G. A., 1972, Primary features of cherty iron-formations: Sedimentary Geology, v. 7, p. 241-261, doi: 10.1016/0037-0738(72)90024-3.

James, H. L., 1954, Sedimentary facies of iron-formation: Economic Geology, v. 49, p. 235-293, doi: 10.2113/gsecongeo.49.3.235.

Kappler, A., C. Pasquero, K. O. Konhauser, and D. K. Newman, 2005, Deposition of banded iron formations by anoxygenic phototrophic Fe (II)-oxidizing bacteria: Geology, v. 33, p. 865-868, doi: 10.1130/G21658.1.

Kimberley, M. M., 1983, Ferriferous ooids, *in* T. M. Peryt, ed., Coated Grains: New York, Springer-Verlag, p. 100-108, doi: 10.1007/978-3-642-68869-0_8.

Klein, C., 2005, Some Precambrian banded iron-formations (BIFs) from around the world: Their age, geologic setting, mineralogy, metamorphism, geochemistry, and origins: American Mineralogist, v. 90, p. 1473-1499, doi: 10.2138/am.2005.1871.

Knox, R. W. O'B., 1970, Chamosite ooliths from the Winter Gill ironstone (Jurassic) of Yorkshire, England: Journal of Sedimentary Research, v. 40, p. 1216-1225, doi: 10.1306/74D7216C-2B21-11D7-8648000102C1865.

Pufahl, P. K., and P. W. Fralick, 2004, Depositional controls on Palaeoproterozoic iron formation accumulation, Gogebic Range, Lake Superior region, USA: Sedimentology, v. 51, p. 791-808, doi: 10.1111/j.1365-3091.2004.00651.x.

Simonson, B. M., 1985, Sedimentological constraints on the origins of Precambrian iron-formations: GSA Bulletin, v. 96, p. 244-252, doi: 10.1130/0016-7606(1985)96<244:SCOTOO>2.0.CO;2.

Taylor, K. G., J. A. Simo, D. Yocum, and D. A, Leckie, 2002, Stratigraphic significance of ooidal ironstones from the Cretaceous Western Interior Seaway: The Peace River Formation, Alberta, Canada, and the Castlegate Sandstone, Utah, USA: Journal of Sedimentary Research, v. 72, p. 316-327, doi: 10.1306/060801720316.

Van Houten, F. B., 1985, Oolitic ironstones and contrasting Ordovician and Jurassic paleogeography: Geology, v. 13, p. 722-724, doi: 10.1130/0091-7613(1985)13<722:OIACOA>2.0.CO;2.

Van Houten, F. B., and D. P. Bhattacharyya, 1982, Phanerozoic oolitic ironstones—geologic record and facies model: Annual Review of Earth and Planetary Sciences, v. 10, p. 441-457, doi: 10.1146/annurev.ea.10.050182.002301.

Young, T. P., 1989, Phanerozoic ironstones: An introduction and review, *in* T. P. Young, and W. E. G. Taylor, eds., Phanerozoic Ironstones: London, GSL Special Publication 46, p. ix-xxv, doi: 10.1144/GSL.SP.1989.046.01.02.

Tuffaceous sedimentary rocks:

Altaner, S. P., and R. E. Grim, 1990, Mineralogy, chemistry, and diagenesis of tuffs in the Sucker Creek Formation (Miocene), eastern Oregon: Clays and Clay Minerals, v. 38, p. 561-572, doi: 10.1346/CCMN.1990.0380601.

Cadigan, R. A, 1963, Tuffaceous sandstones in the Triassic Chinle Formation, Colorado Plateau, *in* Geological Survey Research 1963: Washington, DC, USGS Professional Paper 475-B, p. B48-B51.

Cather, S. M., 1988, Jemez-derived pumice near San Antonio, New Mexico: Depositional processes and implications: New Mexico Geology, v. 10, p. 65-66.

Dickinson, W. R., 1985, Interpreting provenance relations from detrital modes of sandstones, *in* G. G. Zuffa, ed., Provenance of Arenites (NATO Science Series C): Dordrecht, Netherlands, D. Reidel Publishing, p. 333-361, doi: 10.1007/978-94-017-2809-6_15.

Fisher, R. V., 1966, Rocks composed of volcanic fragments and their classification: Earth-Science Reviews, v. 1, p. 287-298, doi: 10.1016/0012-8252(66)90010-9.

Fritz, W. J., and M. F. Howells, 1991, A shallow marine volcaniclastic facies model: an example from sedimentary rocks bounding the subaqueously welded Ordovician Garth Tuff, North Wales, UK: Sedimentary Geology, v. 74, p. 217-240, doi: 10.1016/0037-0738(91)90041-B.

Hay, R. L., and A. Iijima, 1968, Nature and origin of palagonite tuffs of the Honolulu Group on Oahu, Hawaii, *in* R. R. Coats, R. L. Hay, and C. A. Anderson, eds., Studies in Volcanology—A Memoir in Honor of Howel Williams: Boulder, CO, GSA Memoir 116, p. 331-376, doi: 10.1130/MEM116-p331.

Marfil, R., A. Hall, S. Garcia-Gil, and M. G. Stamatakis, 1998, Petrology and geochemistry of diagenetically altered tuffaceous rocks from the Middle Triassic of central Spain: Journal of Sedimentary Research, v. 68, p. 391-403, doi: 10.2110/jsr.68.391.

Pablo-Galan, L. de, and M. D. L. Chavez-Garcia, 1996, Diagenesis of Oligocene vitric tuffs to zeolites, Mexican volcanic belt: Clays and Clay Minerals, v. 44, p. 324-338, doi: 10.1180/minmag.1994.58A.2.93.

Reed, J. K., M. Gipson, Jr., and D. Vass, 1993, Hydrocarbon potential of sandstone reservoirs in the Neogene, East Slovakian Basin. Part 2: Zeolites and clay minerals: Journal of Petroleum Geology, v. 16, p. 223-236, doi: 10.1111/j.1747-5457.1993.tb00108.x.

Schmid, R., 1981, Descriptive nomenclature and classification of pyroclastic deposits and fragments: Geologische Rundschau, v. 70, p. 41-43, doi: 10.1007/BF01822152.

Shane, P. A. R., 1991, Remobilised silicic tuffs in middle Pleistocene fluvial sediments, southern North Island, New Zealand: New Zealand Journal of Geology and Geophysics, v. 34, p. 489-499, doi: 10.1080/00288306.1991.9514485.

Smith, R. L., 1960, Zones and zonal variations in welded ash flows, Washington, DC, USGS Professional Paper 354-F, p. 149-159.

Facing Page: Top – An opal cemented Tertiary sandstone from Texas texturized by P. van Gogh (aka, Peter Scholle). Bottom – A porous sandstone with inspirational insights (also the handiwork of P. van Gogh).

SAND & SANDSTONE TEXTURES

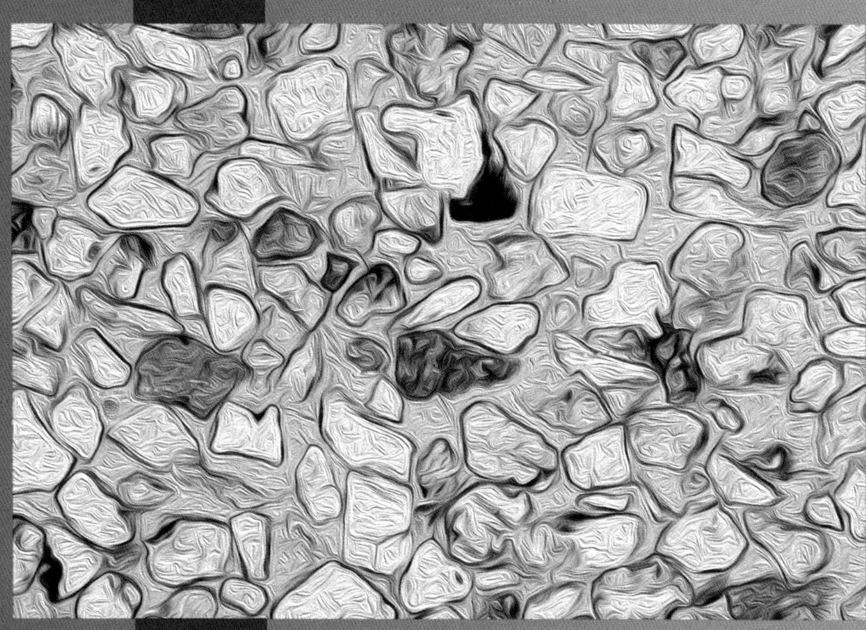

CHAPTER 6

SIZE AND SORTING MATTERS

Texture, Fabric, Matrix, Maturity and Textural Classification

Introduction – The term "texture" encompasses a wide range of attributes of sediments/rocks and their constituent grains, including grain size and sorting, particle morphology (form and sphericity, rounding, and surface texture), grain orientation, imbrication and packing. All of those properties have significance in interpreting transport processes and depositional settings of sedimentary rocks, but they also have economic importance in a wide variety of fields ranging from engineering of construction and road materials to understanding and predicting the porosity and permeability relationships of such materials in petroleum exploration/production or hydrologic contexts.

Most textural properties mentioned above are best measured in unconsolidated or easily disaggregateable materials where grains can be size-sorted by sieving, settling, laser particle analysis or other techniques or where individual grains can be viewed in three dimensions. These properties generally are far more difficult to measure accurately in thin sections of consolidated rocks. To use just one example, the size of a grain in thin section can never exceed the longest axis of the grain but it can easily be shorter because most cuts through grains are tangential or oblique to that axis. So, in most cases, the basic size-, sorting- and shape-related properties of grains are merely estimated in thin sections through the use of visual comparators. Even there, the most useful comparators are ones that have been specifically corrected for thin-section use. Detailed direct measurements on grain size and shape can, of course, be done using thin-section microscopy, especially through point-counting large numbers of grains, but there too, correction factors must be used to overcome, to the degree possible, the two-dimensional (2-D) view of three-dimensional (3-D) grains afforded in thin sections (see Harrell and Eriksson, 1979; Johnson, 1994). Advances in computerized photomicrographic image analysis can make the process of measurement and measurement correction both more accurate and far less time consuming (see, for example, Schäfer and Teyssen, 1987; Seelos and Sirocko, 2005; Syvitski, 2007).

An enormous amount of effort was made in the period from the 1930s to the 1970s to perfect textural measurements and to find reliable statistical measures that could be used to identify specific environments of deposition. That work still finds application in soft and unconsolidated sediments, but it will not be discussed in detail in this book, because it is of lesser applicability to petrographic studies of hard rocks. A number of papers in the bibliography at the close of this section can be used to follow up on unconsolidated sediment studies (especially the excellent overview of statistical measures provided in Folk, 1980) and almost all textbooks on sands and sandstones include discussions of these topics.

Grain size – There is a standard grain size chart for sediments and sedimentary rocks, termed the Wentworth or Udden-Wentworth scale as it is based on the work of Udden (1914) as refined by Wentworth (1922). In the version shown in Table 6.1, a column of phi (ϕ) equivalents (from Krumbein, 1938) is included ($\phi = -\log_2$ grain diameter in mm). The Wentworth classification, which is widely applied in North America and many other areas, uses the measured or estimated average grain size to provide a grain-size term for the entire sediment (shown in the fourth column) or rock (shown in the last column). Thus, for example, an unconsolidated sediment could be called a "very coarse sand" or a "medium silt" whereas lithified rocks of the same grain size would be termed a "very coarse sandstone" or a "medium siltstone". As will be seen in the sorting section, this name can be further modified to include terms that better express the range of sizes in either sediment or rock. But it should be noted again that these are only textural names for terrigenous sediments or rocks—these terms do not reflect composition in any way. There are many compositional classifications, but they are separately covered in Chapter 7.

Table 6.2 shows a second widely-used grain size classification, an International Standards Chart that was designed mainly for soils, but is also used for natural and engineered materials including sediments and rocks (the latter by adding the suffix "stone" to the sediment names). Part of the point of including both charts is to highlight the fact that the clay-silt boundaries differ in the two classifications, as do the specific usages of divisions within sands, silts and gravels. Thus, when using either classification to communicate with others (the basic point of having classifications), it is imperative to state which classification you are using.

The measurement of the average grain size of rocks in thin sections is certainly more laborious than sieving or settling tube analysis of unconsolidated materials. Grain size can be studied under the microscope on a grain-by-grain basis, using a calibrated eyepiece micrometer, a transparent grain size gauge or some other measuring technique. Where accurate measurements are not of great importance, one can simply determine the dimensions of a few grains that visually appear to be of "average size". One also can pick grains at the large and small ends of the spectrum of grains and measure those as well. Those few measurements can easily be corrected for the

Table 6.1: The Udden-Wentworth grain size scale for clastic terrigenous sediments and rocks. It is based on publications of Udden (1918) and Wentworth (1922) with the addition of a phi scale from Krumbein (1938).

Millimeters (mm)		Micrometers (μm)	Phi (ϕ)	Wentworth size class		Rock name
	4096		-12	Boulder	Gravel	Conglomerate
	256		-8	Cobble	Gravel	Conglomerate
	64		-6	Pebble	Gravel	Conglomerate
	4		-2	Granule	Gravel	Conglomerate
	2.00		-1	Very coarse sand	Sand	Sandstone
	1.00	1000	0	Coarse sand	Sand	Sandstone
1/2	0.50	500	1	Medium sand	Sand	Sandstone
1/4	0.25	250	2	Fine sand	Sand	Sandstone
1/8	0.125	125	3	Very fine sand	Sand	Sandstone
1/16	0.0625	62.5	4	Coarse silt	Silt	Siltstone
1/32	0.031	31	5	Medium silt	Silt	Siltstone
1/64	0.0156	15.6	6	Fine silt	Silt	Siltstone
1/128	0.0078	7.8	7	Very fine silt	Silt	Siltstone
1/256	0.0039	3.9	8	Clay	Mud	Claystone
	0.00006	0.06	14			

Table 6.2: The International Standards Chart (BS EN ISO 14688-1:2002) for description of grain sizes in soils and other granular materials.

General names		Detailed names	Size range
Very coarse soil		Large boulder (LBo)	>630 mm
		Boulder (Bo)	200-630 mm
		Cobble (Co)	63-200 mm
Coarse soil	Gravel	Coarse gravel (CGr)	20-63 mm
		Medium gravel (MGr)	6.3 20 mm
		Fine gravel (FGr)	2.0 - 6.3 mm
	Sand	Coarse sand (CSa)	0.63 - 2.0 mm
		Medium sand (MSa)	0.2 - 0.63 mm
		Fine sand (FSa)	0.063 - 0.2 mm
Fine soil	Silt	Coarse silt, (CSi)	0.02 - 0.063 mm
		Medium silt (MSi)	0.0063 - 0.02 mm
		Fine silt (FSi)	0.002 - 0.0063 mm
	Clay	Clay (Cl)	<0.002 mm

2-D/3-D effect noted above by using equations provided in Harrell and Eriksson (1979) or Johnson (1994). For many purposes, such approximations are time and cost effective solutions for estimating grain size. If more accurate, statistically valid grain size distributions are required, then point counting and measurement of hundreds of grains will be needed. Some modern, automated procedures for doing that are outlined in the aforementioned references.

Grain sorting – This is a measure of the distribution (total range and modes) of grain sizes in any given sediment or rock sample—the better the sorting, the narrower the range of size grades in the sample. Sorting can be expressed quantitatively in terms of diameter ratios between the smallest and largest significant grain populations in a sample (or the standard deviation from the mean grain size of a sample). More commonly however, it is verbally described as steps on a scale ranging from very well sorted to very poorly sorted samples. See Table 6.3 for the correspondence between verbal and quantitative categories and see Folk (1980) for a thorough discussion of this topic. Sediments and rocks can have complex sorting patterns, with two or more size modes (termed bimodal or polymodal distributions), or grain size distributions that differ in shape from typical bell-shaped curves. The latter variations are statistically described in terms of sediment skewness and kurtosis. The discussion of these topics is well beyond the scope and intent of this book, but many references, including Folk (1980), cover these topics in great detail.

Sorting can reflect the influence of several factors, most notably the size range of materials supplied from a particular source area, the types of transport and depositional processes, and postdepositional influences that range from grain breakdown to overgrowth during diagenesis. Most commonly, however, sorting is interpreted to reflect transport/depositional processes. For example, high-energy eolian and beach environments generally are associated with well-sorted deposits; low-energy fluvial, alluvial, neritic, lagoonal or swampy environments are most commonly associated with poorly sorted deposits (sometimes termed diamictons). Many glacial deposits (e.g., tills) are very poorly sorted with grains ranging from clay- to boulder-sized. In addition to its value in environmental interpretation, sorting is important from an economic perspective because it is strongly correlated to both porosity and permeability (with lower values of both in more poorly sorted sediments and rocks).

The measurement of sorting on unconsolidated samples is normally done by statistical analysis of data from sieving and other methods of measuring the full range of grain sizes in a sample. But in thin section studies, sorting is normally estimated through the use of a comparison chart, preferably one that has been designed specifically to compensate for the 2-D vs. 3-D problem noted earlier. Such a chart is supplied here as Fig. 6.1—it includes specific references to the Folk (1980 and earlier) verbal sorting categories and quantitative phi (ϕ) standard deviation ranges.

Grain shape – is one of the least commonly studied textural property of terrigenous rocks in normal petrographic work. It refers to the tendency of grains to fall in one of four categories of shapes—discoidal or oblate, bladed or tabular, equant or spherical and prolate or rod-shaped—as shown in Fig. 6.2, following the classification by Zingg (1935). That classification is based on the relative ratios of mutually perpendicular long-, intermediate- and short-axes of particles. Thus, as with several other types of textural analysis, identification of grain shapes requires having loose grains that can be analyzed in 3-D. Furthermore, even in unconsolidated sediment samples, shape analysis is done mainly on pebble-sized and larger particles that are not well represented in most thin sections. Thus, at best, most petrographic studies only note nonquantitative characteristics such as "elongate grains". However, more detailed grain shape analysis has considerable significance for engineering studies or evaluation of construction materials.

Grain rounding – is a measure of the amount of smoothing and abrasion a grain has undergone during transport and deposition. It is one of the most important textural properties that can be measured in terrigenous deposits because it can help in estimating the distance

Table 6.3: Nomenclature and statistical measures of sorting in sediments (Folk 1966 and 1980). To determine the diameter ratio when grain-sizes are measured in millimeters, divide the 16th percentile diameter size by the 84th percentile. The phi standard deviation is utilized for grain sizes measured in ϕ by subtracting the 16th from the 84th percentile and dividing by 2.

Diameter ratio (millimeters)	Phi Standard Deviation	Verbal scale	
1.0	.00		MATURE
		Very well sorted	
1.6	.35		
		Well sorted	
2.0	.50		
		Moderately sorted	SUBMATURE
4.0	1.00		
8.0	1.50	Poorly sorted	
16.0	2.00		
		Very poorly sorted	

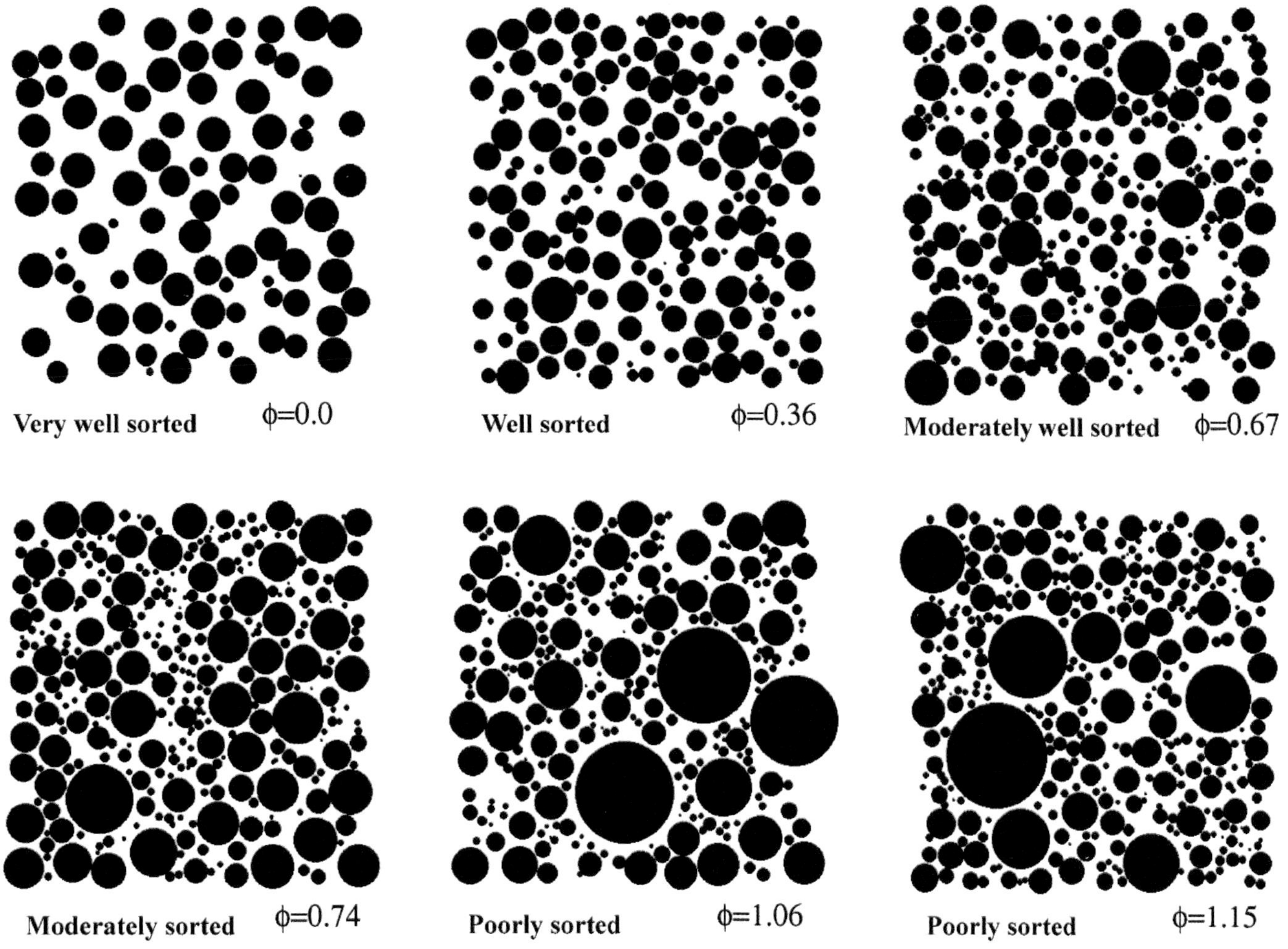

Very well sorted $\phi=0.0$ | Well sorted $\phi=0.36$ | Moderately well sorted $\phi=0.67$

Moderately sorted $\phi=0.74$ | Poorly sorted $\phi=1.06$ | Poorly sorted $\phi=1.15$

Figure 6.1: Six textural comparators for degree of sorting in 2-D. Each comparator is labeled with its sorting description (Folk, 1968) based on the true 3-D sorting value. Adapted from Jerram (2001).

and duration of sediment transport and/or reworking of sand grains from earlier cycle sedimentary rocks.

Although rounding can be described mathematically (by comparing the ratio of the average radius of curvature of the edges/corners to the radius of curvature of the maximum inscribed sphere), in petrographic studies it typically is estimated using a visual comparison chart (almost always that of Powers, 1953, either in its original form as shown in Fig. 6.3 or in a drafted outline version of it). It should be remembered, however, that roundness tends to be better in coarse grains (sands and larger) than in finer grained ones, in part because rates of rounding are higher where grain impacts are more energetic.

As noted above, degree of rounding is very much a function of grain composition because rounding is strongly influenced by differences in the relative hardness and cleavage of detrital minerals. For sediment maturity studies, the degree of roundness of monocrystalline quartz grains should be used as the standard measure; for transport distance studies, on the other hand, comparative rounding values for both durable and nondurable grains are optimal. In its most simplified form, consistently well-rounded quartz results from prolonged

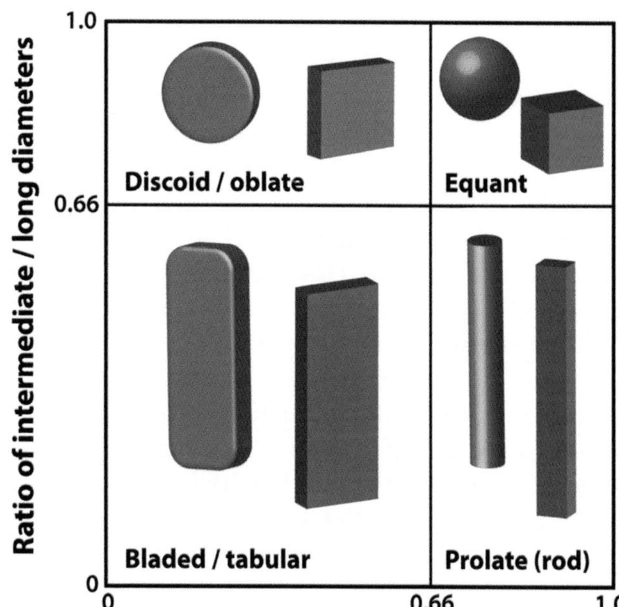

Figure 6.2: Classification of pebble shapes based on the relative lengths of their three mutually perpendicular axes. Adapted from Zingg (1935) and Krumbein (1941).

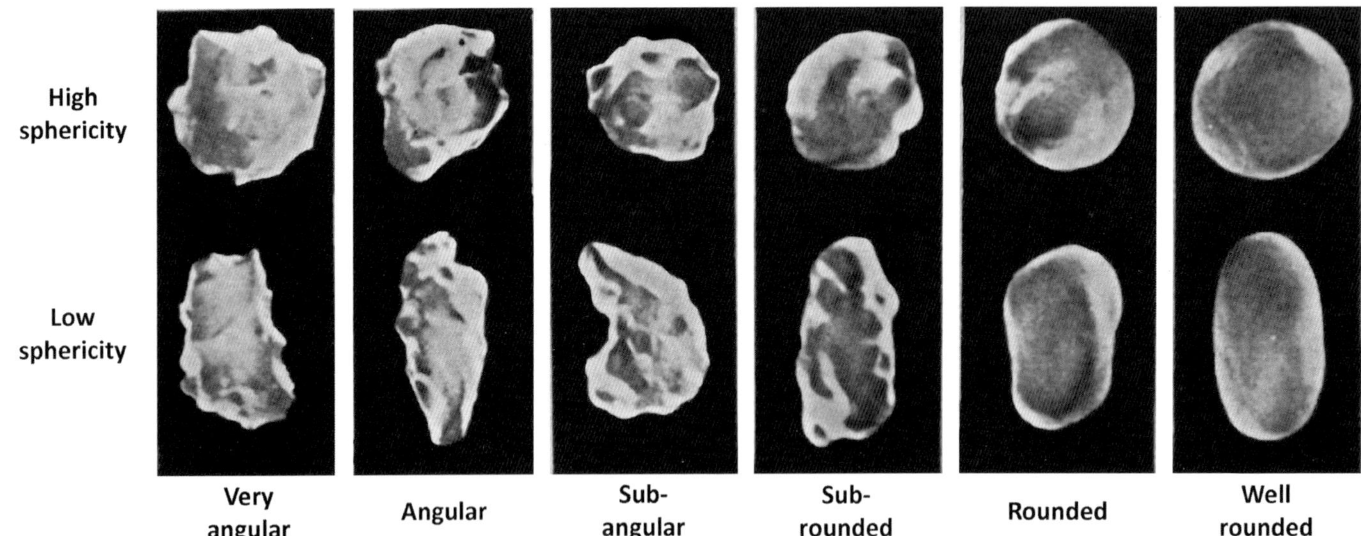

Figure 6.3: A visual comparator showing levels of rounding in high and low sphericity sand grains (from Powers, 1953).

abrasion in high-energy settings (e.g., eolian or beach environments) or is the result of multiple cycles of sedimentation and erosion.

Surface texture – This term refers to the whole range of impact and abrasion marks, dissolution features and other "blemishes" on the surface of detrital grains. These are not features normally discernable in thin section, but rather require examination of whole-grain surfaces in binocular stereoscopes or, more commonly, using scanning electron microscopy. This is a very specialized topic that will not be covered here, although some SEM images of quartz surface textures are shown in Chapter 1 (and it is the surface textures of quartz grains that are most commonly studied). The interested reader is referred to citations at the close of that chapter for further information.

Fabric (grain orientation, imbrication and packing) – Orientation is a measure of the azimuthal direction of the long axis of particles; imbrication refers to the "shingling" or consistent dip of grains, especially platy ones; and packing refers to the more general patterns of spacing and proximity of grains. Packing also includes determination of whether clastic grains touch each other throughout (a grain supported fabric) or whether grains "float" in matrix (a mud- or matrix-supported fabric). Orientation and imbrication are mostly 3-D measures of the grain fabric of rocks and sediments and thus they are normally only looked at in specialized studies. Study of 3-D properties requires multiple, directionally-oriented thin sections or the use of nonpetrographic techniques (such as examination of directional magnetic properties). Nonetheless, preferred orientation of elongate grains or imbricate inclination of grains (typically dipping in an up-current direction) can be seen in thin sections cut with a known orientation relative to bedding and should be noted where visible. Those observations can then be followed by more systematic sampling, section cutting and other studies if warranted. Packing, on the other hand, generally can be evaluated in 2-D space as long as thin sections are cut perpendicular to bedding. Thus, it commonly is examined in studies of compaction and/or the relative timing of compaction and cementation (e.g., McBride, 2012).

Matrix – To this point, the book's focus has been mainly on the mineral grains that form the framework of terrigenous sediments. Interstitial to those framework grains, however, there can be open pore space, cement (that is, minerals precipitated diagenetically into formerly open pores), or matrix (fine-grained detrital material originally deposited between the coarser particles). Any one, two, or all three of these, can occur in any given sample. Cements and pores will be discussed in other chapters, but detrital matrix, because it is a primary mineral constituent that can greatly affect rock texture and fabric, will be discussed here.

The presence of matrix has significance both in terms of defining the original rock composition and in interpreting its environment of deposition. Although interstitial matrix typically consists of clay and silt (within a sandstone framework), some workers restrict the definition of matrix to just clay-sized material (and call it "clay matrix"); others have adopted limits of roughly 30 μm for matrix which encompasses grains up to and including medium silt. Where such matrix is present, it is reflective of the processes of sediment deposition and winnowing. As such, the presence or absence of matrix is a major factor in determining the "maturity" of sediments or sedimentary rocks.

Before proceeding to a discussion of maturity, however, it is important to note that over the past three decades there has been an significant shift in perspective relative to matrix. Where clay and silt were once assumed to be detrital components of many or most sandstones, today the pendulum has swung to considering much clay and silt, especially in litharenites, to be the product of compaction or disintegration of claystone or siltstone clasts (SRFs) or diagenetic precipitates or infiltrates of pore-filling clays (including soil-forming processes involving infiltration and precipitation; e.g., Matlack et al., 1989; Dunn, 1992). The distinction between true matrix and the "pseudomatrix" of deformed SRF or clay precipitates can be quite easily made in some cases and quite difficult in others, but it certainly is critical to the proper interpretation of textural properties of sediments. The photographs and captions throughout this book should help to delineate ways of distinguishing matrix and pseudomatrix that depend largely on the patterns of matrix distribution. For example, precipitated clays commonly have grain-rimming textures; squashed and smeared out SRF are commonly localized or still have recognizable grain outlines; and true matrix is more compositionally uniform throughout the sample.

Textural maturity – the concept of stages of textural maturity (Folk, 1951; Folk, 1956; and Folk, 1980) is essentially the synthesis of all the aspects of sediment or rock texture discussed above. The concept postulates that as sediments suffer a progressively greater input of mechanical energy through the abrasive and sorting action of waves or currents, they pass sequentially through four stages of development as shown in Table 6.4. In essence, the removal of matrix is followed by sediment sorting and sediment rounding at ever greater levels of "environmental energy" within the depositional setting. The following quote may help to clarify the concept:

> *"It is important to realize that the maturity rating of an environment depends on how much mechanical energy is exerted on a sediment after it has been moved essentially to its final resting place, by currents and waves at the final site of deposition; it does not depend on the energy expenditure required to move it from the source area to the site of final deposition. A flood or turbidity current, for example, may expend a tremendous amount of energy while it is transporting sediment, but once the sediment is dropped it is simply buried by more sediment and never suffers any further sorting or winnowing; thus such sediments have low maturity. Waves on a beach, on the contrary, sort and rework the sediment continually..." From Folk (1980, p. 106).*

The appropriate term for the textural maturity of a given deposit can simply be included as part of the rock's overall textural name — for example, a *supermature* coarse sandstone or a *submature* fine sandstone. But it is important to recognize that the concept and terminology of maturity applies mainly to moderately coarse rocks or sediments (primarily granule- and pebble-conglomerates, sandstones, and coarse siltstones). Beyond those ranges, the winnowing, sorting and rounding processes are not entirely comparable to those in the intermediate size ranges. Two additional things should also be noted regarding textural maturity. First, "textural inversions" can sometimes be found where the maturity stage sequence is not easily applied—for example, extremely well-rounded sandstones with bimodal grain distributions that technically mean that the rock is not well sorted. Such inversion can result from extensive bioturbation of layered sediments, storm mixing of sediments, and even some depositional processes (eolian transport for example). Second, in looking at the factors involved in textural maturity (matrix, sorting and rounding) one must strive to recognize depositional fabrics, and learn to subtract diagenetic alteration (overgrowth cements, pseudomatrix and the like).

Textural classification of complex sediments and rocks – Some sediment/rock terminology schemes were presented in the opening section of this chapter (Tables

Table 6.4: Definitions of the four stages of textural maturity of clastic terrigenous sediments or sedimentary rocks (from Folk, 1951, 1956, and 1980).

Stages of Textural Maturity

I. Immature stage - Sediment contains over 5 percent terrigenous clay matrix; sand grains usually poorly sorted and angular.

II. Submature stage - Sediment contains under 5 percent clay, but sand grains are still poorly sorted and are not well rounded.

III. Mature stage - Sediment contains little or no clay, and sand grains are well sorted but still not well rounded.

IV. Supermature stage - Sediment contains no clay, sand grains are well sorted and well rounded. This determination should be made, if possible, on quartz grains of medium and fine sand size.

6.1 and 6.2). Those terms were based on the average grain size of a sample and are thus especially well suited to well-sorted samples. But many terrigenous deposits are far from well sorted, and thus several nomenclatural schemes that reflect that diversity of grain sizes exist. Four of those are shown graphically in Fig. 6.4. Three of the classifications allow plotting of three categories of sediment size, sand silt and clay; the fourth classification combines silt and clay and plots it against sand content. As in previous diagrams, the names shown are for sediments, but the classifications can easily be used for rocks simply by appending "stone" to the end of the name. Such classifications are most easily applied to unconsolidated sediments where quantitative data from sieving, settling tube or other direct measurement techniques is easily plotted. However, for consolidated rocks, it is common for less accurate visual estimates of percentages of sand, silt and clays to be used. The categories shown in the plots in Fig. 6.4 generally suffice to reflect the accuracy of such estimates. For use with unconsolidated sediments, on the other hand, these classification schemes have an additional level of more detailed subdivisions that can further refine descriptions (and those more detailed schemes are available in the citations provided).

As in previous sections, it is important to note that the actual names used in each of these classifications are very similar, but the definitional boundaries vary significantly. Thus, it is necessary to specify which classification scheme was used when applying any such names.

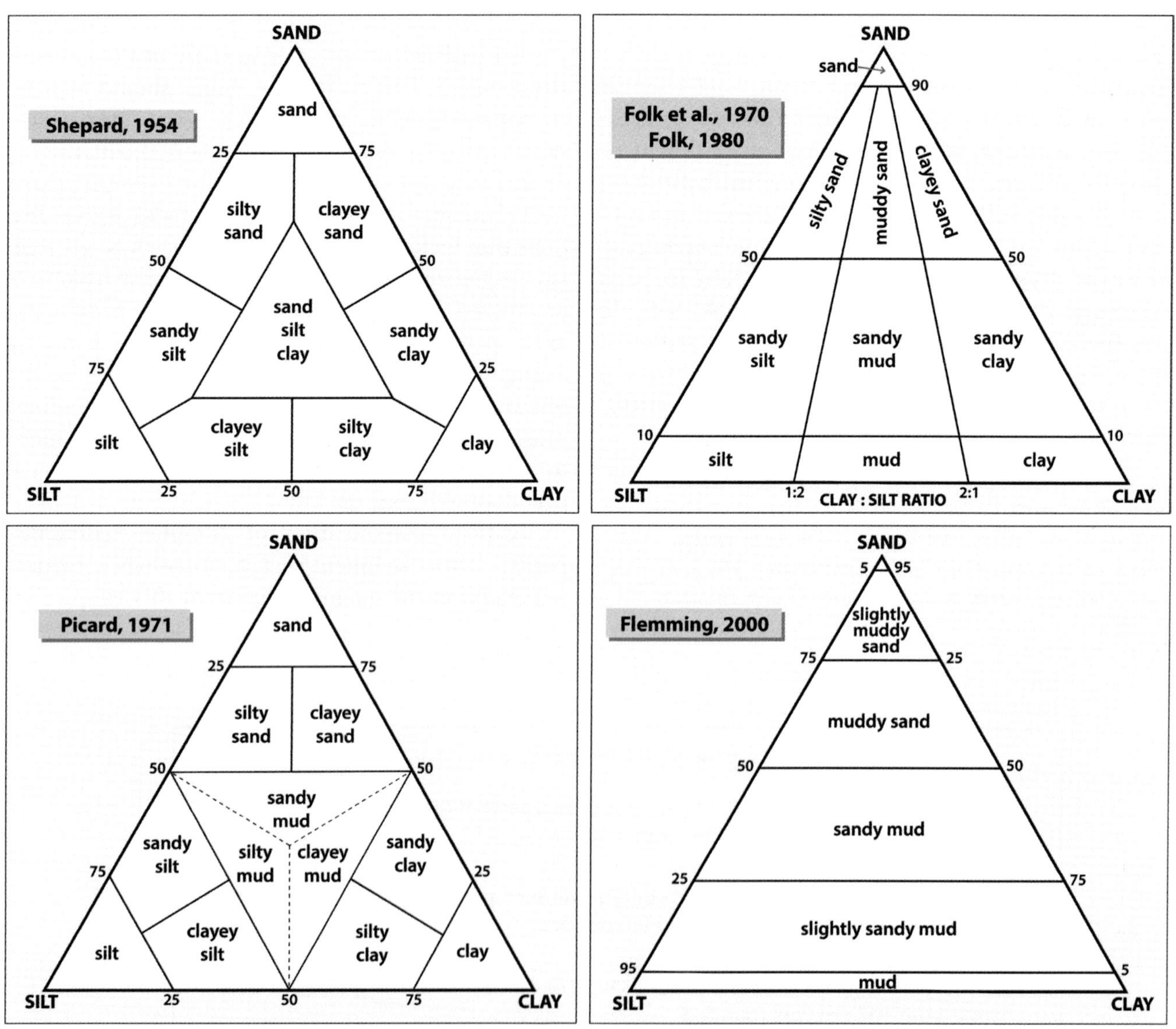

Figure 6.4: Four textural classifications of fine-grained or mixed grain-size clastic terrigenous sediments (adapted from Shepard, 1954; Folk et al., 1970 and Folk,1980; Picard, 1971; and Flemming, 2000).

Mid. Eocene – Lo. Oligocene Spears Gp., Socorro Co., New Mexico

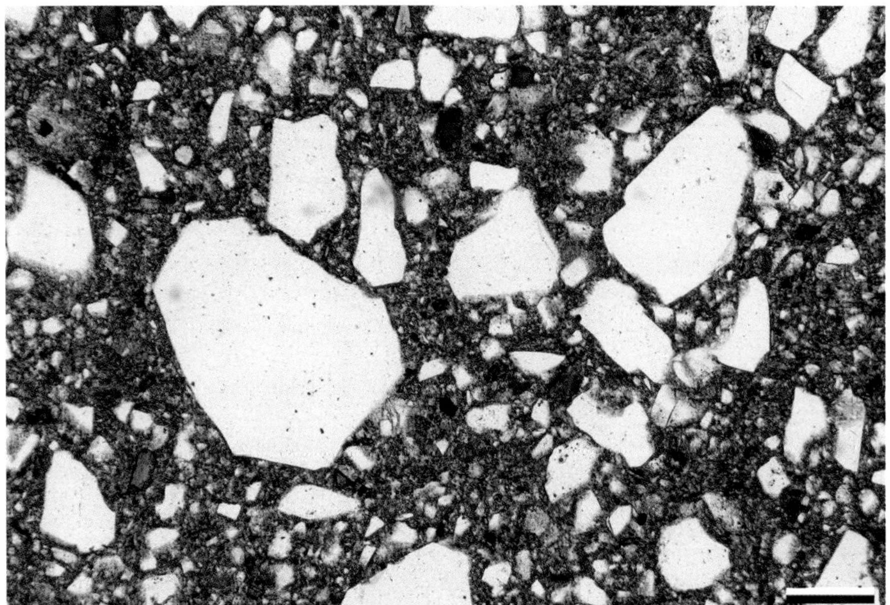

An immature sandstone. This rock is very poorly-sorted, with extensive muddy (clay and silt) matrix. Even the largest grains (the ones most rapidly abraded during transport) show little rounding. The determination that the matrix here is primary is crucial to the naming of this rock; it is relatively easy in this case, however, because of the lack of a grain-supported fabric. Instead, the primary silt and clay matrix is supporting isolated ("floating") sand grains. Such fabrics can form from normal transport processes, but this example is from a fluidized sandstone injection dike.

PPL, Scale bar = 0.10 mm

Paleocene, Fatehgarh Fm., Barmer Basin, Rajasthan, India

Another example of an immature sandstone with muddy (clay and silt) matrix, but in this case, a rock with a largely grain-supported fabric and sand-sized grains with a greater degree of rounding. The homogeneity of matrix size and composition is evidence that this is true primary matrix that was incorporated during the deposition of this rock. This rock represents a soil which formed through mixing of quartz sand grains and silt/clay matrix by pedogenic processes such as illuviation (downward infiltration of fines), burrowing and rooting.

PPL, Scale bar = 0.49 mm

Pennsylvanian Pottsville Fm., Licking Co., Ohio

Another sandstone with a submature texture based on its muddy (silt- and clay-rich) matrix. In this case, however, some of the silt-sized and larger matrix constituents were leached at some point in the diagenetic history. This resulted in secondary porosity development throughout much of the matrix and most of the areas of blue-dyed epoxy mark sites of such dissolution.

PPL, BDI, Scale bar = 0.34 mm

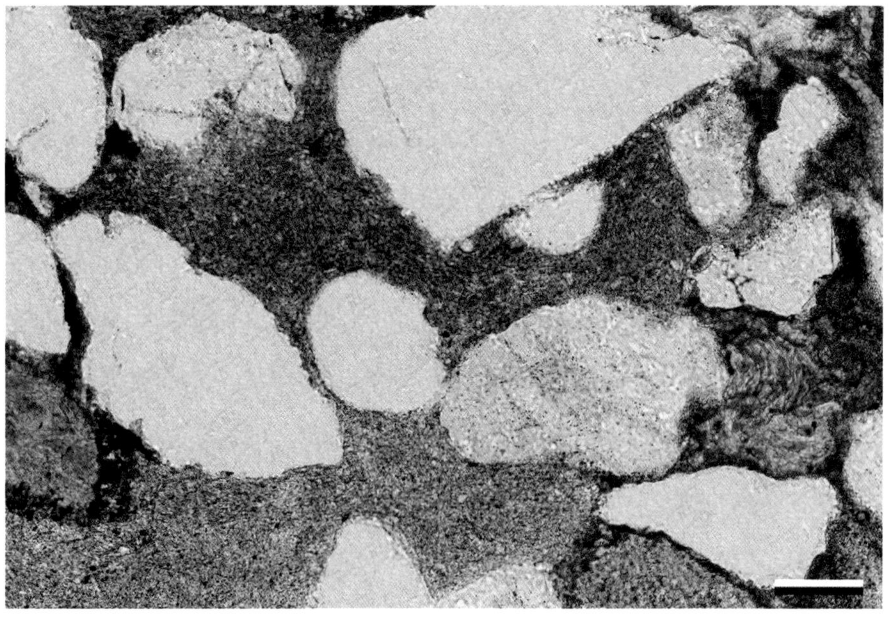

Up. Cambrian Sillery Gp., Gaspé Peninsula, Quebec, Canada

An excellent example of a clay "pseudomatrix" produced by the compaction and flowage of soft sedimentary rock fragments (claystone clasts). This yields a patchy "matrix" with locally and subtly different compositions reflecting original variations in the compositions of individual SRFs (in this example, note the differences in color between the "matrix" in the upper and lower parts of the photo). Failure to recognize pseudomatrix affects not just textural maturity determinations but also compositional determinations—this is a shale arenite only if one recognizes the highly deformed shale clasts.

PPL, Scale bar = 0.10 mm

Silurian Tuscarora Fm., Lebanon Co., Pennsylvania

This rock too has numerous shaly sedimentary rock fragments, and although these have not been strongly deformed, their mere presence gives the appearance of clay "matrix". Thus, careful observation is needed before accepting the mere presence of clays as evidence of primary matrix. The past few decades have seen a growing understanding of the importance of pseudomatrix produced through degradation and deformation of detrital clasts or through precipitation of authigenic clays.

PPL, Scale bar = 0.26 mm

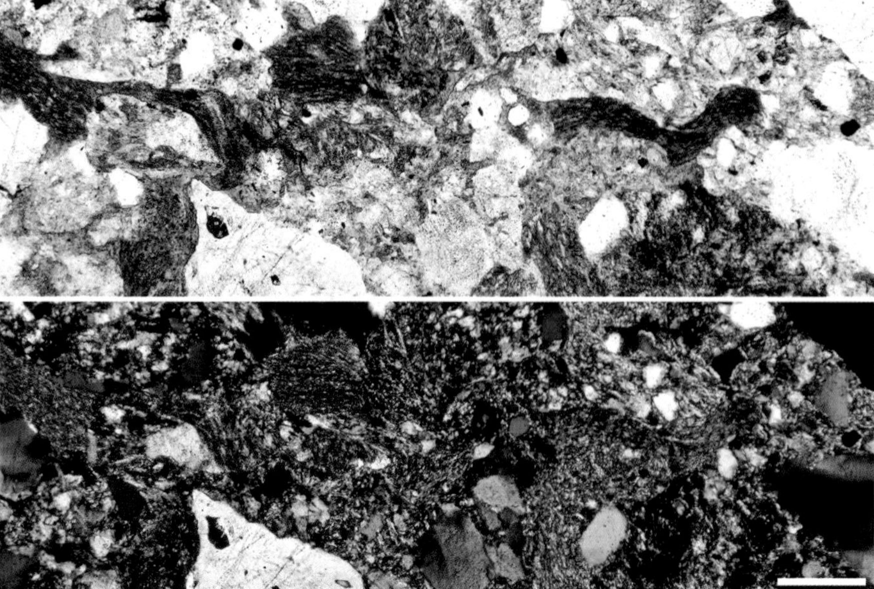

Silurian Tuscarora Sandstone, Lebanon Co., Pennsylvania

An example showing both plane- and cross-polarized views of a sandstone composed mainly of SRFs that have been deformed and fused into what looks like a matrix-dominated rock (but is actually a pseudomatrix-dominated rock). The compositional variations of clay and silt in these clasts has produced a quilt of telltale textural differences that betray the pseudomatrix origin. Had the clasts been of a more uniform lithology, the task or recognizing pseudomatrix would have been much more difficult.

PPL / XPL, Scale bar = 0.26 mm

Lo. Cambrian Unicoi Fm., Shenandoah Valley, Virginia

Another example of squashed sedimentary rock fragments producing pseudomatrix. These now illitic grains have been intensely deformed to the point where they have flowed into adjacent pores and would have fused into a single unrecognizable mass in the absence of the thin iron-stained cement rinds that define the contacts between original grains.

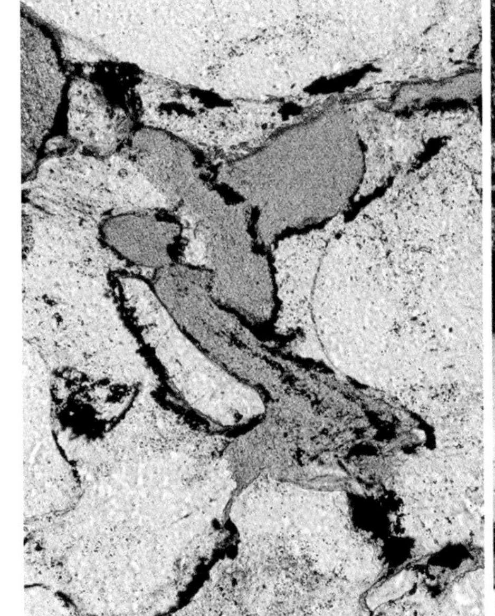

PPL | XPL, Scale bar = 0.10 mm

Up. Triassic New Haven Arkose, New Haven Co., Connecticut

A submature sandstone. At first glance, this might be called an immature sandstone given all the dark-colored material between poorly sorted and largely angular grains. Upon closer examination, however, a significant part of the dark-colored matter consists of hematite and clay cement rims on grains, detrital biotite flakes and small shale or slate clasts rather than a true matrix of detrital clays and fine silt. There also are some pores completely filled with authigenic clay cements. Authigenic clays do not count in maturity determination, because they were not part of the primary deposit.

PPL, Scale bar = 0.51 mm

Up. Triassic New Haven Arkose, New Haven Co., Connecticut

A higher-magnification view, under cross-polarized light, from the same sample shown above. Note the abundance of small, elongate, dark fragments scattered within the birefringent illitic clays— these are fragments of hematite coatings that surrounded now deformed and recrystallized shale or slate clasts. The clays still retain concentration in patchy clumps that reflect the original clast distribution, although recrystallization to coarse illite has obscured finer details. That makes determination of this as pseudomatrix rather than true matrix more difficult, although no less important.

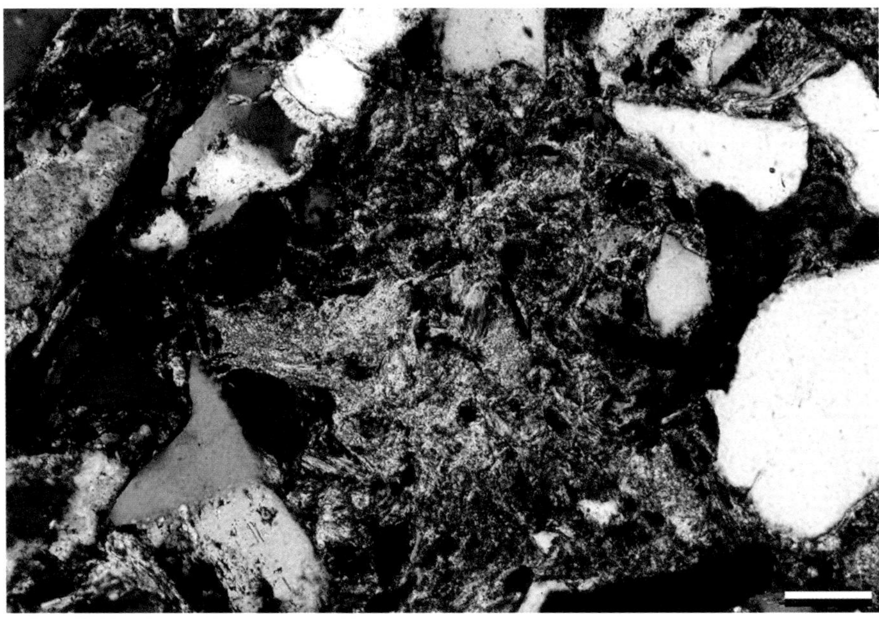

XPL, Scale bar = 0.26 mm

Lo. Permian Cutler Fm., Mesa Co., Colorado

Because the distinction between true matrix and pseudomatrix is both important and sometimes difficult, several examples are provided here. This one is especially complex, because it shows considerable dark material, which might be assumed to be clay, followed by light material filling pores. In fact, the dark materials are detrital biotites and iron oxide coatings, and much of the light-colored interstitial material consists of highly deformed clayey particles (rock fragments and/or altered clay cements but not primary matrix—see next photo).

PPL, Scale bar = 0.51 mm

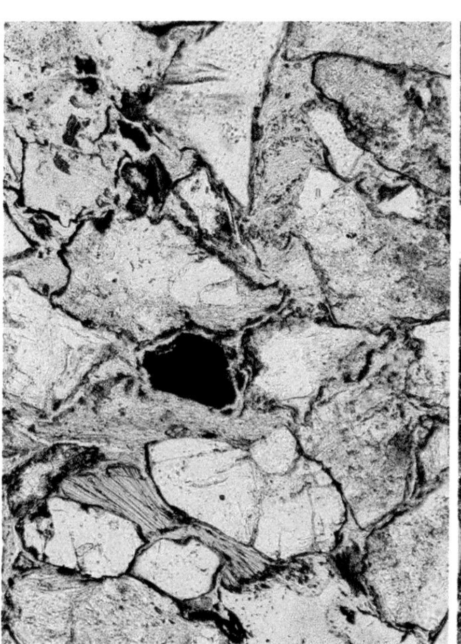

Lo. Permian Cutler Fm., Mesa Co., Colorado

A higher-magnification view from the same sample shown above. Note the grain-coating iron-oxide cements that predate the highly deformed illitic clays interstitial to the recognizable gains. The fact that the clays show multiple domains and do not show grain-coating fabrics indicates that they probably were originally shale clasts that have been deformed and recrystallized. In any case, they are not matrix because they postdate the early grain-coating cements. So these clays represent pseudomatrix. Note also the clear-cut deformation of the detrital mica in the lower part of the photo.

PPL | XPL, Scale bar = 0.10 mm

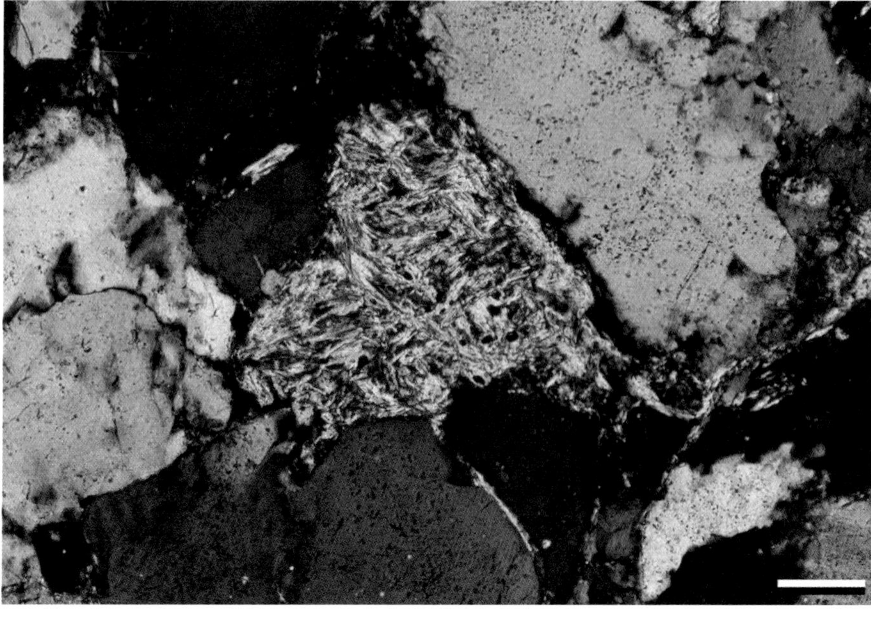

Lo. Triassic Bunter Sandstone, Tyrol, Austria

In sandstones that have undergone substantial burial, the extent and coarseness of clay recrystallization (to illite or even muscovite) can progressively obscure clues to the detrital or authigenic origins of precursor clays. In this example (and the next), patches of coarse illite/muscovite are scattered through an otherwise texturally mature sandstone. The majority of the clays may be the result of illitic replacement (illitization) of a detrital feldspar.

XPL, Scale bar = 0.10 mm

Lo. Triassic Bunter Sandstone, Tyrol, Austria

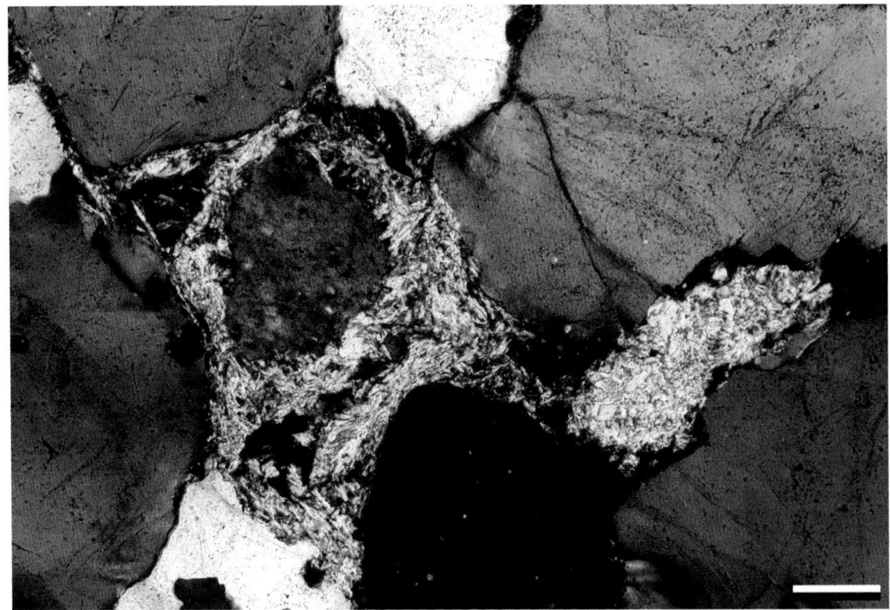

This photo shows a different area of the mature sandstone illustrated in the previous photo. It, too, shows the very patchy distribution of coarse, illitic clays, but here the patches appear to be altered grains partly deformed or "smeared" by deformation. The original grains may have been shale SRFs or illitized detrital feldspars. In any case, although clays are widespread in this rock, none are primary detrital matrix and so none affect the classification of this well sorted and somewhat rounded rock as a mature sandstone.

XPL, Scale bar = 0.26 mm

Up. Triassic Chinle? Fm., Moffat Co., Colorado

A rock that again at first glance could be mistaken for an immature sandstone with substantial matrix. In fact, this is a submature to mature sandstone with no clay matrix but with considerable dark, ferruginous cement and extensive authigenic kaolinite cement in many of the larger pores. The patchy distribution of the dark iron oxides, their grain coating character (visible locally) and the otherwise moderately sorted character of the deposit, all should be clues that stimulate more careful examination of the sample.

PPL, RDI, Scale bar = 0.51 mm

Mid. Permian Bell Canyon Fm., Culberson Co., Texas

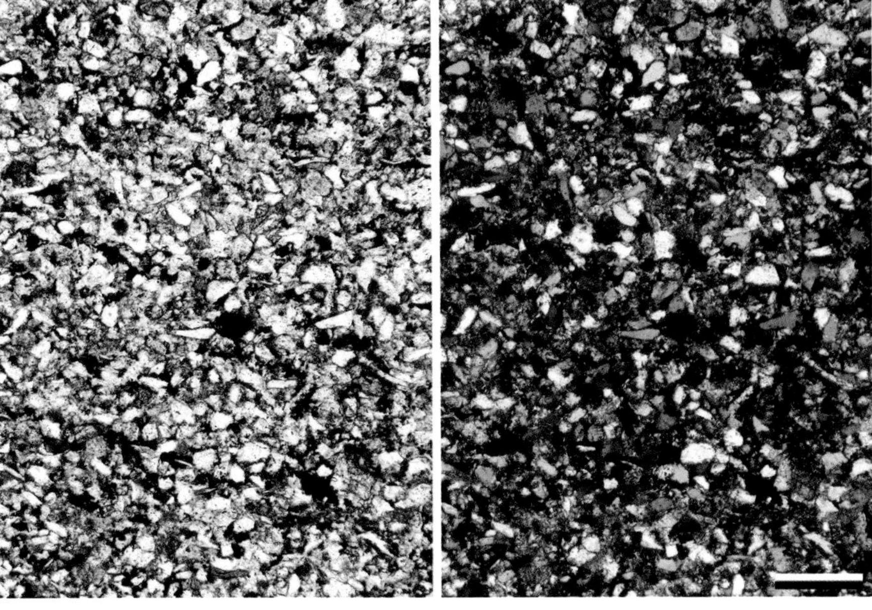

Although Folk's textural maturity classification applies mainly to sandstones, the basic constituent fabric characteristics (matrix, sorting and rounding) can be extended to siltstones as well. In this coarse siltstone/very fine sandstone, one can see little or no true matrix, exceptionally good sorting of grains and poor rounding. This is a deep-water deposit that has retained good sorting from having earlier been transported as eolian material (but silt-sized material is rarely well rounded, even with eolian transport). This illustrates some of the complexities involved in environmental interpretation of textural fabrics.

PPL | XPL, Scale bar = 0.26 mm

Unknown unit, unknown U.S. location

Having covered the textural maturity grades that involve interpretation of matrix, we will now look at the more mature categories. This is submature sandstone; the grains generally are very angular and poorly sorted, and the rock contains substantial primary and minor secondary porosity but no clay matrix. Recognizing textures in uncemented or largely uncemented sandstones is far easier than in cemented ones, especially where porosity is impregnated with dyed resins because the shapes and sizes of grains are starkly silhouetted against a colored backdrop.

PPL, BDI, Scale bar = 0.34 mm

Lo. Permian Abo Sandstone, Rio Arriba Co., New Mexico

Very angular (unrounded) grains and poor sorting in a submature to barely mature sandstone. The rock does not have any significant clay matrix, although it does have some silt-sized grains located in interstitial positions in a grain-supported fabric. Thus, it meets the definition of a submature sandstone (<5% clay; poor rounding, poor sorting). This sandstone also is fully cemented by calcite (recognizable by its high birefringence) which provides good contrast with the mainly quartzose grains, again allowing easy recognition of grain shapes and sizes.

XPL, Scale bar = 0.26 mm

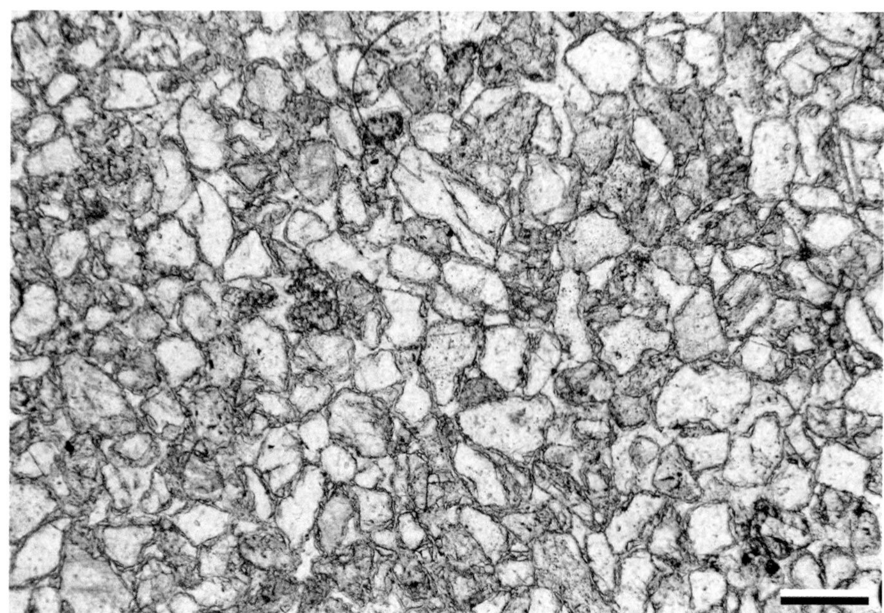

Up. Cretaceous Aachen Sandstone Fm., Liège Province, Belgium

A mature sandstone. Although this rock has considerable clay, all of that clay is of diagenetic origin (seen here as brownish rinds of chlorite that surround virtually all grains and predate chalcedonic quartz cementation of the rest of the rock). The constituent detrital quartz grains are well sorted but are poorly to moderately rounded, completing the definition of a mature sandstone.

PPL, Scale bar = 0.26 mm

Mid. Jurassic Entrada Fm., Moffat Co., Colorado

A sandstone that is borderline between mature and supermature. The grains in this rock are well sorted and fairly well rounded but have achieved neither the full degree of rounding nor the degree of sorting required for a supermature designation. The rock is also compositionally not fully "mature", with a large number of feldspars (the brownish or cloudy grains) interspersed with the clear quartz grains.

PPL, PFS, KFS, BDI, Scale bar = 0.26 mm

Mid. Jurassic Entrada Fm., Uintah Co., Utah

A supermature sandstone that meets all aspects of the definition of that textural term. It lacks any clay matrix and consists of very well-sorted and very well-rounded quartz grains. This sandstone is eolian—a depositional environment which is most commonly, but by no means exclusively, associated with sedimentation of supermature sandstones. As in most such deposits, it is also compositionally mature with virtually nothing but quartz present as a result of mechanical, and perhaps also chemical, degradation of other, more unstable minerals.

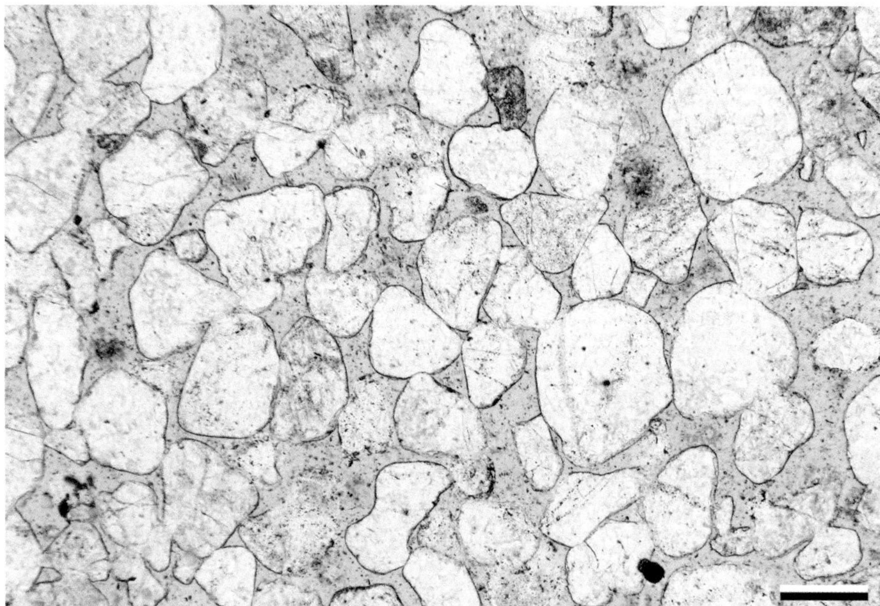

PPL, BDI, Scale bar = 0.26 mm

Mid. Jurassic Entrada Fm., Moffat Co., Colorado

The highest-maturity stages of sandstones are based on levels of grain rounding. What seems as though it should be the easiest of observations is easy only where there has been little diagenetic alteration of the grains. Generally, the most mature sandstones are very quartz rich, and quartz grains, especially in the absence of clay, bitumen or microcrystalline quartz coatings, are prone to overgrowth by syntaxial cements (as seen here). Those overgrowths typically have angular (unrounded) crystal faces, and one needs to evaluate the shapes of the cores (dusty in this example) rather than the overgrowths for maturity determinations.

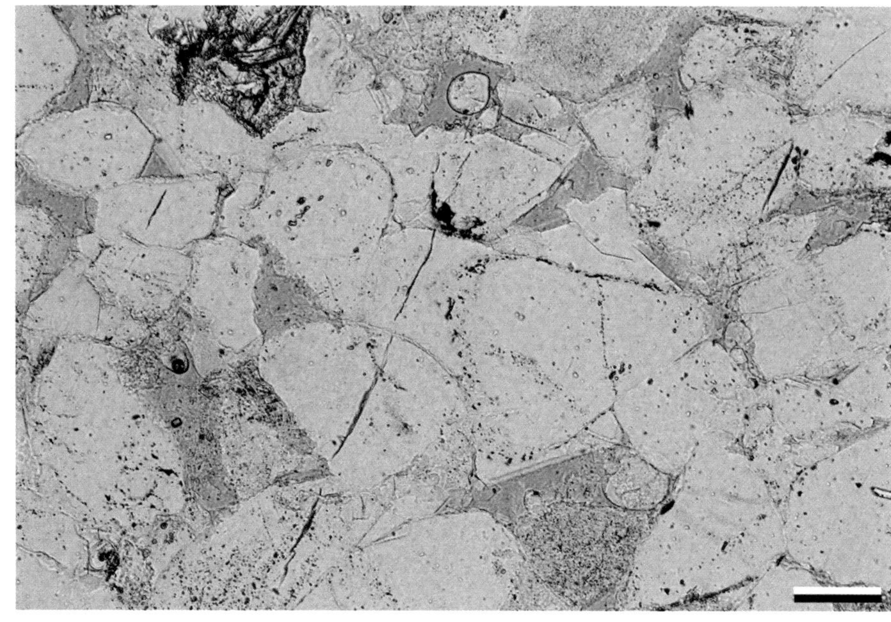

PPL, BDI, Scale bar = 0.08 mm

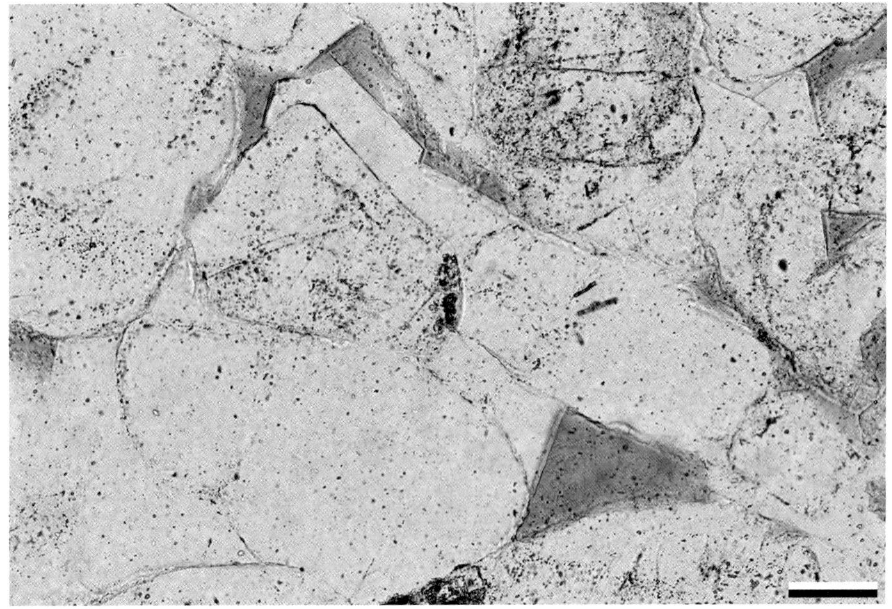

Mid. Jurassic Entrada Fm., Moffat Co., Colorado

This photomicrograph is from the same unit (but not the same sample) as the previous image. It shows, in more detail, the euhedral quartz overgrowths on rounded, detrital quartz cores. Accurate observation of the original (precement) shapes and sizes of detrital grains is necessary in order to obtain correct values for grain size, rounding, sorting and maturity.

PPL, KFS, BDI, Scale bar = 0.09 mm

Silurian Tuscarora Sandstone, Lebanon Co., Pennsylvania

When completely cemented, unlike the previous sample, quartzarenites can look like one enormous mass of quartz and evaluation of primary grain roundness and sorting can be very difficult. The abundance of inclusions in these grain give promise that with close examination, grain cores can be differentiated from interlocking cement overgrowths. Indeed, when looked at carefully, one can see numerous cloudier, rounded forms of fairly consistent size that are the detrital grains as opposed to the clear and more irregular overgrowth cements. Thus, this is a supermature sandstone.

PPL, Scale bar = 0.26 mm

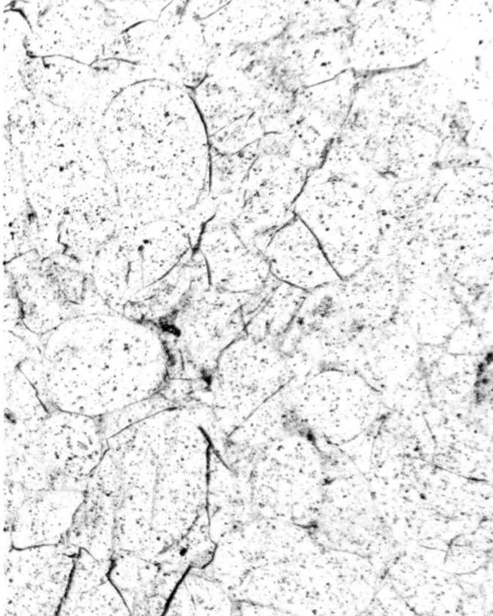

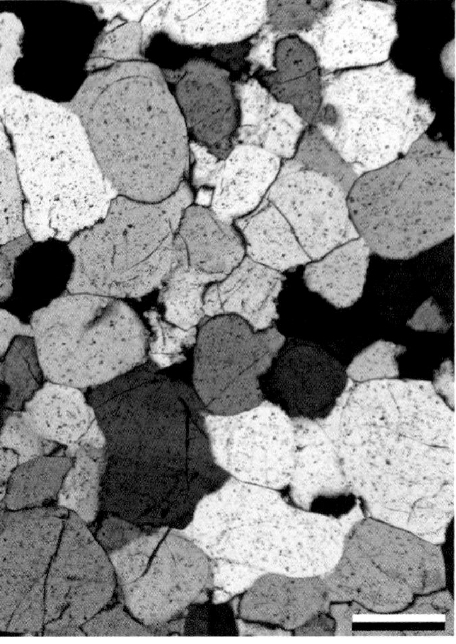

Up. Cambrian Gatesburg Fm., Centre Co., Pennsylvania

Another example of a completely quartz-cemented supermature sandstone. The rock has no detrital clay matrix and contains well-rounded and well-sorted grains, virtually entirely composed of quartz. Thus it is both texturally and compositionally supermature, but the maturity is harder to judge given the degree of overgrowth cementation. Fortunately, there generally is a clear line of micro-inclusions marking the contacts between detrital grain cores and overgrowths allowing assessment of size and rounding. In addition, cross-polarized light allows easier evaluation of sizes because grains show varied extinction positions.

PPL | XPL, Scale bar = 0.26 mm

Mid.–Up. Cambrian Riley Fm., Hickory Sandstone Mbr., Llano Co., Texas

In-situ brittle grain deformation in a supermature quartzarenite, as seen here, also can complicate textural evaluation. The shattering of grains, especially where there is more shear rotation of fragments, can impact estimates of both grain size and sorting. In this case, the use of plane- and cross-polarized light images allows one to see the excellent rounding and fairly good sorting in this area of the sample.

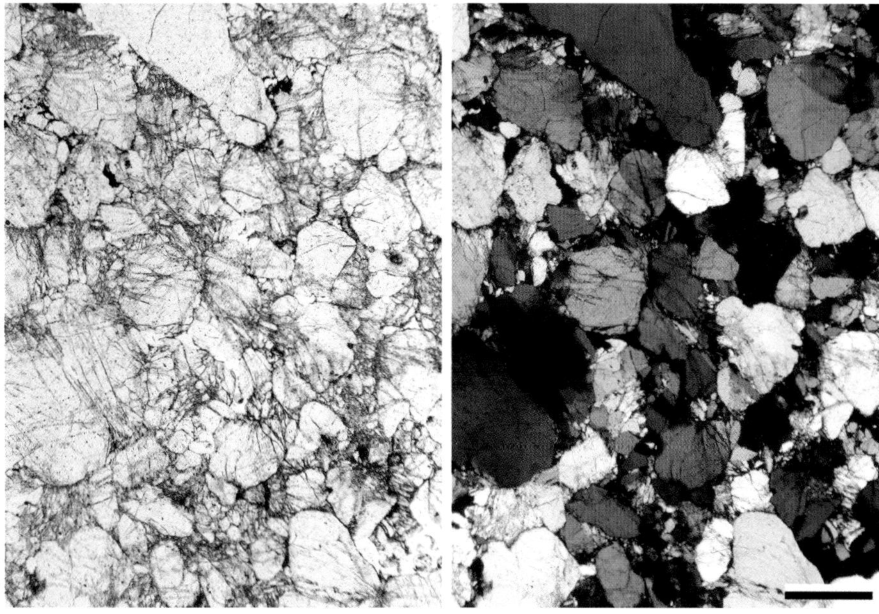

PPL | XPL, Scale bar = 0.26 mm

Silurian Clinton Ironstone, Giles Co., Virginia

An extremely well-rounded sandstone with large and small quartz grains and numerous small sedimentary rock fragments. This bimodal sorting conflicts, in a sense, with the term "supermature"; on the other hand, the rounding is far better than what would normally found in mature rocks. That conundrum is common in sandstones derived from sedimentary sources (second-cycle sands) or in deposits of eolian deflationary flats (Folk, 1968). This extremely well rounded but bimodal texture is termed a "textural inversion", and the rock would be termed "bimodal supermature".

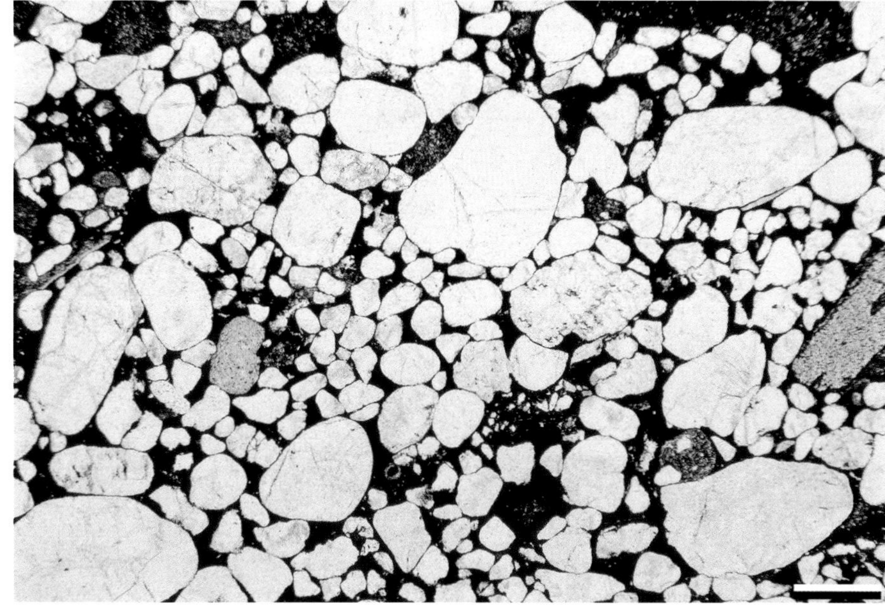

PPL, Scale bar = 0.51 mm

Mid. Cambrian Riley Fm., Hickory Sandstone Mbr., Llano Co., Texas

Another supermature sandstone with a bimodal distribution of its very well-rounded constituent quartz grains. The grains are outlined by early hematite coatings. This is part of a thick, shallow-marine, transgressive lag deposit overlying the "Great Unconformity" atop 1.1 to 1.3 billion year old Precambrian basement. These well-rounded grains were derived, in part, from earlier eolian sandstones and have undergone virtually no quartz overgrowth cementation (McBride, 2012). Because of their uniform rounding and weak cementation, they are now processed as frac sands in central Texas.

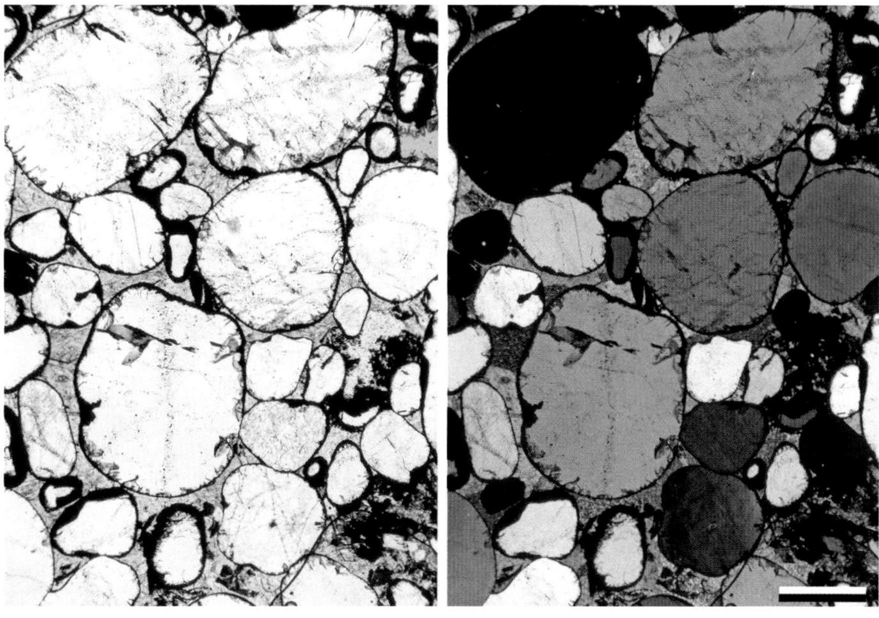

PPL | XPL, BDI, Scale bar = 0.51 mm

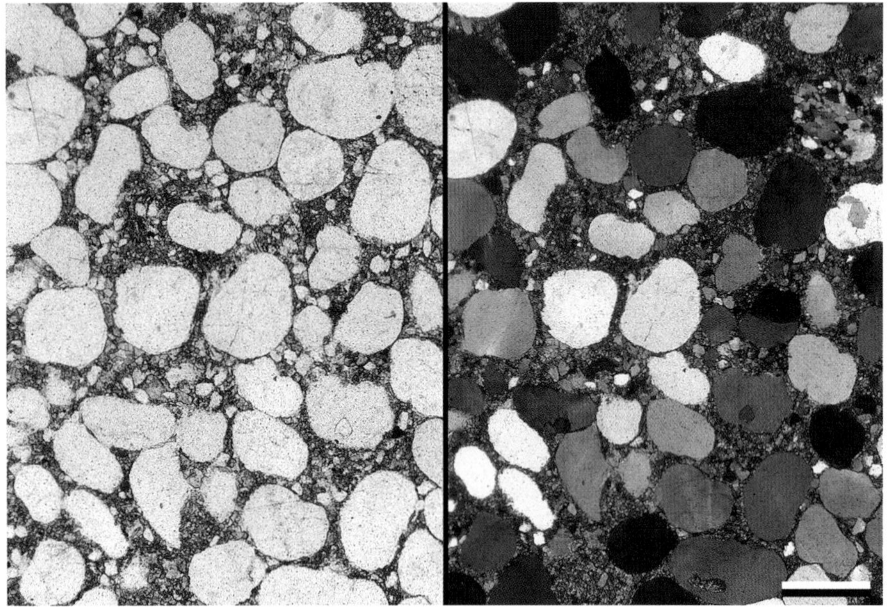

Up. Cambrian Eilean Dubh Fm., Sutherland, NW Scotland, U.K. †

A remarkable example of complete textural inversion. Here, large and well-rounded quartz grains are set in a matrix of much finer and more angular silt. Such profound textural differences typically occur through storm mixing, bioturbation or other processes that combine sediments from disparate setting (for example, washover fans delivering reworked barrier beach or dune sands into a normally, protected silty, lagoonal tidal flat environment). This example formed in a carbonate sabkha environment (supratidal mudflat) onto which eolian dunes migrated during periods of relative sea-level fall.

PPL | XPL, Scale bar = 0.49 mm

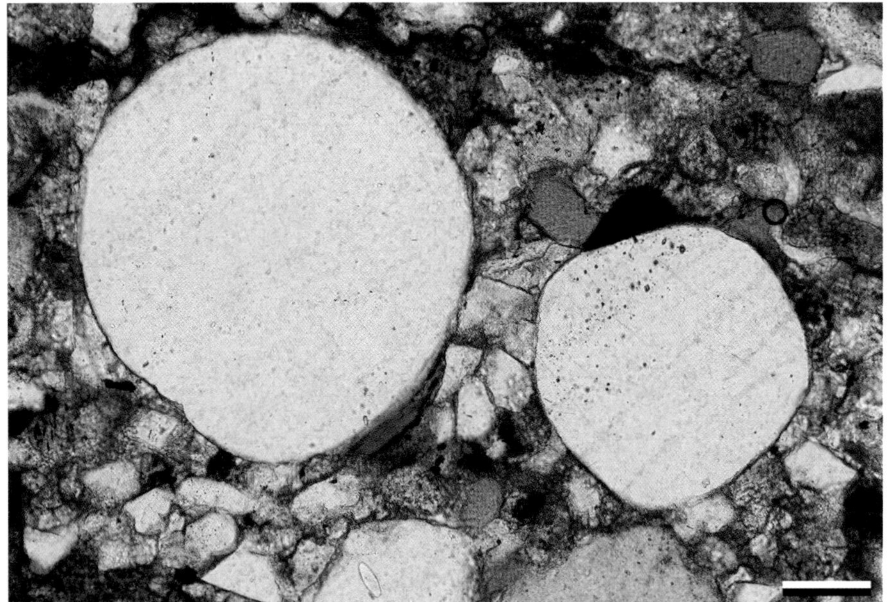

Mid. Jurassic Curtis Fm., Moffat Co., Colorado

An example of large, extremely well-rounded quartz grains scattered in a rock dominated by much smaller, more angular and poorly-sorted grains of quartz, feldspar and glauconite. This bimodal deposit (both in terms of size and rounding) results from sediment mixing by storms, bioturbation, depositional processes or, in part, through the derivation of grains from disparate (first- and second-cycle) sources. This "textural inversion" (partially excellent rounding but poor sorting) does not follow the normal pattern of maturity development and has no unambiguous name in the maturity classification.

PPL, Scale bar = 0.10 mm

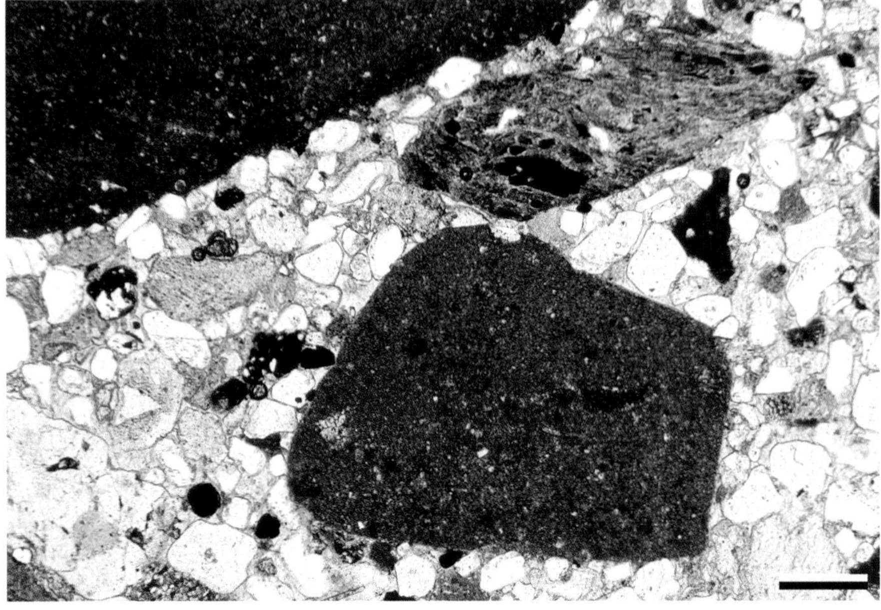

Paleogene Vieja Gp., Presidio Co., Texas

A somewhat unusual bimodal grain size distribution in an SRF-rich sandstone. Generally, the coarsest grains in such rocks are composed of abrasion resistant quartz. Here, the rock consists of small quartz grains mixed with large and abrasionally-weak limestone lithoclasts and some smaller shale clasts. This textural and compositional mismatch probably indicates a very local source of the limestone fragments that therefore had little time to be abraded. The finer-grained, yet well-rounded quartz sands may be second cycle from clastic terrigenous sources or it may simply reflect more distant sources and greater transport-related abrasion.

PPL, Scale bar = 0.51 mm

Cited References and Additional Information Sources

General:

Dennison, J. M., and J. H. Shea, 1966, Reliability of visual estimates of grain abundance: Journal of Sedimentary Research, v. 36, p. 81-89, doi: 10.1306/74D71410-2B21-11D7-8648000102C1865D.

Flemming, B. W., 2000, A revised textural classification of gravel-free muddy sediments on the basis of ternary diagrams: Continental Shelf Research, v. 20, p. 1125-1137, doi: 10.1016/S0278-4343(00)00015-7.

Folk, R. L., 1956, The role of texture and composition in sandstone classification: Discussion: Journal of Sedimentary Research, v. 26, p. 166-171.

Folk, R. L., 1980, Petrology of Sedimentary Rocks: Austin, TX, Hemphill's Book Store, 184 p. (free download at https://www.lib.utexas.edu/geo/folkready/contents.html)

Folk, R. L., P. B. Andrews, and D. W. Lewis, 1970, Detrital sedimentary rock classification and nomenclature for use in New Zealand: New Zealand Journal of Geology and Geophysics, v. 13, p. 937-968, doi: 10.1080/00288306.1970.10418211.

Friedman, G. M., 1961, Distinction between dune, beach, and river sands from their textural characteristics: Journal of Sedimentary Research, v. 31, p. 514-529, doi: 10.1306/74D70BCD-2B21-11D7-8648000102C1865D.

McKinley, J. M., P. M. Atkinson, C. D. Lloyd, A. H. Ruffell, and R. H. Worden, 2011, How porosity and permeability vary spatially with grain size, sorting, cement volume, and mineral dissolution in fluvial Triassic sandstones: The value of geostatistics and local regression: Journal of Sedimentary Research, v. 81, p. 844-858, doi: 10.2110/jsr.2011.71.

Passega, R., 1957, Texture as characteristic of clastic deposition: AAPG Bulletin, v. 41, p. 1952-1984.

Picard, M. D., 1971, Classification of fine-grained sedimentary rocks: Journal of Sedimentary Research, v. 41, p. 179-195, doi: 10.1306/74D7221B-2B21-11D7-8648000102C1865D.

Schäfer, A., and T. Teyssen, 1987, Size, shape and orientation of grains in sands and sandstones—image analysis applied to rock thin-sections: Sedimentary Geology, v. 52, p. 251-271, doi: 10.1016/0037-0738(87)90064-9.

Shepard, F. P., 1954, Nomenclature based on sand-silt-clay ratios: Journal of Sedimentary Research, v. 24, p. 151-158, doi: 10.1306/D4269774-2B26-11D7-8648000102C1865D.

Terry, J. P., and J. Goff, 2014, Megaclasts: Proposed revised nomenclature at the coarse end of the Udden-Wentworth grain-size scale for sedimentary particles: Journal of Sedimentary Research, v. 84, p. 192-197, doi: 10.2110/jsr.2014.19.

Matrix and maturity:

Audley-Charles, M. G., 1967, Greywackes with a primary matrix from the Viqueque Formation (upper Miocene-Pliocene), Timor: Journal of Sedimentary Research, v. 37, p. 5-11, doi: 10.1306/74D7163B-2B21-11D7-8648000102C1865D.

Brenchley, P. J., 1969, Origin of matrix in Ordovician greywackes, Berwyn hills, north Wales: Journal of Sedimentary Research, v. 39, p. 1297-1301, doi: 10.1306/74D71E10-2B21-11D7-8648000102C1865D.

Cox, R., and D. R. Lowe, 1996, Quantification of the effects of secondary matrix on the analysis of sandstone composition, and a petrographic-chemical technique for retrieving original framework grain modes of altered sandstones: Journal of Sedimentary Research, v. 66, p. 548-558, doi: 10.1306/D42683A1-2B26-11D7-8648000102C1865D.

Dickinson, W. R., 1970, Interpreting detrital modes of graywacke and arkose: Journal of Sedimentary Research, v. 40, p. 695-707, doi: 10.1306/74D72018-2B21-11D7-8648000102C1865D.

Dott, R. H., 1964, Wacke, graywacke and matrix; what approach to immature sandstone classification?: Journal of Sedimentary Research, v. 34, p. 625-632, doi: 10.1306/74D71109-2B21-11D7-8648000102C1865D.

Dunn, T. L., 1992, Infiltrated materials in cretaceous volcanogenic sandstones, San Jorge Basin, Argentina, *in* D. W. Houseknecht, and E. D. Pittman, eds., Origin, Diagenesis, and Petrophysics of Clay Minerals in Sandstones: Tulsa, OK, SEPM Special Publication 47, p. 159-174, doi: 10.2110/pec.92.47.0159.

Fischer, G., 1933, Die Petrographie der Grauwacken: Preussische Geologische Landesanstalt, Jahrbuch, v. 54, p. 320-320.

Folk, R. L., 1951, Stages of textural maturity in sedimentary rocks: Journal of Sedimentary Research, v. 21, p. 127-130.

Hathon, L. A., and D. W. Houseknecht, 1992, Origin and diagenesis of clay minerals in the Oligocene Sespe Formation, Ventura basin, *in* D. W. Houseknecht, and E. D. Pittman, eds., Origin, Diagenesis, and Petrophysics of Clay Minerals in Sandstones: Tulsa, OK, SEPM Special Publication 47, p. 185-195, doi: 10.2110/pec.92.47.0185.

Matlack, K. S., D. W. Houseknecht, and K. R. Applin, 1989, Emplacement of clay into sand by infiltration: Journal of Sedimentary Research, v. 59, p. 77-87, doi: 10.1306/212F8F21-2B24-11D7-8648000102C1865D.

Moraes, M. A. S., and L. F. De Ros, 1992, Depositional, infiltrated and authigenic clays in fluvial sandstones of the Jurassic Sergi Formation, Recôncavo Basin, northeastern Brazil, *in* D. W. Houseknecht, and E. D. Pittman, eds., Origin, Diagenesis, and Petrophysics of Clay Minerals in Sandstones: Tulsa, OK, SEPM Special Publication 47, p. 197-208, doi: 10.2110/pec.92.47.0197.

Rahmani, R., A., 1968, Greywackes with a primary matrix from the Viqueque Formation (upper Miocene-Pliocene), Timor: Comment on a paper by MG Audley-Charles: Journal of Sedimentary Research, v. 38, p. 271-273.

Shannon, P. M., 1978, The petrology of some lower Palaeozoic greywackes from South-east Ireland: A clue to the origin of the matrix: Journal of Sedimentary Research, v. 48, p. 1185-1192, doi: 10.1306/212F7626-2B24-11D7-8648000102C1865D.

Whetten, J. T., and J. W. Hawkins, 1970, Diagenetic origin of graywacke matrix minerals: Sedimentology, v. 15, p. 347-361, doi: 10.1111/j.1365-3091.1970.tb02191.x.

Wilson, M. D., and E. D. Pittman, 1977, Authigenic clays in sandstones; recognition and influence on reservoir properties and paleoenvironmental analysis: Journal of Sedimentary Research, v. 47, p. 3-31, doi: 10.1306/212F70E5-2B24-11D7-8648000102C1865D.

Worden, R. H., and S. Morad, 2003, Clay minerals in sandstones: Controls on formation, distribution and evolution, *in* R. H. Worden, and S. Morad, eds., Clay Mineral Cements in Sandstones (IAS Special Publication 34): Oxford, UK, Wiley-Blackwell, p. 1-41, doi: 10.1002/9781444304336.ch1.

Grain size and sorting:

Blair, T. C., and J. G. McPherson, 1999, Grain-size and textural classification of coarse sedimentary particles: Journal of Sedimentary Research, v. 69, p. 6-19, doi: 10.2110/jsr.69.6.

Blott, S. J., and K. Pye, 2001, GRADISTAT: A grain size distribution and statistics package for the analysis of unconsolidated sediments: Earth Surface Processes and Landforms, v. 26, p. 1237-1248, doi: 10.1002/esp.261.

Folk, R. L., 1954, The distinction between grain size and mineral composition in sedimentary-rock nomenclature: Journal of Geology, v. 62, p. 344-359, doi: 10.1086/626171.

Folk, R. L., 1966, A review of grain-size parameters: Sedimentology, v. 6, p. 73-93, doi: 10.1111/j.1365-3091.1966.tb01572.x.

Folk, R. L., 1968, Bimodal supermature sandstones: Product of the desert floor: International Geological Congress, Report of 23rd Session (Prague), Proceedings of Section 8, Genesis and Classification of Sedimentary Rocks, p. 9-32.

Friedman, G. M., 1962, On sorting, sorting coefficients, and the lognormality of the grain-size distribution of sandstones: Journal of Geology, v. 70, p. 737-753.

Gao, S., and M. Collins, 1992, Net sediment transport patterns inferred from grain-size trends, based upon definition of "transport vectors": Sedimentary Geology, v. 81, p. 47-60, doi: 10.1016/0037-0738(92)90055-V.

Harrell, J., 1984, A visual comparator for degree of sorting in thin and plane sections: Journal of Sedimentary Research, v. 54, p. 646-650.

Harrell, J. A., and K. A. Eriksson, 1979, Empirical conversion equations for thin-section and sieve derived size distribution parameters: Journal of Sedimentary Research, v. 49, p. 273-280, doi: 10.1306/212F7711-2B24-11D7-8648000102C1865D.

Jerram, D. A., 2001, Visual comparators for degree of grain-size sorting in two and three-dimensions: Computers & Geosciences, v. 27, p. 485-492, doi: 10.1016/S0098-3004(00)00077-7.

Johnson, M. R., 1994, Thin section grain size analysis revisited: Sedimentology, v. 41, p. 985-999, doi: 10.1111/j.1365-3091.1994.tb01436.x. [see also discussion and reply in Sedimentology 43: 189-191]

Kellerhals, R., J. Shaw, and V. K. Arora, 1973, On grain size from thin-sections: Journal of Geology, v. 83, p. 79-96, doi: 10.1086/628046.

Klovan, J. E., 1966, The use of factor analysis in determining depositional environments from grain-size distributions: Journal of Sedimentary Research, v. 36, p. 115-125, doi: 10.1306/74D7141A-2B21-11D7-8648000102C1865D.

Krumbein, W. C., 1938, Size frequency distributions of sediments and the normal phi curve: Journal of Sedimentary Research, v. 8, p. 84-90, doi: 10.1306/D4269008-2B26-11D7-8648000102C1865D.

Longiaru, S., 1987, Visual comparators for estimating the degree of sorting from plane and thin section: Journal of Sedimentary Research, v. 57, p. 791-794.

McLaren, P., and D. Bowles, 1985, The effects of sediment transport on grain-size distributions: Journal of Sedimentary Research, v. 55, p. 457-470, doi: 10.1306/212F86FC-2B24-11D7-8648000102C1865D.

Moiola, R. J., and D. Weiser, 1968, Textural parameters: An evaluation: Journal of Sedimentary Research, v. 38, p. 45-53, doi: 10.1306/74D718C5-2B21-11D7-8648000102C1865D.

Passega, R., 1964, Grain size representation by CM patterns as a geological tool: Journal of Sedimentary Research, v. 34, p. 830-847.

Seelos, K., and F. Sirocko, 2005, RADIUS–rapid particle analysis of digital images by ultra-high-resolution scanning of thin sections: Sedimentology, v. 52, p. 669-681, doi: 10.1111/j.1365-3091.2005.00715.x.

Sevon, W. D., 1966, Distinction of New Zealand beach, dune, and river sands by their grain size distribution characteristics: New Zealand Journal of Geology and Geophysics, v. 9, p. 212-223, doi: 10.1080/00288306.1966.10422810.

Syvitski, J. P. M., ed., 2007, Principles, Methods and Application of Particle Size Analysis: Cambridge, Cambridge University Press, 388 p.

Udden, J. A., 1914, Mechanical composition of clastic sediments: GSA Bulletin, v. 25, p. 655-744.

Visher, G. S., 1969, Grain size distributions and depositional processes: Journal of Sedimentary Research, v. 39, p. 1074-1106, doi: 10.1306/74D71D9D-2B21-11D7-8648000102C1865D.

Wentworth, C. K., 1922, A scale of grade and class terms for clastic sediments: Journal of Geology, v. 30, p. 377-392.

Shape and roundness:

Benn, D. I., and C. K. Ballantyne, 1993, The description and representation of particle shape: Earth Surface Processes and Landforms, v. 18, p. 665-672, doi: 10.1002/esp.3290180709.

Crook, K. A. W., 1968, Weathering and roundness of quartz sand grains: Sedimentology, v. 11, p. 171-182, doi: 10.1111/j.1365-3091.1968.tb00851.x.

Dott, R. H., Jr., 2003, The importance of eolian abrasion in supermature quartz sandstones and the paradox of weathering on vegetation-free landscapes: Journal of Geology, v. 111, p. 387-405, doi: 10.1086/375286.

Hofmann, H. J., 1994, Grain-shape indices and isometric graphs: Journal of Sedimentary Research, v. 64, p. 916-920. [see also discussion and reply in JSR, v. 65, p. 719-723].

Howard, J. L., 1992, An evaluation of shape indices as palaeoenvironmental indicators using quartzite and metavolcanic clasts in Upper Cretaceous to Palaeogene beach, river and submarine fan conglomerates: Sedimentology, v. 39, p. 471-486, doi: 10.1111/j.1365-3091.1992.tb02128.x.

Khalaf, F. I., and I. M. Gharib, 1985, Roundness parameters of quartz grains of Recent aeolian sand deposits in Kuwait: Sedimentary Geology, v. 45, p. 147-158, doi: 10.1016/0037-0738(85)90028-4.

Krinsley, D. H., and J. C. Doornkamp, 2011, Atlas of Quartz Sand Surface Textures: Cambridge, UK, Cambridge University Press, 102 p.

Krumbein, W. C., 1941, Measurement and geological significance of shape and roundness of sedimentary particles: Journal of Sedimentary Research, v. 11, p. 64-72, doi: 10.1306/D42690F3-2B26-11D7-8648000102C1865D.

MacLeod, N., 2002, Geometric morphometrics and geological shape-classification systems: Earth-Science Reviews, v. 59, p. 27-47, doi: 10.1016/S0012-8252(02)00068-5.

Mazzullo, J., D. Sims, and D. Cunningham, 1986, The effects of eolian sorting and abrasion upon the shapes of fine quartz sand grains: Journal of Sedimentary Research, v. 56, p. 45-56, doi: 10.1306/212F887D-2B24-11D7-8648000102C1865D.

McBride, E. F., 2012, Geology of the Late Cambrian Hickory Sandstone in the Voca frac sand district, western Llano Uplift, Texas: GSA Abstracts with Programs, v. 45, no. 3, p. 85.

McBride, E. F., and M. D. Picard, 1987, Downstream changes in sand composition, roundness, and gravel size in a short-headed, high-gradient stream, northwestern Italy: Journal of Sedimentary Research, v. 57, p. 1018-1026, doi: 10.1306/212F8CD3-2B24-11D7-8648000102C1865D.

McFarlane, M. J., S. H. Coetzee, J. R. Kuhn, C. H. M. Vanderpost, and F. D. Eckardt, 2007, In situ rounding of quartz grains within an African surface weathering profile in North West Ngamiland, Botswana: Zeitschrift für Geomorphologie, v. 51, p. 269-286, doi: 10.1127/0372-8854/2007/0051-0269.

Oakey, R. J., M. Green, P. A. Carling, M. W. E. Lee, D. A. Sear, and J. Warburton, 2005, Grain-shape analysis—A new method for determining representative particle shapes for populations of natural grains: Journal of Sedimentary Research, v. 75, p. 1065-1073, doi: 10.2110/jsr.2005.079.

Powers, M. C., 1953, A new roundness scale for sedimentary particles: Journal of Sedimentary Research, v. 23, p. 117-119.

Sneed, E. D., and R. L. Folk, 1958, Pebbles in the lower Colorado River, Texas A study in particle morphogenesis: Journal of Geology, v. 66, p. 114-150, doi: 10.1086/626490.

Twenhofel, W. H., 1945, The rounding of sand grains: Journal of Sedimentary Research, v. 15, p. 59-71, doi: 10.1306/D4269224-2B26-11D7-8648000102C1865D.

Zingg, T., 1935, Beiträge zur Schotteranalyse: Schweizerische Mineralogische und Petrologische Mitteilungen, v. 15, p. 39-140.

Orientation, imbrication and packing:

Bonham, L. C., and J. H. Spotts, 1971, Measurement of grain orientation, *in* R. E. Carver, ed., Procedures in Sedimentary Petrology: New York, Wiley-Interscience, p. 285-312.

Martini, I. P., 1971, A test of validity of quartz grain orientation as a paleocurrent and paleoenvironmental indicator: Journal of Sedimentary Research, v. 41, p. 60-68, doi: 10.1306/74D721E4-2B21-11D7-8648000102C1865D.

McBride, E. F., 2012, Heterogeneous packing and quartz cementation of the Eureka quartzarenite (Middle Ordovician), Utah and Nevada, USA: Journal of Sedimentary Research, v. 82, p. 664-680, doi: 10.2110/jsr.2012.58.

Onions, D., and G. V. Middleton, 1968, Dimensional grain orientation of Ordovician turbidite graywackes: Journal of Sedimentary Research, v. 38, p. 164-174, doi: 10.1306/74D7190B-2B21-11D7-8648000102C1865D.

Shelton, J. W., and D. E. Mack, 1970, Grain orientation in determination of paleocurrents and sandstone trends: AAPG Bulletin, v. 54, p. 1108-1119.

Spotts, J. H., 1964, Grain orientation and imbrication in Miocene turbidity current sandstones, California: Journal of Sedimentary Research, v. 34, p. 229-253, doi: 10.1306/74D71023-2B21-11D7-8648000102C1865D.

Facing Page: Top – A volcanic arenite from the Up. Cretaceous Ildefonso Fm., Ponce-Coamo area, Puerto Rico (XPL). Bottom – Navigating the terrible torrent of compositional terminology for sandstones.

SANDSTONE CLASSIFICATION

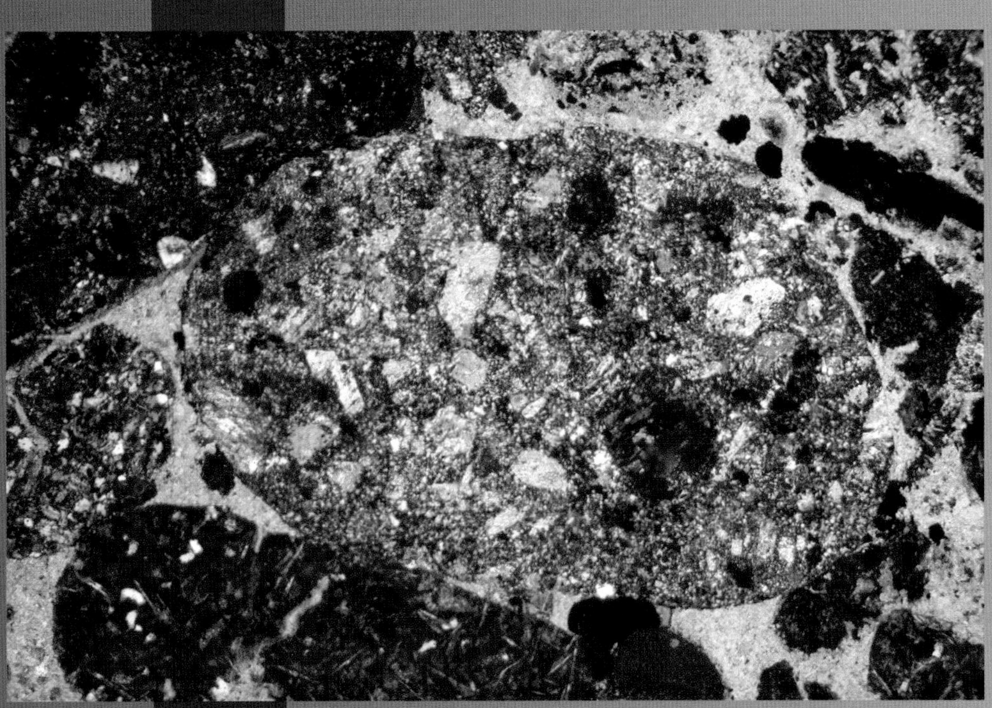

CHAPTER 7

subarkose
graywacke
quartz arenite
labile sandstone
calclithite
quartzite
feldspathic graywacke
high-rank graywacke
arkose
sublabile sandstone
sublitharenite
lithic wacke
volcanic arenite
phyllarenite
orthoquartzite
impure arkose
quartzarenite
labile graywacke
litharenite
Arghh!
sedarenite
feldspathic litharenite
lithic arkose
quartz wacke
feldspathic wacke
arkosite

COMPOSITIONAL CLASSIFICATION OF SANDSTONES

More than 50 different classification schemes for sandstones have been proposed over the past century. Some of those classifications are mainly related to the textural properties of clastic terrigenous deposits (and those are discussed in Chapter 6). Many of the sandstone classifications combine texture and composition, sometimes confusing the important distinction between those two characteristics. Others focus exclusively on composition, although some provide separate terms to describe textures. However, the many classifications typically share a few common characteristics. Most plot compositional data on ternary diagrams, with quartz, feldspar and lithic fragments as the poles—referred to as QFL diagrams. As always, however, the devil is in the details. For example, what qualifies as quartz (monocrystalline quartz, polycrystalline quartz, chert or other quartzose rock fragments) varies from classification to classification, as do the rock names and the percentage boundaries applied in different nomenclatural schemes.

Summary diagrams of several examples from the "golden age" of rampant sandstone classification (the late 1940s through the 1970s) are shown in Figures 7.1 and 7.2, mainly to show the divergence of approaches and terminology used. This comparison alone should suffice to highlight the fact that one must always state which classification is being used when one writes a report or paper, as the names have no rigorously consistent meanings from one classification to the next.

Not surprisingly, with such diverse nomenclatures having been used for many years, the geologic literature is muddled with a chaos of different names (or, worse yet, common names with varied meanings). Fortunately, in the decades since the 1970s, few new sandstone classifications were developed, and most workers have adopted one of the handful of "winners" in the Darwinian struggle for nomenclatural supremacy. The classification of Folk (Folk et al., 1970; Folk, 1980) is the one that will be discussed in detail here; however, the classifications of Dott (1964), McBride (1963) and a few others also retain strong adherents.

Part of the reason why some classifications have gained adherents and others have not is a reflection of the fact that there really are two purposes for classifications. The first is descriptive—to provide a clear, logical and easily usable way of accurately and reproducibly describing the composition of rocks or sediments. The second is interpretive—to have named groupings that provide some meaningful insights into provenance, depositional environments, compositional or diagenetic stability, or other interpretive topics. The classifications that have survived, and the classifications of Folk (1980 and earlier) in particular, are the ones that best achieve those two widely divergent goals.

So let's examine the Folk classification (Figure 7.3). At its heart, it uses the standard QFL ternary diagram. It includes a few specific grain types that are lithic fragments at the Q and F poles, however; metaquartzite is grouped with quartz (in part because metaquartzite is difficult to reliably distinguish from other polycrystalline quartz) and granitic and gneissic rock fragments are grouped with feldspars (because they are identified primarily by their feldspar content and are difficult to distinguish from each other). On the other hand, the Folk classification, unlike many others, groups chert with rock fragments and not with quartz (because it is easily distinguishable from "normal" quartz grains and because it really is a lithic fragment). The generalized names for the rocks close to the three poles are quartzarenite, arkose and litharenite (but as will be seen later those terms are not always the final ones applied). The term "quartzarenite" is used only for rocks with more than 95% quartz, a higher threshold than in most classifications because it helps to distinguish truly exceptionally quartz-rich deposits from the normally quartzose deposits that predominate in the geologic record. Arkose (for feldspar-rich rocks), litharenite (for lithic-fragment-rich rocks) and the other terms used for intermediate compositions have broad boundaries as shown in Figure 7.3.

Unlike in other classifications, however, most of those general terms are supplanted, where possible, by more specific terms as shown in the three subsidiary ternary diagrams in Figure 7.3. Rather than just calling a rock an arkose, one should, if possible, look in more detail at the feldspars and call it a "plagioclase arenite" or "K-feldspar arenite". Likewise, although a rock may be a litharenite, by looking in more detail at the nature of the constituent lithic fragments, one would instead term it a "volcanic arenite", a "phyllarenite" or a "sedarenite" (and if it classed as a sedarenite, one should characterize it in more detail and call it a "chert arenite", a "calclithite", a "sandstone arenite" or a "shale arenite").

The Folk classification also provides separate terms to describe the overall textural maturity of a sample (shown in very abbreviated form in the box in the lower part of Figure 7.3). It takes into account sorting, rounding and matrix content of the deposits, features that otherwise do not enter into the compositional classification. Thus, a rock could be named a "supermature quartzarenite", a "submature plagioclase subarkose", an "immature shale arenite", or any other combination of textural and compositional terms. This differs considerably from other classifications, for example Dott (1964) in which matrix content directly affects primary rock name (arenites versus wackes).

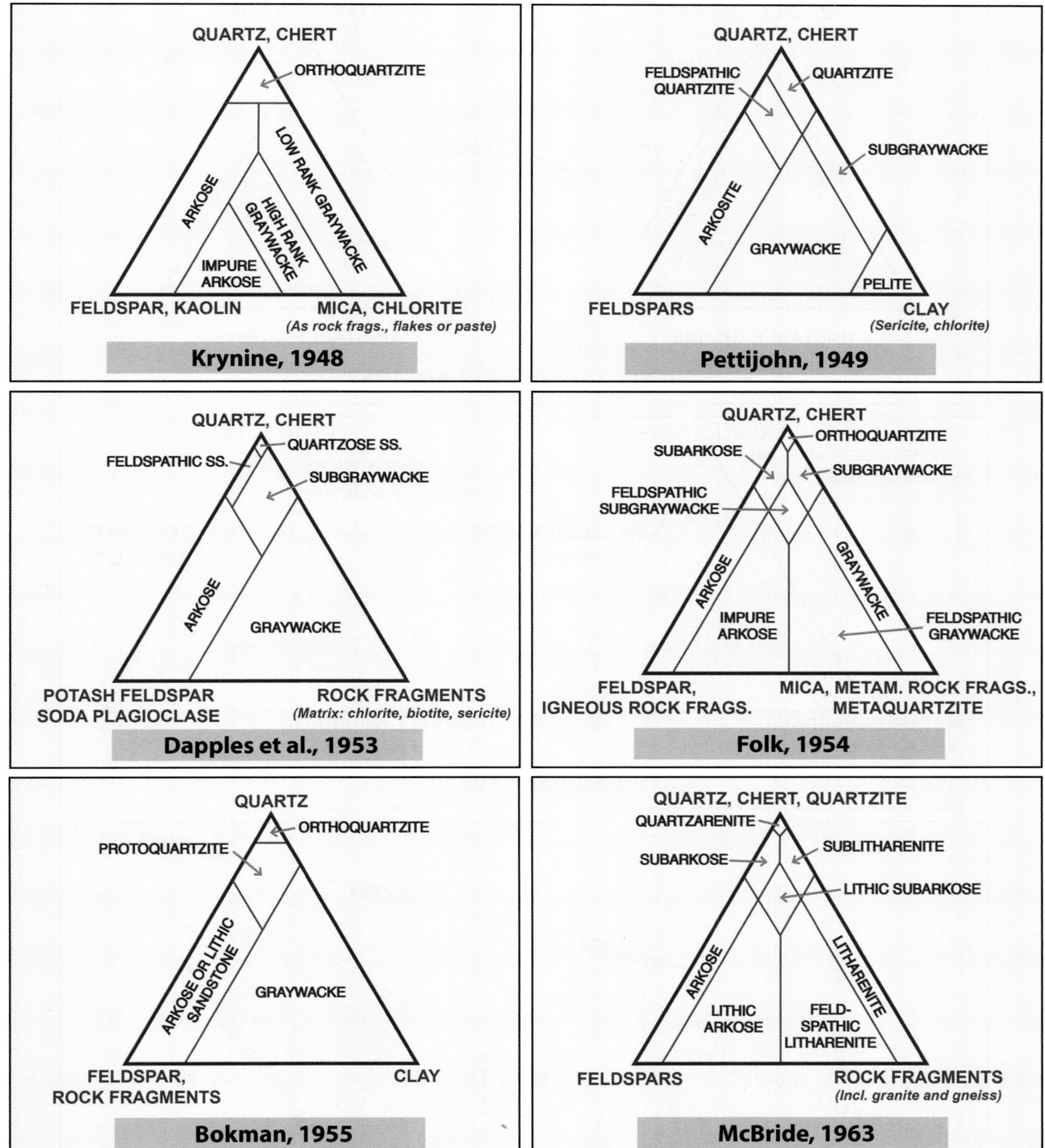

Figure 7.1: Comparative QFL diagrams for six sandstone classifications from the 1948-1963 period showing the variations in terminology and in materials plotted at each pole. Adapted from Krynine (1948), Pettijohn (1949), Dapples et al. (1953), Folk (1954), Bokman (1955), and McBride (1963).

Although the Folk classification sounds quite complex, it has proven to be both a very workable way to accurately classify rocks and sediments and an effective way to convey information about interpretive aspects of those deposits. As gross generalizations, quartzarenites are the products of intense abrasion, weathering, and/or diagenetic removal of less stable minerals; commonly they are derived from stable cratonic, multi-cycle

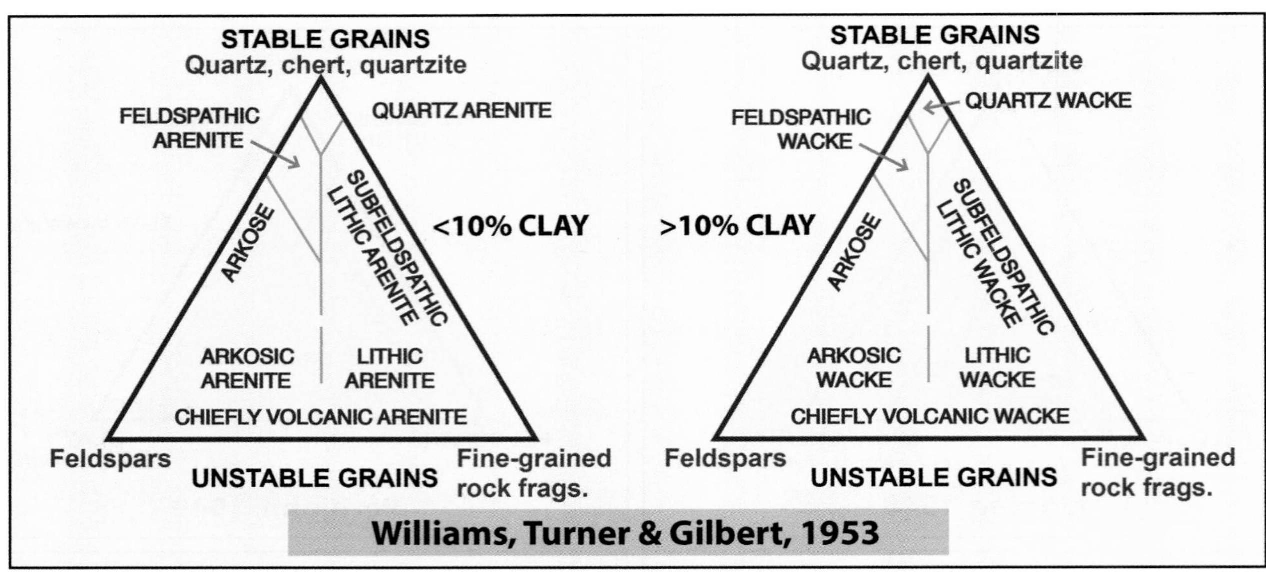
STABLE GRAINS
Quartz, chert, quartzite
FELDSPATHIC ARENITE
QUARTZ ARENITE
ARKOSE
SUBFELDSPATHIC LITHIC ARENITE
<10% CLAY
ARKOSIC ARENITE
LITHIC ARENITE
CHIEFLY VOLCANIC ARENITE
Feldspars
UNSTABLE GRAINS
Fine-grained rock frags.
STABLE GRAINS
Quartz, chert, quartzite
QUARTZ WACKE
FELDSPATHIC WACKE
>10% CLAY
ARKOSE
SUBFELDSPATHIC LITHIC WACKE
ARKOSIC WACKE
LITHIC WACKE
CHIEFLY VOLCANIC WACKE
Feldspars
UNSTABLE GRAINS
Fine-grained rock frags.
Williams, Turner & Gilbert, 1953

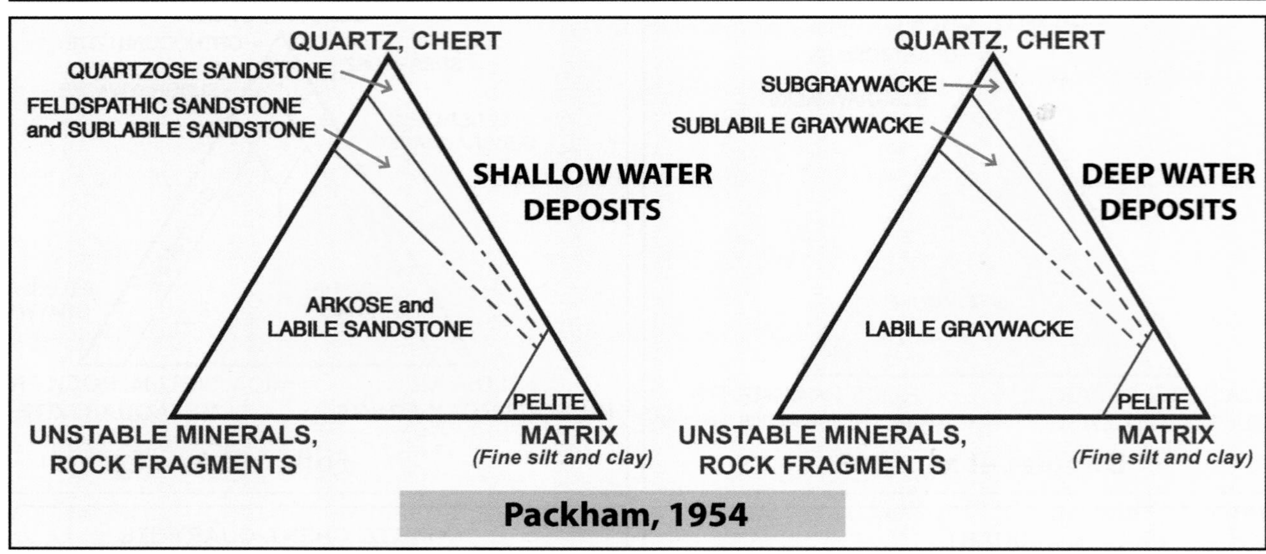
QUARTZ, CHERT
QUARTZOSE SANDSTONE
FELDSPATHIC SANDSTONE and SUBLABILE SANDSTONE
SHALLOW WATER DEPOSITS
ARKOSE and LABILE SANDSTONE
PELITE
UNSTABLE MINERALS, ROCK FRAGMENTS
MATRIX
(Fine silt and clay)
QUARTZ, CHERT
SUBGRAYWACKE
SUBLABILE GRAYWACKE
DEEP WATER DEPOSITS
LABILE GRAYWACKE
PELITE
UNSTABLE MINERALS, ROCK FRAGMENTS
MATRIX
(Fine silt and clay)
Packham, 1954

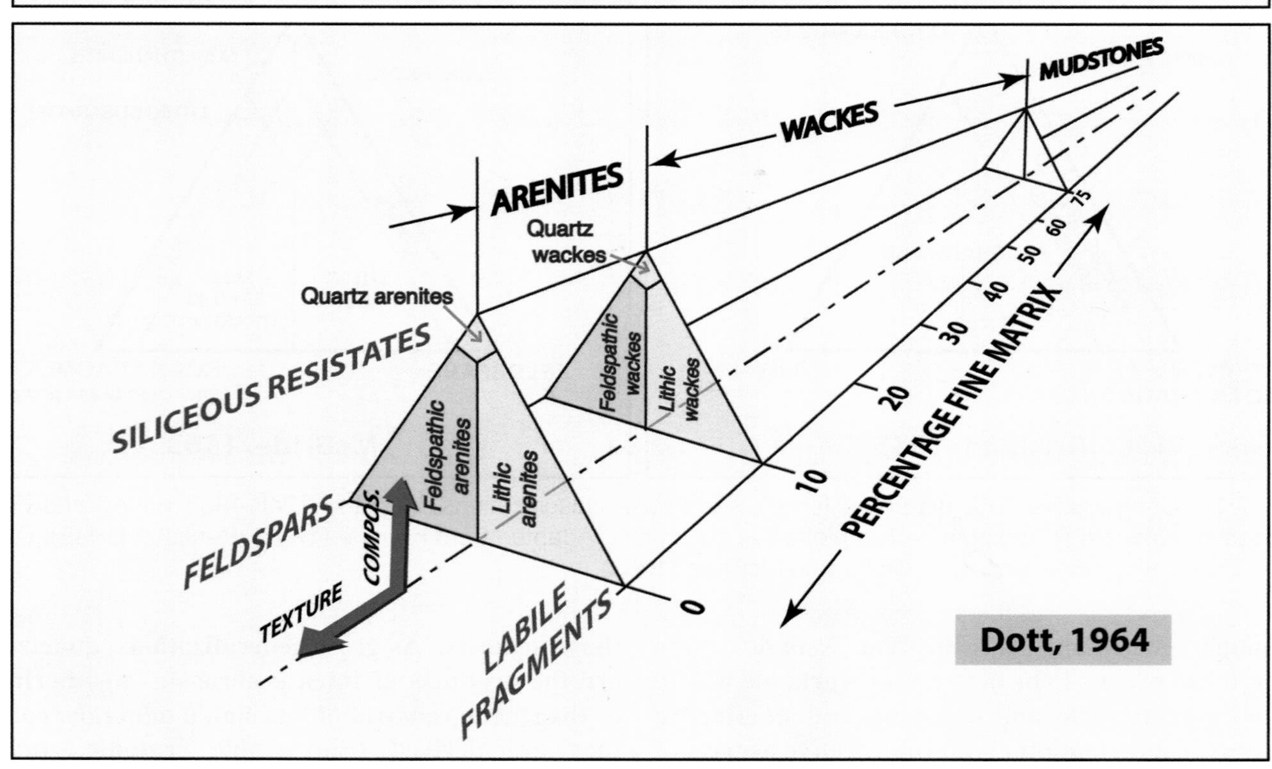
MUDSTONES
WACKES
ARENITES
Quartz wackes
Quartz arenites
SILICEOUS RESISTATES
Feldspathic wackes
Lithic wackes
Feldspathic arenites
Lithic arenites
FELDSPARS
COMPOS.
TEXTURE
LABILE FRAGMENTS
0
10
20
30
40
50
60
75
PERCENTAGE FINE MATRIX
Dott, 1964

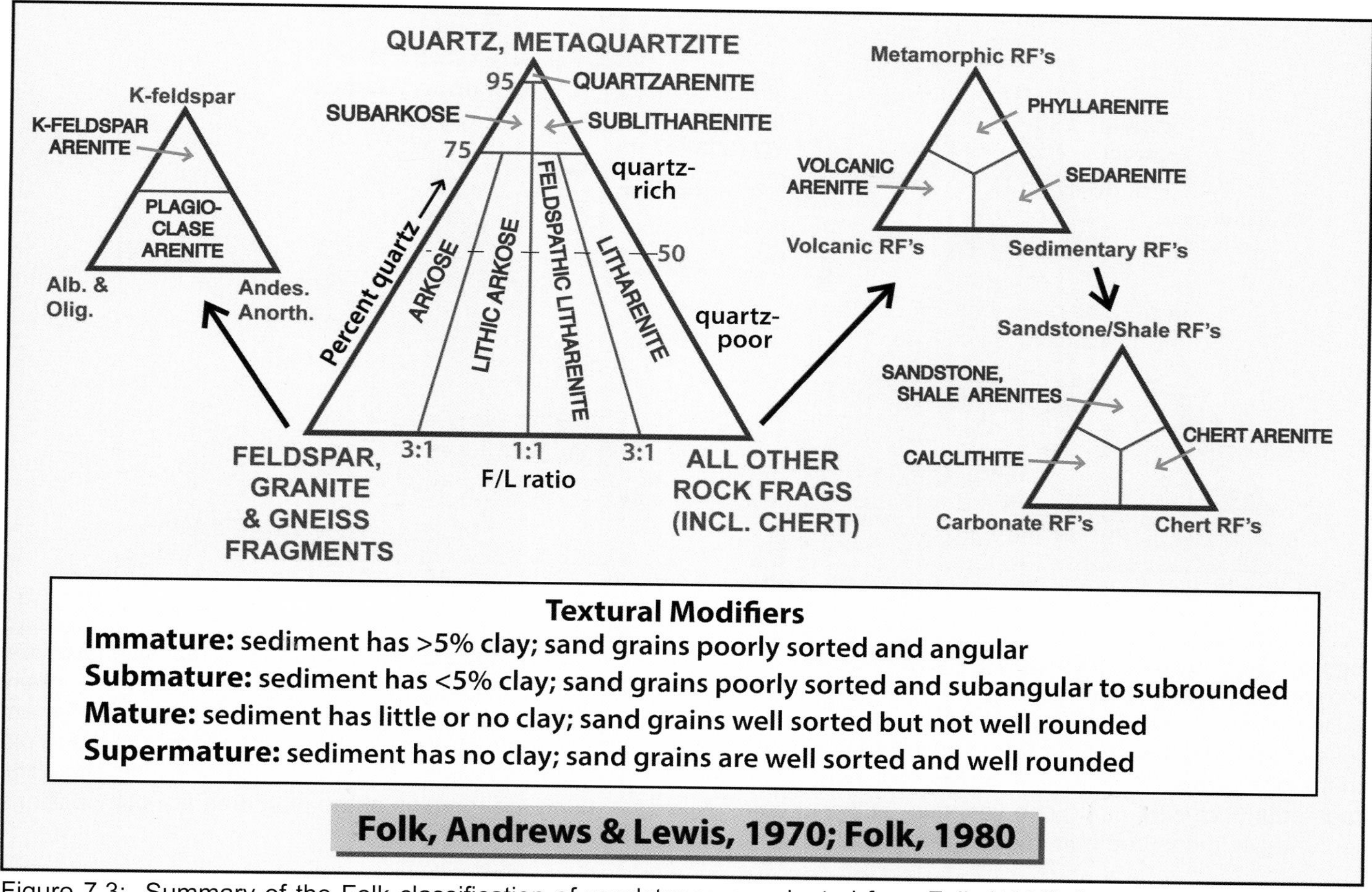

Figure 7.3: Summary of the Folk classification of sandstones as adapted from Folk (1980), based, in part, on earlier versions of the 1980 book as well as Folk et al. (1970).

sedimentation regimes. Arkoses (feldspathic arenites) and many calclithites and chert arenites commonly are derived from block-faulted basement uplifts associated with a variety of tectonic settings; arkoses in particular are more common in arid climates or high-rainfall zones of rapid erosion and short transport. Shale arenites, phyllarenites and some volcanic arenites are commonly associated with collisional tectonics and arc-trench systems.

Dickinson et al. (1983) provided considerably more detailed summaries of sandstone compositions as related to source terranes and tectonic settings and two of the summary diagrams from that paper are redrawn here as Figure 7.4. The two diagrams differ mainly with regard to what is specifically plotted at the Q, F and L poles (see figure caption for details) and both show the generalized compositional boundaries related to known tectonic source regions based on a very extensive data set (roughly 7500 samples). For those wanting to look further into the provenance and its relation to both compositional and textural aspects of classification the references in the provenance section of the bibliography at the end of this chapter are recommended reading. In addition, as noted in Chapter 3, some methods of counting the mineral constituents within lithic grains (specifically the Gazzi-Dickinson method) may yield better provenance results and may yield rock names that more closely match X-ray or bulk geochemical data.

In the final analysis, however, sandstone compositions are the product of a long series of factors stretching from source terranes and climate, to modes of transport, winnowing and deposition and on through diagenetic histories that may have been long and complex. Sandstone composition alone can provide only a very generalized framework for provenance interpretation. That framework must be supplemented with detailed analysis of specific QFL grains, petrographic analysis of detrital heavy mineral suites and geochemical/radiometric analysis of individual mineral separates to provide really useful provenance information.

A few closing words about classifications: the two chapters of this book devoted to textural and

Figure 7.2: Comparative nomenclatural diagrams for three sandstone classifications that integrate clayey matrix as a substantial part of their terminology. Adapted from Williams, Turner and Gilbert (1953), Packham (1954), and Dott (1964).

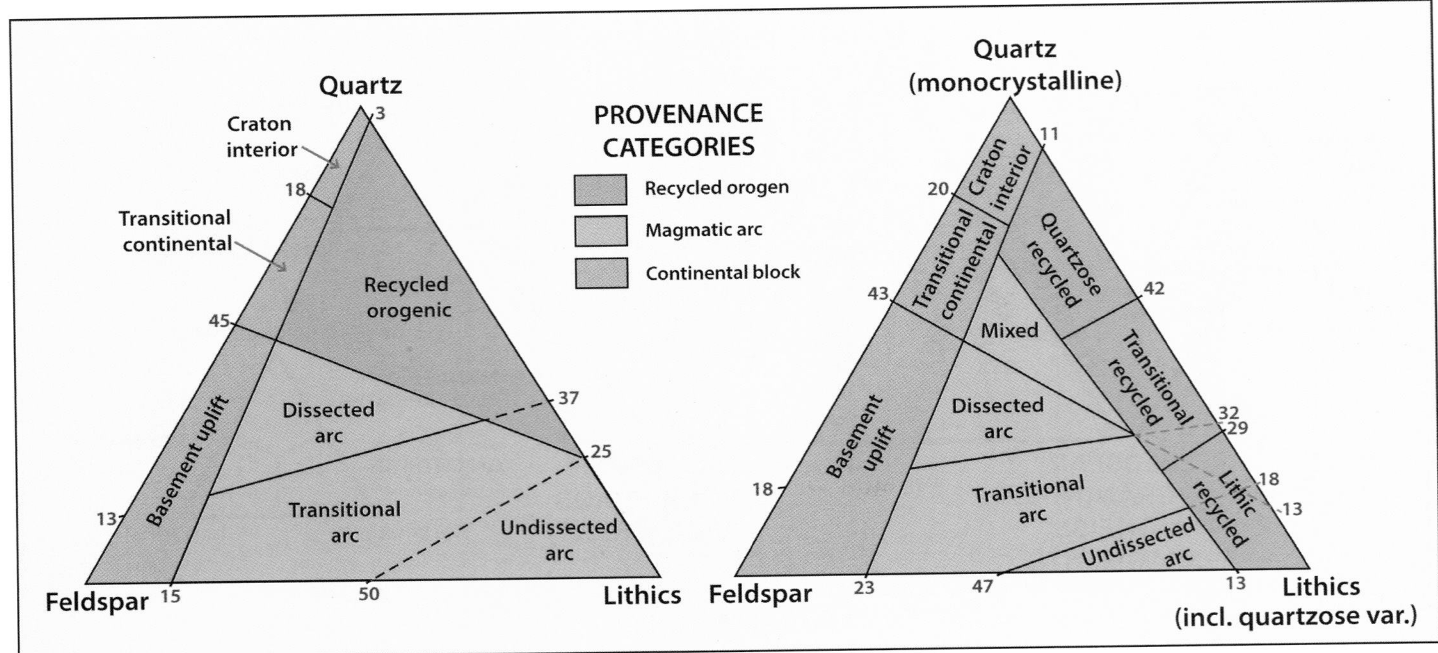

Figure 7.4: Two types of QFL plots showing the relationship between detrital terrigenous sandstone compositional modes and inferred provenance types; from Dickinson et al. (1983). The QFL plot on the left shows total quartzose grains (including polycrystalline grains such as chert and quartzite) at the Q pole, monocrystalline feldspars at the F pole and unstable polycrystalline lithic fragments (igneous, sedimentary or metamorphic) at the L pole. The modified QmFLt plot on the right shows exclusively monocrystalline quartz at the Qm pole, an unchanged F pole and total polycrystalline lithic fragments (including quartzose varieties) at the Lt pole. Numbers along the margins of the diagrams are compositional percentages measured from the nearest apical pole.

compositional terminology have provided you with many technical terms with very specific definitions. However, words are neither communication nor language, although they are the tools for both. You need to craft the terms presented in these chapters into effective communications with your target audience, turning words into sentences, and sentences into insights. "A crumbly green rock that tastes a little like salt" is effective communication (at a third grade school show-and-tell and so is "coarse sandstone: slightly glauconitic, weakly halite cemented, submature feldspathic chert arenite" (in field notes or a technical paper). The latter name puts into 10 words a great deal more information about the rock in question than the first name. It has a grain-size term, some adjectival phrases describing accessory grains and cements, a maturity term that conveys textural information, and two compositional words describing the first and second most abundant grain types in the rock. It is an example of how to put together the words provided in the classification chapters into more inclusive names—but is it effective communication? Well that depends on the audience. In the third grade class, the answer would be an unequivocal "No!" To an audience of sedimentologists the answer probably would be "yes", but a petroleum geologist may well say "I learned nothing about the porosity or permeability of this rock" and a materials engineer might want to know more details about the color and shape of anything that will be incorporated into building materials. However, there is nothing that precludes you from adding porosity terms, color categorizations or more specific shape and rounding/angularity terms into a rock name if that is what your target audience needs. So again, terminology is simply a shorthand device that allows the condensation of an enormous amount of information into a small space. Effective communication, on the other hand, is the use of those shorthand words to get very specific knowledge into the minds of your target audience. Just like great literature, that is where the art lies.

Finally, a few words about how to collect compositional data. For many purposes, visual estimates of grain percentages are both sufficient and cost effective (time being money) and a visual comparison chart is provided in the introduction of this book for that purpose. However, visual estimates can be surprisingly inaccurate (e.g., Allen, 1956; Dennison and Shea, 1966) and point counting of at least 300–500 grains is necessary for quantitative studies. References to books dealing with sedimentary techniques also are provided in the introduction to this volume and most of those sources detail specific methodologies for point counting.

Carboniferous Kulm graywacke, Bavaria, Germany

This is a submature medium-rank phyllarenite according to the Folk classification (Fig 7.3, Folk, 1980; Folk et al., 1970). Most of the matrix within this sample consists of micas and other minerals derived from the breakdown of schistose metamorphic rock fragments. More robust metamorphic rock fragments also comprise the dominant grains. In the McBride (1963) classification (Fig. 7.1), this sample is a litharenite, and in the Dott (1964) classification (Fig. 7.2), it would be classified as a lithic arenite.

PPL | XPL, Scale bar = 0.1 mm

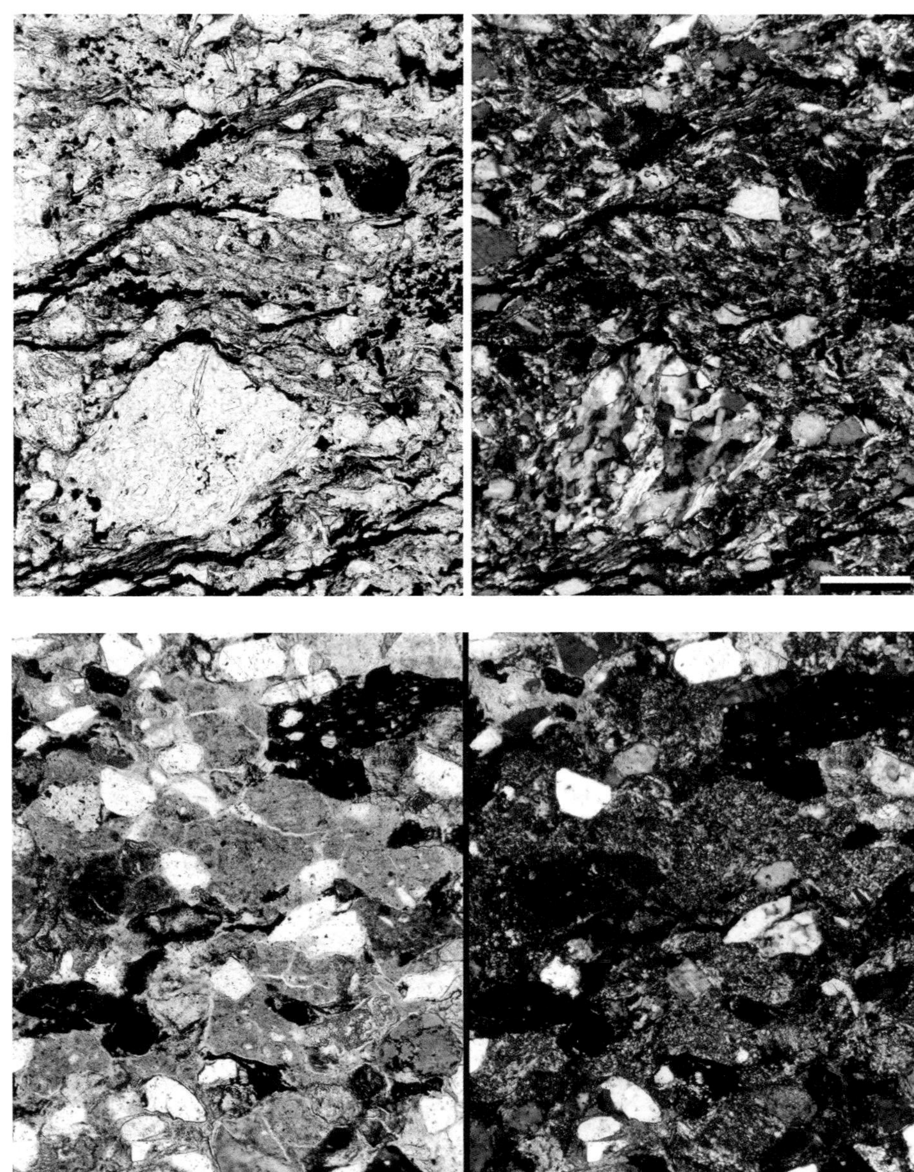

Triassic Chinle Fm., eastern Arizona

This is a mature low-rank phyllarenite (Folk), a litharenite (McBride) or lithic arenite (Dott). The shale clasts within this rock have been strongly compacted; the clays within them show preferential alignment. Many of the quartz grains are polycrystalline grains of metamorphic origin. The provenance of this material is probably quite local from remnant basement uplifts and younger sediments in a relatively stable platform setting.

PPL | XPL, Scale bar = 0.26 mm

Up. Cretaceous Ildefonso Fm., Ponce-Coamo area, Puerto Rico

This sample is composed of well-rounded but poorly-sorted volcaniclastic grains. In the Folk classification, it is a calcite-cemented, mature volcanic arenite. Because of the overall poor grain sorting within this rock, it would not be classified as a supermature rock. In the McBride classification, it would be a litharenite, and it is a lithic arenite in the Dott classification. From a provenance perspective, this rock was derived from an active Caribbean volcanic island arc.

XPL, Scale bar = 0.67 mm

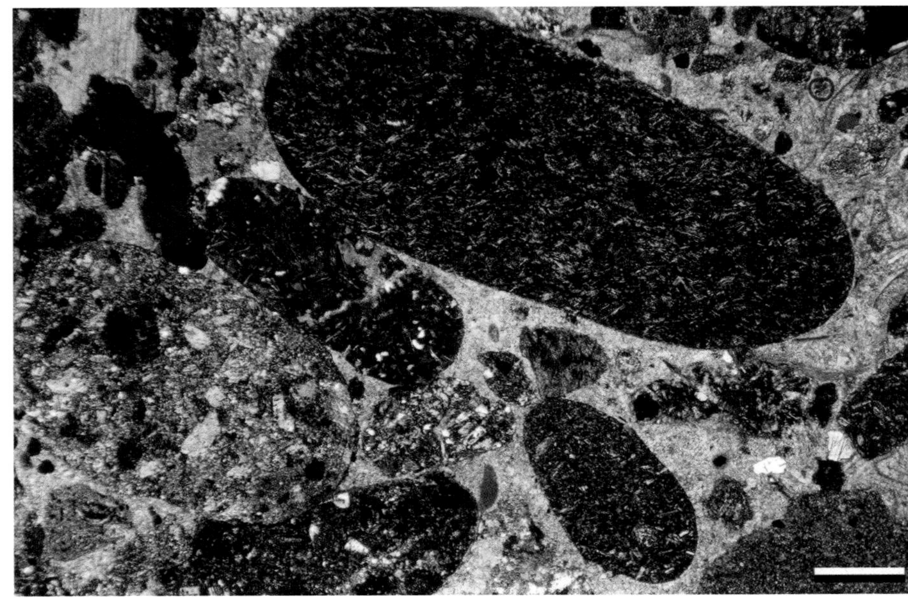

Up. Cretaceous Shumagin Fm., Shumagin Islands, Alaska

This is an immature, poorly-sorted volcanic arenite (Folk). Most of the rock fragments are subangular to subrounded, and the larger volcaniclastic grains are floating in a silt- to clay-sized matrix. In the McBride classification, this rock lies outside the normal sandstone terminology shown in Figure 7.1 and, because of its combination of matrix, rock fragments and feldspars, would be termed a "graywacke". It is a lithic wacke in the Dott classification (Fig. 7.2). The provenance of this material is largely from an active volcanic arc.

PPL, Scale bar = 0.26 mm

Up. Cretaceous Prince Creek Fm., North Slope, Alaska

This is a chert arenite (Folk). The Folk classification groups chert with lithic fragments, but in both the McBride and Dott classifications chert is grouped with quartz grains, making this rock a quartzarenite/quartz arenite in those latter classifications. The grains in this sample are moderately well sorted but are not well rounded; thus, the fuller Folk textural and compositional name for this sample is a "mature chert arenite". This is a nonmarine sandstone deposited in a foreland basin adjacent to a thrust belt.

PPL | XPL, BDI, Scale bar = 0.26 mm

Up. Eocene Oberaudorf Beds, Tirol, Austria

A sample containing angular clasts of chert as well as carbonate rock and fossil fragments in a muddy matrix. This rock would be classified as an immature cherty calclithite (Folk), as a cherty litharenite (McBride) or as a cherty lithic wacke (Dott). This deposit, like the previous one, is from a foreland basin, in this case receiving detritus from the European Alpine thrust belt.

PPL | XPL, Scale bar = 0.51 mm

ᴾliocene Verde Fm., Yavapai Co., ᴬrizona

ʼhis sample includes a variety of sandstone ınd siltstone grains (SRFs) in addition to ounded to subrounded quartz grains. The ample is thus classified as a mature siltstone ırenite (Folk), litharenite (McBride) or a lithic ırenite (Dott). These deposits formed in a lowly subsiding interior basin proximal to a edimentary source terrane containing clastic errigenous rocks.

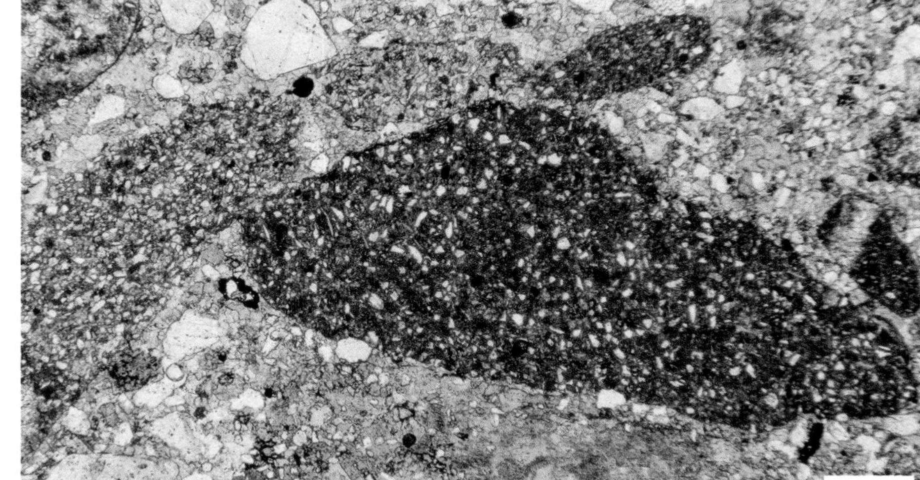

ᴾPL, Scale bar = 0.51 mm

ᵎilurian Tuscarora Sandstone, ˌebanon Co., Pennsylvania

ʼhis rock is a submature sedarenite (Folk). ʼhe grains present include monocrystalline ˌuartz, chert, shale, sandstone and schist rock ragments. Because this sample contains redominantly mixed sedimentary plus ubordinate metamorphic rock fragments, he name is the more generic sedarenite. In he other classifications, it is a litharenite McBride) or a lithic arenite (Dott). This ock too is from a foreland basin that received naterial from an eroding thrust belt, in this ase, one that developed during the Taconic rogeny.

ᴾPL | XPL, Scale bar = 0.34 mm

ᴶp. Cambrian Sillery Gp., Gaspé ᴾeninsula, Quebec, Canada

ʼhis sandstone contains a variety of shale ınd siltstone SRFs in addition to rounded to ubrounded quartz grains. It may have some rimary matrix, but much of the interstitial lay appears to result from crushing of shale ᵎRFs. Thus, in the Folk classification this is ermed an immature to submature shale arenite depending on what percentage of the matrix s considered primary or secondary); in other lassifications it is a litharenite (McBride) or a ithic wacke (Dott).

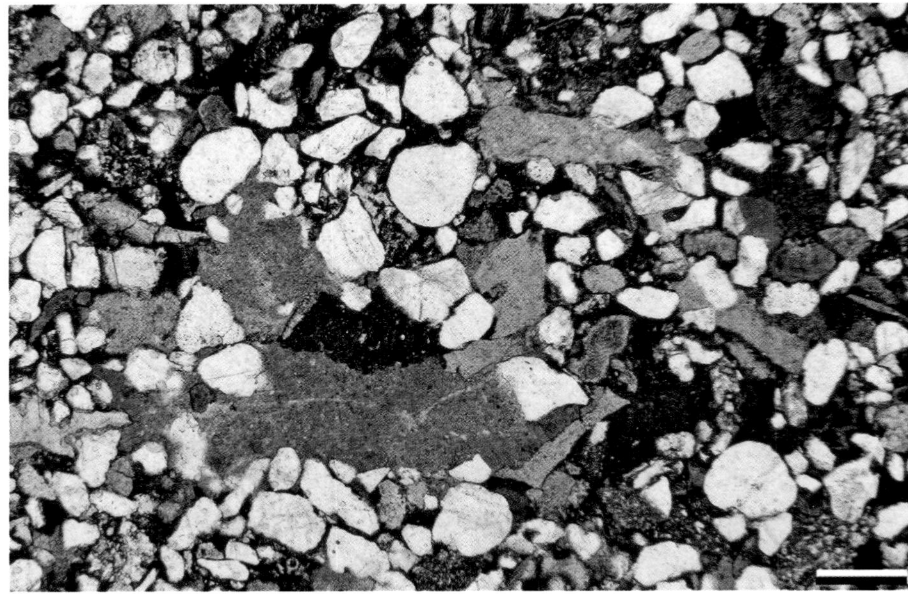

ᴾPL, Scale bar = 0.51 mm

Lo. Oligocene Nile Gp., Kongahu Breccia, Westland, New Zealand

The grains in this sample are quartz, feldspars and carbonate rock or fossil fragments (stained red). All of the grains are encased in blue-stained, highly ferroan calcite cement. This rock would be classified as a submature subarkose in the Folk classification (using the separate textural and composition names), or it is classified as a subarkose (McBride) or a feldspathic arenite (Dott). This marine deposit was shed from uplifts paralleling the Alpine strike-slip fault zone of New Zealand.

PPL | XPL, AFeS, Scale bar = 0.33 mm

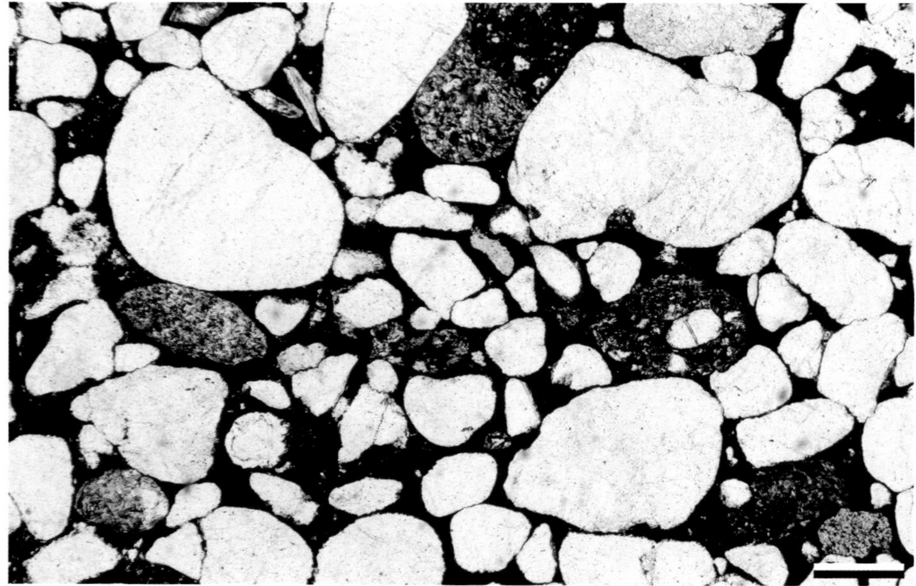

Silurian Clinton Ironstone, Giles Co., Virginia

This is a bimodal, hematite-cemented sandstone consisting of large, well-rounded second-cycle quartz grains and sandstone rock fragments probably eroded from thrusted sedimentary rocks in the Taconic orogenic belt. This rock is classified as a sublitharenite in the Folk and McBride schemes (or a sandstone arenite if those grains exceeded 25% of the grain total in the sample as a whole). It would be a lithic arenite in the Dott classification.

PPL, Scale bar = 0.26 mm

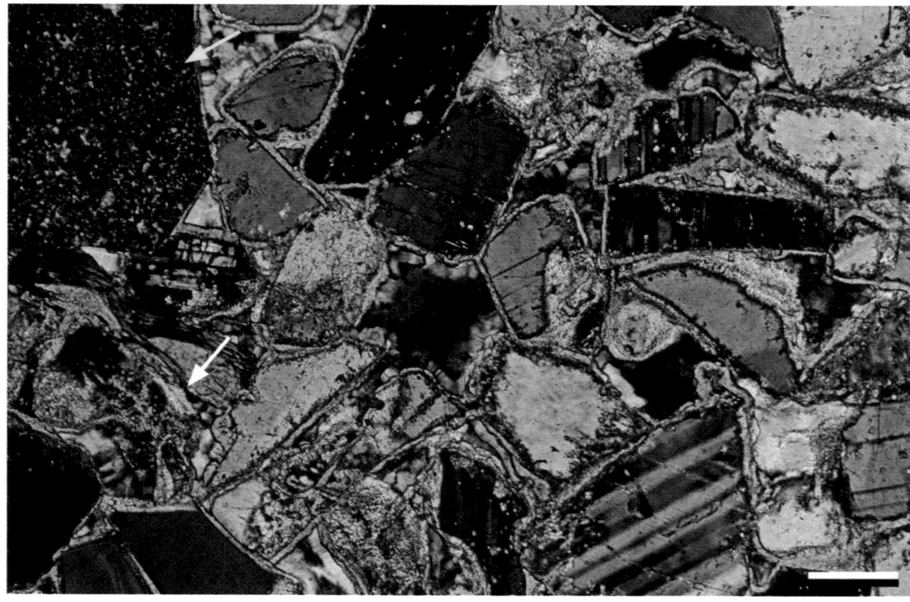

Cretaceous Mowry Fm., Moffat Co., Colorado

The grains in this well-cemented sample include feldspars (both plagioclase and K-feldspars), chert (example at yellow arrow), schist rock fragments (example at white arrow) and quartz. Thus, this is a mature lithic arkose in the Folk classification. It is an lithic arkose in the McBride system and a feldspathic arenite in the Dott classification. This deposit was formed in a shallow-marine foreland basin associated with thrusting and uplift in the associated Sevier orogenic belt.

XPL, Scale bar = 0.1 mm

Mid. Jurassic Entrada Fm., Moffat Co., Colorado

This rock is a laminated, mature arkose (more specifically plagioclase arenite) in the Folk classification. In the other classifications, it is an arkose (McBride) or feldspathic arenite (Dott). The rock has interlaminated quartz-rich and feldspar-rich horizons that also have preferential concentrations of heavy minerals (tourmaline, magnetite and zircon). The rock was deposited under very arid climatic conditions as part of an eolian dune complex in a stable shelf setting.

PPL | XPL, PFS, KFS, Scale bar = 0.51 mm

Paleocene sandstone, Sri Lanka

This rock is a submature arkose (K-feldspar arenite) according to the Folk classification. It contains K-feldspar, plagioclase, rock fragments and quartz. Most of the rock fragments were plastically deformed during compaction. In the other two classifications, this sample would be a arkose (McBride) or a feldspathic arenite (Dott). This sandstone was deposited as a marine turbidite, sourced from a granitic or charnockitic terrane.

PPL, KFS, BDI, Scale bar = 0.49 mm

Mid. Permian Wegener Halvø Fm., Karstryggen area, East Greenland

This sample is composed of K-feldspars and mono- and polycrystalline quartz grains in a silty matrix. This is an immature arkose (K-feldspar arenite) in the Folk classification. In the McBride classification it is an arkose, and in the Dott classification it is a feldspathic wacke. This rock was derived from Hercynian basement exposed in a block-faulted terrane.

XPL, Scale bar = 0.51 mm

Mid. Jurassic Entrada Fm., Moffat Co., Colorado

This is a weakly cemented sandstone with well-sorted and fairly well-rounded grains. It is classed as a mature to supermature subarkose in the Folk scheme; a subarkose using the McBride classification or a feldspathic arenite (Dott). In this rock, the feldspars were stained orange/brown to make them more visible. This sample was deposited under very arid conditions as part of an eolian dune complex on a stable platform.

PPL, KFS, BDI, Scale bar = 0.26 mm

Mid. Jurassic Fulmar Fm., United Kingdom sector, North Sea †

In the Folk classification, this rock is termed an immature to submature subarkose. Using the McBride classification, this rock would fall in the lithic subarkose field. It is a feldspathic wacke in the Dott classification, because of its abundant matrix. However, if most of the "matrix" here was formed by degradation of lithic fragments (the dark area was once a mica), or through formation of diagenetic clay cements, then this rock would actually be classified as a feldspathic arenite (Dott) or a submature to mature subarkose (Folk). The framboidal pyrite and organic fragments suggest that the clay is detrital.

PPL, KFS, AFeS, BDI, Scale bar = 0.07 mm

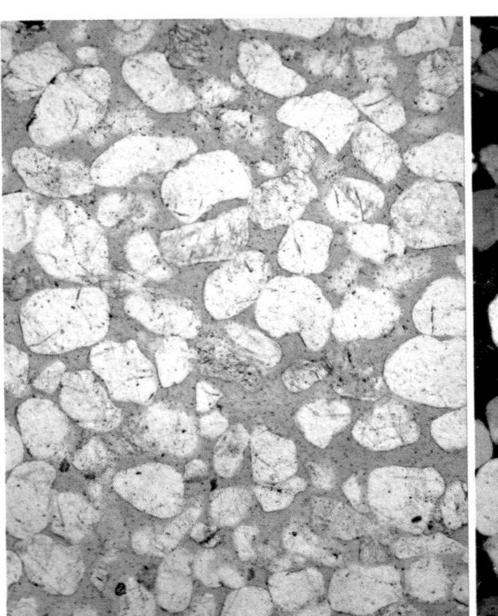

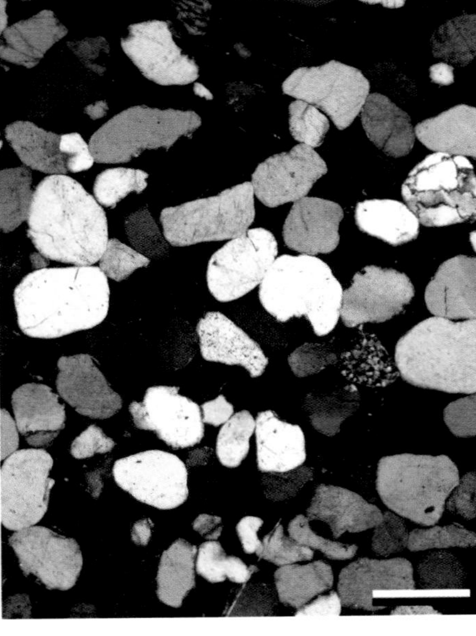

Mid. Jurassic Entrada Fm., Uintah Co., Utah

This sandstone would be classified as a supermature quartzarenite (both Folk and McBride) or as a quartz arenite (Dott). Almost all the grains visible in this thin section are monocrystalline quartz with less than 5% polycrystalline quartz, chert or feldspar grains. For the McBride and Dott classifications, chert grains are grouped with quartz, but the Folk classification groups chert grains with lithic fragments. As with previous examples from the Entrada Fm., this sample was deposited in an arid eolian dune setting in a stable shelf area.

PPL | XPL, BDI, Scale bar = 0.63 mm

Silurian Tuscarora Sandstone, Lebanon Co., Pennsylvania

A supermature quartz-cemented quartzarenite (Folk and McBride) or quartz arenite (Dott). There are two areas with some minor, brightly birefringent clays that formed as cements and thus do not affect the textural maturity name for this rock. This deposit was interpreted to have formed in a single cycle through long-term abrasion in a beach or beach-dune environment within a low-relief setting (Folk, 1960). Quartzarenites, however, also can be formed by near-complete diagenetic removal of unstable minerals (termed "diagenetic quartzarenites"; e.g., McBride, 1985; Cox et al., 2002).

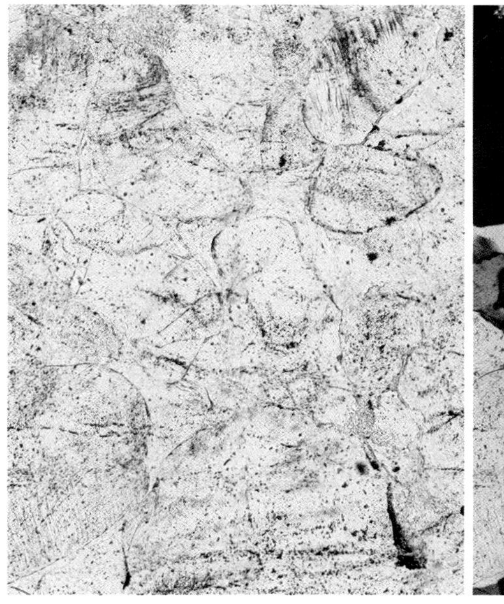

PPL | XPL, Scale bar = 0.17 mm

Cited References and Additional Information Sources

General:

Allen, J., 1956, Estimation of percentages in thin sections—considerations in visual psychology: Journal of Sedimentary Research, v. 26, p. 160-161.

Allen, J. R. L., 1962, Petrology, origin and deposition of the highest Lower Old Red Sandstone of Shropshire, England: Journal of Sedimentary Research, v. 32, p. 657-697, doi: 10.1306/74D70D49-2B21-11D7-8648000102C1865D.

Blatt, H., R. Tracy, and B. Owens, 2006, Petrology: Igneous, Sedimentary, and Metamorphic (3rd Edition): New York, NY, W. H. Freeman, 530 p.

Boggs, S., Jr., 1967, A numerical method for sandstone classification: Journal of Sedimentary Research, v. 37, p. 548-555, doi: 10.1306/74D71717-2B21-11D7-8648000102C1865D.

Bokman, J. W., 1955, Sandstone classification—relation to composition and texture: Journal of Sedimentary Research, v. 25, p. 201-206, doi: 10.1306/74D7044D-2B21-11D7-8648000102C1865D.

Carr, J. R., and M. J. Hibbard, 1991, Open-ended mineralogical/textural rock classification: Computers & Geosciences, v. 17, p. 1409-1463, doi: 10.1016/0098-3004(91)90005-X.

Chen, Pei-Yuan, 1968, A modification of sandstone classification: Journal of Sedimentary Research, v. 38, p. 54-60, doi: 10.1306/74D718CA-2B21-11D7-8648000102C1865D.

Crook, K. A. W., 1960, Classification of arenites: American Journal of Science, v. 258, p. 419-428, doi: 10.2475/ajs.258.6.419.

Dapples, E. C., 1947, Sandstone types and their associated depositional environments: Journal of Sedimentary Research, v. 17, p. 91-100, doi: 10.1306/D42692BA-2B26-11D7-8648000102C1865D.

Dapples, E. C., W. C. Krumbein, and L. L. Sloss, 1953, Petrographic and lithologic attributes of sandstones: Journal of Geology, v. 61, p. 291-317, doi: 10.1086/626098.

Dean, W. E., M. Leinen, and D. A. V. Stow, 1985, Classification of deep-sea, fine-grained sediments: Journal of Sedimentary Research, v. 55, p. 250-256, doi: 10.1306/212F868E-2B24-11D7-8648000102C1865D.

Dennison, J. M., and J. H. Shea, 1966, Reliability of visual estimates of grain abundance: Journal of Sedimentary Research, v. 36, p. 81-89, doi: 10.1306/74D71410-2B21-11D7-8648000102C1865D.

Dott, R. H., 1964, Wacke, graywacke and matrix; what approach to immature sandstone classification?: Journal of Sedimentary Research, v. 34, p. 625-632, doi: 10.1306/74D71109-2B21-11D7-8648000102C1865D.

Ehrlich, R., S. J. Crabtree, K. O. Horkowitz, and J. P. Horkowitz, 1991, Petrography and reservoir physics, I: Objective classification of reservoir porosity: AAPG Bulletin, v. 75, p. 1547-1562.

Fisher, R. V., 1966, Rocks composed of volcanic fragments and their classification: Earth-Science Reviews, v. 1, p. 287-298, doi: 10.1016/0012-8252(66)90010-9.

Flemming, B. W., 2000, A revised textural classification of gravel-free muddy sediments on the basis of ternary diagrams: Continental Shelf Research, v. 20, p. 1125-1137, doi: 10.1016/S0278-4343(00)00015-7.

Folk, R. L., 1954, The distinction between grain size and mineral composition in sedimentary-rock nomenclature: Journal of Geology, v. 62, p. 344-359.

Folk, R. L., 1956, The role of texture and composition in sandstone classification: Discussion: Journal of Sedimentary Research, v. 26, p. 166-171.

Folk, R. L., 1980, Petrology of Sedimentary Rocks: Austin, TX, Hemphill's Book Store, 184 p. (also in numerous earlier editions)

Folk, R. L., P. B. Andrews, and D. W. Lewis, 1970, Detrital sedimentary rock classification and nomenclature for use in New Zealand: New Zealand Journal of Geology and Geophysics, v. 13, p. 937-968, doi: 10.1080/00288306.1970.10418211.

Füchtbauer, H., 1959, Zur Nomenklatur der Sedimentgesteine: Erdöl und Kohle, v. 12, p. 605-613.

Grabau, A. W., 1904, On the classification of sedimentary rocks: American Geologist, v. 33, p. 228-247.

Hallsworth, C. R., and R. Knox, 1999, BGS Rock Classification Scheme. Volume 3, Classification of sediments and sedimentary rocks: Nottingham, UK, British Geological Survey, 44 p.

Haralick, R. M., and K. Shanmugam, 1973, Computer classification of reservoir sandstones: Geoscience Electronics, IEEE Transactions, v. 11, p. 171-177, doi: 10.1109/TGE.1973.294312.

Herron, M. M., 1988, Geochemical classification of terrigenous sands and shales from core or log data: Journal of Sedimentary Research, v. 58, p. 820-829, doi: 10.1306/212F8E77-2B24-11D7-8648000102C1865D.

Huckenholz, H. G., 1963, A contribution to the classification of sandstones: Geologiska Foreningens I Stockholm Forhandlingar, v. 85, Part 1, no. 512, p. 156-172, doi: 10.1080/11035896309448877.

Klein, G., deV., 1963, Analysis and review of sandstone classifications in the North American geological literature, 1940–1960: GSA Bulletin, v. 74, p. 555-576, doi: 10.1130/0016-7606(1963)74[555:AAROSC]2.0.CO;2.

Krumbein, W. C., and L. L. Sloss, 1963, Stratigraphy and Sedimentation (2nd Edition): San Francisco, CA, W.H. Freeman & Co., 660 p.

Krynine, P. D., 1948, The megascopic study and field classification of sedimentary rocks: Journal of Geology, v. 56, p. 130-165, doi: 10.1086/625492.

Lerbekmo, J. F., 1962, Field classification of sandstones: AAPG Bulletin, v. 46, p. 1526-1527.

Mansfield, C. F., and T. S. Ahlbrandt, 1978, A detrital sandstone classification for petrologists: Earth Science Bulletin, v. 11, p. 1-10.

McBride, E. F., 1963, A classification of common sandstones: Journal of Sedimentary Research, v. 33, p. 664-669, doi: 10.1306/74D70EE8-2B21-11D7-8648000102C1865D.

Moncrieff, A. C. M., 1989, Classification of poorly-sorted sedimentary rocks: Sedimentary Geology, v. 65, p. 191-194, doi: 10.1016/0037-0738(89)90015-8.

Mount, J., 1985, Mixed siliciclastic and carbonate sediments: A proposed first-order textural and compositional classification: Sedimentology, v. 32, p. 435-442, doi: 10.1111/j.1365-3091.1985.tb00522.x.

Okada, H., 1971, Classification of sandstone: Analysis and proposal: Journal of Geology, v. 79, p. 509-525, doi: 10.1086/627673.

Packham, G. H., 1954, Sedimentary structures as an important factor in the classification of sandstones: American Journal of Science, v. 252, p. 466-476, doi: 10.2475/ajs.252.8.466.

Pettijohn, F. J., 1954, Classification of sandstones: Journal of Geology, v. 62, p. 360-365.

Pettijohn, F. J., P. E. Potter, and R. Siever, 1987, Sand and Sandstones (2nd Edition): New York, Springer-Verlag, 553 p.

Pingitore, N. E., Jr., and J. D. Shotwell, 1977, Sandstone classification; a view from factor space: Neues Jahrbuch für Geologie und Palaeontologie Abhandlungen, v. 3, p. 177-188.

Shepard, F. P., 1954, Nomenclature based on sand-silt-clay ratios: Journal of Sedimentary Research, v. 24, p. 151-158, doi: 10.1306/D4269774-2B26-11D7-8648000102C1865D.

Tallman, S. L., 1949, Sandstone types: Their abundance and cementing agents: Journal of Geology, v. 57, p. 582-591, doi: 10.1086/625671.

Travis, R. B., 1970, Nomenclature for sedimentary rocks: AAPG Bulletin, v. 54, p. 1095-1107.

Wang, Chao-Siang, 1967, On the occurrence of quartz wacke and its bearing on the problems of sandstone classification: Geological Society of China, Proceedings, v. 10, p. 99-106.

Washburn, A. L., J. E. Sanders, and R. F. Flint, 1963, A convenient nomenclature for poorly sorted sediments: Journal of Sedimentary Research, v. 33, p. 478-480.

Wentworth, C. K., 1922, A scale of grade and class terms for clastic sediments: Journal of Geology, v. 30, p. 377-392, doi: 10.1086/622910.

Williams, H., F. J. Turner, and C. M. Gilbert, 1954, Petrography: An Introduction to the Study of Rocks in Thin Section (1st Edition): San Francisco, CA, W. H. Freeman & Co., 406 p.

Yanov, E. N., 1977, Classification of sandstones and siltstones by composition of grains: Lithology and Mineral Resources, v. 12, p. 466-472.

Zuffa, G. G., 1980, Hybrid arenites: Their composition and classification: Journal of Sedimentary Research, v. 50, p. 21-29, doi: 10.1306/212F7950-2B24-11D7-8648000102C1865D.

Interpretation and provenance analysis:

Allen, J. L., and C. L. Johnson, 2010, Facies control on sandstone composition (and influence of statistical methods on interpretations) in the John Henry Member, Straight Cliffs Formation, Southern Utah, USA: Sedimentary Geology, v. 230, p. 60-76, doi: 10.1016/j.sedgeo.2010.06.023.

Basu, A., 2003, A perspective on quantitative provenance analysis, *in* R. Valloni, and A. Basu, eds., Quantitative Provenance Studies in Italy (Memorie descrittive della carta geologica d'Italia, Vol.61): Roma, Istituto poligrafico e Zecca dello Stato, p. 11 - 22.

Blatt, H., 1967, Provenance determinations and recycling of sediments: Journal of Sedimentary Research, v. 37, p. 1031-1044, doi: 10.1306/74D71825-2B21-11D7-8648000102C1865D.

Blatt, H., 1985, Provenance studies and mudrocks: Journal of Sedimentary Research, v. 55, p. 69-75, doi: 10.1306/212F8611-2B24-11D7-8648000102C1865D.

Blatt, H., and R. L. Jones, 1975, Proportions of exposed igneous, metamorphic, and sedimentary rocks: GSA Bulletin, v. 86, p. 1085-1088, doi: 10.1130/0016-7606(1975)86<1085:POEIMA>2.0.CO;2.

Chandler, F. W., 1988, Quartz arenites: Review and interpretation: Sedimentary Geology, v. 58, p. 105-126, doi: 10.1016/0037-0738(88)90065-6.

Cox, R., E. D. Gutmann, and P. G. Hines, 2002, Diagenetic origin for quartz-pebble conglomerates: Geology, v. 30, p. 323-326, doi: 10.1130/0091-7613(2002)030<0323:DOFQPC>2.0.CO;2.

Cox, R., and D. R. Lowe, 1996, Quantification of the effects of secondary matrix on the analysis of sandstone composition, and a petrographic-chemical technique for retrieving original framework grain modes of altered sandstones: Journal of Sedimentary Research, v. 66, p. 548-558, doi: 10.1306/D42683A1-2B26-11D7-8648000102C1865D.

Dickinson, W. R., 1970, Interpreting detrital modes of graywacke and arkose: Journal of Sedimentary Research, v. 40, p. 695-707, doi: 10.1306/74D72018-2B21-11D7-8648000102C1865D.

Dickinson, W. R., L. S. Beard, G. R. Brakenridge, J. L. Erjavec, R. C. Ferguson, K. F. Inman, R. A. Knepp, F. A. Lindberg, and P. T. Ryberg, 1983, Provenance of North American Phanerozoic sandstones in relation to tectonic setting: GSA Bulletin, v. 94, p. 222-235, doi: 10.1130/0016-7606(1983)94<222:PONAPS>2.0.CO;2.

Dickinson, W. R., 1985, Interpreting provenance relations from detrital modes of sandstones, *in* G. G. Zuffa, ed., Provenance of Arenites (NATO Science Series C): Dordrecht, Netherlands, D. Reidel Publishing, p. 333-361, doi: 10.1007/978-94-017-2809-6_15.

Folk, R. L., 1960, Petrography and origin of the Tuscarora, Rose Hill, and Keefer formations, Lower and Middle Silurian of eastern West Virginia: Journal of Sedimentary Research, v. 30, p. 1-58, doi: 10.1306/74D709C5-2B21-11D7-8648000102C1865D.

Garzanti, E., C. Doglioni, G. Vezzoli, and S. Ando, 2007, Orogenic belts and orogenic sediment provenance: Journal of Geology, v. 115, p. 315-334, doi: 10.1086/512755.

Ingersoll, R. V., T. F. Fullard, R. L. Ford, J. P. Grimm, J. D. Pickle, and S. W. Sares, 1984, The effect of grain size on detrital modes: A test of the Gazzi-Dickinson point-counting method: Journal of Sedimentary Research, v. 54, p. 103-116, doi: 10.1306/212F83B9-2B24-11D7-8648000102C1865D.

Johnsson, M. J., 1993, The system controlling the composition of clastic sediments, *in* M. J. Johnsson, and A. Basu, eds., Processes Controlling the Composition of Clastic Sediments: Boulder, CO, GSA Special Paper 284, p. 1-19, doi: 10.1130/SPE284-p1.

Le Pera, E., J. Arribas, S. Critelli, and A. Tortosa, 2001, The effects of source rocks and chemical weathering on the petrogenesis of siliciclastic sand from the Neto River (Calabria, Italy): Implications for provenance studies: Sedimentology, v. 48, p. 357-378, doi: 10.1046/j.1365-3091.2001.00368.x.

McBride, E. F., 1985, Diagenetic processes that affect provenance determinations in sandstone, *in* G. G. Zuffa, ed., Provenance of Arenites (NATO Science Series C): Dordrecht, Netherlands, D. Reidel Publishing, p. 95-113, doi: 10.1007/978-94-017-2809-6_5.

Milliken, K. L., 1988, Loss of provenance information through subsurface diagenesis in Plio-Pleistocene sandstones, northern Gulf of Mexico: Journal of Sedimentary Research, v. 58, p. 992-1002, doi: 10.1306/212F8EE0-2B24-11D7-8648000102C1865D.

Molinaroli, E., and A. Basu, 1993, Toward quantitative provenance analysis: A brief review and case study, *in* M. J. Johnsson, and A. Basu, eds., Processes Controlling the Composition of Clastic Sediments: Boulder, CO, GSA Special Paper 284, p. 323-334, doi: 10.1130/SPE284-p323.

Roser, B. P., and R. J. Korsch, 1988, Provenance signatures of sandstone-mudstone suites determined using discriminant function analysis of major-element data: Chemical Geology, v. 67, p. 119-139, doi: 10.1016/0009-2541(88)90010-1.

Suttner, L. J., A. Basu, and G. H. Mack, 1981, Climate and the origin of quartz arenites: Journal of Sedimentary Research, v. 51, p. 1235-1246, doi: 10.1306/212F7E73-2B24-11D7-8648000102C1865D.

Suttner, L. J., and P. K. Dutta, 1986, Alluvial sandstone composition and paleoclimate: I. Framework mineralogy: Journal of Sedimentary Research, v. 56, p. 329-345, doi: 10.1306/212F8909-2B24-11D7-8648000102C1865D.

von Eynatten, H., C. Barceló-Vidal, and V. Pawlowsky-Glahn, 2003, Composition and discrimination of sandstones: A statistical evaluation of different analytical methods: Journal of Sedimentary Research, v. 73, p. 47-57, doi: 10.1306/070102730047.

Weltje, G. J., 2002, Quantitative analysis of detrital modes: Statistically rigorous confidence regions in ternary diagrams and their use in sedimentary petrology: Earth-Science Reviews, v. 57, p. 211-253, doi: 10.1016/S0012-8252(01)00076-9.

Weltje, G. J., and H. von Eynatten, 2004, Quantitative provenance analysis of sediments: Review and outlook: Sedimentary Geology, v. 171, p. 1-11, doi: 10.1016/j.sedgeo.2004.05.007.

Zahid, K. M., and D. L. Barbeau, Jr., 2012, Constructing sandstone provenance and classification ternary diagrams using an electronic spreadsheet: Journal of Sedimentary Research, v. 82, p. 131-132, doi: 10.2110/jsr.2012.12.

Zuffa, G. G., 1985, Optical analyses of arenites: Influence of methodology on compositional results, *in* G. G. Zuffa, ed., Provenance of Arenites (NATO Science Series C): Dordrecht, Netherlands, D. Reidel Publishing, p. 165-189, doi: 10.1007/978-94-017-2809-6_8.

Facing Page: Top – Cyclic, meter-scale bedding in Upper Cretaceous shales, Haute Maritime, France. Bottom – A pyritic burrow and pellets from the Pennsylvanian Morrow B Sandstone, Ochiltree Co., Texas (PPL, RDI).

Mudrocks: Siltstones, Mudstones, Claystones & Shales

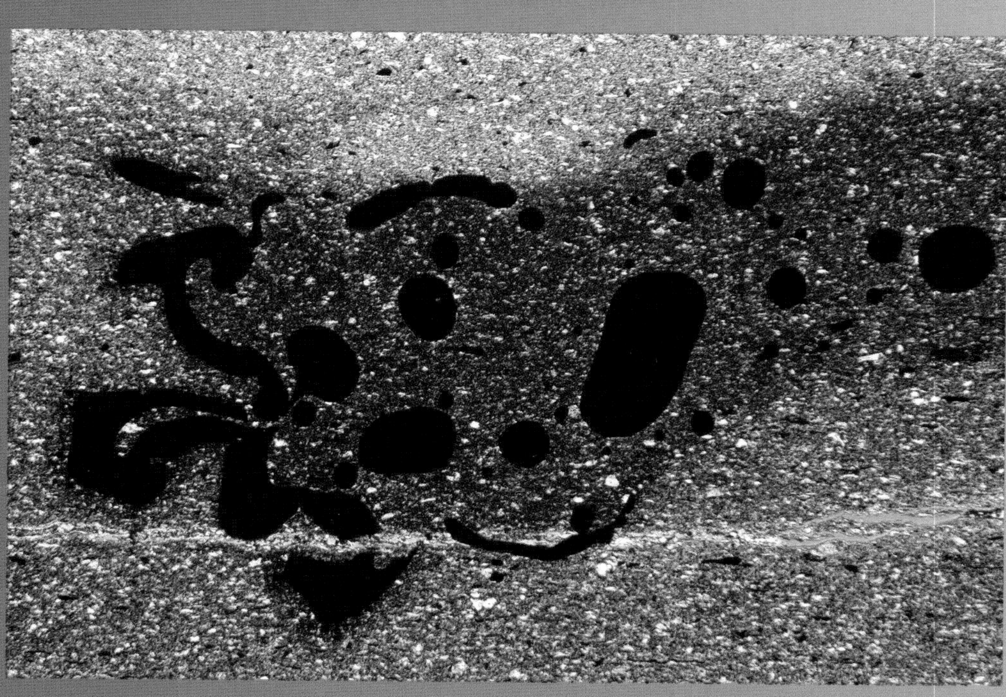

Chapter 8

SILTSTONES, MUDSTONES & SHALES

"A Picture is a Fact" Ludwig Wittgenstein, Tractatus Logico-Philosophicus, 1922

Shale and mudstone are both widely used terms for fine-grained terrigenous clastic rocks (although some use fissility as a requirement for the use of the term "shale"), but there is at present no broadly agreed upon terminology for naming and classifying these rocks (see discussions in Schieber et al., 1998, and Potter et al., 2005). Because in past stratigraphic and sedimentologic studies, the great majority of fine-grained rocks have been designated as shales (such as, for example, the Cretaceous Eagle Ford Shale that is, in many places, a marl or even a limestone) or the Monterey Shale (a diatomite or chert depending on diagenesis), we will use the term shale in this chapter, with the understanding that it includes what some prefer to identify as mudstone. Also, because the most basic definition of shales, that they be dominated by particles smaller than 62.5 μm (e.g., Blatt et al., 1980), implies that shales span the clay-silt boundary, a good many rocks that are identified as siltstones in the literature also qualify as shales (and vice versa).

The compositional variability of the fine-grained deposits commonly included in the word "shale" has both academic and economic importance. Figure 8.1 shows a ternary plot of the average composition for some of the major "unconventional" hydrocarbon plays in North America (along with a few outlines of the variability even within single formations). From that plot, it is clear that the carbonate and siliceous (both quartz and biogenic silica) components play a major role in determining the economic viability of such exploration targets. Both siliceous and carbonate admixtures increase the brittleness of shale and thus their suitability for both natural and artificial fracturing. Not surprisingly, then, the siliceous and carbonate end members on carbonate-silica-clay ternary plots (variously termed chert, diatomite, radiolarite, tripolite, or siliceous mudstone at the silica pole and chalk or carbonate mudstone at the carbonate pole) were among the first "unconventional" reservoirs to be exploited. In the case of the siliceous Monterey Formation, production extends back to 1901; in the case of chalks, significant oil production began in the 1970s.

Historically, a variety of classification schemes exist for fine-grained sedimentary rocks (e.g., Ingram, 1953; Dunbar and Rodgers, 1957; Folk, 1965; Picard, 1971; Lewan, 1978; Blatt et al., 1980; Lundegard and Samuels, 1980; Potter et al., 1980; Macquaker and Adams, 2003), but in spite of the dedicated efforts of their authors, none of them experienced widespread application. If one examines citation statistics, above classification schemes have been cited two to three times per year since publication. In comparison, the widely used carbonate classification schemes by Folk (1959) and Dunham (1962) have an annual citation average of 26 and 55, respectively. The need for a good classification scheme for shale is obvious.

The naming of any rock must on one hand convey a maximum amount of information about the rock, yet at the same time accept incompleteness for the sake of brevity. What would the key properties of a shale be with that purpose in mind? Grain size, for example, while a useful property of sandstones, lacks utility for shale because it is too small to be readily discerned. Mineral composition, though unquestionably useful, is again stymied by small grain size. XRD or whole rock chemistry data would be needed to make it workable. In fact, without advanced instrumentation, the most accessible properties of a shale are probably its color, its relative softness (lithification state), its reaction with hydrochloric acid (is it calcareous?) and textural features such as lamination, bioturbation, etc. Whereas it is not uncommon to see geologists use color charts to describe rock color, simpler qualifiers (gray, red, green, greenish, etc.) are in many instances sufficient. Relative softness can be tested be scratching the sample with a nail, and most geologists also would have some hydrochloric acid handy. So, aside from color, most of what we are able to say about a shale at first encounter is decidedly of qualitative nature, and that circumstance makes textural features very valuable when we set out to describe and categorize shales.

Characteristics of texture and fabric commonly reflect sedimentary and diagenetic processes. They range from millimeter to centimeter scale and are readily observed on polished slabs and in petrographic thin sections. In light of above observations and considering the fact that most shale sequences exhibit much larger variations in texture and fabric than in composition (Nuhfer et al., 1979; Cluff, 1980; Nuhfer, 1981; Schieber, 1989, 1999), it is obvious that texture and fabric are indispensable for meaningful description and classification. This state of affairs is analogous to the situation in carbonate rocks, where within narrow compositional confines the textural variability of carbonate rocks was made interpretable through the frame of reference provided by the carbonate microfacies concept (e.g., Wilson, 1975; Flügel, 1978). Thin-section-scale integration of paleontologic, sedimentologic and petrographic observations is the basis of the microfacies concept as pioneered by Brown (1943) and Cuvillier (1952), and adaptation of this approach to shales (e.g., Schieber, 1989, 1994, 1999) is now becoming accepted practice (Aplin and Macquaker, 2011; Hammes et al., 2011; Hart et al., 2013). In absence of a widely adopted classification scheme, we consider it best practice to record a shale's composition and fabric elements as accurately as possible, thus enabling future readers to understand them in the context of more comprehensive forthcoming classification schemes.

Because we believe that pictures provide the most direct access to the multifaceted entities that are "shale" and convey the circumstances of their origin in the most compact way possible, this chapter is all about pictures. The images shown come from shales that range in age from Precambrian to Cenozoic and were collected on four continents, though most are from North America. We tried to focus on examples that convey the most common textural and compositional aspects of shales but are fully aware that a comprehensive illustration of their full textural and compositional range will require a book in its own right. We organized the illustrations so that we first show the most common textures (laminated, lenticular, homogenized), and then follow these with thematic sets, such as on bioturbation, sedimentary structures reflective of depositional processes, fossil content and diagenesis. In order to convey a mental image of what these features may tell us about processes and conditions of deposition and diagenesis, commentary was added that benefited from in depth study of many of these samples in the context of formation-scale examination of these rocks. We also added a few images from SEM observations on polished thin sections, broken surfaces and argon ion milled surfaces. The SEM study of argon ion-milled shales is emerging as the new method of choice for detailed examination of shale fabrics and diagenesis (Loucks et al., 2009; Schieber, 2013), but it will not end

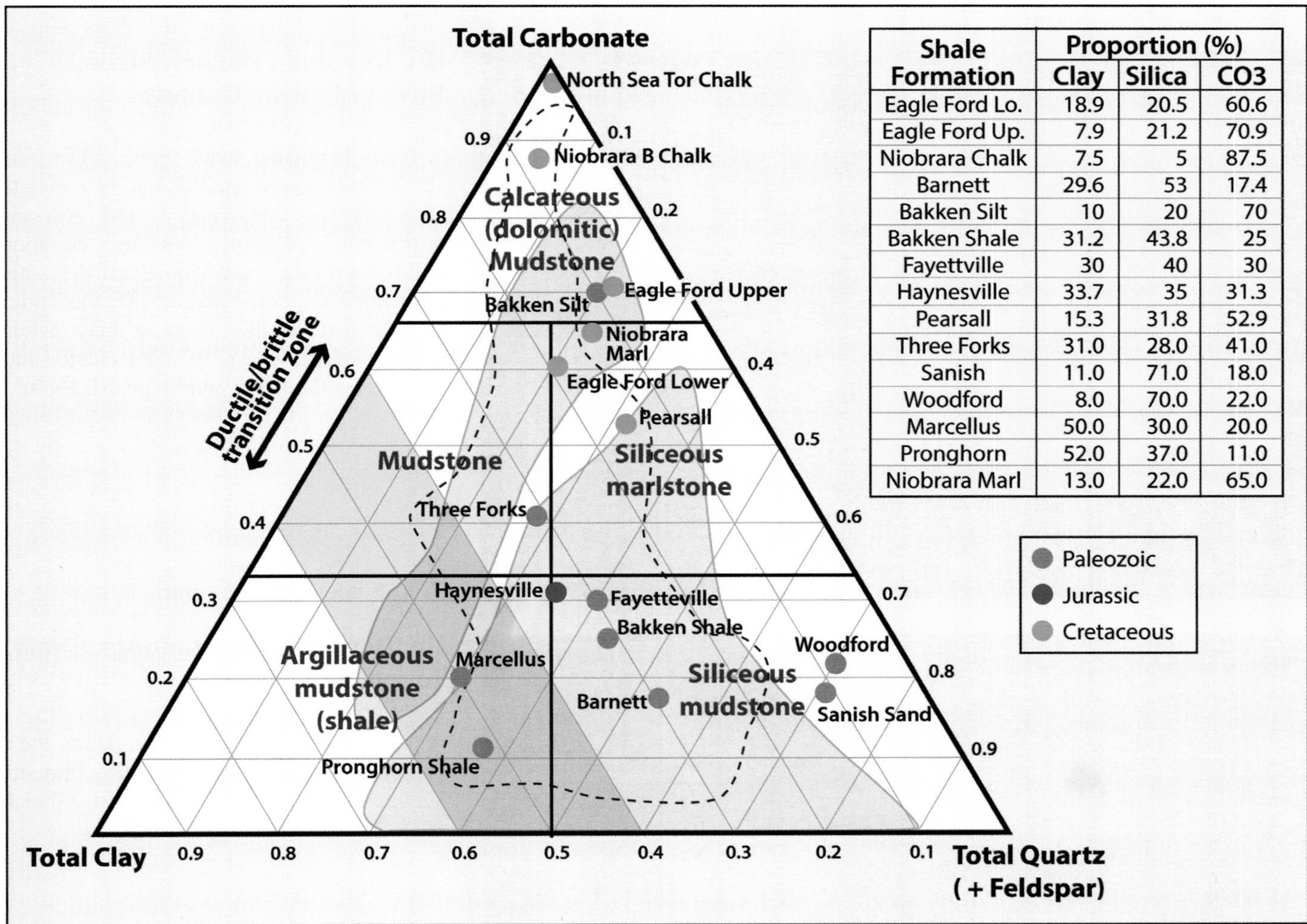

Shale Formation	Proportion (%)		
	Clay	Silica	CO3
Eagle Ford Lo.	18.9	20.5	60.6
Eagle Ford Up.	7.9	21.2	70.9
Niobrara Chalk	7.5	5	87.5
Barnett	29.6	53	17.4
Bakken Silt	10	20	70
Bakken Shale	31.2	43.8	25
Fayettville	30	40	30
Haynesville	33.7	35	31.3
Pearsall	15.3	31.8	52.9
Three Forks	31.0	28.0	41.0
Sanish	11.0	71.0	18.0
Woodford	8.0	70.0	22.0
Marcellus	50.0	30.0	20.0
Pronghorn	52.0	37.0	11.0
Niobrara Marl	13.0	22.0	65.0

Figure 8.1 A ternary diagram showing variations in mineralogical composition in some major North American "shale" exploration plays (and one European chalk play). The colored dots represent regionally averaged compositional values in labeled units; the pale blue overlay shows the range of locality-specific values in the Barnett "Shale"; the pale violet defines similar local compositional data from the Eagle Ford "Shale"; and the dashed black line defines the range of local compositional data from the Marcellus "Shale" in the Appalachian Basin. The main ternary plot and data points are courtesy of Tom Anderson, Energy & Geoscience Institute, University of Utah. The regional play overlays are adapted from Passey et al. (2010) and the Marcellus data range is adapted from Wang and Carr (2013). Most mudstones contain more than just these limited components, but ternary diagrams can be a useful starting point for understanding mudstone formation. The generally high carbonate content of Cretaceous units reflects the Mesozoic evolution of calcareous plankton.

the need for petrographic thin sections. The latter will continue to have much to offer because even at sub-SEM resolutions there is a wealth of information to be accessed, the viewing areas are much larger (better context), the cost of thin section preparation is comparatively low, and access to petrographic microscopes is widespread.

For those that wish to learn more about shales, their nuances, their practical uses and their potential to tell us more about earth history, there is a small number of books that are instructive reading. First among these is the original "Sedimentology of Shales" by Potter et al. (1980), an assessment of the rather poor state of understanding at the time of its publication, but with an excellent catalogue of questions that need to be addressed to make progress. The "Argillaceous Rock Atlas" by O'Brien and Slatt (1990) marks a helpful addition to the shale literature by providing a catalogue of images (mostly using light microscopy) from a wide variety of shale formations and organizing them into perceived sedimentary environments. In 1998, a two volume set "Shales and Mudstones" (Schieber et al., 1998) compiled state of the art research papers on a wide range of topics (including basin studies, sedimentology, sequence stratigraphy, paleontology, geochemistry, economic geology, petrography and geophysics) and took stock on how much progress had been made since 1980. Finally, in 2005, the follow-up to "Sedimentology of Shales", titled "Mud and Mudstones" (Potter et al., 2005), was published. It provides a well-organized overview of the multiple themes that constitute shale research with an in-depth set of references and good questions for future research. It is already ten years out of date, and shale research is expanding rapidly. Whoever choses to write the next update will face a monumental task.

Up. Devonian Bakken Shale, Saskatchewan, Canada

A finely-laminated black shale. The bright silt laminae appear parallel, but may, in detail, show lateral convergence and low-angle downlap relationships. Although commonly believed to be the result of quiescent deposition, experimental studies suggest that this style of well-defined parallel laminae may actually reflect bedload transport of flocculated muds by currents (Schieber and Southard, 2009).

PPL, Scale bar = 2.64 mm

Up. Devonian Chattanooga Shale, Tennessee

A finely-laminated black shale. Here, the silt laminae in places show thickening into small lenses, suggesting a bedload origin (Schieber et al., 2007).

PPL, Scale bar = 2.00 mm

Up. Devonian Chattanooga Shale, Tennessee

A finely-laminated black shale from the same outcrops as the previous image, but at higher magnification. The laminae vary in thickness and diverge and converge across the section. In places, laminae are gently inclined and appear to downlap towards common surfaces (two marked with arrows). These observations suggest that the mud was originally deposited in bedload by bottom currents (e.g., Schieber et al., 2007; Schieber and Southard, 2009). It should be noted that primary dip angles have been lowered by substantial compaction (perhaps as much as 10x) as a result to the initial high water content of these deposits.

PPL, Scale bar = 0.47 mm

Up. Devonian – Lo. Mississippian Woodford Shale, Lincoln Co., Oklahoma

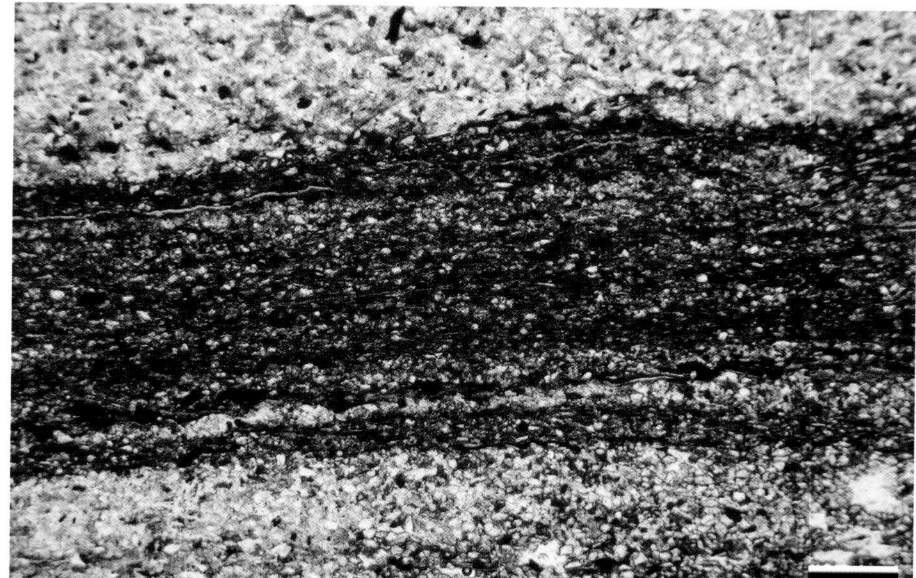

Interbedded siltstone and shale (middle) with small, probably artificial, fractures (blue epoxy fill). The siltstones are event beds and are interpreted as storm or turbidite generated. The fractures follow the planar internal fabric of the rock and most likely formed as the rock expanded after being brought to the surface. Wetting and drying of core also can lead to fractures of this type.

PPL, BDI, Scale bar = 0.22 mm

Up. Cretaceous Monte Antola Fm., Liguria, Italy

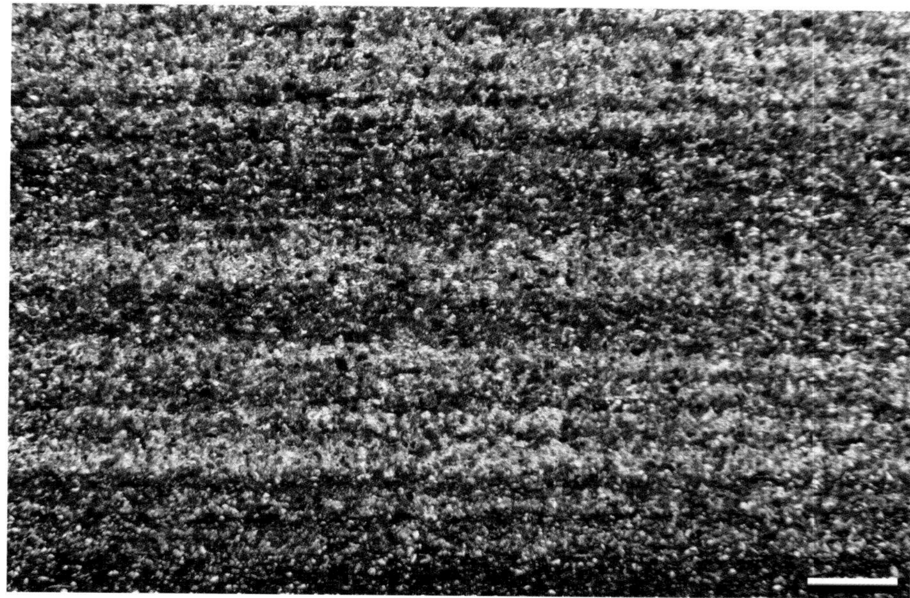

A laminated, carbonate-rich microturbidite siltstone. However, whereas a turbidite origin might be inferred from the overall context and the sedimentary structures of the bed as a whole, the laminae as shown here are rather generic and nondiagnostic.

PPL, Scale bar = 0.51 mm

Up. Cretaceous Monte Antola Fm., Liguria, Italy

A basinal laminated siltstone from a distal turbidite succession. Shows abundant small carbonate clasts (medium gray to brownish) that include pellets or peloids, microfossil fragments and calcareous nannofossils (visible only at much higher magnification; Scholle, 1971).

PPL, Scale bar = 0.26 mm

Mid. Proterozoic Rampur Shale, India

A shale with lenticular fabric. The flattened lenses originated as current rip-ups of soft and water-rich (~80%) seafloor muds that were redeposited, piled up and compacted. This is a common fabric in many shale successions. Similar fabrics can arise when we have densely spaced burrows or transported and redeposited fecal pellets, but a thin section cut parallel to bedding should help to sort out the possibilities (see following image). The dark zones of organic matter and clays may reflect accentuation of primary concentrations by dissolution along microstylolites or wispy solution seams.

PPL, Scale bar = 0.19 mm

Mid. Proterozoic Rampur Shale, India

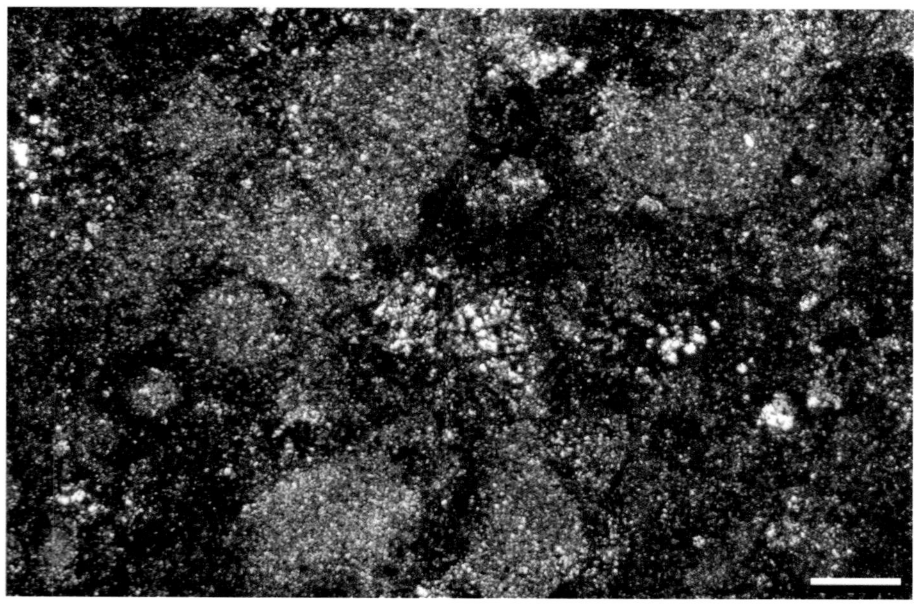

A view of the same sample as in previous image, but here the section is cut parallel to bedding. The lenses from the previous image are discrete bodies with irregular outlines, and thus the lenses may be clasts. Fecal pellets would be more similar in outline and size; burrows would appear as sinuous traces in this view. So, at times it is instructive to make thin sections parallel to bedding. For more detail see Schieber et al. (2010).

PPL, Scale bar = 0.19 mm

Lo. Permian Wolfcamp Shale, Reeves Co., Texas

Wavy-lenticular lamination in very silt-rich shale. The lamination could reflect bedload transport of silt by weak and intermittent currents (otherwise ripples would more likely form). The darker seams are caused by organic matter concentration, perhaps accentuated by dissolution along microstylolites (solution seams).

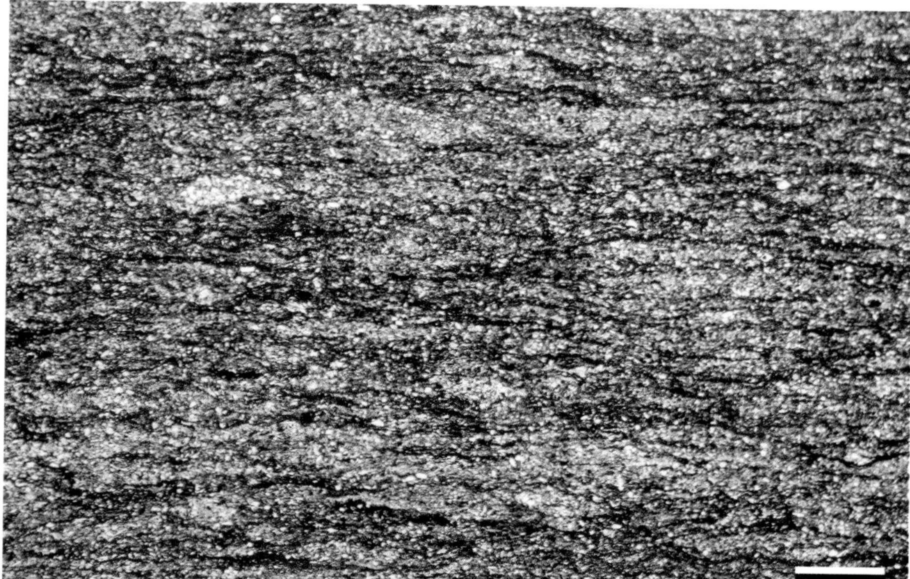

PPL, Scale bar = 0.14 mm

Up. Jurassic (Kimmeridgian) Kimmeridge Clay, Dorset, England, U.K.

A slightly silty claystone with minor glauconite (green grains) and numerous organic fragments. The brown streaks are probably kerogen, and the black particles are diagenetic pyrite (small concretions and framboids).

PPL, Scale bar = 0.51 mm

Cretaceous Mowry Fm., Moffat Co., Colorado

A carbonaceous silty shale with subhorizontal epoxy-filled (pink) induced fractures (related to core expansion). The lower half of the slide shows a shale with abundant pyrite-filled (black) and also calcite-filled (clear) planktonic foraminifera and large kerogen streaks (dark brown, 0.1–0.2 mm long). The upper half of the slide lacks foraminifera and large kerogen streaks. The lower half likely reflects slow deposition with strong pelagic input, and the upper half suggests faster deposition via storms or turbidites.

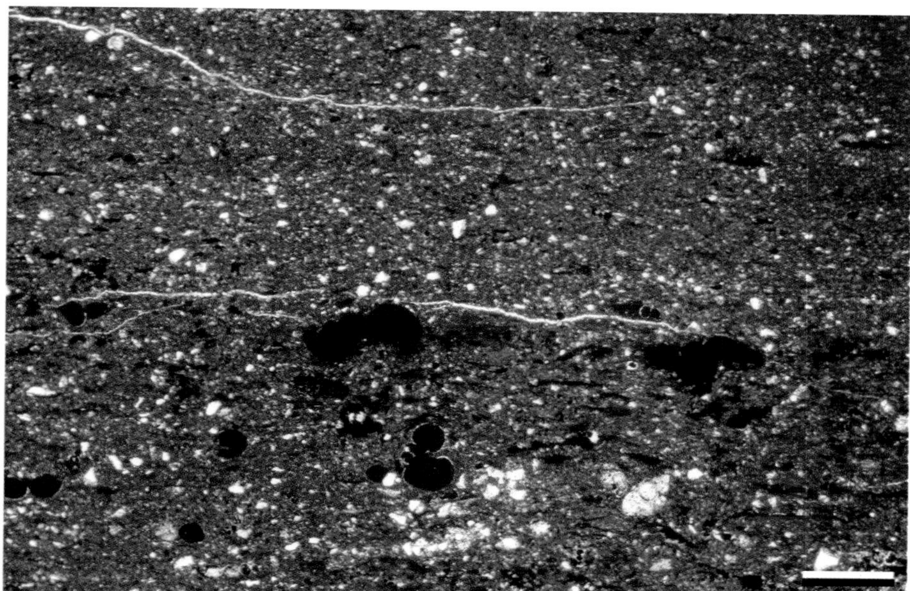

PPL, RDI, Scale bar = 0.34 mm

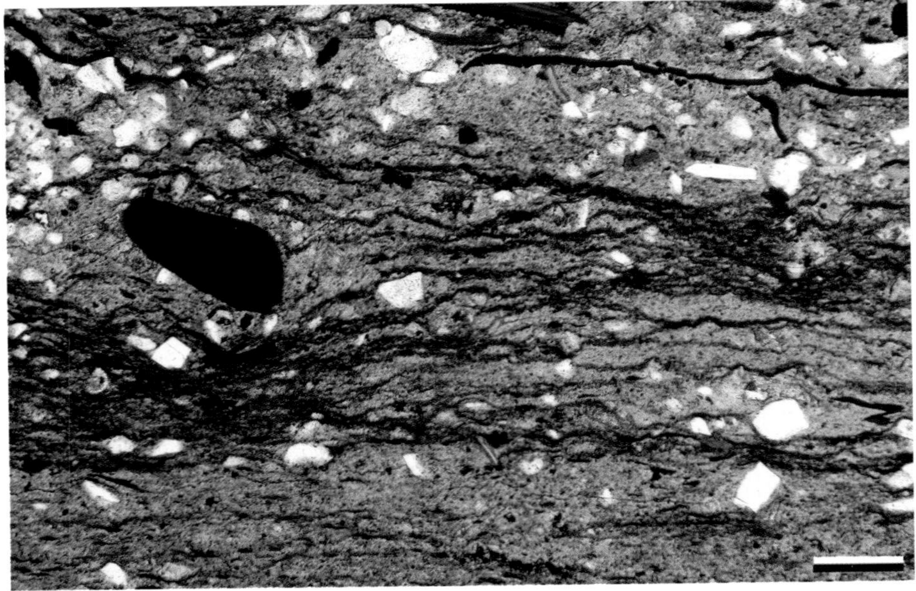

Cretaceous Mowry Fm., Moffat Co., Colorado

A siliceous and carbonaceous shale, where siliceous refers to the abundant presence of diagenetic silica in the matrix (with silica derived from primary biogenic opal in radiolarians or sponge spicules). Volcanic ash (bentonite beds) are another possible source of silica. The red-brown particles are fish debris, and scattered black particles are pyrite. The large rounded black grain is an oil-stained fish bone fragment. Around this grain, we see a deflection of wispy, dark (organic-rich) seams, and these are interpreted as bedding-parallel pressure solution seams (opal dissolution) along which organic matter was concentrated.

PPL, RDI, Scale bar = 0.26 mm

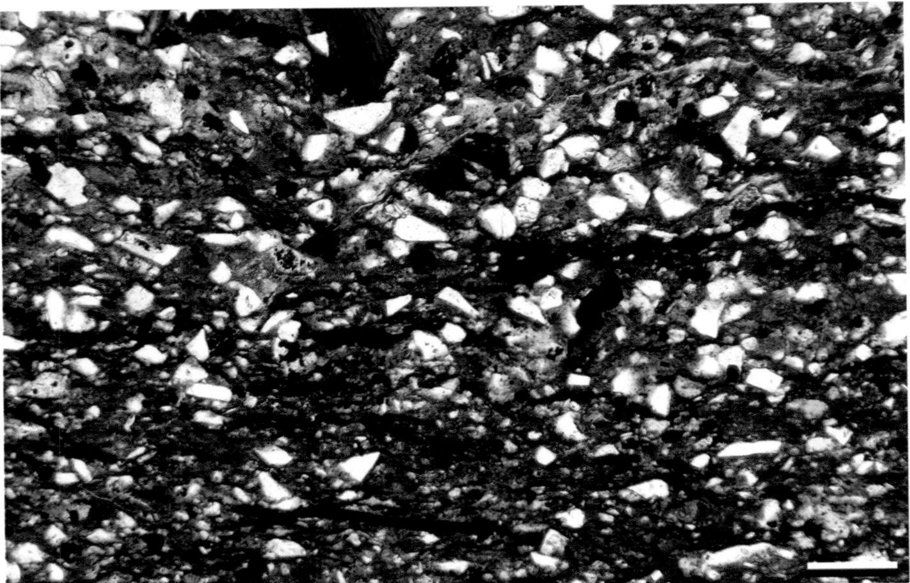

Cretaceous Mowry Fm., Moffat Co., Colorado

A carbonaceous shale with abundant scattered fine-sand grains (variably rounded). That the sand grains are mixed with the clay matrix suggests postdepositional mixing of sand layers/laminae by burrowing organisms. The brown color of the matrix is related to kerogen, and the black streaks are terrestrial plant matter and/or pyrite formation in connection with organic particles. Fractures (pink epoxy) were caused by coring and sample handling.

PPL, RDI, Scale bar = 0.26 mm

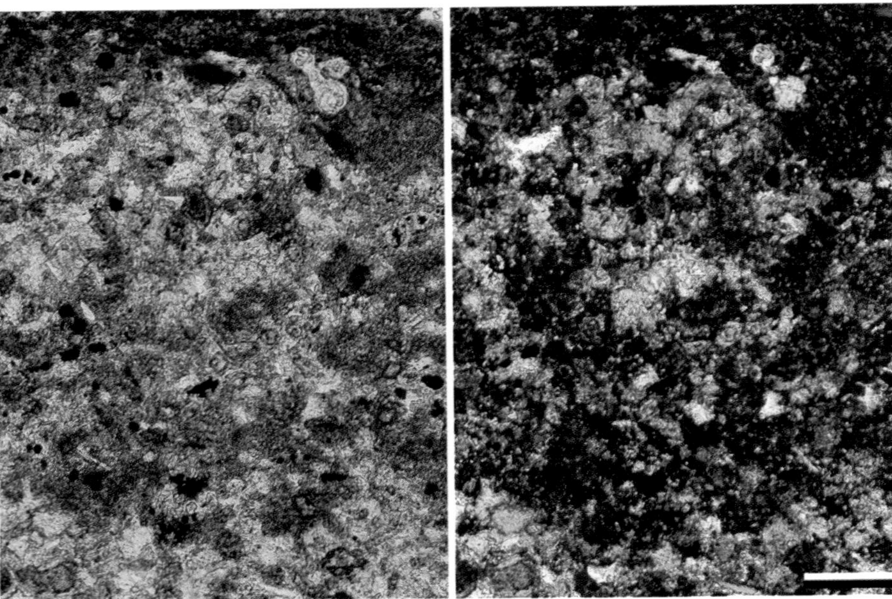

Up. Cretaceous Monte Antola Fm., Liguria, Italy

A siltstone with grains of both clastic (clear quartz grains) and biogenic (gray-brown carbonate grains) origin; in addition, there are some detrital dolomite crystals. In the cross-polarized light image on the right, the carbonates show up as brightly birefringent areas. Some of the carbonate grains have been partially to completely silicified.

PPL | XPL, Scale bar = 0.10 mm

Up. Cretaceous Blackhawk Fm., Kenilworth Mbr., Book Cliffs, Utah

A moderately bioturbated shale from a core slab. Silt-rich layers are lighter in color and may show basal scours, grading and low-angle laminae. These are interpreted as muddy tempestites (event beds, storm deposits). Event beds are disrupted but well visible. On the other hand, intervening muddy beds are more strongly bioturbated. This unit was deposited on a storm-wave-dominated shelf with deltaic sediment input.

PPL, Scale bar = 3.81 mm

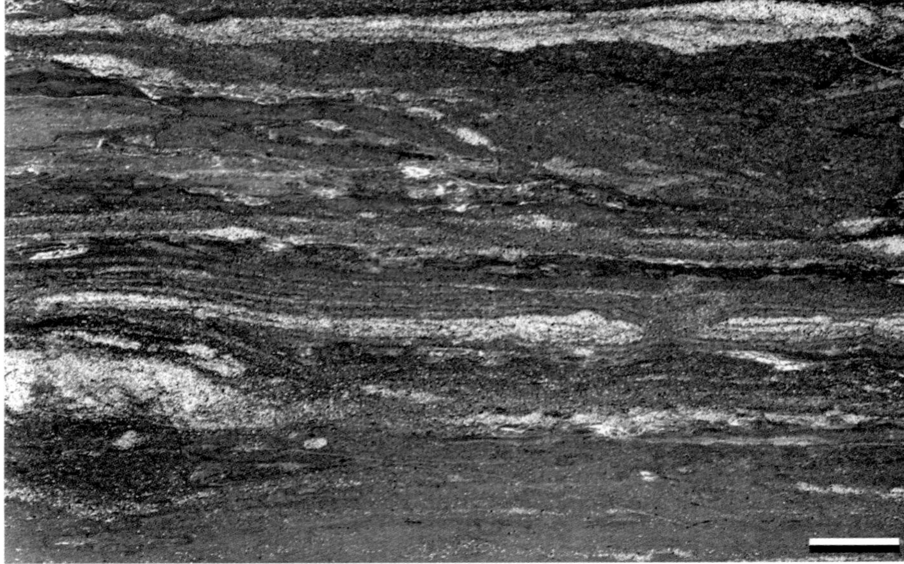

Up. Cretaceous Blackhawk Fm., Kenilworth Mbr., Book Cliffs, Utah

A mostly bioturbated shale. One thicker silt bed is still visible, whereas other silt-rich layers are barely recognizable. Most of the bioturbation is generic, but the small dark burrows are attributed to the ichnogenus *Chondrites*. The vertical burrow that cuts across the silt layer at left shows zoning, and may be attributable to the ichnogenus *Skolithos* (or possibly *Cylindrichnus*, were it not for the discontinuous zoning and the lack of downward tapering). Silt-rich layers may be relict tempestites, similar to those in the image above. Deposition was on a storm-wave-dominated shelf with deltaic sediment input.

PPL, Scale bar = 4.76 mm

Up. Cretaceous Blackhawk Fm., Kenilworth Mbr., Book Cliffs, Utah

A strongly bioturbated to homogenized–churned shale, containing only a few areas with relict bedding. One easily recognizable ichnogenus is *Chondrites*, the small dark burrows. The large semiconcentric feature to the right could be a cross section of a *Scolicia* or an *Asterosoma*-type burrow. Deposition of this unit was on a storm-wave-dominated shelf with deltaic sediment input.

PPL, Scale bar = 4.45 mm

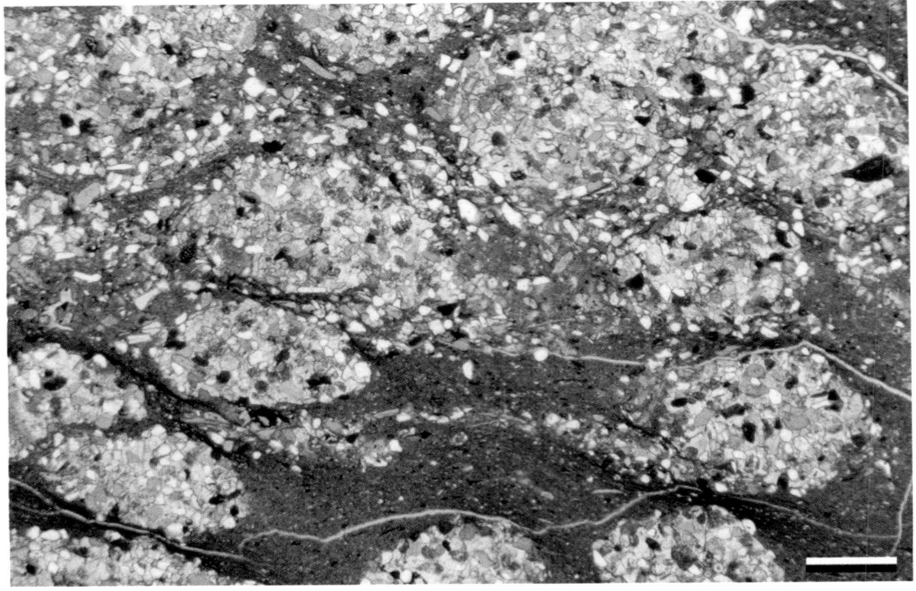

Mid. Jurassic Curtis Fm., Moffat Co., Colorado

Fine-grained sand from an overlying sandstone bed has filled millimeter-scale burrows in a shallow-marine shale. Note the abundant pale green glauconite grains in the burrow fills. There is differential compaction of the shales around the burrows, because the infill was framework supported whereas the water-rich shale matrix was not. That the burrows stayed open to allow sand infiltration suggests that the sediment was semiconsolidated.

PPL, RDI, Scale bar = 0.51 mm

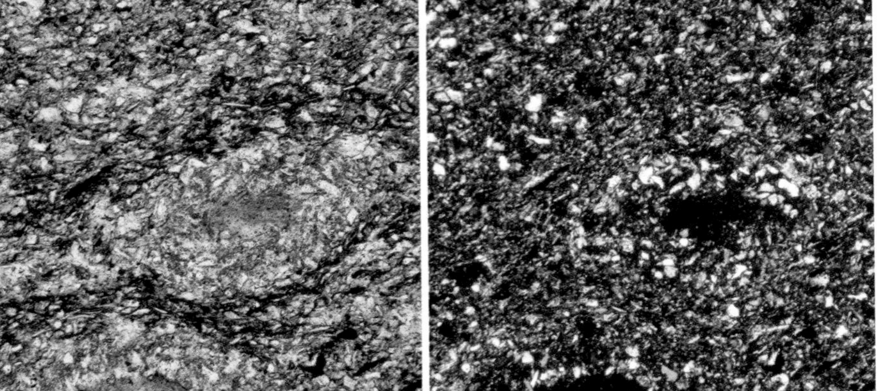

Up. Cretaceous Monte Antola Fm., Liguria, Italy

A silt-rich, bioturbated turbidite-associated shale. This section has phosphate-filled (brownish in PPL, black in XPL) burrows with quartz-lined walls and compactional deformation between burrows. The lined burrows suggest a sediment that was originally of a soft, soupy consistency.

PPL | XPL, Scale bar = 0.26 mm

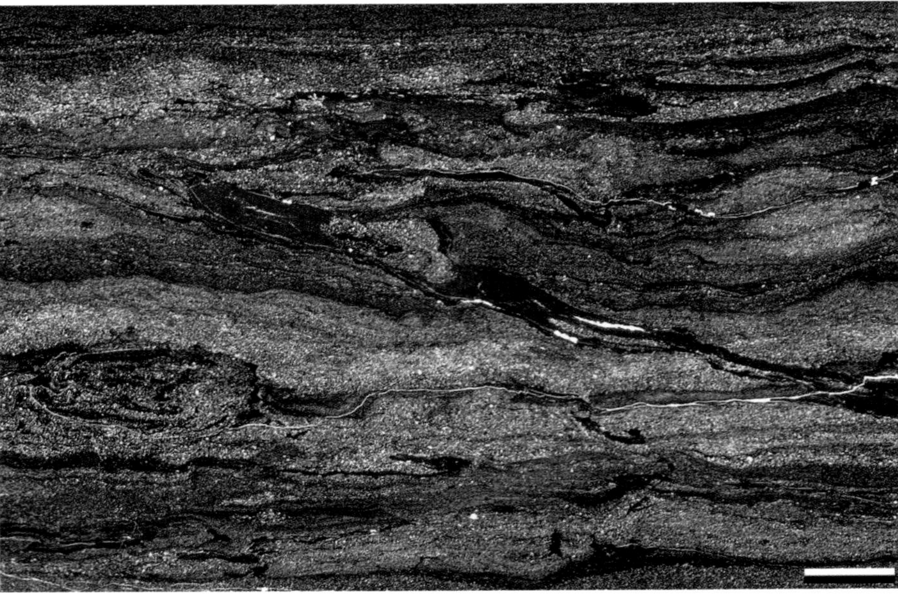

Up. Devonian Sonyea Gp., New York

A thin section from a red-stained (hematitic) immature floodplain soil. The layering, reflecting successive flooding, is still clearly recognizable. Disturbance of the fabric was mostly due to root traces. Flooding occurred too frequently to allow a mature (and homogenized) soil to form. For details see Schieber (1999).

PPL, Scale bar = 2.38 mm

Jp. Devonian Sonyea Gp., New York

A considerably more mature soil from the same section as the sample above. The soil is also reddish because of hematite impregnation, and no bedding is preserved. The original fabric has been completely obliterated by plant roots and seasonal shrinkage and expansion. The silt-rich lighter clots probably reflect old root traces. For more detail see Schieber (1999).

PPL, Scale bar = 2.38 mm

Mid. Proterozoic Mt. Isa Gp., Queensland, Australia

A graded bed in a silt-rich shale. The unit contains numerous graded beds that have small clasts at the base and successively finer silt and clay laminae upwards. These are interpreted as earthquake-triggered muddy turbidites, because deposition occurred near an active fault.

PPL, Scale bar = 2.00 mm

Jp. Cretaceous Blackhawk Fm., Kenilworth Mbr., Book Cliffs, Utah

A variably bioturbated shale. The basal third is strongly bioturbated with sand/silt-filled burrow tubes, and is overlain by two laminated-graded beds (arrows) that are only bioturbated near the top. These appear to be event beds and most likely originated as muddy tempestites. The unit was deposited on a shallow, storm-dominated shelf.

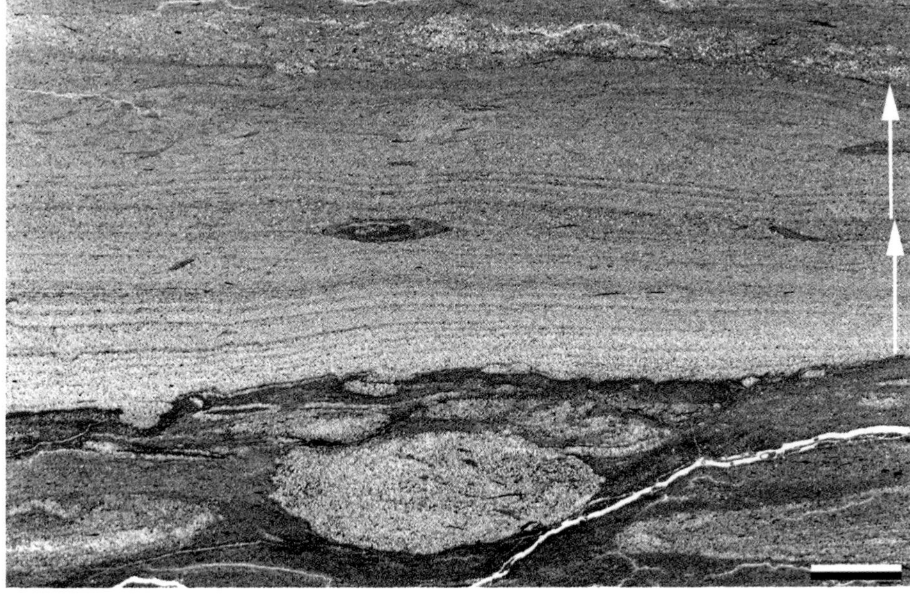

PPL, Scale bar = 3.15 mm

Cretaceous Mowry Fm., Moffat Co., Colorado

A small channel cut into a shale showing that deposition was located in an area were traction currents were flowing across the seafloor. The channel was filled with very fine-grained sand grading upward into siltstones. This graded fill probably represents turbidity current deposition.

PPL, RDI, Scale bar = 0.51 mm

Up. Devonian Sonyea Gp., New York

A cross-laminated and parallel-laminated silt-rich shale. This deposit shows subtle grading upwards from a base with flame structures, overlain by cross-laminae, and topped with parallel laminae that diminish upwards. This unit probably was deposited as a fine grained turbidite (Stow and Shanmugam, 1980) with Bouma C–E succession at the shelf to basin transition (Schieber, 1999). The horizontal fractures may be an artifact of sample handling and thin-section preparation.

PPL, Scale bar = 2.38 mm

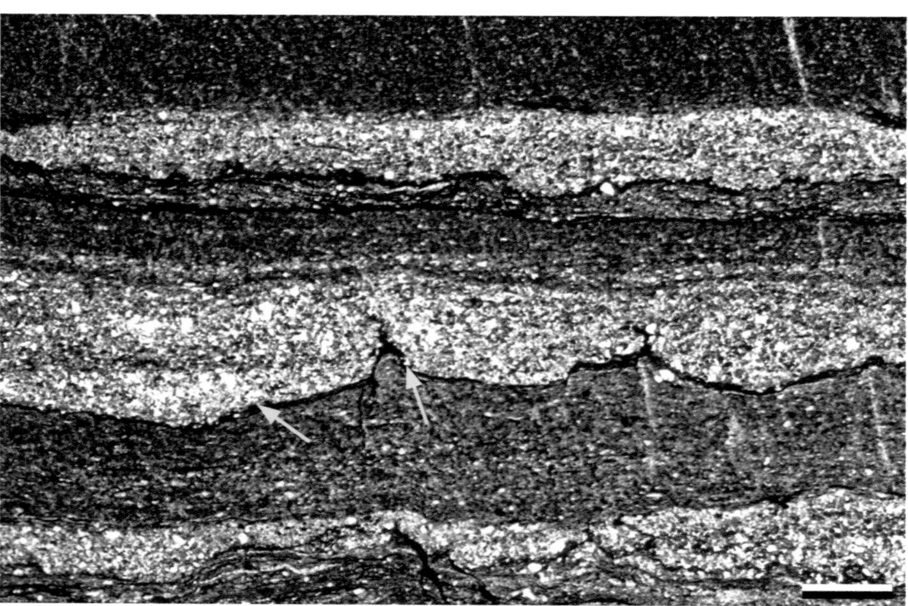

Mid. Proterozoic Belt Supergroup, Newland Fm., Meagher Co., Montana

A clay-rich shale with silt layers. The silt layers all have bulbous features at the base that are small-scale load casts. These formed because the underlying mud was water-rich and could not support the denser silt layer. Fabric collapse in the underlying mud led to downward movement of the clay-silt interface and the formation of a load structure. The yellow arrows show a distinctive carbonaceous biofilm layer that was deformed, but not cut, by the overlying silt layer, clearly demonstrating that these are load structures, not scours. The near-vertical white lines on the right half of the image are scratches on the thin section.

PPL, Scale bar = 0.45 mm

;retaceous Mowry Fm., Moffat Co., ;olorado

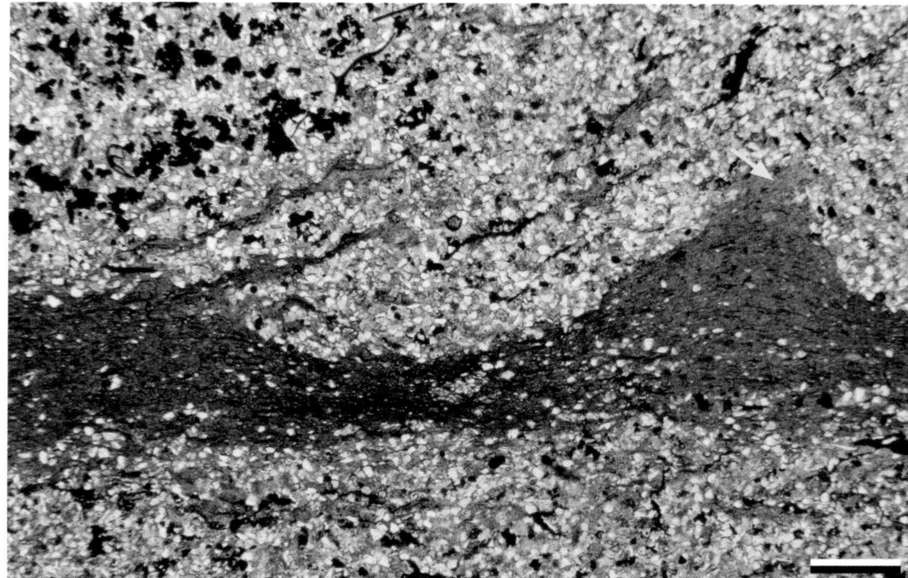

Vhen coarse sediments are dumped onto vater-rich clay substrates, load structures and lame structures can form. These very fine-;rained, cross-bedded sandstones are partially inking into the underlying clays forming ulbous load structures. As the water-rich lays flow due to the overburden loading, they an be partially intruded, as in this case, into the verlying sandstones forming flame structures example at arrow). Some of the clays may lso have been forced into inclined bedding lanes within the sandstones. Replacive pyrite black) crystals accentuate the cross bedding in ne sandstones.

'PL, Scale bar = 0.51 mm

Jp. Devonian Chattanooga Shale, 'ennessee

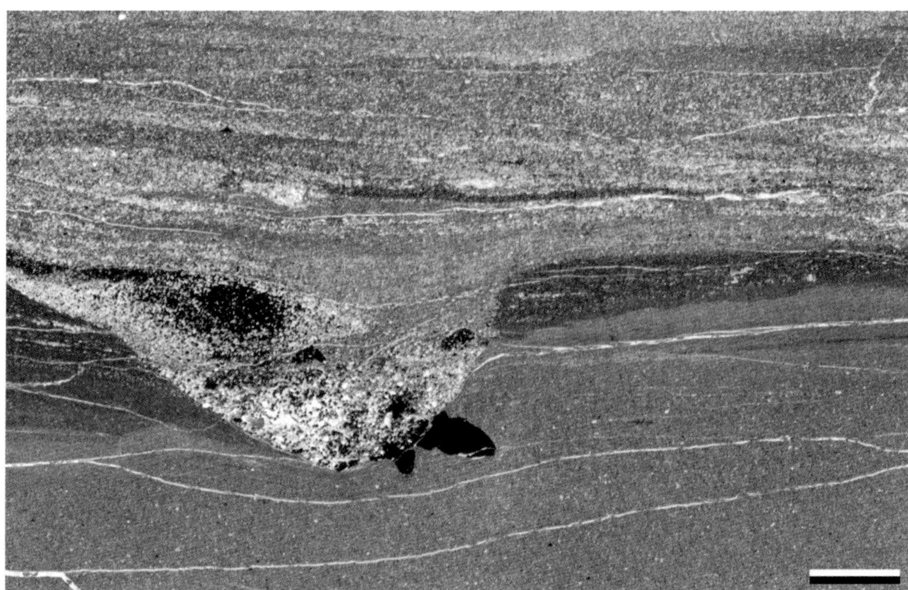

scour at the base of a silty shale bed. The ottom of the scour pit is filled with silt and ome fossil debris, but the upper part of the pit nd the overlying bed are dominated by clay nd fine-grained silt. This is another example f shale-on-shale erosion, a common feature 1 the rock record, but difficult to detect in utcrop and drill core. That it is a scour and ot a load cast is indicated by the fact that the cour pit truncates the internal layering of the nderlying bed.

'PL, Scale bar = 2.96 mm

Jp. Devonian Sonyea Gp., New 'ork

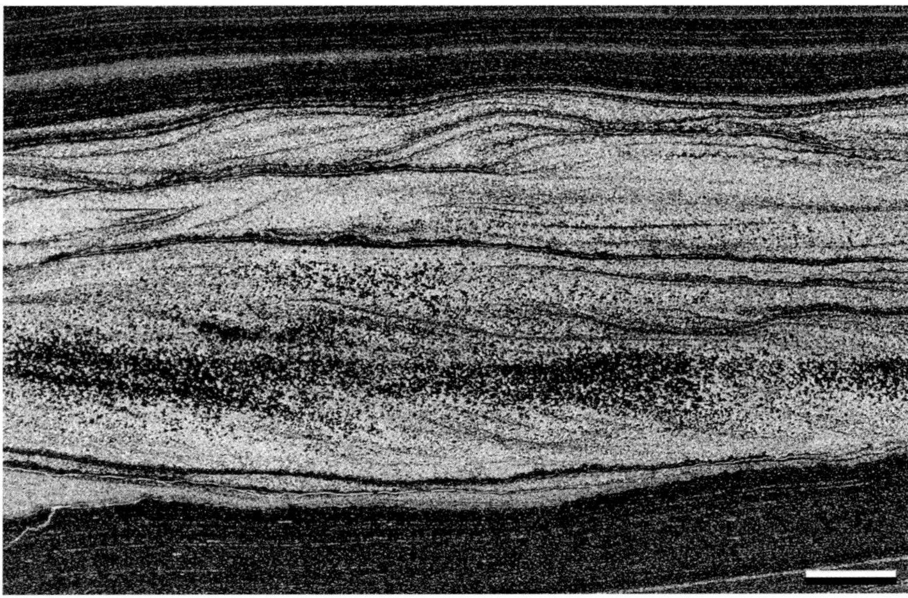

finely laminated black shale with a silt nterval that shows cross lamination and len-cular buildup of silt. The opposing dips of ne foresets and the dark seams between the ilt layers suggest multiple depositional events nd variable current directions. This is a distal eposit within the North American Devonian nland sea, and the silt is interpreted as having een deposited by bottom currents.

'PL, Scale bar = 2.38 mm

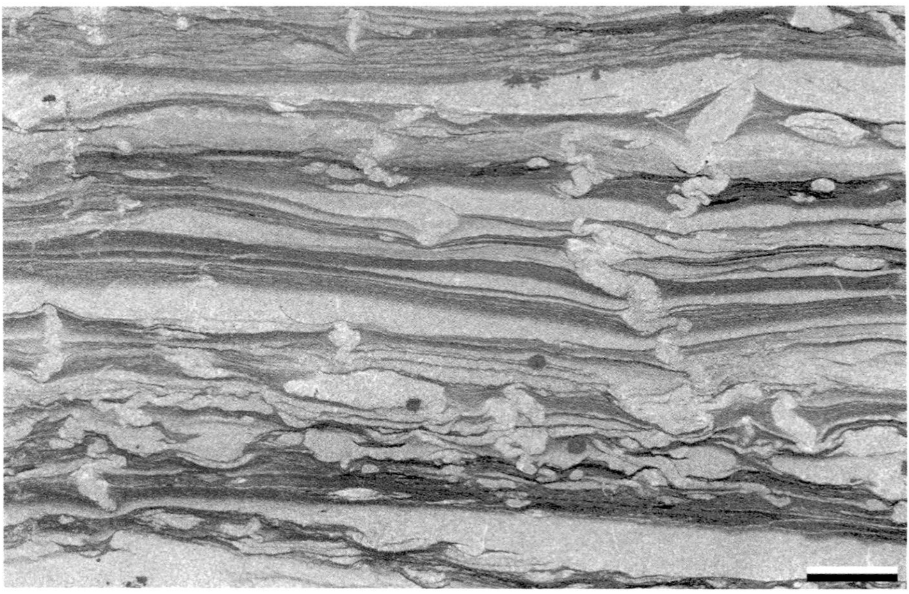

Cambrian Eau Claire Fm., Indiana

A shale with wavy-lenticular layers of fine sand to silt and deformed vertical cracks. The cracks can extend both up and down from sandy layers and show ptygmatic folding. Differential compaction around these cracks, as well as their vertical shortening, indicates that they formed in soft, water-rich sediments. They are interpreted as syneresis (also written synaeresis or synæresis) cracks, rather than desiccation cracks, in part based on bedding-plane observations of the layer from which this sample was taken. The darker gold spots within sand-silt layers are pyrite concretions. These sediments were deposited in a shallow-shelf environment.

Core photograph, Scale bar = 8.00 mm

Cretaceous Mowry Fm., Moffat Co., Colorado

The silt layer in this image fills a downward extending crack that is complexly deformed (ptygmatic folding) as a result of compaction (by a factor of ~2x). The sediment at the time of folding was still rather soft and behaved in plastic manner. The crack is most likely a syneresis crack although, alternatively, it could be a burrow. A bedding-plane surface is needed to make a definitive interpretation.

PPL, RDI, Scale bar = 0.10 mm

Eocene Green River Fm., Wyoming

Mud-filled desiccation cracks in a shale. The arrows mark a large crack that is filled with clays (instead of sand) and is similar in composition and hardness to the wall rock. In outcrop or hand sample, such cracks are not easily seen. Additional clay-filled cracks in the bottom part of the image lack the differential compaction of syneresis cracks suggesting that, unlike the previous examples, these cracks originated as desiccation cracks. Although these structures could be mistaken for burrows, the fact that they remain in the plane of the section, coupled with outcrop observations, confirm that these are indeed cracks.

PPL, Scale bar = 2.99 mm

ᴾermian Wolfcamp Shale, Reeves ᶜo.,Texas

ʼhe subtle recumbent fold structure in this ilty shale reflects soft-sediment flowage on . gentle slope (fold nose marked by arrow). ;uch folds occur very early, at or near the eafloor, before fine-grained sediments have ındergone substantial dewatering. They are ·ommonly found in association with debris lows and other deposits related to sediment iquefaction. Most such structures are larger han thin-section scale, but it is not uncommon ɔ find evidence of such deformation at milli- neter scales as well.

ᴾPL, Scale bar = 0.34 mm

Jp. Devonian Chattanooga Shale, ʼennessee

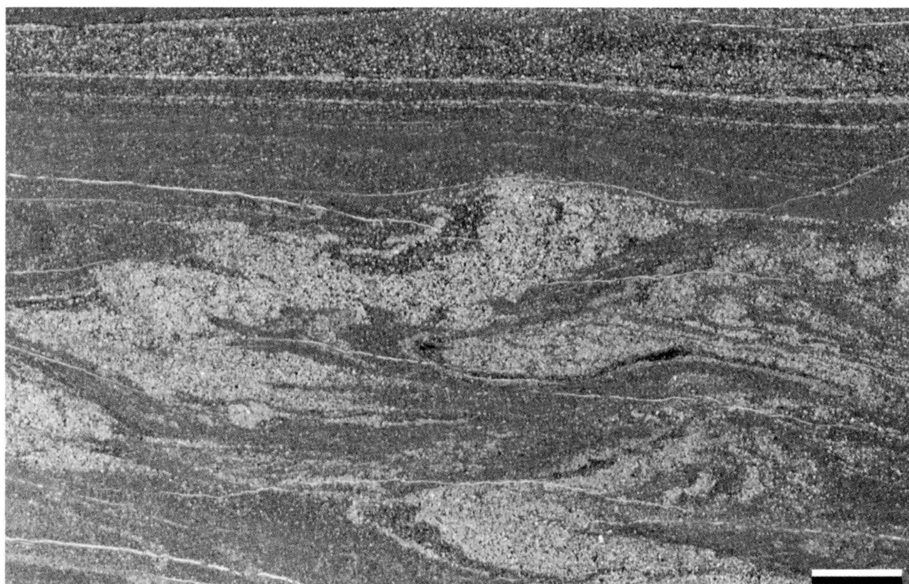

ˢoft sediment deformation in a carbonaceous hale. In the lower two thirds of the image, the ilt layers have been seriously contorted while ıe sediment was still soft. In the upper third f the section, the laminae are continuous and ndeformed. This kind of deformation, that nly affects horizons of some centimeters to ecimeters, can have multiple causes, such as eismic shaking, storm waves and tsunamis.

ᴾPL, Scale bar = 2.00 mm

Jp. Devonian Chattanooga Shale, ʼennessee

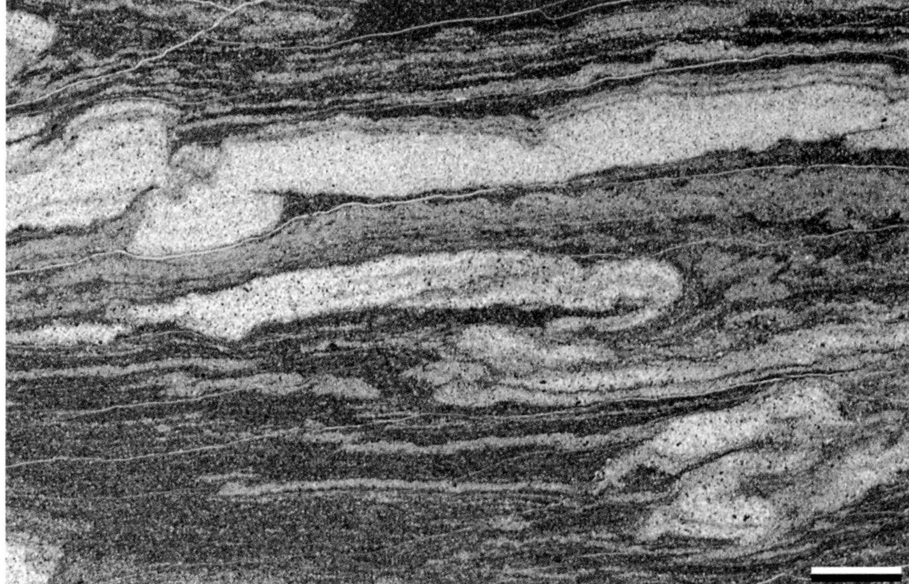

ᴀnother example of soft sediment deformation ı a black shale. Similar to the previous photo- ıicrograph, this example also has faulting off- etting the layers. In contrast to the previous xample, this suggests that the sediment in this ase was more cohesive (more consolidated) ⁄hen deformation occurred.

ᴾPL, Scale bar = 2.00 mm

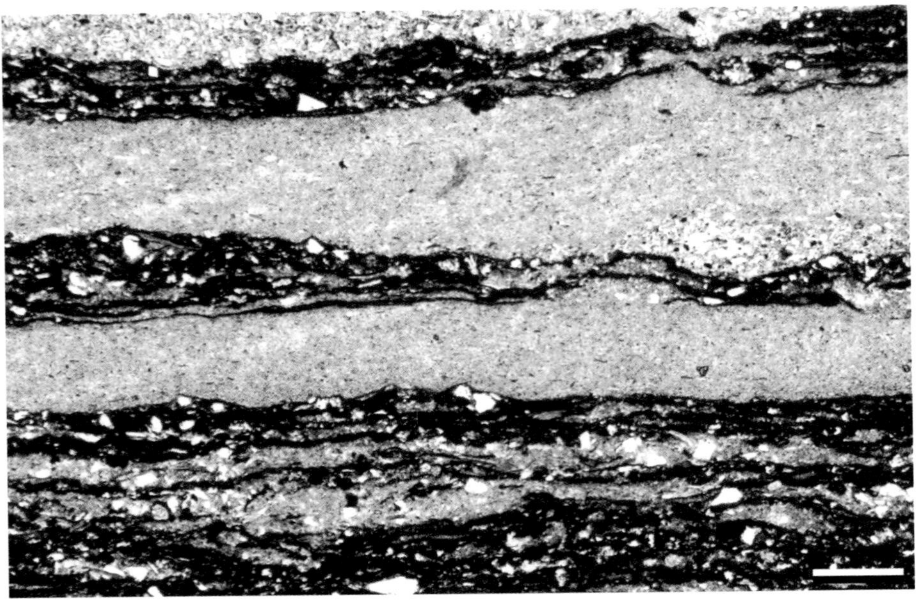

Mid. Proterozoic Belt Supergroup, Newland Fm., Meagher Co., Montana

Benthic microbial mats and distal storm deposits (tempestites). The mats are characterized by dark-colored, wavy–crinkly, carbonaceous laminae, with internal drapes of clay and silt. Mat growth was interrupted by occasional storm deposits (the light-colored layers). The storm deposits have silt at the base and grade upward into clays. For details see Schieber (1986, 2007a).

PPL, Scale bar = 0.51 mm

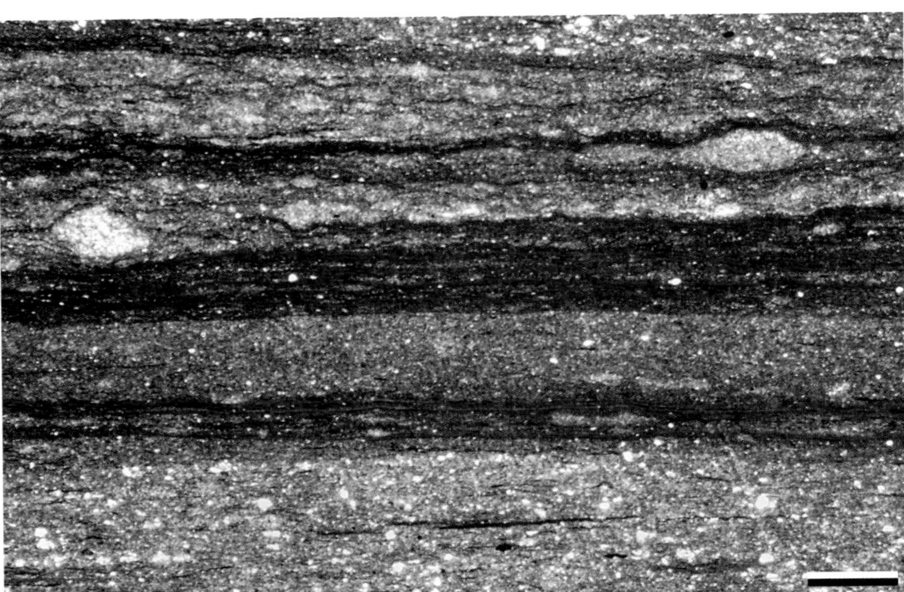

Eocene Green River Fm., Wyoming

Kerogen-rich laminae (reddish-brown) in a "lean" dolomitic oil shale. The kerogen laminae are interpreted as microbial mats on the basis of mechanical behavior (internal cohesiveness) and geochemical considerations. In places, abundant tiny dolomite concretions give rise to a lenticular-wavy fabric. Carbonaceous shreds in non-mat layers are likely mat rip-ups. For more detail see Schieber (2007b).

PPL, Scale bar = 1.21 mm

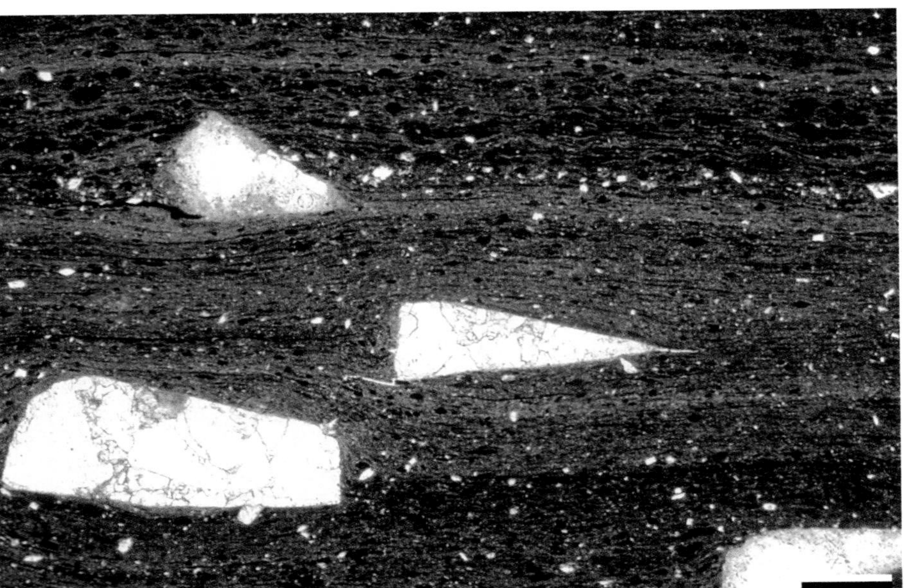

Eocene Green River Fm., Wyoming

A kerogen-rich oil shale layer with kerogen laminae. These laminae have been interpreted as benthic microbial mats that grew at the bottom of a shallow playa lake (Eugster and Hardie, 1975). The clear triangular crystals are shortite ($Na_2Ca_2(CO_3)_3$), a mineral that is common in saline lacustrine deposits.

PPL, Scale bar = 1.21 mm

Jurassic Diaspri Radiolarite, Liguria, Italy

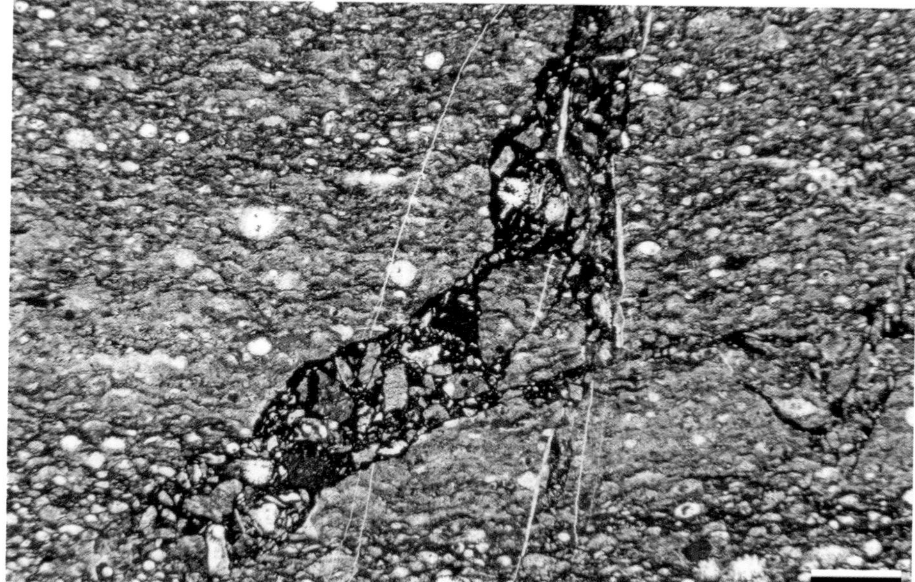

A brecciated radiolarian chert with minor interspersed reddish shale, a rock that falls near the siliceous pole of the ternary compositional diagram for shales (Fig. 8.1). This is part of an ophiolite sequence—the remains of Tethyan ocean-floor deposits. The circular white objects are relatively poorly preserved radiolarians. This rock is highly brittle and is brecciated, here and in most samples, an indication of the influence of siliceous material in decreasing the ductility and increasing the brittleness of shales.

PPL, Scale bar = 0.1 mm

Cretaceous Mowry Fm., Moffat Co., Colorado

A view of an organic-rich, silty, siliceous shale with planar fabric. One can see partially preserved and silica-filled radiolaria (light brown), darker brown pieces of fish debris and horizontal black streaks that are pyrite precipitates associated with organic particles. The silica-filled (chert) remains of radiolaria occur throughout this view. One radiolarian (yellow arrow) has the original skeletal structure preserved as clear margins. There is a hint of an underlying pelletal fabric, probably due to an abundance of original planktonic fecal pellets.

PPL, RDI, Scale bar = 0.10 mm

Cretaceous Mowry Fm., Moffat Co., Colorado

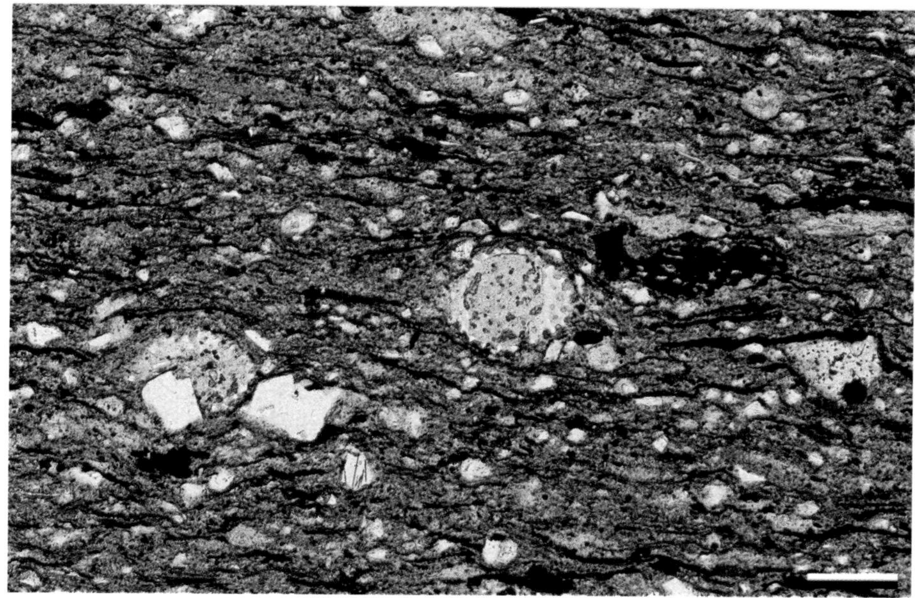

A closer view of a siliceous shale with remains of radiolarians (the rounded perforated features at center and center left). The larger (0.1–0.2 mm) black particles are woody fragments of terrestrial plants. The uncollapsed radiolarians show differential compaction around them, suggesting early infill and replacement by chert.

PPL, RDI, Scale bar = 0.10 mm

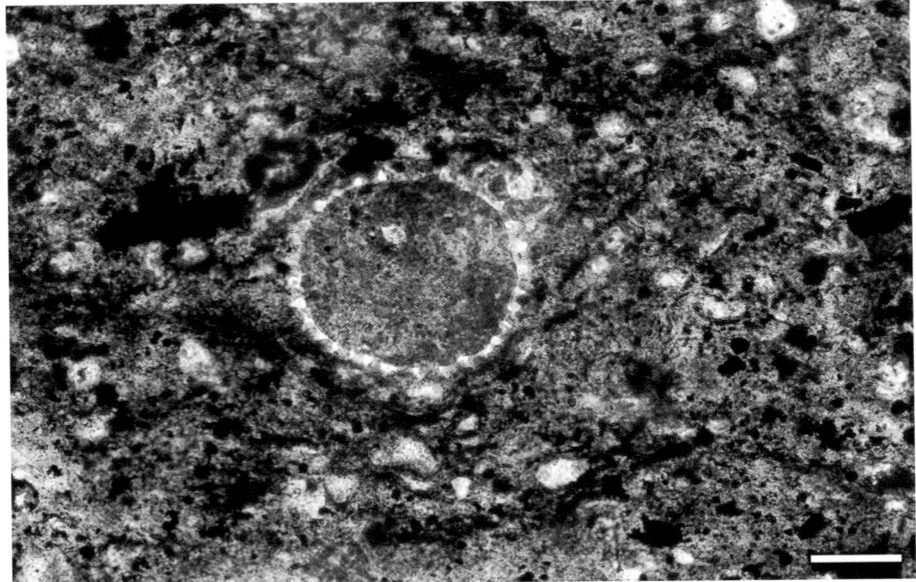

Cretaceous Mowry Fm., Moffat Co., Colorado

A close-up of a cross section of a radiolarian test from the same interval shown in the previous image. The clear wall of the radiolarian is now composed of quartz (after originally opaline silica) and shows regular perforations; the center of the test is infilled with brownish chert (with the color a result of minute organic inclusions).

PPL, RDI, Scale bar = 0.05 mm

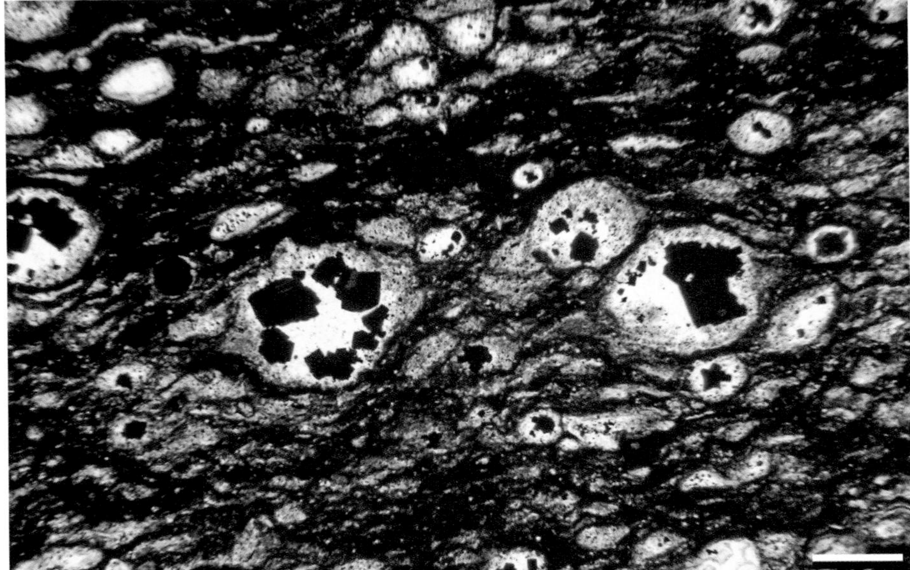

Up. Devonian – Lo. Mississippian Woodford Shale, Lincoln Co., Oklahoma

This example shows a highly siliceous (because of biogenic silica) and carbonaceous shale. Here, the bright spherical areas are radiolaria that have been infilled with diagenetic pyrite (black) and chalcedony/chert cement. The wall structures of the radiolaria have been largely dissolved, although some structure is preserved locally where early chert replacement occurred. Quite a few of the spherical structures look flattened, probably because they were not fully filled with cement prior to compaction.

PPL, Scale bar = 0.14 mm

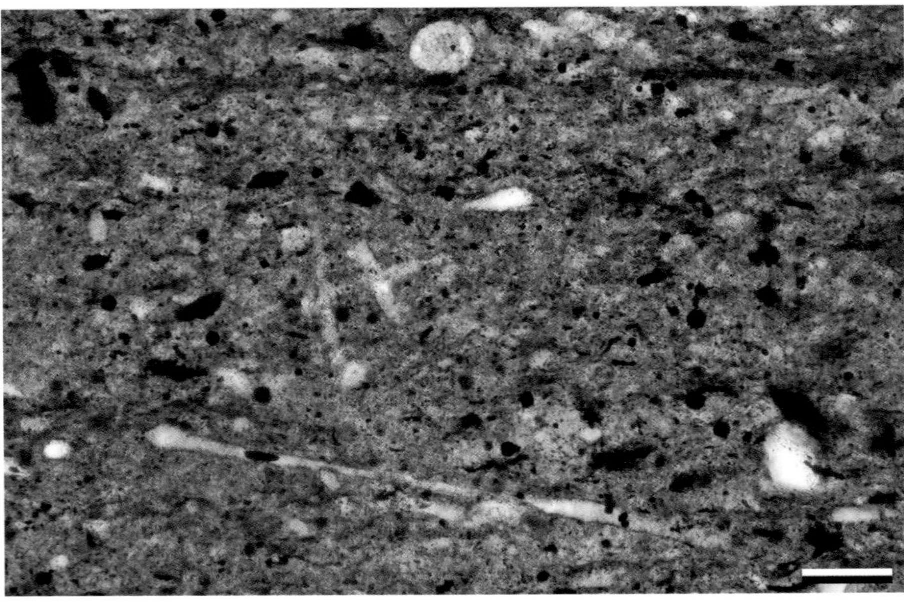

Cretaceous Mowry Fm., Moffat Co., Colorado

This example of an organic-rich, siliceous shale shows biogenic silica components that are mainly sponge spicules (straight, light colored elements), rather than the radiolaria that were seen in several previous images. These spicules come from the breakdown of sponges from any of several groups. Spicules from the class Hyalospongia reflect very deep water sedimentation (generally >1,000 m [3,300 ft]); those from the class Demospongia can originate in much shallower waters, commonly even within the photic zone.

PPL, RDI, Scale bar = 0.05 mm

Lo. Permian Bone Spring Formation, Reeves Co., Texas

A highly fossiliferous shale with abundant siliceous sponge spicules. Some of the sponge spicules still show well preserved central canals and a few even retain intragranular porosity (blue epoxy). The fine-grained shaly carbonate matrix also contains nanoscale porosity not visible at this magnification. The sample is oil stained (brownish colors) and organic rich.

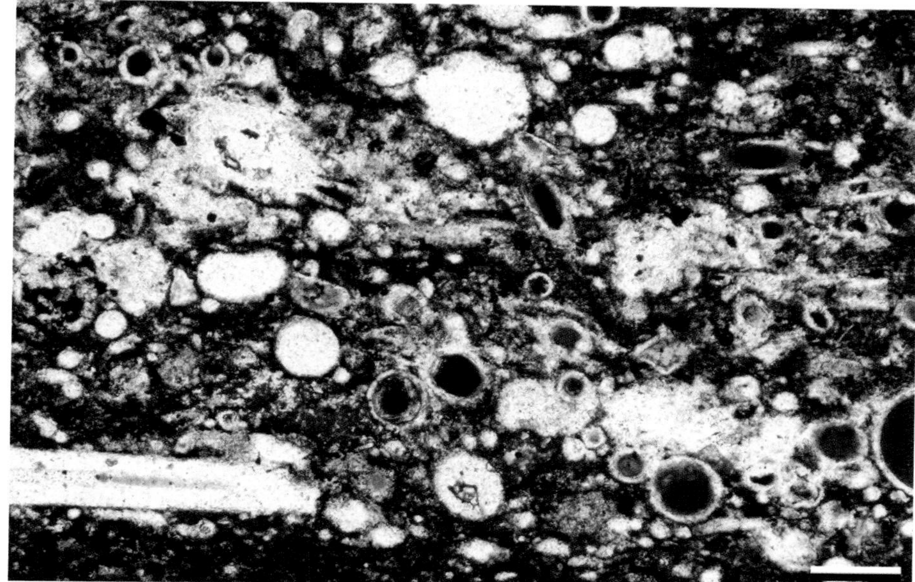

PPL, BDI, Scale bar = 0.10 mm

Lo. Permian Bone Spring Formation, Reeves Co., Texas

A spiculitic shale with carbonate lenses and clasts. The sample is organic rich and contains live oil. In the cross-polarized light image, silica-rich areas are speckled gray and white, whereas carbonate-rich areas show higher-order, pastel birefringence colors.

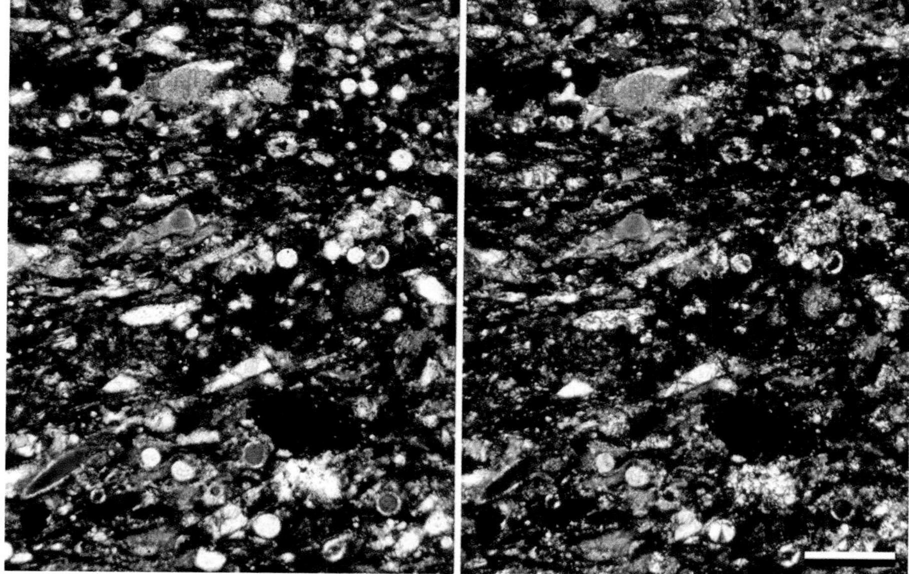

PPL | XPL, Scale bar = 0.26 mm

Eocene Green River Fm., Lincoln Co., Wyoming

A fragmented, phosphatic fish bone in a laminated lime mud. Irregular seams between lime mud laminae could be related to bedload processes (Schieber et al., 2013) or they could reflect slightly deformed original sediment couplets (varves). Note the compactional drape over the upward pointing bone spur. The plane-polarized light view of this photomicrograph is located in Chapter 5 (page 119).

XPL, Scale bar = 0.51 mm

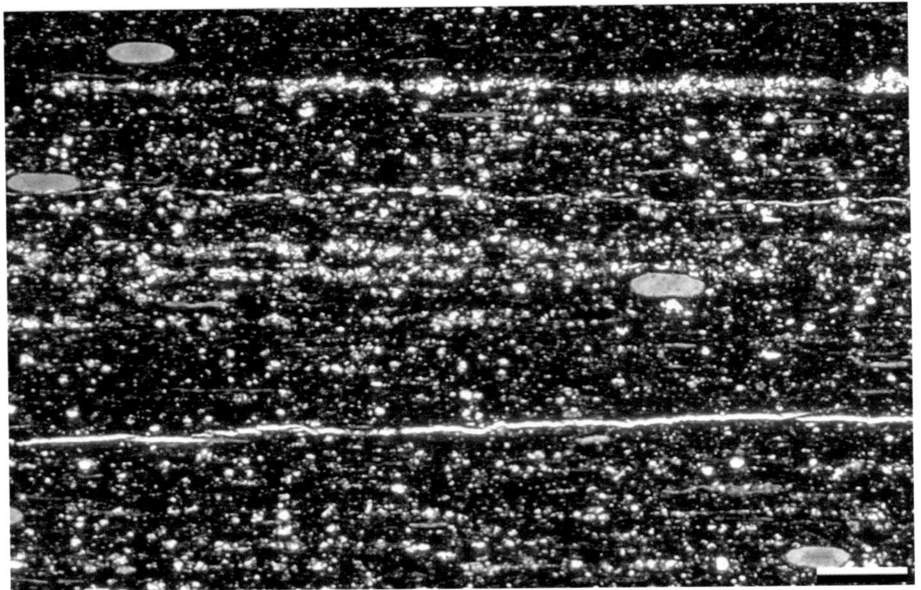

Up. Devonian Chattanooga Shale, Tennessee

Carbonaceous shale (~10% TOC) with fine parallel silt laminae. This sample contains abundant yellowish to brownish streaks and oval bodies that represent collapsed algal cysts and spores. Generically, these cysts/spores are commonly referred to as *Tasmanites*, but one should keep in mind that there around 29 genera and 100 species in the prasinophyte algal family Tasmanacea (Haq and Boersma, 1998). Originally spherical, the hollow *Tasmanites* cysts are flattened by compaction.

PPL, Scale bar = 0.32 mm

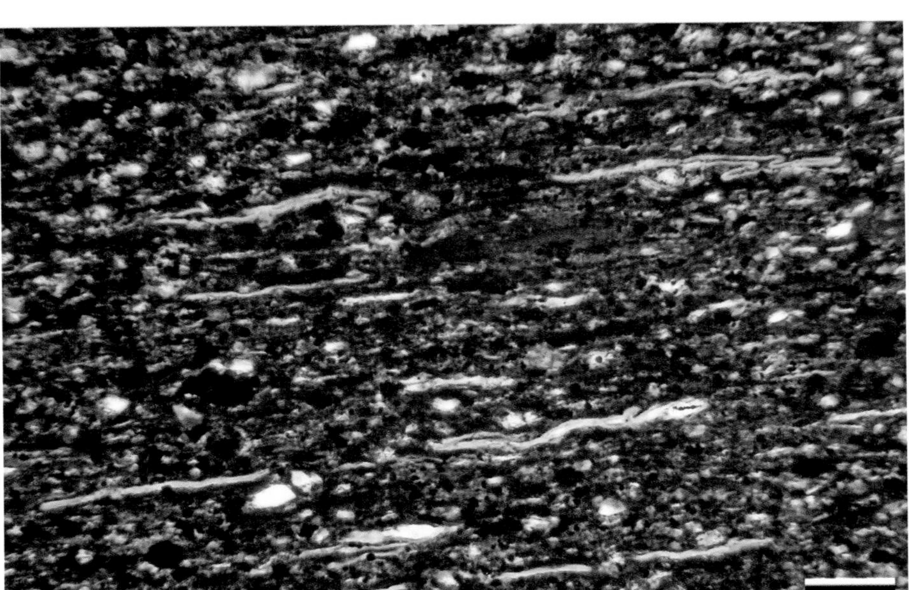

Up. Devonian – Lo. Mississippian Woodford Shale, Lincoln Co., Oklahoma

Flattened *Tasmanites* in a carbonaceous shale with abundant tiny clumps (black) of diagenetic pyrite (framboids, framboid clusters and tiny concretions). Tasmanitids can add considerable organic matter to these types of basinal deposits—organic matter can form coals and rich oil shales.

PPL, Scale bar = 0.34 mm

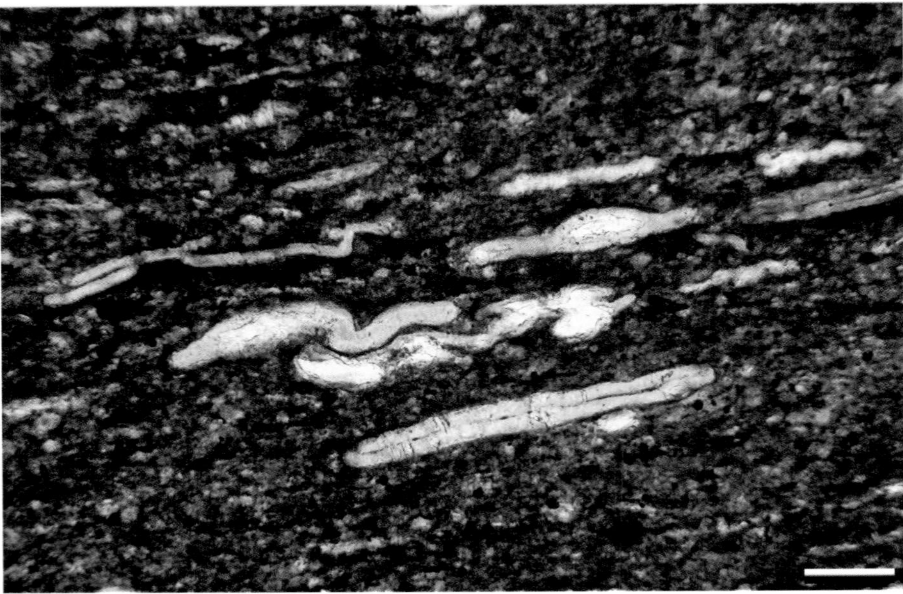

Up. Devonian – Lo. Mississippian Woodford Shale, Lincoln Co., Oklahoma

A close-up of *Tasmanites* in a black shale. The bright interior fills of some cysts are diagenetic quartz (chalcedony and chert) that were precipitated in the cysts before they became fully compacted. The silica was derived from dissolution of radiolaria (Schieber et al., 2000).

PPL, Scale bar = 0.45 mm

Up. Devonian – Lo. Mississippian Woodford Shale, Lincoln Co., Oklahoma

A bedding plane view of a shale with abundant *Tasmanites* cysts or spores. In plan view, the flattened cysts/spores reveal their original spherical shape.

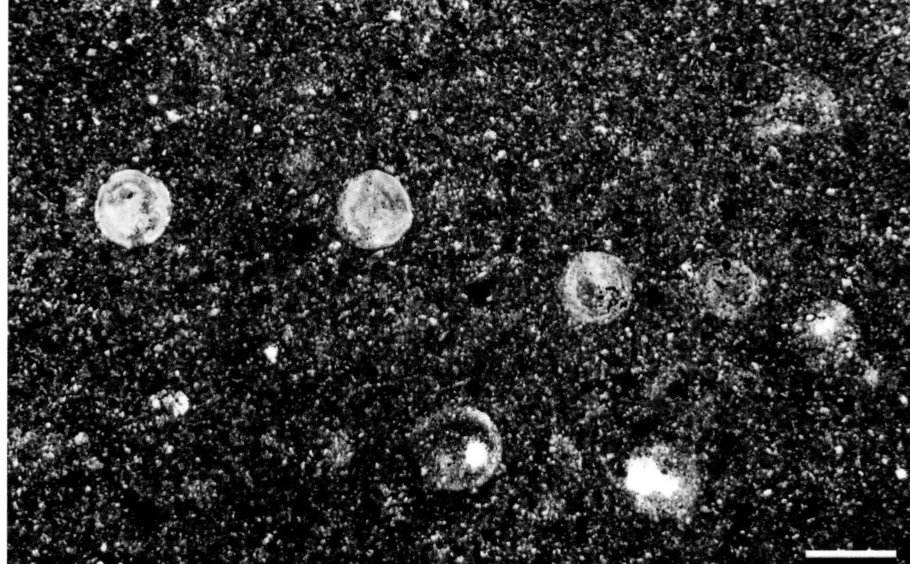

PPL, Scale bar = 0.34 mm

Up. Cretaceous Niobrara Fm., Smoky Hill Mbr., Logan Co., Kansas

An outcrop example of a pure chalk, the carbonate end member of the spectrum of fine-grained sedimentary rocks. Essentially everything in this field of view is composed of calcite and is of biogenic origin. Several species of coccolith shields are visible; the broken and attached spines (yellow arrows) are from genera such as *Prediscosphaera*. Remarkable porosity can be preserved in such rocks where they have not been deeply buried or where overpressuring and/or early hydrocarbon entry prevented or retarded compaction and recrystallization.

SEM, Scale bar = 1.7 μm

Up. Cretaceous Hod Gp. chalk, Danish Sector, North Sea

A hand-colorized SEM image of a less pure chalk that has undergone substantial burial (~2 km [~6,560 ft]). Recognizable coccolith remains are colored green; clays are colored blue; diagenetic nanosilica is colored red. Considerably less porosity is present as compared to the previous sample because of compaction and growth of calcite cements (tan colored). Clays and nanosilica clusters, the latter probably derived from dissolution of radiolaria, occlude pores but may also contribute to the brittleness of these rocks.

Col SEM, Scale bar = 2.2 μm

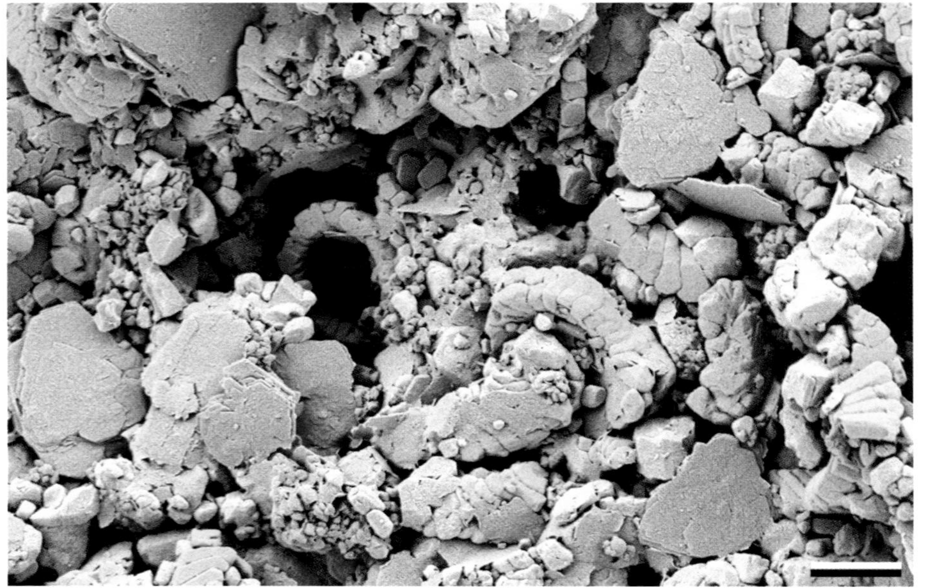

Up. Cretaceous Hod Gp. chalk, Norwegian Sector, North Sea

Another hand-colored SEM image, this one depicting a marly (clay-rich) chalk, a transitional rock type between chalks and shales. The same color scheme is used as in the previous image, and far more clay, mainly kaolinite, is visible here. Despite the clays, which tend to accelerate carbonate dissolution and reprecipitation, considerable porosity remains in this sample at nearly 3-km (9,800-ft) burial depth, largely because of very strong regional overpressure.

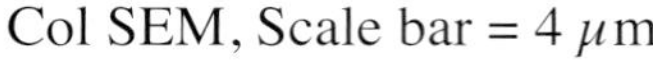

Col SEM, Scale bar = 4 μm

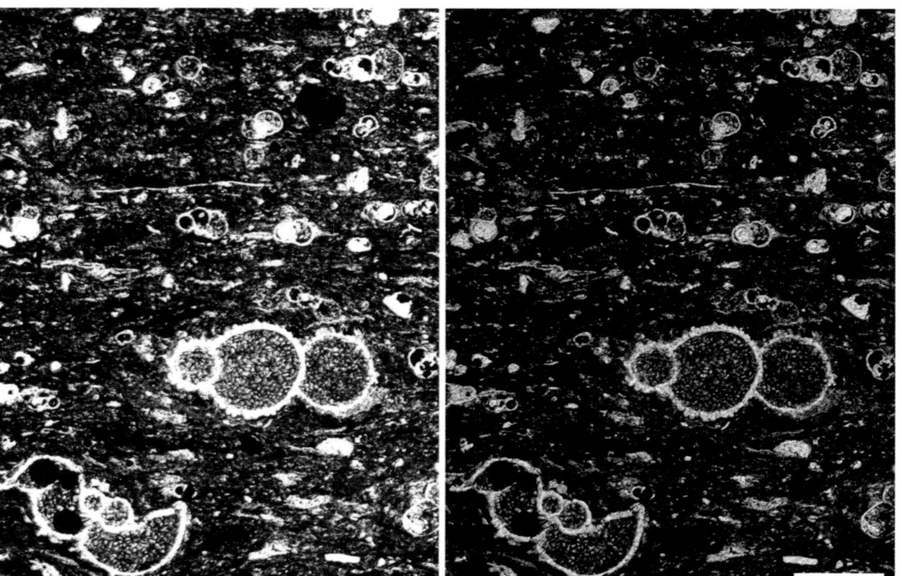

Cretaceous (Cenomanian) Eagle Ford Shale, southwest Texas

Abundant planktonic foraminifera in a carbonaceous shale. The chamber walls of the foraminifera consist of calcite, and they are filled with calcite cement (clear), pyrite framboids (black), and authigenic clays (kaolinite, brownish). The shale matrix consists of nannofossil-rich planktonic fecal pellets, clays, silt and organic matter.

PPL | XPL, Scale bar = 0.06 mm

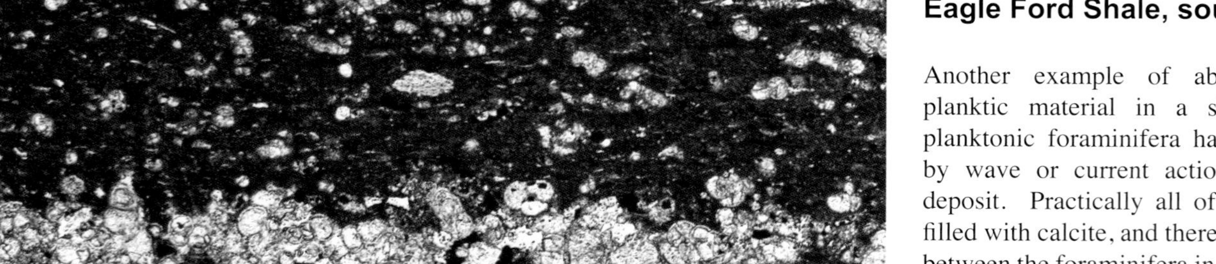

Up. Cretaceous (Cenomanian) Eagle Ford Shale, southwest Texas

Another example of abundant carbonate planktic material in a shale. Here, the planktonic foraminifera have been reworked by wave or current action into a thin lag deposit. Practically all of the chambers are filled with calcite, and there is also calcite spar between the foraminifera in the lag, suggesting that the mud matrix was winnowed out.

PPL, Scale bar = 0.12 mm

Up. Pennsylvanian Graham Fm., Finis Shale, central Texas

The bright streaks in this photomicrograph are the remains of benthic agglutinated foraminifera (Milliken et al., 2007). They stand out because the foraminifera constructed their chamber walls from small, detrital quartz silt grains, and subsequent compaction flattened the chambers (note the medial sutures). Later, the silt clusters were cemented with silica and transformed into chert-like streaks that made it into the rock record. The next photomicrograph shows more detail of the wall structure.

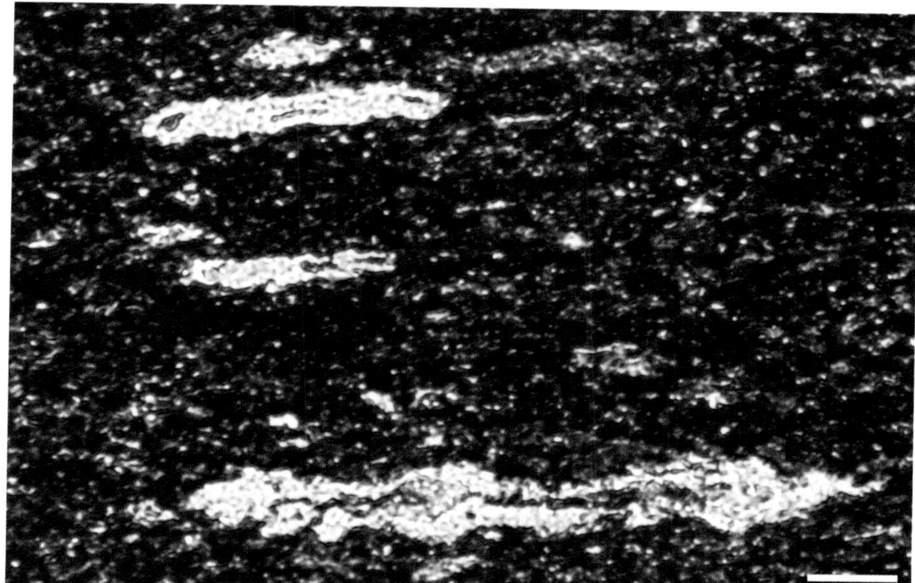

PPL, Scale bar = 0.09 mm

Up. Pennsylvanian Graham Fm., Finis Shale, central Texas

A closer view of an agglutinated foraminifer from the Finis Shale. This is not the normal pairing of two images of the same area; rather it is a continuous image with one part in plane-polarizes and the other part in cross-polarized light. The clear appearance in thin section (right side) is related to a dominance of quartz in the wall structure. When cross-polarized light is used, the individual quartz grains become visible and, because of their small size, give rise to a cherty appearance (left side-XPL). Whether this is indeed chert cannot be decided with a petrographic microscope due to the small grain size.

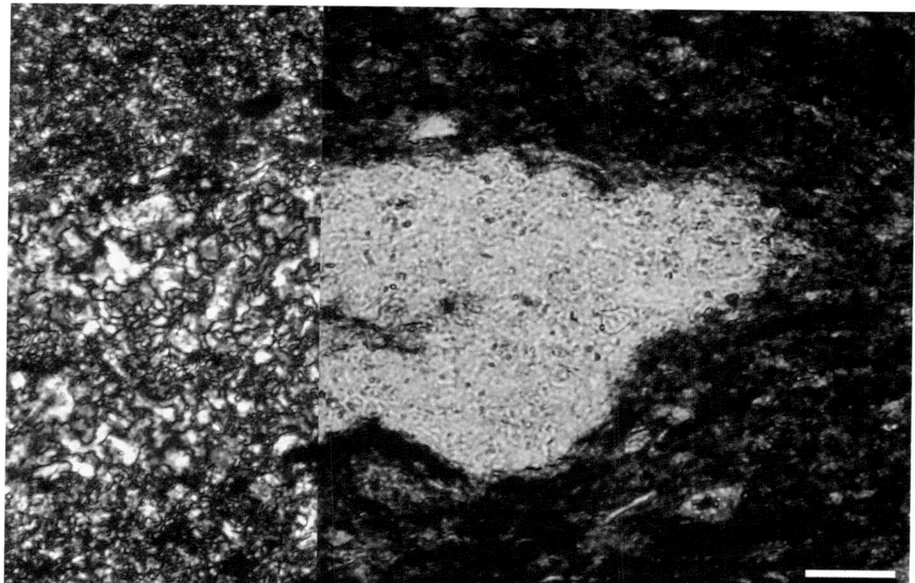

XPL | PPL, Scale bar = 31.2 μm

Up. Devonian Cleveland Shale, Kentucky

A scanned cathodoluminescence image of a shale with abundant agglutinated foraminifera. The colorful (mostly reddish) spots within the "cherty streaks" are tiny grains of detrital quartz, mostly less than 5 μm in size. This well-sorted characteristic (in comparison to the rest of the shale) is related to grain-size selectivity by the foraminifera (Pike and Kemp, 1996). The low-luminescence areas between detrital grains are filled with diagenetic quartz.

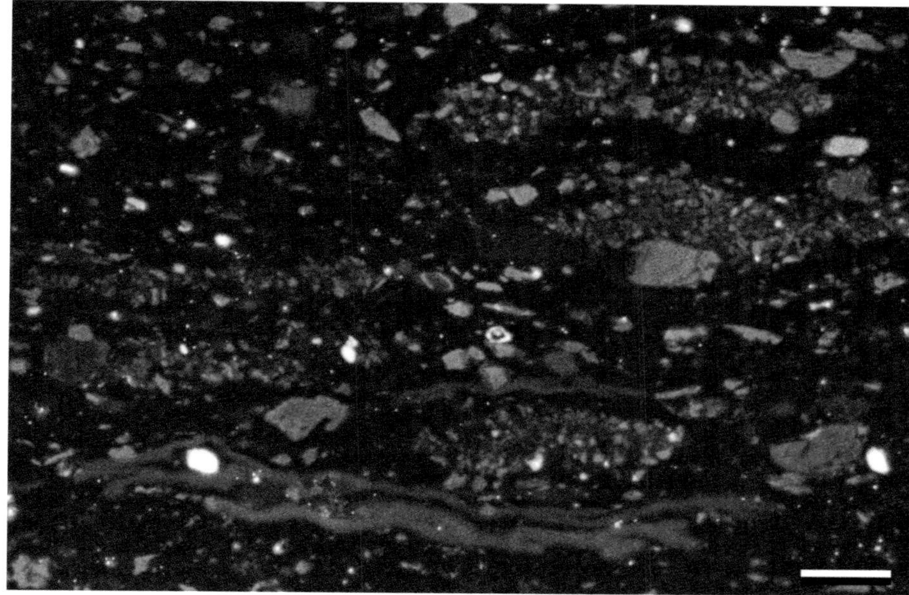

CCL, Scale bar = 25.5 μm

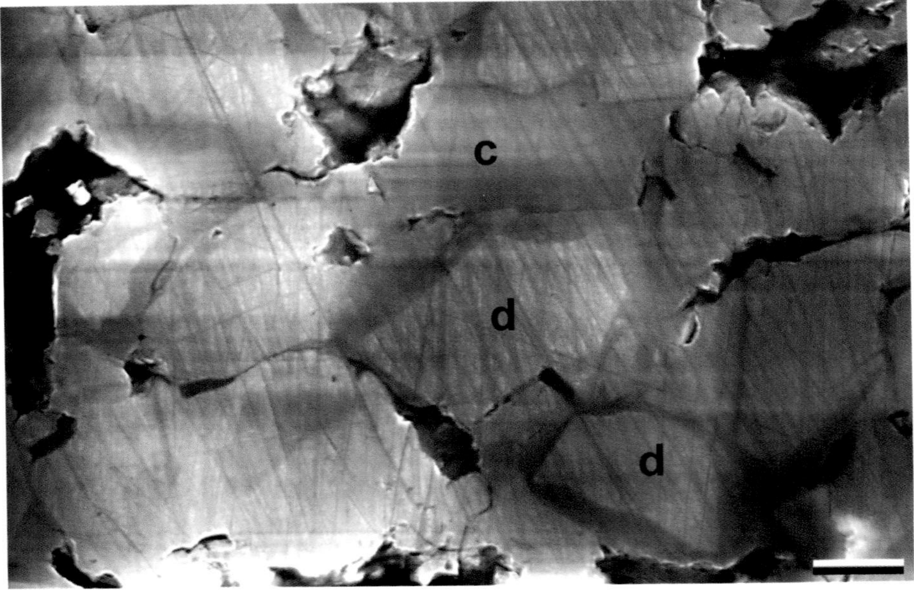

Mississippian Barnett Shale, central Texas

A high-magnification SEM image of an agglutinated foraminifera similar to those from the previous image, but now in charge contrast mode. The detrital grains (d) are subtly less conductive and thus brighter than the diagenetic quartz cement (c) between them. Because of this, the angularity of the detrital quartz grains is clearly visible. The lines across the grains are scratches from polishing the thin section. For more on imaging benthic agglutinated foraminifera, see Schieber (2009).

SE-CCI(SEM), Scale bar = 2.3 μm

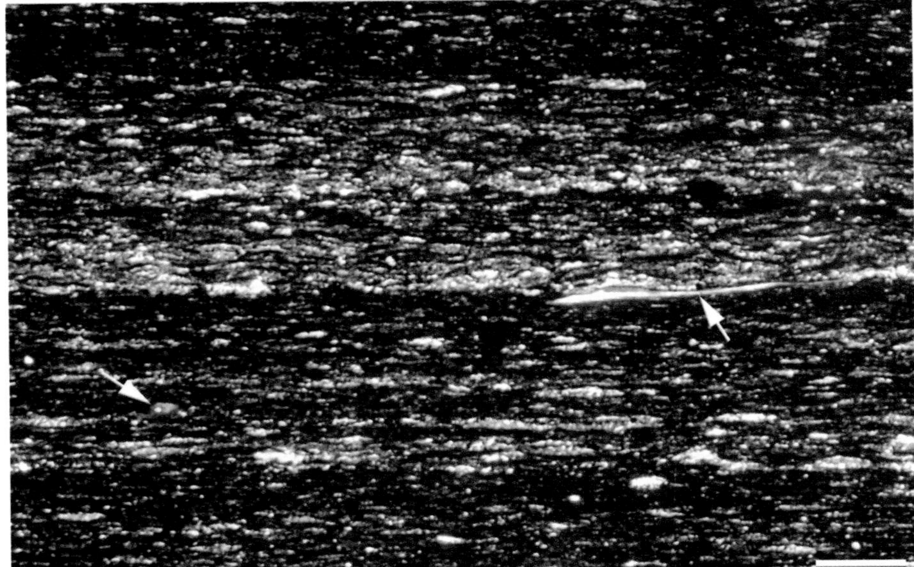

Lo. Jurassic (Toarcian) Posidonia Shale, southwest Germany

A carbonaceous shale with abundant bright lenticular features, giving rise to a lenticular fabric. The bright features are planktonic fecal pellets that have been flattened by compaction. They look bright because they contain abundant calcite debris from planktonic microorganisms (coccoliths and others). The layers with high abundance of pellets have sharp bases, suggesting that these were reworked and winnowed by bottom currents or wave action. The arrows point to pieces of phosphatic fish debris. This unit was deposited in a dysoxic shallow-shelf environment.

PPL, Scale bar = 0.32 mm

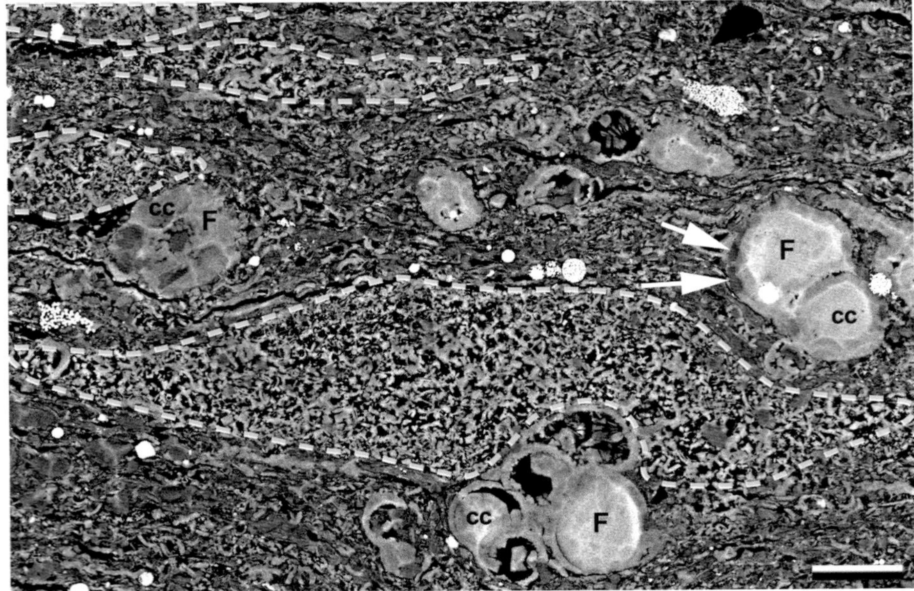

Up. Cretaceous (Cenomanian) Eagle Ford Shale, southwest Texas

Fecal pellets of a planktic organisms in a carbonaceous and calcareous shale. The pellets (marked with yellow dashed lines) have been deformed and flattened during compaction and differ in texture and chemistry (the pellets are calcite-rich) from the surrounding matrix. Other components are foraminifera (F) with calcite cement (CC) fills and kerogen fills (dark gray to black). In places, the chamber walls have been replaced by diagenetic quartz (white arrows).

BSE(SEM), Scale bar = 24.6 μm

Up. Cretaceous (Cenomanian) Eagle Ford Shale, southwest Texas

A detailed view of the fecal pellets of a planktic organism from the previous image. This enlargement shows microfossil and nanno-fossil debris, calcite rhombs and various levels of cementation. In spite of significant burial, there is abundant microporosity preserved in these pellets, an important factor in gas shales of Mesozoic age.

BSE(SEM), Scale bar = 3.0 μm

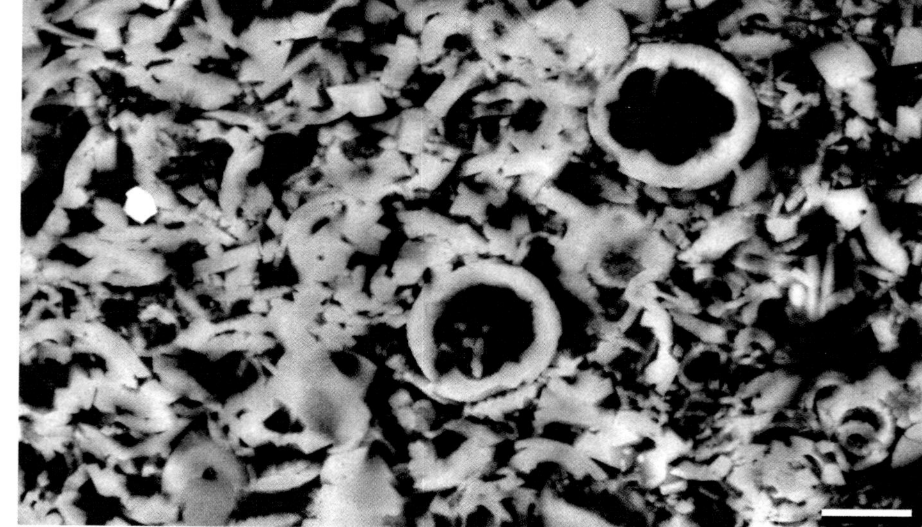

Up. Devonian Chattanooga Shale, Tennessee

A carbonaceous laminated shale with lenticular silt layers and a small phosphate nodule. Differential compaction of shale around the nodule indicates an early diagenetic formation when the water content of the sediment was approximately 80 volume%.

PPL, Scale bar = 2.00 mm

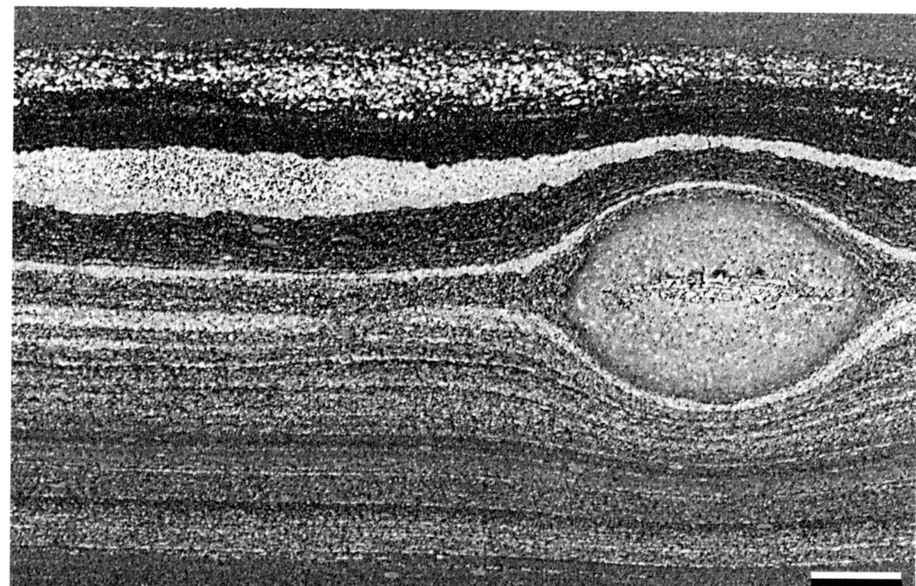

Mid. Permian Bell Canyon Fm., Reeves Co., Texas

A silt-rich shale with brownish deformed clasts. The latter are probably rip-up clasts of contemporaneous overbank muds in this submarine fan system.

PPL, Scale bar = 0.07 mm

Lo. Jurassic (Toarcian) Posidonia Shale, southwest Germany

An interesting diagenetic phenomenon—cone-in-cone calcite growing parallel to bedding. Profuse calcite growth has pushed apart and disjointed the shale fabric. This kind of growth has been interpreted as an indicator of fluid overpressure in other shale successions (Cobbold and Rodrigues, 2007).

PPL, Scale bar = 2.00 mm

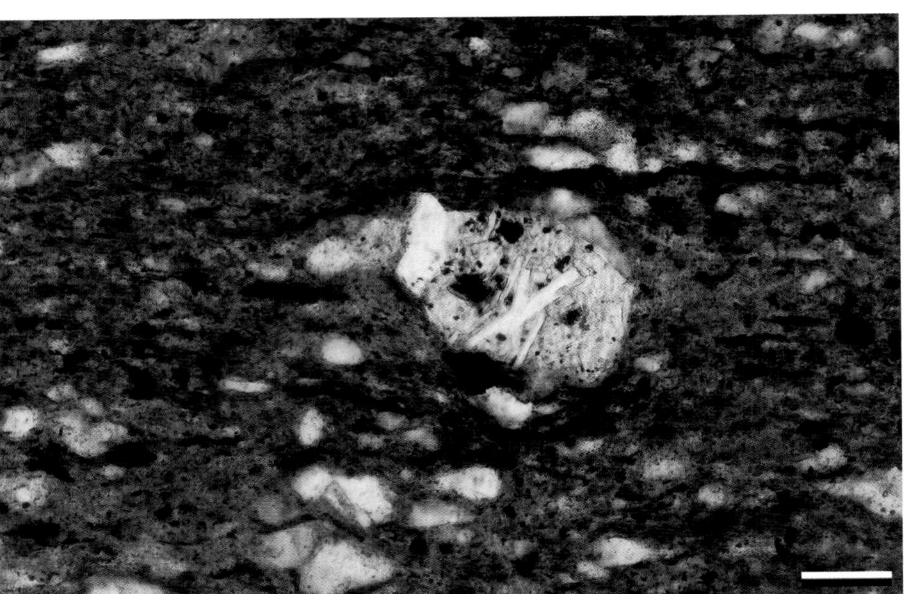

Cretaceous Mowry Fm., Moffat Co., Colorado

A rock fragment in an organic-rich silty and pyritic shale. The large grain in the center shows lath-shaped plagioclase phenocrysts that strongly suggest that this is a volcanic rock fragment. Because recognition of rock fragments becomes more problematic at smaller grain sizes, finding rock fragments is far less common in mudrocks than in coarser-grained deposits.

PPL, RDI, Scale bar = 0.05 mm

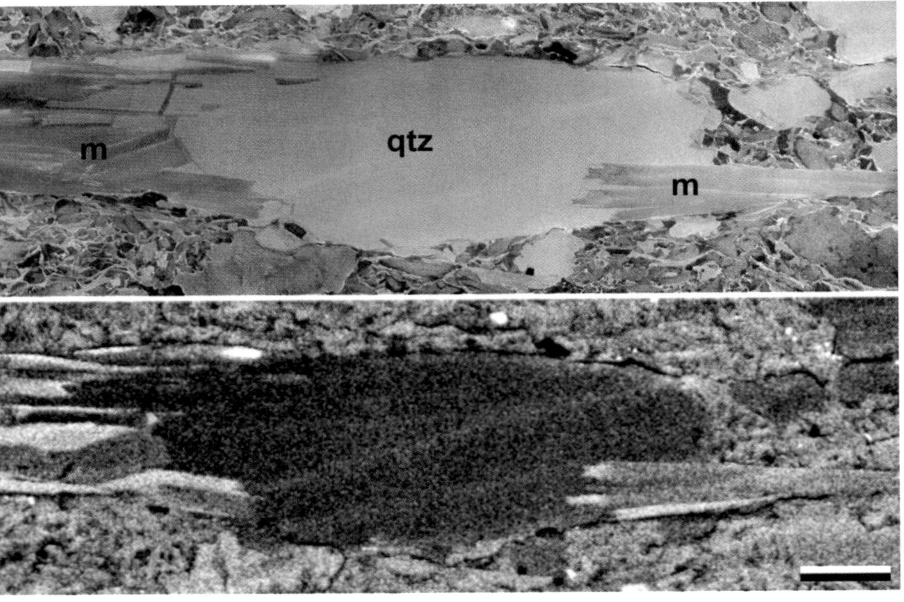

Paleocene Lepsa Fm., eastern Carpathians, Romania

SEM imaging of a rock fragment in an organic-rich calcareous shale (marl). The image shows a quartz grain (qtz) that is intergrown with micas (m). The alignment of the intergrown micas (biotite and muscovite) on both sides of the grain suggests that this grain originated from schistose metamorphic rocks.

SE(SEM) / BSE(SEM), Scale bar = 9.3 μm

Silurian Crab Orchard Fm. – Devonian Ohio Shale, eastern Kentucky

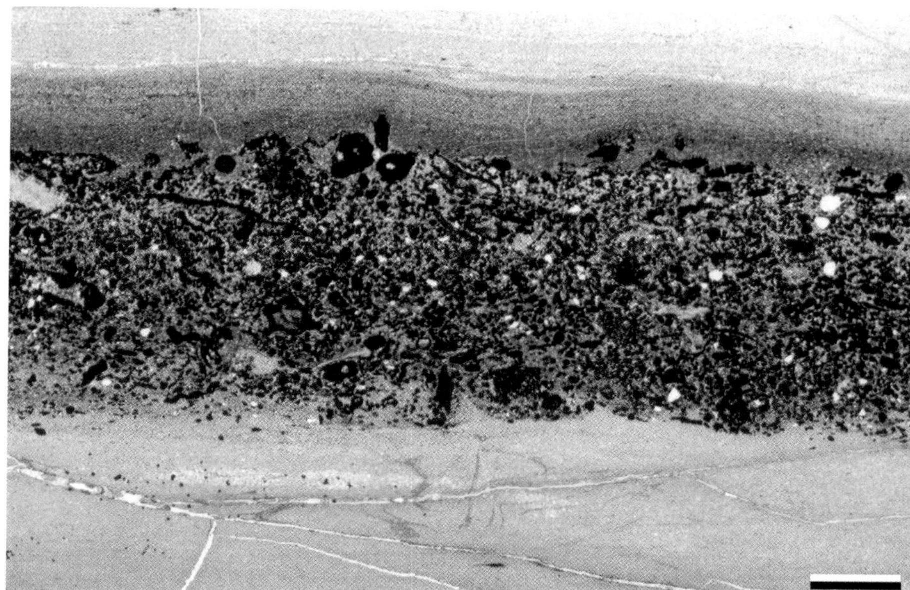

The Silurian-Devonian boundary in eastern Kentucky. The lower gray shale is from the Silurian Crab Orchard Formation, followed by a basal lag with abundant reworked pyrite (black; eroded from concretions, burrow tubes, etc.), conodonts, quartz grains (bright) and shale rip-ups. This basal Devonian lag of the Ohio Shale is overlain by a thin drape of carbonaceous shale (with burrows) that passes upwards into bioturbated gray shales. It is a major sequence boundary, but it is not very conspicuous in either outcrop or core.

PPL, Scale bar = 2.01 mm

Up. Devonian Chattanooga Shale, Dowelltown Mbr., Tennessee

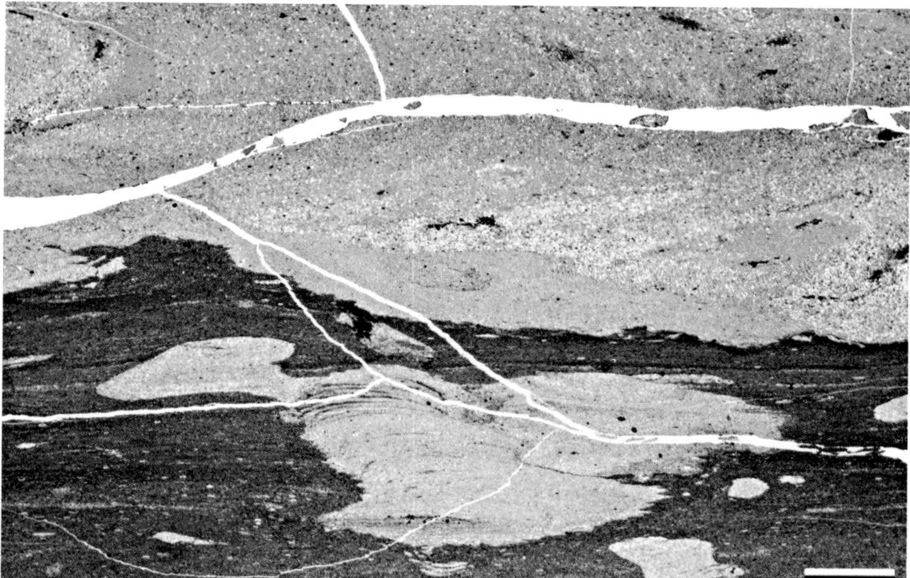

Interbedded carbonaceous (black) and clay-rich (gray) shales. The lower carbonaceous layer contains *Tasmanites* (tiny orange specks) and is perforated by burrows (gray, filled from above). The central burrow, probably *Teichichnus* shows spreiten. The carbonaceous interval has relict layering and fabric disruption (burrowing) from earlier in the depositional history. The upper gray layer is burrow mottled. Sharp black/gray boundary is related to erosion that exposed firm, consolidated muds.

PPL, Scale bar = 3.12 mm

Up. Devonian New Albany Shale, Indiana

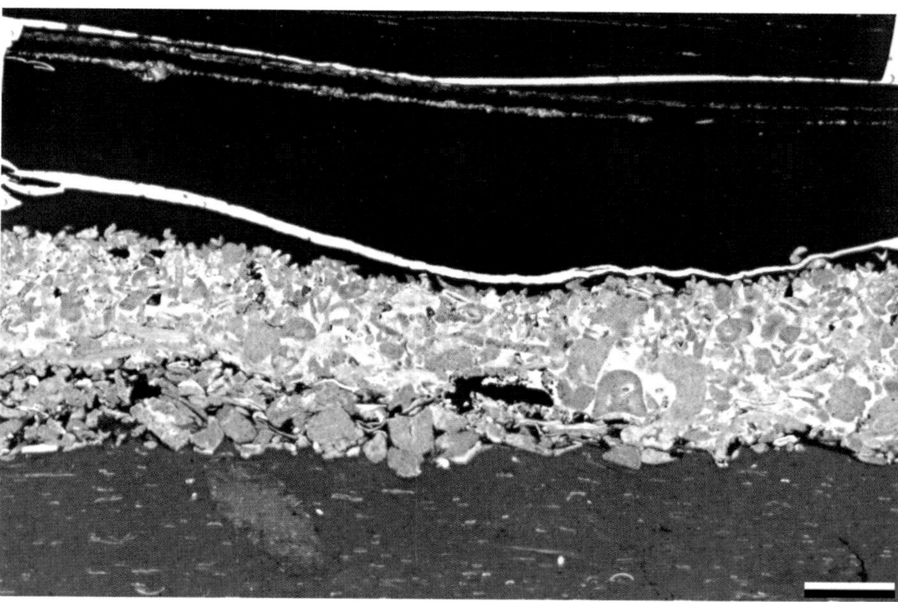

A sequence boundary in a thin section. The brownish-gray shale at the bottom (with shelly debris and bioturbation) is of the Selmier Member (Frasnian) of the New Albany Shale, the black and highly carbonaceous top belongs to the Morgan Trail Member (Famennian). Between these two members is a layer of carbonate debris that was reworked from Ordovician carbonates exposed on the Cincinnati Arch during sea-level fall.

PPL, Scale bar = 1.65 mm

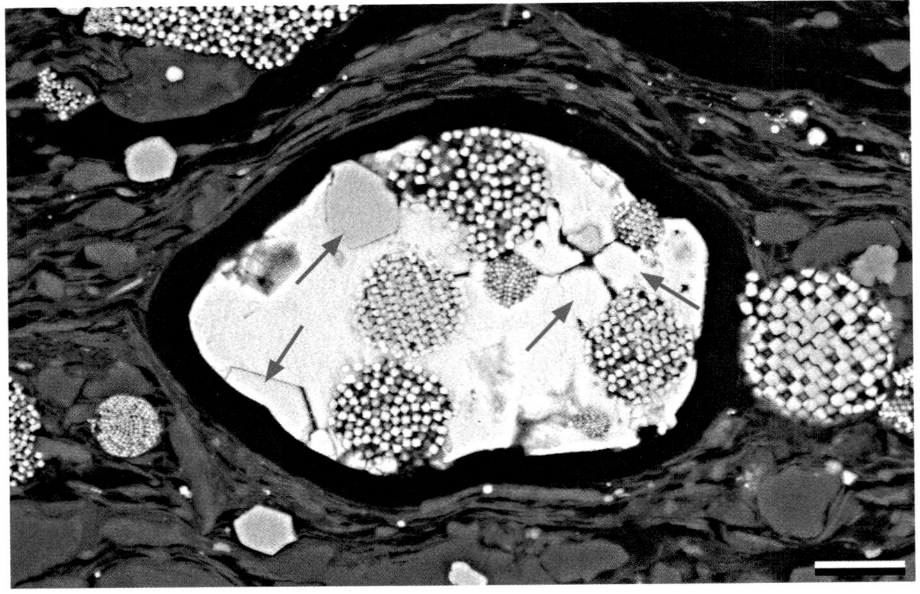

Up. Devonian Chattanooga Shale, Tennessee

Early diagenetic pyrite in a black shale. The rounded objects with abundant internal crystallites are pyrite framboids. In the center, these pyrite framboids are filling an organic cyst (black rim), probably *Tasmanites*. The cement between framboids in the cyst interior is marcasite–the determination is based on distinctive crystal habit (examples at arrows) and electron backscatter diffraction analysis. Differential compaction around the mineralized cyst indicates early diagenetic formation of pyrite and marcasite in a water-rich, soupy substrate. These pyrite fills can be reworked later into pyritic lags (Schieber and Baird, 2000).

BSE(SEM), Scale bar = 6.4 μm

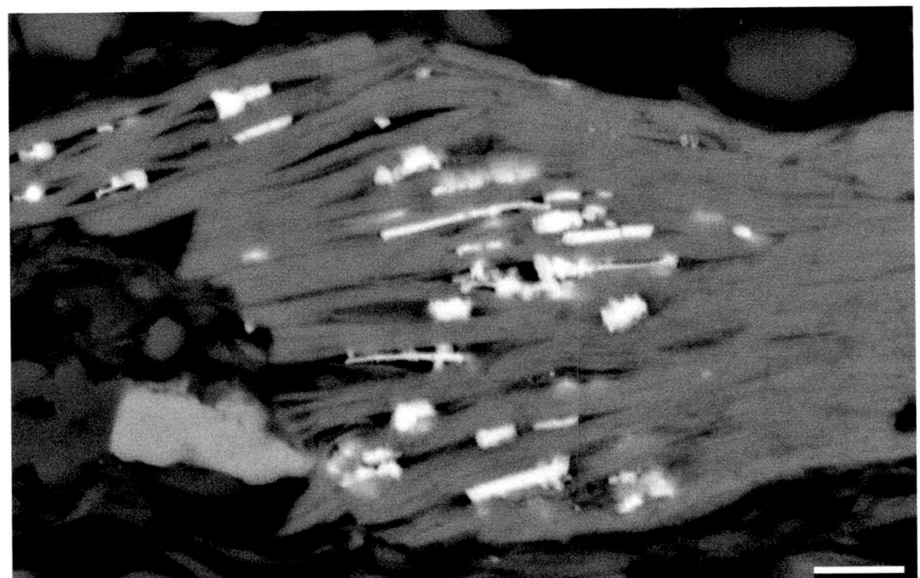

Up. Devonian Cleveland Shale, Kentucky

Later diagenetic pyrite in a black shale. Here, a biotite grain was degraded and defoliated in pore waters, and the iron that was released precipitated as pyrite along cleavage planes. This suggests pore waters had free sulfide. Because the iron release during biotite alteration is comparatively slow, these pyrite crystals are most likely something that happened later in burial, unlike the more abundantly observed formation of early pyrite framboids.

BSE(SEM), Scale bar = 2.3 μm

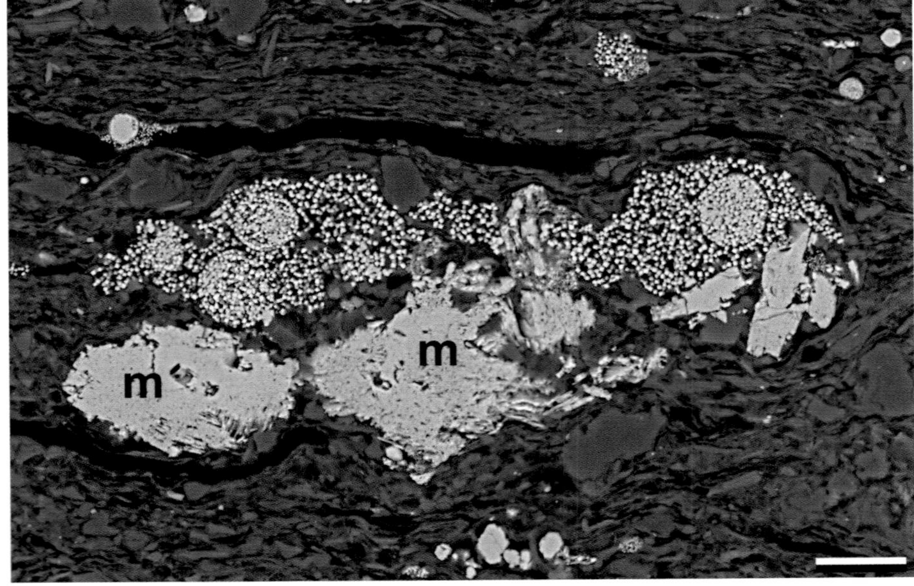

Up. Devonian Cleveland Shale, Kentucky

Marcasite and pyrite in a black shale. The pyrite formed as spherical framboids. Some of the framboids appear to be at various stages of disintegration/decay. The adjacent bright aggregates of bladed crystals consist of marcasite (m) that grew when the sediment was reoxidized and the initial framboids were partially destroyed. Find out more about sedimentary marcasite formation in Schieber (2011).

BSE(SEM), Scale bar = 17.7 μm

Up. Devonian New Albany Shale, Indiana

Quartz cement in a black shale. In the left-hand image (BSE), quartz (medium gray) is dominant in the rock. In the right-hand charge contrast image, the detrital quartz cores (D) are shown to be encased by diagenetic coalescing quartz cement rims (darker gray). These cements increase the brittleness of the shale, a desirable property in shale gas exploration. For more on diagenetic quartz in shales see Schieber et al. (2000) or Thyberg et al. (2012).

BSE(SEM) | CCI(SEM), Scale bar = 2.4 μm

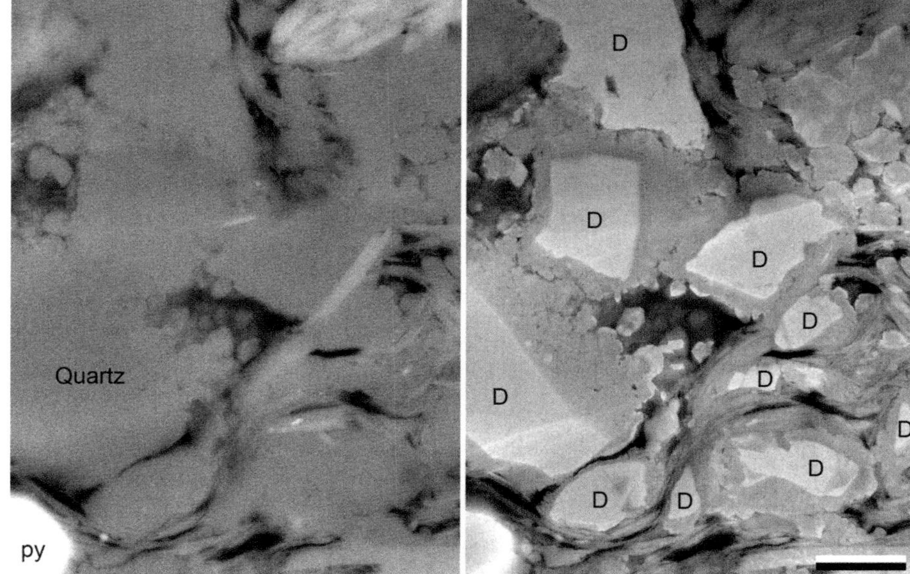

Oligocene Bituminous Marl Fm., Romania

Even though shales are fine grained, electron microscopy allows a close look at their diagenetic history. In this image, a coccolithophore (coccosphere; labeled "co") with interlocking elements is overgrown by diagenetic calcite (cc) and infilled with diagenetic smectite clays (sc). The clays probably formed early in burial history. Late in diagenesis, silica-rich fluids invaded and deposited replacive and pore filling quartz (qtz).

SE(SEM), Scale bar = 1.5 μm

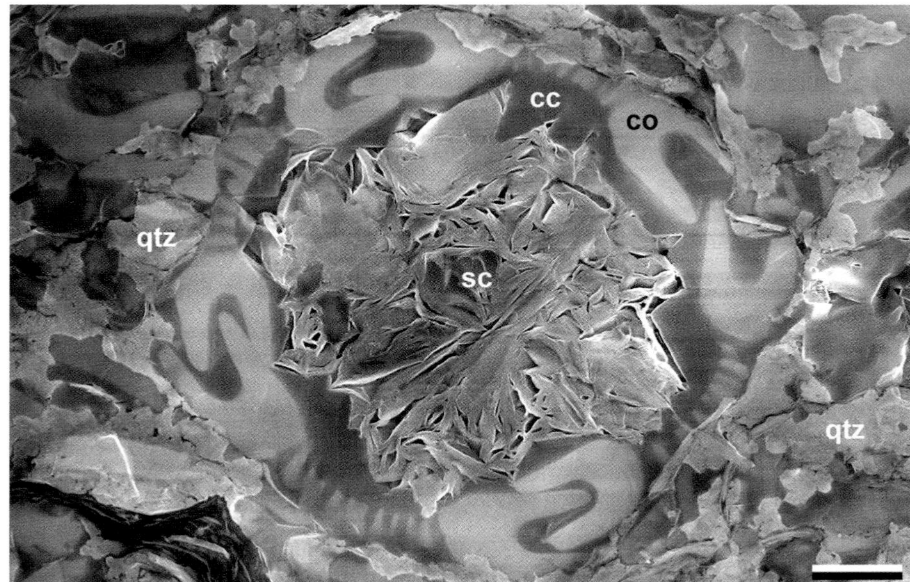

Up. Devonian New Albany Shale, Indiana

Early diagenetic concretions in a black shale. These concretions consist of pyrite (bright), ankerite (medium gray) and iron carbonate. The differential compaction around the concretions indicates early formation in water-rich sediments. The enclosure of pyrite in ankerite records the passage (related to burial) of the sediment from the sulfate reduction zone to the methanogenic (fermentation) zone.

BSE(SEM), Scale bar = 0.47 μm

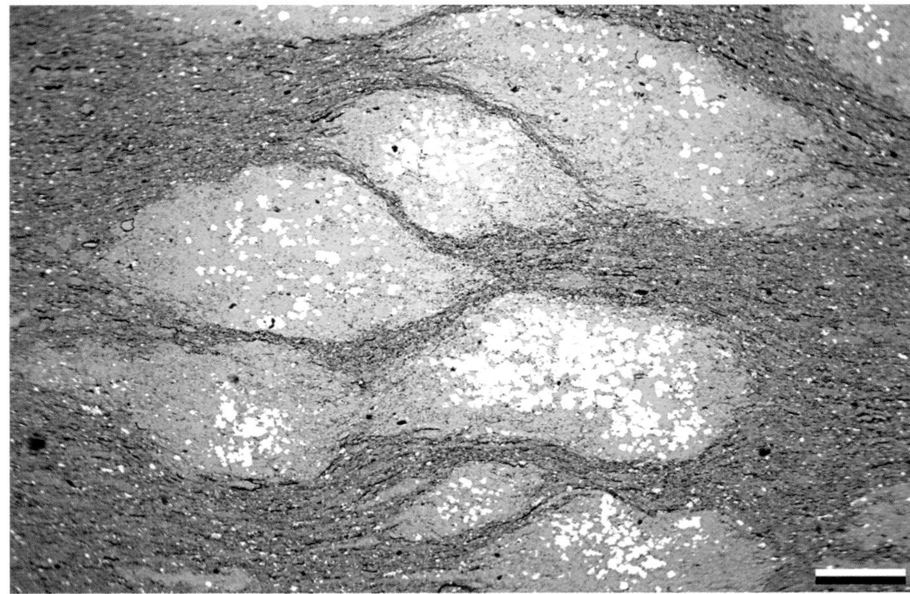

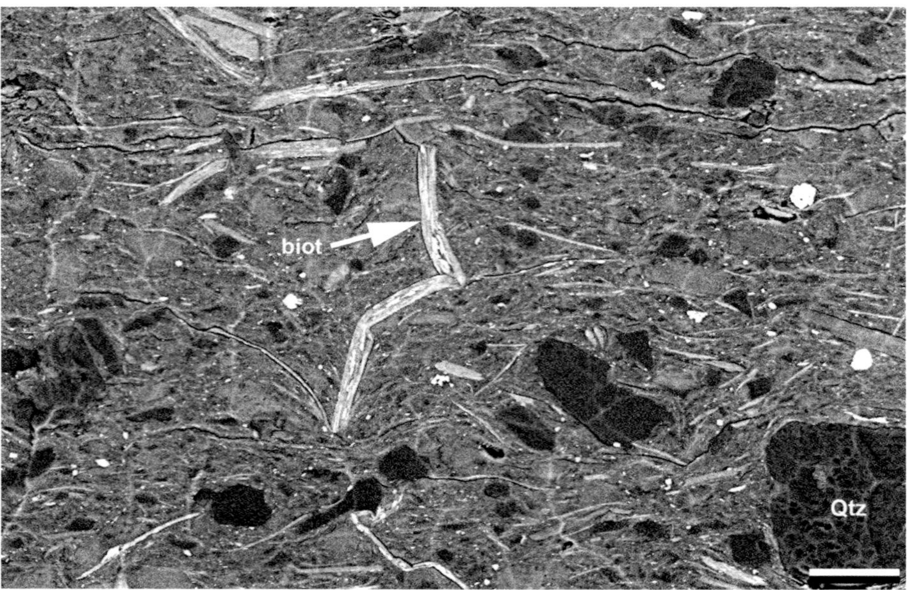

Cambrian Eau Claire Fm., Indiana

An ion milled shale from the subsurface o Indiana. This image shows how much detai is visible once mechanical damage has bee removed via ion milling. Quartz grains (Qtz are dark and may show fluid inclusion holes Mica flakes show vertical compression and dif ferential compaction and alteration to kaolinit and pyrite. The fabric is randomized, probabl by bioturbation.

BSE(SEM), Scale bar = 22.6 μm

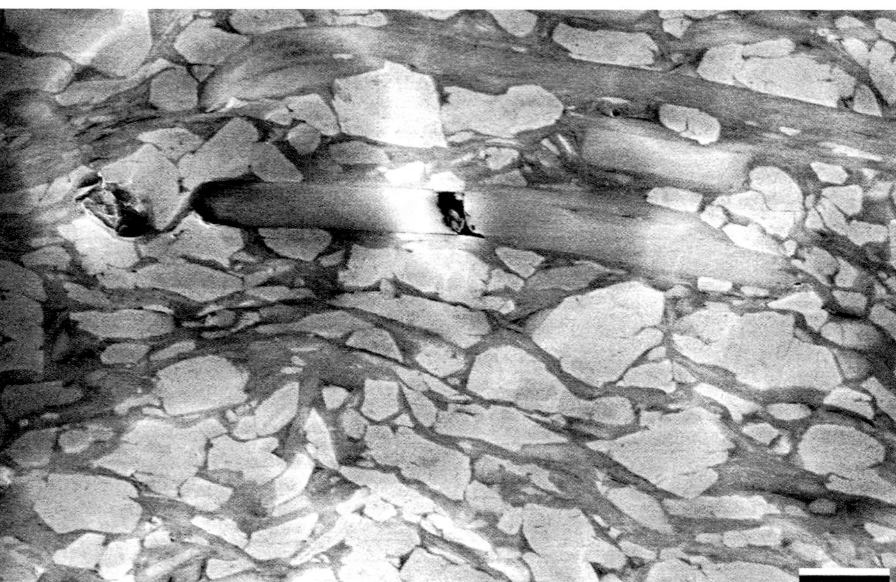

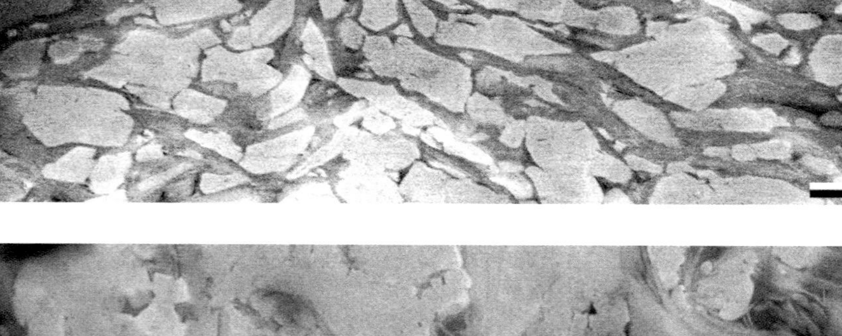

Cambrian Eau Claire Fm., Indiana

Another example of the benefits of ion millin (see also other examples in Loucks et al., 200 or Schieber, 2013). Every single silt grain i clearly visible, even (at higher magnifications those that are only a fraction of micron in size. By removing surface damage, io milling enables one to take advantage of th full resolving power of electron microscopes This secondary electron image gains extr contrast from conductivity differences betwee minerals.

SE-CCI(SEM), Scale bar = 22.6 μm

Devonian New Albany Shale, Illinois

Detail view of an ion milled black shale Here, the grain fabric consists of abundan diagenetic quartz (in part with euhedral crysta terminations) that holds together an earlie fabric of clay and mica flakes. The dark interstitial material is kerogen. This secondary electron image gains extra contrast from the conductivity differences between minerals.

SE-CCI(SEM), Scale bar = 1.0 μm

Devonian New Albany Shale, Illinois

A very high-magnification view of an ion milled black shale. Here, one can examine the fabric of clay flakes and discern the spatial relationships between clay particles that may be as small as 100 nm, or smaller. The dark triangular spaces are actually open pores.

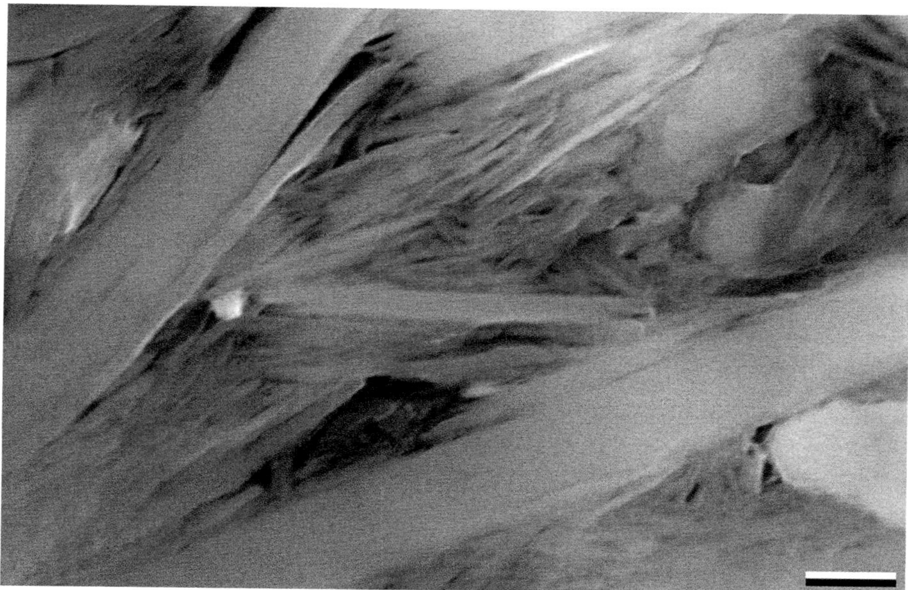

SE-CCI(SEM), Scale bar = 0.39 μm

Cited References and Additional Information Sources

Anderson, T., 2014, Key parameters for liquid-rich unconventional plays: Case studies from North America: AAPG Search and Discovery Article #80354.

Aplin, A. C., A. J. Fleet, and J. H. S. Macquaker, eds., 1999, Muds and Mudstones: Physical and Fluid-flow Properties: London, GSL Special Publication 158, 190 p., doi: 10.1144/GSL.SP.1999.158.01.01.

Aplin, A. C., and J. H. S. Macquaker, 2011, Mudstone diversity: Origin and implications for source, seal, and reservoir properties in petroleum systems: AAPG Bulletin, v. 95, p. 2031-2059, doi: 10.1306/03281110162.

Blatt, H., G. V. Middleton, and R. C. Murray, 1980, Origin of Sedimentary Rocks: Englewood Cliffs, NJ, Prentice Hall Inc., 782 p.

Brown, J. S., 1943, Suggested use of the word microfacies: Economic Geology, v. 38, p. 325, doi: 10.2113/gsecongeo.38.4.325.

Camp, W. K., E. Diaz, and B. Wawak, 2013, Electron Microscopy of Shale Hydrocarbon Reservoirs: Tulsa, OK, AAPG Memoir 102, 260 p.

Cluff, R. M., 1980, Paleoenvironment of the New Albany Shale Group (Devonian-Mississippian) of Illinois: Journal of Sedimentary Research, v. 50, p. 767-780, doi: 10.1306/212F7AE0-2B24-11D7-8648000102C1865D.

Cobbold, P. R., and N. Rodrigues, 2007, Seepage forces, important factors in the formation of horizontal hydraulic fractures and bedding-parallel fibrous veins ('beef' and 'cone-in-cone'): Geofluids, v. 7, p. 313-322, doi: : 10.1111/j.1468-8123.2007.00183.x.

Cook, A. C., and N. R. Sherwood, 1991, Classification of oil shales, coals and other organic-rich rocks: Organic Geochemistry, v. 17, p. 211-222, doi: 10.1016/0146-6380(91)90079-Y.

Cuvillier, J., 1952, Le notion de 'microfacies' et ses applications: VIII Congreso Nazionale di Metano e Petroleo, Sect. I, p. 1-7.

Dean, W. E., M. Leinen, and D. A. V. Stow, 1985, Classification of deep-sea, fine-grained sediments: Journal of Sedimentary Research, v. 55, p. 250-256, doi: 10.1306/212F868E-2B24-11D7-8648000102C1865D.

Dunbar, C. O., and J. Rodgers, 1957, Principles of Stratigraphy: New York, Wiley, 357 p.

Dunham, R. J., 1962, Classification of carbonate rocks according to their depositional texture, *in* W. E. Ham, ed., Classification of Carbonate Rocks: Tulsa, OK, AAPG Memoir 1, p. 108-121.

Eugster, H. P., and L. A. Hardie, 1975, Sedimentation in an ancient playa-lake complex: The Wilkins Peak Member of the Green River Formation of Wyoming: GSA Bulletin, v. 86, p. 319-334, doi: 10.1130/0016-7606(1975)86<319:SIAAPC>2.0.CO;2.

Flügel, E., 1978, Mikrofazielle Untersuchungsmethoden von Kalken: Berlin, Springer, 454 p.

Folk, R. L., 1959, Practical petrographic classification of limestones: AAPG Bulletin, v. 43, p. 1-38.

Folk, R. L., 1965, Some aspects of recrystallization in ancient limestones, *in* L. C. Pray, and R. S. Murray, eds., Dolomitization and Limestone Diagenesis: Tulsa, OK, SEPM Special Publication 13, p. 14-48, doi: 10.2110/pec.65.07.0014.

Hammes, U., H. S. Hamlin, and T. E. Ewing, 2011, Geologic analysis of the Upper Jurassic Haynesville Shale in east Texas and west Louisiana: AAPG Bulletin, v. 95, p. 1643-1666, doi: 10.1306/02141110128.

Haq, B. U., and A. Boersma, 1998, Introduction to Marine Micropaleontology (2nd Edition): Amsterdam, Elsevier, 384 p.

Hart, B. S., J. H. S. Macquaker, and K. G. Taylor, 2013, Mudstone ("shale") depositional and diagenetic processes: Implications for seismic analyses of source-rock reservoirs: Interpretation, v. 1, p. B7-B26, doi: 10.1190/INT-2013-0003.1.

Ingram, R. L., 1953, Fissility of mudrocks: GSA Bulletin, v. 64, p. 869-878, doi: 10.1130/0016-7606(1953)64[869:FOM]2.0.CO;2.

Lewan, M. D., 1978, Laboratory classification of very fine grained sedimentary rocks: Geology, v. 6, p. 745-748, doi: 10.1130/0091-7613(1978)6<745:LCOVFG>2.0.CO;2.

Loucks, R. G., R. M. Reed, S. C. Ruppel, and D. M. Jarvie, 2009, Morphology, genesis, and distribution of nanometer-scale pores in siliceous mudstones of the Mississippian Barnett Shale: Journal of Sedimentary Research, v. 79, p. 848-861, doi: 10.2110/jsr.2009.092.

Lundegard, P. D., and N. D. Samuels, 1980, Field classification of fine-grained sedimentary rocks: Journal of Sedimentary Research, v. 50, p. 781-786, doi: 10.1306/212F7AE5-2B24-11D7-8648000102C1865D.

Macquaker, J. H. S., and A. E. Adams, 2003, Maximizing information from fine-grained sedimentary rocks: An inclusive nomenclature for mudstones: Journal of Sedimentary Research, v. 73, p. 735-744, doi: 10.1306/012203730735.

Macquaker, J. H. S., K. G. Taylor, and R. L. Gawthorpe, 2007, High-resolution facies analyses of mudstones: Implications for paleoenvironmental and sequence stratigraphic interpretations of offshore ancient mud-dominated successions: Journal of Sedimentary Research, v. 77, p. 324-339, doi: 10.2110/jsr.2007.029.

Macquaker, J. H. S., K. G. Taylor, M. Keller, and D. Polya, 2014, Compositional controls on early diagenetic pathways in fine-grained sedimentary rocks: Implications for predicting unconventional reservoir attributes of mudstones: AAPG Bulletin, v. 98, p. 587-603, doi: 10.1306/08201311176.

Milliken, K. L., Suk-Joo Choh, P. Papazis, and J. Schieber, 2007, "Cherty" stringers in the Barnett Shale are agglutinated foraminifera: Sedimentary Geology, v. 198, p. 221-232, doi: 10.1016/j.sedgeo.2006.12.012.

Millot, G., 1970, Geology of Clays: Weathering, Sedimentology, Geochemistry (Translated by W. R. Farrand and H. Paquet): New York, Springer-Verlag, 429 p.

Nuhfer, E. B., 1981, Mudrock fabrics and their significance: Discussion: Journal of Sedimentary Research, v. 51, p. 1027-1029.

Nuhfer, E. B., R. J. Vinopal, and D. S. Klanderman, 1979, X-radiograph Atlas of Lithotypes and other Structures in the Devonian Shale Sequence of West Virginia and Virginia: Springfield, VA, METC/CR-79/27, NTIS, 45 p.

O'Brien, N. R., 1996, Shale lamination and sedimentary processes, *in* A. E. S. Kemp, ed., Palaeoclimatology and Palaeoceanography from Laminated Sediments: London, GSL Special Publication 116, p. 23-36, doi: 10.1144/GSL.SP.1996.116.01.04.

O'Brien, N. R., and R. M. Slatt, 1990, Argillaceous Rock Atlas: New York, Springer-Verlag, 141 p.

Passey, Q. R., K. M. Bohacs, W. L. Esch, R. Klimentidis, and S. Sinha, 2010, From oil-prone source rock to gas-producing shale reservoir—geologic and petrophysical characterization of unconventional shale-gas reservoirs: CPS/SPE International Oil & Gas Conference and Exhibition in China, Beijing, China, 8–10 June 2010, Paper SPE 131350, 29 p., doi: 10.2118/131350-MS.

Pedersen, T. F., and S. E. Calvert, 1990, Anoxia vs. productivity: What controls the formation of organic-carbon-rich sediments and sedimentary rocks: AAPG Bulletin, v. 74, p. 454-466.

Picard, M. D., 1971, Classification of fine-grained sedimentary rocks: Journal of Sedimentary Research, v. 41, p. 179-195, doi: 10.1306/74D7221B-2B21-11D7-8648000102C1865D.

Pike, J., and A. E. S. Kemp, 1996, Silt aggregates in laminated marine sediment produced by agglutinated foraminifera: Journal of Sedimentary Research, v. 66, p. 625-631, doi: 10.1306/D42683C9-2B26-11D7-8648000102C1865D.

Plummer, P. S., and V. A. Gostin, 1981, Shrinkage cracks: Desiccation or synaeresis?: Journal of Sedimentary Research, v. 51, p. 1147-1156, doi: 10.1306/212F7E4B-2B24-11D7-8648000102C1865D.

Potter, P. E., J. B. Maynard, and P. J. Depetris, 2005, Mud and Mudstones: Introduction and Overview: New York, Springer, 297 p.

Potter, P. E., J. B. Maynard, and W. A. Pryor, 1980, Sedimentology of Shale: Study Guide and References Source: New York, Springer-Verlag, 306 p.

Sageman, B. B., A. E. Murphy, J. P. Werne, C. A. Ver Straeten, D. J. Hollander, and T. W. Lyons, 2003, A tale of shales: The relative roles of production, decomposition, and dilution in the accumulation of organic-rich strata, Middle–Upper Devonian, Appalachian basin: Chemical Geology, v. 195, p. 229-273, doi: 10.1016/S0009-2541(02)00397-2.

Schieber, J., 1986, The possible role of benthic microbial mats during the formation of carbonaceous shales in shallow Mid-Proterozoic basins: Sedimentology, v. 33, p. 521-536, doi: 10.1111/j.1365-3091.1986.tb00758.x.

Schieber, J., 1989, Facies and origin of shales from the mid-Proterozoic Newland Formation, Belt Basin, Montana: Sedimentology, v. 36, p. 203-219, doi: 10.1111/j.1365-3091.1989.tb00603.x.

Schieber, J., 1994, Reflection of deep vs shallow water deposition by small scale sedimentary features and microfabrics of the Chattanooga Shale in Tennessee, *in* A. F. Embry, B. Beauchamp, and D. J. Glass, eds., Pangea: Global Environments and Resources: Calgary, Canadian Society of Petroleum Geologists Memoir 17, p. 773-784.

Schieber, J., 1998, Sedimentary features indicating erosion, condensation, and hiatuses in the Chattanooga Shale of central Tennessee: relevance for sedimentary and stratigraphic evolution, *in* J. Schieber, W. Zimmerle, and P. V. Sethi, eds., Shales and Mudstones: Volume I, Basin studies, sedimentology, and paleontology): Stuttgart, E. Schweizerbart'sche Verlagsbuchhandlung, p. 187-215.

Schieber, J., 1999, Distribution and deposition of mudstone facies in Upper Devonian Sonyea Group of New York: Journal Sedimentary Research, v. 69, p. 909-925, doi: 10.2110/jsr.69.909.

Schieber, J., 2007a, Microbial mats on muddy substrates – Examples of possible sedimentary features and underlying processes, *in* J. Schieber, P. K. Bose, P. G. Eriksson, S. Banerjee, S. Sarkar, W. Altermann, and O. Catuneanu, eds., Atlas of Microbial Mat Features Preserved within the Clastic Rock Record: Amsterdam, Elsevier, p. 117-134.

Schieber, J., 2007b, Benthic microbial mats as an oil shale component: Green River Formation (Eocene) of Wyoming and Utah, *in* J. Schieber, P. K. Bose, P. G. Eriksson, S. Banerjee, S. Sarkar, W. Altermann, and O. Catuneanu, eds, Atlas of Microbial Mat Features Preserved within the Clastic Rock Record: Amsterdam, Elsevier, p. 189-197.

Schieber, J., 2009, Discovery of agglutinated benthic foraminifera in Devonian black shales and their relevance for the redox state of ancient seas: Palaeogeography, Palaeoclimatology, Palaeoecology, v. 271, p. 292-300, doi: 10.1016/j.palaeo.2008.10.027.

Schieber, J., 2011, Marcasite in black shales—a mineral proxy for oxygenated bottom waters and intermittent oxidation of carbonaceous muds: Journal of Sedimentary Research, v. 81, p. 447-458, doi: 10.2110/jsr.2011.41.

Schieber, J., 2013, SEM observations on ion-milled samples of Devonian black shales from Indiana and New York: The petrographic context of multiple pore types, *in* W. K. Camp, E. Diaz, and B. Wawak, eds., Electron Microscopy of Shale Hydrocarbon Reservoirs: Tulsa, OK, AAPG Memoir 102, p. 153-172, doi: 10.1306/13391711M1023589.

Schieber, J., and G. Baird, 2001, On the origin and significance of pyrite spheres in Devonian black shales of North America: Journal of Sedimentary Research, v. 71, p. 155-166, doi: 10.1306/051600710155 6.

Schieber, J., D. Krinsley, and L. Riciputi, 2000, Diagenetic origin of quartz silt in mudstones and implications for silica cycling: Nature, v. 406, p. 981-985, doi: 10.1038/35023143.

Schieber, J., and J. B. Southard, 2009, Bedload transport of mud by floccule ripples—Direct observation of ripple migration processes and their implications: Geology, v. 37, p. 483-486, doi: 10.1130/G25319A.1.

Schieber, J., J. B. Southard, and A. Schimmelmann, 2010, Lenticular shale fabrics resulting from intermittent erosion of water-rich muds—interpreting the rock record in the light of recent flume experiments: Journal of Sedimentary Research, v. 80, p. 119-128, doi: 10.2110/jsr.2010.005.

Schieber, J., J. Southard, and K. Thaisen, 2007, Accretion of mudstone beds from migrating floccule ripples: Science, v. 318, p. 1760-1763, doi: 10.1126/science.1147001.

Schieber, J., and W. Zimmerle, 1998, The history and promise of shale research, *in* J. Schieber, W. Zimmerle, and P. V. Sethi, eds., Shales and Mudstones: Volume I, Basin studies, sedimentology, and paleontology): Stuttgart, E. Schweizerbart'sche Verlagsbuchhandlung, p. 1-10.

Schieber, J., W. Zimmerle, and P. V. Sethi, eds., 1998, Shales and Mudstones: Volume I, Basin Studies, Sedimentology, and Paleontology; and Volume 2, Petrography, Petrophysics, Geochemistry, and Economic Geology: Stuttgart, E. Schweizerbart'sche Verlagsbuchhandlung, 384 p and 296 p. respectively.

Scholle, P. A., 1971, Sedimentology of fine-grained deep-water carbonate turbidites, Monte Antola Flysch (Upper Cretaceous), northern Apennines, Italy: GSA Bulletin, v. 82, p. 629-658, doi: 10.1130/0016-7606(1971)82[629:SOFDCT]2.0.CO;2.

Sorby, H. C., 1908, On the application of quantitative methods to the study of the structure and history of rocks: Quarterly Journal of the Geological Society, v. 64, p. 171-233, doi: 10.1144/GSL.JGS.1908.064.01-04.12.

Stow, D. A. V., and D. J. W. Piper, 1984, Deep-water fine-grained sediments: history, methodology and terminology: London, GSL Special Publication 15, p. 3-14, doi: 10.1144/GSL.SP.1984.015.01.01.

Stow, D. A. V., and G. Shanmugam, 1980, Sequence of structures in fine-grained turbidites: Comparison of recent deep-sea and ancient flysch sediments: Sedimentary Geology, v. 25, p. 23-42, doi: 10.1016/0037-0738(80)90052-4.

Thyberg, B., and J. Jahren, 2011, Quartz cementation in mudstones: sheet-like quartz cement from clay mineral reactions during burial: Petroleum Geoscience, v. 17, p. 53-63, doi: 10.1144/1354-079310-028.

Wang, G., and T. R. Carr, 2013, Organic-rich Marcellus Shale lithofacies modeling and distribution pattern analysis in the Appalachian Basin: AAPG Bulletin, v. 97, p. 2173-2205, doi: 10.1306/05141312135.

Weaver, C. E., 1989, Clays, Muds, and Shales (Developments in Sedimentology 44): New York, Elsevier, 819 p.

Wignall, P. B., 1994, Black Shales: Oxford, UK, Oxford Science Publications, 127 p.

Wilson, J. L., 1975, Carbonate Facies in Geologic History: New York, Springer Verlag, 471 p.

Facing Page: Top – Calcareous rhizoliths in a terra rossa overlying a karsted surface, Cenozoic of Malta. Bottom – Core photograph of bioturbated marl with *Zoophycos* burrows showing spreite from Lower Paleocene Ekofisk Fm., Norwegian Sector, North Sea.

DIAGENESIS:
SYNSEDIMENTARY AND SURFICIAL DIAGENETIC FEATURES

CHAPTER 9

GENERAL INTRODUCTION TO DIAGENESIS

The term "diagenesis" refers to essentially everything that happens to sediments and rocks after their deposition but prior to metamorphism. There are a variety of diagenetic processes, biological, chemical and physical, that ultimately convert sediments into sedimentary rocks. The earliest of those events are covered in this chapter on near-surface diagenesis; subsequent chapters cover processes and products that occur primarily during later stages of diagenesis (mainly mesogenesis). Those include mechanical and chemical compaction, cementation, dissolution, replacement and structural deformation. All these processes can profoundly affect the porosity, permeability and hydrocarbon reservoir potential of clastic terrigenous deposits, and most of them are a function of initial sediment composition and the changes in pressure, temperature and water chemistry that accompany progressive burial. Less explicitly covered, but potentially no less important, is diagenesis that can occur during one or more episodes of local or regional uplift and consequent exposure (telogenesis). These events also introduce changes in the pressure/temperature/water chemistry regime of rocks, and thus can cause major diagenetic changes, especially grain dissolution and cementation. In an attempt to address the impacts of the various diagenetic events that rocks may experience, this book includes both a section on the recognition of porosity types as well as one on paragenesis (i.e., the placement of diagenetic events into a temporal sequence related to the burial/uplift history of rocks).

Burial diagenesis is critically important in controlling the porosity of clastic terrigenous rocks and is, in the main, porosity destructive—that is, almost all rocks lose porosity with increased burial depth. Nonetheless several factors can retard or inhibit porosity loss, including early grain-coating cements, that block later overgrowth cementation, regional overpressuring of basins that reduce effective overburden stresses and under some circumstances, hydrocarbon entry that can reduce rock-water interactions. In addition, the processes of dissolution and fracturing may, under the right circumstance, lead to actual increases in subsurface porosity. So the discussion of porosity destruction, preservation and creation pervades all chapters in the diagenesis section, and emphasis is placed on recognition of key features associated with anomalous porosity retention or creation.

SYNSEDIMENTARY AND SURFICIAL DIAGENETIC FEATURES

The earliest diagenetic processes, specifically synsedimentary and shallow subsurface transformations in marine and nonmarine settings, are covered in this chapter. These include biological processes such as burrowing and boring, and biogeochemical processes such as microbially-mediated precipitation of cements or alteration of precursor materials. Also included are diagenetic products that involve winnowing, erosion and reduction of sediment accumulation rates (and thus allow sediments to remain in the zone of surficial diagenesis for longer periods of time). These include concretions, nodules and hardgrounds. Finally, this section includes products of other surficial (but not necessarily synsedimentary) diagenetic processes—mainly telogenetic ones related to weathering and soil formation. Such processes can act for limited durations during active sedimentation, especially in nonmarine settings. However, such processes also can act during later uplift and exposure of sedimentary rocks and act for much longer time periods, producing karst, deeply weathered soil profiles and lateritic or bauxitic deposits.

Burrows and borings — these biogenic structures are produced at or near the sedimentation surface in marine and nonmarine settings. Burrows are produced by a wide variety of organisms that graze on or dig into sediments in the search for food, for protection from predators or to escape from depositional "catastrophic events" (storm erosion, turbidity currents and the like). Crustacean, molluscan, echinoid and worm burrows are common in marine settings, but other mobile organisms also can produce burrows. Burrows can be useful in the determination of oxygenated versus anoxic bottom conditions (especially because burrows are definitively produced *in situ*, unlike shells which can be transported). Burrows range in size from millimeters to meters, so many are not readily visible in thin section (and only one is illustrated in this section because many are shown in Chapter 8).

Borings, on the other hand, are produced by organisms that drill, rasp or dissolve their way into hard substrates, such as shells, hardgrounds or concretions. These can range from micron-scale microbial tubules to millimeter- or centimeter-scale structures produced by higher organisms (sponges, bivalves, barnacles and others). Burrows and borings may be differentiated by their relationship to their substrate—burrowing organisms avoid hard shells or rock fragments in a soft substrate; boring organisms tunnel directly into or through shells, hardgrounds or other hard objects. The main importance of recognizing borings in concretions or hardgrounds is that they provide evidence of extremely early sediment lithification.

Concretions and nodules — concretions are hard, spherical masses formed in soft sedimentary rocks by the precipitation of authigenic minerals, commonly as intergranular cements in the host sediment. Concretions may range from millimeters to several meters in diameter. Carbonate minerals (calcite, dolomite and siderite), phosphates (apatite), pyrite and iron oxides are the most common constituents of concretions and may also be found as disseminated early-stage cements, especially in organic-rich sediments. Silica precipitation also may occur in spherical masses, but these form mainly as replacement features at various stages of burial and thus are termed "silica nodules", "chert nodules" or "flint nodules" rather than concretions. Concretions occur in marine, nonmarine or mixing-zone settings and can form in almost all types of sedimentary rocks, including sandstones, siltstones and shales. Regardless of composition, they typically form early, prior to significant sediment burial, compaction or lithification. Indeed, in many cases, formation begins within centimeters of the sedimentation surface—the evidence for this can come from extensive later-stage compactional drape of unconsolidated sediment around the hard nodules or it can come from erosional reworking as well as biological boring or encrustation of exhumed concretions. Such reworked concretions are termed "hiatus concretions" (Voigt, 1968; Kennedy and Klinger, 1972; Baird, 1976; Zatoń, 2010), and they can be important in recognition of otherwise subtle unconformities. Concretions generally form around a nucleus (commonly decomposing plant or animal remains, burrows and the like). Microbial activity in such organic remains or in organic-rich sediments, along with associated sulfate reduction and decompositional release of phosphorus are often involved in concretion formation (e.g., Curtis and Coleman, 1986; Coleman, 1993). Concretions can have varied growth patterns—concentric growth in some cases and pervasive or disseminated growth in other cases (Mozley, 1996), and the elongation axis of concretions formed in nonmarine settings has been shown to be related to groundwater paleoflow paths (e.g., Johnson, 1989; McBride et al., 1994). Finally, it should be noted that although many concretions start to form early in the burial history of sediments, some concretions (for example septarian ones) have complex, multistage structures and chemistries that indicate formation over extended time periods, perhaps especially in overpressured sections where compaction may be slowed (Sellés-Martínez, 1996).

Hardgrounds — are lithified sediment horizons that were cemented at or near the sediment-water interface. They are commonly an indication of strong winnowing, erosion and slow sediment accumulation rates. Weakly cemented hardgrounds are termed "firmgrounds". Hardgrounds most commonly form in carbonate sediments but also can be found in clastic terrigenous strata. The hiatus interval required to produce hardgrounds in terrigenous deposits, however, may be considerably longer than for carbonate ones (Molenaar et al., 1988; Bjørkum and Walderhaug, 1990). In many cases, hardground formation starts with lithification of burrows and/or formation of concretions; in other cases, entire sediment layers are uniformly cemented into a lithified pavement or are encrusted with laminar to digitate microbial precipitates (carbonate or phosphate). Hardgrounds may be cemented with calcium carbonate, calcium phosphate and even glauconitic minerals. True hardgrounds show no signs of subaerial exposure but do show evidence of lithification in a marine setting (such as borings by marine organisms, attached encrusting fauna, and erosion and reworking into pebble conglomerates that include marine fossils). Although localized hardgrounds may form in many environments, regionally extensive and stratigraphically correlative hardgrounds generally are interpreted as condensed sections related to hiatal intervals or highstand flooding events (e.g., Fürsich et al., 1981 and 1992; Bromley and Gale, 1982; Loutitt et al., 1988; Grimm, 2000).

Soils and weathering products (calcretes, caliche crusts, and bauxites) — weathering surfaces and soils are hiatus markers in nonmarine sections. The time required for their formation can range from hundreds to many millions of years, and they can form on a wide range of sedimentary (and non-sedimentary) rocks. In addition, although exposure-related features can be virtually synsedimentary, they also can form much later, typically during telogenesis. As such, they are an incredibly complex topic that can scarcely be summarized in this setting, and readers are encouraged to look at the books cited in the bibliography at the end of this chapter. Still, recognition of exposure surfaces is so important both in sequence stratigraphy and in understanding patterns of diagenetic alteration in sedimentary deposits that we should at least note some of the diagnostic exposure fabrics potentially visible at the scale of thin sections. The most important and distinctive such features, formed even during short exposure cycles, are: dark, multiply laminated crusts, cemented root traces (rhizoliths), clay cutans, soil pisoids or glaebules (irregular, concentrically-coated grains), and circumgranular cracking (as open or cemented pores). In settings with longer-term exposure, one can find reddened (terra rossa) horizons, ferruginous or aluminous pisolitic deposits (bauxites) and, where associated with carbonate rocks, major paleokarst features. Each of the small-scale features is illustrated in this chapter and is further discussed in the figure captions.

Mid. Permian Wegener Halvø Fm., Karstryggen area, East Greenland

A pelecypod shell fragment cut by large clionid sponge borings. The borings (dark areas) are filled with carbonate mud (micrite) and quartz silt. Many different organisms can bore into hard substrates (rasping, dissolving or chipping their way into hardgrounds, shells or other lithified seafloor materials). In this example, the massive bivalve shell was penetrated by probable boring sponges. Sponge borings usually have large, excavated living chambers (as in this example) with narrow exterior openings on the surface of the shell (not well shown in this sample).

PPL, AFeS, BDI, Scale bar = 0.26 mm

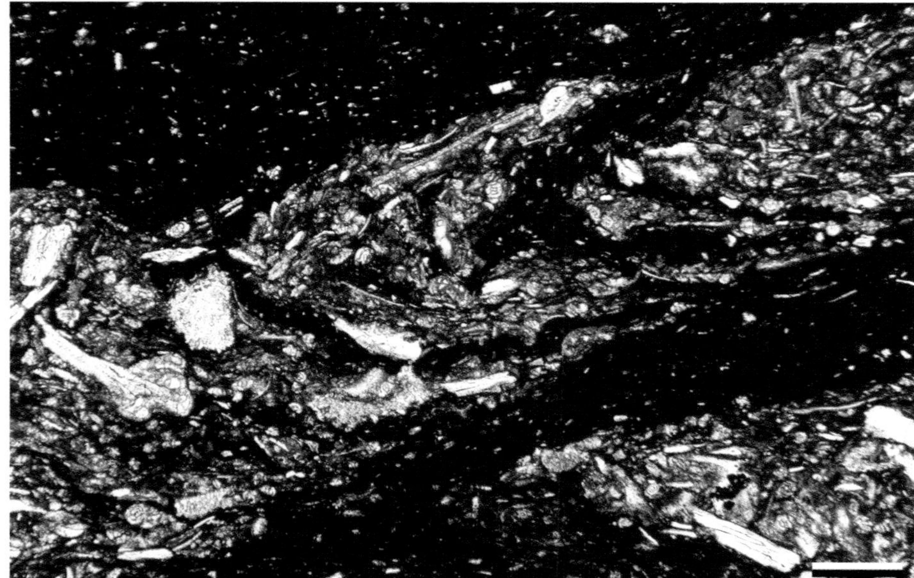

Lo. Pennsylvanian Morrow B Sandstone, Ochiltree Co., Texas

This burrow in a slightly silty shale is filled with carbonate-rich debris. The distinctive curved layers within the burrow, termed spreite or menisca, are the result of back-filling of the tunnel by the burrowing organism. In this case the burrow is introducing coarser material (foraminiferal and other bioclastic debris) from an overlying into the shale. Recognition of the internal fabric is key to distinguishing this as a burrow, and identification of the burrow type can often provide valuable paleoecological information (e.g., Bromley, 1996). Numerous examples of burrows also are depicted in the previous chapter.

PPL, RDI, Scale bar = 0.26 mm

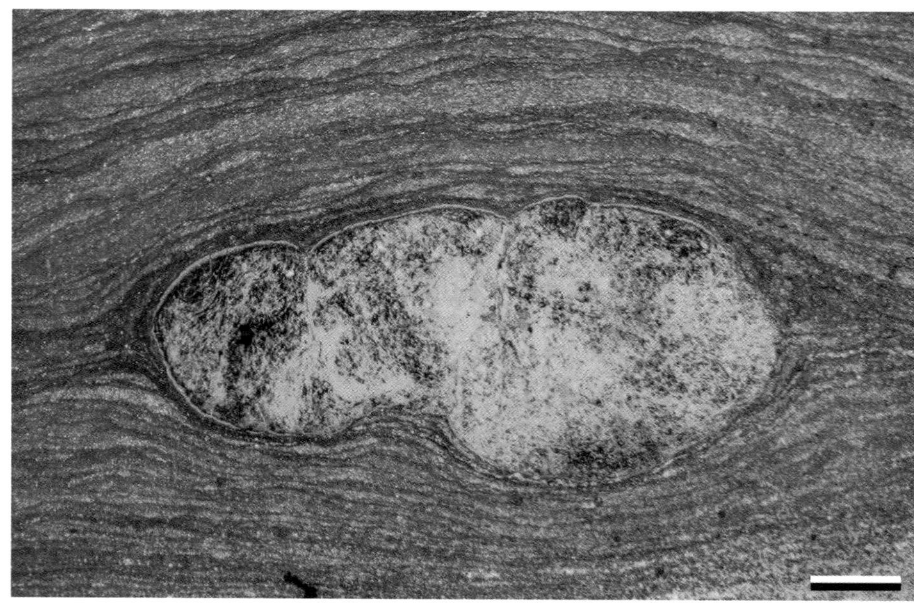

Eocene Green River Fm., Lincoln Co., Wyoming

A phosphatic concretion in lacustrine marls. Such structures are early diagenetic features that form below the sediment/water interface in suboxic to reducing conditions within organic-rich sediments (De las Heras et al., 1989). This sample exhibits differential compaction (compactional drape) around the concretion indicating that it was a hard object prior to substantial burial. Such drape features are especially common in claystones and fine-grained limestones (sediments that can have sea-floor porosities of 70% and more); in such units, flow occurs around hard objects during even minor overburden loading and compaction.

PPL, Scale bar = 0.80 mm

Lo. Cretaceous (Albian) Upper Folkestone Beds, Kent, England, U.K.

A phosphate concretion growing in a glauconitic sandstone. The phosphate precipitated as a cement, surrounding quartz and glauconite grains in the precursor sediment. Phosphate concretions typically form early, prior to burial or other cementation. Both the glauconite grains and the early phosphate concretions indicate deposition in a shallow marine environment with low sediment accumulation rates.

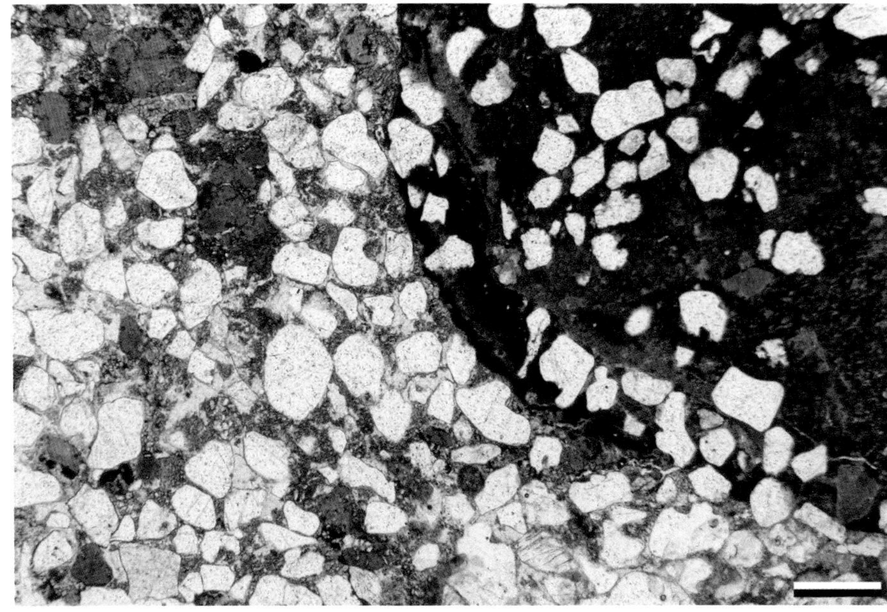

PPL, Scale bar = 0.51 mm

Miocene Santa Fe Gp., Zia Fm., Cerro Conejo Mbr., Sandoval Co., New Mexico

A small portion of a large calcite-cemented concretion within an eolian sandstone. The quartz and lithic grains have loose packing indicating that the calcite cementation was "early" (that is, prior to any significant burial). These calcite cements are poikilotopic and twinned, with a single crystal cementing all the grains in this view. According to Mozley and Davis (2005), this sample is part of a composite concretion that contained multiple nucleation sites for individual poikilotopic calcite crystals. These concretions thus have bumpy surfaces, reflecting those individual poikilotopic masses.

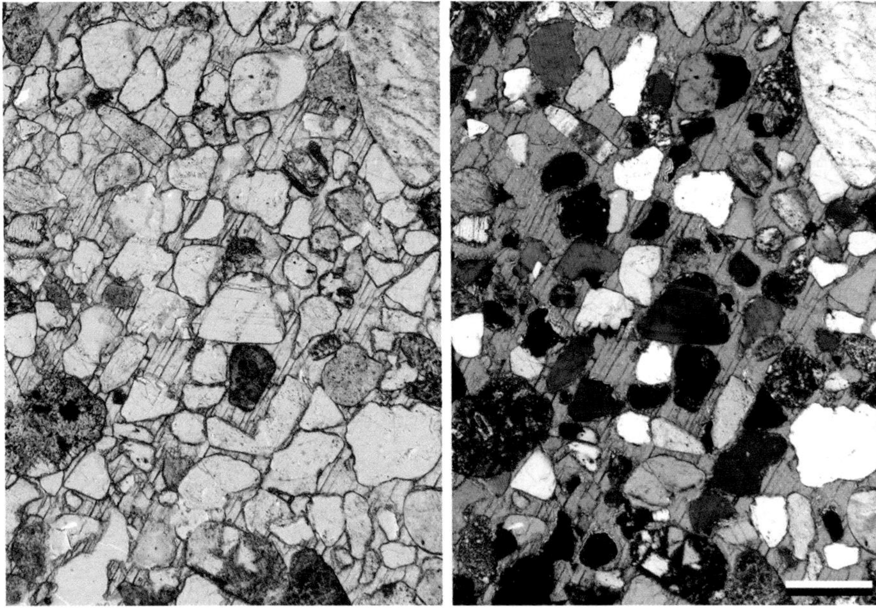

PPL | XPL, Scale bar = 0.26 mm

Neogene Santa Fe Gp., Socorro Co., New Mexico

A concretion that grew in laminated fluvial sediments. The concretion formed by the selective cementation of the finer-grained horizons, likely due to selective capillary retention of water in the finer-grained layers (although there are thin, grain-coating cements in the coarser layers as well). The cements in this area are aphanocrystalline to finely crystalline calcite (Mozley and Davis, 2005). Crystal growth appears to have started simultaneously throughout the concretion and the loose packing of the coarser detrital layers indicates an early onset for concretion growth.

PPL, BDI, Scale bar = 0.51 mm

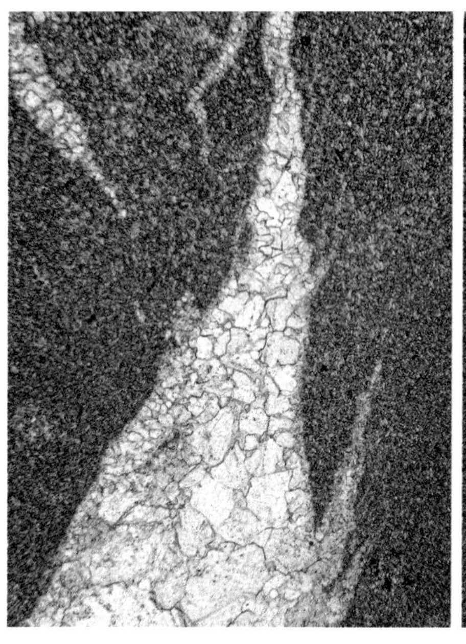
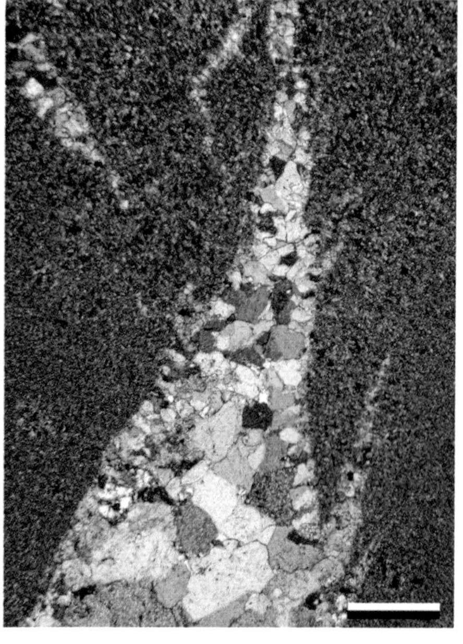

Paleogene Vieja Gp., Presidio Co., Texas

This is a view of a small part of a large carbonate septarian concretion with at least two generations of calcite (early, fracture-rimming fibrous calcite followed by blocky calcite). Septarian nodules are roughly spherical diagenetic structures formed of calcite, siderite, iron oxides or other materials. Their most distinctive characteristic is a complex network of intersecting radial and/or concentric fractures similar to shrinkage cracks. The fractures generally are filled with calcite cements, in many cases consisting of multiple generations of yellow-brown to white, fibrous to bladed calcite.

PPL | XPL, Scale bar = 0.51 mm

Up. Cretaceous (Turonian) Eagle Ford Shale, Dallas Co., Texas

Another view of cements in a fracture within a septarian concretion. Here, multiple generations of cement are visible, starting with an early-stage, nonferroan calcite (pink-stained cement) that formed skeletal crystals or crystals that were later corroded. The second stage of cement formed under reducing conditions and incorporated iron into the calcite lattice (blue stain). The ferroan calcite filled both the skeletal voids and other remaining pore space in the fracture.

PPL, AFeS, Scale bar = 0.51 mm

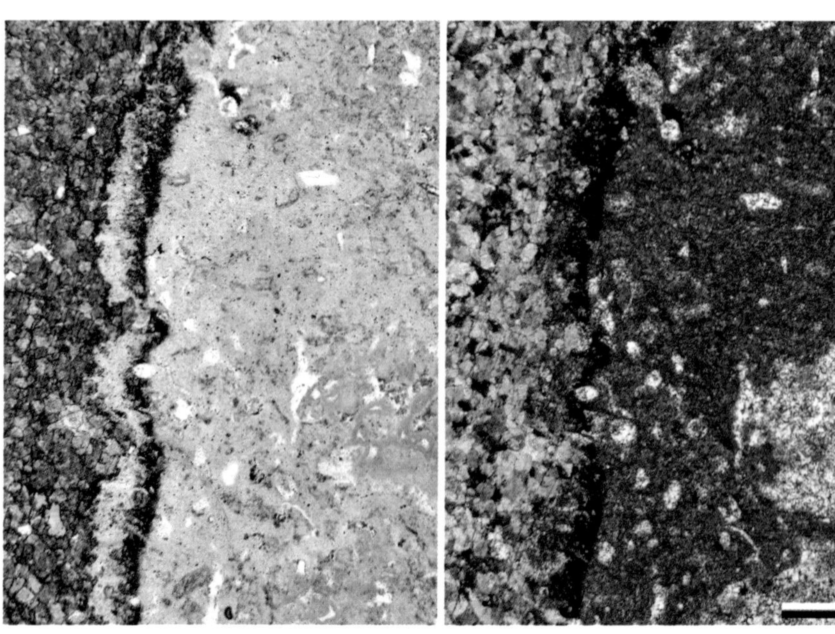

Devonian Jefferson Fm., Bridger Range, central Montana

This chert nodule has replaced a precursor carbonate that contained siliceous sponge spicules. Sponge spicules provide a source of easily remobilized biogenic silica for silicification of the surrounding sediments. The original carbonate rock was later replaced by fabric-destroying dolomite. The lighter colored areas (in PPL) within the chert nodule are more coarsely crystalline quartz after sponge spicules. Those areas are coarser because silica there is a void filling cement within leached spicules rather than a replacement of carbonate matrix.

PPL | XPL, AFeS, Scale bar = 0.51 mm

Up. Cretaceous Hod Gp., Danish sector, North Sea

An example of a diffuse boundary between a host chalk and chert nodule. This fine-grained, foraminiferal limestone was partially to completely replaced by chert. The rock's original fabric partially controlled silicification within the nodule (light colored area in PPL)—the foraminifers and their interiors have been replaced by more coarsely crystalline silica than the matrix. Silicification has been extensive within the nodule and grades into the host rock where the carbonate matrix is preserved, but the foraminifers have been replaced or infilled by silica. The source of the silica may be radiolarians or sponge spicules.

PPL | XPL, Scale bar = 0.51 mm

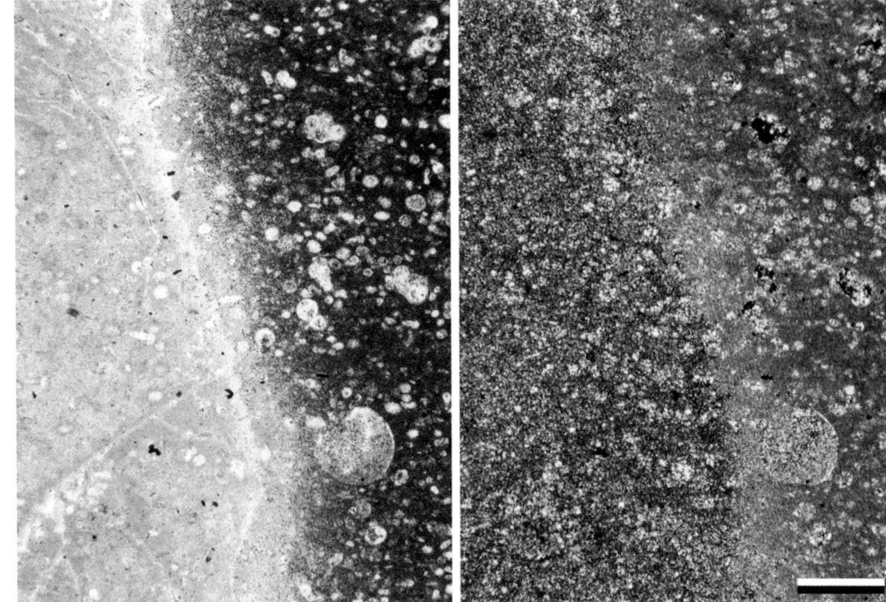

Oligocene – Miocene Otekaike Fm., Oamaru, Otago, New Zealand

Stromatolitic crusts can develop on marine hardgrounds exposed at the seafloor for long time periods (Martín-Algarra and Sánchez-Navas, 1995). Here, the digitate, laminated microbial crusts are largely phosphatic and are yellow-brown in plane-polarized light. Some laminae, particularly the last stages of microbial growth, are stained by iron oxides (dark brown layers), and interstices between microbialite heads are filled with glauconite and quartz sand grains. The glauconite is an indicator, along with phosphatization, of slow sediment accumulation rates during hardground formation.

PPL | XPL, AS, Scale bar = 0.51 mm

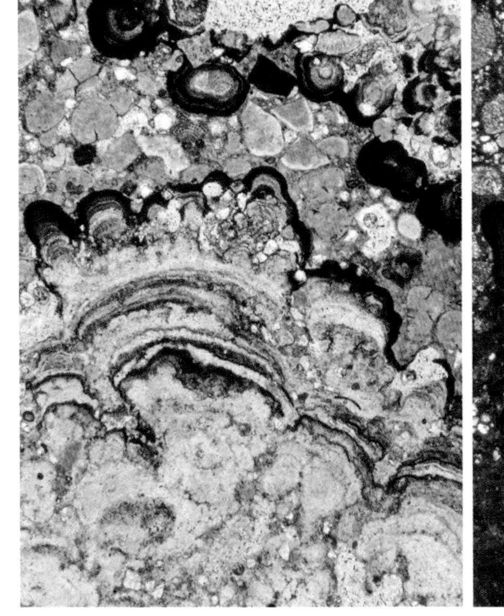

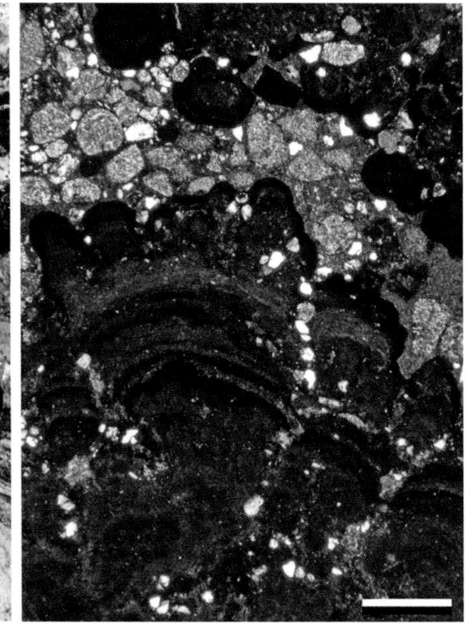

HARDGROUNDS

Eocene Totara Limestone, Otago, New Zealand

This is a clast of a phosphatic stromatolitic crust that was eroded from a submarine hardground (similar to that in the previous photo). It is indicative of at least episodically high-energy conditions and erosion associated with many such hiatal surfaces. In addition to the crust fragment, the rock also contains phosphate nodules and glauconite grains, again indicating very slow net accumulation rates. Most of the bioclasts visible in this photomicrograph are foraminifers and bryozoans.

PPL, Scale bar = 0.51 mm

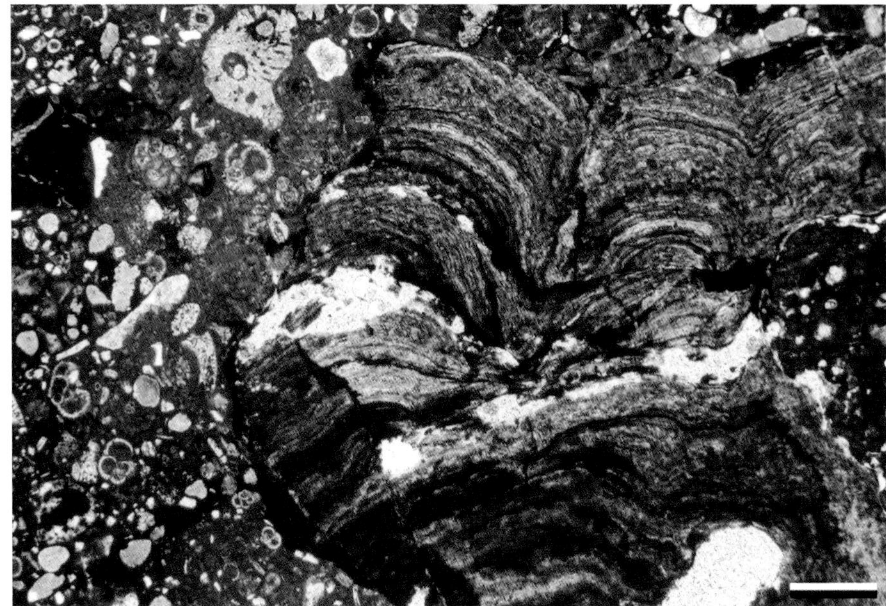

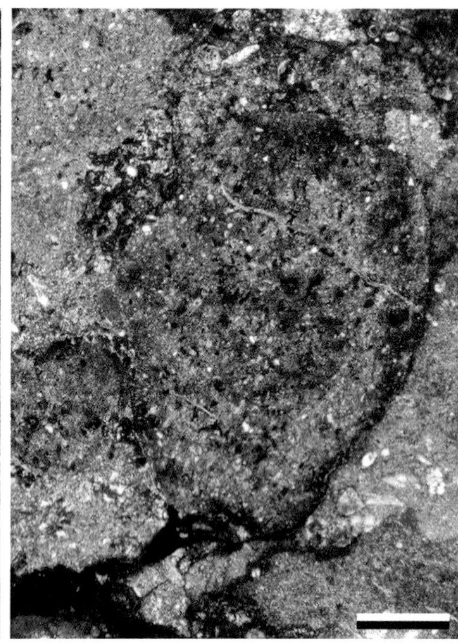

Miocene (Aquitanian – Burdigalian) Globigerina Limestone Fm., Gozo, Malta

Marine hardgrounds, erosion, boring and diagenetic mineralization here too are commonly associated with hiatal surfaces. This sample has numerous rip-up clasts produced by submarine erosion of a hardground surface (Pedley and Bennett, 1985; Rehfeld and Janssen, 1995). The large central grain may be sitting on an underlying, iron-stained, eroded carbonate hardground surface. The clasts, themselves, are heavily impregnated with phosphate, glauconite and iron oxides.

PPL | XPL, Scale bar = 0.51 mm

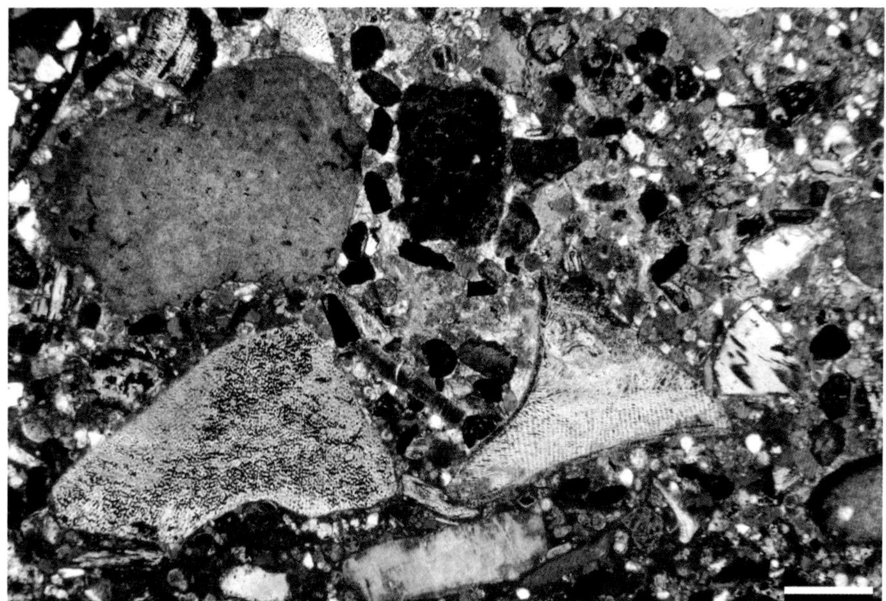

Up. Cretaceous top Eagle Ford Shale – basal Austin Chalk, Dallas Co., Texas

Hiatal surfaces, or surfaces of nondeposition, can be thin and thus difficult to recognize. This is a "bone bed", a surface on top of the Eagle Ford Shale that is littered with abundant phosphatic debris (shark's teeth, fish bone and scales and phosphatic pellets) along with glauconite grains. The sparse terrigenous debris (only a few scattered quartz grains), the dominance of biogenic phosphatic material (from decomposed vertebrate fauna) and the presence of glauconite (some of which may be replaced fecal material) is characteristic of marine hiatus surfaces in low-energy settings.

PPL, Scale bar = 0.51 mm

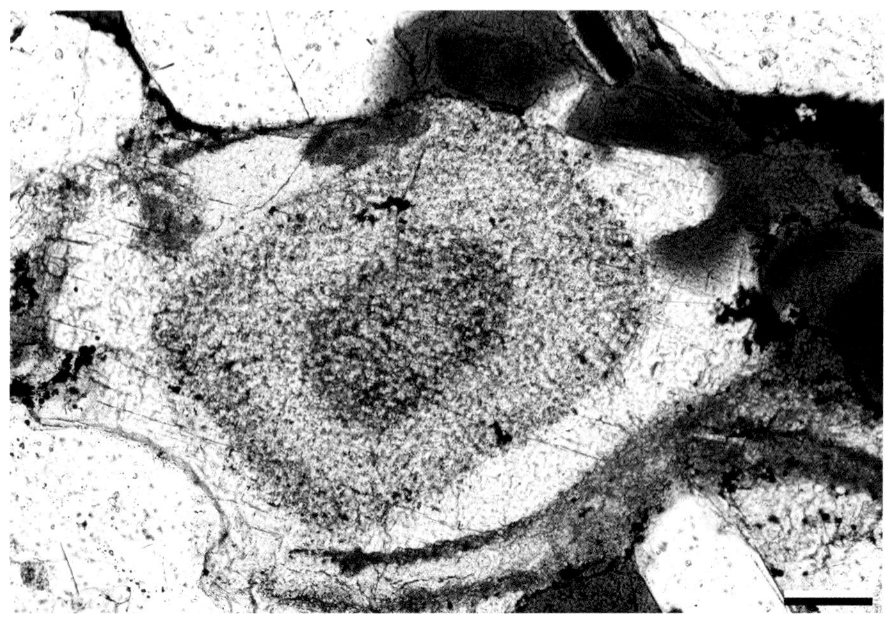

Mid. Cambrian Riley Fm., Lion Mountain Mbr., Burnet Co., Texas

A sample from a probable interval of slow sedimentation that contains glauconite, quartz grains and echinoderm and other shell fragments. The glauconite in this rock formed both as grains and as early marine, pore-filling cements. Here, the original intragranular pores in an echinoderm plate were cemented with greenish glauconite. Seafloor cementation often preferentially impacts internal pores of bioclasts because those are sites of microbial decomposition of organic matter. The echinoderm fragment was later overgrown with coarse, syntaxial calcite that deformed adjacent pelletal glauconite grains.

PPL, Scale bar = 0.1 mm

Neogene caliche, Midland Co., Texas

This mature "High Plains" caliche contains irregularly-shaped coated grains (termed soil pisoids or pisoliths). There are abundant inclusions of detrital terrigenous silt and sand grains that were engulfed by pedogenic calcite. The insoluble quartz grains were concentrated during the net dissolution of underlying strata that characterizes such long-term exposure surfaces (Reeves, 1970 and 1976). A calcrete with little or no evidence of biological activity, such as this one, is termed an "alpha calcrete" by Wright (1991).

PPL | XPL, Scale bar = 0.64 mm

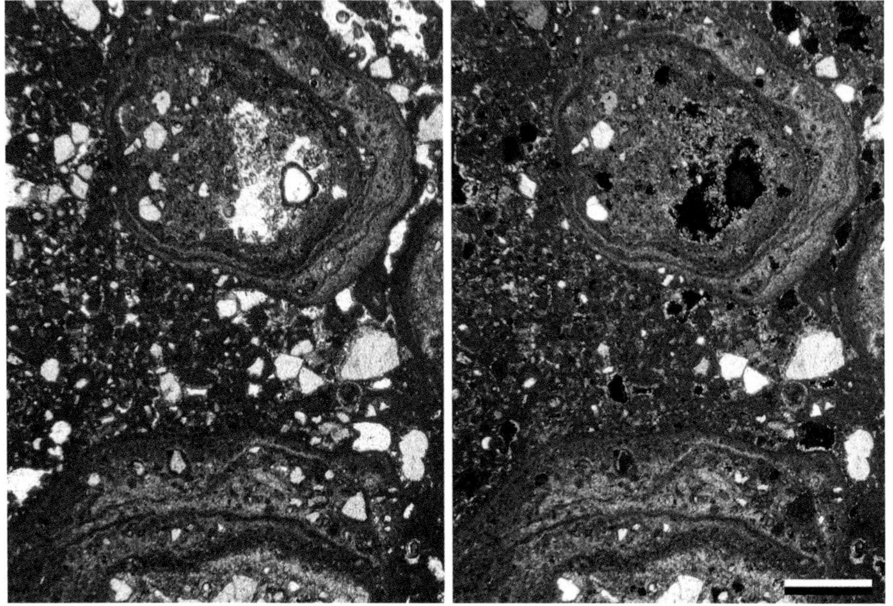

Neogene caliche, Midland Co., Texas

Another view of the same mature "High Plains" caliche. Note the irregular, highly asymmetrical coated grains (pisoids/pisoliths) and abundant inclusions of detrital terrigenous silt and sand. Caliche pisoids grow with preferential downward elongation, but typically are rotated into more random orientations by bioturbation, cryoturbation and displacive crystal growth during the long periods (commonly hundreds of thousands to many millions of years) of exposure over which such deposits form.

PPL, Scale bar = 0.51 mm

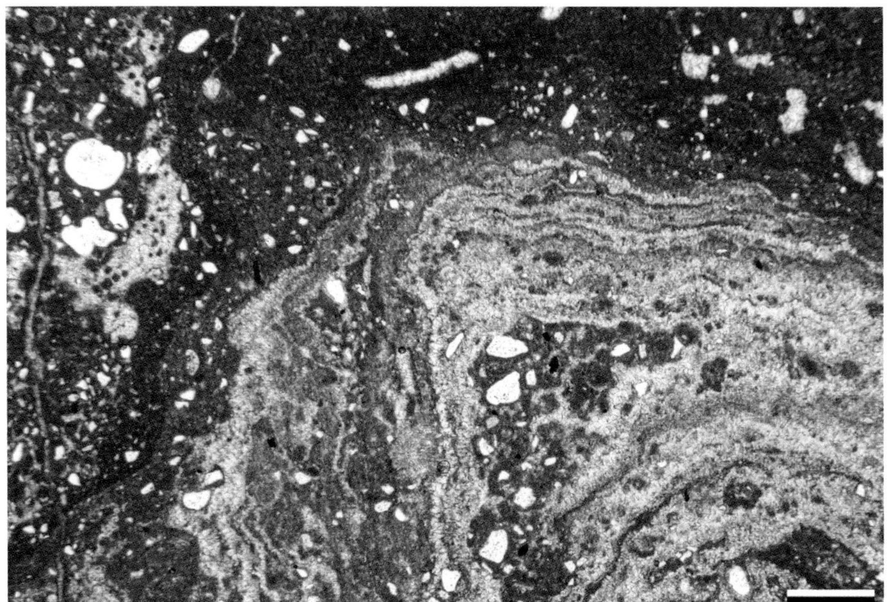

Neogene terra rossa, Malta

Like the Neogene caliches from Texas, this terra rossa contains large well-developed soil pisoids, laminated soil crusts and rhizoliths (lithified root traces). The term terra rossa is Italian for red earth and was first used to describe soils around the Mediterranean that were derived from the weathering of limestones. These features represent long-term exposure (millions of years) and concentration of clays, iron oxides and other insoluble materials within the soils.

PPL, Scale bar = 0.26 mm

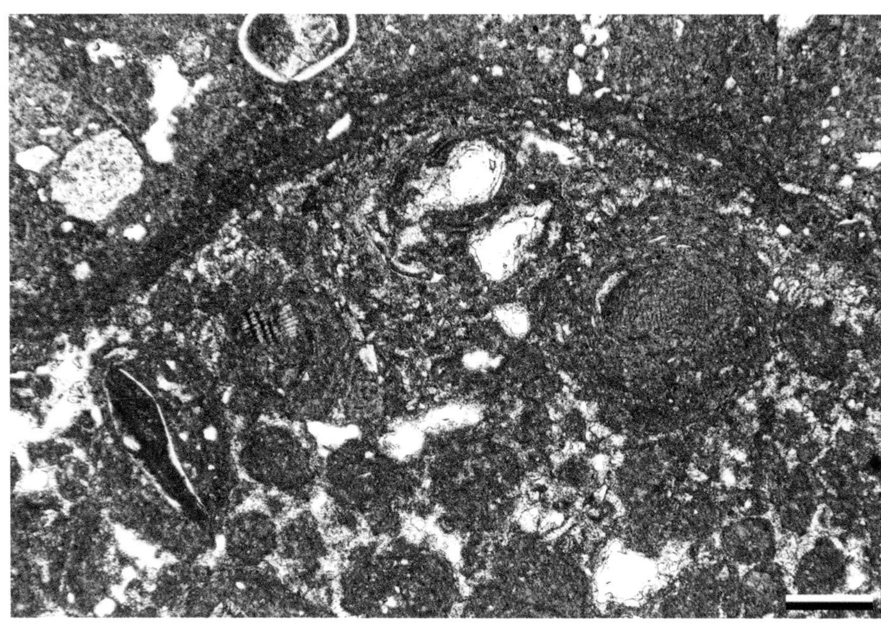

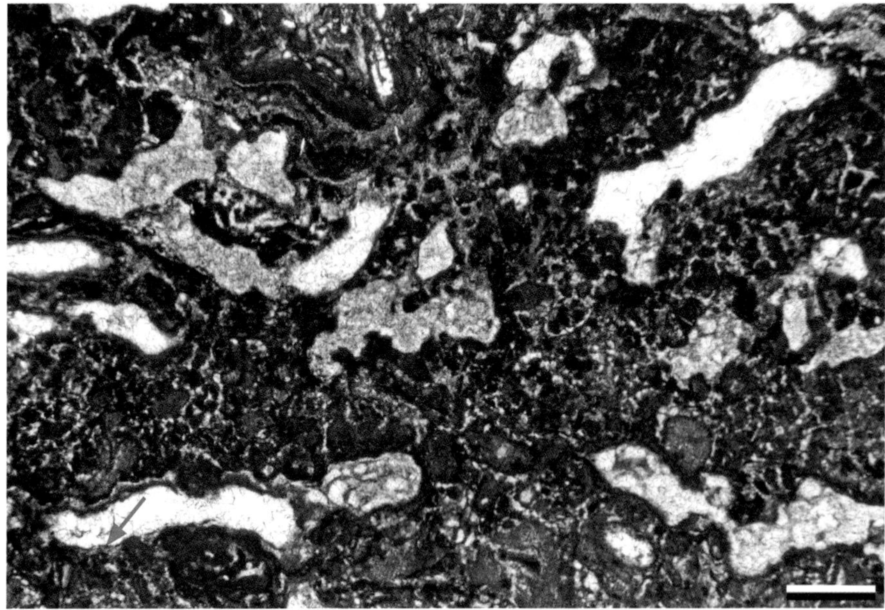

Up. Pennsylvanian (Virgilian) Holder Fm., Otero Co., New Mexico

A view of a lithified paleosol or calcrete deposit. Note the sharply bounded, irregular surface (uppermost part of photomicrograph) underlain by a laminar, somewhat contorted microcrystalline crust. The abundant circular to elliptical tubes with dark rinds are calcified root structures (rhizoliths) and some are lined with lighter-colored clay cutans (red arrow). This Carboniferous example is related to one of many glacioeustatic sea-level drops that resulted in exposure of shelfal strata (Goldstein, 1988). Calcretes with extensive evidence of biological activity, such as this one, are termed "beta calcretes" by Wright (1991).

PPL, Scale bar = 0.51 mm

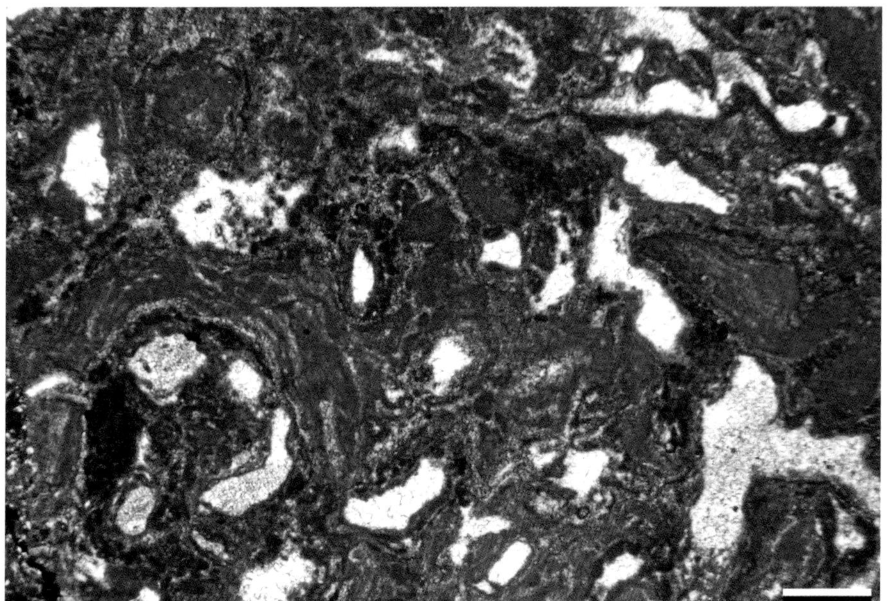

Up. Pennsylvanian (Virgilian) Holder Fm., Otero Co., New Mexico

An example of a calichified crust (calcrete) with several characteristic features—the best-developed ones in this example are incipient soil nodules (glaebules) as well as soil crusts and rhizoliths (the elongate to ovoid calcite spar-filled areas that are the cemented traces of decomposed roots). A small "nodule" in the lower-left corner is partially coated by microcrystalline calcite crusts. Like the previous example, this is a "beta-calcrete".

PPL, Scale bar = 0.51 mm

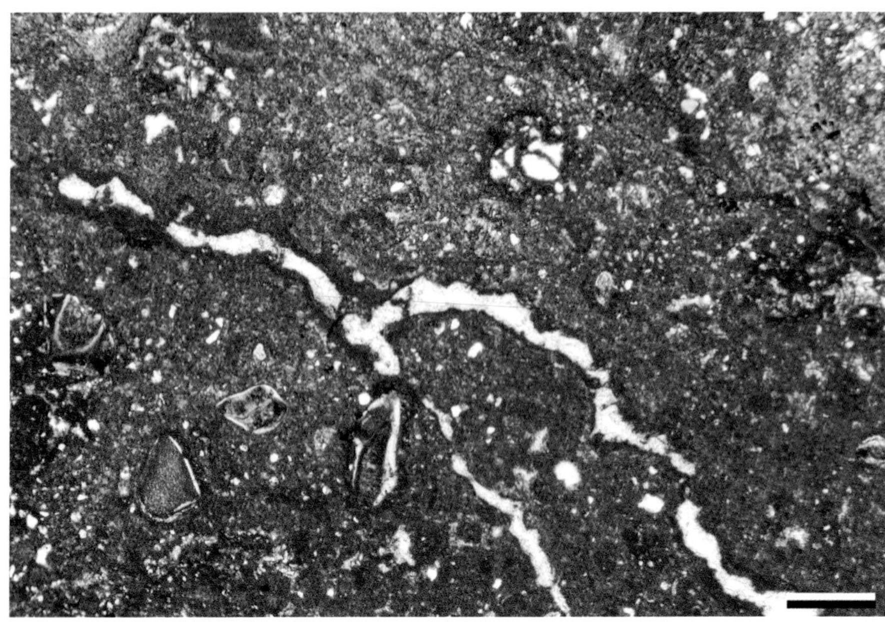

Neogene terra rossa, Malta

An example of a young, laminated, microcrystalline soil crust (calcrete) in which micrite-sized pedogenic calcite has engulfed carbonate and terrigenous clastic detritus (see photo on the title page of this chapter for an outcrop view of this rock). Note the vague laminations and well developed, branching, carbonate-lined root tubules of different sizes (rhizoliths). The dark-brown patchy areas are somewhat more heavily calcified, incipient nodules (glaebules). Note also the remarkable similarity of size and shape of structures and even the degree of fabric preservation between this and the previous (roughly 300 my old) example.

PPL, Scale bar = 0.51 mm

Pleistocene-Holocene coastal alluvium, Canterbury, New Zealand

This sample is from a calcareous soil that was formed on subaerially-exposed coastal sand deposits. This view shows numerous quartz sand grains exhibiting circumgranular cracking (and circumgranular porosity). During exposure and soil formation, circumgranular cracks form around siliciclastic grains and other soil elements due to alternating expansion and shrinkage of soils that causes the grains to be released from the surrounding matrix (Swineford, et al., 1958; Esteban and Klappa, 1983). This is a common feature in soils in general and in caliches in particular.

PPL, AFeS, BDI, Scale bar = 0.10 mm

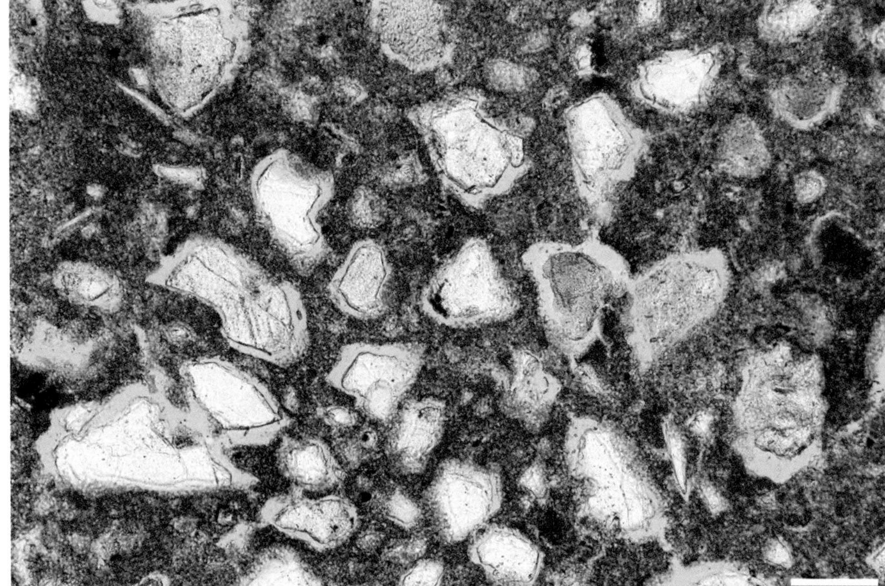

Pleistocene-Holocene coastal alluvium, Canterbury, New Zealand

This sample is from the same soil shown in the previous photomicrograph. This view shows rhizoliths as well as numerous quartz sand grains exhibiting circumgranular cracking (example at yellow arrow). Rhizoliths and rhizocretions are sedimentary structures, produced by replacement or cementation of former roots or root traces. They can be millimeters to meters in length, occur at and below hiatus surfaces and taper along their long axis. Clay cutans (greenish clays within the root traces) also are present.

PPL, BDI, Scale bar = 0.26 mm

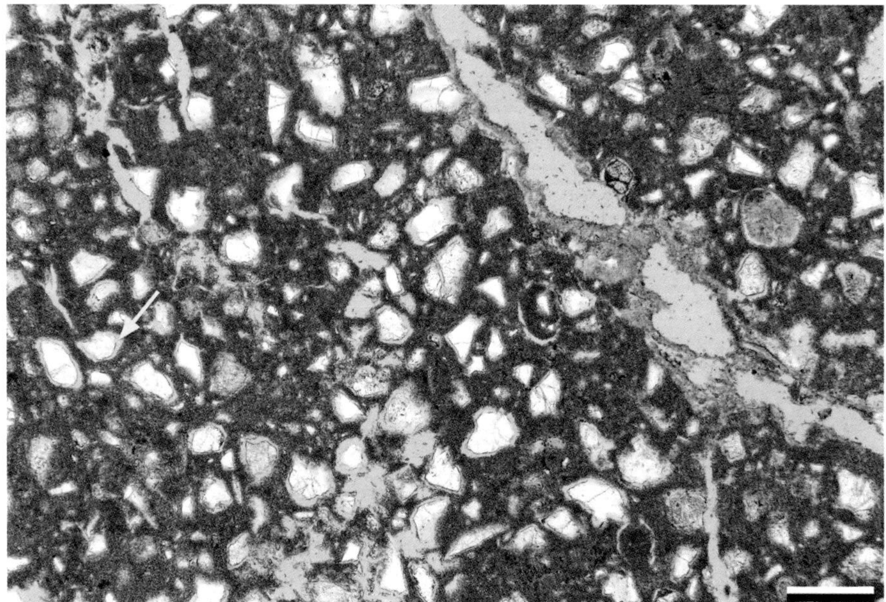

Paleocene Fatehgarh Fm., Rajasthan, India

A rhizolith (fossil root trace) within a paleosol. The rhizolith can be seen crossing the photomicrograph from top left to bottom right. The matrix comprises silt-rich clay with outsized sand grains of highly degraded lava. The root trace is filled by green clays, calcite cements and organic fragments. This is identified as a rhizolith by the presence of organic remains and a possible bifurcation. Identification as a burrow is ruled out since it tapers downwards and has a fill, parallel to the margin that suggests that the root may have shrunk as it decomposed.

PPL, AFeS, Scale bar = 0.12 mm

Paleocene Fatehgarh Fm., Rajasthan, India

Pore-lining and pore-bridging clays (XRD and SEM confirm an illite composition with overlying kaolinite). This was originally deposited as a clay-free arenite in a fluvial channel, and the clay represents infiltration of fine-grained material via illuviation of what was likely to have been smectitic clay from overlying soil horizons. Clay coats such as these can result in preservation of porosity through inhibition of later quartz overgrowth; however, they may also act to compartmentalize the pore system, resulting in lower permeability. The clays are brown in color due to staining from oil.

PPL | XPL, BDI, Scale bar = 0.14 mm

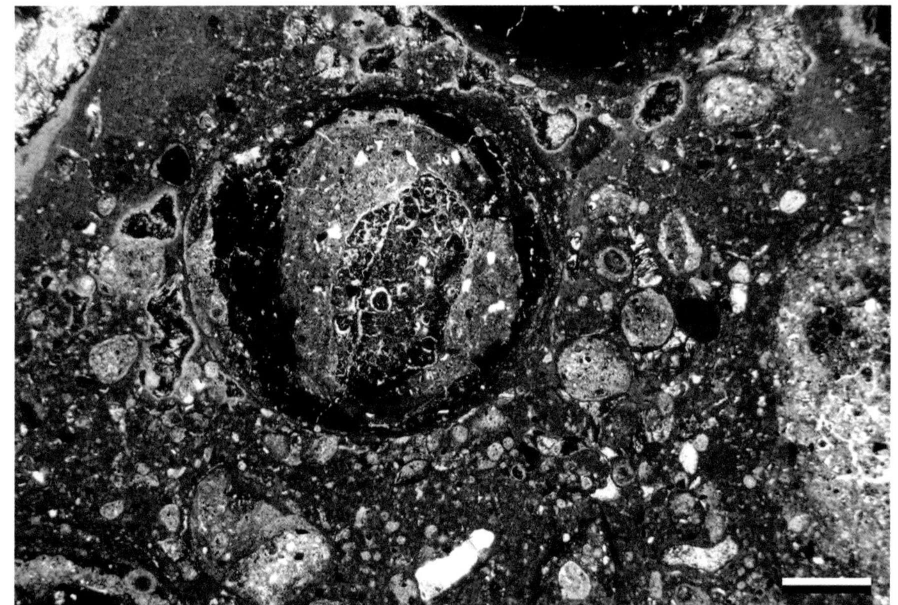

Cretaceous – Tertiary Arkansas bauxite, Pulaski Co., Arkansas

This ferruginous, pisolitic bauxite is an example of long-term weathering and "aluminous soil" formation in a high-rainfall setting. Note the abundant irregular pisoids and circumgranular cracking of finer-grained matrix. Intense chemical weathering has left behind only the most insoluble, aluminous and ferruginous chemical components.

PPL, Scale bar = 0.38 mm

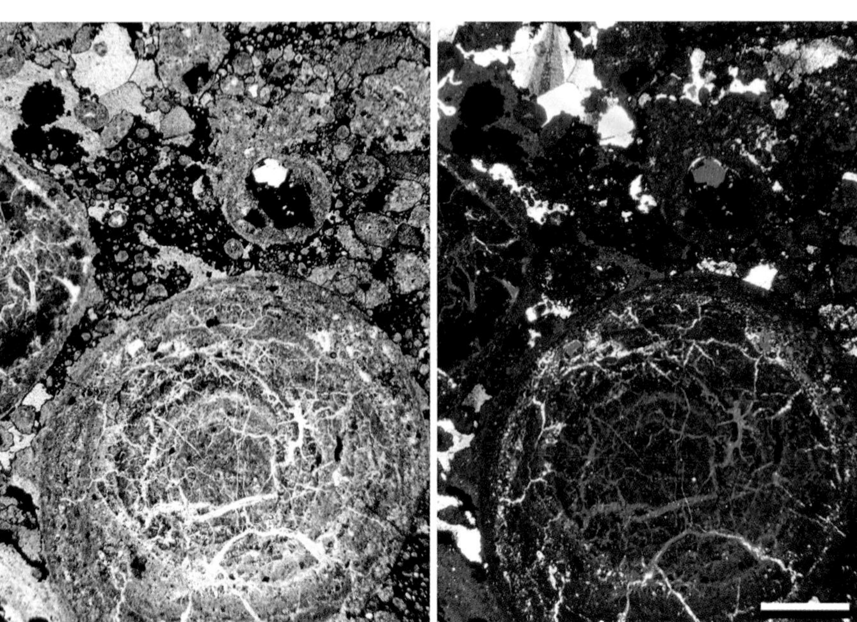

Cretaceous – Tertiary Arkansas bauxite, Bauxite Co., Arkansas

Well-developed soil pisoids within a kaolinitic bauxite. The pisoids are highly fractured due to expansion and desiccation events that also resulted in formation of circumgranular cracks. Note the extremely low birefringence to virtual isotropism of the bauxite and kaolinite. The reddish color in parts of this photomicrograph results from hematite cementation/staining. The development of pisolitic fabric, as shown here, is characteristic of many soil-type deposits including bauxites and caliche crusts.

PPL | XPL, Scale bar = 0.64 mm

Pennsylvanian Cheltenham Fm.?, Franklin Co., Missouri

A kaolinite-rich claystone with abundant ball-like structures. These claystones are sometimes termed "flint clays" because of their habit of breaking with a conchoidal fracture. These deposits partially fill karstic surface topography developed on underlying Mississippian carbonates and appear to be nonmarine (swamp and marsh) deposits in a warm/hot, very humid environment. This organic-rich environment (and the associated organic acids) resulted in the removal of most soluble grains and the concentration of Al-rich clays over a long period of exposure (Keller, 1981).

PPL, Scale bar = 0.26 mm

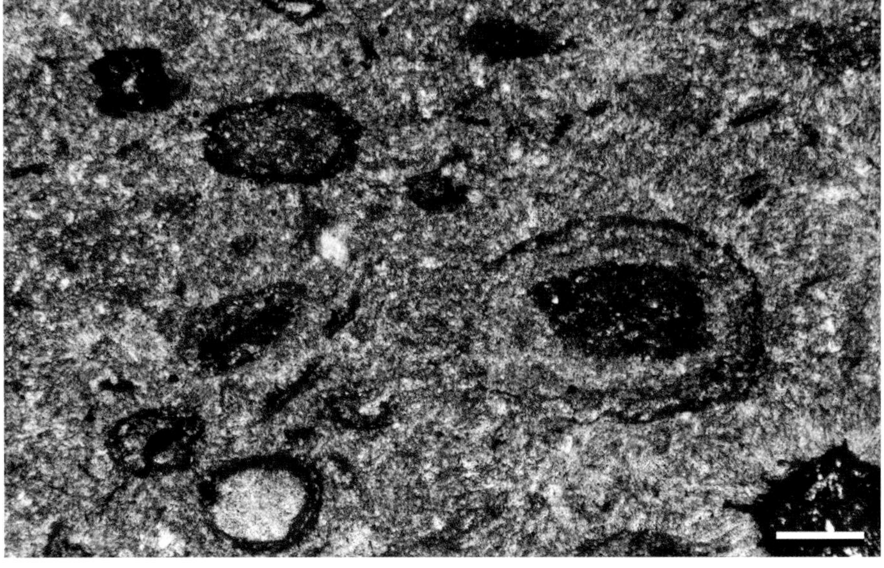

Cited References and Additional Information Sources

Burrows and borings:

Bromley, R. G., 1968, Burrows and borings in hardgrounds: Bulletin of the Geological Society of Denmark, v. 18, p. 247-250.

Bromley, R. G., 1996, Trace Fossils: Biology, Taxonomy and Applications (2nd Edition).: London, Chapman & Hall, 361 p.

Droser, M. L., and D. J. Bottjer, 1986, A semiquantitative classification of ichnofabric: Journal of Sedimentary Research, v. 56, p. 558-569, doi: 10.1306/212F89C2-2B24-11D7-8648000102C1865D.

Johnson, M. E., M. A. Wilson, and J. A. Redden, 2010, Borings in quartzite surf boulders from the Upper Cambrian basal Deadwood Formation, Black Hills of South Dakota: Ichnos, v. 17, p. 48-55, doi: 10.1080/10420941003659618.

Miller, W., III, ed., 2011, Trace Fossils: Concepts, Problems, Prospects: New York, Elsevier Science, 632 p.

Pemberton, S. G., J. A. MacEachern, and R. W. Frey, 1992, Trace fossil facies models: Environmental and allostratigraphic significance, *in* R. G. Walker, and N. P. James, eds., Facies Models: Response to Sea Level Change: Toronto, Geological Association of Canada, p. 47-72.

Concretions and nodules:

Astin, T. R., 1986, Septarian crack formation in carbonate concretions from shales and mudstones: Clay Minerals, v. 21, p. 617-631.

Baird, G. C., 1976, Coral encrusted concretions: A key to recognition of a "shale-on-shale" erosion surface: Lethaia, v. 9, p. 293-302, doi: 10.1111/j.1502-3931.1976.tb01325.x.

Bojanowski, M. J., and E. N. K. Clarkson, 2012, Origin of siderite concretions in microenvironments of methanogenesis developed in a sulfate reduction zone: An exception or a rule?: Journal of Sedimentary Research, v. 82, p. 585-598, doi: 10.2110/jsr.2012.50.

Boles, J. R., C. A. Landis, and P. Dale, 1985, The Moeraki boulders—anatomy of some septarian concretions: Journal of Sedimentary Research, v. 55, p. 398-406, doi: 10.1306/212F86E3-2B24-11D7-8648000102C1865D.

Chan, M. A., J. Ormö, A. J. Park, M. Stich, V. Souza-Egipsy, and G. Komatsu, 2007, Models of iron oxide concretion formation: Field, numerical, and laboratory comparisons: Geofluids, v. 7, p. 356-368, doi: 10.1111/j.1468-8123.2007.00187.x.

Coleman, M. L., 1993, Microbial processes: Controls on the shape and composition of carbonate concretions: Marine Geology, v. 113, p. 127-140, doi: 10.1016/0025-3227(93)90154-N.

Curtis, C. D., and M. L. Coleman, 1986, Controls on the precipitation of early diagenetic calcite, dolomite and siderite concretions in complex depositional sequences, *in* D. L. Gautier, ed., Roles of Organic Matter in Sediment Diagenesis: Tulsa, OK, SEPM Special Publication 38, p. 23-33, doi: 10.2110/pec.86.38.0023.

De las Heras, X., J. O. Grimalt, J. Albaiges, R. Julia, and P. Anadon, 1989, Origin and diagenesis of the organic matter in Miocene freshwater lacustrine phosphates (Cerandya Basin, Eastern Pyrencos): Organic Geochemistry, v. 14, p. 667-677, doi: 10.1016/0146-6380(89)90046-6.

DeCraen, M., R. Swennen, E. M. Keppens, C. I. Macaulay, and K. Kiriakoulakis, 1999, Bacterially mediated formation of carbonate concretions in the Oligocene Boom Clay of northern Belgium: Journal of Sedimentary Research, v. 69, p. 1098-1106, doi: 10.2110/jsr.69.1098.

Gautier, D. L., 1982, Siderite concretions: Indicators of early diagenesis in the Gammon Shale (Cretaceous): Journal of Sedimentary Research, v. 52, p. 859-871, doi: 10.1306/212F8076-2B24-11D7-8648000102C1865D.

Hesse, R., 1989, Silica diagenesis: Origin of inorganic and replacement cherts: Earth-Science Reviews, v. 26, p. 253-284, doi: 10.1016/0012-8252(89)90024-X.

Johnson, M. R., 1989, Paleogeographic significance of oriented calcareous concretions in the Triassic Katberg Formation, South Africa: Journal of Sedimentary Research, v. 59, p. 1008-1010, doi: 10.1306/212F90D9-2B24-11D7-8648000102C1865D.

Kennedy, W. J., and H. C. Klinger, 1972, Hiatus concretions and hardground horizons in the Cretaceous of Zululand: Palaeontology, v. 15, p. 539-549.

Kidder, D. L., 1985, Petrology and origin of phosphate nodules from the Midcontinent Pennsylvanian epicontinental sea: Journal of Sedimentary Research, v. 55, p. 809-816, doi: 10.1306/212F880A-2B24-11D7-8648000102C1865D.

Marshall-Neill, G., and A. Ruffell, 2004, Authigenic phosphate nodules (Late Cretaceous, Northern Ireland) as condensed succession microarchives: Cretaceous Research, v. 25, p. 439-452, doi: 10.1016/j.cretres.2004.03.001.

McBride, E. F., M. D. Picard, and R. L. Folk, 1994, Oriented concretions, Ionian coast, Italy: Evidence of groundwater flow direction: Journal of Sedimentary Research, v. A64, p. 535-540, doi: 10.1306/D4267DFC-2B26-11D7-8648000102C1865D.

Mozley, P. S., and S. J. Burns, 1993, Oxygen and carbon isotopic composition of marine carbonate concretions: An overview: Journal of Sedimentary Research, v. 63, p. 73-83, doi: 10.1306/D4267A91-2B26-11D7-8648000102C1865D.

Mozley, P. S., and J. M. Davis, 2005, Internal structure and mode of growth of elongate calcite concretions: Evidence for small-scale, microbially induced, chemical heterogeneity in groundwater: GSA Bulletin, v. 117, p. 1400-1412, doi: 10.1130/B25618.1.

Parry, W. T., 2011, Composition, nucleation, and growth of iron oxide concretions: Sedimentary Geology, v. 233, p. 53-68, doi: 10.1016/j.sedgeo.2010.10.009.

Raiswell, R., and Q. J. Fisher, 2000, Mudrock-hosted carbonate concretions: A review of growth mechanisms and their influence on chemical and isotopic composition: Journal of the Geological Society, v. 157, p. 239-251, doi: 10.1144/jgs.157.1.239.

Raiswell, R., and Q. J. Fisher, 2004, Rates of carbonate cementation associated with sulphate reduction in DSDP/ODP sediments: Implications for the formation of concretions: Chemical Geology, v. 211, p. 71-85, doi: 10.1016/j.chemgeo.2004.06.020.

Sellés-Martínez, J., 1996, Concretion morphology, classification and genesis: Earth-Science Reviews, v. 41, p. 177-210, doi: 10.1016/S0012-8252(96)00022-0.

Soudry, D., and Z. Lewy, 1990, Omission-surface incipient phosphate crusts on early diagenetic calcareous concretions and their possible origin, Upper Campanian, southern Israel: Sedimentary Geology, v. 66, p. 151-163, doi: 10.1016/0037-0738(90)90012-I.

Voigt, E., 1968, Über Hiatus-Konkretionen (dargestellt an Beispielen aus dem Lias): Geologische Rundschau, v. 58, p. 281-296, doi: 10.1007/BF01820609.

Zatoń, M., 2010, Hiatus concretions: Geology Today, v. 26, p. 186-189, doi: 10.1111/j.1365-2451.2010.00762.x.

Hardgrounds:

Bjørkum, P. A., and O. Walderhaug, 1990, Lateral extent of calcite-cemented zones in shallow marine sandstones, *in* A. T. Buller, E. Berg, O. Hjelmeland, J. Kleppe, O. Torsæter, and J. O. Aasen, eds., North Sea Oil and Gas Reservoirs—II: London, Graham & Trotman, p. 331-336, doi: 10.1007/978-94-009-0791-1_28.

Bromley, R. G., 1975, Trace fossils at omission surfaces, *in* R. W. Frey, ed., The Study of Trace Fossils: New York, Springer-Verlag, p. 399-428, doi: 10.1007/978-3-642-65923-2_18.

Bromley, R. G., and A. S. Gale, 1982, The lithostratigraphy of the English Chalk Rock: Cretaceous Research, v. 3, p. 273-306, doi: 10.1016/0195-6671(82)90030-1.

Fürsich, F. T., W. J. Kennedy, and T. J. Palmer, 1981, Trace fossils at a regional discontinuity surface: The Austin/Taylor (Upper Cretaceous) contact in central Texas: Journal of Paleontology, v. 55, p. 537-551.

Fürsich, F. T., W. Oschmann, I. B. Singh, and A. K. Jaitly, 1992, Hardgrounds, reworked concretion levels and condensed horizons in the Jurassic of western India: Their significance for basin analysis: Journal of the Geological Society, v. 149, p. 313-331, doi: 10.1144/gsjgs.149.3.0313.

Grimm, K., 2000, Stratigraphic condensation and the redeposition of economic phosphorite: Allostratigraphy of Oligo-Miocene shelfal sediments, Baja California Sur, Mexico, *in* C. R. Glenn, L. Prévôt-Lucas, and J. Lucas, eds., Marine Authigenesis: From Global to Microbial: Tulsa, OK, SEPM Special Publication 66, p. 325-347, doi: 10.2110/pec.00.66.0325.

Loutit, T. S., J. Hardenbol, P. R. Vail, and G. R. Baum, 1988, Condensed sections: The key to age dating and correlation of continental margin sequences, *in* C. K. Wilgus, B. S. Hastings, C. G. St. C. Kendall, H. W. Posamentier, C. A. Ross, and J. C. Van Wagoner, eds., Sea-Level Changes: An Integrated Approach: Tulsa, OK, SEPM Special Publication 42, p. 183-213, doi: 10.2110/pec.88.01.0183.

Martín-Algarra, A., and A. Sánchez-Navas, 1995, Phosphate stromatolites from condensed cephalopod limestones, Upper Jurassic, southern Spain: Sedimentology, v. 42, p. 893-919, doi: 10.1111/j.1365-3091.1995.tb00416.x.

Molenaar, N., G. P. Van de Bilt, E. R. Van den Hoek Ostende, and S. D. Nio, 1988, Early diagenetic alteration of shallow-marine mixed sandstones: An example from the Lower Eocene Roda Sandstone Member, Tremp-Graus Basin, Spain: Sedimentary Geology, v. 55, p. 295-318, doi: 10.1016/0037-0738(88)90136-4.

Pedley, H. M., and S. Bennett, 1985, Phosphorites, hardgrounds and syndepositional subsidence structures in the Miocene Globigerina Limestone Formation, Maltese Islands: Sedimentary Geology, v. 45, p. 1-34, doi: 10.1016/0037-0738(85)90022-3.

Rehfeld, U., and A. W. Janssen, 1995, Development of phosphatized hardgrounds in the Miocene Globigerina Limestone of the Maltese Archipelago, including a description of *Gamopleura melitensis* sp. nov. (Gastropoda, Euthecosomata): Facies, v. 33, p. 91-106, doi: 10.1007/BF02537445.

Paleosols, calcretes, and bauxites:

Bhattacharyya, D. P., and P. K. Kakimoto, 1982, Origin of ferriferous ooids: An SEM study of ironstone ooids and bauxite pisoids: Journal of Sedimentary Research, v. 52, p. 849-857, doi: 10.1306/212F8071-2B24-11D7-8648000102C1865D.

Dill, H. G., 2001, The geology of aluminium phosphates and sulphates of the alunite group minerals: A review: Earth-Science Reviews, v. 53, p. 35-93, doi: 10.1016/S0012-8252(00)00035-0.

Esteban, M., and C. F. Klappa, 1983, Subaerial exposure environment, *in* P. A. Scholle, D. G. Bebout, and C. H. Moore, eds., Carbonate Depositional Environments: Tulsa, OK, AAPG Memoir 33, p. 1-54.

Goldstein, R. H., 1988, Paleosols of Late Pennsylvanian cyclic strata, New Mexico: Sedimentology, v. 35, p. 777-804, doi: 10.1111/j.1365-3091.1988.tb01251.x.

Gordon, M., Jr., J. I. Tracey, Jr., and M. W. Ellis, 1958, Geology of the Arkansas bauxite region: Washington, DC, USGS Professional Paper 299, 268 p.

Keller, W. D., 1981, The sedimentology of flint clay: Journal of Sedimentary Research, v. 51, p. 233-244, doi: 10.1306/212F7C57-2B24-11D7-8648000102C1865D.

Kraus, M. J., 1999, Paleosols in clastic sedimentary rocks: Their geologic applications: Earth-Science Reviews, v. 47, p. 41-70, doi: 10.1016/S0012-8252(99)00026-4.

Mack, G. H., W. C. James, and H. C. Monger, 1993, Classification of paleosols: GSA Bulletin, v. 105, p. 129-136, doi: 10.1130/0016-7606(1993)105<0129:COP>2.3.CO;2.

Martini, I. P., ed., 1992, Weathering, Soils and Paleosols: Amsterdam, Elsevier Science Publishers, 618 p.

Mongelli, G., and P. Acquafredda, 1999, Ferruginous concretions in a Late Cretaceous karst bauxite: Composition and conditions of formation: Chemical Geology, v. 158, p. 315-320, doi: 10.1016/S0009-2541(99)00061-3.

Price, G. D., P. J. Valdes, and B. W. Sellwood, 1997, Prediction of modern bauxite occurrence: Implications for climate reconstruction: Palaeogeography, Palaeoclimatology, Palaeoecology, v. 131, p. 1-13, doi: 10.1016/S0031-0182(96)00145-9.

Purvis, K., and V. P. Wright, 1991, Calcretes related to phreatophytic vegetation from the Middle Triassic Otter Sandstone of south west England: Sedimentology, v. 38, p. 539-551, doi: 10.1111/j.1365-3091.1991.tb00366.x.

Reeves, C. C., Jr., 1970, Origin, classification and geologic history of caliche on southern High Plains, Texas and eastern New Mexico: Journal of Geology, v. 78, p. 352-362, doi: 10.1086/627521.

Reeves, C. C., Jr., 1976, Caliche: Origin, Classification, Morphology and Uses: Lubbock, TX, Estacado Books, 233 p.

Reinhardt, J., and W. R. Sigleo, eds., 1988, Paleosols and Weathering Through Geologic Time: Principles and Applications: Boulder, CO, GSA Special Paper 216, 181 p.

Retallack, G. J., 1990, Soils of the Past: An Introduction of Paleopedology: Boston, MA, Unwin Hyman, 520 p.

Swineford, A., A. B. Leonard, and J. C. Frye, 1958, Petrology of the Pliocene pisolitic limestone in the Great Plains: Kansas Geological Survey Bulletin 130, Part 2, p. 97-116.

Wright, V. P., ed., 1986, Paleosols: Their Recognition and Interpretation: Oxford, Blackwell Scientific Publications, 315 p.

Wright, V. P., 1992, Paleosol recognition: a guide to early diagenesis in terrestrial settings, *in* K. H. Wolf, and G. V. Chilingarian, eds., Diagenesis, III: New York, Elsevier Science Publishers, p. 591-629.

Wright, V. P., and M. E. Tucker, 1991, Calcretes: An introduction, *in* V. P. Wright, and M. E. Tucker, eds., Calcretes (IAS Reprint Series 2): Oxford, UK, Blackwell Publishing, p. 1-22, doi: 10.1002/9781444304497.ch.

Wright, V. P., M. S. Turner, J. E. Andrews, and B. Spiro, 1993, Morphology and significance of super-mature calcretes from the Upper Old Red Sandstone of Scotland: Journal of the Geological Society, v. 150, p. 871-883, doi: 10.1144/gsjgs.150.5.0871.

Yaalon, D. H., ed., 1971, Paleopedology: Origin, Nature and Dating of Paleosols: Jerusalem, International Society of Soil Science and Israel Universities Press, 350 p.

Facing Page: Top – Photomicrograph (XPL) of *in-situ* deformation of a plagioclase grain, Up. Triassic New Haven Arkose, Connecticut. Bottom – Compacted tension gashes produced by downslope sliding in marly shales, Ordovician Cowhead Breccia, western Newfoundland. Photo from Noel P. James.

DIAGENESIS:
COMPACTION

CHAPTER 10

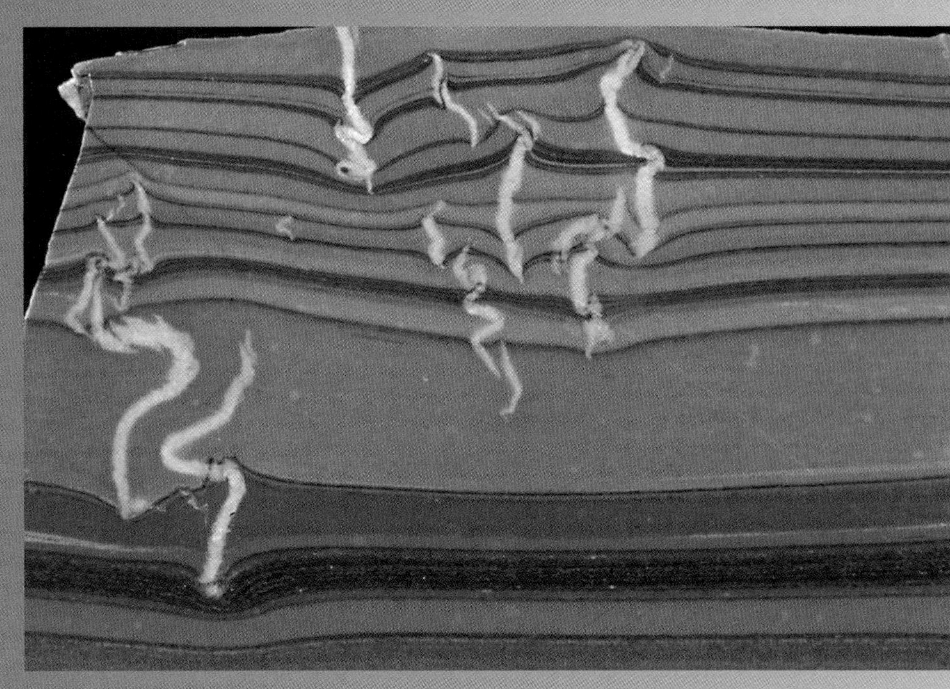

COMPACTION

Compaction is one of the major processes by which sediments lose porosity and begin the transformation to sedimentary rocks. Compaction is driven mainly by overburden loading and involves changes in the packing density of constituent grains. This is accomplished initially through grain reorientation and repacking accompanied by water expulsion from porous sediments. With additional overburden loading, fracturing and cleavage of brittle grains and plastic deformation of ductile grains contribute to increased packing density and concomitant loss of pore space.

Further reduction of intergranular pore space, beyond that produced by "mechanical compaction", results from pressure-solution processes, sometimes termed "chemical compaction". Chemical compaction includes selective dissolution and interpenetration at grain-to-grain contacts, as well as broader dissolution along solution seams and stylolites. Although more common in carbonate rocks, chemical compaction features are widespread in clastic terrigenous deposits as well (e.g., Heald, 1955; Walderhaug and Bjørkum, 2003).

Chemical compaction depends on the relative pressure-related solubilities of the varied grains and rock fragments in clastic terrigenous rocks, along with pressure, temperature and water chemistry conditions present at various stages of burial. Organic matter and clays play additional roles in enhancing pressure dissolution (Oelkers et al., 1992; Bjørkum, 1996). The porosity loss patterns are more complex in chemical compaction as compared with mechanical compaction, because they involve not just deformation, but actual removal of material at the sites of compaction. That removal may or may not be matched or offset by reprecipitation of the solutes as cements at nearby or distant sites.

Localized compaction can also occur in millimeter- to centimeter-thick deformation zones in which compactional (overburden) strains equal or exceed shear strains (e.g., Olsson et al., 2002; Aydin and Ahmadov, 2009). Such layers, termed "compaction bands", are tabular and parallel (or near-parallel) to bedding and form only in porous, medium- to coarse-grained sandstones (Schultz et al., 2012). Compaction bands can significantly increase the heterogeneity of porosity and permeability in sandstone sections.

All sediments undergo compaction-driven porosity loss as a function of burial depth, but the rates of compaction are a function of rock properties such as grain size and composition (Fig. 10.1). In general, fine-grained sediments (shale, marl, chalk and siliceous ooze) have the highest rates of porosity loss, primarily because they have exceptionally high initial water contents (70 to 80% or more in many cases). Despite the relatively low permeabilities of fine-grained sediments, early water expulsion and associated porosity loss are far more rapid than in coarser-grained deposits. Shale and chalk lose roughly half their porosity during the first kilometer of burial; sandstone, limestone and dolomite lose less, but still very significant, amounts of porosity with comparable burial. For sandstone, compaction typically accounts for porosity loss from initial values of 40–50% to values of 25–32% before lithification substantially reduces or terminates further compaction (Houseknecht, 1987; Paxton et al., 2002; Taylor et al., 2010). The rate of burial (sometimes termed the burial trajectory), which ultimately depends on tectonic history of the basin, also can affect the degree of sediment compaction relative to cementation (e.g., Houseknecht and Hathon, 1987).

Although some compaction is inevitable in sediments, there are circumstances in which compaction is reduced or even eliminated. Most obviously, complete or nearly complete near-surface cementation (as in concretions, hardgrounds or other cemented zones) largely precludes compaction in those layers, albeit at the cost of all or most porosity. More effectively, at least from the standpoint of an explorationist interested in porosity preservation, partial early cementation can have similar effects, especially, but not exclusively, where cements are concentrated at or near grain contacts (sometimes termed "spot-welding"). In such cases, especially common in vadose meteoric settings, the rock fabric can be strengthened sufficiently to withstand burial to considerable depths with little compactional porosity loss.

Another common circumstance in which compaction is minimized is regional overpressuring. Overpressuring refers to elevated pore-fluid pressures (that is, nondirectional pressures that exceed the normal hydrostatic gradients and partially offset vertical lithostatic stresses). Overpressuring can result from a variety of mechanisms (Swarbrick and Osborne, 1998), including overburden loading rates that exceed the rate at which water can escape from fine-grained and low-permeability sections, exceptional thermal heating of waters, organic maturation and hydrocarbon generation, gypsum dehydration and others. Viewed in its simplest form, because water is effectively incompressible, sediments and rocks cannot compact in settings where water cannot escape. Overpressuring (with associated reduction in vertical overburden stresses) has been shown to be a widespread and effective porosity preserving mechanism in the North Sea and other exploration provinces (Ramm and Bjørlykke, 1994; Osborne and Swarbrick, 1999; Block et al., 2002). It should be noted, however, that compaction is essentially irreversible and overpressuring can only preserve porosity still present at the time of initiation of

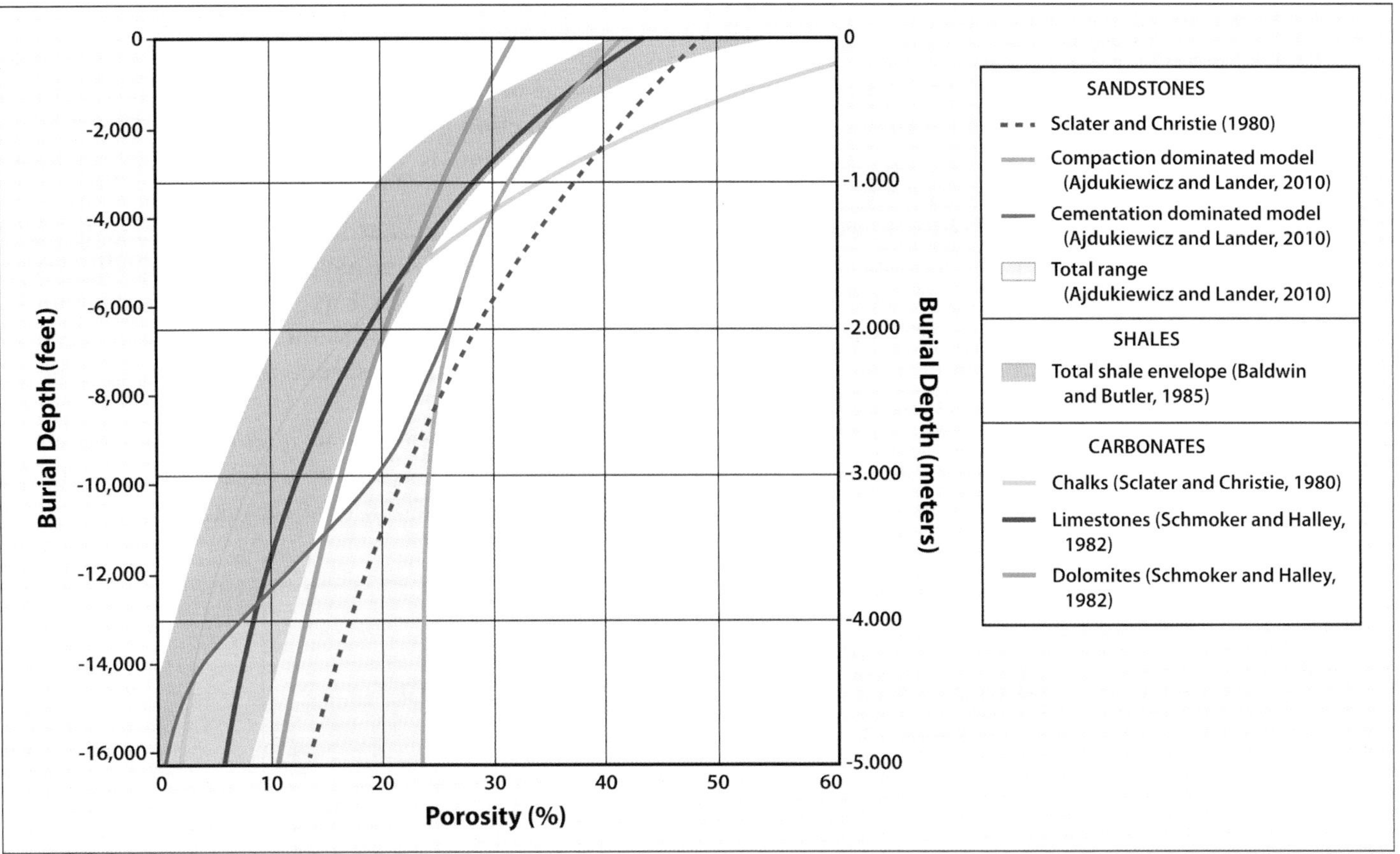

Figure 10.1 Smoothed porosity-depth curves for sandstones and shales with comparative data for three types of carbonate rocks (shallow-marine limestones and dolomites and deeper-water chalks). These curves were constructed primarily through statistical modeling, based on empirical data from cores and geophysical logs. Compaction and cementation components of porosity loss are differentiated only for sandstones, where separate curves are shown for depths below 2 km (6,560 ft). However, compaction is of primary importance in the early stages of burial for the other rock types as well. Data sources are given in the table legend; see Ehrenberg and Nadeau (2005) for a better sense of the enormous diversity in measured porosity-depth data for sandstones and carbonate rocks.

anomalous pressure; thus, early onset of overpressuring (relative to burial) is critical in minimizing compaction.

From a petrographic perspective, recognition of compaction (and evaluation of the relative importance of compaction versus cementation in porosity loss) centers around five major types of observations:

1. Evaluation of the number of grain-to-grain contacts – Uncompacted sand, when viewed in 2-D sections shows few grain-to-grain contacts and numerous "floating grains" (less than 1.6 contacts per grain and 16.6% floating grains as reported by Taylor, 1950; all grains are, of course, supported by additional contacts outside the plane of view). Through progressive burial to 2.8 km (8,300 ft), the average number of grain-to-grain contacts increased to more than 5 in compacted sandstones (Taylor, 1950).
2. Evaluation of the shape of grain contacts – Uncompacted sand generally shows tangential or point contacts; with progressive burial, the grain contacts become more complex. Moderate compaction results in grain rotation, and long and straight contacts become more common; with further compaction and some grain deformation, concavo-convex contacts become increasingly common; and with extensive compaction, sutured (pressure-solution) contacts dominate (Taylor, 1950).
3. Recognition of grain deformation features (brittle or plastic) – These observations include grain-scale fracturing, plastic deformation of soft grains (e.g., glauconite claystone or SRFs) and compactional drape of fine-grained sediments around objects that were hard when they were deposited or were hardened in near-seafloor diagenesis (e.g., fossil remains or concretions).
4. Recognition of grain-scale and larger pressure-dissolution (chemical compaction) features – this overlaps with the sutured grain-to-grain contacts noted in criterion 2, but extends to larger-scale features such as stylolites or solution seams.
5. Measurements of intergranular volume (IGV) and cement volume – this is the most quantitative way to distinguish the relative importance of

compaction versus cementation in porosity loss (Houseknecht, 1987). This technique involves measuring the total IGV of a rock sample (sometimes termed the "minus-cement porosity") which is the combination of current porosity and the volume of primary porosity filled with cement. A starting IGV of 40% is assumed for this calculation based on the average porosity of modern sands. Compaction, both mechanical and chemical, reduces the volume of spaces between grains (the IGV); cementation occludes pores, but does not reduce IGV. In this methodology, the IGV of a samples is plotted against the volume of cement as shown in Figure 10.2 (see Houseknecht, 1987; Paxton et al., 2002), thereby distinguishing porosity loss resulting from compaction versus porosity loss from cementation. Data on a wide range of sandstones (Paxton et al., 2002) show that compaction generally reduces IGV to about 26%; cementation is needed to reduce IGV to lower values.

It should be noted that it is not always easy to make intergranular volume measurements, especially where cements, such as syntaxial quartz overgrowths, can be difficult to differentiate from their substrate grains, and the use of cathodoluminescence microscopy is essential in such cases.

Because textural fabrics were covered in Chapter 6, the illustrations in this chapter will deal mainly with the other criteria for recognition of compaction.

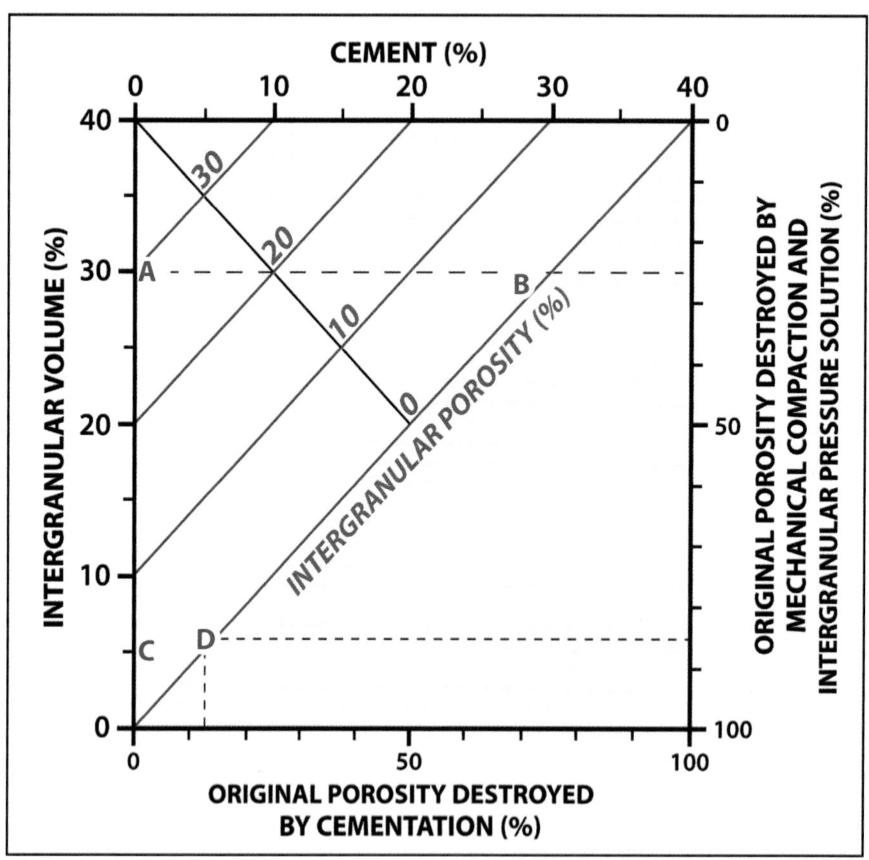

Figure 10.2 Diagram for evaluating the relative importance of compactional processes and cementation to porosity loss (from Houseknecht, 1987; see also discussion by Ehrenberg, 1989). **An initial porosity of 40% is assumed for all samples.** Petrographically measured data are plotted using the upper (Cement) and left-hand (Intergranular volume, IGV) axes. Examples marked A, B, C and D are data plotted for four end-member sandstones illustrated in the Houseknecht paper (p. 636). A = sandstone with 30% IGV and essentially no cement (sample thus retains 30% porosity); B = sandstone with 29% IGV, virtually all filled with cement (and thus with only about 1% remaining porosity); C = sandstone with 5% IGV and essentially no cement (sample thus retains 5% porosity); D = sandstone with 6% IGV most of which is cement filled (and sample thus has only 1-2% remaining porosity). The bottom and right-hand axes show the relative percentages of the original 40% porosity destroyed by cementation and compaction respectively. For sample A, for example, 25% of the assumed original 40% porosity was destroyed by compaction (10% lost relative to the 40% initially available = 25%) and essentially none was affected by cementation as shown by the broadly dashed line; for sample D, 87% of the original porosity was destroyed by compaction and about 12% was filled by cement as shown by the two finely dashed lines.

Up. Cretaceous Monte Antola Fm., Liguria, Italy

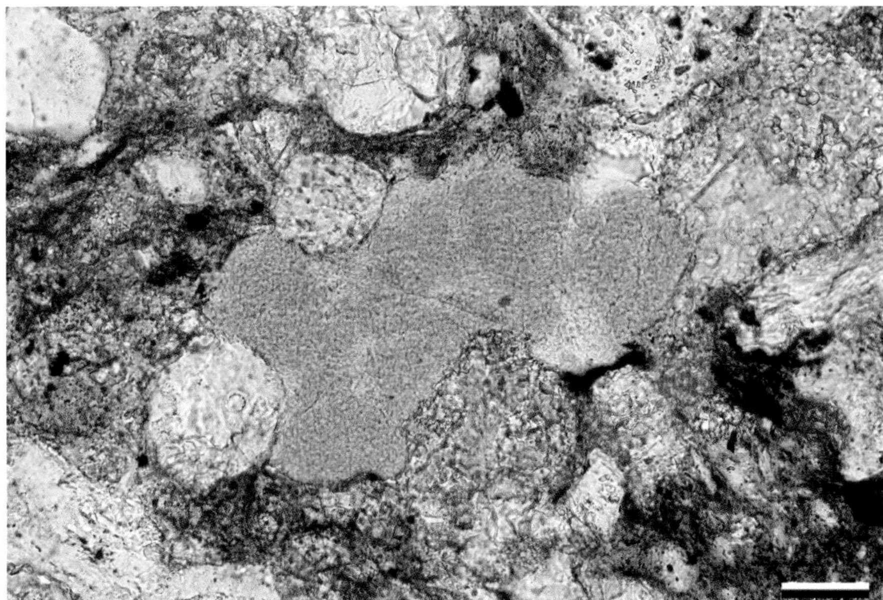

Grain deformation is one of the clearest indicators of compaction, and glauconite grains are among the most easily deformable materials. Here, a glauconite grain has been deeply indented by calcitized sponge spicules and lithic grains during mechanical compaction. The clay-rich clasts (shales) also are deformed extensively due to their softness. Because of such ductile compaction, it is often difficult to determine if the rock contains true matrix or pseudomatrix.

PPL, Scale bar = 0.04 mm

Mid. Cambrian Riley Fm., Lion Mountain Mbr., Burnet Co., Texas

Ductile deformation of glauconite grains within this greensand has destroyed the primary porosity of the rock, at least locally. Individual glauconite grains are still visible because the exterior rims of these Fe-rich grains were oxidized to limonite and/or hematite prior to burial. The fracture porosity that is visible in this photomicrograph is a result of stress release during uplift (or desiccation during sample preparation).

PPL, BDI, Scale bar = 0.26 mm

Lo. Cretaceous (Albian), Nahr Umr Fm., offshore Qatar

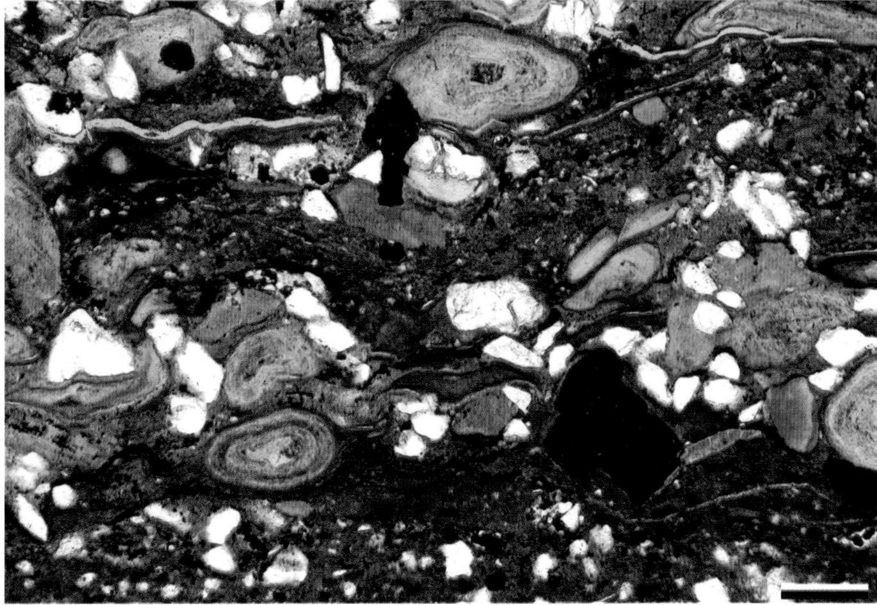

Here, chamosite ooids and glauconite grains have been flattened and deformed during compaction. Quartz grains embayed many of these grains. The original cortices of the ooids are still visible within most of the chamositic grains despite their deformation. Partial pyrite replacement of the matrix (black crystals), including organic material, occurred early as indicated by the observation that both matrix and grains are compactionally deformed around the pyrite crystals. The matrix and other grains also have been partially impregnated by phosphate.

PPL, BDI, Scale bar = 0.26 mm

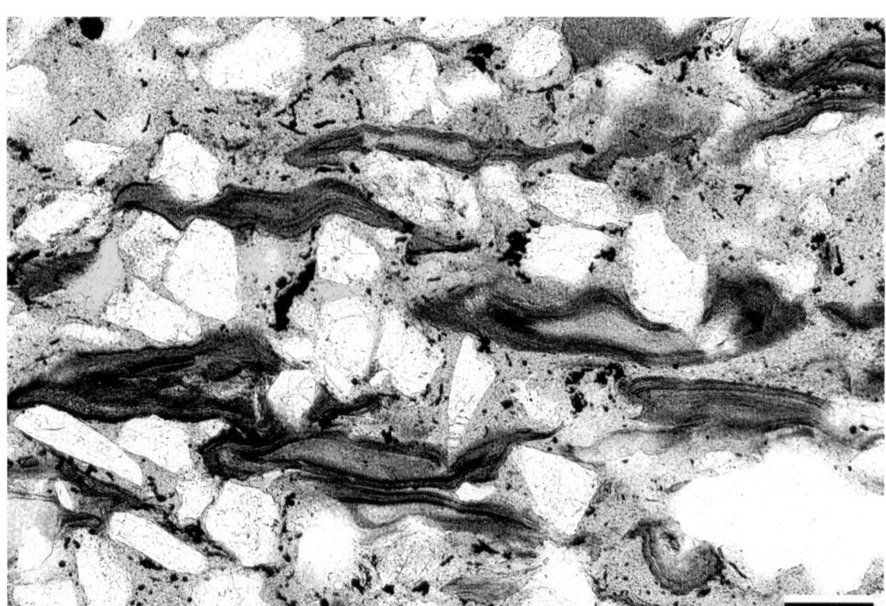

Lo. Cretaceous (Albian), Nahr Umr Fm., offshore Qatar

This is a strongly-compacted, poorly-cemented chamositic sandstone. During compaction, the ductile ooids were nearly completely flattened, smeared out and embayed by adjacent, rigid quartz grains. However, despite the extreme compaction, the cortices are still visible.

PPL, BDI, Scale bar = 0.1 mm

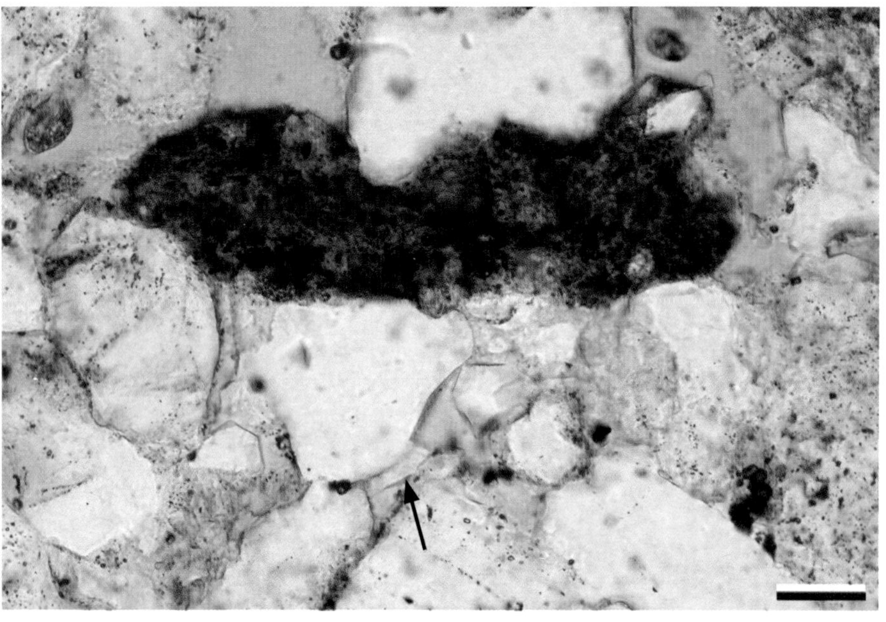

Mississippian Berea Sandstone, Midland Co., Michigan

This photomicrograph shows a shale clast that has been deformed by an adjacent quartz grain during compaction. Where compaction is extensive, grains like this can be severed or intruded into surrounding porosity. This rock may have been stabilized by localized euhedral quartz overgrowth cements that "spot welded" the quartz grains (arrow) in place limiting compaction and porosity loss during burial.

PPL, BDI, Scale bar = 0.04 mm

Lo. Cambrian Unicoi Fm., Shenandoah Valley, Virginia

In this photomicrograph, a shale-rich arenite has been highly compacted. Quartz and feldspar grains have deeply embayed adjacent illite-rich shale rock fragments, and formed concavo-convex boundaries with them. The compaction has forced the ductile shale grains into surrounding pore spaces. The presence of thin iron oxide cement rims on both the quartz and shale clasts provide proof that these clays were originally grains, not matrix.

XPL, Scale bar = 0.10 mm

Permian basinal sandstone, Presidio Co., Texas

This illustration shows how compaction of shale grains within a sandstone can influence the porosity evolution of a rock. Compaction has caused the flowage of the ductile shale clasts into surrounding pore space, producing pseudomatrix and simultaneously destroying whatever primary porosity was present in the original deposit.

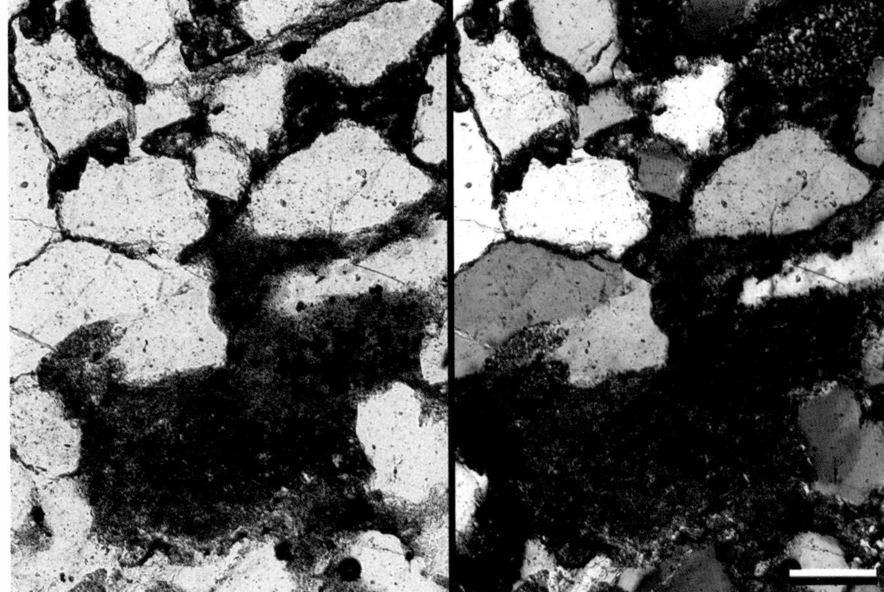

PPL | XPL, Scale bar = 0.10 mm

Up. Jurassic Ile Fm., Norwegian sector, North Sea †

This sample shows varying styles of compaction. The quartz grain at the top right of the image shows brittle fracturing (red arrow), while other quartz grains display point contacts with some indented (pressure dissolved) margins (black arrow). An expanded muscovite mica (center) was compacted in a ductile manner. Grain coating diagenetic clays are chlorite and post date some of the compaction as they terminate at grain contacts. Dissolution of the K-feldspar at right of center is interpreted to postdate compaction because it maintained the now delicate grain shape.

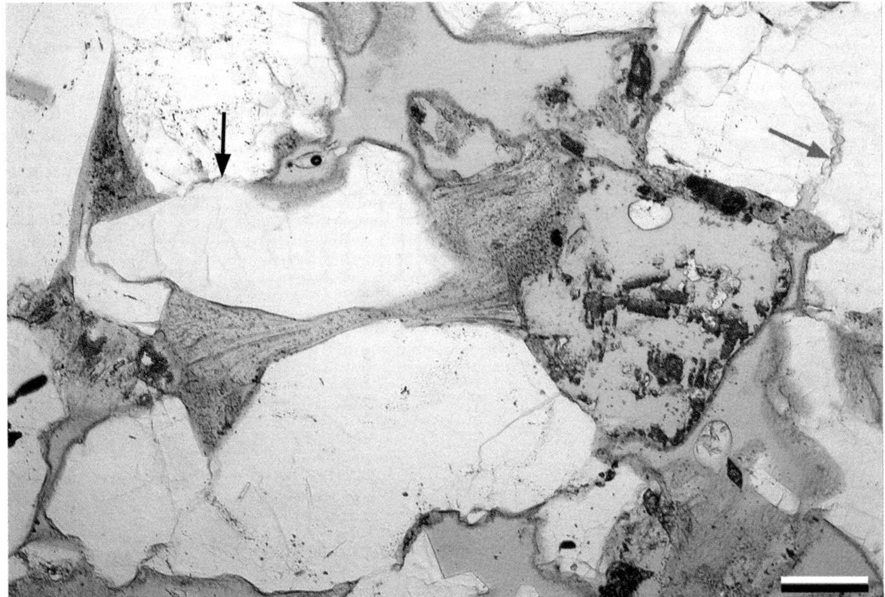

PPL, KFS, BDI, Scale bar = 0.12 mm

Lo. Permian Cutler Fm., Mesa Co., Colorado

Micas are very ductile, and during compaction, they usually bend rather than breaking or shattering. Because the highly foliated phyllosilicates crystals are bent, their birefringence varies as their orientation changes in the thin section. In addition, quartz grains in this sandstone have embayed adjacent micaceous grains.

PPL / XPL, Scale bar = 0.26 mm

Unknown age, formation and location

During mechanical compaction, this biotite (greenish colored grain) was pushed into and folded against a muscovite grain (pink/purple to blue/yellow colors). The muscovite grain deformed, bending around and conforming to the contour of the feldspar grain (note the variations in birefringence within the muscovite as its orientation varies because of the deformation).

XPL, BDI, Scale bar = 0.07 mm

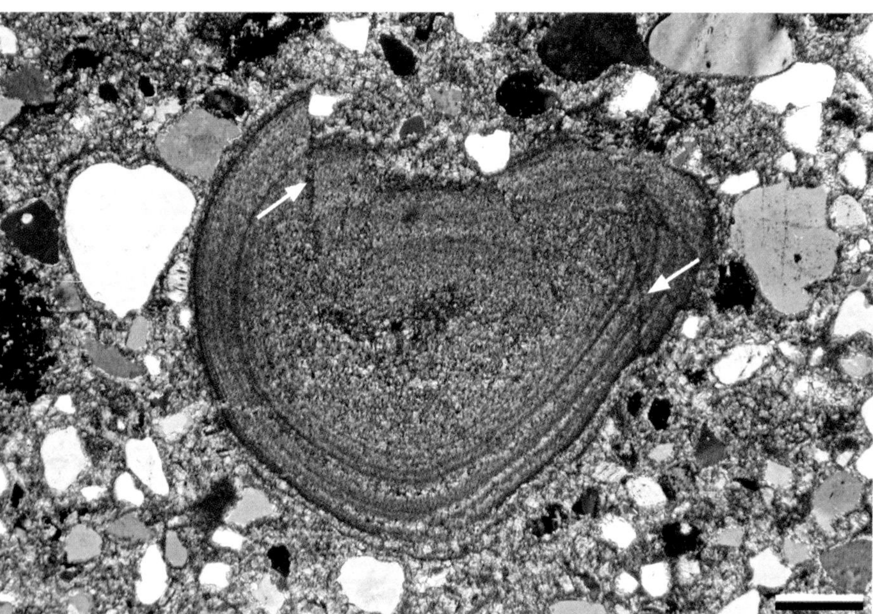

Up. Cambrian Gatesburg Fm., Pennsylvania

This calcitic ooid in a sandy limestone has undergone primarily brittle compactional deformation. The cortices of the ooid were sheared (white arrows), and those pieces have been forced into the core. The rest of this sample may have been partially lithified prior to compaction as the ooid has only partially collapsed. Early dissolution of the original nucleus of this ooid may have facilitated this compaction.

XPL, Scale bar = 0.26 mm

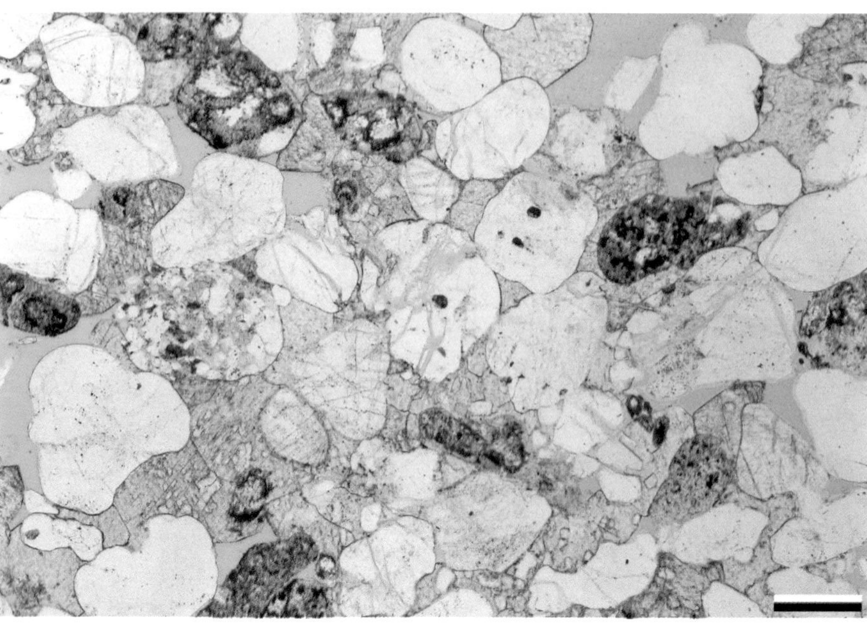

Lo. Permian Enler Gp., Carlane Fm., Co. Tyrone, Northern Ireland, U.K. †

Well-rounded quartz, feldspar and lithic grains in a fluvial sandstone most likely derived from an earlier eolian source. Many of the grains (especially rigid quartz) have been fractured during compaction. Fractures mostly emanate from grain-to-grain contacts. There is non-ferroan calcite cement filling some pores, and this cement also is present within some fractures, indicating that cementation at least partially postdates compaction and fracturing. Nonetheless, it is a sample with a fairly high IGV.

PPL, KFS, AFeS, BDI, Scale bar = 0.32 mm

Up. Cretaceous Mesaverde Gp., McKinley Co., New Mexico

It is common to have both brittle and ductile mechanical deformation in different components within a single rock sample. In this example, the large microcline feldspar grain at center (identified by tartan twinning in cross-polarized light) has been shattered during burial and compaction. These open fractures represent secondary porosity within the rock. The clay-rich grains, on the other hand, have behaved in a ductile manner, compressing and flowing into available pore space, creating pseudomatrix (although some true matrix also is present). Straight grain contacts from grain reorientation also are visible in places.

PPL, BDI, Scale bar = 0.16 mm

Up. Triassic – Lo. Jurassic New Haven Arkose, New Haven Co., Connecticut

During mechanical compaction, brittle grains, such as feldspars, break along zones of weakness (cleavage, compositional lamellae and twin planes). In this example, an orthoclase grain is wedged between polycrystalline quartz grains. The orthoclase has shattered along exsolution lamellae (see examples of perthite in Chapter 2). Quartz also can fail by fracturing, but because it has no cleavage it has more shear strength than feldspars.

XPL, Scale bar = 0.38 mm

Unknown age, formation and location

In this example, a twinned plagioclase has fractured parallel to the twin planes during compaction as the grain was forced against an adjacent quartz grain.

XPL, BDI, Scale bar = 0.07 mm

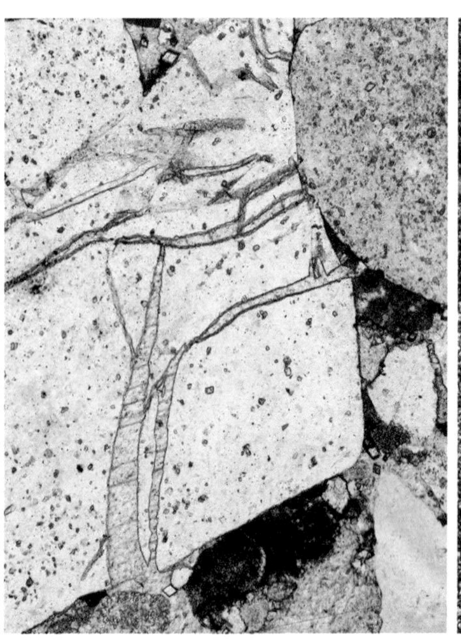

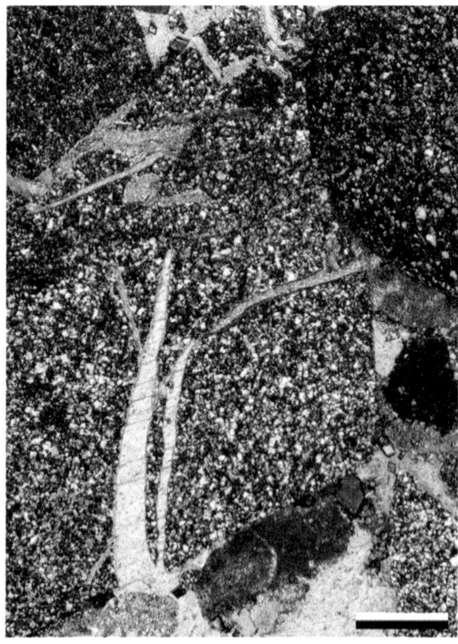

Lo. Permian (Wolfcampian) Hueco Fm., Powwow Conglomerate, Hudspeth Co., Texas

Because of its tightly interlocked microcrystalline fabric, chert usually breaks with a conchoidal fracture, like quartz. In this example, the large colorless chert grain has been compressed against the brownish chert. The rounded, brownish chert acts as a anvil transferring stress to the adjacent grain producing radiating conchoidal fractures in the grain. These fractures were later filled by calcite cement.

PPL | XPL, Scale bar = 0.33 mm

Pennsylvanian Conemaugh Gp., Athens Co., Ohio

Here, a chert grain was compressed between two quartz grains. Two of the grains have behaved brittlely and sheared during compaction. Iron oxides cemented the grains and fractures and thus clearly postdate fracturing.

PPL, BDI, Scale bar = 0.16 mm

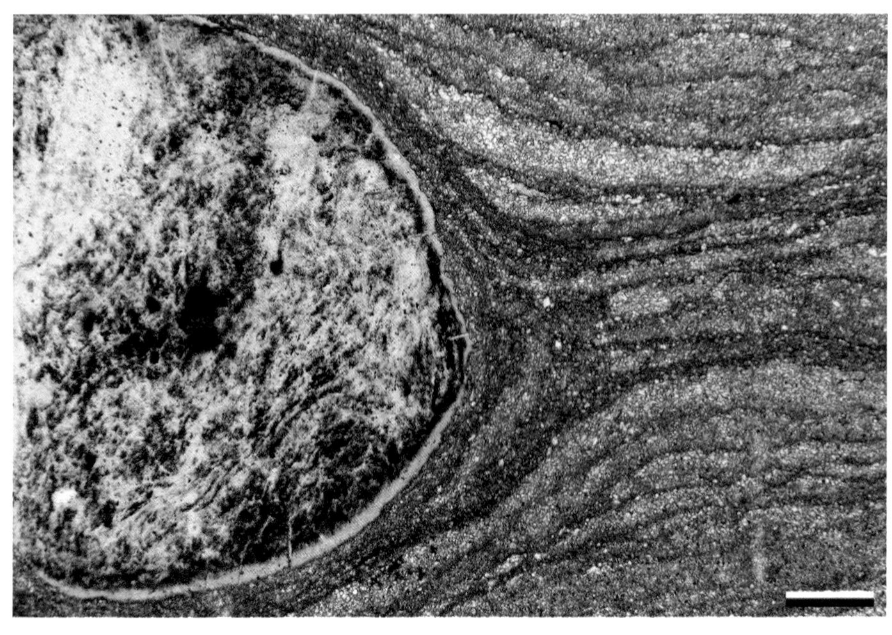

Eocene Green River Fm., Lincoln Co., Wyoming

Physical compaction features are especially common in fine-grained sediments that have escaped pervasive cementation in marine or other shallow-burial settings. With initial seafloor porosities of 70% or more, shales and marls are susceptible to early mechanical dewatering and grain reorientation. This example shows burial-related compactional drape (and possibly also some depositional onlap) of laminated, fine-grained, lacustrine, shaly carbonate muds around a synsedimentary phosphatic nodule. The amount of compaction can be quantified, in some cases, by measuring the deformation around such structures.

PPL, Scale bar = 0.26 mm

Cretaceous Mowry Fm., Moffat Co., Colorado

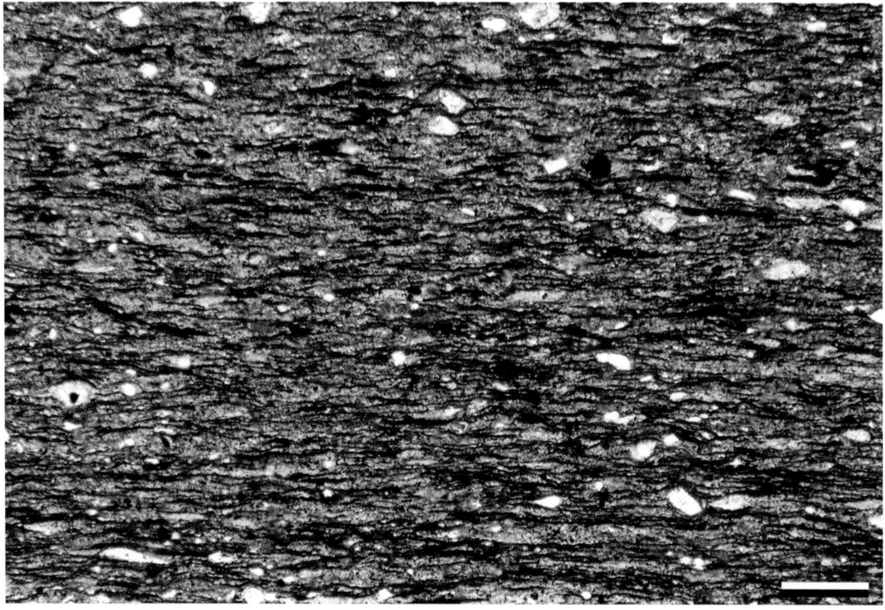

Compaction in these shales and marls is both mechanical and chemical. Mechanical compaction has resulted in bedding-parallel realignment of grains and the development of some fissility through deformation of organic matter. Additional chemical compaction here is manifested by the concentration of insoluble materials (clays, organics and iron oxides in wispy, microstylolitic seams. However, it is difficult in this example to quantify the relative importance of bedding-parallel reorientation of grains versus concentration of insoluble materials by pressure-solution.

PPL, Scale bar = 0.51 mm

Lo. Permian Bone Spring Fm., Reeves Co., Texas

Inhomogeneities within the sediments can influence pressure solution during burial and compaction. This sample contains lens-shaped areas that formed as a result of early cementation of burrows (based on consistent size and shapes) prior to onset of pressure solution. The coarser burrow fills were more resistant to compaction than surrounding material, resulting in the burrows showing less evidence of compaction, whereas surrounding areas contain abundant bedding-parallel lines of clays and organic matter. These materials can be further concentrated by microstylolites or wispy pressure solution seams.

PPL, Scale bar = 0.01 mm

Lo. Ordovician Phycodes Gp., Bavaria, Germany

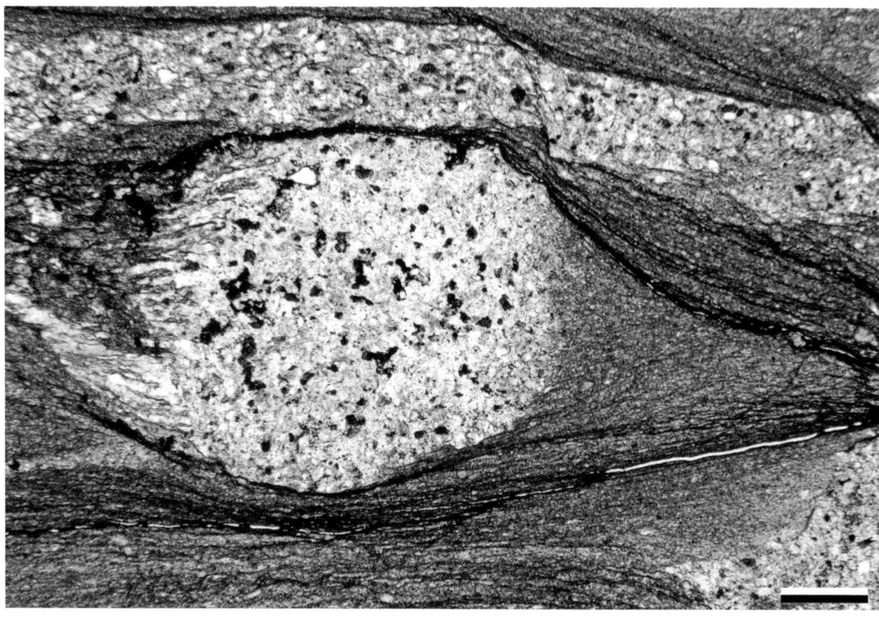

The structures in this siltstone resulted from mechanical and chemical compaction (compactional drape) around an early-cemented, sand-filled burrow. Compaction and shear here involved formation of numerous pressure solution seams (brownish, irregular streaks) in clay- and silt-sized sediments not strongly cemented during early diagenesis. Solution seams are more planar than stylolites and involve less dissolution along any single surface, but they can occur in such numbers that, in aggregate, they accomplish extensive volume loss; these swarms of solution surfaces are sometimes called "horsetail seams".

PPL, Scale bar = 0.51 mm

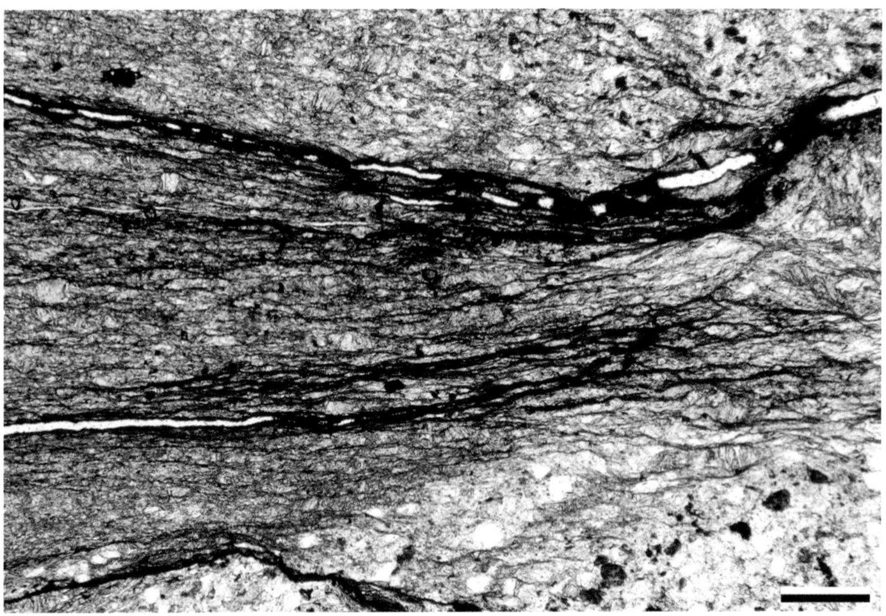

Lo. Ordovician Phycodes Gp., Bavaria, Germany

A closeup view of wispy pressure solution seams or "horsetail seams" shown in the previous example. The concentrated dark material within these seams is a mixture of insoluble materials, including clays, organics and iron oxides and/or sulfides. The swarms of seams encase lenses of more resistant (cemented or coarser-grained) material.

PPL, Scale bar = 0.26 mm

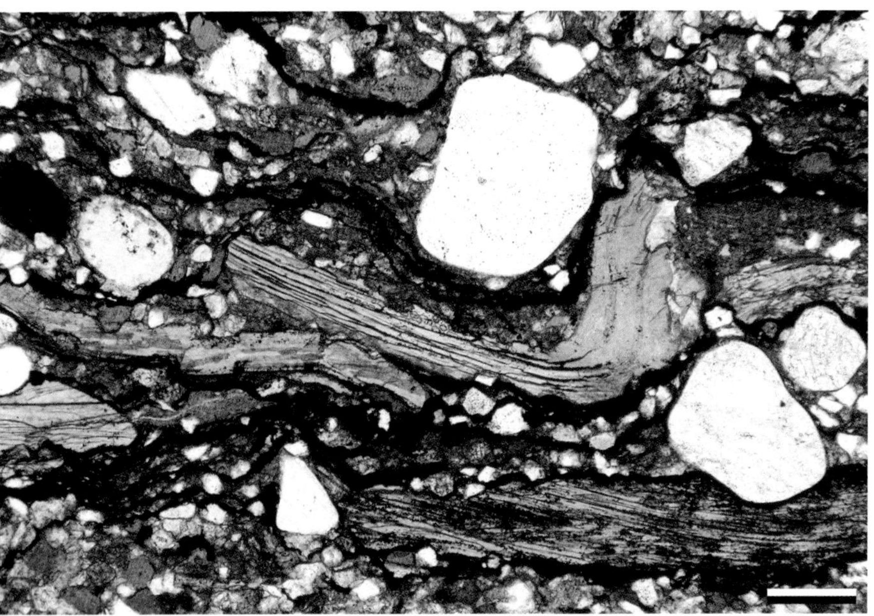

Mid. Jurassic Curtis Fm., Moffat Co., Colorado

Well-developed solution seams within a sandy, glauconitic carbonate shale. Compaction, both mechanical and chemical, has increased packing density and produced numerous wispy pressure solution (microstylolitic) seams. These pressure solution seams consist of residual clays, organics and pyrite. Carbonate bioclasts were fractured and have also been embayed by harder, less soluble grains, especially quartz, during compaction.

PPL, Scale bar = 0.26 mm

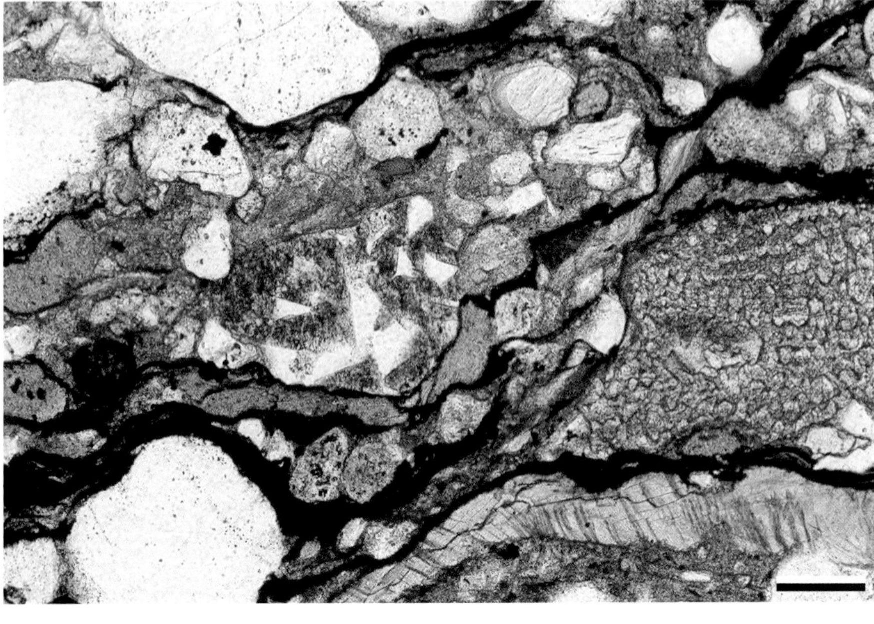

Mid. Jurassic Curtis Fm., Moffat Co., Colorado

A close-up view of the grain boundaries from a different area of the previous sample showing well-developed horsetail or wispy pressure solution seams. Glauconite grains have been squashed between a graphic granite clast (center) and nearby quartz grains and carbonate bioclasts. The spongy grain on the right side of photomicrograph is an echinoid fragment, and the grain below it is a bivalve shell. Both of these carbonate grains were deformed and partially dissolved during compaction. The dissolved materials go into solution and eventually may result in nearby cementation.

PPL, Scale bar = 0.10 mm

Up. Jurassic Fulmar Fm., United Kingdom sector, North Sea †

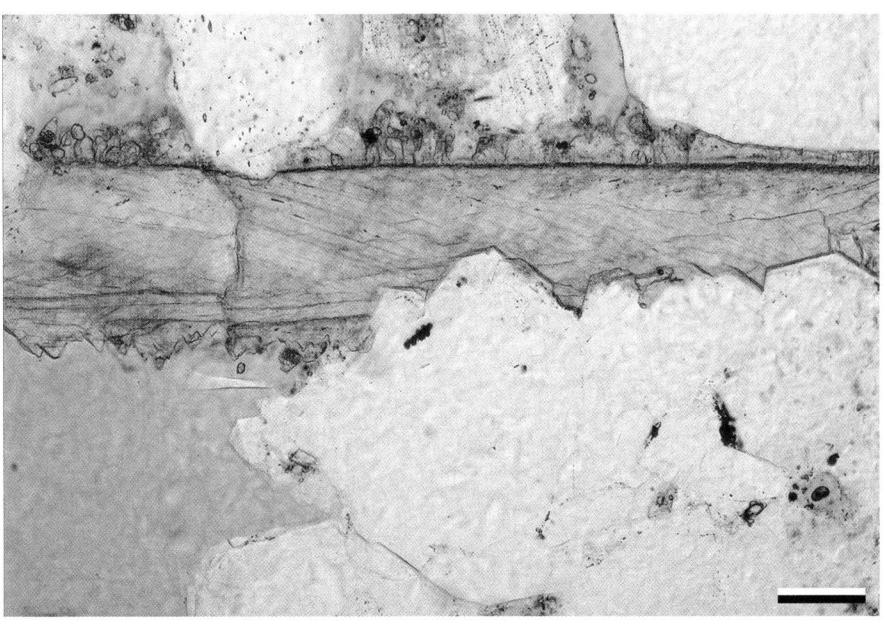

This bivalve shell is fractured and has undergone pressure solution by the surrounding quartz grains. The quartz grains in the bottom half of the image display euhedral overgrowths and so the relative timing of compaction can be determined. The blocky crystals of calcite on the bivalve shell grew first, followed by the quartz overgrowths and finally compaction (or continued compaction) occurred. Alternatively, the quartz overgrowths could have replaced adjacent calcite grains either before or after compaction. However, the absence of calcitic inclusions in the quartz makes that interpretation less likely.

PPL, AFeS, BDI, Scale bar = 0.07 mm

Up. Permian Cutler Fm., Mesa Co., Colorado

During burial, grains realign in response to lithostatic stresses until the grains are tightly packed. Once the grains can no longer move or deform substantially, chemical compaction processes (pressure solution) take over. In this example, a pressure-solution contact (arrow) has developed between two adjacent feldspar grains, forming a serrated or sutured grain boundary. The boundary marked by insoluble materials that may be remnants of clay matrix or clay coatings trapped between the adjacent grains.

PPL, Scale bar = 0.10 mm

Silurian Clinton Ironstone, Giles Co., Virginia

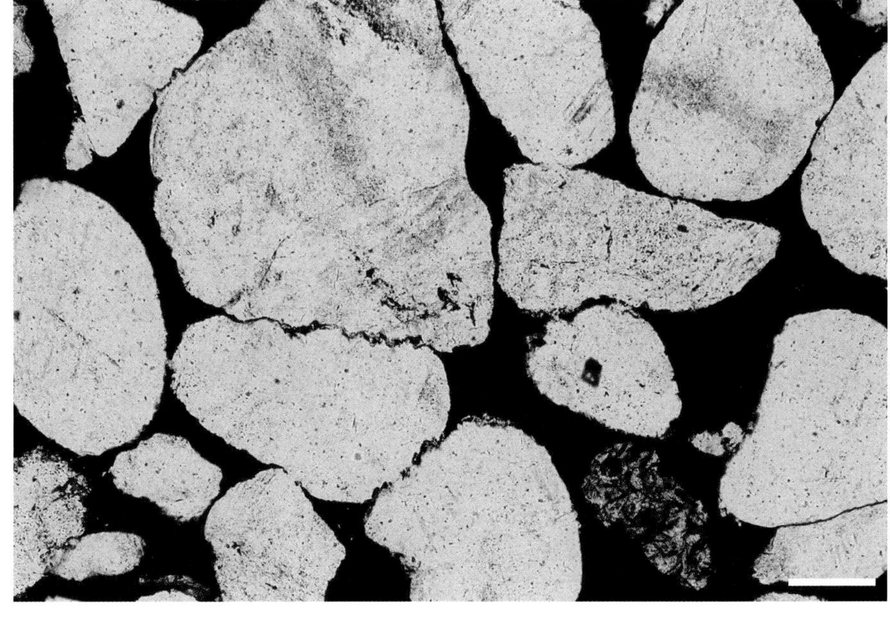

An example of intergranular pressure solution. During burial compaction, grains in contact with each other can undergo grain dissolution and suturing, as in the quartz grains in the center of the photomicrograph. Note the jagged boundaries and the insoluble materials trapped along the intergranular pressure-solution contacts—the clays in those contact areas may play a significant role in the dissolution (Bjørkum, 1996). The rock was later cemented by iron oxides that probably stained some of the materials in the intergranular sutures making those features more prominent than usual.

PPL, Scale bar = 0.10 mm

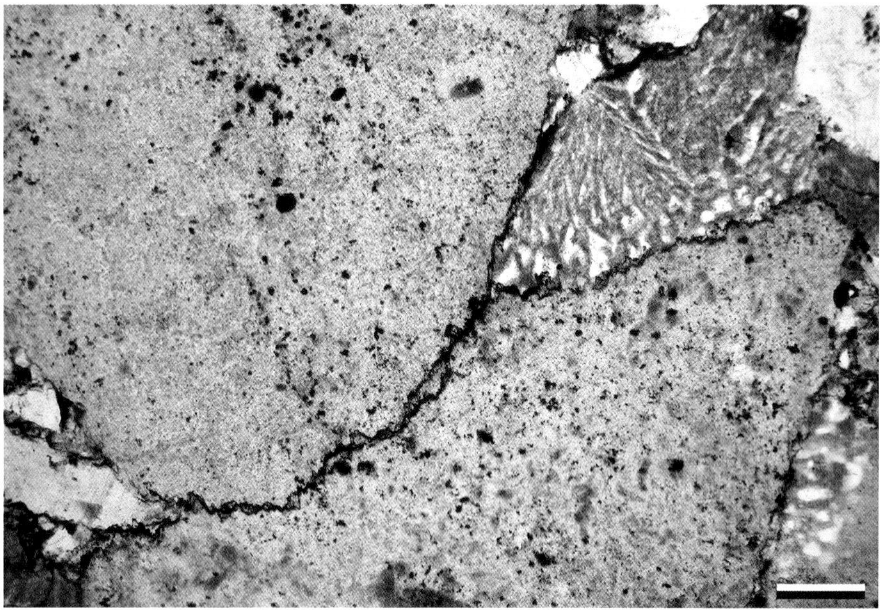

Up. Pennsylvanian – Lo. Permian Granite Wash, southern Oklahoma

This lithic sandstone is highly compacted and intergranular microstylolites are preser between most of the grains. In this photomi crograph, chert and plutonic rock fragment (including graphic granite) have low-relie sutured boundaries where the grains hav dissolved into each other leaving behind a zon of insoluble residues. In general, styloliti sutures in siliciclastic sediments have les relief than in limestones. This is an exceller example of a rock dominated by compac tion-related rather than cementation-relate porosity destruction (see Houseknecht, 1987)

PPL, Scale bar = 0.09 mm

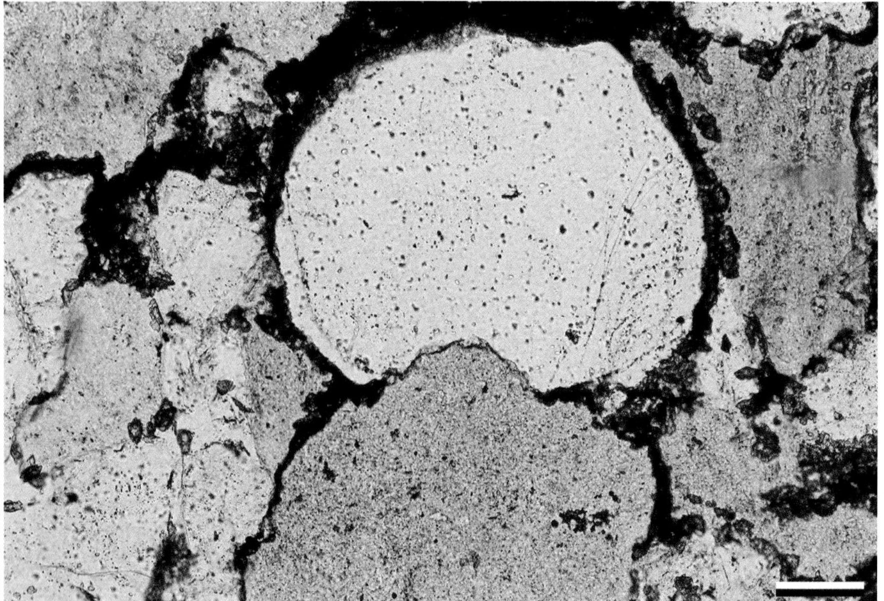

Triassic Ivishak Fm., North Slope, Alaska

In this photomicrograph, a probable second generation quartz grain (note the abrade quartz overgrowths) has been solution penetrated by a chert clast during compactior The quartz grain here has, somewha unusually, undergone preferential dissolutio relative to the chert grain. The reason that thi chert is apparently less soluble than the quart may have something to do with anisotropy o defects within the quartz crystal lattice. I addition, some crystal faces (planes) are mor susceptible to pressure solution than other (e.g., Hicks et al., 1986).

PPL, BDI, Scale bar = 0.10 mm

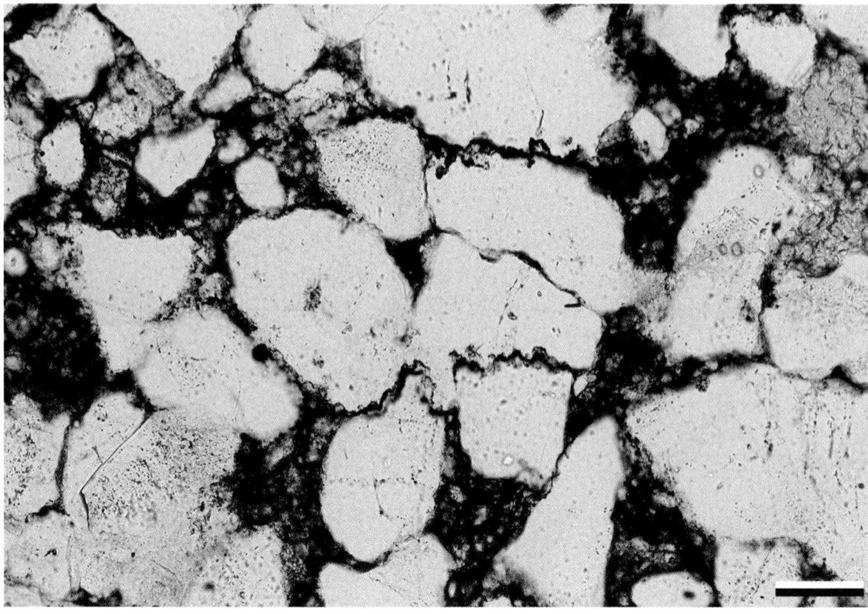

Unknown age and formation, Moffat Co., Colorado

This sandstone is highly compacted wit stylolitic boundaries between many of th quartz grains. Clay-rich siltstone clasts hav been completely crushed to fill the primar pores and kaolinite cements fill most of th remnant intergranular porosity. Within th solution seams, clay and silt are trappe between the grains. These microstylolites, i places, have 0.05 mm or more relief on them representing at least that amount of dissolutio between the quartz grains. Such volume los can provide a source of silica for diageneti precipitates in nearby strata.

PPL, RDI, Scale bar = 0.10 mm

Lo. Cretaceous Brisi Sandstone, Bavaria, Germany

A well-developed stylolitic dissolution surface in a sandstone. The dark zone extending across the slide is the stylolitic seam that contains insoluble residues, such as clays and organic matter. It cross-cuts numerous detrital grains and the amplitude of the stylolite provides a minimum estimate of material removal (in this case at least 2 mm). Stylolites typically are more widely spaced in clean sandstones and more closely spaced in calcareous or argillaceous sandstones (Heald, 1955), and the localized presence of detrital or authigenic clays may accelerate stylolite formation (Heald, 1959; Renard et al., 1997).

PPL, Scale bar = 0.51 mm

Up. Cretaceous Mesaverde Gp., Piceance Creek Basin, Colorado

This pressure solution surface may have originally started out as a clay-rich lamina, but it has further concentrated clays and organic matter during dissolution. Small, unfilled fractures are visible within the stylolite. They are formed during load release that was either related to natural conditions during uplift or artificial conditions during core recovery or thin section preparation.

PPL, BDI, Scale bar = 0.10 mm

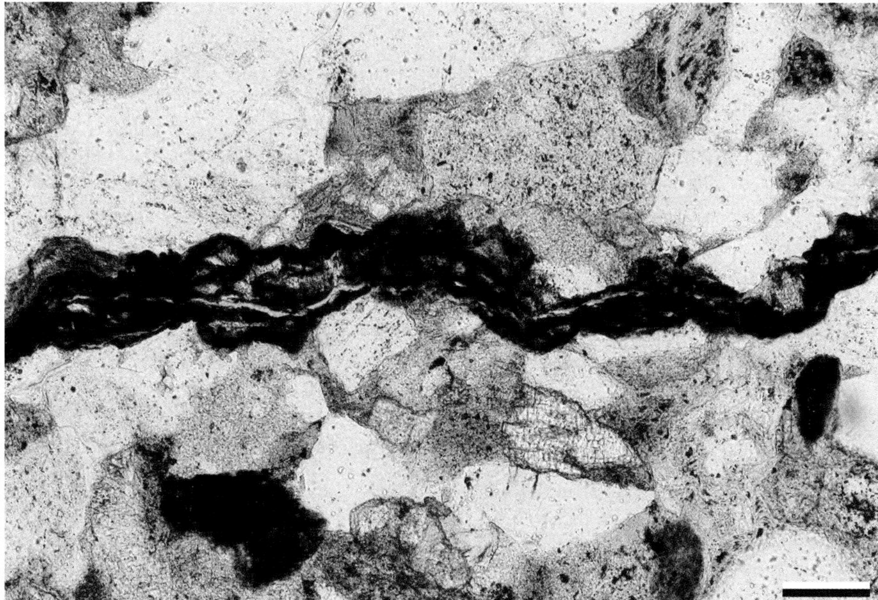

Up. Cretaceous Mesaverde Gp., Piceance Basin, Colorado

In this sample, many of the grain boundaries are marked by microstylolites and squashed clay-rich grains. The clay minerals have been embayed and forced into surrounding pore spaces by harder quartz and feldspar grains. In the areas with fewer clay minerals, secondary porosity also has developed by both the fracturing of grains and leaching of feldspars. The initial presence of clay minerals in this rock may have accelerated formation of microstylolites, as noted earlier.

PPL, BDI, Scale bar = 0.16 mm

Lo. Cambrian Unicoi Fm., Shenandoah Valley, Virginia

A highly-compacted sandstone with numerous pressure solution seams surrounding grains throughout the photomicrograph. The shale clasts have been crushed and show flowage into surrounding open pores, with some of the clays forming insoluble residues within the stylolite seams. In addition, the quartz grains have undulose extinction that may, in part, relate to compactional stresses. This is an excellent example of a rock in which almost all porosity loss resulted from compaction and almost none was from cementation (see intergranular volume discussions in Houseknecht, 1987 and Paxton, et al., 2002).

XPL, Scale bar = 0.26 mm

Mid. Jurassic Entrada Fm., Moffat Co., Colorado

A well-cemented and compacted quartz sandstone with numerous stylolitic contacts between quartz grains. In most cases, sandstones with this type of texture originally had grain-coating clays (commonly detrital, infiltrated clays that may appear similar to clay cements). Those clays inhibited overgrowth nucleation and promoted intergranular pressure solution. The clays recrystallized to illite during deep burial. This is another example of a rock that has effectively no porosity and in which virtually all porosity loss was the result of compaction and essentially none was related to cementation.

XPL, Scale bar = 0.10 mm

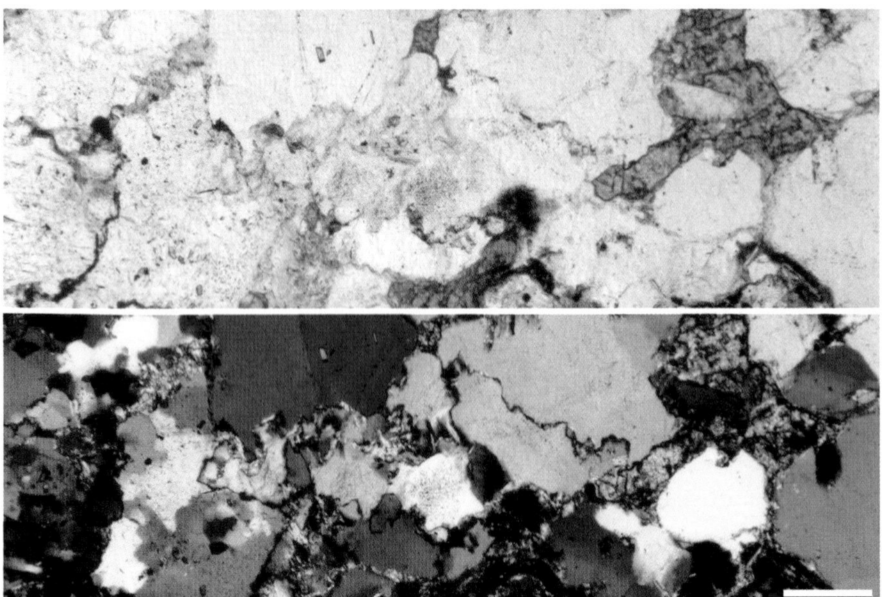

Lo. Carboniferous Bundoran Shale Fm., Dowra Sandstone Mbr., Co. Fermanagh, Northern Ireland, U.K.†

Stylolites within a strongly compacted sandstone. The primary pores were lost to compaction, and secondary pores were filled by ferroan and nonferroan calcite cements with local ankerite replacement. These stylolites have little to distinguish them in plane-polarized light, where there is no concentration of insoluble residue. They are readily seen in cross-polarized light, however. The sample is from 1,040 m (3,412 ft) depth, but this level of compaction suggests that the rock was once buried to a greater depth.

PPL / XPL, AFeS, BDI, Scale bar = 0.17 mm

Cited References and Additional Information Sources

Ajdukiewicz, J. M., and R. H. Lander, 2010, Sandstone reservoir quality prediction: The state of the art: AAPG Bulletin, v. 94, p. 1083-1091, doi: 10.1306/intro060110.

Aplin, A. C., I. F. Matenaar, D. K. McCarty, and B. A. van Der Pluijm, 2006, Influence of mechanical compaction and clay mineral diagenesis on the microfabric and pore-scale properties of deep-water Gulf of Mexico mudstones: Clays and Clay Minerals, v. 54, p. 500-514, doi: 10.1346/CCMN.2006.0540411.

Aydin, A., and R. Ahmadov, 2009, Bed-parallel compaction bands in aeolian sandstone: Their identification, characterization and implications: Tectonophysics, v. 479, p. 277-284, doi: 10.1016/j.tecto.2009.08.033.

Baldwin, B., 1971, Ways of deciphering compacted sediments: Journal of Sedimentary Research, v. 41, p. 293-301, doi: 10.1306/74D7224D-2B21-11D7-8648000102C1865D.

Baldwin, B., and C. O. Butler, 1985, Compaction curves: AAPG Bulletin, v. 69, p. 622-626.

Bjørkum, P. A., 1996, How important is pressure in causing dissolution of quartz in sandstones?: Journal of Sedimentary Research, v. 66, p. 147-154, doi: 10.1306/D42682DE-2B26-11D7-8648000102C1865D.

Bjørkum, P. A., E. H. Oelkers, P. H. Nadeau, O. Walderhaug, and W. M. Murphy, 1998, Porosity prediction in quartzose sandstones as a function of time, temperature, depth, stylolite frequency, and hydrocarbon saturation: AAPG Bulletin, v. 82, p. 637-648.

Bjørlykke, K., 1988, Sandstone diagenesis in relation to preservation, destruction and creation of porosity, *in* G. V. Chilingarian, and K. H. Wolf, eds., Diagenesis: New York, Elsevier Science Publ., p. 555-588, doi: 10.1016/S0070-4571(08)70180-8.

Bjørlykke, K., M. Ramm, and G. C. Saigal, 1989, Sandstone diagenesis and porosity modification during basin evolution: Geologische Rundschau, v. 78, p. 243-268, doi: 10.1007/BF01988363.

Bloch, S., R. H. Lander, and L. Bonnell, 2002, Anomalously high porosity and permeability in deeply buried sandstone reservoirs: Origin and predictability: AAPG Bulletin, v. 86, p. 301-325, doi: 10.1306/61EEDABC-173E-11D7-8645000102C1865D.

Chuhan, F. A., A. Kjeldstad, K. Bjørlykke, and K. Høeg, 2002, Porosity loss in sand by grain crushing—Experimental evidence and relevance to reservoir quality: Marine and Petroleum Geology, v. 19, p. 39-53, doi: 10.1016/S0264-8172(01)00049-6.

Dewers, T., and P. Ortoleva, 1990, A coupled reaction/transport/mechanical model for intergranular pressure solution, stylolites, and differential compaction and cementation in clean sandstones: Geochimica et Cosmochimica Acta, v. 54, p. 1609-1625, doi: 10.1016/0016-7037(90)90395-2.

Dewers, T., and P. Ortoleva, 1991, Influences of clay minerals on sandstone cementation and pressure solution: Geology, v. 19, p. 1045-1048, doi: 10.1130/0091-7613(1991)019<1045:IOCMOS>2.3.CO;2.

Dickinson, W. W., and K. L. Milliken, 1995, The diagenetic role of brittle deformation in compaction and pressure solution, Etjo sandstone, Namibia: Journal of Geology, v. 103, p. 339-347, doi: 10.1086/629751.

Dutta, T., G. Mavko, T. Mukerji, and T. Lane, 2009, Compaction trends for shale and clean sandstone in shallow sediments, Gulf of Mexico: The Leading Edge, v. 28, p. 590-596, doi: 10.1190/1.3124935.

Dutton, S. P., 1997, Timing of compaction and quartz cementation from integrated petrographic and burial-history analyses, Lower Cretaceous Fall River Formation, Wyoming and South Dakota: Journal of Sedimentary Research, v. 67, p. 186-196, doi: 10.1306/D426852C-2B26-11D7-8648000102C1865D.

Dzevanshir, R. D., L. A. Buryakovskiy, and G. V. Chilingarian, 1986, Simple quantitative evaluation of porosity of argillaceous sediments at various depths of burial: Sedimentary Geology, v. 46, p. 169-175, doi: 10.1016/0037-0738(86)90057-6.

Ehrenberg, S. N., 1989, Assessing the relative importance of compaction processes and cementation to reduction of porosity in sandstones: Discussion: AAPG Bulletin, v. 73, p. 1274-1276.

Ehrenberg, S. N., 1995, Measuring sandstone compaction from modal analyses of thin sections: How to do it and what the results mean: Journal of Sedimentary Research, v. 65, p. 369-379, doi: 10.1306/D42680C7-2B26-11D7-8648000102C1865D.

Ehrenberg, S. N., and P. H. Nadeau, 2005, Sandstone vs. carbonate petroleum reservoirs: A global perspective on porosity-depth and porosity-permeability relationships: AAPG Bulletin, v. 89, p. 435-445, doi: 10.1306/11230404071.

Fisher, Q. J., R. J. Knipe, M. Casey, and M. B. Clennell, 1999, Mechanical compaction of deeply buried sandstones of the North Sea: Marine and Petroleum Geology, v. 16, p. 605-618, doi: 10.1016/S0264-8172(99)00044-6.

Heald, M. T., 1955, Stylolites in sandstones: Journal of Geology, v. 63, p. 101-114, doi: 10.1086/626237.

Heald, M. T., 1959, Significance of stylolites in permeable sandstones: Journal of Sedimentary Research, v. 29, p. 251-253, doi: 10.1306/74D708F3-2B21-11D7-8648000102C1865D.

Hicks, B. D., K. R. Applin, and D. W. Houseknecht, 1986, Crystallographic influences on intergranular pressure solution in a quartzose sandstone: Journal of Sedimentary Research, v. 56, p. 784-787, doi: 10.1306/212F8A49-2B24-11D7-8648000102C1865D.

Houseknecht, D. W., 1984, Influence of grain size and temperature on intergranular pressure solution, quartz cementation, and porosity in a quartzose sandstone: Journal of Sedimentary Research, v. 54, p. 348-361, doi: 10.1306/212F8418-2B24-11D7-8648000102C1865D.

Houseknecht, D. W., 1987, Assessing the relative importance of compaction processes and cementation to reduction of porosity in sandstones: AAPG Bulletin, v. 71, p. 633-642.

Houseknecht, D. W., 1988, Intergranular pressure solution in four quartzose sandstones: Journal of Sedimentary Research, v. 58, p. 228-246, doi: 10.1306/212F8D64-2B24-11D7-8648000102C1865D.

Houseknecht, D. W., and L. A. Hathon, 1987, Petrographic constraints on models of intergranular pressure solution in quartzose sandstones: Applied Geochemistry, v. 2, p. 507-521, doi: 10.1306/212F8D64-2B24-11D7-8648000102C1865D.

Kristiansen, K., M. Valtiner, G. W. Greene, J. R. Boles, and J. N. Israelachvili, 2011, Pressure solution–The importance of the electrochemical surface potentials: Geochimica et Cosmochimica Acta, v. 75, p. 6882-6892, doi: 10.1016/j.gca.2011.09.019.

Lander, R. H., and O. Walderhaug, 1999, Predicting porosity through simulating sandstone compaction and quartz cementation: AAPG Bulletin, v. 83, p. 433-449.

Lehner, F., and Y. Leroy, 2004, Sandstone compaction by intergranular pressure solution, *in* Y. Guéguen, and M. Boutéca, eds., Mechanics of Fluid-saturated Rocks: Amsterdam, Elsevier-Academic Press, p. 115-170, doi: 10.1016/S0074-6142(03)80019-0.

Lundegard, P. D., 1992, Sandstone porosity loss—a "big picture" view of the importance of compaction: Journal of Sedimentary Research, v. 62, p. 250-260, doi: 10.1306/D42678D4-2B26-11D7-8648000102C1865D.

Makowitz, A., and K. L. Milliken, 2003, Quantification of brittle deformation in burial compaction, Frio and Mount Simon Formation sandstones: Journal of Sedimentary Research, v. 73, p. 1007-1021, doi: 10.1306/051003731007.

Mitra, S., and W. C. Beard, 1980, Theoretical models of porosity reduction by pressure solution for well-sorted sandstones: Journal of Sedimentary Research, v. 50, p. 1347-1360, doi: 10.1306/212F7BEE-2B24-11D7-8648000102C1865D.

Mørk, M. B. E., and K. Moen, 2007, Compaction microstructures in quartz grains and quartz cement in deeply buried reservoir sandstones using combined petrography and EBSD analysis: Journal of Structural Geology, v. 29, p. 1843-1854, doi: 10.1016/j.jsg.2007.08.004.

Nagtegaal, P. J. C., 1978, Sandstone-framework instability as a function of burial diagenesis: Journal of the Geological Society, v. 135, p. 101-105, doi: 10.1144/gsjgs.135.1.0101.

Oelkers, E. H., P. A. Bjørkum, and W. M. Murphy, 1996, A petrographic and computational investigation of quartz cementation and porosity reduction in North Sea sandstones: American Journal of Science, v. 296, p. 420-452, doi: 10.2475/ajs.296.4.420.

Olsson, W. A., D. J. Holcomb, and J. W. Rudnicki, 2002, Compaction localization in porous sandstone: Implications for reservoir mechanics: Oil & Gas Science and Technology – Revue d'IFP Energies nouvelles, v. 57, p. 591-599, doi: 10.2516/ogst:2002040.

Osborne, M. J., and R. E. Swarbrick, 1999, Diagenesis in North Sea HPHT clastic reservoirs—consequences for porosity and overpressure prediction: Marine and Petroleum Geology, v. 16, p. 337-353, doi: 10.1016/S0264-8172(98)00043-9.

Paxton, S. T., J. O. Szabo, J. M. Ajdukiewicz, and R. E. Klimentidis, 2002, Construction of an intergranular volume compaction curve for evaluating and predicting compaction and porosity loss in rigid-grain sandstone reservoirs: AAPG Bulletin, v. 86, p. 2047-2067, doi: 10.1306/61EEDDFA-173E-11D7-8645000102C1865D.

Peltonen, C., Ø. Marcussen, K. Bjørlykke, and J. Jahren, 2008, Mineralogical control on mudstone compaction: A study of Late Cretaceous to Early Tertiary mudstones of the Vøring and Møre basins, Norwegian Sea: Petroleum Geoscience, v. 14, p. 127-138, doi: 10.1144/1354-079308-758.

Pittman, E. D., and R. E. Larese, 1991, Compaction of lithic sands: Experimental results and applications: AAPG Bulletin, v. 75, p. 1279-1299.

Ramm, M., and K. Bjørlykke, 1994, Porosity/depth trends in reservoir sandstones: assessing the quantitative effects of varying pore-pressure, temperature history and mineralogy, Norwegian Shelf data: Clay Minerals, v. 29, p. 475-490, doi: 10.1180/claymin.1994.029.4.07.

Renard, F., P. Ortoleva, and J.-P. Gratier, 1997, Pressure solution in sandstones: Influence of clays and dependence on temperature and stress: Tectonophysics, v. 280, p. 257-266, doi: 10.1016/S0040-1951(97)00039-5.

Renard, F., A. Park, P. Ortoleva, and J.-P. Gratier, 1999, An integrated model for transitional pressure solution in sandstones: Tectonophysics, v. 312, p. 97-115, doi: 10.1016/S0040-1951(99)00202-4.

Rittenhouse, G., 1971a, Pore-space reduction by solution and cementation: AAPG Bulletin, v. 55, p. 80-91.

Rittenhouse, G., 1971b, Mechanical compaction of sands containing different percentages of ductile grains: A theoretical approach: AAPG Bulletin, v. 55, p. 92-96.

Robinson, A., and J. Gluyas, 1992, Model calculations of loss of porosity in sandstones as a result of compaction and quartz cementation: Marine and Petroleum Geology, v. 9, p. 319-323, doi: 10.1016/0264-8172(92)90080-X.

Sathar, S., R. H. Worden, D. R. Faulkner, and P. C. Smalley, 2012, The effect of oil saturation on the mechanism of compaction in granular materials: Higher oil saturations lead to more grain fracturing and less pressure solution: Journal of Sedimentary Research, v. 82, p. 571-584, doi: 10.2110/jsr.2012.44.

Schmoker, J. W., and R. B. Halley, 1982, Carbonate porosity versus depth: a predictable relation for south Florida: AAPG Bulletin, v. 66, p. 2561-2570.

Schultz, R. A., C. H. Okubo, and H. Fossen, 2010, Porosity and grain size controls on compaction band formation in Jurassic Navajo Sandstone: Geophysical Research Letters, v. 37, L22306, doi: 10.1029/2010GL044909.

Sclater, J. G., and P. A. F. Christie, 1980, Continental stretching: An explanation of the post-mid-Cretaceous subsidence of the central North Sea Basin: Journal of Geophysical Research, v. 85, p. 3711-3739, doi: 10.1029/JB085iB07p03711.

Sheldon, H. A., J. Wheeler, R. H. Worden, and M. J. Cheadle, 2003, An analysis of the roles of stress, temperature, and pH in chemical compaction of sandstones: Journal of Sedimentary Research, v. 73, p. 64-71, doi: 10.1306/070802730064.

Sibley, D. F., and H. Blatt, 1976, Intergranular pressure solution and cementation of the Tuscarora Quartzite: Journal of Sedimentary Research, v. 46, p. 881-896, doi: 10.1306/212F7081-2B24-11D7-8648000102C1865D.

Sloss, L. L., and D. E. Feray, 1948, Microstylolites in sandstone: Journal of Sedimentary Research, v. 18, p. 3-13, doi: 10.1306/D42692D8-2B26-11D7-8648000102C1865D.

Souque, C., Q. J. Fisher, M. Casey, and P. Bentham, 2010, Structural controls on mechanical compaction within sandstones: An example from the Apsheron Peninsula, Azerbaijan: Marine and Petroleum Geology, v. 27, p. 1713-1724, doi: 10.1016/j.marpetgeo.2010.06.010.

Stone, W. N., and R. Siever, 1996, Quantifying compaction, pressure solution and quartz cementation in moderately-and deeply-buried quartzose sandstones from the Greater Green River Basin, Wyoming, *in* L. J. Crossey, R. Loucks, and M. W. Totten, eds., Siliciclastic Diagenesis and Fluid Flow: Concepts and Applications: Tulsa, OK, SEPM Special Publication 55, p. 129-150, doi: 10.2110/pec.96.55.0129.

Swarbrick, R. E., and M. J. Osborne, 1998, Mechanisms that generate abnormal pressures: An overview, *in* B. E. Law, G. F. Ulmishek, and V. I. Slavin, eds., Abnormal Pressures in Hydrocarbon Environments: Tulsa, OK, AAPG Memoir 70, p. 13-34.

Tada, R., R. Maliva, and R. Siever, 1987, A new mechanism for pressure solution in porous quartzose sandstone: Geochimica et Cosmochimica Acta, v. 51, p. 2295-2301, doi: 10.1016/0016-7037(87)90282-1.

Tada, R., and R. Siever, 1989, Pressure solution during diagenesis: Annual Review of Earth and Planetary Sciences, v. 17, p. 89-118 doi: 10.1146/annurev.ea.17.050189.000513.

Taylor, J. M., 1950, Pore-space reduction in sandstones: AAPG Bulletin, v. 34, p. 701-716.

Taylor, T. R., M. R. Giles, L. A. Hathon, T. N. Diggs, N. R. Braunsdorf, G. V. Birbiglia, M. G. Kittridge, C. I. Macaulay, and I. S. Espejo, 2010, Sandstone diagenesis and reservoir quality prediction: Models, myths, and reality: AAPG Bulletin, v. 94, p. 1093-1132, doi: 10.1306/04211009123.

Trurnit, P., 1968, Pressure solution phenomena in detrital rocks: Sedimentary Geology, v. 2, p. 89-114, doi: 10.1016/0037-0738(68)90030-4.

Walderhaug, O., and P. A. Bjørkum, 2003, The effect of stylolite spacing on quartz cementation in the Lower Jurassic Stø Formation, southern Barents Sea: Journal of Sedimentary Research, v. 73, p. 146-156, doi: 10.1306/090502730146.

Wilson, J. C., and E. F. McBride, 1988, Compaction and porosity evolution of Pliocene sandstones, Ventura Basin, California: AAPG Bulletin, v. 72, p. 664-681.

Wolf, K. H., and G. V. Chilingarian, 1976, Diagenesis of sandstones and compaction, *in* G. V. Chilingarian, and K. H. Wolf, eds., Compaction of Coarse-Grained Sediments, II (Developments in Sedimentology 18B): New York, Elsevier Scientific Publishing, p. 69-444, doi: 10.1016/S0070-4571(08)71100-2.

Wong, Teng-Fong, 1990, Mechanical compaction and the brittle–ductile transition in porous sandstones, *in* R. J. Knipe, and E. H. Rutter, eds., Deformation Mechanisms, Rheology and Tectonics: London, GSL Special Publication 54, p. 111-122, doi: 10.1144/GSL.SP.1990.054.01.12.

Worden, R. H., M. Mayall, and I. J. Evans, 2000, The effect of ductile-lithic sand grains and quartz cement on porosity and permeability in Oligocene and lower Miocene clastics, South China Sea: Prediction of reservoir quality: AAPG Bulletin, v. 84, p. 345-359, doi: 10.1306/C9EBCDE7-1735-11D7-8645000102C1865D.

Yasuhara, H., D. Elsworth, and A. Polak, 2004, Compaction and diagenesis of sandstones—the role of pressure solution, *in* O. Stephanson, ed., Coupled Thermo-Hydro-Mechanical-Chemical Processes in Geo-Systems Fundamentals—Modelling, Experiments and Applications: New York, Elsevier Geo-Engineering Book Series 2, p. 733-738, doi: 10.1016/S1571-9960(04)80126-5.

Facing Page: Top – Chalcedony and chert fracture-fill cements, Eocene Inglis Fm., Marion Co., Florida (PPL). Bottom – Euhedral overgrowths on inclusion-rich quartz, Oligocene – Miocene Fontainebleau Sandstone, Paris Basin, France (PPL | XPL).

DIAGENESIS: INTRODUCTION AND QUARTZ & SILICA CEMENTS

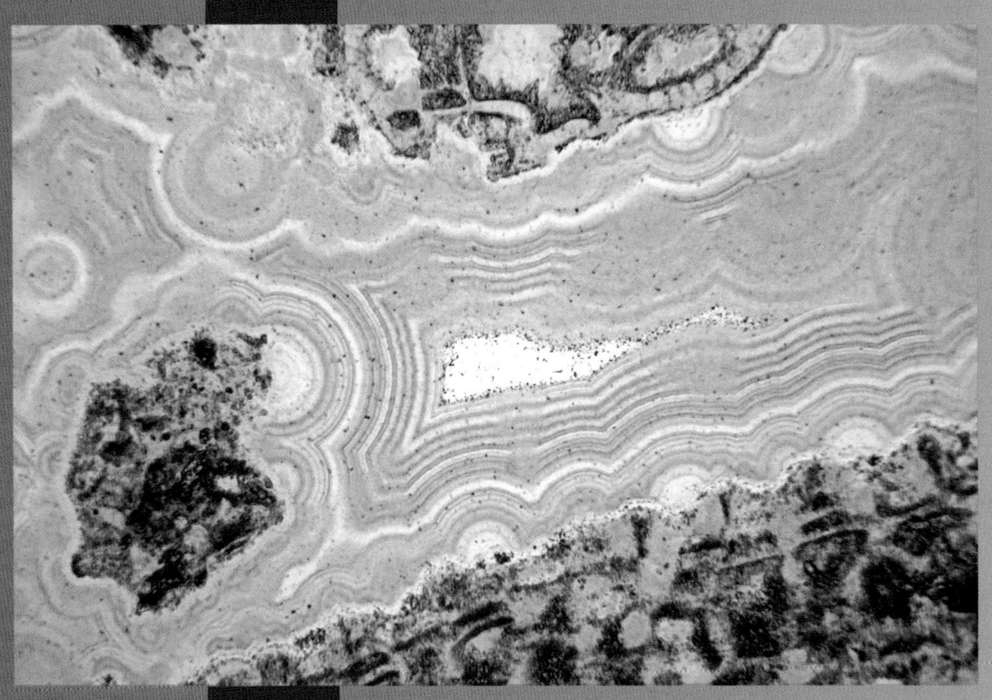

CHAPTER 11

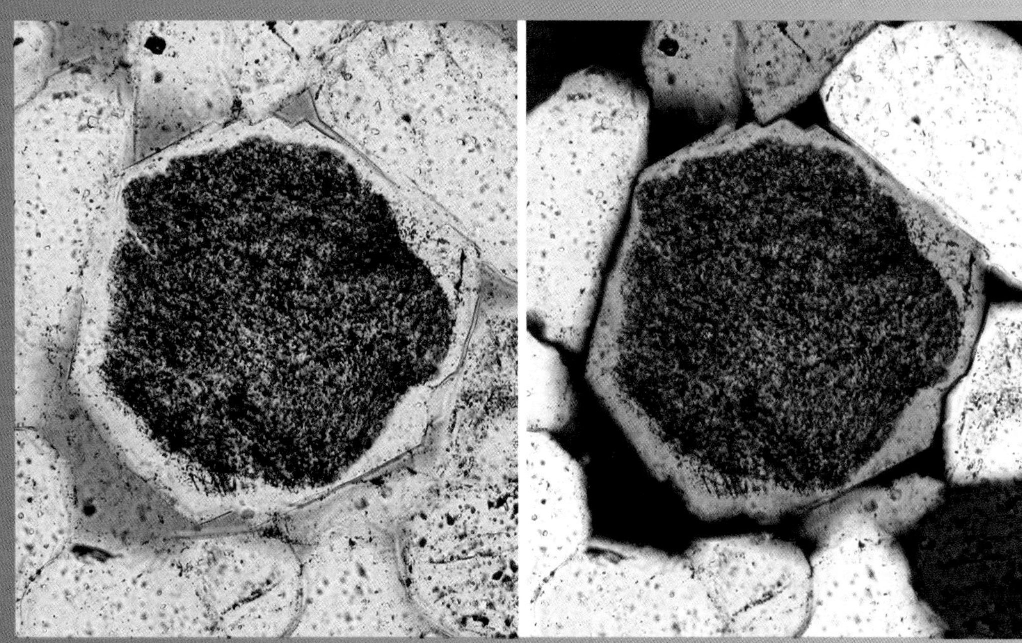

INTRODUCTION TO CEMENTATION

Cementation, the authigenic precipitation of minerals in pore spaces within rocks, is one of the most important processes in the lithification of clastic terrigenous deposits (we also include displacive, authigenic mineral precipitates within this general term). Cements can have a wide range of crystal sizes (terminology shown in Table 11.1) and fabrics. They can form throughout the history of sedimentary deposits, starting with surficial (eogenetic) processes in marine and nonmarine settings and continuing through all stages of burial (mesogenetic) diagenesis as well as uplift-associated (telogenetic) diagenesis. Hundreds of different minerals are found as cements in the panoply of different sandstones and mudrocks, and some of the more common ones are shown in Table 11.2. However, most typical sandstones and mudrocks contain perhaps one to five cementing minerals, making identification far less complex than it might appear from the table.

All cements form by precipitation of materials from aqueous solution, and variations in subsurface fluid temperatures, pressures and chemistries (pH, salinity, specific ionic abundances, etc.) are the major controls on which minerals are precipitated or dissolved. Solutes can be derived from many sources. Some may come directly from seawater or via influx of meteoric waters; others may come from reflux of evaporitic brines. Additional solutes come from circulating basinal fluids, through chemical dissolution of soluble minerals, pressure solution along stylolites and solution seams, maturation of organic matter or dehydration of gypsum beds. These subsurface fluids can be moved through the sedimentary section at basinal scales via compactional dewatering or thermal convection, commonly aided by permeability "highways" created by fractures and faults. In all settings, cements may form through local dissolution or alteration of unstable minerals, relatively small-scale diffusive transport and nearby reprecipitation of more stable minerals. In very low-permeability settings, that may be the only viable mechanism for cementation. On the other hand, the near-complete exclusion of water (or water contact with grains), as in some hydrocarbon reservoirs, may inhibit both water movement and cement formation.

It should be noted that cements form in all types of pores: primary as well as secondary (dissolved), and intragranular as well as intergranular. That makes the distinction between cementation and mineral replacement difficult in some cases. Indeed, in some classifications the term "replacement" includes the "solution-fill" process in which a mineral is dissolved and the resulting void space is partially or completely refilled at a later date by another mineral or minerals. The gap between solution and fill can be days or many millions of years, blurring the distinction between cementation and replacement. Nonetheless, we have separate chapters for those two processes and will guide readers through the distinction where possible. We also will provide examples showing temporal successions of cements through time, termed "paragenetic sequences". Recognition of such sequences is especially important because they are critical to understanding the history of pore filling, and thus of porosity change and reservoir potential through time as well as linking diagenesis to burial history.

Further consideration should be given to the connections between cementation and porosity. At first glance, the relationship seems simple—by definition, filling a pore with cement represents a loss of porosity. When considered from a broader areal or temporal perspective, however, the relationship can be far more complex. If the solutes for the precipitated cement are locally derived, the porosity loss related to cement formation may be balanced by nearby porosity gain through removal of less stable grains (or earlier cements), making the overall process effectively porosity neutral. Likewise, from a temporal perspective, the immediate loss of pore space through cementation may be offset ultimately by having a stronger rock, less susceptible to long-term compactional porosity reduction. Alternatively, the precipitation of cements that can be dissolved during later diagenesis (especially carbonate, sulfate and halide cements) also may preclude compaction in the short term and allow porosity resurrection and late-stage hydrocarbon reservoir development in the longer term. So cementation should be looked at as part of a much broader complex of diagenetic phenomena, including

Table 11.1: Size terminology for carbonate cement crystals (adapted from Folk, 1965). It can, however, be applied to cements of any composition.

Size (mm)	Size (µm)	Name
		extremely coarsely crystalline
4.0	4000	
		very coarsely crystalline
1.0	1000	
		coarsely crystalline
0.25	250	
		medium crystalline
0.062	62	
		finely crystalline
0.016	16	
		very finely crystalline
0.004	4	
		aphanocrystalline
0.001	1	

Table 11.2 Some relatively common cement and replacement minerals in sandstones and shales along with their generalized chemical formulas. Many of these minerals, especially the clays and zeolites, have complex and varied structures with extensive elemental substitutions — thus, the formulas given are the most common or most basic ones. Data adapted from many sources, especially Ali et al., 2010.

Carbonates

Ankerite: $Ca(Fe,Mg,Mn)(CO_3)_2$

Calcite: $CaCO_3$

Dolomite: $CaMg(CO_3)_2$

Siderite: $FeCO_3$

Clays

Chamosite: ~ $(Fe^{2+},Mg,Fe^{3+})_5Al(Si_3Al)O_{10}(O,OH)_8$

Chlorite: $(Fe,Mg,Li)_5(Si, Al)_4O_{10}(OH)_8$

Glauconite: $(K,Na)Fe^{3+},Al,Mg)_2(Si, Al)_4O_{10}(OH)_2$

Illite: $(K,H_3O)(Al,Mg,Fe)_2(Si,Al)_4O_{10}(OH)_2 \cdot H_2O$

Kaolinite/dickite: $Al_2Si_2O_5(OH)_4$

Smectite: $(Na,Ca)_{0.33}(Al,Mg)_2Si_4O_{10}(OH)_2 \cdot nH_2O$

Feldspars

Orthoclase/microcline: $KAlSi_3O_8$

Plagioclase: mainly albite $Na(AlSi_3O_8)$
some with very minor incorporation of Ca

Halides

Fluorite: CaF_2

Halite: NaCl

Sylvite: KCl (uncommon as a cement)

Oxides/hydroxides

Anatase/rutile/brookite: TiO_3

Goethite: FeO(OH)

Hematite: Fe_2O_3

Limonite: $Fe_2O_3 \cdot n(H_2O)$

Phosphates

Fluorapatite: $Ca_5(PO_4)_3F$

Francolite (collophane): $Ca(PO_4,CO_3)_3(F,OH,Cl)$

Silica

Chalcedony, chert and quartz: SiO_2

Opal: $SiO_2 \cdot nH_2O$

Sulfates

Anhydrite: $CaSO_4$

Barite: $BaSO_4$

Gypsum: $CaSO_4 \cdot 2H_2O$

Sulfides

Marcasite: FeS_2 (with trace Cu and As)

Pyrite: FeS_2

Sphalerite: ZnS to (Zn,Fe)S

Zeolites

Analcime: $NaAlSi_2O_6 \cdot H_2O$

Chabazite: $CaAl_2Si_4O_{12} \cdot 6H_2O$

Erionite: $(Na_2K_2Ca)\ 2Al_4Si_{14}O_{36} \cdot 15H_2O$

Heulandite/clinoptilolite:
$(Ca,Na)_{2\text{-}3}Al_3(Al,Si)_2Si_{13}O_{36} \cdot 12H_2O$

Laumontite: $Ca(AlSi_2O_6)_2 \cdot 6H_2O$

Natrolite: $Na_2Al_2Si_3O_{10} \cdot 2H_2O$

Mordenite: $(Ca,Na_2,K_2)Al_2Si_{10}O_{24} \cdot 7H_2O$

Phillipsite: $(Ca,K,Na)_{1\text{-}2}(Si,Al)_8O_{16} \cdot 6H_2O$

Scolecite: $CaAl_2Si_3O_{10} \cdot 3H_2O$

Thomsonite: $NaCa_2Al_5Si_5O_{20} \cdot 6H_2O$

replacement, dissolution and large-scale patterns of solute transport.

In this extensive section of the book, we have divided cements into eight chapters, each with its own introduction that will provide keys to mineral recognition and more details on settings of cement formation, including timing, geochemical constraints and additional controls (temperature, pressure and others). The eight chapters are: quartz/silica, feldspars, clay minerals, zeolites, carbonates, sulfates and halides, iron oxides/hydroxides and sulfides and other cements. Hydrocarbons also are included in the final section. Although not technically cements, hydrocarbons are located in pores and, especially when occurring in relatively solid (tarry or asphaltic) forms, they play much the same role as mineral cements. It should be clearly stated, however,

that cements in several of these categories, especially clays and zeolites, are commonly so finely crystalline that individual grains cannot be resolved even with the best petrographic microscopes. Conventional petrography can provide clues about which minerals are likely involved, but the use of SEM, XRD, microprobe or other analytical instrumentation is essential for precise identifications. For that reason, we have included SEM and BSE images in many places in this chapter.

Cited References and Additional Information Sources

Ajdukiewicz, J. M., and R. H. Lander, 2010, Sandstone reservoir quality prediction: The state of the art: AAPG Bulletin, v. 94, p. 1083-1091, doi: 10.1306/intro060110.

Ali, S. A., W. J. Clark, W. R. Moore, and J. R. Dribus, 2010, Diagenesis and reservoir quality: Oilfield Review, v. 22, no. 2, p. 14-27.

Barclay, S. A., and R. H. Worden, 2000, Geochemical modelling of diagenetic reactions in a sub-arkosic sandstone: Clay Minerals, v. 35, p. 57-67, doi: 10.1180/000985500546729.

Bjørlykke, K., 1983, Diagenetic reactions in sandstones, *in* A. Parker, and B. W. Sellwood, eds., Sediment Diagenesis (NATO ASI Series Volume 115): Dordrecht, Netherlands, D. Reidel Publishing, p. 169-213, doi: 10.1007/978-94-009-7259-9_3.

Bjørlykke, K, 1988, Sandstone diagenesis in relation to preservation, destruction and creation of porosity, *in* G. V. Chilingarian, and K. H. Wolf eds., Diagenesis: New York, Elsevier Science, p. 555-588, doi: 10.1016/S0070-4571(08)70180-8.

Bjørlykke, K., and J. Jahren, 2012, Open or closed geochemical systems during diagenesis in sedimentary basins: Constraints on mass transfer during diagenesis and the prediction of porosity in sandstone and carbonate reservoirs: AAPG Bulletin, v. 96, p. 2193-2214, doi: 10.1306/04301211139.

Bloch, S., R. H. Lander, and L. Bonnell, 2002, Anomalously high porosity and permeability in deeply buried sandstone reservoirs: Origin and predictability: AAPG Bulletin, v. 86, p. 301-328, doi: 10.1306/61EEDABC-173E-11D7-8645000102C1865D.

Caracciolo, L., J. Arribas, R. V. Ingersoll, and S. Critelli, 2014, The diagenetic destruction of porosity in plutoniclastic petrofacies: The Miocene Diligencia and Eocene Maniobra formations, Orocopia Mountains, southern California, USA, *in* R. A. Scott, H. R. Smyth, A. C. Morton, and N. Richardson, eds., Sediment Provenance Studies in Hydrocarbon Exploration and Production: London, GSL Special Publication 386, p. 49-62, doi: 10.1144/SP386.9.

Chuhan, F. A., K. Bjørlykke, and C. J. Lowrey, 2001, Closed-system burial diagenesis in reservoir sandstones: Examples from the Garn Formation at Haltenbanken area, offshore mid-Norway: Journal of Sedimentary Research, v. 71, p. 15-26, doi: 10.1306/041100710015.

Curtis, C. D., 1978, Possible links between sandstone diagenesis and depth-related geochemical reactions occurring in enclosing mudstones: Journal of the Geological Society, v. 135, p. 107-117, doi: 10.1144/gsjgs.135.1.0107.

Dapples, E. C., 1967, Chapter 3. Diagenesis of sandstones, *in* G. Larsen, and G. V. Chilingar, eds., Diagenesis in Sediments: Amsterdam, Elsevier, p. 91-125.

Folk, R. L., 1965, Some aspects of recrystallization in ancient limestones, *in* L. C. Pray, and R. S. Murray, eds., Dolomitization and Limestone Diagenesis: Tulsa, OK, SEPM Special Publication 13, p. 14-48, doi: 10.2110/pec.65.07.0014.

Gluyas, J. G., S. M. Grant, and A. G. Robinson, 1993, Geochemical evidence for a temporal control on sandstone cementation, *in* A. D. Horbury, and A. G. Robinson, eds., Diagenesis and Basin Development: Tulsa, OK, AAPG Studies in Geology 36, p. 23-33.

Haszeldine, R. S., C. I. Macaulay, A. Marchand, M. Wilkinson, C. M. Graham, A. Cavanagh, A. E. Fallick, and G. D. Couples, 2000, Sandstone cementation and fluids in hydrocarbon basins: Journal of Geochemical Exploration, v. 69/70, p. 195-200, doi: 10.1016/S0375-6742(00)00126-6.

Hayes, J. B., 1979, Sandstone diagenesis—the hole truth, *in* P. A. Scholle, and P. R. Schluger, eds., Aspects of Diagenesis: Tulsa, OK, SEPM Special Publication 26, p. 127-139, doi: 10.2110/pec.79.26.0127.

Kordi, M., B. Turner, and A. M. K. Salem, 2011, Linking diagenesis to sequence stratigraphy in fluvial and shallow marine sandstones: Evidence from the Cambrian–Ordovician lower sandstone unit in southwestern Sinai, Egypt: Marine and Petroleum Geology, v. 28, p. 1554-1571, doi: 10.1016/j.marpetgeo.2011.05.003.

Mackenzie, F. T., ed., 2005, Sediments, Diagenesis, and Sedimentary Rocks (Treatise on Geochemistry, vol. 7): New York, Elsevier, 446 p.

Morad, S., J. M. Ketzer, and L. F. De Ros, 2000, Spatial and temporal distribution of diagenetic alterations in siliciclastic rocks: Implications for mass transfer in sedimentary basins: Sedimentology, v. 47, p. 95-120, doi: 10.1046/j.1365-3091.2000.00007.x.

Pittman, E. D., 1979, Recent advances in sandstone diagenesis: Annual Review of Earth and Planetary Sciences, v. 7, p. 39, doi: 10.1146/annurev.ea.07.050179.000351.

Primmer, T. J., C. A. Cade, J. Evans, J. G. Gluyas, M. S. Hopkins, N. H. Oxtoby, P. C. Smalley, E. A. Warren, and R. H. Worden, 1997, Global patterns in sandstone diagenesis: Their application to reservoir quality prediction for petroleum exploration, *in* J. A. Kupecz, J. Gluyas, and S. Bloch, eds., Reservoir Quality Prediction in Sandstones and Carbonates: Tulsa, OK, AAPG Memoir 69, p. 61-77.

Ramm, M., and K. Bjørlykke, 1994, Porosity/depth trends in reservoir sandstones: Assessing the quantitative effects of varying pore-pressure, temperature history and mineralogy, Norwegian Shelf data: Clay Minerals, v. 29, p. 475-490, doi: 10.1180/claymin.1994.029.4.07.

Taylor, T. R., M. R. Giles, L. A. Hathon, T. N. Diggs, N. R. Braunsdorf, G. V. Birbiglia, M. G. Kittridge, C. I. Macaulay, and I. S. Espejo, 2010, Sandstone diagenesis and reservoir quality prediction: Models, myths, and reality: AAPG Bulletin, v. 94, p. 1093-1132, doi: 10.1306/04211009123.

Tobin, R. C., and D. Schwarzer, 2014, Effects of sandstone provenance on reservoir quality preservation in the deep subsurface: Experimental modelling of deep-water sand in the Gulf of Mexico, *in* R. A. Scott, H. R. Smyth, A. C. Morton, and N. Richardson, eds., Sediment Provenance Studies in Hydrocarbon Exploration and Production: London, GSL Special Publication 386, p. 27-47, doi: 10.1144/SP386.17.

Turner, P., S. D. Burley, D. Rey, and J. Prosser, 1995, Burial history of the Penrith Sandstone (Lower Permian) deduced from the combined study of fluid inclusion and palaeomagnetic data, *in* P. Turner, and A. Turner, eds., Palaeomagnetic Applications in Hydrocarbon Exploration and Production: London, GSL Special Publication 98, p. 43-78, doi: 10.1144/GSL.SP.1995.098.01.04.

Wilkinson, M., and R. S. Haszeldine, 2011, Oil charge preserves exceptional porosity in deeply buried, overpressured, sandstones: Central North Sea, UK: Journal of the Geological Society, v. 168, p. 1285-1295, doi: 10.1144/0016-76492011-007.

Worden, R. H., and S. D. Burley, 2003, Sandstone diagenesis: The evolution of sand to stone, *in* S. D. Burley, and R. H. Worden, eds., Sandstone Diagenesis: Recent and Ancient (IAS Reprint Series 4): Oxford, UK, Publishing Ltd., p. 3-44, doi: 10.1002/9781444304459.ch/summary.

INTRODUCTION TO QUARTZ AND SILICA CEMENTS

Quartz and silica are some of the most abundant cements in sandstones. Quartz cement generally occurs as single-crystal overgrowths (or apparent single-crystal overgrowths) on the surfaces of detrital quartz grains. Other forms of silica can fill pore space, regardless of substrate mineralogy, and include microcrystalline quartz (chert), chalcedony and opal. Silica cements, and quartz in particular, can have a major effect on porosity loss, but they also can act to stabilize the rock against further compaction. Reservoir quality is thus, in part, a complex interplay between habit and quantity of the quartz or silica cement. Diagenetic quartz also can occur as small crystals within mudrocks (Schieber et al., 2000; Thyberg and Jahren, 2011; Milliken, 2013).

Optical properties and recognition of quartz and silica cements:

Syntaxial quartz overgrowths – monocrystalline quartz overgrowths are the most commonly observed type of silica cement, but it is sometimes difficult to distinguish such overgrowths as they are typically in optical continuity with their substrates. Several features may help to differentiate overgrowths from underlying detrital grains:

1. Lines of impurities or dust rims outlining the original surfaces of the detrital grain;
2. Inclusion clouds, vacuoles or other features within the quartz grain terminate against the usually clear, more inclusion-poor quartz overgrowth (although the reverse situation, with clear cores and cloudy overgrowths, also can occur); and
3. The presence of euhedral terminations where quartz crystals grew into adjacent pore spaces.

Cement abundances are usually quantified by point counting. However, where one or more of the above characteristics are not present, or where the percentage of overgrowths is low, cathodoluminescence can be used to aid recognition and more reliably determine cement abundances.

Microcrystalline quartz (microgranular quartz) – consists of mosaics of minute (< 5–20 μm diameter) subequidimensional crystals. In some cases, the bipyramidal shape of quartz can be seen. Microcrystalline quartz normally has a dark gray, anomalous birefringence and some microporosity. These cements are best recognized in SEM or can sometimes be inferred from XRD of the clay-sized fraction. Microcrystalline quartz can occur as a result of opal-CT alteration and is common where spiculitic or diatomaceous remains have served as sources of silica. Such cement is common within both sandstones and mudrocks. Chalcedony and moganite, fibrous forms of microcrystalline quartz, also can be found as cements and large void fills.

Opal – is a hydrous silica containing significant amounts of water (generally 10% or less, but in some cases considerably more) in submicroscopic pores (see Chapter 1). Opal-A is amorphous and is isotropic in thin section, but it can be identified using XRD; opal-CT is nearly isotropic (i.e., very weakly birefringent) in thin section and finely crystalline. Opal-CT can transform to microcrystalline quartz and chalcedony, and so opal is rarely found in sedimentary rocks older than the Cenozoic.

Controls on formation:

Quartz overgrowths start as numerous small crystals nucleating on detrital quartz grains. Favorably oriented crystals may eventually coalesce into a single large crystal that is in optical continuity with its host grain. Formation of overgrowths is strongly affected by the substrate grains. Nonundulatory monocrystalline quartz grains are more likely to host overgrowths than undulose or polycrystalline grains (James et al., 1986). Quartz overgrowths form mainly during mesogenetic burial, typically at temperatures above 60–80°C, with maximum formation commonly in the temperature range of 90–165°C (Bjørlykke and Egeberg, 1993; Walderhaug, 1994; Walderhaug, 2000). Although rare, the presence of nonsyntaxial and pseudopoikilotopic quartz cement is diagnostic of high temperature diagenesis in some settings (see Spötl et al. , 1996).

Although most quartz cementation takes place in mesogenetic environments, eogenetic quartz and opaline silica cementation may occur in water table silcretes, especially in microbially-influenced settings in arid to semi-arid regions (e.g., Abdel-Wahab et al., 1998; McBride et al., 2012). In addition, eogenetic or telogenetic quartz cements can form during deep weathering under humid, tropical conditions.

There are many potential sources of silica for quartz and opal cementation. There is no chemical method to recognize solute sources in precipitated cements, but it sometimes is possible to infer sources from observations on cement distribution. Common sources include:

1. Pressure dissolution at grain contacts or stylolites;
2. Breakdown of unstable grains such as feldspar and volcanic glass;
3. Dissolution of biogenic silica from radiolaria, diatoms or sponge spicules, or glass from volcanic ash beds; and
4. Detrital quartz dissolution, for example during carbonate replacement.

Additional sources can include diagenetic products derived from associated mudrocks, including pressure dissolution of silt-sized quartz or opaline skeletons and possibly the transformation of smectite to illite during burial diagenesis.

There are many factors that may inhibit quartz overgrowth formation, mainly by isolating detrital quartz substrates or minimizing fluid flow. These include:

1. **Lack of suitable substrates in sandstones with quartz-poor initial composition.**
2. **Extensive detrital clay matrix;**
3. **Clay coatings inferred to have been inherited (Wilson, 1992) or to have formed during infiltration of muddy fluids shortly following deposition of the sediment (Dunn, 1992);**
4. **Formation of grain-coating diagenetic clays, in particular chlorite (Berger et al., 2009; Ajdukiewicz and Larese, 2012), or formation of early microcrystalline quartz cement coatings (Aase et al., 1996; French et al., 2012);**
5. **High oil or gas saturations, oil-wet grain surfaces or bitumen impregnation (Worden et al., 1998; Wilkinson and Haszeldine, 2011); and**
6. **Overpressurization causing reduction of vertical effective stress (e.g., Ramm and Bjørlykke, 1994; Bloch et al., 2002).**

Quartz Overgrowths

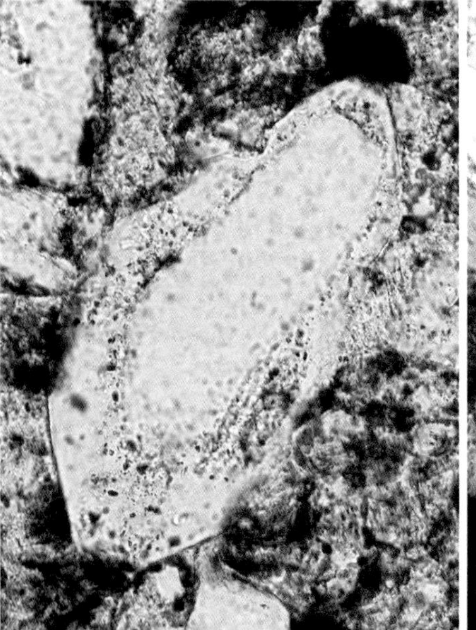

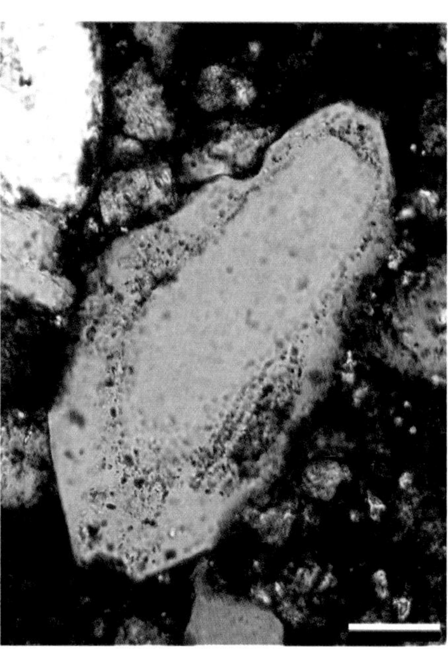

Paleogene Vieja Gp., Presidio Co., Texas

A euhedral, syntaxial quartz overgrowth on a detrital grain. Syntaxial overgrowths, by definition, are in optical continuity with the underlying detrital grains. Authigenic overgrowths can be difficult to differentiate from the underlying grain where both are compositionally nearly identical. In this example, however, mineral and fluid inclusions pervade the outer (overgrowth) part of the crystal and are scarce in the detrital core, clearly distinguishing the detrital core from the authigenic overgrowth. The sharp, bipyramidal crystal outlines formed by the overgrowths are further indication of authigenesis.

PPL | XPL, Scale bar = 0.06 mm

Oligocene – Miocene Fontainebleau Sandstone, Paris Basin, France

Within this porous sandstone, all the detrital quartz grains have euhedral overgrowths. Quartz overgrowths can be highly zoned, although this zoning is only rarely visible using traditional petrographic techniques. Thus, cathodoluminescence (see later examples) and backscatter electron imaging generally are necessary to see the extent of the zonation. Haddad et al. (2006) showed that these overgrowths are strongly zoned, are composed of quartz and poorly-ordered silica and can display a systematic rotation of the crystallographic axis causing the overgrowths to not always be syntaxial with their detrital grains.

PPL, BDI, Scale bar = 0.26 mm

Paleogene Vieja Gp., Presidio Co., Texas

Euhedral, hexagonal syntaxial quartz overgrowths here have formed on several fine sand sized detrital quartz grains. The mineral inclusions within the overgrowths were probably incorporated from the surrounding fine-grained sediment as the quartz crystals only partially displaced that matrix.

PPL, Scale bar = 0.05 mm

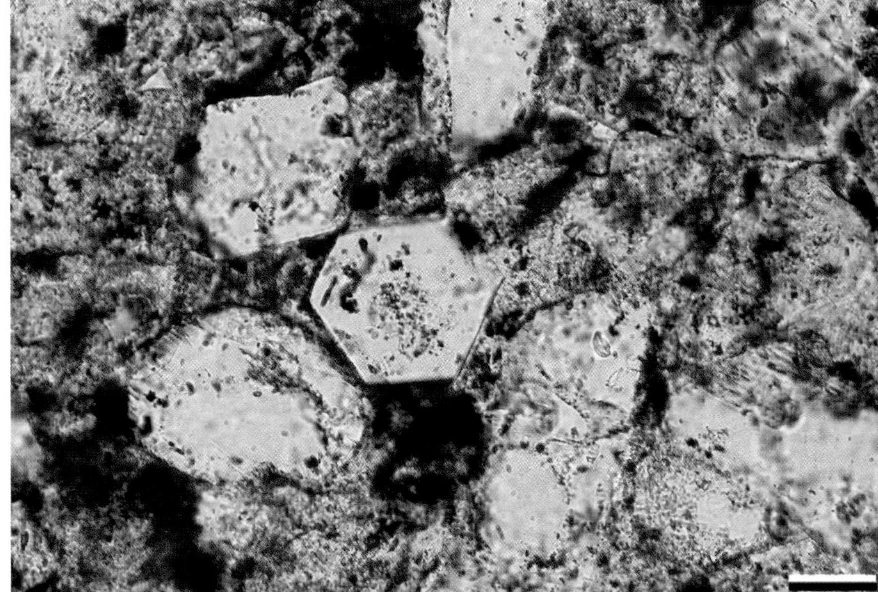

Mid. Permian Brushy Canyon Fm., Culberson Co., Texas

An authigenic overgrowth on a detrital quartz grain. The well-rounded nucleus is outlined by a thin layer of inclusions (mostly small carbonate inclusions) on its surface. The authigenic overgrowth shows euhedral crystal shape (hexagonal) where fully developed and is in optical continuity with the underlying quartz grain. Such euhedral outlines produced by overgrowths must be carefully distinguished from the primary euhedral outlines of some detrital quartz grains derived from volcanic source rocks. The carbonate cements and quartz overgrowths in this example are likely to be, in part, coeval. The remaining porosity was largely filled with calcite cement.

XPL, Scale bar = 0.06 mm

Mid. Jurassic Entrada Fm., Moffat Co., Colorado

A large euhedral quartz overgrowth (near center) occludes part of the rock's primary porosity. It and the other visible overgrowths are inclusion poor, which is common, especially in overgrowths growing into open pores. Not all of the quartz grains have overgrowths, and not all overgrowths are euhedral. Zoned dolomite cements formed before, during and after quartz overgrowths as shown by the absence of quartz overgrowths in zones with extensive dolomite cementation (black arrow) as well as dolomite encasing the lower edge of the overgrowth on the large euhedral grain near the photo center (red arrow).

PPL, AFeS, BDI, Scale bar = 0.10 mm

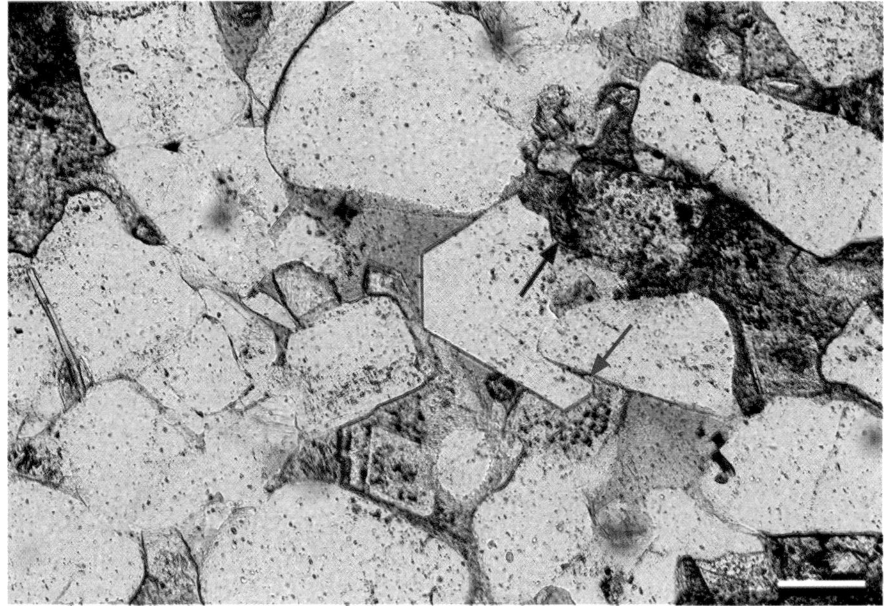

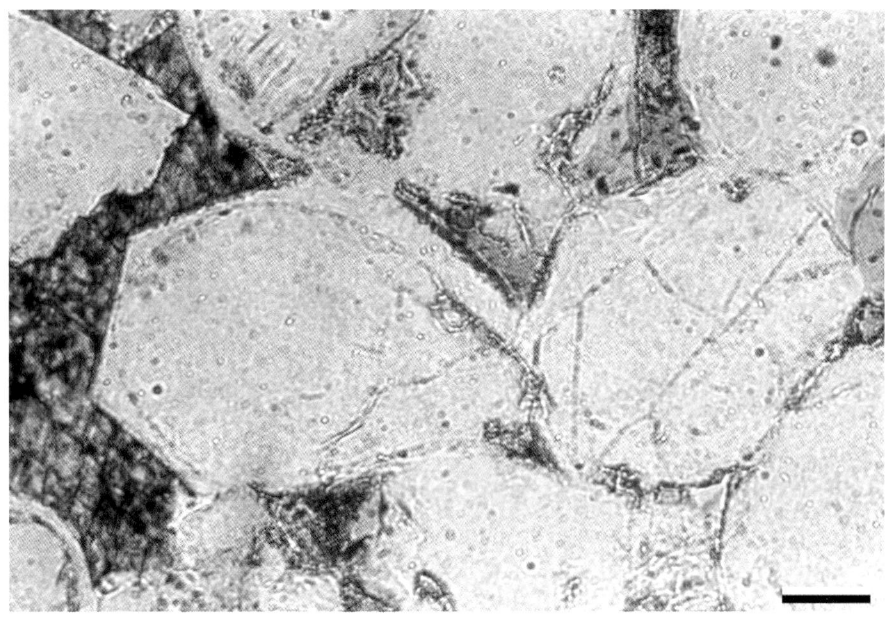

Mid. Pennsylvanian Strawn Gp., Gray Sandstone, Texas

An intermediate to advanced stage of quartz overgrowth cementation in which a significant amount of primary pore space was filled. Note euhedral crystal faces where overgrowths did not interfere with each other and the enhanced visibility of overgrowths due to inclusions or "dust rims" at overgrowth contacts with the detrital grain cores. Nonferroan (weakly pink stained) and later ferroan (blue stained) calcite cements filled all remaining pores. The nonferroan calcite cementation was partially coeval with quartz cementation; the ferroan calcites, however, clearly postdate quartz overgrowths. Photo from Shirley P. Dutton.

PPL, AFeS, Scale bar = 0.01 mm

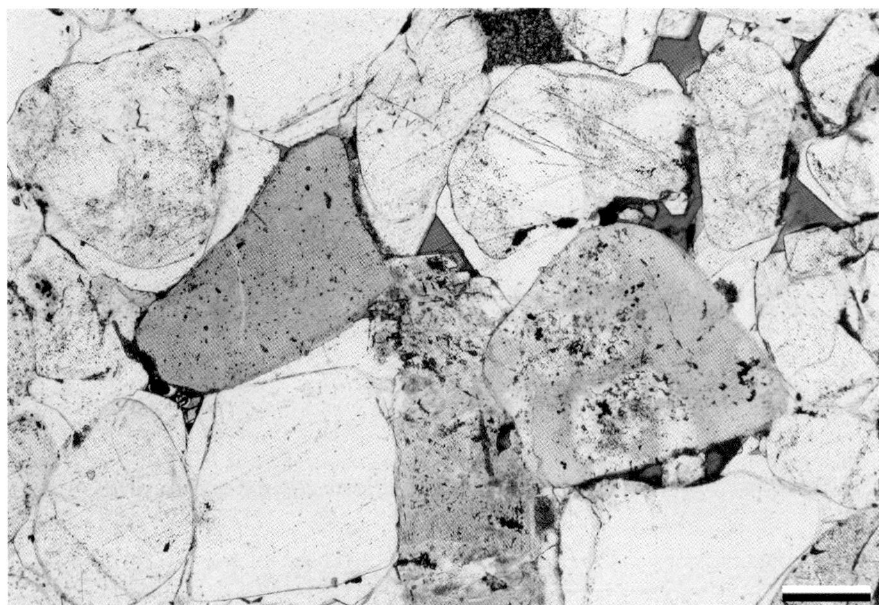

Cretaceous Dakota Gp., Moffat Co., Colorado

Syntaxial quartz overgrowths on detrital quartz grains occlude much of the primary porosity within this sample. Overgrowths have not formed on the chert grains (pink due to dye injection into microporosity resulting from partial dissolution) because they have diverse microcrystal orientations. Nonetheless, they are partially encased within quartz overgrowths that grew on nearby detrital monocrystalline quartz-grain substrates. The feldspar (dusty brown grain, lower center) has a thin authigenic feldspar cement overgrowth that may have formed at about the same time as the quartz overgrowths.

PPL, RDI, Scale bar = 0.21 mm

Cretaceous Dakota Gp., Moffat Co., Colorado

Syntaxial quartz overgrowths fill most of the primary pores in areas dominated by monocrystalline quartz grains. Dusty rims on many of the quartz grains outline the original grain shapes. In areas with more lithic grains, there is less porosity loss by quartz cementation. In addition, many of lithic grains have been partially leached. The lithic grain at the arrow, in particular, has been almost completely dissolved, with only remnants of silt-sized, original constituent particles and iron oxide coatings marking the exterior outline of the now nearly vanished grain.

PPL, RDI, Scale bar = 0.26 mm

Mid. Jurassic Entrada Fm., Moffat Co., Colorado

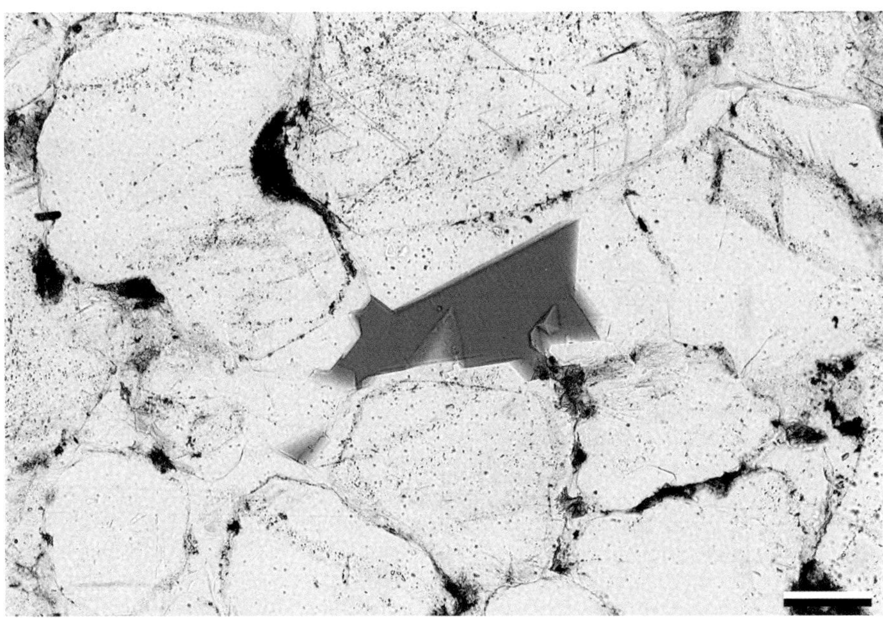

Here, syntaxial quartz overgrowths have formed on several quartz grains and coalesced along a series of compromise crystal boundaries. The cements growing into open pores have euhedral outlines and are somewhat inclusion poor (clear) relative to the host grains. Most of the detrital grains had early, partial iron oxide and clay coatings that failed to prevent later overgrowth nucleation because of their patchy distribution. Where the coatings are thin or lacking, it is difficult to see where the detrital cores end and the authigenic overgrowths begin.

PPL, RDI, Scale bar = 0.10 mm

Lo. Pennsylvanian Morrow B Sandstone, Ochiltree Co., Texas

The contacts between monocrystalline quartz grains and their syntaxial overgrowths can be difficult to see in thin section, as in this case. The quartz grain in the center of this image has a euhedral outline and some irregular projections that would not have survived transport, but there is little visible distinction between the grain and its overgrowth. These unrounded features and the fact that most other quartz grains in the sample have readily visible, well-delineated overgrowths, has to suffice for identification. Cathodoluminescence also can help to distinguish compositional differences between overgrowths and detrital grain cores.

PPL, RDI, Scale bar = 0.16 mm

Lo. Pennsylvanian Morrow B Sandstone, Ochiltree Co., Texas

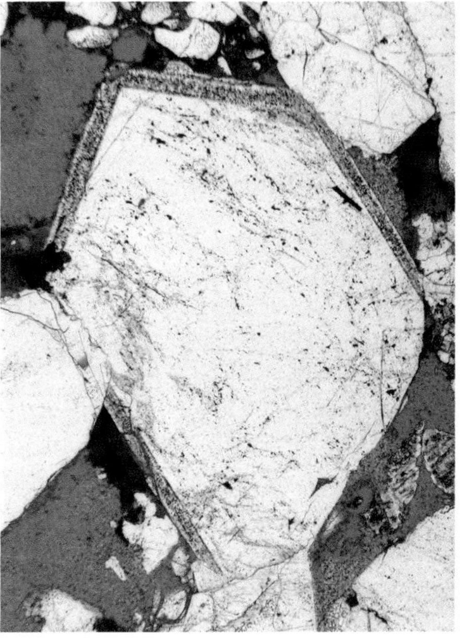

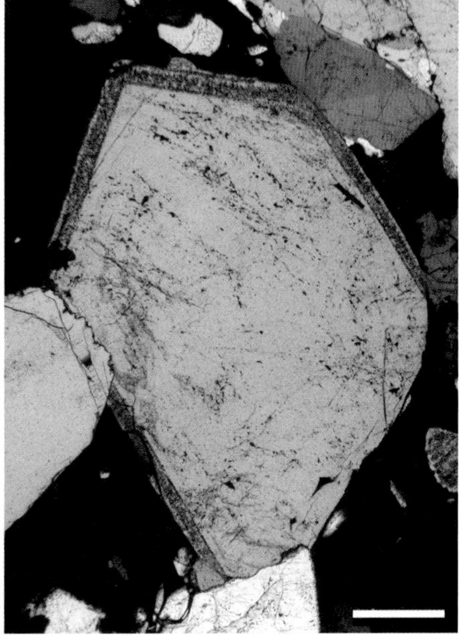

This quartz appears to have a remarkably inclusion-rich overgrowth around most of its margins. However, although many quartz grains in this sample have overgrowths, the rest are relatively inclusion-poor. That makes it likely that this is either primary growth zonation in a quartz grain derived from a granitic or pegmatitic source, or it is a little-rounded, second-cycle grain reworked from a sandstone source. The image should be a reminder that not everything that looks like an authigenic overgrowth is actually that. Note also the complex pressure-dissolved margins with adjacent grains.

PPL | XPL, RDI, Scale bar = 0.51 mm

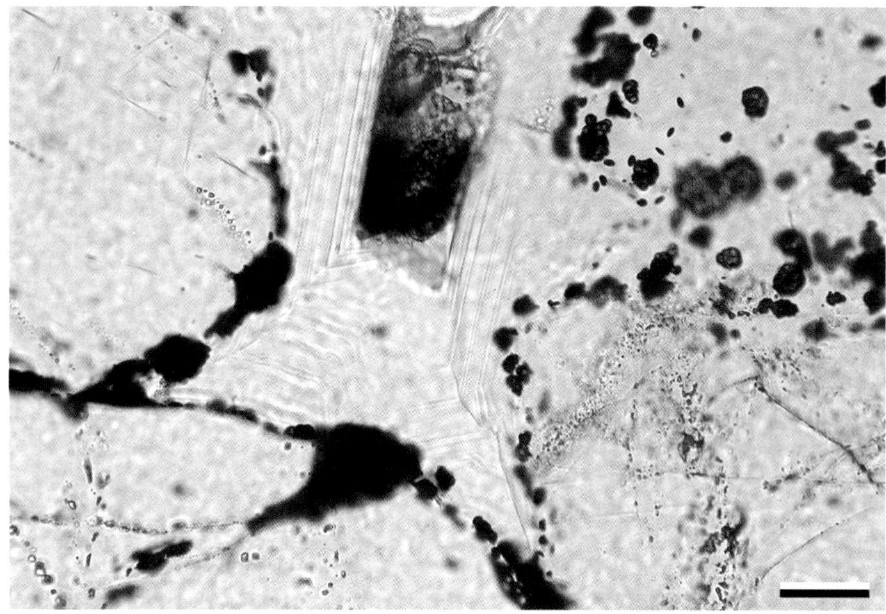

Quaternary? Thar desert, Rajasthan, India

The sample is a type of duricrust (part ferricrete and part silcrete) from a desert region outcrop. Small aggregates of iron oxyhydroxides mark the boundary between the detrital grains and the quartz cement. Also, note the growth lines within the quartz cement. The iron and quartz cements represent precipitation during different climatic conditions. Silcretes are commonly found in arid and semi-arid environments; ferricretes, on the other hand, originate in humid tropical environments (Mares, 1999). Silcrete formation temperatures (derived from oxygen isotopes) can be as low as 10–30°C (Kelly et al., 2007).

PPL, BDI, Scale bar = 0.07 mm

Silurian Tuscarora Sandstone, Lebanon Co., Pennsylvania

When quartz overgrowths are almost ubiquitous in quartzarenites, they can interlock and substantially obscure the original fabric of the rock, as in this example. In particular, even well-rounded grains can appear to be angular due to the overgrowths. In this example, indeed, the original detrital quartz grains were well rounded, as shown by the thin "dust rims" that help to distinguish detrital cores from authigenic overgrowths.

XPL, Scale bar = 0.10 mm

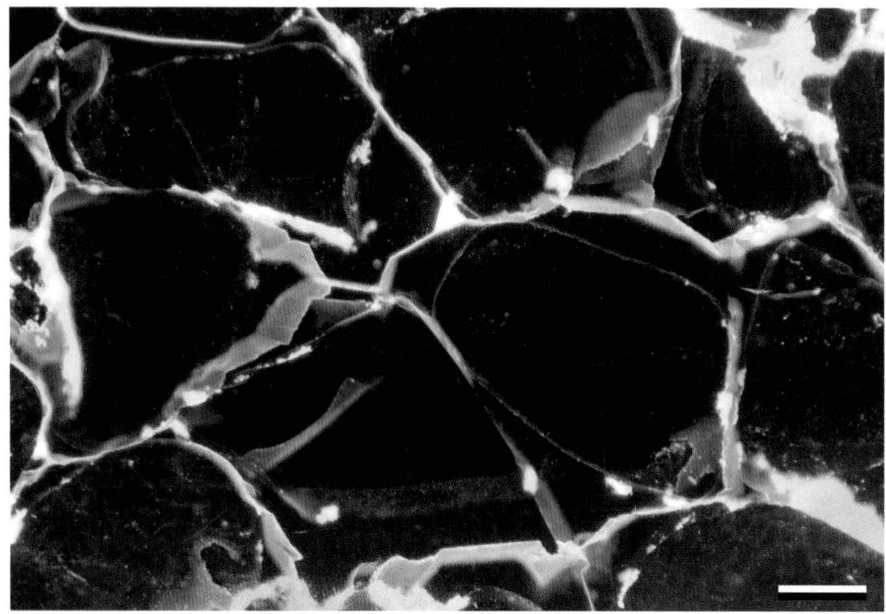

Cretaceous Dakota Gp., Moffat Co., Colorado

In addition to conventional petrography, fluorescence and cathodoluminescence microscopy can be utilized to distinguish quartz grains and their overgrowths, and those techniques are especially useful in fully cemented samples (such as the previous example). By impregnating this sample with fluorescent epoxy, the microporosity within the dust rims surrounding the grains fluoresces orange, highlighting the contacts between the grains and their overgrowths. The outer contacts of the overgrowths are marked by brighter yellow fluorescence. The orange fluorescence shows that the detrital quartz consists of well-rounded to subrounded grains.

FL 470, RDI, Scale bar = 0.10 mm

Devonian Hoing Sandstone Mbr., Cedar Valley Limestone, Illinois

Where "dust rims" and recognizable crystal terminations are scarce or absent, quartz overgrowths can be very difficult to recognize in thin section. In this example, the cross-polarized image on the left shows sutured boundaries between grains that might be interpreted as being related to compaction rather than cementation. The cathodoluminescence image on the right, however, allows clear distinction of luminescent detrital cores and interlocking-nonluminescent overgrowths with some euhedral crystal outlines. Photomicrographs by R. F. Sippel (from Sippel, 1968).

XPL | CL, Scale bar = 0.013 mm

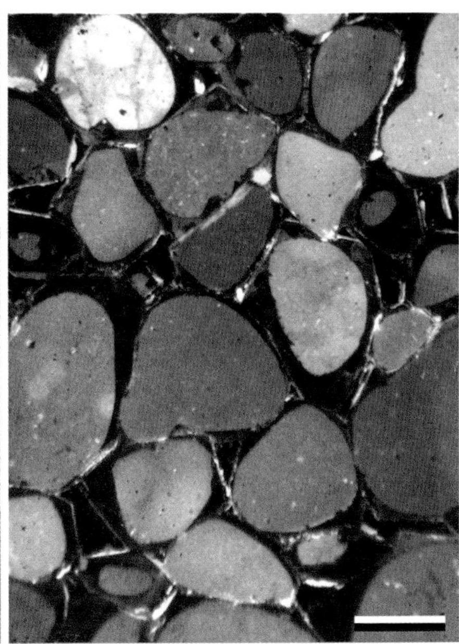

Lo. Jurassic Navajo Sandstone, Arizona

A quartz arenite with detrital quartz grains of metamorphic derivation (see Chapter 1). In scanned color CL, the grain on the left shows a dark, low temperature overgrowth (white arrows) that was rounded by transport and abrasion, indicating that this grain was likely recycled from an older sandstone source. The pore space of most of the sandstone is filled with dark green to non-luminescent epoxy, and the bright yellow-greenish spots are from abrasive grit that got trapped in the epoxy that was used to stabilize the sample.

CCL, Scale bar = 18 μm

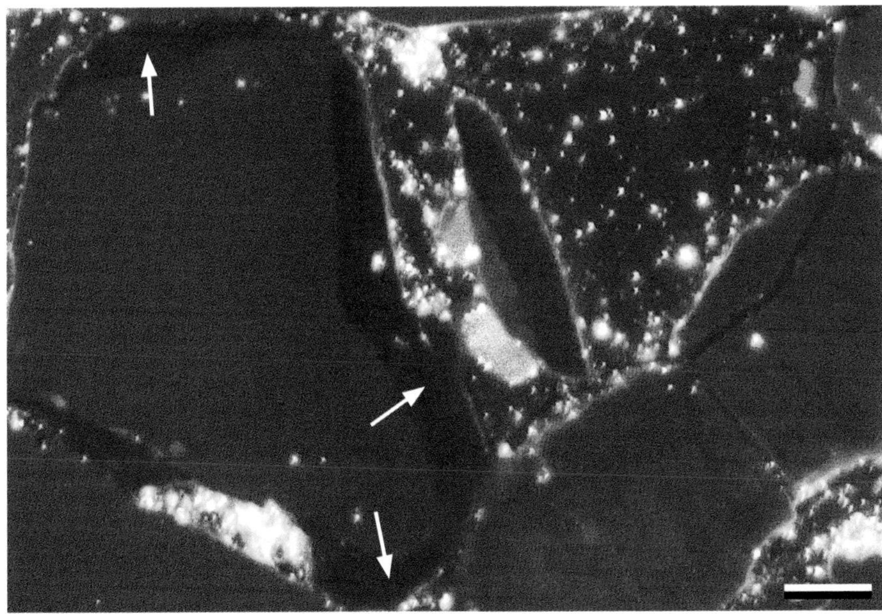

Mid. Proterozoic Belt Supergroup, Neihart Quartzite, Little Belt Mountains, Montana

A scanned color CL image of a quartz-cemented quartzarenite. The bimodal sand grains are well rounded, probably multicycle, and judging from their bluish colors are most likely of plutonic origin. The reddish discolorations are probably a sign of hydrothermal alteration. The cement has bluish and reddish colors. The blues probably reflect relatively low-temperature, shallow-burial diagenesis; the reds record elevated temperatures during late diagenesis.

CCL, Scale bar = 76 μm

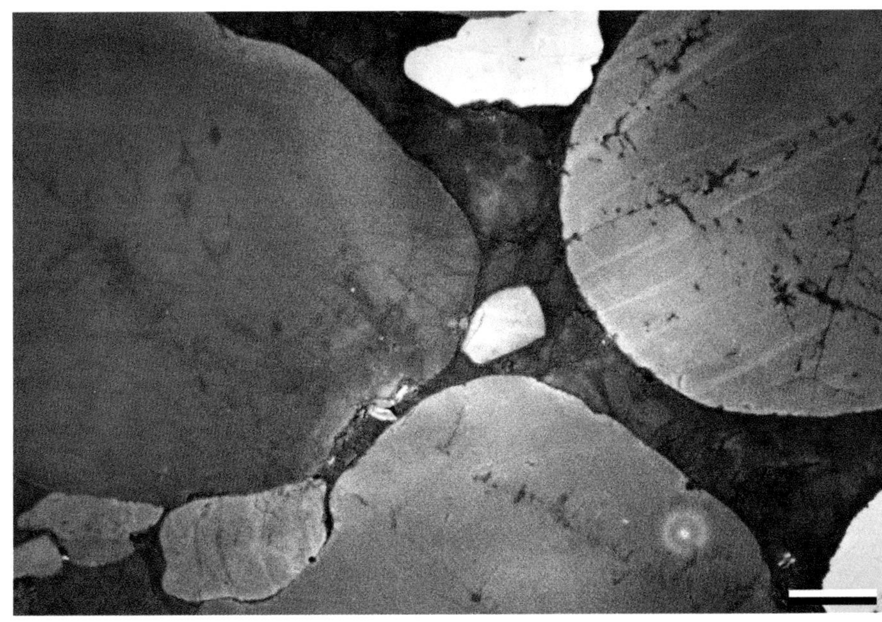

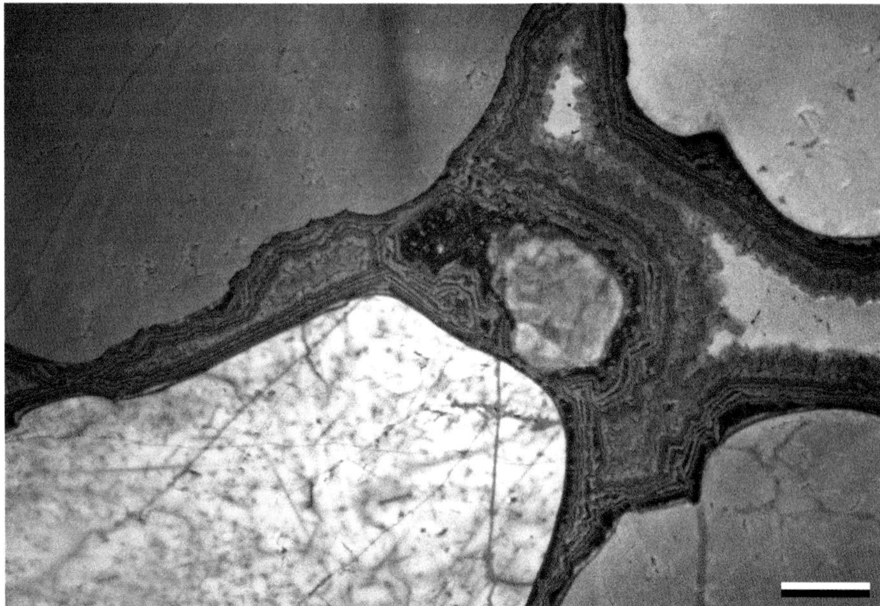

Lo. Cretaceous "Basement sands" (Maxon Fm. equivalent), west Texas

A quartz arenite that was cemented by syntaxial quartz overgrowths. In a scanned color CL, the quartz grains are plutonic (bluish) and metamorphic (reddish) in origin. The most remarkable feature here is the growth history displayed within the pore-filling quartz cement. The banding delineates a cement stratigraphy analogous to what is customarily seen in carbonate cements. The greenish colors are from epoxy that infills residual porosity.

CCL, Scale bar = 0.12 mm

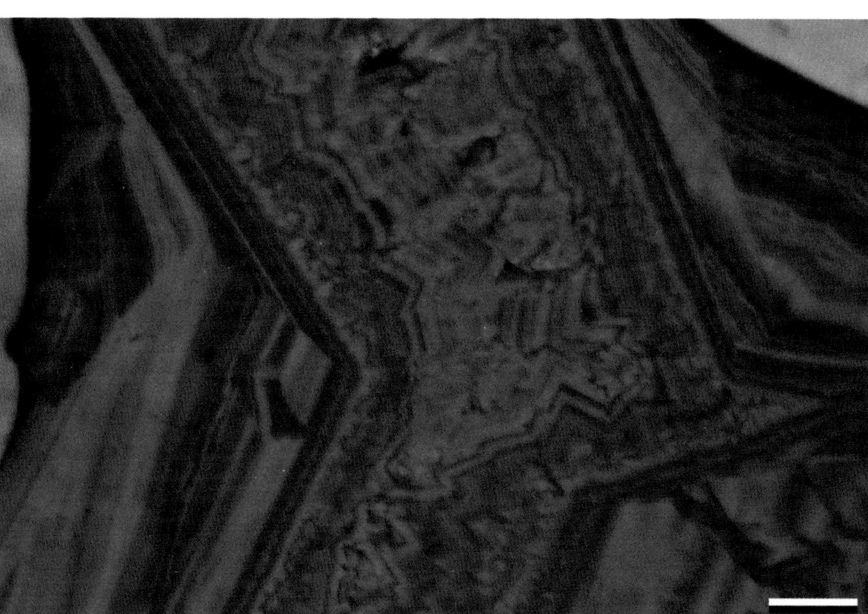

Lo. Cretaceous "Basement sands" (Maxon Fm. equivalent), west Texas

A detailed view of the quartz cement from the previous sample. In scanned color CL, submicron-scale growth banding in the quartz overgrowths is visible. These euhedral overgrowths have distinct growth periods, as well as episodes where growth was interrupted; the more subtle banding may reflect fluctuation in pore fluid chemistry.

CCL, Scale bar = 18.8 μm

Proterozoic Siyeh Fm., Belt Basin, Alberta, Canada

A quartz arenite with variably colored detrital quartz grains in scanned color CL. The grains can be interpreted to come from plutonic (bluish) and metamorphic (reddish-brown) sources. There are two distinguishable quartz cement generations visible. The first cement generation is dark and almost non-luminescent (suggestive of low-temperature formation) and shows dark bluish growth zones. A later reddish cement infills the final voids and fractures within the darker first generation cement. This geometric relationship, as well as the color, suggests later formation from higher temperature fluids.

CCL, Scale bar = 80 μm

Up. Jurassic Ula Fm., Norwegian sector, North Sea †

Grain-coating, euhedral, microcrystalline quartz (microquartz) growing on detrital quartz grains, but not in optical continuity with their substrate. The formation of such thin, microcrystalline quartz cements is thought to inhibit growth of larger, euhedral, monocrystalline quartz overgrowths during burial (Aase et al., 1996), and the source of the silica is most commonly attributed to dissolution of opaline biota (primarily sponge spicules or radiolarian tests). There are minor strands and flakes of illite on top of the microcrystalline quartz.

SEM, Scale bar = 18 μm

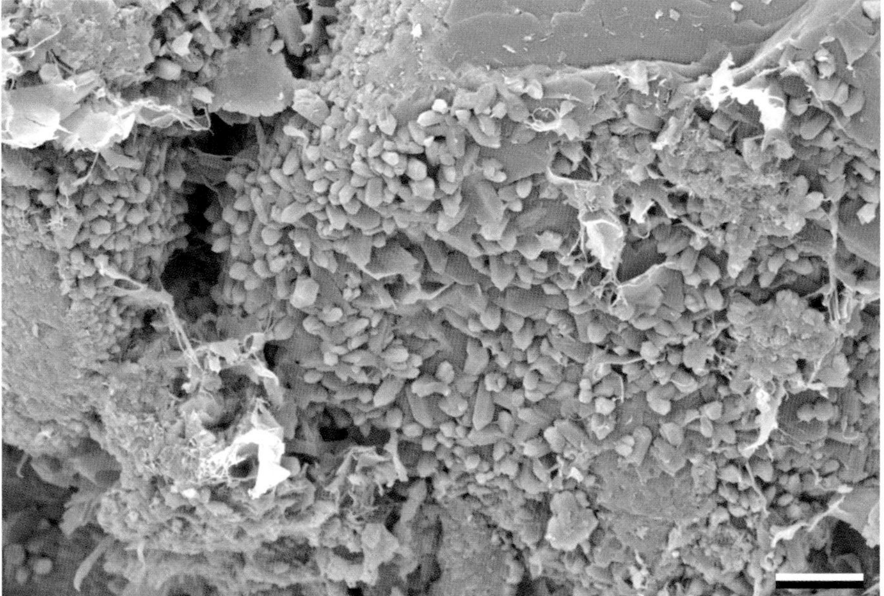

Up. Jurassic Spekk Fm., Møre Basin, Norwegian Sea †

Numerous quartz crystals forming a grain overgrowth. One will normally see this style of overgrowth on a polycrystalline grain. In this case, it appears that the grain had a clay component, and it may be a metamorphic rock fragment. The clay could also be a detrital grain coat or poorly-ordered diagenetic clay. The grains in the top left and bottom right of the image are feldspars showing partial dissolution.

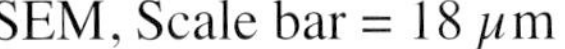

SEM, Scale bar = 18 μm

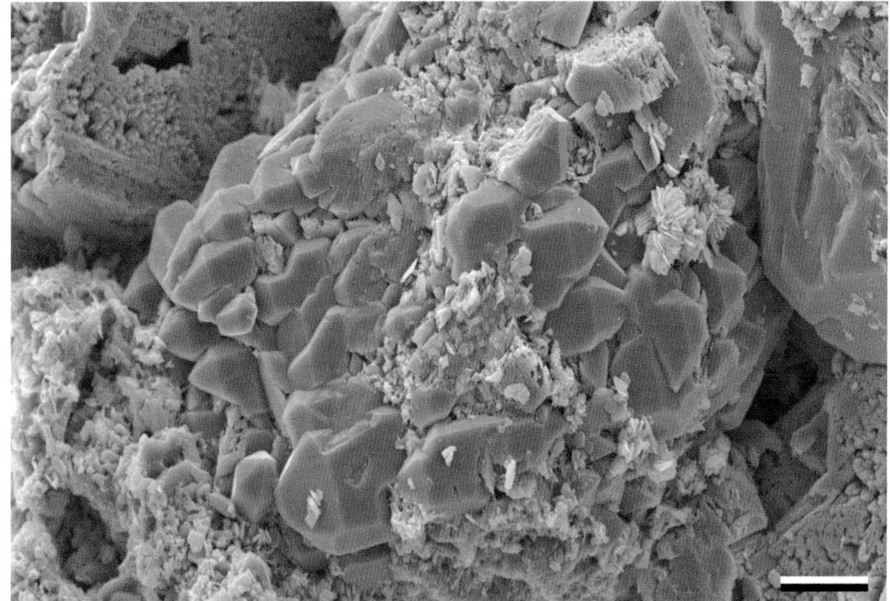

Lo. Jurassic Tilje Fm., Norwegian sector, North Sea †

This image shows large, euhedral, interlocking quartz crystals filling a pore. The surrounding grains have chlorite grain coats (examples near white arrows) that may prevent widespread quartz overgrowth formation by isolating the grains. These quartz cements have nucleated on breaks in the chlorite coatings. Along with quartz and chlorite cements, there also are hexagonal booklets of kaolinite (or possibly dickite) partially filling the porosity and especially clearly visible along the right edge of the image.

SEM, Scale bar = 13 μm

Lo. Jurassic Tilje Fm., Norwegian sector, North Sea †

Well-formed quartz overgrowths in an otherwise chlorite-cemented sandstone. The imprint of the chlorite coatings from an adjacent grain can be seen on the quartz overgrowth; the grain was separated during sample preparation. The growth of thin clay coatings, as seen here and in the previous image, can completely envelop grains. That ultimately can help to prevent further cementation, thus preserving the porosity and permeability of a sandstone (Ehrenberg, 1993). Thicker cement coatings, however, can negatively impact reservoir properties.

SEM, Scale bar = 17 μm

Lo. Cretaceous Travis Peak Fm., Eastland Co., Texas

A chert arenite cemented by radial megaquartz. Each detrital chert grain has a radially-oriented rim of bladed to equant megaquartz (drusy quartz). The crystal size of these cements coarsens into the pore. The monocrystalline quartz grain in this photomicrograph does not have an overgrowth on it. Megaquartz (as opposed to microquartz) is defined as having crystals larger than 20 μm.

XPL, Scale bar = 0.10 mm

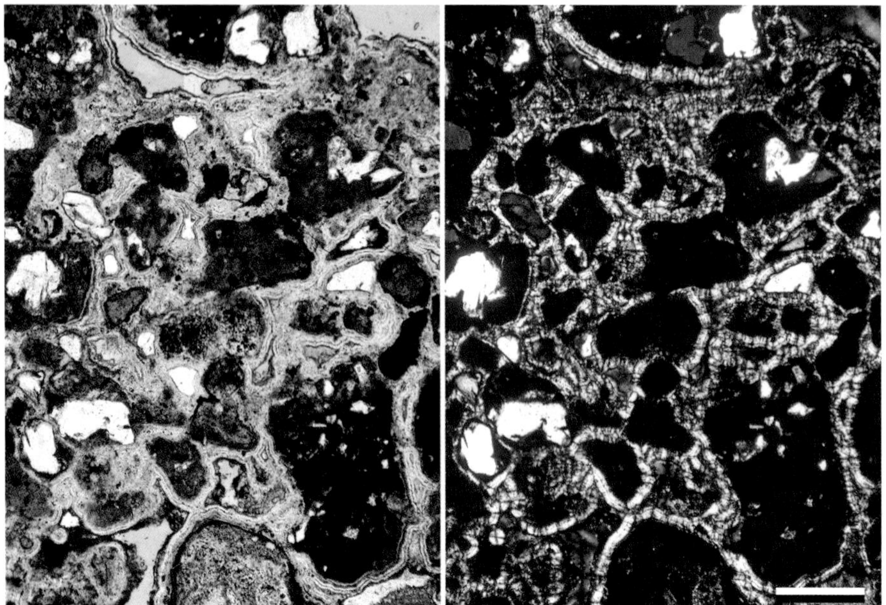

Pliocene – Pleistocene silcrete, east coast, Aruba

An incipient silcrete developed on a limestone terrace topped by alluvium containing abundant igneous rock fragments (the dark grains in this image) and quartz sand (the white grains in PPL) that may have acted as a source of silica. In this case, silica (mainly in the form of fibrous chalcedony) has produced layered crusts that surround and bind the lithic clasts. Thinner iron oxide/hydroxide cement layers also are present. This duricrust may have formed over a long time period (millions of years) in a semiarid setting—despite being a Caribbean island, Aruba averages less than 50 cm (20 in) of rain per year.

PPL | XPL, BDI, Scale bar = 0.26 mm

Paleogene Vieja Gp., Presidio Co., Texas

Several generations of vug-filling chalcedony in a brecciated volcaniclastic deposit are delineated by zones of iron-oxides (hematite) and trapped water-filled inclusions. The fibrous character of the chalcedony is clearly visible in the photomicrograph on the right. Most chalcedony consists of length-fast and/or length-slow fibrous quartz and moganite (also length-slow). Here, the initial cement is length-fast chalcedony (brown zones); it was followed by length-slow chalcedony (colorless bands shown by arrow). In the final phase, there are various cuts through the fibers, ranging from parallel to perpendicular to the fiber axes.

PPL | XPL | GP, Scale bar = 0.26 mm

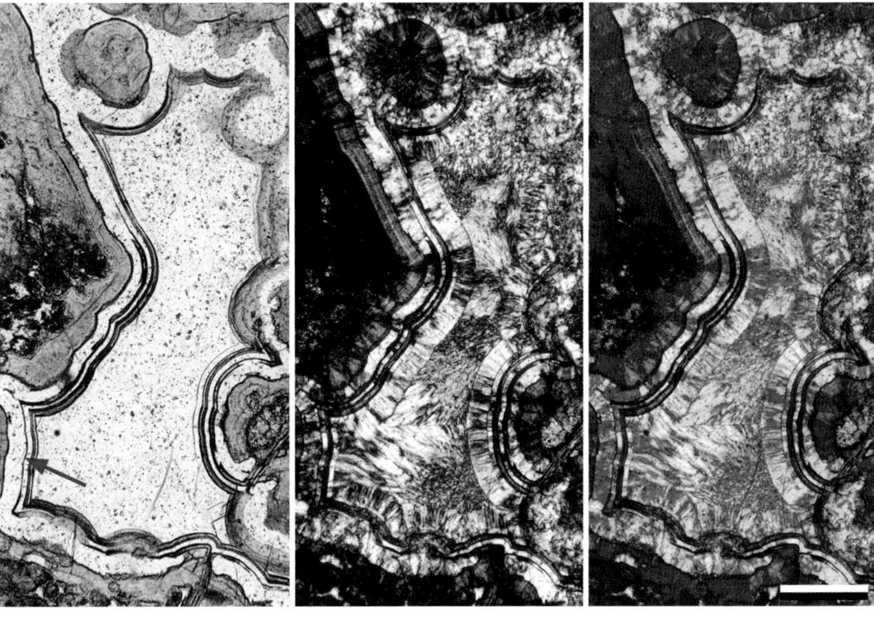

Paleogene Vieja Gp., Presidio Co., Texas

A series of laminated cherts filling a large void located between breccia fragments within a volcaniclastic rock. The void was first lined with several generations of chalcedony separated by hematitic bands. The final filling of the void was accomplished by a sequence of graded cherts. Note how the geopetal chert fills within the pore are mirrored within the open fracture (lower right). Although the origin of this type of graded chert filling is unclear, the texture is not uncommon.

PPL | XPL, Scale bar = 0.51 mm

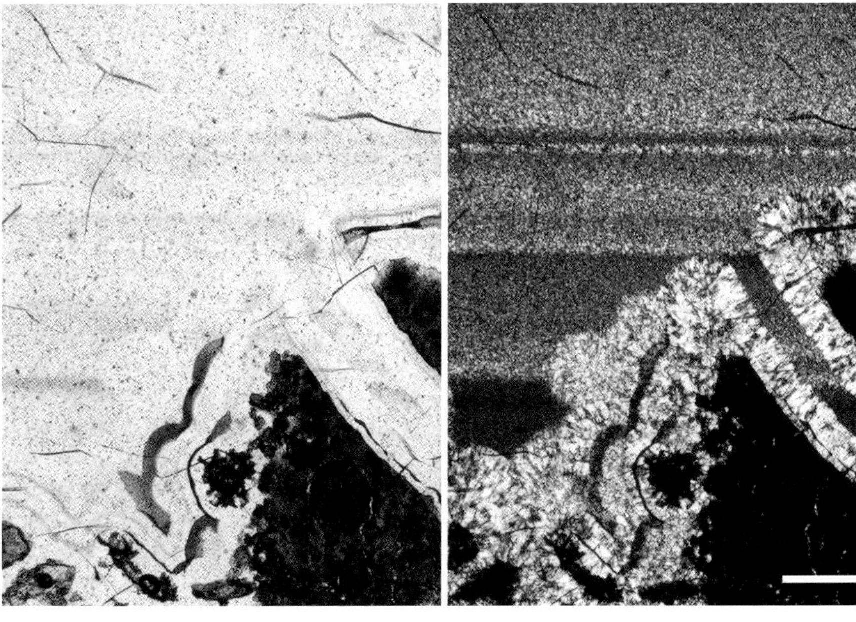

Paleogene Vieja Gp., Presidio Co., Texas

The earliest generation of chalcedony (the widespread, inclusion-rich brownish phase) is length-fast. Length-fast chalcedony has the crystallographic c-axis (slow axis) perpendicular to the length of the fibers. To determine if the chalcedony is length-fast or -slow, one must insert an oriented mineral plate (z, α = slow or x, γ = fast) into the microscope. Here, a length-fast gypsum plate was inserted from the SE quadrant. The birefringence colors for the brown chalcedony increase in the NW and SE quadrants (blues) and decrease in the NE and SW quadrants (yellows), thus indicating length-fast chalcedony. The opposite occurs in the clear, length-slow chalcedony areas.

PPL | GP, Scale bar = 0.10 mm

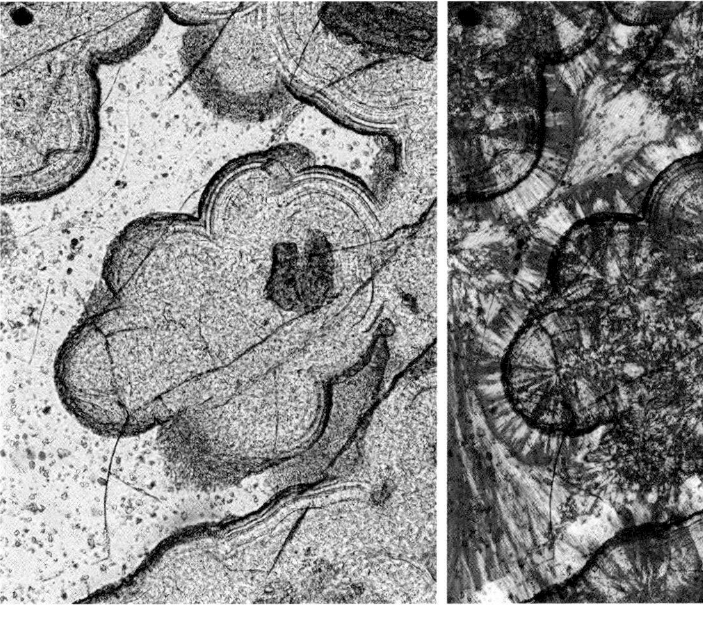

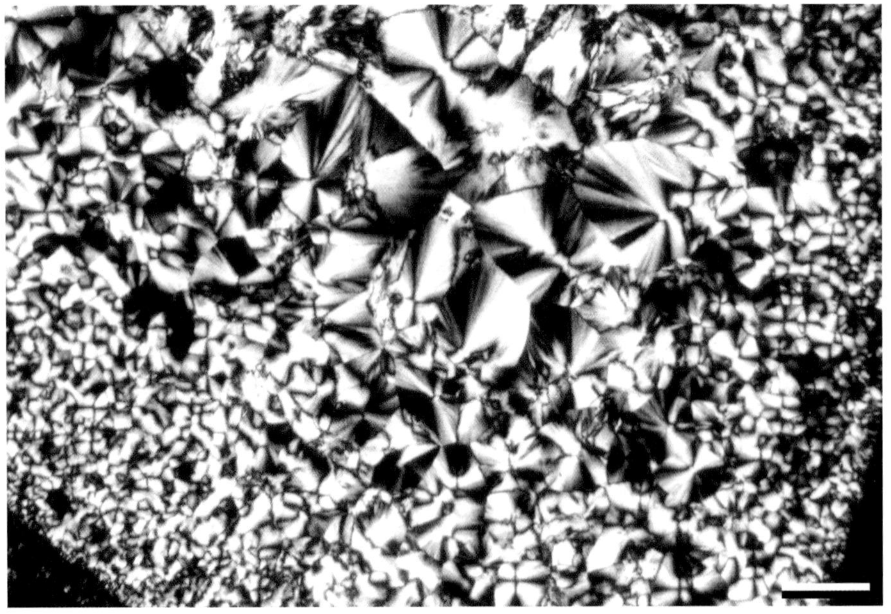

Up. Jurassic Radiolariti, northern Italy

Chalcedonic quartz infilling of a cavity. Commonly, as in this example, chalcedony occurs as radiating bundles of fibers. In this case, the bundles increase in size from the margins of the cavity to the center. Chalcedony, like most other forms of microquartz, contains numerous fluid-filled microinclusions. These inclusions are responsible for the relatively low refractive index of chalcedony.

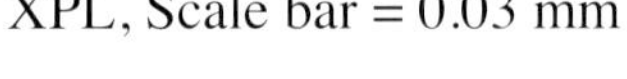

XPL, Scale bar = 0.03 mm

Up. Jurassic Radiolariti, northern Italy

In this photomicrograph, zebraic chalcedony lines a cavity. Zebraic chalcedony is a banded, fibrous form of microquartz. The fibers are alternately light and dark (as viewed along the fiber elongation direction under cross-polarized light) as a result of twisting of the crystal axes. McBride and Folk (1977) have described the association of zebraic chalcedony with evaporite minerals (such as halite, gypsum and anhydrite), but it has also been described in deep marine strata with no known associated evaporites (Keene, 1983), an interpretation more likely for this sample.

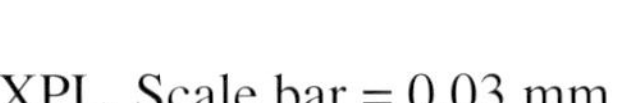

XPL, Scale bar = 0.03 mm

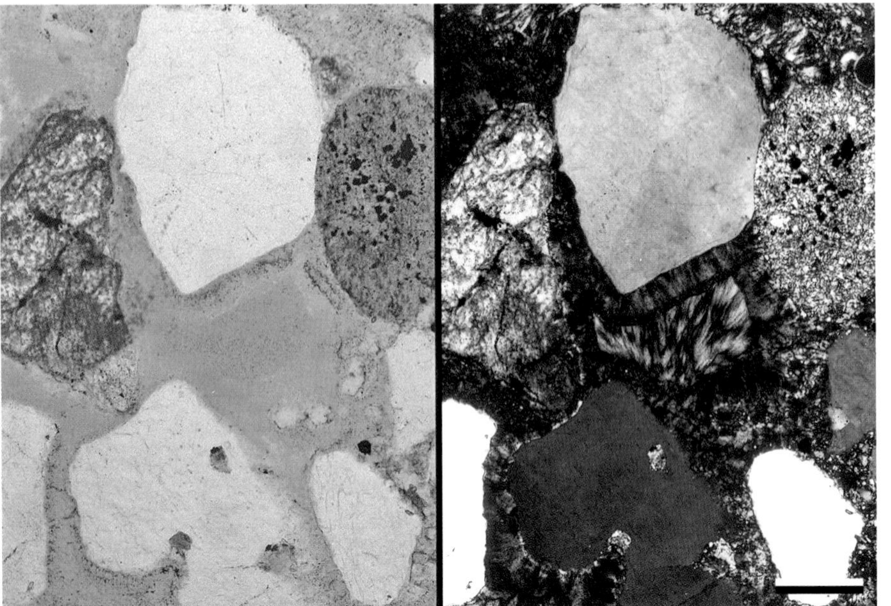

Mid. Jurassic Fulmar Fm., United Kingdom sector, North Sea †

Sandstone with chalcedony cement. The early diagenetic nature of the cement is evidenced by the loosely packed detrital grains. Other sandstones in this succession contain evidence of altered sponge spicules, a likely source of the silica. Some of the oversized pores may have contained spicular remains, but now contain a mixture of microcrystalline quartz with overlying chalcedony. Both silica cements here are highly porous and show impregnation by the blue resin.

PPL | XPL, KFS, BDI, Scale bar = 0.12 mm

Mid. Jurassic Fulmar Fm., United Kingdom sector, North Sea †

This image comes from a sponge spicule-rich sandstone. The cement consists of microcrystalline quartz which is arranged in a botryoidal habit. The quartz crystals are easily visible at the margins of the cement. In this case, the microporosity may have been further cemented by quartz, making the individual crystals difficult to distinguish.

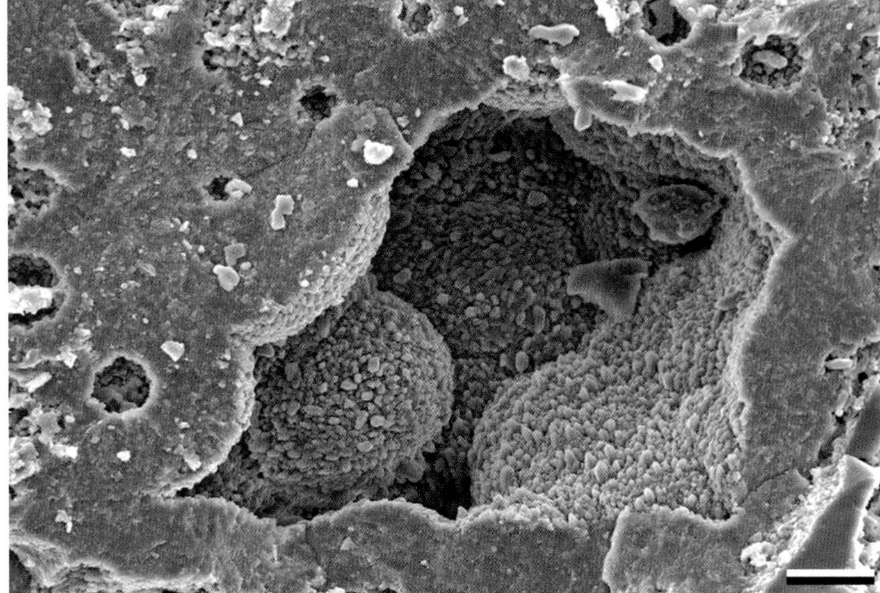

SEM, Scale bar = 14 μm

Mid. Eocene Jackson Gp., Carlos Sandstone, Bastrop Co., Texas

This is an opal- and quartz-cemented sandstone. Opal constitutes the first cement that formed within the pores. It is colorless, has low relief and is laminated. The opal cement is followed by a thin pore-rimming zone of chalcedony cement that has a slight brown coloration. The final stage of cementation is colorless chalcedony. These are volcanic-rich sediments, and the alteration of lithic fragments and siliceous ash is the source for the silica cements. Alteration and cementation most likely occurred in meteoric waters, and such deposits have been termed "groundwater silcretes" (McBride et al., 2012).

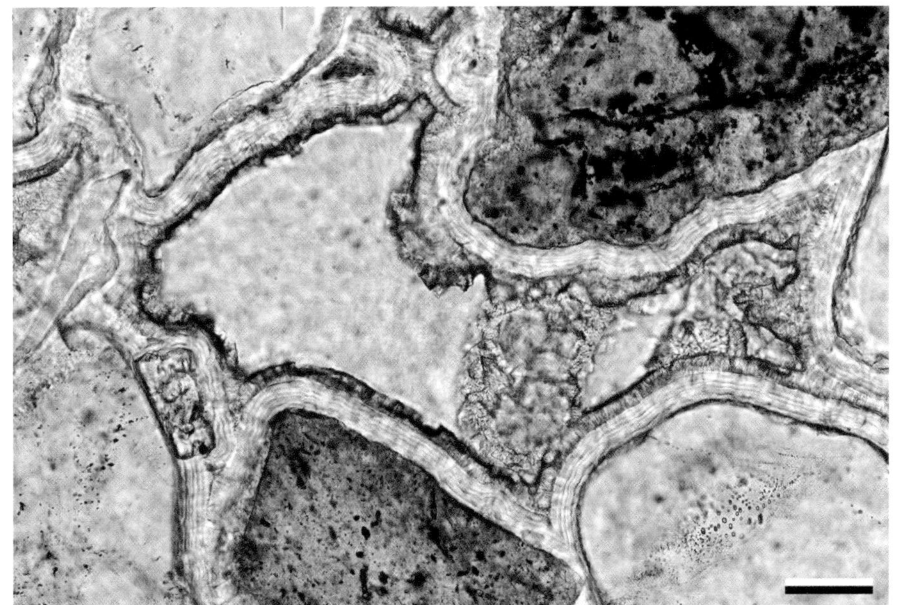

PPL, Scale bar = 0.05 mm

Mid. Eocene Jackson Gp., Carlos Sandstone, Bastrop Co., Texas

Within this pore, two separate generations of opal and chalcedony cementation are visible. A first-generation opal cement is followed by iron-stained chalcedony, then another zone of opal cement and then yet another generation of iron-stained chalcedony. The amorphous opal cements are isotropic (black bands) in cross-polarized light. The final filling of the pore is by zebraic chalcedony.

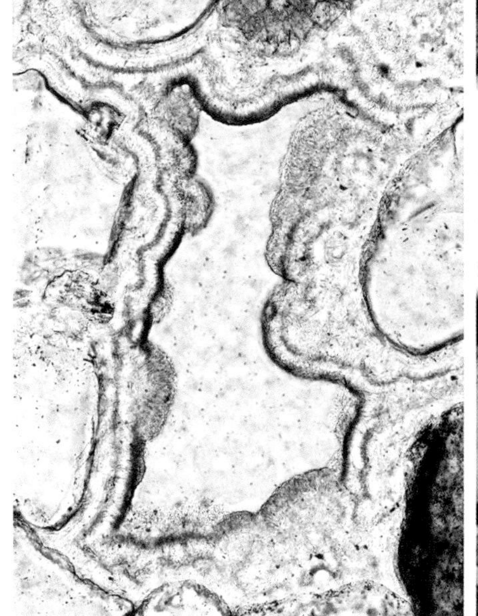

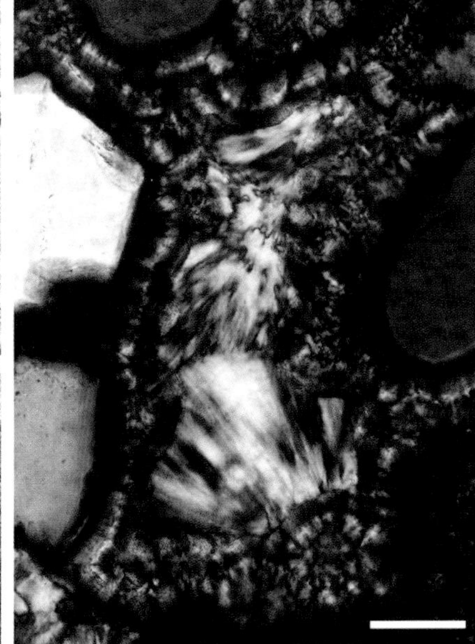

PPL | XPL, Scale bar = 0.06 mm

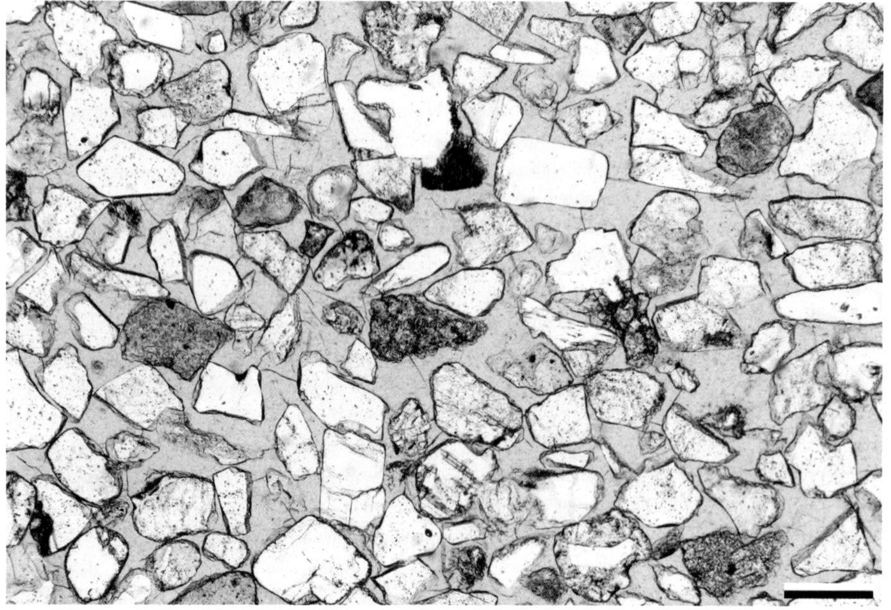

Oligocene Catahoula Fm., Fayette Co., Texas

An opal-cemented sandstone deposited in a probable fluvial environment. The opal cements are brownish and have lower relief than the grains. Opal generally is brownish because it contains abundant water-filled inclusions. Opal cement is completely isotropic under cross-polarized light (see next photo), because it is an amorphous, hydrous form of silica and thus is sometimes termed a mineraloid (a mineral-like material that does not demonstrate crystallinity).

PPL, Scale bar = 0.26 mm

Oligocene Catahoula Fm., Fayette Co., Texas

Here, isotropic opal cements surround quartz, feldspar, and other detrital grains. Opal cements commonly are associated with volcaniclastic (siliceous) sediments, because the source of the silica is from the dissolution of volcanic glass. Opal cements are chemically metastable and will, in time, dissolve or convert to more stable quartz cement.

PPL | XPL, Scale bar = 0.1 mm

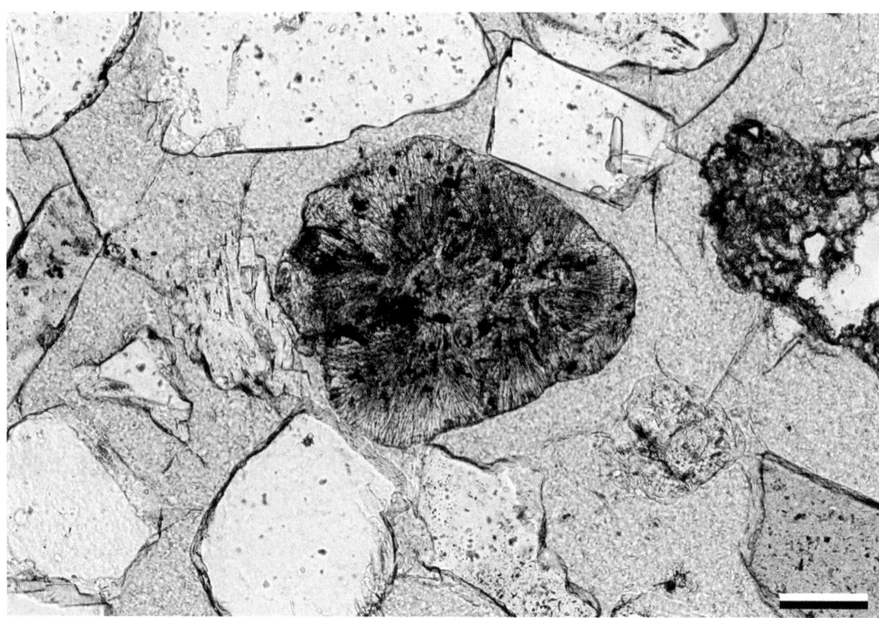

Oligocene Catahoula Fm., Fayette Co., Texas

An example of detrital volcanic rock fragments associated with the opal-cemented unit shown in the all three images on this page. The spherulitic volcanic grain in the center of the image may be a devitrified glass fragment. The needle-like crystals are probably plagioclase or tridymite (a high-temperature polymorph of quartz). During devitrification, glass converts to more stable quartz and releases silica into solution.

PPL, Scale bar = 0.1 mm

Up. Miocene Virgin Valley Fm., Harney Co., Oregon

These freshwater diatoms were cemented by opal (pale brown cements). As in previous examples, the opal is amorphous and isotropic in cross-polarized light. Here the source of opal, however was biogenic (diatoms) rather than volcanic. Because the opaline diatoms tests were unstable, they were either dissolved (with the voids later infilled) or they were directly replaced and partially filled by chalcedony. The interiors of the diatoms also are filled by chalcedony. A more detailed view of these diatoms and their fills can be seen in Chapter 5 (page 114).

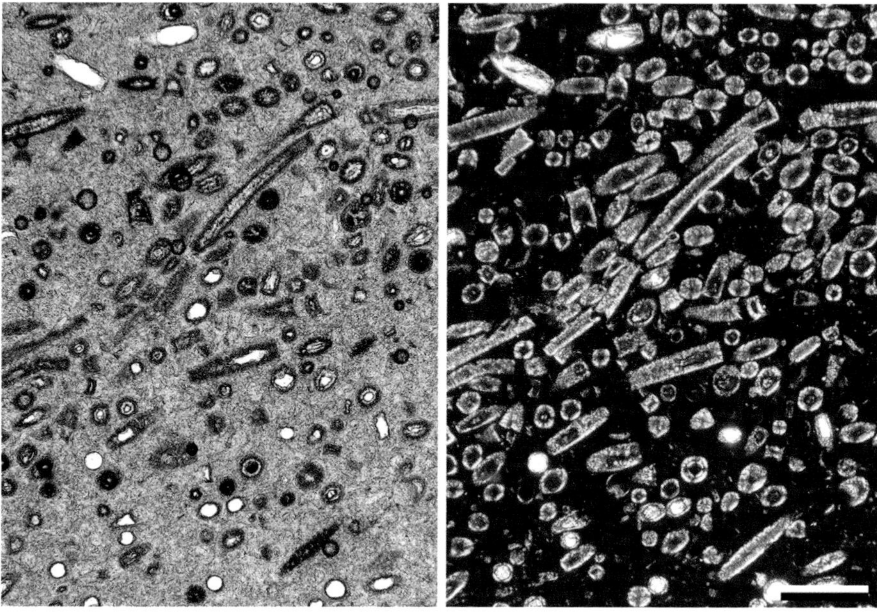

PPL | XPL, Scale bar = 0.26 mm

Cited References and Additional Information Sources

Aase, N. E., P. A. Bjørkum, and P. H. Nadeau, 1996, The effect of grain-coating microquartz on preservation of reservoir porosity: AAPG Bulletin, v. 80, p. 1654-1673.

Abdel-Wahab, A., A. M. K. Salem, and E. F. McBride, 1998, Quartz cement of meteoric origin in silcrete and nonsilcrete sandstones, Lower Carboniferous, western Sinai, Egypt: Journal of African Earth Sciences, v. 27, p. 277-290, doi: 10.1016/S0899-5362(98)00061-X.

Ajdukiewicz, J. M., and R. E. Larese, 2012, How clay grain coats inhibit quartz cement and preserve porosity in deeply buried sandstones: Observations and experiments: AAPG Bulletin, v. 96, p. 2091-2119, doi: 10.1306/02211211075.

Aplin, A. C., E. A. Warren, S. M. Grant, and A. G. Robinson, 1993, Mechanisms of quartz cementation in North Sea reservoir sandstones: Constraints from fluid compositions, *in* A. D. Horbury, and A. G. Robinson, eds., Diagenesis and Basin Development: Tulsa, OK, AAPG Studies in Geology 36, p. 5-22.

Berger, A., S. Gier, and P. Krois, 2009, Porosity-preserving chlorite cements in shallow-marine volcaniclastic sandstones: Evidence from Cretaceous sandstones of the Sawan gas field, Pakistan: AAPG Bulletin, v. 93, p. 595-615, doi: 10.1306/01300908096.

Bjørlykke, K., and P. K. Egeberg, 1993, Quartz cementation in sedimentary basins: AAPG Bulletin, v. 77, p. 1538-1548.

Bloch, S., R. H. Lander, and L. Bonnell, 2002, Anomalously high porosity and permeability in deeply buried sandstone reservoirs: Origin and predictability: AAPG Bulletin, v. 86, p. 301-328, doi: 10.1306/61EEDABC-173E-11D7-8645000102C1865D.

Cook, J. E., L. B. Goodwin, and D. F. Boutt, 2011, Systematic diagenetic changes in the grain-scale morphology and permeability of a quartz-cemented quartz arenite: AAPG Bulletin, v. 95, p. 1067-1088, doi: 10.1306/11151010009.

Dapples, E. C., 1967, Silica as an agent in diagenesis, *in* G. Larsen, and G. V. Chilingar, eds., Diagenesis in Sediments: New York, Elsevier Publishing Co., p. 323-342, doi: 10.1016/S0070-4571(08)70845-8.

Dunn, T. L., 1992, Infiltrated materials in Cretaceous volcanogenic sandstones, San Jorge Basin, Argentina, *in* D. W. Houseknecht, and E. D. Pittman, eds., Origin, Diagenesis, and Petrophysics of Clay Minerals in Sandstones: Tulsa, OK, SEPM Special Publication 47, p. 159-174, doi: 10.2110/pec.92.47.0159.

Dutton, S. P., 1997, Timing of compaction and quartz cementation from integrated petrographic and burial-history analyses, Lower Cretaceous Fall River Formation, Wyoming and South Dakota: Journal of Sedimentary Research, v. 67, p. 186-196, doi: 10.1306/D426852C-2B26-11D7-8648000102C1865D.

Dutton, S. P., and T. N. Diggs, 1990, History of quartz cementation in the Lower Cretaceous Travis Peak Formation, east Texas: Journal of Sedimentary Research, v. 60, p. 191-202, doi: 10.1306/212F914C-2B24-11D7-8648000102C1865D.

Ehrenberg, S. N., 1993, Preservation of anomalously high porosity in deeply buried sandstones by grain-coating chlorite: Examples from the Norwegian continental shelf: AAPG Bulletin, v. 77, p. 1260-1286.

Evans, J., A. J. C. Hogg, M. S. Hopkins, and R. J. Howarth, 1994, Quantification of quartz cements using combined SEM, CL, and image analysis: Journal of Sedimentary Research, v. 64, p. 334-338, doi: 10.1306/D4267D93-2B26-11D7-8648000102C1865D.

French, M. W., R. H. Worden, E. Mariani, R. E. Larese, R. R. Mueller, and C. E. Kliewer, 2012, Microcrystalline quartz generation and the preservation of porosity in sandstones: Evidence from the Upper Cretaceous of the Subhercynian Basin, Germany: Journal of Sedimentary Research, v. 82, p. 422-434, doi: 10.2110/jsr.2012.39.

Giles, M. R., S. L. Indrelid, G. V. Beynon, and J. Amthor, 2000, The origin of large-scale quartz cementation: Evidence from large data sets and coupled heat–fluid mass transport modelling, *in* R. H. Worden, and S. Morad, eds., Quartz Cementation in Sandstones (IAS Special Publication 29): Oxford, UK, Wiley-Blackwell, p. 21-38, doi: 10.1002/9781444304237.ch2.

Goldstein, R. H., and C. Rossi, 2002, Recrystallization in quartz overgrowths: Journal of Sedimentary Research, v. 72, p. 432-440, doi: 10.1306/110201720432.

Götze, J., 2012, Mineralogy, geochemistry and cathodoluminescence of authigenic quartz from different sedimentary rocks, *in* J. Götze, and R. Möckel, eds., Quartz: Deposits, Mineralogy and Analytics: New York, Springer, p. 287-306, doi: 10.1007/978-3-642-22161-3_13.

Grant, S. M., and N. H. Oxtoby, 1992, The timing of quartz cementation in Mesozoic sandstones from Haltenbanken, offshore mid-Norway: Fluid inclusion evidence: Journal of the Geological Society, v. 149, p. 479-482, doi: 10.1144/gsjgs.149.4.0479.

Haddad, S. C., R. H. Worden, D. J. Prior, and P. C. Smalley, 2006, Quartz cement in the Fontainebleau Sandstone, Paris Basin, France: Crystallography and implications for mechanisms of cement growth: Journal of Sedimentary Research, v. 76, p. 244-256, doi: 10.2110/jsr.2006.024.

Harwood, J., A. C. Aplin, C. I. Fialips, J. E. Iliffe, R. Kozdon, T. Ushikubo, and J. W. Valley, 2013, Quartz cementation history of sandstones revealed by high-resolution Sims oxygen isotope analysis: Journal of Sedimentary Research, v. 83, p. 522-530, doi: 10.2110/jsr.2013.29.

Hendry, J. P., and N. H. Trewin, 1995, Authigenic quartz microfabrics in Cretaceous turbidites: Evidence for silica transformation processes in sandstones: Journal of Sedimentary Research, v. 65, p. 380-392, doi: 10.1306/D42680CC-2B26-11D7-8648000102C1865D.

James, W. C., G. C. Wilmar, and B. G. Davidson, 1986, Role of quartz type and grain size in silica diagenesis, Nugget Sandstone, south-central

Wyoming: Journal of Sedimentary Research, v. 56, p. 657-662, doi: 10.1306/212F8A03-2B24-11D7-8648000102C1865D.

Keene, J. B., 1983, Chalcedonic quartz and occurrence of quartzine (length-slow chalcedony) in pelagic sediments: Sedimentology, v. 30, p. 449-454, doi: 10.1111/j.1365-3091.1983.tb00683.x.

Kelly, J. L., B. Fu, N. T. Kita, and J. W. Valley, 2007, Optically continuous silcrete quartz cements of the St. Peter Sandstone: High precision oxygen isotope analysis by ion microprobe: Geochimica et Cosmochimica Acta, v. 71, p. 3812-3832, doi: 10.1016/j.gca.2007.05.014.

Khalaf, F. I., 1988, Petrography and diagenesis of silcrete from Kuwait, Arabian Gulf: Journal of Sedimentary Research, v. 58, p. 1014-1022, doi: 10.1306/212F8EEA-2B24-11D7-8648000102C1865D.

Lander, R. H., R. E. Larese, and L. M. Bonnell, 2008, Toward more accurate quartz cement models: The importance of euhedral versus noneuhedral growth rates: AAPG Bulletin, v. 92, p. 1537-1563, doi: 10.1306/07160808037.

Lander, R. H., and O. Walderhaug, 1999, Predicting porosity through simulating sandstone compaction and quartz cementation: AAPG Bulletin, v. 83, p. 433-449.

Makowitz, A., R. H. Lander, and K. L. Milliken, 2006, Diagenetic modeling to assess the relative timing of quartz cementation and brittle grain processes during compaction: AAPG Bulletin, v. 90, p. 873-885, doi: 10.1306/12190505044.

Marchand, A. M. E., A. E. Fallick, R. S. Haszeldine, C. I. Macaulay, and R. Swennen, 2000, Quartz cementation inhibited by crestal oil charge: Miller deep water sandstone, UK North Sea: Clay Minerals, v. 35, p. 201-210, doi: 10.1180/000985500546585.

Mares, M. A., 1999, Encyclopedia of Deserts: Norman, OK, University of Oklahoma Press, 654 p.

McBride, E. F., 1989, Quartz cement in sandstones: A review: Earth-Science Reviews, v. 26, p. 69-112, doi: 10.1016/0012-8252(89)90019-6.

McBride, E. F., and R. L. Folk, 1977, The Caballos Novaculite revisited: Part II: Chert and shale members and synthesis: Journal of Sedimentary Research, v. 47, p. 1261-1286, doi: 10.1306/212F731A-2B24-11D7-8648000102C1865D.

McBride, E. F., R. L. Folk, and T. E. Yancey, 2012, Silica-cemented sandstones (groundwater silcretes) in outcrops of the Jackson Group, Texas Coastal Plain: Gulf Coast Association of Geological Societies Transactions, v. 62, p. 273-285.

Milliken, K. L., 2013, SEM-based cathodoluminescence imaging for discriminating quartz types in mudrocks: Unconventional Resources Technology Conference (Denver, CO, 12-14 August 2013), Paper 1582467, 10 p.

Mizutani, S., 1970, Silica minerals in the early stages of diagenesis: Sedimentology, v. 15, p. 415-436, doi: 10.1111/j.1365-3091.1970.tb02193.x.

Oelkers, E. H., P. A. Bjørkum, and W. M. Murphy, 1992, The mechanism of porosity reduction, stylolite development and quartz cementation in North Sea sandstones, *in* Y. K. Kharaka, and A. S. Maest, eds., Water-Rock Interaction 2, Proceedings of the Water-Rock Interaction Conference 7 (Park City Utah): Rotterdam, Balkema, p. 1183-1186.

Peltonen, C., Ø. Marcussen, K. Bjørlykke, and J. Jahren, 2009, Clay mineral diagenesis and quartz cementation in mudstones: The effects of smectite to illite reaction on rock properties: Marine and Petroleum Geology, v. 26, p. 887-898, doi: 10.1016/j.marpetgeo.2008.01.021.

Ramm, M., and K. Bjørlykke, 1994, Porosity/depth trends in reservoir sandstones: Assessing the quantitative effects of varying pore-pressure, temperature history and mineralogy, Norwegian Shelf data: Clay Minerals, v. 29, p. 475-490, doi: 10.1180/claymin.1994.029.4.07.

Robinson, A., and J. Gluyas, 1992, Duration of quartz cementation in sandstones, North Sea and Haltenbanken Basins: Marine and Petroleum Geology, v. 9, p. 324-327, doi: 10.1016/0264-8172(92)90081-O.

Schieber, J., D. Krinsley, and L. Riciputi, 2000, Diagenetic origin of quartz silt in mudstones and implications for silica cycling: Nature, v. 406, p. 981-985, doi: 10.1038/35023143.

Siever, R., 1959, Petrology and geochemistry of silica cementation in some Pennsylvanian sandstones, *in* H. A. Ireland, ed., Silica in Sediments: Tulsa, OK, SEPM Special Publication 7, p. 55-79, doi: 10.2110/pec.59.01.0055.

Siever, R., 1986, Burial diagenesis of sandstones, *in* F. A. Mumpton, ed., Studies in Diagenesis: Washington, DC, USGS Bulletin 1578, p. 237-248.

Sippel, R. F., 1968, Sandstone petrology, evidence from luminescence petrography: Journal of Sedimentary Research, v. 38, p. 530-554, doi: 10.1306/74D719DD-2B21-11D7-8648000102C1865D.

Spötl, C., D. W. Houseknecht, and S. J. Burns, 1996, Diagenesis of an 'overmature' gas reservoir: The Spiro sand of the Arkoma Basin, USA: Marine and Petroleum Geology, v. 13, p. 25-40, doi: 10.1016/0264-8172(95)00037-2.

Stone, W. N., and R. Siever, 1996, Quantifying compaction, pressure solution and quartz cementation in moderately-and deeply-buried quartzose sandstones from the Greater Green River Basin, Wyoming, *in* L. J. Crossey, R. Loucks, and M. W. Totten, eds., Siliciclastic Diagenesis and Fluid Flow: Concepts and Applications: Tulsa, OK, SEPM Special Publication 55, p. 129-150, doi: 10.2110/pec.96.55.0129.

Summerfield, M. A., 1983, Silcrete, *in* A. S. Goudie, and K. Pye, eds., Chemical Sediments and Geomorphology: Precipitates and Residua in the Near-surface Environment: London, Academic Press, p. 59-92.

Thyberg, B., and J. Jahren, 2011, Quartz cementation in mudstones: Sheet-like quartz cement from clay mineral reactions during burial: Petroleum Geoscience, v. 17, p. 53-63, doi: 10.1144/1354-079310-028.

Vagle, G. B., A. Hurst, and H. Dypvik, 1994, Origin of quartz cements in some sandstones from the Jurassic of the Inner Moray Firth (UK): Sedimentology, v. 41, p. 363-377, doi: 10.1111/j.1365-3091.1995.tb02107.x.

Walderhaug, O., 1990, A fluid inclusion study of quartz-cemented sandstones from offshore mid-Norway—possible evidence for continued quartz cementation during oil emplacement: Journal of Sedimentary Research, v. 60, p. 203-210, doi: 10.1306/212F9151-2B24-11D7-8648000102C1865D.

Walderhaug, O., 1994, Precipitation rates for quartz cement in sandstones determined by fluid-inclusion microthermometry and temperature-history modeling: Journal of Sedimentary Research, v. 64, p. 324-333, doi: 10.2110/jsr.64.324.

Walderhaug, O., 1996, Kinetic modeling of quartz cementation and porosity loss in deeply buried sandstone reservoirs: AAPG Bulletin, v. 80, p. 731-745.

Walderhaug, O., 2000, Modeling quartz cementation and porosity in Middle Jurassic Brent Group sandstones of the Kvitebjørn field, northern North Sea: AAPG Bulletin, v. 84, p. 1325-1339, doi: 10.1306/A9673E96-1738-11D7-8645000102C1865D.

Walderhaug, O., and P. A. Bjørkum, 2003, The effect of stylolite spacing on quartz cementation in the Lower Jurassic Stø Formation, southern Barents Sea: Journal of Sedimentary Research, v. 73, p. 146-156, doi: 10.1306/090502730146.

Wangen, M., 1999, Modelling quartz cementation of quartzose sandstones: Basin Research, v. 11, p. 113-126, doi: 10.1046/j.1365-2117.1999.00091.x.

Wilkinson, M., and R. S. Haszeldine, 2011, Oil charge preserves exceptional porosity in deeply buried, overpressured, sandstones: Central North Sea, UK: Journal of the Geological Society, v. 168, p. 1285-1295, doi: 10.1144/0016-76492011-007.

Williams, L. A., and D. A. Crerar, 1985, Silica diagenesis, II. General mechanisms: Journal of Sedimentary Research, v. 55, p. 312-321, doi: 10.1306/212F86B1-2B24-11D7-8648000102C1865D.

Wilson, M. D., 1992, Inherited grain-rimming clays in sandstones from eolian and shelf environments: Their origin and control on reservoir properties, *in* D. W. Houseknecht, and E. D. Pittman, eds., Origin, Diagenesis, and Petrophysics of Clay Minerals in Sandstones: Tulsa, OK, SEPM Special Publication 47, p. 209-225, doi: 10.2110/pec.92.47.0209.

Worden, R. H., and S. Morad, 2000, Quartz cementation in oil field sandstones: A review of the key controversies, *in* R. H. Worden, and S. Morad, eds., Quartz Cementation in Sandstones (IAS Special Publication 29): Oxford, UK, Blackwell Publishing Ltd., p. 1-20, doi: 10.1002/9781444304237.ch1

Worden, R. H., N. H. Oxtoby, and P. C. Smalley, 1998, Can oil emplacement prevent quartz cementation in sandstones?: Petroleum Geoscience, v. 4, p 129-137, doi: 10.1144/petgeo.4.2.129.

Facing Page: Top – Orthoclase overgrowths on rounded microcline grains, Cretaceous Prince Creek Fm., North Slope, Alaska (PPL | XPL). Bottom – Orthoclase overgrowth on a rounded microcline grain Permian Rush Springs Fm., Caddo Co., Oklahoma (XPL).

DIAGENESIS:
FELDSPAR CEMENTS

CHAPTER 12

INTRODUCTION TO FELDSPAR CEMENTS

Detrital feldspars most commonly undergo dissolution or alteration and replacement in subsurface settings, yet in many sandstones one also can find examples of authigenic feldspar cements. Such cements, in almost all cases, are either albite (the sodic end-member feldspar) or orthoclase (the K-spar end member)—see Chapter 2 for more details. Such feldspar cements commonly occur as thin or irregular overgrowths (not always in optical continuity with their detrital host grain) or as fills of microfractures within feldspar grains; however, they can form major pore-occluding cements in some clastic terrigenous rocks. The overgrowths can be monocrystalline or polycrystalline (sometimes even consisting of a mosaic of micron-sized rhombs—Worden and Rushton, 1992; De Ros et al., 1994). Despite their generally minor role in cementation of sandstones, the recognition of authigenic feldspars is important in understanding burial-related pore-fluid chemistry variations through time, especially because such overgrowths (K-feldspar overgrowths in particular) can be radiometrically dated using K–Ar and $^{40}Ar/^{39}Ar$ methods (e.g., Hagen et al., 2001; Mark et al., 2005 and 2008). In this regard especially, it should be noted that feldspar compositional zonation and overgrowth can be inherited from igneous, metamorphic or sedimentary source rocks, and such "overgrowths" must be carefully distinguished from authigenic cement precipitation during burial (just as recycled quartz overgrowths must be distinguished from ones produced *in situ* during burial).

Optical properties and recognition of feldspar cements:

The first step in identifying feldspar overgrowths is the characterization of the substrate feldspars on which they are growing (note that overgrowths are not always compositionally identical to their substrates). See Chapter 2 for guidelines to feldspar identification (including twinning, cleavage, zonation and alteration; however, staining is the most reliable way to identify feldspars under the microscope). The differentiation of overgrowths from detrital host grains is the next step in the process. Overgrowths generally are optically clear, especially when compared with the vacuolized or sericitized remains of detrital feldspars, although many exceptions can be found to this criterion. Albite overgrowths, like quartz overgrowths, are commonly set off from their detrital host grains by a zone of mineral and/or fluid inclusions. Saigal et al. (1988) and Chowdhury and Noble (1993) listed additional criteria for recognition of authigenic albite:

1. Homogeneous and pure albite composition (>99% Ab);
2. Euhedral habit of albite crystals with smooth crystal surfaces and sharp edges and corners;
3. General lack of twinning in overgrowths;
4. Visible growth of the replacement albite into adjacent pores;
5. Delicate skeletal texture in partially dissolved areas of albitized feldspar;
6. Occurrence of ankerite cement inclusions in authigenic albite; and
7. Absence of albite cements and replacements in zones with early carbonate cementation.

Cathodoluminescence petrography is another technique used to distinguish authigenic albite and orthoclase from detrital grains (Saigal, et al., 1988; Lee and Parsons, 2003); authigenic feldspars typically show little or no luminescence. Finally, because feldspar overgrowths commonly are not compositionally identical to their substrates, one can find stable feldspar overgrowths surrounding what is now secondary porosity. Compositionally selective dissolution yields skeletal grains with internal porosity formed where an unstable detrital feldspar was dissolved subsequent to stable overgrowth formation. In some cases, an additional generation of authigenic feldspar can precipitate within such skeletal feldspar grains. Such intragranular precipitation would be termed "replacement" by many workers, despite a potentially large time gap between dissolution and pore filling (e.g., Milliken, 1989).

Controls on formation:

Authigenic albite overgrowths have been reported to be contemporaneous precipitates with quartz overgrowths in the North Sea at temperatures of around 110°C and 2.5-km (8,200-ft) burial depth (even in the presence of some hydrocarbons as indicated by trapped oil inclusions; Nedkvitne et al., 1993). Schmid et al. (2004) reported a 90°C temperature for the onset of albite cementation in a basin west of Ireland, but also noted 105–110°C as the temperature of maximum albite authigenesis. Both of the latter temperature determinations were based on fluid inclusion microthermometry. Authigenic K-feldspar has also been reported to form early in the diagenetic history, "preceding most other common cementing minerals including calcite, laumontite and quartz" (Milliken, 1989).

There are many potential sources of K^+ or Na^+ for feldspar cementation. Common internal sources are dissolution of detrital feldspars or volcanic rock fragments and clay mineral transformations. External sources include diagenetic products derived from associated mudrocks, such as pressure dissolution of silt-sized feldspars, brines generated from dewatering of evaporites and possibly also through transformation of smectite to illite during burial diagenesis.

Up. Pennsylvanian Alamitos Fm., San Miguel Co., New Mexico

An altered albite grain that has been vacuolized, sericitized and iron-stained. The inclusion-poor overgrowth, also albite, was either chemically more stable than the detrital core or the grain alteration occurred prior to cementation. Note the euhedral terminations on the overgrowth that was encased by later calcite cement.

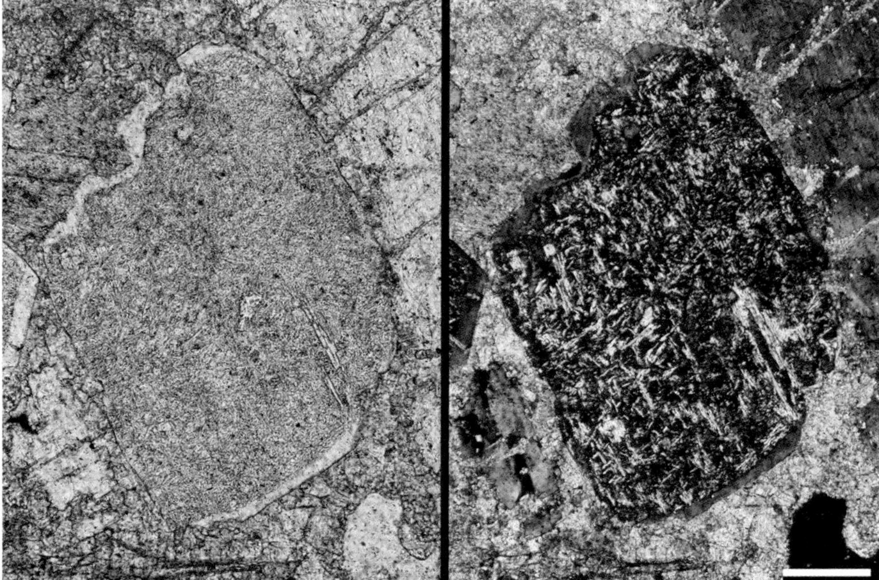

PPL | XPL, Scale bar = 0.12 mm

Mid. Ordovician Newtown Gneiss, New Haven Co., Connecticut

This photomicrograph looks similar to the feldspar overgrowth in the previous photomicrograph, but this grain is within a gneiss. The crystal is a zoned plagioclase in which the core and outer rim are clearly of different compositions. The core shows considerable alteration (mainly vacuolization and sericitization), whereas the rim is largely unaltered. This example illustrates that alteration can take place in source rocks and is highly composition dependent. It also shows that some primary zonation could easily be confused with overgrowths in reworked grains.

XPL, Scale bar = 0.08 mm

Oligocene Up. Spears Gp., Socorro Co., New Mexico

In this volcaniclastic sandstone, a rounded detrital plagioclase grain hosts an overgrowth of indeterminate origin. The slight rounding and sharp truncation of the overgrowth could indicate that it is also detrital. Regardless of origin, it was clearly chemically more stable than the grain it encloses. The dissolution of the original plagioclase grain occurred preferentially along the twin planes. Authigenic growth of plagioclase cements that occurred after core dissolution is evidenced by the crystals extending from the inner surface of the overgrowth into the void space within the leached grain (examples at black arrows).

PPL | XPL, BDI, Scale bar = 0.14 mm

Up. Cretaceous Nise Fm., Vøring Basin, Norwegian Sea †

A feldspar grain that has been partially dissolved. The remnants of the original feldspar are the long thin strands of undissolved lamellae that cross much of the partial void. The smaller, more irregularly-oriented crystals in the upper part of the grain (arrow) appear to have grown from that margin into the pore created by earlier dissolution. EDX spectra show that these crystals are albite, whereas most of the grains in this rock are either calcic plagioclase or K-feldspar. This further suggests a diagenetic origin. The dissolved grain is coated by a shell of clay and microcrystalline quartz which marks the outlines of the original grain.

SEM, Scale bar = 13 μm

Up. Cretaceous Prince Creek Fm., North Slope, Alaska

In the center of this photomicrograph is a microcline crystal with its characteristic tartan twinning. The original grain is rounded, with a feldspar overgrowth (white arrow) partially filling the adjacent open pore. Overgrowths on microcline are rarely twinned (unlike the detrital grain cores) and commonly have an orthoclase composition.

PPL | XPL, BDI, Scale bar = 0.1 mm

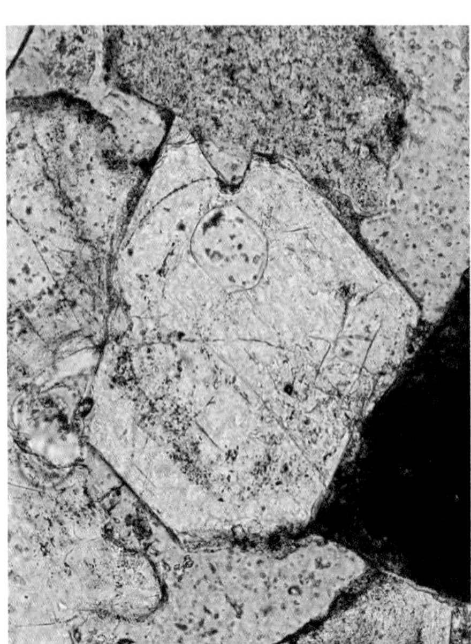

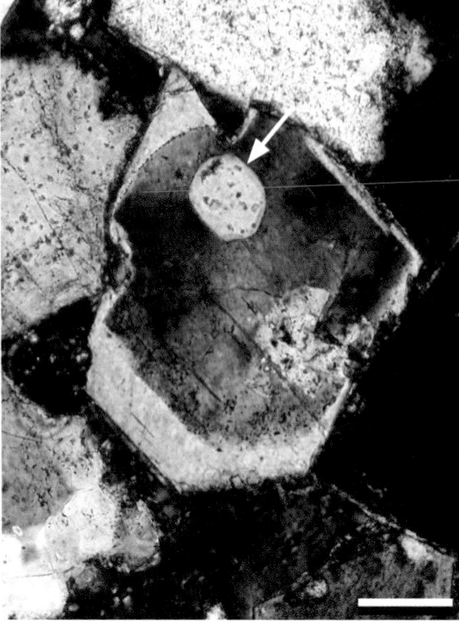

Up. Cretaceous Prince Creek Fm., North Slope, Alaska

Unlike quartz overgrowths, feldspar overgrowths usually have some compositional difference between the detrital grain core and its overgrowth. Here, a rounded orthoclase grain has an inclusion-poor, euhedral and clearly authigenic overgrowth. Because the overgrowth, likely albite, is compositionally different from the core grain, the birefringence is also different. The orthoclase grain also contains a large, euhedral quartz inclusion (white arrow).

PPL | XPL, BDI, Scale bar = 0.08 mm

Up. Cretaceous Prince Creek Fm., North Slope, Alaska

The partially altered detrital orthoclase core in the center of this image can be readily distinguished by its vacuoles (parallel to the cleavage). The core is surrounded by clear, unaltered overgrowths which formed in optical continuity with the detrital grain. The overgrowths are definitely authigenic, because they interlock with surrounding grains and other nearby overgrowths. When feldspar grains are inclusion-free, their overgrowths can be mistaken for quartz overgrowths in some cases.

PPL | XPL, BDI, Scale bar = 0.05mm

Mid. Permian Rush Springs Fm., Caddo Co., Oklahoma

This detrital orthoclase grain has authigenic overgrowths covering most of its surface. The grain is dark-colored (because it contains abundant inclusions) and has cleavage traces cutting across it. The orthoclase cements are clear, have euhedral terminations that extend into the porosity and show twinning (near the white arrow on the XPL image) that indicates that they are not in optical continuity with their host grain. Calcite cements (stained red) fill the remaining porosity.

PPL | XPL, AS, Scale bar = 0.05 mm

Lo. Pennsylvanian Morrow B Sandstone, Ochiltree Co., Texas

This inclusion-rich K-feldspar grain has a euhedral, syntaxial, inclusion-poor K-feldspar overgrowth. The feldspar overgrowth formed after quartz overgrowths, based on the character of their contacts—that is, the feldspar overgrowths precipitated around the euhedral quartz overgrowths (especially well shown along the lower margin of the feldspar). The timing of the ankerite cement at upper left is more problematic. The ankerite clearly precipitated after the quartz, but possibly simultaneously with the feldspar cements, based on the complex relations between those minerals near the upper tips of the feldspar.

PPL | XPL, RDI, Scale bar = 0.10 mm

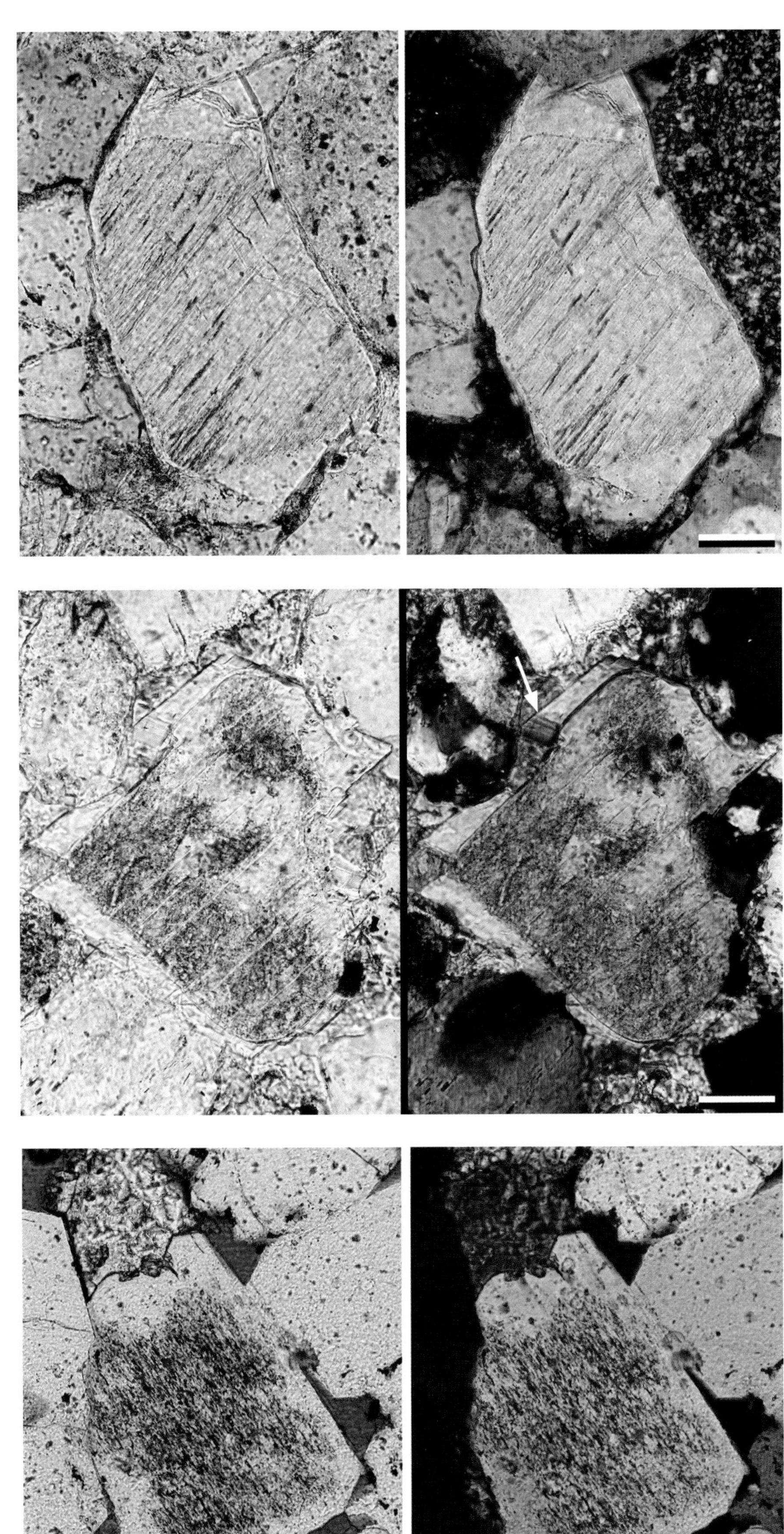

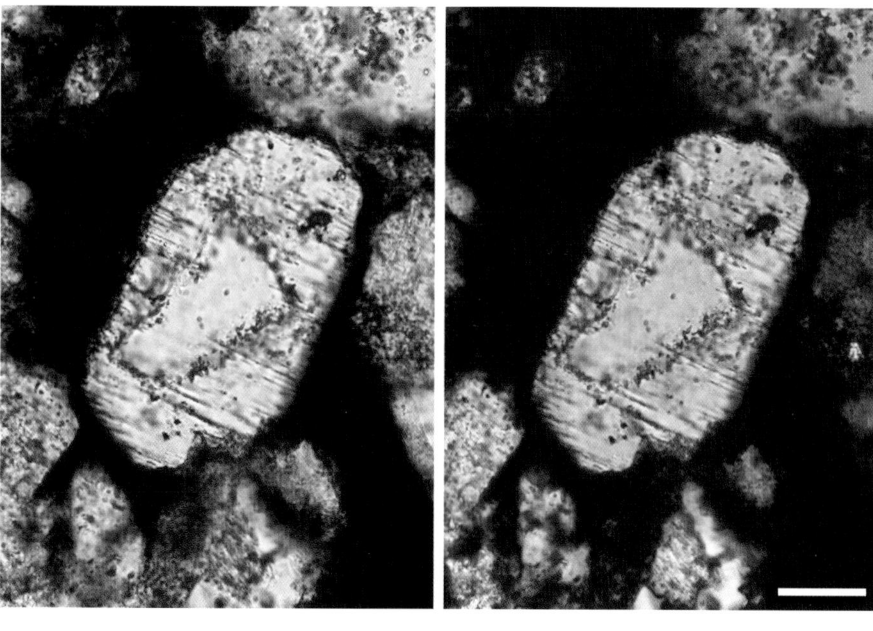

Paleogene Vieja Gp., Presidio Co., Texas

Not all feldspar grains are more inclusion rich than their overgrowths. In this example, an inclusion-free orthoclase grain is encased by an inclusion-rich overgrowth. The inclusions are aligned along the twin planes of the feldspar, but they do not pass into the host grain, clearly marking the outline of the host. Alternatively, however, the clear core could represent a host grain dissolved and replaced by albite after overgrowth formation.

PPL | XPL, Scale bar = 0.05 mm

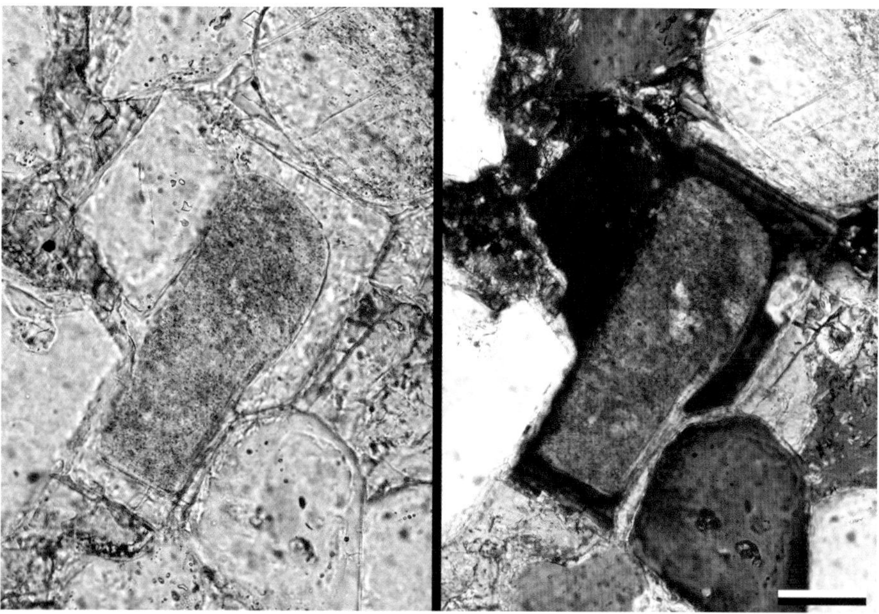

Mid. Permian Rush Springs Fm., Caddo Co., Oklahoma

The detrital orthoclase grain in the center of this image is brownish due to the presence of abundant inclusions. The core grain is moderately rounded and has large overgrowths that extend outward into the surrounding porosity. The difference in birefringence between the host grain and the overgrowth indicate that they are not in optically continuity. Normally not considered an important pore filling cement, this example shows that feldspar cements can significantly reduce local porosity and, in some cases, even overall reservoir porosity. Calcite cements (stained red) fill all remaining porosity in this case.

PPL | XPL, AS, Scale bar = 0.05 mm

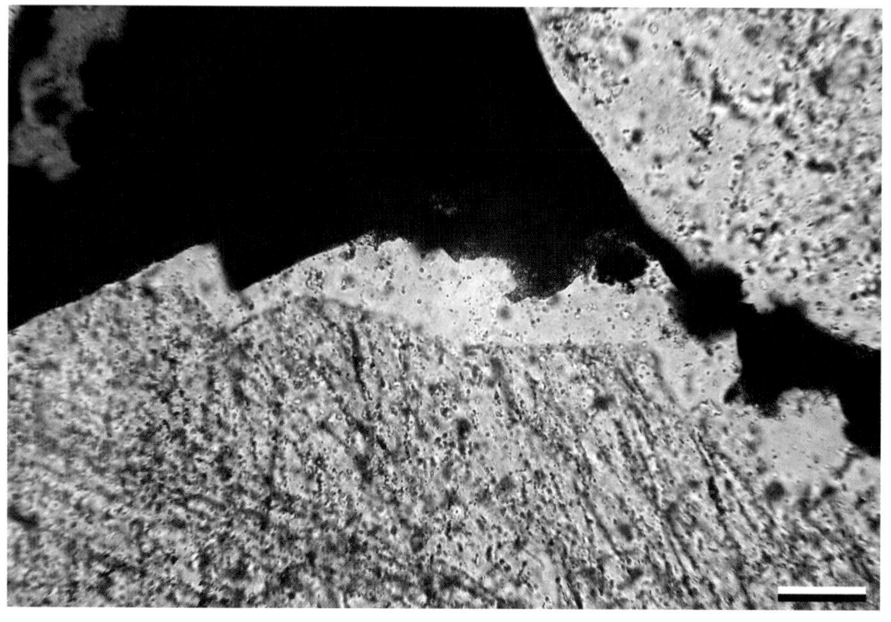

Pennsylvanian Conemaugh Gp., Athens Co., Ohio

This K-feldspar grain has extensive euhedral overgrowths. The feldspar grain is inclusion-rich, but the overgrowths are inclusion-poor, creating a clear demarcation between the grain and the overgrowth. The overgrowths are displacing iron-stained infiltrated clays associated with paleosols.

PPL, BDI, Scale bar = 0.04 mm

Lo. Pennsylvanian Morrow B Sandstone, Ochiltree Co., Texas

This K-feldspar grain underwent extensive leaching with only a few scattered, cloudy (inclusion-rich) remnants of the original crystal remaining (best seen in the PPL image). Iron oxide/clay cements outline the shape of the original grain as well as the fractures that cut through it prior to dissolution. The intragranular or moldic porosity that was produced by the dissolution was reduced by later albite overgrowths (white, inclusion-poor areas in the PPL view and near white first-order colors in XPL). Minor ankerite was the final phase of cementation within some of the secondary pore space.

PPL | XPL, RDI, Scale bar = 0.51 mm

Up. Cretaceous Prince Creek Fm., North Slope, Alaska

Feldspars commonly undergo dissolution in subsurface environments. This example shows a relatively early stage in which there has been dissolution of the original feldspar grain leaving behind the chemically more stable zones (especially the outer rim) and overgrowths (the most prominent one shown by red arrow). Note the strong control of grain dissolution patterns along preferred crystallographic directions within the mineral.

PPL, BDI, Scale bar = 0.04 mm

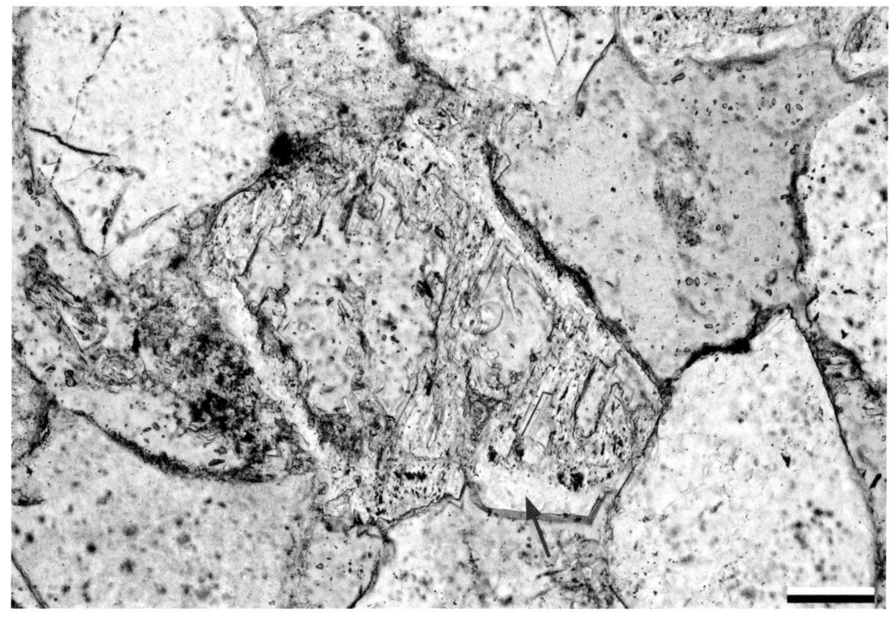

Cretaceous Dakota Gp., Moffat Co., Colorado

Conditions within a reservoir vary during the diagenetic history of the unit. A grain may be stable for a long period of time, then become unstable because of changing water chemistry. In this photomicrograph, the outline of an inclusion-rich (brown) orthoclase grain is visible. At some point, fluids moving through the unit dissolved the less stable feldspar (when compared to quartz). Later, orthoclase overgrowths (relatively inclusion free) formed inside the dissolution-generated secondary pore.

PPL | XPL, RDI, Scale bar = 0.12 mm

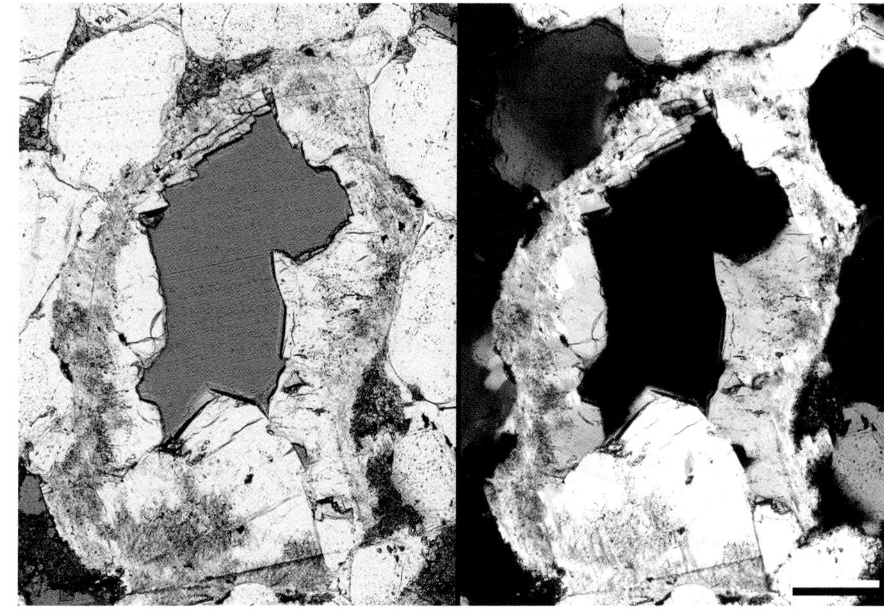

Up. Jurassic Fulmar Fm., United Kingdom sector, North Sea †

An SEM view of K-feldspar overgrowths. The detrital K-feldspar grains are partially dissolved (note especially the grain on the right side), but the cements are not. Where as the overgrowths may have formed after the dissolution event, it is also possible that the overgrowths precipitated prior to dissolution but have a composition different from the grains. This compositional variation may have made the cements less susceptible to dissolution during diagenesis.

SEM, Scale bar = 15 μm

Cited References and Additional Information Sources

Ali, A. D., and P. Tuner, 1982, Authigenic K-feldspar in the Bromsgrove Sandstone Formation (Triassic) of central England: Journal of Sedimentary Research, v. 52, p. 187-197.

Baskin, Y., 1956, A study of authigenic feldspars: Journal of Geology, v. 64, p. 132-155, doi: 10.1086/626330.

Chowdhury, A. H., and J. P. A. Noble, 1993, Feldspar albitization and feldspar cementation in the Albert Formation reservoir sandstones, New Brunswick, Canada: Marine and Petroleum Geology, v. 10, p. 394-402, doi: 10.1016/0264-8172(93)90083-5.

De Ros, L. F., G. N. C. Sgarbi, and S. Morad, 1994, Multiple authigenesis of K-feldspar in sandstones: Evidence from the Cretaceous Areado Formation, São Francisco Basin, central Brazil: Journal of Sedimentary Research, v. 64, p. 778-787, doi: 10.1306/D4267EBF-2B26-11D7-8648000102C1865D.

Glasmann, J. R., 1992, The fate of feldspar in Brent Group reservoirs, North Sea: A regional synthesis of diagenesis in shallow, intermediate, and deep burial environments, *in* A. C. Morton, R. S. Haszeldine, M. R. Giles, and S. Brown, eds., Geology of the Brent Group: London, GSL Special Publication 61, p. 329-350, doi: 10.1144/GSL.SP.1992.061.01.17.

Gold, P. B., 1987, Textures and geochemistry of authigenic albite from Miocene sandstones, Louisiana Gulf Coast: Journal of Sedimentary Research, v. 57, p. 353-362, doi: 10.1306/212F8B2A-2B24-11D7-8648000102C1865D.

González-Acebrón, L., J. Arribas, and R. Mas, 2010, Role of sandstone provenance in the diagenetic albitization of feldspars: A case study of the Jurassic Tera Group sandstones (Cameros Basin, NE Spain): Sedimentary Geology, v. 229, p. 53-63, doi: 0.1016/j.sedgeo.2010.06.005.

Götze, J., M. R. Krbetschek, D. Habermann, and D. Wolf, 2000, High-resolution cathodoluminescence studies of feldspar minerals, *in* M. Pagel, V. Barbin, P. Blanc, and D. Ohnenstetter, eds., Cathodoluminescence in Geosciences: Berlin, Springer-Verlag, p. 245-270, doi: 10.1007/978-3-662-04086-7_10.

Hagen, E., S. P. Kelley, H. Dypvik, O. Nilsen, and B. Kjølhamar, 2001, Direct dating of authigenic K-feldspar overgrowths from the Kilombero Rift of Tanzania: Journal of the Geological Society, v. 158, p. 801-807, doi: 10.1144/jgs.158.5.801.

Larsen, D., J. L. Loomis, and L. J. Crossey, 2000, Diagenesis in the Point Lookout Sandstone, San Juan Basin, New Mexico and Colorado: Influence of depositional conditions, cyclic stratigraphy, and changing hydrologic regime: The Mountain Geologist, v. 37, p. 109-133.

Lee, M. R., and I. Parsons, 2003, Microtextures of authigenic Or-rich feldspar in the Upper Jurassic Humber Group, UK North Sea: Sedimentology, v. 50, p. 597-608, doi: 10.1046/j.1365-3091.2003.00567.x.

Mark, D. F., S. P. Kelley, M. R. Lee, J. Parnell, S. C. Sherlock, and D. J. Brown, 2008, Ar-Ar dating of authigenic K-feldspar: Quantitative modelling of radiogenic argon-loss through subgrain boundary networks: Geochimica et Cosmochimica Acta, v. 72, p. 2695-2710, doi: 10.1016/j.gca.2008.03.018.

Mark, D. F., J. Parnell, S. P. Kelley, M. Lee, S. C. Sherlock, and A. Carr, 2005, Dating of multistage fluid flow in sandstones: Science, v. 309, p. 2048-2051, doi: 10.1126/science.1116034.

Milliken, K. L., 1989, Petrography and composition of authigenic feldspars, Oligocene Frio Formation, South Texas: Journal of Sedimentary Research, v. 59, p. 361-374, doi: 10.1306/212F8F94-2B24-11D7-8648000102C1865D.

Nedkvitne, T., D. A. Karlsen, K. Bjørlykke, and S. Larter, 1993, Relationship between reservoir diagenetic evolution and petroleum emplacement in the Ula Field, North Sea: Marine and Petroleum Geology, v. 10, p. 255-270, doi: 10.1016/0264-8172(93)90108-5.

Saigal, G. C., S. Morad, K. Bjørlykke, P. K. Egeberg, and P. Aagaard, 1988, Diagenetic albitization of detrital K-feldspar in Jurassic, Lower Cretaceous, and Tertiary clastic reservoir rocks from offshore Norway, I. Textures and origin: Journal of Sedimentary Research, v. 58, p. 1003-1013, doi: 10.1306/212F8EE5-2B24-11D7-8648000102C1865D.

Schmid, S., R. H. Worden, and Q. J. Fisher, 2004, Diagenesis and reservoir quality of the Sherwood Sandstone (Triassic), Corrib Field, Slyne Basin, west of Ireland: Marine and Petroleum Geology, v. 21, p. 299-315, doi: 10.1016/j.marpetgeo.2003.11.015.

Sherlock, S. C., T. Lucks, S. P. Kelley, and A. Barnicoat, 2005, A high resolution record of multiple diagenetic events: Ultraviolet laser microprobe Ar/Ar analysis of zoned K-feldspar overgrowths: Earth and Planetary Science Letters, v. 238, p. 329-341, doi: 10.1016/j.epsl.2005.07.018.

Worden, R. H., and J. C. Rushton, 1992, Diagenetic K-feldspar textures: A TEM study and model for diagenetic feldspar growth: Journal of Sedimentary Research, v. 62, p. 779-789, doi: 10.1306/D42679D8-2B26-11D7-8648000102C1865D.

Facing Page: Top – Illite/smectite clays lining a pore, Cretaceous Prince Creek Fm., North Slope, Alaska (PPL). Bottom – Kaolinite booklets formed after the first generation of quartz overgrowths and followed by a second generation of quartz cementation, Cretaceous Dakota Gp., Moffat Co., Colorado (PPL).

DIAGENESIS:
Clay Cements

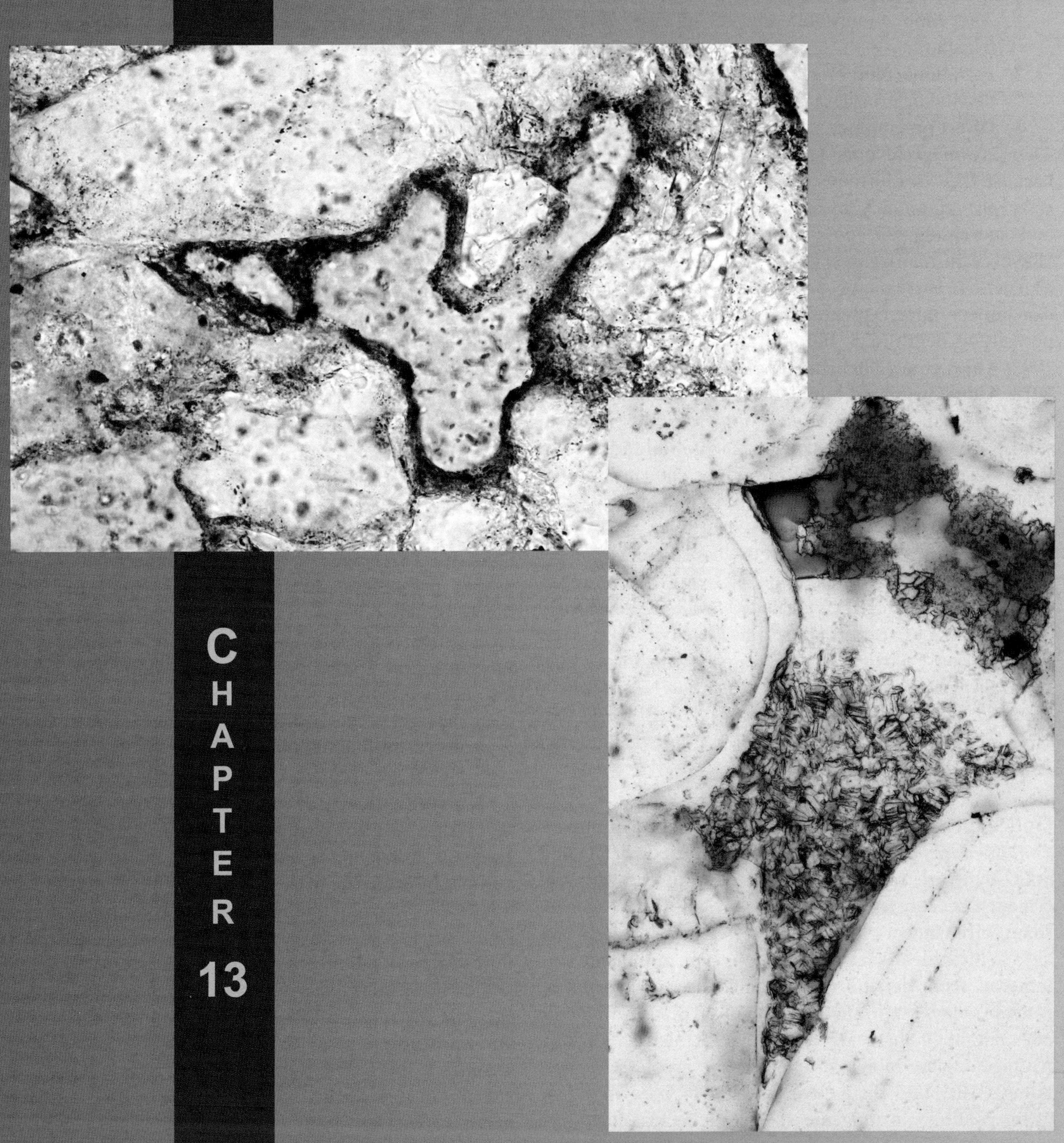

CHAPTER 13

INTRODUCTION TO CLAY CEMENTS

Clay minerals are a complex family of aluminosilicates, and generalized chemical formulas for them can be found in Table 11.2 (page 247). They can be either platy or fibrous with a high degree of chemical substitution. The word "clay" also has grain size connotations, so when referring to clay cements, it is best to call them "clay minerals". Clay minerals occur in siliciclastic rocks as detrital and/or diagenetic components, which are commonly difficult to differentiate. Diagenetic clay minerals, the focus of this chapter, form in several ways:

1. Alteration of unstable silicate minerals, such as feldspars;
2. Pseudomorphic or neomorphic transformation of detrital or precursor diagenetic clays; or
3. Direct precipitates.

They are of great interest to the oil and gas industry, because they can have a significant impact on sandstone reservoirs, commonly lowering porosity and permeability and increasing the possibility of formation damage. However, in many cases, clays (especially chlorite) form eogenetic to early mesogenetic grain-coating cements that can impede later cementation and preserve exceptional porosity at depth (e.g., Pittman et al., 1992; Ehrenberg, 1993; Anjos et al., 2003; Berger et al., 2009; Gould et al., 2010; Ajdukiewicz and Larese, 2012).

The main diagenetic clay minerals observed in sandstones are kaolinite, chlorite, smectite, illite and mixed layer varieties, such as illite/smectite or chlorite/smectite (corrensite). Some clay minerals have polymorphs (e.g., kaolinite and dickite). Other clays, such as palygorskite or berthierine, are rare and are not discussed in detail. Clay minerals are best identified using XRD analysis; however, the clay fraction in XRD typically is considered to be the $< 4\ \mu m$ material. This can create some issues with clay XRD identifications, particularly in samples with larger clay mineral species, such as kaolinite which can be $> 20\ \mu m$ in diameter and form elongate "verms". SEM/EDS and thin section analysis are required to differentiate detrital and authigenic phases where possible.

Optical properties and recognition of clay cements:

Optical properties, such as refractive index (RI) and optic sign, are rarely used in identifying clay mineral cements because of their small crystal sizes. Instead, color, birefringence and multi-crystal morphology are keys to their identification. Distinguishing authigenic clay minerals from detrital ones is aided by a few features. Detrital clays typically are smaller than authigenic ones, although there are exceptions to this relationship. Authigenic clay minerals tend to have greater chemical homogeneity and distinctive patterns of pore filling or grain coating. Authigenic clays can be classified into several fabrics, using the terminology of Neasham (1977) and Wilson and Pittman (1977):

1. Discrete (not intergrown) crystals;
2. Intergrown crystal pore linings or pore-fillings;
3. Pore bridging crystals;
4. Fracture fillings; and
5. Pseudomorphic replacements.

Diagenetic clay minerals can, in some cases, contribute to "diagenetic pseudomatrix" during burial and compaction (Fröhlich et al., 2010), although most pseudomatrix is derived from compressed sedimentary clay-rich rock fragments. The specific characteristics of the more commonly encountered clay minerals are given below.

Chamosite is a diagenetic Fe-rich chlorite. It is normally finely crystalline and green (brown when weathered). It commonly is associated with ooids, grain coats and, in some cases, also forms pore-filling cement. Because of its finely crystalline nature, chamosite normally displays dark gray anomalous birefringence. Chamosite is also commonly associated with other iron-rich minerals, such as siderite, and may, in part, be an alteration product of earlier iron-bearing clays such as berthierine (e.g., Hornibrook and Longstaffe, 1996).

Glauconite is a broad term for greenish peloids that include a range of clay minerals composed of glauconitic smectite to glauconitic mica. Glauconite typically is an early diagenetic precipitate associated with marine pore fluids. In most cases, glauconite is finely crystalline, has a bright-green color in plane-polarized light and shows an anomalous, speckled birefringence in cross-polarized light. It is rarely observed as pore-filling cement although it can precipitate within fractures, within carbonate bioclasts and borings. Thus, it is most common as a replacement, especially in fecal pellets. Glauconitic pellets are susceptible to compaction and are commonly squeezed into pores forming a pseudomatrix that can be confused with deformed chamosite. Studies have shown that it typically forms in outer-shelf settings with from 50–150 m (165–500 ft) water depth.

Kaolinite/dickite is an aluminosilicate clay mineral that typically forms discrete (patchy), pore-filling cement. Kaolinite commonly is associated with feldspar alteration and is intermixed with authigenic illite/smectite clay minerals. During modal analyses, it is important to count whether it is in primary or secondary pores. In thin section, kaolinite can occur as cloudy white masses and/or individual blocky crystals or vermicular clusters. It is microporous and has a dark gray birefringence with an appearance similar to chert. Kaolinite can occur in secondary, oversized pores and can form through the alteration of volcanic ash by organic acids. In SEM, kaolinite displays pseudohexagonal plates and commonly forms stacks of crystals with a book- or worm-like (vermicular) habit. Dickite is thought to be the high temperature form of kaolinite (Beaufort et al., 1998) and forms fracture-fills associated with high-temperature dewatering of deep basins (Houseknecht and Ross, 1992). However, dickite has also been reported to occur in soils and shallow reservoirs (Anovitz et al., 1991; De Bona et al., 2008). Dickite tends to form larger crystals with thicker "plates", but it can be definitively identified only through XRD.

Smectite group clays form as predominantly eogenetic cements under oxidizing conditions in alkaline waters (Jeans, 1978) and in soils in semi-arid settings. Smectites are not very stable during burial and transform to mixed-layer illite/smectite and ultimately to illite with increasing temperatures (e.g., Lynch et al., 1997; McKinley et al., 2003). In thin section, smectite can occur as flakes or as poorly formed grain coats. In SEM, smectite commonly exhibits a web-like, grain-coating texture, with an intricate boxwork pattern. Smectites can have a wide range of compositions; therefore, distinguishing between different minerals that make up the smectite group requires study beyond thin section and SEM/EDS. The birefringence of smectite is very similar to that of illite. Illite, illite/smectite and smectite have similar habit and optical properties in thin section and only can be distinguished by careful use of XRD techniques that test for expandability. Smectitic clays can swell when in contact with fresh water and thus can be partly lost during thin-section preparation.

Illite is primarily a grain-coating and pore-bridging clay, which in some cases, completely fills pores. The crystals are normally so thin that they do not display correct optical properties. In many instances though, a high birefringence can be seen. Illite typically exhibits a fibrous morphology, with crystals up to 10 μm long, 0.1–1 μm wide, but only a few nm thick (as seen in SEM). Illite forms grain-coating precipitates, with thin strands and hair-like crystals extending perpendicular to the grain. Illite commonly is associated with the alteration of kaolinite and smectite or degradation of feldspars or micas. In rocks exposed to deeper burial and elevated temperatures, illite can be coarse enough to observe under the petrographic microscope, where it resembles muscovite. K–Ar and, more recently, $^{40}Ar/^{39}Ar$ radiometric dating of clay mineral cements (Hamilton, 2003; Clauer et al., 2012), especially illites and mixed layer illite/smectites, has proven valuable in providing radiometric ages for diagenetic events. However, reliable dating requires high-resolution *in-situ* analysis or effective separation of specific authigenic phases from other clays, especially detrital ones.

Chlorite is green in thin section and, in some cases, individual crystals are large enough to be visible under the microscope. Chlorite group clays have a characteristic platy, grain coating morphology and anomalous blue birefringence. The optical properties are not easily seen (even the larger crystals are still too small for such observations), so chlorites are recognized by color (green), platy to vermicular habit and anomalous birefringence. Chlorite coatings have been shown globally to be very effective in inhibiting the growth of quartz cements (Ehrenberg, 1993). Eogenetic chlorites generally form under reducing conditions from marine waters; mesogenetic chlorite forms from alteration of precursor green marine clays, trioctaedral smectites, and ferromagnesian grains through a reported range of temperatures from 100-200°C (see Table 13.1 and Spötl et al., 1994 and 1996).

Controls on formation:

During early diagenesis, the formation of clay minerals is controlled by depositional facies, sediment composition and climate. Common early diagenetic clays include kaolinite, berthierine, glauconite and smectite (along with some mixed-layer clays and palygorskite). Mg-smectites and palygorskite form as fibers in evaporative lacustrine, fluvial and eolian sediments and are commonly associated with paleosols or with gypsum, anhydrite or halite deposits. They are rare in sediments subjected to considerable burial (see McKinley et al., 2003 for diagenetic controls on smectite distribution). Smectite and illite/smectite form from the breakdown of volcanic ash (producing bentonites—e.g., Altaner and Grim, 1990; Jeans et al., 2000) as well as volcanic rock fragments and micas and so may be associated in thin section with degraded grains and zeolites. Smectites and other clays that were infiltrated by groundwater to form clay cutans typically are arranged tangential to host grains (e.g., Walker et al., 1978); authigenic smectites, on the other hand, generally are oriented perpendicular to their substrates.

Some green clay minerals such as berthierine, odinite (serpentinite group clays) and glauconite form as grain coats, pellets, ooids and void fillings or as replacements of detrital grains (Odin and Matter, 1981). They are common in marginal marine, deltaic and shelf settings (Hornibrook and Longstaffe, 1996) and are commonly associated with phosphate (suggesting nutrient-rich waters) (Worden and Morad, 2003). Berthierine and/or odinite are interpreted to be the precursors of some burial diagenetic chlorite (Ehrenberg, 1993; Ryan and Hillier, 2002).

Early kaolinite generally forms in terrestrial sediments deposited in humid climates, through the action of acidic pore waters on feldspars and micas (Emery et al., 1990). Albite and plagioclase are more prone to alteration than K-feldspars and break down to yield kaolinite, silica and Na and Ca ions (Worden and Morad, 2003).

During burial diagenesis, temperature, pressure and pore water chemistry all influence the precipitation of clay minerals, although temperature is the key governing

control. Those burial-related transformations, including subsurface clay-producing alterations of feldspars and lithic grains, are summarized in Table 13.1. In general, there is a wholesale diagenetic shift in the clay mineral constituents of sandstones during progressive burial from early-formed green marine clays, smectite-group clays and kaolinite, to a higher temperature and pressure assemblage, including dickite, chlorite and illite.

Table 13.1 Tabulation of common diagenetic clay mineral reactions and transformations in sandstones including reaction products, approximate temperature ranges of substantial mineral development and typical reaction byproducts. Note overall clay mineralogical shift from early-formed (eogenetic) green marine clays, kaolinite and smectite to later (higher temperature/mesogenetic) chlorite and illite. Key to numbered data sources (1) Aagaard et al., 2000; (2) Altaner and Grim, 1990; (3) Beaufort et al., 1998; (4) Bjørkum and Gjelsvik, 1988; (5) Bjørlykke et al., 1995; (6) Boles and Franks, 1979; (7) De Ros et al., 1994; (8) Ehrenberg and Nadeau, 1989; (9) Ehrenberg et al.,1993; (10) Emery et al., 1990; (11) Hillier, 1994; (12) Ketzer et al., 2003a; (13) Lander and Bonnell, 2010; (14) Lanson et al., 2002; (15) Leone et al., 1983; (16) McAulay et al., 1993; (17) McKinley et al., 2003; (18) Milliken, 2003; (19) Odin and Matter, 1981; (20) Osborne et al., 1994; (21) Ryan and Reynolds,1996; (22) Ryan et al., 1998; (23) Worden and Morad, 2003.

Starting materials	Temperature (degrees centigrade) 30 60 90 120 150 180	Other products (where known or deduced)	Source
illite, muscovite or kaolinite	glauconite		12, 19
feldspars or micas and/or low-pH meteoric waters	kaolinite	either silicic acid and K ions released or Ca ions	10, 18, 20
volcanic ash	smectite	Na, Ca, and bicarbonate ions	2, 15
dioctahedral smectite + K and Al ions	smectite-illite illite-smectite illite	Na, Ca, Fe, Mg, Si ions released with water	6, 17
trioctahedral smectite	mixed layer clays chlorite	quartz	11, 22
berthierine/odinite?	chlorite		1, 6, 9, 21
ferromagnesian grains	chlorite		7, 23
kaolinite + K-feldspar	illite	quartz and possibly albite	4, 5, 8, 13, 14
kaolinite	dickite		3, 9, 16
feldspar + acidic waters	illite	quartz	16, 23
kaolinite + Fe, Mg ions	chlorite		6, 9
dickite	illite		23

Chamosite Cements

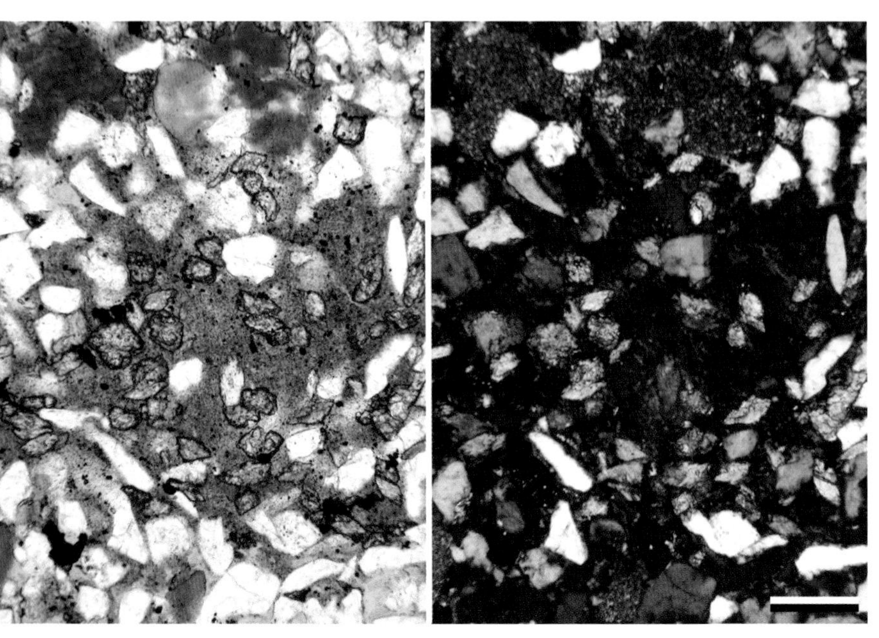

Lo. Cretaceous (Albian) Nahr Umr Fm., offshore Qatar

This shallow-marine sediment from a slow-sediment accumulation interval shows a common assemblage of near-seafloor to early burial cements. Chamosite (center) comprises the cloudy, pale greenish areas of extremely finely-crystalline cements (perhaps after berthierine precursors). The chamosite is accompanied by glauconite (mainly as peloidal grains), siderite (the irregular, flattened rhombs) and calcium phosphate (the darker brown areas).

PPL | XPL, BDI, Scale bar = 0.16 mm

Lo. Jurassic Tilje Fm., Norwegian sector, North Sea †

Chamosite can be the dominant cement in some rocks. It is an iron-rich chlorite and commonly is replaced by iron hydroxides or carbonates to produce ironstones. Some of the chamosite here forms coats around detrital grains. Kaolinite and siderite postdate chamosite, suggesting an early origin for the chamosite. Chamosite, siderite and pyrite are commonly found together in marine sandstones. Their formation involves the presence of reducing conditions, bacterial oxidation and, depending on the amount of organic matter, generation of post-oxic, sulfidic and/or methanic environments (Mücke, 2006).

PPL, BDI, Scale bar = 0.14 mm

Lo. Cretaceous (Albian) Nahr Umr Fm., offshore Qatar

Chamositic cement and small, authigenic pyrite framboids in a shallow-marine sandstone. The weak cementation by these minerals has allowed considerable compaction (especially deformation of soft, chamositic and glauconitic grains) despite relatively little overburden loading. Chamosite cements can have berthierine precursors that form during very early sulfate reduction in Fe-rich, organic-rich, phosphatic and glauconitic sediments and both of those early cements can be precursors for later, grain-rimming chlorite cements (Pe-Piper and Weir-Murphy, 2008).

PPL, BDI, Scale bar = 0.08 mm

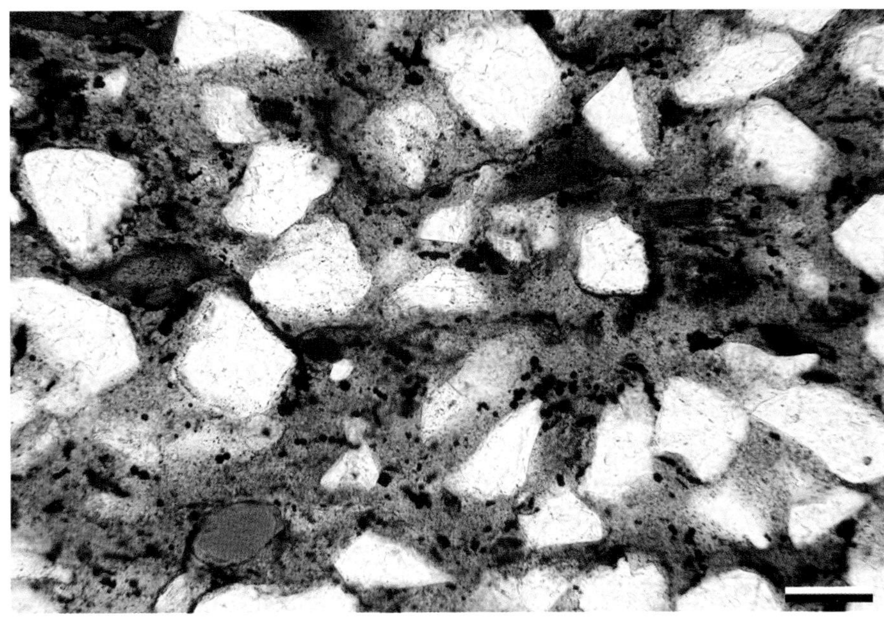

Mid. Jurassic Curtis Fm., Moffat Co., Colorado

Although glauconite is most common in sandstones as pellets or altered clays, it is also found as an early diagenetic cement, especially in association with carbonate bioclasts. In this example, intragranular porosity within a bryozoan fragment was completely filled by authigenic glauconite. Foraminiferal tests, echinoderm plates and other porous skeletal fragments commonly show such fills in glauconitic sandstones.

PPL, BDI, Scale bar = 0.10 mm

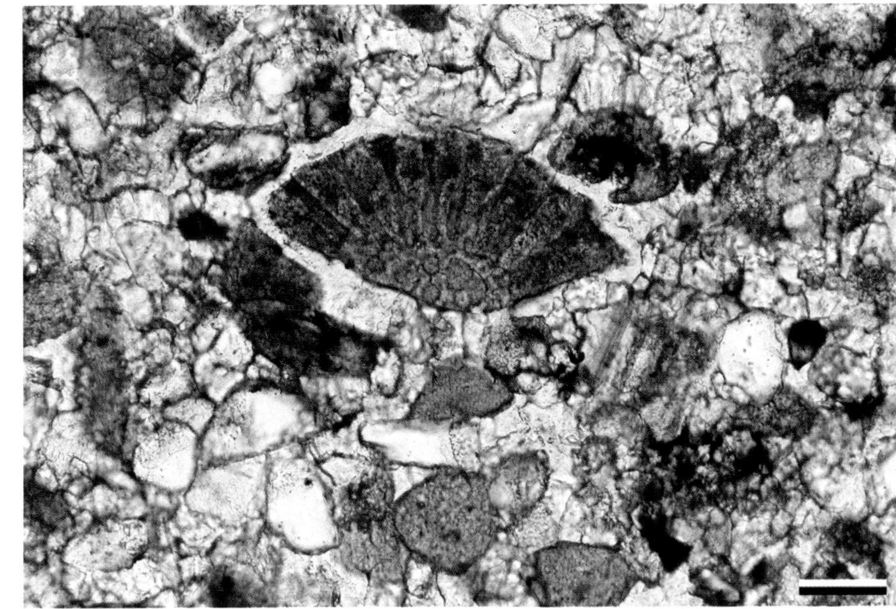

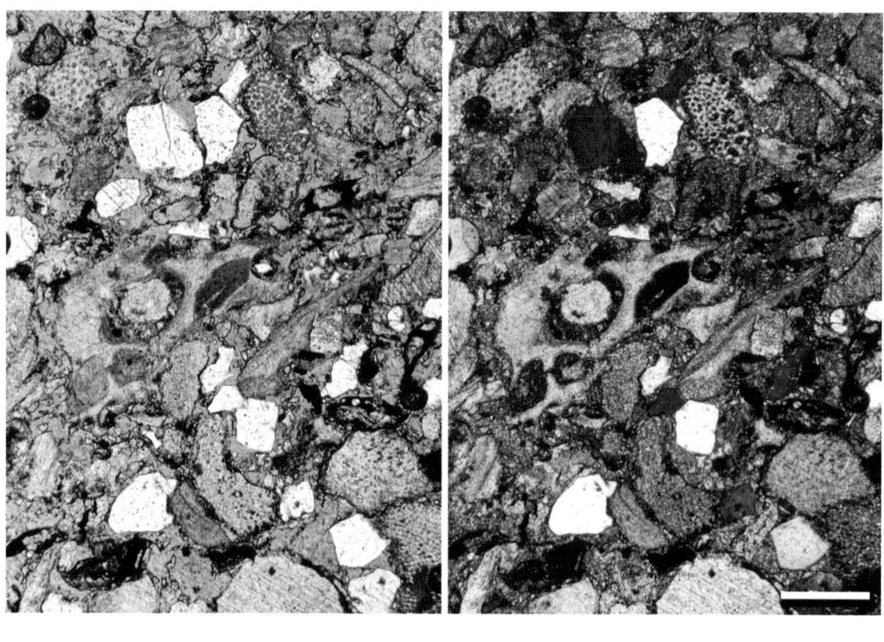

Pennsylvanian – Permian Nuka Fm., North Slope, Alaska

A glauconite-cemented calcareous sandstone. The grains with high birefringence are calcitic echinoderm and bryozoan fragments; quartz grains with low birefringence colors also are present. The greenish interstitial material is glauconite which is only rarely found as a substantial intergranular cement (although it commonly fills small intragranular voids such as those shown here and in the previous photograph). Glauconite cements are largely restricted to marine units and are inferred to form very early, probably at or near the seafloor.

PPL | XPL, Scale bar = 0.33 mm

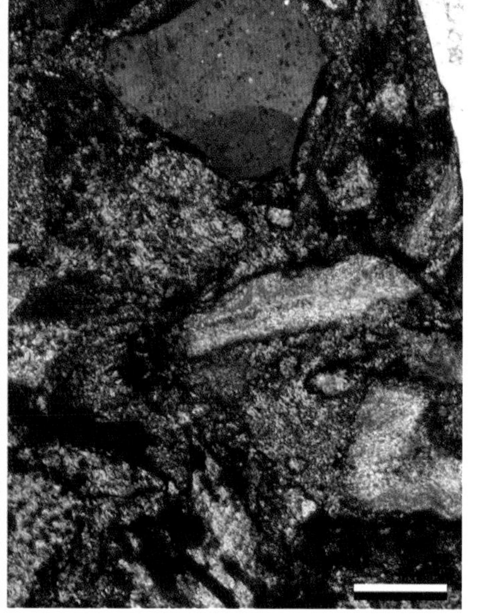

Pennsylvanian – Permian Nuka Fm., North Slope, Alaska

A higher magnification image of the glauconitic and calcareous sandstone shown above. Although most of the pore-filling glauconite consists of squashed peloidal glauconitic grains, there also is some additional growth of glauconite cement in this sandstone that completes the porosity occlusion. This slide also contains abundant carbonate skeletal fragments, some of which have pores infilled with precipitated glauconite cement (at bottom left corner, for example). The glauconite cements show the characteristic dark green, speckled appearance of this mineral in both plane- and cross-polarized light views.

PPL | XPL, Scale bar = 0.10 mm

Lo. Jurassic Tilje Fm., Norwegian sector, North Sea †

Aggregates of blocky to locally vermicular (booklet-like) kaolinite within a primary pore. The plates consist of euhedral, subhexagonal crystals, which display high degrees of elongation along the a- or b-axis direction. The elongation of kaolinite appears to result from changes during deeper burial (Beaufort et al. 1998). Loosely packed kaolinite such as this is prone to movement within a hydrocarbon reservoir and thus can cause "formation damage".

SEM, Scale bar = 15.9 μm

Cretaceous Lange Fm., Norwegian sector, North Sea †

This image shows grain-coating flakes of chlorite and microcrystalline quartz partly enveloped within quartz overgrowths (as at yellow arrow). The crystalline aggregate on the left of the image is most likely pore-filling kaolinite/dickite. Some of the coarser dickite crystals are associated with smaller and more, probable kaolinite, suggesting that the dickite may have resulted from alteration of kaolinite during burial.

SEM, Scale bar = 7.4 μm

Mid. Permian Wegener Halvø Fm., Karstryggen area, East Greenland

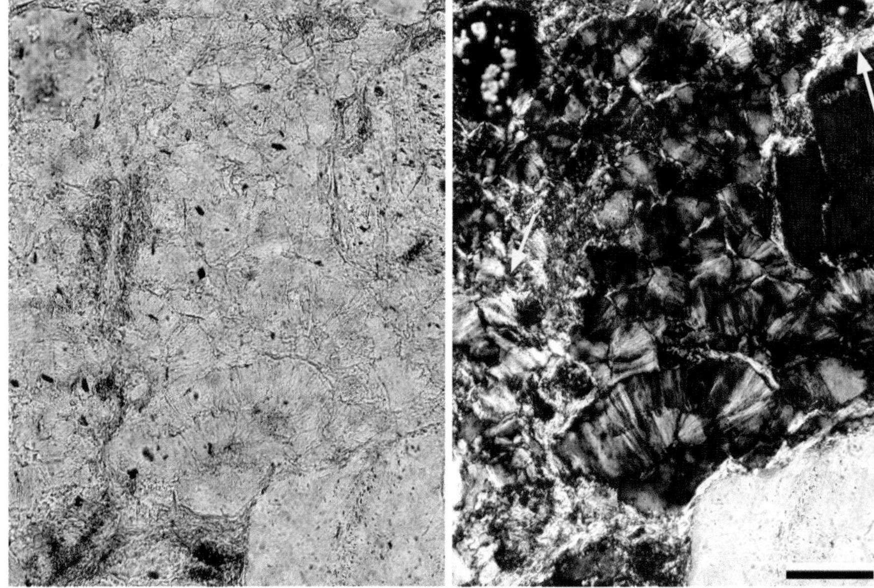

Authigenic kaolinite/dickite cements grew as booklets or stacks of plates in this sandstone (best seen in the cross-polarized light image). Considerable porosity remains in interstices between crystal stacks, as shown by blue-dyed epoxy in the plane-polarized light image. Petrographic distinction of kaolinite from its polymorph, dickite, is difficult and unreliable in thin section. Kaolinite is more common as an early cement (<50°C); dickite predominates in higher temperature settings, but exceptions occur. Minor illite (bright white to yellow birefringence) is also present in places (examples near yellow arrows).

PPL | XPL, AFeS, BDI, Scale bar = 0.05 mm

Cretaceous unknown unit, Moffat Co., Colorado

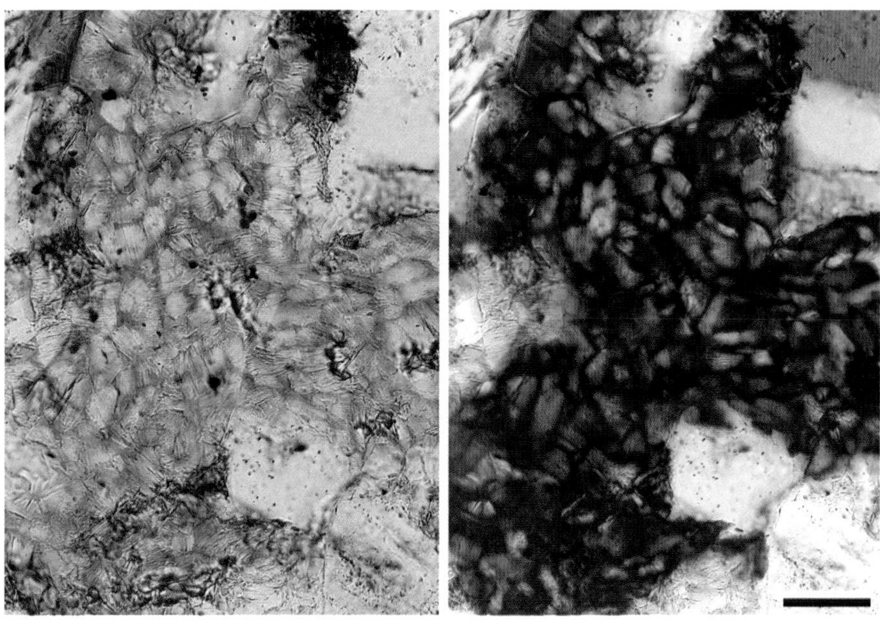

Another example of authigenic vermicular kaolinite cement booklets in the pore space of a sandstone. Here again, considerable microporosity remains in the interstices between crystal stacks as shown by the red-dyed epoxy in the plane-polarized light image; however, such cements clearly reduce local permeability. These are the earliest cements in this unit and may well have formed through relatively early influx of low-pH, meteoric pore fluids. Sardini et al. (2009) discuss methodologies to map and quantify microporosity in such clay cement aggregates.

PPL | XPL, RDI, Scale bar = 0.05 mm

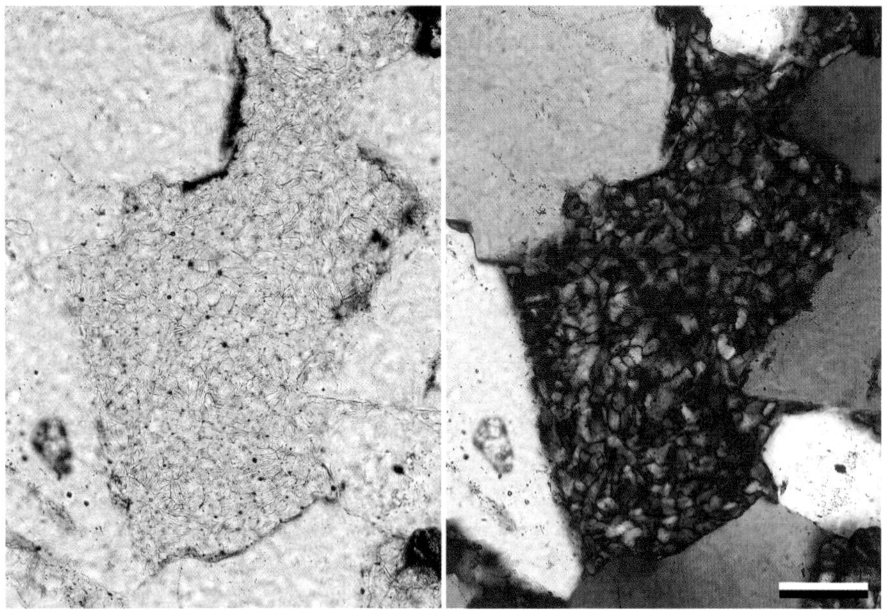

Mid. Jurassic Ile Fm., Møre Basin, Norwegian Sea †

An image that shows a microporous mass of authigenic, blocky kaolinite within a pore. This pore is larger than others surrounding it, and both the pore and its kaolinite fill may be the result of dissolution of a feldspar grain. Normally, when this is the case, there is some hint of a grain outline, but that is not the case in this example.

PPL | XPL, BDI, Scale bar = 0.07 mm

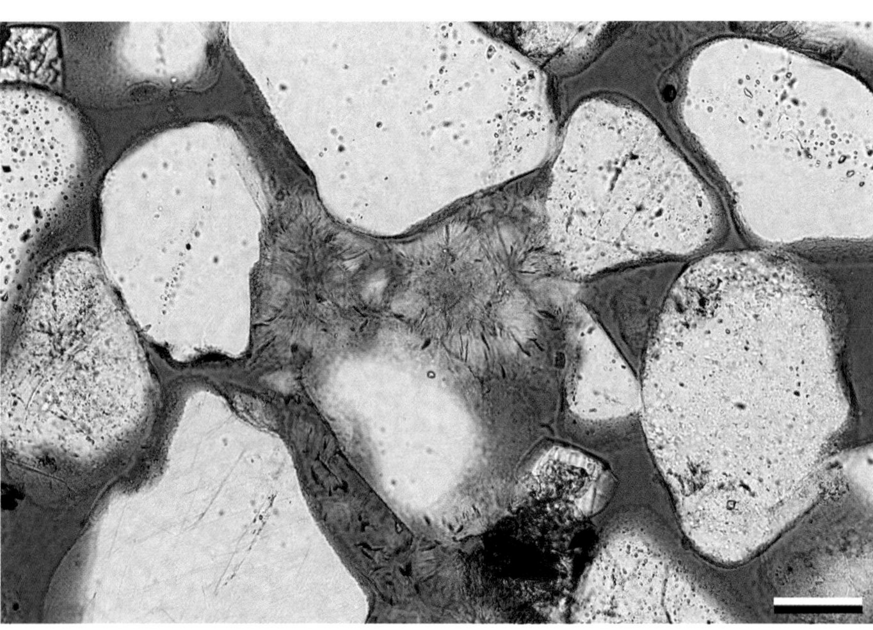

Mid. Jurassic Entrada Fm., Emery Co., Utah

A characteristic of kaolinite cement is that it commonly is localized in specific areas within the host rock. In this example, kaolinite booklets have substantially filled the pore at the center of the image but are entirely absent in the surrounding pores. That localization of precipitation may be related to local sources of solutes needed to precipitate kaolinite. *In-situ* dissolution of feldspar is a commonly cited example of such a source, but local microbial influences have also been suggested (see discussion in Milliken, 2003).

PPL, RDI, Scale bar = 0.05 mm

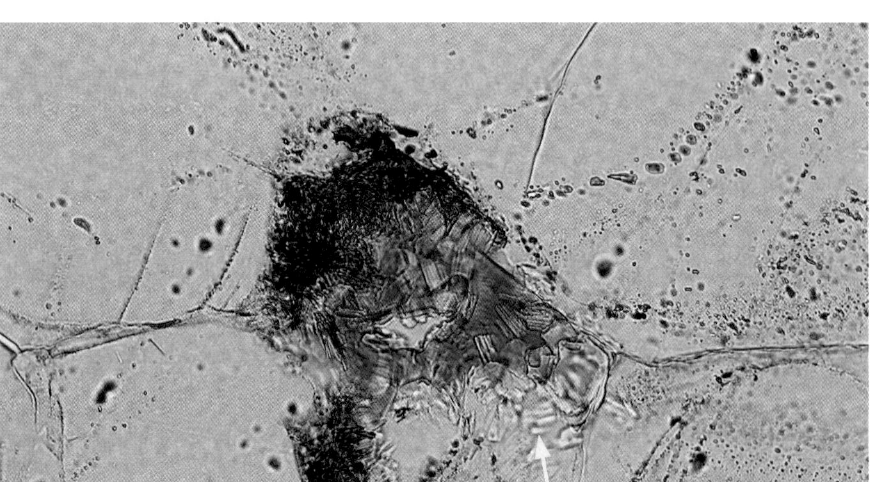

Cretaceous unknown formation, Moffat Co., Colorado

As seen in previous photographs, kaolinite may fill pores, but it typically does not coat grain surfaces nor does it fully occlude pores. Thus, later cements can grow into the unfilled intercrystalline micropores within the kaolinite, engulfing the precursor kaolinite books. Here one can see quartz overgrowths that have incorporated some of the loosely-packed kaolinite in the pore (arrow). The remaining void shows substantial intercrystalline microporosity (filled with pink epoxy) in the purely kaolinite-cemented areas.

PPL, RDI, Scale bar = 0.04 mm

Up. Pennsylvanian Bursum Fm., Socorro Co., New Mexico

This is an example of nearly complete encasement of vermicular kaolinite within a "micropoikilotopic" calcite cement crystal. As in the previous example, vermicular stacks of kaolinite partially filled pore space in a sandstone. Here, however, subsequent formation of coarsely-crystalline calcite cement led to the incorporation of the loosely-packed kaolinite stacks as inclusions within the calcite. This is a common occurrence in many calcareous sandstones.

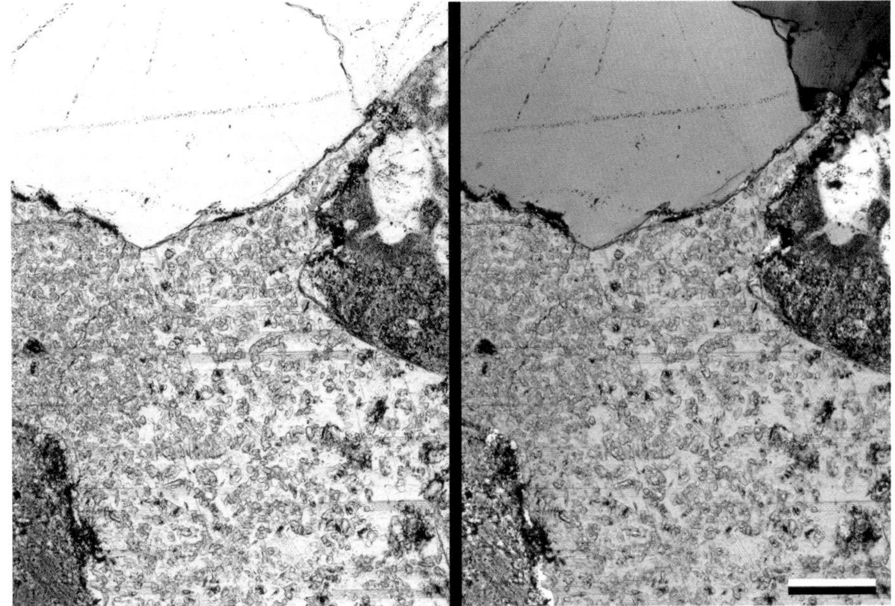

PPL | XPL, BDI, Scale bar = 0.14 mm

Up. Pennsylvanian Bursum Fm., Socorro Co., New Mexico

Another view of coarsely-crystalline calcite cement with kaolinite inclusions from the same unit as the previous image. In this case, the kaolinite booklets are larger and more scattered, so it is easier to see the characteristic vermicular stacks of plates that make up these crystals and crystal aggregates. The throughgoing cleavage of the calcite crystal shows that all the kaolinite stacks are incorporated within a single large crystal that filled this pore.

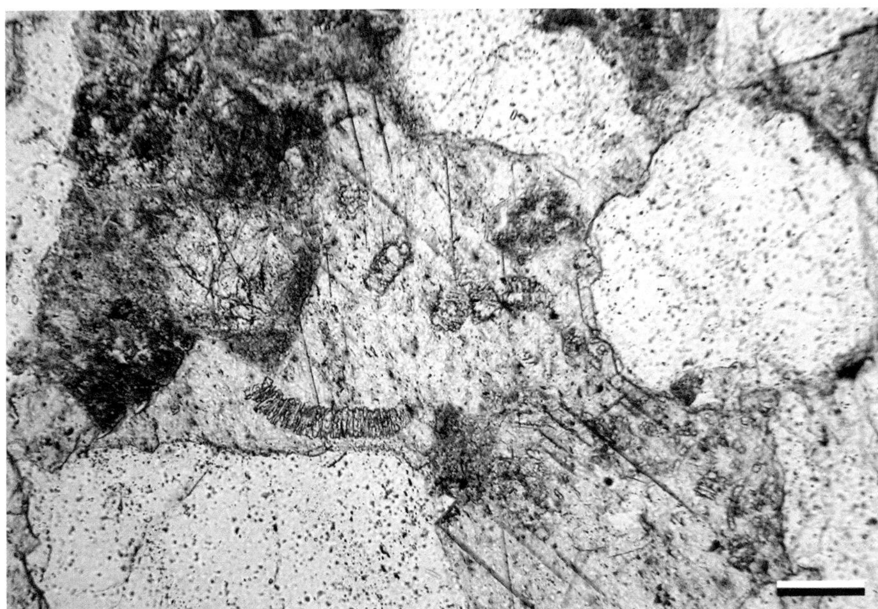

PPL, BDI, Scale bar = 0.14 mm

Mid. Jurassic Entrada Fm., Moffat Co., Colorado

Not all kaolinite (or especially dickite) cement forms during early diagenesis. In this example, probable dickite occurs in a large fracture that cuts what was an already well-cemented sandstone. Within the fracture, the dickite postdates initial dolomite cement (yellow arrows) as well as "dogtooth" or scalenohedral calcite crystals (red arrows) that were partially corroded and stained by migrated hydrocarbons. The central, pale pink area contains the final fill—packed kaolinite/dickite clusters with virtually no porosity (see next image for a closeup view of the kaolinite/dickite fill).

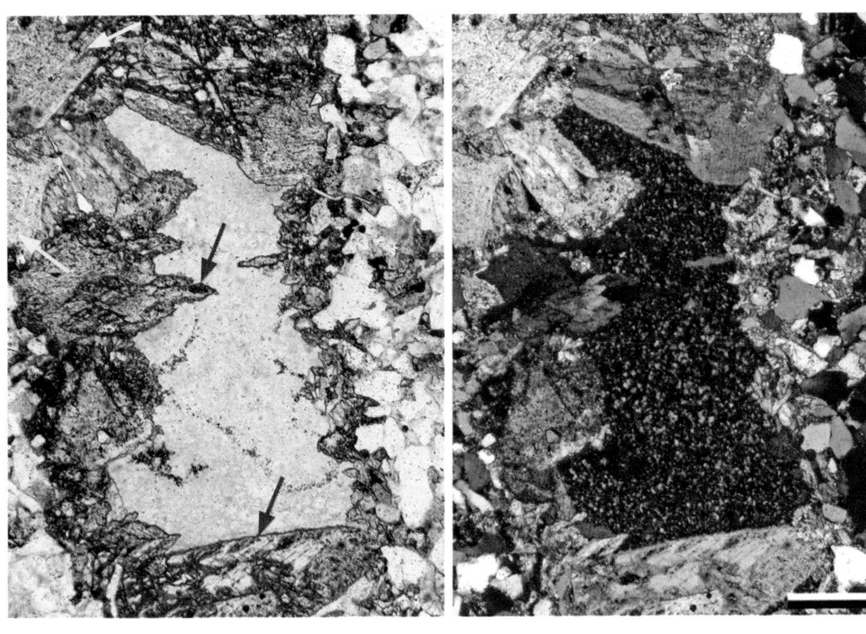

PPL | XPL, PFS, BDI, Scale bar = 0.26 mm

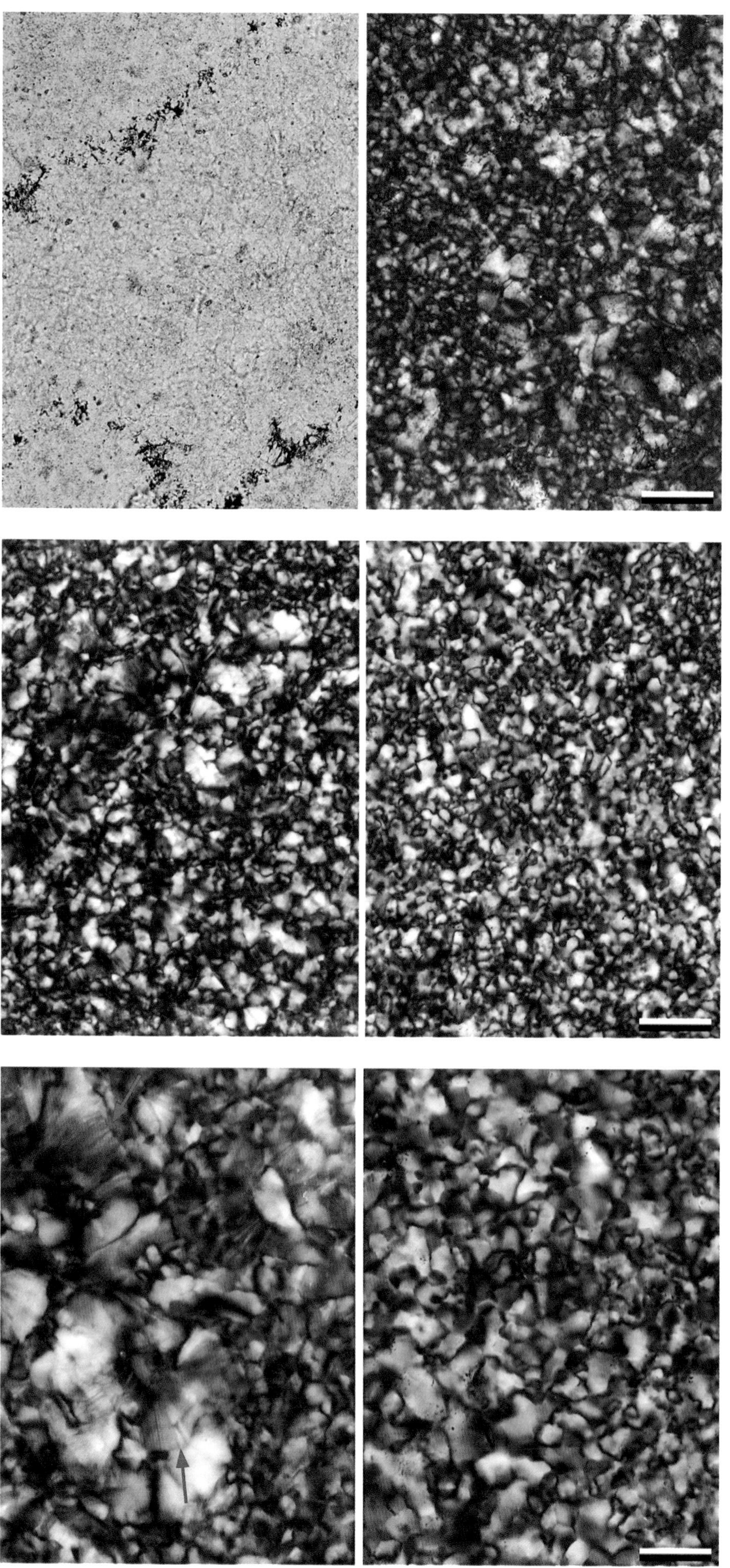

Mid. Jurassic Entrada Fm., Moffat Co., Colorado

A detailed view of the previous sample showing a late-stage kaolinite/dickite fill of a fracture. It is difficult to distinguish the individual crystal packets in the plane-polarized light image, because they are tightly packed and have effectively no visible porosity. The cross-polarized light image better shows the outlines of crystal packets. These fairly "blocky" or uniform crystal packets may indicate that this is the dickite polymorph (although XRD is required to confirm that observation). This view also shows how closely kaolinite/dickite cements can resemble chert. The slight pink color in the kaolinite is an artifact of feldspar staining.

PPL | XPL, PFS, BDI, Scale bar = 0.05 mm

Mid. Jurassic Entrada Fm., Moffat Co., Colorado and Proterozoic Lorrain Fm., Ontario, Canada

This composite image allows comparison of the kaolinite/dickite fracture-fill cement shown in previous photos (left) and a typical chert (right) at the same moderately high magnification. Both have comparable birefringence along with similar ranges of constituent crystal sizes and overall texture—all features that make these two deposits appear quite similar at this magnification (and even more similar at lower magnifications). The next slide shows the same paired rock types at twice the magnification.

XPL | XPL, Scale bar = 0.05 mm

Mid. Jurassic Entrada Fm., Moffat Co., Colorado and Proterozoic Lorrain Fm., Ontario, Canada

This composite image again pairs the kaolinite/dickite cement shown in previous images (left) and a Precambrian chert (right), this time at high magnification. Both still have comparable birefringence and overall texture. But at this high magnification, the probable dickite shows traces of blocky, non-interlocked, striated vermicular crystal stacks (examples at arrows), whereas the chert shows complex and irregular crystal boundaries and a tightly interlocked crystal fabric. Both textures are typical for these rock types and show the level of observation needed for identification.

XPL | XPL, Scale bar = 0.025 mm

Lo. Cretaceous Dakota Gp., Muddy Sandstone, Wyoming

Kaolinite/dickite booklets may form early in the diagenetic history of sediments, especially where meteoric waters invade underlying marine strata. Thus, where such sediments undergo early erosion or redeposition, individual kaolinite booklets may be reworked as detrital grains, although the delicate nature of such particles generally implies short and/or gentle transport. These grains are inferred to be reworked because they appear fragmented and are roughly the same size as the nearby detrital quartz and feldspar grains.

PPL, BDI, Scale bar = 0.014 mm

Pennsylvanian Pottsville Fm., Licking Co., Ohio

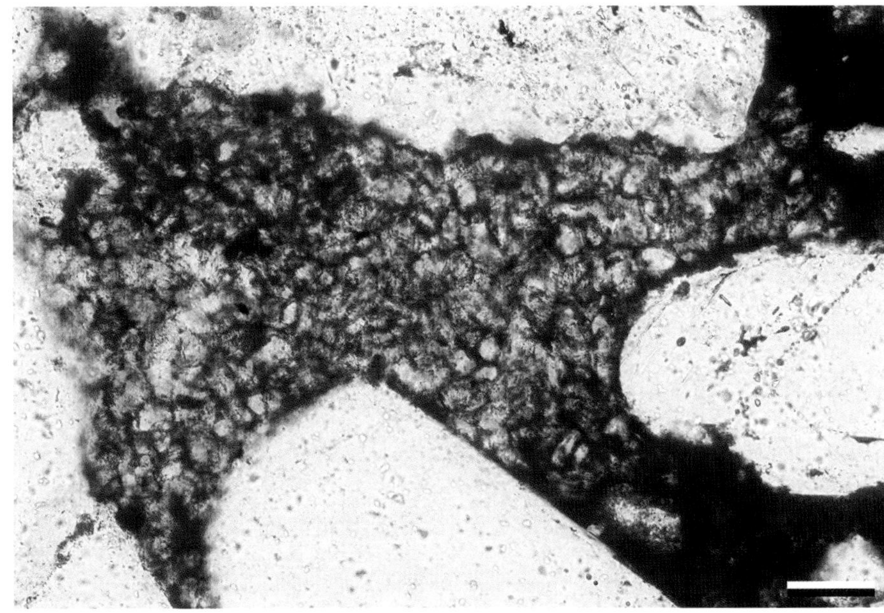

The eogenetic to early mesogenetic kaolinitic cement seen here retained extensive intercrystalline porosity and was later coated and stained by iron oxide. The combination of clay and iron oxide cements largely prevented formation of quartz overgrowths that might have otherwise have far more drastically reduced pore space (although some small quartz overgrowths had begun to grow on the grain at right center). It is possible that some of the brown staining is also the result of later-stage asphaltic hydrocarbon residues, but that was not definitively determined in this sample.

PPL, BDI, Scale bar = 0.08 mm

Cretaceous unknown formation, Moffat Co., Colorado

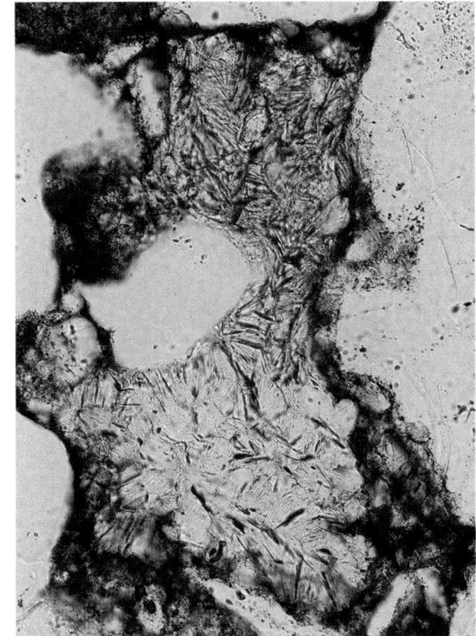

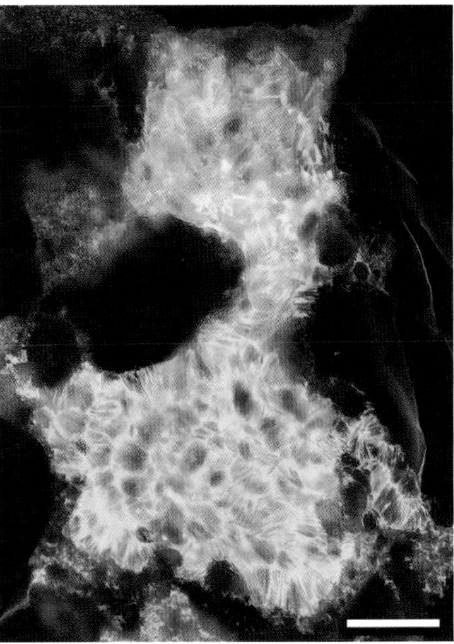

Even where kaolinite/dickite is fairly densely packed within intergranular pores, it can still retain substantial micro- and nanoporosity. Here the red epoxy impregnation shows variations in preserved porosity in kaolinite/dickite cements across a single pore in the left-hand image. The image at right, taken with fluorescent illumination, however, shows oil presence (the greenish fluorescence) in both the loosely and more tightly packed areas of kaolinite cement.

PPL | FL 470, RDI, Scale bar = 0.06 mm

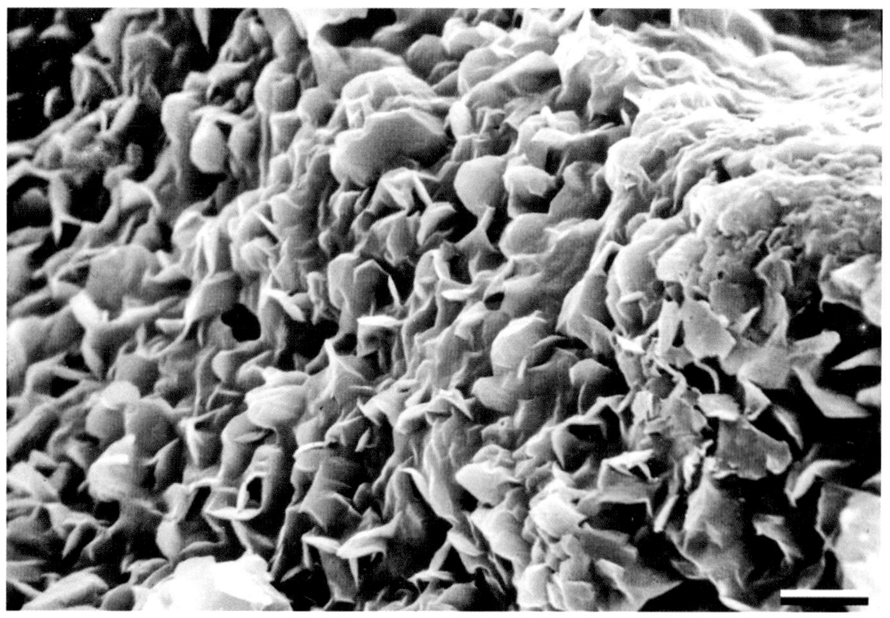

Up. Cretaceous Frontier Fm., Wyoming

An SEM view of authigenic smectite as a cementing agent in a sandstone. The highly crenulate, honeycombed, interlocking crystals are typical of smectite. The smooth intergrowth or fusing of adjacent crystals serves to distinguish these crystals from chlorite, but XRD should be used for more definitive determinations. The high degree of crystallinity shown in these clays is a strong indicator of an authigenic origin and this criterion is valid for other clay minerals as well. Photo from Edward D. Pittman.

SEM, Scale bar = 0.33 μm

Paleocene Tang Fm., Møre Basin, Norwegian Sea †

A thin layer of web-like authigenic smectite on a K-feldspar. The clay flakes here lie parallel to the feldspar surface, and the honeycomb pattern reflects the upturned edges of the crystals. The weak crenulation displayed by the smectite is also common in illite/smectite—thus, EDX spectra and clay fraction XRD are needed to reliably determine the exact mineralogy. The EDX will penetrate the whole clay crystal and thus can read underlying or nearby minerals, so multiple readings should be taken (with an appropriate spot size). The spectrum for this specimen shows Si, Al, Ca and Mg, in deceasing levels, with a low K peak.

SEM, Scale bar = 8.8 μm

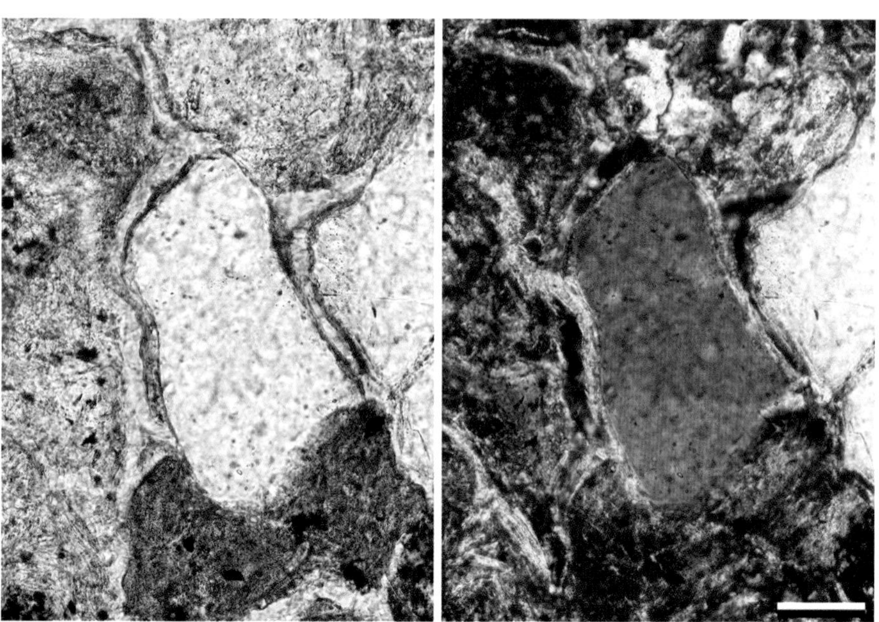

Up. Miocene San Pablo Gp., Neroly Sandstone, Contra Costa Co., California

Thick, grain-coating clay cement in a volcaniclastic sandstone. These clays were identified using XRD by Lerbekmo (1957) as a mixture of beidellite (an aluminous montmorillonite) and nontronite (an iron-rich smectite), both members of the smectite group of clays. The greenish-brown color and moderate birefringence are common for smectitic clays. Note the absence of the grain-coating clay where the quartz grain was in original contact with the SRF (bottom); but it is present where the SRF has deformed around the quartz, indicating some compaction prior to cement formation.

PPL | XPL, BDI, Scale bar = 0.08 mm

Up. Miocene San Pablo Gp., Neroly Sandstone, Contra Costa Co., California

A higher-magnification view of thick, grain-coating smectitic clay cement from the same sample shown in the previous photograph. The moderately high birefringence and brown to green color are typical of smectitic clays (especially for iron-rich smectites or smectites with chlorite interlayers). Such grain-coating cements can be quite effective in isolating the framework grains from pore waters and thus in precluding later formation of overgrowth cements on quartz, feldspar and other detrital grains (e.g., Ajdukiewicz and Larese, 2012).

PPL | XPL, BDI, Scale bar = 0.04 mm

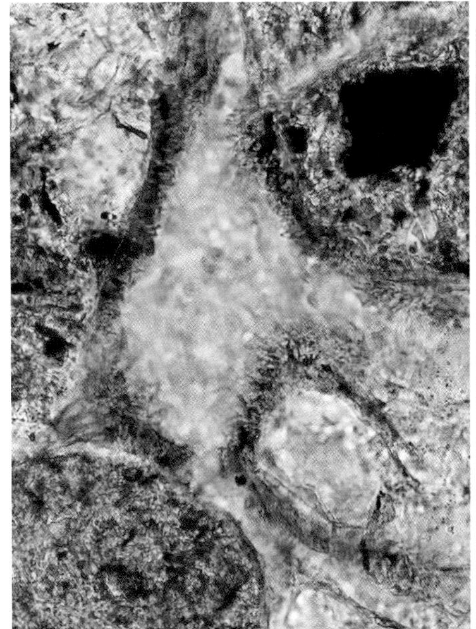

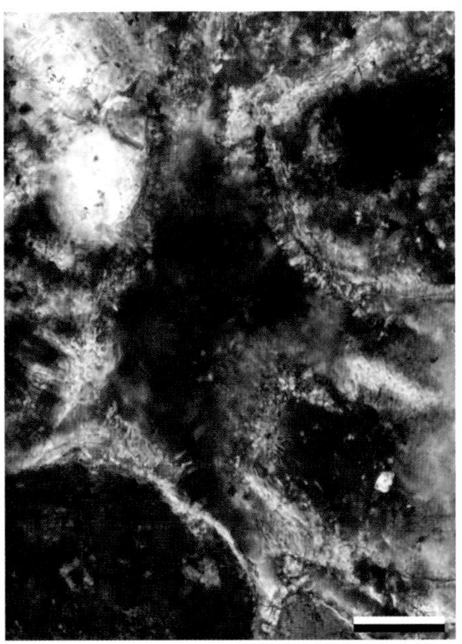

Cretaceous Lange Fm., Norwegian sector, North Sea †

Pore-filling illite/smectite (confirmed by XRD and EDX analysis). The illite/smectite consists of flaky, ragged crystals with some filamentous margins. There is a loose resemblance to a possible precursor web-like, or boxwork habit. The quartz grain on the right edge has developed an overgrowth, the formation of which was interrupted by growth of the illite/smectite. Some continued formation of minute, very localized quartz overgrowths after clay coating may account for the irregular clay-quartz contact.

SEM, Scale bar = 3.7 μm

Up. Cretaceous Prince Creek Fm., North Slope, Alaska

A view of pore-lining clay coatings surrounding mainly detrital chert and quartz grains. Clays (mainly illite/smectite) here formed fairly complete coatings with irregular thickness around detrital grains. Note the brownish color and moderately high birefringence of the clay films. Although it is possible that these are precipitated clay cements, it is also possible that these are infiltrated detrital clays (clay cutans).

PPL | XPL, BDI, Scale bar = 0.10 mm

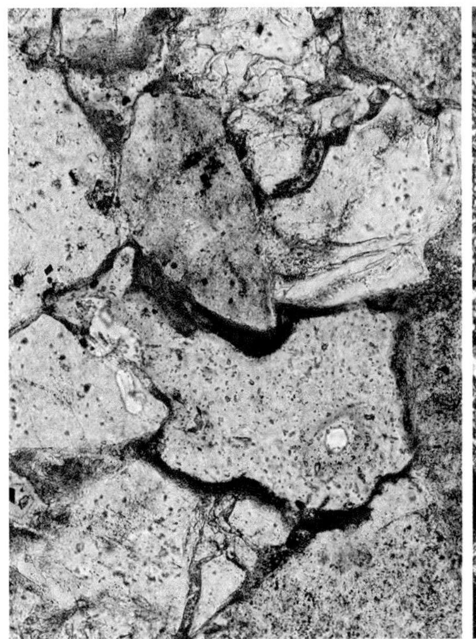

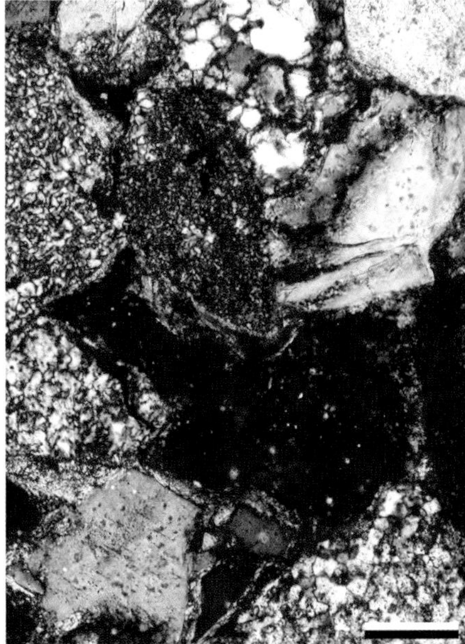

Lo. Permian Cutler Fm., Mesa Co., Colorado

This rock shows hematite-stained, clay-coating cement that formed around virtually all the detrital grains in this lithic and feldspathic arenite. Those grain-coats of probable illite-smectite were effective in preventing overgrowth cementation—so effective that the rock did not have the strength to resist overburden stress-induced compaction. Thus, this rock has an "overcompacted" fabric with both ductile and brittle grain deformation and some grain interpenetration. However, after such substantial iron staining and deformation it is difficult to determine if these clays are of detrital or authigenic origin.

PPL, Scale bar = 0.1 mm

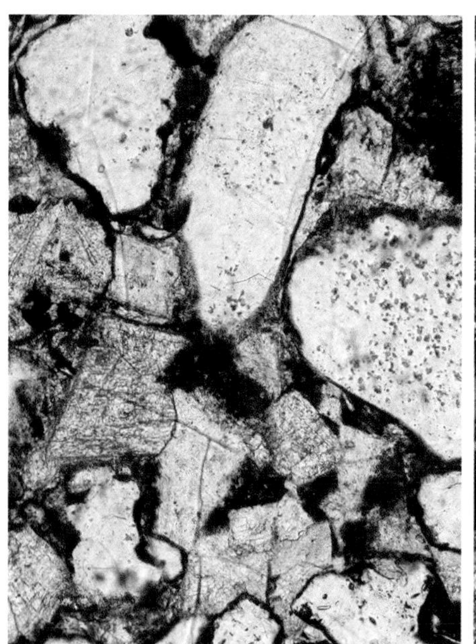

Triassic Sherwood Sandstone Gp., Co. Antrim, Northern Ireland, U.K.

Another example of a rock in which grain coating (and hematite stained) illite/smectite cement isolated grains from overgrowth cementation. In this example, however, pore filling dolomite cement effectively occluded all pore space, preventing compaction but destroying the reservoir potential of this rock. Although clay coats can prevent syntaxial overgrowths, they do not prevent other, non-overgrowth cements. Good reservoirs need just enough cementation to retain compaction-resistant strength, but not so much as to drastically reduce reservoir capacity.

PPL | XPL, Scale bar = 0.10 mm

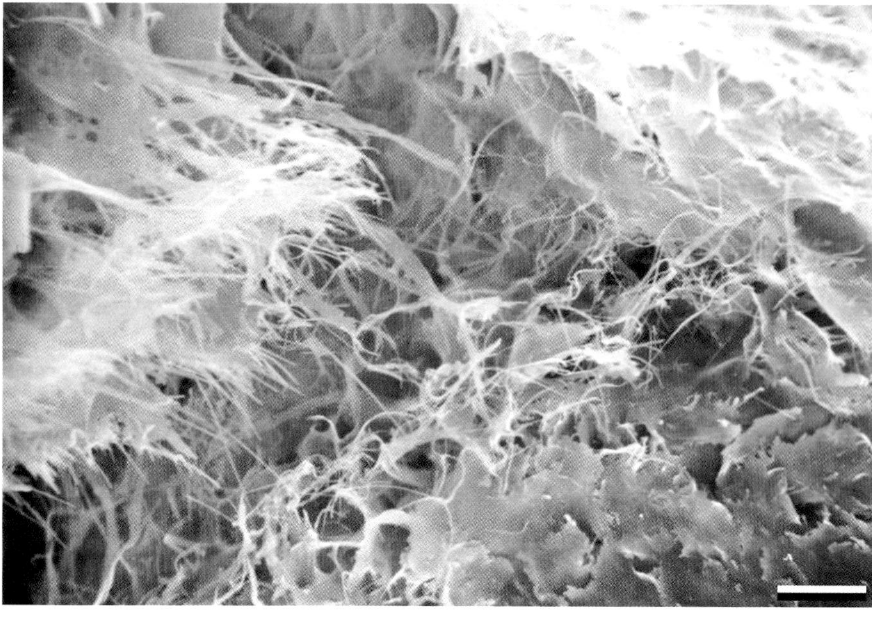

Lo. Permian Rotliegend Sandstone, United Kingdom sector, North Sea

SEM view of authigenic illite cement in a sandstone. This delicate growth form, consisting of sheets with wispy to hair-like terminations, is the most common habit for illitic cement in sandstones. Pure illites can be confused with mixed-layer illite/smectite cement and, in some cases, with sericite as well. Identifications should, therefore, be checked with XRD. Photo by Edward D. Pittman.

SEM, Scale bar = 0.7 μm

Permian Unayzah Fm., Qatar †

Platy (pseudo-boxwork) illite within a pore. This morphology is not specific to any single clay species and is perhaps more typical of smectite, illite/smectite and illite/chlorite. This illite is possibly a replacement of smectite. The plates of illite here consist of some overlayered ribbon-like crystals, and the margins of the plates display a frayed "cockscomb" margin.

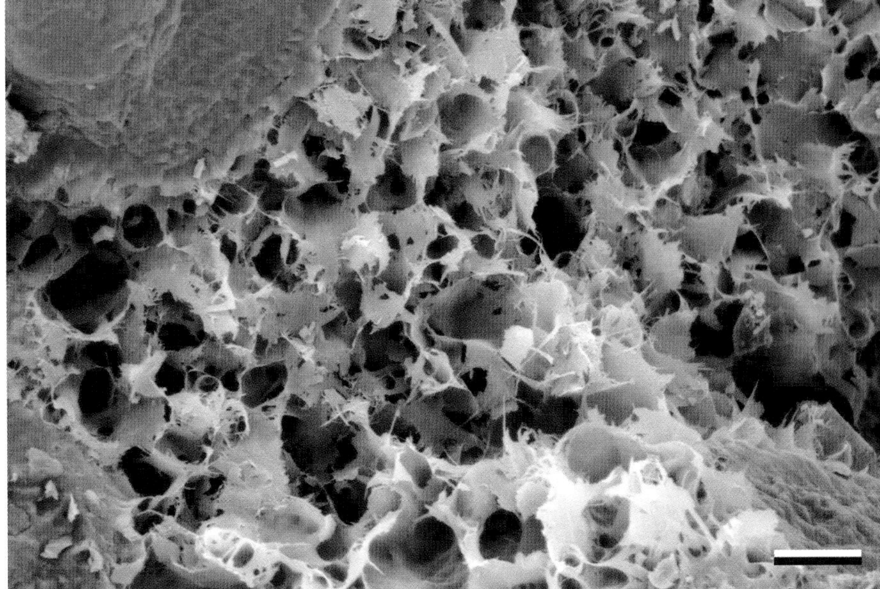

SEM, Scale bar = 25 μm

Permian Unayzah Fm., Qatar †

Pore-bridging illite here consist of sheets of crystals with "cockscomb" margins that give rise to ribbons. These ribbons are folded over on themselves but also bridge the pore between two grains. On closer inspection, the sheet-like morphology comprises a mat of crisscrossing filaments. It should be noted, however, that the appearance of illite in SEM can be affected significantly by sample handling, especially sample drying (e.g., McHardy et al., 1982).

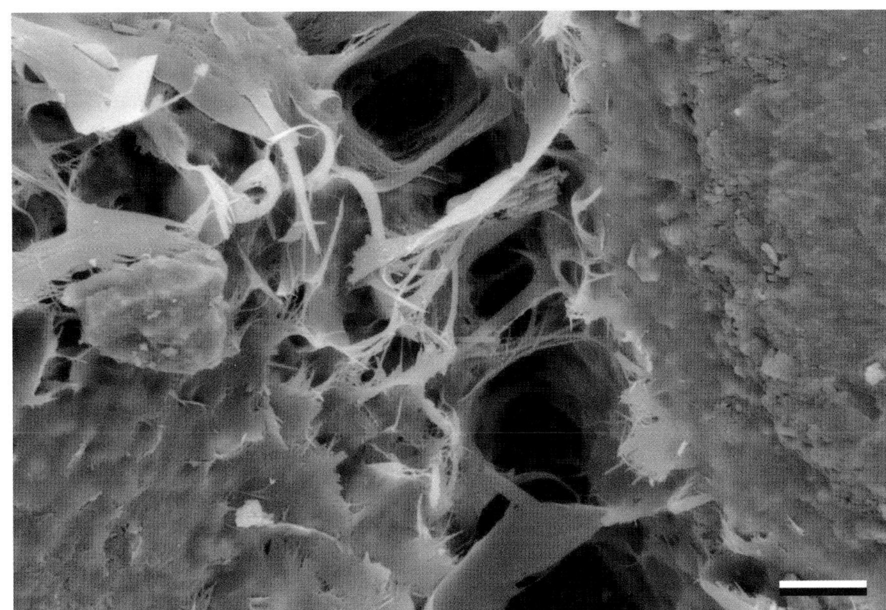

SEM, Scale bar = 12.7 μm

Devonian Jauf Fm., Qatar †

A higher-magnification SEM view of illite within a pore showing pseudo-boxwork texture. This image shows a mixture of sheets comprising matted ribbons and hairy, filamentous strands projecting from the ends of the sheets.

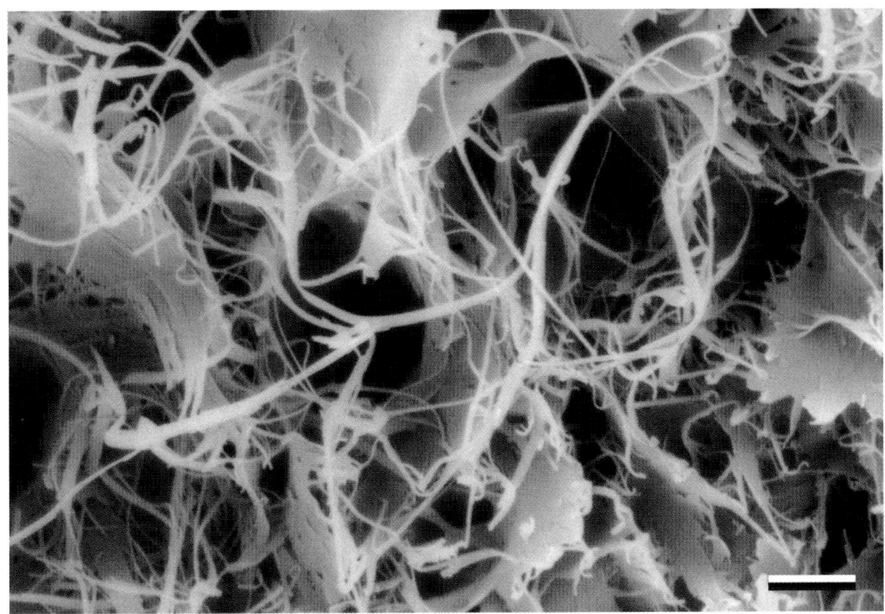

SEM, Scale bar = 3.1 μm

Paleocene Fatehgarh Fm., Rajasthan, India

An SEM view of grains that are completely coated by diagenetic clay. These clays are mostly illite with some kaolinite and, unlike those in the preceding images, they are not well crystallized. These clays are pore bridging in some places and represent cutans that formed through the infiltration of clay into a sandstone (see subsequent photo). In addition, the clays have been partly recrystallized.

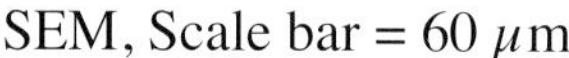

SEM, Scale bar = 60 μm

Paleocene Fatehgarh Fm., Rajasthan, India

A thin section view of the same sample shown in the SEM above. The clay is finely crystalline and individual crystals/particles cannot be distinguished. The brown color in this case is, at least partly, the result of staining by hydrocarbons. Some illite naturally has a brown color or can be stained by iron oxides, so color is not always an indication of oil staining. The pore bridging nature of the cutans serves to compartmentalize some of the pore space, which would therefore result in a lower permeability. However, clay coats such as this also reduce sites for potential quartz overgrowth and thus can help maintain porosity.

PPL, BDI, Scale bar = 0.04 mm

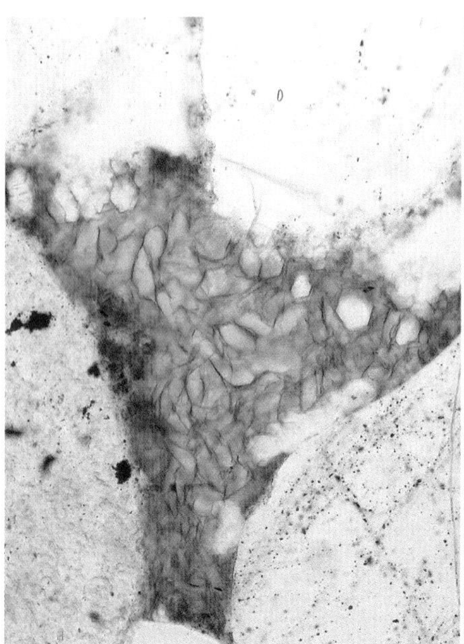

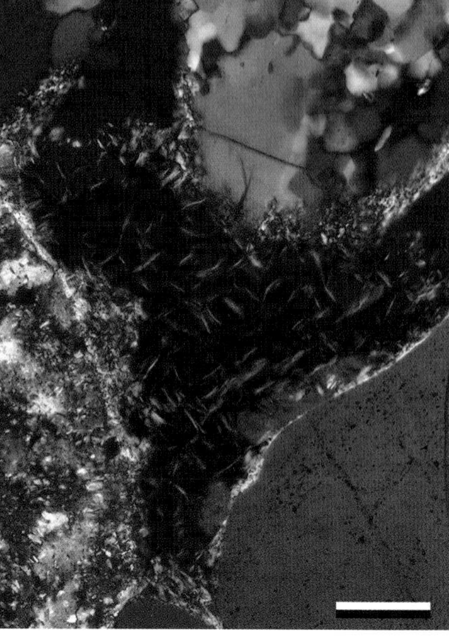

Precambrian unknown unit, Bihar, India

Illite and euhedral diagenetic quartz filling a pore. The illite either grew from the surrounding grains or, more likely, formed through recrystallization of an illite/smectite precursor. Sometimes networks of illite, such as these, fill secondary pores within feldspar gains or are associated with extreme expansion and alteration of micas.

PPL | XPL, BDI, Scale bar = 0.07 mm

Proterozoic (Huronian) Cobalt Gp., Lorrain Fm., Ontario, Canada

Coarsely-crystalline illitic and/or sericitic cement postdates quartz overgrowths in this sandstone. Both of these mica-type clay minerals have near-muscovite compositions, but sericite forms mainly in high-temperature diagenetic to incipient metamorphic settings. All three (muscovite, sericite and illite) share common features: largely white to colorless in plane-polarized light and bright second-order to low third-order birefringence in cross-polarized light. However, muscovite generally is far coarser than either of the clays. XRD is essential for precise identifications.

XPL, Scale bar = 0.16 mm

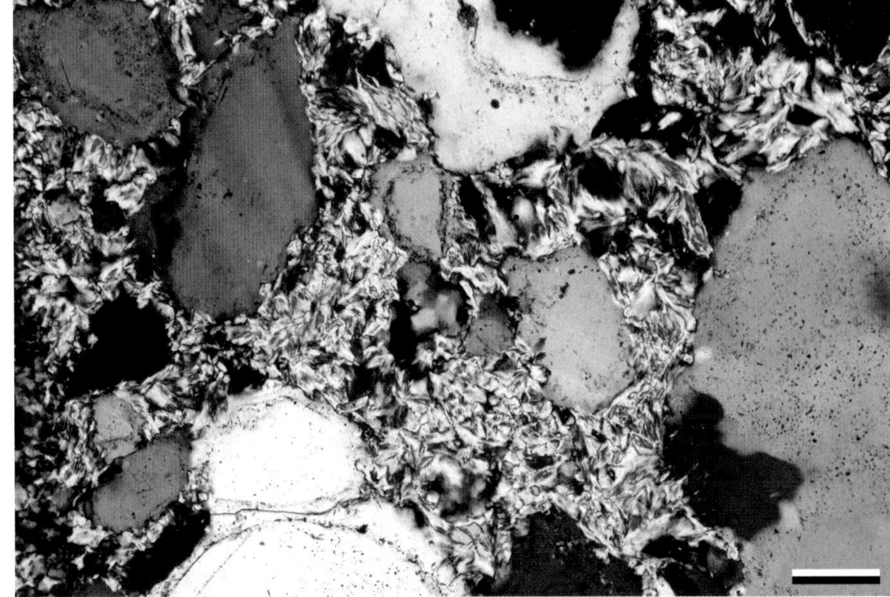

Proterozoic (Huronian) Cobalt Gp., Lorrain Fm., Ontario, Canada

A view of an area in the previous sample completely filled by the coarsely-crystalline illite/sericite cement. Note how the cements form radial crystal bundles within the pore space. Some intercrystalline porosity is still present here, but the clay cements have greatly reduced the overall permeability of the unit. The metamorphic rock fragment section of Chapter 3 includes some images of muscovite-rich grains that show the similar birefringence but coarser crystal size of muscovite in metamorphic rocks.

PPL | XPL, Scale bar = 0.10 mm

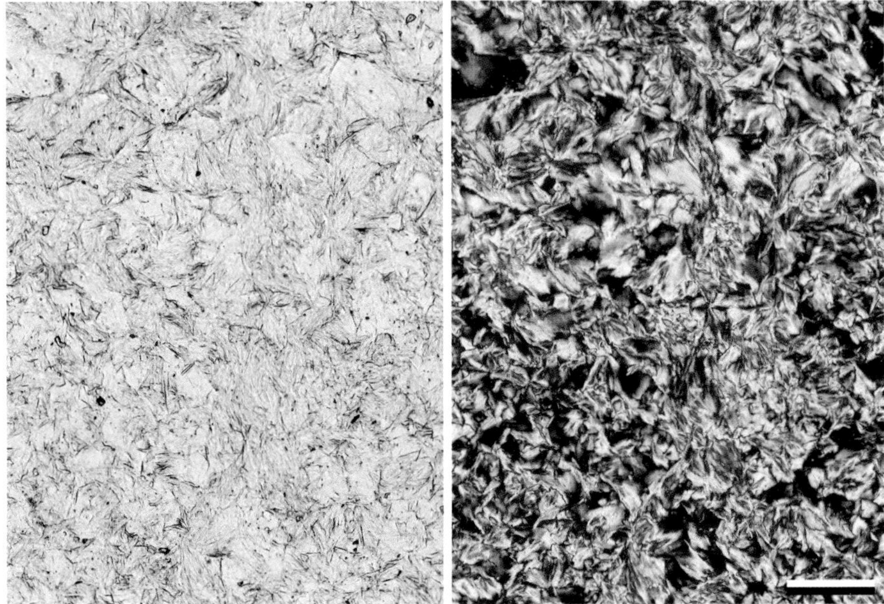

Up. Triassic – Lo. Jurassic Huizachal Fm., Tamaulipas, Mexico

Illite cement rimming grains in an arkosic sandstone. Although largely illite now, these grain coatings may have originated as smectite or illite/smectite coatings (possibly infiltrated, although more likely authigenic) but with increasing depth and temperature they have transformed to predominantly illitic clay.

XPL, Scale bar = 0.10 mm

Up. Cretaceous Tuscaloosa Fm., Louisiana

Authigenic chlorite cement can have a wide variety of crystal morphologies. In this example, chlorite occurs as individual idiomorphic crystals which are plate-like and are attached to the detrital sand grains along their thin edge. This is perhaps the most common morphology for authigenic chlorite. Photo by G. W. Smith.

SEM, Scale bar = 0.35 μm

Lo. Jurassic Tilje Fm., Norwegian sector, North Sea †

Unlike the previous image of monomineralic chlorite, in some cases the clays observed under SEM are a mixture. Although more complicated, that can help to decipher the relative timing of cements (discussed more fully in Chapter 23 - Paragenesis). Here the chlorite plates preceded the quartz (center). Both the chlorite and, in places, the quartz are overlain by ribbons of illite (for example, above and to the left of the largest quartz crystal).

SEM, Scale bar = 2.2 μm

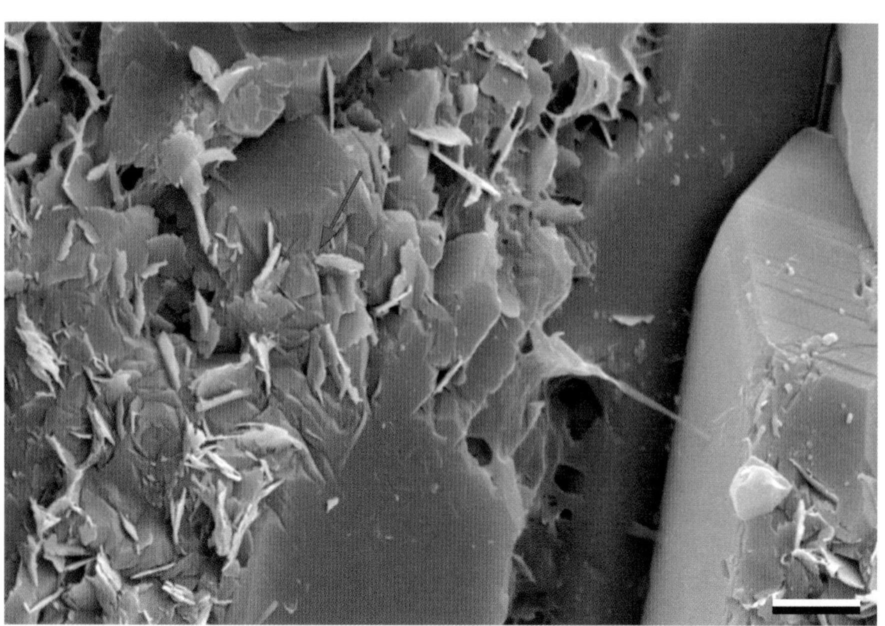

Mid. Jurassic Ile Fm., Norwegian sector, North Sea †

Isolated chlorite plates and minor illite, both of which lie on a detrital grain, have been substantially enclosed by quartz overgrowths. Where the grain has been broken, the boundary between the clay-free quartz and the quartz enclosing the chlorite can be seen (an example shown by red arrow).

SEM, Scale bar = 3.2 μm

Cretaceous Lange Fm., Norwegian sector, North Sea †

An image showing ragged-edged chlorite platelets stacked face-to-face in a beehive-like structure. The chlorite is overlain by euhedral quartz crystals and larger quartz overgrowths enclose other areas of chlorite.

SEM, Scale bar = 7.3 μm

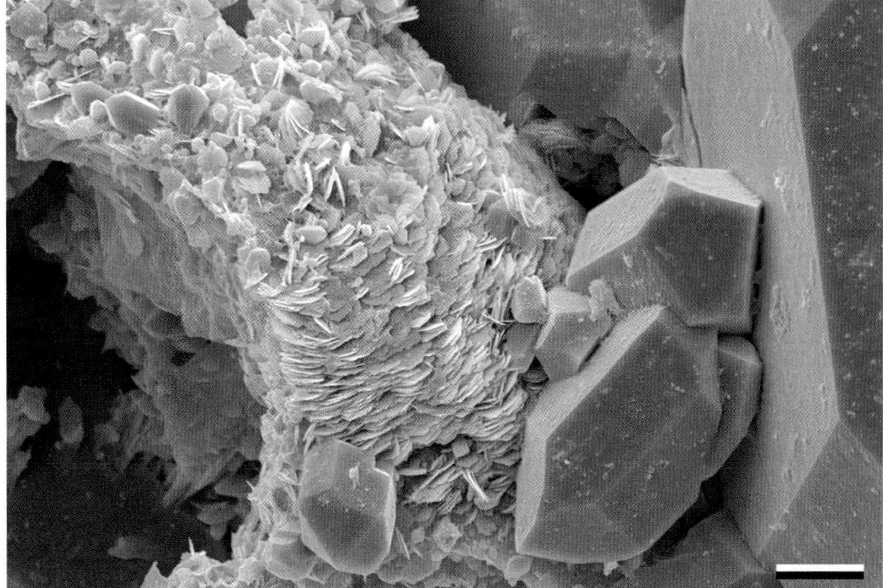

Lo. Jurassic Tilje Fm., Norwegian sector, North Sea †

A view of grain-coating authigenic chlorite that surrounded a detrital grain that was subsequently largely dissolved, yielding a hollow clay shell. This process produced a secondary pore that was isolated from the main porosity network before the sample was broken for SEM examination.

SEM, Scale bar = 6.3 μm

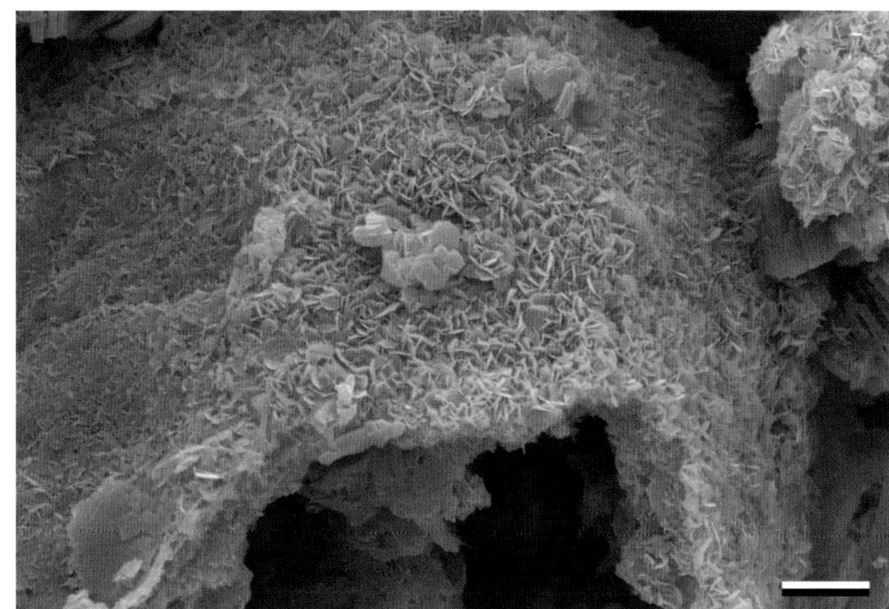

Up. Triassic Snadd Fm., Barents Sea †

Views of chlorite in thin section are rarely as impressive as those in SEM. In thin section, chlorite commonly is seen as thin rinds of radially-arranged, greenish clays that formed on grain surfaces. Here there is some evidence that the chlorite may have precipitated prior to the principal period of compaction because the clay coating continues around the grains even at grain-to-grain contacts.

PPL, BDI, Scale bar = 0.03 mm

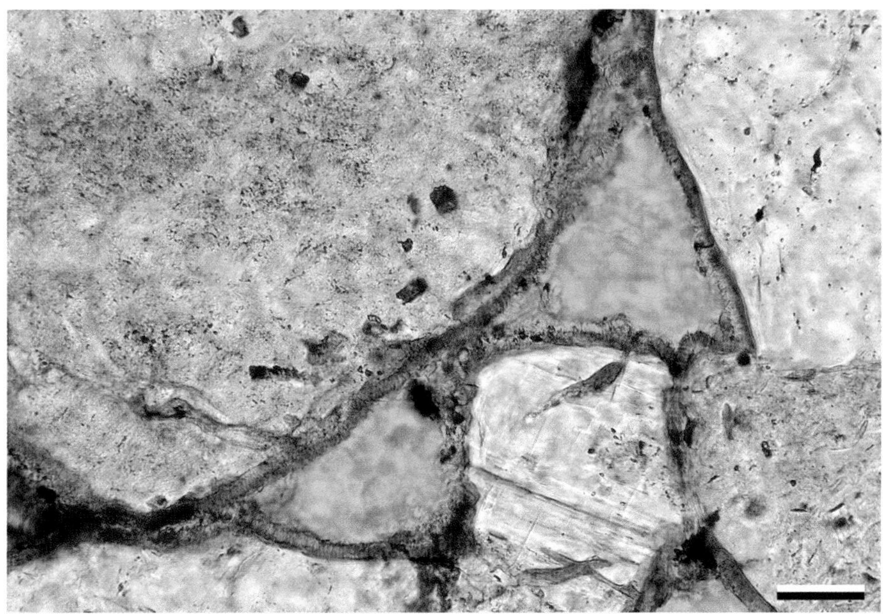

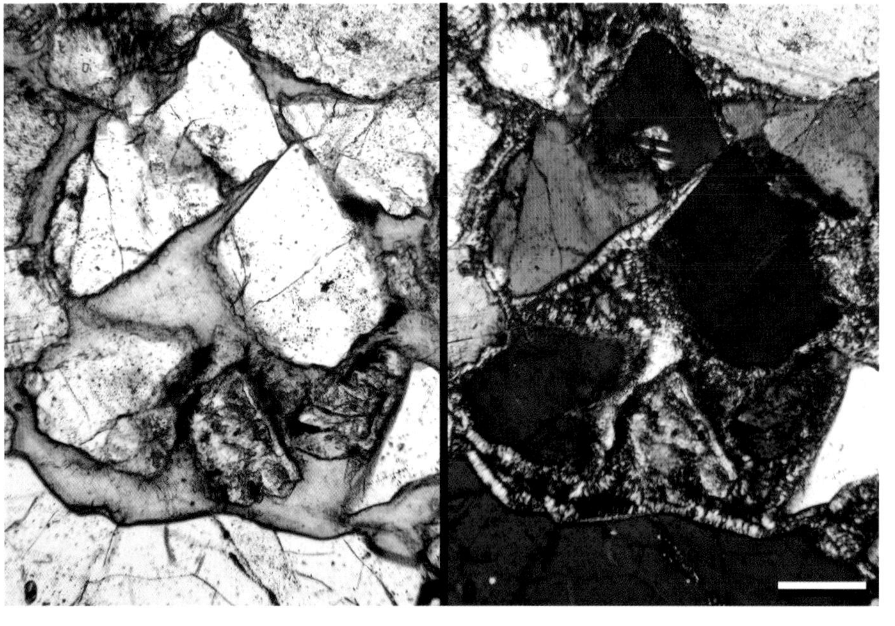

Lo. Cretaceous Patula Arkose, Coahuila, Mexico

A low-magnification view of two stages of chlorite cementation in an arkosic sandstone. Here, authigenic clay formation has extended well beyond the "grain coating" stage to completely obliterate porosity. The light olive-green color is characteristic of chlorite in plain-polarized light as is the "anomalous blue" color under cross-polarized light. The next photograph more clearly shows the details of the two stages of cementation present here.

PPL | XPL, Scale bar = 0.08 mm

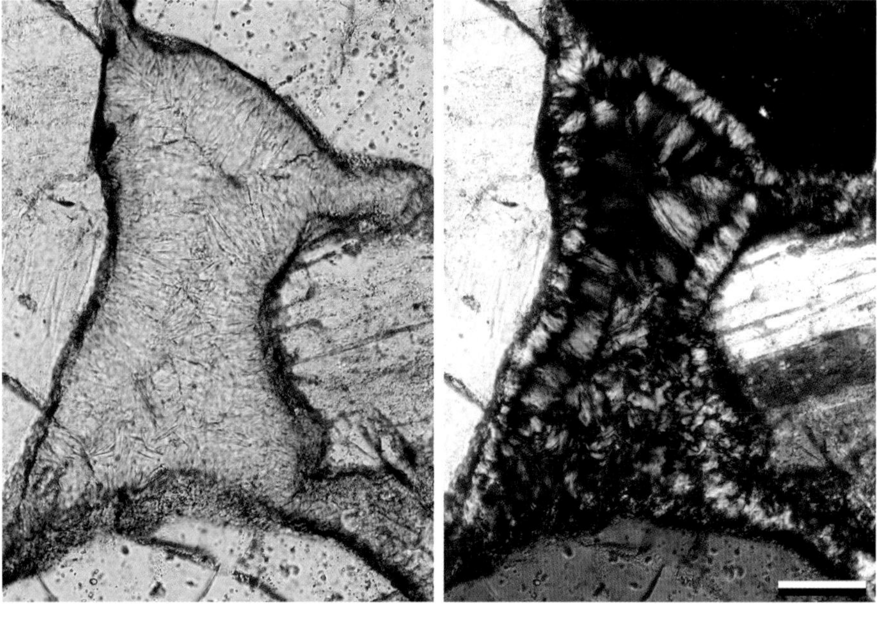

Lo. Cretaceous Patula Arkose, Coahuila, Mexico

A closeup view of two stages of chlorite cement in an arkosic sandstone from the same sample as above. In the cross-polarized light image, one can clearly see the radial crystals of the initial, grain-coating chlorite cement crusts. That is followed by coarser fans of pore-filling chlorite. Note the "ultrablue" anomalous birefringence colors characteristic of chlorite that are especially well developed in the first-stage chlorites. This difference may reflect compositional variations between the earlier- and later-stage chlorites as they may have formed under very different conditions (see Table 13.1).

PPL | XPL, Scale bar = 0.1 mm

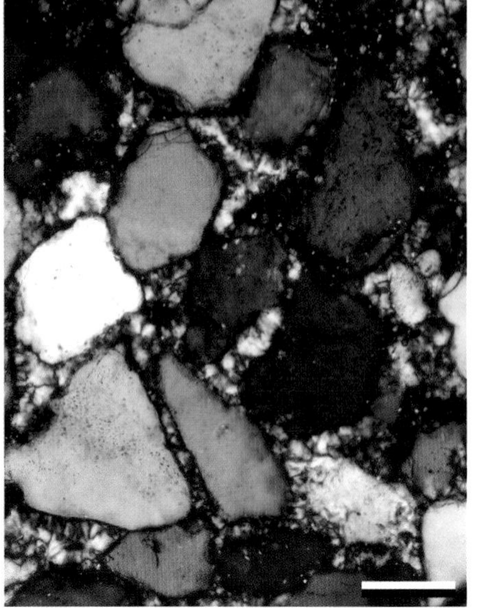

Up. Cretaceous (Campanian) sandstone, Steenteveld, Belgium

Authigenic chlorite cements here are thin, very pale brownish green coatings visible on the detrital quartz grains in the plane-polarized light view. In the cross-polarized light view, the crystal orientation of the chlorite cements is visible (especially at higher magnification), forming radial rims around the grains. The chlorite cements have low first-order birefringence (to anomalous) colors. After the chlorite cements, chalcedony cements fill the remaining porosity.

PPL | XPL, Scale bar = 0.1 mm

Ordovician Griffel Schiefer, Frankenwald, Bavaria, Germany

These knots of authigenic chlorite (showing typical "anomalous blue" or "ultrablue" birefringence) have grown displacively in the matrix of an organic carbon rich shale that underwent burial and near-metamorphic diagenesis during the Variscan orogeny. The growth of chlorite in sedimentary rocks is a common phenomenon in the entire range of temperatures and pressures from moderate sedimentary burial through greenschist metamorphism. Chlorite is most readily recognized by its birefringence and morphology, but XRD is needed for detailed mineralogical studies.

PPL | XPL, Scale bar = 0.10 mm

Up. Triassic New Haven Arkose, New Haven Co., Connecticut

Authigenic vermicular chlorite as fracture-filling cements in a strongly deformed sandstone. Chlorite (as well as kaolinite and vermiculite) commonly, but by no means always, assumes this vermicular growth form as "books" or "worms" of clay plates. Note the greenish color in plane-polarized light and "ultrablue" extinction colors that are typical of chlorite. Minor authigenic illite/sericite is also visible in this sample (marked by white to yellow birefringence).

PPL | XPL, Scale bar = 0.01 mm

Unknown age, formation and location

Chlorite may form vermicular cements in rocks, but vermicular chlorite also can form by the alteration of kaolinite or dickite. To determine their origin, further analytical work is necessary to determine if these dark green vermicular chlorites are replacements of earlier kaolinite cements or first generation precipitates.

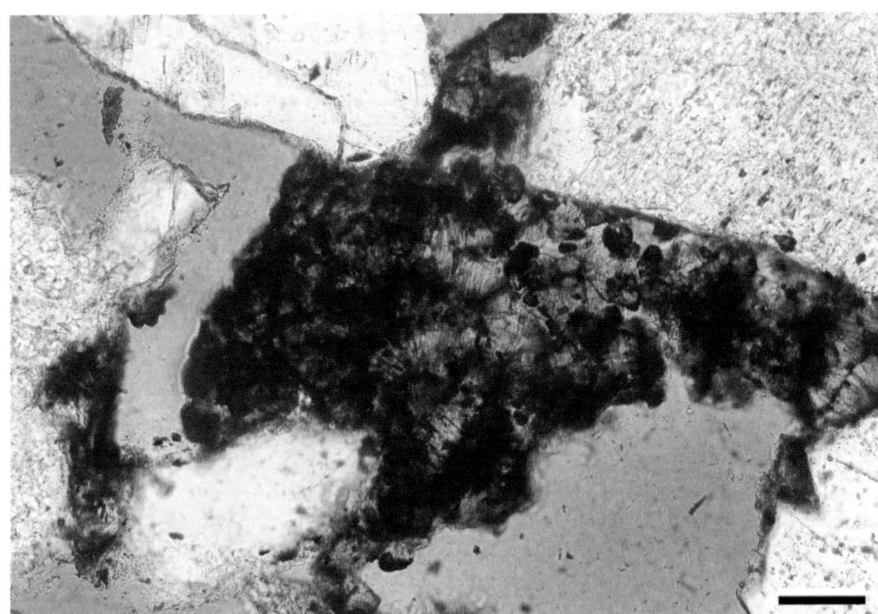

PPL, BDI, Scale bar = 0.08 mm

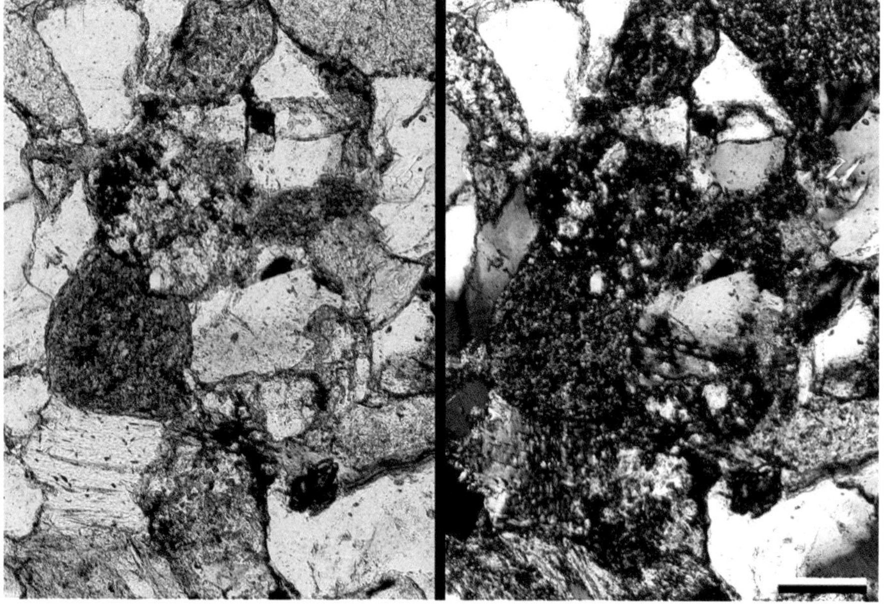

Up. Triassic – Lo. Jurassic Huizachal Fm., Nuevo Leon, Mexico

This lithic arenite is cemented by both chlorite and mixed-layer clays. These cements form tangential growths that partially to completely rim the grains. The chlorite cements are the greenish phases in plane-polarized light, whereas the brownish, highly birefringent cements are the mixed-layer cements. These cements may represent a progressive change with increasing temperature and burial depth. Details of the temperature induced gradations in chlorite polytypes and crystal habits are documented in Spötl et al. (1994).

PPL | XPL, Scale bar = 0.10 mm

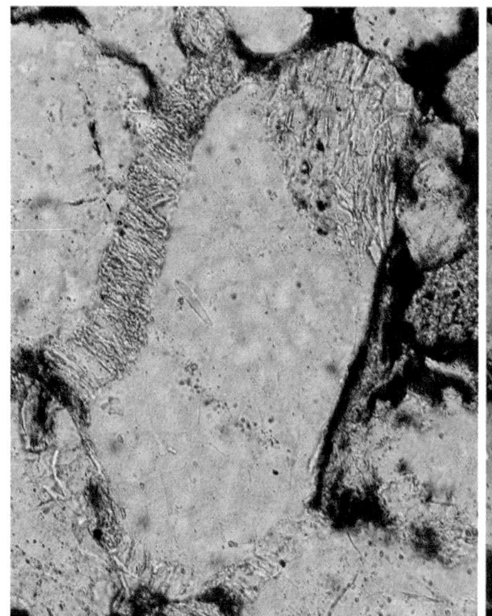

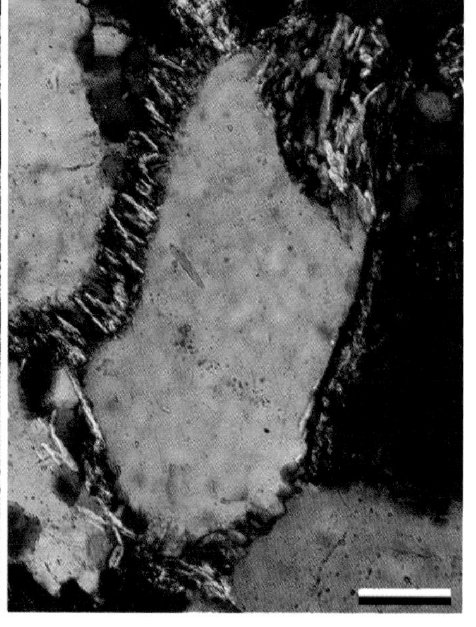

Up. Silurian Bloomsburg Fm., Warren Co., New Jersey

Many of the quartz grains in this sample have radial chlorite-illite cements which were later encased in quartz. Like the previous sample, the chlorite cements are pale green in plane-polarized light, and the illite cements are brightly birefringent in cross-polarized light. The cements have not collapsed during compaction because of the quartz overgrowths that surround them. These quartz cements appear to be spot welding the grains, providing rigid structural support to the rock.

PPL | XPL, Scale bar = 0.10 mm

Cited References and Additional Information Sources

Aagaard, P., and J. S. Jahren, 1992, Diagenetic illite-chlorite assemblages in arenites. II. Thermodynamic relations: Clays and Clay Minerals, v. 40, p. 547-547, doi: 10.1346/CCMN.1992.0400508.

Aagaard, P., J. S. Jahren, A. O. Harstad, O. Nilsen, and M. Ramm, 2000, Formation of grain-coating chlorite in sandstones. Laboratory synthesized vs. natural occurrences: Clay Minerals, v. 35, p. 261-269, doi: 10.1180/000985500546639.

Ajdukiewicz, J. M., and R. E. Larese, 2012, How clay grain coats inhibit quartz cement and preserve porosity in deeply buried sandstones: Observations and experiments: AAPG Bulletin, v. 96, p. 2091-2119, doi: 10.1306/02211211075.

Altaner, S. P., and R. E. Grim, 1990, Mineralogy, chemistry, and diagenesis of tuffs in the Sucker Creek Formation (Miocene), Eastern Oregon: Clays and Clay Minerals, v. 38, p. 561-572, doi: 10.1346/CCMN.1990.0380601.

Anjos, S. M. C., L. F. De Ros, and C. M. A. Silva, 2003, Chlorite authigenesis and porosity preservation in the Upper Cretaceous marine sandstones of the Santos Basin, offshore eastern Brazil, *in* R. H. Worden, and S. Morad, eds., Clay Mineral Cements in Sandstones (IAS Special Publication 34): Oxford, Wiley-Blackwell, p. 291-316, doi: 10.1002/9781444304336.ch13.

Anovitz, L. M., D. Perkins, and E. J. Essene, 1991, Metastability in near-surface rocks of minerals in the system Al_2O_3–SiO_2–H_2O: Clays and Clay Minerals, v. 39, p. 225-233, doi: 10.1346/CCMN.1991.0390301.

Aoyagi, K., and T. Kazama, 1980, Transformational changes of clay minerals, zeolites and silica minerals during diagenesis: Sedimentology, v. 27, p. 179-188, doi: 10.1111/j.1365-3091.1980.tb01168.x.

Awwiller, D. N., 1993, Illite/smectite formation and potassium mass transfer during burial diagenesis of mudrocks: A study from the Texas Gulf Coast Paleocene-Eocene: Journal of Sedimentary Research, v. 63, p. 501-512, doi: 10.1306/D4267B3B-2B26-11D7-8648000102C1865D.

Barnes, D. A., J.-P. Girard, and J. L. Aronson, 1992, K-Ar dating of illite diagenesis in the Middle Ordovician St. Peter Sandstone, Central Michigan Basin, USA: Implications for thermal history, *in* D. W. Houseknecht, and E. D. Pittman, eds., Origin, Diagenesis, and Petrophysics of Clay Minerals in Sandstones: Tulsa, OK, SEPM Special Publication 47, p. 35-48, doi: 10.2110/pec.92.47.0035.

Beaufort, D., A. Cassagnabere, S. Petit, B. Lanson, G. Berger, J. C. Lacharpagne, and H. Johansen, 1998, Kaolinite-to-dickite reaction in sandstone reservoirs: Clay Minerals, v. 33, p. 297-316, doi: 10.1180/000985598545499.

Berger, A., S. Gier, and P. Krois, 2009, Porosity-preserving chlorite cements in shallow-marine volcaniclastic sandstones: Evidence from Cretaceous sandstones of the Sawan gas field, Pakistan: AAPG Bulletin, v. 93, p. 595-615, doi: 10.1306/01300908096.

Berger, G., B. Velde, and T. Aigouy, 1999, Potassium sources and illitization in Texas Gulf Coast shale diagenesis: Journal of Sedimentary Research, v. 69, p. 151-157, doi: 10.2110/jsr.69.151.

Billault, V., D. Beaufort, A. Baronnet, and J.-C. Lacharpagne, 2003, A nanopetrographic and textural study of grain-coating chlorites in sandstone reservoirs: Clay Minerals, v. 38, p. 315-328, doi: 10.1180/0009855033830098.

Bjørkum, P. A., and N. Gjelsvik, 1988, An isochemical model for formation of authigenic kaolinite, K-feldspar, and illite in sediments: Journal of Sedimentary Research, v. 58, p. 506-511, doi: 10.1306/212F8DD2-2B24-11D7-8648000102C1865D.

Bjørlykke, K., and P. Aagaard, 1992, Clay minerals in North Sea sandstones, *in* D. W. Houseknecht, and E. D. Pittman, eds., Origin, Diagenesis and Petrophysics of Clay Minerals in Sandstones: Tulsa, OK, SEPM Special Publication 47, p. 145-157, doi: 10.2110/pec.92.47.006.

Bjørlykke, K., P. Aagaard, P. K. Egeberg, and S. P. Simmons, 1995, Geochemical constraints from formation water analyses from the North Sea and the Gulf Coast basins on quartz, feldspar and illite precipitation in reservoir rocks, *in* J. M. Cubitt, and W. A. England, eds., The Geochemistry of Reservoirs: London, GSL Special Publication 86, p. 33-50, doi: 10.1144/GSL.SP.1995.086.01.03.

Boles, J. R., and S. G. Franks, 1979, Clay diagenesis in Wilcox sandstones of southwest Texas; implications of smectite diagenesis on sandstone cementation: Journal of Sedimentary Research, v. 49, p. 55-70, doi: 10.1306/212F76BC-2B24-11D7-8648000102C1865D.

Clauer, N., and N. Liewig, 2013, Episodic and simultaneous illitization in oil-bearing Brent Group and Fulmar Formation sandstones from the northern and southern North Sea based on illite K-Ar dating: AAPG Bulletin, v. 97, p. 2149-2171, doi: 10.1306/04021312122.

Clauer, N., H. Zwingmann, N. Liewig, and R. Wendling, 2012, Comparative $^{40}Ar/^{39}Ar$ and K-Ar dating of illite-type clay minerals: A tentative explanation for age identities and differences: Earth-Science Reviews, v. 115, p. 76-96, doi: 10.1016/j.earscirev.2012.07.003.

De Bona, J., N. Dani, J. M. Ketzer, and L. F. De Ros, 2008, Dickite in shallow oil reservoirs from Recôncavo Basin, Brazil: Diagenetic implications for basin evolution: Clay Minerals, v. 43, p. 213-233, doi: 10.1180/claymin.2008.043.2.06.

de Caritat, P., I. Hutcheon, and J. L. Walshe, 1993, Chlorite geothermometry: a review: Clays and Clay Minerals, v. 41, p. 219-239, doi: 10.1346/CCMN.1993.0410210.

De Ros, L. F., G. N. C. Sgarbi, and S. Morad, 1994, Multiple authigenesis of K-feldspar in sandstones: Evidence from the Cretaceous Areado Formation, São Francisco Basin, central Brazil: Journal of Sedimentary Research, v. 64, p. 778-787, doi: 10.1306/D4267EBF-2B26-11D7-8648000102C1865D.

Dowey, P. J., D. M. Hodgson, and R. H. Worden, 2012, Pre-requisites, processes, and prediction of chlorite grain coatings in petroleum reservoirs: A review of subsurface examples: Marine and Petroleum Geology, v. 32, p. 63-75, doi: 10.1016/j.marpetgeo.2011.11.007.

Ehrenberg, S. N., 1993, Preservation of anomalously high porosity in deeply buried sandstones by grain-coating chlorite: Examples from the Norwegian continental shelf: AAPG Bulletin, v. 77, p. 1260-1286.

Ehrenberg, S. N., P. Aagaard, M. J. Wilson, and A. R. Fraser, D. M. L., 1993, Depth-dependent transformation of kaolinite to dickite in sandstones of the Norwegian continental shelf: Clay Minerals, v. 28, p. 325-352, doi: 10.1180/claymin.1993.028.3.01.

Ehrenberg, S. N., and P. H. Nadeau, 1989, Formation of diagenetic illite in sandstones of the Garn Formation, Haltenbanken area, Mid-Norwegian continental shelf: Clay Minerals, v. 24, p. 233-253, doi: 10.1180/claymin.1989.024.2.09.

Emery, D., R. J. Myers, and R. Young, 1990, Ancient subaerial exposure and freshwater leaching in sandstones: Geology, v. 18, p. 382-406, doi: 10.1130/0091-7613(1990)018<1178:ASEAFL>2.3.CO;2.

Franks, S. G., and H. Zwingmann, 2010, Origin and timing of late diagenetic illite in the Permian–Carboniferous Unayzah sandstone reservoirs of Saudi Arabia: AAPG Bulletin, v. 94, p. 1133-1159, doi: 10.1306/04211009142.

Friis, H., N. Molenaar, and T. Varming, 2013, Chlorite meniscus cement–implications for diagenetic mineral growth after oil emplacement: Terra Nova, v. 26, p. 14-21, doi: 10.1111/ter.12061.

Fröhlich, S., J. Redfern, L. Petitpierre, J. D. Marshall, M. Power, and P. V. Grech, 2010, Diagenetic evolution of incised channel sandstones: Implications for reservoir characterisation of the Lower Carboniferous Marar Formation, Ghadames Basin, western Libya: Journal of Petroleum Geology, v. 33, p. 3-18, doi: 10.1111/j.1747-5457.2010.00461.x.

Gould, K., G. Pe-Piper, and D. J. W. Piper, 2010, Relationship of diagenetic chlorite rims to depositional facies in Lower Cretaceous reservoir sandstones of the Scotian Basin: Sedimentology, v. 57, p. 587-610, doi: 10.1111/j.1365-3091.2009.01106.x.

Hamilton, P. J., 2003, A review of radiometric dating techniques for clay mineral cements in sandstones, *in* R. H. Worden, and S. Morad, eds., Clay Mineral Cements in Sandstones (IAS Special Publication 34): Oxford, Wiley-Blackwell, p. 253-287, doi: 10.1002/9781444304336.ch12.

Hillier, S., 1994, Pore-lining chlorites in siliciclastic reservoir sandstones: Electron microprobe, SEM and XRD data, and implications for their origin: Clay Minerals, v. 29, p. 665-680, doi: 10.1180/claymin.1994.029.4.20.

Hornibrook, E. R. C., and F. J. Longstaffe, 1996, Berthierine from the Lower Cretaceous Clearwater Formation, Alberta, Canada: Clays and Clay Minerals, v. 44, p. 1-21, doi: 10.1346/CCMN.1996.0440101.

Humphreys, B., S. J. Kemp, G. K. Lott, D. Bermanto, D. A. Dharmayanti, and I. Samsori, 1994, Origin of grain-coating chlorite by smectite transformation: An example from Miocene sandstones, North Sumatra back-arc basin, Indonesia: Clay Minerals, v. 29, p. 681-692, doi: 10.1180/claymin.1994.029.4.21.

Jeans, C. V., 1978, The origin of the Triassic clay assemblages of Europe with special reference to the Keuper Marl and Rhaetic of parts of England: Philosophical Transactions of the Royal Society of London, Series A, Mathematical and Physical Sciences, v. 289, p. 549-636, doi: 10.1098/rsta.1978.0068.

Jeans, C. V., D. S. Wray, R. J. Merriman, and M. J. Fisher, 2000, Volcanogenic clays in Jurassic and Cretaceous strata of England and the North Sea Basin: Clay Minerals, v. 35, p. 25-55, doi: 10.1180/000985500546710.

Ketzer, J. M., S. Morad, and A. Amorosi, 2003a, Predictive diagenetic clay-mineral distribution in siliciclastic rocks within a sequence stratigraphic framework, *in* R. H. Worden, and S. Morad, eds., Clay Mineral Cements in Sandstones (IAS Special Publication 34): Oxford, Wiley-Blackwell, p. 43-61, doi: 10.1002/9781444304336.ch2.

Ketzer, J. M., S. Morad, J. P. Nystuen, and L. F. De Ros, 2003b, The role of the Cimmerian unconformity (Early Cretaceous) in the kaolinitization and related reservoir-quality evolution in Triassic sandstones of the Snorre Field, North Sea, *in* R. H. Worden, and S. Morad, eds., Clay Mineral Cements in Sandstones (IAS Special Publication 34): Oxford, Wiley-Blackwell, p. 361-382, doi: 10.1002/9781444304336.ch13.

Lander, R. H., and L. M. Bonnell, 2010, A model for fibrous illite nucleation and growth in sandstones: AAPG Bulletin, v. 94, p. 1161-1187, doi: 10.1306/04211009121.

Lanson, B., D. Beaufort, G. Berger, A. Bauer, A. Cassagnabere, and A. Meunier, 2002, Authigenic kaolin and illitic minerals during burial diagenesis of sandstones: A review: Clay Minerals, v. 37, p. 1-22, doi: 10.1180/000985502371.

Lemon, N. M., and C. J. Cubitt, 2003, Illite fluorescence microscopy: A new technique in the study of illite in the Merrimelia Formation, Cooper Basin, Australia, *in* R. H. Worden, and S. Morad, eds., Clay Mineral Cements in Sandstones (IAS Special Publication 34): Oxford, Wiley-Blackwell, p. 409-424, doi: 10.1002/9781444304336.ch18.

Leone, G., E. Reyes, G. Cortecci, A. Pochini, and J. Linares, 1983, Genesis of bentonites from Cabo de Gata, Almeria, Spain: A stable isotope study: Clay Minerals, v. 18, p. 227-238, doi: 10.1180/claymin.1983.018.3.01.

Lerbekmo, J. F., 1957, Authigenic montmorillonoid cement in andesitic sandstones of central California: Journal of Sedimentary Research, v. 27,

p. 298-305, doi: 10.1306/74D706CD-2B21-11D7-8648000102C1865D.

Lynch, F. L., 1997, Frio shale mineralogy and the stoichiometry of the smectite-to-illite reaction: The most important reaction in clastic sedimentary diagenesis: Clays and Clay Minerals, v. 45, p. 618-631, doi: 10.1346/CCMN.1997.0450502.

Lynch, F. L., L. E. Mack, and L. S. Land, 1997, Burial diagenesis of illite/smectite in shales and the origins of authigenic quartz and secondary porosity in sandstones: Geochimica et Cosmochimica Acta, v. 61, p. 1995-2006, doi: 10.1016/S0016-7037(97)00066-5.

Macchi, L., 1987, A review of sandstone illite cements and aspects of their significance to hydrocarbon exploration and development: Geological Journal, v. 22, p. 333-345, doi: 10.1002/gj.3350220406.

McAulay, G. E., S. D. Burley, and L. H. Johnes, 1993, Silicate mineral authigenesis in the Hutton and NW Hutton fields: Implications for subsurface porosity development, *in* J. R. Parker, ed., Petroleum Geology of Northwest Europe: Proceedings of the 4th Conference: London, GSL, p. 1377-1394, doi: 10.1144/0041377.

McHardy, W. J., M. J. Wilson, and J. M. Tait, 1982, Electron microscope and X-ray diffraction studies of filamentous illitic clay from sandstones of the Magnus Field: Clay Minerals, v. 17, p. 23-40, doi: 10.1180/claymin.1982.017.1.04.

McKinley, J. M., R. H. Worden, and A. H. Ruffell, 2003, Smectite in sandstones: A review of the controls on occurrence and behaviour during diagenesis, *in* R. H. Worden, and S. Morad, eds., Clay Mineral Cements in Sandstones (IAS Special Publication 34): Oxford, Wiley-Blackwell, p. 109-128, doi: 10.1002/9781444304336.ch5.

Milliken, K. L., 2003, Microscale distribution of kaolinite in Breathitt Formation sandstones (Middle Pennsylvanian): Implications for mass balance, *in* R. H. Worden, and S. Morad, eds., Clay Mineral Cements in Sandstones (IAS Special Publication 34): Oxford, Wiley-Blackwell, p. 343-360, doi: 10.1002/9781444304336.ch15.

Morad, S., 1990, Mica alteration reactions in Jurassic reservoir sandstones from the Haltenbanken area, offshore Norway: Clays and Clay Minerals, v. 38, p. 584-590, doi: 10.1346/CCMN.1990.0380603.

Morad, S., R. H. Worden, and J. M. Ketzer, 2003, Oxygen and hydrogen isotopic composition of diagenetic clay minerals in sandstones: A review of the data and controls, *in* R. H. Worden, and S. Morad, eds., Clay Mineral Cements in Sandstones (IAS Special Publication 34): Oxford, Wiley-Blackwell, p. 63-91, doi: 10.1002/9781444304336.ch3.

Mücke, A., 2006, Chamosite, siderite and the environmental conditions of their formation in chamosite-type Phanerozoic ooidal ironstones: Ore Geology Reviews, v. 28, p. 235-249, doi: 10.1016/j.oregeorev.2005.03.004.

Neasham, J. W., 1977, The morphology of dispersed clay in sandstone reservoirs and its effect on sandstone shaliness, pore space and fluid flow properties: Society of Petroleum Engineers Annual Technical Conference and Exhibition, Denver, Oct. 9-12., Paper SPE 6858, doi: 10.2118/6858-MS.

Odin, G. S., and A. Matter, 1981, De glauconarium origine: Sedimentology, v. 28, p. 611-641, doi: 10.1111/j.1365-3091.1981.tb01925.x.

Osborne, M., R. S. Haszeldine, and A. E. Fallick, 1994, Variation in kaolinite morphology with growth temperature in isotopically mixed pore-fluids, Brent Group, UK North Sea: Clay Minerals, v. 29, p. 591-608, doi: 10.1180/claymin.1994.029.4.15.

Pe-Piper, G., and S. Weir-Murphy, 2008, Early diagenesis of inner-shelf phosphorite and iron-silicate minerals, Lower Cretaceous of the Orpheus graben, southeastern Canada: Implications for the origin of chlorite rims: AAPG Bulletin, v. 92, p. 1153-1168, doi: 10.1306/05050807118.

Pittman, E. D., R. E. Larese, and M. T. Heald, 1992, Clay coats: Occurrence and relevance to preservation of porosity in sandstones, *in* D. W. Houseknecht, and E. D. Pittman, eds., Origin, Diagenesis, and Petrophysics of Clay Minerals in Sandstones: Tulsa, OK, SEPM Special Publication 47, p. 241-255, doi: 10.2110/pec.92.47.0241.

Ryan, P. C., M. E. Conrad, K. Brown, and C. P. Chamberlain, R. C., Jr., 1998, Oxygen isotope compositions of mixed layer serpentine–chlorite and illite–smectite in the Tuscaloosa Formation (U.S. Gulf Coast): Implications for pore fluids and mineralogic reactions: Clays and Clay Minerals, v. 46, p. 357-368, doi: 10.1346/CCMN.1998.0460401.

Ryan, P. C., and S. Hillier, 2002, Berthierine/chamosite, corrensite, and discrete chlorite from evolved verdine and evaporite-associated facies in the Jurassic Sundance Formation, Wyoming: American Mineralogist, v. 87, p. 1607-1615.

Ryan, P. C., and R. C. Reynolds, 1996, The origin and diagenesis of grain-coating serpentine-chlorite in Tuscaloosa Formation sandstones, US Gulf Coast: American Mineralogist, v. 81, p. 213-225.

Sanjuan, B., J.-P. Girard, S. Lanini, A. Bourguignon, and E. Brosse, 2003, Geochemical modelling of diagenetic illite and quartz cement formation in Brent Sandstone reservoirs: Example of the Hild Field, Norwegian North Sea, *in* R. H. Worden, and S. Morad, eds., Clay Mineral Cements in Sandstones (IAS Special Publication 34): Oxford, Wiley-Blackwell, p. 425-452, doi: 10.1002/9781444304336.ch19.

Sardini, P., A. El Albani, D. Pret, S. Gaboreau, M. Siitari-Kauppi, and D. Beaufort, 2009, Mapping and quantifying the clay aggregate microporosity in medium- to coarse-grained sandstones using the ^{14}C-PMMA method: Journal of Sedimentary Research, v. 79, p. 584-592, doi: 10.2110/jsr.2009.063.

Shaw, H. F., and D. M. Conybeare, 2003, Patterns of clay mineral diagenesis in interbedded mudrocks and sandstones: An example from the Palaeocene of the North Sea, *in* R. H. Worden, and S. Morad, eds., Clay Mineral Cements in Sandstones (IAS Special Publication 34): Oxford, Wiley-Blackwell, p. 129-145, doi: 10.1002/9781444304336.ch6.

Spötl, C., D. W. Houseknecht, and S. J. Burns, 1996, Diagenesis of an 'overmature' gas reservoir: The Spiro sand of the Arkoma Basin, USA: Marine and Petroleum Geology, v. 13, p. 25-40, doi: 10.1016/0264-8172(95)00037-2.

Spötl, C., D. W. Houseknecht, and F. J. Longstaffe, 1994, Authigenic chlorites in sandstones as indicators of high-temperature diagenesis, Arkoma foreland basin, USA: Journal of Sedimentary Research, v. 64A, p. 553-566, doi: 10.1306/D4267E06-2B26-11D7-8648000102C1865D.

Środoń, J., 1999a, Use of clay minerals in reconstructing geological processes: Recent advances and some perspectives: Clay Minerals, v. 34, p. 27-37, doi: 10.1180/000985599546046.

Środoń, J., 1999b, Nature of mixed-layer clays and mechanisms of their formation and alteration: Annual Review of Earth and Planetary Sciences, v. 27, p. 19-53, doi: 10.1146/annurev.earth.27.1.19.

Storvoll, V., K. Bjørlykke, D. Karlsen, and G. Saigal, 2002, Porosity preservation in reservoir sandstones due to grain-coating illite: A study of the Jurassic Garn Formation from the Kristin and Lavrans fields, offshore Mid-Norway: Marine and Petroleum Geology, v. 19, p. 767-781, doi: 10.1016/S0264-8172(02)00035-1.

Walker, T. R., B. Waugh, and A. J. Crone, 1978, Diagenesis in first-cycle desert alluvium of Cenozoic age, southwestern United States and northwestern Mexico: GSA Bulletin, v. 89, p. 19-32, doi: 10.1130/0016-7606(1978)89<19:DIFDAO>2.0.CO;2.

Wilkinson, M., R. S. Haszeldine, and A. Fallick, 2006, Jurassic and Cretaceous clays of the northern and central North Sea hydrocarbon reservoirs reviewed: Clay Minerals, v. 41, p. 151-186, doi: 10.1180/0009855064110197.

Wilson, M. D., and E. D. Pittman, 1977, Authigenic clays in sandstones: Recognition and influence on reservoir properties and paleoenvironmental analysis: Journal of Sedimentary Research, v. 47, p. 3-31, doi: 10.1306/212F70E5-2B24-11D7-8648000102C1865D.

Wilson, M. J., 2013, Rock-Forming Minerals, Volume 3C, Sheet Silicates: Clay Minerals (2nd Edition): London, GSL, 736 p.

Worden, R. H., and S. Morad, 2003, Clay minerals in sandstones: Controls on formation, distribution and evolution, *in* R. H. Worden, and S. Morad, eds., Clay Mineral Cements in Sandstones (IAS Special Publication 34): Oxford, Wiley-Blackwell, p. 1-41, doi: 10.1002/9781444304336.ch1.

Facing Page: Top — Phillipsite filling vesicles in a basalt porphyry, Oligocene Deborah Volcanic Fm., Otago, New Zealand (XPL). Bottom — SEM view of clinoptilolite, Miocene Barstow Fm., California.

DIAGENESIS:
ZEOLITE CEMENTS

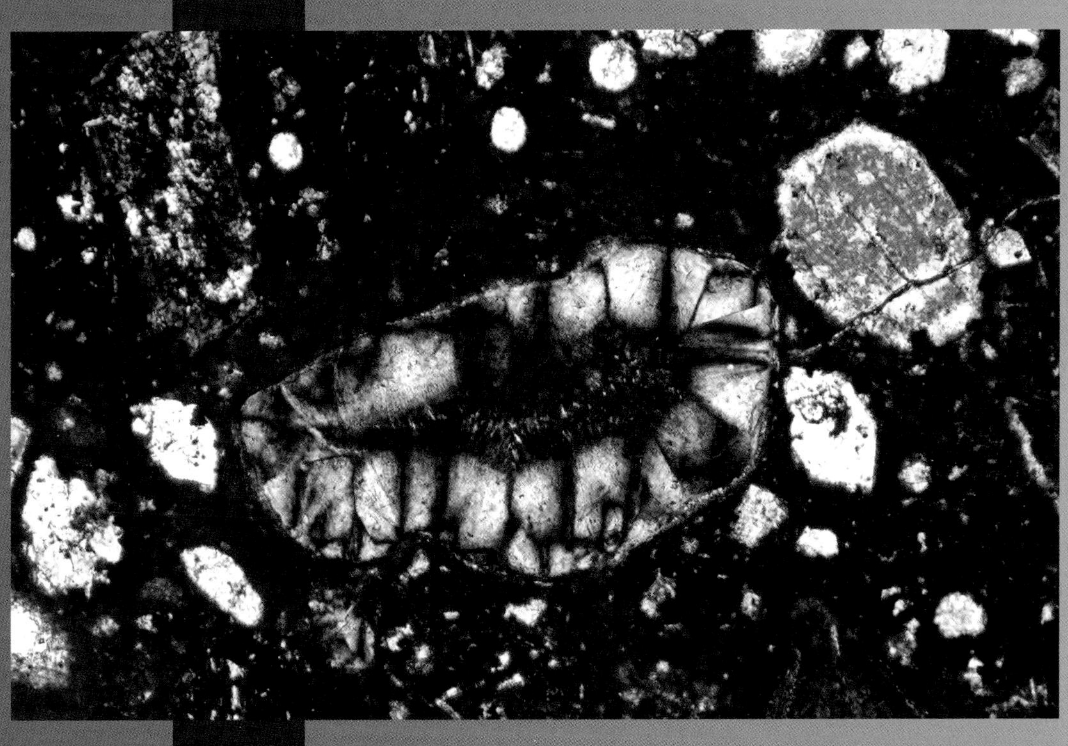

CHAPTER 14

Zeolite Cements

Zeolites are a large and complex group of hydrated aluminosilicate minerals that are among the most abundant authigenic silicates in sedimentary deposits. They are so complex that only a superficial summary can be given in this short introduction, and readers are urged to consult the vastly more detailed references provided at the end of the chapter.

Both water molecules and cations can be substituted or replaced in most zeolites without disrupting the crystal structure, accounting for their great compositional diversity as well as their widespread use in ion-exchange applications. Their classification and nomenclature has undergone, and continues to undergo, extensive review and revision (e.g., Coombs et al. 1998; McCusker et al., 2001). Although more than 30 different zeolites have been noted from sediments and sedimentary rock, only five are common (analcime, clinoptilolite, heulandite, laumontite and phillipsite) and three others (chabazite, erionite and mordenite) are relatively common (Gottardi and Galli, 1985; Hay and Sheppard, 2001). The generalized or simplified formulas for these minerals are given in Table 11.2 (page 247).

Most authigenic sedimentary zeolites are found as displacement, replacement or cementing minerals and also are commonly found as vug or fracture fills. Zeolites generally are formed from volcanic materials, clay minerals or precursor zeolites, typically in settings with neutral to alkaline water chemistry (Hall, 1998). More specifically, volcanic glass, quartz, plagioclase, pyroxene, leucite, nepheline and montmorillonite are commonly involved in early zeolite formation (Hay, 1966). Sites of zeolite formation include tephra-rich saline and alkaline lake deposits, volcaniclastic strata, soils and other surficial deposits and some deep-sea sediments (especially clay-rich, volcaniclastic or biosiliceous ones; e.g., Kastner and Stonecipher, 1978). Silica-rich zeolites predominate in glass-rich sediments, and Fe/Mg-rich zeolites predominate in mafic VRF-rich sandstones. Early-formed phillipsite and clinoptilolite dominate deep-sea systems; clinoptilolite, phillipsite, erionite, chabazite and analcime are common in young, tephra-rich lacustrine deposits; and analcime, phillipsite, clinoptilolite, chabazite and natrolite are found in young, alkaline soils (Hay and Sheppard, 2001). Laumontite occurs as a burial diagenetic cement in feldspathic sandstones, particularly those rich in plagioclase, where it is commonly associated with regional albitization (Helmold and van de Kamp, 1984).

Zeolite minerals also are influenced strongly by diagenetic processes. Changes in temperature, pressure and water chemistry commonly result in alteration of early-formed zeolites to later-stage forms. Iijima and Utada (1966) and Iijima (1988) proposed a scheme of generalized depth-related zonation of zeolites and associated minerals in progressively buried marine, silicic, volcaniclastic deposits that remains largely valid:

Stage 1: Volcanic glass

Stage 2: Transformation of glass to clinoptilolite and/or mordenite at temperatures of roughly 41°–55°C

Stage 3: Reaction of clinoptilolite and mordenite to form analcime at temperatures of about 84°–91°C; heulandite/analcime and deeper laumontite/analcime associations also are commonly formed in this zone (with laumontite persisting even into early metamorphic stages; Coombs et al., 1959)

Stage 4: Analcime reacts to form albite at temperatures of 120°–124°C

Sedimentary zeolite identification:

Identification of specific zeolite minerals in thin section generally is difficult because of the low birefringence common to this mineral group and the variability of relief even within single minerals because of their substantial compositional variability. Coarsely crystalline zeolites (for example cuboctahedral crystals of analcime, tabular crystals of clinoptilolite or radiating sprays of phillipsite crystals) can be identified optically by their crystal morphologies, but refined optical identification typically is done by examination of mineral separates in carefully calibrated immersion oils (e.g., Mason and Sand, 1960).

Far better accuracy of zeolite identification typically can be achieved using SEM, again based on crystal morphology (Mumpton and Ormsby, 1976; Welton, 1984) and such SEM-derived morphologic descriptions are given below. However, further study using X-ray diffraction (XRD), Atomic Force Microscopy (Fuoco, 2012) and other analytical techniques is essential for accurate identification of zeolite minerals and especially for determination of their specific compositions. Whole-rock XRD generally suffices for identification of zeolites composing 10% or more of a sample; XRD analysis of concentrated mineral separates generally is required to identify lower-abundance zeolites (Hay, 1966).

References especially useful for comparative SEM morphological identification of sedimentary zeolites are Mumpton and Ormsby (1976) and Welton (1984), and references cited therein. The descriptions given below are adapted largely from those two works.

Analcime is crystallographically complex (with cubic or pseudocubic habit), has poor cleavage and subconchoidal fracture. In thin section, it is colorless and isotropic to very weakly birefringent. It is recognized in SEM by its characteristic cuboctahedral and trapezohedral habits, with crystals commonly 10-30 μm

in diameter. The cubic habit of analcime is recognizable even in subhedral fragments or in crystals that are partially coated with clays.

Chabazite is triclinic and pseudohexagonal with one distinct cleavage direction and uneven fracture. It is colorless and biaxial (+ or –) or uniaxial in thin section, with very low birefringence. In SEM, it generally shows cube-like rhombohedra (commonly 1–10 μm across) and clusters of intergrown cubes or rhombs; it also is found as platy, discoidal crystals and intergrown clusters or rosettes of such discs.

Clinoptilolite is monoclinic with one perfect cleavage. It is colorless and biaxial (+ or –) in thin section with very low birefringence. In SEM, it is found as small (1–5 μm) tabular plates or laths, often displaying a characteristic monoclinic symmetry. Clinoptilolite also occurs as distinctive "coffin-shaped" crystals.

Erionite is a group of hexagonal minerals that generally show one distinct cleavage. They are colorless and uniaxial (+) in thin section and are slightly more birefringent than most of the siliceous, alkali-rich zeolites (but still have low birefringence). They are primarily found as rod- or needle-shaped crystals. In SEM, they commonly appear as closely-packed (sometime woven mat-like) fibers. Also found as clusters or barrel-shaped bundles of acicular needles (10–>100 μm in length) and locally as needles with broom-like "split ends".

Heulandite is monoclinic and has one perfect cleavage and uneven or subconchoidal fracture. It is colorless and biaxial (+) in thin section and has very low birefringence. In SEM, it is found as monoclinic, blocky to "coffin-shaped" crystals that are similar to those of clinoptilolite (thus requiring XRD or other analysis for reliable determinations).

Laumontite is monoclinic and has two perfect cleavages and uneven fracture. It is colorless and biaxial (-) in thin section, with very low birefringence. Laumontite does not have distinctive SEM characteristics but commonly forms blocky pore-filing fabrics similar to calcite.

Mordenite is orthorhombic and has one perfect and one distinct cleavage as well as uneven fracture. It is colorless and biaxial (+ or -) in thin section, with very low birefringence. In SEM, it occurs as long, extremely delicate (submicron scale) fibrous crystals, as curved fibers, as cottony fiber-aggregates, as needles, and as short, stubby crystals. Mordenite commonly is difficult to differentiate in SEM from erionite, although mordenite generally has thinner fibers than erionite.

Phillipsite is monoclinic with one good and one less distinct cleavage. It is colorless in thin section and biaxial (+) with very low birefringence. In SEM, it occurs as thick prisms and stubby laths, roughly 3–30 μm in length and 0.3–3 μm thick. The pseudo-orthorhombic symmetry generally is evident, especially where crystals are capped by two-sided "domes". In addition, phillipsite commonly occurs as rosettes or bundles of spherically radiating crystals (Mumpton and Ormsby, 1976).

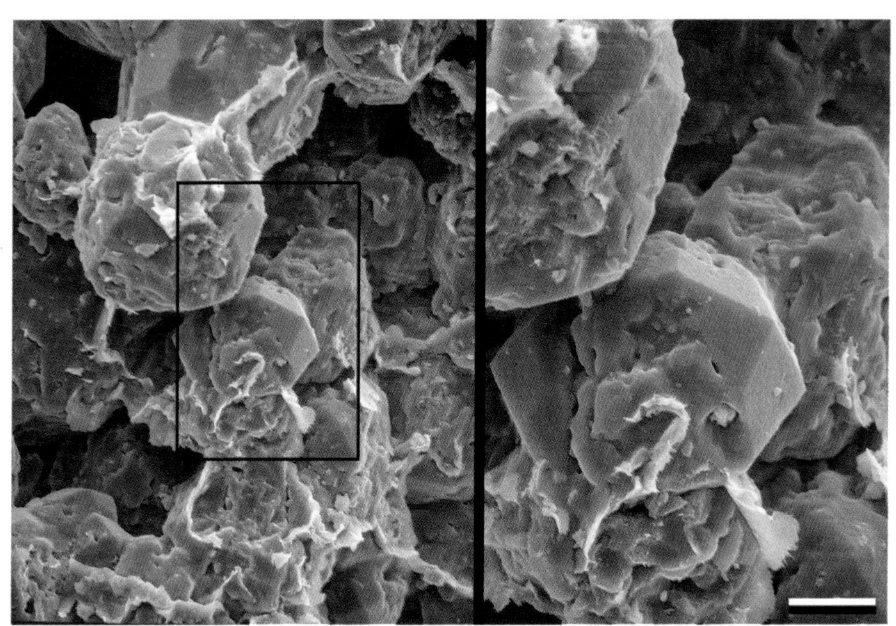

Pliocene Big Sandy Fm., Mohave Co., Arizona

Analcime is one of the most abundant zeolites in sedimentary rocks, especially in older sedimentary units. Although analcime is grouped in the zeolites, it also is similar to the feldspathoids, and it is one of the few zeolites that may occur in rocks lacking glassy material. Analcime forms cements in sedimentary rocks and also replaces clays and pyroclastics in lake deposits or precipitates within vesicles in igneous rocks. Even though the analcime crystals in this SEM image are coated with smectite/illite clays, the trapezohedral crystal structure of analcime is still visible (Welton, 1984). Photomicrographs from Joann E. Welton.

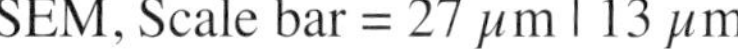

SEM, Scale bar = 27 μm | 13 μm

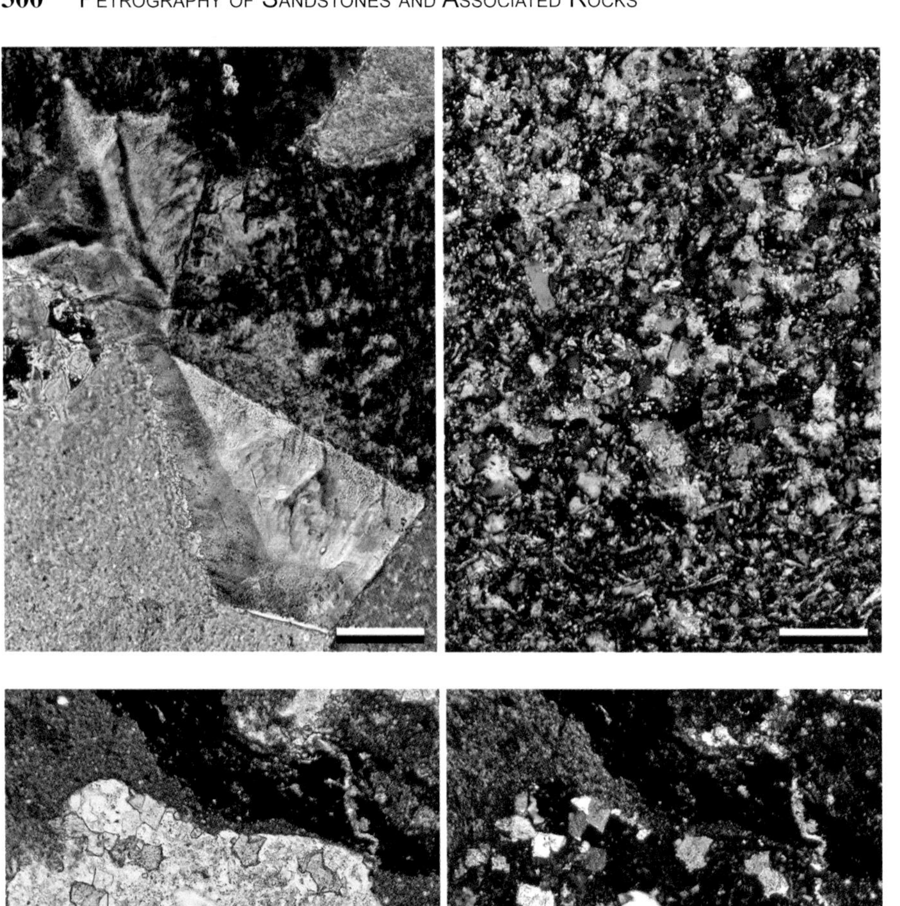

Up. Triassic Lockatong Fm., Hunterdon Co., New Jersey

This analcime is from a rift-basin where it occurs mainly as an eogenetic displacive mineral within lacustrine shales. The example at left shows a cut through a trapezohedral crystal of analcime (note the hourglass shape) that is surrounded by eogenetic dolomite (El Tabakh and Schreiber, 1994). Analcime commonly forms euhedral to subhedral crystals, and its birefringence is very low to isotropic. The second view shows disseminated analcime partially replaced by albite—such albite alteration is mesogenetic and occurs at temperatures above 120°C (Iijima, 1988), in this case in the vicinity of basaltic dikes (Van Houten, 1962).

XPL | XPL, Scale bar = 0.05 | 0.26 mm

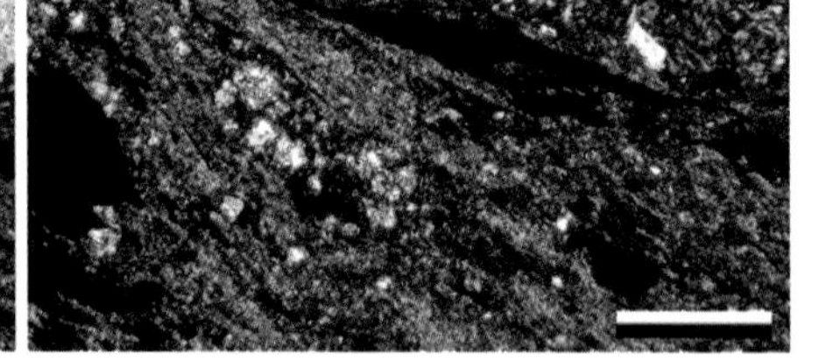

Up. Triassic Lockatong Fm., Hunterdon Co., New Jersey

Zeolitic nodules are common within these pyrite-rich, lacustrine mudstone deposits. This vug is lined with trapezohedral crystals of analcime followed by dolomite. The analcime may be replacive after gypsum, but it more commonly forms primary precipitates, so these nodules most likely originated as displacive analcime. The crystals contain brown-colored, primarily fluid-filled inclusions and have a very low (near isotropic) birefringence.

PPL | XPL, Scale bar = 0.10 mm

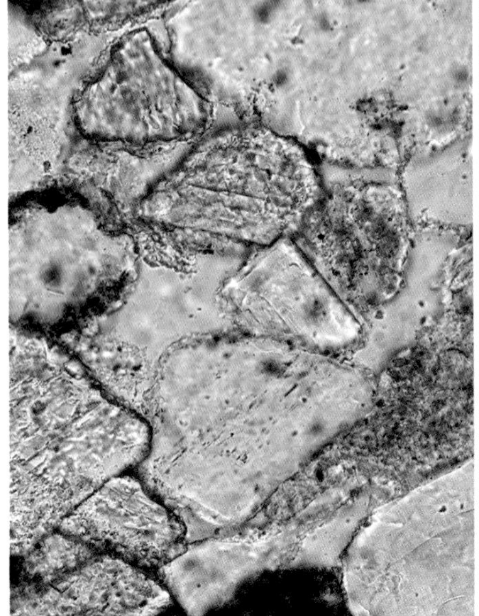

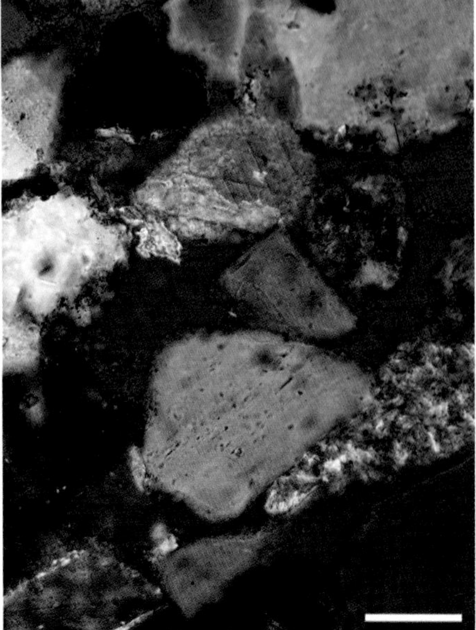

Cretaceous (Cenomanian-Turonian) Seabee Fm., North Slope, Alaska

Poikilotopic analcime fills intergranular pores in a lithic sandstone containing numerous detrital dolomite grains. The isotropic nature and low relief of analcime makes it one of the easiest zeolites to identify in thin section, but makes it more difficult to recognize poikilotopic crystals. No crystal boundaries are present in the large patch of analcime cement suggesting it is one large crystal. The use of partially crossed polarizers results in a dark gray extinction color rather than complete extinction. Analcime in this sample was derived from alteration of volcanic glass shards in adjacent tuffaceous laminae. Photomicrograph from Kenneth P. Helmold.

PPL | PXPL, BDI, Scale bar = 0.03 mm

Pliocene – Pleistocene unknown formation, Graham Co., Arizona

Chabazite, a well-crystallized zeolite, is moderately common in Neogene lacustrine, tuffaceous units and altered volcanic glasses (Welton, 1984). Chabazite commonly occurs as euhedral rhombohedral crystals that make it quite distinctive from most other zeolites. In this SEM image, a crystal of erionite (the large cluster of crystal laths at center) is surrounded and partially coated by chabazite. The closeup view shows linear strands of rhombohedral chabazite crystals. Chabazite also forms disc-shaped crystal arrays, shown in the lower magnification view. Photomicrographs from Joann E. Welton.

SEM, Scale bar = 7.5 μm | 0.67 μm

Pliocene – Pleistocene unknown formation, Graham Co., Arizona

These large discs of chabazite have euhedral rhombohedral chabazite crystals coating them (Welton, 1984). In addition to the chabazite discs, linear strands of rhombohedral chabazite crystals coat large crystals of erionite in the background. Photomicrograph from Joann E. Welton.

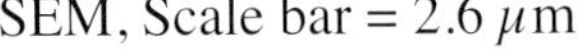

SEM, Scale bar = 2.6 μm

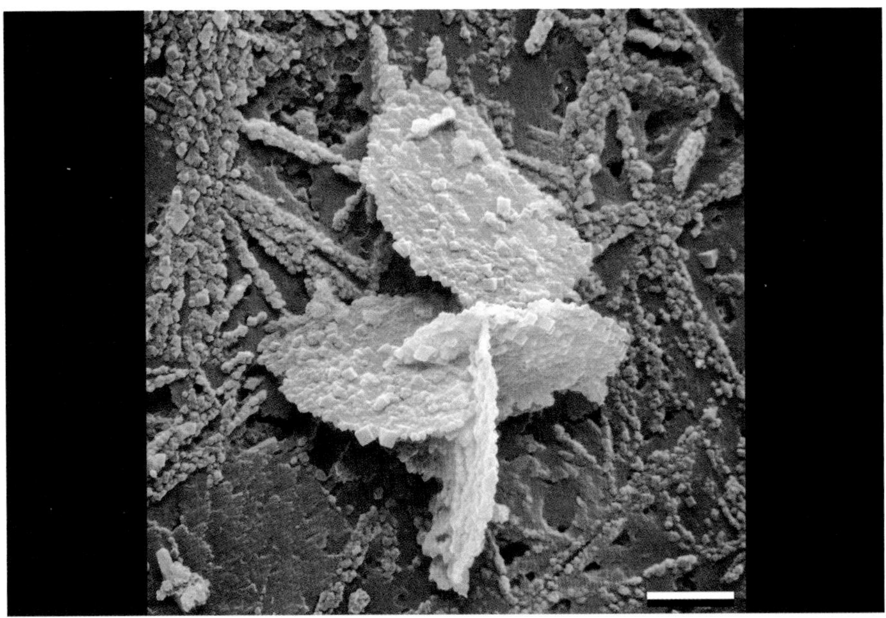

Pliocene – Pleistocene unknown formation, Graham Co., Arizona

A low magnification view of zeolite-replaced volcanic ash (Welton, 1984). Much of the original fabric of the ash, shards and glass bubbles is still visible despite having been replaced and cemented by chabazite. In this instance, the chabazite forms crystal rosettes of cement that coat the ash fragments. Photomicrograph from Joann E. Welton.

SEM, Scale bar = 77.4 μm

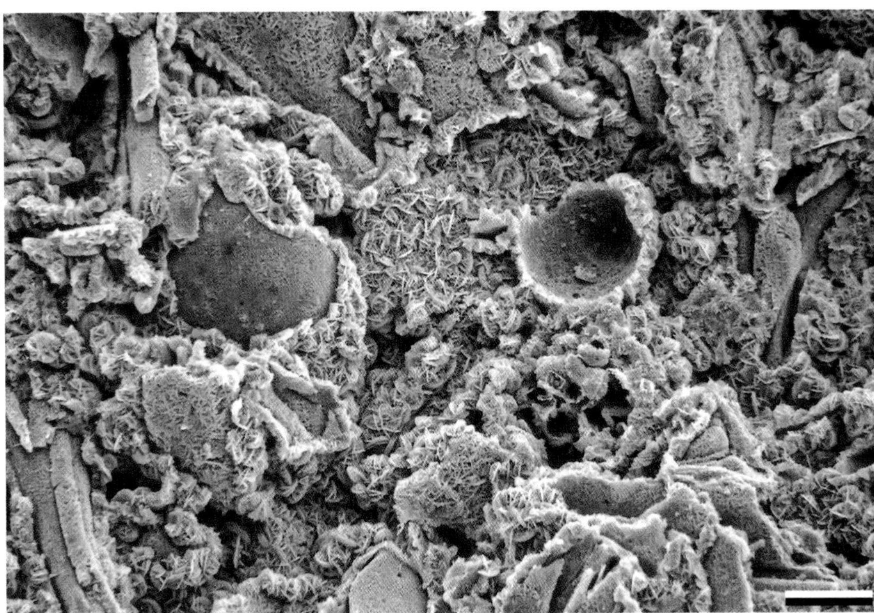

Cenozoic unknown formation, San Bernardino Co., California

In SEM, clinoptilolite commonly appears as jumbled stacks of platy crystals, as in this example of pore filling cements (Welton, 1984). It is a member of the heulandite group of zeolites and thus is crystallographically very similar to heulandite. In the close-up view on the right (an enlargement of the box on the left image), a few of the tabular clinoptilolite crystals show a distinctive "coffin" shape and scattered clay coats bridge several of the crystals. Because clinoptilolite and heulandite are difficult to distinguish, microprobe analysis is the most definitive method for their identification. Photomicrographs from Joann E. Welton.

SEM, Scale bar = 11.8 μm | 2.4 μm

Mid. Eocene – Lo. Oligocene Spears Gp., Catron Co., New Mexico

Because clinoptilolite and heulandite are difficult to tell apart, they are commonly identified petrographically as clinoptilolite-heulandite. There is a solid solution series between them, with clinoptilolite containing higher Si/Al ratios (> 4), as determined by microprobe analysis. In this view, stacks of platy coffin-shaped crystals of microprobe-analyzed clinoptilolite partially fill the porosity within a volcanic arenite. Photomicrograph from Steven M. Cather.

SEM, Scale bar = 36 μm

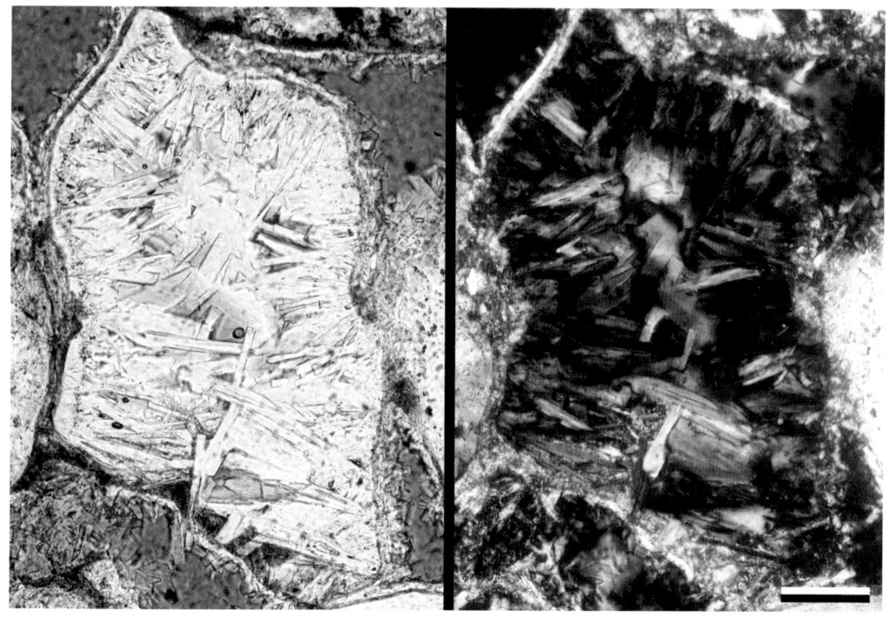

Pliocene – Pleistocene Gila Gp., New Mexico

In thin-section, clinoptilolite is colorless, biaxial (+ or -) and has very low birefringence, and these needle-like, tabular and "coffin-shaped" crystals are typical for clinoptilolite-heulandite. Here a leached grain, possibly a feldspar or volcanic rock fragment, was infilled by medium to coarsely crystalline clinoptilolite. Smaller crystals of clinoptilolite coat the exterior of the grain over illite/smectite cements (the high birefringence phase). This rock contains both primary intergranular and secondary intragranular porosity.

PPL | XPL, BDI, Scale bar = 0.11 mm

Mid. Eocene – Lo. Oligocene Spears Gp., Catron Co., New Mexico

Here, clinoptilolite, as determined by microprobe analysis, occurs as a pore-filling cement in volcanic arenite (Cather, 1986). The clinoptilolite cements, which formed over thin, grain-coating clay cements, appear as low-relief, tabular crystals.

PPL, BDI, Scale bar = 0.04 mm

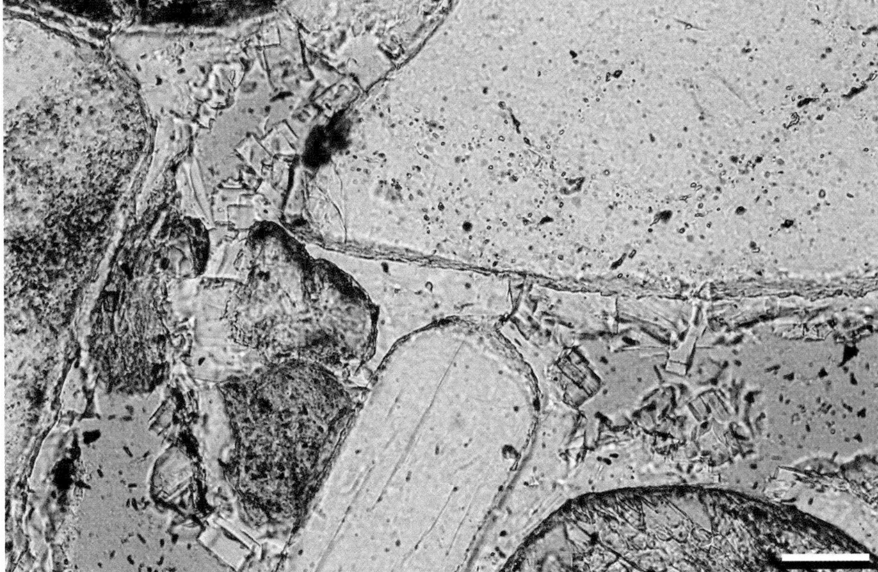

Up. Miocene "Rome Beds", Malheur Co., Oregon

Erionite, along with many other zeolites, commonly occurs as an alteration and cement in volcanic tuffs. Here are two different magnifications of a sample that contains devitrified glass shards. The shards are both replaced by woven networks of erionite and are cemented by acicular erionite crystals (Welton, 1984). The center of the left image is magnified in the right image; the box in the right image outlines the region of even greater magnification shown in the left image below. Photomicrographs from Joann E. Welton.

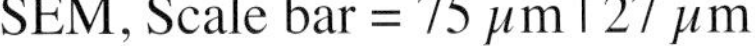

SEM, Scale bar = 75 μm | 27 μm

Up. Miocene "Rome Beds", Malheur Co., Oregon

Typically, erionite forms acicular to "woolly" masses of hexagonal rods and fibers, woven surfaces or bundles of crystals or large blocky to barrel-shaped crystals. These two images are from the previous sample of devitrified tuff (Welton, 1984). The left image is from the box in the previous image and demonstrates the acicular woolly fabric. The bundles and woven textures on the right image are from the surface of an altered glass shard just outside the area of the lowest magnification image. In thin section, erionite is uniaxial (+) and has very low birefringence. Photomicrographs from Joann E. Welton.

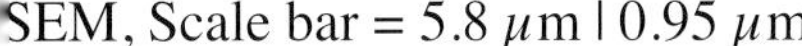

SEM, Scale bar = 5.8 μm | 0.95 μm

Unknown age and formation, Canada Creek, Nova Scotia, Canada

Heulandite, like its close relative clinoptilolite, commonly shows coffin-shaped to blocky crystals in SEM (Welton, 1984). Heulandite has one perfect cleavage, is colorless, biaxial (+) and has very low birefringence. Because it can be easily confused with other zeolite minerals, especially clinoptilolite, XRD and quantitative geochemistry (microprobe analysis) are necessary for definitive identification. Photomicrograph from Joann E. Welton.

SEM, Scale bar = 4.0 μm

Cenozoic unknown formation, California

This pore in a feldspathic sandstone is filled with blocky, coarsely-crystalline laumontite cement (Welton, 1984). In thin section, laumontite is colorless, biaxial (-) and has very low birefringence. In SEM, laumontite can be confused with carbonate cements (unless EDS data is obtained). However, in thin section, those two mineral are easily distinguishable because of their birefringence differences (low for laumontite and high for carbonates). Photomicrograph from Joann E. Welton.

SEM, Scale bar = 8.5 μm

Paleocene Fatehgarh Fm., Rajasthan, India

A grain coated by chlorite flakes that are overlain and partly enclosed by laumontite crystals. The laumontite here has formed as short, euhedral prisms. Both chlorite and zeolites such as laumontite are common in sandstones that are rich in basaltic grains such as this one. Formation of laumontite can be promoted by the degradation of plagioclase and the chlorite can result from the alteration of ferromagnesian minerals such as pyroxene. A thin-section view of the same sample is shown in the next image.

SEM, Scale bar = 9 μm

Paleocene Fatehgarh Fm., Rajasthan, India

Laumontite cement within a volcaniclastic sandstone that contains altered basaltic grains. Such grains display microporosity and now are composed of chlorite and albite. Laumontite and quartz have similar birefringence color, but the laumontite differs from quartz by showing a strong cleavage. In addition, it shows negative relief compared to the surrounding quartz because of laumontite's lower refractive index. The brightly birefringent lozenges are authigenic titanite (sphene).

PPL | XPL, BDI, Scale bar = 0.04 mm

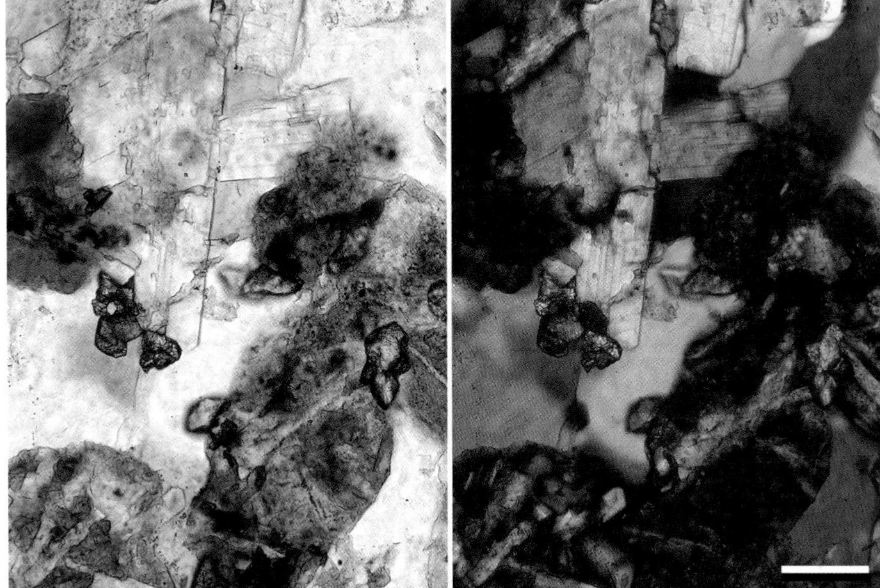

Up. Jurassic Naknek Fm., Cook Inlet, Alaska

Intergranular pores in a feldspathic sandstone filled by two generations of cement—an initial coating of pore-lining chlorite, followed by a later occurrence of pore-filling laumontite. Each pore is filled by a single crystal of laumontite (with higher than normal birefringence colors due to partially cross-polarized illumination). The prominent cleavage helps to distinguish it from other zeolites. Laumontite also partially replaces adjacent grains of albitized plagioclase. Laumontite is a by-product of albitization where detrital grains of calcic-plagioclase are replaced by the Na-feldspar albite. Photomicrographs from Kenneth P. Helmold.

PPL | PXPL, Scale bar = 0.09 mm

Miocene Obispo Fm., San Luis Obispo Co., California

In these SEM images, mordenite has formed coatings within cristobalite-lined cavities (Welton, 1984). The mordenite occurs as delicate, acicular, radial arrays of crystals. Mordenite can be difficult to distinguish from erionite in SEM; therefore, XRD is usually needed to make a reliable identification. In thin section, mordenite is biaxial (+ or -), colorless and has very low birefringence. Photomicrographs from Joann E. Welton.

SEM, Scale bar = 11.2 μm | 2.4 μm

Miocene Obispo Fm., San Luis Obispo Co., California

In this SEM, mordenite occurs as masses of delicate, curved crystals growing within a porcellanite. This type of growth has been called a "rat's nest" of interfingering crystals (Welton, 1984). Photomicrograph from Joann E. Welton.

SEM, Scale bar = 10 μm

Pleistocene sediments, Lake Tecopa, Inyo Co., California

Phillipsite commonly forms radiating bundles (or rosettes) of crystals that are visible in either SEM or thin section. These phillipsite rosettes formed in tuffaceous sediments within a saline lake (Welton, 1984). Photomicrograph from Joann E. Welton.

SEM, Scale bar = 3.1 μm

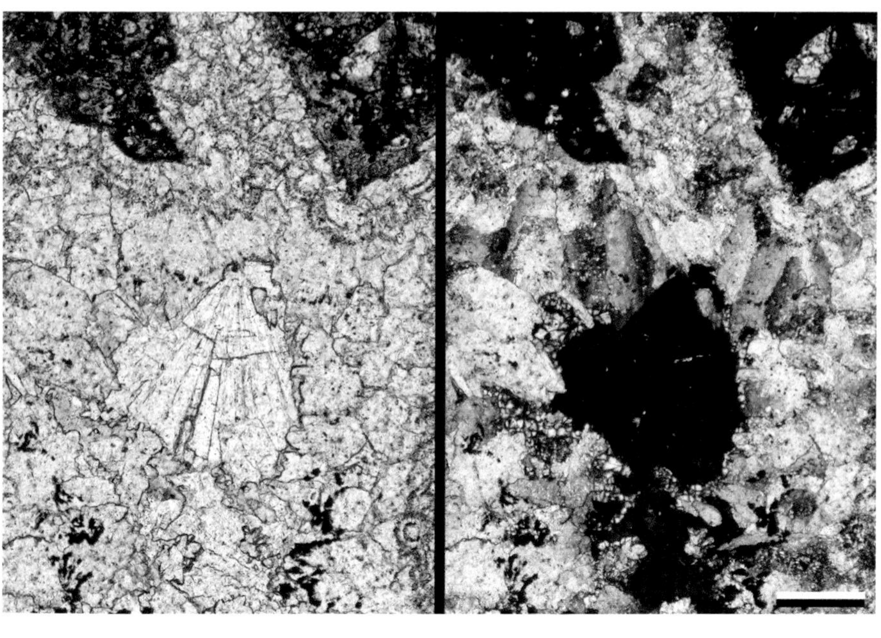

Lo. Oligocene Deborah Volcanic Fm., Otago, New Zealand

In these marine pillow basalt breccias, phillipsite forms fan-like clusters of crystals that were later enclosed in zoned, marine calcite cements. Note the very low birefringence of phillipsite; it is easy to overlook or misidentify zeolites in thin section, especially because they can resemble holes in the slide.

PPL | XPL, AS, Scale bar = 0.26 mm

Lo. Oligocene Deborah Volcanic Fm., Otago, New Zealand

The vesicles in this basalt porphyry are filled with at least two generations of zeolites, likely phillipsite. The fills can easily be confused with chalcedony, except for the fact that their birefringence is slightly lower than that of silica. The bright yellow phase around the vesicles and fractures is palagonite alteration of the basalt groundmass.

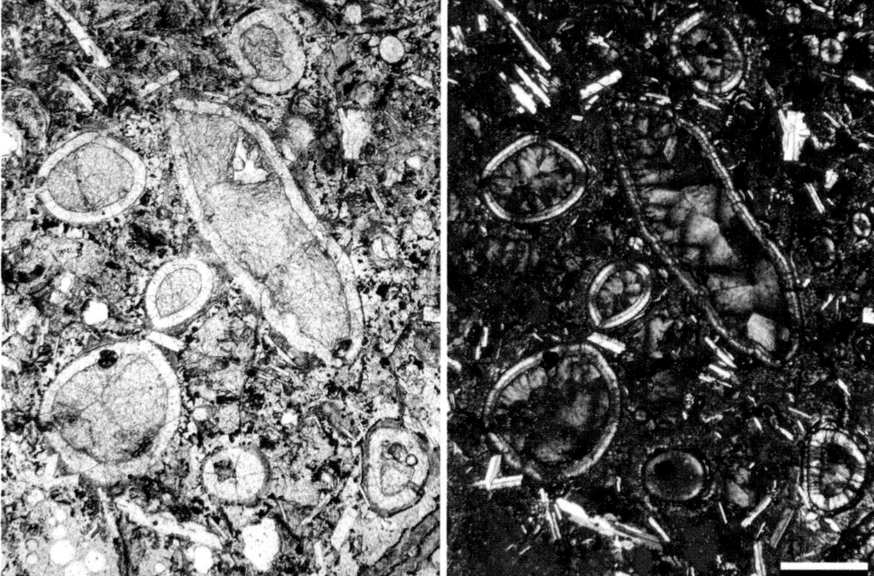

PPL | XPL, Scale bar = 0.64 mm

Pleistocene sediments, Lake Tecopa, Inyo Co., California

Although searlesite is not a zeolite (it belongs to the phyllosilicate group), it commonly forms in the same environments and with zeolite minerals, especially phillipsite and analcime. In this example, searlesite fills large voids and replaces glass shards in a tuffaceous sediment. Searlesite, a monoclinic mineral, has higher birefringence than most zeolites (with first-order colors similar to quartz), commonly has prismatic crystals and typically forms relatively late in the paragenetic sequence of zeolites.

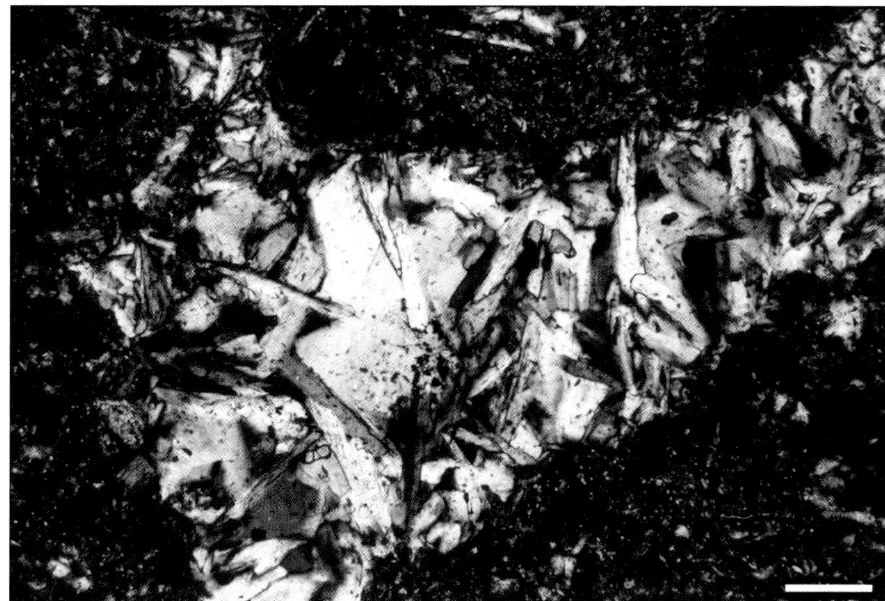

XPL, Scale bar = 0.01 mm

Cited References and Additional Information Sources

Bernoulli, D., R. E. Garrison, and F. Melieres, 1978, Phillipsite cementation in a foraminiferal sandstone at Hole 373A and "The case of the violated foram", *in* Initial Reports of the Deep Sea Drilling Project, vol. 42: Washington, DC, U.S. Government Printing Office, p. 478-482.

Bish, D. L., and J. M. Boak, 2001, Clinoptilolite-heulandite nomenclature, *in* D. L. Bish, and D. W. Ming, eds., Natural Zeolites: Occurrence, Properties, Applications (Reviews in Mineralogy and Geochemistry 45): Chantilly, VA, Mineralogical Society of America, p. 207-216, doi: 10.2138/rmg.2001.45.5.

Boles, J. R., and D. S. Coombs, 1977, Zeolite facies alteration of sandstones in the Southland Syncline, New Zealand: American Journal of Science, v. 277, p. 982-1012, doi: 10.2475/ajs.277.8.982.

Cather, S.M., 1986, Volcano-sedimentary evolution and tectonic implications of the Datil Group (latest Eocene-early Oligocene), west-central New Mexico: Ph.D. Dissertation, The University of Texas at Austin, 486 p.

Cochemé, J. J., A. C. Lassauvagerie, J. Gonzalez-Sandoval, E. Perez-Segura, and P. Münch, 1996, Characterisation and potential economic interest of authigenic zeolites in continental sediments from NW Mexico: Mineralium Deposita, v. 31, p. 482-491, doi: 10.1007/BF00196129.

Coombs, D. S., A. Alberti, T. Armbruster, G. Artioli, C. Colella, E. Galli, J. D. Grice, F. Liebau, J. A. Mandarino, H. Minato, E. H. Nickel, E. Passaglia, D. R. Peacor, S. Quartieri, R. Rinaldi, M. Ross, R. A. Sheppard, E. Tillmanns, and G. Vezzalini, 1998, Recommended nomenclature for zeolite minerals: Report of the Subcommittee on Zeolites of the International Mineralogical Association, Commission on New Minerals and Mineral Names: Mineralogical Magazine, v. 62, p. 533-571, doi: 10.1180/002646198547800.

Coombs, D. S., A. S. Ellis, W. S. Fyfe, and A. M. Taylor, 1959, The zeolite facies, with comments on the interpretation of hydrothermal syntheses: Geochimica et Cosmochimica Acta, v. 17, p. 53-107, doi: 10.1016/0016-7037(59)90079-1.

Deffeyes, K. S., 1959, Zeolites in sedimentary rocks: Journal of Sedimentary Research, v. 29, p. 602-609, doi: 10.1306/74D709AC-2B21-11D7-8648000102C1865D.

El Tabakh, M., and B. C. Schreiber, 1994, Lithologies and diagenesis of the lacustrine sediments of the Lockatong Formation (Upper Triassic) in the Newark Rift Basin, *in* A. J. Lomando, B. C. Schreiber, and P. M. Harris, eds., Lacustrine Reservoirs and Depositional Systems: Tulsa, OK, SEPM, Core Workshop Notes #19, p. 239-295, doi: 10.2110/cor.94.01.0239.

Fuoco, D., 2012, A new method for characterization of natural zeolites and organic nanostructure using Atomic Force Microscopy: Nanomaterials, v. 2, p. 79-91, doi: 10.3390/nano2010079.

Gilbert, C. M., and M. G. McAndrews, 1948, Authigenic heulandite in sandstone, Santa Cruz County, California: Journal of Sedimentary Research, v. 18, p. 91-99, doi: 10.1306/D4269323-2B26-11D7-8648000102C1865D.

Gottardi, G., 1989, The genesis of zeolites: European Journal of Mineralogy, v. 1, p. 479-487.

Gottardi, G., and E. Galli, 1985, Natural Zeolites: Berlin, Springer-Verlag, 409 p.

Gude, A. J., and R. A. Sheppard, 1981, Woolly erionite from the Reese River zeolite deposit, Lander County, Nevada, and its relationship to other erionites: Clays and Clay Minerals, v. 29, p. 378-384, doi: 10.1346/CCMN.1981.0290507.

Hall, A., 1998, Zeolitization of volcaniclastic sediments: The role of temperature and pH: Journal of Sedimentary Research, v. 68, p. 739-745, doi: 10.2110/jsr.68.739.

Hay, R. L., 1966, Zeolites and zeolitic reactions in sedimentary rocks: Boulder, CO, GSA Special Paper 85, 130 p., doi: 10.1130/SPE85-p1.

Hay, R. L., and R. A. Sheppard, 2001, Occurrence of zeolites in sedimentary rocks: An overview, *in* D. L. Bish, and D. W. Ming, eds., Natural Zeolites: Occurrence, Properties, Applications (Reviews in Mineralogy and Geochemistry 45): Chantilly, VA, Mineralogical Society of America, p. 217-234, doi: 10.2138/rmg.2001.45.6.

Helmold, K. P., and P. C. van de Kamp, 1984, Diagenetic mineralogy and controls on albitization and laumontite formation in Paleogene arkoses, Santa Ynez Mountains, California, *in* D. A. McDonald and R. C. Surdam, eds., Clastic Diagenesis: AAPG Memoir 37, p. 239-276.

Iijima, A., 1980, Geology of natural zeolites and zeolitic rocks: Pure and Applied Chemistry, v. 52, p. 2115-2130, doi: 10.1351/pac198052092115.

Iijima, A., 1988, Diagenetic transformation of minerals as exemplified by zeolites and silica minerals—a Japanese view, *in* G. V. Chilingarian, and K. H. Wolf, eds., Diagenesis II, Developments in Sedimentology 43: Amsterdam, Elsevier Science Publishers, p. 147-211.

Iijima, A., 2001, Zeolites in petroleum and natural gas reservoirs, *in* D. L. Bish, and D. W. Ming, eds., Natural Zeolites: Occurrence, Properties, Applications (Reviews in Mineralogy and Geochemistry 45): Chantilly, VA, Mineralogical Society of America, p. 347-402, doi: 10.2138/rmg.2001.45.12.

Iijima, A., and M. Utada, 1966, Zeolites in sedimentary rocks, with reference to the depositional environments and zonal distribution: Sedimentology, v. 7, p. 327-357, doi: 10.1111/j.1365-3091.1966.tb01299.x.

Kastner, M., 1979, Zeolites, *in* R. G. Burns, ed., Marine Minerals: Washington, DC, Mineralogical Society of America Short Course Notes 6, p. 111-122.

Kastner, M., and S. A. Stonecipher, 1978, Zeolites in pelagic sediments of the Atlantic, Pacific, and Indian Oceans, *in* L. B. Sand, and F. A. Mumpton, eds., Natural Zeolites: Occurrence, Properties, Use: New York, Pergamon Press, p. 199-220.

Madsen, B. M., and K. J. Murata, 1970, Occurrence of laumontite in the Tertiary sandstones of the central Coast Ranges of California: Washington, DC, USGS Professional Paper 700-D, p. 188-195.

Mason, B. H., and L. B. Sand, 1960, Clinoptilolite from Patagonia: The relationship between clinoptilolite and heulandite: American Mineralogist, v. 45, p. 341-350.

McCusker, L. B., F. Liebau, and G. Engelhardt, 2001, Nomenclature of structural and compositional characteristics of ordered microporous and mesoporous materials with inorganic hosts (IUPAC Recommendations 2001): Pure and Applied Chemistry, v. 73, p. 381-394, doi: 10.1351/pac200173020381.

Moncure, G. K., R. C. Surdam, and H. McKague, 1981, Zeolite diagenesis below Pahute Mesa, Nevada Test Site: Clays and Clay Minerals, v. 29, p. 385-396, doi: 10.1346/CCMN.1981.0290508.

Mumpton, F. A., and W. C. Ormsby, 1976, Morphology of zeolites in sedimentary rocks by scanning electron microscopy: Clays and Clay Minerals, v. 24, p. 1-23, doi: 10.1346/CCMN.1976.0240101.

Noh, J. H., 1998, Geochemistry and paragenesis of heulandite cements in a Miocene marine fan-delta system of the Pohang Basin, Republic of Korea: Clays and Clay Minerals, v. 46, p. 204-214, doi: 10.1346/CCMN.1998.0460211.

Noh, J. H., and J. R. Boles, 1993, Origin of zeolite cements in the Miocene sandstones, North Tejon oil fields, California: Journal of Sedimentary Research, v. 63, p. 248-260, doi: 10.1306/D4267AD2-2B26-11D7-8648000102C1865D.

Passaglia, E., and R. A. Sheppard, 2001, The crystal chemistry of zeolites, *in* D. L. Bish and D. W. Ming, eds., Natural Zeolites: Occurrence, Properties, Applications (Reviews in Mineralogy and Geochemistry 45): Chantilly, VA, Mineralogical Society of America, p. 69-116, doi: 10.2138/rmg.2001.45.2.

Reed, J. K., M. Gipson, Jr., and D. Vass, 1993, Hydrocarbon potential of sandstone reservoirs in the Neogene, East Slovakian Basin. Part I: Zeolites and clay minerals: Journal of Petroleum Geology, v. 16, p. 223-236, doi: 10.1111/j.1747-5457.1993.tb00108.x.

Sheppard, R. A., and A. J. Gude, III, 1973, Zeolites and associated authigenic silicate minerals in tuffaceous rocks of the Big Sandy Formation, Mojave County, Arizona: Washington, DC, USGS Professional Paper 830, 36 p.

Sheppard, R. A., and R. L. Hay, 2001, Formation of zeolites in open hydrologic systems, *in* D. L. Bish, and D. W. Ming, eds., Natural Zeolites: Occurrence, Properties, Applications (Reviews in Mineralogy and Geochemistry 45): Chantilly, VA, Mineralogical Society of America, p. 261-275, doi: 10.2138/rmg.2001.45.8.

Surdam, R. C., and J. R. Boles, 1979, Diagenesis of volcanic sandstones, *in* P. A. Scholle, and P. Schluger, eds., Aspects of Diagenesis: Tulsa, OK, SEPM Special Publication 26, p. 227-242, doi: 10.2110/pec.79.26.0227.

Surdam, R. C., and R. A. Sheppard, 1978, Zeolites in saline, alkaline-lake deposits, *in* L. B. Sand and F. A. Mumpton, eds., Natural Zeolites: Occurrence, Properties, Use: New York, Pergamon Press, p. 145-174.

Tang, Zhaohui, J. Parnell, and F. J. Longstaffe, 1997, Diagenesis of analcime-bearing reservoir sandstones: The Upper Permian Pingdiquan Formation, Junggar Basin, northwest China: Journal of Sedimentary Research, v. 67, p. 486-498, doi: 10.1306/D42685A4-2B26-11D7-8648000102C1865D.

Taylor, M. W., and R. C. Surdam, 1971, Zeolitic reactions in the tuffaceous sediments at Teels Marsh, Nevada: Clays and Clay Minerals, v. 29, p. 341-352, doi: 10.1346/CCMN.1981.0290504.

Tschernich, R. W., 1992, Zeolites of the World: Phoenix, AZ, Geoscience Press, 563 p.

Utada, M., 2001a, Zeolites in burial diagenesis and low-grade metamorphic rocks, *in* D. L. Bish, and D. W. Ming, eds., Natural Zeolites: Occurrence, Properties, Applications (Reviews in Mineralogy and Geochemistry 45): Chantilly, VA, Mineralogical Society of America, p. 277-304, doi: 10.2138/rmg.2001.45.9.

Utada, M., 2001b, Zeolites in hydrothermally altered rocks, *in* D. L. Bish, and D. W. Ming, eds., Natural Zeolites: Occurrence, Properties, Applications (Reviews in Mineralogy and Geochemistry 45): Chantilly, VA, Mineralogical Society of America, p. 305-322, doi: 10.2138/rmg.2001.45.10.

Van Houten, F. B., 1962, Cyclic sedimentation and the origin of analcime-rich Upper Triassic Lockatong, west-central New Jersey and adjacent Pennsylvania: American Journal of Science, v. 260, p. 561-576, doi: 10.2475/ajs.260.8.561.

Vavra, C. L., 1989, Mineral reactions and controls on zeolite-facies alteration in sandstone of the central Transantarctic Mountains, Antarctica: Journal of Sedimentary Research, v. 59, p. 688-703, doi: 10.1306/212F904D-2B24-11D7-8648000102C1865D.

Walton, A. W., 1975, Zeolitic diagenesis in Oligocene volcanic sediments, Trans-Pecos Texas: GSA Bulletin, v. 86, p. 615-624, doi: 10.1130/0016-7606(1975)86<615:ZDIOVS>2.0.CO;2.

Welton, J. E., 1984, SEM Petrology Atlas: Tulsa, OK, AAPG Methods in Exploration Series No. 4, 237 p.

Wise, W. S., and R. W. Tschernich, 1978, Habits, crystal forms and composition of thomsonite: Canadian Mineralogist, v. 16, p. 487-493.

Worden, R. H., 2006, Dawsonite cement in the Triassic Lam Formation Shabwa Basin, Yemen: A natural analogue for a potential mineral product of subsurface CO_2 storage for greenhouse gas reduction: Marine and Petroleum Geology, v. 23, p. 61-77, doi: 10.1016/j.marpetgeo.2005.07.001

Zhang, Lifei, 1992, Origin of laumontite and conditions for its formation in the Triassic Yanchang Series, north Shaanxi: Journal of Geochemical Exploration, v. 69-70, p. 145-152.

Facing Page: Top — Aragonite cements in a Pleistocene-Holocene pisolitic caliche from Abu Dhabi, United Arab Emirates (XPL). Bottom — Fracture filling calcite with abundant water- and oil-filled inclusions from the Jurassic Carmel Fm., Emery Co., Utah (PPL, RDI).

DIAGENESIS: Carbonate Cements and Authigenic Precipitates

CHAPTER 15

CARBONATE CEMENTS

Carbonate minerals, particularly calcite, dolomite, ankerite and siderite, are important as cements and replacements in sandstones and mudstones. Their compositions (or compositional ranges) are shown in Table 11.2 (page 247) and Figure 15.1. Perhaps surprisingly, carbonate cements can be difficult to work with in terms of quantifying conditions of formation (including time, temperature and water chemistry). There are two reasons for this. First, some carbonate minerals (especially dolomite in the context of cements in clastic terrigenous deposits) can recrystallize at one or more stages, resetting their diagnostic fabrics and elemental and isotopic composition. Second, all carbonate cements are relatively soft minerals with multiple cleavages. That means that fluid inclusions, when subjected to temperatures higher than those at which they formed, can stretch or leak, resetting apparent temperatures of formation (Goldstein, 1986 and 2001).

That said, there is abundant evidence that carbonate cements form at virtually any stage of diagenesis, from synsedimentary to deep burial (see overview in Morad, 1998). Early-formed marine precipitates (and many biogenic grains) are composed mainly of aragonite or high-Mg calcite, at least in modern, tropical to temperate oceans. These marine carbonate cements occur primarily in calcareous sands in localized settings including hardgrounds, beachrocks and hot and cold sea-floor hydrocarbon seeps. Early carbonate cements in sandstones also appear to form preferentially in highstand systems tracts, especially below parasequence boundaries or maximum flooding surfaces because of the availability of carbonate bioclasts and the extended residence time in such zones for potential diagenetic alteration (Morad et al., 2010). However, the relatively unstable aragonite and high-Mg calcite cement most commonly produced in those settings, unless isolated from water during burial, will dissolve, invert or ultimately recrystallize to low-Mg calcite or other carbonate cements (e.g., Molenaar, 1990).

Low-Mg calcite, siderite and dolomite cements also are precipitated in a range of near-surface settings, especially in subaerial exposure surfaces (caliche crusts, calcretes and dolocretes; see Goudie, 1983; Wright and Tucker, 1991), areas of shallow meteoric water circulation, sites of synsedimentary marine precipitation, as halos around oil or gas seeps, or in areas of near-synsedimentary, microbially-influenced diagenesis. Such early carbonate cements generally form under oxidizing conditions, but some precipitate under reducing conditions in fine-grained, organic-rich marine deposits (where otherwise rare carbonate cements such as rhodochrosite may also form—e.g., Minoura et al., 1991; Calvert and Pedersen, 1996).

Mesogenetic (burial diagenetic) settings also are common places to form carbonate cements, with calcite, ferroan calcite, dolomite, ferroan dolomite, ankerite and siderite all being reported from burial environments. Many mesogenetic settings are sites of dissolution or alteration of earlier, unstable carbonate cements as well as carbonate bioclasts or lithoclasts. Such carbonate dissolution, along with dissolution of calcic feldspars and some clay mineral transformations, provides the main source of material for mesogenetic carbonate cements. Other types of mesogenetic carbonate diagenesis are found as well. Early-formed dolomite typically converts to better-ordered and more stable dolomite at elevated temperatures and pressures (Mazzullo, 1992); ferroan carbonates (ankerite, siderite, ferroan calcite and ferroan dolomite) can precipitate, and all tend to be more coarsely crystalline than early-formed equivalents. Some especially coarsely-crystalline (poikilotopic) calcites have been reported from mesogenetic settings (Saigal and Bjørlykke, 1987; Girard, 1998), although such calcite cements also are known to precipitate in near-surface concretions (e.g., Beckner and Mozley, 1998).

Most carbonate cementation in sandstones tends to be concentrated in nodules, lenses or relatively thin and discrete layers, albeit ones that can have lateral extents of tens of kilometers (Kantorowicz et al., 1987; Bjørkum and Walderhaug, 1990; Taylor et al., 2000). The formation of such carbonate cements can drastically reduce sandstone porosity, but the solubility of carbonates under many conditions can lead to extensive and well-connected secondary porosity development during burial diagenesis.

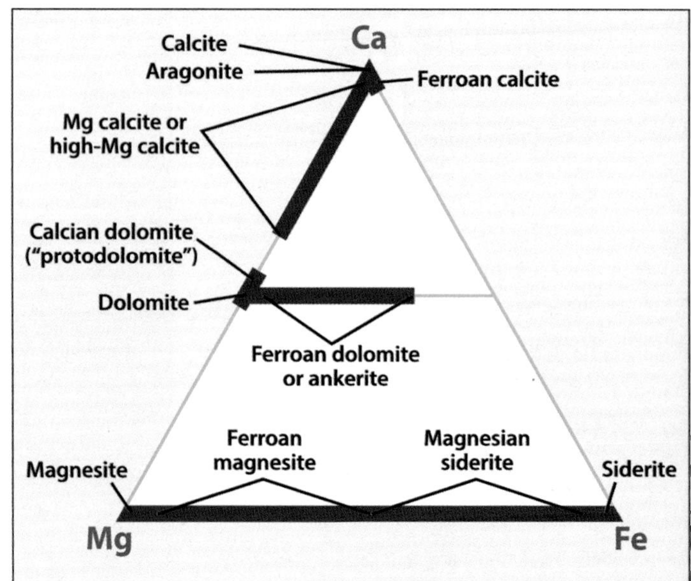

Figure 15.1 A diagram showing the general compositions of carbonate minerals found in modern and ancient rocks.

In addition, small volumes of early carbonate cement (especially dolomite) can retard sandstone compaction and associated porosity loss (e.g., Burley, 1984; Souza et al., 1995) by strengthening the grain framework. On the other hand, larger volumes of early carbonate cements, unless leached at later stages, typically are detrimental to reservoir properties (e.g., Amthor and Okkerman, 1998).

Optical properties and conditions of formation:

All the rhombohedral carbonates have high relief (relief that changes with stage rotation) and high birefringence; they are most easily and effectively distinguished from each other by staining (see Dickson, 1966) but generally also can be identified using standard petrographic techniques.

Aragonite is a polymorph of calcite and is orthorhombic with acicular crystals. It is found in carbonate grains as well as cements, especially during icehouse climatic periods (Wilkinson and Given, 1986). As a cement in sandstones, aragonite is relatively rare, except in beachrocks, marine hardgrounds and marine hydrocarbon seeps, and most such aragonite cements have a radial-fibrous morphology. However, aragonite, either in grains or cements, is not stable in meteoric waters or during burial. Therefore, it is rarely found in pre-Cenozoic rocks, although exceptions are found, mainly in rocks that exclude significant pore-water movement (e.g., concretions, tar sands and impermeable shales). Aragonite can be identified in rock samples by XRD analyses or in thin sections by staining using Feigl's solution (Friedman, 1959). Aragonitic grains (ooids, peloids or bioclasts) normally dissolve to produce secondary pores.

High-Mg calcite (calcite variously defined as containing more than 1, 2 or 5 mol percent Mg^{2+}) occurs as a marine cement, primarily in modern carbonate-rich deposits, and it typically has a very finely crystalline fabric (often misnamed a "micritic" fabric). It, too, is unstable during diagenesis, but high-Mg calcites convert to normal or low-Mg calcite without forming macro-pores. High-Mg calcite can be identified in samples by XRD analyses and by Titan-yellow staining of thin sections (Friedman, 1959; Choquette and Trusell, 1978).

Low-Mg calcite (normally just called calcite) is the predominant $CaCO_3$ cement, especially in nonmarine and burial-diagenetic settings. Calcite cements in sandstones commonly form as microcrystalline (micritic to microspar size) precipitates, although in large pores they can consist of blocky spar crystals that generally increase in size from the pore margins inward. Overgrowths of bioclasts, and especially syntaxial overgrowths of echinoderm fragments, commonly result in localization of calcite cementation. In other cases, calcite is precipitated as exceptionally large single crystals that fill multiple pores and engulf multiple grains. These are termed poikilotopic calcites and, as noted earlier, they can form in both near-surface and burial settings. Calcite crystals can be ferroan or nonferroan, with ferroan varieties normally indicating formation under reducing conditions, most commonly during mesogenesis, but also in some eogenetic situations. Although optically indistinguishable, ferroan and nonferroan calcites are readily differentiated using a combined alizarin red-S and potassium ferricyanide stain, although the exact stain mixture and length of immersion must be carefully standardized (Dickson, 1966). Nonferroan calcite generally stains pink, ferroan calcite stains purplish blue or mauve, and such staining can reveal even small-scale compositional zonation in calcite and other carbonate cements. Cathodoluminescence also can be useful in highlighting compositional zonation in calcite that otherwise is not visible with standard petrography (Meyers, 1991). Calcite is most reliably distinguished from dolomite using stains or XRD, but in the absence of those aids, calcite crystals can be recognized by the fact that they typically do not show the well-formed rhombic outlines of dolomite/ankerite (or siderite) but do show twinning more commonly than dolomite/ankerite or siderite.

Siderite, in plane-polarized light, is pale brownish to deep yellow or reddish-brown and has higher relief than calcite; it commonly shows surficial alteration with iron oxide or hydroxide coatings. It does not normally take a stain in the solutions mentioned above (although in some cases it can take a very slight stain). Its crystal habit is rhombic, but typically the rhombs are irregular or highly flattened (lozenge-shaped) and, in some cases, crystals are almost needle-or hair-like. Siderite forms a solid solution series with magnesite and rhodochrosite, and so siderite crystals commonly contain considerable Mg^{2+} and/or Mn^{2+}. A microspherulitic variety, termed sphaerosiderite, is formed from reducing groundwaters in swampy paleosols and is characterized by sweeping or cruciform extinction (Ludvigson et al., 2013).

Siderite forms during eogenesis in nonmarine to shallow-marine deposits, where it is precipitated from suboxic, nonsulfidic, meteoric or marine pore waters (Mozley and Wersin, 1992; Morad, 1998). Such siderite typically is associated with zones of microbial methanogenesis, even in deeper water sediments (Ellwood et al., 1988). Although petrographically indistinguishable, eogenetic siderites that precipitated from meteoric or marine pore waters may be differentiated by their elemental composition—those precipitated in marine pore waters generally have higher Mg^{2+}/Ca^{2+} ratios and contain less Mn^{2+} and Fe^{2+} and more Ca^{2+} and Mg^{2+} than those formed from meteoric waters (Mozley, 1989). Because siderite can form during very early diagenesis, it (and, less commonly, ankerite) may occur as rounded "rip-up" clasts in sandstone. These clasts are especially common in coastal plain, deltaic, and shallow marine deposits. In addition, siderite cements commonly can precipitate during mesogenesis (Macaulay et al., 1993; Rossi et al., 2001) and can be significant in occluding reservoir porosity. In both surficial and deep burial settings, siderite formation may be associated with mica alteration (Morad, 1990).

Dolomite is similar in optical properties to calcite, but dolomite has slightly higher relief, almost always shows rhombic crystal outlines, rarely displays twinning and commonly has visible compositional zonation. Dolomite has more equant rhombs than the more lozenge-shaped crystals typical of siderite. Dolomite, like calcite, can have significant incorporation of other elements, especially Fe^{2+} and Mn^{2+} in its lattice, and dolomite has a continuous solid-solution series with ferroan dolomite and ankerite (Fig. 15.1). The easiest way to distinguish calcite from dolomite and to identify iron-rich and iron-poor dolomites (as with calcites) is though the use of alizarin red-S/potassium ferricyanide staining (nonferroan dolomite takes no stain; iron-bearing dolomite stains pale turquoise blue). Higher levels of Fe^{2+} and Mn^{2+} substitution are discussed later under the "ankerite" heading.

Dolomite forms as aphanocrystalline to microcrystalline precipitates or early carbonate replacements in a wide variety of eogenetic settings including hypersaline lakes and salinas, tidal flats or sabkhas, exposure surfaces, stratified lagoons with anoxic bottom waters, coastal water mixing zones, and in fine-grained and organic-rich sediments in deep-marine settings. The specific conditions interpreted to favor dolomite precipitation in these widely varied settings are equally diverse, but include: reduced ionic concentrations, increased Mg/Ca ratios, reduced sulfate concentrations, microbial mediation, elevated temperature and others (see Morad, 1992; Vasconcelos and McKenzie, 1997; Morad, 1998; and Machel, 2004). In most such eogenetic settings, dolomite is formed primarily as a cement or a displacement/replacement mineral in carbonate-rich deposits, but dolomites also form in associated sandstones, siltstones and shales, albeit less abundantly. And as noted earlier, the eogenetic formation of moderate amounts of dolomite cement, because it is the most compaction-resistant of the carbonate minerals (see Figure 10.1), can help sandstones to retain porosity during burial.

Dolomite also forms in mesogenetic settings, both as a directly precipitated pore-filling cement and as a replacement of earlier, less stable forms of dolomite and other carbonate cements. Indeed, carbonate cementation during mesogenesis is dominated by ferroan dolomite (along with related ferroan carbonates, ankerite and siderite), which grows at similar or higher temperatures as illite and quartz cements (roughly 100°C). A distinctive form of dolomite, termed saddle (or baroque) dolomite, also occurs as sandstone cements and fracture fills in mesogenetic settings. It is characterized by curved crystal faces and cleavage surfaces, along with sweeping (undulatory) extinction (Radke and Mathis, 1980; Machel, 1987). The many saddle dolomites analyzed to date have a wide range of fluid-inclusion homogenization temperatures (60-160°C), but all share freezing temperatures indicative of formation from saline brines, demonstrating "that saddle dolomite is a reliable indicator of rock-brine interactions at temperatures that coincide largely with the liquid hydrocarbon 'window' and extend well into the dry-gas zone" (Spötl and Pitman, 1998, p. 437).

Ankerite (and ferroan dolomite) Ankerite is similar to dolomite in composition and appearance, but it has substantial Fe^{2+} and typically also Mn^{2+} substitution for Mg^{2+} in its crystal lattice. Commonly, especially in early-formed ankerites, excess Ca^{2+} is found in Mg^{2+} lattice sites as well (Boles, 1978; Hendry et al., 2000, Hendry, 2002). The exact definition of ankerite versus ferroan dolomite is confusing because usage varies. Many workers follow Deer et al. (1992) in using the term "ferroan dolomites" for crystals in which the Mg:Fe ratio is higher than 4 and restrict "ankerite" to crystals in which this ratio is lower than 4. Others draw no distinction between ferroan dolomite and ankerite and use the two terms interchangeably (e.g., Reeder and Dollase, 1989).

In thin section, ankerite is similar to dolomite, with fairly high relief, high birefringence and equant, rhombic crystal outlines. However, ankerite crystals, can have a "rusty" appearance related to iron oxide and hydroxide staining or alteration rims (a characteristic they share with the more lozenge-shaped siderite crystals), especially in outcrop samples. Ankerite commonly forms rhombohedral overgrowths on authigenic and detrital dolomite. Ankerite crystals take on a definitive deep turquoise color when stained with potassium ferricyanide. XRD, microprobe or other analytical techniques can provide more definitive determination and more exact compositions for this family of carbonates.

Ankerite cements and/or replacements can form in a variety of settings. Ankerite is found in eogenetic mudrock-hosted concretions occur in low sulfate, anoxic settings undergoing microbial methanogenesis, commonly in association with siderite (Morad, 1998). Most ankerite cements in sandstones, however, occur as mesogenetic precipitates, typically formed over a wide range of temperatures (from 30 to 120°C and above), sometimes in association with hydrocarbon emplacement (Boles, 1978; Kantorowicz, 1985; Burley et al., 1989; Hendry et al., 2000).

Pleistocene – Holocene pisolitic caliche Abu Dhabi, United Arab Emirates

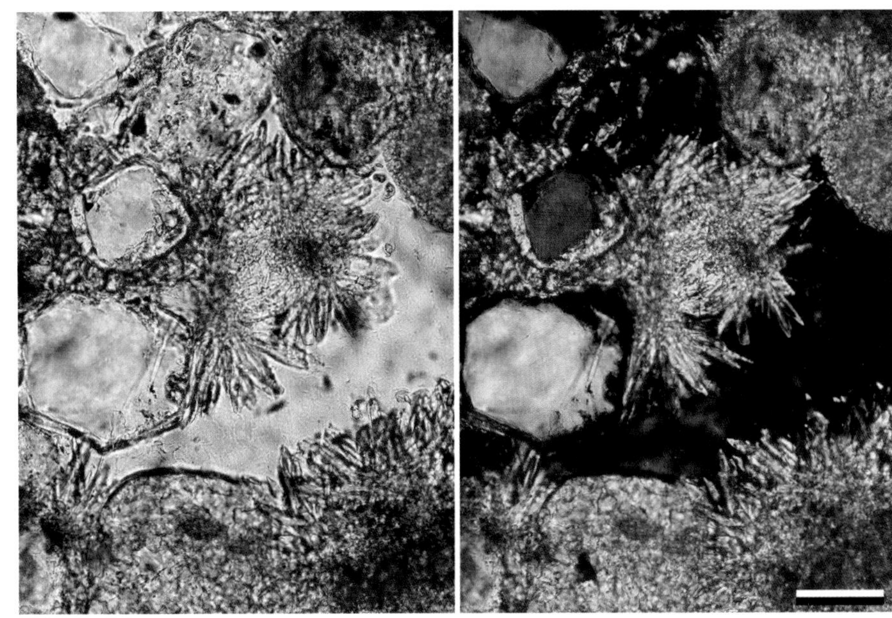

In this sample, carbonate and quartz sand grains are cemented by needles of aragonite. These precipitates are the most common marine cements forming today in temperate to tropical settings, and they generally are the dominant cements that form during icehouse periods (as opposed to greenhouse intervals during which low-Mg marine calcites predominate; Wilkinson and Given, 1986). Aragonite is not a geologically stable mineral, and with burial, influx of meteoric waters and time, aragonite typically is either dissolved or replaced by calcite.

PPL | XPL, Scale bar = 0.05 mm

Pleistocene – Holocene marine sediment, Abu Dhabi, United Arab Emirates

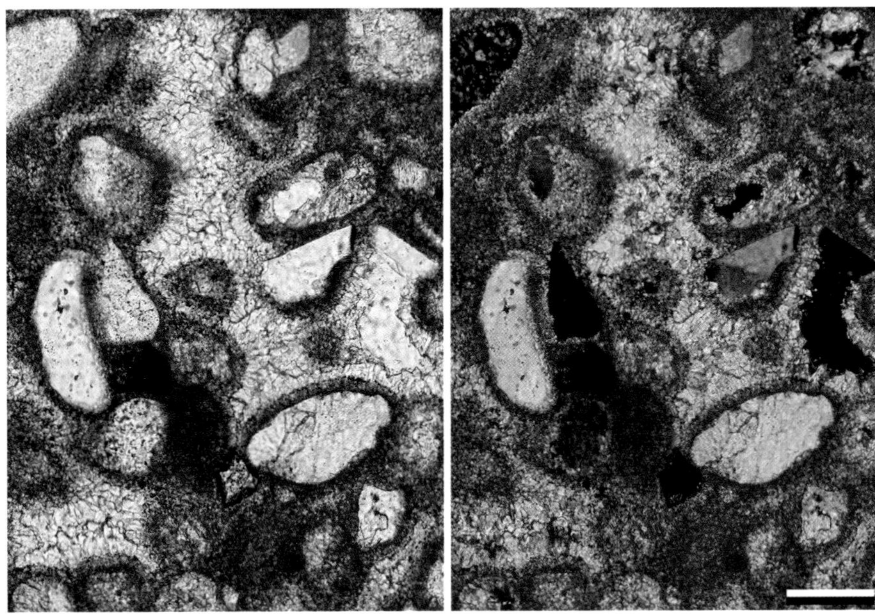

This calcareous sandstone has multiple generations of eogenetic carbonate cement. The first generation is a dark, fibrous, isopachous, aragonite cement that coats both quartz and carbonate grains. It was precipitated as a beachrock cement in a marginal marine setting. This is followed by clear, isopachous bladed to blocky calcite cements, probably formed in a meteoric to marginal marine phreatic environment. A marine to meteoric transition in carbonate cements is common in coastal marine settings.

PPL | XPL, Scale bar = 0.10 mm

Up. Cretaceous Pierre Shale, Pueblo Co., Colorado

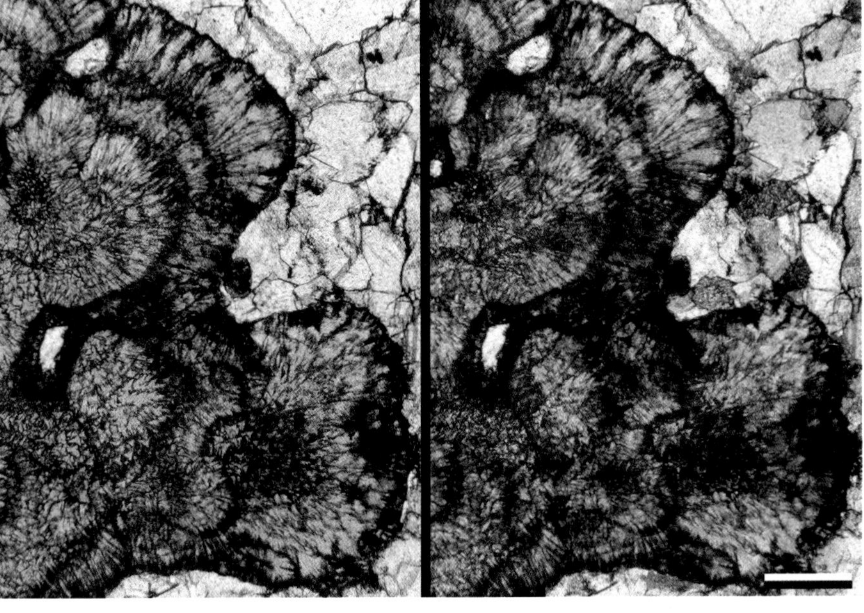

Seafloor hydrocarbon cold- and warm-seeps are another common site of marine carbonate cementation (Hovland et al., 1987; Roberts et al., 1993), and in most such settings, microbes play a major role in carbonate precipitation. This example shows small fans or botryoids of brownish calcite that formed seafloor mounds in organic-rich shales that underwent biogenic methane generation. The brown color reflects incorporation of microbial organic matter as inclusions within the calcite. The mounds are largely devoid of other organisms within the vent zone, but show abundant organic remains in surrounding areas (Kauffman et al., 1996).

PPL | XPL, Scale bar = 0.51 mm

Up. Cretaceous Pierre Shale, Pueblo Co., Colorado

Another view of seafloor methane seep cements from the Pierre Shale, in this case cut perpendicular to the growth surface. Fans of fibrous calcite crystals radiate outward and have dark and light bands that show variations in the incorporation of microbial organic matter (and perhaps also reflect variations in rates of precipitation). The formation of such seafloor cements provided firm substrates in an area of otherwise soupy muds and thus allowed growth of hundreds of isolated but sizable carbonate mounds (termed tepee mounds in this area) in an otherwise carbonate-poor environment (Kauffman et al., 1996).

PPL, Scale bar = 0.51 mm

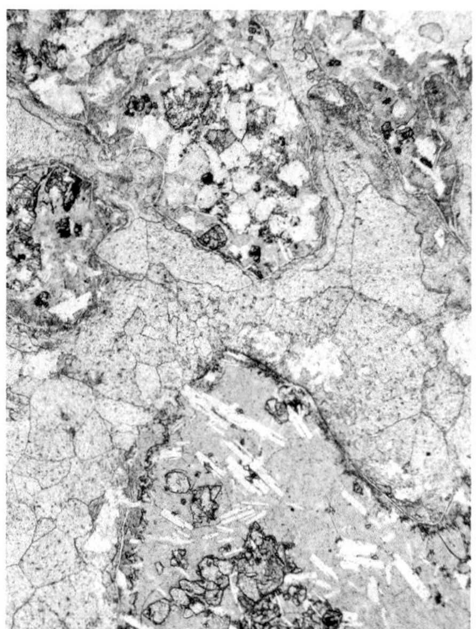

Oligocene Deborah Volcanic Fm., north Otago, New Zealand

Eogenetic marine carbonate cements also are found commonly associated with hot seeps and lava flows. These volcanic rock fragments (derived from submarine pillow basalts) were rimmed by palagonite alteration (yellow) and then cemented by zeolites (mainly phillipsite) and calcite. The calcites are nonferroan, microfibrous, marine precipitates and can be inorganic or microbially influenced. In cross-polarized light, these calcites show strongly undulatory or sweeping extinction. They could possibly be mistaken for saddle dolomite, but staining is an inexpensive reliable way to distinguish those two cement types.

PPL | XPL, Scale bar = 0.51 mm

Oligocene Deborah Volcanic Fm., north Otago, New Zealand

A closer view of calcite cement between basaltic pillows. The cements formed in warm to hot waters within active submarine flows. These fabrics are commonly termed "feathery" or "plumose" and give a sweeping extinction pattern with stage rotation under cross-polarized light.

PPL | XPL, Scale bar = 0.26 mm

Oligocene Deborah Volcanic Fm., north Otago, New Zealand

Another view of calcite cement that formed between pillow basalts. Clear, relatively inclusion-poor, botryoidal, fibrous, and feathery calcite crystals form the first generation of cement and that is followed by a consistently dark, organic-inclusion-rich zone that is transitional to another inclusion-poor zone formed of feathery crystals with scalenohedral terminations. These variations may reflect episodic changes of water temperatures in the intra-pillow pores.

XPL, BDI, Scale bar = 0.51 mm

Oligocene Deborah Volcanic Fm., north Otago, New Zealand

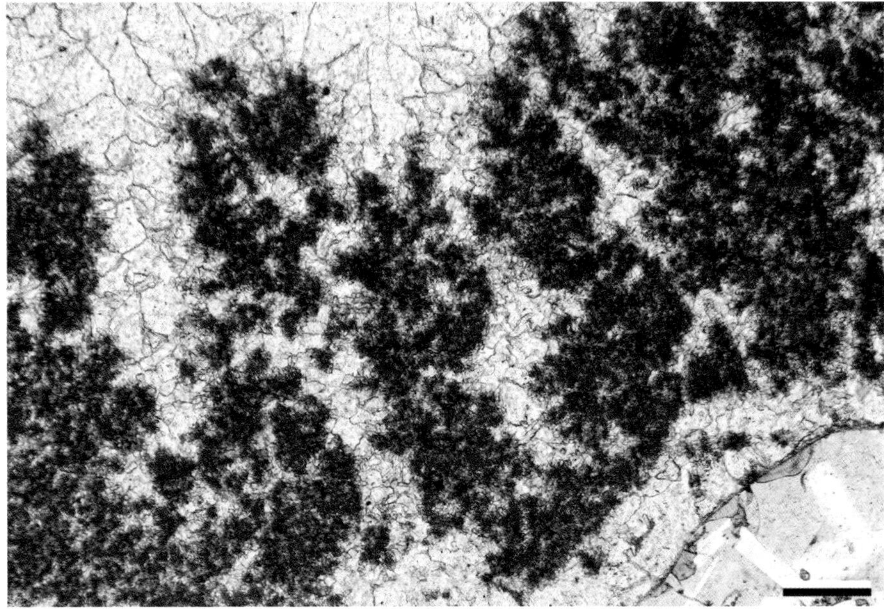

The darker and/or peloidal layers in many marine cements are associated with incorporation of organic matter of probable microbial origin. In this area of interpillow calcite cementation, the evidence for microbially-influenced precipitation is clearer because the dark, peloidal crystal aggregates are in the form of clearly recognizable "shrubs" known from modern marine as well as nonmarine warm- and hot-springs deposits (e.g., Pentecost, 1990; Chafetz and Guidry, 1999). Later, blocky calcite cements have enclosed and preserved these shrubs; commonly such shrubs fall apart into individual peloids.

PPL, Scale bar = 0.26 mm

Quaternary travertine, Bonneville Co., Idaho

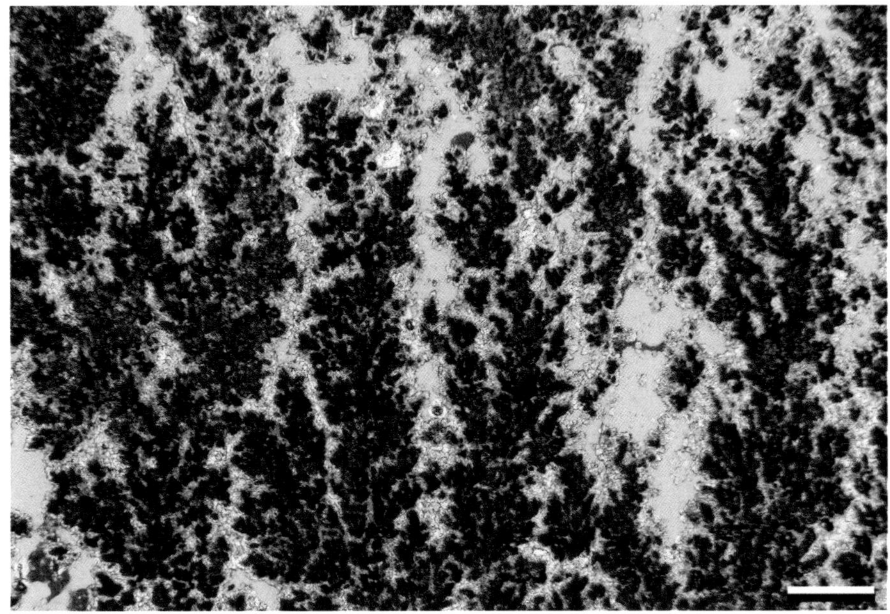

Peloidal microbial shrubs are even more common as eogenetic precipitates in nonmarine warm- and hot-spring travertine deposits. In this example, large microbial shrubs show typical peloidal fabric; there has been little additional calcite cementation and considerable porosity remains. As can be seen in comparison with the previous photograph, marine and nonmarine microbial shrubs are remarkably similar and therefore must generally be differentiated by examining their depositional contexts or geochemical signatures.

PPL, BDI, Scale bar = 0.51 mm

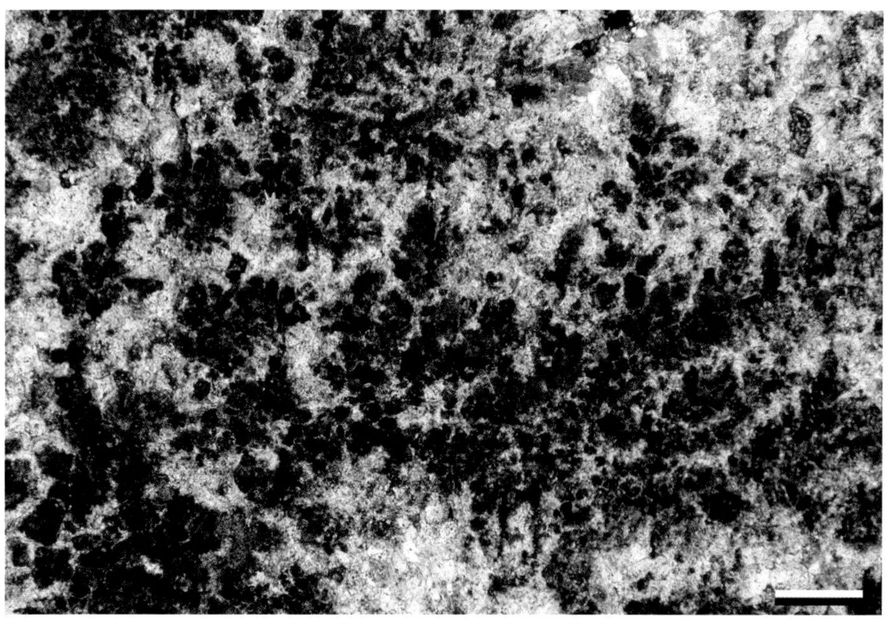

Quaternary travertine, Bonneville Co., Idaho

Another example of a nonmarine microbial travertine with peloidal microbial shrubs. In this case, however, the shrubs have been encased in blocky calcite cement that greatly enhances the likelihood of preservation of the shrub structure. In the absence of such secondary cementation, the shrubs can disaggregate and form geopetal layers of peloids in individual pores and larger cavities.

XPL, BDI, Scale bar = 0.26 mm

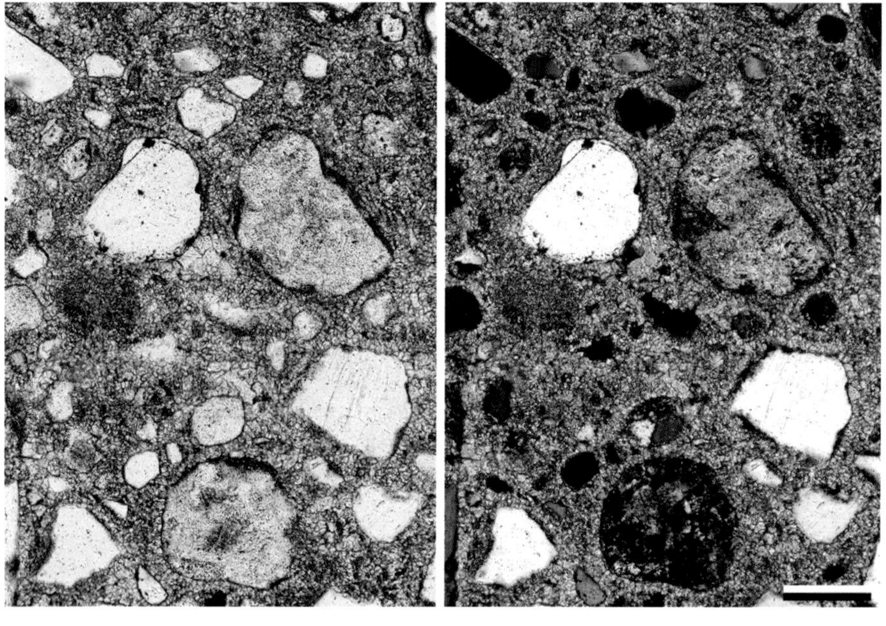

Mid. Eocene Baca Fm., Socorro Co., New Mexico

This alluvial sandstone has been completely cemented in eogenetic to mesogenetic settings with what at first glance appears to be a simple mosaic of finely-crystalline calcite. When examined more carefully, however, it is apparent that there is a range of cement sizes (see enlarged view of a similar area in the same sample in next photo), with the most finely crystalline cements in direct grain contact and the coarsest ones in the centers of former pores. Such cement size transitions are found in both marine and nonmarine rocks, although the diversity of fabric seen here, as noted below, is especially common in nonmarine strata.

PPL | XPL, Scale bar = 0.10 mm

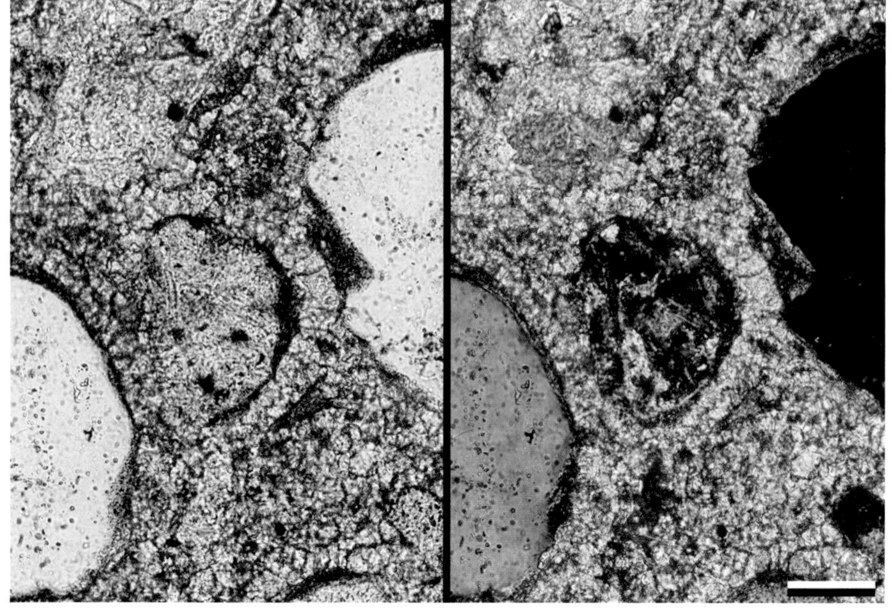

Mid. Eocene Baca Fm., Socorro Co., New Mexico

This alluvial sandstone shows a variety of calcite cement morphologies. An irregular coating of aphanocrystalline calcite forms a patchy coating on most grains and probably reflects very early pedogenic processes. A second generation of bladed calcite crystals surrounds many grains and is followed by complete filling of remaining pores by very finely crystalline, equant, calcite spar, both formed in deep vadose or phreatic settings. Multiple carbonate cement fabrics are a common feature in nonmarine deposits and can be attributed to several causes (see Chafetz et al., 1985).

PPL | XPL, Scale bar = 0.05 mm

Neogene Santa Fe Gp., Socorro Co., New Mexico

Grain size can play an important role in rock cementation, especially in vadose meteoric settings. In this example, a concretion that grew in laminated fluvial sediments has formed by the selective calcite cementation of the finer-grained horizons, likely related to selective capillary retention of water in the finer-grained horizons. There are cements in the coarser layers as well, but most of those cements have initiated at grain contacts, again an indication of formation under vadose conditions (with selective capillary water retention). See Mozley and Davis (2005) for more details on cementation in this unit.

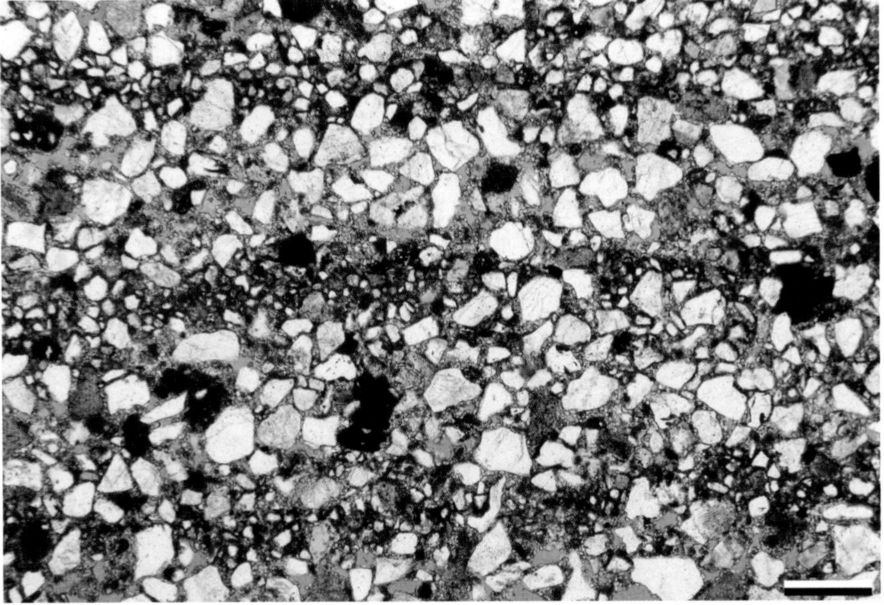

PPL, BDI, Scale bar = 0.51 mm

Mid. Eocene – Lo. Oligocene, Spears Gp., Socorro Co., New Mexico

Blocky, anhedral calcite cement has completely occluded porosity in this volcanic arenite that is part of a thick alluvial unit that has undergone purely terrestrial eogenetic to mesogenetic diagenesis. The uniformity of crystal-size distribution indicates that, unlike the prior two samples, cementation occurred under more uniform phreatic conditions.

XPL, Scale bar = 0.26 mm

Eocene Galisteo Fm., Sandoval Co., New Mexico

Nonmarine calcite cement in sandstones exposed to hydrothermal fluids during eogenesis or mesogenesis can sometimes show unusual and diagnostic fabrics. In this example, radial crusts and feathery or flamboyant calcite crystal aggregates are present in an extremely loosely packed grain framework that may have been expanded by dispersive water pressures and/or crystal growth forces (Watts, 1978). The inferred hydrothermal waters would have come from Oligocene eruptive centers in the nearby Ortiz Mountains and the Cerrillos Hills (S. Cather, 2014, personal communication).

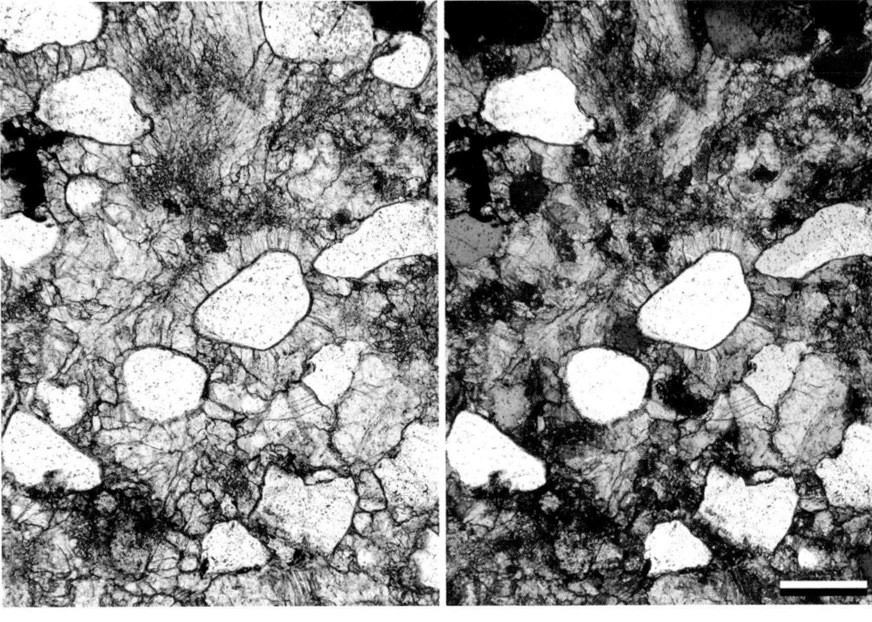

PPL | XPL, BDI, Scale bar = 0.26 mm

Lo.–Mid. Eocene Cub Mountain Fm., Lincoln Co., New Mexico

Here, even more dramatically flamboyant and partially displacive calcite cements characterize cementation in a sandstone that was affected by Oligocene or younger hydrothermal fluid flow associated with the nearby Sierra Blanca intrusive complex. Such spectacularly radiating calcite crystal fabrics (sometimes termed calcite feathers) have been described as common precipitates in hot springs travertines and associated rocks (see Guo and Riding, 1998).

XPL, Scale bar = 0.51 mm

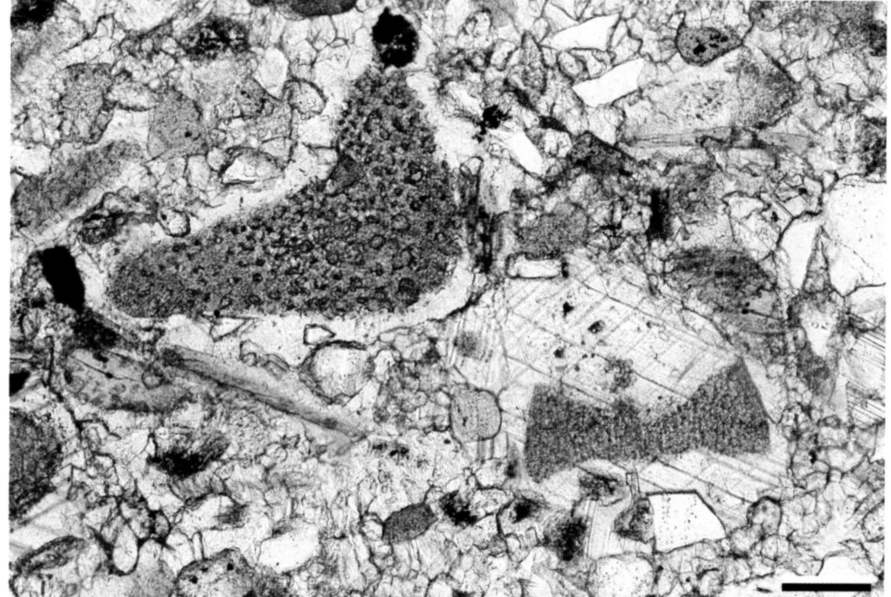

Mid. Jurassic Curtis Fm., Moffat Co., Colorado

Sandstones, either marine or nonmarine, may contain carbonate bioclasts that can host cement overgrowths. If the bioclasts are marine echinoid or crinoid fragments, they are especially susceptible to eogenetic or mesogenetic calcite cementation, because each fragment is a single calcite crystal. Syntaxial overgrowths, optically continuous with the bioclast, then can encase nearby grains. Note the faint twinning (cross-hatching) within the larger syntaxial overgrowth—this is typical for calcite cements. The phosphate cement in the crinoid pores and green glauconite grains are also present..

PPL, RDI, Scale bar = 0.26 mm

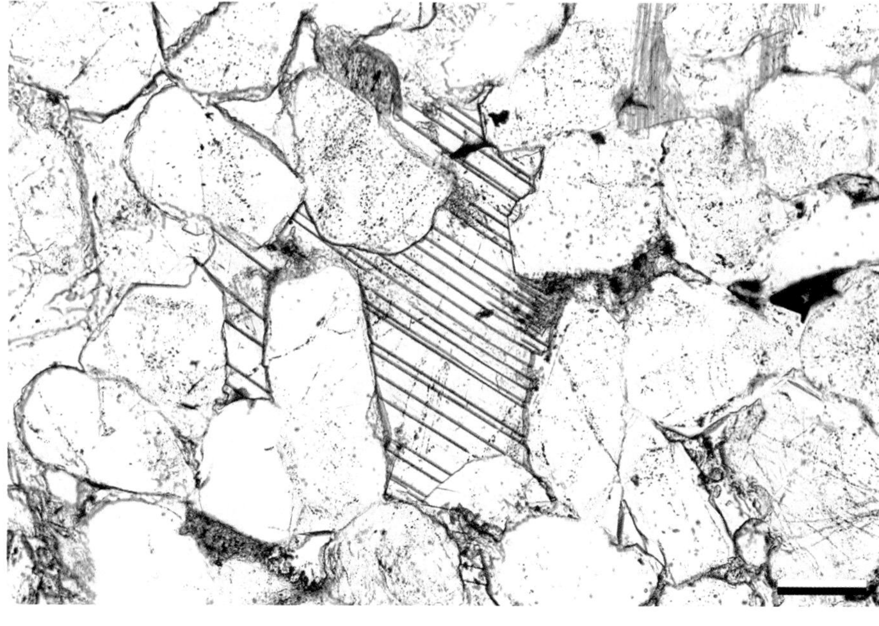

Mid. Jurassic Carmel Fm., Emery Co., Utah

This quartz sandstone was first cemented by quartz (note the "dust" rims and inclusion-poor overgrowth cements on some of the quartz grains). That was followed by mesogenetic, very coarsely crystalline, calcite cement. The calcite cement has much higher relief than the quartz grains and cements. In this photomicrograph, the linear features cutting across the cement are well-developed calcite twins. This type of stress-induced twinning is diagnostic of deformed calcite but is rare in other carbonate minerals.

PPL, RDI, Scale bar = 0.10 mm

Mid. Permian Rush Springs Fm., Caddo Co., Oklahoma

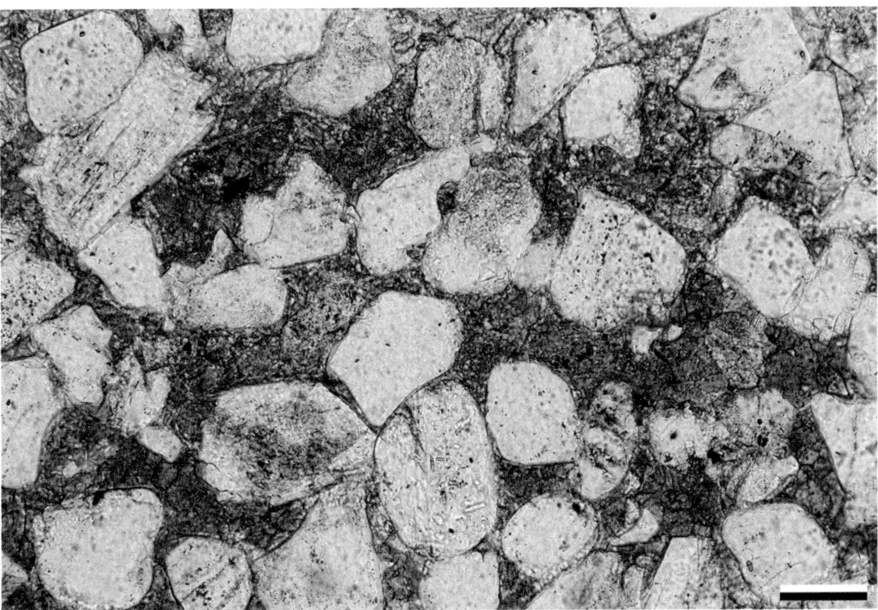

In this example, calcite and dolomite cements formed in sandstones in a relatively near-surface environment, primarily in areas overlying the Cement oil field. The large, red-stained, poikilotopic calcite cements, in particular, contain isotopically light carbon, indicating that they formed in a setting undergoing oxidation of hydrocarbons in fluids as they leaked from underlying reservoirs during eogenesis or telogenesis (Donovan, 1974; Kirkland et al., 1995). By using stains, such as alizarin red-S, one can identify and quantify (using image processing programs) the amount of calcite cement in thin sections.

PPL, AS, Scale bar = 0.10 mm

Miocene Santa Fe Gp., Zia Fm., Cerro Conejo Mbr., Sandoval Co., New Mexico

These poikilotopic calcite cements formed within a concretion in eolian sediments and originated through growth on a preferential nucleation site (typically carbonate bioclasts or carbonate rock fragments). This example shows calcite's very high birefringence and relief (as compared to quartz). At least two-thirds of this field of view is part of a single crystal of calcite cement that encases numerous detrital quartz and feldspar grains and rock fragments.

XPL, Scale bar = 0.26 mm

Up. Permian Schuchert Dal Fm., Jameson Land, East Greenland

Normally, compositional variations within calcite cements are not petrographically visible. By using a combined alizarin red-S and potassium ferricyanide stain, however, even modest variations in the amount of iron in the calcite lattice (a result of changes in water chemistry through time) can be inferred from variations in the color of the stain (Dickson, 1966). In this photomicrograph, the earliest cements are quartz overgrowths; they are followed by moderately ferroan (purple) calcite and finally very ferroan (deep blue) calcite that is associated with traces of asphaltic hydrocarbons (black).

PPL, AFeS, BDI, Scale bar = 0.26 mm

Mid. Permian Huledal Fm., Karstryggen area, East Greenland

Carbonate staining can provide information about the composition of fluids responsible for calcite precipitation and thus the timing of such cementation. This calcite-cemented lithic arenite is from the margins of an intrusive dike. The earliest calcite cement is nonferroan (pink-stained) and that grades into more extensively zoned, moderately ferroan (purple-stained) calcite. Finally most of the remaining porosity was filled by very ferroan (blue-stained) calcite. Because iron is incorporated into calcite mainly under reducing conditions, one can infer that pore fluids became increasingly reducing with time and burial.

PPL, AFeS, BDI, Scale bar = 0.26 mm

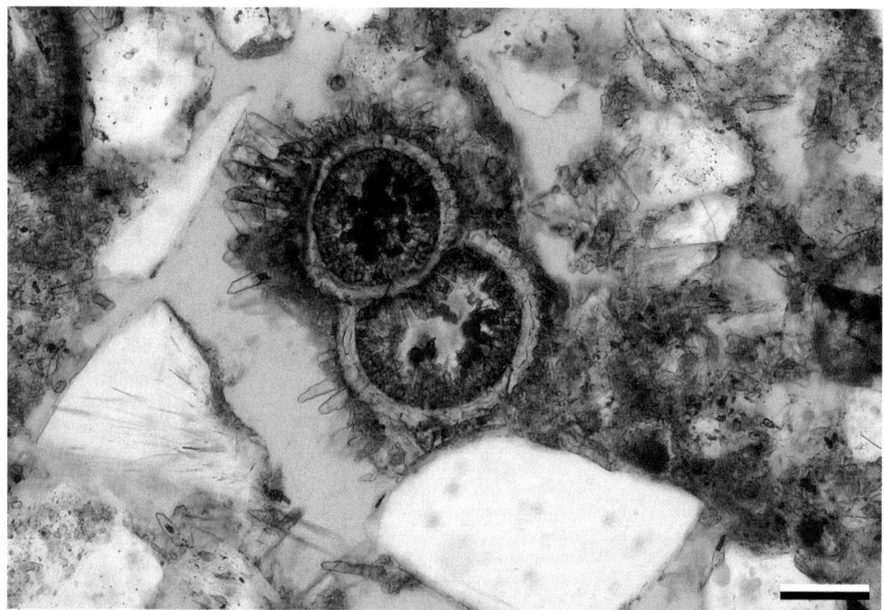

Miocene (Tortonian), Tagus Basin, Setúbal Peninsula, Portugal †

A foraminiferal test with a nonferroan calcite wall (stained pink) surrounded by fringing crystals of ferroan calcite cement (stained dark blue). In the absence of staining, the fibrous to bladed calcite cements might well be assumed to be seafloor (synsedimentary) precipitates; the incorporation of iron, however, implies at least somewhat later, subsurface, reducing conditions of formation. The spherical ferruginous cements in the center of the test probably are outcrop-weathered pyrite framboids (now iron oxyhydroxides).

PPL, AFeS, KFS, BDI, Scale bar = 0.07 mm

Up. Cretaceous Blålange Fm., Vøring Basin, Norwegian Sea †

Slightly poikilotopic pore-filling ferroan calcite. The stain color that ferroan calcite takes variable, based on the Fe concentration. Dolomite tends to take a more patchy stain with a different shade of blue. The grains coated by a green clay along with the large open pores and few grain contacts suggest that there was little compaction prior to cementation—thus, the cement most likely formed during early subsurface diagenesis. The yellowish grains are stained K-feldspars.

PPL, AFeS, KFS, Scale bar = 0.12 mm

Paleogene Vieja Gp., Presidio Co., Texas

Concretions are formed by localized carbonate precipitation, that can span eogenetic meteoric to burial (mesogenetic) diagenetic environments. This example shows a small part of a large septarian nodule. Septarian nodules are characterized by angular cracks that decrease in size and abundance towards the edge of the nodule. The cracks visible here are filled with an early, finely crystalline calcite cement that is followed by more coarsely crystalline calcite.

XPL, Scale bar = 0.51 mm

Up. Cretaceous Eagle Ford Shale, Dallas Co., Texas

Septarian nodules commonly are cemented by multiple generations of calcite that can be clear or white to strongly-colored (yellow to dark brown). In this example, the first cements are nonferroan and are followed by highly ferroan calcites (with blue stain visible even in this cross-polarized light image). The staining indicates precipitation from very different pore fluids. The initial cements formed in oxidizing waters (conditions in which iron does not go into the calcite lattice); later cements formed from reducing pore fluids with available iron (where Fe^{2+} can go into the calcite lattice; Lindholm, 1974).

XPL, AFeS, Scale bar = 0.51 mm

Up. Cretaceous Eagle Ford Shale, Dallas Co., Texas

A comparison of plane-polarized and epifluorescent light views of the same sample as the previous photomicrograph. The early nonferroan (pink-stained and skeletal) cements in the septarian fractures fluoresce strongly because of the incorporation of abundant hydrocarbon microinclusions within the calcites (visible only at higher magnifications). The later ferroan cements are nonfluorescent, because they precipitated from pore fluids that apparently did not contain hydrocarbons. It also shows that calcite does not have mineral fluorescence at the 470 nm wavelength used.

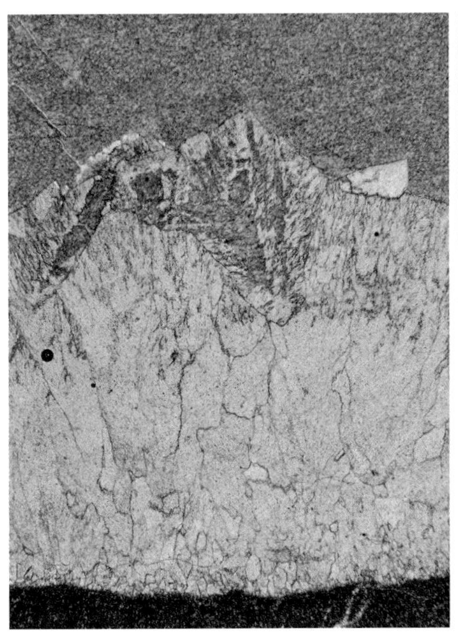

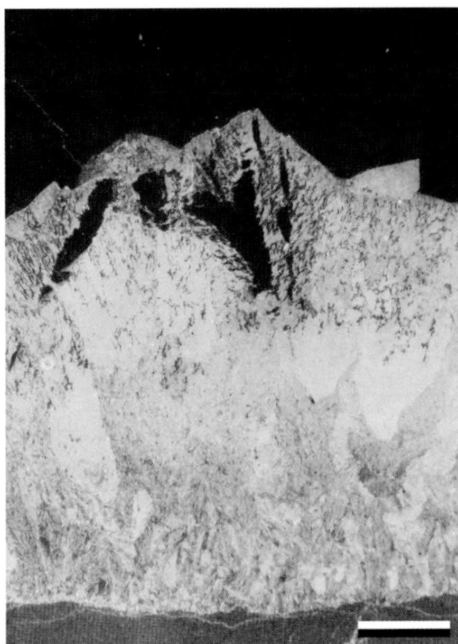

PPL | FL 470, AFeS, Scale bar = 0.64 mm

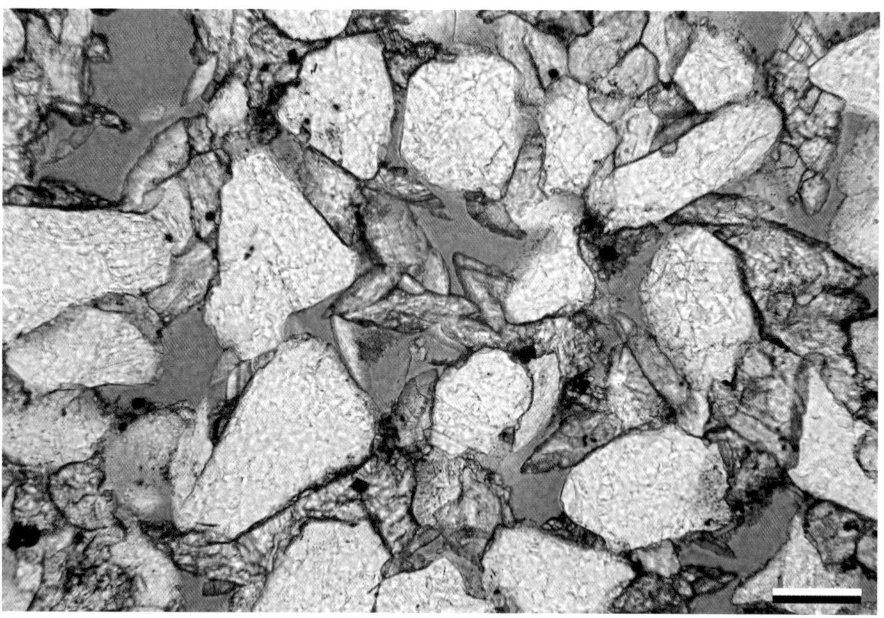

Lo. Cretaceous (Albian) Nahr Umr Fm., offshore Qatar

Siderite commonly occurs as yellowish or reddish-brown lozenge-shaped crystals. This is an example of a siderite-cemented shallow-marine quartzarenite. The siderite cements are clear to very pale brown and many of the crystals display typical lozenge or flattened rhomb shapes. Siderite cements commonly form in eogenetic meteoric to shallow-marine environments that are suboxic and sulfide-poor (Morad, 1998); they also can form during later burial (Macaulay et al., 1993; Rossi et al., 2001). The loose packing in this sample indicates that the siderite was probably early and stabilized the rock during burial.

PPL, BDI, Scale bar = 0.08 mm

Lo. Cretaceous (Albian) Nahr Umr Fm., offshore Qatar

An example of a quartzarenite with much of its primary porosity destroyed by siderite cements. The well-developed lozenge-shaped crystals are easily seen here. Many of the crystals also have some iron-staining around their exterior surfaces which is typical for both siderite and ankerite. The pyrite crystals (opaque) may be a later replacement during burial.

PPL | XPL, BDI, Scale bar = 0.12 mm

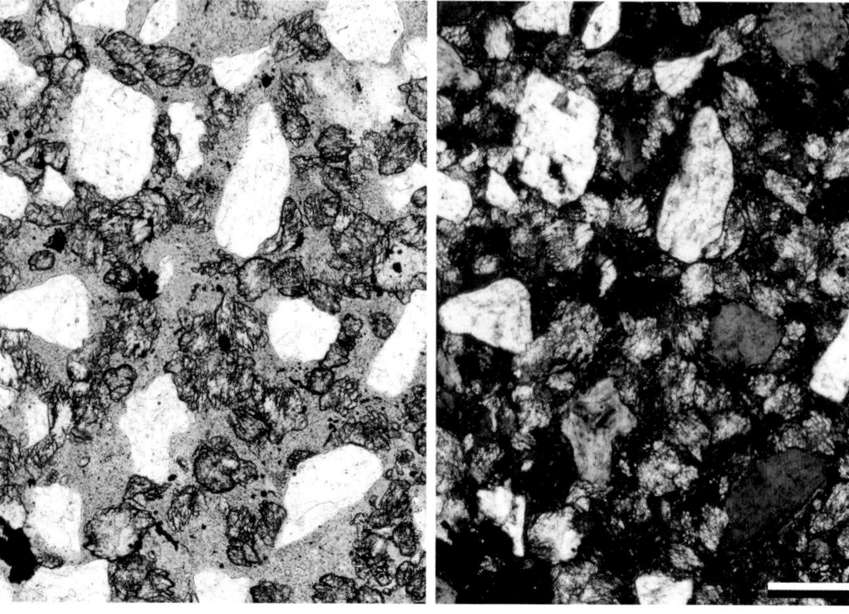

Lo. Cretaceous (Albian) Nahr Umr Fm., offshore Qatar

Siderite cements commonly have iron oxide or hydroxide coatings related to later diagenetic partial dissolution and oxidation of the iron from the siderite. The rhombic siderite appears to be intergrown with porous chamosite cements in this quartzarenite. The siderite cements in this sample show more equant (albeit irregular) rhombs and fewer lozenge-shaped crystals than were seen in the previous photomicrographs from the same unit.

PPL | XPL, BDI, Scale bar = 0.14 mm

Mid. Jurassic Fensfjord Fm., Norwegian sector, North Sea †

Siderite cements in marine sandstones commonly can be associated with degraded grains, as seen here. Lozenges of siderite are interspersed between layers of an expanded mica; where not constrained by other detrital grains the mica has fanned out into open pore space. Less well-formed crystals also are associated with a degraded grain of unknown origin, which now consists of quartz and diagenetic replacive clay. The siderite has taken a slight pink stain in this case.

PPL, AFeS, KFS, BDI, Scale bar = 0.12 mm

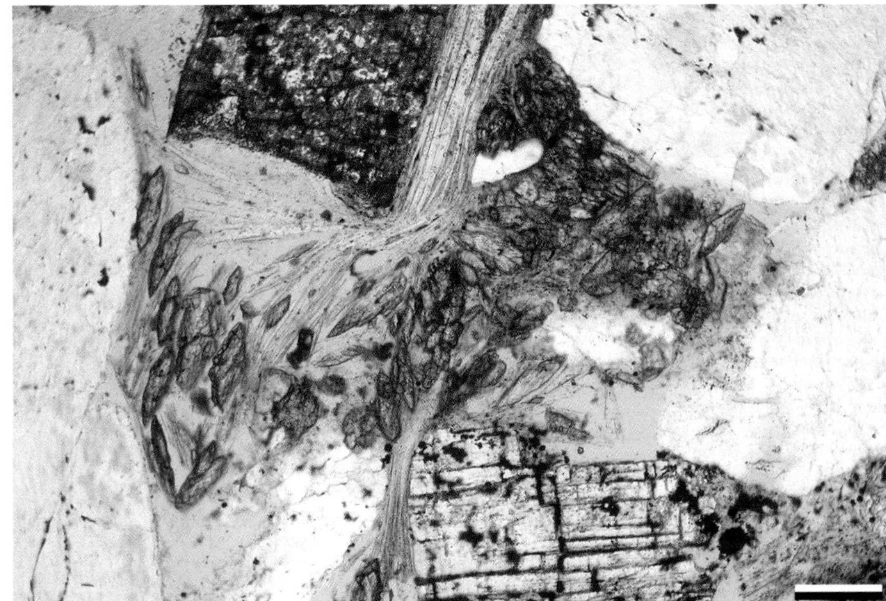

Paleocene Tang Fm., Vøring Basin, Norwegian Sea †

Cubic to granular siderite crystals apparently associated with a highly altered grain. The siderite is found in conjunction with strongly ferroan dolomite/ankerite (blue-stained, rhombic crystals along top edge). Although this siderite crystal habit is not typical, the crystals have a fairly characteristic very pale yellow/brown color and have not taken a stain. The siderite clearly has a higher relief than the nearby ferroan dolomite/ankerite, as shown by the difference in focus.

PPL, AFeS, KFS, BDI, Scale bar = 0.03 mm

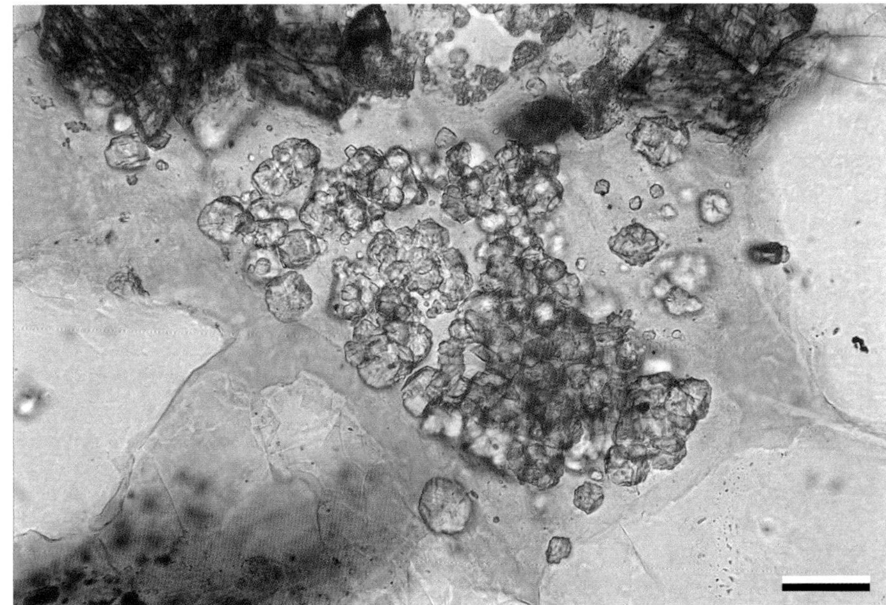

Lo. Triassic Ivishak Fm., North Slope, Alaska

In this chert arenite, siderite formed grain-coating cements on chert clasts and quartz grains. These cements are rhombic to slightly lozenge-shaped and have a strong reddish-brown hue. Geochemical studies on the Ivishak Formation revealed that much of this siderite cement formed in a fresh-water environment (based on high iron, but low calcium and magnesium concentrations), but siderite formation also continued during deeper burial (Harun, 1996).

PPL, BDI, Scale bar = 0.10 mm

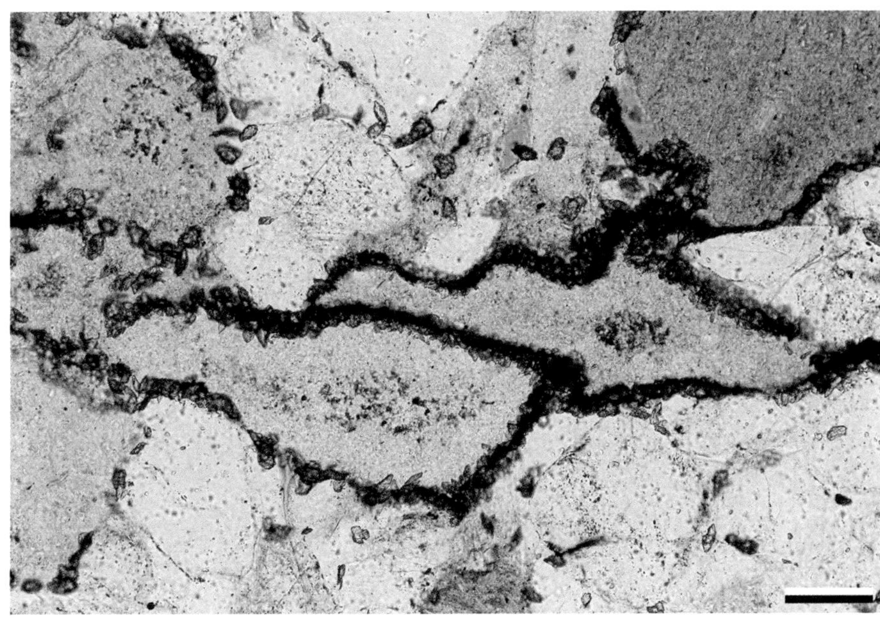

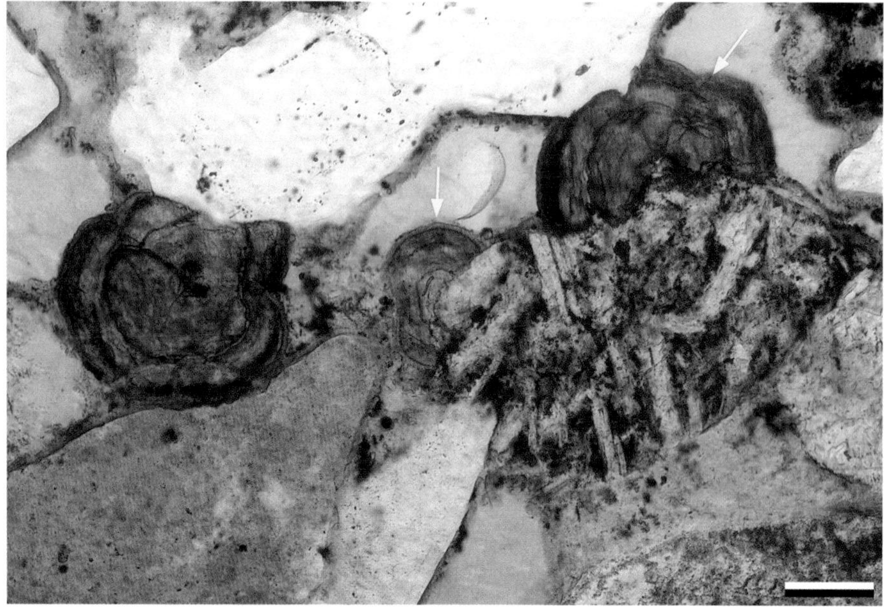

Up. Triassic Snadd Fm., Barents Sea †

The siderite cements (examples at arrows) have an unusual nodular habit and have overgrown detrital grains. Other nodular siderites (mainly sphaerosiderite) show sweeping extinction and typically are restricted to water-saturated freshwater sediments. The siderite has, in places, taken a weak blue color (probably from injection of blue dye along microfractures), and the thin chlorite grain-coating cement in this sample appears to be absent where the siderite cement has grown on the grain, suggesting that siderite predates chlorite. The grain at center-right is a degraded volcanic rock fragment with lath-shaped feldspars.

PPL, AFeS, KFS, BDI, Scale bar = 0.12 mm

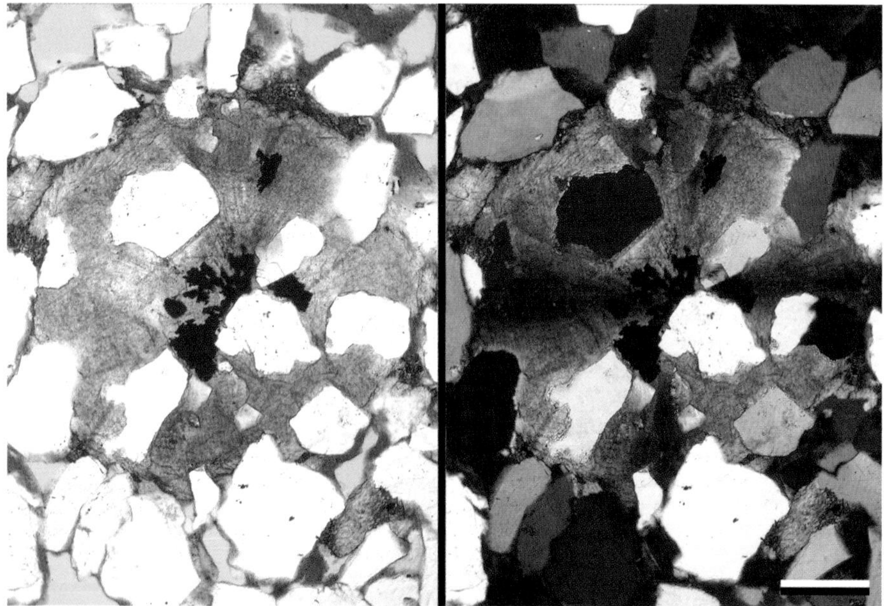

Paleocene Fatehgarh Fm., Rajasthan, India

A fluvial sandstone with a sphaerosiderite nodule. The "cruciform" sweeping extinction is apparent in the XPL image, and it also displays some pleochroism. Sphaerosiderite is more common in mudrocks, but here it has engulfed a number of quartz grains, suggesting that it forms passively rather than displacively. The nodule shows a slight blue stain (from dye injection) and has an ankerite overgrowth. The opaque mineral in the center is pyrite and suggests that there were minor amounts of sulfate present. When this was consumed, methanogenesis resulted in iron being used to form siderite (e.g., Ellwood et al., 1988).

PPL | XPL, AFeS, BDI, Scale Bar = 0.06 mm

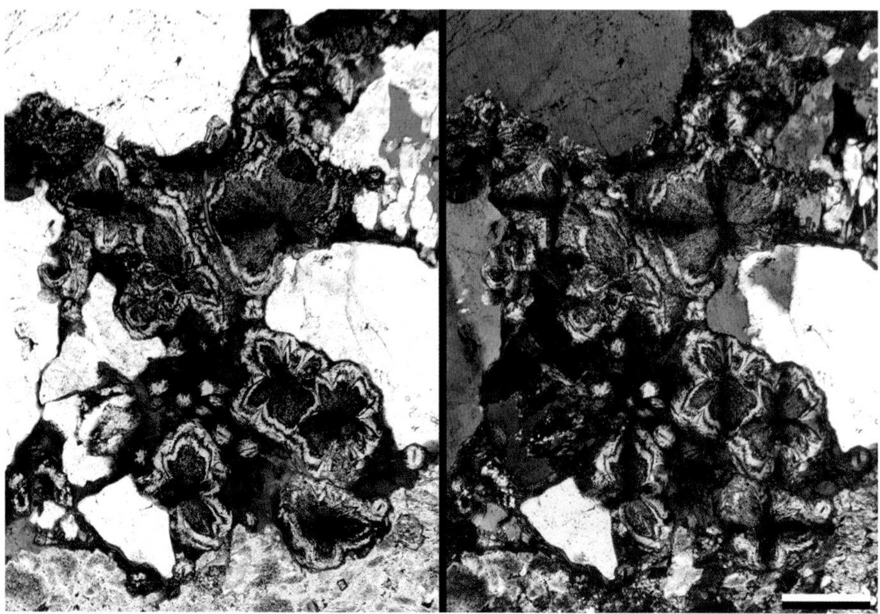

Lo. Pennsylvanian Morrow B Sandstone, Ochiltree Co., Texas

These large sphaerosiderite cements occur within inferred paleosol horizons in nonmarine sandstones. The sphaerosiderite is highly zoned in plane-polarized light and has sweeping and spherulitic (cruciform) extinction patterns in cross-polarized light. Sphaerosiderite is characteristic of continental wetland and paleosol environments and has been utilized to help reconstruct regional paleohydrology and paleoclimatology utilizing $\delta^{18}O$ analysis (Ludvigson et al., 2013).

PPL | XPL, RDI, Scale bar = 0.26 mm

Lo. Pennsylvanian Morrow B Sandstone, Ochiltree Co., Texas

This sphaerosiderite cement is from the same interval as the previous example, but it has undergone partial leaching during burial diagenesis. However, the characteristic sweeping extinction and the lozenge-shaped rhombs of the partly corroded siderite crystal remain. The siderite cement is accompanied by authigenic quartz and probable illite/smectite or smectite/chlorite cements (the brownish clays around the exterior of the crystal). Photomicrographs from Sara Gallagher.

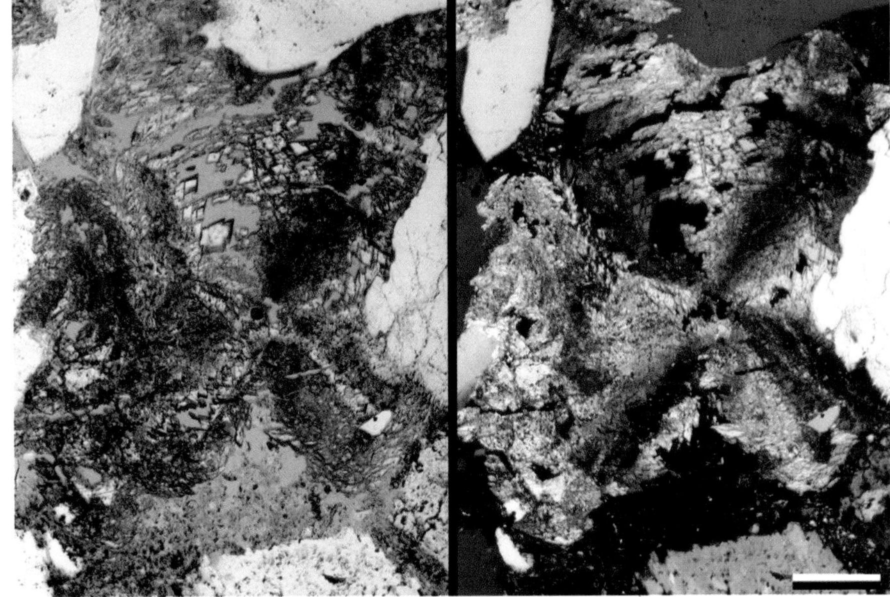

PPL | XPL, RDI, Scale bar = 0.10 mm

Lo. Pennsylvanian Morrow B Sandstone, Ochiltree Co., Texas

This back-scattered electron image of a sphaerosiderite shows compositional zoning within the crystal. The bright interior of the crystal contains a higher percentage of $FeCO_3$ (92 mol%) than the exterior parts (74 mol%; Gallagher, 2014). The iron-rich cores were interpreted to have resulted from precipitation from nonmarine pore fluids, whereas precipitation of the exterior of the crystal reflected mixing of nonmarine and marine pore fluids. Later dissolution preferentially affected the more Mg-rich outer parts of the crystal, yielding a blocky compositional mosaic in those areas. BSE photograph from Sara Gallagher.

BSE, Scale bar = 4.6 μm

Up. Jurassic Fulmar Fm., United Kingdom sector, North Sea †

An SEM view of diagenetic siderite cement that is surrounded by microcrystalline quartz and clay. The crystal shows the typical rhombohedral cleavage and crystal habit of siderite on the exposed surface.

SEM, Scale bar = 7.3 μm

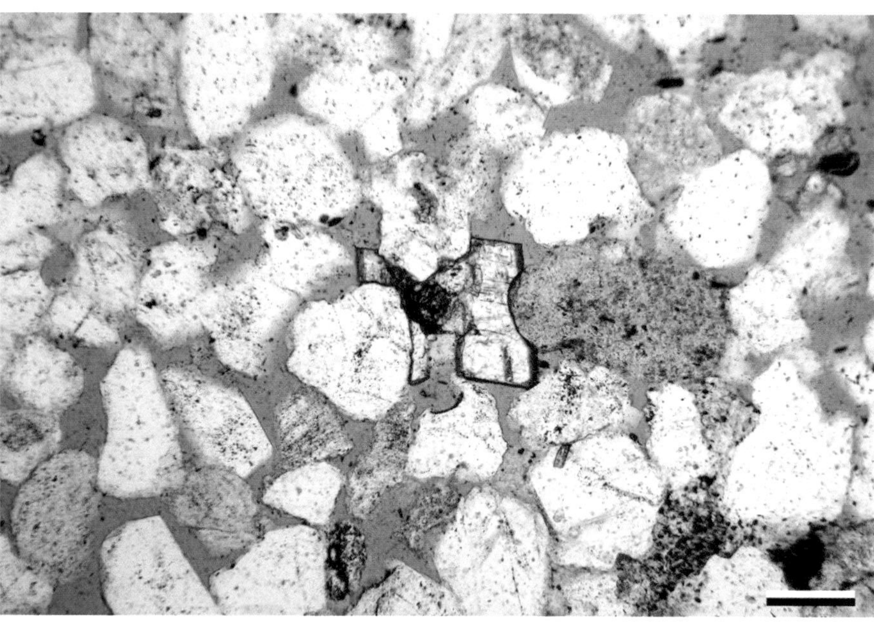

Lo. Mississippian Berea Sandstone, Ohio

In this photomicrograph, a single euhedral crystal of dolomite has cemented several detrital grains together. The crystal is zoned, with a cloudy core, followed by a clear zone and a final thin brown rim that reflects later oxidation of iron in the outermost (ankeritic?) zone of the dolomite. Dolomite rarely exhibits twinning, and the lack of twinning can help distinguish it from calcite, especially where the crystals are not euhedral. However, staining is the easiest and most reliable way to distinguish calcite from dolomite. The embayment within the crystal is probably related to an unstable grain having been dissolved.

PPL, BDI, Scale bar = 0.16 mm

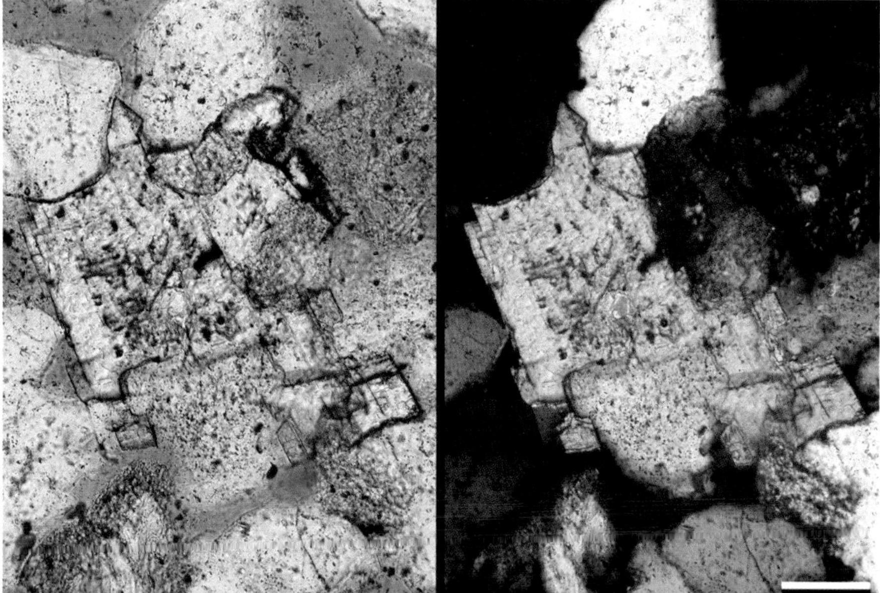

Lo. Mississippian Berea Sandstone, Ohio

In this example, a large and complexly-shaped single crystal of authigenic dolomite fills a large pore between detrital grains. The crystal is slightly zoned, with more inclusions in the core than in the margins. The crystal also has a thin iron oxide rim related to later oxidation of iron in the outermost, possibly ankeritic, zones. Later dissolution has partially to completely removed some SRFs (for example at upper right).

PPL | XPL, BDI, Scale bar = 0.10 mm

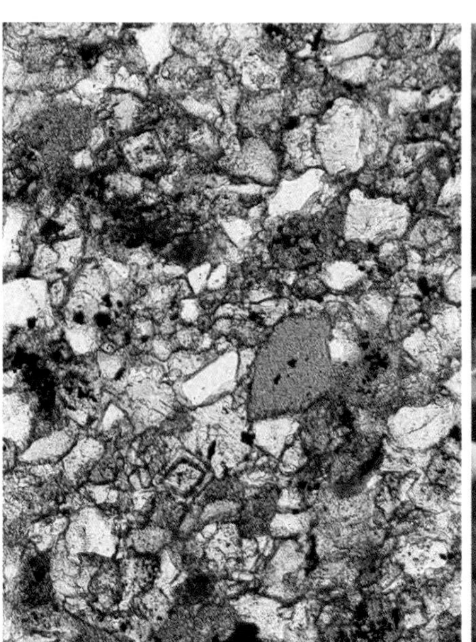

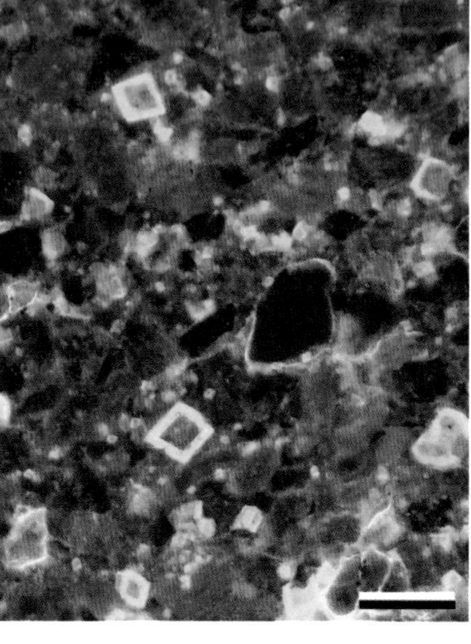

Mid. Jurassic Curtis Fm., Moffat Co., Colorado

Euhedral zoned dolomite has grown displacively and as cement in this shaly phosphate- and glauconite-rich siltstone. In the fluorescent view, the cores of the dolomite crystals are largely nonfluorescent, whereas the rims are highly fluorescent. During initial dolomite cementation, there were no hydrocarbons in the pore fluids, but later dolomite precipitation occurred in the presence of hydrocarbons, yielding the bright green fluorescence. In addition, the use of fluorescent epoxies has highlighted micropores that might not be visible otherwise (bright orange areas in the right-hand image).

PPL | FL 470, RDI, Scale bar = 0.10 mm

Mid. Jurassic Entrada Fm., Moffat Co., Colorado

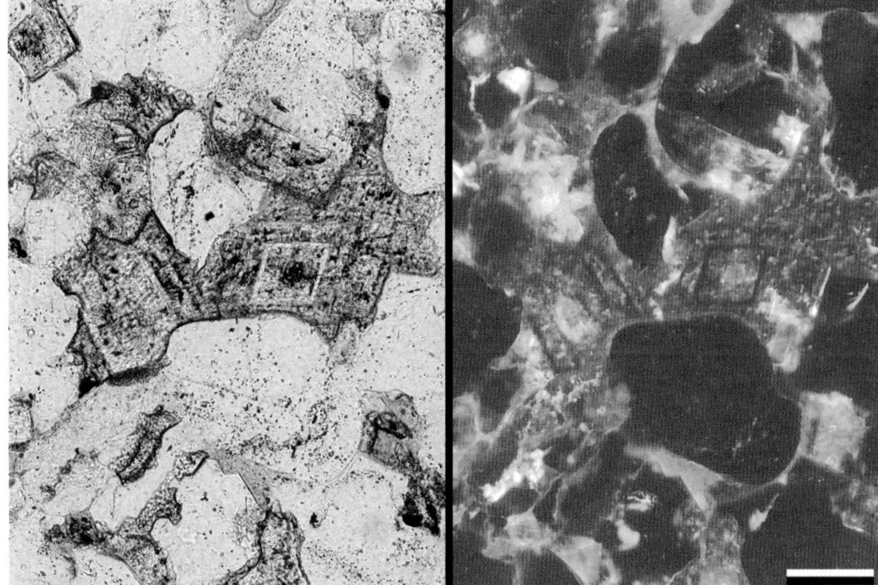

This sandstone contains medium- to coarsely-crystalline, zoned dolomite cement (and strong zoning is an important feature for dolomite recognition). Here, quartz cement does not occur where dolomite is in contact with the grains, and the quartz cement laps on top of the dolomite cements. This indicates that dolomite precipitated prior to the quartz cement. In fluorescence, the inclusion-rich dolomite zones contain hydrocarbons, whereas the limpid (inclusion-poor) zones do not. The early-formed, less-ordered dolomite cores contain microporosity (orange fluorescence) probably related to later dissolution.

PPL | FL 470, AS, BDI, Scale bar = 0.10 mm

Lo. Permian Hueco Fm., Powwow Conglomerate, Hudspeth Co., Texas

This chert arenite has two different carbonate cements. Nonferroan dolomite (unstained) precipitated first in the available pore space. The dolomite cements are euhedral and slightly zoned with the cores being darker and more inclusion-rich than the later zones. The dolomite crystals (but not the detrital grains) are coated with iron oxides probably due to later alteration of an outermost iron-rich (dolomite/ankerite) zone. The iron oxides were then followed by nonferroan (pink stained) to iron-rich (purple to blue stained) calcite cements which fill most remaining pores.

PPL, AFeS, BDI, Scale bar = 0.12 mm

Up. Jurassic Fulmar Fm., United Kingdom sector, North Sea †

Several examples of nonferroan dolomite crystals with thin, ferroan margins are visible in the center and left of this image. An additional and more uniformly ferroan dolomite crystal is present in the lower part of the image. The K-feldspars in this sample lack overgrowths where they are in contact with the dolomite (as in the area near red arrow), indicating that dolomite formation preceded feldspar overgrowth.

PPL, AFeS, KFS, BDI, Scale bar = 0.07 mm

Lo. Cretaceous, Lusitanian Basin, Portugal †

A sandstone that is cemented by rhombic ferroan dolomite. Zoning such as this commonly is better seen in cathodoluminescence, but because of the strong color distinctions between zones in this example, the zonation is easily visible with conventional petrography. The dark brown areas are slightly ferroan dolomite or ankerite that has been altered during outcrop exposure of this sample (which is more weathered than the sample seen in the next image).

PPL, AFeS, KFS, BDI, Scale bar = 0.49 mm

Lo. Cretaceous, Lusitanian Basin, Portugal †

An image of a sandstone sample from the same outcrop as the preceding image. The ferroan dolomite here has a core of ankerite, which is darker blue in color, reflecting a higher iron content.

PPL, AFeS, KFS, BDI, Scale bar = 0.49 mm

Eocene Galisteo Fm., Sandoval Co., New Mexico

The range of compositions and variations of solubilities of carbonate minerals in different settings commonly leads to complex paragenetic relations between such minerals. Here mesogenetic ankerite cements (the dark areas in this image) were overgrown by non-ferroan dolomite cement (white). The ankerite was later dissolved or altered, leaving iron oxide and hydroxide residues. The alteration most likely occurred during telogenesis (uplift and exposure of this outcrop sample).

PPL, BDI, Scale bar = 0.10 mm

Up. Pennsylvanian – Lo. Permian Granite Wash, southern Oklahoma

Mesogenetic diagenesis can include the formation of saddle or baroque dolomites. Saddle dolomites have curved crystal faces and cleavages and exhibit sweeping or undulose extinction (see the cross-polarized light view). They typically have >60°C precipitation temperatures. Overall, saddle dolomites form from saline brines at variable temperatures and depths, often associated with hydrocarbon maturation and migration (Machel; 1987; Spötl and Pitman, 1998).

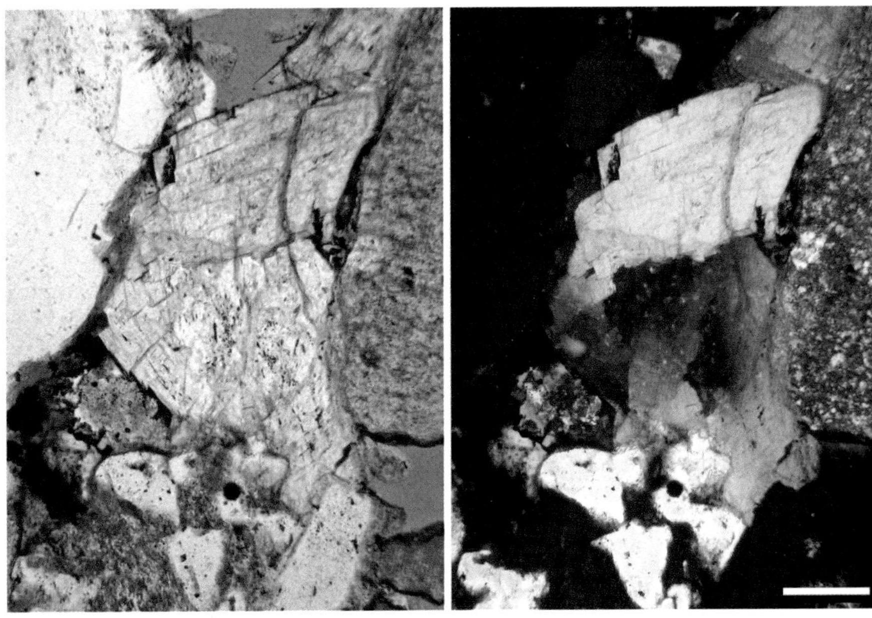

PPL | XPL, BDI, Scale bar = 0.09 mm

Up. Pennsylvanian – Lo. Permian Granite Wash, southern Oklahoma

In this sample, the main cements are saddle dolomite (also termed baroque dolomite). Note the highly curved crystal face on the dolomite in the center and the strongly undulatory extinction in all the dolomite crystals. Because later-formed dolomite, including ankerite and some saddle dolomite, can have high iron concentrations, it is common to see iron oxide staining where such crystals have reacted with meteoric pore fluids in surface (telogenetic) or subsurface (mesogenetic) settings.

XPL, BDI, Scale bar = 0.14 mm

Lo. Mississippian Berea Sandstone, Midland Co., Michigan

Ankerite, an iron-rich (and sometimes also magnesium-rich) carbonate, forms a solid solution series with dolomite. It is common, as in this example, for ankerite to form as overgrowth zones on precursor authigenic or detrital dolomite crystals. Dolomite and ankerite look very similar, but ankerite crystals usually have a stronger rust-colored stain due to the higher iron content of the mineral. The dolomites are the cloudy gray, rhombic crystals and the ankerites are the iron-stained rhombic overgrowths encasing the dolomites. Ankerite forms in a wide variety of settings and temperatures in the burial diagenetic realm.

PPL, BDI, Scale bar = 0.10 mm

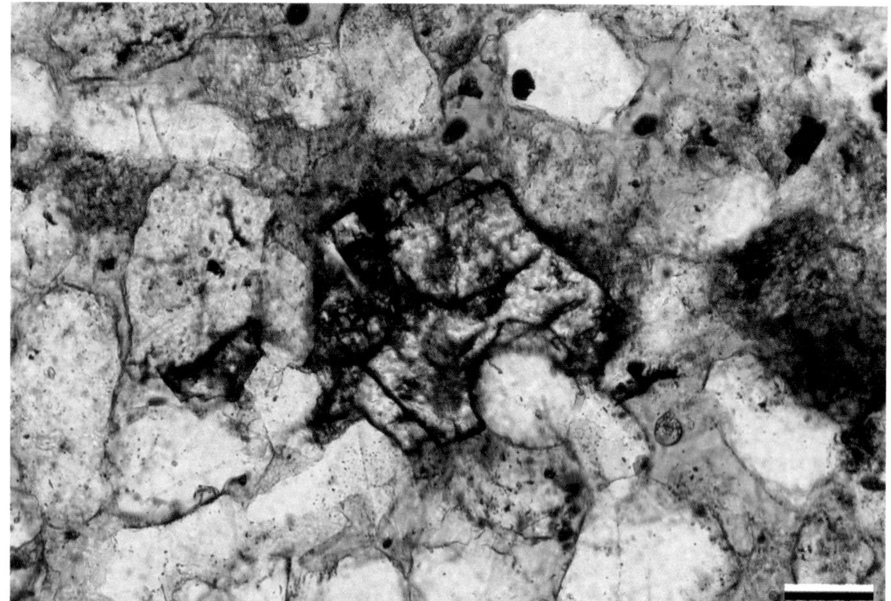

Lo. Mississippian Berea Sandstone, Midland Co., Michigan

Another example of carbonate cements with a probable dolomite core and thinner ankerite rim. The iron oxide and hydroxide staining on the ankerites are oxidation rims formed by reaction with the surrounding pore fluids. Like dolomite, ankerite crystals typically are rhombic.

PPL, BDI, Scale bar = 0.06 mm

Mid. Ordovician Martinsburg Fm., Berks Co., Pennsylvania

Ankerite cements in this lithic arenite range in size from very finely to medium crystalline. The euhedral rhombs grew displacively in this rock and have a crust of iron oxide or hydroxide on them related to oxidation by uplift-associated meteoric pore fluids. The cores of these crystals may well be closer to dolomite than to ankerite in composition, but that is impossible to prove without further analytical work (XRD, microprobe or other quantitative techniques).

PPL, Scale bar = 0.10 mm

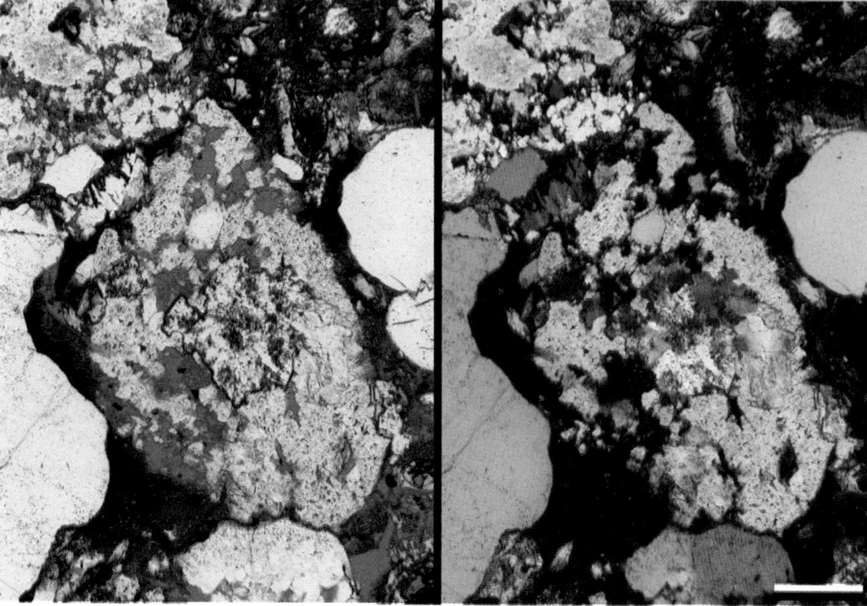

Lo. Pennsylvanian Morrow B Sandstone, Ochiltree Co., Texas

Ankerite cements here are infilling a partially leached feldspar grain. These ankerite crystals are not heavily stained by iron oxides like those in previous examples, because this sample is from a subsurface core, so these cements probably were never exposed to meteoric waters. Microprobe data on similar crystals from a nearby well (ca. 300 m [1,000 ft] away) show that these cements are indeed ankerite and not dolomite. The ankerite crystals have a complex sweeping or undulatory extinction in cross-polarized light that probably reflect many lattice defects.

PPL | XPL, RDI, Scale bar = 0.26 mm

Lo. Pennsylvanian Morrow B Sandstone, Ochiltree Co., Texas

Another example of ankerite filling porosity probably created by feldspar dissolution (widespread in this sample). The ankerite cements formed during deep, late-stage burial, after quartz overgrowth formation, kaolinite precipitation and dissolution of feldspars. As in the previous photomicrograph, these cements can be easily confused with dolomite (very high to high relief, very high birefringence, rhombohedral crystals and sweeping extinction). Although staining can help in differentiation, microprobe or XRD data are best used to resolve the exact mineralogy of the cements. Photomicrographs from Sara Gallagher.

PPL | XPL, RDI, Scale bar = 0.10 mm

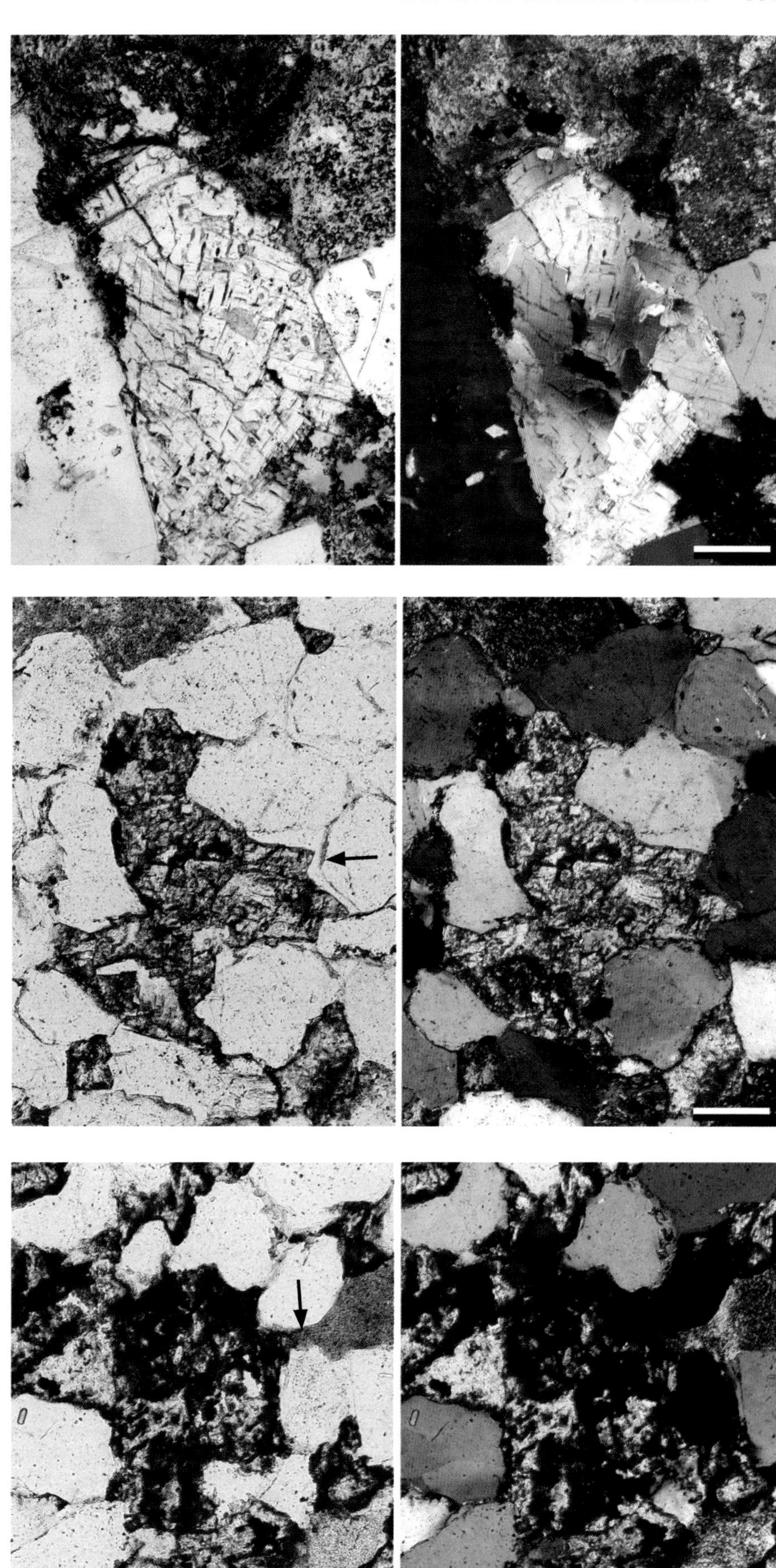

Permian basinal sandstone, Presidio Co., Texas

This quartz-rich sandstone is primarily cemented with subhedral to euhedral ankerite (strongly birefringent, rhombic crystals that have a "rusty" appearance due to weathering and iron oxide and hydroxide staining). At least two cements precede the ankerite—thin, illite/smectite clay coatings surround most of the detrital grains (example at black arrow) and are followed by quartz overgrowths. As a precipitate that fills the remaining porosity, the authigenic ankerite is clearly a late-stage mesogenetic product.

PPL | XPL, Scale bar = 0.10 mm

Permian basinal sandstone, Presidio Co., Texas

This sample is similar to the previous one. It shows some strongly compacted shale clasts that flowed into adjacent pores (example at black arrow), but most porosity loss was due to postcompactional ankerite cementation. Numerous small, round, dark bodies are visible within the ankerite cement—these probably are microcrystalline dolomite or ankerite clusters formed in association with organic matter by microbial processes, probably during early diagenesis. Those clusters may then have acted as nuclei for much later precipitation of the coarse, mesogenetic ankerite cement.

PPL | XPL, Scale bar = 0.10 mm

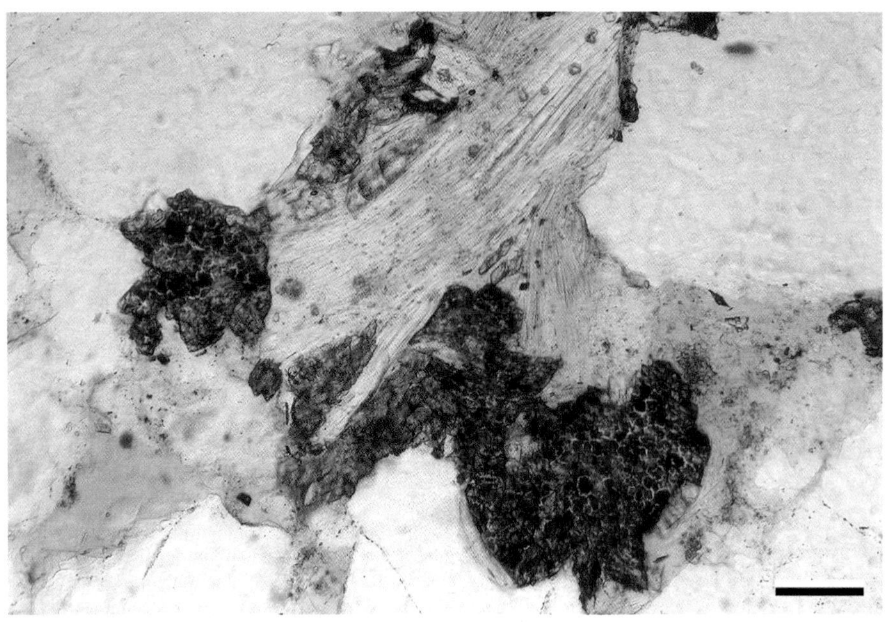

Mid. Jurassic Ness Fm., Norwegian sector, North Sea †

Lozenges of siderite here are associated with a partly expanded muscovite mica. The siderite is, in places, enclosed by a dark blue-stained carbonate, which is probably ankerite. It can be difficult, even in stained sections, to differentiate between ankerite and highly ferroan calcite because of the intensity of their staining. The rhombic shape and lack of twinning are two characteristics that make it more likely that these crystals are ankerite.

PPL, AFeS, KFS, BDI, Scale bar = 0.07 mm

Cited References and Additional Information Sources

Amthor, J. E., and J. Okkerman, 1998, Influence of early diagenesis on reservoir quality of Rotliegende sandstones, northern Netherlands: AAPG Bulletin, v. 82, p. 2246-2265.

Baker, J. C., J. Kassan, and P. J. Hamilton, 1996, Early diagenetic siderite as an indicator of depositional environment in the Triassic Rewan Group, southern Bowen Basin, eastern Australia: Sedimentology, v. 43, p. 77-88, doi: 10.1111/j.1365-3091.1996.tb01461.x.

Beckner, J. R., and P. S. Mozley, 1998, Origin and spatial distribution of early vadose and phreatic calcite cements in the Zia Formation, Albuquerque Basin, New Mexico, USA, *in* S. Morad, ed., Carbonate Cementation in Sandstones: Distribution Patterns and Geochemical Evolution (IAS Special Publication 26): Oxford, UK, Blackwell Publishing, p. 27-51, doi: 10.1002/9781444304893.ch2.

Bjørkum, P. A., and O. Walderhaug, 1990, Geometrical arrangement of calcite cementation within shallow marine sandstones: Earth-Science Reviews, v. 29, p. 145-161, doi: 10.1016/0012-8252(0)90033-R.

Boles, J. R., 1978, Active ankerite cementation in the subsurface Eocene of southwest Texas: Contributions to Mineralogy and Petrology, v. 68, p. 13-22, doi: 10.1007/BF00375443.

Boles, J. R., 1998, Carbonate cementation in Tertiary sandstones, San Joaquin basin, California, *in* S. Morad, ed., Carbonate Cementation in Sandstones: Distribution Patterns and Geochemical Evolution (IAS Special Publication 26): Oxford, UK, Blackwell Publishing, p. 261-284, doi: 10.1002/9781444304893.ch12.

Browne, G. H., and D. M. Kingston, 1993, Early diagenetic spherulitic siderites from Pennsylvanian palaeosols in the Boss Point Formation, Maritime Canada: Sedimentology, v. 40, p. 467-474, doi: 10.1111/j.1365-3091.1993.tb01346.x.

Burley, S. D., 1984, Patterns of diagenesis in the Sherwood Sandstone Group (Triassic), United Kingdom: Clay Minerals, v. 19, p. 403-440, doi: 10.1180/claymin.1984.019.3.11.

Burley, S. D., J. Mullis, and A. Matter, 1989, Timing diagenesis in the Tartan Reservoir (UK North Sea): Constraints from combined cathodoluminescence microscopy and fluid inclusion studies: Marine and Petroleum Geology, v. 6, p. 98-120, doi: 10.1016/0264-8172(89)90014-7.

Calvert, S. E., and T. F. Pedersen, 1996, Sedimentary geochemistry of manganese: Implications for the environment of formation of manganiferous black shales: Economic Geology, v. 91, p. 36-47, doi: 10.2113/gsecongeo.91.1.36.

Chafetz, H. S., and S. A. Guidry, 1999, Bacterial shrubs, crystal shrubs, and ray-crystal shrubs: Bacterial vs. abiotic precipitation: Sedimentary Geology, v. 126, p. 57-74, doi: 10.1016/S0037-0738(99)00032-9.

Chafetz, H. S., B. H. Wilkinson, and K. M. Love, 1985, Morphology and composition of non-marine carbonate cements in near-surface settings, *in* N. Schneidermann, and P. M. Harris, eds., Carbonate Cements: Tulsa, OK, SEPM Special Publication 36, p. 337-347, doi: 10.2110/pec.85.36.0337.

Choquette, P. W., and F. C. Trusell, 1978, A procedure for making the Titan-yellow stain for Mg calcite permanent: Journal of Sedimentary Research, v. 48, p. 639-641, doi: 10.1306/212F74FF-2B24-11D7-8648000102C1865D.

Conybeare, D. M., and H. F. Shaw, 2000, Fracturing, overpressure release and carbonate cementation in the Everest Complex, North Sea: Clay Minerals, v. 35, p. 135-149, doi: 10.1180/000985500546530.

Dale, A., C. M. John, P. S. Mozley, P. C. Smalley, and A. H. Muggeridge, 2014, Time-capsule concretions: Unlocking burial diagenetic processes in the Mancos Shale using carbonate clumped isotopes: Earth and Planetary Science Letters, v. 394, p. 30-37, doi: 10.1016/j.epsl.2014.03.004.

Dapples, E. C., 1971, Physical classification of carbonate cement in quartzose sandstones: Journal of Sedimentary Research, v. 41, p. 196-204, doi: 10.1306/74D72220-2B21-11D7-8648000102C1865D.

Deer, W. A., R. A. Howie, and J. Zussman, 1992, An Introduction to the Rock-forming Minerals (2nd Edition): London, Longman Group Ltd., 696 p.

Dickson, J. A. D., 1966, Carbonate identification and genesis as revealed by staining: Journal of Sedimentary Research, v. 36, p. 491-505, doi: 10.1306/74D714F6-2B21-11D7-8648000102C1865D.

Donovan, T. J., 1974, Petroleum microseepage at Cement, Oklahoma: Evidence and mechanism: American Association of Petroleum Geologists Bulletin, v. 58, p. 429-446.

Dutton, S. P., 2008, Calcite cement in Permian deep-water sandstones, Delaware Basin, west Texas: Origin, distribution, and effect on reservoir properties: AAPG Bulletin, v. 92, p. 765-787, doi: 10.1306/01280807107.

Dutton, S. P., H. S. Hamlin, R. L. Folk, and S. J. Clift, 1996, Early siderite cementation as a control on reservoir quality in submarine fan sandstones, Sonora Canyon gas play, Val Verde Basin, Texas, *in* L. J. Crossey, R. Loucks, and M. W. Totten, eds., Siliciclastic Diagenesis and Fluid Flow: Concepts and Applications: Tulsa, OK, SEPM Special Publication 55, p. 115-127, doi: 10.2110/pec.96.55.0115.

Dutton, S. P., C. D. White, B. J. Willis, and D. Novakovic, 2002, Calcite cement distribution and its effect on fluid flow in a deltaic sandstone, Frontier Formation, Wyoming: AAPG Bulletin, v. 86, p. 2007-2021, doi: 10.1306/61EEDDE6-173E-11D7-8645000102C1865D.

Ellwood, B. B., T. H. Chrzanowski, F. Hrouda, G. J. Long, and M. L. Buhl, 1988, Siderite formation in anoxic deep-sea sediments:

A synergetic bacterially controlled process with important implications in paleomagnetism: Geology, v. 16, p. 980-982, doi: 10.1130/0091-7613(1988)016<0980:SFIADS>2.3.CO;2.

Fleming, N., 1993, Calcium carbonate cementation of sandstones: Geology Today, v. 9, p. 223-226.

Folk, R. L., H. S. Chafetz, and P. A. Tiezzi, 1985, Bizarre forms of depositional and diagenetic calcite in hot-spring travertines, central Italy, *in* N. Schneidermann, and P. M. Harris, eds., Carbonate Cements: Tulsa, OK, SEPM Special Publication 36, p. 349-369, doi: 10.2110/pec.85.36.0349.

Friedman, G. M., 1959, Identification of carbonate minerals by staining methods: Journal of Sedimentary Research, v. 29, p. 87-97, doi: 10.1306/74D70894-2B21-11D7-8648000102C1865D.

Gallagher, S., 2014, Depositional and diagenetic controls on reservoir heterogeneity: Upper Morrow Sandstone, Farnsworth Unit, Ochiltree County, Texas: Master's thesis, New Mexico Institute of Mining and Technology, Socorro, 214 p.

Girard, J., 1998, Carbonate cementation in the Middle Jurassic Oseberg reservoir sandstone, Oseberg Field, Norway: A case of deep burial-high temperature poikilotopic calcite, *in* S. Morad, ed., Carbonate Cementation in Sandstones: Distribution Patterns and Geochemical Evolution (IAS Special Publication 26): Oxford, UK, Wiley-Blackwell, p. 285-307, doi: 10.1002/9781444304893.ch13.

Goldstein, R. H., 1986, Reequilibration of fluid inclusions in low-temperature calcium-carbonate cement: Geology, v. 14, p. 792-795, doi: 10.1130/0091-7613(1986)14<792:ROFIIL>2.0.CO;2.

Goldstein, R. H., 2001, Fluid inclusions in sedimentary and diagenetic systems: Lithos, v. 55, p. 159-193, doi: 10.1016/S0024-4937(00)00044-X.

Goudie, A. S., 1983, Calcrete, *in* A. S. Goudie, and K. Pye, eds., Chemical Sediments and Geomorphology: Precipitates and Residua in the Near-surface Environment: London, Academic Press, p. 93-132.

Guo, L., and R. Riding, 1998, Hot-spring travertine facies and sequences, Late Pleistocene, Rapolano Terme, Italy: Sedimentology, v. 45, p. 163-180, doi: 10.1046/j.1365-3091.1998.00141.x.

Hall, J. S., P. Mozley, J. M. Davis, and N. D. Roy, 2004, Environments of formation and controls on spatial distribution of calcite cementation in Plio-Pleistocene fluvial deposits, New Mexico, USA: Journal of Sedimentary Research, v. 74, p. 643-653, doi: 10.1306/020904740643.

Harun, N. T., 1996, Siderite in the Ivishak Sandstone, Prudhoe Bay, Alaska: Is it an indicator of an early burial environment?: AAPG Annual Meeting Abstracts, Search and Discover Article #91019.

Hendry, J. P., 2002, Geochemical trends and palaeohydrological significance of shallow burial calcite and ankerite cements in Middle Jurassic strata on the East Midlands Shelf (onshore UK): Sedimentary Geology, v. 151, p. 149-176, doi: 10.1016/S0037-0738(01)00236-6.

Hendry, J. P., N. H. Trein, and A. E. Fallick, 1996, Low-Mg calcite marine cement in Cretaceous turbidites: Origin, spatial distribution and relationship to seawater chemistry: Sedimentology, v. 43, p. 877-900, doi: 10.1002/9781444304459.ch5.

Hendry, J. P., M. Wilkinson, A. E. Fallick, and R. S. Haszeldine, 2000, Ankerite cementation in deeply buried Jurassic sandstone reservoirs of the central North Sea: Journal of Sedimentary Research, v. 70, p. 227-239, doi: 10.1306/2DC4090D-0E47-11D7-8643000102C1865D.

Hendry, J. P., M. Wilkinson, A. E. Fallick, and N. H. Trewin, 2000, Disseminated 'jigsaw piece' dolomite in Upper Jurassic shelf sandstones, Central North Sea: An example of cement growth during bioturbation?: Sedimentology, v. 47, p. 631-644, doi: 10.1046/j.1365-3091.2000.00319.x.

Hovland, M., M. R. Talbot, H. Qvale, S. Olaussen, and L. Aasberg, 1987, Methane-related carbonate cements in pockmarks of the North Sea: Journal of Sedimentary Research, v. 57, p. 881-892, doi: 10.1306/212F8C92-2B24-11D7-8648000102C1865D.

Kantorowicz, J. D., 1985, The origin of authigenic ankerite from the Ninian Field, North Sea: Nature, v. 315, p. 214-216, doi: 10.1038/315214a0.

Kantorowicz, J. D., I. D. Bryant, and J. M. Dawans, 1987, Controls on the geometry and distribution of carbonate cements in Jurassic sandstones: Bridport Sands, southern England and Viking Group, Troll Field, Norway, *in* J. D. Marshall, ed., Diagenesis of Sedimentary Sequences: London, GSL Special Publication 36, p. 103-118, doi: 10.1144/GSL.SP.1987.036.01.09.

Kauffman, E. G., M. A. Arthur, B. Howe, and P. A. Scholle, 1996, Widespread venting of methane-rich fluids in Late Cretaceous (Campanian) submarine springs (Tepee Buttes), Western Interior seaway, U.S.A.: Geology, v. 24, p. 799-802, doi: 10.1130/0091-7613(1996)024<0799:WVOMRF>2.3.CO;2.

Kirkland, D. W., R. E. Denison, and M. A. Rooney, 1995, Diagenetic alteration of Permian strata at oil fields of south central Oklahoma, USA: Marine and Petroleum Geology, v. 12, p. 629-644, doi: 10.1016/0264-8172(95)98089-N.

Lindholm, R. C., 1974, Fabric and chemistry of pore filling calcite in septarian veins: Models for limestone cementation: Journal of Sedimentary Research, v. 44, p. 428-440, doi: 10.1306/74D72A4A-2B21-11D7-8648000102C1865D.

Ludvigson, G. A., D. A. Fowle, J. A. Roberts, S. G. Driese, M. A. Villareal, J. J. Smith, and M. B. Suarez, 2013, Paleoclimatic applications and modern process studies of pedogenic siderite, *in* S. G. Driese, and L. C. Nordt, eds., New Frontiers in Paleopedology and Terrestrial Paleoclimatology: Tulsa, OK, SEPM Special Publication 104, p. 79-87, doi: 10.2110/sepmsp.104.01.

Lynch, F. L., and L. S. Land, 1996, Diagenesis of calcite cement in Frio Formation sandstones and its relationship to formation water chemistry: Journal of Sedimentary Research, v. 66, p. 439-446, doi: 10.1306/D426836A-2B26-11D7-8648000102C1865D.

Macaulay, C. I., A. E. Fallick, R. S. Haszeldine, and G. E. McAulay, 2000, Oil migration makes the difference: Regional distribution of carbonate cement $\delta^{13}C$ in northern North Sea Tertiary sandstones: Clay Minerals, v. 35, p. 69-76, doi: 10.1180/000985500546738.

Macaulay, C. I., R. S. Haszeldine, and A. E. Fallick, 1993, Distribution, chemistry, isotopic composition and origin of diagenetic carbonates: Magnus Sandstone, North Sea: Journal of Sedimentary Research, v. 63, p. 33-43, doi: 10.1306/D4267A82-2B26-11D7-8648000102C1865D.

Machel, H. G., 1987, Saddle dolomite as a by-product of chemical compaction and thermochemical sulfate reduction: Geology, v. 15, p. 936-940, doi: 10.1130/0091-7613(1987)15<936:SDAABO>2.0.CO;2.

Machel, H. G., 2004, Concepts and models of dolomitization: A critical reappraisal, *in* C. J. R. Braithwaite, G. Rizzi, and G. Darke, eds., The Geometry and Petrogenesis of Dolomite Hydrocarbon Reservoirs: London, GSL Special Publication 235, p. 7-63, doi: 10.1144/GSL.SP.2004.235.01.02.

Machemer, S. D., and I. Hutcheon, 1988, Geochemistry of early carbonate cements in the Cardium Formation, central Alberta: Journal of Sedimentary Research, v. 58, p. 136-147, doi: 10.1306/212F8D37-2B24-11D7-8648000102C1865D.

Mazzullo, S. J., 1992, Geochemical and neomorphic alteration of dolomite: A review: Carbonates and Evaporites, v. 7, p. 21-37, doi: 10.1007/BF03175390.

McKay, J. L., F. J. Longstaffe, and A. G. Plint, 1995, Early diagenesis and its relationship to depositional environment and relative sea-level fluctuations (Upper Cretaceous Marshybank Formation, Alberta and British Columbia): Sedimentology, v. 42, p. 161-190, doi: 10.1111/j.1365-3091.1995.tb01276.x.

Meyers, W. J., 1991, Calcite cement stratigraphy: An overview, *in* C. E. Barker, and O. C. Kopp, eds., Luminescence Microscopy and Spectroscopy: Qualitative and Quantitative Applications: Tulsa, OK, SEPM Short Course 25, p. 133-148, doi: 10.2110/scn.91.25.0133.

Milliken, K. L., 2002, Petrography of ankerite cement, grain replacement, and fracture fill in foreland sandstones of the central Rocky Mountains: AAPG Annual Meeting Abstracts 2002 AAPG Search and Discovery Article #90007©2002.

Minoura, K., S. Nakaya, and A. Takemura, 1991, Origin of manganese carbonates in Jurassic red shale, central Japan: Sedimentology, v. 38, p. 137-152, doi: 10.1111/j.1365-3091.1991.tb01859.x.

Molenaar, N., 1990, Calcite cementation in shallow marine Eocene sandstones and constraints of early diagenesis: Journal of the Geological

Society, v. 147, p. 759-768, doi: 10.1144/gsjgs.147.5.0759.

Morad, S., 1990, Mica alteration reactions in Jurassic reservoir sandstones from the Haltenbanken area, offshore Norway: Clays and Clay Minerals, v. 38, p. 584-590, doi: 10.1346/CCMN.1990.0380603.

Morad, S., 1992, The role of mixing-zone dolomitization in sandstone cementation: Evidence from the Triassic Buntsandstein, the Iberian Range, Spain: Sedimentary Geology, v. 80, p. 53-65, doi: 10.1016/0037-0738(92)90031-L.

Morad, S., 1998, Carbonate cementation in sandstones: Controls on distribution patterns and geochemical evolution, *in* S. Morad, ed., Carbonate Cementation in Sandstones: Distribution Patterns and Geochemical Evolution (IAS Special Publication 26): Oxford, UK, Wiley-Blackwell, p. 1-26, doi: 10.1002/9781444304893.ch1.

Morad, S., K. Al-Ramadan, J. M. Ketzer, and L. F. De Ros, 2010, The impact of diagenesis on the heterogeneity of sandstone reservoirs: A review of the role of depositional facies and sequence stratigraphy: AAPG Bulletin, v. 94, p. 1267-1309, doi: 10.1306/04211009178.

Mozley, P. S., 1989, Relation between depositional environment and the elemental composition of early diagenetic siderite: Geology, v. 17, p. 704-706, doi: 10.1130/0091-7613(1989)017<0704:RBDEAT>2.3.CO;2.

Mozley, P. S., and J. M. Davis, 2005, Internal structure and mode of growth of elongate calcite concretions: Evidence for small-scale, microbially induced, chemical heterogeneity in groundwater: GSA Bulletin, v. 117, p. 1400-1412, doi: 10.1130/B25618.1.

Mozley, P. S., and K. Hoernle, 1990, Geochemistry of carbonate cements in the Sag River and Shublik Formations (Triassic/Jurassic), North Slope, Alaska: Implications for the geochemical evolution of formation waters: Sedimentology, v. 37, p. 817-836, doi: 10.1111/j.1365-3091.1990.tb01827.x.

Mozley, P. S., and P. Wersin, 1992, Isotopic composition of siderite as an indicator of depositional environment: Geology, v. 20, p. 817-820, doi: 10.1130/0091-7613(1992)020<0817:ICOSAA>2.3.CO;2.

Nash, A. J., and E. D. Pittman, 1975, Ferro-magnesian calcite cement in sandstones: Journal of Sedimentary Research, v. 45, p. 258-265, doi: 10.1306/212F6D2A-2B24-11D7-8648000102C1865D.

Pentecost, A., 1990, The formation of travertine shrubs: Mammoth Hot Spring, Wyoming: Geological Magazine, v. 127, p. 159-168, doi: 10.1017/S0016756800013844.

Pentecost, A., 2005, Travertine: Berlin, Springer, 445 p.

Pitman, J. K., and C. Spötl, 1996, Origin and timing of carbonate cements in the St. Peter Sandstone, Illinois Basin: Evidence for a genetic link to Mississippi Valley-type mineralization, *in* L. J. Crossey, R. Loucks, and M. W. Totten, eds., Siliciclastic Diagenesis and Fluid Flow: Tulsa, OK, SEPM Special Publication 55, p. 187-203, doi: 10.2110/pec.96.55.0187.

Prosser, D. J., J. A. Daws, A. E. Fallick, and B. P. J. Williams, 1993, Geochemistry and diagenesis of stratabound calcite cement layers within the Rannoch Formation of the Brent Group, Murchison Field, North Viking Graben (northern North Sea): Sedimentary Geology, v. 87, p. 139-164, doi: 10.1016/0037-0738(93)90002-M.

Pye, K., and D. H. Krinsley, 1986, Diagenetic carbonate and evaporite minerals in Rotliegend aeolian sandstones of the Southern North Sea: Their nature and relationship to secondary porosity development: Clay Minerals, v. 21, p. 443-457, doi: 10.1180/claymin.1986.021.4.03.

Radke, B. M and R. L. Mathis, 1980, On the formation and occurrence of saddle dolomite: Journal of Sedimentary Research, v. 50, p. 1149-1168, doi: 10.1306/212F7B9E-2B24-11D7-8648000102C1865D.

Reeder, R. J., and W. A. Dollase, 1989, Structural variation in the dolomite-ankerite solid-solution series: An X-ray, Mössbauer, and TEM study: American Mineralogist, v. 74, p. 1159-1167.

Roberts, H. H., P. Aharon, and M. M. Walsh, 1993, Cold-seep carbonates of the Louisiana continental slope-to-basin floor, *in* R. Rezak, and D. L. Lavoie, eds., Carbonate Microfabrics: New York: Springer-Verlag, p. 95-104, doi: 10.1007/978-1-4684-9421-1_7.

Rossi, C., R. Marfil, K. Ramseyer, and A. Permanyer, 2001, Facies-related diagenesis and multiphase siderite cementation and dissolution in the reservoir sandstones of the Khatatba Formation, Egypt's Western Desert: Journal of Sedimentary Research, v. 71, p. 459-472, doi: 10.1306/2DC40955-0E47-11D7-8643000102C1865D.

Saigal, G. C., and K. Bjørlykke, 1987, Carbonate cements in clastic reservoir rocks from offshore Norway—relationships between isotopic composition, textural development and burial depth, *in* J. D. Marshall, ed., Diagenesis of Sedimentary Sequences: London, GSL Special Publication 36, p. 313-324, doi: 10.1144/GSL.SP.1987.036.01.22.

Sample, J. C., M. R. Reid, H. J. Tobin, and J. C. Moore, 1993, Carbonate cements indicate channeled fluid flow along a zone of vertical faults at the deformation front of the Cascadia accretionary wedge (northwest U.S. coast): Geology, v. 21, p. 507-510, doi: 10.1130/0091-7613(1993)021<0507:CCICFF>2.3.CO;2.

Searl, A., 1989, Saddle dolomite: A new view of its nature and origin: Mineralogical Magazine, v. 53, p. 547-555, doi: 10.1180/minmag.1989.053.373.05.

Souza, R. S. de, L. F. De Ros, and S. Morad, 1995, Dolomite diagenesis and porosity preservation in lithic reservoirs: Carmópolis Member, Sergipe-Alagoas Basin, northeastern Brazil: AAPG Bulletin, v. 79, p. 725-748.

Spötl, C., and J. K. Pitman, 1998, Saddle (baroque) dolomite in carbonates and sandstones: A reappraisal of a burial-diagenetic concept, *in* S. Morad, ed., Carbonate Cementation in Sandstones. (IAS Special Publication 26): Oxford, UK, Wiley-Blackwell, p. 437-460, doi: 10.1002/9781444304893.ch19.

Taylor, K. G., and R. L. Gawthorpe, 2003, Basin-scale dolomite cementation of shoreface sandstones in response to sea-level fall: GSA Bulletin, v. 115, p. 1218-1229, doi: 10.1130/B25227.1.

Taylor, K. G., R. L. Gawthorpe, C. D. Curtis, J. D. Marshall, and D. N. Awwiller, 2000, Carbonate cementation in a sequence-stratigraphic framework: Upper Cretaceous sandstones, Book Cliffs, Utah-Colorado: Journal of Sedimentary Research, v. 70, p. 360-372, doi: 10.1306/2DC40916-0E47-11D7-8643000102C1865D.

Vasconcelos, C., and J. A. McKenzie, 1997, Microbial mediation of modern dolomite precipitation and diagenesis under anoxic conditions (Lagoa Vermelha, Rio De Janeiro, Brazil): Journal of Sedimentary Research, v. 67, p. 378-390, doi: 10.1306/D4268577-2B26-11D7-8648000102C1865D.

Walderhaug, O., and P. A. Bjørkum, 1998, Calcite cement in shallow marine sandstones: Growth mechanisms and geometry, *in* S. Morad, ed., Carbonate Cementation in Sandstones: Distribution Patterns and Geochemical Evolution (IAS Special Publication 26): Oxford, UK, Wiley-Blackwell, p. 179-192, doi: 10.1002/9781444304893.ch8.

Watson, R. S., N. H. Trewin, and A. E. Fallick, 1995, The formation of carbonate cements in the Forth and Balmoral fields, northern North Sea: A case for biodegradation, carbonate cementation and oil leakage during early burial, *in* A. Hartle, and J. Prosser, eds., Characterization of Deep Marine Clastic Systems: London, GSL Special Publication 94, p. 177-200, doi: 10.1144/GSL.SP.1995.094.01.13.

Watts, N. L., 1978, Displacive calcite: Evidence from recent and ancient calcretes: Geology, v. 6, p. 699-703, doi: 10.1130/0091-7613(1978)6<699:DCEFRA>2.0.CO;2.

Wilkinson, B. H., and R. K. Given, 1986, Secular variation in abiotic marine carbonates: Constraints on Phanerozoic atmospheric carbon dioxide contents and oceanic Mg/Ca ratios: Journal of Geology, v. 94, p. 321-333, doi: 10.1086/629032.

Wilkinson, M., R. S. Haszeldine, A. E. Fallick, and M. J. Osborne, 2000, Siderite zonation within the Brent Group: Microbial influence or aquifer flow?: Clay Minerals, v. 35, p. 107-117, doi: 10.1180/000985500546512.

Wright, V. P., and M. E. Tucker, 1991, Calcretes: An introduction, *in* V. P. Wright, and M. E. Tucker, eds., Calcretes (International Association of Sedimentologists Reprint Series 2): Oxford, UK, Blackwell Publishing Ltd., p. 1-22, doi: 10.1002/9781444304497.ch.

Facing Page: Top — Bedding-plane view of halite crystal casts, Permian Yeso Fm., Socorro Co., New Mexico. Bottom — Anhydrite nodule, Triassic Sherwood Sandstone Gp., 1,174m (3,853 ft) depth, County Antrim, Northern Ireland (XPL).

DIAGENESIS: Sulfate & Halide Cements and Authigenic Precipitates

CHAPTER 16

INTRODUCTION TO SULFATE AND HALIDE CEMENTS

Sulfates and halides can form significant cements in sandstones (Table 11.2, page 247) but typically only in very specific settings. Early formed gypsum, anhydrite and halite are common in arid, highly evaporitic settings and form not only cements but also bedded deposits and displacive or replacive precipitates. Additional sulfates and halides, such as glauberite (a sodium calcium sulfate), barite ($BaSO_4$), celestine ($SrSO_4$), sylvite (KCl) and many others, can form in surficial settings, but most such minerals are rare and may not survive subsequent diagenetic alterations because of their generally high solubility. Eogenetic sulfates and halides are especially common precipitates in arid-region playa lakes, dunes and interdune flats, as well as continental and coastal sabkhas (e.g., Kinsman, 1969; Benison and Goldstein, 2000; Warren, 2006). Gypsum is the most common of these minerals, and extensive gypsum cementation, mostly within 10 m (33 ft) of the ground surface, can produce gypsiferous soils, gypsum crusts and fully cemented gypcretes (Watson, 1985; Nettleton, 1991; Hartley and May, 1998). Gypsum cements in terrigenous strata can consist of coarsely poikilotopic crystals (encompassing numerous detrital sand grains), isolated lenticular crystals or crystal clusters (desert roses), gypsarenites with graded cement sizes and alabastrine gypsum with individual crystallites that average less than 50 μm in diameter (Watson, 1985). Although gypsum is the major precipitate in such surficial deposits, in settings with especially hot, arid, evaporitic and low humidity conditions, anhydrite (anhydrous calcium sulfate) and even halite can be found cementing surficial and near-surficial sands and muds, especially in sabkhas and saline pans (Kinsman, 1969; Casas and Lowenstein, 1989; Warren, 2006).

Complete cementation, even with minerals as soluble as halite or sylvite, can occur in such settings with infiltration of surficial brines to depths between 10 and 40 m (33-132 ft). Below those depths, evaporite cements are unlikely to be dissolved in arid settings; thus, ancient evaporite-cemented deposits from such settings are not uncommon (Casas and Lowenstein, 1989; Benison and Goldstein, 2000). Deeper-water settings can also be sites of eogenetic sulfate mineral precipitation, with barite and celestine formation reported as nodules, crystal clusters and cements in surficial to shallowly buried marine sediments (e.g., Baker and Bloomer, 1988; Paytan et al., 2002). In this context, it should be noted that gypsum and anhydrite are easily, in some cases repeatedly, transformed with changes in temperature and water chemistry. Thus, what may be initially formed as gypsum cement, may transform to anhydrite, in surficial settings and especially during burial. Conversely, anhydrite can convert to gypsum during wetter climatic periods or telogenetic uplift.

Considerable cementation by sulfate and halide minerals can take place in the transition between eogenesis and mesogenesis, mainly through the process of brine reflux. This process involves density-driven brines from mesosaline and hypersaline surface water bodies (ranging in size from small salinas to the dessicated Late Miocene Mediterranean Sea) percolating laterally and downward through connected pore networks, fractures and faults. Such fluids can precipitate a wide variety of diagenetic minerals, especially dolomite, but also gypsum, anhydrite and halite (Jones and Xiao, 2005). Those precipitates may take the form of cements, displacive nodules or mineral replacements in sandstones as well as limestones through which the brines pass.

Under mesogenetic conditions, patterns of cementation by sulfates and halides are both more diverse and more complex, and some additional minerals, such as fluorite (CaF_2), can be introduced in that setting. Because of their solubility, eogenetic evaporite minerals (whether as early cements or bedded evaporite deposits) are subject to alteration and remobilization during burial. Therefore, considerable mesogenetic sulfate and halide cementation takes place in sandstones proximal to evaporite-rich carbonate rocks or bedded evaporites. However, basin-scale water movements and hydrothermal processes can transport sulfate- or chloride-rich fluids into virtually any clastic terrigenous deposits, regardless of location. In that context, a very major part of subsurface remobilization of sulfate is water expulsion associated with the dehydration of gypsum to anhydrite as a function of increased temperatures and pressures at depth. Predicting the depth of that transformation is complex, however, because heat flow, tectonic factors and thermal conductivity of surrounding units all can play major roles; thus, the depth can vary from as little as 400 m (1,300 ft) to as much as 4 km (13,000 ft) (Jowett et al., 1993).

Subsurface diagenesis involving local dissolution, aqueous transport and subsequent reprecipitation produces porosity "winners" and "losers", with some rocks gaining porosity from removal of earlier cements (e.g., Strong, 1993; Binh et al., 2007) and others losing porosity from precipitation of additional sulfate or halide cements (e.g., McBride et al., 1987). Those processes, both dissolution and precipitation, can take place in either shallow or deep settings. Relatively early precipitation of pore-filling anhydrite cements was seen by James (1992) in Pennsylvanian units in the Rocky Mountains; on the other hand, late precipitation of evaporite cements was seen in the Triassic of the North Sea (Binh, et al., 2013) and the Jurassic of the Gulf Coast (McBride et al., 1987). The latter study demonstrated anhydrite cementation at temperatures of 90° and 100°C (approximately 2.3 km [7,500 ft] depth) and halite cementation at temperatures below 120°C (< 3 km [10,000 ft]). On the dissolution side, relatively early removal of carbonate and evaporite cements was inferred in the study by Strong (1993), whereas very late-stage dissolution of halite was seen in the study by Binh et al. (2013). It is important to note in this diversity of results that evaporite cements, although porosity destructive when they form, can have positive impacts on ultimate reservoir performance. By blocking other pore-filling cements from forming, these minerals allow the possibility of later cement removal and porosity resurrection during mesogenesis or telogenesis.

Finally, hydrothermal alteration (focused fluid flow associated with high salinity and/or metal-rich brines generally at elevated temperatures) is important in some subsurface settings. Many sulfate and halide minerals, especially barite, celestine and fluorite, can be precipitated through such processes (e.g., Breit et al., 1990; Sullivan et al., 1994; Hartley and May, 1998; Williams-Jones et al., 2000; Paytan et al., 2002). None of the three above-named minerals is an especially common cement in sandstones or mudstones, but each can be important locally, and their recognition can be important in identifying hydrothermal activity (although not all barite or celestine is of hydrothermal origin). They can occur as disseminated cements as well as vug- or vein-filling precipitates.

Optical properties

Many of the minerals discussed here, especially anhydrite, halite and other evaporite minerals are so soluble that special thin-section preparation techniques (primarily cutting and grinding in oil) are necessary to retain those minerals in thin sections.

Gypsum is most commonly found as large, generally displacive, discoidal or twinned crystals or as intergrown, anhedral, poikilotopic cements that incorporate or engulf multiple detrital grains. Gypsum is also found as nodular masses. It is monoclinic, has pronounced cleavage, is typically colorless in thin section, has low birefringence (just slightly higher than quartz), low relief, and is biaxial (+). It can show somewhat mottled birefringence where partial dissolution or selective "erosion" has taken place during thin section preparation.

Anhydrite typically is found as displacive, nodular masses of tiny cleavage fragments, as coarser crystalline, equant to lath-like pore-filling crystals, and as poikilotopic cements. It is orthorhombic, has two near-perfect cleavages and one good cleavage, is typically colorless in thin section, has high (third-order) birefringence, low relief, and is biaxial (+).

Barite (or baryte) often is found in displacive nodules or rosettes but also forms bedded deposits (Brodtkorb et al., 1982), replacements, vein fillings and cements both in early- and late-diagenetic settings. It is orthorhombic, has three cleavages (one perfect, one imperfect and one poor), is typically colorless in thin section, has low birefringence (first-order, but slightly higher than quartz or gypsum), moderate relief, and is biaxial (+).

Celestine (formerly celestite) often is found in displacive nodules or rosettes but also forms bedded deposits, replacements and cements. It is found in shallow- and deeper-water deposits especially in evaporative or brine-reflux settings with Sr derived directly from seawater or from dissolution of aragonite or other Sr-bearing minerals (Scholle et al., 1990; Hanor, 2000 and 2004). It is most common in carbonate and evaporite strata but can be found cementing clastic terrigenous deposits as well (e.g., Dutton and Land, 1985). It is orthorhombic, has three cleavages (one perfect, one good and one poor), is typically colorless in thin section, has low birefringence (first-order, essentially identical to quartz and slightly lower than gypsum), moderate relief, and is biaxial (+). It often occurs as tabular to lath-like crystals or rosettes with radiating crystal structure.

Halite is common, but not abundant, as a sandstone cement and is somewhat under-recognized because of extensive or even complete dissolution unless special care is taken with sample and thin-section preparation. It is normally colorless in thin section and is a cubic (and thus isotropic) mineral; both of those factors contribute to its difficulties in recognition. It has low relief and is usually recognizable by the presence of perfect cleavage lines and fluid inclusion trains in what might otherwise be taken for open pore space. Impregnation of thin sections with dyed resins helps in identification, because actual porosity should be strongly colored where dyes are present.

Fluorite is a fairly rare cement in sandstones, but recognition is important as an indicator of hydrothermal activity. Fluorite, like halite, is a cubic, and thus an isotropic mineral (although it can, in rare cases, exhibit weak, anomalous birefringence). It has moderate relief and perfect cleavages, ranges from colorless to fairly strongly colored (purple, yellow or green) and sometimes can show color zonation. Unlike halite, fluorite is not highly soluble and thus is not likely to be lost during sample preparation. Like halite, however, it is most easily recognized by the presence of cleavage and fluid inclusion traces along with its cubic crystal outlines and possible coloration.

Mid. Permian Queen Fm., Lea Co., New Mexico

Poikilotopic gypsum cement fills the porosity within this sandstone. These gypsum crystals have low negative relief and low birefringence (gray to white). In cross-polarized light, they can have a mottled appearance due to dissolution during thin section preparation. Because of gypsum's solubility, reservoirs that are cemented early have a higher potential for porosity resurrection. In this rock, there may have already been some dissolution of the gypsum (and the gypsum in this outcrop sample almost certainly was anhydrite prior to uplift and erosion).

PPL | XPL, BDI, Scale bar = 0.11 mm

Cretaceous Mowry Fm., Moffat Co., Colorado

This feldspathic arenite is cemented by both poikilotopic gypsum and calcite cements. The calcite cements appear to be forming at the expense of the gypsum cements during diagenesis. If the poikilotopic gypsum is later dissolved, secondary porosity is created. This type porosity easily can be misidentified later as primary porosity if no remnants of the cements are left.

XPL, RDI, Scale bar = 0.26 mm

Mid. Jurassic Carmel Fm., Emery Co., Utah

A sandstone with poikilotopic gypsum cements (the gray areas with distinct cleavage, low relief and numerous inclusions or surface corrosion). The gypsum cements postdate earlier authigenic quartz overgrowths and are either contemporaneous with or predate coarse calcite cements (best seen at top and right). The source of the sulfate for these gypsum cements is thought to be reflux of brines from marginal marine sabkhas or salinas in overlying strata.

XPL, RDI, Scale bar = 0.10 mm

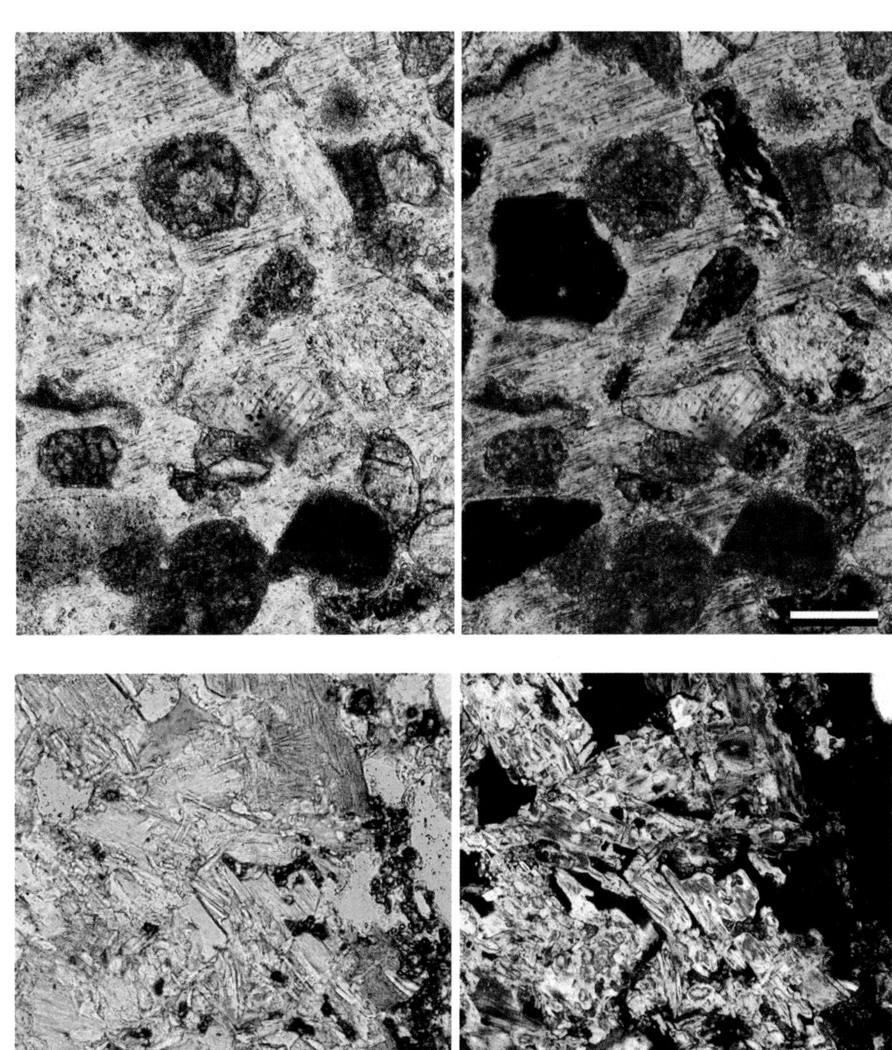

Holocene sabkha sands, Abu Dhabi, United Arab Emirates

Sandstones on the landward margins of supratidal sabkhas in Abu Dhabi are widely cemented by poikilotopic gypsum. The yellow-red birefringent cement is a single crystal of gypsum (the birefringence is slightly high due to the >30 μm thickness of this thin section). The original evaporites formed as a result of storm washover and evaporation on the low-relief sabkha flats and were later dissolved by groundwater and reprecipitated as coarse poikilotopic cements (Kinsman, 1969 and 1974).

PPL | XPL, Scale bar = 0.10 mm

Recent hot spring tufa, Park Co., Wyoming

A calcium sulfate-cemented lithic sandstone associated with a hot-spring. The sulfate here was generated by sulfidic gases (H_2S and SO_2) interacting with Ca-rich waters to precipitate gypsum and anhydrite (both appear present in these high-temperature deposits). The yellow areas are a sulfur-stained isotropic material (opal?). Hot springs and fumaroles provide an alternative to evaporite-derived sulfates, at least for localized cementation. A similar process is often invoked for large-scale carbonate dissolution (cave formation) and gypsum precipitation using H_2S leakage from hydrocarbon-rich basins (e.g., Hill, 1995).

PPL | XPL, BDI, Scale bar = 0.10 mm

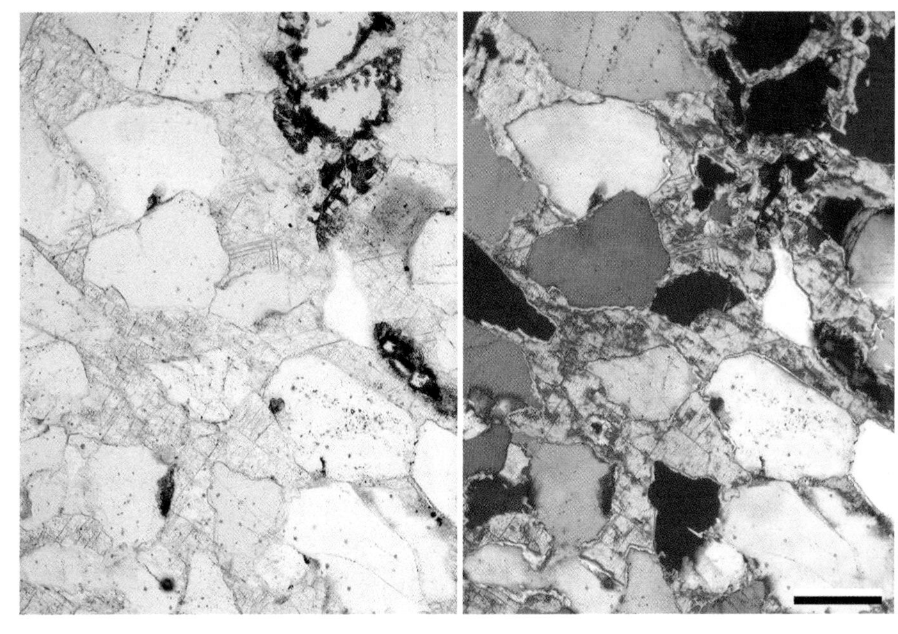

Devonian Jubah Fm., Qatar †

Large crystals of anhydrite with typical bright second-order blue birefringence. Another crystal (top left) shows lower, gray to yellow, birefringence. The range of birefringence will vary with the orientation and thickness of the mineral under the microscope. The edges of the blue anhydrite show this gradation well, as they thin against detrital grains. In this sample the two cleavages at 90° are prominent, and narrow twins, similar to those in calcite, also are visible. The corroded margins of the quartz grains may suggest that there was a precursor calcite cement, although anhydrite embayment of quartz is possible (Schenk and Richardson, 1985).

PPL | XPL, KFS, BDI, Scale bar = 0.06 mm

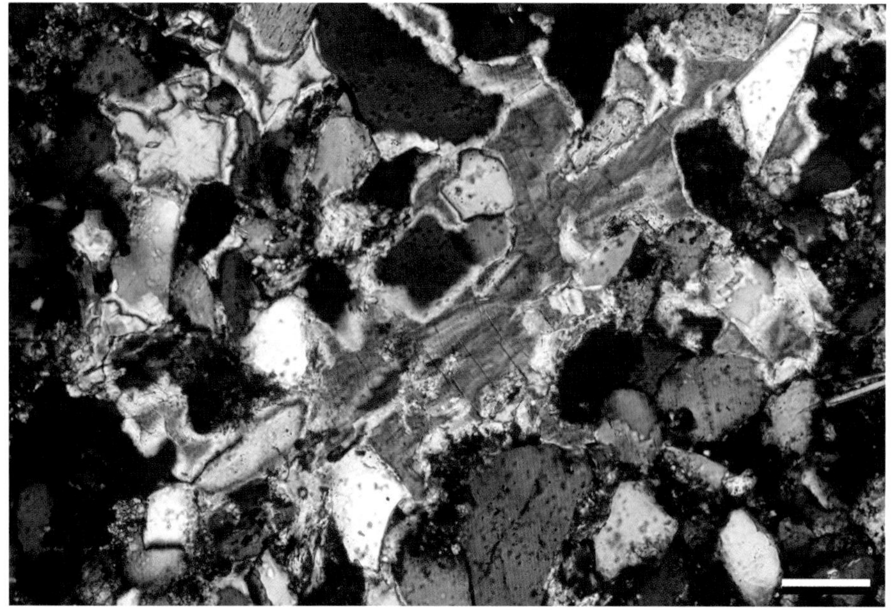

Mid. Permian Queen Fm., Lea Co., New Mexico

Most of the porosity within this sandstone is occluded by poikilotopic anhydrite cement. Anhydrite is easy to distinguish from gypsum because anhydrite has higher birefringence (usually second- to third-order colors) and gypsum has low (first-order) birefringence. Anhydrite may exhibit pseudo-cubic cleavages; in this sample, however, only one cleavage direction is visible. Even more than for gypsum, preservation of anhydrite requires thin section preparation using oil. Creation of secondary porosity by anhydrite dissolution during later diagenesis is common because of its solubility.

XPL, Scale bar = 0.11 mm

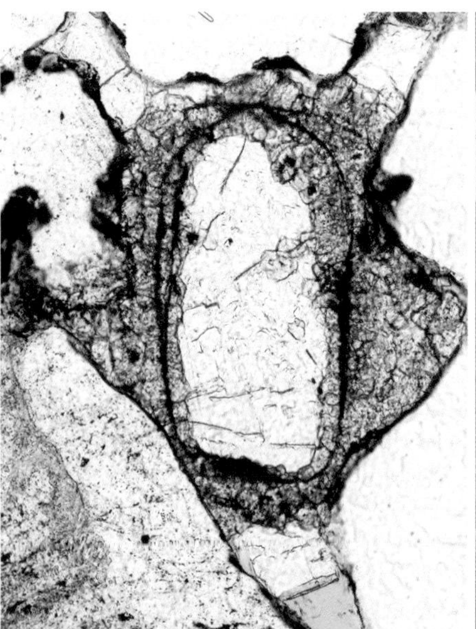

Triassic Sherwood Sandstone Gp., Co. Antrim, Northern Ireland, U.K. †

An example of anhydrite filling primary and secondary dissolution porosity. All the detrital grains are outlined by a hematite dust line. The grain at the center of the image most likely was replaced along its edges by nonferroan dolomite that also filled some surrounding pore space. The rest of the central grain was later dissolved and then filled by anhydrite cement (with additional anhydrite precipitating in remnant primary pores). This is a subsurface sample; finding anhydrite in outcrop samples is rare unless the material is from very arid areas.

PPL | XPL, AFeS, KFS, Scale bar = 0.06 mm

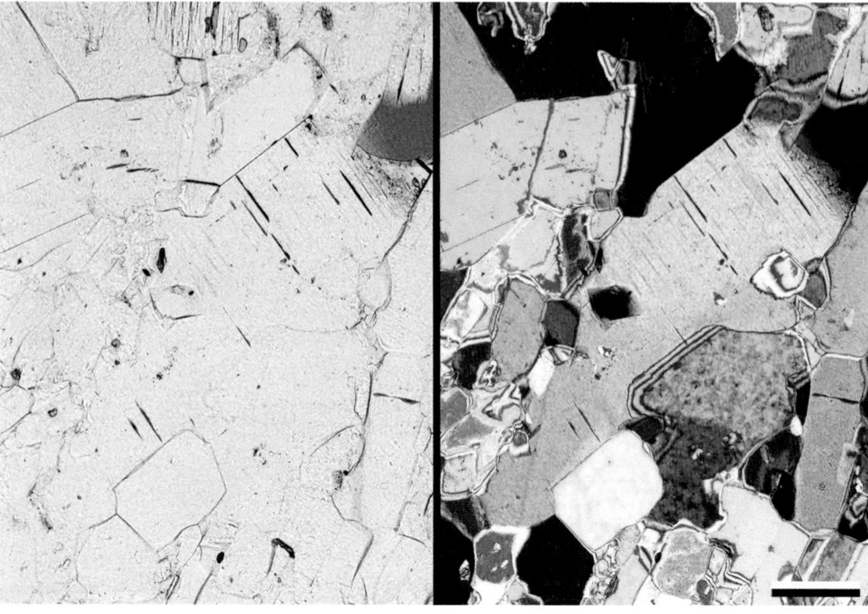

Mid. Permian Queen Fm., Lea Co., New Mexico

This fracture fill contains both anhydrite and halite. The anhydrite crystals are birefringent, whereas halite is isotropic (no birefringence). The cleavages within the anhydrite are clearly visible because they are marked by the presence of numerous minute, dark inclusions. Like gypsum, anhydrite commonly has a mottled appearance in cross-polarized light due to some dissolution during thin-section grinding. The source of these cements was undoubtedly reflux of brines from the sabkhas and salinas that produced massive bedded evaporites in immediately overlying units.

PPL | XPL, BDI, Scale bar = 0.14 mm

Lo. Cretaceous Marnes Bleues, Alpes-de-Haute-Provence, France

This early barite concretion is from an organic-rich, basinal black shale. The felted mass of barite crystals can easily be confused with gypsum or celestine. Barite has higher relief, a smaller 2V and parallel extinction, whereas gypsum has an inclined extinction and celestine has lower relief. Barite is common in anoxic marine sediments, and the occurrence of barite layers may mark changes in redox conditions and/or interruptions in sedimentation.

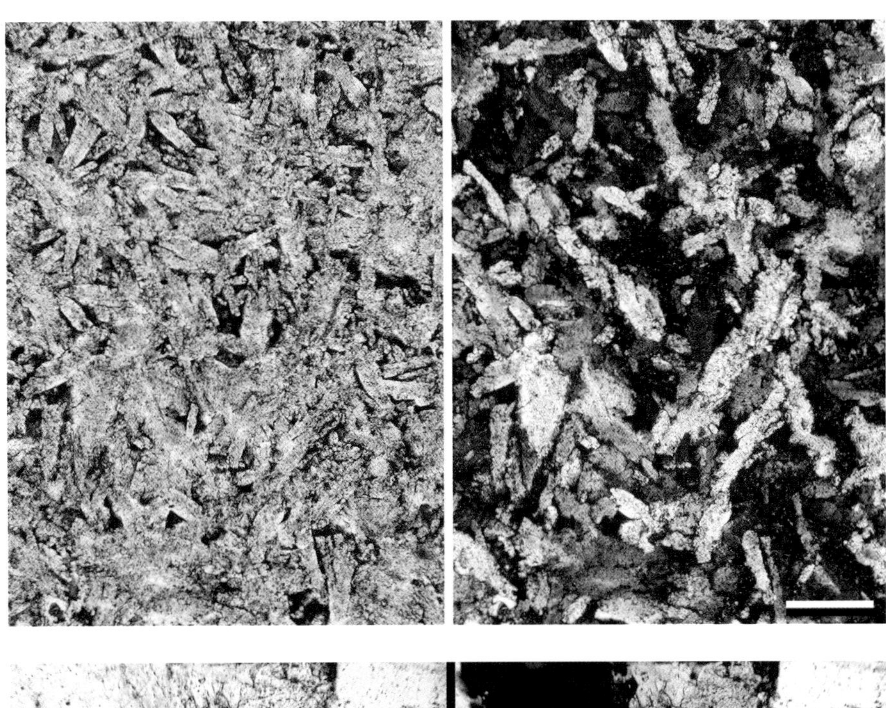

PPL | XPL, BDI, Scale bar = 0.26 mm

Up. Pennsylvanian Bursum Fm., Socorro Co., New Mexico

This example shows a feldspathic arenite that is cemented by a combination of barite and quartz. The barite cement has higher relief and is more inclusion-rich than the quartz cement. Relative timing of the cements appear to be that the quartz overgrowths formed first and were followed by the barite. Barite can form during early to late diagenesis in a variety of environments. In this sample, the barite cements likely formed as part of a low- to moderate-temperature hydrothermal system.

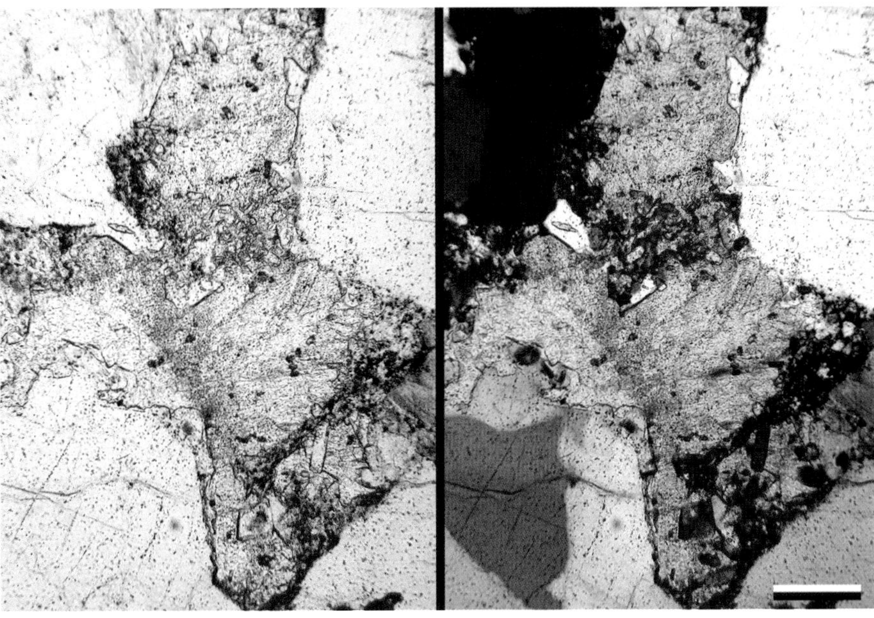

PPL | XPL, BDI, Scale bar = 0.09 mm

Up. Jurassic Ula Fm., Norwegian sector, North Sea †

Barite can originate from cold seep waters, but in most sandstone reservoirs it suggests the presence of warmer fluids. The barite here is in the form of relatively large, poikilotopic crystals that enclose quartz grains. The large area of porosity into which the barite precipitated may be the result of dilation along fractures or perhaps reflects grain dispersal through crystallization forces during relatively early diagenesis (especially given the absence of other cements). Barite shows a range of dull gray birefringence, high relief, (which distinguishes it from gypsum) some twinning (as in this example) and two cleavages.

XPL, KFS, Scale bar = 0.12 mm

Mississippian Dartry Limestone Fm., Co. Sligo, Ireland

A large fracture filled by elongate splays of barite crystals. Barite can be confused with quartz in thin section because of their similar birefringences and the fact that both can show hexagonal crystal morphologies. However, barite is biaxial, and quartz is uniaxial. Elongate crystals like the ones in this photomicrograph can also be misidentified as celestine and x-ray examination is recommended for absolute certainty in identification. These barite crystals formed in a hydrothermal vein in a limestone.

XPL, Scale bar = 0.51 mm

Mid. Permian Wegener Halvø Fm., Jameson Land, East Greenland

A view of the edge of a fracture in a lithic sandstone. Barite and saddle dolomite are the main precipitates in this vein, but galena and fluorite are found as cements or replacements in nearby strata, making a hydrothermal origin of these cements likely. The barite here is a mix of large crystals forming rosette clusters and much smaller, equant, anhedral crystals or crystal fragments.

XPL, BDI, Scale bar = 0.51 mm

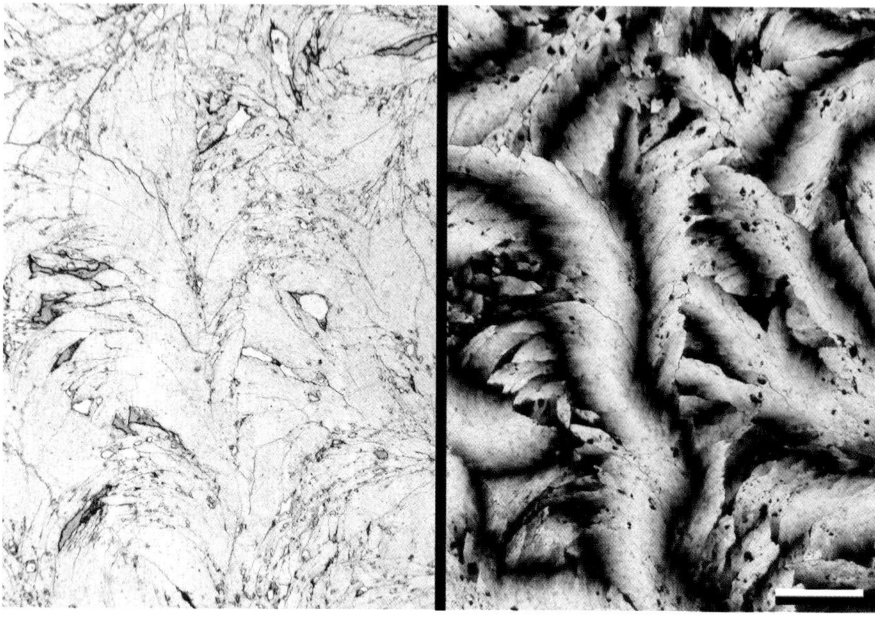

Mid. Permian Wegener Halvø Fm., Jameson Land, East Greenland

A fracture-filling barite cement with a distinctive feathery or plumose crystal pattern. The crystals have a strongly undulose or sweeping extinction. Although the specific significance of this texture is unclear, this is part of the same hydrothermal deposit discussed in the previous image.

PPL | XPL, BDI, Scale bar = 0.51 mm

Devonian Jauf Fm., Qatar †

A view of pore-filling barite that formed of a cluster of small, subhedral crystals. The barite is closely associated with, and is overlain by, illite clay. The image was taken with a mix of SE and BSE detectors, and so the barite is not as bright as it can look under purely back-scatter imaging (normally almost white). The illite overlying it has also acted to lessen the brightness.

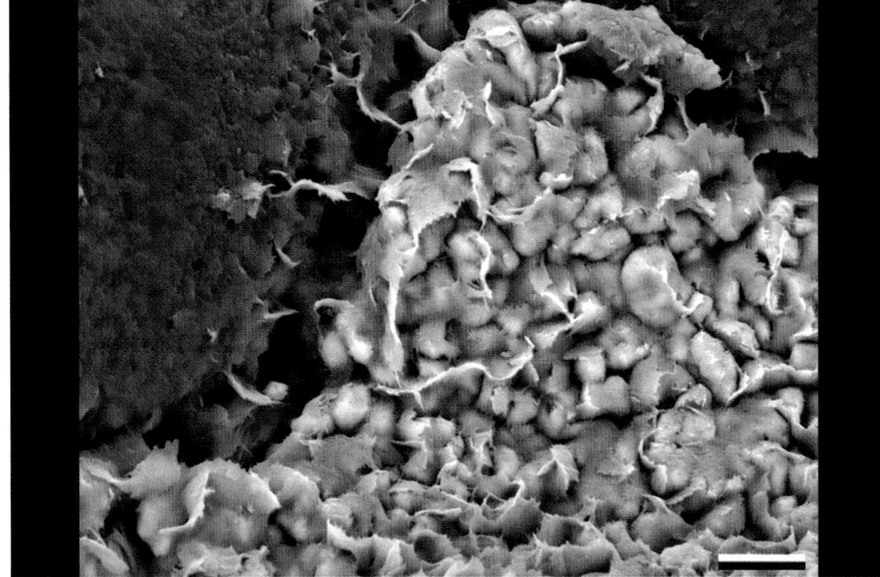

SEM+BSE, Scale bar = 13 μm

Mid. Permian Karstryggen Fm., Jameson Land, East Greenland

A nodule with scattered, radiating bundles of celestine. It has moderate relief, low birefringence and is biaxial positive. Celestine can be mistaken for quartz (uniaxial), gypsum (low negative relief) and barite (lower 2V and higher relief). These nodules grew within pockets formed during a Permian karstification event. Later diagenetic fluids migrating through the sediments precipitated celestine. Based on the radiogenic Sr-isotopic signature, the strontium was sourced by surface to near-surface fluids dissolving feldspars in arkosic redbeds and the adjacent highlands. The sulfur probably came from a nearby gypsum basin.

PPL | XPL, BDI, Scale bar = 0.26 mm

Celestine Cements

Mid. Permian Queen Fm., Lea Co., New Mexico

A fracture in a carbonate-rich sandstone that was filled first with anhydrite (seen along bottom margin) and then with halite. The halite has cubic cleavage (visible in plane-polarized light). Halite also commonly contains both fluid and mineral inclusions. Trails of fluid inclusions can help one to see the otherwise nearly invisible halite in plane-polarized light, although mineral inclusions are usually more easily seen in cross-polarized light (as in this sample). Because of its high solubility, halite can be very mobile in subsurface settings, dissolving and reprecipitating repeatedly (e.g., Glennie et al., 1978; Burley, 1984).

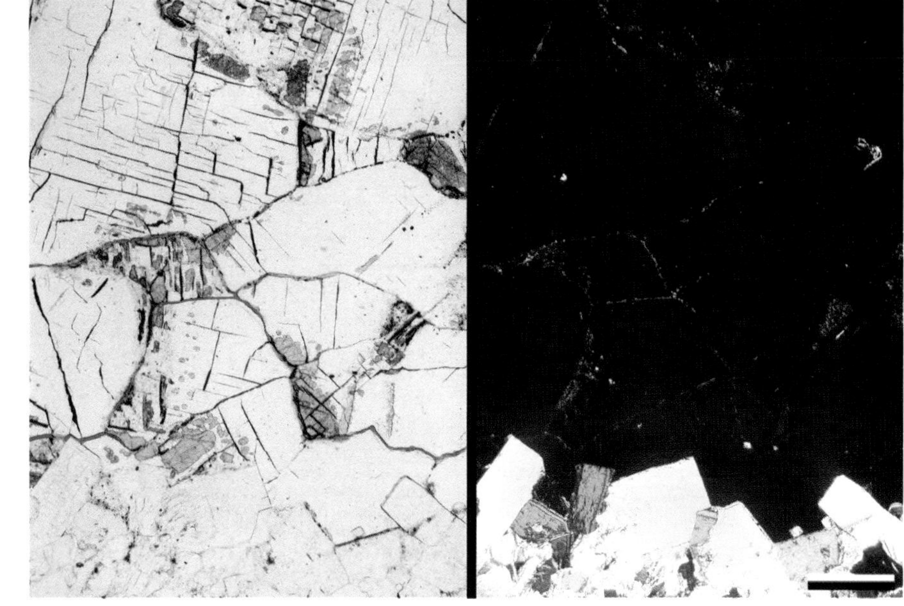

PPL | XPL, BDI, Scale bar = 0.34 mm

Halite Cements

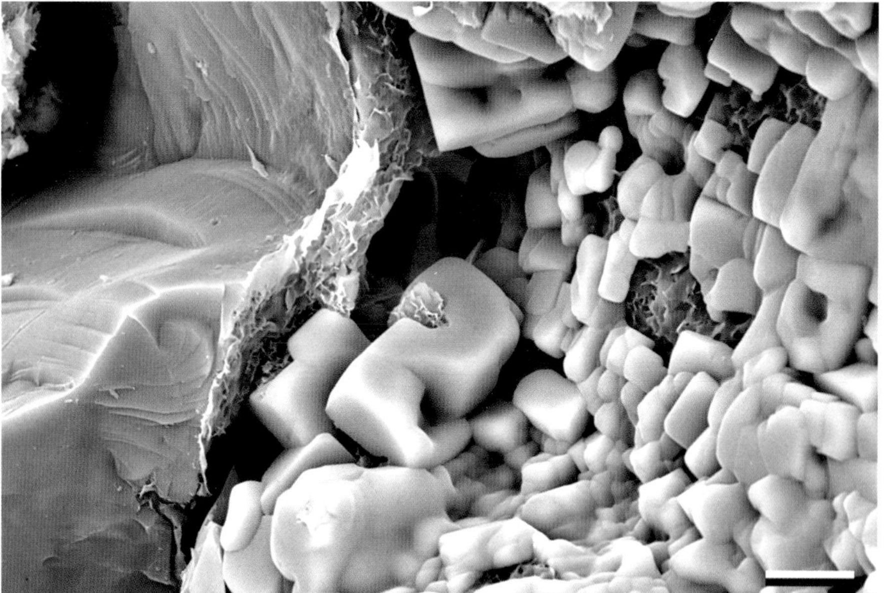

Permian Unayzah Fm., Qatar †

Because halite is prone to dissolution and reprecipitation, it's presence within a sandstone may result from a number of processes. In a succession with evaporites, it is likely to be an actual diagenetic mineral. However, in many cases, it is an artifact of poor sample handling where either saline pore waters dry out in the sample or there is contamination from sea-water or drilling fluid. Halite shows up brightly in BSE imaging and forms cubic crystals, which are rarely perfect due to dissolution. Here, halite overlies and encloses illite within a primary pore.

SEM+BSE, Scale bar = 16 μm

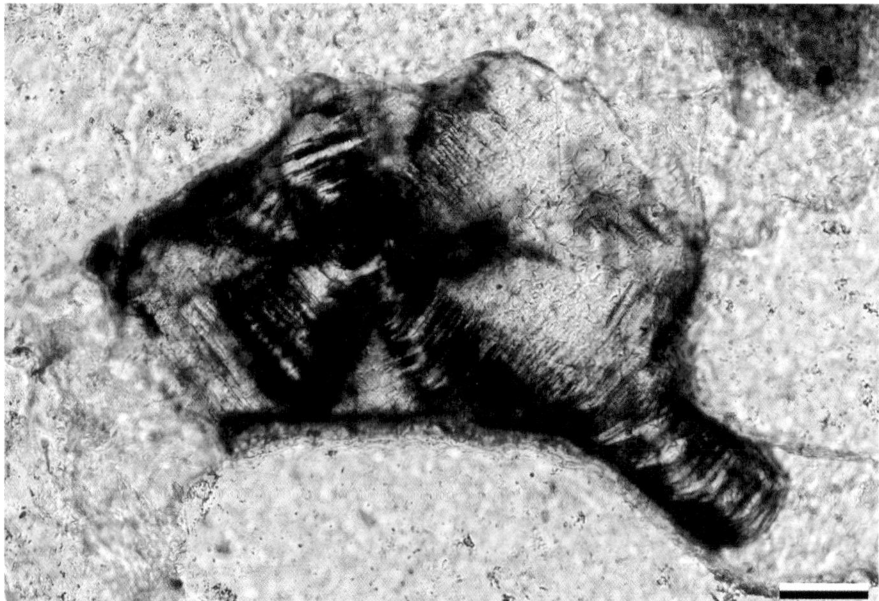

Mid. Permian Wegener Halvø Fm., Karstryggen area, East Greenland

Fluorite is relatively rare as a cement in sandstones. However, when it does occur, fluorite can provide valuable information on the thermal history of the rock because it forms from hydrothermal fluids (e.g., Williams-Jones et al., 2000). This sector-zoned crystal shows an unusually intense blue/purple color in thin section. Normally, at least in thin section, fluorite is colorless to pale pink, blue or purple. The color of fluorites can be influenced by radiation damage, thermal history, rare earth elements and/or other trace elements.

PPL, BDI, Scale bar = 0.05 mm

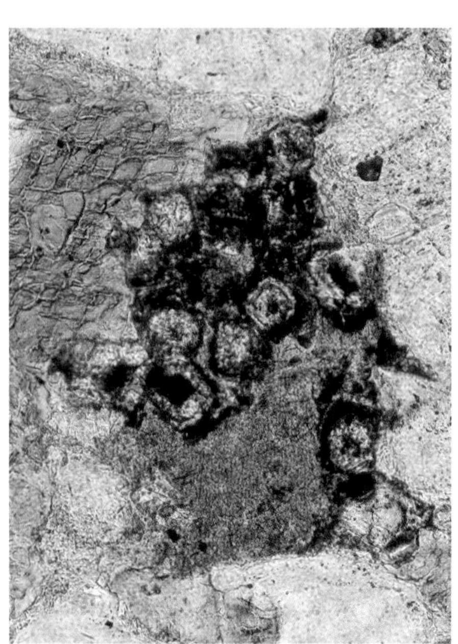

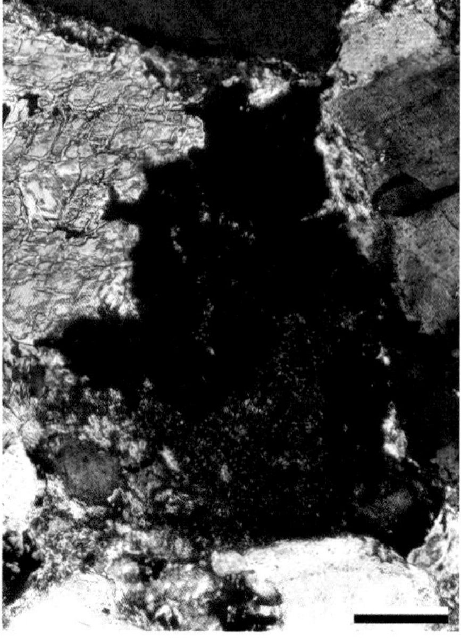

Mid. Permian Wegener Halvø Fm., Karstryggen area, East Greenland

Another view of isotropic fluorite cements partially filling a void in a lithic arenite that also contains stained calcite cements (the brightly birefringent mineral) and clay cements. Like halite, fluorite is cubic, but it also can assume a variety of other crystal forms such as octahedra and dodecahedra. All these forms are nonbirefringent (isotropic) in cross-polarized light. In this rock, the highly zoned, cubic crystals of fluorite precipitated after the carbonate cements. However, their timing relative to the clay cements is more problematic.

PPL | XPL, AFeS, BDI, Scale bar = 0.05 mm

Cited References and Additional Information Sources

Baker, P. A., and S. H. Bloomer, 1988, The origin of celestite in deep-sea carbonate sediments: Geochimica et Cosmochimica Acta, v. 52, p. 335-339, doi: 10.1016/0016-7037(88)90088-9.

Barone, M., R. Dominici, and S. Lugli, 2007, Interpreting gypsarenites in the Rossano basin (Calabria, Italy): A contribution to the characterization of the Messinian salinity crisis in the Mediterranean, *in* J. Arribas, S. Critelli, and M. J. Johnsson, eds., Sedimentary Provenance and Petrogenesis: Perspectives from Petrography and Geochemistry: Boulder, CO, GSA Special Paper 420, p. 135-148, doi: 10.1130/2006.2420(09).

Bein, A., and A. R. Dutton, 1993, Origin, distribution and movement of brine in the Permian Basin (U.S.A.): A model for displacement of connate brine: GSA Bulletin, v. 105, p. 695-707, doi: 10.1130/0016-7606(1993)105<0695:ODAMOB>2.3.CO;2.

Benison, K. C., and R. H. Goldstein, 2000, Sedimentology of ancient saline pans: An example from the Permian Opeche Shale, Williston Basin, North Dakota, U.S.A.: Journal of Sedimentary Research, v. 70, p. 159-169, doi: 10.1306/2DC40907-0E47-11D7-8643000102C1865D.

Benison, K. C., and R. H. Goldstein, 2001, Evaporites and siliciclastics of the Permian Nippewalla Group of Kansas, USA: A case for non-marine deposition in saline lakes and saline pans: Sedimentology, v. 48, p. 165-188, doi: 10.1046/j.1365-3091.2001.00362.x.

Binh, T. T. N., S. J. Jones, N. R. Goulty, A. J. Middleton, N. Grant, A. Ferguson, and L. Bowen, 2013, The role of fluid pressure and diagenetic cements for porosity preservation in Triassic fluvial reservoirs of the Central Graben, North Sea: AAPG Bulletin, v. 97, p. 1273-1302, doi: 10.1306/01151311163.

Boyer, R. E., 1961, Occurrence of radioactive fluoritic sandstone, Wet Mountains, Colorado: Economic Geology, v. 56, p. 780-783, doi: 10.2113/gsecongeo.56.4.780.

Breit, G. N., M. B. Goldhaber, D. R. Shawe, and E. C. Simmons, 1990, Authigenic barite as an indicator of fluid movement through sandstones within the Colorado Plateau: Journal of Sedimentary Research, v. 60, p. 884-896, doi: 10.1306/D426763B-2B26-11D7-8648000102C1865D.

Brodtkorb, M. K. de, V. Ramos, M. Barbieri, and S. Ametrano, 1982, The evaporitic celestite-barite deposits of Neuquen, Argentina: Mineralium Deposita (Berlin), v. 17, p. 423-436, doi: 10.1016/j.marpetgeo.2010.06.010.

Burley, S. D., 1984, Patterns of diagenesis in the Sherwood Sandstone Group (Triassic), United Kingdom: Clay Minerals, v. 19, p. 403-440, doi: 10.1180/claymin.1984.019.3.11.

Butler, G. P., P. M. Harris, and C. G. St. C. Kendall, 1982, Recent evaporites from the Abu Dhabi coast flats, *in* C. R. Handford, R. G. Loucks, and G. R. Davies, eds., Depositional and Diagenetic Spectra of Evaporites—A Core Workshop (Calgary, 1982): Tulsa, OK, SEPM Core Workshop No. 3, p. 33-64, doi: 10.2110/cor.82.01.0033.

Casas, E., and T. K. Lowenstein, 1989, Diagenesis of saline pan halite: Comparison of petrographic features of modern, Quaternary and Permian halites: Journal of Sedimentary Research, v. 59, p. 724-739, doi: 10.1306/212F905C-2B24-11D7-8648000102C1865D.

Dixon, S. A., D. M. Summers, and R. C. Surdam, 1989, Diagenesis and preservation of porosity in Norphlet Formation (Upper Jurassic), southern Alabama: AAPG Bulletin , v. 73, p. 707-728.

Dutton, S. P., and L. S. Land, 1985, Meteoric burial diagenesis of Pennsylvanian arkosic sandstones, southwestern Anadarko Basin, Texas: AAPG Bulletin, v. 69, p. 22-38.

Dworkin, S. I., and L. S. Land, 1994, Petrographic and geochemical constraints on the formation and diagenesis of anhydrite cements, Smackover sandstones, Gulf of Mexico: Journal of Sedimentary Research, v. 64, p. 339-348, doi: 10.1306/D4267D98-2B26-11D7-8648000102C1865D.

El Tabakh, M., R. Riccioni, and B. C. Schreiber, 1997, Evolution of Late Triassic rift basin evaporites (Passaic Formation)—Newark Basin, eastern North America: Sedimentology, v. 44, p. 767-790, doi: 10.1046/j.1365-3091.1997.d01-47.x.

Fishman, N. S., 1997, Basin-wide fluid movement in a Cambrian paleoaquifer: Evidence from the Mt. Simon Sandstone, Illinois and Indiana, *in* I. P. Montañez, J. M. Gregg, and K. L. Shelton, eds., Basin-wide Diagenetic Patterns: Tulsa, OK, SEPM Special Publication 57, p. 221-234, doi: 10.2110/pec.97.57.0221.

Füchtbauer, H., 1983, Facies controls on sandstone diagenesis, *in* A. Parker, and B. W. Sellwood, eds., Sandstone Diagenesis (NATO ASI Series Volume 115): Dordrecht, Netherlands, D. Reidel Publishing Co., p. 269-288, doi: 10.1007/978-94-009-7259-9_5.

Gaupp, R., and J. A. Okkerman, 2011, Diagenesis and reservoir quality of Rotliegend sandstones in the northern Netherlands—A review, *in* J. Grötsch, and R. Gaupp, eds., The Permian Rotliegend of the Netherlands: Tulsa, OK, SEPM Special Publication 98, p. 193-226, doi: 10.2110/pec.11.98.0193.

Glennie, K. W., G. C. Mudd, and P. J. C. Nagtegaal, 1978, Depositional environment and diagenesis of Permian Rotliegendes sandstones in Leman Bank and Sole Pit areas of the UK southern North Sea: Journal of the Geological Society, v. 135, p. 25-34, doi: 10.1144/gsjgs.135.1.0025.

Gurova, Y. P., and A. A. Val'ter, 1970, Temperatures of formation of fluorite from Lower Cambrian sandstones of Podolia: International Geology Review, v. 12, p. 62-64, doi: 10.1080/00206817009475208.

Handford, C. R., 1991, Marginal marine halite: Sabkhas and salinas, *in* J. L. Melvin, ed., Evaporites: Petroleum and Mineral Resources (Developments in Sedimentology 50): New York, Elsevier Science Publishers, p. 1-66.

Hanor, J. S., 2000, Barite–celestine geochemistry and environments of formation, *in* C. N. Alpers, J. L. Jambor, and D. K. Nordstrom, eds., Sulfate Minerals: Crystallography, Geochemistry, and Environmental Significance (Reviews in Mineralogy and Geochemistry 40): Chantilly, VA, Mineralogical Society of America, p. 193-275, doi: 10.2138/rmg.2000.40.4.

Hanor, J. S., 2004, A model for the origin of large carbonate-and evaporite-hosted celestine ($SrSO_4$) deposits: Journal of Sedimentary Research, v. 74, p. 168-175, doi: 10.1306/092203740168.

Hartley, A. J., and G. May, 1998, Miocene gypcretes from the Calama Basin, northern Chile: Sedimentology, v. 45, p. 351-364, doi: 10.1046/j.1365-3091.1998.0166e.x.

Hill, C. A., 1995, Sulfur redox reactions, native sulfur, Mississippi Valley-type deposits and sulfuric acid karst, Delaware Basin, New Mexico and Texas: Environmental Geology, v. 25, p. 16-23, doi: 10.1016/0264-8172(95)98089-N.

Hoholick, J. D., T. Metarko, and P. E. Potter, 1984, Regional variations of porosity and cement: St. Peter and Mount Simon sandstones in Illinois Basin: AAPG Bulletin, v. 68, p. 753-764.

James, W. C., 1992, Sandstone diagenesis in mixed siliciclastic-carbonate sequences: Quadrant and Tensleep Formations (Pennsylvanian), northern Rocky Mountains: Journal of Sedimentary Research, v. 62, p. 810-824, doi: 10.1306/D42679E7-2B26-11D7-8648000102C1865D.

Jones, G. D., and Y. Xiao, 2005, Dolomitization, anhydrite cementation, and porosity evolution in a reflux system: Insights from reactive transport models: AAPG Bulletin, v. 89, p. 577-601, doi: 10.1306/12010404078.

Jowett, E. C., L. M. Cathles, III, and B. W. Davis, 1993, Predicting depths of gypsum dehydration in evaporitic sedimentary basins: AAPG Bulletin, v. 77, p. 402-413.

Kinsman, D. J. J., 1969, Modes of formation, sedimentary associations, and diagnostic features of shallow water and supratidal evaporites: AAPG Bulletin, v. 53, p. 830-840.

Kinsman, D. J. J., 1974, Calcium sulfate minerals of evaporite deposits: their primary mineralogy, *in* A. J. Coogan, ed., Fourth Symposium on Salt: Cleveland, OH, Northern Ohio Geological Society, p. 343-348.

Kirkland, D. W., and R. Evans, eds., 1973, Marine Evaporites, Origin, Diagenesis and Geochemistry: Stroudsburg, PA, Dowden, Hutchinson and Ross, 426 p.

Kopnin, V. I., 1993, Solikamsky potash basin, *in* B. I. Chuashov, and A. E. Nairn, eds., International Congress on the Permian System of the World (Perm, U.S.S.R.), Guides to Geological Excursions, Excursion II: Central

Urals: Columbia, SC, Urals Branch, USSR Academy of Sciences and Earth Sciences and Resources Institute, University of South Carolina, Occasional Publications ESRI, New Series, No. 10, p. 208-225.

Laier, T., and B. L. Nielsen, 1989, Cementing halite in Triassic Bunter Sandstone (Tonder, southwest Denmark) as a result of hyperfiltration of brines: Chemical Geology, v. 76, p. 353-363, doi: 10.1016/0009-2541(89)90103-4.

Levandowski, D. W., M. E. Kaley, S. R. Silverman, and R. G. Smalley, 1973, Cementation in Lyons Sandstone and its role in oil accumulation, Denver Basin, Colorado: AAPG Bulletin, v. 57, p. 2217-2244.

Makhnach, A. A., 2008, Diagenetic gypsum, anhydrite and halite in non-evaporite deposits of Belarus: Baltica, v. 21, p. 25-39.

McBride, E. F., L. S. Land, and L. E. Mack, 1987, Diagenesis of eolian and fluvial feldspathic sandstones, Norphlet Formation (Upper Jurassic), Rankin County, Mississippi, and Mobile County, Alabama: AAPG Bulletin, v. 71, p. 1019-1034.

McManus, K. M., and J. S. Hanor, 1993, Diagenetic evidence for massive evaporite dissolution, fluid flow, and mass transfer in the Louisiana Gulf Coast: Geology, v. 21, p. 727-730, doi: 10.1130/0091-7613(1993)021<0727:DEFMED>2.3.CO;2.

McNeil, B., H. F. Shaw, and A. H. Rankin, 1998, The timing of cementation in the Rotliegend sandstones of the southern North Sea: A petrological and fluid inclusion study of cements: Journal of Petroleum Geology, v. 21, p. 311-328, doi: 10.1111/j.1747-5457.1998.tb00784.x.

Möller, P., S. Schulz, and K. H. Jacob, 1980-81, Formation of fluorite in sedimentary basins: Chemical Geology, v. 31, p. 97-117, doi: 10.1016/0009-2541(80)90070-4.

Moser, M. R., A. H. Rankin, and H. J. Milledge, 1992, Hydrocarbon-bearing fluid inclusions in fluorite associated with the Windy Knoll bitumen deposit, UK: Geochimica et Cosmochimica Acta, v. 56, p. 155-168, doi: 10.1016/0016-7037(92)90123-Z.

Nettleton, W. D., 1991, Occurrence, characteristics, and genesis of carbonate, gypsum, and silica accumulations in soils: Lincoln, NE, USDA National Soil Survey Laboratory, SSSA Special Publication 26, 16 p.

Paytan, A., S. Mearon, K. Cobb, and M. Kastner, 2002, Origin of marine barite deposits: Sr and S isotope characterization: Geology, v. 30, p. 747-750, doi: 10.1130/0091-7613(2002)030<0.

Purvis, K., 1992, Lower Permian Rotliegend sandstones, southern North Sea: A case study of sandstone diagenesis in evaporite-associated sequences: Sedimentary Geology, v. 77, p. 155-171, doi: 10.1016/0037-0738(92)90123-9.

Purvis, K., and J. A. Okkerman, 1996, Inversion of reservoir quality by early diagenesis: An example from the Triassic Buntsandstein, offshore the Netherlands, *in* H. E. Rondeel, D. A. J. Batjes, and W. H. Nieuwenhuijs, eds., Geology of Gas and Oil under the Netherlands: Berlin, Springer, p. 179-189, doi: 10.1007/978-94-009-0121-6_16.

Pye, K., and D. H. Krinsley, 1986, Diagenetic carbonate and evaporite minerals in Rotliegend aeolian sandstones of the southern North Sea: Their nature and relationship to secondary porosity development: Clay Minerals, v. 21, p. 443-457, doi: 10.1180/claymin.1986.021.4.03.

Schenk, C. J., and R. W. Richardson, 1985, Recognition of interstitial anhydrite dissolution: A cause of secondary porosity, San Andres Limestone, New Mexico and upper Minnelusa Formation, Wyoming: AAPG Bulletin, v. 69, p. 1064-1076.

Scholle, P. A., L. Stemmerik, and O. Harpøth, 1990, Origin of major karst-associated celestite mineralization in Karstryggen, central East Greenland: Journal of Sedimentary Research, v. 60, p. 397-410, doi: 10.1306/212F91A6-2B24-11D7-8648000102C1865D.

Schreiber, B. C., 1986, Arid shorelines and evaporites, *in* H. G. Reading, ed., Sedimentary Environments and Facies (2nd edition): Oxford: Blackwell Scientific Publications, 189-228.

Schreiber, B. C., ed., 1988, Evaporites and Hydrocarbons: New York, Columbia University Press, 475 p.

Schreiber, B. C., and M. El Tabakh, 2000, Deposition and early alteration of evaporites: Sedimentology, v. 47, p. 215-239, doi: 10.1046/j.1365-3091.2000.00002.x.

Strong, G. E., 1993, Diagenesis of Triassic Sherwood Sandstone Group rocks, Preston, Lancashire, UK: A possible evaporitic cement precursor to secondary porosity?, *in* C. P. North, and D. J. Prosser, eds., Characterization of Fluvial and Aeolian Reservoirs: London, GSL Special Publication 73, p. 279-289, doi: 10.1144/GSL.SP.1993.073.01.17.

Sullivan, M. D., A. E. Fallick, R. S. Haszeldine, A. J. Boyce, and G. Rogers, 1994, Late anhydrite cements mark basin inversion: Isotopic and formation water evidence, Rotliegend Sandstone, North Sea: Marine and Petroleum Geology, v. 11, p. 46-54, doi: 10.1016/0264-8172(94)90008-6.

Tieman, D. J., and O. C. Kopp, 1983, Deformation fabrics in barite and fluorite of the Del Rio District: Their implications for the relative timing of mineralization and deformation: Economic Geology, v. 78, p. 91-104, doi: 10.2113/gsecongeo.78.1.91.

Waldschmidt, W. A., 1958, Halite as cementing mineral in sandstones: AAPG Bulletin, v. 42, p. 871-875.

Warren, J. K., 1989, Evaporite Sedimentology: Importance in Hydrocarbon Accumulation: Englewood Cliffs, NJ, Prentice Hall Advanced Reference Series, 285 p.

Warren, J. K., 2006, Chapter 1. Interpreting evaporite texture, *in* Evaporites: Sediments, Resources and Hydrocarbons: Berlin, Springer, p. 1-57, doi: 10.1007/3-540-32344-9_1.

Watson, A., 1985, Structure, chemistry and origins of gypsum crusts in southern Tunisia and the central Namib Desert: Sedimentology, v. 32, p. 855-875, doi: 10.1111/j.1365-3091.1985.tb00737.x.

Williams-Jones, A. E., I. M. Samson, and G. R. Olivo, 2000, The genesis of hydrothermal fluorite-REE deposits in the Gallinas Mountains, New Mexico: Economic Geology, v. 95, p. 327-341, doi: 10.2113/gsecongeo.95.2.327.

Facing Page: Top — Hematite-cemented sandstone, Cretaceous Woodbine Fm., Anderson Co., Texas (XPL). Bottom — Spearhead twins in marcasite cement. Up. Devonian New Albany Shale, Camp Run Mbr., Kentucky (RL).

DIAGENESIS: IRON SULFIDE, OXIDE & HYDROXIDE CEMENTS

CHAPTER 17

INTRODUCTION TO IRON CEMENTS

The type and abundance of minerals that we observe in the earth's crust and its sedimentary cover is governed by elemental abundances and thermodynamic mineral equilibria. Iron is fourth in abundance (e.g., Mason, 1966) by weight (~5%) after oxygen, silica, and aluminum (~82.5% cumulatively), and whereas that relationship readily explains the preponderance of silica and clay minerals as cements in sedimentary rocks, the story for iron is a bit more complicated. Due to its multiple redox states, iron can form (or be part of) minerals in oxidizing as well as reducing environments, and the main "sinks" in the sedimentary rock record are Precambrian banded iron formations, Phanerozoic ironstones and continental red beds.

Iron is generally supplied to sedimentary basins in the form of iron hydroxide coatings on fine particles (Carroll, 1958) and as iron silicates in the sand fraction (Walker, 1967). In fluvial sediments, the commonly oxidizing pore water conditions result in intrastratal alteration of detrital iron silicates (pyroxene, hornblende, biotite) and in the precipitation of iron hydroxides (limonite, goethite; yellow-brownish color) within pore spaces. Over time, iron hydroxides in both fine and coarse sediments are converted to hematite that gives the rocks their characteristic red color (Walker, 1967). The timing of hematite formation can, under some circumstance, be dated using paleomagnetic information (e.g., Lu et al., 1994) or geochronologic data (e.g., Reiners et al., 2014).

Under the microscope, iron oxides either seemed to have formed either disseminated brownish-reddish species in fine-grained rocks or grain-coating and pore-filling cements of iron hydroxides (limonite, goethite) and/or hematite. In transmitted light, hematite is opaque (or reddish at very thin crystal edges), whereas the hydroxides have a brownish-reddish appearance. If such sediments (muds and sands) are deposited in a marine setting where co-deposited organic matter renders pore fluids anoxic early in the diagenetic history (Berner, 1981), detrital iron hydroxides dissolve and release ferrous iron (Fe^{2+}) to the pore waters. Depending on the prevailing conditions, there are several potential pathways for the formation of iron minerals.

During shallow burial (eogenesis), sulfate reducing bacteria participate in organic matter degradation. Associated production of hydrogen sulfide will precipitate dissolved iron in the form of iron monosulfides that will further react to form pyrite (Berner, 1970). In fine-grained rocks, early diagenetic pyrite is overwhelmingly in the form of micron-size framboidal pyrite (Wilkin and Barnes, 1997), whereas in sandy sediments, it may form coarser-grained, blocky pyrite cements. Marcasite, a dimorph of pyrite, has been observed as a cement mineral in situations where earlier-formed diagenetic pyrite underwent partial oxidation (Schieber, 2007, 2011). Persistent reducing conditions may occur in the near-surface under unusual circumstances, such as in chimneys of hydrocarbon leakage (oil or gas seeps). In such settings, iron oxide and hydroxide coatings are removed leaving "bleached" zones in red-bed sediments. In addition, pyrite, marcasite, calcite and dolomite cements may be precipitated, acting as markers for underlying hydrocarbon deposits (Reynolds et al., 1990; Kirkland et al., 1995).

In the case of reducing pore waters that do not contain sulfide, dissolved iron is readily incorporated into carbonate minerals (ferroan calcite, ferroan dolomite, ankerite, siderite). Sulfide formation does not take place in the absence of sulfate (no microbial sulfate reduction is possible in that situation), and thus iron-bearing carbonates are to be expected in freshwater-dominated and nonmarine reducing settings (estuarine and lacustrine successions). However, iron-bearing carbonates can also form in marine sediments. For example, below the zone of microbial sulfate reduction, in the realm of methanogenic fermenting microbes (Curtis et al., 1977; Berner, 1981; Burdige, 1993), organic matter degradation reactions produce bicarbonate (Froelich et al., 1979). Thus, carbonate precipitation is favored in that situation and may be manifested in the form of siderite concretions or other iron-enriched carbonate minerals (Curtis et al., 2000).

Under mildly reducing pore water conditions and at low rates of sedimentation, iron-bearing phyllosilicates, such as berthierine, chamosite and glauconite may form in surficial sediments (Odin, 1988). These precipitate preferentially in the reducing microenvironments of fecal pellets (Porrenga, 1966), as well as in the interior of fossil tests, and as pore filling cements in sandy layers. These kinds of minerals are typical for shallow marine successions and are in many instances associated with hiatal surfaces (Ketzer et al., 2003). Under favorable circumstances, stratigraphic iron enrichment may lead to economic accumulations of iron minerals. The associated strata are known as ironstones (Van Houten and Bhattacharyya, 1982) and are mined in multiple locations as a source of iron ore. The iron-bearing clays are not covered in this chapter, however, as they were previously discussed in Chapters 5 and 13.

Up. Devonian Geneseo Shale, New York

This siltstone contains abundant pyrite (FeS_2) cements. In plane-polarized light, metallic sulfide and some oxide minerals are opaque, and are best studied in polished sections on a reflected light microscope. In this example, pyrite occurs as cubic crystals that are highly reflective (pale yellow to white). Organic matter, bacteria and iron hydroxides on detrital grains can interact to create an environment suitable for the precipitation of pyrite as cements or replacements within the rock.

RL, Scale bar = 0.13 mm

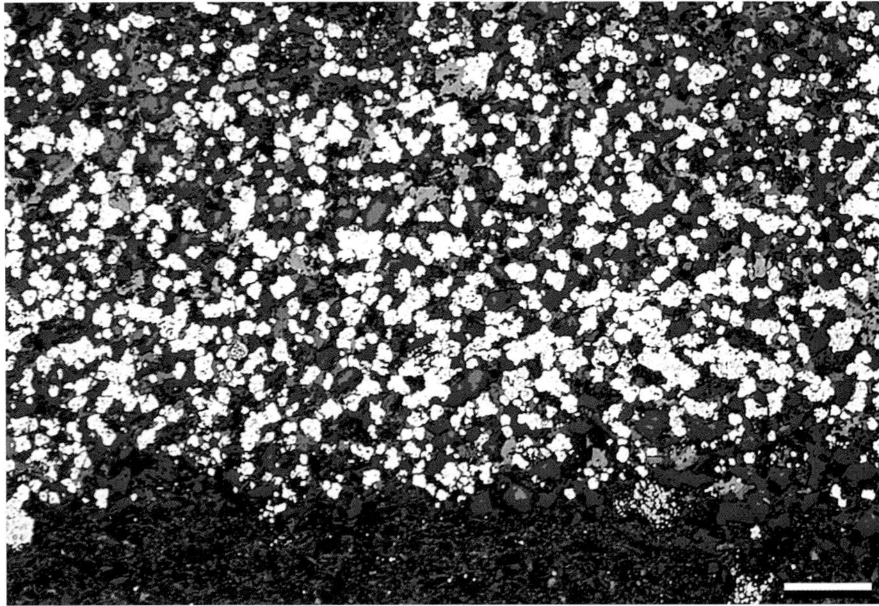

Mid. Jurassic Entrada Fm., Slick Rock Mbr., Emery Co., Utah

In this iron-cemented sandstone, scattered remnants of earlier pyrite cements are found within the goethite cements that now dominate the rock. The pyrite in this sample is pale yellow in reflected light, whereas the goethite is gray. Pyrite precipitated from waters that were reducing; later, possibly during uplift, oxidizing pore fluids migrated through these rocks converting the pyrite to goethite. The identification of the goethite is also supported by XRD data (Radhu, 2013).

RL, RDI, Scale bar = 0.09 mm

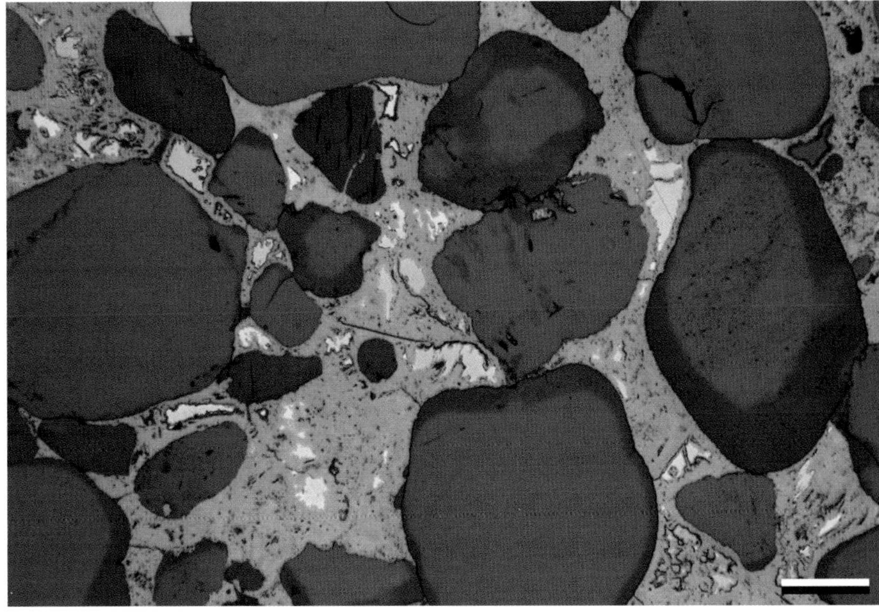

Mid. Permian Rush Springs Fm., Caddo Co., Oklahoma

These pyrite cements formed as a result of reducing conditions in hydrocarbon seeps over the Cement oil field (Reynolds et al., 1990). Pyrite is opaque and essentially black in plane-polarized light (left) so positive identification requires reflected light. Because reflected light microscopes are not always available and polished thin sections are more expensive than unpolished ones, oblique reflected light (in this case using a high-intensity, flexible fiber-optic light) is often used to identify iron sulfides and oxides. In oblique reflected light (right side), pyrite appears golden, just as it does in hand specimen.

PPL | ORL, Scale bar = 0.10 mm

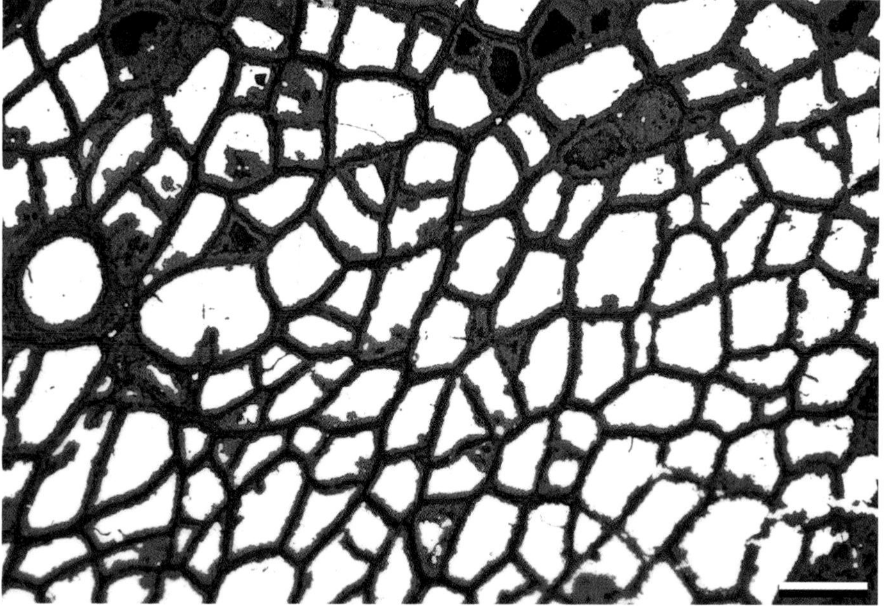

Up. Devonian New Albany Shale, Camp Run Mbr., Kentucky

The zooecia of this coral fragment are infilled by pyrite cements (white in reflected light). The coral fragment is part of a pyritized lag deposit associated with a sequence boundary containing quartz grains, reworked pyrite, fish debris and miscellaneous bioclasts. The red coloration along thin edges of the pyrite is due to alteration of pyrite to iron oxides/hydroxides, probably during uplift and exposure.

RL, Scale bar = 32 μm

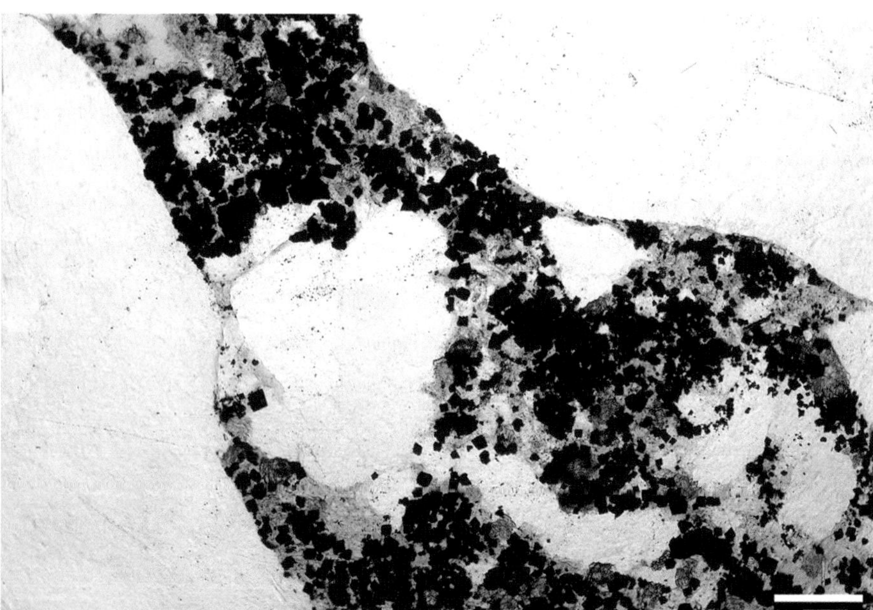

Lo. Jurassic Tilje Fm., Norwegian sector, North Sea †

Here, clusters of cubic pyrite (black) are seen within pores and in association with detrital clay. Although pyrite is abundant in this sample, it typically represents less than 2% of most terrigenous rocks—it is more common in marine rocks and generally less common in terrestrial deposits. It is seen in thin section primarily as opaque cubes and framboids. When small crystals of pyrite are visible, they are most commonly of early diagenetic origin and are associated with the presence of organic material.

PPL, KFS, BDI, Scale bar = 0.07 mm

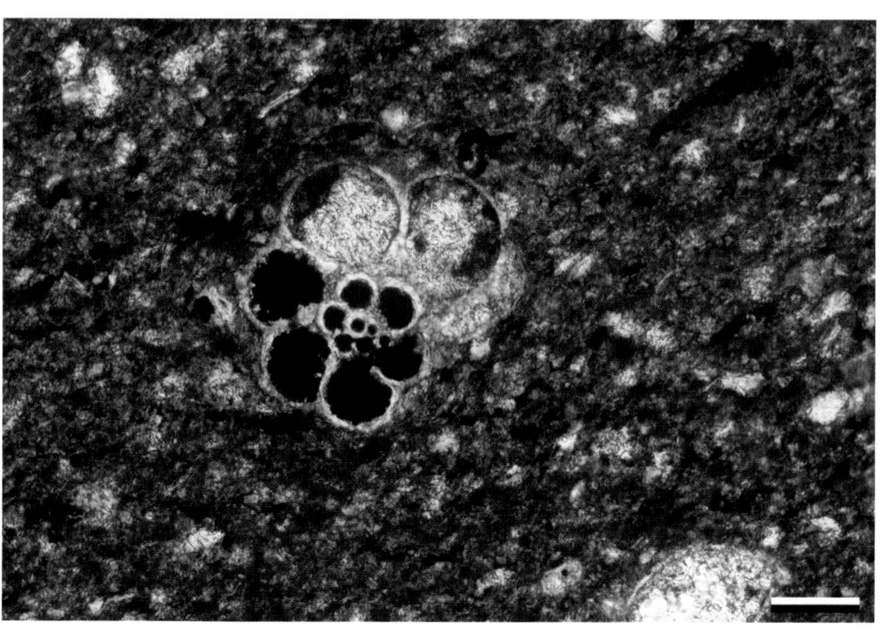

Cretaceous Mowry Fm., Moffat Co., Colorado

The chambers of this foraminifer are filled by both pyrite (black opaque) and calcite (colorless) cements. It is common for pyrite to form within the living chambers of organisms because decaying organic matter can produce micro-reducing environments that are conducive to pyrite formation.

PPL, Scale bar = 0.07 mm

Up. Devonian (Famennian) Ohio Shale, Knox Co., Kentucky

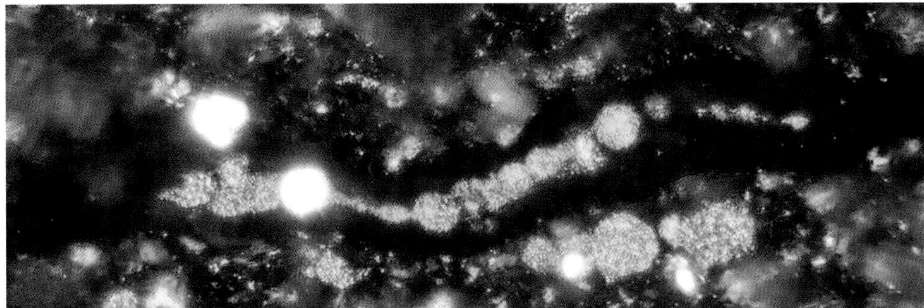

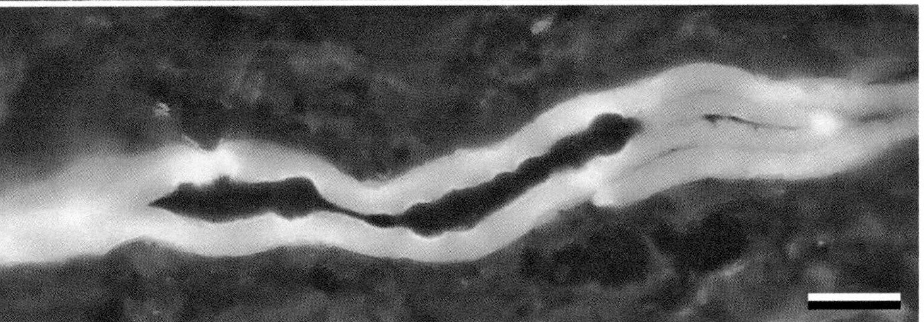

In this view, a *Tasmanites* cyst from the Ohio Shale is dark brown in oblique reflected light (top) and appears bright in fluorescent illumination (bottom). The brightly reflecting gold-colored framboidal pyrite cements (in oblique-reflected light) formed both within the cysts and in the surrounding sediment prior to compaction. These framboids consists of numerous tetrahedral crystals of pyrite in spherical aggregates. Photomicrograph from Cortland Eble.

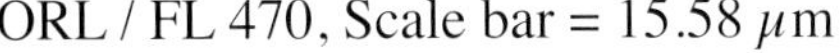

ORL / FL 470, Scale bar = 15.58 μm

Up. Devonian (Famennian) Ohio Shale, Knox Co., Kentucky

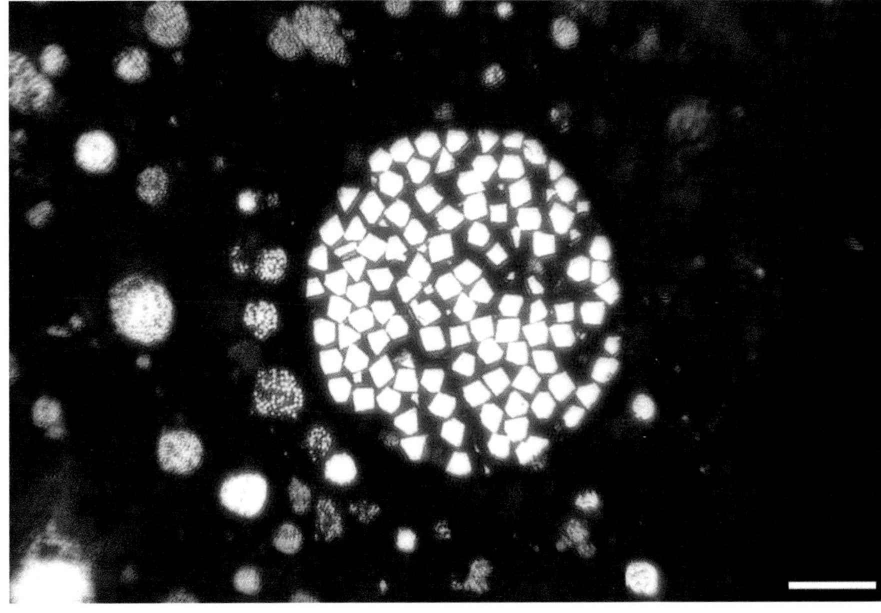

Large and small pyrite framboids, such as the ones seen here, are ubiquitous in the Ohio Shale and are the principal contributor to its characteristic high sulfur content (commonly 3 to 6%). The abundance of these pyrite framboids in the sediment attests to deposition in dysaerobic to anoxic conditions, which also promoted the preservation of organic matter. Photomicrograph from Cortland Eble.

RL, Scale bar = 12.28 μm

Up. Cretaceous Nise Fm., Vøring Basin, Norwegian Sea †

These framboids of pyrite have a wide range of sizes and are somewhat larger than average, with some reaching nearly 30 μm in diameter (although most are less than 10 μm). The component crystals of the framboids are rarely seen in thin section and so confirmation that these are framboids generally requires SEM imaging. However, the identification as pyrite is easily done with reflected light.

PPL, KFS, BDI, Scale bar = 0.07 mm

Up. Cretaceous (Campanian) Hod Gp., Danish sector, North Sea

An SEM view into the breached chambers of a planktic foraminifer showing an authigenic pyrite framboid in the bottom chamber (black box shows area enlarged in the right-hand image). The framboid and isolated pyrite crystals are partially overgrown with later-stage illitic clays (see next image). Also visible in the low magnification image are doubly-terminated, authigenic quartz crystals. Framboids (a term that comes from the French word for raspberry) consist of clusters of euhedral octahedral pyrite crystals.

SEM | SEM, Scale bar = 20 μm | 3.3 μm

Up. Cretaceous (Campanian) Hod Gp., Danish sector, North Sea

A higher magnification, artificially colorized SEM view of a central portion of the framboid from the previous image showing, in more detail, the octahedral pyrite crystals of the framboid (shown in an arbitrarily selected yellow color) along with the illite/smectite clay minerals (blue) that have grown in interstices on the outer surface of the framboid. Such framboidal pyrite clusters have been interpreted as replacements of precursor framboids of greigite (Fe_3S_4) crystals (Wilkin and Barnes, 1997).

Col SEM, Scale bar = 1.25 μm

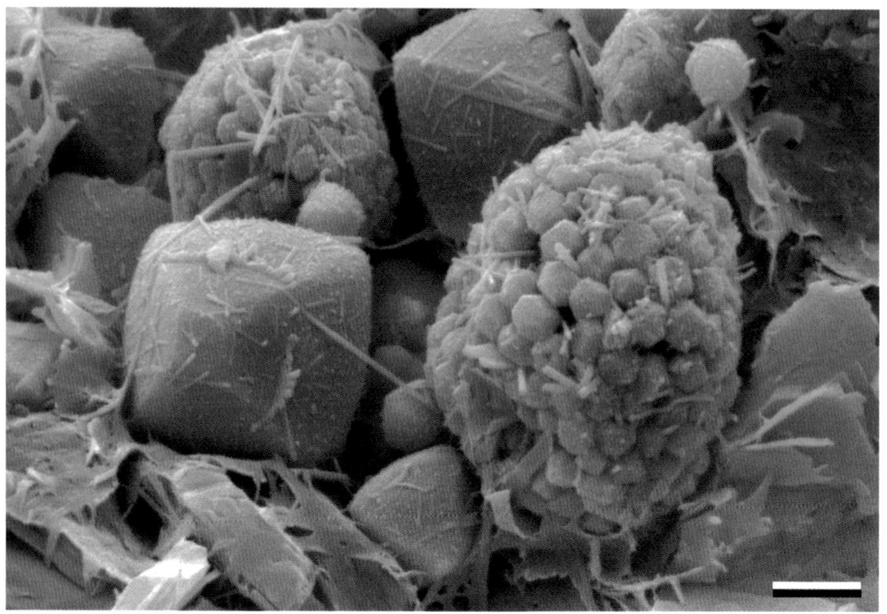

Lo. Jurassic Tilje Fm., Norwegian sector, North Sea †

This image shows pyrite in two crystal habits—euhedral octahedrons and spherical framboids. The framboids are composed of numerous small dodecahedrons (pyritohedrons). Pyrite framboids can also consist of cubic crystals of pyrite. The pyrite is overlain by thin, filamentous to platy illite crystals.

SEM, Scale bar = 2 μm

Up. Devonian New Albany Shale, Camp Run Mbr., Kentucky

The cements in this lag deposit are mainly a mix of pyrite and marcasite (FeS_2). Although chemically identical, pyrite and marcasite (both white in this image) differ in their optical properties. In reflected light, marcasite behaves pleochroically (unlike pyrite), and with crossed polarizers, marcasite (anisotropic) will show changing colors, whereas pyrite (isotropic) will stay uniformly dark. The lag deposit marks a sequence boundary and is composed of quartz grains, fossil fragments, fish debris and reworked pyrite crystals (Schieber, 2007). Pyrite concretions and pyritized bone and burrow-fills within the lag are reworked from underlying strata.

RL, Scale bar = 0.13 mm

Up. Devonian New Albany Shale, Camp Run Mbr., Kentucky

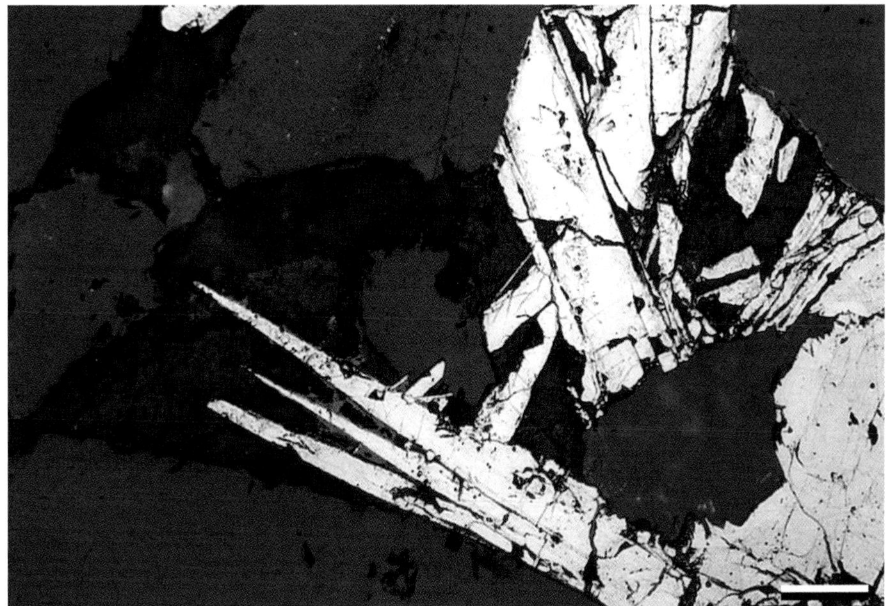

These pores in the same lag deposit as the previous photomicrograph are partially filled by marcasite cement. Spearhead twins are characteristic of marcasite. Marcasite exhibits coloration in cross-polarized light and is pleochroic in reflected light, with slight variations of color depending on the crystallographic orientation of the crystals. Schieber (2007) ascribed marcasite formation to the oxidation and dissolution of pyrite in sediments lowering the pH and increasing the iron concentration. That was followed by an influx of hydrogen sulfide from the decay of organics, leading to marcasite precipitation.

RL, Scale bar = 32 μm

Up. Jurassic Fulmar Fm., United Kingdom sector, North Sea †

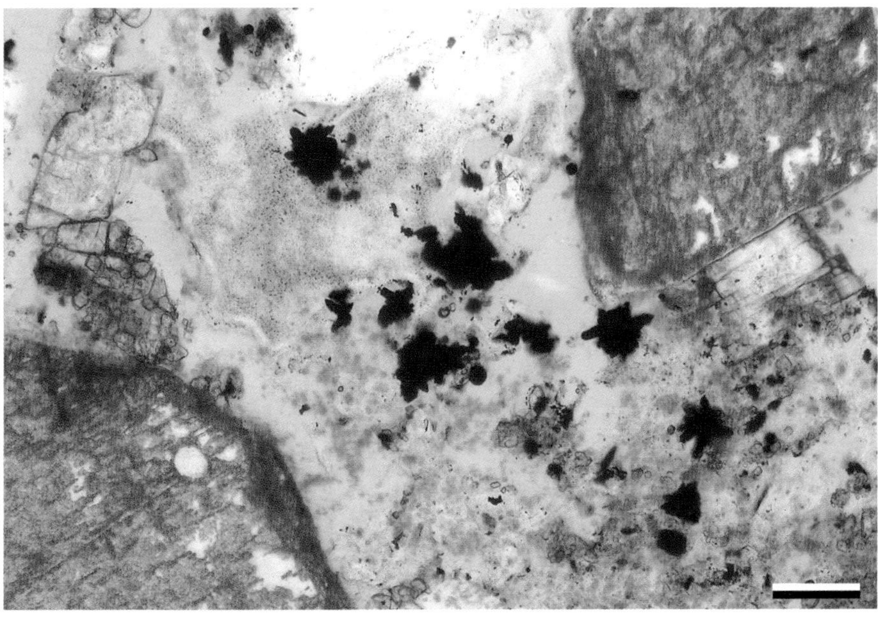

Radially arranged clusters of pyrite crystals which, on a small scale, resemble the crystal habit typical of marcasite and may, in fact, include some marcasite (especially the elongate crystals). Differentiation between the two requires reflected light. The pyrite crystals are located within a partially dissolved plagioclase and they have grown within, or were partially encompassed by, chalcedony cement. The chalcedony is microporous and has taken some color from blue resin infiltration.

PPL, AFeS, KFS, BDI, Scale bar = 0.07 mm

Up. Jurassic Fulmar Fm., United Kingdom sector, North Sea †

A view of crystal clusters of marcasite with distinctive crystal habit that appear nearly white under backscatter imaging. Nearby are quartz overgrowths, rhombic calcite crystals, oil coated authigenic illite and detrital clay. Next to the radiating marcasite crystals are poorly preserved pyrite framboids which may have acted as nuclei for the later marcasite growth.

SEM+BSE, Scale bar = 9 μm

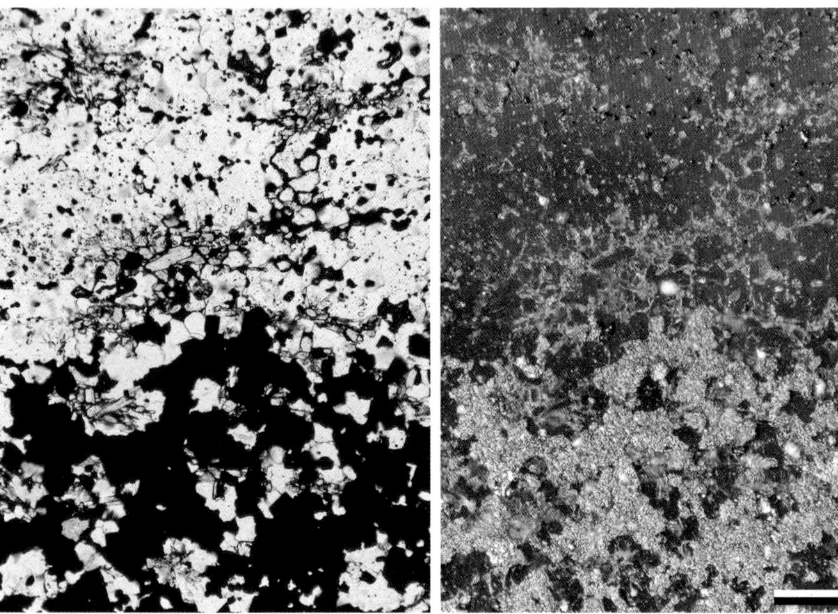

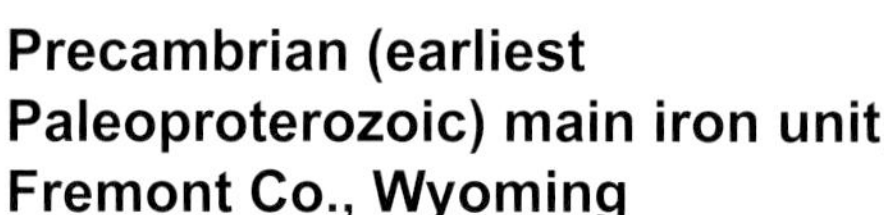

Precambrian (earliest Paleoproterozoic) main iron unit, Fremont Co., Wyoming

The opaque minerals in the plane-polarized light view (left) are a combination of mostly magnetite ($Fe^{3+}_2Fe^{2+}O_4$) along with lesser amounts of hematite (Fe_2O_3) and limonite ($FeO(OH) \cdot nH_2O$). The magnetite is highly reflective and silvery gray in oblique-reflected light. The yellow colors mark "limonite" staining and the red oxide staining is hematite. Magnetite is generally resistant to weathering, but can oxidize to hematite and hydrate to limonite.

PPL | ORL, Scale bar = 0.26 mm

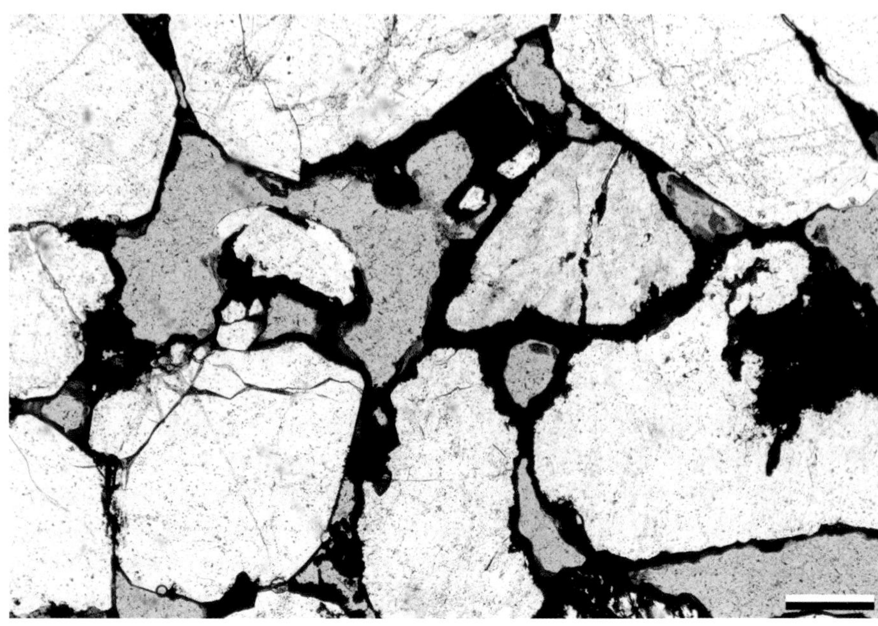

Pennsylvanian Conemaugh Gp., Athens Co., Ohio

Clay cutans are soil features that are commonly preserved in the rock record. They form from clays migrating downward within a soil, ultimately lining pores. With time, cutans can thicken and produce rounded pores. Cutans are commonly stained by hematite cements, as in this example. The hematite generally is derived from the dissolution of detrital iron-rich minerals. Hematite is opaque in plane-polarized light (except along thin edges where it can appear red). It is generally red in oblique-reflected light and pale bluish gray in incident-reflected light.

PPL, BDI, Scale bar = 0.26 mm

Silurian Clinton Ironstone, Giles Co., Virginia

This is an example of a hematite-rich oolitic ironstone deposit that is widely distributed through the eastern United States. The larger ovoid lumps of opaque material are deformed ooids scattered amidst well-rounded, bimodal quartz grains in this marginal marine sandstone. Hematite is the main cement and probably precipitated early in the diagenetic history of the rock.

PPL, Scale bar = 0.51 mm

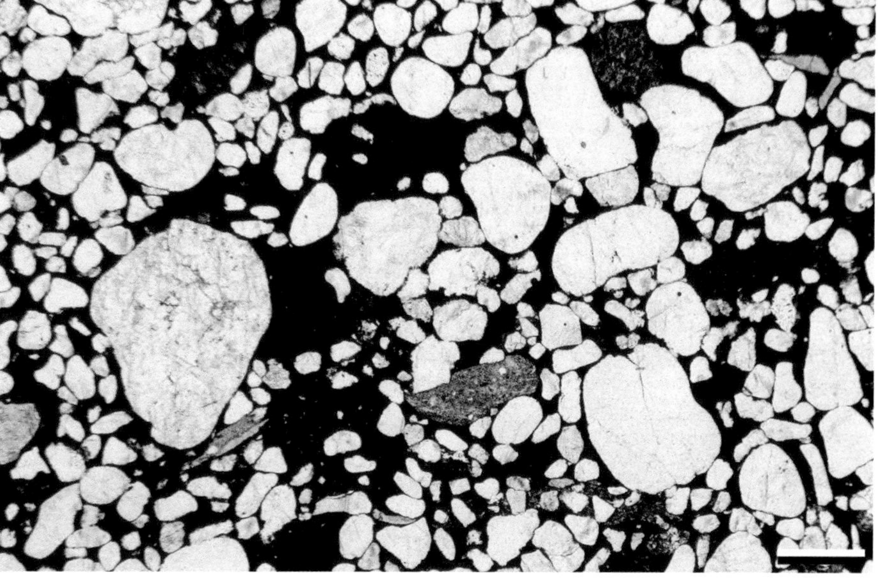

Miocene Hayner Ranch Fm., south-central New Mexico

The hornblende grain in this view was partially corroded to form hematite cement in nearby pores. Dissolution of ferromagnesian minerals can be an important source of iron for diagenetic minerals. That is especially prevalent in sediments deposited in arid conditions where such unstable detrital grains were not removed during initial weathering. Pores and translucent grains both appear pale blue to white in this photo. Photomicrograph by T. R. Walker.

PPL+ORL, Scale bar = 0.07 mm

Mid.–Up. Eocene Baca Fm., Socorro Co., New Mexico

A typical arid-region, continental "red bed". The iron in the reddish staining was probably derived from the eogenetic weathering and *in situ* alteration of iron-rich detrital minerals (augite, amphibole, olivine, biotite and others; e.g., previous photo) in these alluvial deposits. The precipitation of hematite grain coatings, and hematite staining of clay coatings, is inferred to have occurred over a long period of time in the thick vadose zone typically found in alluvial fans and other sub-humid to arid, basin-margin settings (Walker, 1976).

PPL, BDI, Scale bar = 0.1 mm

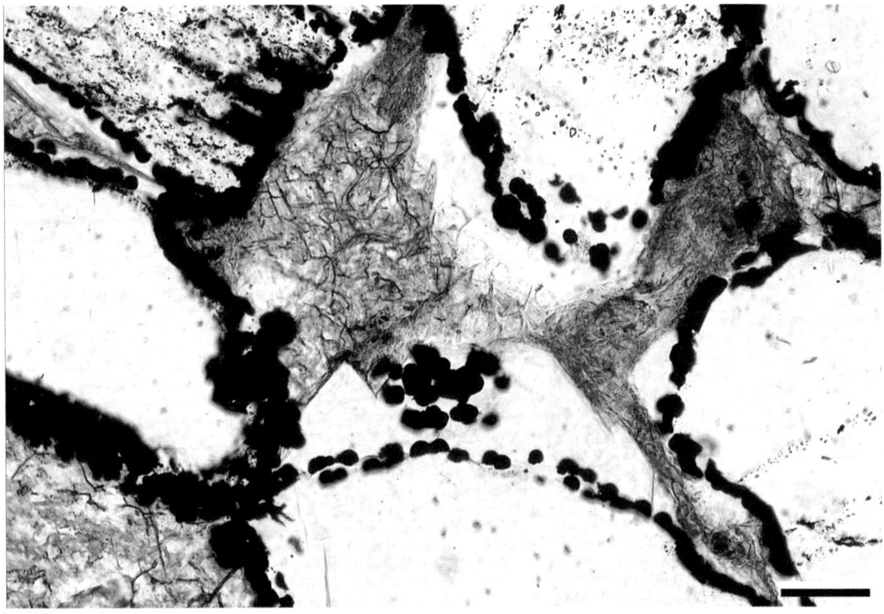

Paleocene (with Quaternary? cementation), Rajasthan, India

This image displays a sample from the Thar Desert. What was once a friable sandstone is now cemented by hematite clusters (possibly replaced pyrite or marcasite framboids); in some examples the iron cement is more abundant, constituting a ferricrete. Here, however, the thin layer of hematite clusters is overlain by quartz cement and in turn by non-ferroan calcite and fibrous clay. Deciphering whether diagenetic cements are formed in burial or surficial settings is sometimes difficult, but in this case the iron minerals are known to be concentrated at the present desert surface and so the overlying cements are clearly very young.

PPL, AFeS, BDI, Scale bar = 0.07 mm

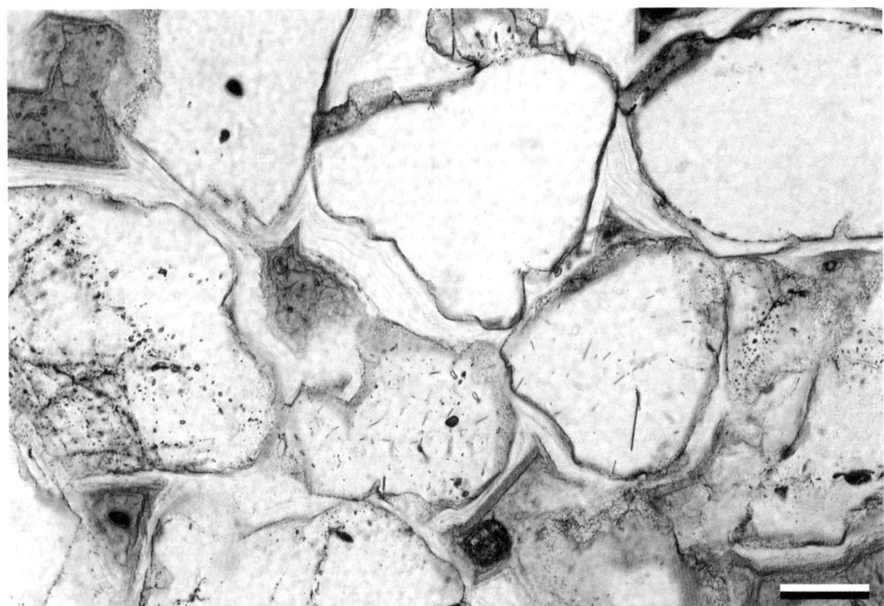

Paleocene (with Quaternary? cementation), Rajasthan, India

This is the same sandstone silcrete shown above, probably formed under meteoric phreatic conditions. It displays prominent ferruginous "dust rims" around the detrital grains as opposed to the discrete hematite clusters seen in the previous image. The rims are red-brown in color and are overlain by banded quartz overgrowths that were not impeded by the earlier iron oxides. The overgrowths show growth lines that parallel the initial grain surface. Quartz is not a common silcrete mineral; usually opal or chalcedony is precipitated in silcretes.

PPL, BDI, Scale bar = 0.07 mm

Mid. Jurassic Entrada Fm., Slick Rock Mbr., Emery Co., Utah

This sandstone was originally cemented by pyrite (inclusions are still present elsewhere in the slide), but during later diagenesis, the pyrite was oxidized to a combination of hematite and goethite. The hematite is the brighter reflecting phase and goethite is the brownish-gray phase; the dark blobs are dead hydrocarbons. The hematite and goethite may have formed from meteoric fluids that entered after basinal brines that were associated with hydrocarbon migration.

RL, RDI, Scale bar = 0.09 mm

Lo. Eocene Rockdale Fm., Simsboro Mbr., Bastrop Co., Texas

This oblique-reflected light image highlights the thick hematite and limonite cements in this sandstone. The hematite ranges from a very dark red to orange, and limonite is the yellow phase. Limonite, an iron hydroxide, was once considered to be a mineral, but is now classified as a mineraloid—a mixture of iron hydroxide and oxide minerals, such as goethite, jarosite, hematite and others. The sandstone was part of a meandering river and marsh complex that contains scattered terrestrial fossils. The hematite and limonite probably formed in a lateritic soil.

ORL, Scale bar = 0.15 mm

Mid. Jurassic Entrada Fm., Slick Rock Mbr., Emery Co., Utah

This quartzarenite is from an eolian dune complex and is completely cemented by goethite (FeO(OH)). Goethite usually appears deep red in both plane- and cross-polarized light, and unlike many iron oxides and sulfides, is not opaque. These isopachous and fibrous goethite cements filled the primary porosity. They probably formed during burial diagenesis (see also next image) and do not represent paleosol formation.

PPL+ORL, RDI, Scale bar = 0.10 mm

Mid. Jurassic Entrada Fm., Slick Rock Mbr., Emery Co., Utah

Goethite (red) and hematite (opaque black) here partially filled an open fracture associated with a deformation band (the shattered, sheared and compacted zone along the lower edge of the photomicrograph). These fracture fills are highly zoned, with crystal shapes in the early stages that are more representative of pyrite, not goethite or hematite. That has led to an interpretation that the goethite may have partially formed by oxidation of pyrite precursors (S. Radhu, personal communication), although the later stages most likely were primary goethite precipitates.

PPL+ORL, RDI, Scale bar = 0.26 mm

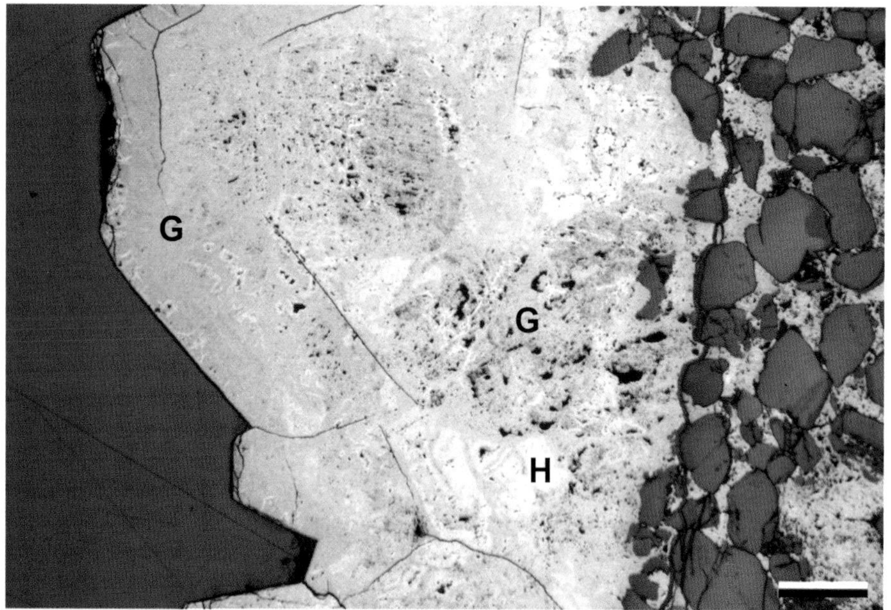

Mid. Jurassic Entrada Fm., Slick Rock Mbr., Emery Co., Utah

This reflected light image shows a different area of the fracture illustrated in the previous photomicrograph (center and left) along with detrital grains of the host sediment at right. The filling is composed of goethite (darker gray phase; G) and hematite (light gray phase; H). Zoning within the crystals close to the wall of the fracture indicates that they may be a pseudomorphous replacement of precursor pyrite. With influx of oxidizing, sulfide-poor pore fluids, goethite (and to a lesser degree hematite) may have replaced pyrite and formed additional cement crystals near the center of the fracture (left side of image).

RL, RDI, Scale bar = 0.09 mm

Lo. Pennsylvanian Morrow B Sandstone, Ochiltree Co., Texas

This fracture in a poorly-sorted quartzarenite is filled by coarsely crystalline goethite that also partially replaces the surrounding rock. The fill is composed of goethite, but the shape of the fill appears to be pseudomorphous after hematite.

PPL | XPL, RDI, Scale bar = 0.51 mm

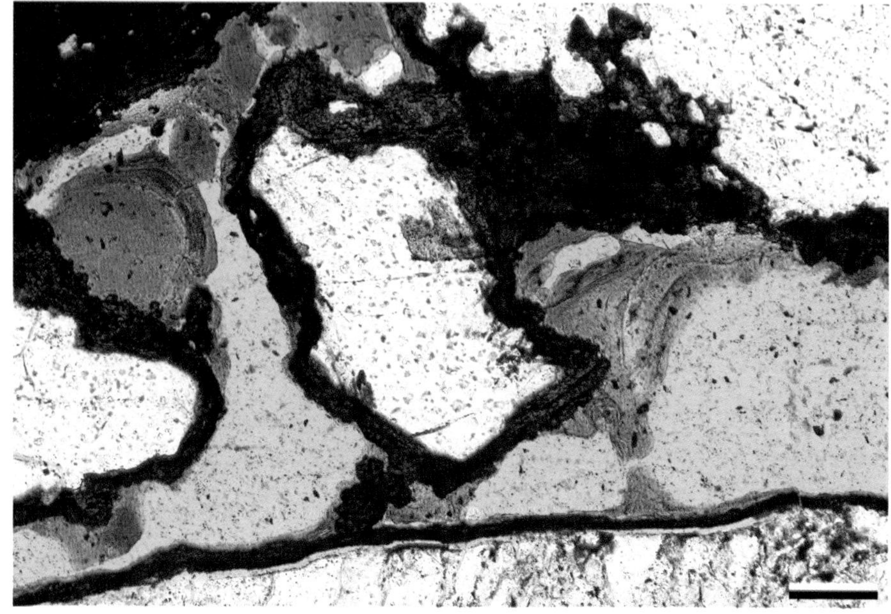

Lo. Cretaceous Patuxent Fm., Cumberland Co., North Carolina

This thin section contains two fabrics that are indicative of having formed in a paleosol—ferruginous crusts and clay cutans. The ferruginous crusts formed first and were probably originally hematite. Later, detrital clays (brown) were deposited by colloidal transport of material from higher in the soil zone into this layer. The clays were deposited in meniscus coatings (termed cutans), thereby also forming correspondingly rounded pores. Here, hematite (nearly opaque) was hydrated to goethite (dark red), possibly during uplift-related diagenesis (telogenesis).

PPL, BDI, Scale bar = 0.10 mm

Lo. Cretaceous Patuxent Fm., Cumberland Co., North Carolina

Soil crusts cement this quartz sandstone. The hematitic crusts rim the quartz grains and fossilized root traces (the circular structure near the center was likely such a rhizolith). The ferruginous crusts probably started out as hematite (opaque) but were partially replaced by goethite (red translucent phase).

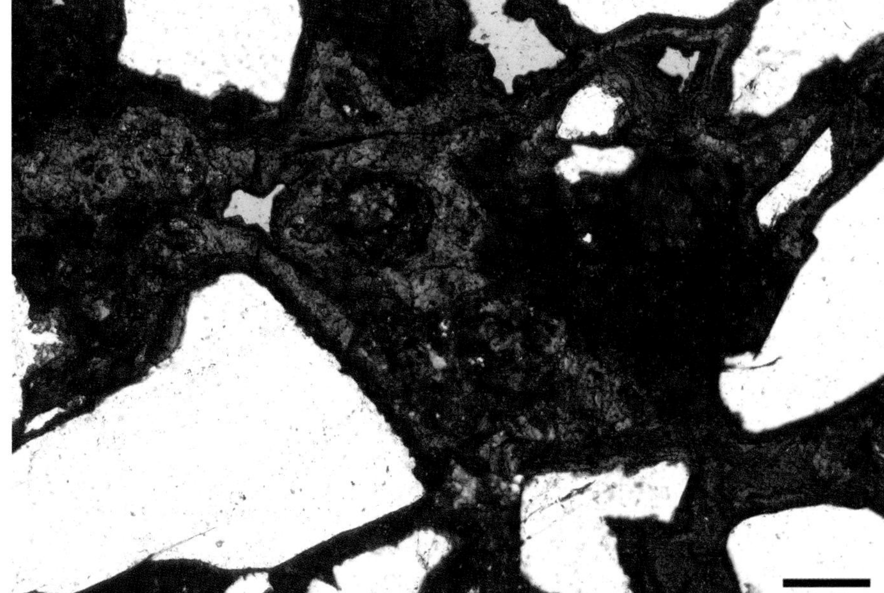

PPL, BDI, Scale bar = 0.07 mm

Mid. Eocene Reklaw Fm., Newby Mbr., Bastrop Co., Texas

This sandstone is cemented by goethite (and the goethite is probably after hematite). The iron oxides and hydroxides have been interpreted to have formed as one of several discrete ferricretes layers with iron derived from the alteration and leaching of "glauconite" and associated eogenetic iron minerals from overlying beds (Sharp and Sartin, 1988).

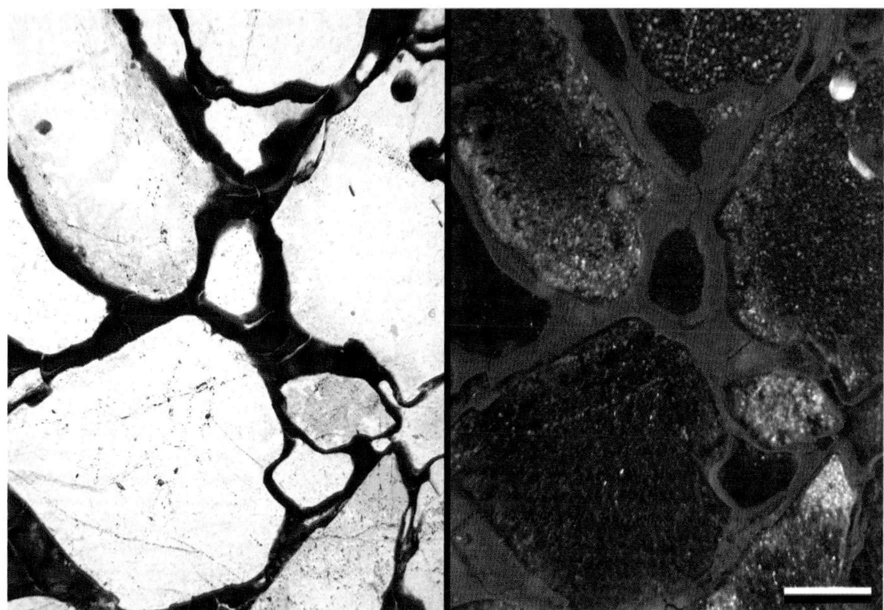

PPL | ORL, Scale bar = 0.20 mm

Cited References and Additional Information Sources

Berner, R. A., 1970, Sedimentary pyrite formation: American Journal of Science, v. 268, p. 1-24, doi: 10.2475/ajs.268.1.1.

Berner, R. A., 1981, A new geochemical classification of sedimentary environments: Journal of Sedimentary Research, v. 51, p. 359-365, doi: 10.1306/212F7C7F-2B24-11D7-8648000102C1865D.

Burdige, D. J., 1993, The biogeochemistry of manganese and iron reduction in marine sediments: Earth-Science Reviews, v. 35, p. 249-284, doi: 10.1016/0012-8252(93)90040-E.

Butler, I. B., and D. Rickard, 2000, Framboidal pyrite formation via the oxidation of iron (II) monosulfide by hydrogen sulphide: Geochimica et Cosmochimica Acta, v. 64, p. 2665-2672, doi: 10.1016/S0016-7037(00)00387-2.

Carroll, D., 1958, Role of clay minerals in the transportation of iron: Geochimica et Cosmochimica Acta: v. 14, p. 1-28, doi: 10.1016/0016-7037(58)90090-5.

Chan, M. A., J. Ormö, A. J. Park, M. Stich, V. Souza-Egipsy, and G. Komatsu, 2007, Models of iron oxide concretion formation: Field, numerical, and laboratory comparisons: Geofluids, v. 7, p. 356-368, doi: 10.1111/j.1468-8123.2007.00187.x.

Chan, M. A., W. T. Parry, and J. R. Bowman, 2000, Diagenetic hematite and manganese oxides and fault-related fluid flow in Jurassic sandstones, southeastern Utah: AAPG Bulletin, v. 84, p. 1281-1310, doi: 10.1306/A9673E82-1738-11D7-8645000102C1865D.

Cornell, R. M., and U. Schwertmann, 2003, The Iron Oxides: Structure, Properties, Reactions, Occurrences and Uses (2nd Edition): Weinheim, Wiley-VCH, 703 p.

Curtis, C. D., R. G. Burns, and J. V. Smith, 1977, Sedimentary geochemistry: Environments and processes dominated by involvement of an aqueous phase: Philosophical Transactions of the Royal Society of London, Series A, Mathematical and Physical Sciences, v. 286, p. 352-372, doi: 10.1098/rsta.1977.0123.

Curtis, C. D., J. C. W. Cope, D. Plant, and J. H. S. Macquaker, 2000, 'Instantaneous' sedimentation, early microbial sediment strengthening and a lengthy record of chemical diagenesis preserved in Lower Jurassic ammonitiferous concretions from Dorset: Journal of the Geological Society, v. 157, p. 165-172, doi: 10.1144/jgs.157.1.165.

Eichhubl, P., W. L. Taylor, D. D. Pollard, and A. Aydin, 2004, Paleo-fluid flow and deformation in the Aztec Sandstone at the Valley of Fire, Nevada—Evidence for the coupling of hydrogeologic, diagenetic, and tectonic processes: GSA Bulletin, v. 116, p. 1120-1136, doi: 10.1130/

B25446.1.

Folk, R. L., 1976, Reddening of desert sands: Simpson Desert, N.T., Australia: Journal of Sedimentary Research, v. 46, p. 604-615, doi: 10.1306/212F6FFF-2B24-11D7-8648000102C1865D.

Froelich, P. N., G. P. Klinkhammer, M. L. Bender, N. A. Luedtke, G. R. Heath, D. Cullen, P. Dauphin, D. Hammond, B. Hartman, and V. Maynard, 1979, Early oxidation of organic matter in pelagic sediments of the eastern equatorial Atlantic: Suboxic diagenesis: Geochimica et Cosmochimica Acta, v. 43, p. 1075-1090, doi: 10.1016/0016-7037(79)90095-4.

Ixer, R. A., P. Turner, and B. Waugh, 1979, Authigenic iron and titanium oxides in Triassic red beds (St. Bees Sandstone), Cumbria, northern England: Geological Journal, v. 14, p. 179-192, doi: 10.1002/gj.3350140214.

Ketzer, J. M., S. Morad, and A. Amorosi, 2003, Predictive diagenetic clay-mineral distribution in siliciclastic rocks within a sequence stratigraphic framework, *in* R. H. Worden, and S. Morad, eds., Clay Mineral Cements in Sandstones (IAS Special Publication 34): Oxford: Wiley-Blackwell, p. 43-61, doi: 10.1002/9781444304336.ch2.

Kirkland, D. W., R. E. Denison, and M. A. Rooney, 1995, Diagenetic alteration of Permian strata at oil fields of south central Oklahoma, USA: Marine and Petroleum Geology, v. 12, p. 629-644, doi: 10.1016/0264-8172(95)98089-N.

Lindsley, D. H., ed., 1991, Oxide Minerals: Petrologic and Magnetic Significance (Reviews in Mineralogy 25): Chantilly, VA, Mineralogical Society of America, 509 p.

Love, L. G., 1971, Early diagenetic polyframboidal pyrite, primary and redeposited, from the Wenlockian Denbigh Grit Group, Conway, North Wales, U.K.: Journal of Sedimentary Petrology, v. 41, p. 1038-1044, doi: 10.1306/74D723EC-2B21-11D7-8648000102C1865D.

Lu, G., C. McCabe, D. J. Henry, and A. Schedl, 1994, Origin of hematite carrying a Late Paleozoic remagnetization in a quartz sandstone bed from the Silurian Rose Hill Formation, Virginia, USA: Earth and Planetary Science Letters, v. 126, p. 235-246, doi: 10.1016/0012-821X(94)90109-0.

Mason, B. H., 1966, Principles of Geochemistry (3rd Edition): New York, NY, Wiley & Sons, 329 p.

McMahon, P. B., F. H. Chapelle, W. F. Falls, and P. M. Bradley, 1992, Role of microbial processes in linking sandstone diagenesis with organic-rich clays: Journal of Sedimentary Research, v. 62, p. 1-10, doi: 10.1306/D4267870-2B26-11D7-8648000102C1865D.

McManus, K. M., and J. S. Hanor, 1988, Calcite and iron sulfide cementation of Miocene sediments flanking the West Hackberry salt dome, southwest Louisiana, USA: Chemical Geology, v. 74, p. 99-112, doi: 10.1016/0009-2541(88)90148-9.

Mücke, A., 1994, Postdiagenetic ferruginization of sedimentary rocks (sandstones, oolitic ironstones, kaolins and bauxites)—including a comparative study of the reddening of red beds, *in* K. H. Wolf, and G. V. Chilingarian, eds., Diagenesis IV (Developments in Sedimentology 51): Amsterdam, Elsevier, p. 361-395, doi: 10.1016/S0070-4571(08)70444-8.

Odin, G. S., ed., 1988, Green Marine Clays: Oolitic Ironstone Facies, Verdine Facies, Glaucony Facies and Celadonite-Bearing Rock Facies—A Comparative Study (Developments in Sedimentology 45): Amsterdam, Elsevier Science Publ. Co., 445 p.

Parry, W. T., 2011, Composition, nucleation, and growth of iron oxide concretions: Sedimentary Geology, v. 233, p. 53-68, doi: 10.1016/j.sedgeo.2010.10.009.

Porrenga, D. H., 1966, Clay minerals in recent sediments of the Niger delta: Clays and Clay Minerals, v. 14, p. 221-233, doi: 10.1346/CCMN.1966.0140119.

Prosser, D. J., J. A. Daws, A. E. Fallick, and B. P. J. Williams, 1994, The occurrence and $\delta^{34}S$ of authigenic pyrite in Middle Jurassic Brent Group sediments: Journal of Petroleum Geology, v. 17, p. 407-428, doi: 10.1111/j.1747-5457.1994.tb00148.x.

Reiners, P. W., M. A. Chan, and N. S. Evenson, 2014, (U-Th)/He geochronology and chemical compositions of diagenetic cement, concretions, and fracture-filling oxide minerals in Mesozoic sandstones of the Colorado Plateau: GSA Bulletin, v. 126, p. 1363-1383, doi: 10.1130/B30983.1.

Reynolds, R. L., N. S. Fishman, R. B. Wanty, and M. B. Goldhaber, 1990, Iron sulfide minerals at Cement oil field, Oklahoma: Implications for magnetic detection of oil fields: GSA Bulletin, v. 102, p. 368-380, doi: 10.1130/0016-7606(1990)1022.3.CO;2.

Salem, A. M. K., A. Abdel-Wahab, and E. F. McBride, 1998, Diagenesis of shallowly buried cratonic sandstones, southwest Sinai, Egypt: Sedimentary Geology, v. 119, p. 311-335, doi: 10.1016/S0037-0738(98)00056-6.

Schieber, J., 2007, Oxidation of detrital pyrite as a cause for marcasite formation in marine lag deposits from the Devonian of the eastern US: Deep Sea Research, Part II: Topical Studies in Oceanography, v. 54, p. 1312-1326, doi: 10.1016/j.dsr2.2007.04.005.

Schieber, J., 2011, Marcasite in black shales—a mineral proxy for oxygenated bottom waters and intermittent oxidation of carbonaceous muds: Journal of Sedimentary Research, v. 81, p. 447-458, doi: 10.2110/jsr.2011.41.

Schieber, J., and G. Baird, 2001, On the origin and significance of pyrite spheres in Devonian black shales of North America: Journal of Sedimentary Research, v. 71, p. 155-166, doi: 10.1306/051600710155.

Sharp, P. S., and A. A. Sartin, 1988, The Claiborne Group of East Texas, *in* O. T. Hayward, ed., South-Central Section of the Geological Society of America, Decade of North American Geology Centennial Field Guide, Volume 4: Boulder, CO, GSA, p. 335-338.

Stel, H., 2009, Diagenetic crystallization and oxidation of siderite in red bed (Buntsandstein) sediments from the Central Iberian Chain, Spain: Sedimentary Geology, v. 213, p. 89-96, doi: 10.1016/j.sedgeo.2008.12.001.

Surdam, R. C., Z. S. Jiao, and D. B. MacGowan, 1993, Redox reactions involving hydrocarbons and mineral oxidants: A mechanism for significant porosity enhancement in sandstones: AAPG Bulletin, v. 77, p. 1509-1518.

Van Houten, F. B., 1973, Origin of red beds, a review—1961-1972: Annual Review of Earth and Planetary Sciences, v. 1, p. 39-61, doi: 10.1146/annurev.ea.01.050173.000351.

Van Houten, F. B., and D. P. Bhattacharyya, 1982, Phanerozoic oolitic ironstones—geologic record and facies model: Annual Review of Earth and Planetary Sciences, v. 10, p. 441-457, doi: 10.1146/annurev.ea.10.050182.002301.

Vaughan, D. J., ed., 2006, Sulfide Mineralogy and Geochemistry (Reviews in Mineralogy 61): Chantilly, VA, Mineralogical Society of America, 714 p.

Walker, T. R., 1967, Formation of red beds in modern and ancient deserts: GSA Bulletin, v. 78, p. 353-368, doi: 10.1130/0016-7606(1967)78[353:FORBIM]2.0.CO;2.

Walker, T. R., 1976, Diagenetic origin of continental red beds, *in* H. Falke, ed., The Continental Permian in Central West and South Europe: Dordrecht (Netherlands), D. Reidel Publishing Co., p. 240-282, doi: 10.1007/978-94-010-1461-8_20.

Walker, T. R., E. E. Larson, and R. P. Hoblitt, 1981, Nature and origin of hematite in the Moenkopi Formation (Triassic), Colorado Plateau: A contribution to the origin of magnetism in red beds: Journal of Geophysical Research: Solid Earth, v. 86, p. 317-333, doi: 10.1029/JB086iB01p00317.

Wilkin, R. T., and H. L. Barnes, 1997, Formation processes of framboidal pyrite: Geochimica et Cosmochimica Acta, v. 61, p. 323-339, doi: 10.1016/S0016-7037(96)00320-1.

Facing Page: Top — Phosphate-cemented sandstone containing quartz grains and phosphate clasts, Pliocene Bone Valley Fm., Polk Co., Florida (XPL). Bottom — Oil-stained zones in a calcite cemented sandstone above an oil seep. Permian Rush Springs Fm., Caddo Co., Oklahoma (PPL | XPL).

DIAGENESIS:
OTHER CEMENTS

CHAPTER 18

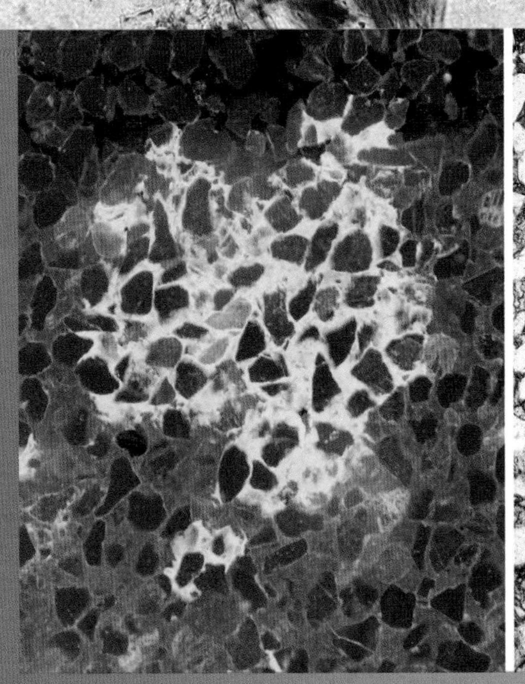

Introduction to Other Cements

This last section on cementation provides images that feature some of the less common mineral cements that were not covered in previous chapters. It is by no means an all-encompassing chapter because there are simply too many minor detrital mineral overgrowths or volumetrically insignificant, fully authigenic precipitates to cover in detail.

Phosphates are probably the most common and volumetrically important minerals that will be illustrated here. These include a variety of apatite group minerals (mainly carbonate-fluorapatite or hydroxylapatite) as well as less common minerals such as wavellite ($Al_3(PO_4)_2(OH,F)_3 \cdot 5H_2O$). Apatite cements tend to form in eogenetic, marine, diagenetic environments in nodules, hardgrounds and bedded sedimentary phosphorites—such occurrences were discussed and illustrated in Chapters 5 and 9 and so will not be elaborated on in this introduction. The less common phosphate cements, such as wavellite, generally form in nonmarine or hydrothermal settings, especially during surficial weathering of phosphatic sediments and rocks (Blanchard and Denahan, 1967).

A number of detrital heavy minerals, including tourmaline, brookite, titanite (formerly termed sphene), rutile, anatase and zircon typically form either authigenic overgrowths or fully authigenic precipitates during burial (Alty, 1933; Awasthi, 1961; Morad, 1986 and 1988; Henry et al., 1994; Milliken, 2007; Hay and Dempster, 2009; Dill, 2010). Other minerals, especially metallic sulfides such sphalerite ($(Zn,Fe)S$) and galena (PbS), can grow as fully authigenic cements or replacements, especially in hydrothermal settings. Other minerals, including prehnite (an inosilicate commonly associated with zeolites), can occur in cavities and veins, particularly in association with volcaniclastic sediments. See the individual illustrations in this chapter for mineral-specific recognition features.

The last "cement" covered in this chapter is solid to semisolid hydrocarbons. Although neither a mineral nor a cement in the conventional sense of those terms, hydrocarbons that enter pores as fluids can lose volatiles and harden during diagenesis to form pore-filling, rock-solidifying and permeability-reducing materials that thus are the equivalents of cements (Lomando, 1992). In addition, the presence of heavy oils, just as with normal oils, can substantially inhibit the formation of "true" mineral cements in many circumstances (e.g., Worden et al., 1998).

Altered hydrocarbons form a spectrum spanning semisolids such as heavy oils, tars and waxes (like ozocerite) that can be partially recovered through the application of heat or chemical solvents, as well as near-solid asphalts such as uintaite (gilsonite) that form not just intergranular cements, but also larger vein and vug fillings (Monson and Parnell, 1992). The hardening of fluid oils to highly viscous, semisolid material may occur through thermal chemical maturation, thermochemical sulfate reduction, biodegradation (microbial decomposition) and other processes (Kelemen et al., 2010). The biodegradation can be associated with meteoric water influx, but recent work has shown that it also proceeds under normal burial conditions at temperatures up to about 80°C, especially at or near oil/water contacts (Head et al., 2003; Larter et al., 2006; Hein et al., 2013).

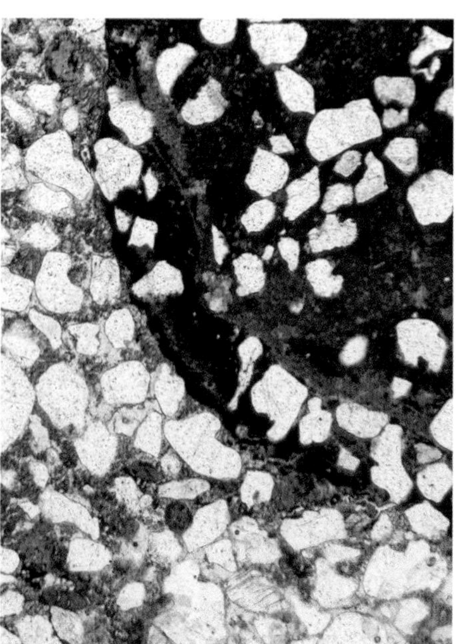

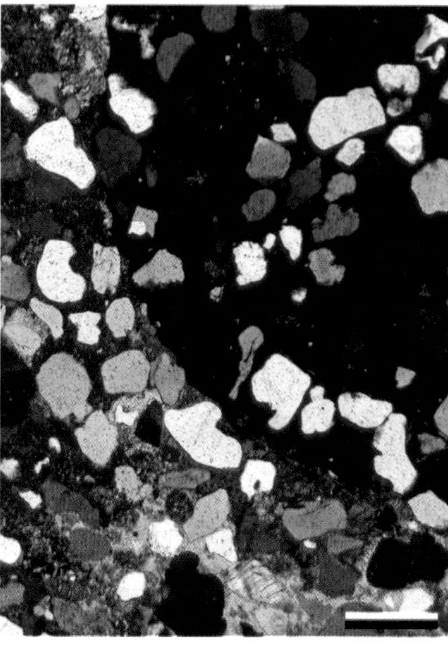

Lo. Cretaceous (Albian) Upper Folkestone Beds, Kent, England, U.K.

Phosphate precipitated as a cement in this concretion, surrounding quartz and glauconite grains in the precursor sediment. Phosphate concretions typically form early, prior to substantial burial or other cementation. Indeed, many phosphatic nodules are formed so early that they are reworked as pebbles or are bored by organisms at hiatus surfaces.

PPL | XPL, Scale bar = 0.51 mm

Mid. Cambrian Fucoid Mbr., An t-Sròn Fm., northwest Scotland, U.K. †

A sandstone cemented entirely by phosphate in a unit that was formerly quarried for fertilizer. The phosphate is microcrystalline and brownish in plane-polarized light. It is effectively isotropic in cross-polarized light when microcrystalline, which is a key distinguishing character from most other cements. The phosphate also has a higher relief than the enclosed quartz grains, which therefore appear less prominent. The "floating grains" with few point contacts suggest that the phosphate is either an early cement or a phosphate replacement of a precursor matrix.

PPL | XPL, Scale bar = 0.12 mm

Lo. Cretaceous (Albian) Gault Clay, Kent, England, U.K.

This phosphate-cemented sandstone contains sand-sized grains of quartz and glauconite that are encased in a pore-filling phosphatic (calcium fluorapatite) cement that completely occluded porosity, at least locally. This cement has patchy, concretionary distribution localized within specific bedding-parallel intervals. These cements are inferred to have formed in a moderately deep shelf setting during a transgression.

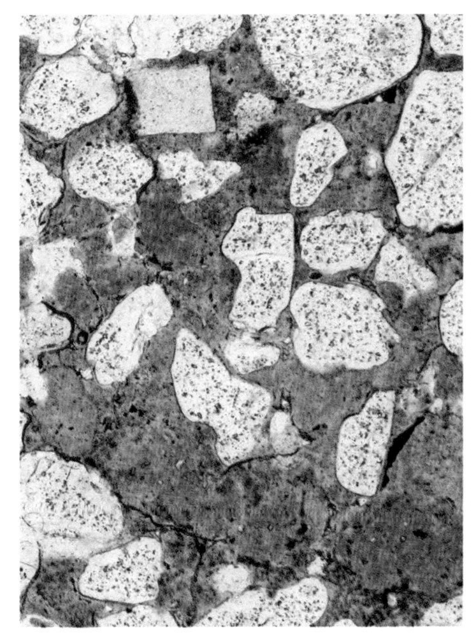

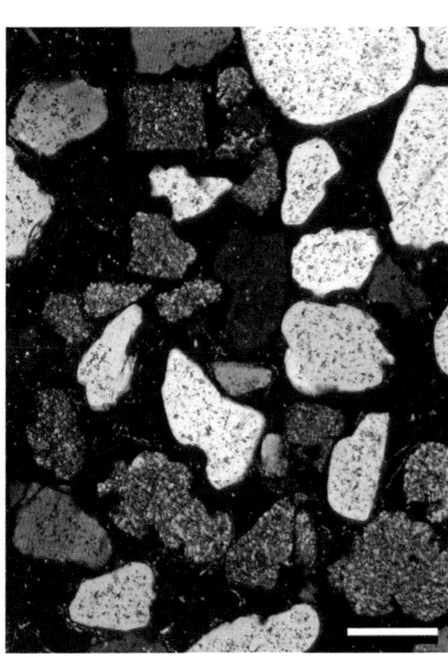

PPL | XPL, Scale bar = 0.51 mm

Lo. Permian Phosphoria Fm., Meade Peak Mbr., Daggett Co., Utah

This sample comprises quartz grains, phosphatic peloids, bone fragments, and scattered oolitically-coated phosphate grains that have been cemented by phosphate (note that in the cross-polarized light view that only the quartz is birefringent). The brownish color is typical of phosphate (carbonate fluorapatite, hydroxylapatite or other apatite minerals). Phosphate deposits of this type commonly mark zones of upwelling of cool, phosphate-saturated, oceanic waters on ancient continental margins, other zones of high biological productivity or hiatus intervals.

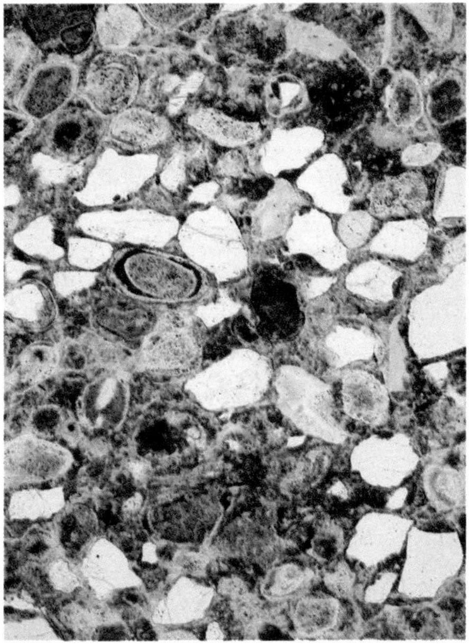

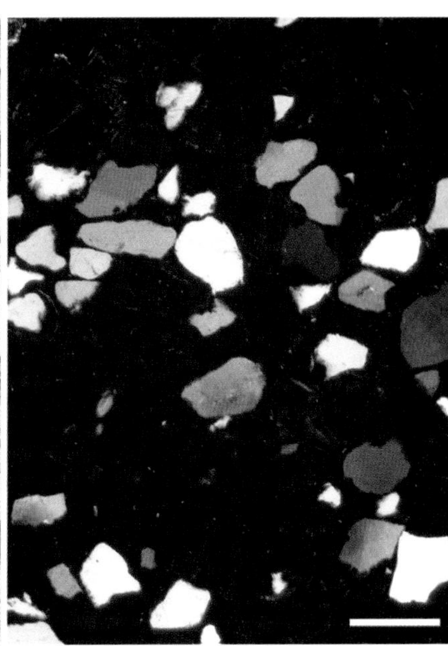

PPL | XPL, BDI, Scale bar = 0.38 mm

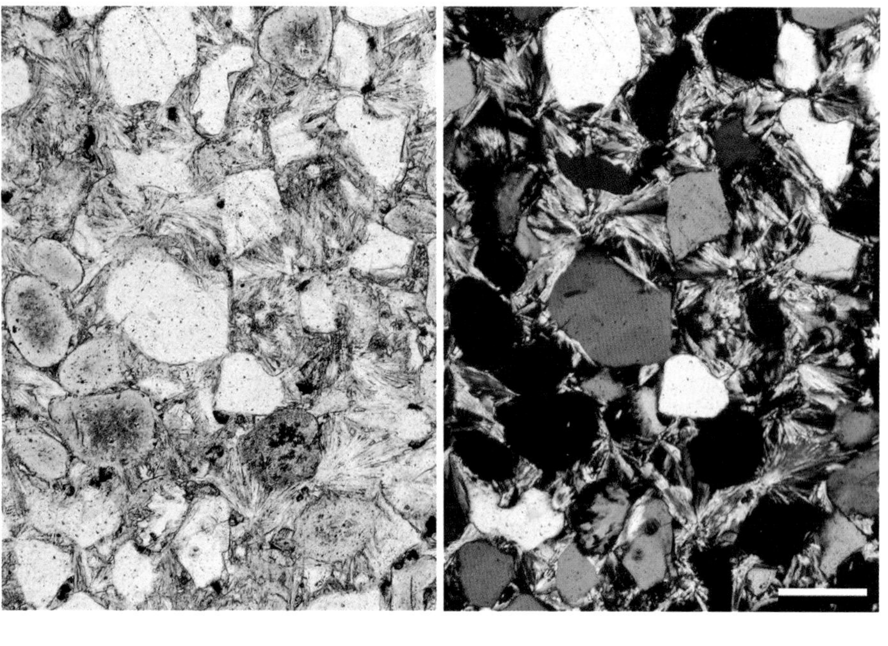

Pliocene Bone Valley Fm., Polk Co., Florida

A phosphate-cemented sandstone containing quartz grains and phosphate peloids or "pellets". The detrital phosphate grains are completely isotropic or have extremely low birefringence. The extensive phosphatic cement, composed primarily of wavellite, has moderate birefringence and radiating, fibrous crystal structure. Phosphate cements are most common in sediments with abundant detrital phosphate or in association with hiatus intervals.

PPL | XPL, Scale bar = 0.26 mm

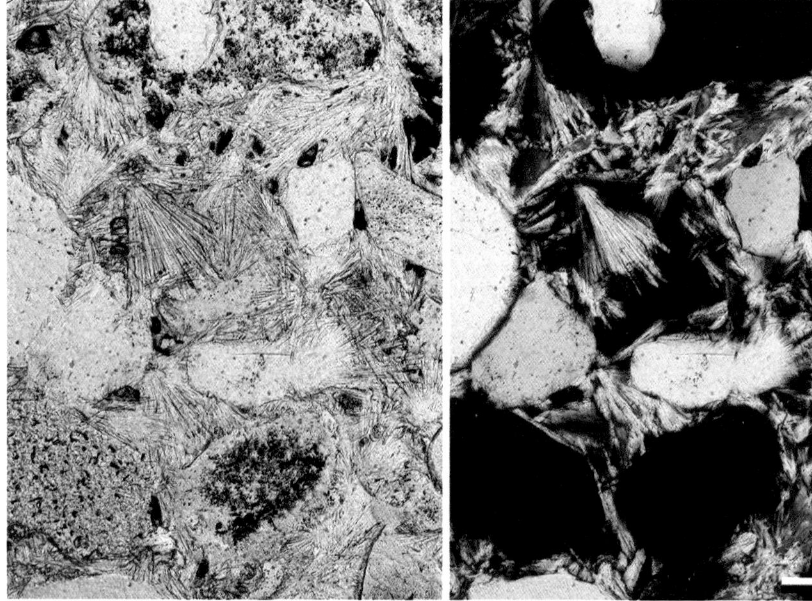

Pliocene Bone Valley Fm., Polk Co., Florida

This higher magnification view of the previous sample shows the quartz and nonbirefringent phosphatic grains and more clearly shows the wavellite structure. The wavellite occurs in radiating to botryoidal masses of fibrous crystals that are biaxial positive and have low relief and high first-order to low second-order birefringence. These cements typically form in near-surface to shallow-burial environments where groundwaters leach phosphate and other unstable grains, including feldspars and clays (Bergendahl, 1955; Blanchard and Denahan, 1967; Klemic and Mrose, 1972).

PPL | XPL, Scale bar = 0.12 mm

Paleocene Fatehgarh Fm., Rajasthan, India

A cluster of diagenetic titanite crystals (formerly termed sphene) that range from subhedral to euhedral rhombs. In this case, they may have crystallized from available silica and titanium ions, with calcium derived from the albitization of calcic-plagioclase and with possible concurrent precipitation of laumontite. A similar process, involving calcium from calcite dissolution, was suggested by Morad (1988). Titanite is also common in sandstones that have undergone extensive intergranular pressure solution, as titanium liberated by quartz dissolution has low mobility and typically precipitates locally (relative to the site of titanium liberation).

PPL | XPL, BDI, Scale bar = 0.07 mm

Mid. Jurassic Entrada Fm., Moffat Co., Colorado

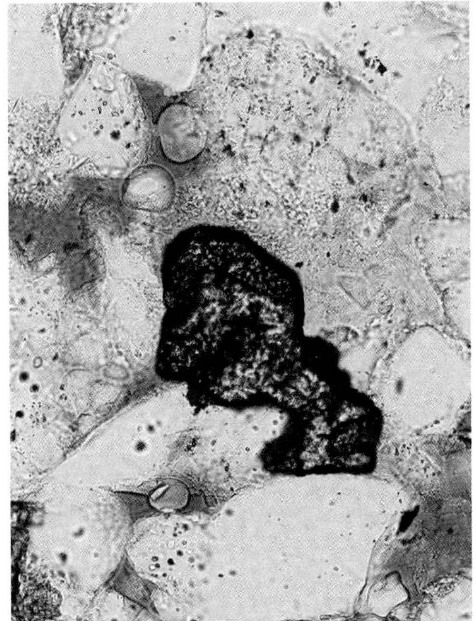

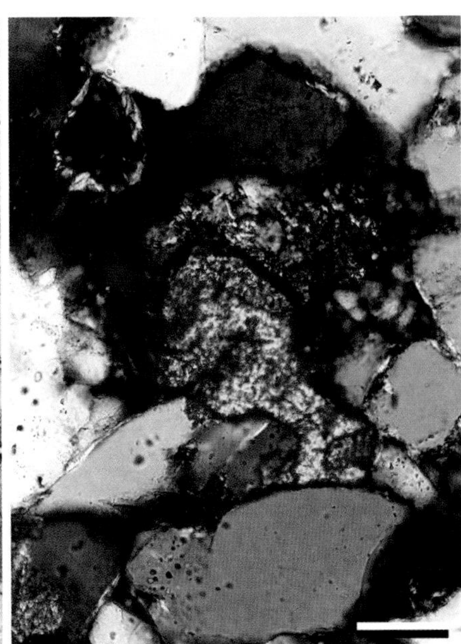

In this sandstone, minor amounts (<1%) of titanite occur as authigenic cements. Titanite cements are uncommon, but they provide information about the unit's burial diagenesis as they form from pore fluids at elevated temperatures. The source of the major elements for titanite is the dissolution of titanium-bearing minerals and rock fragments (e.g., pyroxenes, amphiboles, volcanic rock fragments), regional albitization, the dissolution of calcium-rich feldspars, carbonate grains and/or cements, and clay diagenesis. In this case, the correspondence of the titanite grain boundaries with nearby grains suggest the presence of overgrowths on a detrital titanite core.

PPL | XPL, RDI, Scale bar = 0.05 mm

Paleocene Tang Fm., Vøring Basin, Norwegian Sea †

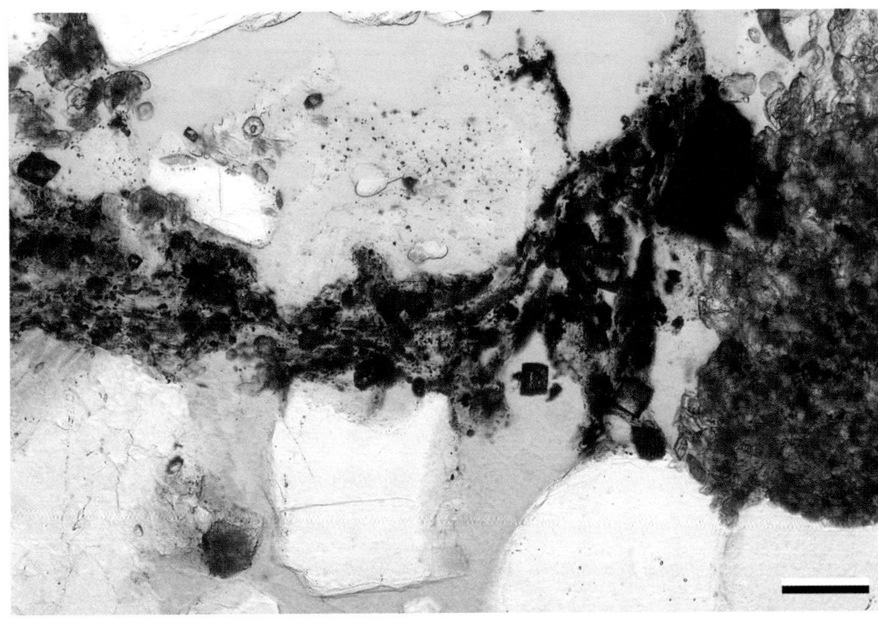

Cubes of anatase associated with detrital pseudomatrix, probably a degraded biotite grain—a likely the source of the titanium in the anatase. The pseudomatrix appears to have undergone ductile deformation by what is now kaolinite, but it was likely a rigid grain at the time of deformation, possibly a feldspar. Siderite cement occurs along the right side of the image. Anatase is similar to rutile in having very high relief. It has high birefringence (but slightly lower birefringence than rutile) and is uniaxial (-).

PPL, BDI, Scale bar = 0.07 mm

Paleocene Fatehgarh Fm., Rajasthan, India

Anatase is a common diagenetic mineral in sandstones, although it generally is minor in abundance. It tends to be associated with the degradation of biotite and rutile grains, which provide a local source of titanium ions. Anatase is difficult to photograph, because of its generally small size and high relief. The crystals here are well formed, show a typical greenish brown color and are surrounded by kaolinite. Diagenetic anatase is normally recognizable by its euhedral rectangular shape.

PPL, BDI, Scale bar = 0.07 mm

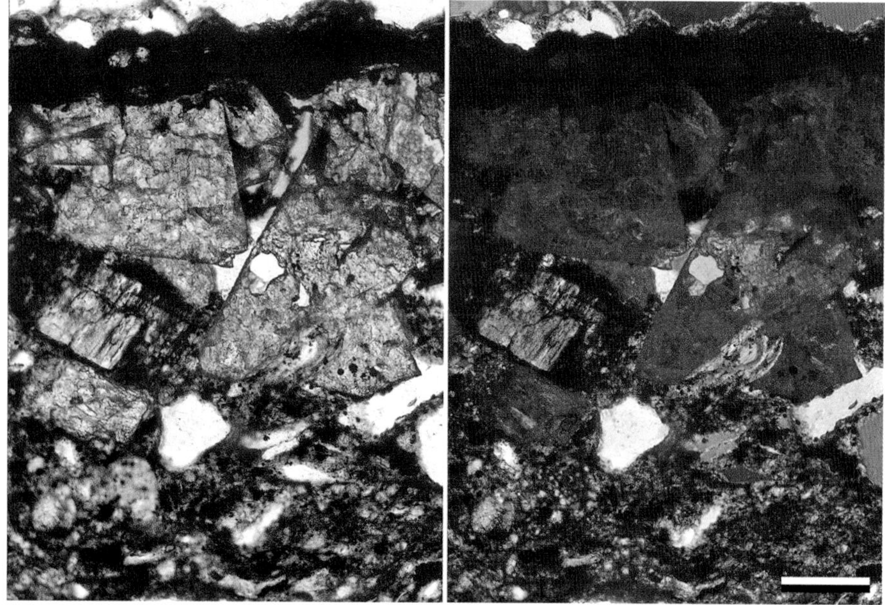

Up. Jurassic Fulmar Fm., United Kingdom sector, North Sea †

Sphalerite within a fractured sandstone. Sphalerite is a rare but distinctive cement; it is normally pale brown, but can be pale yellow or colorless. These crystals range from colorless to brown and display some cleavage. In plane-polarized light, they show high relief and are virtually isotropic under cross-polarized light. The sphalerite here has grown into a fracture. Sphalerite is normally a hydrothermal sulfide, and its occurrence in sandstones generally indicates the influence of hot brines.

PPL | XPL, BDI, Scale bar = 0.12 mm

Mid. Jurassic Entrada Fm., Moffat Co., Colorado

After early, but incomplete, quartz cementation of this lithic arenite (that includes a detrital blue tourmaline grain at upper center), hydrothermal fluids migrating through the unit locally precipitated sphalerite cement. Note how the sphalerite crystal (dark area in lower center) conforms to both adjacent grains and precursor cement boundaries.

PPL, Scale bar = 0.05 mm

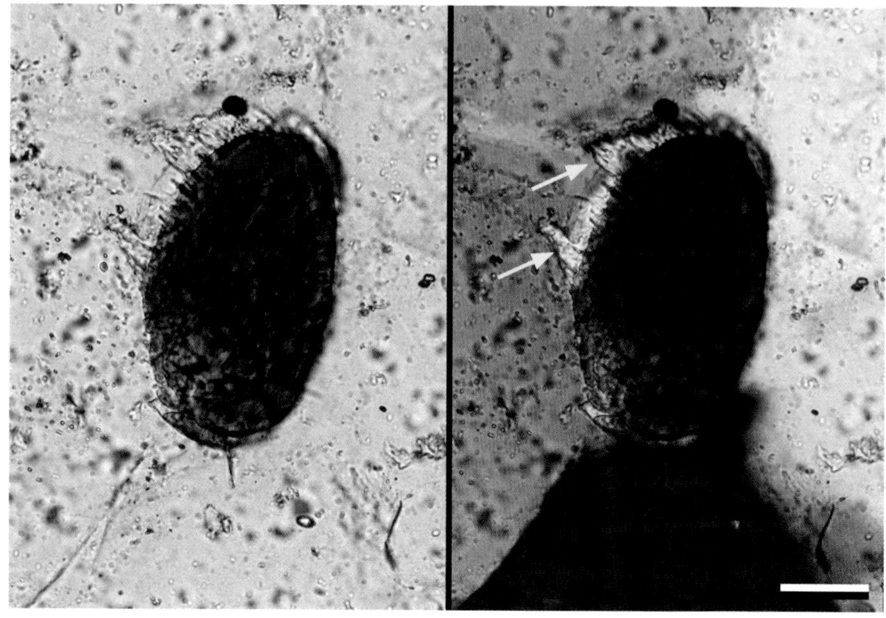

Up. Cambrian Gatesburg Fm., Centre Co., Pennsylvania

A rounded, detrital tourmaline with authigenic overgrowths. The tourmaline grain shows very poor cleavage, high relief, green/brown color with extreme pleochroism (from medium green to completely black with stage rotation), and oriented, ragged overgrowths (yellow arrows). Tourmaline is derived from acidic igneous rocks as well as regional and contact metamorphic rocks. Authigenic overgrowths on tourmaline grains exhibit an unusual characteristic of growing asymmetrically because the overgrowths form preferentially on one end of the c-axis (Alty, 1933; Henry et al., 1994).

PPL | XPL, Scale bar = 0.05 mm

Paleocene lava, Rajasthan, India

Prehnite, an inosilicate ($Ca_2Al_2Si_3O_{10}(OH)_2$), commonly forms curved fiber masses. It is a biaxial (+) (with a 2V = 67-71°) mineral with moderate relief. The crystals display up to second-order birefringence. The mass of fibrous crystals displays sweeping extinction. The prehnite here occurs in an amygdule as a secondary alteration mineral. It also occurs in sandstones which have experienced temperatures above 200°C (Cho et al., 1986).

PPL | XPL, BDI, Scale bar = 0.25 mm

Triassic Skagerrak Fm., United Kingdom sector, North Sea

This vug is partially filled with bladed calcite cements that are followed by zoned, euhedral, baroque dolomite crystals. The remaining porosity was filled by hydrocarbons. Later, those hydrocarbons were degraded to bitumen or "dead" oil coatings on the crystals within that pore as well as many nearby smaller pores. Based on the pore shape, the vug may have started out as an evaporite nodule that was leached and later partially infilled by the carbonate cements and hydrocarbons.

PPL, BDI, Scale bar = 0.09 mm

Triassic Skagerrak Fm., United Kingdom sector, North Sea

This fracture was partially filled with blocky calcite cements. Prior to complete cementation of the fracture, hydrocarbons moved into the system, essentially shutting down further diagenesis in the rock. The hydrocarbons form a thin brownish ("dead" oil) stain on the surface of the carbonate cements. This sample is from the same thin section as the previous photomicrograph, but notice the color difference in the blue-dyed epoxy. Here the epoxy is yellowish-green because "live" (fluid) oil within the fracture mixed with epoxy during sample preparation.

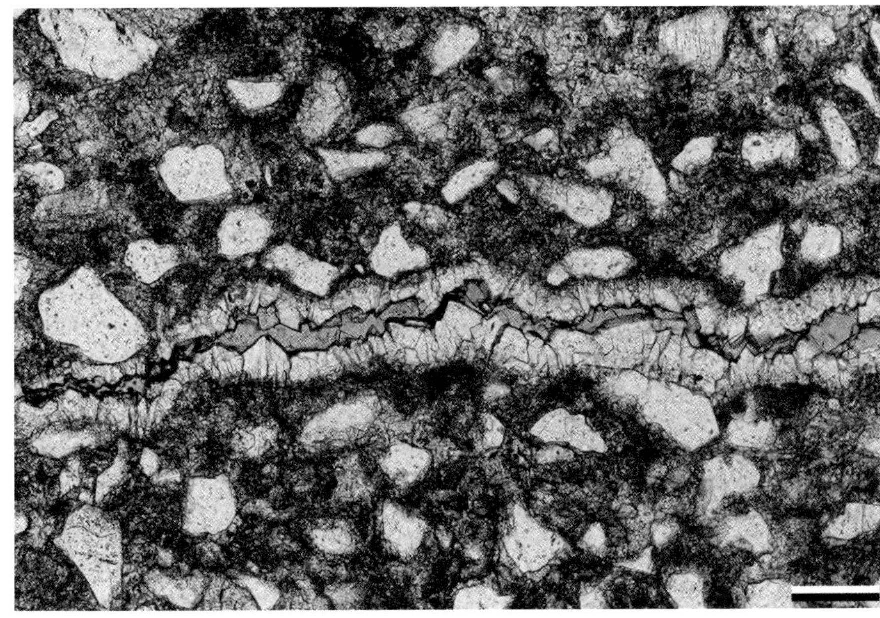

PPL, BDI, Scale bar = 0.14 mm

Pliocene Up. Monterey Fm., Santa Barbara Co., California

Hydrocarbons such as oil and asphaltic residues can effectively fill the pore space in sandstones. In this tar sand, amorphous, brownish-to-black, semisolid petroleum hydrocarbons completely fill pore space. These hydrocarbons can be very effective in reducing the permeability of the unit by blocking pores. Hydrocarbons and clays (especially iron-stained clays) can be difficult to distinguish from each other. The presence of hydrocarbon staining on clays can be confirmed by fluorescence with ultraviolet light (for "live oil") or by extraction with solvents.

PPL, Scale bar = 0.10 mm

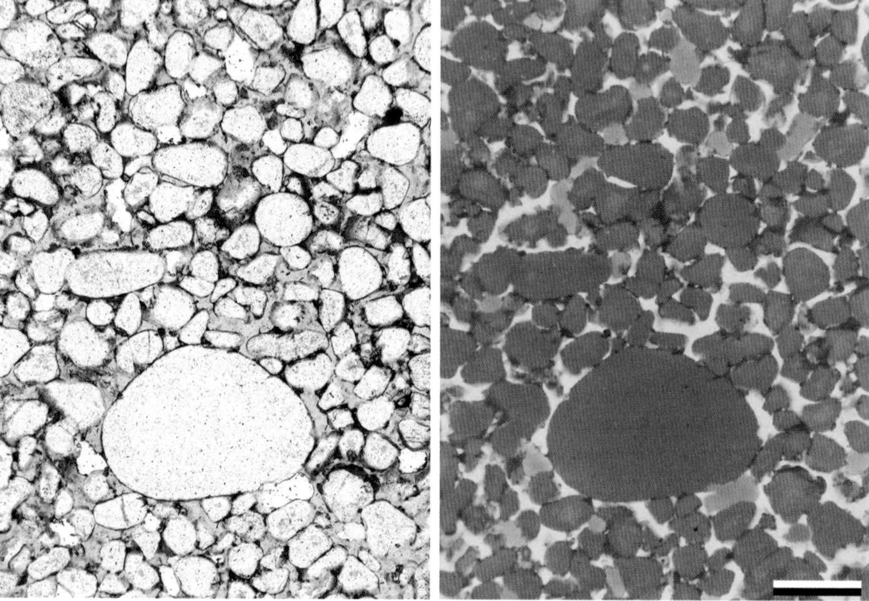

Mid.–Up. Ordovician Simpson Gp., Oil Creek Fm., Murray Co., Oklahoma

This well-rounded quartzarenite has all the intergranular porosity filled by a combination of "live" oil and minor "dead" oil or bitumen (darker brown material in plane-polarized light and dark areas between grains in fluorescent light). Fluorescence microscopy is a relatively easy and inexpensive way to see both the presence and type of hydrocarbons. It also can aid in recognition of microporosity if thin sections have been impregnated with fluorescent epoxy.

PPL | FL 470, Scale bar = 0.26 mm

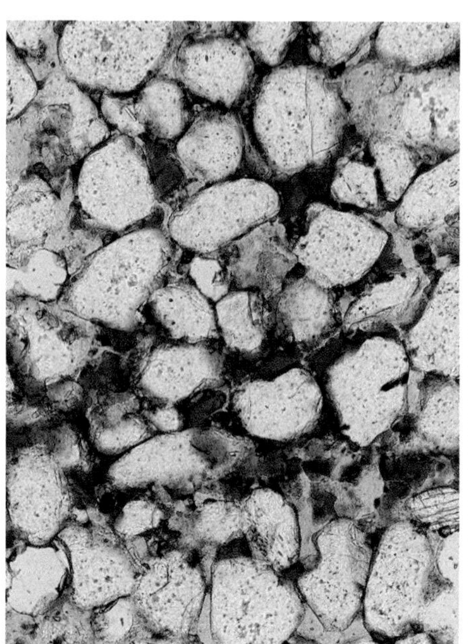

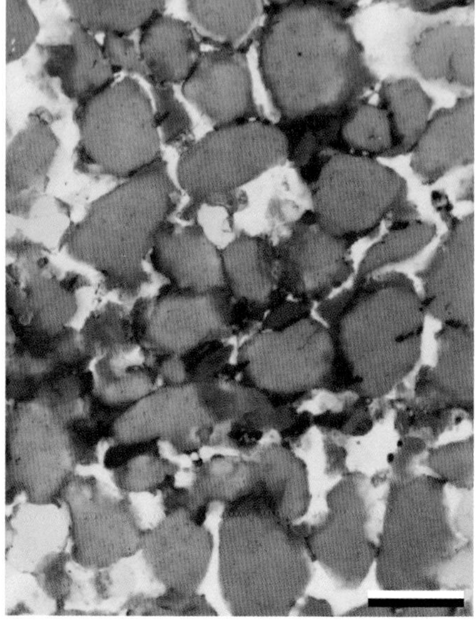

Mid.–Up. Ordovician Simpson Gp., Oil Creek Fm., Murray Co., Oklahoma

A close-up view of the previous thin section showing a zone with significant "dead" oil. The "dead" oil is non-luminescent because it lacks volatiles that are found in "live" hydrocarbons. Bitumen ("dead" oil) potentially can disrupt the units permeability by blocking pore throats. In this example, many of the pores are filled with bitumen, thus preventing fluid migration through these zones. Significant live oil saturation also can impact fluid flow of other phases (specifically water).

PPL | FL 470, Scale bar = 0.10 mm

Eocene Thumbli Fm., Rajasthan, India

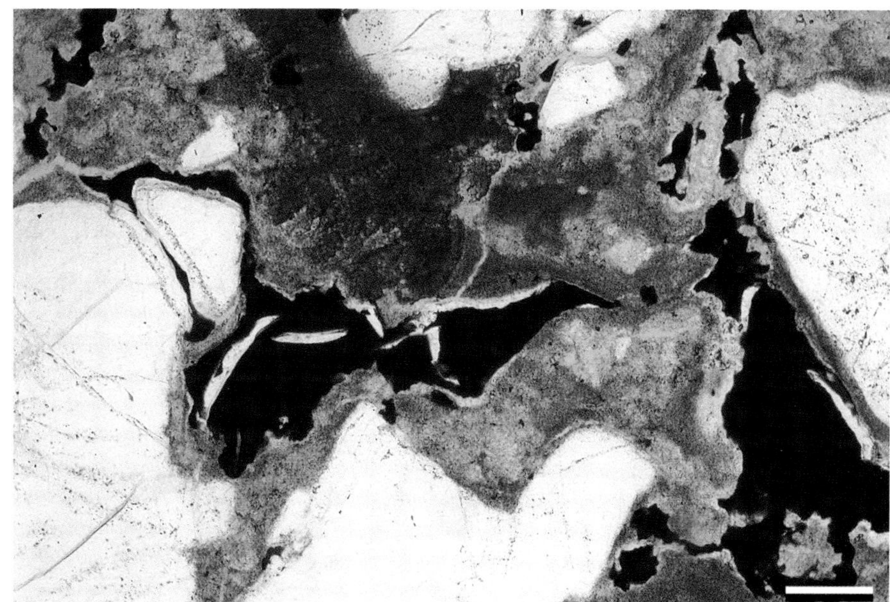

Bituminous hydrocarbons in pore spaces. The pores are partly filled by kaolinite, which the hydrocarbons have not invaded, as seen in the top right. Other areas are surrounded by microcrystalline siderite. The hydrocarbons have contracted away from the edges of the pores and display white curved shapes—these are cracks not filled by blue resin, possibly because the oil shrank after resin impregnation. The shrinkage cracks and brown color are characteristic. Most thin sections are cleaned with solvent, so only the heavier hydrocarbons remain. Generally, oils are seen as thin films around pore margins.

PPL, BDI, Scale bar = 0.49 mm

Paleocene Fatehgarh Fm., Rajasthan, India

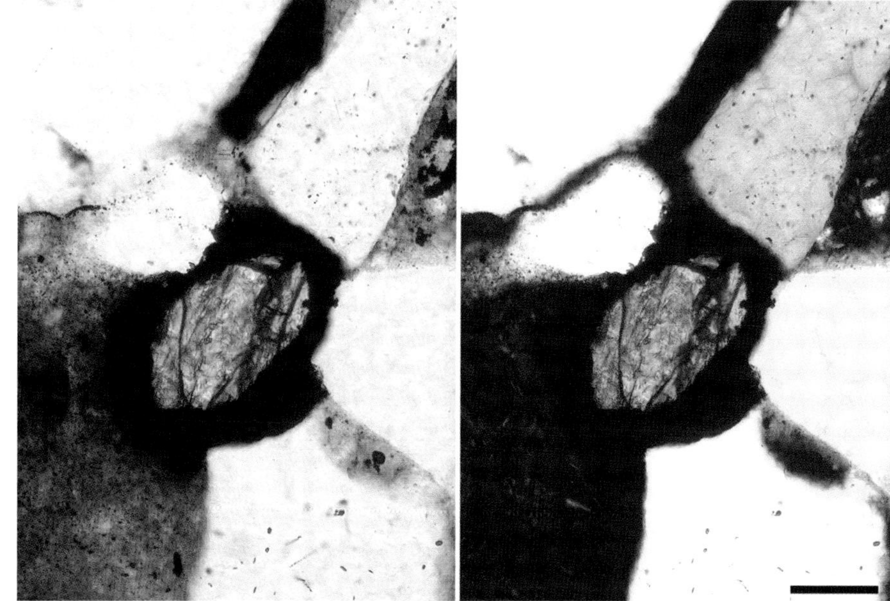

A subrounded detrital zircon grain enveloped by hydrocarbons. The association of degraded oils with zircon and monazite grains is a common one. This happens because most zircons and monazite grains are radioactive and the radiation matures and degrades the oil. In this case, the oil was within microporous detrital clay surrounding the grain.

PPL | XPL, BDI, Scale bar = 0.07 mm

Up. Jurassic Spekk Fm., Vøring Basin, Norwegian Sea †

A strongly compacted and quartz-cemented sample in which, as above, oil in an isolated pore has been degraded because of its proximity to a zircon grain. The radiation is likely to have degraded oil into an asphaltic residue. In this example, quartz cementation did not stop after oil entered the pore. The surface of the quartz likely remained water wet, but the quartz overgrowths do contain clearly visible hydrocarbon inclusions (examples at red arrows).

PPL, BDI, Scale bar = 0.03 mm

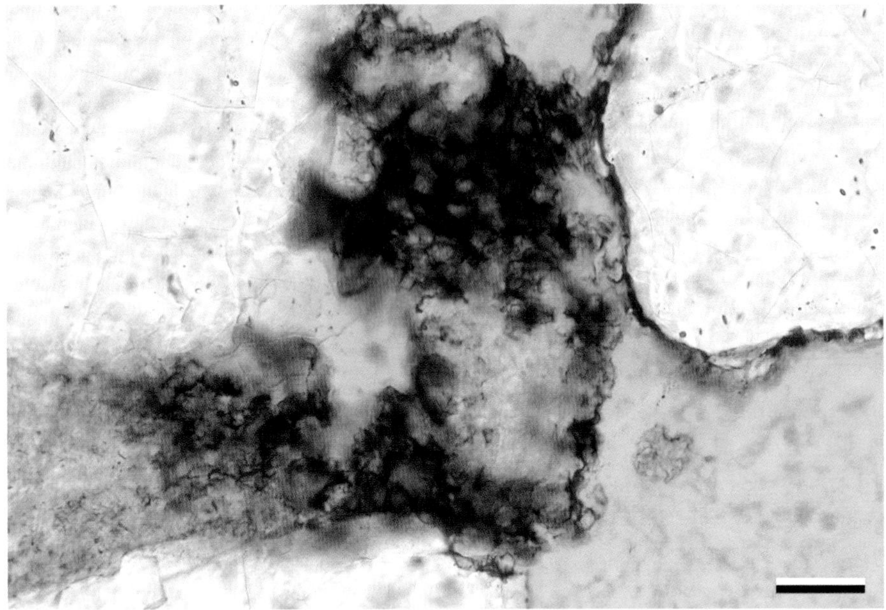

Paleocene Fatehgarh Fm., Rajasthan, India

Hydrocarbons within microporosity between vermicular kaolinite books. Within oil-wet sandstones, it is known that kaolinite is preferentially wetted by oil, while illite and other minerals are preferentially wetted by water (Fassi-Fihri et al., 1995). This means that in many thin sections from oil reservoirs, kaolinitic areas can be selectively associated with oil staining.

PPL, BDI, Scale bar = 0.03 mm

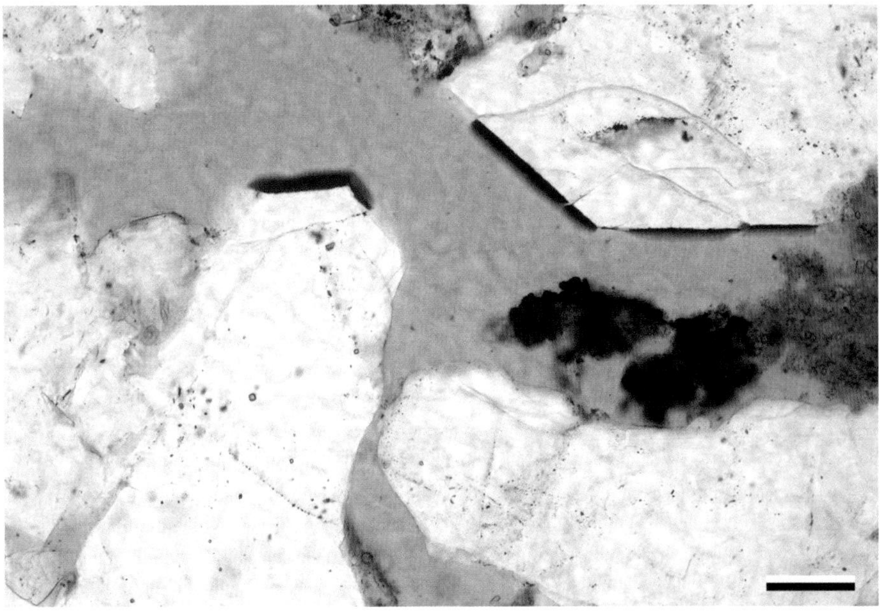

Up. Jurassic Fulmar Fm., United Kingdom sector, North Sea †

Brown hydrocarbon patches within primary pore space. The brown patches are not, in themselves, evidence of hydrocarbons because they could also either be organic fragments or oil-stained clays. Quite often, when hydrocarbons are very dark, it is worth checking that the material is not pyrite by using reflected light. The clay to the right is also brownish and likely to be oil stained. The strongest evidence for oil in these pore spaces is the thin black coating on the quartz overgrowths.

PPL, KFS, BDI, Scale bar = 0.07 mm

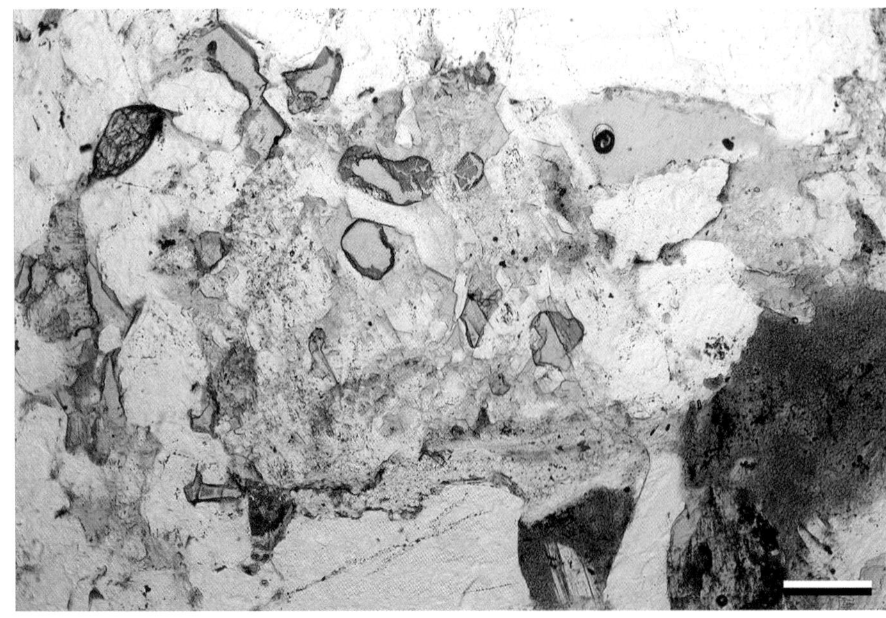

Mid. Jurassic Ile Fm., Vøring Basin, Norwegian Sea †

Oil stain within secondary pores in a dissolved plagioclase feldspar grain. The hydrocarbons are lighter brown than those in the previous picture and probably represent a less bituminous oil. A patch of kaolinite below the degraded grain, possibly an altered and deformed mica, also has an oil stain around its margins. The oil charge clearly occurred after dissolution of the plagioclase feldspar (center).

PPL, KFS, BDI, Scale bar = 0.12 mm

Mid. Permian Rush Springs Fm., Caddo Co., Oklahoma

Not all hydrocarbons are easily visible utilizing standard petrographic techniques. In this example from sandstones overlying the Cement oil field, hydrocarbons have seeped upward and contributed to calcite cement formation (e.g., Donovan et al., 1974; Kirkland et al., 1995). In plane-polarized light, there is little or no evidence of hydrocarbons being present, but in fluorescent light, the extent of hydrocarbon impregnation of clay-rich areas is visible (bright yellow areas). Also more easily seen in fluorescent light is the patchy nature of the carbonate cements (dull, nonfluorescent areas).

PPL | FL 470, Scale bar = 0.26 mm

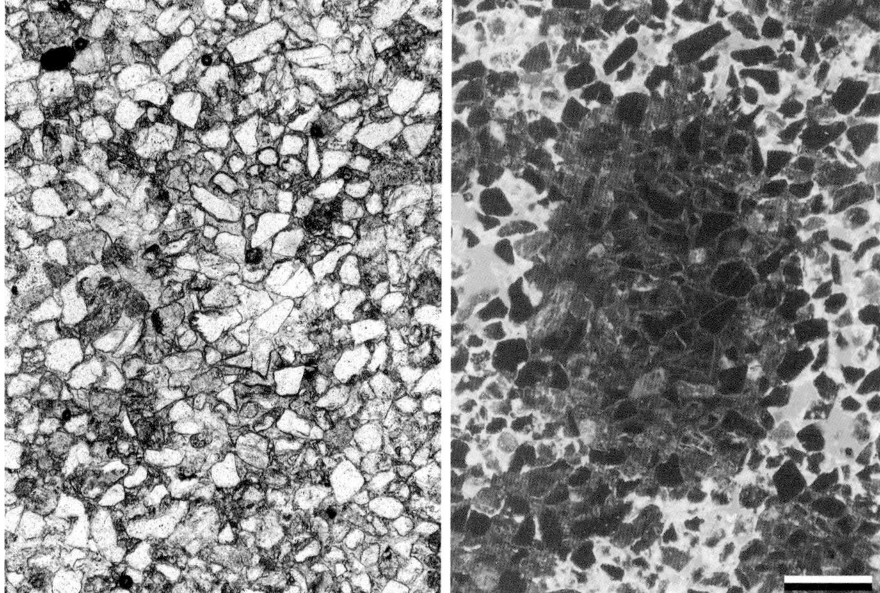

Up. Jurassic Fulmar Fm., United Kingdom sector, North Sea †

A drape of residual oil in a sandstone. The sample was cleaned with solvent, so this represents the heavier, less soluble hydrocarbons. Sometimes this bitumen represents an earlier oil charge that was bacterially degraded. The drapes of hydrocarbons (examples at yellow arrows) have shrunk slightly, and it can be seen that it is only a thin film. In samples coated with gold/palladium rather than carbon, the oil composition can be partially determined by EDX analysis. But on thin drapes such as this, some electrons make it through the hydrocarbons and read the underlying mineral.

SEM, Scale bar = 7 μm

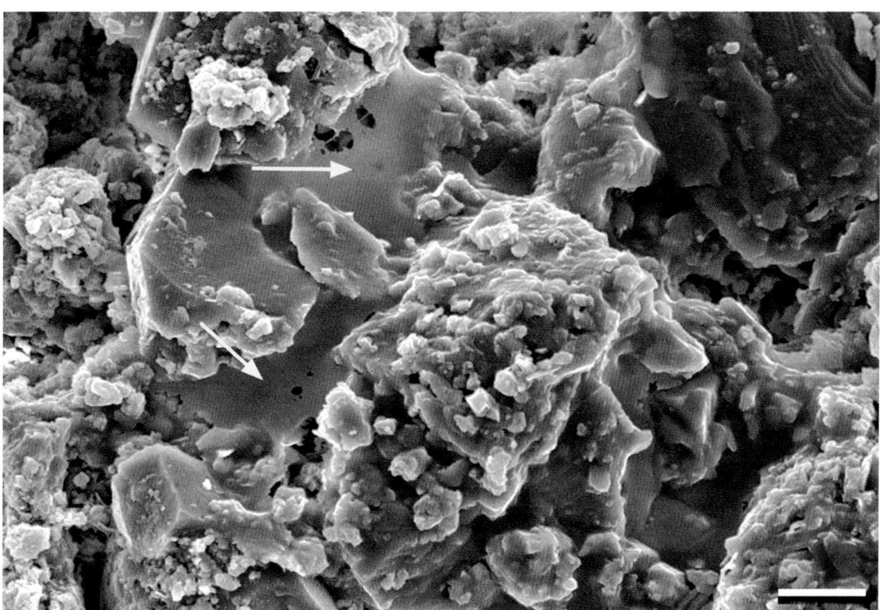

Cited References and Additional Information Sources

Alty, S. W., 1933, Some properties of authigenic tourmaline from Lower Devonian sediments: American Mineralogist, v. 18, p. 351-355.

Awasthi, N., 1961, Authigenic tourmaline and zircon in the Vindhyan formations of Sone Valley, Mirzapur District, Uttar Pradesh, India: Journal of Sedimentary Research, v. 31, p. 482-484.

Bergendahl, M. H., 1955, Wavellite spherulites in the Bone Valley Formation of central Florida: American Mineralogist, v. 40, p. 497-504.

Blanchard, F. N., and S. A. Denahan, 1967, Wavellite-cemented sandstones from northern Florida: Florida Scientist, v. 29, p. 248-256.

Braithwaite, C. J. R., and C. D. Gribble, 1998, Phosphatic microbial biofilms cementing gravels in a vadose environment: Journal of Coastal Research, v. 14, p. 1422-1425.

Cathcart, J. B., and D. L. Schmidt, 1977, Middle Paleozoic sedimentary phosphate in the Pensacola Mountains, Antarctica: Washington, DC, USGS Professional Paper 456-E, 17 p.

Cho, M., J. G. Liou, and S. Maruyama, 1986, Transition from the zeolite to prehnite-pumpellyite facies in the Karmutsen metabasites, Vancouver Island, British Columbia: Journal of Petrology, v. 27, p. 467-494, doi: 10.1093/petrology/27.2.467.

Clarke, O. M., Jr., and W. D. Keller, 1984, A gibbsite-cemented quartz sandstone: Journal of Sedimentary Research, v. 54, p. 154-158, doi: 10.1306/212F83D2-2B24-11D7-8648000102C1865D.

Cornelius, C. D., 1987, Classification of natural bitumen: A physical and chemical approach, *in* R. F. Meyer, ed., Exploration for Heavy Crude Oil and Natural Bitumen: Tulsa, OK, American Association of Petroleum Geologists Studies in Geology 25, p. 165-174.

Dill, H. G., 2010, Authigenic heavy minerals a clue to unravel supergene and hypogene alteration of marine and continental sediments of Triassic to Cretaceous age (SE Germany): Sedimentary Geology, v. 228, p. 61-76, doi: 10.1016/j.sedgeo.2010.04.006.

Donovan, T. J., I. Friedman, and J. D. Gleason, 1974, Recognition of petroleum-bearing traps by unusual isotopic compositions of carbonate-cemented surface rocks: Geology, v. 2, p. 351-354, doi: 10.1130/0091-7613(1974)2<351:ROPTBU>2.0.CO;2.

Ehrenberg, S. N., I. Skjevrak, and A. E. Gilje, 1995, Asphaltene-rich residues in sandstone reservoirs of Haltenbanken province, mid-Norwegian continental shelf: Marine and Petroleum Geology, v. 12, p. 53-69, doi: 10.1016/0264-8172(95)90387-Y.

Fassi-Fihri, O., M. Robin, and E. Rosenberg, 1995, Wettability studies at the pore level: A new approach by the use of cryo-scanning electron microscopy: SPE Formation Evaluation, v. 10, p. 11-19, http://dx.doi.org/10.2118/22596-PA.

Garden, I. R., S. C. Guscott, S. D. Burley, K. A. Foxford, J. J. Walsh, and J. Marshall, 2001, An exhumed palaeo-hydrocarbon migration fairway in a faulted carrier system, Entrada Sandstone of SE Utah, USA: Geofluids, v. 1, p. 195-213, doi: 10.1046/j.1468-8123.2001.00018.x.

Gautier, D. L., 1979, Preliminary report of authigenic, euhedral tourmaline crystals in a productive gas reservoir of the Tiger Ridge Field, north-central Montana: Journal of Sedimentary Research, v. 49, p. 911-916, doi: 10.1306/212F7874-2B24-11D7-8648000102C1865D.

George, S. G., 1993, Black sandstones in the Midland Valley of Scotland: Thermally metamorphosed hydrocarbon reservoirs?: Transactions of the Royal Society of Edinburgh: Earth Sciences, v. 84, p. 61-72, doi: 10.1017/S0263593300005927.

Gentzis, T., and F. Goodarzi, 1990, A review of the use of bitumen reflectance in hydrocarbon exploration with examples from Melville Island, Arctic Canada, *in* V. F. Nuccio, and C. E. Barker, eds., Applications of Thermal Maturity Studies to Energy Exploration: Denver, CO, Rocky Mountain Section SEPM, p. 23-36.

Hay, D. C., and T. J. Dempster, 2009, Zircon alteration, formation and preservation in sandstones: Sedimentology, v. 56, p. 2175-2191, doi: 10.1111/j.1365-3091.2009.01075.x.

Head, I. M, D. M. Jones, and S. R. Larter, 2003, Biological activity in the deep subsurface and the origin of heavy oil: Nature, v. 426, p. 344-352, doi: 10.1038/nature02134.

Hein, F. J., D. Leckie, S. Larter and J. Suter, 2013, Heavy-oil and Oil-sand Petroleum Systems in Alberta and Beyond: Tulsa, OK, AAPG Studies in Geology 64, 131 p.

Henry, D. J., G. Lu, and C. McCabe, 1994, Epigenetic tourmaline in sedimentary red-beds; an example from the Silurian Rose Hill Formation, Virginia: The Canadian Mineralogist, v. 32, p. 599-605.

Huc, A. Y., P. Nederlof, R. Debarre, B. Carpentier, M. Boussafir, F. Laggoun-Défarge, A. Lenail-Chouteau, and N. Bordas-Le Floch, 2000, Pyrobitumen occurrence and formation in a Cambro–Ordovician sandstone reservoir, Fahud Salt Basin, north Oman: Chemical Geology, v. 168, p. 99-112, doi: 10.1016/S0009-2541(00)00190-X.

Kelemen, S. R., C. C. Walters, P. J. Kwiatek, M. Afeworki, M. Sansone, H. Freund, R. J. Pottorf, H. G. Machel, T. Zhang, G. S. Ellis, Y. Tang, and K. E. Peters, 2008, Distinguishing solid bitumens formed by thermochemical sulfate reduction and thermal chemical alteration: Organic Geochemistry, v. 39, p. 1137-1143, doi: 10.1016/j.orggeochem.2008.04.007.

Kelemen, S. R., C. C. Walters, P. J. Kwiatek, H. Freund, M. Afeworki, M. Sansone, W. A. Lamberti, R. J. Pottorf, H. G. Machel, K. E. Peters, and T. Bolin, 2010, Characterization of solid bitumens originating from thermal chemical alteration and thermochemical sulfate reduction: Geochimica et Cosmochimica Acta, v. 74, p. 5305-5332, doi: 10.1016/j.gca.2010.06.013.

Kirkland, D. W., R. E. Denison, and M. A. Rooney, 1995, Diagenetic alteration of Permian strata at oil fields of south central Oklahoma, USA: Marine and Petroleum Geology, v. 12, p. 629-644, doi: 10.1016/0264-8172(95)98089-N.

Klemic, H., and M. E. Mrose, 1972, Geologic relations and x-ray crystallography of wavellite from Jackson County, Wisconsin, and their geologic implications: Washington, DC, USGS Professional Paper 800-C, p. C53-C62.

Larter, S., H. Huang, J. Adams, B. Bennett, O. Jokanola, T. Oldenburg, M. Jones, I. Head, C. Riediger, and M. Fowler, 2006, The controls on the composition of biodegraded oils in the deep subsurface: Part II—Geological controls on subsurface biodegradation fluxes and constraints on reservoir-fluid property prediction: AAPG Bulletin, v. 90, p. 921-938, doi: 10.1306/01270605130.

Lomando, A. J., 1992, The influence of solid reservoir bitumen on reservoir quality: AAPG Bulletin, v. 76, p. 1137-1152.

Milliken, K. L., 2007, Provenance and diagenesis of heavy minerals, Cenozoic units of the northwestern Gulf of Mexico sedimentary basin, *in* M. A. Mange, and D. T. Wright, eds., Heavy Minerals in Use (Developments in Sedimentology 58): New York, Elsevier, p. 247-261, doi: 10.1016/S0070-4571(07)58008-8.

Monson, B., and J. Parnell, 1992, The origin of gilsonite vein deposits in the Uinta Basin, Utah, *in* T. D. Fouch, V. F. Nuccio, and T. C. Chidsey, Jr., eds., Hydrocarbon and Mineral Resources of the Uinta Basin, Utah and Colorado: Salt Lake City, UT, Utah Geological Association Publication 20, p. 257-270.

Morad, S., 1986, SEM study of authigenic rutile, anatase and brookite in Proterozoic sandstones from Sweden: Sedimentary Geology, v. 46, p. 77-89, doi: 10.1130/0016-7606(1986)97<567:AODFOI>2.0.CO;2.

Morad, S., 1988, Diagenesis of titaniferous minerals in Jurassic sandstones from the Norwegian Sea: Sedimentary Geology, v. 57, p. 17-40, doi: 10.1016/0037-0738(88)90016-4.

Morad, S., and A. A. Aldahan, 1982, Authigenesis of titanium minerals in two Proterozoic sedimentary rocks from southern and central Sweden: Journal of Sedimentary Research, v. 52, p. 1295-1305, doi: 10.1306/212F8120-2B24-11D7-8648000102C1865D.

Morton, A. C., G. Borg, P. L. Hansley, P. D. W. Haughton, D. H. Krinsley, and P. Trusty, 1989, The origin of faceted garnets in sandstones: Dissolution or overgrowth?: Sedimentology, v. 36, p. 927-942, doi: 10.1111/j.1365-3091.1989.tb01754.x.

Mossman, D. J., and B. Nagy, 1996, Solid bitumens: an assessment of their characteristics, genesis, and role in geological processes: Terra Nova, v. 8, p. 114-128, doi: 10.1111/j.1365-3121.1996.tb00736.x.

Parnell, J., 1994, Hydrocarbons and other fluids: Paragenesis, interactions and exploration potential inferred from petrographic studies, *in* J. Parnell, ed., Geofluids: Origin, Migration and Evolution of Fluids in Sedimentary Basins: London, GSL Special Publication 78, p. 275-291, doi: 10.1144/GSL.SP.1994.078.01.19.

Pe-Piper, G., A. Karim, and D. J. W. Piper, 2011, Authigenesis of titania minerals and the mobility of Ti: New evidence from pro-deltaic sandstones, Cretaceous Scotian Basin, Canada: Journal of Sedimentary Research, v. 81, p. 762-773, doi: 10.2110/jsr.2011.63.

Ricketts, B. D., 1978, Authigenic tourmaline from the Middle Precambrian Belcher Group, Northwest Territories, Canada: Bulletin of Canadian Petroleum Geology, v. 26, p. 543-550.

Saxena, S. K., 1966, Evolution of zircons in sedimentary and metamorphic rocks: Sedimentology, v. 6, p. 1-33, doi: 10.1111/j.1365-3091.1966.tb01568.x.

Smith, R. D. A., 1987, Early diagenetic phosphate cements in a turbidite basin, *in* J. D. Marshall, ed., Diagenesis of Sedimentary Sequences: London, GSL Special Publication 36, p. 141-156, doi: 10.1144/GSL.SP.

Southgate, P. N., 1986, Cambrian phoscrete profiles, coated grains, and microbial processes in phosphogenesis: Georgina Basin, Australia: Journal of Sedimentary Research, v. 56, p. 429-441, doi: 10.1306/212F893B-2B24-11D7-8648000102C1865D.

Spötl, C., 1990, Authigenic aluminium phosphate-sulphates in sandstones of the Mitterberg Formation, Northern Calcareous Alps, Austria: Sedimentology, v. 37, p. 837-845, doi: 10.1111/j.1365-3091.1990.tb01828.x.

Wilhelms, A., and S. R. Larter, 1994, Origin of tar mats in petroleum reservoirs. Part II: Formation mechanisms for tar mats: Marine and Petroleum Geology, v. 11, p. 442-456, doi: 10.1016/0264-8172(94)90078-7.

Worden, R. H., N. H. Oxtoby, and P. C. Smalley, 1998, Can oil emplacement prevent quartz cementation in sandstones?: Petroleum Geoscience, v. 4, p. 129-137, doi: 10.1144/petgeo.4.2.129.

Facing Page: Top — Outcrop photograph of evaporite nodules that were partially silicified along their margins and then dissolved to create moldic pores, Permian Bell Canyon Fm., Culberson Co., Texas. Bottom — Partially dissolved volcanic rock fragments (secondary pores) in an opal-cemented volcanic arenite, Pliocene-Pleistocene Gila Gp., New Mexico (PPL).

DIAGENESIS:

DISSOLUTION

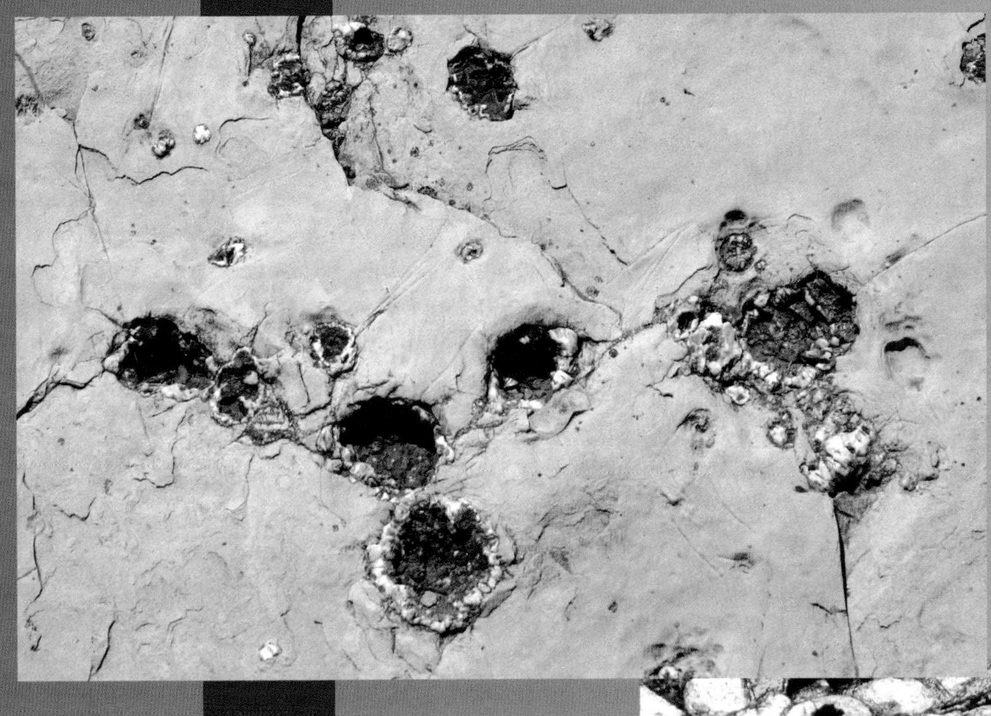

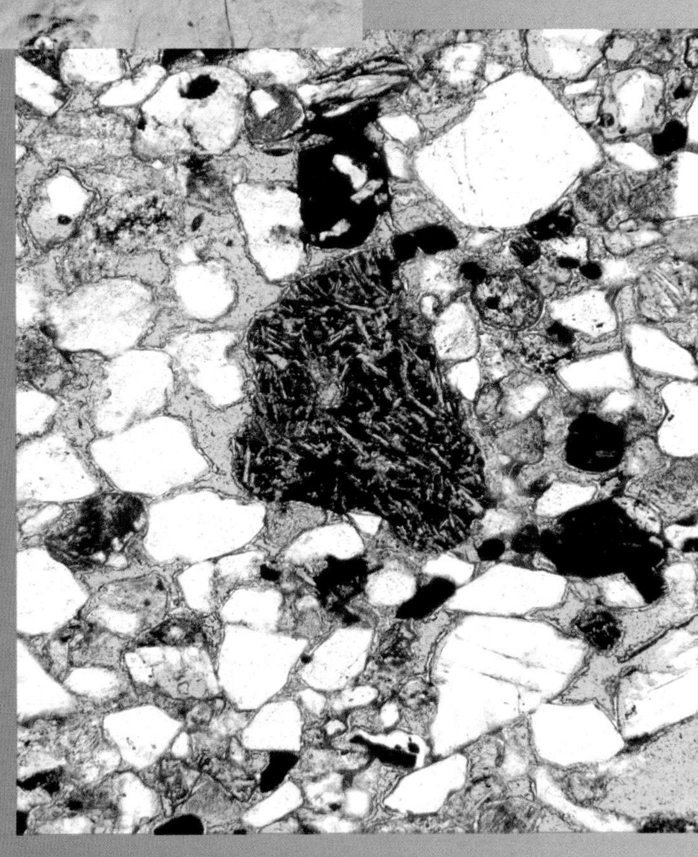

CHAPTER 19

INTRODUCTION TO DISSOLUTION

This chapter deals with the removal of minerals from sandstones and mudrocks during any stage of diagenesis by pore waters undersaturated with respect to one or more minerals. Dissolution has long been known to be important in "more soluble" rocks, especially carbonates and evaporites, but the past five decades have seen a growing understanding that it is an important process in clastic terrigenous deposits as well. It should not have been a surprise—sandstones and mudrocks can comprise many different minerals, detrital as well as authigenic, that were formed under a wide range of conditions. Subsequent eogenetic, mesogenetic and telogenetic diagenetic environments subject minerals to varied temperatures and pressures and especially to diverse water chemistries that potentially can range from nearly pure meteoric water to hypersaline brines and from acidic to alkaline. As a result, one can find instances of dissolution of almost any mineral. Widespread dissolution of common grains and cements may have a significant impact on the ultimate composition of terrigenous rocks, their porosity and their reservoir potential.

Prime among the detrital minerals that commonly undergo dissolution are K-feldspar and calcic plagioclase feldspar (An_{10} and greater; e.g., McBride, 1985; Milliken et al., 1989), whether as isolated grains or as constituents of rock fragments (especially VRFs). Also important for generating secondary porosity is the dissolution of calcium carbonate (aragonite or calcite, either as grains or cements—e.g., Taylor, 1990; Gibling et al., 2000), and any of the common primary or secondary evaporite minerals (gypsum, anhydrite, halite and others; e.g., Schenk and Richardson, 1985). Dissolution of quartz and silica (including glass, opal, chert and chalcedony; Pye and Krinsley, 1985; Shanmugam and Higgins, 1988; Bjørkum, 1996), siderite and dolomite (Burley and Kantorowicz, 1986) and zeolites (Tang et al., 1997; Bernet and Gaupp, 2005) may be locally significant. Finally, many heavy minerals are known to undergo subsurface dissolution (Blatt and Sutherland, 1969; Morad and Aldahan, 1986; Milliken and Mack, 1990; Morton and Hallsworth, 2007; Walderhaug and Porten, 2007), although in most cases, these alterations are not volumetrically important.

The recognition of dissolution fabrics is easy in some instances and quite difficult in others. Schmidt and Macdonald (1979a, 1979b and earlier studies cited in McBride, 1977) outlined several major ways to recognize dissolution. Those criteria are shown in modified form in Figure 19.1 and in the more inclusive list below:

1. Patchy carbonate (or evaporite) cement that is the corroded remnant of once more extensive, now partially dissolved cement;
2. Normal to unusually large moldic pores formed where unstable or calcite-replaced framework grains have been;
3. Unusually long pores resulting from replacement along grain boundaries and subsequent dissolution;
4. "Skeletal" or honeycombed feldspar grains where calcite has replaced part of the feldspar and was later dissolved, or simple moldic pores from dissolution of feldspar;
5. Fossil or other grain molds;
6. Porosity larger than expected for a sandstone that shows strong compaction (indicated by sutured grains, tight packing); also exceptional inhomogeneity of packing;
7. Porosity distributions of samples that have other than an arithmetic normal distribution;
8. Corroded grains, whose margins were replaced by previous cement (for example, rectangular and stair-step cement reentrants associated with evaporite removal); and
9. Broken silicate framework grains with fractures that formed during intense compaction that followed rapid removal of cement.

From this listing, it should also be clear that dissolution porosity may occur in a variety of textural settings: within partially dissolved grains (intragranular porosity, in completely dissolved grains (as moldic pores), between grains (as intergranular pores or as oversized pores and vugs) and in fractures (as solution-enlarged fracture porosity). It should also be pointed out that the chemical dissolution described above does not include the common pressure solution that occurs at grain contacts, solution seams and stylolites (more extensively discussed in Chapters 10 and 21).

By definition, subsurface chemical dissolution produces secondary pores, and most secondary porosity (other than fracture porosity) is produced by dissolution. As a consequence, there normally is an overall depth- and age-related increase in secondary porosity in most clastic terrigenous deposits. But it is not accurate to say that subsurface dissolution always creates increased overall porosity in sandstones or mudstones—that depends on whether a) the dissolved materials recombine locally to produce an equivalent volume of new cements in either primary or secondary pores (reflecting a balance between dissolution and precipitation); or b) whether those solutes are carried out of the local system by large-scale fluid flow, leaving the local primary and/or secondary pores open and substantially uncemented (and also mainly uncompacted). The literature is filled with contrasting conclusions on this topic. Enhanced porosity related

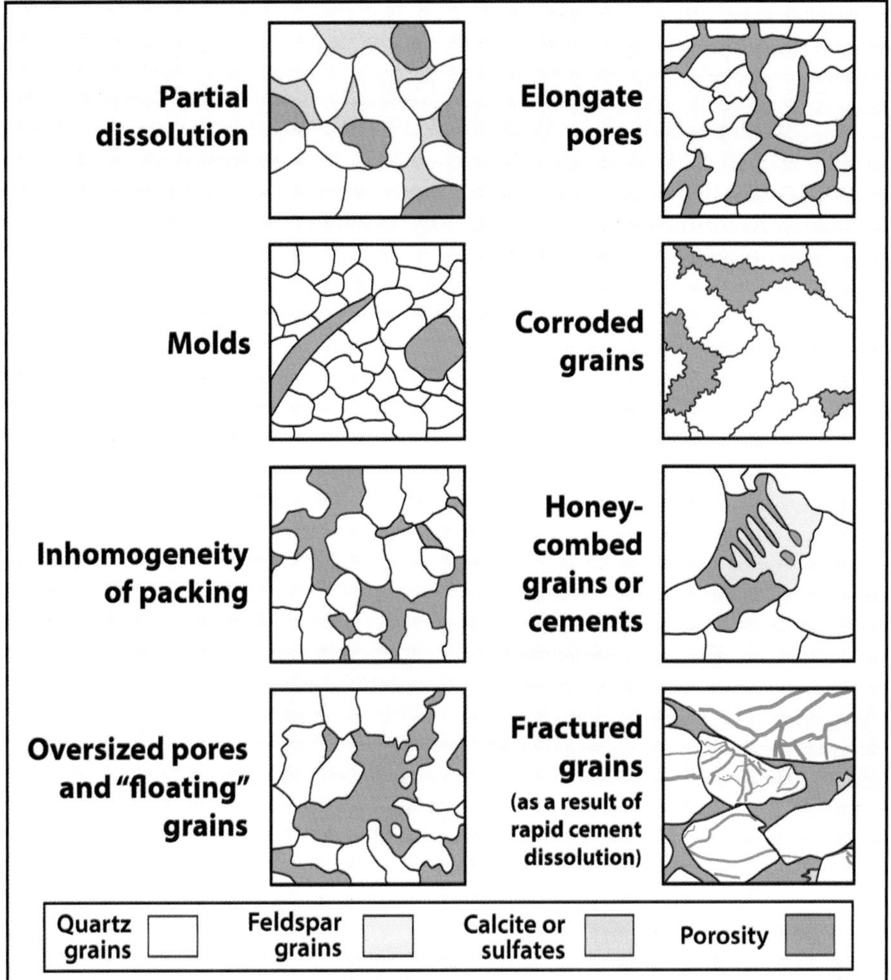

Figure 19.1 Summary of petrographic criteria for recognition of secondary porosity in sandstones (modified from Schmidt and McDonald, 1979b, p. 216).

to subsurface dissolution was reported, for example, by McBride (1977); Land et al. (1987); Bloch and Franks (1993) and Loucks (2005). On the other hand, substantial dissolution with little net porosity increase due to associated precipitation (termed "distributional secondary porosity" by Giles and De Boer, 1990) was noted or modeled by Bjørlykke (1984), Harris (1989), Glasmann (1992), Taylor et al. (2010), Bjørlykke and Jahren (2012) and many others.

In general, major sites of dissolution, and especially of net porosity increases, are in settings where there is strong regional or focused fluid flow. Such settings include areas with eogenetic or telogenetic meteoric water influx driven by hydrostatic head (Emery et al., 1990; Bloch, 1994; Al-Ramadan et al., 2012). Mesogenetic settings are more diverse, and include areas of basin-scale dewatering associated with compaction or clay-mineral transformations (kaolinite-illite or smectite-illite, in particular); zones of rapid, generally localized, loss of overpressures (Wilkinson et al., 1997); fluid mixing zones (Gibling et al., 2000); zones of hydrothermal fluid flow (Bernet and Gaupp, 2005), and areas with influx of acidic, oxalate-rich waters (commonly CO_2-rich fluids associated with organic maturation, hydrocarbon generation—e.g., Welch and Ullman, 1996; Ehrenberg and Jakobsen, 2001). Far less net porosity gain is realized where localized diffusive transport of solutes dominates because pore waters are buffered by rock-water interactions and geochemical gradients are low. In this case, solutes are not transported significant distances, resulting in their reprecipitation in nearby porosity as more stable authigenic minerals.

Regardless of whether or not there is a net gain of pore space through subsurface dissolution, the development of secondary pores (or mixed primary and secondary pore networks) probably leads to changes in overall permeability, and in porosity/permeability relationships, although no consistent patterns have been established to date (Giles and De Boer, 1990; Loucks, 2005; McKinley et al., 2011). In part, that reflects the open- and closed-system relationships discussed above. In near-surface settings in which meteoric dissolution occurs and solutes are easily removed, permeability may be substantially increased (e.g., Shanmugam and Higgins, 1988). In

deeper settings with slower fluid flow and/or more complex rock-water interactions, permeability changes are more variable and difficult to predict.

In closing, we want to emphasize that in studies of dissolution and porosity, perhaps more than in any other area of petrography, it is imperative to pay close attention to sample and thin section preparation techniques. In the cutting of thin-section rock chips and in the grinding of the thin sections, grains may be plucked and grains or cements may be dissolved. Proper preparation (see Pittman, 1992), including care in sample cutting, high-pressure injection with colored impregnation materials, cutting and grinding in oil may minimize such effects and can help in distinguishing artifacts of preparation from preexisting primary or secondary pores.

Mid. Eocene – Lo. Oligocene Lower Spears Gp., Socorro Co., New Mexico

This zoned plagioclase feldspar grain in an immature feldspathic litharenite shows how subtle variations in the crystal's composition influenced later secondary pore development. Here the core was stable and shows little evidence of dissolution, whereas several of the later growth zones were susceptible to dissolution. This selective grain removal has locally created secondary intragranular porosity.

PPL, BDI, Scale bar = 0.34 mm

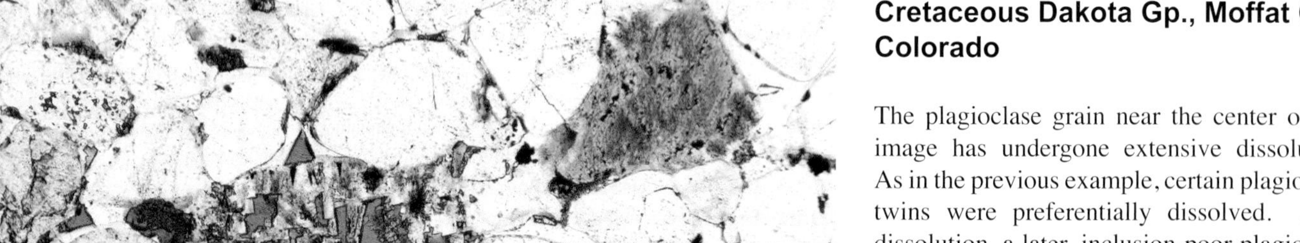

Cretaceous Dakota Gp., Moffat Co., Colorado

The plagioclase grain near the center of this image has undergone extensive dissolution. As in the previous example, certain plagioclase twins were preferentially dissolved. After dissolution, a later, inclusion-poor plagioclase cement (likely albite) grew inward from the grain walls (examples at blue arrows). This sandstone is partially cemented by syntaxial quartz overgrowths which stabilized the rock prior to feldspar dissolution (preventing compaction of the secondary pores). Clay cements formed after quartz cementation and feldspar dissolution (note clays lining the plagioclase remnants in the pore).

PPL, RDI, Scale bar = 0.21 mm

Lo.–Mid. Eocene Cub Mountain Fm., Lincoln Co., New Mexico

Not all feldspars behave the same during diagenesis as a result of compositional variations. On the left is a detrital orthoclase, and on the right is a plagioclase. The plagioclase grain has been almost completely dissolved (note the influence of the twinning), whereas the orthoclase grain underwent far less dissolution, and subsequent precipitation of inclusion-poor cement (black arrow) has filled most of the porosity produced. Later cement (probably albite) also filled some of the pores created within the plagioclase grain. In addition to the feldspar cements, this unit has been lithified by quartz, kaolinite and other authigenic clays.

PPL, KFS, BDI, Scale bar = 0.14 mm

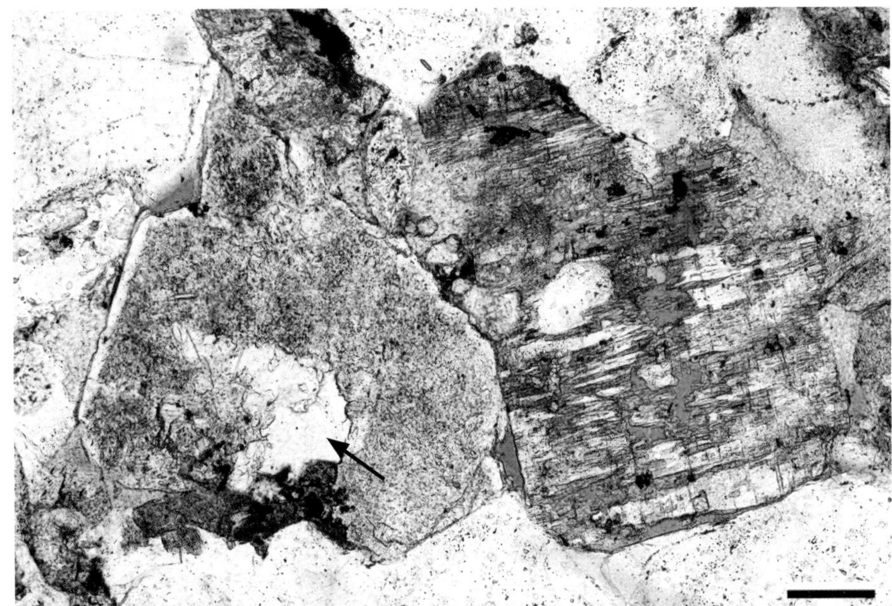

Mid.–Up. Jurassic Melke Fm., Vøring Basin, Norwegian Sea †

Skeletal plagioclase feldspar grains in which dissolution has exploited the cleavage planes. There is a small area of potassium feldspar attached (red arrow), which may be an overgrowth. The delicate character of these grains and their lack of deformation at contacts with each other and more rigid nearby detrital grains suggests that the dissolution postdates compaction. It also shows that the framework was sufficiently rigid at the time of grain dissolution to prevent compaction of the relict, dissolved grains.

PPL, KFS, BDI, Scale bar = 0.07 mm

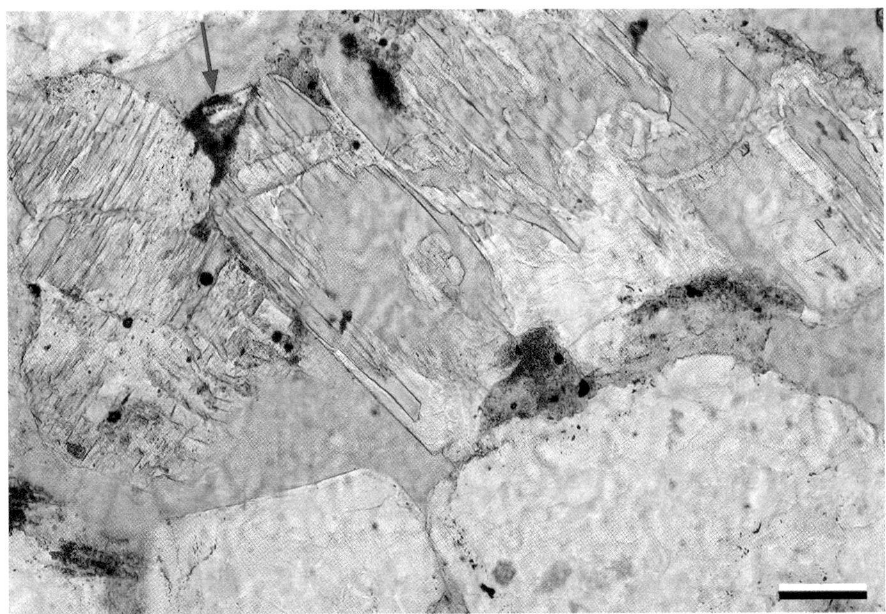

Mid. Jurassic Entrada Fm., Slick Rock Mbr., Emery Co., Utah

The grain at the center of this image has been partially dissolved but is still recognizable as a plagioclase based on the uncollapsed, parallel traces of remnant feldspar and insoluble materials. Nearby quartz overgrowth cements helped to lithify the rock, minimize compaction and ultimately preserve significant porosity within this unit.

PPL | XPL, RDI, Scale bar = 0.05 mm

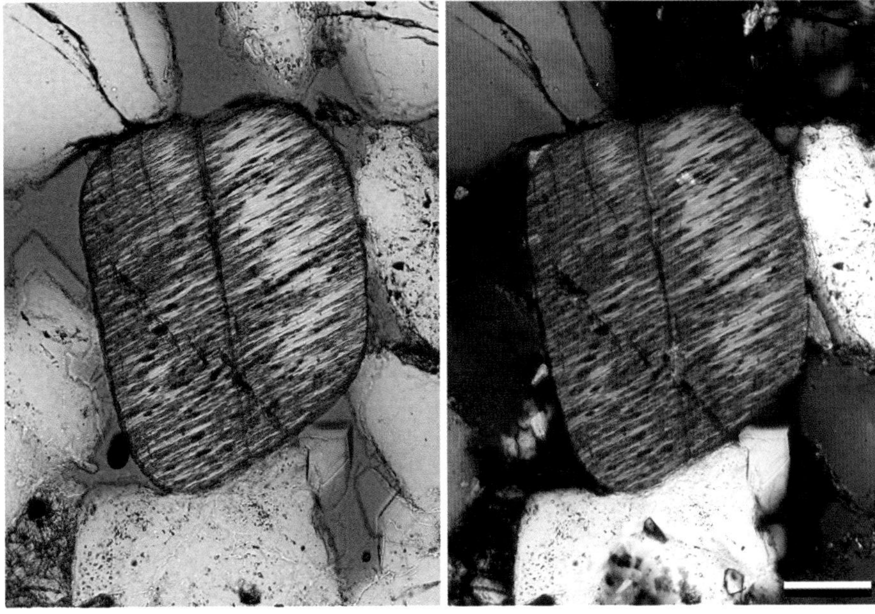

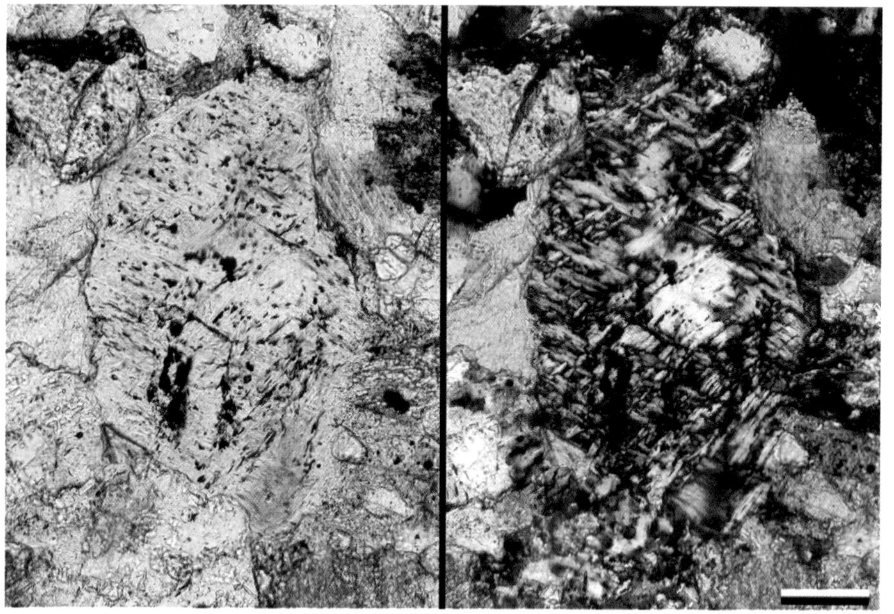

Miocene Upper Marine Molasse, Switzerland

This detrital microcline feldspar has been partially dissolved. The microcline twinning is still visible because one set of twins is better preserved than the other set. The rock was cemented by dolomite prior to feldspar alteration, and subsequently the dolomite also underwent minor dissolution, mainly along crystal boundaries.

PPL | XPL, BDI, Scale bar = 0.10 mm

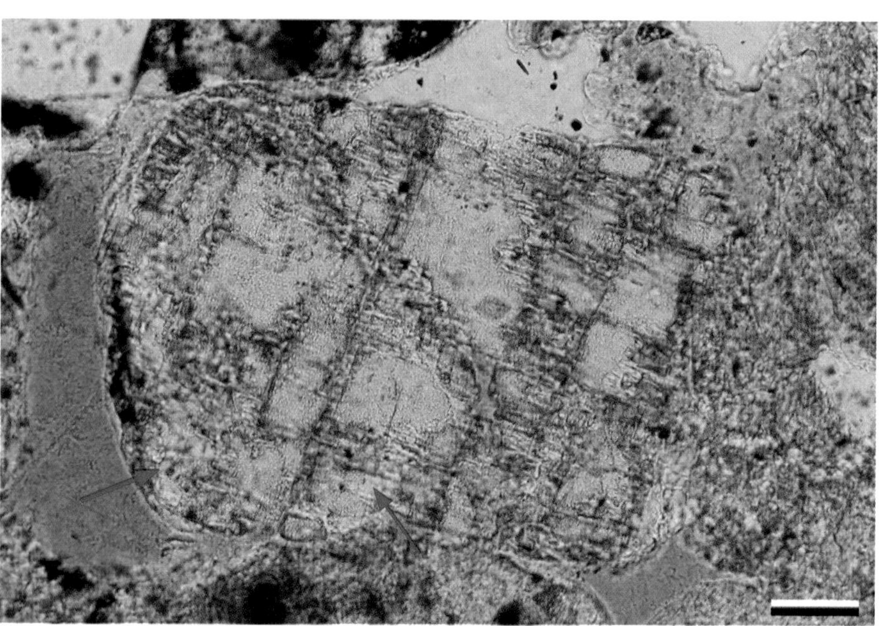

Lo.–Mid. Eocene Cub Mountain Fm., Lincoln Co., New Mexico

The alteration of the orthoclase grain at the center of this image was controlled by cleavage and fracture planes within the grain. Partial dissolution of the grain has produced well-connected secondary intragranular pores. Small, possibly albitic, unstained overgrowths have locally filled minor amounts of the porosity within and around the grain (examples at red arrows).

PPL, KFS, BDI, Scale bar = 0.06 mm

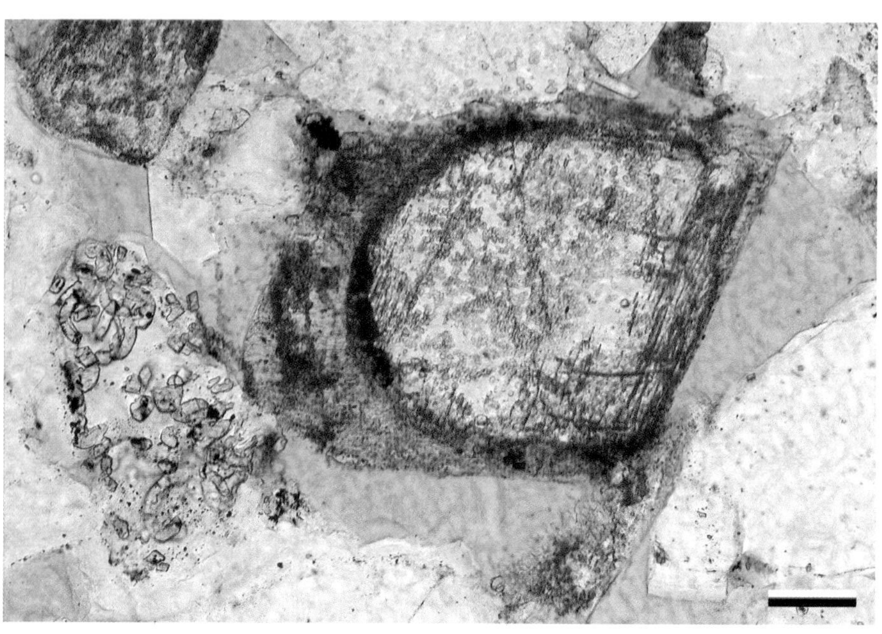

Up. Jurassic Fulmar Fm., United Kingdom sector, North Sea †

Selective dissolution of a potassium feldspar overgrowth. The stain shows that the grain is also potassium feldspar, but there is almost certainly some compositional difference between the two, which clearly affected the dissolution. A quartz grain with vermicular chlorite inclusions is visible to the left of the feldspar.

PPL, KFS, BDI, Scale bar = 0.07 mm

Triassic Sherwood Sandstone Gp., Co. Antrim, Northern Ireland, U.K. †

Dissolution of feldspar grains normally results in a skeletal texture, which comprises remnants of the original crystal along the cleavage planes. Here the unstained, more euhedral crystals are probably albite intragranular cements that grew during or after dissolution. The grain also displays potassium feldspar overgrowths, and these are undegraded. The staining of feldspars here provides some compositional information, but it is no substitute for quantitative analytical studies using x-ray diffraction, microprobe examination or back-scattered electron imaging.

PPL | XPL, KFS, BDI, Scale bar = 0.12 mm

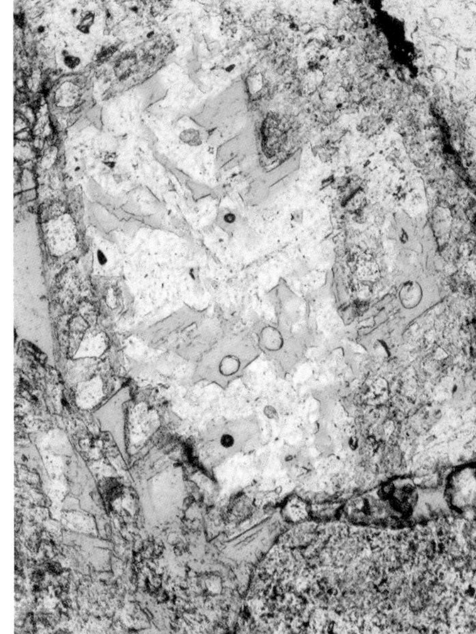

Up. Cretaceous Prince Creek Fm., North Slope, Alaska

Detrital feldspar grains and their authigenic cement overgrowths typically have different compositions, with overgrowths normally being a pure and compositionally uniform end-member precipitate (e.g., Saigal et al., 1988). These compositional distinctions can strongly affect chemical stability during diagenesis. Here, the overgrowth was more stable than the grain core, allowing near-complete dissolution of the core. That dissolution occurred prior to formation of the grain-coating clay cements based on the observation that the clays also occur within the secondary porosity.

PPL, BDI, Scale bar = 0.10 mm

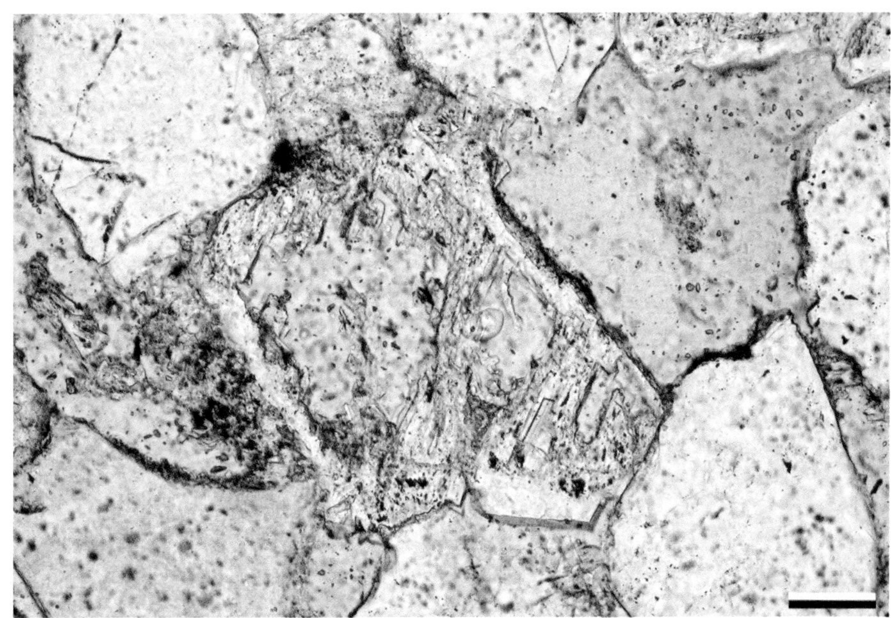

Up. Jurassic Fulmar Fm., United Kingdom sector, North Sea †

Identification of dissolved grains can be quite difficult when little remains of the original grain. For example, there is so little of this grain remaining that identification as potassium feldspar is entirely reliant on the staining. The margin (overgrowth?) remains and was partly disrupted by compaction against an adjacent, rigid quartz grain. Otherwise the grain was not subjected to further compaction. The dissolution thus is largely a post-compaction feature, or the rock was stabilized against compaction by carbonate cements, of which some are observed along the top margin (red arrow).

PPL, AFeS, KFS, BDI, Scale bar = 0.07 mm

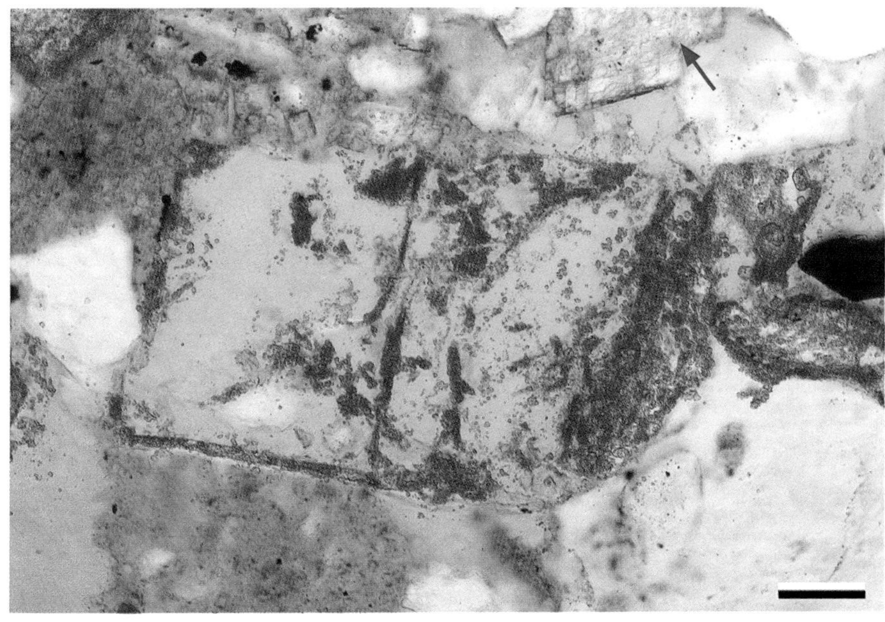

Up. Jurassic Ula Fm., Norwegian sector, North Sea †

An SEM view of a secondary pore generated by extensive dissolution of a potassium feldspar grain. The observed preferred orientations of the remnants suggests that dissolution of the grain is both crystallographically and grain-fracture controlled. Identification of the grain as K-feldspar is based primarily on EDX analysis of the corroded remnants. On the right, the original grain margin (red arrow) is largely intact and separates the corroded grain from a microporous overgrowth. The grain is surrounded by microcrystalline quartz and illite (the latter shown at yellow arrow).

SEM, Scale bar = 43 μm

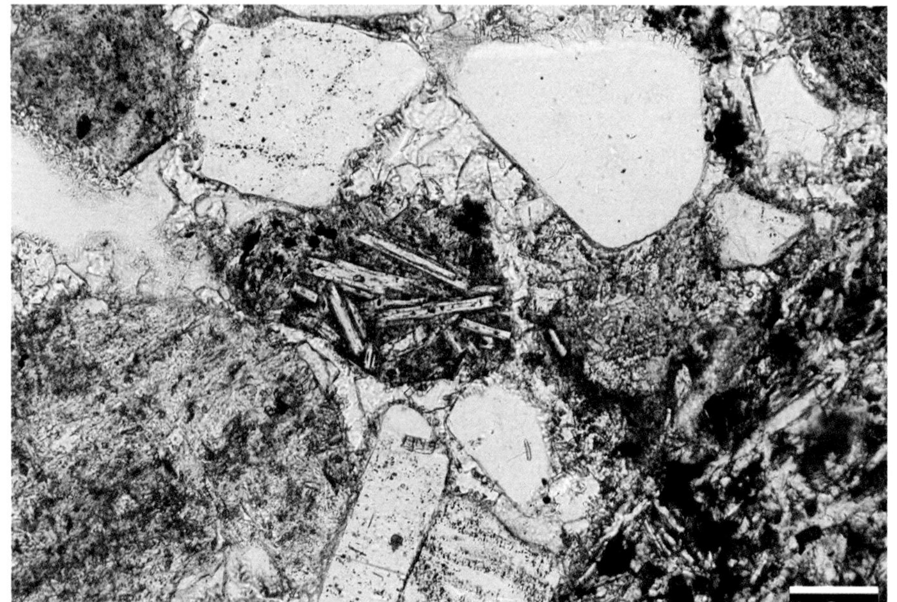

Neogene Santa Fe Gp., Valencia Co., New Mexico

In this submature litharenite (volcanic arenite), the plagioclase phenocrysts within a volcanic rock fragment were selectively altered. In this case, the cores of the feldspars were dissolved but thin remnant (undissolved) rims are still mostly intact. Although this dissolution clearly creates intragranular pores, without further dissolution of other phases, such pores can remain largely unconnected, and thus may not significantly improve reservoir performance.

PPL, KFS, BDI, Scale bar = 0.10 mm

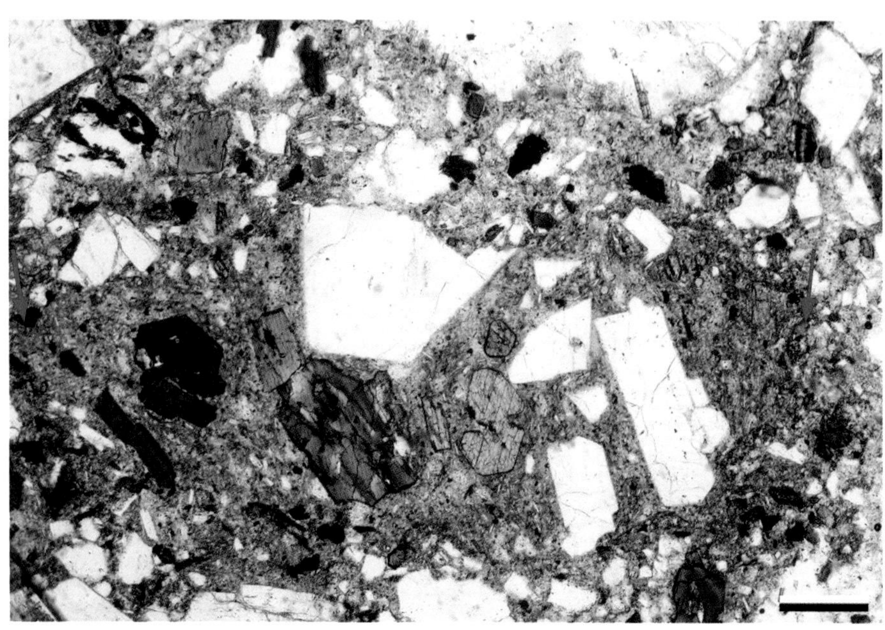

Mid. Eocene – Lo. Oligocene Lower Spears Gp., Socorro Co., New Mexico

This volcanic rock fragment (the large composite particle that extends between the two red arrows), part of a poorly-sorted volcanic arenite, has undergone substantial dissolution. The clast is an andesitic porphyry containing mainly amphibole and feldspar phenocrysts. The finely-crystalline groundmass of the rock fragment was preferentially dissolved, but the phenocrysts are relatively unaltered. Dissolution here has produced intragranular pores. However, the finer-grained matrix of the sandstone itself has undergone only minor dissolution.

PPL, BDI, Scale bar = 0.34 mm

Miocene Lower Santa Fe Gp., Taos Co., New Mexico

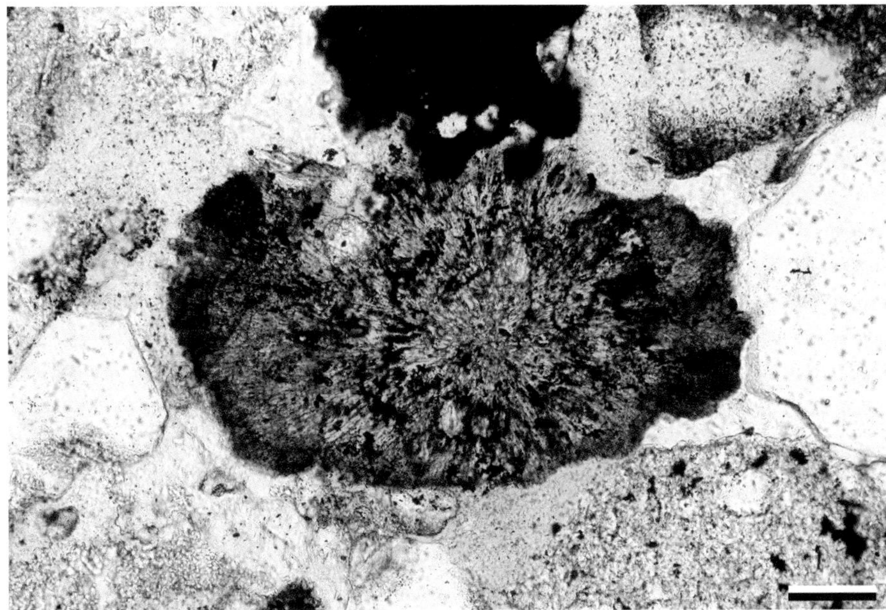

This volcanic rock fragment is a spherulite from a glassy acidic melt in which the radiating crystals were intergrowths of feldspars, silica or other minerals. The texture of the spherulite has been preserved despite extensive dissolution, possibly indicating that a more soluble mineral (e.g., plagioclase) was intergrown with a less soluble mineral (e.g., quartz). Dissolution has produced intragranular pores in this rock fragment as well as many others in the sample. Kaolinite cements, a possible byproduct of this dissolution, precipitated both in the primary pores and within some of the rock fragments (see upper right corner).

PPL, BDI, Scale bar = 0.10 mm

Mid. Jurassic Entrada Fm., Emery Co., Utah

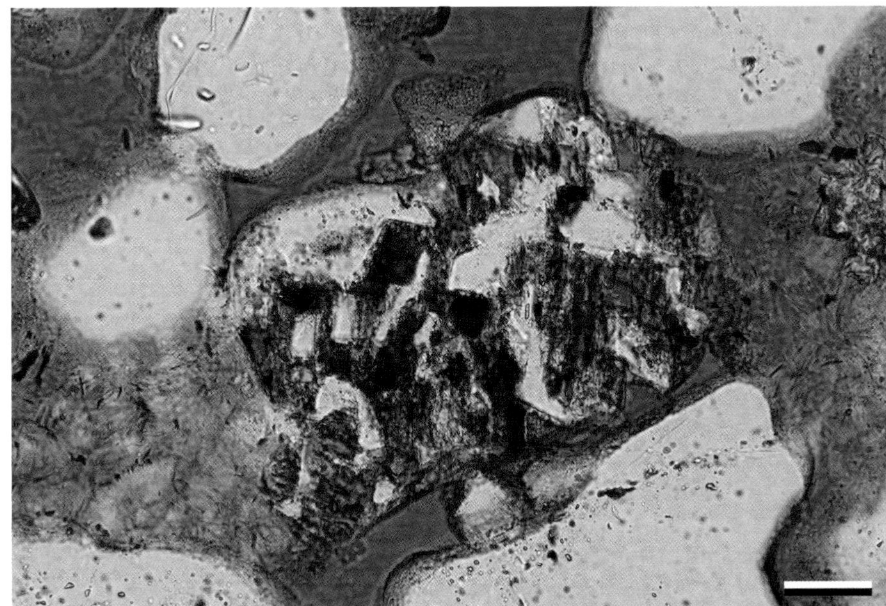

A quartzarenite with a selectively dissolved graphic granite clast. The feldspars within the clast were largely removed leaving mainly the quartz intergrowths. Dissolution likely occurred after most compaction, otherwise the framework of the grain would have undergone some deformation. Vermicular kaolinite cements partially fill the original porosity of the rock.

PPL, RDI, Scale bar = 0.04 mm

Miocene Hayner Ranch Fm., south-central New Mexico

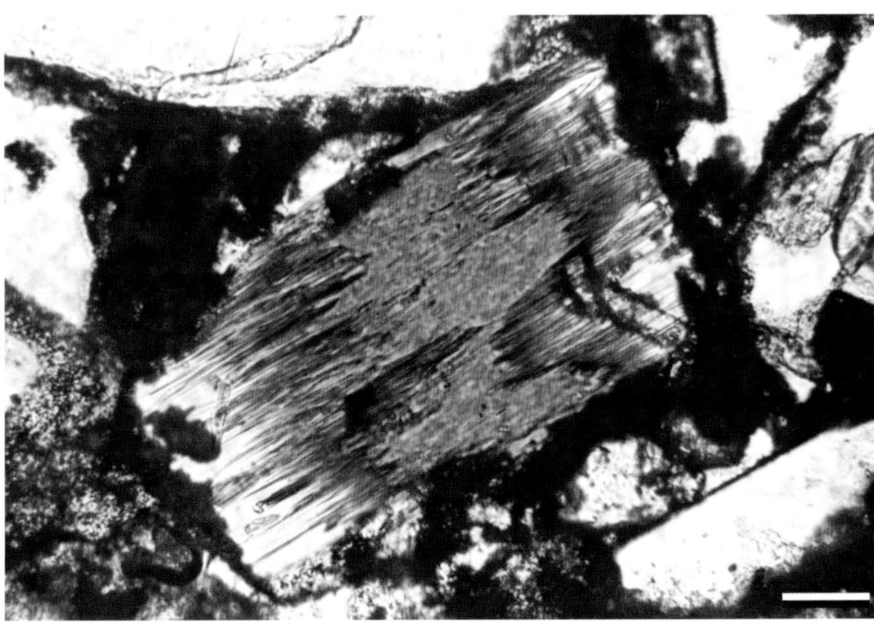

Although feldspars, volcanic rock fragments, and carbonate and evaporite grains or cements are the most commonly dissolved materials in sandstones, detrital heavy minerals also are widely affected. In this example, a detrital hornblende (center of photo) shows corroded margins with thin terminations that clearly could not have survived transport. The white area between greenish hornblende and red-brown authigenic hematite cement is pore space of secondary origin. Photograph by T. R. Walker.

PPL, Scale bar = 0.06 mm

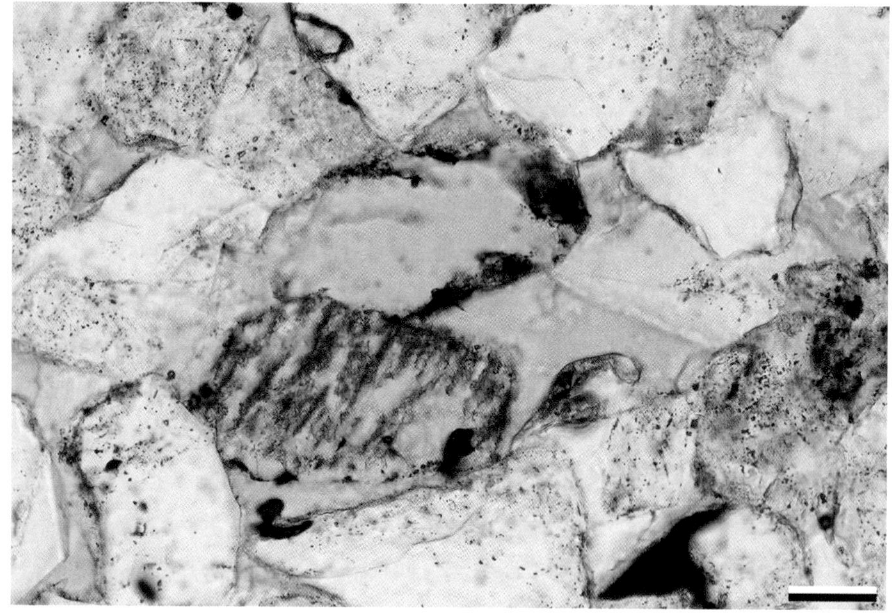

Mississippian Berea Sandstone, Midland Co., Michigan

In this example, the two grains in the center of the photograph have undergone extensive dissolution creating moldic porosity. The lower grain was likely a plagioclase feldspar based on the linear traces of insoluble remains. The upper grain may have been a feldspar, but there are insufficient remnants of the original grain to definitively identify it. Prior to the dissolution of these grains, the rock was cemented by quartz overgrowths that stabilized the framework and prevented compaction from destroying the newly created secondary porosity.

PPL, PFS, BDI, Scale bar = 0.04 mm

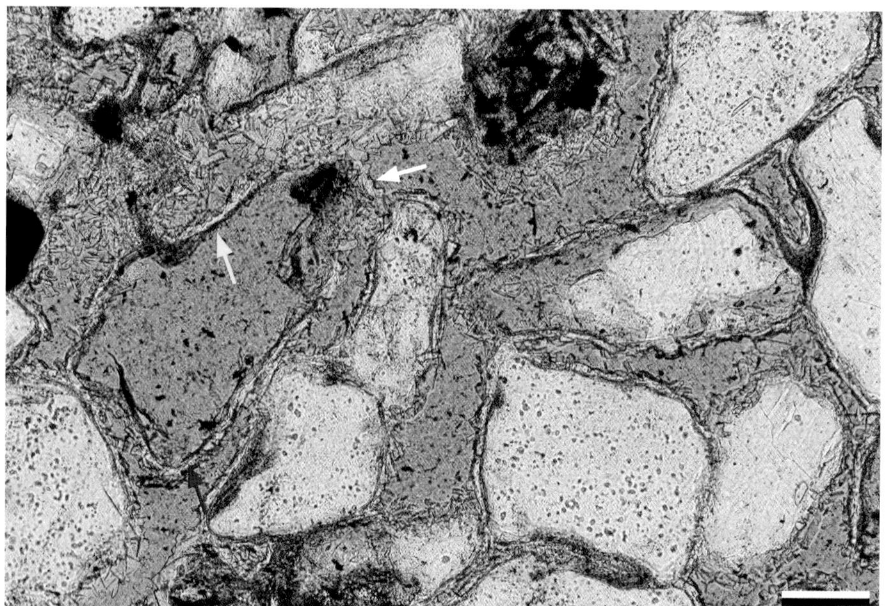

Pliocene – Pleistocene Gila Gp., southern New Mexico

A example of partial to complete dissolution of multiple grains. Some of the dissolved grains are still recognizable as feldspars; others were so completely dissolved that they are visible only because of relict coatings that predated dissolution. In this case, those coatings are complex, involving authigenic clays (yellow arrow) followed by zeolites (the white band indicated by the red arrow); both must predate dissolution because they are found only external to the dissolved grain molds. A generation of more transparent zeolite crystals (white arrow) formed later and is found both inside and outside the grain molds.

PPL, BDI, Scale bar = 0.10 mm

Up. Cretaceous Tuscaloosa Fm., subsurface, Louisiana

A sample etched in dilute hydrochloric acid for about 20 seconds to remove the last remnants of an original carbonate detrital grain (center). This has left a thin, but apparently sturdy shell of authigenic chlorite cement that coated the grain. Such shells are sufficiently durable to survive natural diagenetic dissolution of core grains under subsurface conditions and produce thin rinds that mark the outline of many secondary dissolution pores such as in the previous illustration. These shells also can act as substrates for later cement formation, also shown above.

SEM, Scale bar = 5 μm

Up. Jurassic Fulmar Fm., United Kingdom sector, North Sea †

Secondary pores from the dissolution of sponge spicules such as *Rhaxella* spp.—siliceous spicules that are common in the Jurassic of the North Sea. Although considerable porosity can be created by spicule dissolution, it tends to lead to porosity inversion (through silica reprecipitation) and can greatly lower overall permeability. One spicule can be seen cut along its long axis and the central canal (arrow) is filled by chalcedony, whereas the opaline spicule itself has dissolved. The pores surrounding the spicules also are filled with chalcedony. One secondary pore cut transverse to the spicule axis has been partly occluded by ferroan dolomite (lower right).

PPL, AFeS, KFS, BDI, Scale bar = 0.07 mm

Mid. Eocene Green River Fm., Laney Mbr., Fremont Co., Wyoming

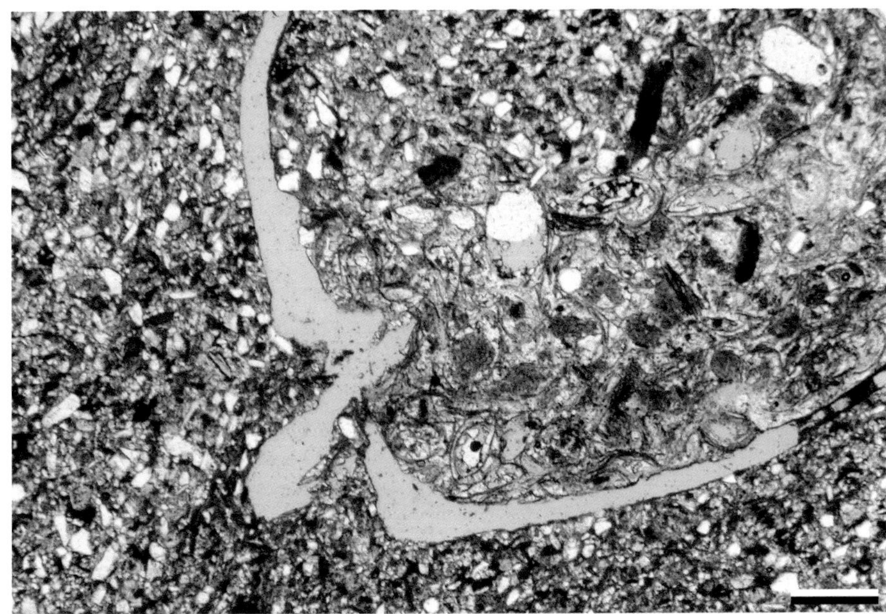

This poorly-sorted sandstone contains numerous dissolved carbonate bioclasts. The large fossil mold here is a cross section through the hinge-line of both valves of a bivalve (mollusk). Most bivalves are composed of aragonite (although some are partially or completely composed of low-Mg calcite) and are commonly dissolved to form biomoldic porosity (e.g., Cherns et al., 2011). The smaller bivalves here have undergone more variable dissolution. Dissolution of bioclasts can be an important source of carbonate for cementation in surrounding sediments.

PPL, BDI, Scale bar = 0.26 mm

Mid. Jurassic Entrada Fm., Slick Rock Mbr., Emery Co., Utah

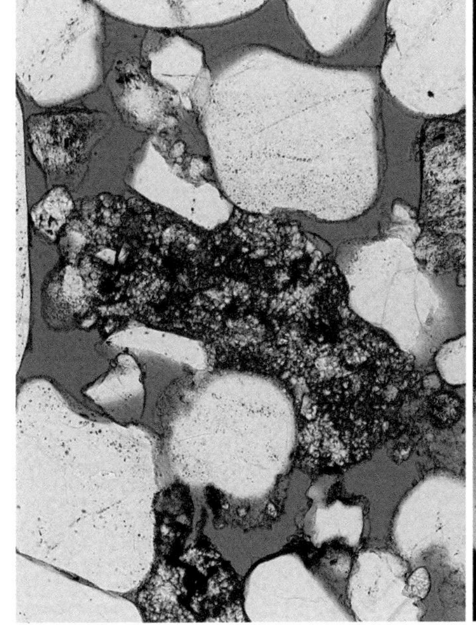

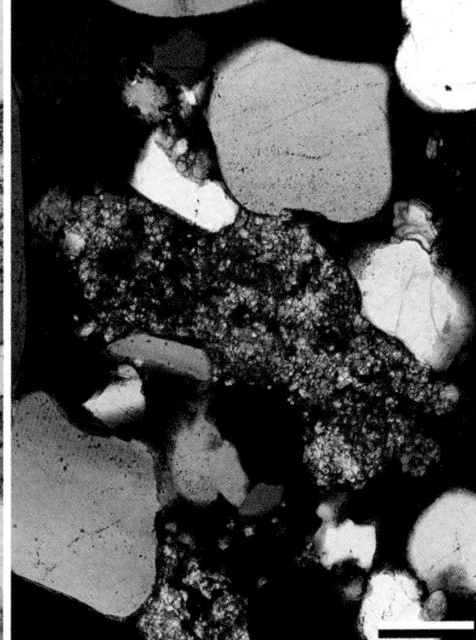

In this example, a carbonate rock fragment in a mature quartz-rich sandstone has been partly dissolved, and that dissolution weakened the rock fragment, contributing to its deformation during burial and compaction. The carbonate clast was originally a dolomitic limestone, based on the presence of rhombic-shaped ghosts within the grain. The dolomite was partially calcitized as well as dissolved.

PPL | XPL, RDI, Scale bar = 0.10 mm

Mid. Eocene Green River Fm., Laney Mbr., Fremont Co., Wyoming

This sedimentary rock fragment (a detrital quartzarenite clast) contains intragranular porosity produced by a combination of fracturing and dissolution of the original cements that held the sandstone clast together during weathering, erosion and deposition. It seems likely that this grain would not have survived as a rock fragment with the amount of cement it currently has, implying some diagenetic dissolution.

PPL | XPL, BDI, Scale bar = 0.26 mm

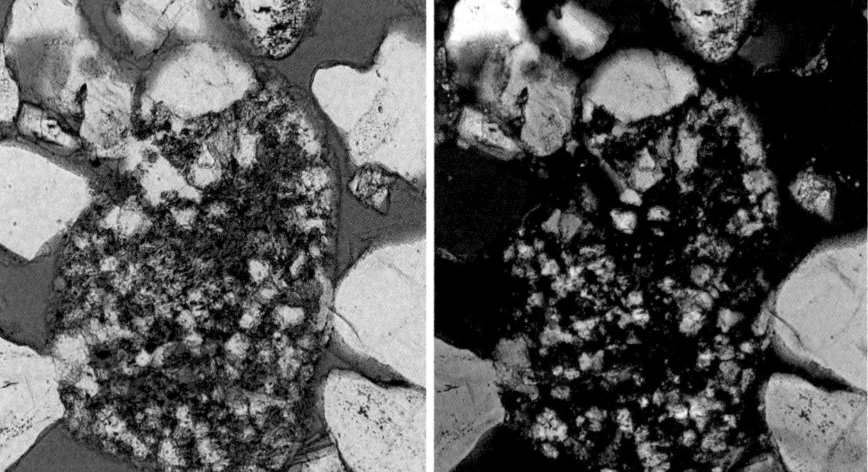

Mid. Jurassic Entrada Fm., Slick Rock Mbr., Emery Co., Utah

An example of a siltstone sedimentary rock fragment that was partially dissolved, yielding intragranular porosity. The clast is composed of grains of quartz silt interspersed with residual clays. In all probability, the material dissolved in this rock fragment was intergranular carbonate cement and possibly some constituent feldspar grains. Despite the dissolution, this clast exhibits almost no sign of having undergone any compaction. This lack of compaction may reflect late-stage, uplift-related leaching (telogenetic) of the grain by meteoric fluids.

PPL | XPL, RDI, Scale bar = 0.10 mm

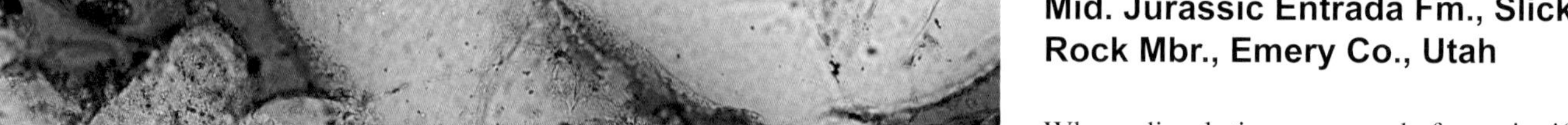

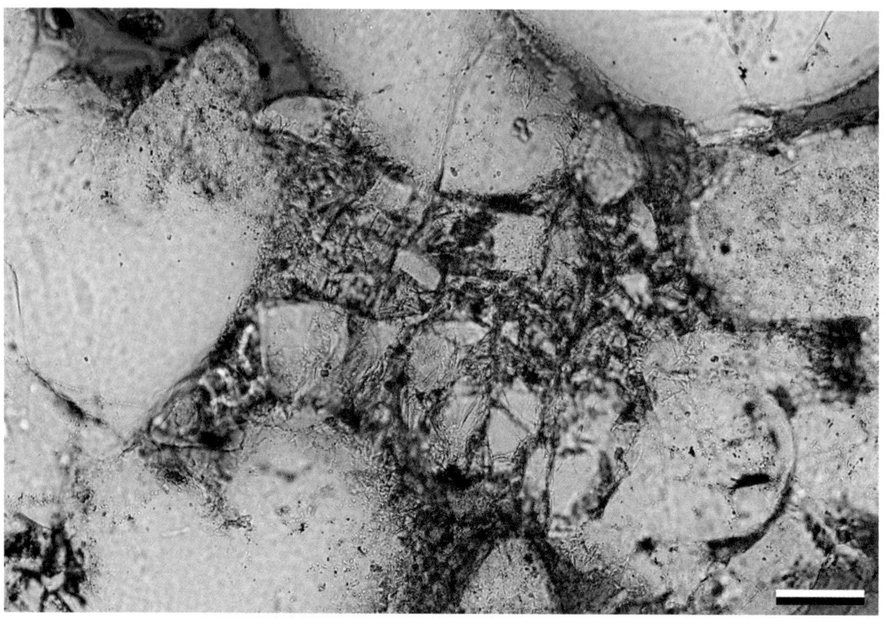

Mid. Jurassic Entrada Fm., Slick Rock Mbr., Emery Co., Utah

When dissolution occurs before significant overburden and compaction, the remnants of leached grains can collapse and fill surrounding porosity. In this example, the material that is now filling this porosity may have been a dissolved feldspar or a sedimentary rock fragment. However, because of the extensive dissolution and deformation, it is difficult to determine the precise identity of the original grain.

PPL, RDI, Scale bar = 0.05 mm

Lo. Triassic Ivishak Fm., North Slope, Alaska

This sedimentary rock fragment consisted of chertified matrix surrounding euhedral dolomite crystals. During burial, the dolomite crystals were dissolved creating dolomoldic porosity, whereas the intercrystalline chert framework remained unaltered. Lozenge-shaped siderite cements surround many of the grains and they precipitated prior to dissolution (based on the lack of siderite cements within the secondary porosity) and also prior to compaction of the unit. The more ductile grains have deformed and flattened during compaction.

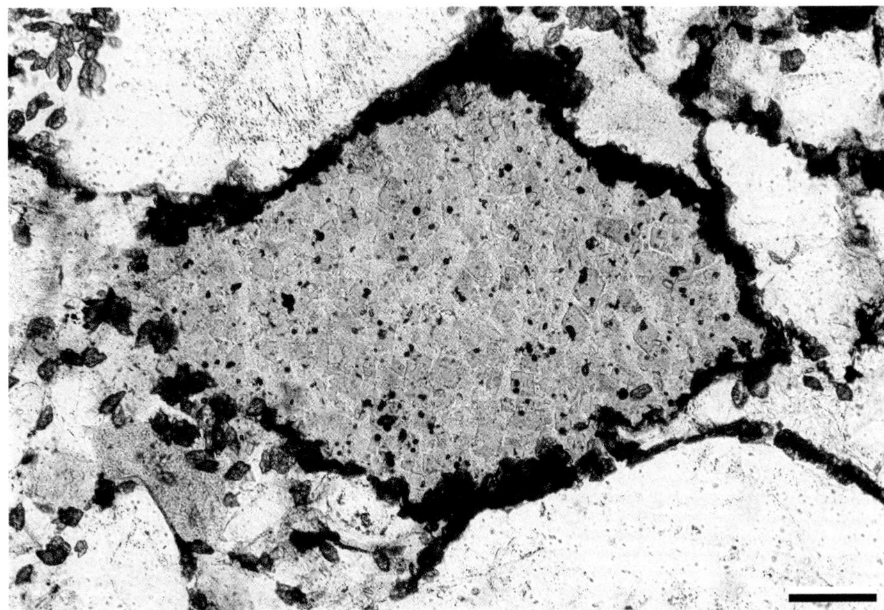

PPL, BDI, Scale bar = 0.10 mm

Lo. Triassic Ivishak Fm., North Slope, Alaska

This litharenite (chert arenite) contains numerous chert clasts that were deformed in their source terrane prior to erosion, transport, and deposition. Some of those clasts contain large fractures filled with fibrous chalcedony/moganite. In the plane-polarized light view, microporosity in the chert grains is clearly visible because of blue-stained epoxy impregnation. The chalcedonic fracture fills show little evidence of dissolution indicating that they were less soluble than the microcrystalline silica in the cherts. As in the previous photomicrograph, the siderite cements formed prior to compaction and dissolution.

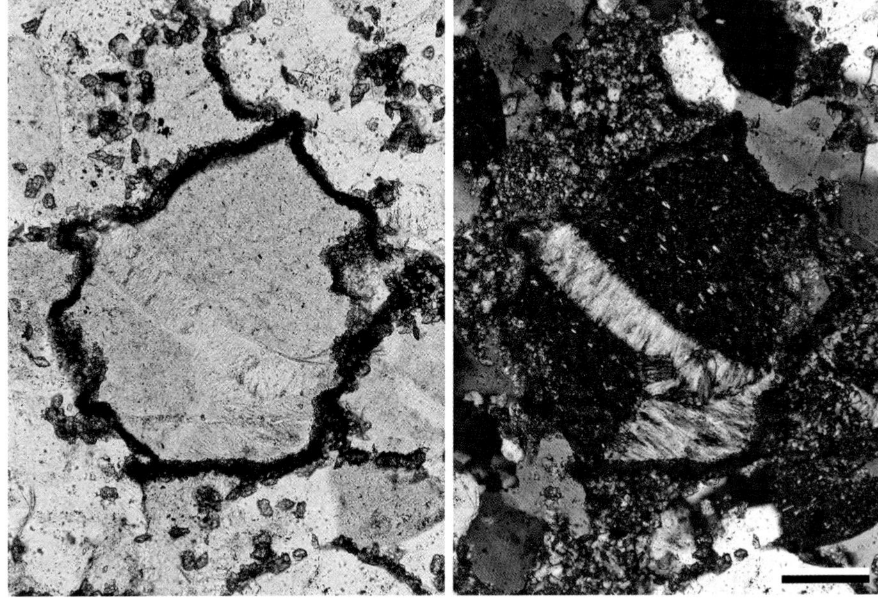

PPL | XPL, BDI, Scale bar = 0.12 mm

Mid. Jurassic Entrada Fm., Slick Rock Mbr., Emery Co., Utah

The large, rounded chert clast in the center of both images was partially dissolved, creating intragranular porosity. Chert porosity can be formed by exposure-related meteoric dissolution (Shanmugam and Higgins, 1988) and, because of its microcrystalline fabric, the porosity produced consists of micro- and nanopores (using Loucks et al., 2012 terminology; see Fig. 22.1). The larger pores in this chert clast are probably related to dissolution of sponge spicules or other siliceous biogenic remains.

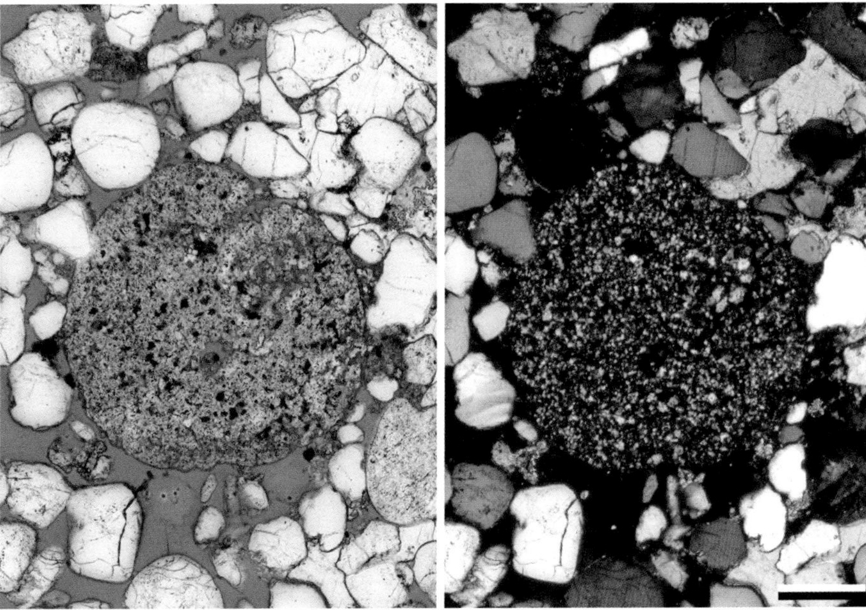

PPL | XPL, RDI, Scale bar = 0.10 mm

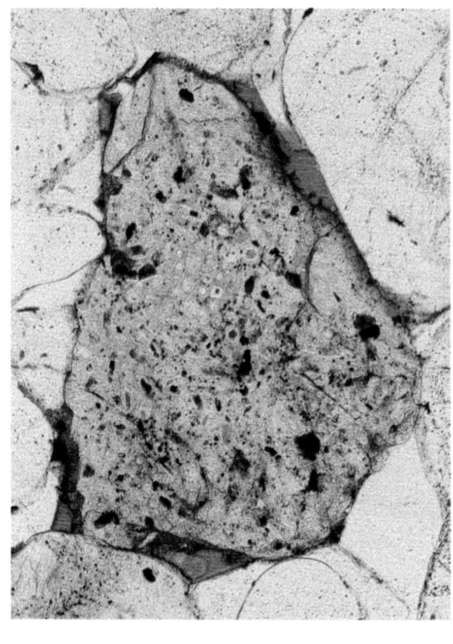
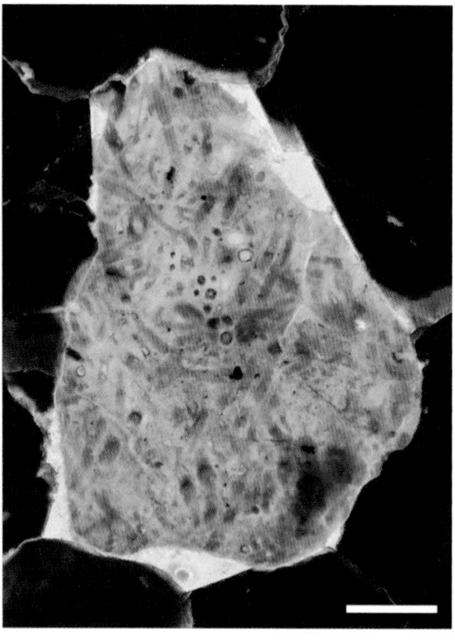

Cretaceous Dakota Gp., Moffat, Co., Colorado

Here, a rounded spiculitic chert clast was partially dissolved to create secondary intragranular porosity. The circular and longitudinal cross sections with central canals are sponge spicules that are still visible despite dissolution. In the fluorescent light view, the green coloration is related to oil impregnation within the grain's micro- and nanoporosity. The yellow orange fluorescence in both the grain and the primary porosity surrounding the grain, however, is the result of fluorescent epoxy injection, and indicates no live oil in those areas. The monocrystalline quartz grains nearly all have syntaxial overgrowths.

PPL | FL 470, RDI, Scale bar = 0.12 mm

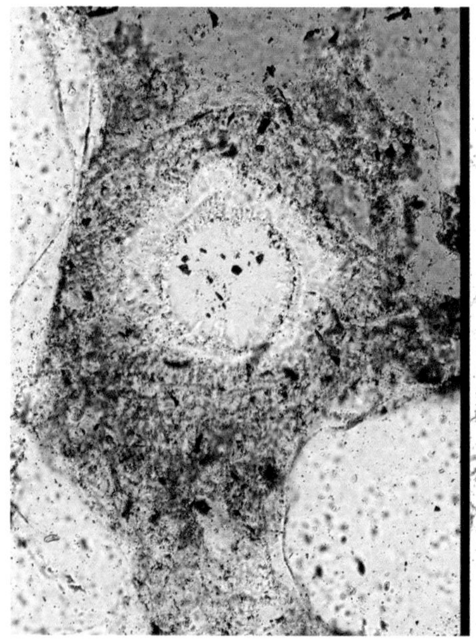
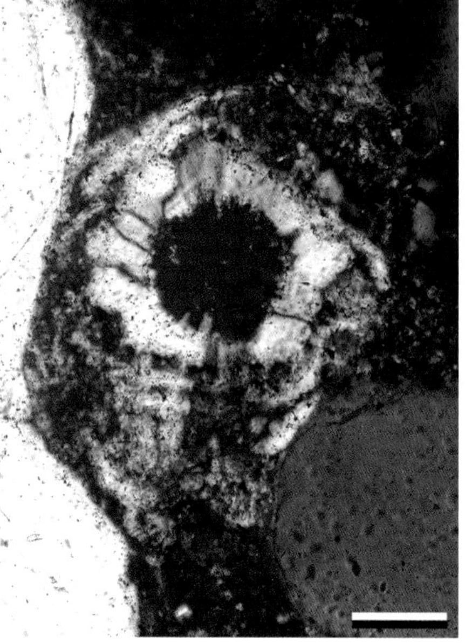

Cretaceous Dakota Gp., Socorro Co., New Mexico

This detrital grain (a reworked fibrous silica spherulite) comprises chalcedony and moganite. The relative proportions of the two minerals in a grain affects its stability, with increased moganite increasing the grain's solubility. In this chalcedony/moganite grain, the monocrystalline quartz core has a chalcedony-rich layer followed by layers that alternated between moganite-rich (dissolved) and chalcedony-rich (preserved).

PPL | XPL, BDI, Scale bar = 0.05 mm

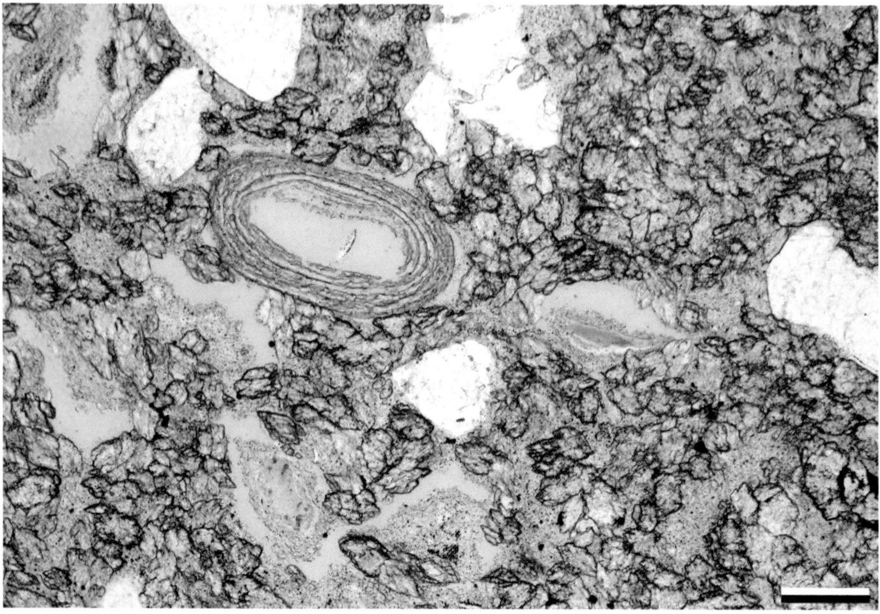

Lo. Cretaceous (Albian) Nahr Umr Fm., offshore Qatar

The "chamosite" ooids in this siderite-cemented sandstone were partially to completely dissolved. The ooid, near the center of the photomicrograph, still has some of the cortices preserved (a common occurrence in compositionally complex and varied "chamositic grains"). While that ooid has been somewhat flattened during burial and compaction, the ooids with only scant remnants of chamosite (and possibly glauconite) have collapsed almost completely. Most of the siderite cements in this sample exhibit a typical lozenge shape.

PPL, BDI, Scale bar = 0.10 mm

Lo. Cretaceous (Albian) Nahr Umr Fm., offshore Qatar

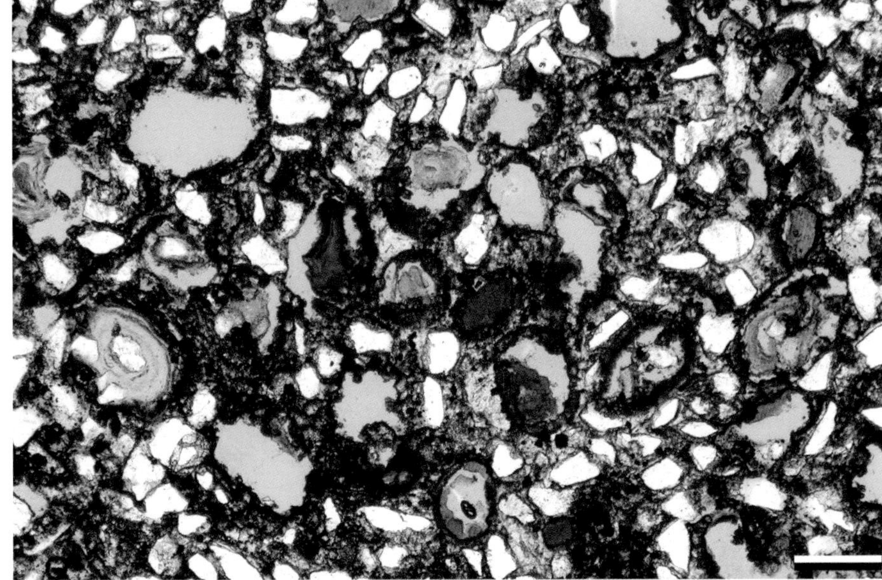

Another example of partially to completely dissolved "chamosite" ooids as well as glauconite grains. Unlike the previous sample, these moldic pores have maintained their shape during compaction. This is likely related to this unit having been more solidly cemented by siderite prior to dissolution of the grains. See Chapter 5 for a discussion of the compositional complexities of "chamosite ooids".

PPL, AFeS, BDI, Scale bar = 0.26 mm

Mesozoic unknown formation, Moffat Co., Colorado

This immature sandstone contains vugs (defined as dissolution pores larger than the individual grains within the rock) that here are partially fabric selective (elongate parallel to bedding and stylolitic surfaces). This secondary pore may have formed by the dissolution of several adjacent unstable grains. The lack of compaction or collapse of surrounding smaller grains into the pore implies substantial rock lithification prior to dissolution. Similar vugs also can be the result of poor sample handling and rip-out of material prior to impregnation and sectioning, although that is not the case here (as shown by the pink-stained epoxy impregnation).

PPL, RDI, Scale bar = 0.26 mm

Miocene Santa Fe Gp., Zia Fm., Cerro Conejo Mbr., Sandoval Co., New Mexico

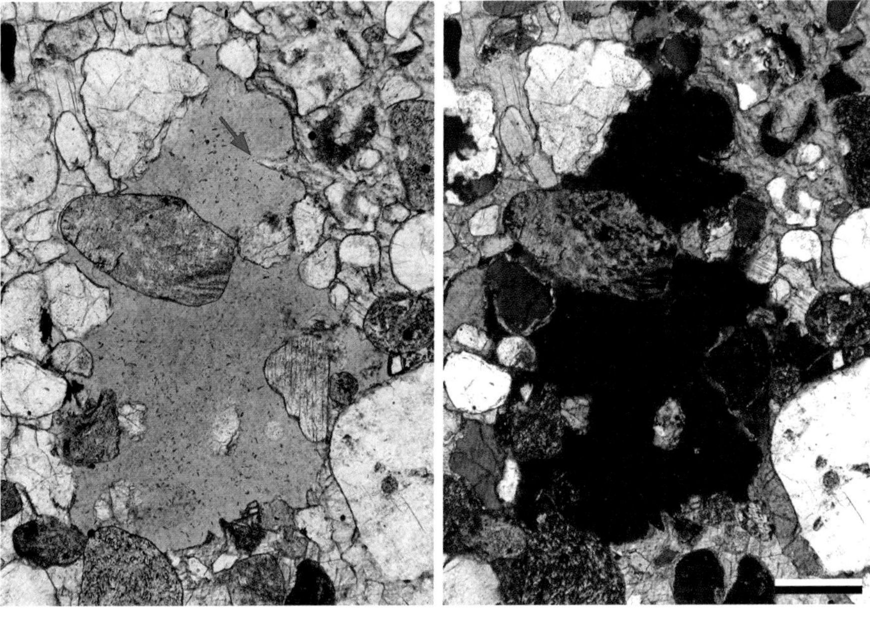

These large, irregular pores are vugs that are only partly fabric selective. This vuggy porosity was a byproduct of concretion formation in a meteoric environment with fluids transporting solutes to cement some areas of the concretion while dissolving grains in other areas. Dissolution also expanded fluid conduits into the interiors of the concretion (Mozley and Davis, 2005). Evidence for a solution origin of this vug is visible near the red arrow where remnants of calcite cement outline parts of at least two grains that were dissolved to form the upper part of this compound structure.

PPL | XPL, BDI, Scale bar = 0.26 mm

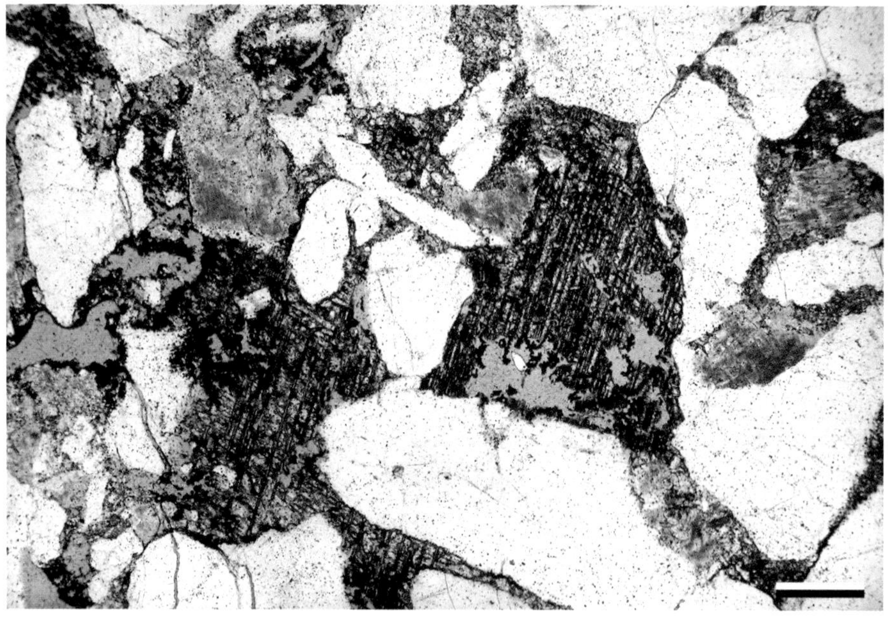

Up. Pennsylvanian Bursum Fm., Socorro Co., New Mexico

An example of partially dissolved calcite cement that, at one time, largely obliterated the primary intergranular pores within large areas of this arkosic sandstone. The remnants of the calcite cements still show parallel twin planes that are consistently oriented throughout the image, evidence that these were poikilotopic cements. Dissolution of calcites most likely occurred as this unit was uplifted (telogenesis), and acidic meteoric fluids circulated through the rock. The black materials within the secondary porosity are iron oxides that formed during or after calcite dissolution.

PPL, BDI, Scale bar = 0.38 mm

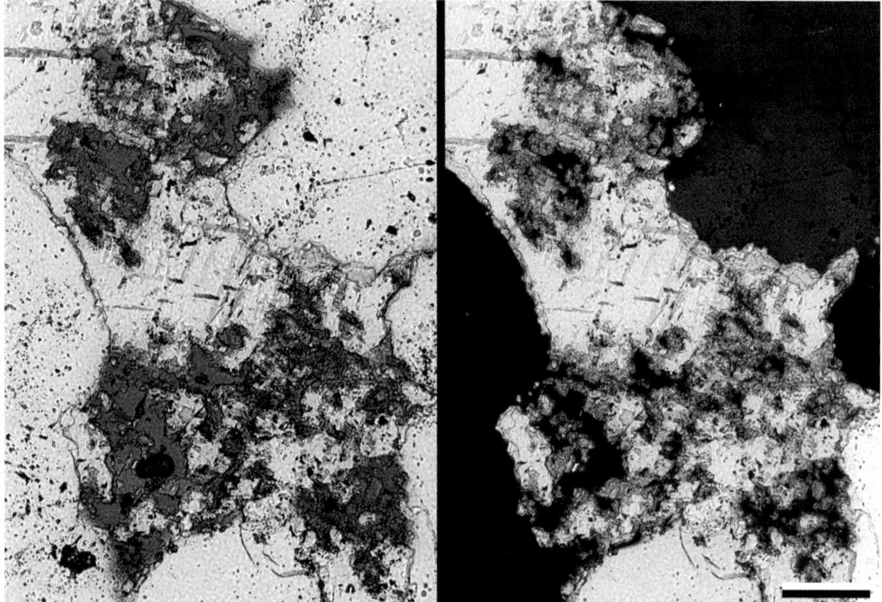

Lo. Pennsylvanian Morrow B Sandstone, Ochiltree Co., Texas

Patchy dissolution of pore-filling cements in this sandstone created cement-dissolution porosity. The cement is a poikilotopic calcite, and cement corrosion has produced very irregular pore margins. If dissolution had continued, all traces of cements in the rock could have been removed "resurrecting" much of the unit's porosity. Cement-dissolution porosity can play a significant role in porosity and permeability development in rocks, but it can be difficult to recognize when all or most of the original cement has been removed.

PPL | XPL, RDI, Scale bar = 0.10 mm

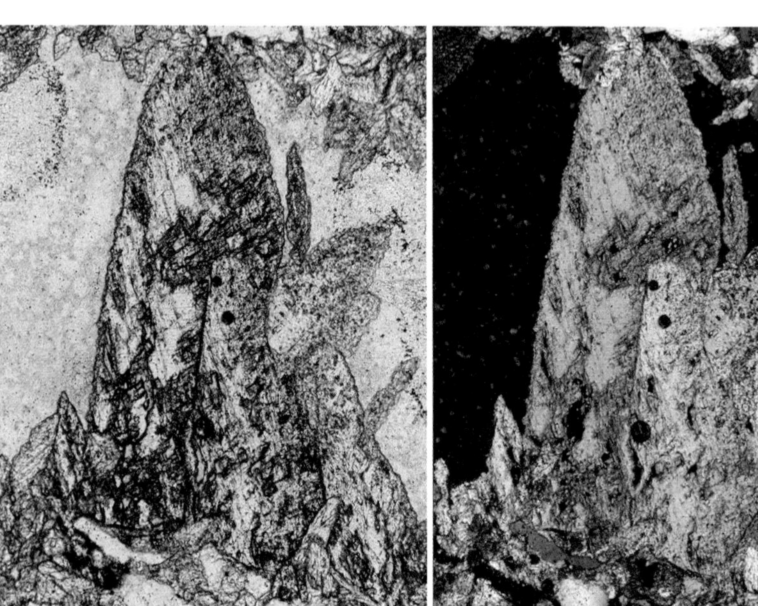

Mid. Jurassic Entrada Fm., Moffat Co., Colorado

In this example, a large fracture within a sandstone was filled with a combination of dogtooth spar calcite crystals, dolomite (not visible in this view) and densely packed kaolinite (the very pale pink stained areas at left and right that show very low to no birefringence). The calcite cements were later partially corroded and now contain intracrystalline porosity (filled with blue-dyed epoxy).

PPL | XPL, AFeS, BDI, Scale bar = 0.10 mm

Mid. Jurassic Entrada Fm., Emery Co., Utah

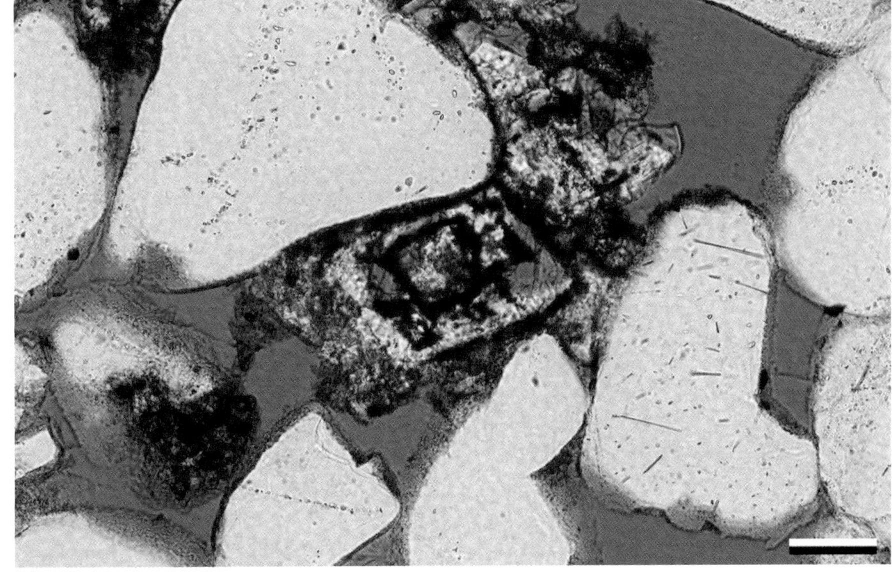

Dolomite cements commonly are compositionally zoned and thus can be susceptible to selective dissolution of less stable zones. The ferroan dolomite cements in this quartzarenite were partially dissolved, creating secondary dolomoldic pores. Originally, the dolomites were zoned with an inner core and outer layers separated by a more ferruginous zone that has subsequently converted to iron oxides.

PPL, RDI, Scale bar = 0.04 mm

Eocene Galisteo Fm., Sandoval Co., New Mexico

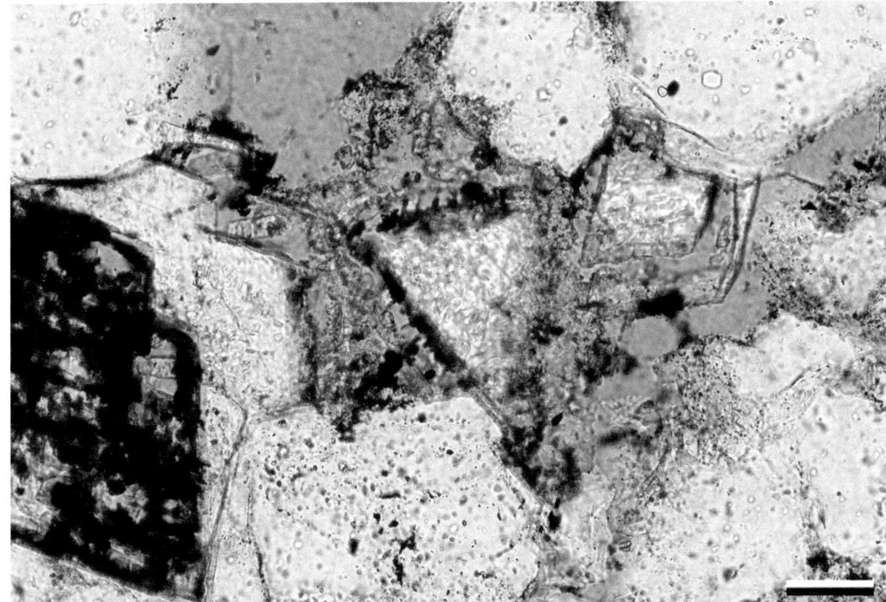

On the left edge of this image, the central zone of a ferroan dolomite or ankerite crystal was selectively dissolved leaving a residue of iron oxides within the crystal-moldic pores. In the other dolomite crystals (near the center of the image), core zones of iron-poor dolomite were followed by a chemically unstable zone that was dissolved, followed by a thin rind of more stable dolomite. The discrepant zonations may reflect different times of formation or may be a result of off-center cuts through .some of the crystals. Where consistent histories are found, however, they can provide a useful record of fluid compositions through time.

PPL, BDI, Scale bar = 0.05 mm

Eocene Galisteo Fm., Sandoval Co., New Mexico

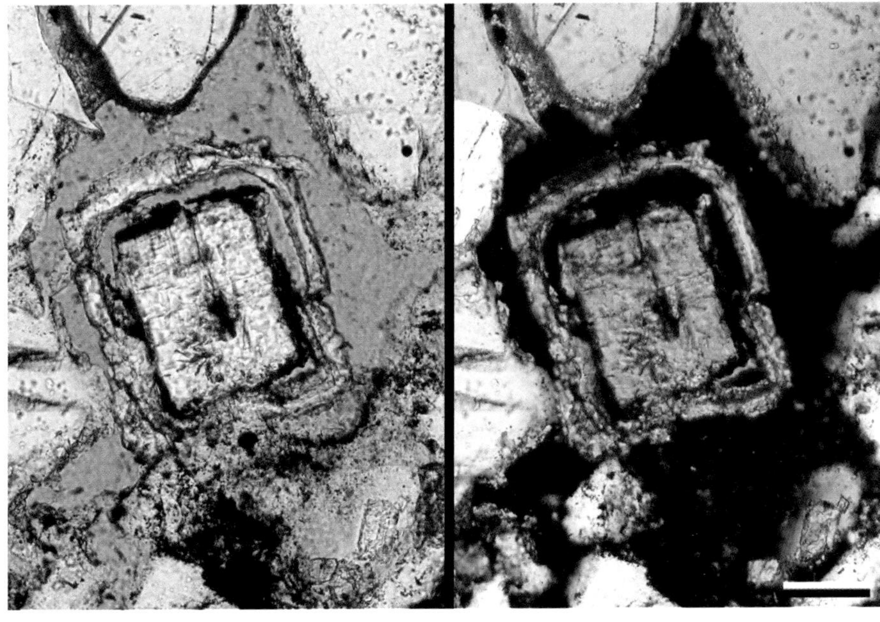

As noted above, not every crystal starts precipitating at the same time and not every cut completely transects crystals. In this photomicrograph from the same sample as the previous slide, a ferroan dolomite/ankerite core is barely visible in this cement crystal. The dark dot in the center of the core zone represents that first growth stage, but only a small corner of that core is intersected by this slice. However, the other zones of dolomite are more apparent and are consistent with the previous example.

PPL, BDI, Scale bar = 0.03 mm

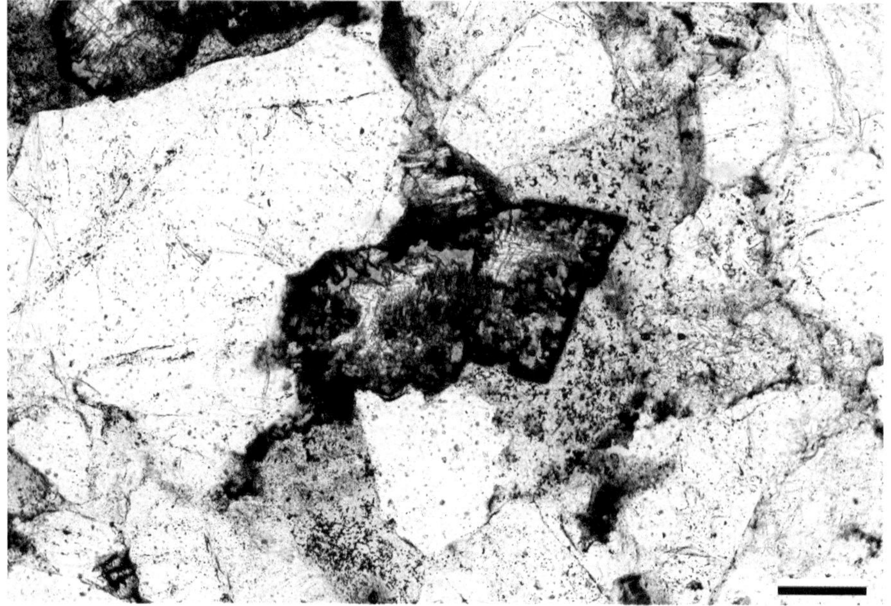

Mid. Eocene Baca Fm., Socorro Co., New Mexico

This example of moldic porosity resulted from the partial dissolution of ferroan dolomite/ankerite cements (or possibly saddle dolomite given the slightly curved crystal faces). Iron oxides coat the pores created during dissolution of the iron-rich dolomites. The adjacent kaolinite cements formed after the dolomite cements but before the dissolution.

PPL, BDI, Scale bar = 0.14 mm

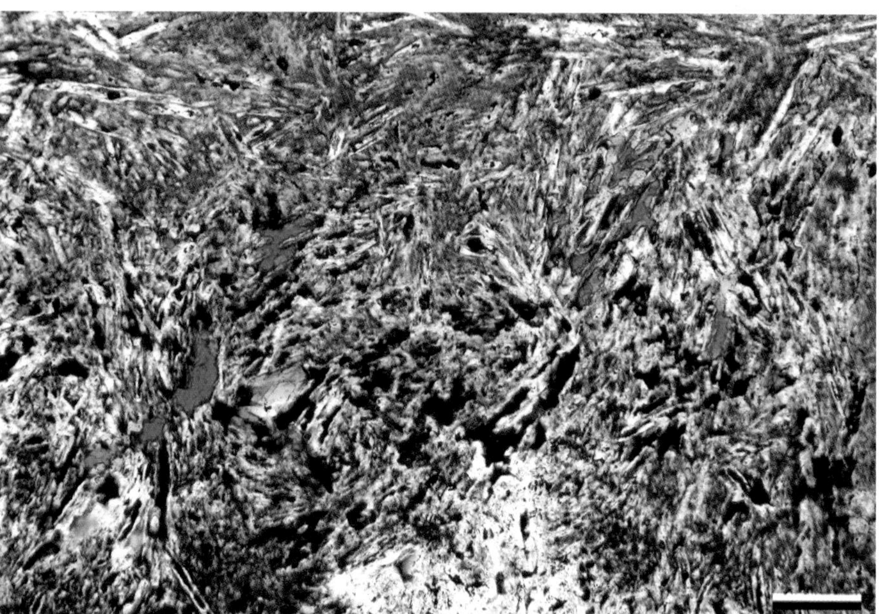

Permian Park City Fm., Tosi Chert Mbr., Park Co., Wyoming

These chert nodules formed through the replacement of nodular evaporites by microcrystalline chert, chalcedony and megaquartz. In the more-finely crystalline areas, the original morphology of the felted anhydrite crystal laths are visible within the silica replacements. Although most of the anhydrite was replaced by silica, unreplaced inclusions of anhydrite were later dissolved producing crystal-moldic pores. This dissolution event may have been associated with migration of hydrocarbons as evidenced by the extensive staining and filling of the porosity by bitumen and live oil.

PPL, BDI, Scale bar = 0.23 mm

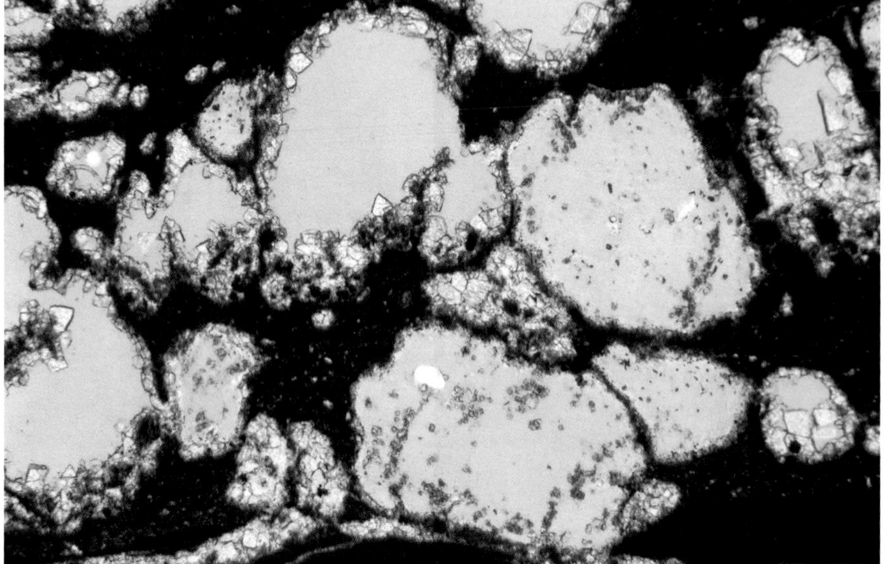

Mid.? Permian (former Ufimian) Up. Solikamskaya Horizon, near Solikamsk, Perm Region, Russia

Large moldic pores resulting from dissolution of evaporite nodules in a dolomitic shale. The original, displacive evaporites (probably anhydrite and/or gypsum) grew syndepositionally in soft muds forming what is often known as a "chicken-wire" fabric. Later dissolution of the evaporites created these large moldic pores, and was followed by precipitation of euhedral dolomite cements rims and fillings in some of these secondary pores. The resultant "cellular" fabric is characteristic of evaporite dissolution, even where no remnants of original minerals remain.

PPL, BDI, Scale bar = 0.51 mm

Mid.? Permian (former Ufimian) Up. Solikamskaya Horizon, near Solikamsk, Perm Region, Russia

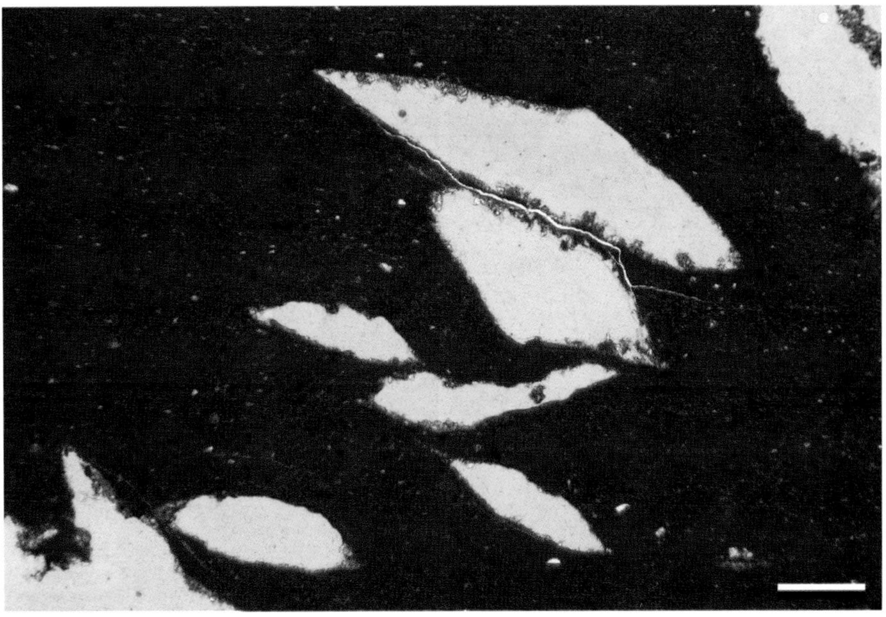

Evaporite minerals in this dolomitic shale were completely dissolved to form crystal-moldic pores. The well-formed, discoidal or lenticular crystal molds indicate that the precursor evaporite mineral was probably gypsum. The absence of any later cements within the molds indicates that dissolution probably occurred at very late stage of telogenesis, probably at or near outcrop.

PPL, AS, BDI, Scale bar = 0.51 mm

Cretaceous Dakota Gp., Socorro Co., New Mexico

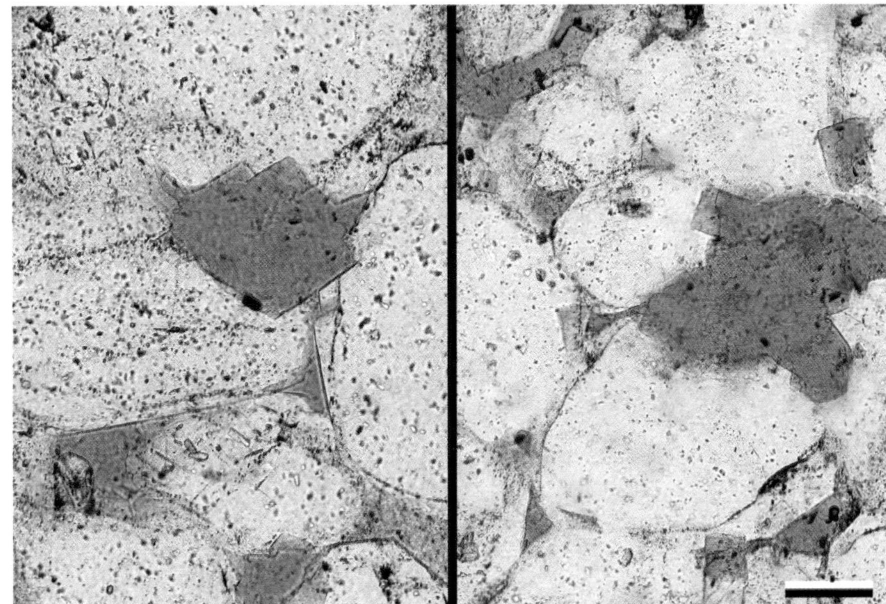

These are two views from different areas in the same thin section showing moldic pores that have square to rectangular outlines. The mineral that filled these pores most likely formed after syntaxial quartz overgrowths, partially replacing both quartz cements and quartz grains. Based on the morphology of the secondary pores, the replacement/cementing mineral was likely anhydrite or halite. While not common, such replacements have been reported in other units (Schenk and Richardson, 1985). Both anhydrite and halite are highly soluble and dissolved during later diagenesis to produce secondary crystal-moldic porosity.

PPL | PPL, BDI, Scale bar = 0.51 mm

Mid. Jurassic Carmel Fm., Crystal Creek Mbr., Cane Co., Utah

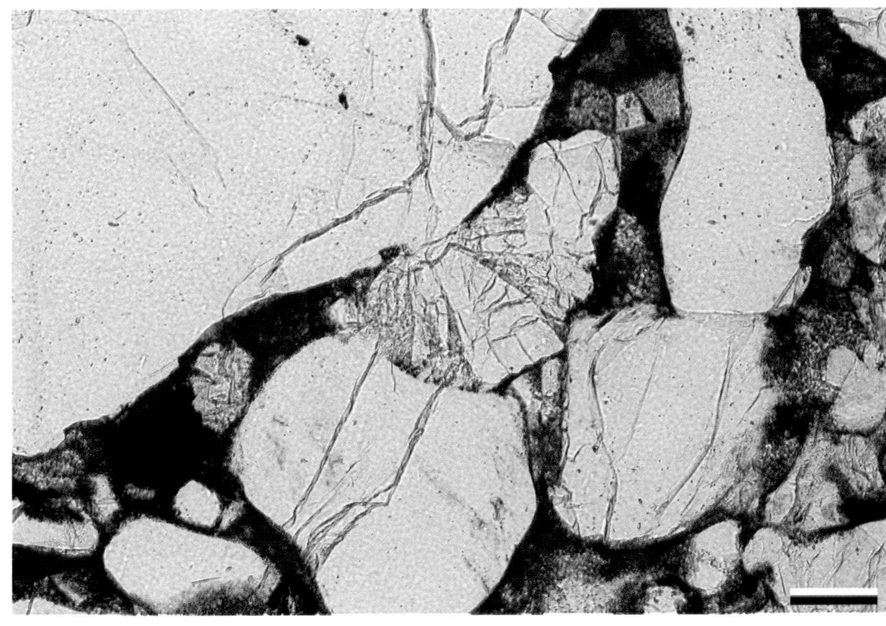

Pervasive shattering of quartz grains by overburden loading. Normally, this occurs in just a few grains; in this rock it affects nearly all grains. Therefore, it is inferred to have involved rapid dissolution of calcium sulfate cements at depth, producing a sudden increase in point-contact load stresses on framework grains that had previously been cushioned by cement (criterion 9 for the recognition of dissolution in the list in the introduction to this chapter). Although no remnants of the inferred cements remain here, the brownish clays may be, in part, an insoluble residue from such dissolved materials (see also next image).

PPL, BDI, Scale bar = 0.26 mm

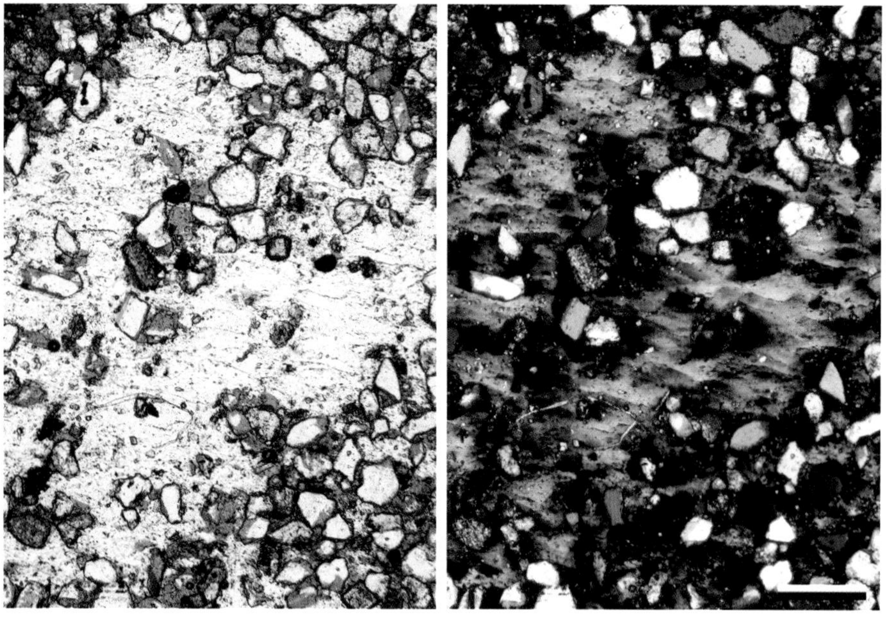

Mid. Jurassic Carmel Fm., Cane Co., Utah

A different interval in the Carmel Formation from that shown in the previous image. Here the terrigenous grains are dispersed in still-preserved gypsum cement in the central part of each view (white in PPL; dark gray in XPL). The gypsum has been partially corroded, possibly during thin-section preparation. This is the type of cement inferred to have been dissolved in a subsurface setting in the previous example.

PPL | XPL, BDI, Scale bar = 0.26 mm

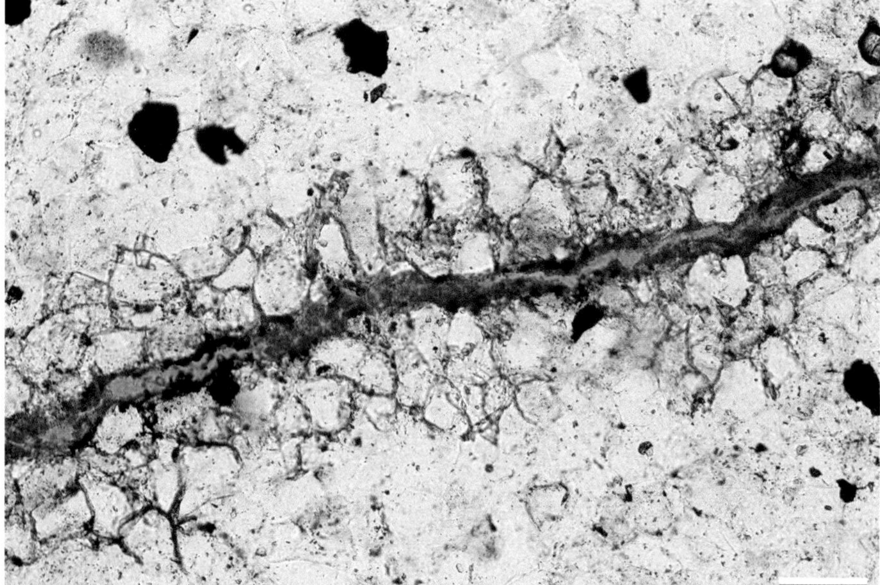

Mississippian Paradise Fm., Hildalgo Co., New Mexico

The fractures in this quartzarenite were conduits for fluid flow. The corrosive fluids that moved through these conduits dissolved parts of quartz grains and quartz cements on either side of the fracture creating additional secondary dissolution porosity associated with the fracture porosity. Solution enlargement of fractures is a common phenomenon in carbonate rocks, but it also occurs, albeit with lesser frequency, in clastic terrigenous units. Micritic carbonate cements were a later-stage precipitate within this fracture, slightly reducing the fracture porosity.

PPL, BDI, Scale bar = 0.10 mm

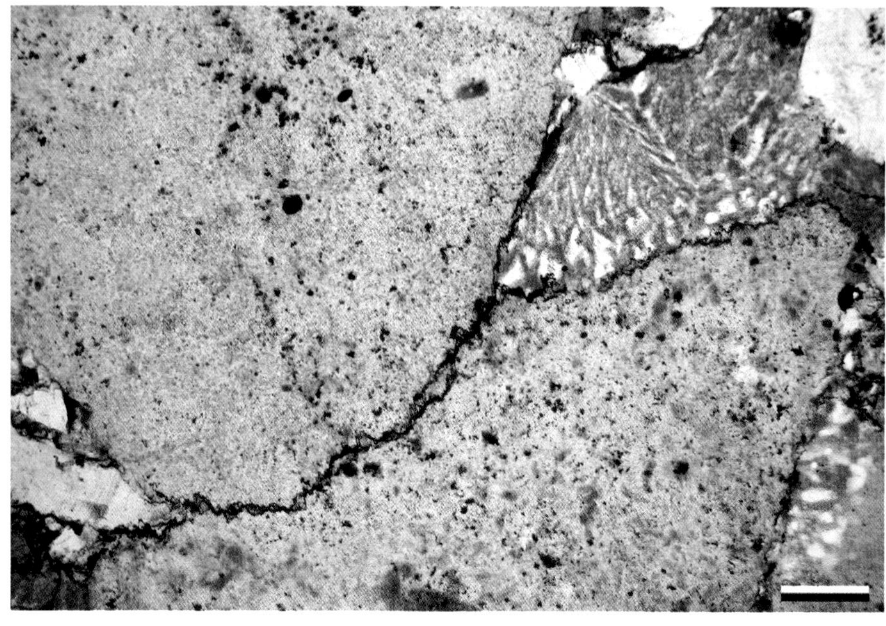

Up. Pennsylvanian – Lo. Permian Granite Wash, southern Oklahoma

Stylolitization results from pressure dissolution of minerals and rocks with an overall loss of volume and little or no porosity formation, unlike most of the previous examples where chemical dissolution resulted in porosity creation. Here, stylolitization produced thin, serrated zones of insoluble residues (commonly clays, iron minerals and/or organic matter). Most solution seams and circumgranular dissolution features are associated with compaction and so are covered in more detail in Chapter 10) and to a lesser degree in Chapter 21. See also photos in Chapter 11 (p. 255) that show CL imaging of quartz overgrowths that look like compaction features.

PPL, BDI, Scale bar = 0.16 mm

Cited References and Additional Information Sources

Al-Ramadan, K., S. Morad, A. K. Norton, and M. Hulver, 2012, Linking diagenesis and porosity preservation versus destruction to sequence stratigraphy of gas condensate reservoir sandstones: The Jauf Formation (Lower to Middle Devonian), eastern Saudi Arabia, *in* S. Morad, J. M. Ketzer, and L. F. De Ros, eds., Linking Diagenesis to Sequence Stratigraphy (IAS Special Publication 45): Oxford, UK, Wiley-Blackwell, p. 297-336, doi: 10.1002/9781118485347.ch13.

Bernet, M., and R. Gaupp, 2005, Diagenetic history of Triassic sandstone from the Beacon Supergroup in central Victoria Land, Antarctica: New Zealand Journal of Geology and Geophysics, v. 48, p. 447-458, doi: 10.1080/00288306.2005.9515125.

Bjørkum, P. A., 1996, How important is pressure in causing dissolution of quartz in sandstones?: Journal of Sedimentary Research, v. 66, p. 147-154, doi: 10.1306/D42682DE-2B26-11D7-8648000102C1865D.

Bjørlykke, K., 1984, Formation of secondary porosity: How important is it?, *in* D. A. McDonald, and R. C. Surdam, eds., Clastic Diagenesis: Tulsa, OK, AAPG Memoir 37, p. 277-286.

Bjørlykke, K., 1988, Sandstone diagenesis in relation to preservation, destruction and creation of porosity, *in* G. V. Chilingarian, and K. H. Wolf, eds., Diagenesis: New York, Elsevier Science Publ., p. 555-588, doi: 10.1016/S0070-4571(08)70180-8.

Bjørlykke, K., and J. Jahren, 2012, Open or closed geochemical systems during diagenesis in sedimentary basins: Constraints on mass transfer during diagenesis and the prediction of porosity in sandstone and carbonate reservoirs: AAPG Bulletin, v. 96, p. 2193-2214, doi: 10.1306/04301211139.

Blatt, H., and B. Sutherland, 1969, Intrastratal solution and non-opaque heavy minerals in shales: Journal of Sedimentary Research, v. 39, p. 591-600, doi: 10.1306/74D71CDA-2B21-11D7-8648000102C1865D.

Bloch, S., 1994, Secondary porosity in sandstones: Significance, origin, relationship to subaerial unconformities, and effect on predrill reservoir quality prediction, *in* M. D. Wilson, ed., Reservoir Quality Assessment and Prediction in Clastic Rocks: Tulsa, OK, SEPM Short Course Notes 30, p. 137-159, doi: 10.2110/scn.94.30.0137.

Bloch, S., and S. G. Franks, 1993, Preservation of shallow plagioclase dissolution porosity during burial: Implications for porosity prediction and aluminum mass balance: AAPG Bulletin, v. 77, p. 1488-1501.

Burley, S. D., and J. D. Kantorowicz, 1986, Thin section and S.E.M. textural criteria for the recognition of cement-dissolution porosity in sandstones: Sedimentology, v. 33, p. 587-604, doi: 10.1111/j.1365-3091.1986.tb00764.x.

Cherns, L., J. R. Wheeley, and V. P. Wright, 2011, Taphonomic bias in shelly faunas through time: Early aragonitic dissolution and its implications for the fossil record, *in* P. A. Allison, and D. J. Bottjer, eds., Taphonomy: Process and Bias Through Time (Topics in Geobiology 32): Dordrecht, Springer Netherlands, p. 79-105, doi: 10.1007/978-90-481-8643-3_3.

Chowdhury, A. H., and J. P. A. Noble, 1992, Porosity evolution in the Albert Formation of the Stoney Creek oil and gas field, Moncton Subbasin, New Brunswick, Canada: AAPG Bulletin, v. 76, p. 1325-1343.

Ehrenberg, S. N., and K. G. Jakobsen, 2001, Plagioclase dissolution related to biodegradation of oil in Brent Group sandstones (Middle Jurassic) of Gullfaks Field, northern North Sea: Sedimentology, v. 48, p. 703-721, doi: 10.1046/j.1365-3091.2001.00387.x.

Emery, D., R. J. Myers, and R. Young, 1990, Ancient subaerial exposure and freshwater leaching in sandstones: Geology, v. 18, p. 382-406, doi: 10.1130/0091-7613(1990)018<1178:ASEAFL>2.3.CO;2.

Gibling, M. R., A. T. Martel, M. H. Nguyen, A. M. Kennedy, J. Shimeld, F. Baechler, S. Forgeron, and B. Mackenzie, 2000, Fluid evolution and diagenesis of a Carboniferous channel sandstone in the Prince Colliery, Nova Scotia, Canada: Bulletin of Canadian Petroleum Geology, v. 48, p. 95-115, doi: 10.2113/48.2.95.

Giles, M. R., and R. B. De Boer, 1990, Origin and significance of redistributional secondary porosity: Marine and Petroleum Geology, v. 7, p. 378-397, doi: 10.1016/0264-8172(90)90016-A.

Glasmann, J. R., 1992, The fate of feldspar in Brent Group reservoirs, North Sea: A regional synthesis of diagenesis in shallow, intermediate, and deep burial environments, *in* A. C. Morton, R. S. Haszeldine, M. R. Giles, and S. Brown, eds., Geology of the Brent Group: London, GSL Special Publication 61, p. 329-350, doi: 10.1144/GSL.SP.1992.061.01.17.

Harris, N. B., 1989, Diagenetic quartzarenite and destruction of secondary porosity: An example from the Middle Jurassic Brent sandstone of northwest Europe: Geology, v. 17, p. 361-364, doi: 10.1130/0091-7613(1989)017<0361:DQADOS>2.3.CO;2. [see also discussions and replies in v. 18, p. 287-288 and 799-800].

Hayes, M. J., and J. R. Boles, 1992, Volumetric relations between dissolved plagioclase and kaolinite in sandstones: Implications for aluminum mass transfer in the San Joaquin basin, California, *in* D. W. Houseknecht, and E. D. Pitman, eds., Origin, Diagenesis, and Petrophysics of Clay Minerals in Sandstones: Tulsa, OK, SEPM Special Publication 47, p. 111-123, doi: 10.2110/pec.92.47.0111.

Helmold, K. P., 1985, Provenance of feldspathic sandstones—the effect of diagenesis on provenance interpretations: A review, *in* G. G. Zuffa, ed., Provenance of Arenites (NATO Science Series C): Dordrecht, Netherlands, D. Reidel Publishing, p. 139-164, doi: 10.1007/978-94-017-2809-6_7.

Huang, W.-L., and J. M. Longo, 1992, The effect of organics on feldspar dissolution and the development of secondary porosity: Chemical Geology, v. 98, p. 271-292, doi: 10.1016/0009-2541(92)90189-C.

Hurst, A. R., 1981, A scale of dissolution for quartz and its implications for diagenetic processes in sandstones: Sedimentology: v. 28, p. 451-459, doi: 10.1111/j.1365-3091.1981.tb01694.x.

Land, L. S., and K. L. Milliken, 1981, Feldspar diagenesis in the Frio Formation, Brazoria County, Texas Gulf Coast: Geology, v. 9, p. 314-318, doi: 10.1130/0091-7613(1981)9<314:FDITFF>2.0.CO;2.

Land, L. S., K. L. Milliken, and E. F. McBride, 1987, Diagenetic evolution of Cenozoic sandstones, Gulf of Mexico sedimentary basin: Sedimentary Geology, v. 50, p. 195-225, doi: 10.1016/0037-0738(87)90033-9.

Loucks, R. G., 2005, Revisiting the importance of secondary dissolution pores in Tertiary sandstones along the Texas Gulf Coast: Gulf Coast Association of Geological Societies Transactions, v. 55, p. 447-455.

Loucks, R. G., M. M. Dodge, and W. E. Galloway, 1979, Importance of secondary leached porosity in Lower Tertiary sandstone reservoirs along the Texas Gulf Coast: Gulf Coast Association of Geological Societies Transactions, v. 29, p. 164-171.

McBride, E. F., 1977, Secondary porosity—importance in sandstone reservoirs in Texas: Gulf Coast Association of Geological Societies Transactions, v. 27, p. 121-122.

McBride, E. F., 1985, Diagenetic processes that affect provenance determinations in sandstone, *in* G. G. Zuffa, ed., Provenance of Arenites (NATO Science Series C): Dordrecht, Netherlands, D. Reidel Publishing, p. 95-113, doi: 10.1007/978-94-017-2809-6_5.

McKinley, J. M., P. M. Atkinson, C. D. Lloyd, A. H. Ruffell, and R. H. Worden, 2011, How porosity and permeability vary spatially with grain size, sorting, cement volume, and mineral dissolution in fluvial Triassic sandstones: The value of geostatistics and local regression: Journal of Sedimentary Research, v. 81, p. 844-858, doi: 10.2110/jsr.2011.71.

Milliken, K. L., 1992, Chemical behavior of detrital feldspars in mudrocks versus sandstones, Frio Formation (Oligocene), South Texas: Journal of Sedimentary Research, v. 62, p. 790-801, doi: 10.1306/D42679DD-2B26-11D7-8648000102C1865D.

Milliken, K. L., 2007, Provenance and diagenesis of heavy minerals, Cenozoic units of the northwestern Gulf of Mexico sedimentary basin, *in* M. A. Mange, and D. T. Wright, eds., Heavy Minerals in Use (Developments in Sedimentology 58): New York, Elsevier, p. 247-261, doi: 10.1016/S0070-4571(07)58008-8.

Milliken, K. L., and L. E. Mack, 1990, Subsurface dissolution of heavy minerals, Frio Formation sandstones of the ancestral Rio Grande Province, South Texas: Sedimentary Geology, v. 68, p. 187-199, doi: 10.1016/0037-0738(90)90111-6.

Milliken, K. L., E. F. McBride, and L. S. Land, 1989, Numerical assessment of dissolution versus replacement in the subsurface destruction of detrital feldspars, Oligocene Frio Formation, South Texas: Journal of Sedimentary Research, v. 59, p. 740-757, doi: 10.1306/212F9061-2B24-11D7-8648000102C1865D.

Morad, S., and A. A. Aldahan, 1986, Alteration of detrital Fe-Ti oxides in sedimentary rocks: GSA Bulletin, v. 97, p. 567-578, doi: 10.1130/0016-7606(1986)97<567:AODFOI>2.0.CO;2.

Morton, A. C., G. Borg, P. L. Hansley, P. D. W. Haughton, D. H. Krinsley, and P. Trusty, 1989, The origin of faceted garnets in sandstones: Dissolution or overgrowth?: Sedimentology, v. 36, p. 927-942, doi: 10.1111/j.1365-3091.1989.tb01754.x.

Morton, A. C., and C. R. Hallsworth, 2007, Stability of detrital heavy minerals during burial diagenesis, *in* M. A. Mange, and D. T. Wright, eds., Heavy Minerals in Use (Developments in Sedimentology 58): Amsterdam, Elsevier, p. 215-245, doi: 10.1130/2006.2420(19).

Mozley, P. S., and J. M. Davis, 2005, Internal structure and mode of growth of elongate calcite concretions: Evidence for small-scale, microbially induced, chemical heterogeneity in groundwater: GSA Bulletin, v. 117, p. 1400-1412, doi: 10.1130/B25618.1.

Nagtegaal, P. J. C., 1978, Sandstone-framework instability as a function of burial diagenesis: Journal of the Geological Society, v. 135, p. 101-105, doi: 10.1144/gsjgs.135.1.0101.

Parsons, I., P. Thompson, M. R. Lee, and N. Cayzer, 2005, Alkali feldspar microtextures as provenance indicators in siliciclastic rocks and their role in feldspar dissolution during transport and diagenesis: Journal of Sedimentary Research, v. 75, p. 921-942, doi: 10.2110/jsr.2005.071.

Pittman, E. D., 1992, Artifact porosity in thin sections of sandstones: Journal of Sedimentary Research, v. 62, p. 734-737.

Pye, K., and D. H. Krinsley, 1985, Formation of secondary porosity in sandstones by quartz framework grain dissolution: Nature, v. 317, p. 54-56, doi: 10.1038/317054a0.

Saigal, G. C., S. Morad, K. Bjørlykke, P. K. Egeberg, and P. Aagaard, 1988, Diagenetic albitization of detrital K-feldspar in Jurassic, Lower Cretaceous, and Tertiary clastic reservoir rocks from offshore Norway, I. Textures and origin: Journal of Sedimentary Research, v. 58, p. 1003-1013, doi: 10.1306/212F8EE5-2B24-11D7-8648000102C1865D.

Schenk, C. J., and R. W. Richardson, 1985, Recognition of interstitial anhydrite dissolution: A cause of secondary porosity, San Andres Limestone, New Mexico and upper Minnelusa Formation, Wyoming: AAPG Bulletin, v. 69, p. 1064-1076.

Schmidt, V., and D. A. McDonald, 1979a, The role of secondary porosity in the course of sandstone diagenesis, *in* P. A. Scholle, and P. Schluger, eds., Aspects of Diagenesis: Tulsa, OK, SEPM Special Publication 26, p. 175-207, doi: 10.2110/pec.79.26.0175.

Schmidt, V., and D. A. McDonald, 1979b, Texture and recognition of secondary porosity in sandstones, *in* P. A. Scholle, and P. Schluger, eds., Aspects of Diagenesis: Tulsa, OK, SEPM Special Publication 26, p. 209-225, doi: 10.2110/pec.79.26.0209.

Shanmugam, G., and J. B. Higgins, 1988, Porosity enhancement from chert dissolution beneath Neocomian unconformity, Ivishak Formation, North Slope, Alaska: AAPG Bulletin, v. 72, p. 523-535.

Siebert, R. M., G. K. Moncure, and R. W. Lahann, 1984, A theory of framework grain dissolution, *in* D. A. McDonald, and R. C. Surdam, eds., Clastic Diagenesis: Tulsa, OK, AAPG Memoir 37, p. 163-176.

Sprunt, E. S., and A. Nur, 1977, Destruction of porosity through pressure solution: Geophysics, v. 42, p. 726-741, doi: 10.1190/1.1440742.

Stoessell, R. K., 1987, Mass transport in sandstones around dissolving plagioclase grains: Geology, v. 15, p. 295-298, doi: 10.1130/0091-7613(1987)15<295:MTISAD>2.0.CO;2.

Stoessell, R. K., and E. D. Pittman, 1990, Secondary porosity revisited: The chemistry of feldspar dissolution by carboxylic acids and anions: AAPG Bulletin, v. 74, p. 1795-1805.

Surdam, R. C., S. W. Boese, and L. J. Crossey, 1984, The chemistry of secondary porosity, *in* D. A. McDonald, and R. C. Surdam, eds., Clastic Diagenesis: Tulsa, OK, AAPG Memoir 37, p. 127-150.

Surdam, R. C., L. J. Crossey, G. Eglinton, B. Durand, J. D. Pigott, R. Raiswell, and R. A. Berner, 1985, Organic-inorganic reactions during progressive burial: Key to porosity and permeability enhancement and preservation [and discussion]: Philosophical Transactions of the Royal Society of London. Series A, Mathematical and Physical Sciences, v. 315, p. 135-156, doi: 10.1098/rsta.1985.0034.

Surdam, R. C., Z. S. Jiao, and D. B. Macgowan, 1993, Redox reactions involving hydrocarbons and mineral oxidants: A mechanism for significant porosity enhancement in sandstones: AAPG Bulletin, v. 77, p. 1509-1518.

Tang, Z., J. Parnell, and F. J. Longstaffe, 1997, Diagenesis of analcime-bearing reservoir sandstones: The Upper Permian Pingdiquan Formation, Junggar Basin, northwest China: Journal of Sedimentary Research, v. 67, p. 486-498, doi: 10.1306/D42685A4-2B26-11D7-8648000102C1865D.

Taylor, T. R., 1990, The influence of calcite dissolution on reservoir porosity in Miocene sandstones, Picaroon Field, offshore Texas Gulf Coast: Journal of Sedimentary Research, v. 60, p. 322-334, doi: 10.1306/212F9192-2B24-11D7-8648000102C1865D.

Taylor, T. R., M. R. Giles, L. A. Hathon, T. N. Diggs, N. R. Braunsdorf, G. V. Birbiglia, M. G. Kittridge, C. I. Macaulay, and I. S. Espejo, 2010, Sandstone diagenesis and reservoir quality prediction: Models, myths, and reality: AAPG Bulletin, v. 94, p. 1093-1132, doi: 10.1306/04211009123.

van de Kamp, P. C., 2008, Smectite-illite-muscovite transformations, quartz dissolution, and silica release in shales: Clays and Clay Minerals, v. 56, p. 66-81, doi: 10.1346/CCMN.2008.0560106.

Walderhaug, O., and K. W. Porten, 2007, Stability of detrital heavy minerals on the Norwegian continental shelf as a function of depth and temperature: Journal of Sedimentary Research, v. 77, p. 992-1002, doi: 10.2110/jsr.2007.091.

Weibel, R., and H. Friis, 2007, Alteration of opaque heavy minerals as a reflection of the geochemical conditions in depositional and diagenetic environments, *in* M. A. Mange, and D. T. Wright, eds., Heavy Minerals in Use (Developments in Sedimentology 58): Amsterdam, Elsevier, p. 277-303, doi: 10.1016/S0070-4571(07)58010-6.

Welch, S. A., and W. J. Ullman, 1996, Feldspar dissolution in acidic and organic solutions: Compositional and pH dependence of dissolution rate: Geochimica et Cosmochimica Acta, v. 60, p. 2939-2948, doi: 10.1016/0016-7037(96)00134-2.

Wilkinson, M., D. Darby, R. S. Haszeldine, and G. D. Couples, 1997, Secondary porosity generation during deep burial associated with overpressure leak-off: Fulmar Formation, UK Central Graben: AAPG Bulletin, v. 81, p. 803-813.

Wilkinson, M., K. L. Milliken, and R. S. Haszeldine, 2001, Systematic destruction of K-feldspar in deeply buried rift and passive margin sandstones: Journal of the Geological Society, v. 158, p. 675-683, doi: 10.1144/jgs.158.4.675.

Zhu, C., A. E. Blum, and D. R. Veblen, 2004, Feldspar dissolution rates and clay precipitation in the Navajo aquifer at Black Mesa, Arizona, USA, *in* R. B. Wanty, and R. R. I. Seal, eds., 11th International Symposium on Water-Rock Interaction WRI-11, Vol. 2: Rotterdam, A. A. Balkema, p. 895-899.

Facing Page: Top – Calcite infill of a partially dissolved feldspar, Pennsylvanian Morrow B Sandstone, Ochiltree Co., Texas (XPL). Bottom – Length-slow chalcedony from Eocene Green River Fm., Laney Mbr., Sweetwater Co., Wyoming (GP).

DIAGENESIS:
REPLACEMENT & RECRYSTALLIZATION

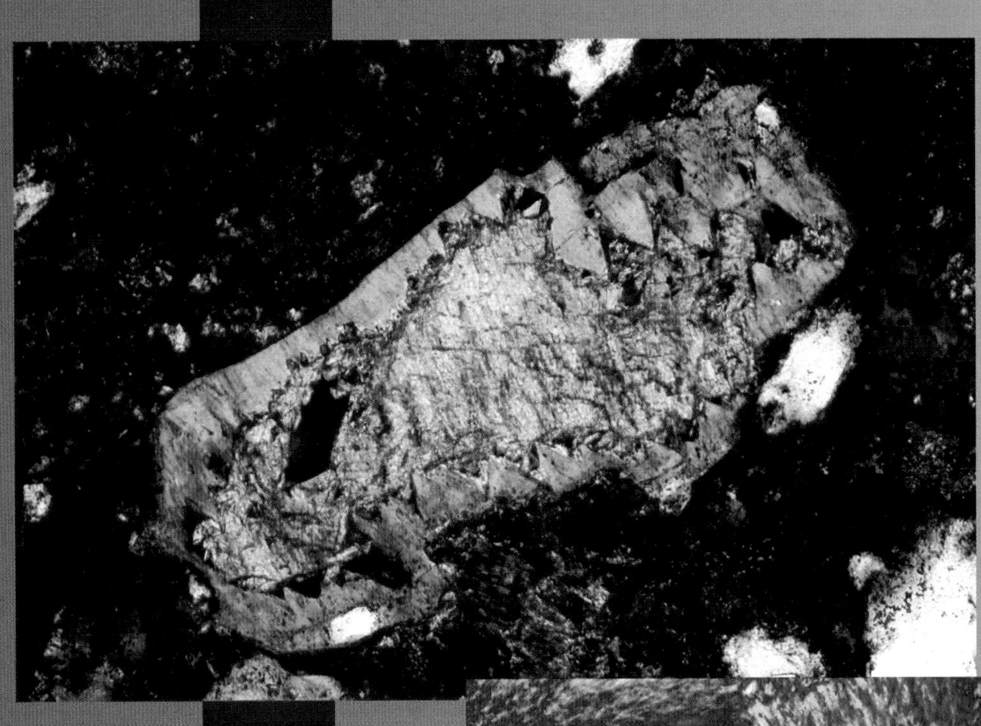

CHAPTER 20

REPLACEMENT AND RECRYSTALLIZATION

Replacement refers to the process of precipitation of minerals that take the place of preexisting minerals, either in detrital grains (monomineralic grains or rock fragments) or previously formed cements. It can even affect nonminerals, as in the case of petrified wood. Replacement, as strictly defined, proceeds by dissolution of the precursor mineral and essentially simultaneous *in-situ* precipitation of the replacement mineral. Typically, that takes place along a solution front or film that can be extremely thin (micrometer to nanometer scales) and replacement can start at one or many small, mineralogically susceptible nucleation sites throughout the host crystal. The general concept of replacement, however, also includes mineral dissolution that is followed by temporally unrelated cementation of the newly-created secondary pores perhaps thousands to many millions of years later. This situation, sometimes termed "solution-cavity fill" (SCF)(see Folk, 1965), is clearly not precisely synonymous with "replacement" as used in its strictest sense. Nonetheless, it should be noted that it commonly is not possible to unequivocally distinguish strictly-defined replacement from SCF, especially where the replacement is complete, with essentially no remnants of the original mineral (but it is possible to tell them apart if relicts of the host mineral are present or other features of the host are preserved within the new phase). Therefore, in this book we use the term replacement in its general sense for either type of alteration. We show examples in this chapter of both near-simultaneous replacements, as narrowly defined, and SCF replacements, despite the fact that the separate processes of dissolution and cementation have already been covered in other chapters.

An almost limitless number of potential replacement pairs exist in sedimentary rocks given the range of detrital minerals present, the possible range of diagenetic precipitates and the enormous variety of temperature, pressure and water chemistry conditions found in near-surface to deep-subsurface settings. We have therefore restricted this chapter mainly to some common, volumetrically significant or paragenetically interesting replacements.

Many other processes also can be included in the term "replacement", although they do not meet the strict definition of that word. The transformation of a mineral to its polymorph (e.g., aragonite to calcite or kaolinite to dickite), for example, is better referred to as "neomorphism" or "inversion" although it often is called "recrystallization". Strain-induced alterations or other changes in crystal size and shape without mineralogical change also are better encompassed by the term "recrystallization" than by the oft-used or misused "replacement". In such cases, recrystallization can be coupled with modifiers to produce terms such as "strain recrystallization" or "degrading crystallization" to describe more exactly the inferred process of formation (Folk, 1965).

To add to the terminological confusion, many true replacements (the substitution of one mineral for another) occur as burial-associated, progressive compositional and mineralogical changes in materials that generally are too finely crystalline to effectively study with light microscopy and thus require XRD, microprobe or other quantitative analytical studies. Such replacements are commonly termed "recrystallization" or mineral "transformation" even though they really involve microscale mineral replacements. Burial-related clay and zeolite mineral changes (the progressive illite/smectite, smectite/chlorite or precursor zeolite to laumontite transformations, for example) are complex and prolonged in duration (e.g., Iijima, 1988; Jahren and Aagaard, 1992; McKinley et al., 2003; Clauer and Liewig, 2013), and they occur at such small scales that petrographic studies typically have little value in their study. Thus, they will not be substantially covered here despite their undoubted importance to sandstone and mudstone diagenesis (although many of those reactions are covered in Chapter 13). Likewise, some of the progressive silica transformations (opal to opal-CT and then to chert or quartz) will not be covered here, because they were touched upon in Chapters 1, 5 and 11.

Recognition of replacement:

The criteria shown below can be useful in identifying replacement in thin section (and can be supplemented with geochemical data as noted in the last criterion):

1. Replacement can affect all or part of a detrital grain or cement, and the most clear-cut recognition of replacement occurs where it has progressed only to the point of partial consumption of its "host". Thus, grains or cements where the original material is still present and only selected areas have been altered to the replacement mineral allow identification of the process, the precursor mineral and the successor mineral. Feldspars are the most common minerals in which one sees such partial alteration (by clays, calcite or other minerals), in part because detrital feldspars commonly arrive in the sediment as inhomogeneous grains, with significant compositional zoning or twinning that later can lead to complex and highly selective replacement of mineralogically susceptible areas (see Gold, 1987; Saigal et al., 1988; Morad et al., 1989; Ramseyer et al.,

1992; Chowdhury and Noble, 1993; Hirt et al., 1993). However, it should be remembered that, especially in feldspars, considerable grain-scale replacement can occur within source rocks and thus predate sedimentation—such alteration can often be difficult to distinguish from mesogenetic sedimentary alteration (e.g., Lee and Parsons, 1998).

2. Even where entire grains or areas of cement have been replaced, it is common to see small relicts of the precursor minerals (or preserved mineral inclusions that are compositionally different from the successor minerals) as solid inclusions scattered through the successor crystals. Indeed, the presence of such disseminated inclusions is one of the best ways to distinguish true replacement from SCF—with SCF, those inclusions might or might not be visible as a residue on the floor of the cavity, but they would not be "suspended" within the cavity-filling cement except as possible projections from the walls of the cavity in the third dimension. A word of caution should also be added about using unreplaced inclusions within replaced grains to identify the original host mineral—in many cases, what remains in the replaced mineral are inclusions that contrasted in composition with the original host grain. Such inclusions commonly are not replaced precisely because they were not the same as their former host.
3. Selective preservation of "unstable" minerals only in zones of extensive early cementation and lack of such preservation in areas without such early cements. Likewise a decrease of "unstable" minerals with depth, temperature and/or relative ease of access to circulating pore fluids, provided the facies and provenance are relatively uniform, also can indicate progressive replacement or dissolution of grains.
4. Mineralogically distinctive grain or crystal shapes occupied by a different mineral can indicate replacement (e.g., a carbonate skeletal fragment or ooid now composed of silica; a bipyramidal, hexagonal quartz crystal outline filled with calcite; or quartz cement with a cubic habit after halite). Such crystal shapes commonly are most easily visible because of "outlining" by early cement coatings (clay, hematite, and others) or if they occur in mudstones. While such grain/crystal shape observations can be reasonably diagnostic of replacement in the loose sense of the word, they commonly do not allow determination of relative timing of the dissolution versus the pore filling and thus the distinction between strict replacement or SCF. The patterns of the fills of the precursor grains/cements sometimes can help in this regard, especially where multiple crystals fill the original single crystal outline. Such crystals that have grown and enlarged inward from the former grain/crystal margins (or the stable, early cements that outlined those margins) generally can be taken as evidence that a void existed that was filled at some later date by the successor cement.
5. Recognition of crystal fabrics that occur only in specific mineral transformations can be important in some cases, such as in the silicification of sulfate evaporites (see details below).
6. Cathodoluminescence (and sometimes staining of carbonate minerals) as well as microprobe and other *in-situ* quantitative analytical techniques can aid in identifying replacement minerals. In general, mesogenetic replacement minerals, especially albitized feldspars, are compositionally uniform, as reflected both in a scarcity of CL zonation, diminished CL intensity and in homogeneous chemical analyses (Morad, et al., 1990; González-Acebrón et al., 2010 and 2012). In that regard, replacement minerals commonly differ from precipitates into open pores that typically show CL banding and chemical zonation.

Mesogenetic feldspar albitization, one of the most important subsurface replacement processes, provides a good example of how several of the recognition criteria listed above can and should be combined to increase reliability of identification. Here are the specific fabric criteria proposed by Saigal et al. (1988) for recognition of albitization:

"1) euhedral habit of albite crystals with sharp edges and corners and smooth crystal faces;
2) generally untwinned albitized grains, mostly riddled with minute brownish inclusions;
3) lack of cathodoluminescence in albite;
4) homogeneous and pure albite composition (> 99% Ab);
5) absence of albitized grains in carbonate-cemented zones; and
6) increase in the percentage and degree (partial to complete) of albitized K-feldspar with depth".

Specific replacements and their settings:

Replacement of feldspars – Feldspar alteration is arguably the most important replacement process in sandstones and can involve a variety of minerals: other feldspars, especially albite (Saigal et al., 1988), calcite (Walker, 1960), chlorite (Morad and Aldahan, 1987b), illite/sericite (Morad and Aldahan, 1987a), kaolinite/dickite (Morad and Aldahan, 1987a; De Bona et al., 2008), quartz (Wallace, 1976; Morad and Aldahan, 1987c), ankerite (Milliken, 2002) and others. Recent studies (e.g., Wilkinson et al., 2001) have shown that up to 15% of the solid rock volume is removed by feldspar dissolution or replacement in arkosic sandstones by 4.5 km (14,800 ft) of burial. This can drastically change the mineralogic character of rocks, converting many arkosic sandstones to "diagenetic quartzarenites" (McBride, 1985) at depth and, in many cases, greatly increasing the relative proportion of secondary to primary porosity.

Albitization is the most widespread of the feldspar replacements and "albitization of plagioclase more calcic than An_{10} and of K-feldspar should be expected for all deeply buried sandstones" (McBride, 1985). Therefore, we provide some additional information about this process, and readers should consult the appended citations for details and examples of the other replacements. Albitization of K-feldspar is a largely temperature-controlled process (Perez and Boles, 2005) and starts at about 70–80° C, roughly contemporaneous with the onset of smectite illitization and significant reactions in kerogen (Aagaard et al., 1990). The zone of maximum K-feldspar albitization is in the 120–150°C range (Baccar et al., 1993), although albitization may continue to far higher temperatures. Albitization of plagioclase, on the other hand, has been shown to decrease with increasing temperature and occurs primarily in the 60° to 100°C range in low pCO_2 conditions and extends to higher temperature only in high pCO_2 conditions (Baccar et al., 1993). The quoted conditions above are for sandstones; both K-feldspar and Ca-plagioclase survive to far higher temperatures and greater depths in mudstones (Milliken, 1992). It should be noted that provenance as well as microtextures within detrital feldspars have been shown to substantially influence their susceptibility to alteration (see Parsons et al., 2005). Finally, the process of replacement may be complicated by multiple replacements or multiple replacement-dissolution-cementation events. This is especially true where relatively soluble minerals such as calcite or anhydrite replace feldspars and are themselves later dissolved or replaced.

Replacement of carbonate minerals – Replacement of carbonate material in clastic terrigenous rocks can occur in carbonate bioclasts, other carbonate grains, and carbonate cements. In grains and bioclasts, early transformations may include alteration of high-Mg calcite to low-Mg calcite as well as inversion or solution-cavity-fill of aragonite. Aphanocrystalline to finely-crystalline, fabric-selective and fabric-preserving dolomitization is also common in a variety of eogenetic settings; siderite and ankerite are more common as cements at that stage, but also can form replacements.

Mesogenetic carbonate replacements can be especially difficult to recognize. Carbonate grains do, of course, continue to undergo recognizable replacement, including silicification (chert, chalcedony or megaquartz replacement) or dolomitization, either of which can occur at many stages of diagenesis from eogenetic through mesogenetic and even during telogenetic uplift. Calcite cement alteration more commonly involves dolomite or ankerite "replacement" of precursor calcite and it often can be difficult to recognize because carbonate-on-carbonate replacement commonly leaves few inclusions. Nonetheless, mesogenetic ankerite replacement of calcite cement (in addition to ankerite cementation) has been noted in some sandstone studies (e.g., Boles and Franks, 1979; Krajewski, 2002; Milliken, 2002). Such studies consistently put the temperatures of ankerite formation at 120°C or higher, with smectite alteration and breakdown for ferromagnesian heavy minerals commonly cited as the source of the cations substituting for calcium.

Dolomite also is a common replacement mineral in mesogenetic settings. Such dolomite typically is coarser, more anhedral (with nonplanar crystal boundaries) and less fabric-preserving than eogenetic dolomite (Sibley and Gregg, 1987). Mesogenetic saddle (baroque) dolomite, with its distinctive undulatory extinction and curved crystal faces, can occur as either cement or replacement (typically of earlier carbonate minerals) and it is consistently associated with moderate- to high-temperature diagenesis associated with thermochemical sulfate reduction and/or brine advection (Machel, 2004).

Replacement of sulfate evaporites – The other group of minerals that almost invariably undergo petrographically-visible diagenetic alterations are the sulfate evaporite minerals. The gypsum-anhydrite transformations were already discussed in Chapter 16 and only the more compositionally significant, but not as ubiquitous, replacements (silicification and calcitization) will be discussed here.

Numerous studies have described specific mineralogies and crystal fabrics in former evaporite deposits that have been partially or completely replaced by silica and quartz and these are well summarized in Milliken (1979). They include zebraic chalcedony, fortification zoning in silica, megaquartz with euhedral terminations as well as undulatory and radiating extinction (also termed flamboyant extinction), and spherules and bands of moganite and length-slow chalcedony (quartzine), although not all are exclusive to silicified evaporites. Studies of silicified evaporites (e.g., Milliken, 1979; Ulmer and Laury, 1984; Ulmer-Scholle and Scholle, 1994) have concluded that silica with these fabrics was a relatively early replacement of bedded to nodular anhydrite (based on mineral inclusions), was influenced by influx of meteoric water, and occurred at temperatures below 40°C. A different type of silicification, also involving

replacement of gypsum and anhydrite, consists of relatively clear, bipyramidal crystals of quartz (sometimes termed Herkimer diamonds). These crystals may have hydrocarbon as well as evaporite mineral inclusions and, based on isotopic geochemistry, are of mesogenetic to telogenetic origin with formation temperatures of 50–85°C (Ulmer-Scholle et al., 1993).

Calcitization of sulfate evaporites is also a common process. It generally consists of coarsely- to very coarsely-crystalline calcite with scattered inclusions of the host evaporite that forms fillings of former anhydrite or gypsum nodules. In many cases, however, the inclusions, if present at all, have fallen to the bottom of the nodule cavity, indicating a solution-cavity fill process rather than direct replacement (Scholle et al., 1992). Sulfate calcitization typically occurs relatively early (in eogenetic settings) or relatively late (in telogenetic settings), in either case typically involving meteoric fluids or the near-surface activity of hypogene fluids derived from expulsion of basinal brines (Stafford et al., 2008).

Other replacements – Innumerable other replacements occur, and several are illustrated below, but space limitations require that discussion of these be limited to their picture captions.

Up. Pennsylvanian Bursum Fm., Socorro Co., New Mexico

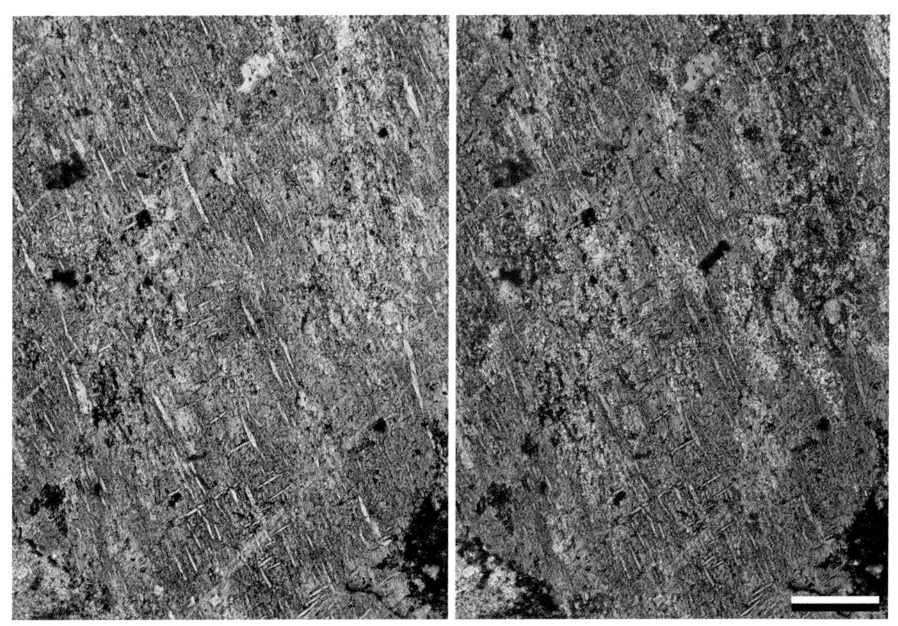

Feldspar alteration is nearly ubiquitous during diagenesis. In this example, the K-feldspar grains are replaced by albite and calcite. The albite preserves much of the original fabric of the K-feldspar grains, including two directions of exsolution lamellae. The hematite staining (orange color) causes the grain to still look like K-feldspar in hand specimen. The calcite may be a replacement, or it may be a void fill after orthoclase dissolution. Because albitization commonly preserves much relict fabric, it is important to do careful petrography and integrate it with other techniques, such as XRD and microprobe analysis.

PPL | XPL, Scale bar = 0.26 mm

Oligocene – Miocene tuffaceous sandstone, Jackson, Co., Oregon

This twinned plagioclase has been partially replaced by albite and illite. In these two cross-polarized light views, albite (the lighter gray phase in each view) can be seen to have partially replaced the more calcic plagioclase grain (darker gray phase with twins). The diagenetic albite does not contain any twins, but it does contain illite clays that formed during alteration of the original grain. Illitization of feldspars may occur in many settings, including eogenetic or telogenetic meteoric diagenesis (particularly in semi-arid to arid environments, Bétard et al., 2009).

XPL | XPL, Scale bar = 0.10 mm

Oligocene – Miocene tuffaceous sandstone, Jackson, Co., Oregon

Despite being substantially replaced by illite (best seen in the image on the left side), the original fabric of the feldspars can still be seen (especially in the image on the right). In this example, the polysynthetic twins are either nearly invisible or visible, depending on the rotation of the stage using cross-polarized light. As in the previous example, it is difficult to determine from a single example if this replacement occurred diagenetically or was inherited from the source terrane.

XPL | XPL, Scale bar = 0.10 mm

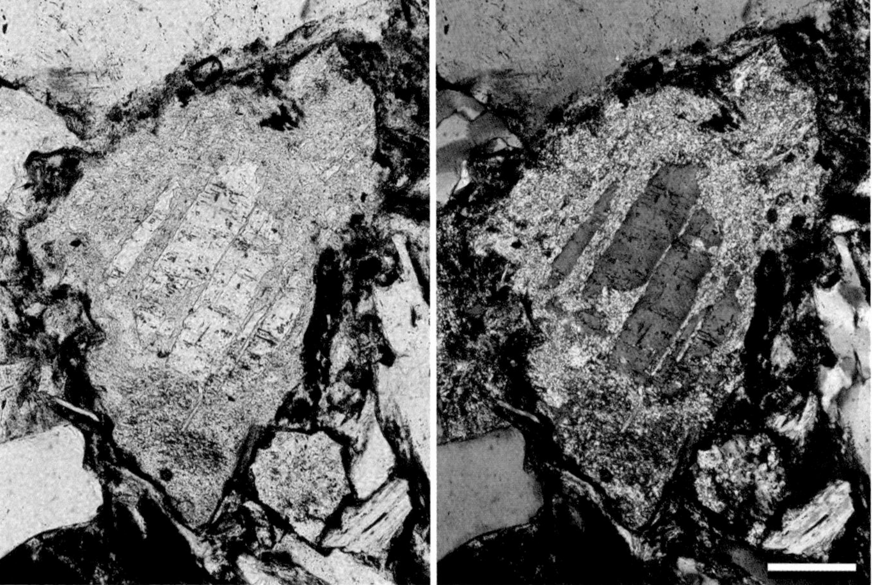

Up. Triassic New Haven Arkose, New Haven Co., Connecticut

This plagioclase feldspar has been extensively replaced by illite/sericite. The replacement has not consumed all of the grain, but rather has selectively attacked the original material along compositional zonation or twin lines that clearly were more susceptible to alteration. The selective replacement is most likely an *in-situ* diagenetic alteration in this case.

PPL | XPL, Scale bar = 0.10 mm

Tertiary intrusive, Clarke Co., Nevada

This intensely altered, compositionally zoned plagioclase feldspar phenocryst is shown in its source rock. Selected zones were preferentially altered to illite or sericite, probably related to hydrothermal alteration. Alteration of feldspars may occur either in source rocks and/or after erosion and deposition of grains in terrigenous clastic deposits. Both types of alteration may have similar appearance and form under similar conditions. Therefore, the presence of alteration can rarely be used as a provenance indicator. On the other hand, alteration commonly cannot be definitively identified as diagenetic, unless one argues that an altered grain is less likely survive transport.

XPL, Scale bar = 0.51 mm

Devonian Old Red Sandstone Gp., Co. Antrim, Northern Ireland, U.K.

A plagioclase feldspar with multiple types of alteration. This grain has been partly replaced by clay minerals (illite) as well as calcite. The calcite is interpreted to have directly replaced the grain (that is not forming as a void filling cement after grain dissolution) based on the observation that the calcite preserves the original fabric (twinning) of the plagioclase grain.

XPL, Scale bar = 0.10 mm

Lo. Pennsylvanian Morrow B Sandstone, Ochiltree Co., Texas

It can be difficult to tell the difference between replacement and the partial or complete dissolution of a grain followed by later filling of the void. The feldspar within this graphic granite clast was replaced by calcite. The calcite replaced the original feldspar and not did not occur as a void fill, as shown by the fact that there are numerous quartz inclusions "floating" in the calcite crystal. If the calcite formed after dissolution of the original grain, then the quartz inclusions (or at least the smaller inclusions that presumably are not interconnected in the third dimension) would not have remained "suspended" in open pore space.

XPL, RDI, Scale bar = 0.51 mm

Up. Triassic – Lo. Jurassic Huizachal Fm., Tamaulipas, Mexico

This plagioclase grain is completely replaced by calcite. What is unusual here is the excellent preservation of the polysynthetic albite twins from the precursor plagioclase grain. Calcite crystals can contain deformation-induced twin lamellae, but calcite twins generally appear as thin lines because they form as a result of shearing of the crystal lattice. These broader, alternating bands of birefringent and nonbirefringent material, on the other hand, reflect different orientations of the crystal lattices, as in albite twinning (see introduction in Chapter 2).

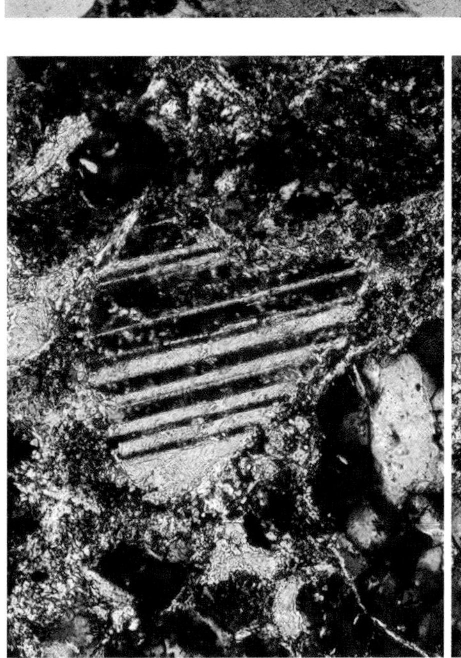

XPL | XPL, Scale bar = 0.05 mm

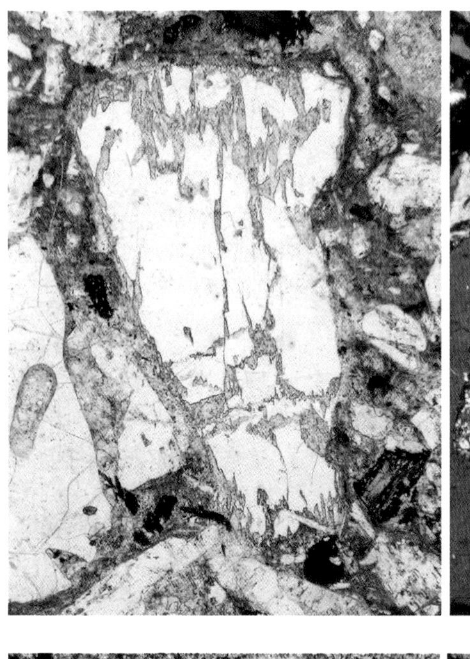

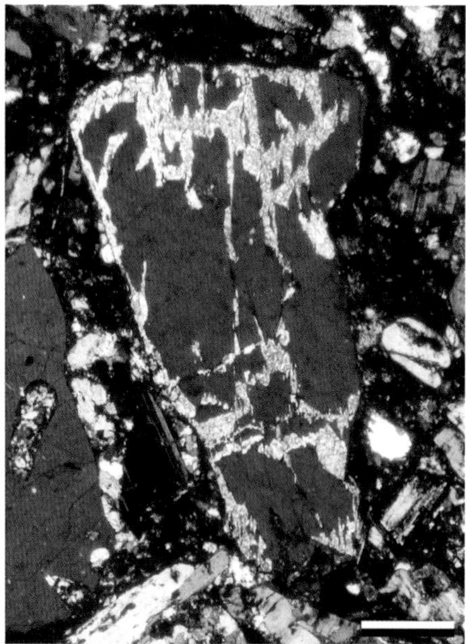

Mid. Eocene – Lo. Oligocene Spears Gp., Socorro Co., New Mexico

Replacement of feldspars by calcite can start early, prior to significant burial, especially in meteoric environments. In this example, a plagioclase grain has been partially replaced by calcite. The replacement occurs along fractures in the grain that were accessible to pore fluids and all of the replacement is a single crystal of calcite (all areas are in optical continuity and show unit extinction). These rocks have never been buried deeply, but may have been exposed to hydrothermal waters associated with nearby volcanism.

PPL | XPL, Scale bar = 0.51 mm

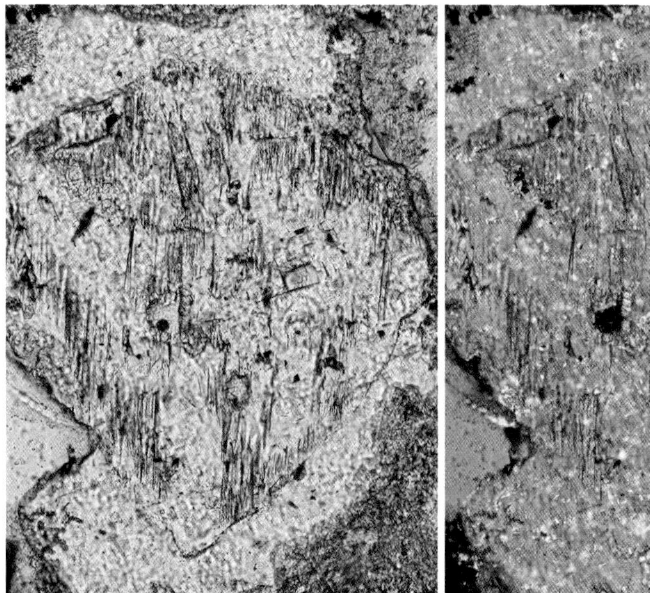

Miocene Lower Santa Fe Gp., Taos Co., New Mexico

The vague, final remnants of a feldspar grain are encased in poikilotopic calcite in this example. Since the replaced part of the crystal contains inclusions of other minerals and structures retained from the original feldspar grain, it is possible that the calcite is a replacement. However, it is also possible that it is a solution-cavity fill of a feldspar "skeletonized" by partial dissolution (see examples in Chapter 19). In either case, the calcite crystal continued to grow into the areas surrounding the grain as a pore-filling cement.

PPL | XPL, BDI, Scale bar = 0.10 mm

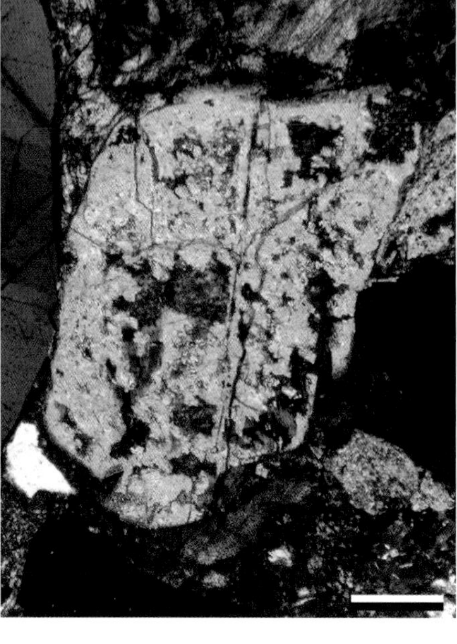

Lo. Pennsylvanian Morrow B Sandstone, Ochiltree Co., Texas

In this example, a single crystal of ankerite (confirmed by microprobe data) has replaced a feldspar grain. Whether the ankerite is a direct replacement or a product of separate processes of feldspar dissolution to produce a void with skeletal crystal remnants coupled with a later infill of that void is difficult to determine in a 2-D section. Inclusions of unleached feldspar and kaolinite "float" in the ankerite, but could be connected to each other or to a less dissolved part of the feldspar outside the plane of this section.

PPL | XPL, RDI, Scale bar = 0.26 mm

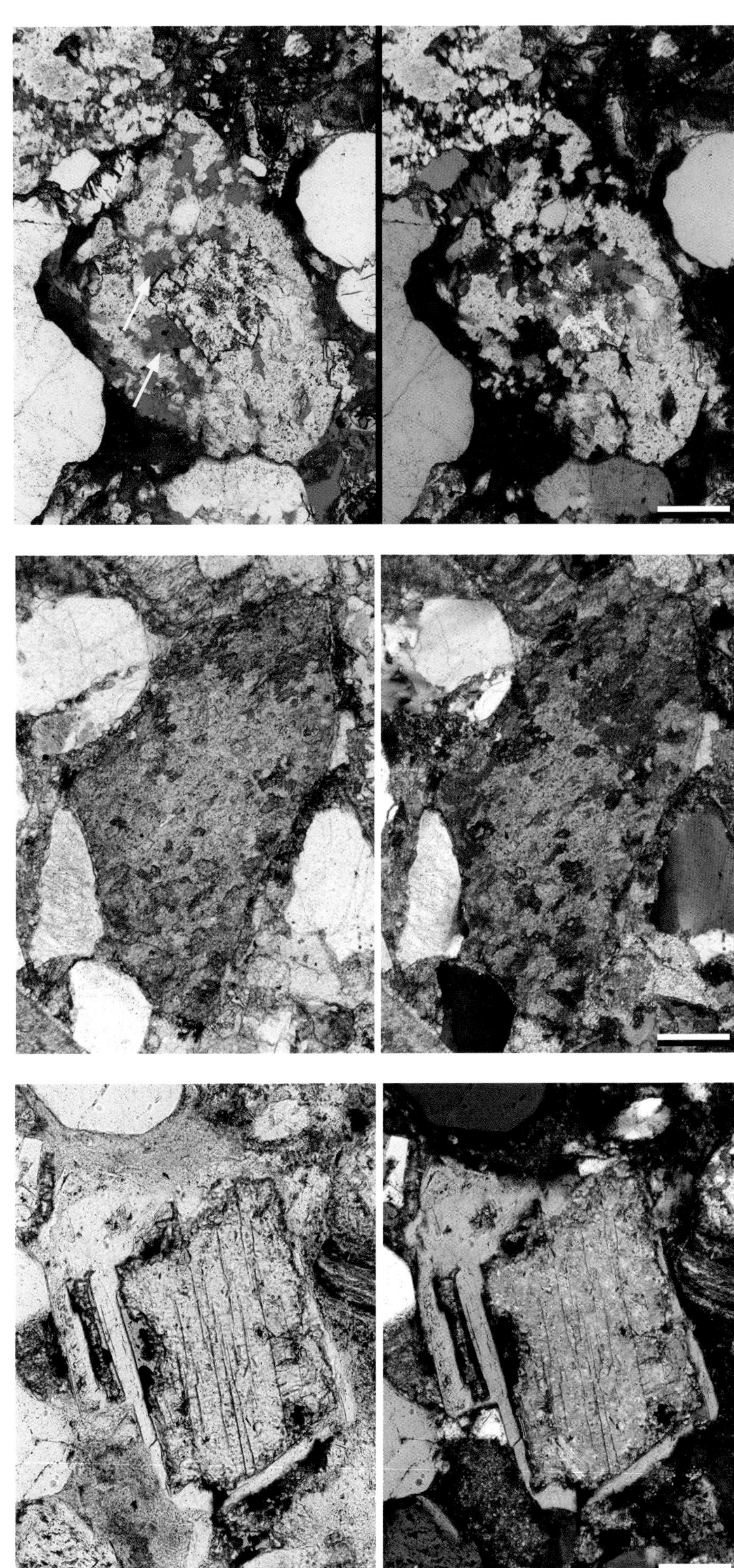

Lo. Pennsylvanian Morrow B Sandstone, Ochiltree Co., Texas

This is an example of a feldspar (probably albite) that has undergone extensive and complex dissolution and replacement. The clear white areas are albite overgrowths that formed within a partially dissolved grain. They are followed by ankerite crystals (highly birefringent crystals in center of grain) that probably replaced parts of the feldspar. The ankerite, in turn, may have undergone dissolution, leaving a second generation of rhombic moldic pores within the crystal (examples at arrows). Further evidence for ankerite dissolution comes from corroded intergranular ankerite cements in other areas of the section.

PPL | XPL, RDI, Scale bar = 0.26 mm

Up. Permian Schuchert Dal Fm., Jameson Land, East Greenland

This orthoclase grain was partially replaced by calcite (the high-birefringence mineral in cross-polarized light). The calcite preserves the exterior morphology of the grain, replaces mineral inclusions within the orthoclase and consists of multiple crystals that are not in optical continuity with each other. In the cross-polarized light view, Carlsbad twinning within the extensive remnants of the original orthoclase grain is clearly visible (the low birefringence dark and light gray areas near the center of the grain).

PPL | XPL, BDI, Scale bar = 0.26 mm

Mid. Eocene – Lo. Oligocene Spears Gp., Socorro Co., New Mexico

This is an example of indirect replacement of a feldspar by calcite via solution-cavity fill. In other words, the detrital plagioclase feldspar grain was partially dissolved, leaving a large rectangular mold surrounded by undissolved feldspar (the rim was probably a compositionally distinct overgrowth on the original grain). The moldic pore was filled with a single calcite crystal (now twinned) at some later date. The calcite crystal margin is ragged, most likely related to later dissolution of the edges of the cement or compromised growth around other mineral inclusions within in the secondary porosity.

PPL | XPL, BDI, Scale bar = 0.10 mm

Lo. Pennsylvanian Morrow B Sandstone, Ochiltree Co., Texas

An example of quartz replacement of parts of an illitized feldspar grain. The source of the silica for the quartz replacement or void fill may have been pressure solution of quartz grains elsewhere in the unit or the dissolution of feldspars and associated clay transformations.

XPL, RDI, Scale bar = 0.26 mm

Up. Jurassic Naknek Fm., Cook Inlet, Alaska

Detrital plagioclase that has been albitized and partially replaced by laumontite (light-colored area in plane-polarized light and low-order birefringence colors in partially cross-polarized light). The dark portions of the plagioclase result from abundant vacuoles created during the albitization process. Note adjacent elongate pore (near yellow arrow) that is lined by chlorite and filled with laumontite cement. Photomicrographs from Kenneth P. Helmold.

PPL | PXPL, Scale bar = 0.20 mm

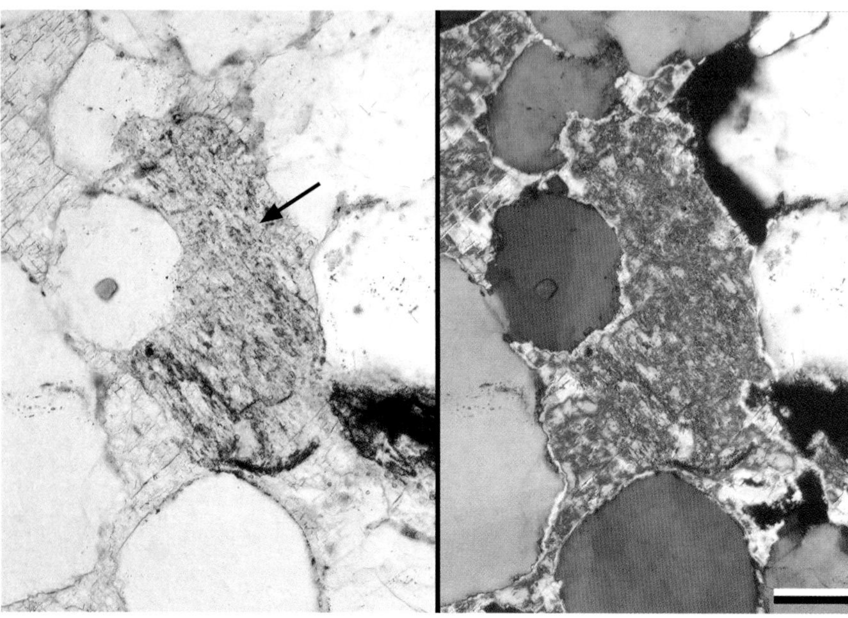

Devonian Jubah Fm., Qatar †

Anhydrite replacing a potassium feldspar (shown with black arrow). The pale yellow stain highlights the fact that there is a significant amount of the original feldspar grain remaining. Thus, the anhydrite replacement is more a filling of all available secondary microporosity in the feldspar (essentially a microscale solution-cavity fill) rather than simple grain replacement. The single crystal of anhydrite also filled surrounding primary intergranular porosity.

PPL | XPL, AFeS, KFS, BDI, Scale bar = 0.04 mm

Mid. Permian Wegener Halvø Fm., central East Greenland

A partly silicified brachiopod from a bed rich in reworked biota from a nearby bioherm. Much of patchy replacement is brown because of abundant calcite inclusions. Those inclusions help to retain the internal fabric of the shell and indicate that replacement occurred along thin solution films (Maliva and Siever, 1988); the later stages of replacement are clear (largely inclusion-free) and have well developed crystal outlines. The source of silica for replacement in this case is almost certainly siliceous sponge spicules based on the fact that spicule remains are still found throughout these rocks.

PPL / XPL, AFeS, BDI, Scale bar = 0.51 mm

Up. Jurassic Ula Fm., Norwegian sector, North Sea †

Shells (probably of bivalves) that were replaced by ferroan dolomite. The ferroan dolomite is composed of a mosaic of large crystals, which have obliterated the original texture of the precursor shell. This sample also contains saddle dolomite and sphalerite, suggesting precipitation of cements and replacement of the shells under hydrothermal conditions with warm to hot briny fluids.

PPL, AFeS, KFS, BDI, Scale bar = 0.49 mm

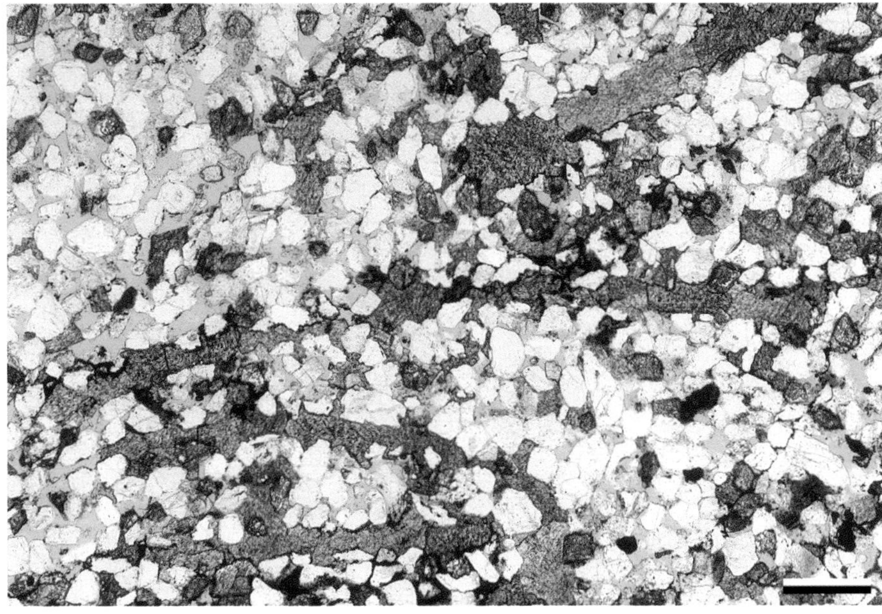

Mid. Jurassic Carmel Fm., Winsor Mbr., Emery Co., Utah

Replacement of a carbonate mineral by another carbonate is common, even in terrigenous rocks. These medium- to coarsely-crystalline rhombs with dusty cores were originally dolomite, but have all been calcitized (sometimes referred to as dedolomitization and determined, in this case, by staining and microprobe analysis). Calcitization of dolomite may occur under a variety of diagenetic conditions. It is especially common in eogenetic and telogenetic settings where dissolution of calcium sulfate minerals increases calcium concentrations in pore fluids and in mesogenetic settings associated with hydrocarbon migration and bacterial or thermochemical sulfate reduction.

PPL | XPL, RDI, Scale bar = 0.10 mm

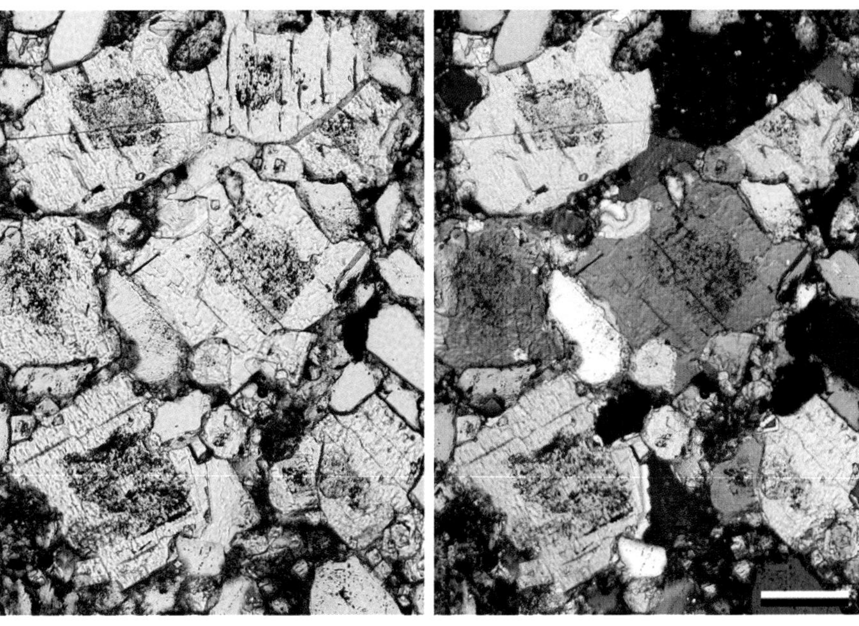

Mid. Jurassic Carmel Fm., Co-op Creek Mbr., Emery Co., Utah

In the center of this image, one can see the outlines of a partially dissolved, euhedral dolomite crystal. The limpid dolomite rim is preserved, but the core of the dolomite has been partially dissolved (leaving collapsed remnants of the original darker residue from core behind; see black arrow). The resultant intracrystal pore was later infilled by poikilotopic calcite cement (with twin bands) that also fills surrounding porosity. Therefore, this is an example of solution-cavity fill rather than direct replacement of dolomite and recognition of the corroded and collapsed original material is key to that determination.

XPL, Scale bar = 0.05 mm

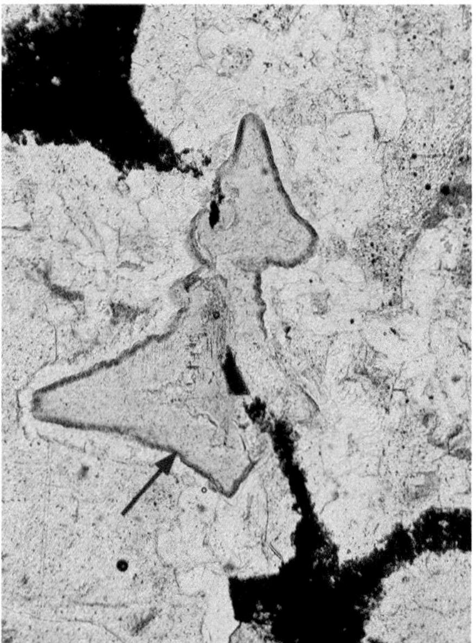

Mid.? Permian (former Ufimian) Solikamskaya Horizon, near Berezniki, Perm region, Russia

Evaporite minerals are highly susceptible to dissolution and/or replacement during diagenesis. In this example, halite (black areas in cross-polarized light) is partially replaced by multiple phases of silica and chalcedony followed by megaquartz. The shape of the chalcedony outline (arrow) reflects the original skeletal halite hopper crystal structure (see halites in Chapter 5). This section is thicker than the standard 30 μm and thus the megaquartz shows anomalously high birefringence colors.

PPL | XPL, BDI, Scale bar = 0.26 mm

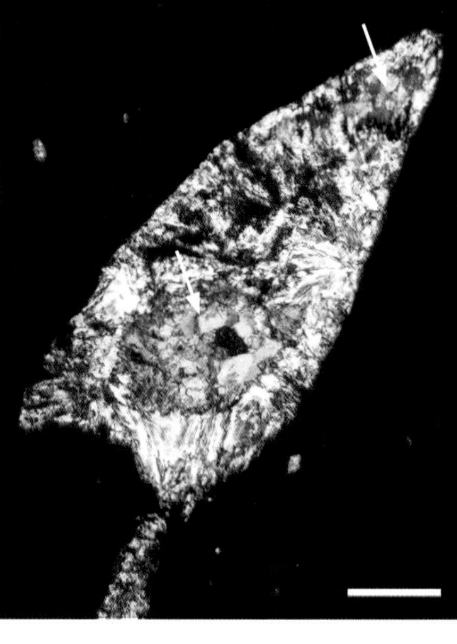

Mid.? Permian (former Ufimian) Solikamskaya Horizon, near Solikamsk, Perm region, Russia

Two views of the same geologic unit showing varied alteration of evaporites. From the shapes of the crystal outlines, both rocks contained displacive, probably synsedimentary, discoidal gypsum. The example on the left shows leaching with no infill of the molds. The example on the right shows complex alteration with fibrous (chalcedonic) silica along the margins and calcite in more central areas (white arrows). The presence of unreplaced dark blebs (probably largely organic matter) indicates that this was a direct replacement, not solution-cavity fill.

PPL | XPL, BDI, Scale bar = 0.53 mm | 0.57 mm

Mississippian Arroyo Peñasco Gp., Terrero Fm., Macho Mbr., San Miguel Co., New Mexico

This is a chalcedony and megaquartz "nodule" showing "fortification zoning"—pseudocubic zoning within non-cubic minerals. Folk and Pittman (1971) considered this fabric indicative of silica replacement of halite. The chalcedony here is mostly length-slow, indicating formation from sulfate-rich pore fluids. There are no solid inclusions or remnants remaining to identify the mineral that was replaced, but this is inferred to have been a replacement of halite or anhydrite.

PPL | XPL, AFeS, BDI, Scale bar = 0.26 mm

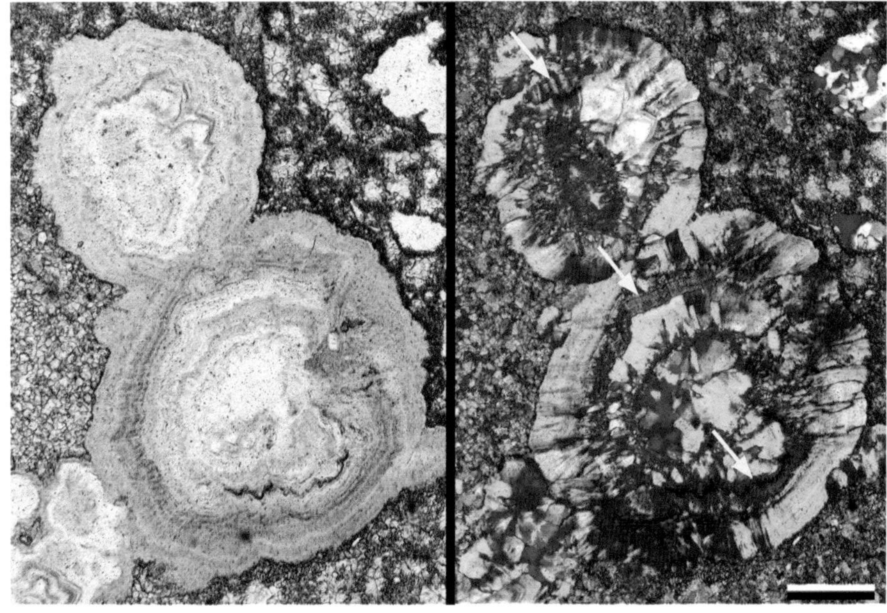

Mississippian Arroyo Peñasco Gp., Terrero Fm., Macho Mbr., San Miguel Co., New Mexico

Mixed length-fast and length-slow chalcedony in nodules from a breccia that formed as a result of evaporite dissolution and karstification. Because length-slow chalcedony is attributed to sulfate-rich fluids, determination of the chalcedony's orientation can provide information on variations in fluid compositions. Here, a length-slow gypsum plate was inserted under cross-polarized light. In the NW-SE quadrants of the nodules, the length-slow chalcedony is blue green in color, whereas in the same quadrant, the length-fast chalcedony is white to reddish-orange (arrows).

PPL | GP, Scale bar = 0.13 mm

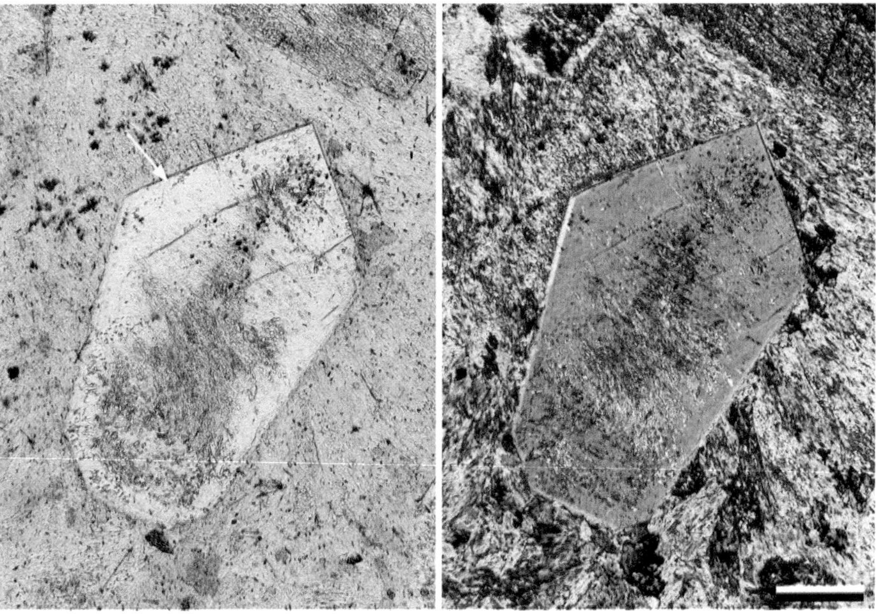

Mid.? Permian (former Ufimian) Solikamskaya Horizon, near Berezniki, Perm region, Russia

This photomicrograph shows anhydrite that has been partially replaced by euhedral megaquartz (yellow arrow). The quartz has anomalous birefringence color because the slide is thicker than 30 μm. The large quartz crystal contains numerous remnant anhydrite inclusions (the dusty areas in plane-polarized light) that were not replaced during silicification. Silicification can occur in near-surface diagenetic environments as well as deeper, higher temperature settings associated with hydrocarbon migration and bacterial sulfate reduction.

PPL | XPL, Scale bar = 0.26 mm

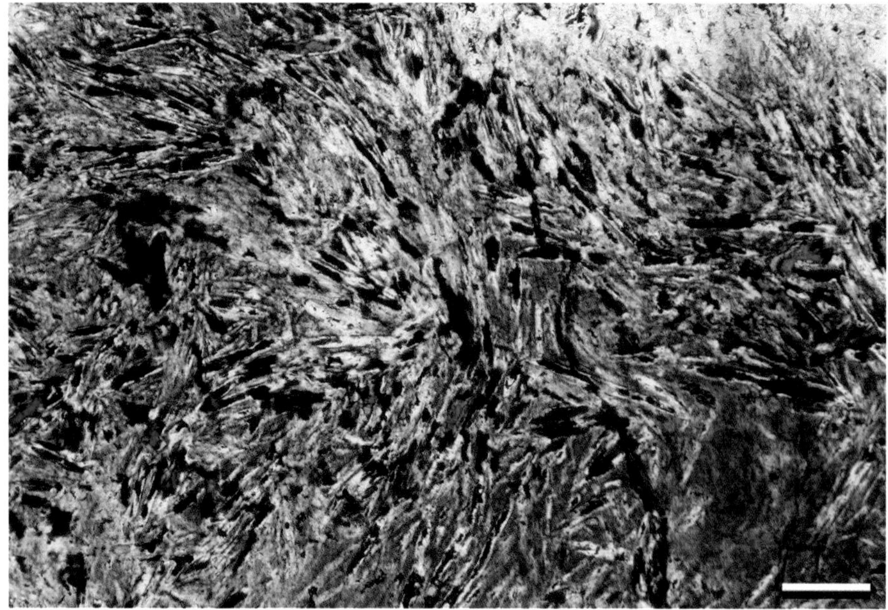

Permian Park City Fm., Tosi Chert Mbr., Park Co., Wyoming

Within the deep-shelf facies of the Tosi Chert, horizons of anhydrite nodules are partially replaced by chert, chalcedony and minor megaquartz. The anhydrite nodules likely formed as displacive precipitates from refluxing brines on the Phosphoria shelf (Ulmer-Scholle and Scholle, 1994). This example shows the outer margin of a nodule that was replaced by chert and chalcedony and contains anhydrite crystal lath pseudomorphs, including skeletal anhydrite crystals. The brown staining is related to oil entry that was coeval with silicification. Residual anhydrite inclusions were leached later, probably on outcrop.

PPL, AFeS, BDI, Scale bar = 0.26 mm

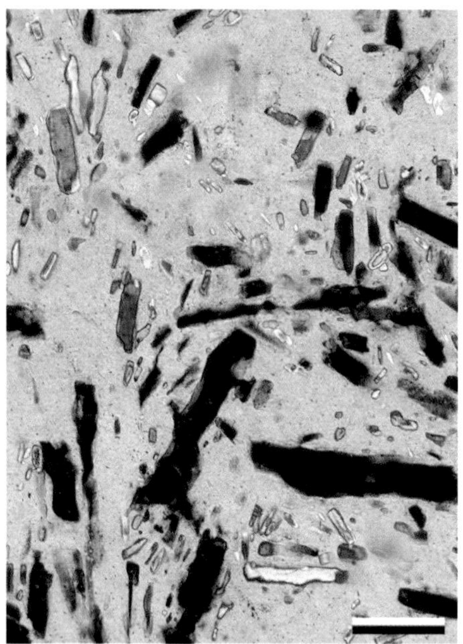

Permian Park City Fm., Tosi Chert Mbr., Park Co., Wyoming

The last stage of silicification within the nodules shown in the previous photomicrograph was euhedral megaquartz. Numerous solid inclusions of anhydrite (second-order birefringence in cross-polarized light) are scattered throughout the replacive megaquartz. Most of the nodules underwent later dissolution which removed relict anhydrite and created large pores within the cores of the nodules. These cores are empty or have bitumen and/or calcite fills. In this view, a few of the molds were filled by hydrocarbons (examples at arrows) and one was filled by slightly ferroan calcite (purple stain in plane-polarized light).

PPL | XPL, AFeS, BDI, Scale bar = 0.10 mm

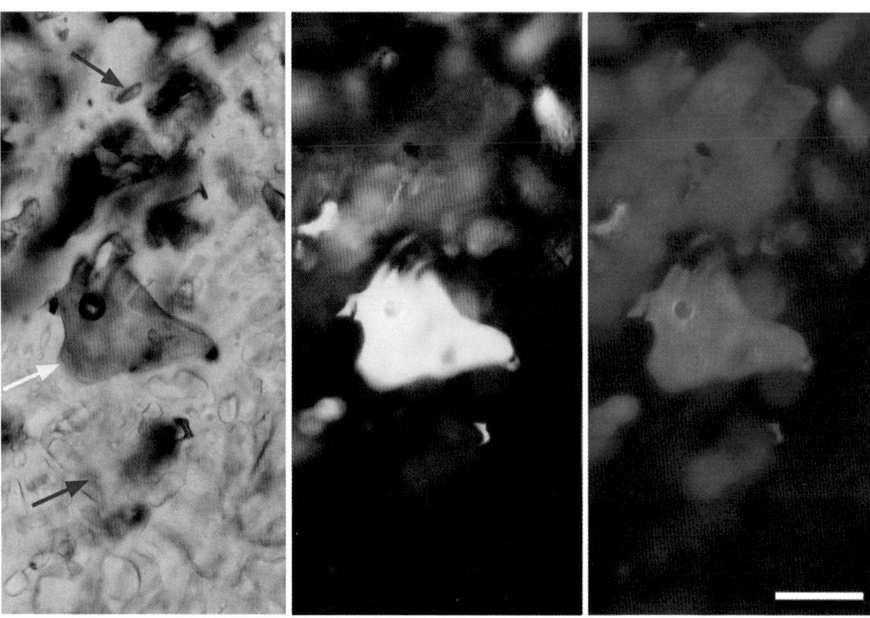

Mid. Permian Yates Fm., Eddy Co., New Mexico

This image shows hydrocarbon (yellow arrow), aqueous (red arrow) and anhydrite (blue arrow) inclusions within authigenic, euhedral quartz replacements of synsedimentary anhydrite nodules formed in silty, sabkha deposits. All the inclusions are primary. The oil inclusion is fluorescent in both wavelengths because the hydrocarbons have a high volatile content. Organic acids associated with the hydrocarbons dissolved terrigenous grains in this unit and that provided a source of silica for the replacement quartz. Fluid inclusion data indicate that silica replacement occurred essentially at maximum burial temperatures (Ulmer-Scholle et al., 1993).

PPL | FL470 | FL365, Scale bar = 0.05 mm

Permian Park City Fm., Tosi Chert Mbr., Big Horn Co., Wyoming

In this sample, cubic crystals of fluorite (high relief, non-birefringent) are associated with evaporite replacement by quartz and calcite. Fluorite typically is a hydrothermal mineral, and in this instance probably formed during thermo-chemical sulfate reduction where nodules of anhydrite were replaced by silica and fluorite (Ulmer-Scholle and Scholle, 1994). The calcite (stained red) most likely was a later void-filling cement.

PPL | XPL, AFeS, BDI, Scale bar = 0.26 mm

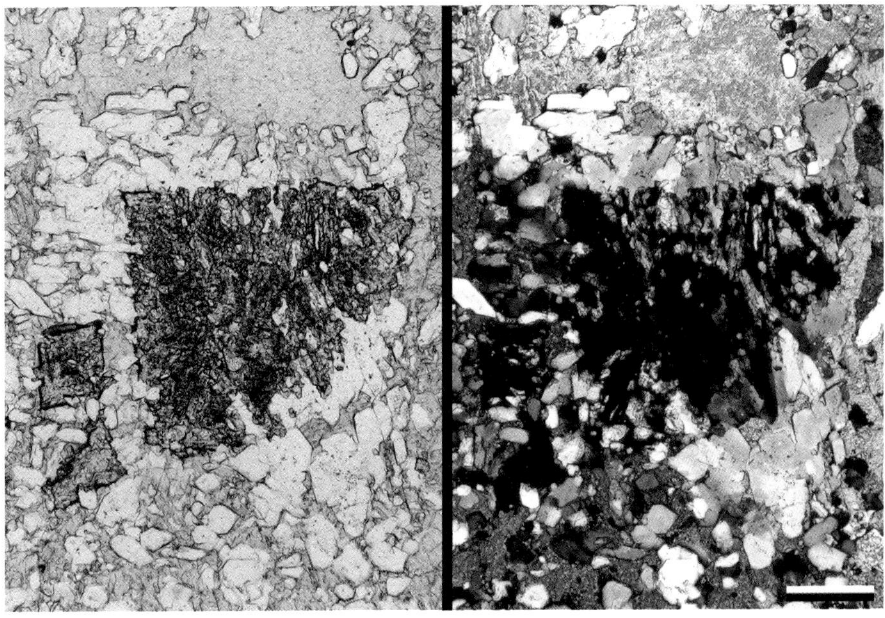

Up. Permian Karstryggen Fm., Jameson Land, East Greenland

This evaporite nodule was replaced by celestine and calcite. Celestine and barite both have elongate crystal morphologies and can be difficult to tell apart. They also form a solid solution series, so XRD analyses may be needed to make a definitive identification. Note the abundant anhydrite inclusions within the very elongate celestine crystals. The celestine is a result of later diagenesis where pore waters became enriched in strontium as they passed through underlying arkosic red beds and reacted with the calcium sulfates to form the celestine (Scholle et al., 1990).

PPL | XPL, Scale bar = 0.26 mm

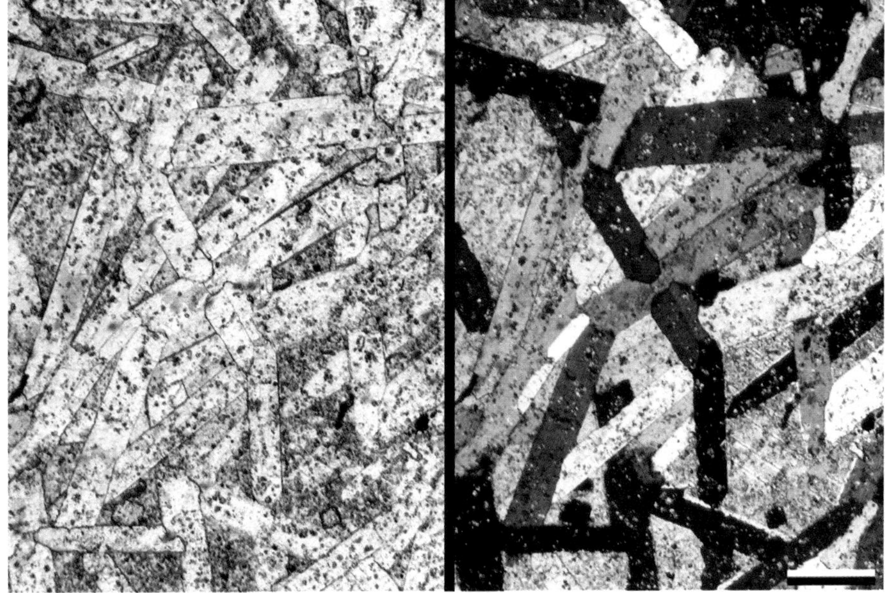

Cretaceous Mowry Fm., Moffat Co., Colorado

In this feldspathic litharenite, the grains appear to be cemented first by a thin rind of poikilotopic calcite (arrow), and then by two generations of chalcedony (numbered) that filled remaining porosity. The problem with this sequence is—how does a thin zone of poikilotopic cement form without filling the entire pore? It is likely that the calcite cements replaced or filled elongate pores left by leaching of another phase or mineral after chalcedony cementation. In these siliceous strata, opal cements are a possible early precipitate; aragonite or high-Mg calcites also are possible early grain-coating cements.

PPL | XPL, RDI, Scale bar = 0.05 mm

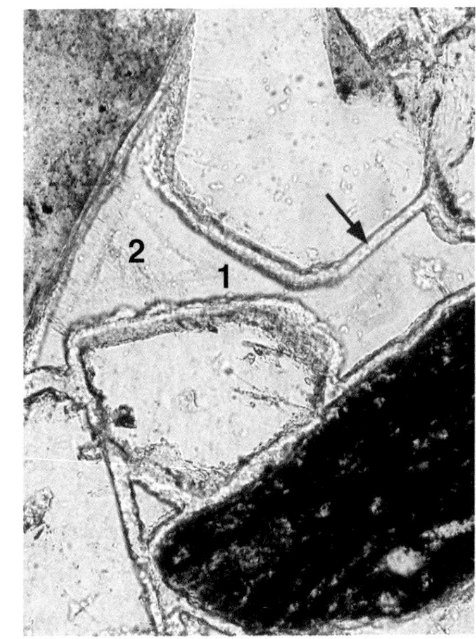

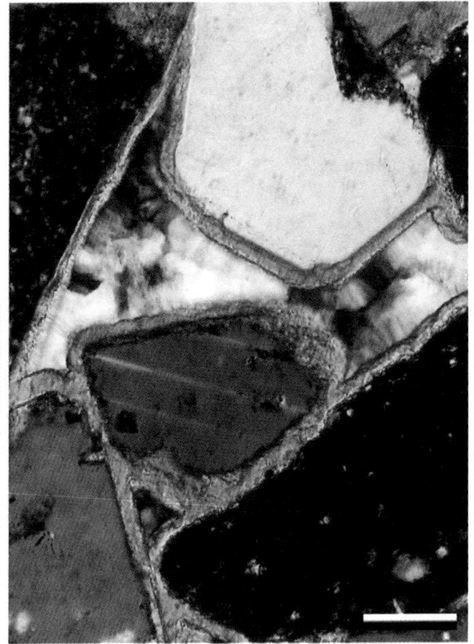

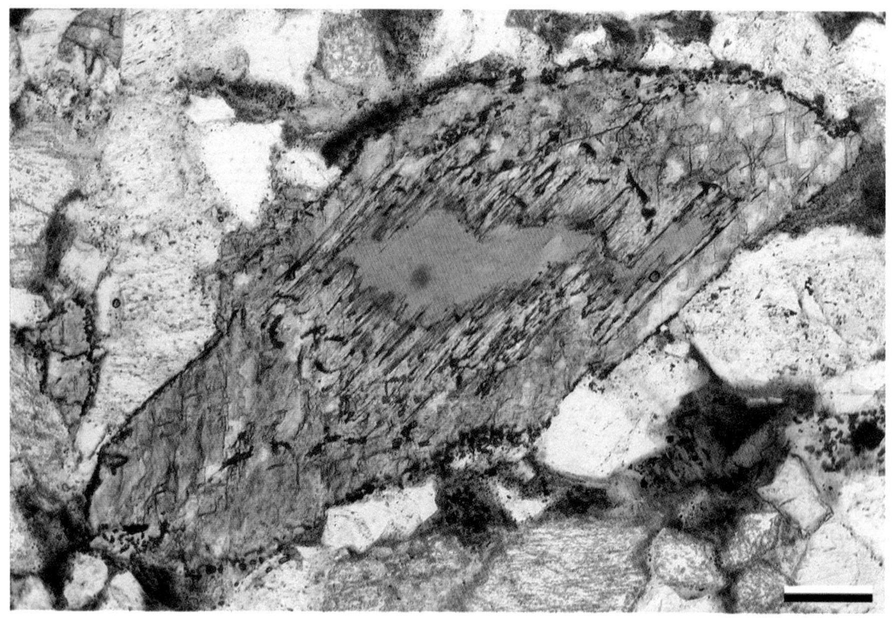

Paleocene unknown unit, Sri Lanka

This partially dissolved hornblende grain was probably compositionally zoned. The non-ferroan calcite (pink and showing cleavage traces of the original skeletal amphibole grain preserved within it) selectively replaced an internal, and especially soluble, zone of the hornblende. The outer zone of the hornblende was apparently dissolved later and that secondary pore space was filled with slightly ferroan calcite (pale purple).

PPL, AFeS, KFS, BDI, Scale bar = 0.12 mm

Up. Pennsylvanian Alamitos Fm., San Miguel Co., New Mexico

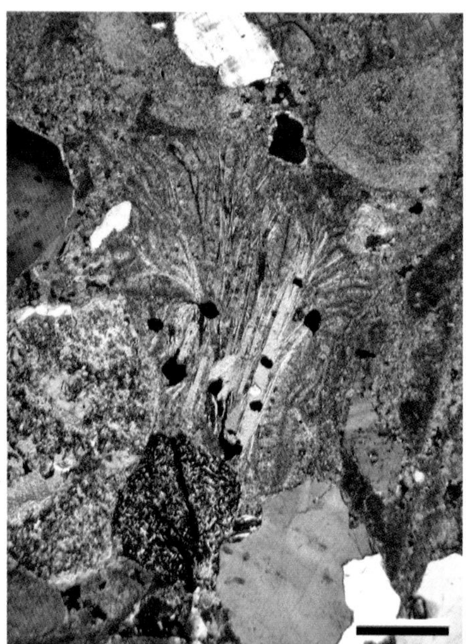

Some of the muscovite micas in this schistose MRF were partially replaced by slightly ferroan calcite. The growth of calcite between the flakes of muscovite caused the grain to expand and likely provided more surface area for further replacement of the micas.

PPL | XPL, AFeS, BDI, Scale bar = 0.38 mm

Up. Cretaceous (Maastrichtian) Tinton Fm., Monmouth Co., New Jersey

This large, brown, vermicular grain is glauconite pseudomorphing a precursor mineral such as biotite or vermiculite. Glauconite commonly forms as an alteration product of biotite or clay minerals in igneous and sedimentary rocks. In addition, these grains have undergone significant outcrop weathering. The greenish material (arrow) is oxidized vivianite; it quickly alters from a bright blue to green (even within this covered thin section).

PPL, Scale bar = 0.08 mm

Up. Cretaceous Blålange Fm., Vøring Basin, Norwegian Sea †

This glauconite has replaced biotite and is highly expanded. Throughout the sample, other biotites are, in part, also replaced by glauconite so this is a consistent alteration. There may be a vermiculite intermediate stage, although that cannot be demonstrated here. Because the glauconite in this example is not microcrystalline, it gives a good indication of the true birefringence of that mineral.

PPL | XPL, AFeS, KFS, BDI, Scale bar = 0.06 mm

Lo. Miocene Arikaree Fm., Platte Co., Wyoming

Celadonite is a mica group mineral that can form as a replacement of ferromagnesian minerals during hydrothermal alteration and/or low-grade metamorphism (zeolite facies). In this example, celadonite occurs as olive green to blue-green platelets that may be replacing a pyroxene. The initial cement in this sandstone is a thin rind (red arrow) of a non-birefringent to very low-birefringent mineral, such as opal or a zeolite.

PPL | XPL, Scale bar = 0.10 mm

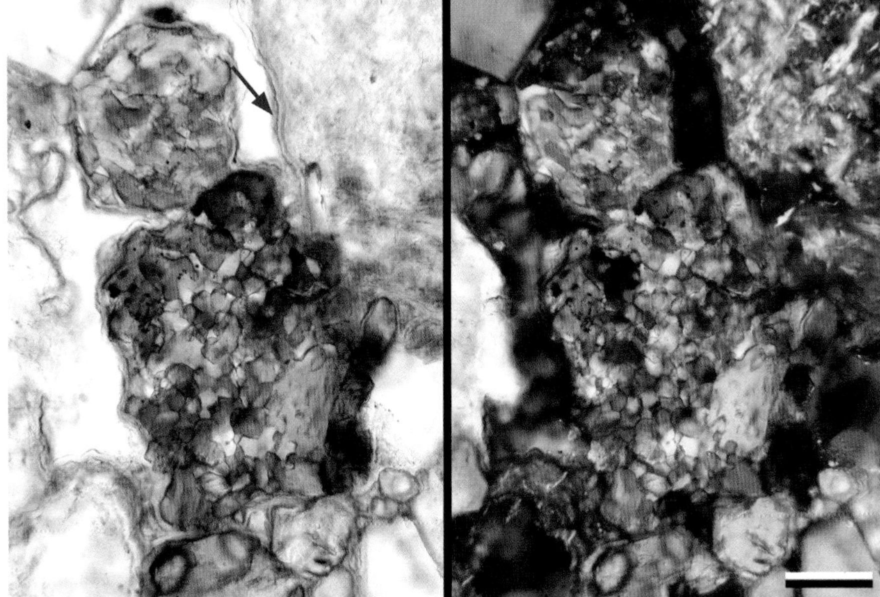

Mid.? Permian Huledal Fm., Jameson Land, East Greenland

This example shows a large volcanic rock fragment from a conglomeratic basal lag deposit on Devonian basement. This VRF contains phenocrysts of pyroxene that were altered to bright green chlorite and plagioclase that were altered to sericite. These types of replacements are typical of hydrothermal alteration.

PPL, BDI, Scale bar = 0.26 mm

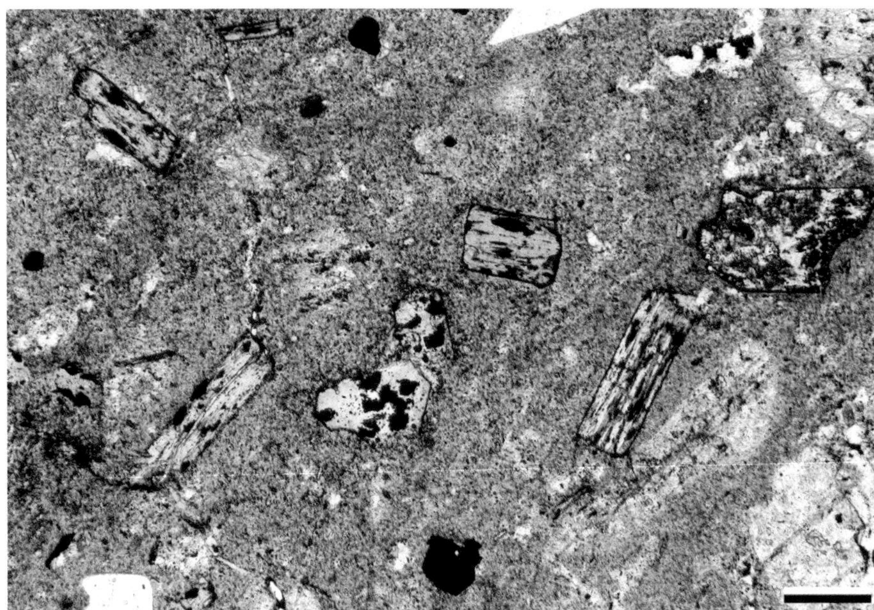

Up. Triassic – Lo. Jurassic Brunswick Fm., Hunterdon Co., New Jersey

This authigenic pyrite crystal formed replacively in a shale. Most pyrite in sedimentary rocks is of diagenetic origin, forming under reducing conditions. Crystals tend to form cubes, pyritohedrons or octohedrons and occur as cements or displacive and replacive crystals and masses. The lack of deformation in surrounding sediment makes replacement more likely than displacement in this lacustrine deposit.

PPL | ORL, Scale bar = 0.26 mm

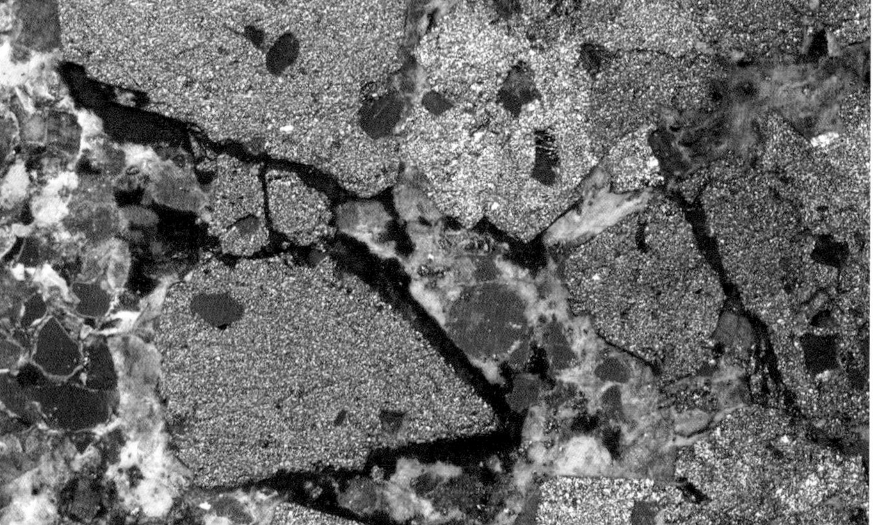

Unknown age and unit, Rajasthan, India

In this example, pyrite is both a replacement and a cement in sandstone. Some of the pyrite crystals have euhedral, cubic terminations, and in oblique reflected light, they appear silvery gold to gold. The reddish brown color surrounding the pyrite was produced by oxidation of the margins to hematite.

ORL, BDI, Scale bar = 0.49 mm

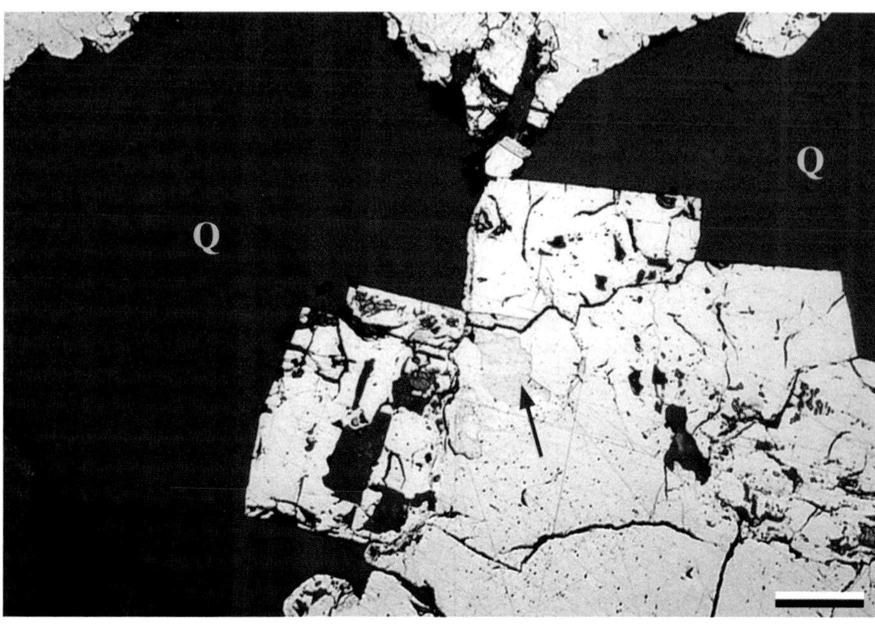

Up. Devonian New Albany Shale, Camp Run Mbr., Kentucky

Sediments within this lag deposit are cemented by a mixture of pyrite and marcasite (FeS_2). The lag deposit marks a sequence boundary and comprises quartz grains, fossil fragments, fish debris and reworked pyrite crystals. In this example, grains of quartz (Q) have been replaced by cubic crystals of pyrite during late diagenesis. The pleochroism (arrow) within the iron sulfides is due to the alteration of pyrite to marcasite (Schieber, 2007).

RL, Scale bar = 32.1 μm

Mid. Jurassic Entrada Fm., Slick Rock Mbr., Emery Co., Utah

This sandstone was cemented by euhedral pyrite (pyritohedron visible at red arrow) that precipitated under reducing conditions and is associated with iron-filled fractures. As the unit was brought back to the surface during telogenesis, the sulfide cements were oxidized to hematite that contains minute relict pyrite inclusions (examples at blue arrows). Inclusions within the original pyrite were inherited by the hematite and preserve the pyrite crystal shape.

RL, RDI, Scale bar = 0.05 mm

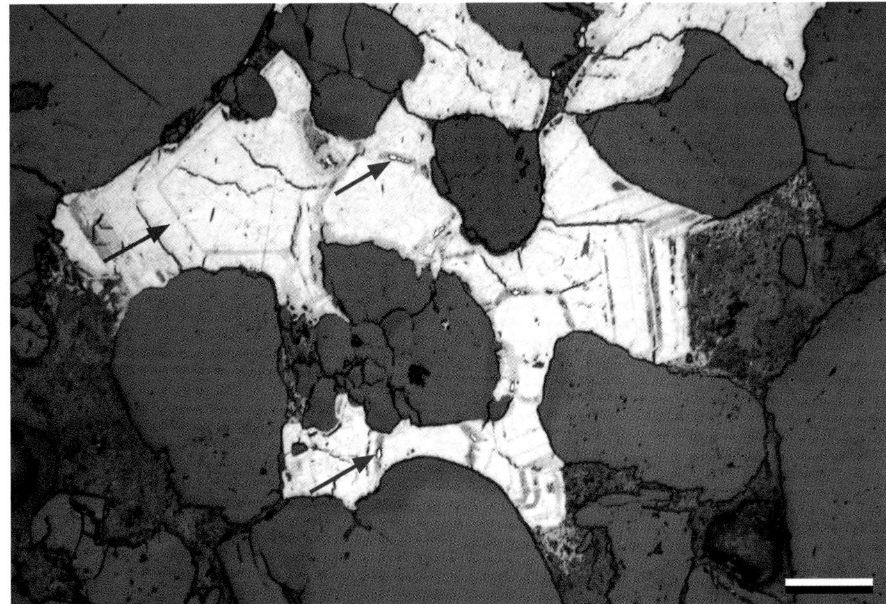

Lo.–Mid. Eocene Cub Mountain Fm., Lincoln Co., New Mexico

This probable amphibole phenocryst in a VRF was largely replaced by hematite. Unlike the other examples, this probably started out as an oxide and not a sulfide because this unit was never buried deeply and probably never was in a reducing environment. Magnetite or iron within the associated volcanic glass may have been the source of the iron for the hematite replacement. Later, any amphibole remnants were leached or replaced by chert.

PPL | XPL, KFS, Scale bar = 0.10 mm

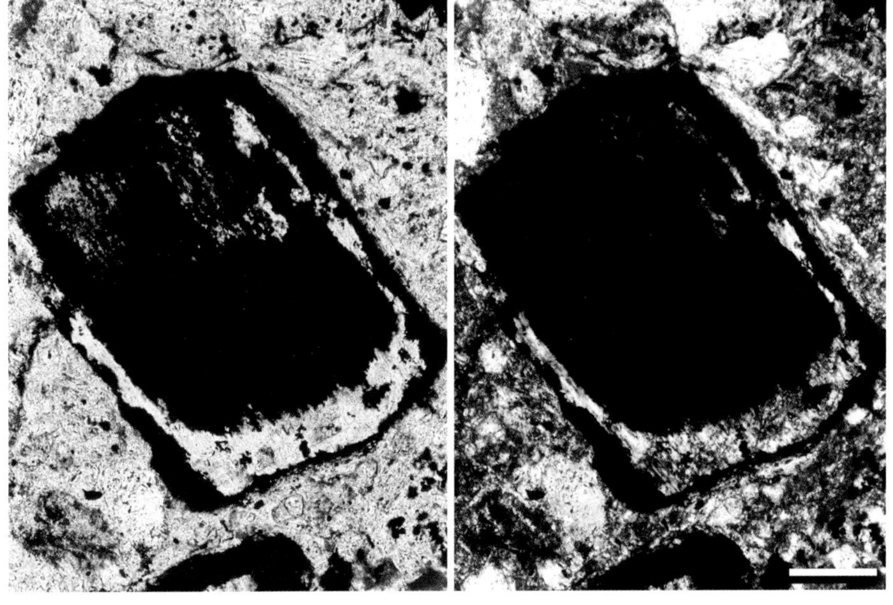

Lo. Cretaceous (Albian) Nahr Umr Fm., offshore Qatar

The replacement shown in these transmitted and oblique-reflected light photomicrographs takes the form of rosettes of needle-like crystals of pyrite or marcasite. Pyrite and marcasite can be difficult to tell apart without polished thin sections or XRD analyses. These extensive iron sulfide replacements occurred in a reducing, organic-rich environment, most likely during early diagenesis near the sediment-water interface.

PPL | ORL, BDI, Scale bar = 0.16 mm

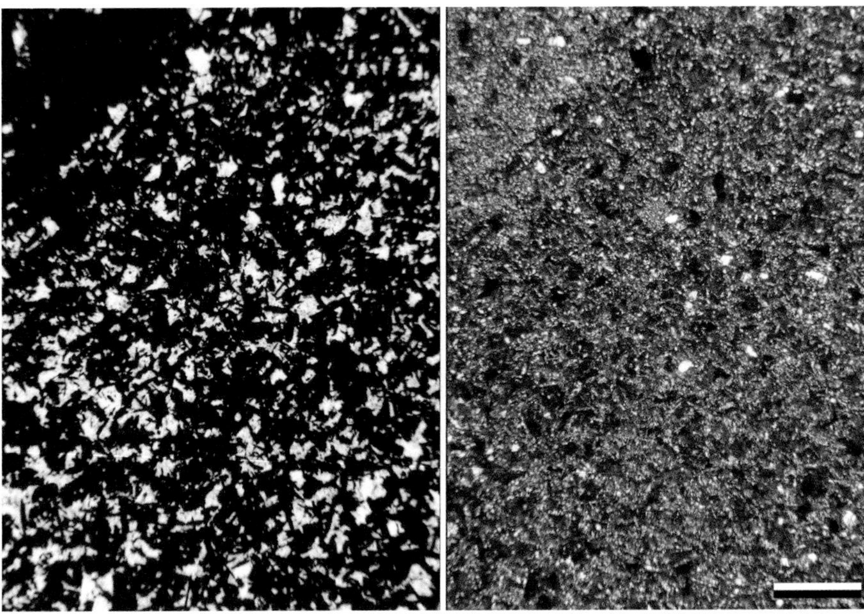

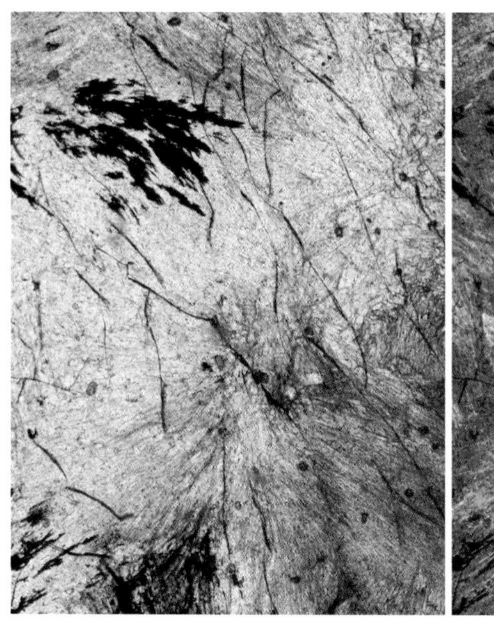
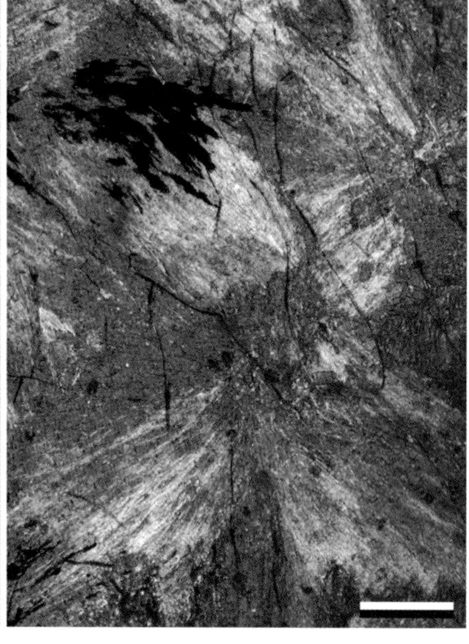

Tertiary ignimbrite, Nye Co., Nevada

The glass in this tuff has been replaced by needle-like crystals of a zeolite mineral, probably mordenite. Zeolites have a wide range of compositions, textures, and optical properties (see Chapter 14). Because of their generally finely crystalline nature and low birefringence, they are commonly overlooked or misidentified with conventional petrography. X-ray and/or geochemical analysis generally is needed for definitive identifications.

PPL | XPL, Scale bar = 0.26 mm

Cretaceous Mowry Fm., Moffat Co., Colorado

The feldspars in this view have been partially replaced by phosphate (brown phase). These sediments were deposited on a distally-steepened shelf edge where upwelling, nutrient-rich waters moved onto the shelf. During periods of slow sedimentation, these waters were responsible for phosphatizing many grains, even, and quite unusually, these feldspars.

PPL | XPL, RDI, Scale bar = 0.10 mm

Cambrian Potsdam Sandstone, St. Lawrence Co., New York

This lithic-rich sandstone was partially replaced by tourmaline. The tourmalines are subhedral to euhedral, hexagonal, pleochroic in colors ranging from pale brown, pale green/ brown to black and are a variety termed schorl. The tourmalines cut across quartz and other grains indicating they are replacing those grains and are not heavy mineral placer deposits. The Potsdam Sandstone sits atop Grenville basement, and hydrothermal fluids, likely enriched in metals and rare earths, moved out of the Grenville units and reacted with these sandstones.

PPL, Scale bar = 0.26 mm

Up. Cretaceous – Paleocene unknown unit, Sri Lanka

Diagenetic replacements involving titanium minerals are common (Morad and Aldahan, 1982 and 1987d; Morad, 1986; Pe-Piper et al., 2005). In this example, rhombic titanite crystals have partially replaced a detrital grain of rutile. Most of the remaining grain was dissolved and the resultant intragranular pore space was filled by nonferroan calcite. The surrounding intragranular pores, however, were mainly filled with slightly more ferroan calcite cement.

PPL | XPL, AFeS, KFS, BDI, Scale bar = 0.06 mm

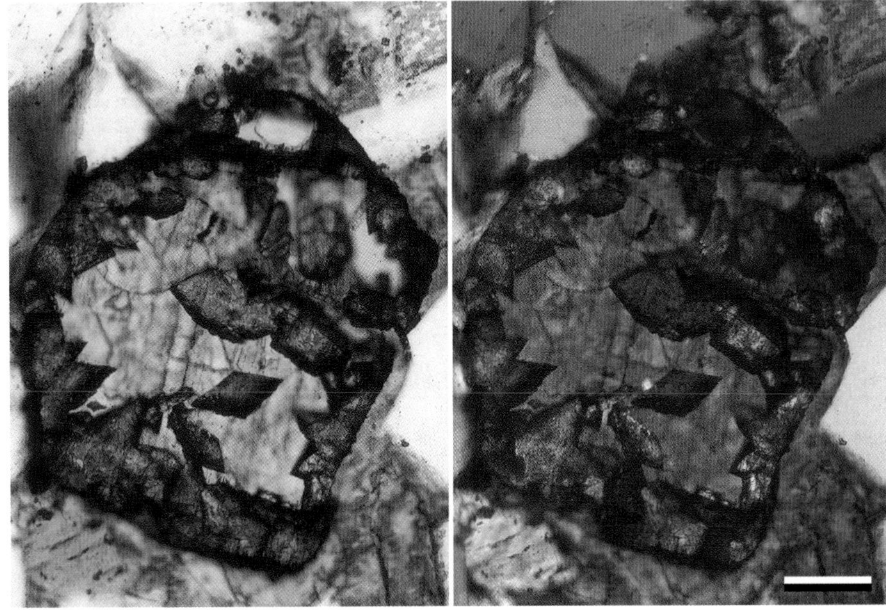

Mid. Permian Wegener Halvø Fm., Karstryggen Plateau, East Greenland

An example of anatase that partially replaced a grain that was later dissolved and filled by calcite. No relict material remains from the original grain which makes it impossible to identify it. In other areas, titanium minerals have been found to replace quartz and feldspar grains (Morad and Aldahan, 1982, 1987a). Anatase can form through hydrothermal alteration of titanium-rich minerals such as magnetite, rutile, titanite, ilmenite and biotite. A hydrothermal origin is supported by the presence of other hydrothermal minerals (fluorite, galena, barite and others) in this area.

PPL | XPL, BDI, Scale bar = 0.05 mm

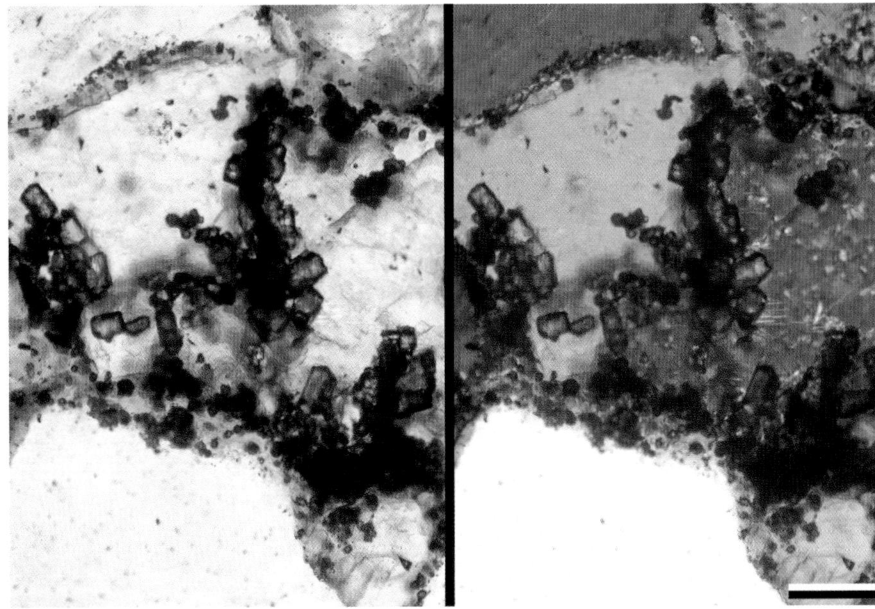

Lo. Pennsylvanian Morrow B Sandstone, Ochiltree Co., Texas

Multiple detrital grains in a placer layer replaced by titanium oxides. The replacements formed through alteration of less stable titanium-bearing precursor grains, such as rutile, ilmenite or magnetite, of which no remnants remain. The outlines of the original grains are preserved, however, and indicate that replacement occurred prior to burial-related compaction (note the grain fracturing) and before the poikilotopic calcite filling of both the secondary pores within the grains and adjacent intergranular pores. Based on their morphology and birefringence, these grains are now brookite or anatase (Morad and Aldahan, 1982).

PPL / XPL, RDI, Scale bar = 0.26 mm

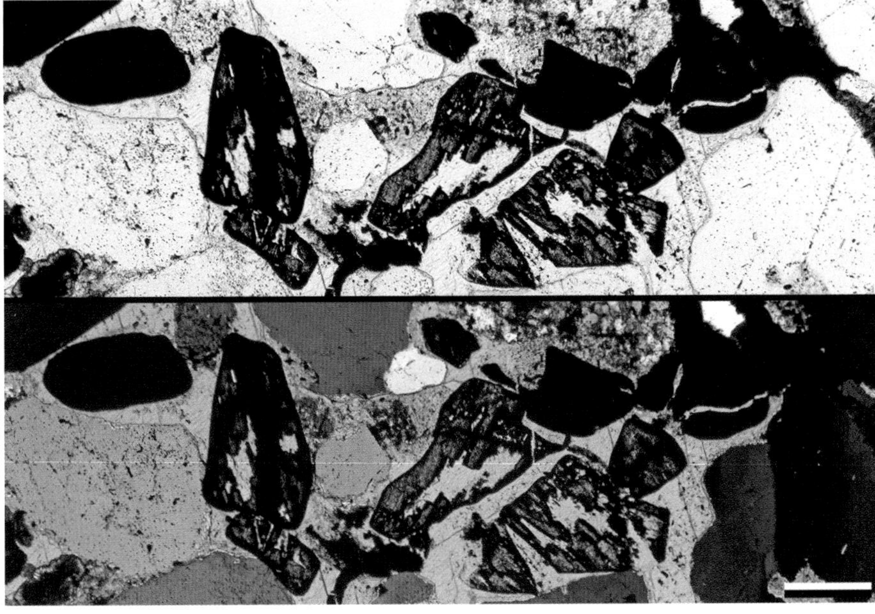

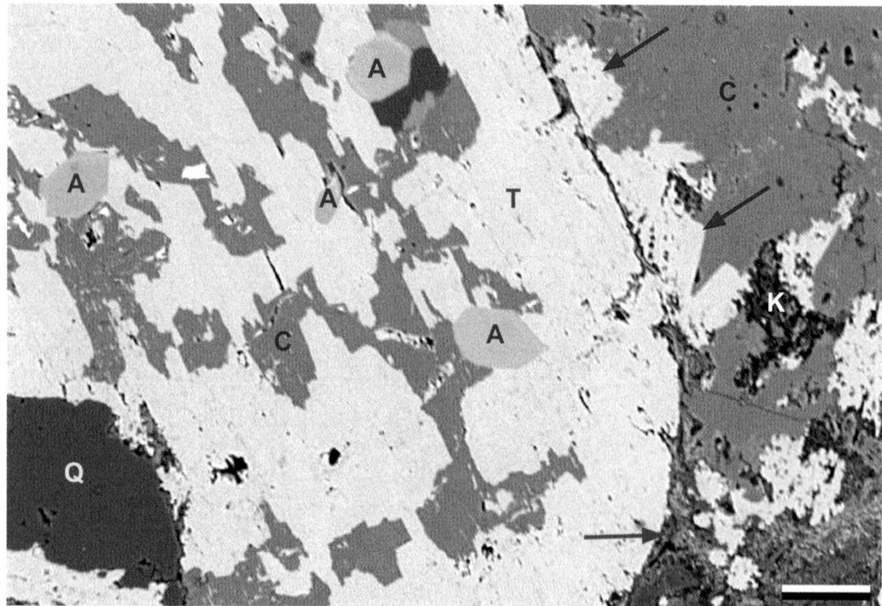

Lo. Pennsylvanian Morrow B Sandstone, Ochiltree Co., Texas

A detrital grain, probably an amphibole (containing apatite (A) and quartz (Q) inclusions) that was replaced by titanium oxide (T). The outline of the original grain is preserved and surrounded by a thin clay coat. Where there are breaks in that coating, titanium oxide cements (red arrows) have grown into the pore space. Other titanium oxide cements also partially fill the primary porosity along with illite (green arrow) and kaolinite (K). After dissolution of any remaining amphibole, poikilotopic calcite (C) filled all the primary and secondary pores in this area. As in the previous example, the titanium oxide is likely anatase or brookite.

BSE(MP), Scale bar = 30 μm

Lo. Oligocene Deborah Volcanic Fm., northern Otago, New Zealand

These fragments of an eroded pillow basalt have thick alteration (replacement) rims of bright yellow palagonite followed by a thin rim of phillipsite (red arrows in PPL) a zeolite cement. Volcanic glass is unstable in most near-surface settings and normally undergoes rapid replacement. Palagonite, a heterogeneous mixture of clay minerals dominated by smectite, here formed botryoidal, fibrous masses of crystals (note the pseudouniaxial crosses in cross-polarized light, example at green arrow). Microfibrous, spherulitic calcite cement followed the zeolite cement and filled all remaining porosity.

PPL | XPL, Scale bar = 0.26 mm

Cited References and Additional Information Sources

Aagaard, P., P. K. Egeberg, G. C. Saigal, S. Morad, and K. Bjørlykke, 1990, Diagenetic albitization of detrital K-feldspars in Jurassic, Lower Cretaceous and Tertiary clastic reservoir rocks from offshore Norway: II. Formation water chemistry and kinetic considerations: Journal of Sedimentary Research, v. 60, p. 575-581, doi: 10.1306/212F91EC-2B24-11D7-8648000102C1865D.

Aplin, A. C., I. F. Matenaar, D. K. McCarty, and B. A. van Der Pluijm, 2006, Influence of mechanical compaction and clay mineral diagenesis on the microfabric and pore-scale properties of deep-water Gulf of Mexico mudstones: Clays and Clay Minerals, v. 54, p. 500-514, doi: 10.1346/CCMN.2006.0540411.

Awwiller, D. N., 1993, Illite/smectite formation and potassium mass transfer during burial diagenesis of mudrocks: A study from the Texas Gulf Coast Paleocene-Eocene: Journal of Sedimentary Research, v. 63, p. 501-512, doi: 10.1306/D4267B3B-2B26-11D7-8648000102C1865D.

Baccar, M. B., B. Fritz, and B. Made, 1993, Diagenetic albitization of K-feldspar and plagioclase in sandstone reservoirs: Thermodynamic and kinetic modeling: Journal of Sedimentary Research, v. 63, p. 1100-1109, doi: 10.1306/D4267CB2-2B26-11D7-8648000102C1865D.

Berger, G., B. Velde, and T. Aigouy, 1999, Potassium sources and illitization in Texas Gulf Coast shale diagenesis: Journal of Sedimentary Research, v. 69, p. 151-157, doi: 10.2110/jsr.69.151.

Bétard, F., L. Caner, Y. Gunnell, and G. Bourgeon, 2009, Illite neoformation in plagioclase during weathering: Evidence from semi-arid northeast Brazil: Geoderma, v. 152, p. 53-62, doi: 10.1016/j.geoderma.2009.05.016.

Bjørkum, P. A., O. Walderhaug, and N. E. Aase, 1993, A model for the effect of illitization on porosity and quartz cementation of sandstones: Journal of Sedimentary Research, v. 63, p. 1089-1091, doi: 10.2110/jsr.63.1089.

Boles, J. R., and S. G. Franks, 1979, Clay diagenesis in Wilcox sandstones of southwest Texas: Implications of smectite diagenesis on sandstone cementation: Journal of Sedimentary Research, v. 49, p. 55-70, doi: 10.1306/212F76BC-2B24-11D7-8648000102C1865D.

Chandler, F. W., 1988, Diagenesis of sabkha-related, sulphate nodules in the early Proterozoic Gordon Lake Formation, Ontario, Canada: Carbonates and Evaporites, v. 3, p. 75-94, doi: 10.1007/BF03174414.

Chowdhury, A. H., and J. P. A. Noble, 1993, Feldspar albitization and feldspar cementation in the Albert Formation reservoir sandstones, New Brunswick, Canada: Marine and Petroleum Geology, v. 10, p. 394-402, doi: 10.1016/0264-8172(93)90083-5.

Clauer, N., and S. Chaudhuri, 1996, Inter-basinal comparison of the diagenetic evolution of illite/smectite minerals in buried shales on the basis of K-Ar systematics: Clays and Clay Minerals, v. 44, p. 818-824, doi: 10.1346/CCMN.1996.0440613.

Clauer, N., and N. Liewig, 2013, Episodic and simultaneous illitization in oil-bearing Brent Group and Fulmar Formation sandstones from the northern and southern North Sea based on illite K-Ar dating: AAPG Bulletin, v. 97, p. 2149-2171, doi: 10.1306/04021312122.

Cook, P. J., 1970, Repeated diagenetic calcitization, phosphatization, and silicification in the Phosphoria Formation: GSA Bulletin, v. 81, p. 2107-2116, doi: 10.1130/0016-7606(1970)81[2107:RDCPAS]2.0.CO;2.

De Bona, J., N. Dani, J. M. Ketzer, and L. F. De Ros, 2008, Dickite in shallow oil reservoirs from Recôncavo Basin, Brazil: Diagenetic implications for basin evolution: Clay Minerals, v. 43, p. 213-233, doi: 10.1180/claymin.2008.043.2.06.

De Ros, L. F., G. N. C. Sgarbi, and S. Morad, 1994, Multiple authigenesis of K-feldspar in sandstones: Evidence from the Cretaceous Areado Formation, São Francisco Basin, central Brazil: Journal of Sedimentary Research, v. 64, p. 778-787, doi: 10.1306/D4267EBF-2B26-11D7-8648000102C1865D.

Dunoyer de Segonzac, G., 1970, The transformation of clay minerals during diagenesis and low-grade metamorphism: A review: Sedimentology, v. 15, p. 281-346, doi: 10.1111/j.1365-3091.1970.tb02190.x.

Elliott, W. C., and G. Matisoff, 1996, Evaluation of kinetic models for the smectite to illite transformation: Clays and Clay Minerals, v. 44, p. 77-87, doi: 10.1346/CCMN.1996.0440107.

Folk, R. L., 1965, Some aspects of recrystallization in ancient limestones, *in* L. C. Pray, and R. S. Murray, eds., Dolomitization and Limestone Diagenesis: Tulsa, OK, SEPM Special Publication 13, p. 14-48, doi: 10.2110/pec.65.07.0014.

Folk, R. L., and J. S. Pittman, 1971, Length-slow chalcedony: A new testament for vanished evaporites: Journal of Sedimentary Research, v. 41, p. 1045-1058, doi: 10.1306/74D723F1-2B21-11D7-8648000102C1865D.

Gold, P. B., 1987, Textures and geochemistry of authigenic albite from Miocene sandstones, Louisiana Gulf Coast: Journal of Sedimentary Research, v. 57, p. 353-362, doi: 10.1306/212F8B2A-2B24-11D7-8648000102C1865D.

González-Acebrón, L., J. Arribas, and R. Mas, 2010, Role of sandstone provenance in the diagenetic albitization of feldspars: A case study of the Jurassic Tera Group sandstones (Cameros Basin, NE Spain): Sedimentary Geology, v. 229, p. 53-63, doi: 10.1016/j.sedgeo.2010.06.005.

González-Acebrón, L., J. Götze, D. Barca, J. Arribas, R. Mas, and C. Pérez-Garrido, 2012, Diagenetic albitization in the Tera Group, Cameros Basin (NE Spain) recorded by trace elements and spectral cathodoluminescence: Chemical Geology, v. 312, p. 148-162, doi: 10.1016/j.chemgeo.2012.04.012.

Helmold, K. P., 1985, Provenance of feldspathic sandstones—the effect of diagenesis on provenance interpretations: A review, *in* G. G. Zuffa, ed., Provenance of Arenites (NATO Science Series C): Dordrecht, Netherlands, D. Reidel Publishing, p. 139-164, doi: 10.1007/978-94-017-2809-6_7.

Hesse, R., 1989, Silica diagenesis: Origin of inorganic and replacement cherts: Earth-Science Reviews, v. 26, p. 253-284, doi: 10.1016/0012-8252(89)90024-X.

Hirt, W. G., H.-R. Wenk, and J. R. Boles, 1993, Albitization of plagioclase crystals in the Stevens sandstone (Miocene), San Joaquin Basin, California, and the Frio Formation (Oligocene), Gulf Coast, Texas: A TEM/AEM study: GSA Bulletin, v. 105, p. 708-714, doi: 10.1130/0016-7606(1993)105<0708:AOPCIT>2.3.CO;2.

Iijima, A., 1988, Diagenetic transformation of minerals as exemplified by zeolites and silica minerals—a Japanese view, *in* G. V. Chilingarian, and K. H. Wolf, eds., Diagenesis II, (Developments in Sedimentology 43): Amsterdam, Elsevier Science Publishers, p. 147-211.

Isaacs, C. M., 1982, Influence of rock composition on kinetics of silica phase changes in the Monterey Formation, Santa Barbara area, California: Geology, v. 10, p. 304-308, doi: 10.1130/0091-7613(1982)10<304:IORCOK>2.0.CO;2.

Jahren, J. S., and P. Aagaard, 1992, Diagenetic illite-chlorite assemblages in arenites. I. Chemical evolution: Clays and Clay Minerals, v. 40, p. 540-546, doi: 10.1346/CCMN.1992.0400507.

Krajewski, K. P., 2002, Catagenic ankerite replacing biogenic calcite in the Marhøgda Bed (Jurassic), Sassenfjorden, Spitsbergen: Polish Polar Research, v. 23, p. 85-99.

Lanson, B., B. A. Sakharov, F. Claret, and V. A. Drits, 2009, Diagenetic smectite-to-illite transition in clay-rich sediments: A reappraisal of X-ray diffraction results using the multi-specimen method: American Journal of Science, v. 309, p. 476-516, doi: 10.2475/06.2009.03.

Lee, M. R., and I. Parsons, 1998, Microtextural controls of diagenetic alteration of detrital alkali feldspars: A case study of the Shap conglomerate (Lower Carboniferous), northwest England: Journal of Sedimentary Research, v. 68, p. 198-211, doi: 10.2110/jsr.68.198.

Loucks, R. G., D. G. Bebout, and W. E. Galloway, 1977, Relationship of porosity formation and preservation to sandstone consolidation history—Gulf Coast Lower Tertiary Frio Formation: Gulf Coast Association of Geological Societies Transactions, v. 27, p. 109-120.

Machel, H. G., 2004, Concepts and models of dolomitization: A critical reappraisal, *in* C. J. R. Braithwaite, G. Rizzi, and G. Darke, eds., The Geometry and Petrogenesis of Dolomite Hydrocarbon Reservoirs: London, GSL Special Publication 235, p. 7-63, doi: 10.1144/GSL.SP.2004.235.01.02.

Maliva, R. G., T. M. Missimer, and J. A. D. Dickson, 2000, Skeletal aragonite neomorphism in Plio-Pleistocene sandy limestones and sandstones, Hollywood, Florida, USA: Sedimentary Geology, v. 136, p. 147-154, doi: 10.1016/S0037-0738(00)00102-0.

Maliva, R. G., and R. Siever, 1988, Mechanism and controls of silicification of fossils in limestones: Journal of Geology, v. 96, p. 387-398, doi: 10.1086/629235.

McBride, E. F., 1985, Diagenetic processes that affect provenance determinations in sandstone, *in* G. G. Zuffa, ed., Provenance of Arenites (NATO Science Series C): Dordrecht, Netherlands, D. Reidel Publishing, p. 95-113, doi: 10.1007/978-94-017-2809-6_5.

McKinley, J. M., R. H. Worden, and A. H. Ruffell, 2003, Smectite in sandstones: A review of the controls on occurrence and behaviour during diagenesis, *in* R. H. Worden, and S. Morad, eds., Clay Mineral Cements in Sandstones (IAS Special Publication 34): Oxford, Wiley-Blackwell, p. 109-128, doi: 10.1002/9781444304336.ch5.

Milliken, K. L., 1979, The silicified evaporite syndrome—two aspects of silicification history of former evaporite nodules from southern Kentucky and northern Tennessee: Journal of Sedimentary Research: v. 49, p. 245-256, doi: 10.1306/212F7707-2B24-11D7-8648000102C1865D.

Milliken, K. L., 1989, Petrography and composition of authigenic feldspars, Oligocene Frio Formation, South Texas: Journal of Sedimentary Research, v. 59, p. 361-374, doi: 10.1306/212F8F94-2B24-11D7-8648000102C1865D.

Milliken, K. L., 1992, Chemical behavior of detrital feldspars in mudrocks versus sandstones, Frio Formation (Oligocene), South Texas: Journal of Sedimentary Research, v. 62, p. 790-801, doi: 10.1306/D42679DD-2B26-11D7-8648000102C1865D.

Milliken, K. L., 2002, Petrography of ankerite cement, grain replacement, and fracture fill in foreland sandstones of the central Rocky Mountains: AAPG Annual Meeting Abstracts, AAPG Search and Discovery Article #90007©2002.

Milliken, K. L., E. F. McBride, and L. S. Land, 1989, Numerical assessment of dissolution versus replacement in the subsurface destruction of detrital feldspars, Oligocene Frio Formation, South Texas: Journal of Sedimentary Research, v. 59, p. 740-757, doi: 10.1306/212F9061-2B24-11D7-8648000102C1865D.

Morad, S., 1986, SEM study of authigenic rutile, anatase and brookite in Proterozoic sandstones from Sweden: Sedimentary Geology, v. 46, p. 77-89, doi: 10.1130/0016-7606(1986)97<567:AODFOI>2.0.CO;2.

Morad, S., 1988, Albitized microcline grains of post-depositional and probable detrital origins in Brøttum Formation sandstones (Upper Proterozoic), Sparagmite Region of southern Norway: Geological Magazine, v. 125, p. 229-239, doi: 10.1017/S0016756800010177.

Morad, S., and A. A. Aldahan, 1982, Authigenesis of titanium minerals in two Proterozoic sedimentary rocks from southern and central Sweden: Journal of Sedimentary Research, v. 52, p. 1295-1305, doi: 10.1306/212F8120-2B24-11D7-8648000102C1865D.

Morad, S., and A. A. Aldahan, 1987a, A SEM study of diagenetic kaolinization and illitization of detrital feldspars in sandstones: Clay Minerals, v. 22, p. 237-243, doi: 10.1180/claymin.1987.022.2.12.

Morad, S., and A. A. Aldahan, 1987b, Diagenetic chloritization of feldspars in sandstones: Sedimentary Geology, v. 51, p. 155-164, doi: 10.1016/0037-0738(87)90046-7.

Morad, S., and A. A. Aldahan, 1987c, Diagenetic replacement of feldspars by quartz in sandstones: Journal of Sedimentary Research, v. 57, p. 488-493, doi: 10.1306/212F8B70-2B24-11D7-8648000102C1865D.

Morad, S., and A. A. Aldahan, 1987d, Diagenetic "replacement" of feldspars by titanium oxides in sandstones: Sedimentary Geology, v. 51, p. 147-153, doi: 10.1016/0037-0738(87)90045-5.

Morad, S., M. Bergan, R. Knarud, and J. P. Nystuen, 1990, Albitization of detrital plagioclase in Triassic reservoir sandstones from the Snorre Field, Norwegian North Sea: Journal of Sedimentary Research, v. 60, p. 411-425, doi: 10.1306/212F91AB-2B24-11D7-8648000102C1865D.

Morad, S., A. Bhattacharyya, I. S. Al-Aasm, and K. Ramseyer, 1991, Diagenesis of quartz in the Upper Proterozoic Kaimur Sandstones, Son Valley, central India: Sedimentary Geology, v. 73, p. 209-225, doi: 10.1016/0037-0738(91)90085-R.

Morad, S., R. Márfil, and J. A. Peña, 1989, Diagenetic K-feldspar pseudomorphs in the Triassic Buntsandstein sandstones of the Iberian Range, Spain: Sedimentology, v. 36, p. 635-650, doi: 10.1111/j.1365-3091.1989.tb02090.x.

Parsons, I., and M. R. Lee, 2009, Mutual replacement reactions in alkali feldspars I: Microtextures and mechanisms: Contributions to Mineralogy and Petrology, v. 157, p. 641-661, doi: 10.1007/s00410-008-0355-4.

Parsons, I., P. Thompson, M. R. Lee, and N. Cayzer, 2005, Alkali feldspar microtextures as provenance indicators in siliciclastic rocks and their role in feldspar dissolution during transport and diagenesis: Journal of Sedimentary Research, v. 75, p. 921-942, doi: 10.2110/jsr.2005.071.

Pe-Piper, G., D. J. W. Piper, and L. Dolansky, 2005, Alteration of ilmenite in the Cretaceous sandstones of Nova Scotia, southeastern Canada: Clays and Clay Minerals, v. 53, p. 490-510, doi: 10.1346/CCMN.2005.0530506.

Perez, R. J., and J. R. Boles, 2005, An empirically derived kinetic model for albitization of detrital plagioclase: American Journal of Science, v. 305, p. 312-343, doi: 10.2475/ajs.305.4.312.

Pytte, A. M., and R. C. Reynolds, 1989, The thermal transformation of smectite to illite, *in* N. D. Naeser, and T. H. McCulloh, eds., Thermal History of Sedimentary Basins: Methods and Case Histories: New York, Springer-Verlag, p. 133-140, doi: 10.1007/978-1-4612-3492-0_8.

Ramseyer, K., J. R. Boles, and P. C. Lichtner, 1992, Mechanism of plagioclase albitization: Journal of Sedimentary Research, v. 62, p. 349-356, doi: 10.1306/D42678FC-2B26-11D7-8648000102C1865D.

Saigal, G. C., S. Morad, K. Bjørlykke, P. K. Egeberg, and P. Aagaard, 1988, Diagenetic albitization of detrital K-feldspar in Jurassic, Lower Cretaceous, and Tertiary clastic reservoir rocks from offshore Norway, I. Textures and origin: Journal of Sedimentary Research, v. 58, p. 1003-1013, doi: 10.1306/212F8EE5-2B24-11D7-8648000102C1865D.

Schieber, J., 2007, Oxidation of detrital pyrite as a cause for marcasite formation in marine lag deposits from the Devonian of the eastern US: Deep Sea Research Part II: Topical Studies in Oceanography, v. 54, p. 1312-1326, doi: 10.1016/j.dsr2.2007.04.005.

Scholle, P. A., L. Stemmerik, and O. Harpøth, 1990, Origin of major karst-associated celestite mineralization in Karstryggen, central East Greenland: Journal of Sedimentary Petrology, v. 60, p. 397-410, doi: 10.1306/212F91A6-2B24-11D7-8648000102C1865D.

Scholle, P. A., D. S. Ulmer, and L. A. Melim, 1992, Late-stage calcites in the Permian Capitan Formation and its equivalents, Delaware Basin margin, west Texas and New Mexico: Evidence for replacement of precursor evaporites: Sedimentology, v. 39, p. 207-234, doi: 10.1111/j.1365-3091.1992.tb01035.x.

Sibley, D. F., and J. M. Gregg, 1987, Classification of dolomite rock textures: Journal of Sedimentary Research, v. 57, p. 967-975, doi: 10.1306/212F8CBA-2B24-11D7-8648000102C1865D.

Stafford, K. W., L. Land, A. B. Klimchouk, and M. O. Gary, eds., 2009, The Pecos River hypogene speleogenetic province: A basin-scale karst paradigm for eastern New Mexico and west Texas, USA: Carlsbad, NM, National Cave and Karst Research Institute Symposium 1, 15 p.

Stroncik, N. A., and H.-U. Schmincke, 2002, Palagonite–a review: International Journal of Earth Sciences, v. 91, p. 680-697, doi: 10.1007/s00531-001-0238-7.

Surdam, R. C., and J. R. Boles, 1979, Diagenesis of volcanic sandstones, *in* P. A. Scholle, and P. Schluger, eds., Aspects of Diagenesis: Tulsa, OK, SEPM Special Publication 26, p. 227-242, doi: 10.2110/pec.79.26.0227.

Ulmer, D. S., and R. L. Laury, 1984, Diagenesis of the Mississippian Arroyo Peñasco Group, north-central New Mexico, *in* W. S. Baldridge, P. W. Dickerson, R. E. Riecker, and J. Zidek, eds., Rio Grande Rift: Northern New Mexico: Albuquerque, NM, New Mexico Geological Society 35th Field Conference Guidebook, p. 91-100.

Ulmer-Scholle, D. S., and P. A. Scholle, 1994, Replacement of evaporites within the Permian Park City Formation, Bighorn Basin, Wyoming, U.S.A.: Sedimentology, v. 41, p. 1203-1222, doi: 10.1111/j.1365-3091.1994.tb01449.x.

Ulmer-Scholle, D. S., P. A. Scholle, and P. V. Brady, 1993, Silicification of evaporites in Permian (Guadalupian) back-reef carbonates of the Delaware Basin, west Texas and New Mexico: Journal of Sedimentary Petrology, v. 63, p. 955-965, doi: 10.1306/D4267C53-2B26-11D7-8648000102C1865D.

Velde, B., and E. Nicot, 1985, Diagenetic clay mineral composition as a function of pressure, temperature, and chemical activity: Journal of Sedimentary Research, v. 55, p. 541-547, doi: 10.1306/212F8724-2B24-11D7-8648000102C1865D.

Velde, B., and G. Vasseur, 1992, Estimation of the diagenetic smectite to illite transformation in time-temperature space: American Mineralogist, v. 77, p. 967-976.

Walker, T. R., 1960, Carbonate replacement of detrital crystalline silicate minerals as a source of authigenic silica in sedimentary rocks: GSA Bulletin, v. 71, p. 145-151, doi: 10.1130/0016-7606(1960)71[145:CRODCS]2.0.CO;2.

Walker, T. R., 1984, Diagenetic albitization of potassium feldspar in arkosic sandstones: Journal of Sedimentary Research, v. 54, p. 3-16, doi: 10.1306/212F8391-2B24-11D7-8648000102C1865D.

Wallace, C. A., 1976, Diagenetic replacement of feldspar by quartz in the Uinta Mountain Group, Utah and its geochemical implications: Journal of Sedimentary Research, v. 46, p. 847-861, doi: 10.1306/212F7072-2B24-11D7-8648000102C1865D.

Walton, A. W., 1975, Zeolitic diagenesis in Oligocene volcanic sediments, Trans-Pecos Texas: GSA Bulletin, v. 86, p. 615-624, doi: 10.1130/0016-7606(1975)86<615:ZDIOVS>2.0.CO;2.

Weibel, R., and H. Friis, 2007, Alteration of opaque heavy minerals as a reflection of the geochemical conditions in depositional and diagenetic environments, *in* M. A. Mange, and D. T. Wright, eds., Heavy Minerals in Use (Developments in Sedimentology 58): Amsterdam, Elsevier, p. 277-303, doi: 10.1016/S0070-4571(07)58010-6.

Wescott, W. A., 1983, Diagenesis of Cotton Valley sandstone (Upper Jurassic), east Texas: Implications for tight gas formation pay recognition: AAPG Bulletin, v. 67, p. 1002-1013.

Wilkinson, M., K. L. Milliken, and R. S. Haszeldine, 2001, Systematic destruction of K-feldspar in deeply buried rift and passive margin sandstones: Journal of the Geological Society, v. 158, p. 675-683, doi: 10.1144/jgs.158.4.675.

Worden, R. H., and J. C. Rushton, 1992, Diagenetic K-feldspar textures: A TEM study and model for diagenetic feldspar growth: Journal of Sedimentary Research, v. 62, p. 779-789, doi: 10.1306/D42679D8-2B26-11D7-8648000102C1865D.

Facing Page: Top – Outcrop example of brittle deformation—multi-stage fractures in shaly limestone, Pennsylvanian Madera Fm., Socorro Co., New Mexico. Bottom – A classic example of ductile deformation of a duck tile.

DIAGENESIS:

DEFORMATION FEATURES

CHAPTER 21

Deformation Features

This chapter covers a range of structural features that can occur at thin-section scale in terrigenous rocks, including natural fractures, deformation bands, grain-scale strain features, stylolites and cone-in-cone structures. However, soft-sediment deformation features are not substantially covered here, because they are visible at thin-section level primarily in very fine-grained strata, and thus were previously illustrated in Chapter 8); likewise, compaction-related deformation features were largely covered in Chapter 10.

The structures discussed in this chapter, especially fractures and deformation bands, play a major role in porosity-permeability relationships and hydrocarbon reservoir performance of many sandstones and mudstones. Although tectonic processes are commonly inferred to be the causes of brittle deformation structures, elevated pore-fluid pressures (and rapid or episodic loss of such overpressures) also have been shown to play a major role in formation of many natural fractures and veins (e.g., Cartwright, 1994; Miller, 1995; Conybeare and Shaw, 2000; Cosgrove, 2001; Cobbold et al., 2013). Natural overpressures can be generated in a variety of settings as a result of rapid burial of low-permeability strata, hydrocarbon generation, expulsion of water from evaporites and other processes (e.g., Burrus, 1998; Swarbrick and Osborne, 1998), and artificial generation of elevated fluid pressures is the basis of modern induced hydraulic fracing.

Fractures, faults and deformation bands – There are many classifications of structural features (see Schultz and Fossen, 2008 and references cited therein), but we will simply use fractures (or microfaults) to denote small, sharply-bounded, partially-filled or unfilled cracks as opposed to shear bands that generally occur in more porous media and have broader zones of deformation that show reorientation and cataclastic features including grain sliding, rotation and fracturing (Aydin, 1978; Fossen et al., 2007; Onasch et al., 2009). The term fracture is used for tensile features with no lateral offset; faults or microfaults are features with lateral shear and offset. Cement-filled fractures or microfaults generally are referred to as veins.

Deformation features occur in virtually all subsurface sandstones and mudstones and play some role, and commonly a major one, in overall fluid movements as well as hydrocarbon reservoir performance. Although, they may look like small and insignificant features in a 2-D thin section view, it should be remembered that these are 3-D structures that may have considerable extent outside the field of view. However, the reservoir impacts of such features can be highly varied (positive or negative), depending on size, abundance, orientation and whether they are open and unfilled or are partially to completely cemented and/or gouge-filled. Clearly, unfilled open fractures can have high porosity and permeability; however, the lack of any propping cements can lead to undesirable fracture closure. At the other end of the spectrum, extensive cementation or other filling may substantially reduce overall rock porosity and permeability, which in turn, may reduce reservoir homogeneity and connectivity. Extensive or complete filling of fractures is quite common, however, because fractures are excellent conduits for diagenetic fluid movement and, in many cases, expose fresh (uncoated) crystal surfaces for cementation. In this context, reservoir-related fractures were divided into four types by Nelson (1985, but cited as 2001 second edition):

Type 1: Fractures that provide the porosity and permeability essential to the rock being a reservoir,

Type 2: Fractures that provide the permeability essential to the rock being a reservoir,

Type 3: Fractures that provide a permeability assist to an already producible reservoir, and

Type 4: Fractures that provide no positive improvements to reservoir quality, but instead increase reservoir anisotropy and inhomogeneity and perhaps create fluid-flow barriers and compartmentalization.

Fractures can have long and complex histories of formation and cementation. Larger fractures can form through coalescence of precursor microcracks (Kranz, 1983) or can grow through multiple episodes of widening and bridging, yielding potentially very complex crack-seal textures (Laubach et al., 2004 a and b).

From a petrographic perspective, there are numerous valuable observations that can be made on the character of fractures, veins or deformation bands and their positive or negative impacts on reservoir properties. These include:

- Determination of the nature of the strain localization structures (deformation bands versus tensile fractures or shear-related microfaults);
- Measurements of the relative orientations, dimensions (widths and, where possible, lengths), and morphologies of fracture sets;
- Observations of temporal relationships (based on cross-cutting relationships) between multiple generations of fractures/bands and the relative timing of fracturing within the overall diagenetic history;
- Determination of porosity enhancement or reduction in host rock adjacent to such deformation structures (including solution enlargement and channel or vug formation along fractures);

- Determination of the amount and composition of cements precipitated within fractures or the amount and nature of the deformed material in or adjacent to deformation bands. Many minerals occur as cements in fractures, especially various forms of silica or quartz and carbonate minerals (calcite, dolomite and ankerite); in some settings iron sulfides, oxides or hydroxides, gypsum or anhydrite, and/or halide minerals are abundant. Additionally, a diverse suite of fracture-filling minerals (including barite, fluorite, galena and others) can form in hydrothermal settings;
- Observations on the presence or absence of hydrocarbon residues can help to define the timing of fractures and their fillings or to establish timing and pathways of hydrocarbon migration;
- More specialized petrographic or geochemical studies (fluid-inclusion geothermometry or isotopic investigations) on fracture and vein fills can help to understand the fluid history and timing of cementation in such structures (e.g., Kelly et al., 2000; Parris et al., 2003).

Stylolites – These features were already discussed and illustrated in Chapter 10 (Compaction). They are mentioned again here mainly because not all stylolites are formed by overburden loading, and thus not all stylolites are bedding-parallel structures. Compressive stresses with any orientation relative to bedding can produce stylolites that cross-cut bedding as well as earlier bedding-parallel stylolites. Stylolites are recognized as irregular, sutured surfaces with concentrations of relatively insoluble materials (especially clays, organic matter, and iron minerals). They also occur as irregular, circumgranular surfaces of dissolution, especially at points of intergranular contact. The clays and other insolubles found along stylolitic surfaces may act mainly to enhance diffusive transport of solutes away from stylolites, thus promoting continued dissolution (e.g., Heald, 1959; Weyl, 1959), or they may have a more active role in localizing and promoting stylolitic dissolution (Lerbekmo and Platt, 1962; Bayly, 1986; Bjørkum, 1996; Renard et al., 1997; Walderhaug and Bjørkum, 2003). In either case, primary concentrations of clays, organic matter and iron minerals may be important in localizing the inception of dissolution surfaces, and such inhomogeneities may be further accentuated by accumulation of insoluble materials during dissolution.

Stylolitic accumulation of insoluble materials into seams that may reach thicknesses of millimeters to centimeters is not inherently a process beneficial for reservoir permeability or homogeneity. However, stylolites can have associated fractures, some natural and some created during coring and sample preparation (Nelson, 1981). It is important, but generally not easy, to distinguish between natural and artificially induced stylolite-associated fractures.

Deformation lamellae and intragranular fracturing – These are stress-related fractures or strain lamellae and inclusion trains within sedimentary grains (most commonly in quartz grains; Fairbairn, 1941). Although some strain lamellae may form *in situ* in tectonically deformed sedimentary rocks, most are inherited from crystalline basement source rocks. Intragranular fractures more commonly are formed *in situ* and can substantially influence porosity distribution and quartz cementation because they increase grain surface area and provide fresh (uncoated) surfaces on which quartz cements can precipitate (Dickinson and Milliken, 1995; Fisher et al., 2000). Growing interest and awareness of the importance of bolide impacts in Earth history (and, in some cases, to hydrocarbon reservoir generation) led to more detailed investigations of deformation lamellae and has enabled distinction of structures produced by hypervelocity impacts from those produced by other forms of deformation (Stöffler and Langenhorst, 1994; Grieve et al., 1996).

Cone-in-cone structures, fibrous veins ("beef") and horizontal hydraulic (or dilatant) fractures – These structures are relatively uncommon and have been controversial in origin. Fibrous veins here refer to veins consisting of fibrous calcite with the fibers oriented generally perpendicular to the vein walls—fibrous gypsum veins (sometimes called satin-spar veins) have similar fabrics and may have comparable origins (e.g., Shearman et al., 1972; Gustavson et al., 1994). Both calcite and gypsum veins generally have a bedding-parallel orientation. Cone-in-cone structures, associated in many cases with fibrous veins, have more complex conical or fan-shaped arrangements of calcite crystals, but also are commonly bedding-parallel. Fibrous and cone-in-cone veins, along with associated, smaller and more irregular dilatant or hydraulic fractures and veins, are found in a variety of lithologies, but are most common in mudrocks and calcareous mudrocks.

Many origins have been proposed for these clearly diagenetic features (see Tarney and Schreiber, 1976) including the force of authigenic cement crystallization in generating and widening fractures (e.g., Woodland, 1964) and pedogenic processes (Aassoumi et al., 1992). Although not all fibrous, cone-in-cone and horizontal hydraulic fractures necessarily have a single origin, there is growing recognition that elevated pore-fluid pressures (overpressures or hydropressures) are the likely cause of many such features (e.g., Shearman et al., 1972; Stoneley, 1983; Machel, 1985; Cobbold and Rodrigues, 2007; Rodrigues et al., 2009; Cobbold et al., 2013).

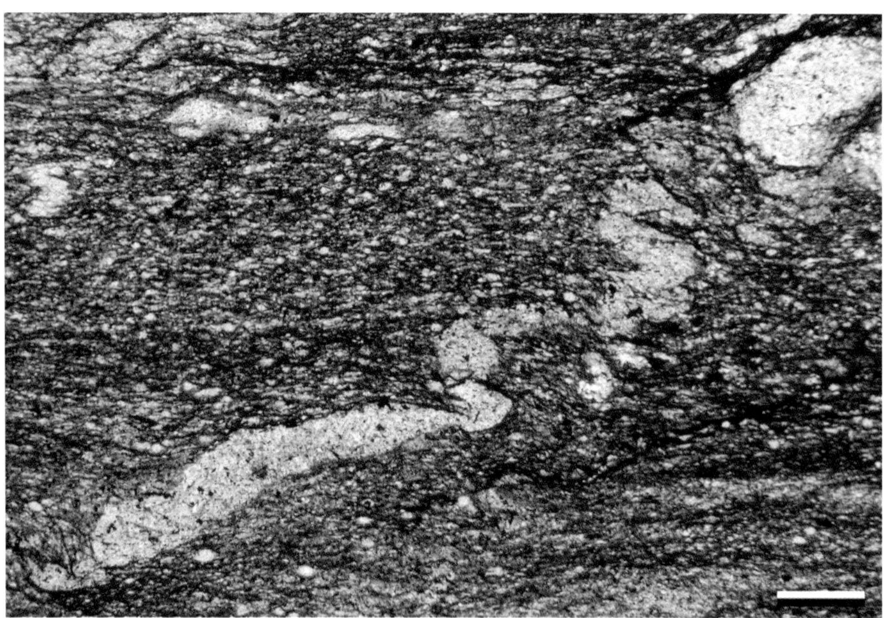

Cretaceous Mowry Fm., Moffat Co., Colorado

An example of ductile, soft-sediment deformation of siltstone-filled burrows in a silty shale. Compaction and soft-sediment deformation are widely seen, especially in fine-grained sediments. Because they were illustrated and discussed in the sections on mudrocks (Chapter 8) and compaction (Chapter 10), they will be mentioned only briefly here. Such deformation features typically start to form very early (virtually from the sediment-water interface) and may continue to grow over long periods of dewatering and compaction.

PPL, RDI, Scale bar = 0.51 mm

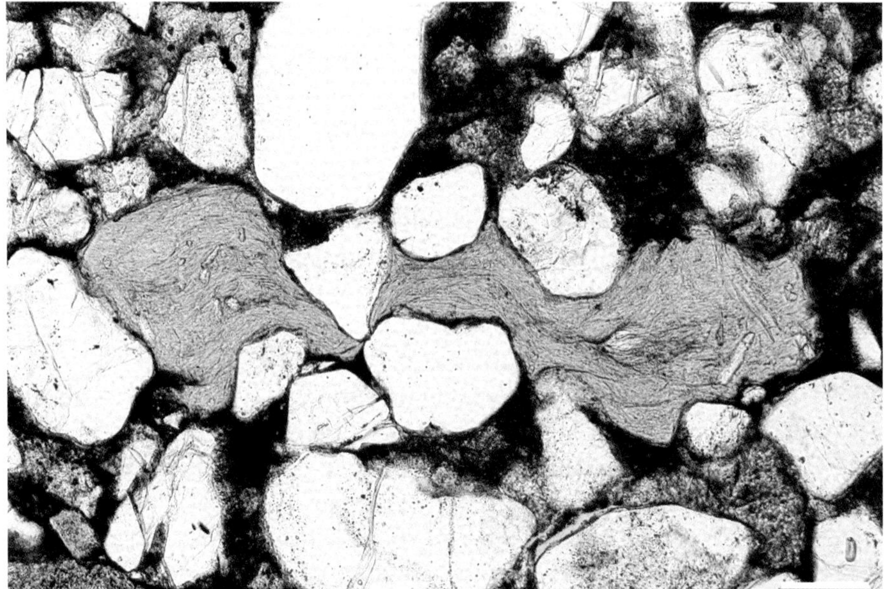

Mid. Jurassic Carmel Fm., Crystal Creek Mbr., Cane Co., Utah

Ductile deformation of a chlorite-illite shale SRF. Several adjacent and more rigid quartz grains have deeply embayed and deformed this rock fragment, presumably as a result of overburden loading. The extent of deformation is facilitated by the lack of substantial cement in this sample or, in this case, by the possible subsurface removal of former calcium sulfate cements. Little evidence of that is visible here, except for the fact that many of the quartz grains in this sample (and a few in this image) have been shattered and contain open fractures, a feature not present in adjacent, still gypsum-cemented samples.

PPL, BDI, Scale bar = 0.10 mm

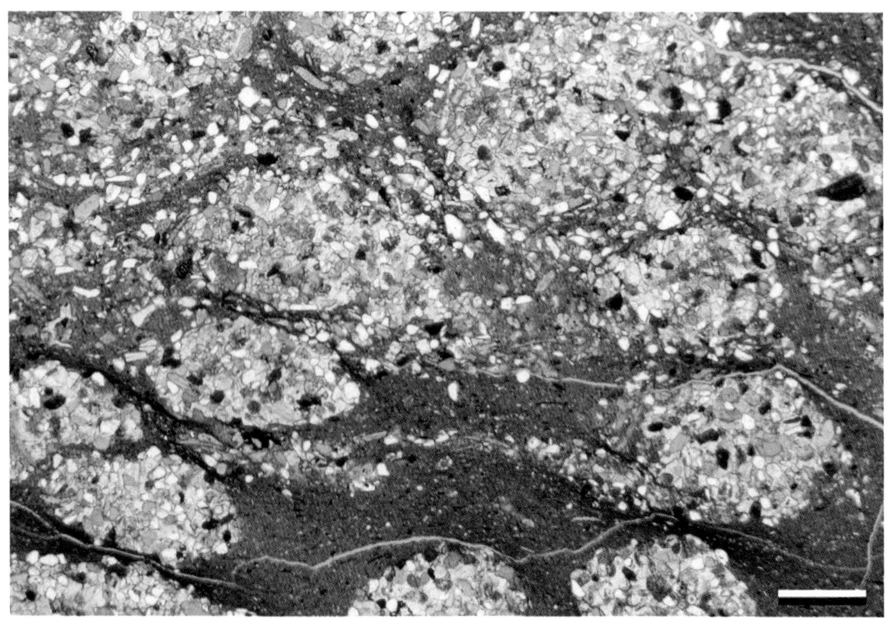

Mid. Jurassic Curtis Fm., Moffat Co., Colorado

Cross-sections of numerous silt-filled, near-horizontal burrows in a slightly silty shale. Most of the burrows have ovoid shapes produced by compactional deformation. Burrow deformation would be much more pronounced if the burrows had been filled with finer-grained material. The silt fills here resisted compaction but considerable deformation and incipient solution seam formation is visible in the surrounding shales. The small open fractures filled with red-dyed epoxy most likely are artifacts of coring or thin-section preparation.

PPL, RDI, Scale bar = 0.10 mm

Cretaceous Mowry Fm., Moffat Co., Colorado

An organic-rich, silty, calcareous mudstone with numerous thin, elongate solution seams (microstylolites). The wrinkly brownish seams typically contain elevated amounts of clay, organic matter and pyrite and most commonly are parallel to bedding, reflecting the influence of overburden loading during burial. At higher magnification, one can see remnants of partly dissolved carbonate grains. Radiolaria, planktonic foraminifers and fragments of organic matter occur throughout this shale and, in many areas, show partial or complete flattening from overburden loading. These features are more fully discussed in the section on compaction (Chapter 10).

PPL, RDI, Scale bar = 0.26 mm

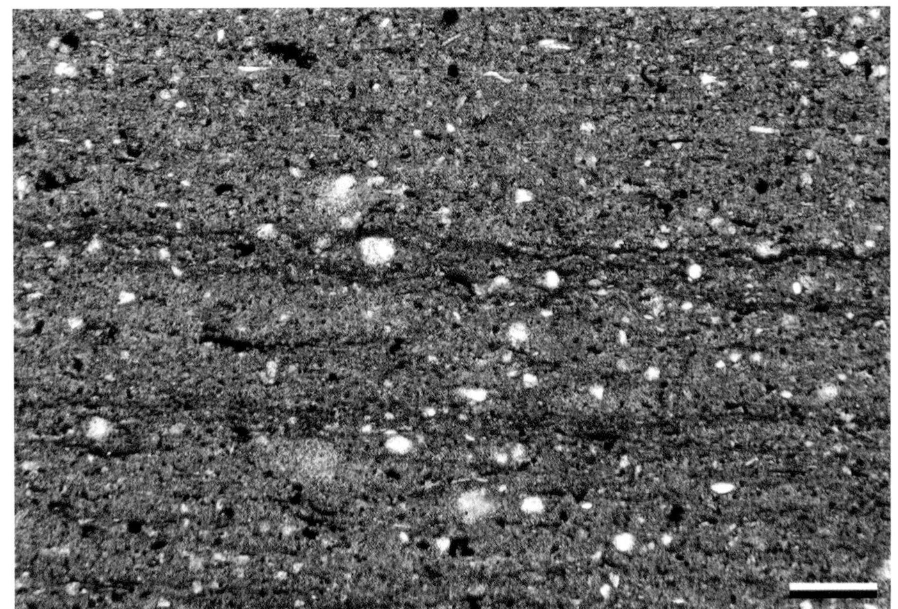

Cretaceous Mowry Fm., Moffat Co., Colorado

Although overburden-driven compaction is the most common cause of stylolite formation, not all stylolites are related to overburden loading. Where rocks are subjected to lateral compressive stresses, stylolites and solution seams that are generally not bedding-parallel can form. Here, an irregular, sutured solution seam or stylolite has formed with an orientation essentially perpendicular to horizontal bedding (marked by compaction-accentuated grain orientation and slight lamination).

PPL, RDI, Scale bar = 0.51 mm

Lo. Ordovician Phycodes Gp., Bavaria, Germany

In this immature arenite, the diagonal to vertical zone in the center of the image is a quartz-cemented sand-filled burrow that has been deformed by compaction. Subsequently, the burrow and adjacent areas were cut by two solution seams (marked with yellow arrows). The seams are at least as well developed where they transect the nearly pure quartz burrow fills as they are in adjacent, more clay-rich areas.

PPL, Scale bar = 0.51 mm

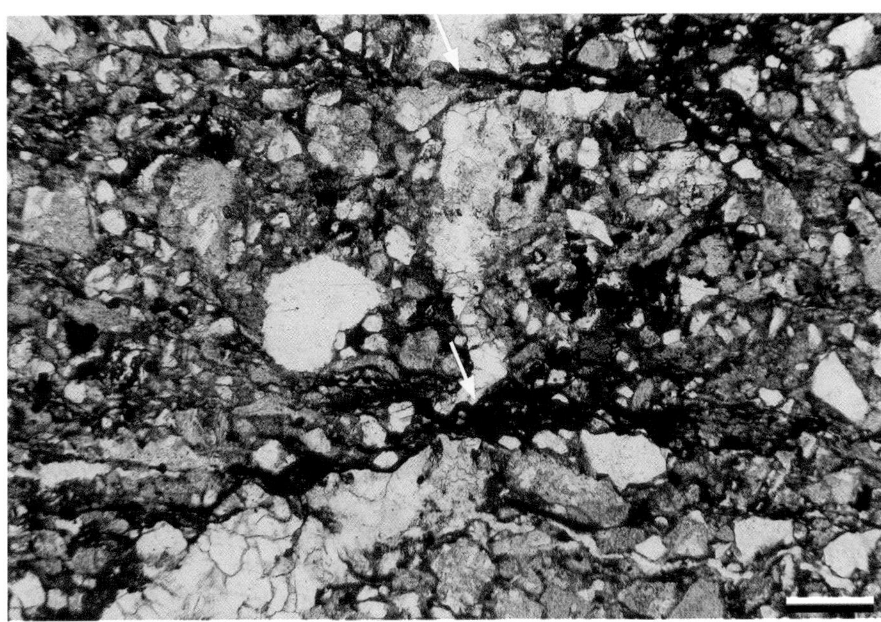

Lo. Pennsylvanian Morrow B Sandstone, Ochiltree Co., Texas

In coarser-grained rocks, the solution seams typical of shales and silty shales sometimes are manifest as thicker seams and/or higher amplitude stylolites that are most commonly oriented parallel to bedding but also can occur in other orientations depending on the direction of maximum compressive stress. This is an example of a bedding-parallel solution seam with a significant concentration of insoluble materials that both aid in seam formation and accumulate there due to localized dissolution. The lack of well-defined "teeth" make this a solution seam rather than a stylolite, although it could be called by either term.

PPL, RDI, Scale bar = 0.51 mm

Lo. Pennsylvanian Morrow B Sandstone, Ochiltree Co., Texas

Pressure solution, whether from overburden loading or tectonic stresses, can be manifest in the form of circumgranular dissolution effects—essentially stylolite swarms that occur between all grains across a broad zone. In this example, most of the detrital quartz grains have lost at least a third of their original volume to dissolution and show interfingering ("tooth-like") contact surfaces with intervening residues of earlier clay cements or matrix that may have been critical to the dissolution process (Bjørkum, 1996).

PPL, RDI, Scale bar = 0.26 mm

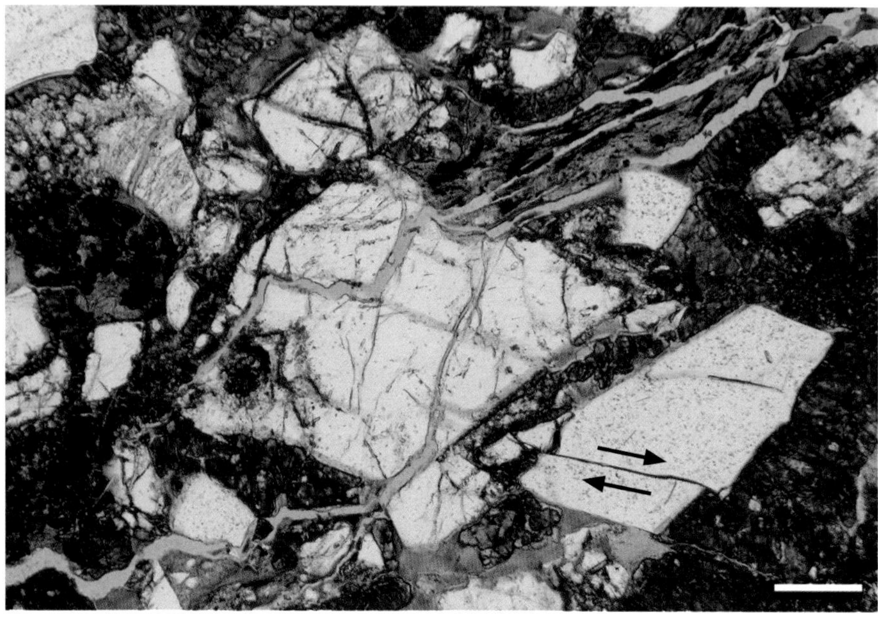

Lo.–Mid. Eocene Cub Mountain Fm., Lincoln Co., New Mexico

Fractures typically are the most significant and most easily recognizable deformation features in sedimentary rocks. These unfilled fractures cut across matrix, cement and grains of differing competency and so, despite being essentially unfilled with cement, are probably natural fractures and not artifacts of sampling. However, it is always difficult to prove unequivocally that unfilled fractures are natural rather than induced. The fractures in this sample that have fillings that are unquestionably diagenetic and the one marked by two black arrows shows grain offset produced by lateral shear. Thus, that fracture is properly termed a microfault.

PPL, AFeS, BDI, Scale bar = 0.26 mm

Lo. Miocene Freshwater Molasse, Switzerland

These circumgranular fractures in a lithic arenite illustrate the difficulty in distinguishing natural versus anthropogenic origins of unfilled fractures. The lack of significant cement in this compacted rock makes circumgranular cracking very likely if and when the rock is deformed. In this example, some of the fractures also cut grains, so a natural origin for those is likely. But similar fabrics can be produced with poor sample handling and even due to poor dye injection practices.

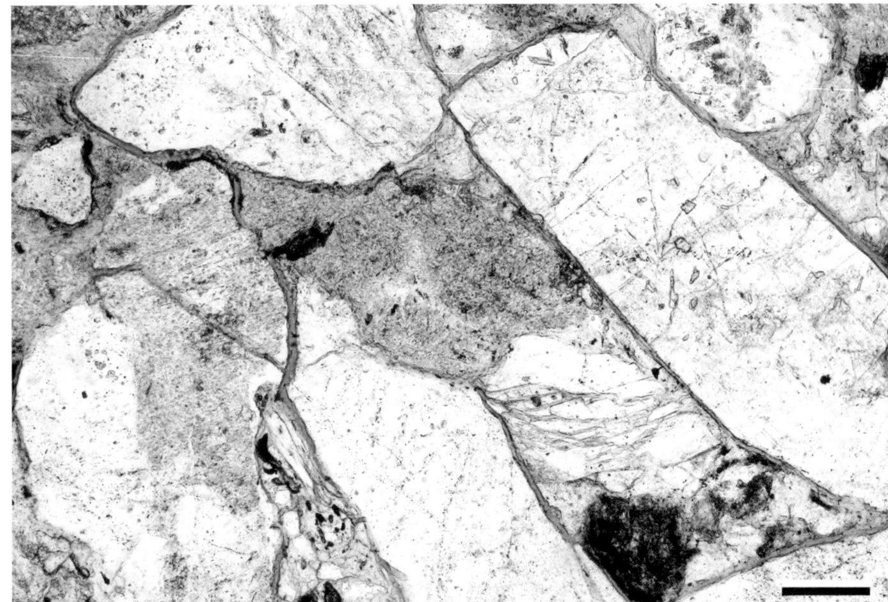

PPL, BDI, Scale bar = 0.16 mm

Unknown Mesozoic unit, Moffat Co., Colorado

These small, essentially unfilled fractures have been impregnated with fluorescent red-dyed epoxy. They are probably natural based on the observation that they are present only in the interior of the sample (making it less likely that they are due to sample handling or desiccation); in addition, fluorescence imaging shows partial filling with liquid hydrocarbons (between the two white arrows). The greenish color marks hydrocarbons that have mingled with the epoxy; the yellow-orange areas are simply fluorescent epoxy.

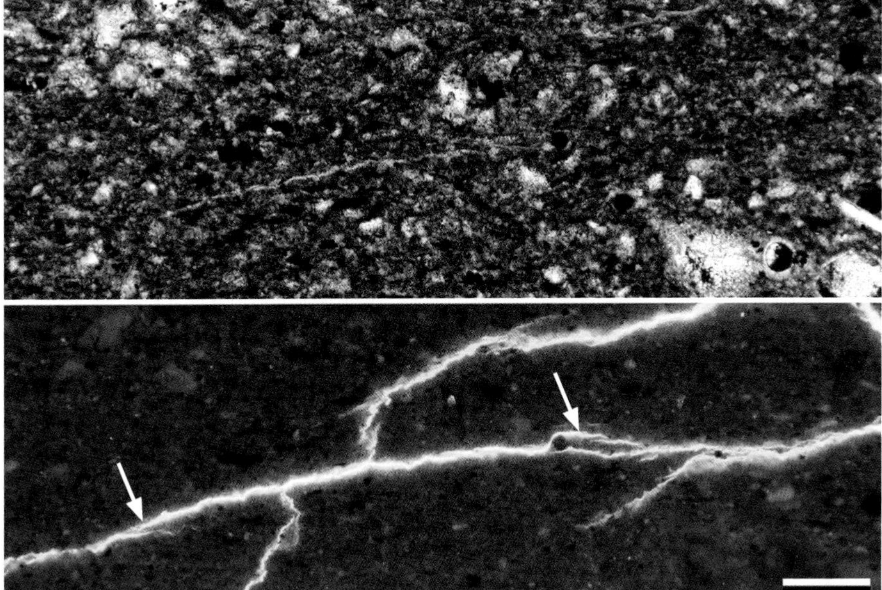

PPL / FL470, RDI, Scale bar = 0.10 mm

Carboniferous Wetzstein Quartzite, Bavaria, Germany

Where (or even if) fractures form in shales and mudstones is very much a function of the rock composition that affects brittleness or ductility (see Figure 8.1). The upper half of this sample is a slightly silty shale; the darker colored, lower half of the sample is a siliceous shale. The extensive, now calcite-filled, vertical fractures that formed in the brittle siliceous shale terminate abruptly upon reaching the more ductile, clay-rich shale layer.

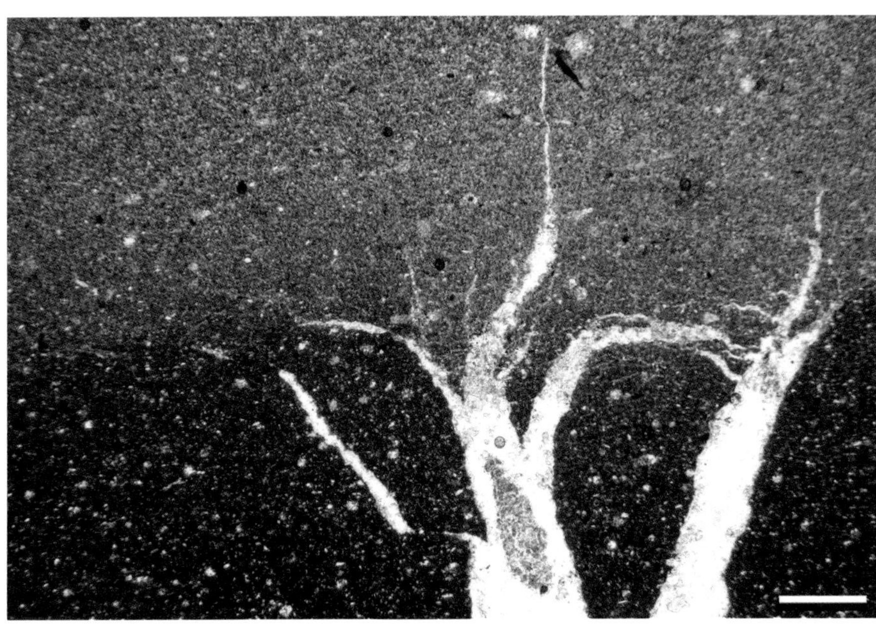

PPL, Scale bar = 0.51 mm

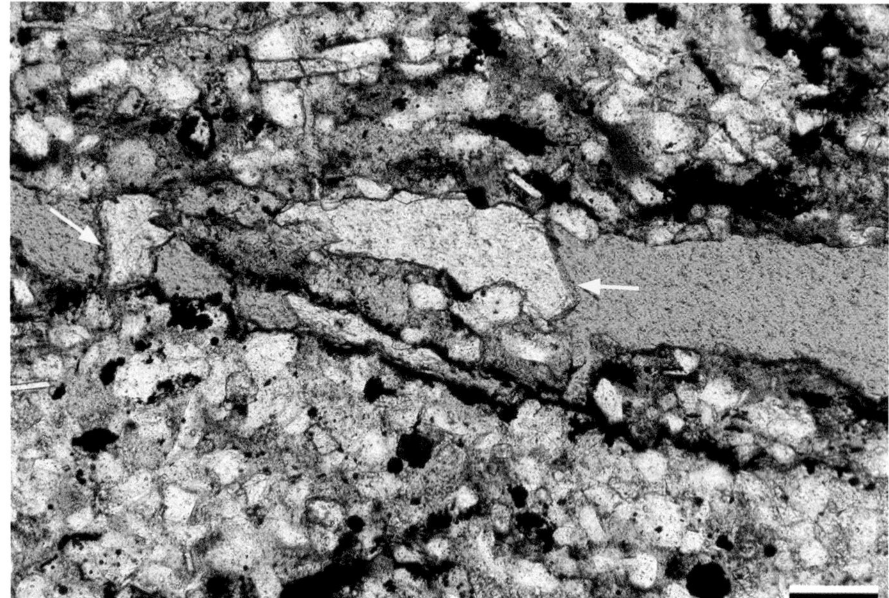

Cretaceous Mowry Fm., Moffat Co., Colorado

As noted in the introduction, the fillings of fractures, or the absence of such fillings, is of critical importance in reservoir performance. In this example, the fracture in the center of the image has had only a few crystals of calcite precipitated in it, but in some cases, isolated single crystals completely span the fracture width (as near the yellow arrows). These crystals may reduce porosity and permeability locally, but they also serve as natural proppants to inhibit closing of the fracture.

PPL, RDI, Scale bar = 0.10 mm

Up. Cretaceous Mesaverde Gp., Piceance Basin, Colorado

This is another example of partial filling of a fracture by calcite crystals. Fractures this wide can have substantial fillings and still be highly conductive for fluids, including hydrocarbons. Here, bladed crystals of calcite, coarser dolomite and some authigenic quartz (e.g., the white crystal near the red arrow) have lined both sides of the fracture. The dark coatings that follow those cements are asphaltic hydrocarbon residues indicating that this fracture continued to act as a fluid conduit during hydrocarbon migration.

PPL, BDI, Scale bar = 0.51 mm

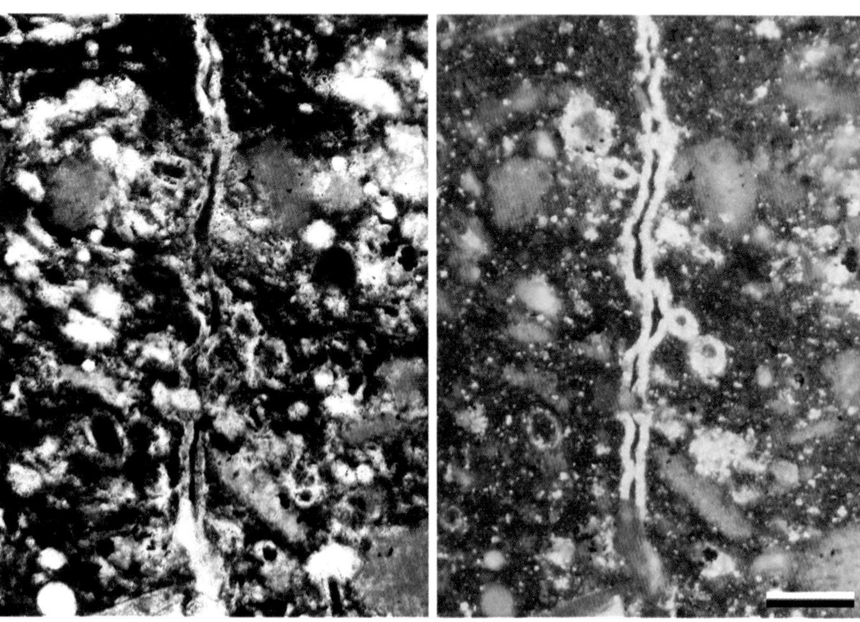

Lo. Permian Bone Spring Fm., Reeves Co., Texas

An example of a partially filled fracture in a shaly siltstone. No fluorescent epoxy was used on this sample so all the visible fluorescence is related to liquid hydrocarbon inclusions trapped in the calcite cements lining both fracture walls (and also visible in matrix pores throughout the rock). The dark, nonfluorescent band in the center of the fracture consists of dead oil (oil that was degraded to asphalt).

PPL | FL470, Scale bar = 0.10 mm

Cretaceous Mowry Fm., Moffat Co., Colorado

Fractures can be partially or fully occluded with material other than cements. This obliquely vertical fracture (note the inclined orientation of bedding in this photo) has a complex history. As with many fractures in fine-grained, relatively poorly consolidated sediment, the initial fracture porosity was probably completely occluded by inflow of "fines"—especially small particles, mainly clays, transported by water inflow from surrounding sediment into the fracture. Later deformation generated new fracture porosity on either side of the earlier fracture fill.

PPL, RDI, Scale bar = 0.10 mm

Paleogene Vieja Gp., Presidio Co., Texas

Completely cement-filled fractures (veins) are the norm in many heavily deformed or deeply buried sandstones. These extensional joints cut large chert and carbonate lithoclasts (SRFs) as well as the adjacent finer-grained sandy matrix and show no sign of shear or lateral offset. The fractures are filled with blocky to bladed calcite crystals that completely span the original fracture width. The fact that the fractures cut grains even in the sandy "matrix" indicates that fracturing took place after substantial matrix cementation.

PPL | XPL, Scale bar = 0.51 mm

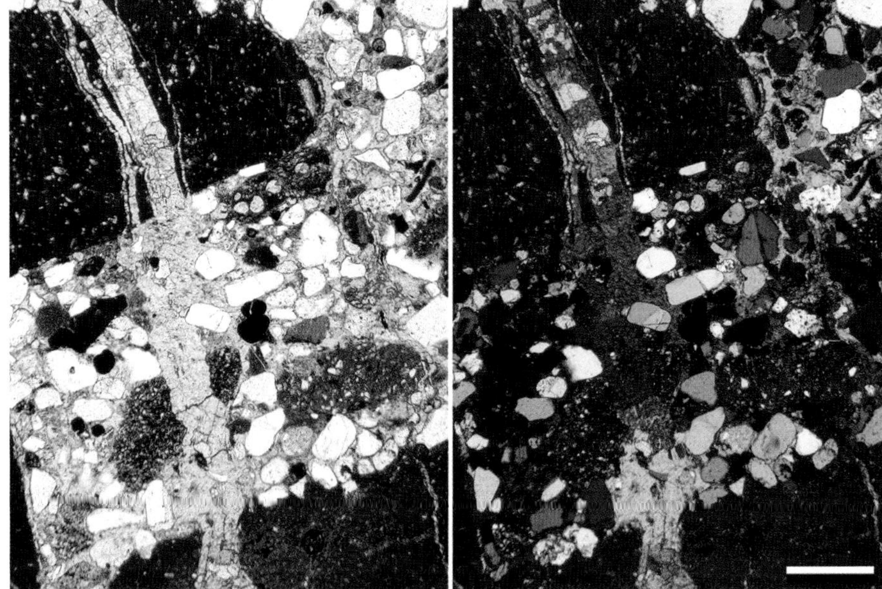

Mid. Jurassic Carmel Fm., Co-op Creek Mbr., Emery Co., Utah

A view of one wall (top) of a several millimeter-wide fracture showing zonation of calcite cements. There is a "stratigraphy" of four cement generations: (1) clear, (2) cloudy scalenohedral, (3) clear, and (4) dark (hydrocarbon inclusion-rich) layers. This zonation can be seen consistently in fractures throughout the sample, and with supplemental fluid inclusion or isotopic (C, O, Sr) studies, one might be able to tie the cement compositional zones to specific markers in the burial/uplift, temperature or hydrocarbon generation/migration histories in this basin.

PPL, RDI, Scale bar = 0.51 mm

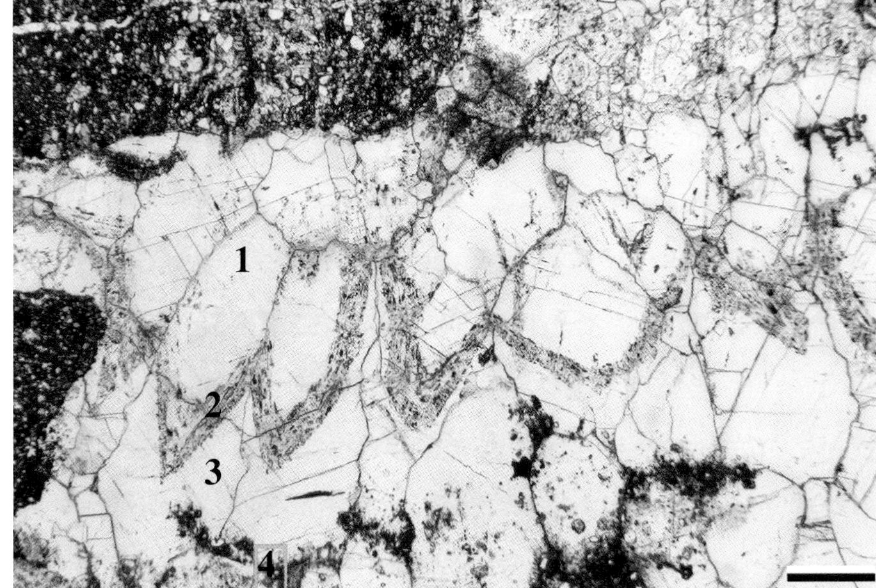

Up. Cretaceous Rio Ancho Fm., La Guajira Department, Colombia

The relative stratigraphies of fractures and fracture fills can be established by cross-cutting relations. The vertical calcite-filled fault here clearly cuts and apparently slightly offsets an earlier calcitic vein with near-horizontal orientation in this very brittle radiolarian chert. The crystallographic similarity of the two calcite vein fills may indicate that there is little temporal difference between them (they may even be part of a conjugate set of related joints). However, there probably is a much greater time difference between these calcitic fractures and earlier quartz-filled fractures that are not well shown here but are illustrated on page 431.

XPL, Scale bar = 0.26 mm

Mid. Permian Wegener Halvø Fm., Karstryggen area, East Greenland

Another example of complex cross-cutting relationships between multiple generations of calcite-filled veins. In this case, the sample was stained to provide some basic geochemical information. The sandstone was cemented by ferroan calcite (purple) in a reducing environment during burial. Grains and cements were cut by a later fracture that is nonferroan, indicating an influx of oxidizing, possibly meteoric waters associated with fracturing. A small, sinuous fracture that cuts grains and the earlier fracture is filled with highly-ferroan calcite, indicating a return to reducing conditions.

PPL, AFeS, BDI, Scale bar = 0.26 mm

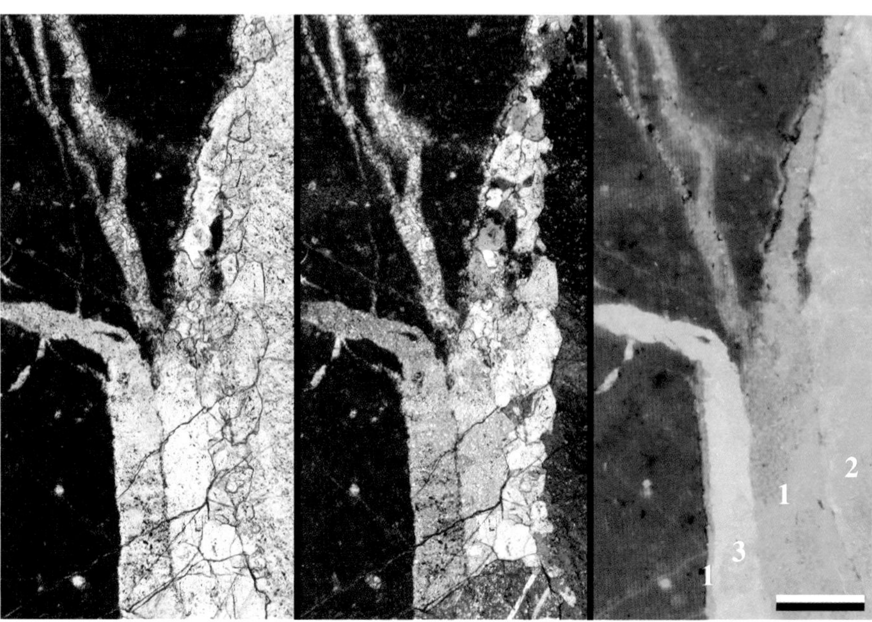

Mid. Permian Cutoff Fm., Culberson Co., Texas

Three views of multi-stage calcitic fracture fills in a basinal, organic-rich calcareous shale. The first stage (#1) of calcite (the clear, branching fracture at the center of this vein) contains few inclusions and no hydrocarbons. A later generation of fracturing and cementation (#2) shows some bands of probable secondary hydrocarbons (incorporated diagenetically along microfractures or twin planes). The final fracture and cementation event (#3) shows slightly brownish calcite with strong yellow fluorescence related to abundant oil inclusions. At the far left there is a thin remnant of the stage-one fracture fill.

PPL | XPL | FL470, Scale bar = 0.26 mm

Mid. Permian Cutoff Fm., Culberson Co., Texas

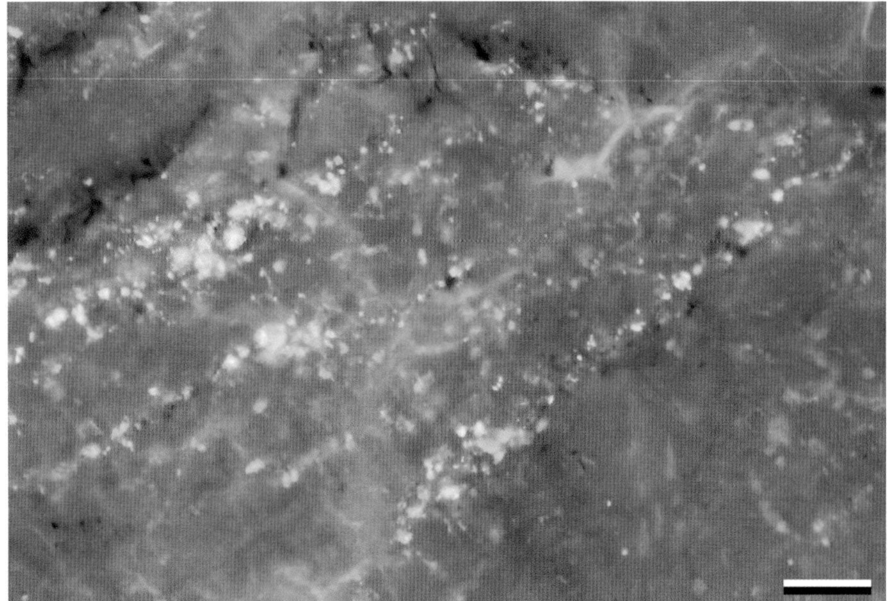

A higher magnification view of the stage 2 cement shown in the previous image. The brightly luminescent secondary fluid-inclusion trains contain live hydrocarbons and cut across individual crystal boundaries. This reflects deformation and hydrocarbon migration that postdated the formation of the stage 2 cements and entrapment of the hydrocarbons along very small microfractures created either during or after stage 3 deformation. Paragenetic sequences such as this can often be directly tied to basin analysis data (see Barker and Halley, 1986).

FL470, Scale bar = 0.05 mm

Paleogene Vieja Gp., Presidio Co., Texas

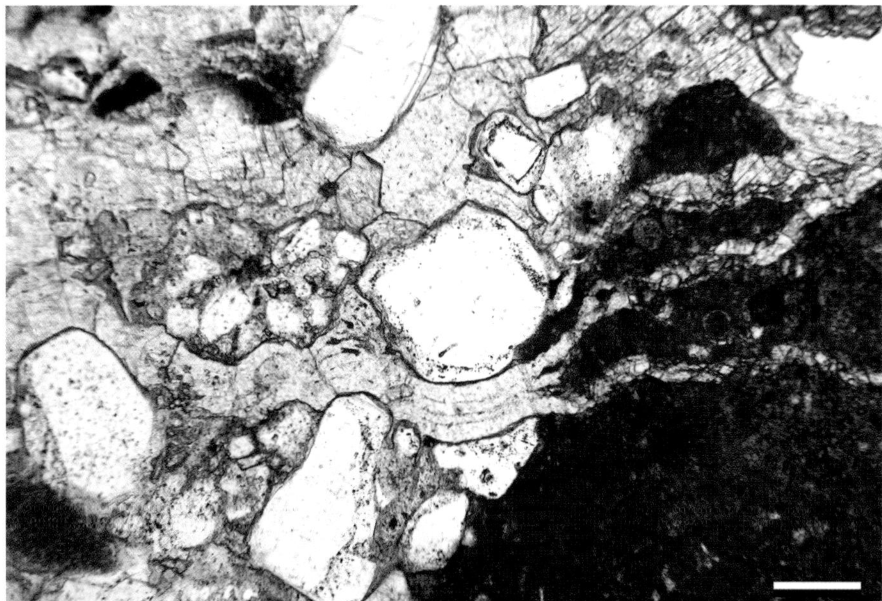

Fracture and vein patterns can be complex, not just because of multiple deformations but also because of differential susceptibilities to fracturing amongst the varied detrital and authigenic components of a terrigenous rock. Here, multiple stages of deformation involving now calcite-filled fractures show how fractures propagated between large quartz grains that were set in a softer matrix of finer grains and clays and were deformed during compaction. Some generations of fractures cut through the softer carbonate lithoclast on the right and others did not.

PPL, Scale bar = 0.09 mm

Cretaceous Mowry Fm., Moffat Co., Colorado

Fracture zones, even when converted into fully cemented veins, can still act as zones of weakness during later deformation. In this example, one can see undulose extinction and small shear planes that reflect post-cementation deformation of this calcite-filled fracture. A small amount of new porosity, marked by areas of red/pink dye impregnation, was created by the second generation of deformation.

XPL, RDI, Scale bar = 0.26 mm

Lo. Cretaceous Palombino Fm., Liguria, Italy

A calcite-filled vein in a siltstone that has undergone further deformation after fracture cementation. The fracture shows an initial stage of left-oblique opening and deformation (white arrows); that was followed by orthogonal opening and further calcite precipitation (red arrows). Shear also led to extensive twinning of the calcite fracture-fill crystals (the thin, parallel dark lines within the crystals).

PPL | XPL, Scale bar = 0.51 mm

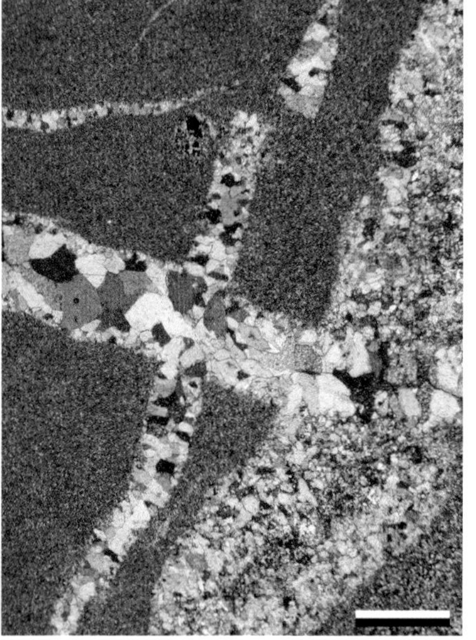

Paleogene Vieja Gp., Presidio Co., Texas

Not all fractures are related to tectonic activity or the influences of elevated pore pressures. In this example from a septarian concretion, the calcite cements are filling shrinkage cracks. Taken out of context, however, the cross-cutting generations of fracture formation and filling could be mistaken for products of compactional, hydraulic or tectonic deformation.

PPL | XPL, Scale bar = 0.51 mm

Eocene? volcaniclastic sandstone, Jackson Co., Oregon

Fractures can be filled by many minerals other than calcite, and quartz is certainly the foremost amongst such cements. These quartz-filled fractures (veins) crosscut glassy VRFs and cherty SRFs along with surrounding sediment. Siliceous grains and silicified matrix are exceptionally brittle, and thus especially susceptible to intense fracturing such as that seen here. In addition, those materials commonly are good sources of silica for quartz cementation through dissolution of remnants of volcanic glass, opal or cristobalite.

XPL, Scale bar = 0.38 mm

Eocene? volcaniclastic sandstone, Jackson Co., Oregon

A higher-magnification view of the multiple quartz-filled veins shown in the center of the previous image but with a slight clockwise rotation of the image. Quartz crystal nucleation took place at just a few preferred locations and the crystals grew into the fracture until they reached the opposite wall or started to interfere with adjacent cement crystals. The result is a mosaic of fairly large, irregularly-shaped crystals that interlock and completely occlude the fracture porosity.

XPL, Scale bar = 0.10 mm

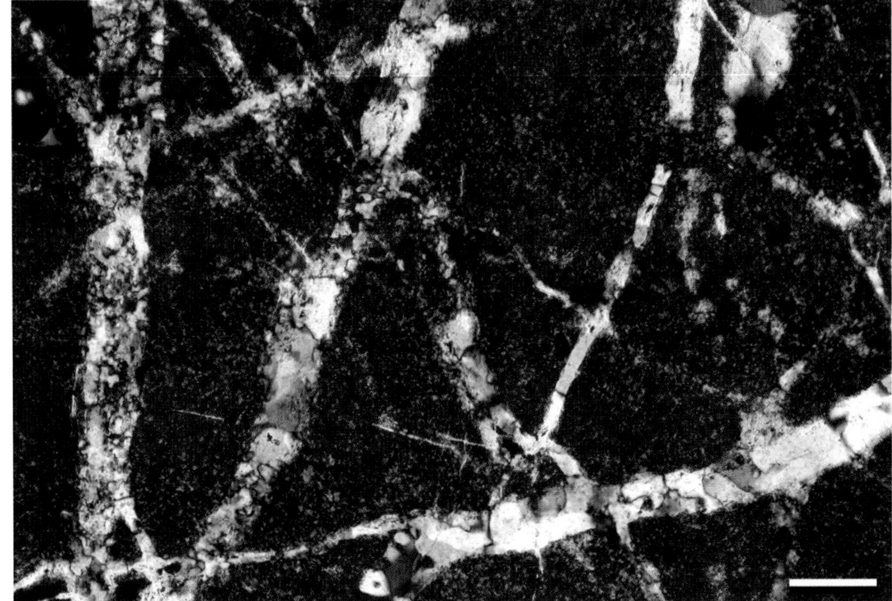

Up. Cretaceous Rio Ancho Fm., La Guajira Department, Colombia

Here multiple generations of cement in a radiolarian chert show different cement fills. Early generations are cemented with bladed quartz, whereas the later fractures are filled with coarsely crystalline calcite. The formation of fractures is common in brittle cherts, and quartz cementation of early generation of fractures is especially common, because of the early and local availability of silica from dissolution of remnants of radiolarian tests and sponge spicules. The later calcitic veins clearly cross-cut the quartz veins, and the calcite is probably derived from later dissolution of carbonate fossils in nearby strata.

XPL, Scale bar = 0.26 mm

Up. Cambrian Conococheague Fm., Giles Co., Virginia

This image shows fracturing in a dolomitic chert. The large, euhedral zoned dolomite crystals within the chert clearly predated fracturing and subsequent cementation by equant megaquartz as shown by the unequivocal cross-cutting relationships. Such temporal relationships can help in understanding the evolution of pore-fluid compositions and other diagenetic parameters through time.

XPL, Scale bar = 0.14 mm

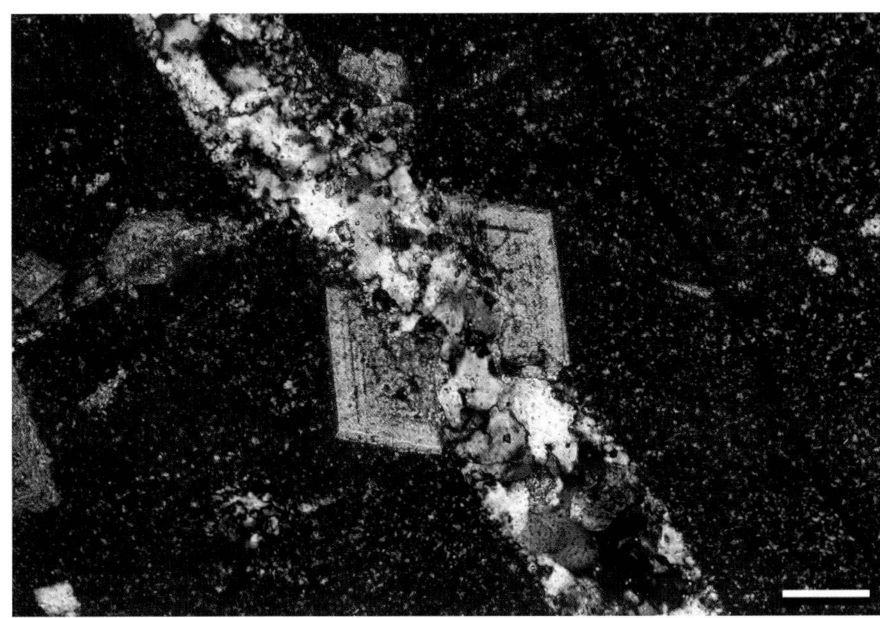

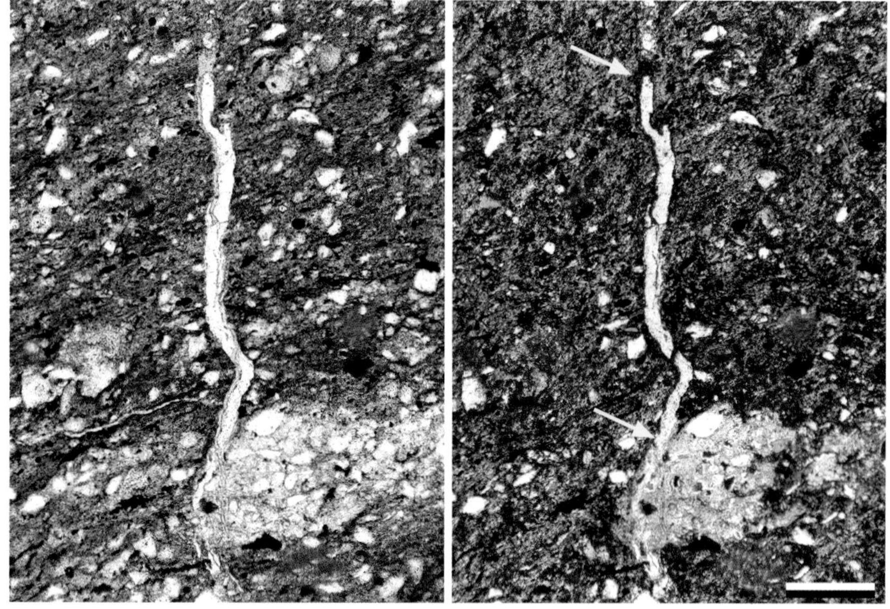

Cretaceous Mowry Fm., Moffat Co., Colorado

This fracture was filled by gypsum in some places (such as the interval between the yellow arrows) and by calcite in other areas. The calcitic cements occur mainly in the vicinity of concentrations of carbonate materials in the matrix of this silty, calcareous shale. Those zones may have supplied solutes for the localized calcite cementation and influx of calcium sulfates apparently allowed later gypsum cementation of the otherwise uncemented stretches of the fracture. Prior to the gypsum cementation, this fracture would still have had substantial porosity.

PPL | XPL, RDI, Scale bar = 0.10 mm

Jurassic Diaspri Radiolarite, Liguria, Italy

A brecciated radiolarian chert from a deep-marine section. These cherts are extremely brittle and fracture readily in an active tectonic setting. Lying directly atop basaltic seafloor, they also are rich in metallic elements, especially manganese and iron. Hence, the hematitic impregnation of the matrix rock as well as ferruginous fillings of the fractures in this shear zone (fault).

PPL, Scale bar = 0.10 mm

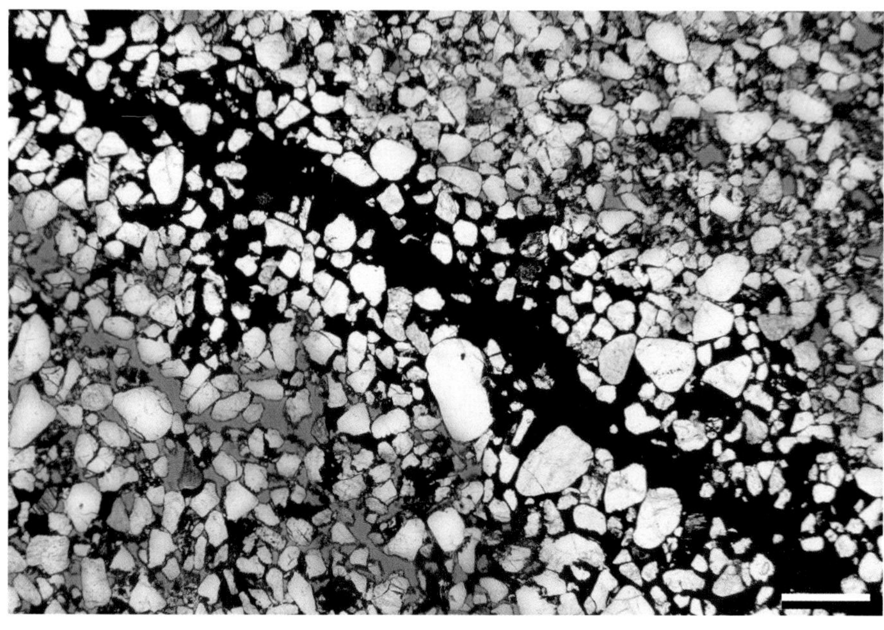

Mid. Jurassic Entrada Fm., Slick Rock Mbr., Emery Co., Utah

This is a hematite/limonite-cemented fracture zone that has substantially reduced porosity in an otherwise porous sandstone. The ferruginous precipitates extend beyond the main fracture into the immediately adjacent rock and have obscured the details of the original structural feature. The cements in this outcrop sample are probably young, telogenetic (uplift-related) precipitates associated with surface-derived meteoric fluids (Chan et al., 2000).

PPL, RDI, Scale bar = 0.51 mm

Mid. Permian Wegener Halvø Fm., Jameson Land, East Greenland

The edge of a large fracture in a shaly siltstone that was cemented by flamboyant crystals of barite and minor saddle dolomite. Fractures commonly act as major conduits for basinal fluid flow during mesogenetic diagenesis, allowing not just cementation of the fractures but also cementation and replacement or mineralization of the host rocks. In this area, those hydrothermal, fracture-associated precipitates include galena and fluorite along with barite, highly ferroan calcite and saddle dolomite.

PPL | XPL, BDI, Scale bar = 0.51 mm

Mid. Jurassic Entrada Fm., Slick Rock Mbr., Emery Co., Utah

A low-porosity deformation band in an otherwise high porosity feldspathic quartzarenite. Note how shearing within the deformation band has shattered grains (the latter especially evident in the XPL view), producing both a much smaller average grain size and much reduced porosity and likely permeability. Although this texture could be mistaken for bedding lamination, these shear zones typically are not oriented parallel to bedding and clearly cross-cut primary sedimentary fabrics.

PPL / XPL, RDI, Scale bar = 0.51 mm

Mid. Jurassic Entrada Fm., Slick Rock Mbr., Emery Co., Utah

A closer view of a deformation band in the same porous eolian sandstone shown above. Note the well-rounded quartz grains and hints of the original bimodal size distribution of grains that is characteristic of many eolian deposits. Within the deformation band, however, these grains have been partially deformed and substantially sheared and fragmented to produce a low-porosity zone that can act as a significant permeability barrier.

PPL | XPL, RDI, Scale bar = 0.30 mm

Mid. Jurassic Entrada Fm., Slick Rock Mbr., Emery Co., Utah

A closer view of a low-porosity, cataclastic deformation band from the same eolian sandstone with well rounded grains shown in the two prior images. In this detailed view, one can see the complete shattering of material along the shear zone and the low porosity of the resultant fill of the shear zone. A few large grains have survived, probably by rotating rather than fracturing.

XPL, RDI, Scale bar = 0.10 mm

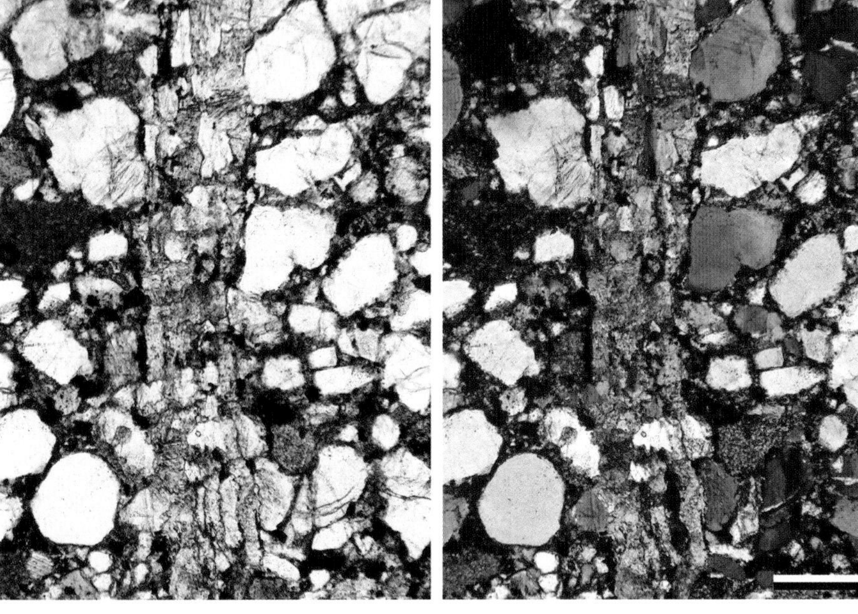

Mid. Ordovician Martinsburg Fm., Berks Co., Pennsylvania

A zone with multiple parallel veins. Note the sheared, rotated and fractured quartz grains scattered through the calcite cements in the veins. Deformation here has developed a strongly preferred grain orientation parallel to the vein walls. However, this dilational deformation zone appears to have formed in already substantially lithified rocks, as opposed to those shown in the three previous illustrations of deformation bands which form only in weakly lithified deposits.

PPL | XPL, Scale bar = 0.26 mm

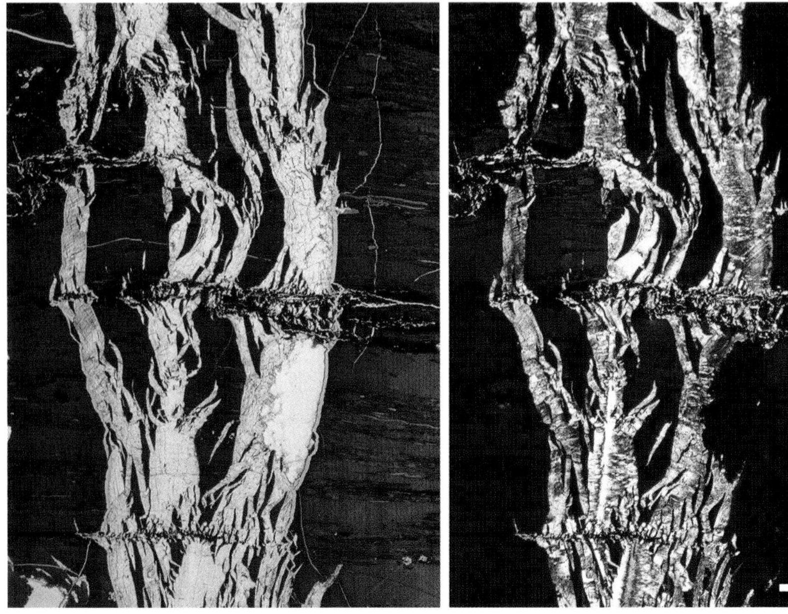

Lo. Pennsylvanian Morrow B Sandstone, Ochiltree Co., Texas

These tendril-like fractures in a thin coal seam within a sandstone unit are termed endogenetic cleats. They typically form perpendicular to bedding, as in this example, in response to physical changes in the coal during maturation. Compaction, dewatering, degassing and shrinkage of organic matter all contribute to their formation and these features can be very important both in coal mining and in coal-bed methane production (Laubach et al., 1998; Rodrigues et al., 2014). In this case, the cleats have been cemented by fibrous calcite with crystal long-axes oriented perpendicular to the cleat walls; calcium sulfate cements also are present locally. Photo from Steven M. Cather.

PPL | XPL, RDI, Scale bar = 0.29 mm

Lo. Cretaceous (Albian) Skull Creek Shale, Niobrara Co., Wyoming

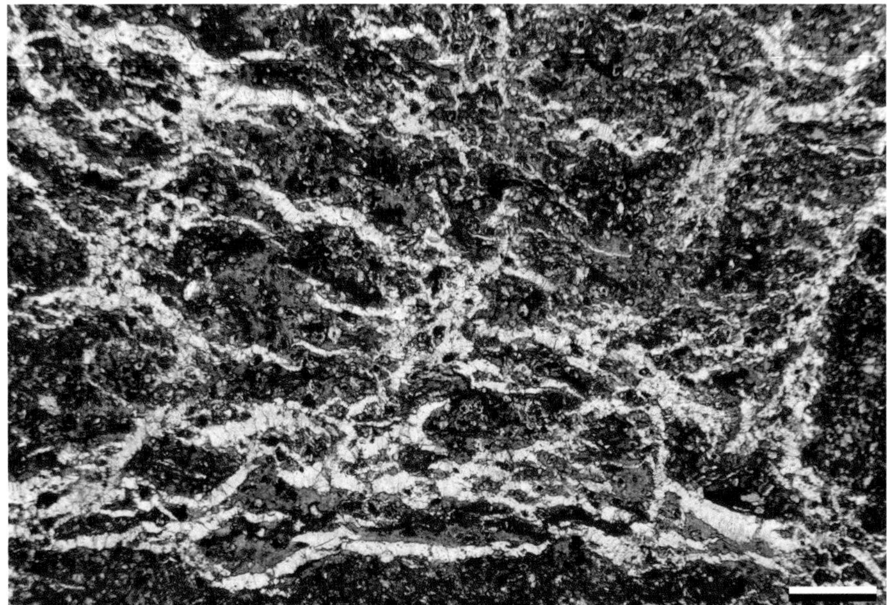

A dense network of small, irregularly anastomosing, but predominantly bedding-parallel veins in a calcareous shale from about 3,100 m (10,200 ft depth). These structures are completely filled with fibrous calcite cements. Such structures have been termed "horizontal hydraulic fractures" and have been related to sediment dilation associated with elevated pore-fluid pressures and associated seepage forces (e.g., Cobbold and Rodrigues, 2007). Overpressures can be associated with hydrocarbon generation and hydrocarbon inclusions are present in some hydraulic fracture-filling cements.

PPL, BDI, Scale bar = 0.51 mm

Lo. Cretaceous (Albian) Skull Creek Shale, Niobrara Co., Wyoming

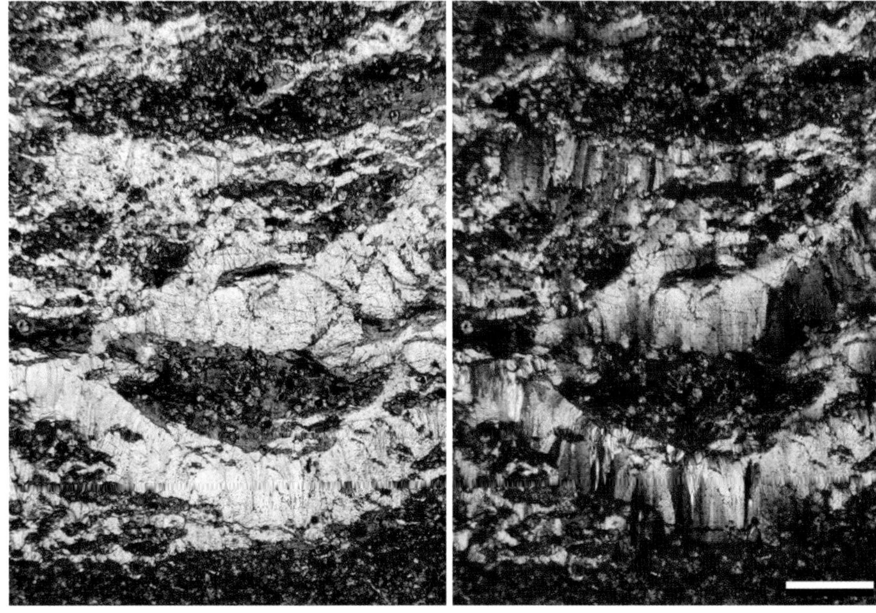

A mix of thick and thin bedding-parallel "hydraulic fractures" from the same sample shown in the previous photo. Here the bedding-perpendicular structure of the fibrous calcite cements is clearly visible, especially in cross-polarized light. Although calcite is the most common filling of such structures, gypsum is another common cementing mineral (distinguished by its much lower birefringence). The calcite cements here have high (third order) birefringence which yields very pale colors.

PPL | XPL, BDI, Scale bar = 0.26 mm

Mid. Jurassic Carmel Fm., Co-op Creek Mbr., Emery Co., Utah

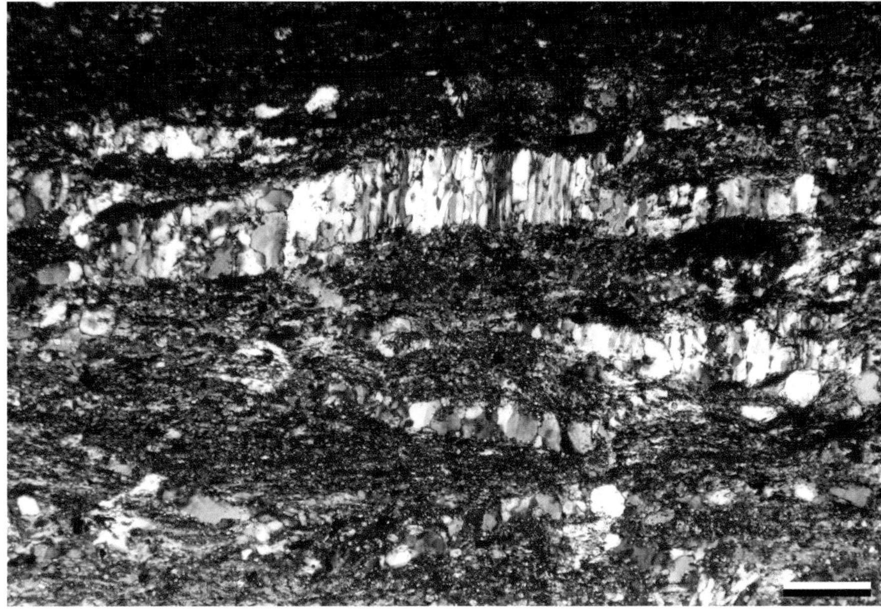

Multiple discontinuous and anastomosing, but predominantly bedding-parallel, veins in a silty mudstone. In this example from a section laden with primary or early diagenetic evaporites, the veins are completely filled with fibrous to bladed gypsum cement. Studies on similar but thicker fibrous veins ("beef"), gypsum has been shown to be the predominant cement at temperatures up to 60°C, calcite is the main cement at intermediate temperatures (70-120°C) and quartz predominates at temperatures above 200°C (Cobbold et al., 2013).

XPL, Scale bar = 0.28 mm

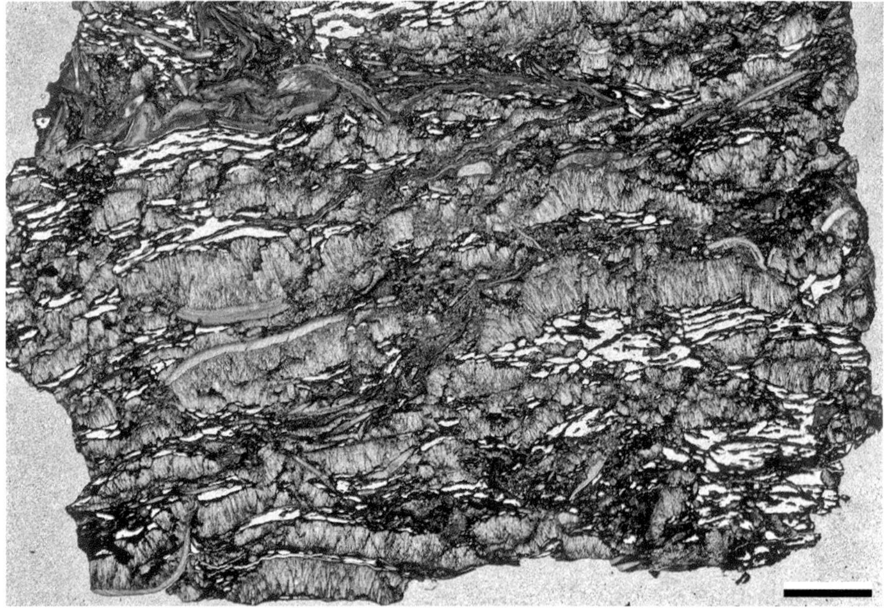

Up. Mississippian Heath Fm., Fergus Co., Montana

This low-magnification image of an entire thin section shows thicker bedding-parallel veins ("beef") in a shaly limestone. The vein-filling cements are fibrous calcite and some were nucleated on brachiopod shells. The calcite crystal fracture fills are oriented predominantly perpendicular to the vein margins, although in some places they grade into cone-in-cone structure with more complex crystal orientations (see next three images). Photo from Julie Dumoulin.

PPL, Scale bar = 2.4 mm

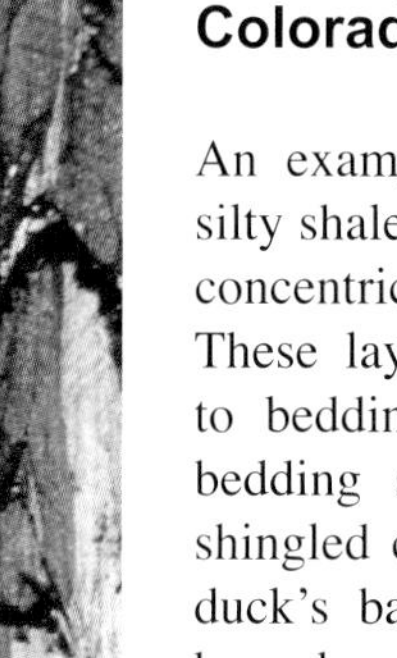

Cretaceous Mowry Fm., Moffat Co., Colorado

An example of cone-in-cone structure in a silty shale. These features consist of layers of concentrically nested cones of calcite crystals. These layers of cones are typically parallel to bedding and when cut perpendicular to bedding show the texture seen here, with shingled crystals that look like feathers on a duck's back. Although many origins have been hypothesized for these features, recent work has linked them to fluid overpressure and hydrocarbon generation, especially where they are associated with bedding-parallel fibrous veins (Rodrigues et al., 2009; Cobbold et al., 2013).

PPL | XPL, RDI, Scale bar = 0.51 mm

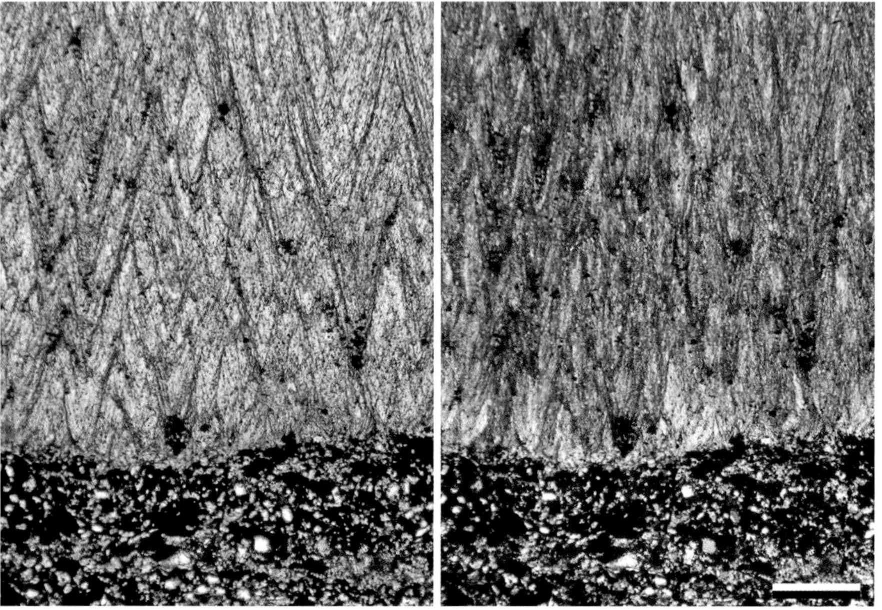

Eocene Scotland Fm., St. Joseph Parish, Barbados

Another example of bedding-parallel cone-in-cone in a shaly siltstone. This example is from a turbidite unit in a strongly deformed, submarine fan deposit from a deep-marine trench. This is a unit that has been described as being associated with hydrocarbon-bearing mud diapirs and other evidence of elevated pore-fluid pressures (Pudsey and Reading, 1982) and so certainly fits the hypothesis that these structures are associated with overpressured sections (see previous caption).

PPL | XPL, Scale bar = 0.26 mm

Eocene Scotland Fm., St. Joseph Parish, Barbados

Although most cone-in-cone features are fairly large (on the order of centimeters in thickness), they also can form as thinner veins (in some cases in direct association with fibrous veins). This example shows a roughly 1-mm-thick cone-in-cone seam from the same turbidite siltstone seen in the previous image.

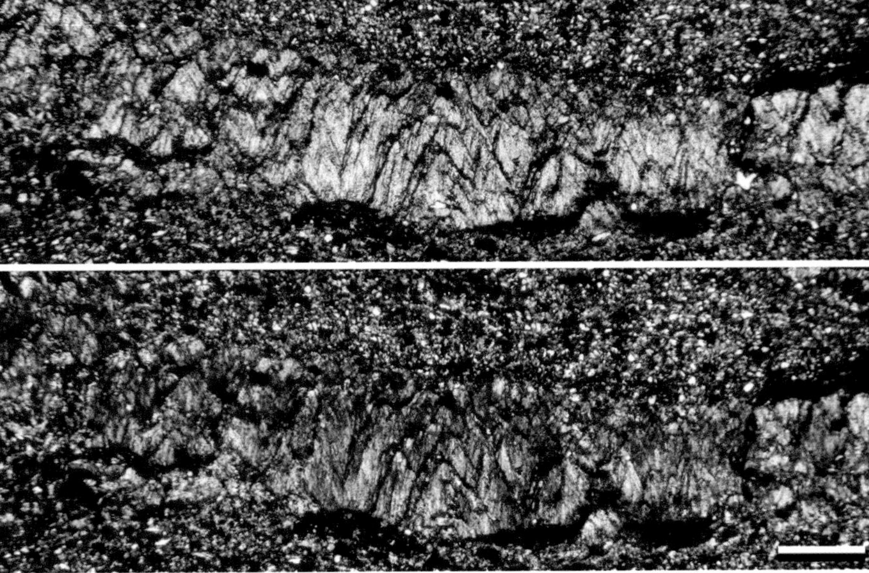

PPL / XPL, BDI, Scale bar = 0.51 mm

Lo. Permian Hueco Fm., Powwow Conglomerate, Hudspeth Co., Texas

Fractures commonly are localized in the more brittle grains in a sandstone, and do not extend through adjacent grains that are less brittle. This is the case here in the compactional fracturing of a chert grain. The thin end of the chert rock fragment shattered where it was in contact with an adjacent detrital quartz grain. The fractures later were filled with ferroan calcite, after the surrounding porosity was partially cemented with dolomite (the large, unstained rhombic crystals especially well seen at upper left).

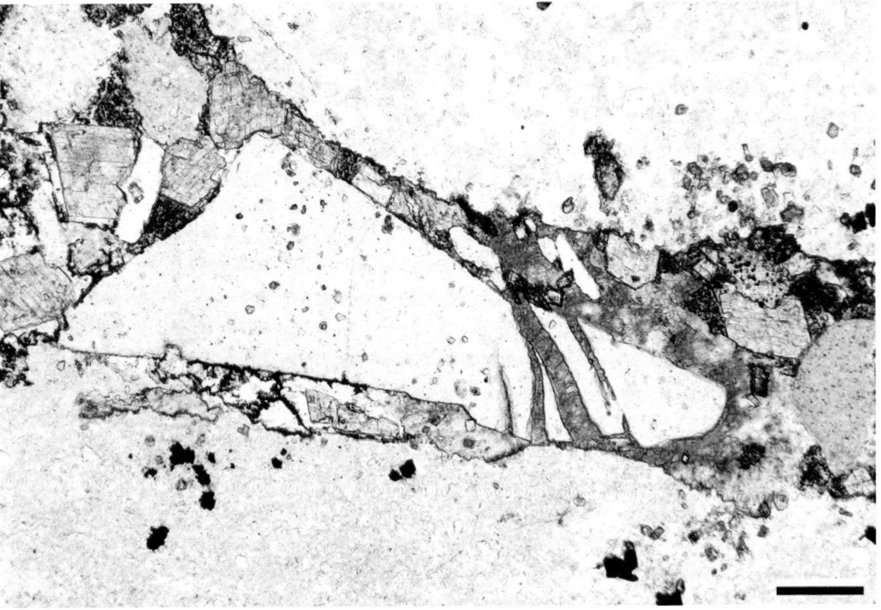

PPL, AFeS, BDI, Scale bar = 0.40 mm

Lo. Permian Hueco Fm., Powwow Conglomerate, Hudspeth Co., Texas

With grain-scale fracturing, one must carefully distinguish between *in-situ* fractures and inherited fractures and fills. In this case, the chert SRF at center has a chalcedony-filled fracture that was formed in the source terrane and was part of the detrital grain (as opposed to the previous example, taken from the same sample, where fracturing and cementation took place in the sandstone). Here, none of the cements that affect the rock as a whole (non-ferroan and ferroan calcites and dolomite) is found in the grain fracture.

XPL, AFeS, BDI, Scale bar = 0.26 mm

Mid. Jurassic Carmel Fm., Crystal Creek Mbr., Cane Co., Utah

An example of pervasive grain-scale fracturing not associated with deformation bands. More than 80% of the grains in this rock are riddled with fractures that have a predominantly bedding-perpendicular trend (although many radiate out from grain contacts reflecting direct stress points). Because normal rates of overburden loading do not usually produce such intense fracturing, this fabric is interpreted to have resulted from a rapid increase in intergranular stress related to abrupt dissolution of evaporite cements (gypsum/anhydrite) in the subsurface (see additional discussion and photos in Chapter 19).

PPL, BDI, Scale bar = 0.26 mm

Mid. Jurassic Carmel Fm., Crystal Creek Mbr., Cane Co., Utah

A more detailed view of the grain-to-grain contact fracturing from the same thin section shown in the previous photo. This formation includes both bedded evaporites and other sandstone and siltstone layers that still have gypsum cements. So the interpretation that the intense fracturing seen here is the result of subsurface evaporite cement removal appears plausible and indeed the kind of fracturing seen here is one of the criteria proposed by Schmidt and McDonald (1979) for the recognition of dissolution and secondary porosity development. Rapid loss of overpressuring is another possible cause of such fracturing, but that is less likely in this case.

PPL, BDI, Scale bar = 0.10 mm

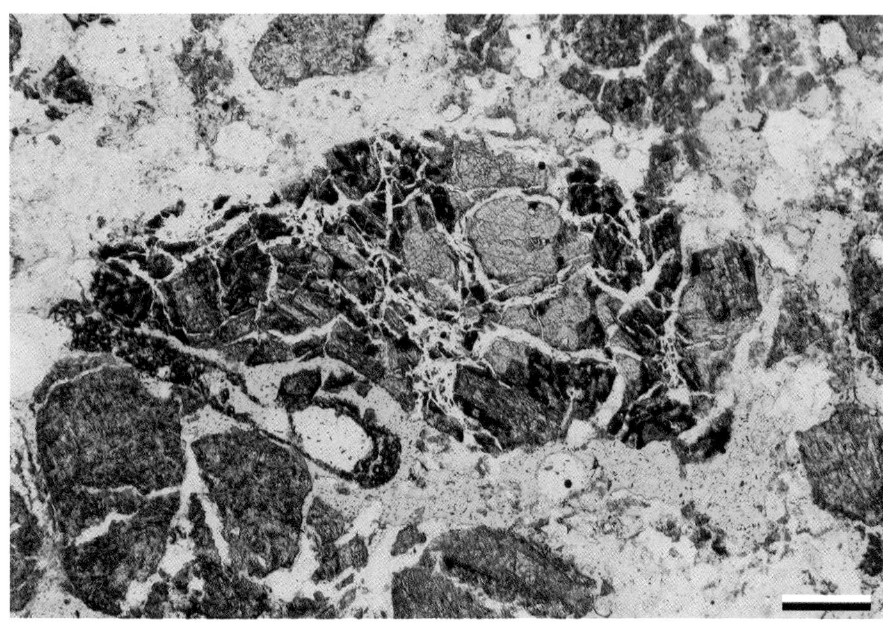

Recent hot-spring tufa, Park Co., Wyoming

Intense intragranular fracturing also can occur in near-surface settings that have had virtually no overburden loading. These shattered VRF grains are from a sulfur-cemented hot-spring tufa. Presumably a combination of rapid thermal stressing and pore-fluid pressure variations contribute to the fracturing, which is common in such deposits. Here, the fractures have been healed by a variety of cements (silica, calcium sulfate and even, in other parts of the section, native sulfur).

PPL, BDI, Scale bar = 0.26 mm

Up. Cambrian Riley Fm., Hickory Sandstone Mbr., Llano Co., Texas

An example of brittle, pervasive grain-scale deformation with microfractures radiating out from point contacts between adjacent grains. There has been sufficient shear and dispersal of crystal fragments to effectively blur many grain boundaries. Such intense deformation, generally is localized in shear bands or fault zones, and it can greatly complicate determination of original fabric and texture, including even such basic properties as grain size and sorting.

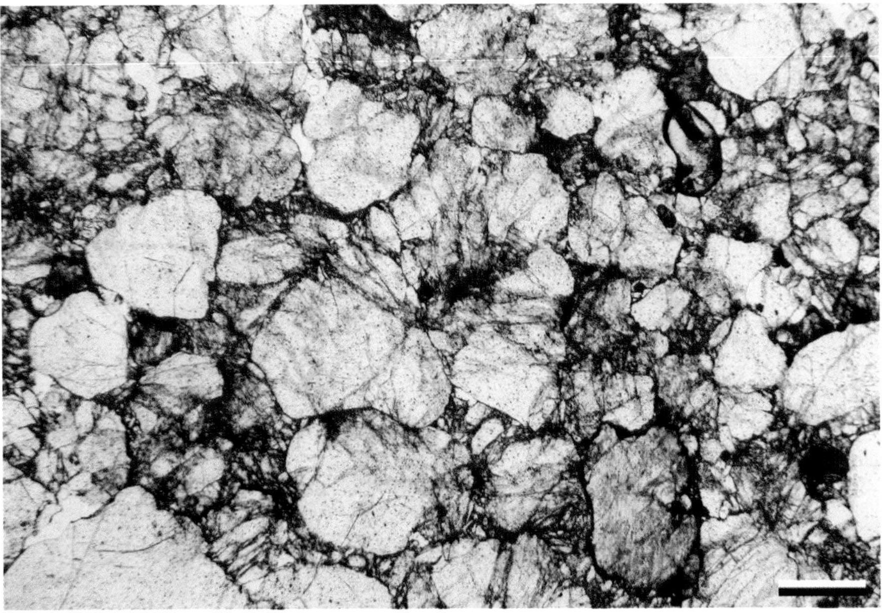

PPL, Scale bar = 0.26 mm

Up. Cambrian Riley Fm., Hickory Sandstone Mbr., Llano Co., Texas

Another view of the fracturing of quartz grains shown in the previous image. Here the consistent fracture patterns can be seen extending through adjacent grains. This was originally a bimodal sandstone comprising larger, well-rounded quartz grains and smaller, more angular grains (also mainly quartz). Despite the intense brittle deformation, that general fabric is still recognizable, at least in this area of the rock.

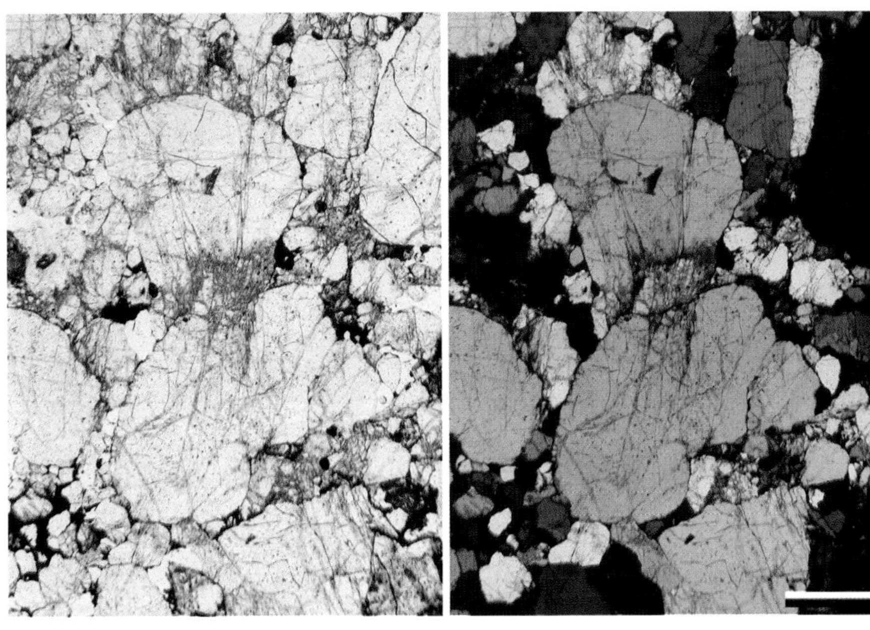

PPL | XPL, Scale bar = 0.32 mm

Silurian Tuscarora Sandstone, Lebanon Co., Pennsylvania

These quartz grains show deformation lamellae (the dark sub-parallel planes) that mark traces of microshearing within the crystals. The lamellae are visible because they are marked by lines of tiny vacuoles (Christie and Ardelle, 1974). Deformation lamellae in individual detrital quartz grains can be inherited from their source terrane; however, when most of the quartz grains show such features they most likely reflect *in-situ* tectonic deformation of the sandstone. Such tectonic deformation lamellae must be distinguished from planar deformation features produced by bolide impacts (e.g., Hamers and Drury, 2011).

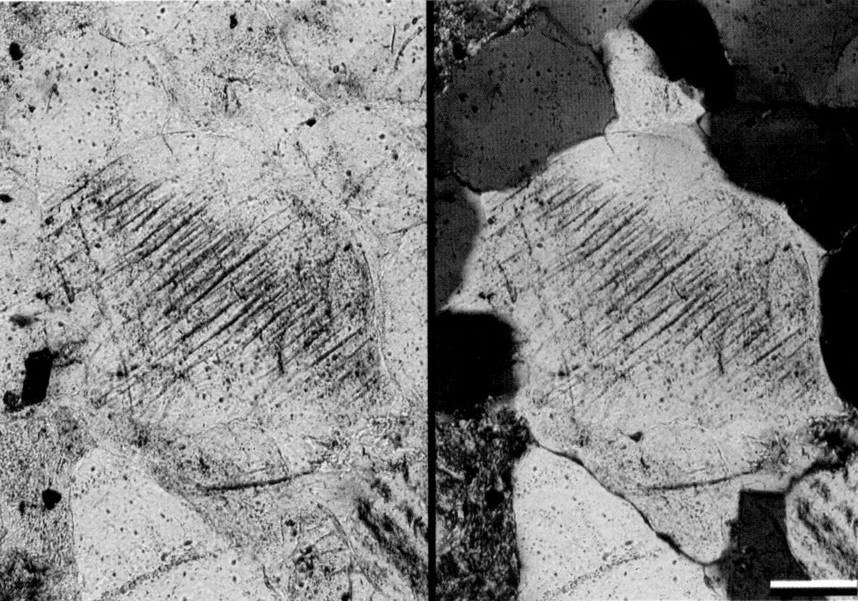

PPL | XPL, Scale bar = 0.10 mm

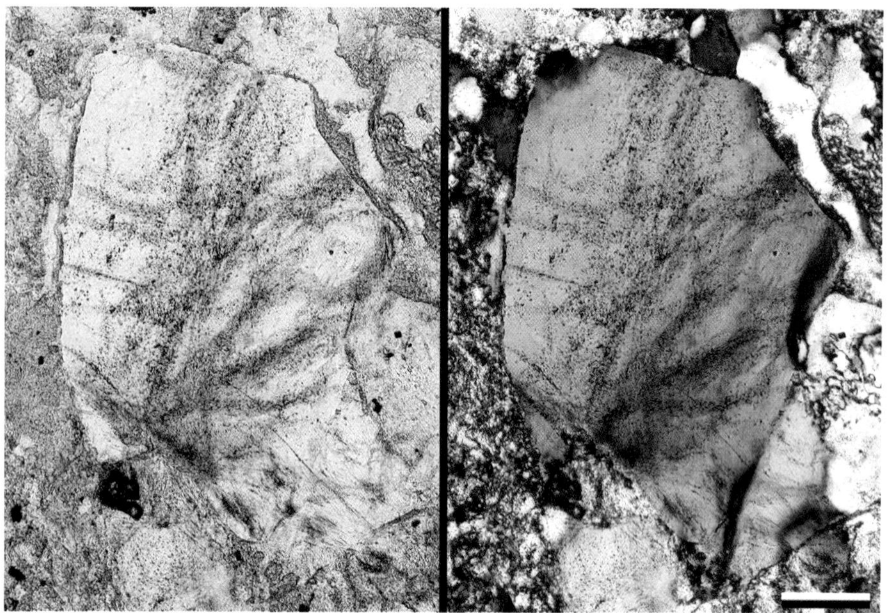

Silurian Tuscarora Sandstone, Lebanon Co., Pennsylvania

Another example of deformation lamellae in quartz grains. These generally are termed Boehm lamellae (and the two terms are often incorrectly used interchangeably). However, Boehm lamellae are defined primarily by broader bands of dusty bubble inclusions in quartz that may be relicts of deformation lamellae (see Fairbairn, 1941). Boehm lamellae in detrital grains can be derived from their source terranes or can be formed *in situ*. In this example, the lamellae probably were formed *in situ* because they appear in the majority of the quartz grains in the rock.

PPL | XPL, Scale bar = 0.09 mm

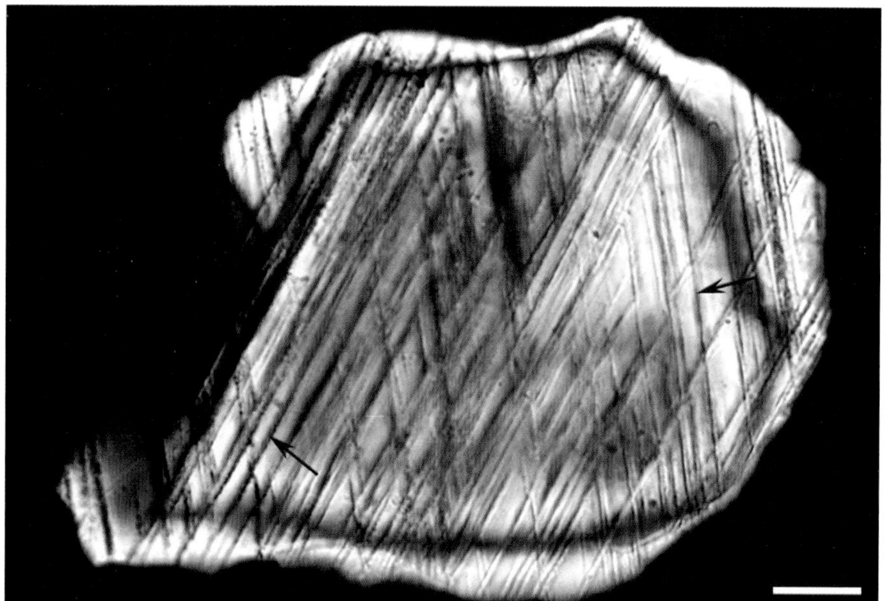

Cretaceous-Paleogene boundary clay, top Hell Creek Fm., Mud Buttes, Bowman Co., North Dakota

An example of impact-shocked quartz. This whole (not thin-sectioned) detrital quartz grain shows two sets of planar, shock lamellae in this orientation (black arrows point to examples of each set; Izett, unpublished data). The grain is from a K/Pg boundary clay layer, and it reflects atmospheric fallout of hypervelocity bolide impact debris from Chicxulub or one of the other boundary impact events. Shocked quartz commonly is used as a definitive marker of such events (Morgan et al., 2006). Photo taken using oil-immersion cell of a Wilcox spindle stage. Photo and caption data from Glen Izett.

XPL, Scale bar = 0.03 mm

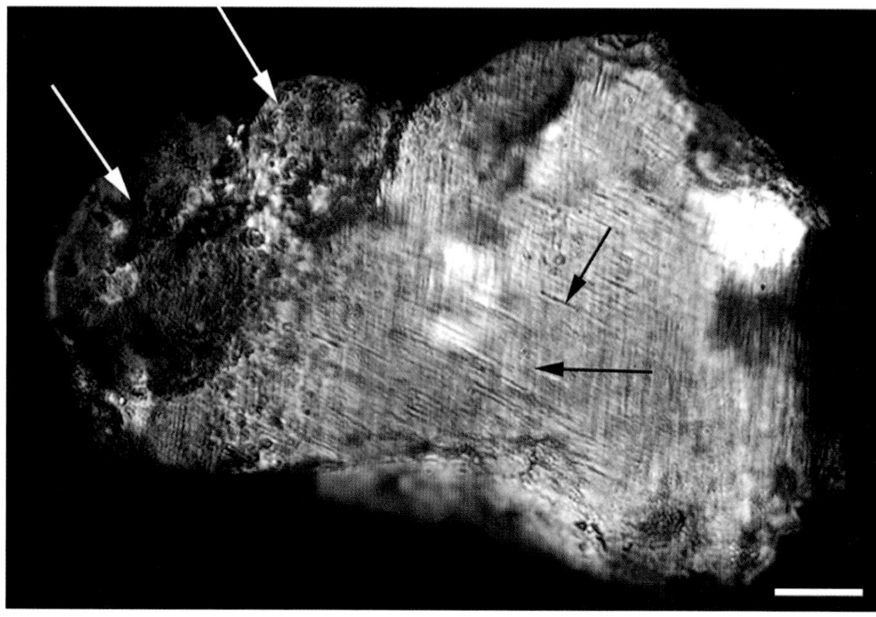

Cretaceous-Paleogene (K/T boundary) upper claystone bed, Las Animas Co., Colorado

A detrital composite quartz-grain mount (not a thin section) showing shocked quartz with two visible sets of shock lamellae (out of five sets present) marked by tips of black arrows. Two spherical areas (white arrows) consist of probable diaplectic or molten glass later devitrified to some form of SiO_2 that, along with the lamellae, indicate that this grain was subjected to hypervelocity shock waves (>30 GPa), partially melted and ejected from the Chicxulub impact crater (Izett, 1990). Photo taken as noted for previous image and provided by Glen Izett.

XPL, Scale bar = 0.027 mm

Cited References and Additional Information Sources

Aassoumi, H., J. Broutin, M. El Wartiti, P. Freytet, J.-C. Koeniguer, C. Quesada, F. Simancas, and N. Toutin-Morin, 1992, Pedological nodules with cone in cone structure in the Permian of Sierra Morena (Spain) and central Morocco: Carbonates and Evaporites, v. 7, p. 140-149, doi: 10.1007/BF03175628.

Aguilera, R., 1995, Naturally Fractured Reservoirs (2nd Edition): Tulsa, OK, PennWell Books, 521 p.

Alexopoulos, J. S., R. A. F. Grieve, and P. B. Robertson, 1988, Microscopic lamellar deformation features in quartz: Discriminative characteristics of shock-generated varieties: Geology, v. 16, p. 796-799, doi: 10.1130/0091-7613(1988)016<0796:MLDFIQ>2.3.CO;2.

Aydin, A., 1978, Small faults formed as deformation bands in sandstone: Pure and Applied Geophysics, v. 116, p. 913-930, doi: 10.1007/BF00876546.

Barker, C. E., and R. B. Halley, 1986, Fluid inclusion, stable isotope, and vitrinite reflectance evidence for the thermal history of the Bone Spring Limestone, southern Guadalupe Mountains, Texas, *in* D. L. Gautier, ed., Roles of Organic Matter in Sediment Diagenesis: Tulsa, OK, SEPM Special Publication 38, p. 189-203, doi: 10.2110/pec.86.38.0189.

Bayly, B., 1986, A mechanism for development of stylolites: Journal of Geology, v. 94, p. 431-435, doi: 10.1086/629041.

Bjørkum, P. A., 1996, How important is pressure in causing dissolution of quartz in sandstones?: Journal of Sedimentary Research, v. 66, p. 147-154, doi: 10.1306/D42682DE-2B26-11D7-8648000102C1865D.

Bjørkum, P. A., E. H. Oelkers, P. H. Nadeau, O. Walderhaug, and W. M. Murphy, 1998, Porosity prediction in quartzose sandstones as a function of time, temperature, depth, stylolite frequency, and hydrocarbon saturation: AAPG Bulletin, v. 82, p. 637-648.

Burrus, J., 1998, Overpressure models for clastic rocks, their relation to hydrocarbon expulsion: A critical reevaluation, *in* B. E. Law, G. F. Ulmishek, and V. I. Slavin, eds., Abnormal Pressures in Hydrocarbon Environments: Tulsa, OK, AAPG Memoir 70, p. 35-63.

Cartwright, J. A., 1994, Episodic basin-wide hydrofracturing of overpressured Early Cenozoic mudrock sequences in the North Sea Basin: Marine and Petroleum Geology, v. 11, p. 587-607, doi: 10.1016/0264-8172(94)90070-1.

Chan, M. A., W. T. Parry, and J. R. Bowman, 2000, Diagenetic hematite and manganese oxides and fault-related fluid flow in Jurassic sandstones, southeastern Utah: AAPG Bulletin, v. 84, p. 1281-1310, doi: 10.1306/A9673E82-1738-11D7-8645000102C1865D.

Christie, J. M., and A. J. Ardell, 1974, Substructures of deformation lamellae in quartz: Geology, v. 2, p. 405-408, doi: 10.1130/0091-7613(1974)2<405:SODLIQ>2.0.CO;2.

Christie, J. M., and C. B. Raleigh, 1959, The origin of deformation lamellae in quartz: American Journal of Science, v. 257, p. 385-407, doi: 10.2475/ajs.257.6.385.

Cobbold, P. R., and N. Rodrigues, 2007, Seepage forces, important factors in the formation of horizontal hydraulic fractures and bedding-parallel fibrous veins ('beef' and 'cone-in-cone'): Geofluids, v. 7, p. 313-322, doi: 10.1111/j.1468-8123.2007.00183.x.

Cobbold, P. R., A. Zanella, N. Rodrigues, and H. Løseth, 2013, Bedding-parallel fibrous veins (beef and cone-in-cone): Worldwide occurrence and possible significance in terms of fluid overpressure, hydrocarbon generation and mineralization: Marine and Petroleum Geology, v. 43, p. 1-20, doi: 10.1016/j.marpetgeo.2013.01.010.

Conybeare, D. M., and H. F. Shaw, 2000, Fracturing, overpressure release and carbonate cementation in the Everest Complex, North Sea: Clay Minerals, v. 35, p. 135-149, doi: 10.1180/000985500546530.

Cosgrove, J. W., 2001, Hydraulic fracturing during the formation and deformation of a basin: A factor in the dewatering of low-permeability sediments: AAPG Bulletin, v. 85, p. 737-748, doi: 10.1306/8626C997-173B-11D7-8645000102C1865D.

Dickinson, W. W., and K. L. Milliken, 1995, The diagenetic role of brittle deformation in compaction and pressure solution, Etjo Sandstone, Namibia: Journal of Geology, v. 103, p. 339-347, http://www.jstor.org/stable/30071226.

Dunn, D. E., L. J. LaFountain, and R. E. Jackson, 1973, Porosity dependence and mechanism of brittle fracture in sandstones: Journal of Geophysical Research, v. 78, p. 2403-2417, doi: 10.1029/JB078i014p02403.

Eichhubl, P., J. N. Hooker, and S. E. Laubach, 2010, Pure and shear-enhanced compaction bands in Aztec Sandstone: Journal of Structural Geology, v. 32, p. 1873-1886, doi: 10.1016/j.jsg.2010.02.004.

Fairbairn, H. W., 1941, Deformation lamellae in quartz from the Ajibik Formation, Michigan: GSA Bulletin, v. 52, p. 1265-1277, doi: 10.1130/GSAB-52-1265.

Fisher, Q. J., R. J. Knipe, and R. H. Worden, 2000, Microstructures of deformed and non-deformed sandstones from the North Sea: Implications for the origins of quartz cement in sandstones, *in* R. H. Worden, and S. Morad, eds., Quartz Cementation in Sandstones (IAS Special Publication 29): Oxford, UK, Blackwell Publishing, p. 129-146, doi: 10.1002/9781444304237.ch10.

Fossen, H., R. A. Schultz, Z. K. Shipton, and K. Mair, 2007, Deformation bands in sandstone: A review: Journal of the Geological Society, v. 164, p. 755-769, doi: 10.1144/0016-76492006-036.

Fowles, J., and S. Burley, 1994, Textural and permeability characteristics of faulted, high porosity sandstones: Marine and Petroleum Geology, v. 11, p. 608-623, doi: 10.1016/0264-8172(94)90071-X.

Franks, P. C., 1969, Nature, origin and significance of cone-in-cone structures in the Kiowa Formation (Early Cretaceous), north-central Kansas: Journal of Sedimentary Research, v. 39, p. 1438-1454, doi: 10.1306/74D71E51-2B21-11D7-8648000102C1865D.

Gale, J. F. W., S. E. Laubach, J. E. Olson, P. Eichhubl, and A. Fall, 2014, Natural fractures in shale: A review and new observations: AAPG Bulletin, v. 98, p. 2165-2216, doi: 10.1306/08121413151.

Grieve, R. A. F., F. Langenhorst, and D. Stöffler, 1996, Shock metamorphism of quartz in nature and experiment: II. Significance in geoscience: Meteoritics & Planetary Science, v. 31, p. 6-35, doi: 10.1111/j.1945-5100.1996.tb02049.x.

Gustavson, T. C., S. D. Hovorka, and A. R. Dutton, 1994, Origin of satin spar veins in evaporite basins: Journal of Sedimentary Research, v. 64, p. 88-94, doi: 10.1306/D4267D1B-2B26-11D7-8648000102C1865D.

Hamers, M. F., and M. R. Drury, 2011, Scanning electron microscope-cathodoluminescence (SEM-CL) imaging of planar deformation features and tectonic deformation lamellae in quartz: Meteoritics & Planetary Science, v. 46, p. 1814-1831, doi: 10.1111/j.1945-5100.2011.01295.x.

Heald, M. T., 1959, Significance of stylolites in permeable sandstones: Journal of Sedimentary Research, v. 29, p. 251-253, doi: 10.1306/74D708F3-2B21-11D7-8648000102C1865D.

Izett, G. A., 1990, The Cretaceous/Tertiary boundary interval, Raton Basin, Colorado and New Mexico, and its content of shock-metamorphosed minerals: Evidence relevant to the K/T boundary impact-extinction theory: Boulder, CO, GSA Special Paper 249, 100 p., doi: 10.1130/SPE249-p1.

Kelly, J., J. Parnell, and H. H. Chen, 2000, Application of fluid inclusions to studies of fractured sandstone reservoirs: Journal of Geochemical Exploration, v. 69, p. 705-709, doi: 10.1016/S0375-6742(00)00096-0.

Knipe, R. J., 1993, The influence of fault zone processes and diagenesis on fluid flow, *in* A. D. Horbury, and A. G. Robinson, eds., Diagenesis and Basin Development: Tulsa, OK, AAPG Studies in Geology 36, p. 135-151.

Knipe, R. J., and H. Rutter, eds., 1990, Deformation Mechanisms, Rheology and Tectonics: London, GSL Special Publication 54, 522 p.

Kranz, R. L., 1983, Microcracks in rocks: A review: Tectonophysics, v. 100, p. 449-480, doi: 10.1016/0040-1951(83)90198-1.

Laubach, S. E., 2003, Practical approaches to identifying sealed and open fractures: AAPG Bulletin, v. 87, p. 561-579, doi: 10.1306/11060201106.

Laubach, S. E., R. H. Lander, L. M. Bonnell, J. E. Olson, and R. M. Reed, 2004a, Opening histories of fractures in sandstone, *in* J. W. Cosgrove, and T. Engelder, eds., The Initiation, Propagation, and Arrest of Joints and Other Fractures: London, GSL Special Publication 231, p. 1-9, doi: 10.1144/GSL.SP.2004.231.01.01.

Laubach, S. E., R. A. Marrett, J. E. Olson, and A. R. Scott, 1998, Characteristics and origins of coal cleat: A review: International Journal of Coal Geology, v. 35, p. 175-207, doi: 10.1016/S0166-5162(97)00012-8.

Laubach, S. E., R. M. Reed, J. E. Olson, R. H. Lander, and L. M. Bonnell, 2004b, Coevolution of crack-seal texture and fracture porosity in sedimentary rocks: Cathodoluminescence observations of regional fractures: Journal of Structural Geology, v. 26, p. 967-982, doi: 10.1016/j.jsg.2003.08.019.

Law, R. D., 1990, Crystallographic fabrics: A selective review of their applications to research in structural geology, *in* R. J. Knipe, and E. H. Rutter, eds., Deformation Mechanisms, Rheology and Tectonics: London, GSL Special Publication 54, p. 335-352, doi: 10.1144/GSL.SP.1990.054.01.30.

Lerbekmo, J. F., and R. L. Platt, 1962, Promotion of pressure-solution of silica in sandstones: Journal of Sedimentary Research, v. 32, p. 397-414, doi: 10.1306/74D70D03-2B21-11D7-8648000102C1865D.

Lianbo, Zeng, and Li Xiang-Yang, 2009, Fractures in sandstone reservoirs with ultra-low permeability: A case study of the Upper Triassic Yanchang Formation in the Ordos Basin, China: AAPG Bulletin, v. 93, p. 461-477, doi: 10.1306/09240808047.

Machel, H. G., 1985, Fibrous gypsum and anhydrite in veins: Sedimentology, v. 32, p. 443-454, doi: 10.1111/j.1365-3091.1985.tb00523.x.

Makowitz, A., R. H. Lander, and K. L. Milliken, 2006, Diagenetic modeling to assess the relative timing of quartz cementation and brittle grain processes during compaction: AAPG Bulletin, v. 90, p. 873-885, doi: 10.1306/12190505044.

Makowitz, A., and K. L. Milliken, 2003, Quantification of brittle deformation in burial compaction, Frio and Mount Simon Formation sandstones: Journal of Sedimentary Research, v. 73, p. 1007-1021, doi: 10.1306/051003731007.

Miller, T. W., 1995, New insights on natural hydraulic fractures induced by abnormally high pore pressures: AAPG Bulletin, v. 79, p. 1005-1018.

Milliken, K. L., and S. E. Laubach, 2000, Brittle deformation in sandstone diagenesis as revealed by scanned cathodoluminescence imaging with application to characterization of fractured reservoirs, *in* M. Pagel, V. Barbin, P. Blanc, and D. Ohnenstetter, eds., Cathodoluminescence in Geosciences: Berlin, Springer, p. 225-243, doi: 10.1007/978-3-662-04086-7_9.

Milliken, K. L., R. M. Reed, and S. E. Laubach, 2005, Quantifying compaction and cementation in deformation bands in porous sandstones, *in* R. Sorkhabi, and Y. Tsuji, eds., Faults, Fluid Flow, and Petroleum Traps: Tulsa, OK, AAPG Memoir 85, p. 237-249, doi: 10.1306/1033726M85252.

Morgan, J. V., C. Lana, A. Kersley, B. Coles, C. Belcher, S. Montanari, E. Diaz-Martinez, A. Barbosa, and V. Neumann, 2006, Analyses of shocked quartz at the global K-P boundary indicate an origin from a single, high-angle, oblique impact at Chicxulub: Earth and Planetary Science Letters, v. 251, p. 264-279, doi: 10.1016/j.epsl.2006.09.009.

Mørk, M. B. E., and K. Moen, 2007, Compaction microstructures in quartz grains and quartz cement in deeply buried reservoir sandstones using combined petrography and EBSD analysis: Journal of Structural Geology, v. 29, p. 1843-1854, doi: 10.1016/j.jsg.2007.08.004.

Mousavi, M. A., and S. L. Bryant, 2013, Geometric models of porosity reduction by ductile grain compaction and cementation: AAPG Bulletin, v. 97, p. 2129-2148, doi: 10.1306/05171311165.

Nelson, R. A., 1981, Significance of fracture sets associated with stylolite zones: AAPG Bulletin, v. 65, p. 2417-2425.

Nelson, R. A., 2001, Geologic Analysis of Naturally Fractured Reservoirs (2nd Edition): Woburn, MA, Butterworth-Heinemann (GPP), 352 p.

Onasch, C. M., W. M. Dunne, J. E. Cook, and A. O'Kane, 2009, The effect of fluid composition on the behavior of well cemented, quartz-rich sandstone during faulting: Journal of Structural Geology, v. 31, p. 960-971, doi: 10.1016/j.jsg.2008.10.010.

Parnell, J., and P. F. Carey, 1995, Emplacement of bitumen (asphaltite) veins in the Neuquén Basin, Argentina: AAPG Bulletin, v. 79, p. 1798-1815.

Parris, T. M., R. C. Burruss, and P. B. O'Sullivan, 2003, Deformation and the timing of gas generation and migration in the eastern Brooks Range foothills, Arctic National Wildlife Refuge, Alaska: AAPG Bulletin, v. 87, p. 1823-1846, doi: 10.1306/07100301111.

Pitman, J. K., and E. S. Sprunt, 1986, Origin and distribution of fractures in Lower Tertiary and Upper Cretaceous rocks, Piceance basin, Colorado, and their relation to the occurrence of hydrocarbons, *in* C. W. Spencer, and R. F. Mast, eds., Geology of Tight Gas Reservoirs: Tulsa, OK, AAPG Studies in Geology Series 24, p. 221-234.

Pudsey, C. J., and H. G. Reading, 1982, Sedimentology and structure of the Scotland Group, Barbados, *in* J. K. Leggett, ed., Trench-Forearc Geology: Sedimentation and tectonics on modern and ancient active plate margins: London, GSL Special Publication 10, p. 291-308, doi: 10.1144/GSL.SP.1982.010.01.19.

Rahl, J. M, M. T. Brandon, H. Deckert, U. Ring, and N. Mortimer, 2011, Tectonic significance of ductile deformation in low-grade sandstones in the Mesozoic Otago subduction wedge, New Zealand: American Journal of Science, v. 311, p. 27-62, doi: 10.2475/01.2011.02.

Renard, F., P. Ortoleva, and J.-P. Gratier, 1997, Pressure solution in sandstones: Influence of clays and dependence on temperature and stress: Tectonophysics, v. 280, p. 257-266, doi: 10.1016/S0040-1951(97)00039-5.

Rodrigues, N., P. R. Cobbold, H. Loseth, and G. Ruffet, 2009, Widespread bedding-parallel veins of fibrous calcite ('beef') in a mature source rock (Vaca Muerta Fm, Neuquén Basin, Argentina): Evidence for overpressure and horizontal compression: Journal of the Geological Society, v. 166, p. 695-709, doi: 10.1144/0016-76492008-111.

Rodrigues, C. F., C. Laiginhas, M. Fernandes, M. J. Lemos de Sousa, and M. A. P. Dinis, 2014. The coal cleat system: A new approach to its study: Journal of Rock Mechanics and Geotechnical Engineering, v. 6, p. 208-218, doi: 10.1016/j.jrmge.2014.03.005.

Schmidt, V., and D. A. McDonald, 1979, Texture and recognition of secondary porosity in sandstones, *in* P. A. Scholle, and P. Schluger, eds., Aspects of Diagenesis: Tulsa, OK, SEPM Special Publication 26, p. 209-225, doi: 10.2110/pec.79.26.0209.

Schultz, R. A., and H. Fossen, 2008, Terminology for structural discontinuities: AAPG Bulletin, v. 92, p. 853-867, doi: 10.1306/02200807065.

Schultz, R. A., and R. Siddharthan, 2005, A general framework for the occurrence and faulting of deformation bands in porous granular rocks: Tectonophysics, v. 411, p. 1-18, doi: 10.1016/j.tecto.2005.07.008.

Shearman, D. J., G. Mossop, H. Dunsmore, and M. Martin, 1972, Origin of gypsum veins by hydraulic fracture: Transactions of the Institute of Mining and Metallurgy Transaction, Section B: Applied Earth Science, v. 81, p. B149-B155.

Sloss, L. L., and D. E. Feray, 1948, Microstylolites in sandstone: Journal of Sedimentary Research, v. 18, p. 3-13, doi: 10.1306/D42692D8-2B26-11D7-8648000102C1865D.

Stearns, D. W., and M. Friedman, 1972, Reservoirs in fractured rock: Geologic exploration methods, *in* H. R. Gould, ed., Stratigraphic Oil and Gas Fields—Classification, Exploration Methods and Case Histories: Tulsa, OK, AAPG Memoir 16, p. 82-100.

Stöffler, D., and F. Langenhorst, 1994, Shock metamorphism of quartz in nature and experiment: I. Basic observation and theory: Meteoritics, v. 29, p. 155-181, doi: 10.1111/j.1945-5100.1994.tb00670.x.

Stone, W. N., and R. Siever, 1996, Quantifying compaction, pressure solution and quartz cementation in moderately-and deeply-buried quartzose sandstones from the Greater Green River Basin, Wyoming, *in* L. J. Crossey, R. Loucks, and M. W. Totten, eds., Siliciclastic Diagenesis and Fluid Flow: Concepts and Applications: Tulsa, OK, SEPM Special Publication 55, p. 129-150, doi: 10.2110/pec.96.55.0129.

Stoneley, R., 1983, Fibrous calcite veins, overpressures, and primary oil migration: AAPG Bulletin, v. 67, p. 1427-1428.

Swarbrick, R. E., and M. J. Osborne, 1998, Mechanisms that generate abnormal pressures: An overview, *in* B. E. Law, G. F. Ulmishek, and V. I. Slavin, eds., Abnormal Pressures in Hydrocarbon Environments: Tulsa, OK, AAPG Memoir 70, p. 13-34.

Tada, R., and R. Siever, 1989, Pressure solution during diagenesis: Annual Review of Earth and Planetary Sciences, v. 17, p. 89-118, doi: 10.1146/annurev.ea.17.050189.000513.

Tarney, J., and B. C. Schreiber, 1976, Cone-in-cone and beef-in-shale textures from DSDP Site 330, Falkland Plateau, South Atlantic, *in* Initial Reports of the Deep Sea Drilling Project, v. 36: College Station, TX, Ocean Drilling Program, p. 865-870.

Walderhaug, O., and P. A. Bjørkum, 2003, The effect of stylolite spacing on quartz cementation in the Lower Jurassic Stø Formation, southern Barents Sea: Journal of Sedimentary Research, v. 73, p. 146-156, doi: 10.1306/090502730146.

Waldschmidt, W. A., P. E. Fitzgerald, and C. L. Lunsford, 1956, Classification of porosity and fractures in reservoir rocks: AAPG Bulletin, v. 40, p. 953-974.

Weyl, P. K., 1959, Pressure solution and the force of crystallization—a phenomenological theory: Journal of Geophysical Research, v. 64, p. 2001-2025, doi: 10.1029/JZ064i011p02001.

Woodland, B. G., 1964, The nature and origin of "cone-in-cone" structure: Fieldiana Geology, v. 13, no. 4, p. 185-305.

Facing Page: Top – Sandstone with primary porosity, Jurassic Entrada Fm., Emery Co., Utah (PPL). Bottom – Primary and secondary feldspar dissolution porosity, Plio–Pleistocene Gila Gp., New Mexico (PPL).

Pore Types and Porosity

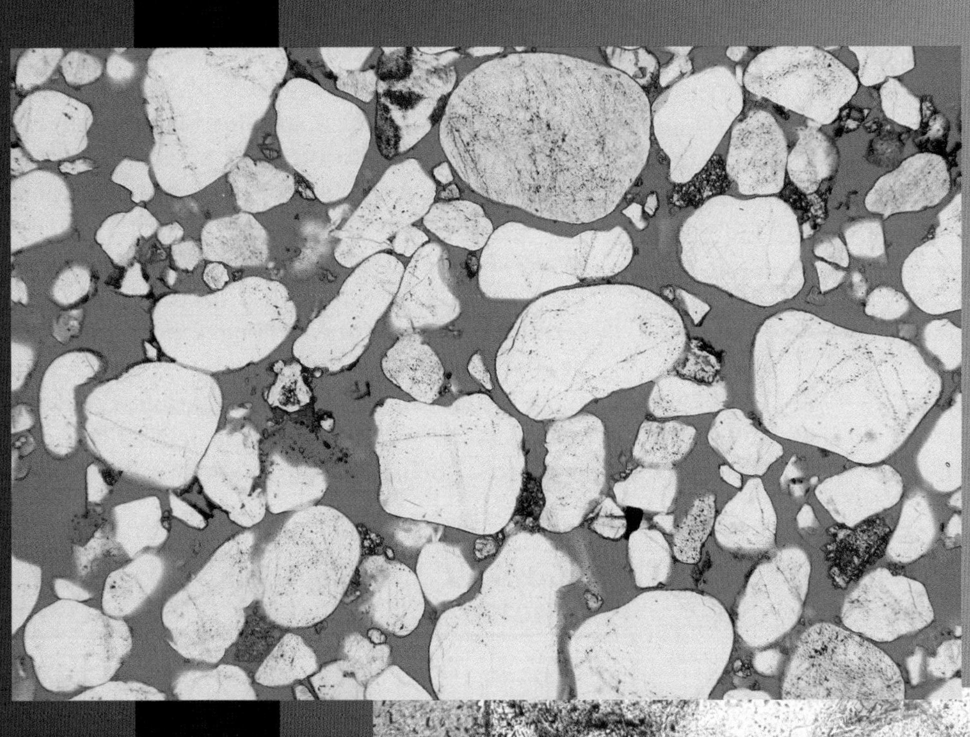

CHAPTER 22

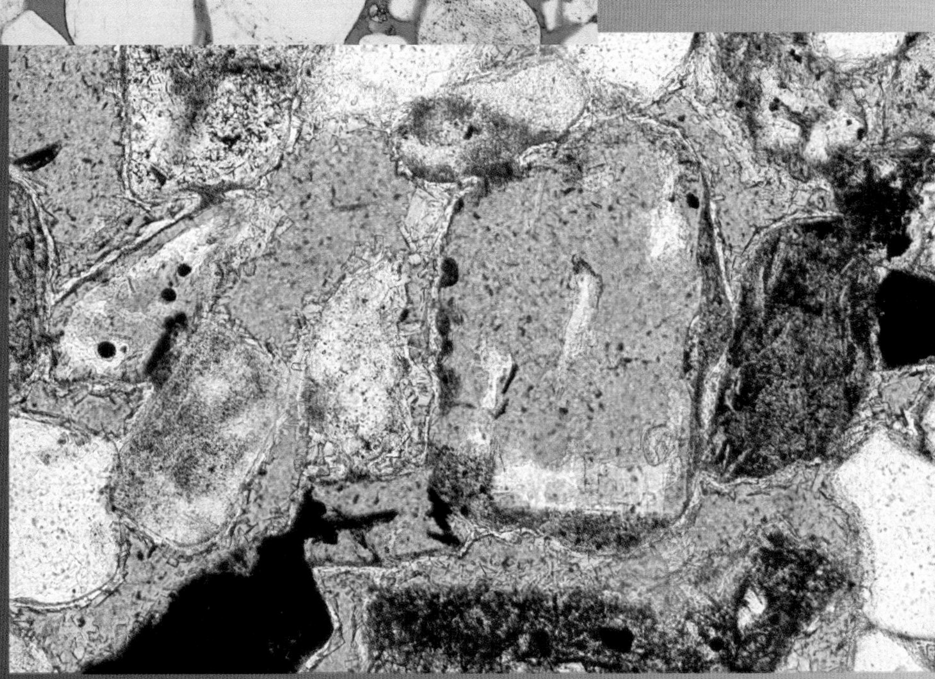

INTRODUCTION TO PORES AND POROSITY

Pores refer to the void spaces in a rock; porosity is a measurement that refers to the percentage of void spaces in a rock (the ratio of the volume of void spaces to the total volume of the rock sample). From an engineering perspective, porosity commonly is subdivided into "total porosity" (all pores, regardless of whether they are interconnected or isolated) versus "effective porosity" (porosity that is interconnected plus, depending on usage, pores associated with clay-bound water). For hydrocarbon explorationists, porosity (especially effective porosity) and permeability are probably the most important properties of any rock. Pores, therefore, are discussed throughout this book, and this relatively short chapter will serve mainly to give examples of the range of pore types in the context of their classification. It should be noted that accurate petrographic identification of the amount of porosity and the types of pores in sedimentary rocks requires careful thin-section preparation (Pittman, 1992a) and is aided by the injection of colored and/or fluorescent dyes into pores prior to thin sectioning (e.g., Ruzyla and Jezek, 1987).

Pore classification – Although several classifications have been proposed for pores in sandstones (e.g., Waldschmidt et al., 1956; Pittman, 1979 and 1984; Ehrlich et al., 1991), none has been accepted universally. The very simple scheme of Pittman (1979) consisting of only four categories (intergranular, dissolution, micro- and fracture pores) has been used by some, but it provides little descriptive range. The widely adopted classification for pores in carbonate rocks (Choquette and Pray, 1970) is far more detailed and it has been adapted by some geologists for use in sandstones.

Adding to pore classification issues in clastic terrigenous deposits is the dramatic change in what sizes of pores are considered important. In the not-so-very-bygone days of "conventional" reservoir exploration, only relatively large pores (typically ones visible with a light microscope or larger) were considered relevant. Today, with fine-grained "unconventional" reservoirs playing a prominent role in exploration, pores considerably smaller have been found to play a significant role, especially in gas production. This wider range of pores of interest is reflected in changes in pore size terminology (Figure 22.1; Loucks et al., 2012) and has led to numerous papers dealing specifically with mudstone pores (see Mudstone/Shale Porosity section of the bibliography for this chapter).

For this book, we have adapted a version of the Choquette and Pray (1970) classification of pores in carbonate rocks and have paired that with elements of the mudstone pore terminologies of Loucks et al. (2012) and Schieber (2013). An overview of that combined terminology, suitable for sandstones and mudrocks, is presented in Figures 22.2 and 22.3. We also use the updated terminology for pore sizes advocated in Loucks et al. (2012), as opposed to the earlier scales of Choquette and Pray (1970) and Rouquerol et al. (1994). However, the variations in the definitions of, in some cases, identical terms used in the three schemes (as shown in Figure 22.1) should alert users to the need to clearly state which size scale they are using in order to avoid confusion.

The genetic classification of pores is considerably more complex than simple size measurement. At its basic level, it begins with a division into primary and secondary pore types. Primary pores are ones formed at the time of sedimentation; secondary pores are those produced after sedimentation, during eogenetic, mesogenetic or telogenetic diagenesis, by a variety of processes, primarily dissolution and fracturing. At

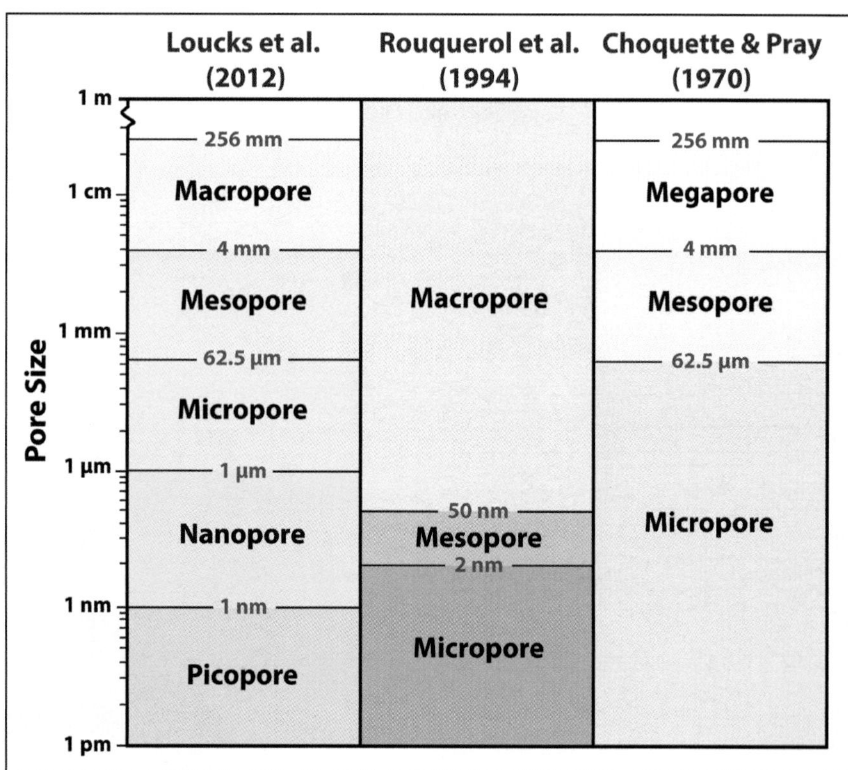

Figure 22.1: A comparison of three terminological schemes for pore sizes in sedimentary rocks and other porous solids. Sizes measurements are typically based on the diameter of circular pores or the width of elongate pores. Adapted from Loucks et al. (2012).

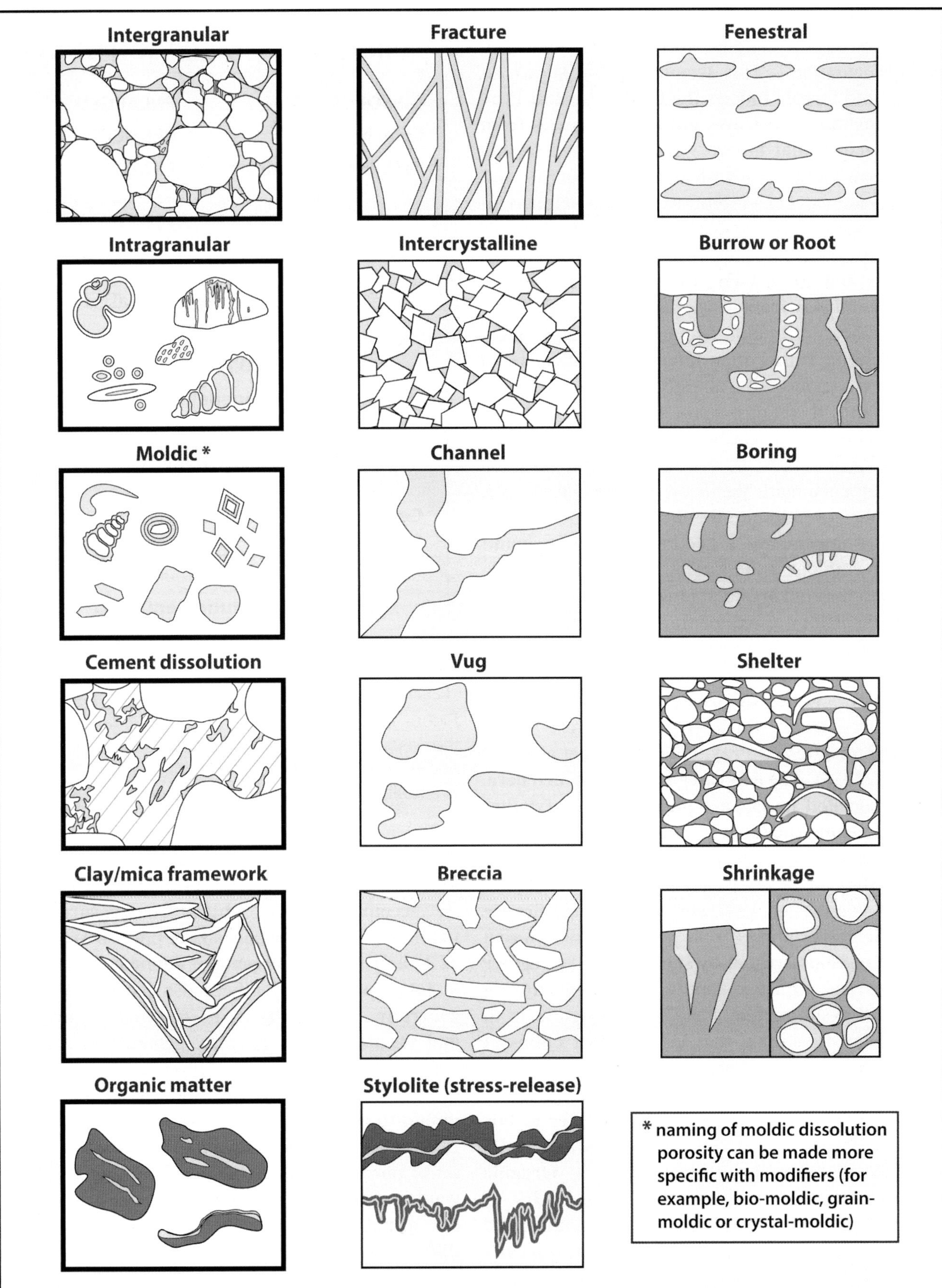

Figure 22.2: An adaptation of the Choquette and Pray (1970) carbonate pore classification for use in clastic terrigenous rocks. A number of categories have been dropped and some new ones added, including clay framework and organic matter pore types that are especially important in mudrocks (Loucks et al., 2012 and Schieber, 2013). Although all these pore types can occur in clastic terrigenous and associated rocks, those outlined with a darker box are most the common ones.

the next level, pores are commonly categorized by two characteristics: relationship to overall rock fabric and process of formation. This terminology is shown in Figure 22.2 and largely follows the Choquette and Pray (1970) nomenclature. It leaves out two categories from the original paper that are very rare in noncarbonate rocks (caverns and growth-framework pores) and adds additional categories that were not included in the original work (stylolite-associated pores, cement-dissolution pores) and two pore types intended primarily, but certainly not exclusively, for mudrocks (organic matter pores and clay framework pores).

This classification modifies some names (e.g., inter- and intragranular rather than inter- and intraparticle) because those terms are already widely used for sandstones, and it retains some categories that are uncommon in most siliciclastic rocks but are included because they can occur in included grains, such as phosphatic or carbonate particles, or associated rock types. Furthermore, despite the fact that some of these pore types are uncommon, a good classification should provide the terminology to describe even relatively rare pore types that may aid in understanding the depositional and diagenetic history of sedimentary deposits.

All the pore types are schematically illustrated in Figure 22.2, and most are shown in the photomicrographs in this chapter. Fuller discussion is available in the original papers from which the terms were taken (primarily Choquette and Pray, 1970, Loucks et al., 2012 and Schieber, 2013). The new categories are defined here. "Stylolite-associated pores" are formed along stylolites and solution seams (bedding-parallel or other) during and after overburden removal or other stress release. They are important in some deep carbonate reservoirs (Carozzi and Von Bergen, 1987) and can have significance in siliciclastic reservoirs as well. "Cement-dissolution pores" refer to voids created by the localized, irregular corrosion or even complete dissolution of cements during diagenesis. "Crystal-moldic pores" is a better term to refer to cement dissolution that leaves intact molds of the crystals; "cement-dissolution pores" better describes irregular crosscutting pores or larger-scale or more extensive removal of cements as illustrated later in this chapter. The porosity types that are included especially for mudrocks are largely self-explanatory. Organic matter pores are found within detrital particles of organic material (with subtypes discussed in Loucks and Reed, 2014) and clay/mica framework pores are typically micron-scale or smaller openings within the assemblages of clay particles that form the matrix of sandstones or that can make up the main part of mudrocks.

Pore classification provides words and phrases that can be used for concise and effective communication amongst geoscientists—it is a vocabulary that allows great flexibility in information transmission. To expand that flexibility, the pore terms shown in Figure 22.2 can be accompanied by objective (measurement-based) modifiers for average pore size or pore-size ranges (Figure 22.1) and for the percentage abundances of pore types or total porosity. Modifiers also can be used to express generally more subjective information about timing and mechanisms of porosity generation, where known (Figure 22.3). How long and complex one makes such descriptions will vary with the level of information one wishes to convey and the technical level of the audience. In some cases, one might want to describe just a particular set of pores in a single picture. In other cases one may want to summarize the porosity of an entire rock—clearly one would use different levels of terminology for those different tasks. For some purposes, just a short phrase such as "a sandstone with 15% primary intergranular macro- and mesopores" will suffice to convey the message that this is a rock with moderate levels of original porosity in the micron to millimeter size range; it also indicates little diagenetic porosity enhancement as the remaining porosity is predominantly primary. A more complex

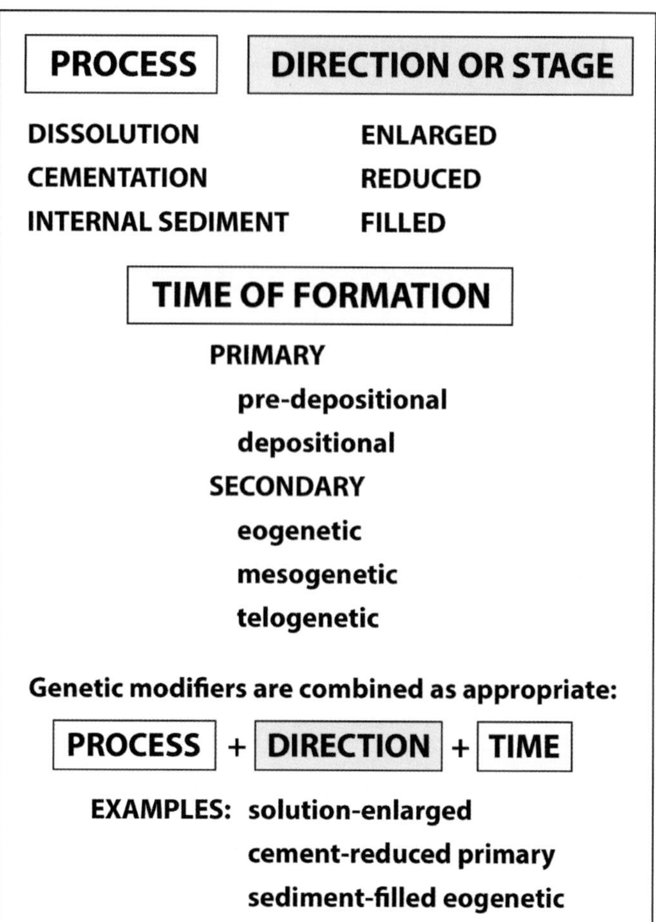

Figure 22.3: A summary of supplementary terminology for the description of pores in sedimentary rocks; from Choquette and Pray (1970). Although designed for carbonate rocks, it is equally applicable to clastic terrigenous deposits.

message is relayed with a phrase such as "a sandstone with 7% primary intergranular macro- and mesopores and 8% telogenetic moldic and fracture pores". In this case, although the total porosity is the same, the message sent is that diagenesis plays a substantial role in the origin of the present-day porosity and that much of it was developed during uplift and perhaps subaerial exposure.

Clearly, a near-infinite number of such phrases can be developed with the available terms, and no two users will emphasize exactly the same points or create exactly the same phrases. But no matter how the rock is described, if the observations are accurate, and the terms are used as defined, they will provide far more information for the reader than a simple set of core- or log-derived porosity and permeability numbers, and can provide valuable insights on the origin of reservoirs and the prediction of additional exploration prospects. Both the interpretative visual descriptions from light or electron microscope and the quantitative data from core or log analysis are useful, but typically they are more informative when used together. That said, petrographic observations of pores and porosity (and even calculations of permeability) can be quantified, systematized and automated using pore casting (Wardlaw, 1976) or through rigorous image analysis techniques (Halley, 1978; Ehrlich et al., 1984; Ehrlich et al., 1991; Gerard et al., 1992; Coskun and Wardlaw, 1993). Petrographic methods can be especially useful for small or friable samples that are difficult to study using conventional core analysis.

Some comments on the porosity of mudstone and shale should be added to this discussion. Fundamentally, fine-grained terrigenous rocks are not different from their coarser relatives, either in basic constituents or in porosity types, although the tools necessary to image them differ significantly. In mudstones, the grains and pores are smaller, the content of clays and organic matter generally are higher, the intergranular attractive or repulsive forces are more dominant, and the permeabilities are markedly lower in the absence of fractures. Nevertheless, if one takes a smaller-scale view of detrital clay and silt as framework grains (along with the remains of planktic or benthic organisms, even nannofossils), then the basic particle-pore relationships seen in coarser deposits also are present at this smaller scale. Primary intergranular or framework pores can be seen between clays or between silt-sized terrigenous or biogenic grains; there can be primary shelter pores where larger particles (platy detrital micas, shell fragments, and others) block influx of smaller particles; and there can be primary intragranular pores especially within organic remains (e.g., radiolarian tests, foraminiferal chambers, coccospheres and coccolith plates, as well as plant remains and other particles of organic matter). One also can observe secondary moldic or intragranular pores from dissolution of unstable detrital grains such as feldspars or heavy minerals (even though such dissolution commonly is less pronounced in shales than in sandstones (e.g., Blatt and Sutherland, 1969; Blatt, 1985; Milliken, 1992). In addition, there are secondary fracture pores created by differential compaction, by tectonic movements, by overpressuring and overpressure release and by organic maturation. Finally, intercrystalline pores can be found between small cement crystals and within clays and other minerals that have undergone subsurface diagenetic transformation. In short, all the major pore types seen in sandstones also are present in mudstones, and the same terminology can be used for both. However, it is worth distinguishing the smaller-sized pores with modifiers noting their nano- and picopore sizes mainly because they have such different fluid transmission properties and play such a large role in the economic viability of unconventional hydrocarbon plays.

Porosity-permeability patterns – The relationship between porosity and permeability is complex. At the most general level, rocks dominated by intergranular and intercrystalline pores tend to have good pore connectivity and thus have high permeabilities relative to porosity; rocks with intragranular, moldic and vuggy pores typically have lower connectivity and correspondingly lower permeabilities for any given porosity (Pittman, 1984). However, there is far more complexity to these relationships than those generalities might imply. Unfortunately for petrographers, the 2-D fabrics observed in thin sections can be difficult to relate to actual 3-D pore networks, and the size and geometric relationships between pores and pore throats that are a major control on pore connectivity are not easily quantified in thin sections (Wardlaw and Cassan, 1978). Although engineering tests such as mercury injection (MICP) data or nuclear magnetic resonance testing are commonly used to address this issue (e.g., Pittman, 1992b), modern 3-D imaging techniques now allow direct visual evaluations of pore systems at petrographic and larger scales. This, in turn, allows integration of porosity and permeability variations with the causative controls determined through conventional petrography. The main techniques now used for pore-system visualization are X-ray computed tomography, mainly at macro- to micropore scales, and a combination of ion-milling and a variety of electron microscopic techniques at the nano- to picopore levels. Together, these techniques have revolutionized the ability to examine, identify and measure pore geometries in a visual context (e.g., Lindquist and Venkatarangan, 1999; Appoloni et al., 2007; Loucks et al., 2009; Milner et al., 2010; Driskill et al., 2013; Schieber, 2013; Ruppert et al., 2013). Some of these procedures are described and illustrated in Chapter 24 (Techniques).

Porosity trends (controls and predictive models) – It is clearly beyond the scope of a book on petrography to delve deeply into controls and models of porosity preservation and generation. Nonetheless, a few words on this topic seem appropriate because many of the controls on porosity loss or creation can only be recognized with petrographic techniques.

As a starting point, it is worthwhile to look at a plot of actual (that is measured, not modeled) worldwide porosity-depth data for sandstone petroleum reservoirs (Fig. 22.4). Although it shows a moderately clear trend of porosity loss with depth (marked by the 90th, 50th and 10th percentile lines), there is, overall, a remarkable diversity in the data. There are so many factors that can affect porosity retention, destruction or creation that virtually any porosity is possible at almost any depth. In the words of Ajdukiewicz and Lander (2010, p. 1089):

> *"WHAT HAS BEEN DESCRIBED AS ANOMALOUS POROSITY IS IN FACT THE HIGH END OF THE RANGE OF POSSIBLE OUTCOMES, WHERE A PARTICULAR COMBINATION OF GRAIN SIZE, SORTING, COMPOSITION, EARLY DIAGENESIS, AND BURIAL HISTORY HAVE ACTED TOGETHER TO MINIMIZE THE EFFECTS OF COMPACTION AND CEMENTATION AND PRESERVE THE GREATEST AMOUNT OF POROSITY AND PERMEABILITY AT DEPTH."*

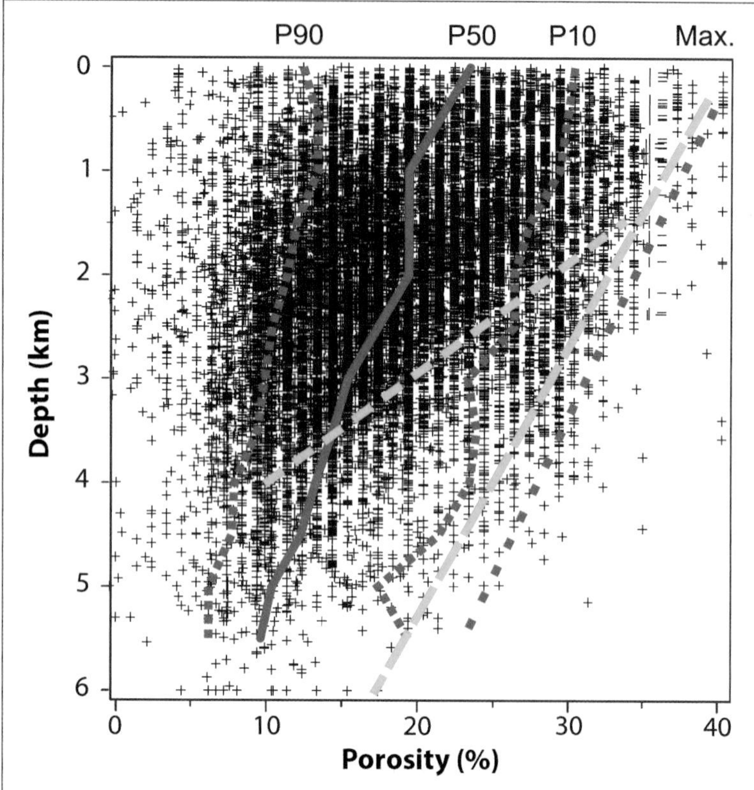

Figure 22.4: Average porosity versus top depth for global petroleum reservoirs composed of sandstone. A few points with values outside the plot axes were plotted at the limiting values. The P90 line denotes points at which 90% of reservoirs have porosity greater than this value, P50 denotes median values, and P10 denotes points at which 10% of reservoirs have porosity greater than this value. The long-dashed green line is the porosity-depth trend for Tertiary sands of south Louisiana (Atwater and Miller, 1965; using data plotted in Blatt et al., 1980, p. 419), an example of quartzose sandstones buried with low geothermal gradient (18°C/km) and widespread overpressures. The short-dashed green line is the porosity-depth trend for the Middle Jurassic Garn Formation, offshore mid-Norway (Ehrenberg, 1990), an example of quartzose sandstones buried with moderate geothermal gradient (35°C/km). Adapted from Ehrenberg and Nadeau (2005, p. 439).

Many of these factors have been discussed in chapters on compaction, cementation and dissolution (Chapters 10-19), but it may be worth reiterating the major processes than can help in porosity preservation (adapted mainly from Bloch et al., 2002; Ehrenberg and Nadeau, 2005; and Taylor et al., 2010).

1. **Sediment composition (quartzose, feldspathic or lithic-rich) is important, because it influences the relative diagenetic potential of strata (especially potential for resisting compaction or for development of secondary porosity);**
2. **Authigenic grain coating or rimming cements such as chlorite, other clays, microquartz, iron oxides or other minerals can retard quartz overgrowth cementation (especially in detrital quartz-rich sandstones);**
3. **Early hydrocarbon emplacement (especially with high hydrocarbon saturations and/or with oil as the wetting phase) can retard further diagenetic modifications;**
4. **Early (relatively shallow) development of fluid overpressure can retard fluid flow, compaction and cementation in quartz-rich as well as quartz-poor lithologies;**
5. **Burial history (burial trajectory) is important because slow burial is more likely to allow formation of cements that can retard compaction as compared with rapid burial;**
6. **Low thermal gradients can retard chemical diagenesis, especially in quartzarenites, because quartz cementation increases exponentially with increasing temperature.**

Although not all of these factors can be recognized in petrographic studies, this list reinforces the fact that petrographic information can and should be integrated with more general depositional modeling and basin analysis (burial history, thermal history, hydrologic models and the like) for maximum utility.

Mid. Jurassic Entrada Fm., Moffat Co., Colorado

An example of a feldspathic quartzarenite with mainly primary intergranular porosity (approximately 24% total porosity based on image analysis). The primary porosity has been reduced slightly by quartz overgrowths and minor kaolinite and dolomite cementation along with minor compaction (note close grain packing in some areas).

PPL, KFS, BDI, Scale bar = 0.26 mm

Mid. Jurassic Entrada Fm., Moffat Co., Colorado

This photomicrograph shows a different area of the same thin section from which the previous image was taken. Note the much greater extent of quartz-overgrowth cementation here, and thus the far greater reduction of primary intergranular porosity (about 10% total porosity in this area based on image analysis, essentially all of it primary). This variability highlights the difficulty of getting meaningful porosity data from single images or even single thin sections. Nonetheless, petrographic study of such variability can yield valuable insights on the controls on porosity in units of interest.

PPL, KFS, BDI, Scale bar = 0.26 mm

Lo. Permian Park City Fm., Daggett Co., Utah

Intergranular pores in a sandstone comprising mainly quartz grains and chert rock fragments. Some minor quartz overgrowths and thin clay cements, along with compaction (evinced by grain breakage), have reduced porosity. Nonetheless, image analysis indicates that nearly 20% porosity remains (at least in the area of this image), with nearly all of it primary.

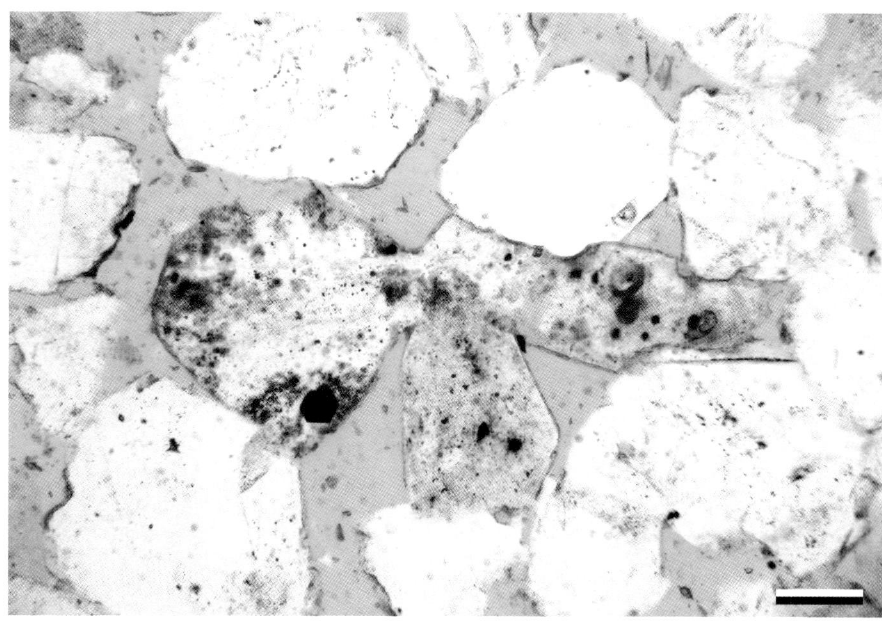

PPL, BDI, Scale bar = 0.08mm

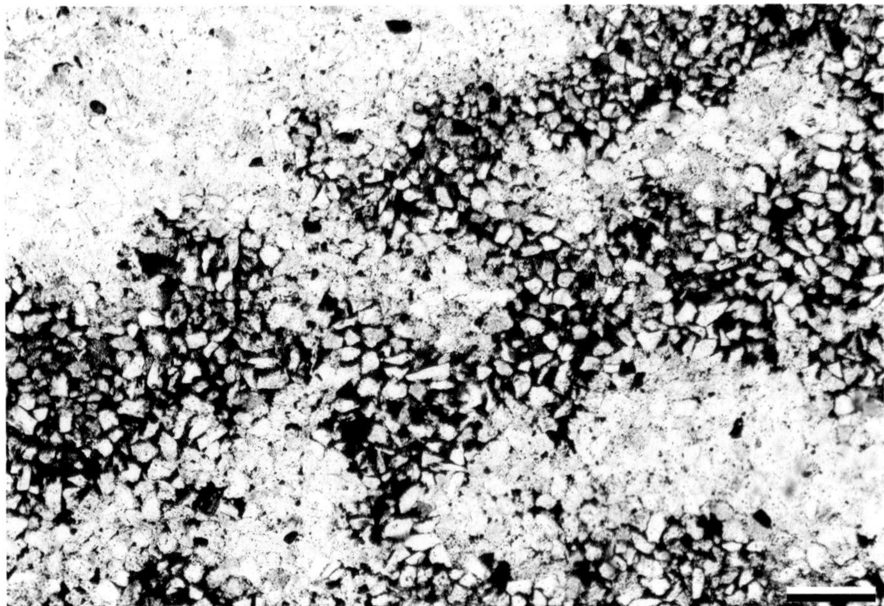

Up. Mississippian Paradise Fm., Hidalgo Co., New Mexico

An example of a complex porosity history in a strongly bioturbated, essentially nonporous quartz arenite. The original intergranular porosity has been occluded by cementation with some areas cemented by hematite and limonite and others by somewhat later quartz overgrowths. At some stage in this porosity-loss scenario, there was selective retention of porosity in the areas that are now quartz cemented and the rock could have had reservoir potential at that time.

PPL, Scale bar = 0.26 mm

Oligocene – Miocene Otekaike Fm., Canterbury, New Zealand

This is an example of a relatively uncommon type of primary porosity in sandstones. Termed intragranular porosity, it is especially common within carbonate bioclasts, but it also can occur in poorly cemented rock fragments and other grains. In this sandy glauconite (greensand), a probable scaphopod shell has retained a large pore in its interior. It has been preserved despite some compaction of the surrounding glauconitic sediment (and some deformation of the shell itself). Note the chemical embayment of the carbonate grain exterior by the physically much softer glauconite grains.

PPL, AFeS, BDI, Scale bar = 0.34 mm

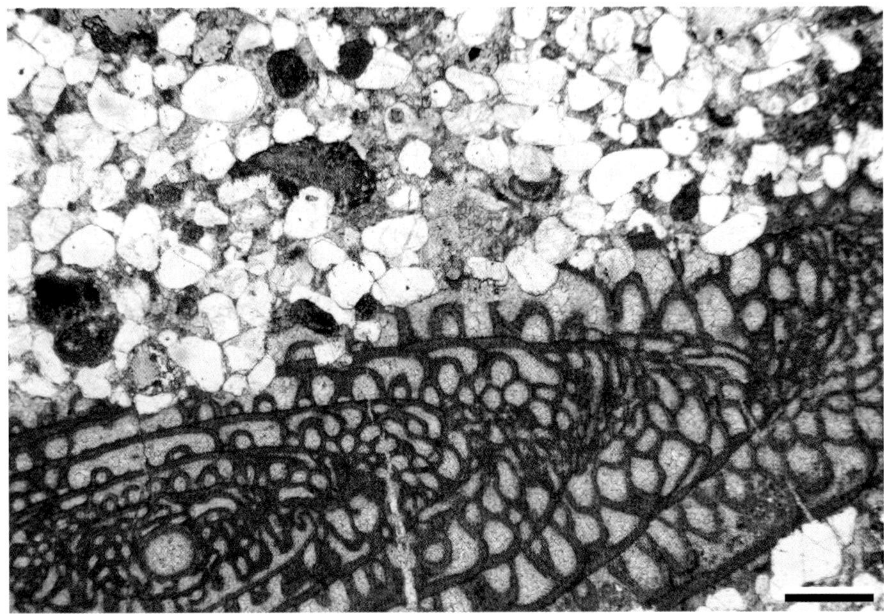

Mid. Permian Brushy Canyon Fm., Culberson Co., Texas

An example of completely cement-reduced primary intragranular porosity. Here, a large fusulinid foraminifer had numerous unfilled living chambers (the gray-white areas within the test) and thus considerable intragranular porosity when originally deposited. Subsequently, all those pores, along with all the intergranular pores in the adjacent sandstone, were filled with fine- to medium-crystalline calcite cement. A likely source of the some calcite cement is the fusulinid test itself, as its outer margin has been extensively dissolved by pressure solution embayment from adjacent quartz grains.

PPL, Scale bar = 0.51 mm

Lo. Oligocene Nile Gp., Westland, New Zealand

An example of calcite cement-filled primary shelter porosity in a sandy carbonate rock. A large bivalve shell (subsequently dissolved, calcite infilled, and then fractured) provided the shelter that prevented sediment infiltration to the area below it. Most shelter porosity, at least on this large a scale, is provided by large bioclasts. The odd coloration of this sample results from carbonate staining of only the right side of the sample. It shows early filling of the void by nonferroan calcite (pink) with the latest (cavity center) calcite having a slightly more ferroan composition (pale purple).

PPL, AFeS, Scale bar = 0.51 mm

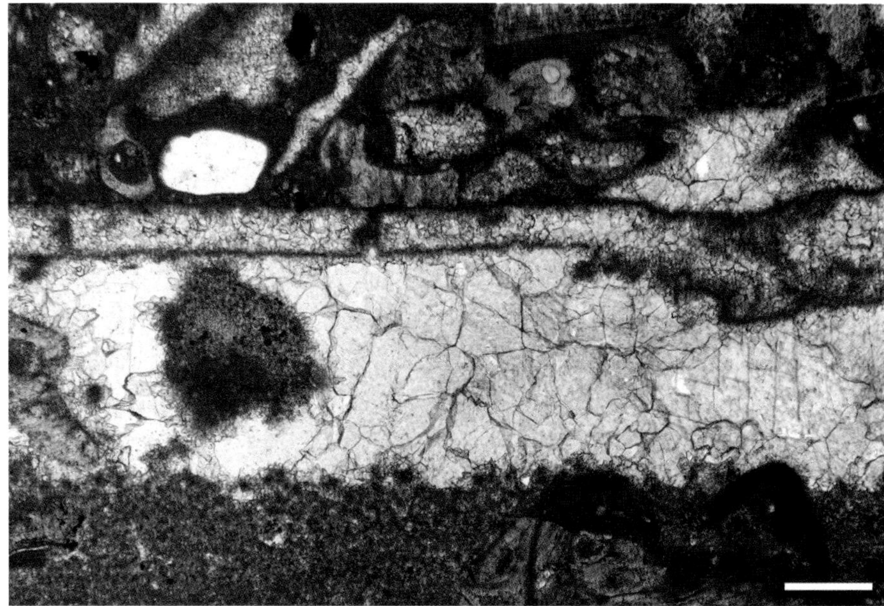

Lo. Oligocene Nile Gp., Westland, New Zealand

Another example of cement-filled primary shelter porosity in a sandy limestone. An echinoid test fragment (the perforated grain that crosses the horizontal center of the image) here precluded infiltration of overlying sediment and thus protected a large shelter pore. The shelter pore was later filled with two stages of carbonate cement. The initial cement rim (perhaps a fibrous high-Mg calcite) was later dissolved to yield about 3% secondary cement dissolution porosity.

PPL, BDI, Scale bar = 0.51 mm

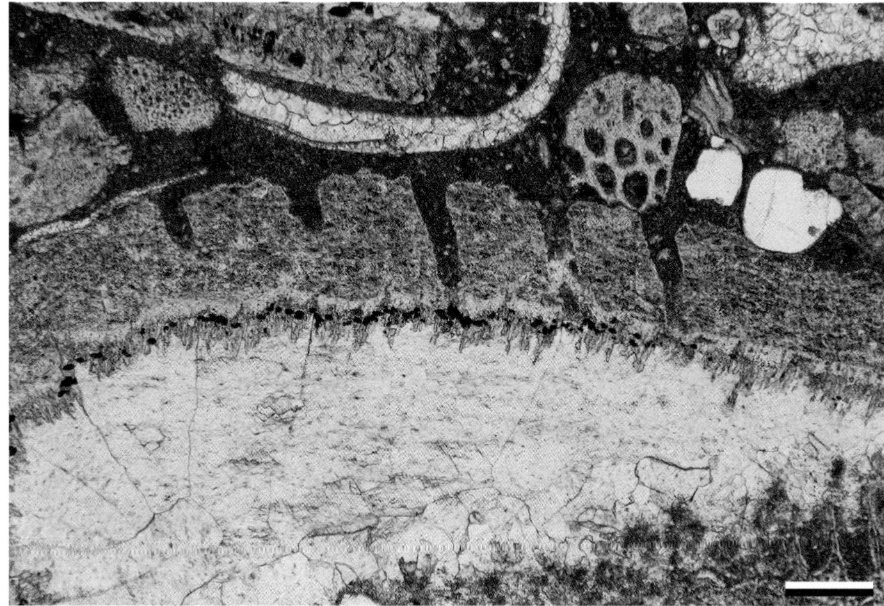

Lo. Cretaceous (Albian) Nahr Umr Fm., subsurface eastern Qatar

An example of intergranular porosity that has been substantially reduced by siderite cementation (now about 12% pore space). Because the cements are so widespread and retain considerable space between the crystals, it is equally appropriate to call this preserved intercrystalline porosity.

PPL | XPL, BDI, Scale bar = 0.05 mm

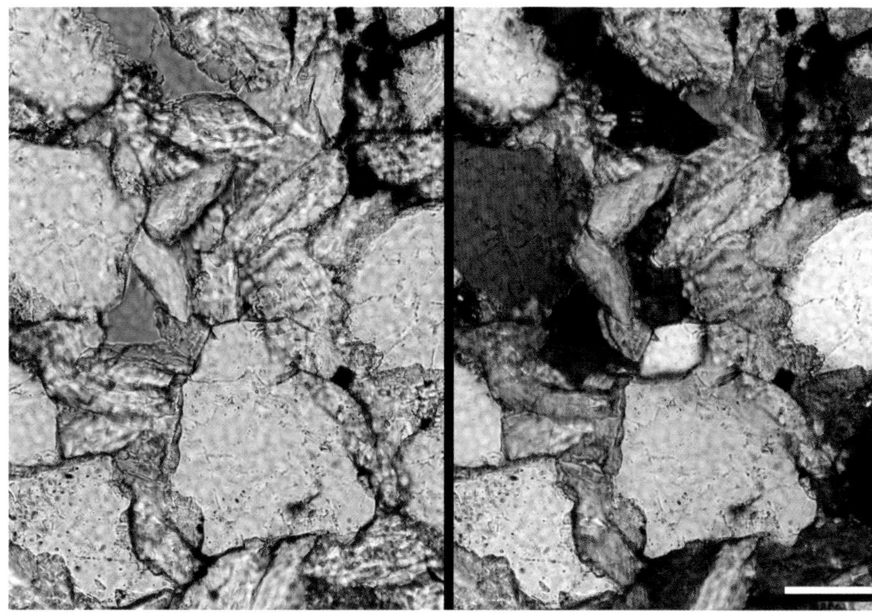

Eocene unknown unit, unknown locality

An example of primary burrow porosity in a strongly bioturbated crevasse-splay siltstone. Here, the burrow fill has more porosity than the surrounding sediment. Burrows can have differential (greater or lesser) porosity relative to their surroundings for a variety of reasons including piping down of coarser or finer sediment, differential sediment packing (in many cases by the burrowing organism) or chemical differences related to levels of organic matter inside or along the walls of the burrows. Slight horizontal lamination in the burrow indicates that it may have been back-filled by the organism (possibly *Taenidium*).

PPL, KFS, BDI, Scale bar = 0.49 mm

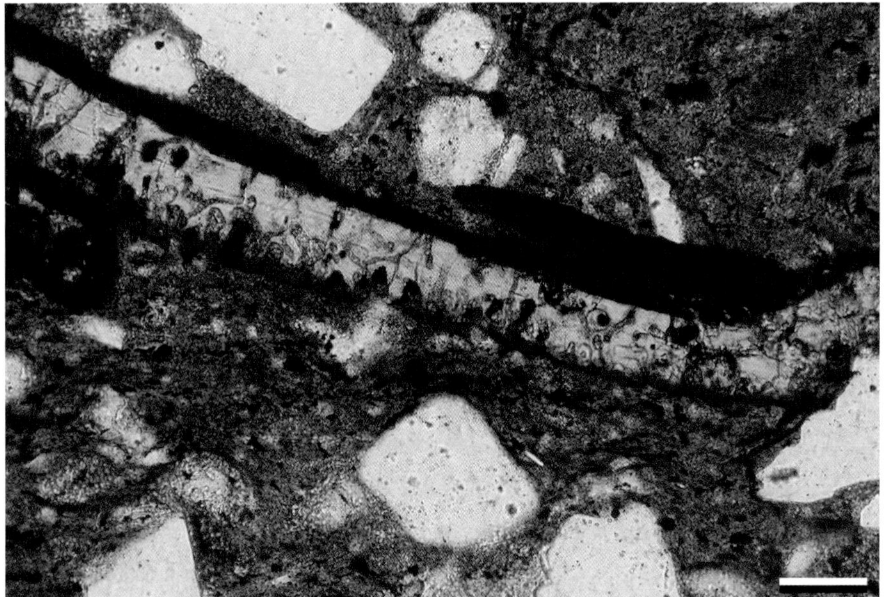

Cretaceous Mowry Fm., Moffat Co., Colorado

Borings (along with filled and unfilled boring porosity) in a phosphatic brachiopod. These small borings show varied diagenetic histories. A few were filled with surrounding sediment, many were filled with precipitated iron sulfide cements (mostly pyrite, but with some altered to iron oxides or hydroxides), and many still contain unfilled primary porosity (marked by red-dyed epoxy fillings). Borings rarely contribute substantially to overall rock porosity, although they can be fairly extensive in slowly deposited sediments associated with hiatal intervals.

PPL, RDI, Scale bar = 0.05 mm

Mid. Jurassic Curtis Fm., Moffat Co., Colorado

Borings within a bivalve shell in a glauconitic sandstone. These are termed borings rather than burrows because they are drilled or dissolved into hard materials rather than soft sediment. Initially these represented open, boring porosity but those pores were subsequently filled, probably very early by adjacent sediment. Many such borings, especially ones produced by boring sponges, have very small surface openings and large interior chambers and those are less likely to become filled with surrounding sediment.

PPL, RDI, Scale bar = 0.10 mm

Pleistocene – Holocene alluvium, Canterbury, New Zealand

Shrinkage and root pores—exposure-related (eogenetic) pore types from a calcareous soil nodule in exposed terrigenous sands. An open root trace (rhizolith) extends diagonally across the view and is partially occluded with cloudy-looking clay and carbonate cement coatings. The largest pore in the image may be a cross-section of a larger root. In addition, many grains were dissolved to form moldic pores; others grains are not dissolved but show circumgranular shrinkage porosity and additional shrinkage cracks are present. This combination of features is characteristic of subaerial exposure (e.g., Tandon and Friend, 1989). Total porosity is 18%.

PPL, BDI, Scale bar = 0.51 mm

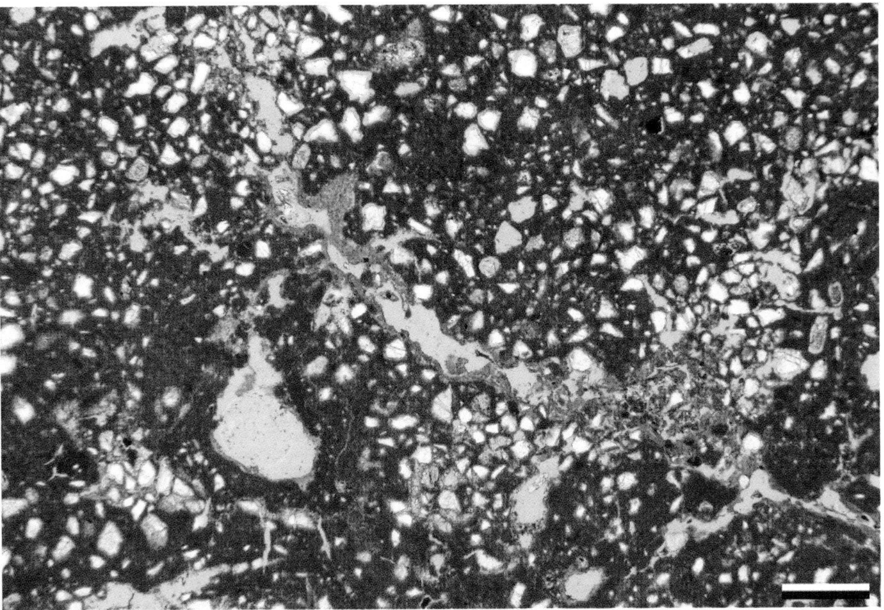

Pleistocene – Holocene coastal alluvium, Canterbury, New Zealand

An example of shrinkage porosity related to circumgranular cracking in calcareous soil nodules in the same sample as above. This type of fabric typically is produced by repeated wetting and drying (with associated shrinkage and expansion) in soils and other surficial materials so it is technically secondary, but essentially synsedimentary. Note also the presence of matrix cracks that also are a form of shrinkage porosity (arrow) and possible root traces in the upper center (root porosity) that may help to connect the otherwise isolated circumgranular shrinkage pores. Total porosity in this area, as measured by image analysis, is about 14%.

PPL, BDI, Scale bar = 0.10 mm

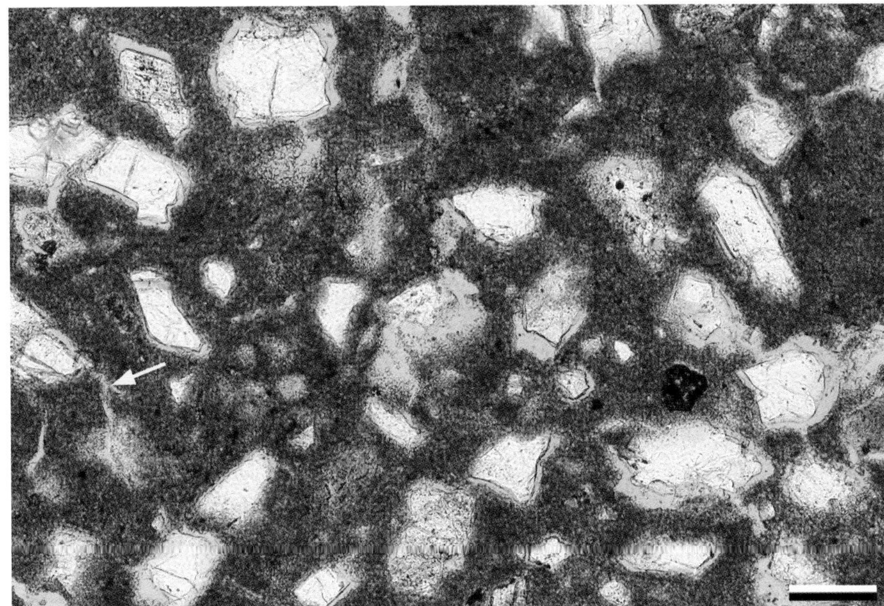

Miocene Lower Santa Fe Gp., Taos Co., New Mexico

Secondary porosity formed by the dissolution of detrital grains (most commonly feldspars, heavy minerals or unstable rock fragments) contributes to total porosity in many diagenetically altered sandstones. In this example, two rock fragments have undergone significant removal of material—interstitial carbonate (left) and constituent feldspars (right). Although most of the secondary porosity remains unfilled, a significant amount of the original primary porosity was filled with kaolinite (the cloudy and patchy white material in the image) which itself has considerable intercrystalline porosity.

PPL, BDI, Scale bar = 0.10 mm

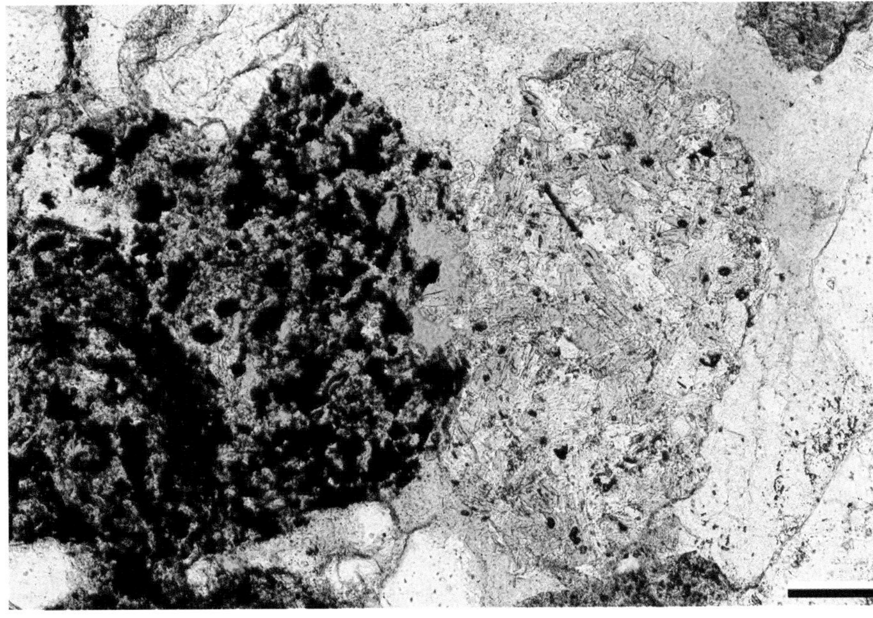

Eocene Green River Fm., Laney Mbr., Fremont Co., Wyoming

Grain dissolution has added some moldic pores to the moderate intergranular porosity in this immature (poorly-sorted and poorly-rounded) sandstone. Because these molds have neither any remnants of the original grains nor any distinctive crystal shapes, it is not possible to determine what materials have been leached. Nonetheless, hydrologists and petroleum geologists can still be pleased with the added pore space. The absence of any cements in the largest mold, however, always raises concern that such pores could be artifacts of grain plucking prior to sample impregnation.

PPL, BDI, Scale bar = 0.26 mm

Lo.–Mid. Eocene Cub Mountain Fm., Lincoln Co., New Mexico

In many arkosic sandstones, feldspar dissolution is the most important source of secondary porosity. In this example, the yellow-stained (sodium cobaltinitrite) K-feldspars have undergone incipient, selective dissolution along compositional zones, cleavage planes and fractures. Some other dissolution pores are present in the rock, along with minor remnants of primary intergranular porosity. However, the majority of the original intergranular porosity has been destroyed by compaction, at least in this area of the sample.

PPL, KFS, BDI, Scale bar = 0.26 mm

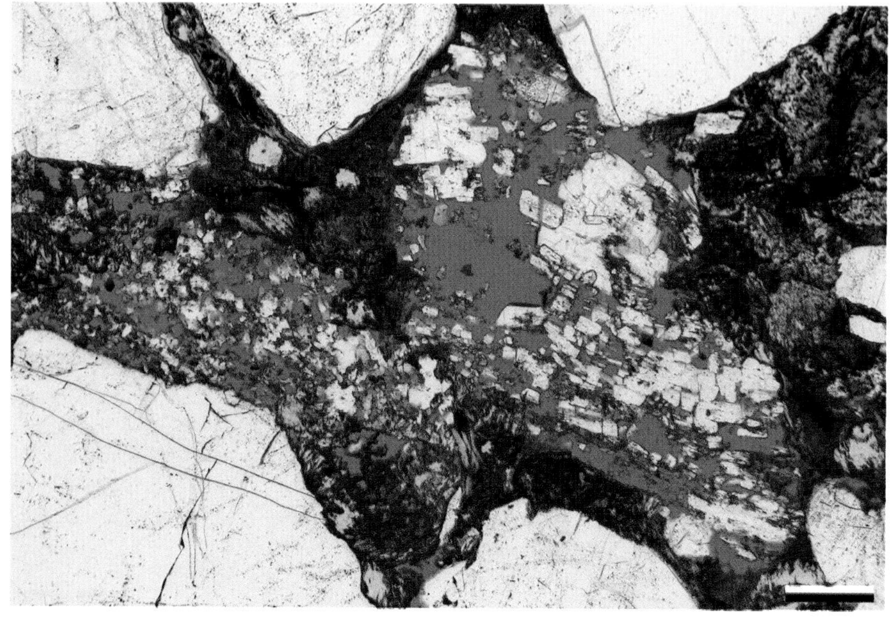

Lo. Pennsylvanian Morrow B Sandstone, Ochiltree Co., Texas

An example of a rock with almost no primary porosity retention and with substantial secondary porosity development. The primary pores here were filled with eogenetic sphaerosiderite (probably of paleosol origin). Subsequent extensive dissolution of the detrital plagioclase feldspar grains at image center produced moldic pores that were reduced slightly by compaction and authigenic cements. Minor, secondary pores also were created through later-stage quartz fracturing; it is volumetrically minor, but could be important in providing permeability pathways between otherwise isolated moldic pores.

PPL, RDI, Scale bar = 0.26 mm

Up. Cretaceous Prince Creek Fm., North Slope, Alaska

The inclusion-rich orthoclase feldspar (center) has a euhedral, inclusion-poor overgrowth. Later partial dissolution of less stable areas created secondary moldic pores within the feldspar grain. Using graphics programs, one can artificially color specific pore types (such as the red color in secondary pores in this image). Utilizing image processing programs (e.g., NIH's free ImageJ), one can determine the percentages of each color. Here, total porosity is 38% (23% primary and 15% secondary). Such analysis depends on the accuracy of pore type identifications and generally does not accurately reflect small micropores and nanopores.

PPL | COL, BDI, Scale bar = 0.04 mm

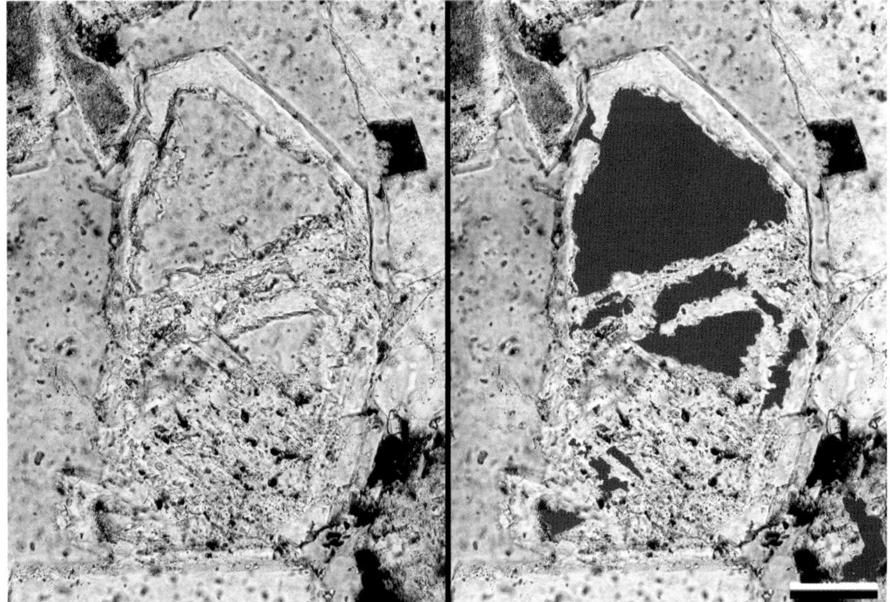

Lo. Pennsylvanian Morrow B Sandstone, Ochiltree Co., Texas

Secondary moldic (feldspar dissolution) pores commonly can become occluded by cement. In this case kaolinite cements (the granular pink and black areas) filled both primary and secondary pores. Thus, the main pores now are intercrystalline, within the pore-filling, vermicular kaolinite booklets. Kaolinite cement typically is very porous (as shown by the red dye impregnation and the dark, residual hydrocarbon staining) and gives the pores a characteristic granular to cloudy appearance. Total image analysis porosity for this sample is 20.2%, but such analysis is less accurate with intercrystalline microporosity.

PPL, RDI, Scale bar = 0.26 mm

Eocene Cub Mountain Fm., Lincoln Co., New Mexico

This high-magnification view shows vermicular kaolinite filling a mix of primary intergranular and feldspar dissolution (moldic) porosity. The individual booklets, and the associated intracrystalline porosity, are clearly visible at this scale. Total porosity in this sample is 25.5% based on image analysis, but such analysis is difficult with the diffuse boundaries between pores and clay packets in kaolinitic areas.

PPL, BDI, Scale bar = 0.07 mm

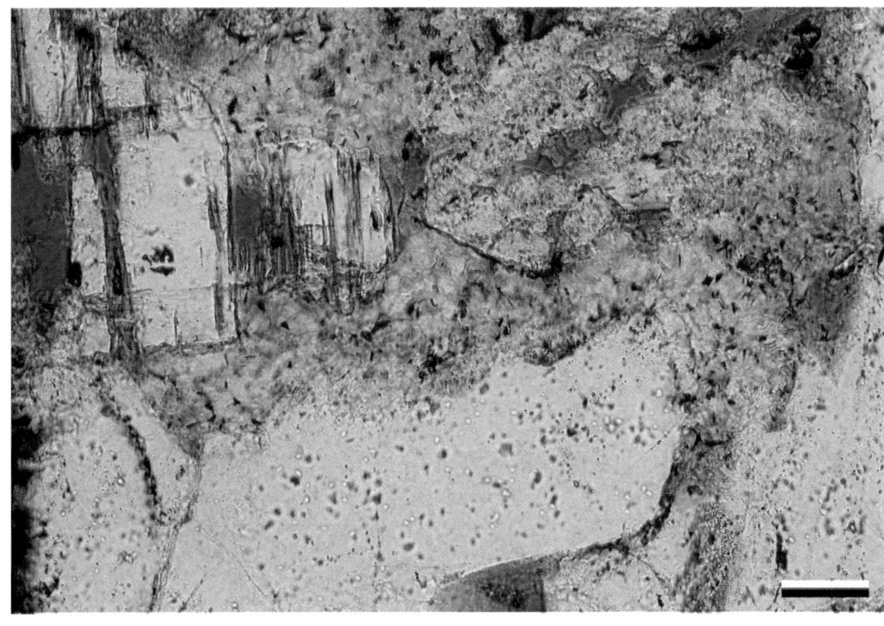

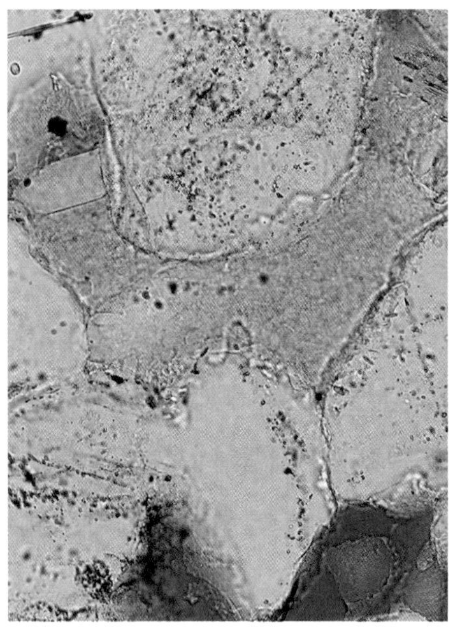

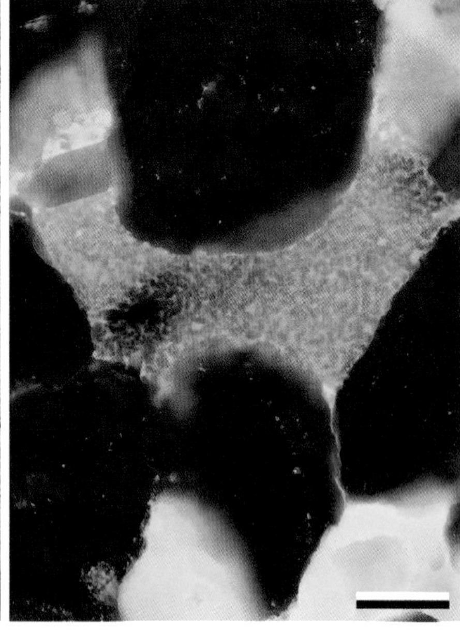

Jurassic Morrison Fm., Moffat Co., Colorado

In many lower magnification views, kaolinite fillings of primary or secondary pores just appear cloudy. Imaging with fluorescence microscopy, as in the view on the right, sometimes can help to resolve the individual booklets and the associated intercrystalline porosity. It is the fluorescent dye in the epoxy that is luminescent here.

PPL | FL470, RDI, Scale bar = 0.51 mm

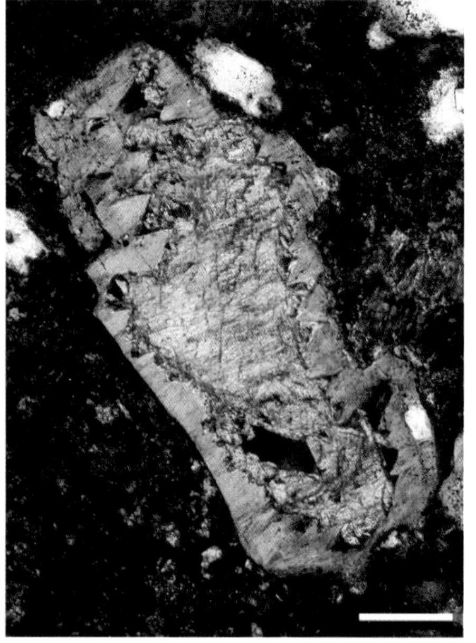

Mid. Eocene – Lo. Oligocene Spears Gp., Socorro Co., New Mexico

Secondary (moldic) porosity resulting from feldspar dissolution can be partially reduced or completely occluded by a variety of cements other than kaolinite. In this example, the exterior of the grain was albitized and the rest of the feldspar was dissolved (based on the absence of feldspar remnants in the grain interior). Subsequently, albite crystals grew inward from the rim, and later, much of the remaining large void was cemented by coarsely crystalline calcite. The calcite was later partially corroded to produce a second stage of dissolution porosity.

PPL | XPL, BDI, Scale bar = 0.07 mm

Lo. Cretaceous (Albian) Nahr Umr Fm., subsurface eastern Qatar

Many grains other than feldspars or rock fragments can undergo dissolution to create secondary (moldic) pores. This is an example of mixed pore types in a siderite-cemented, chamosite ooid-rich, glauconitic quartz arenite. The chamosite ooids probably had berthierine interlayers that were selectively leached to form secondary moldic (oomoldic) porosity that was later reduced by compaction. Some primary porosity has been retained in this sample, although it has been reduced both by compaction and by extensive siderite precipitation (purple-stained crystals).

PPL, AFeS, BDI, Scale bar = 0.10 mm

Mid.? Permian (former Ufimian) Solikamskaya Horizon, near Solikamsk, Perm region, Russia

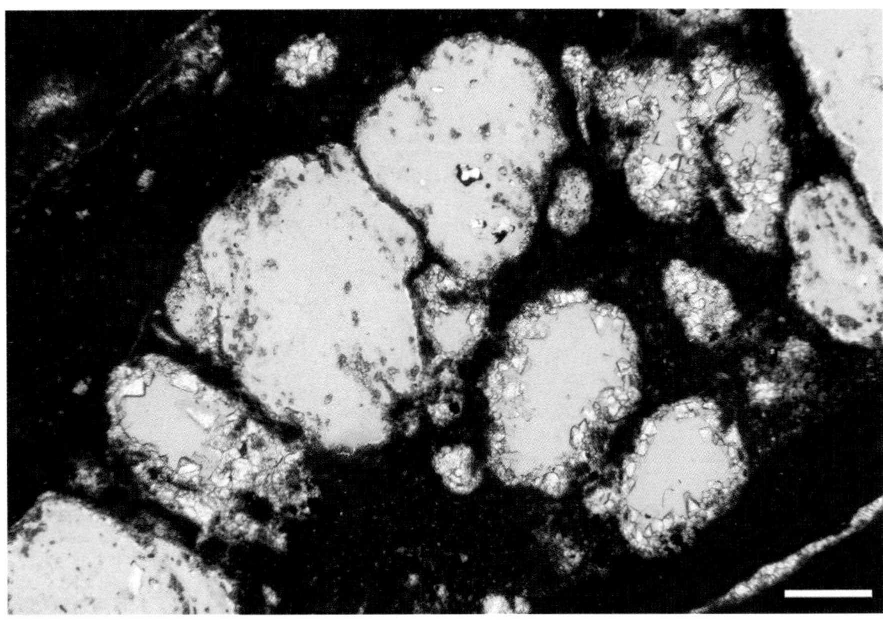

An example of large (although somewhat isolated) moldic pores in a dolomitic shale. In this case, the pores are related to the dissolution of evaporite nodules (typically composites of thousands of small crystal fragments rather than a single crystal). Because there are neither remnants of the original mineral nor distinctive crystal outlines, one can only surmise that these were anhydrite nodules. These moldic pores were slightly reduced by dolomite precipitation around the void margins, but this area of the sample still retains about 30% secondary porosity.

PPL, BDI, Scale bar = 0.51 mm

Cretaceous Dakota Gp., Moffat Co., Colorado

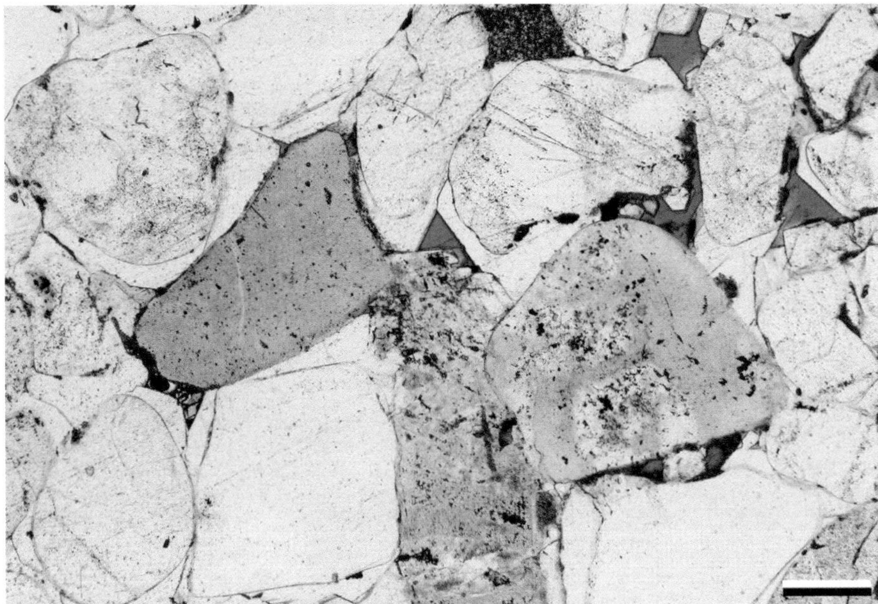

In this example, detrital chert grains (left and right of center in this image) underwent dissolution but did not produce large secondary pores. Instead, they developed disseminated micropores commensurate with their micron-scale crystal structure (as reflected by the pink color). Chert dissolution can occur in association with exposure-related meteoric fluid influx (Shanmugam and Higgins, 1988). The secondary micropores makes the grains easy to confuse with kaolinite cemented molds. Extensive quartz overgrowths greatly reduced primary intergranular porosity here, yielding a total of only 5% total porosity in this sample.

PPL, RDI, Scale bar = 0.26 mm

Mid. Jurassic Entrada Fm., Moffat Co., Colorado

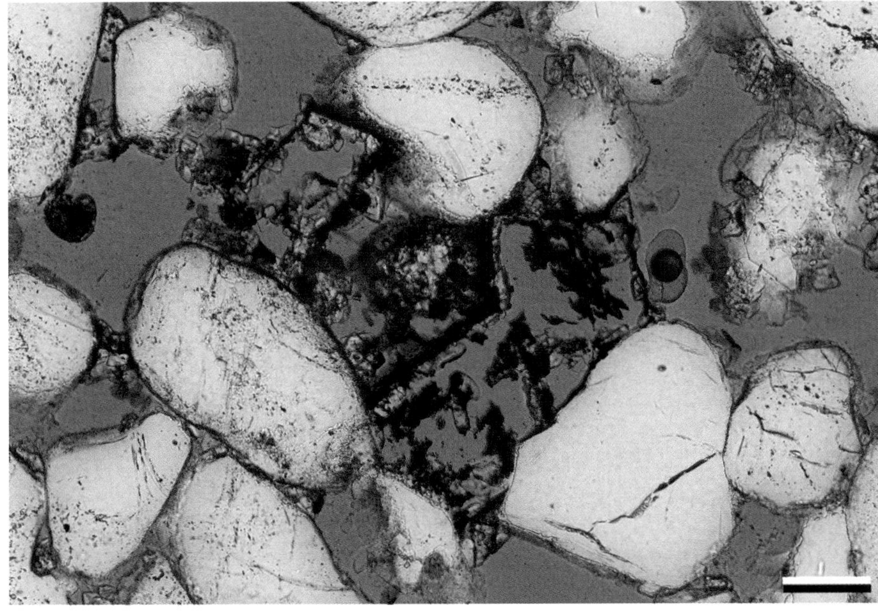

Mixed primary and crystal-moldic secondary pores are present in this porous sandstone (36% measured porosity in this image). The primary intergranular pores were reduced by minor quartz overgrowths and significant precipitation of zoned dolomite and/or ankerite cements. The carbonate crystals subsequently were dissolved almost completely, leaving behind remnants of the chemically most stable zones and iron oxides, both of which preserve the euhedral outlines of the precursor crystals. Thus, these are best termed dolomite/ankerite crystal-moldic pores rather than less specific cement-dissolution pores.

PPL, RDI, Scale bar = 0.04 mm

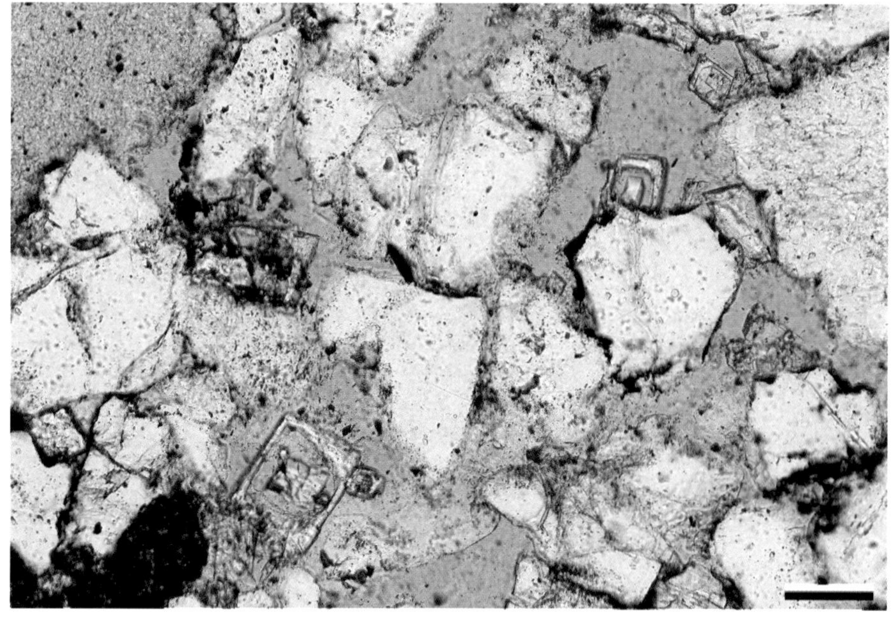

Eocene Galisteo Fm., Sandoval Co., New Mexico

An example of several pore types including primary intergranular and secondary crystal moldic. The crystal-moldic pores are the ones of special interest here and they reflect selective dissolution of particular zones of dolomite cements that had partially occluded primary intragranular pores. The cores and outermost zones of the dolomite were compositionally more stable than the intervening zones during the dissolution event. The fact that both the crystal outline and some of the original crystal remain allow these to be called "crystal-moldic pores" rather than "cement dissolution pores".

PPL, BDI, Scale bar = 0.05 mm

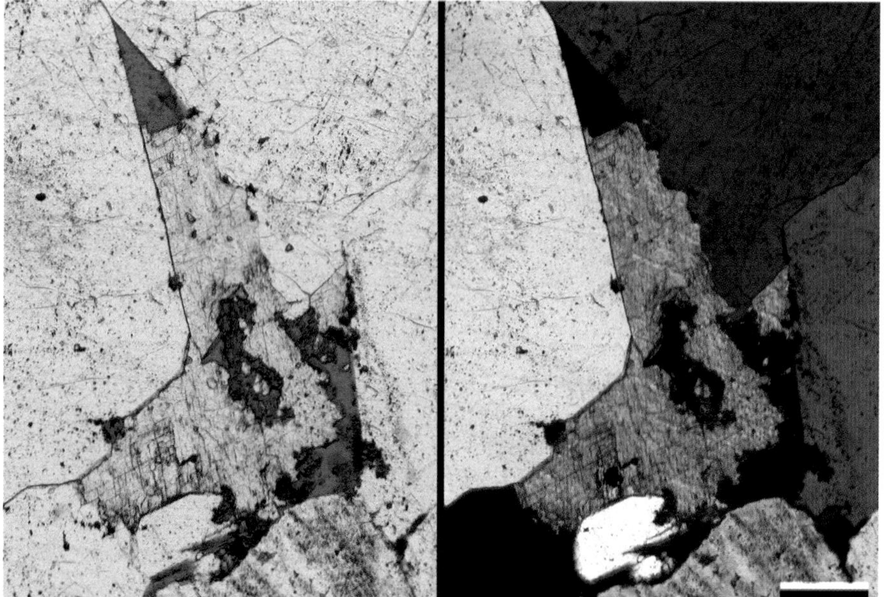

Pennsylvanian Morrow B Sandstone, Ochiltree Co., Texas

Unlike the crystal-moldic porosity shown in the previous two examples, this image and the next show more irregular dissolution of cements that, if carried to completion, would leave little evidence of the former existence of carbonate cements. In this case, the corrosion is attacking coarsely crystalline ankerite cements that postdate small quartz overgrowths. In the context of the entire sample, cement dissolution pores are a significant part of the total porosity.

PPL | XPL, RDI, Scale bar = 0.26 mm

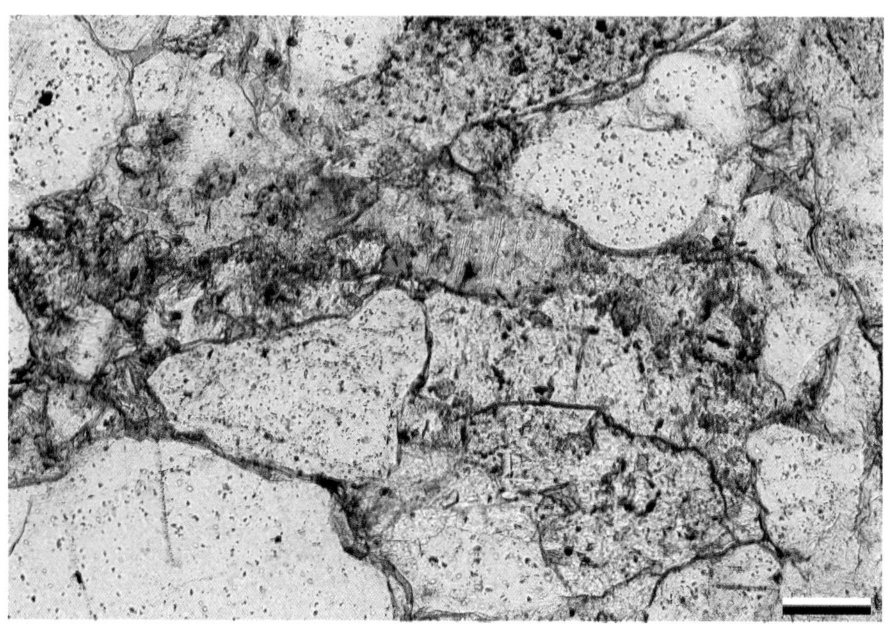

Eocene Galisteo Fm., Sandoval Co. New Mexico

An example of extensive corrosion of calcite cements in a lithic arenite. Here, irregular patches of dissolution have given a Swiss cheese (Emmentaler/Jarlsberg) appearance to the calcites. Many of the dissolution pores have rectilinear outlines, perhaps following cleavage lines or reflecting selective dissolution of former small dolomite inclusions within the calcite. However, the overall pore pattern created by such patchy leaching is best described as "cement dissolution pores".

PPL, BDI, Scale bar = 0.10 mm

Miocene Santa Fe Gp., Zia Fm., Cerro Conejo Mbr., Sandoval Co., New Mexico

Patchy cementation in a concretion led to localized fluid-flow in this alluvial sandstone; associated dissolution, probably of unstable rock fragments, resulted in formation of secondary vuggy porosity. A vug is loosely defined as a cavity larger than the surrounding grains (and thus, one that had to have formed both by the dissolution of multiple particles and after sufficient lithification of the surrounding material to support such a large pore). In this case, the vug also appears to have associated solution-enlarged fluid pathways (termed channel porosity).

PPL, BDI, Scale bar = 0.51 mm

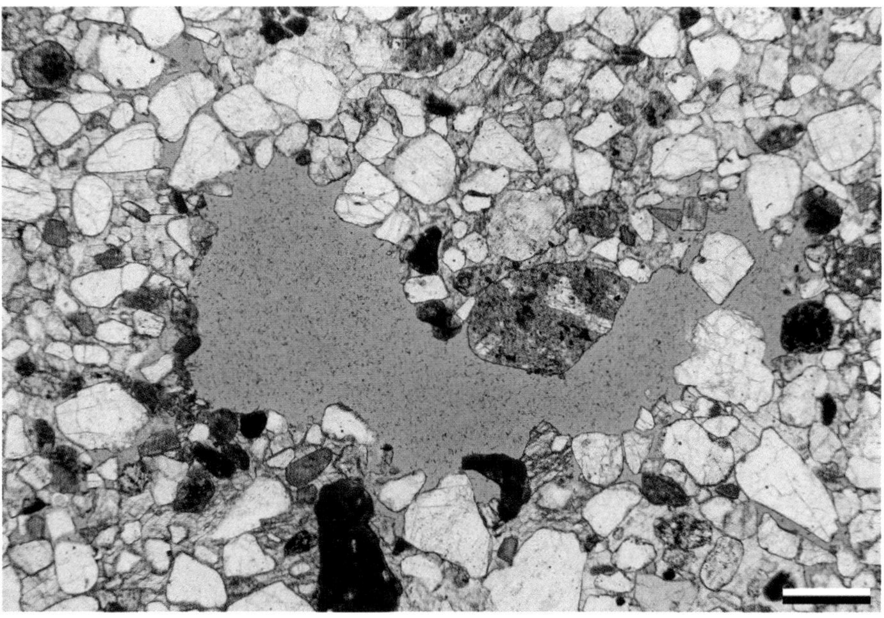

Up. Cretaceous Mesaverde Gp., McKinley Co., New Mexico

Fracture porosity is the one form of secondary porosity that, initially at least, produces an increase in overall porosity (as opposed to dissolution porosity which may be offset by cementation in other pores). Fractures can occur either within shattered grains, as in this example, or as larger, through-going features that transect broader areas. Here, the porosity within shattered grains is the main pore type, but it is supplemented by minor grain dissolution and some circumgranular fracturing.

PPL, BDI, Scale bar = 0.15 mm

Up. Permian Schuchert Dal Fm., Jameson Land, East Greenland

An example of small-scale, circumgranular fracture development around a large polycrystalline quartz grain (center) as well as many others. Such fractures can be developed in sheared rocks or in deformation bands and can be significant contributors to rock permeability. Such unhealed fractures also can be produced by poor sample collection and handling. The completely uncemented fractures here postdate late-stage, highly ferroan calcite in matrix pores and so are either produced during uplift (telogenesis) or by sampling/preparation.

PPL, AFeS, BDI, Scale bar = 0.51 mm

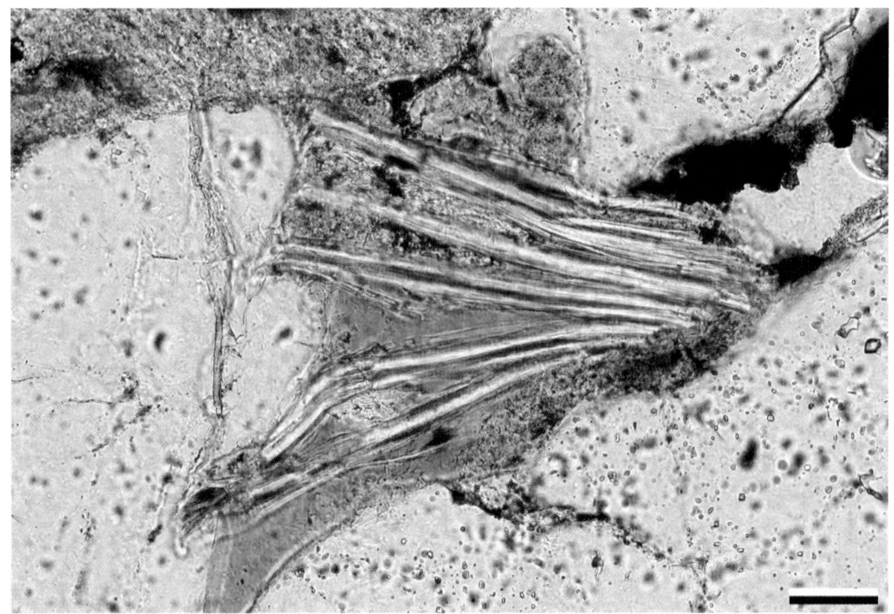

Lo.–Mid. Eocene Cub Mountain Fm., Lincoln Co., New Mexico

Compaction or other compressive deformation can lead to secondary intragranular pore development within micas in a surprising number of cases. There is probably little net addition of porosity, however, as the micas can only expand into whatever nearby primary or secondary pores existed at the time of deformation.

PPL, BDI, Scale bar = 0.05 mm

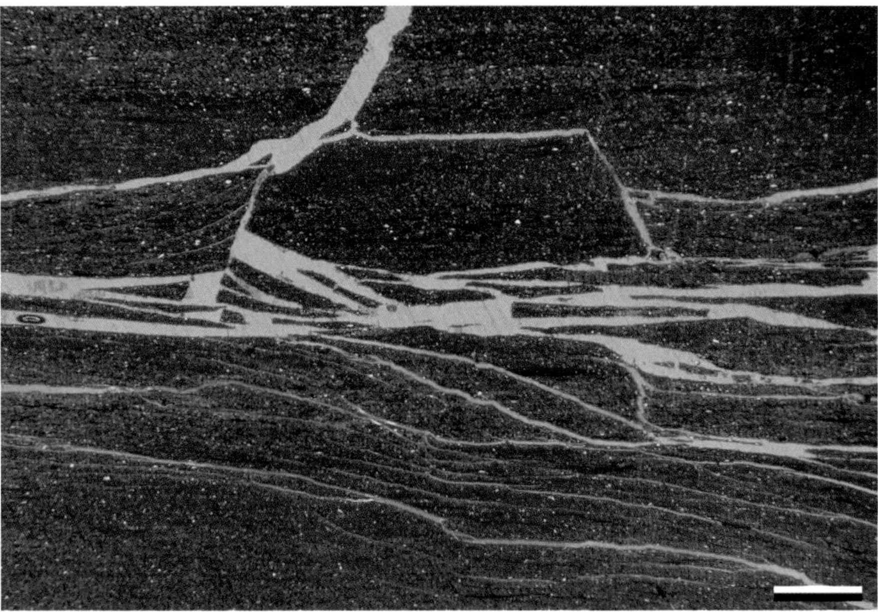

Lo. Pennsylvanian Morrow B Sandstone, Ochiltree Co., Texas

An example of multiple bedding-parallel fractures that are common in laminated or fissile mudrocks (here constituting about 15% pore space). In the absence of any traces of cementation or hydrocarbon staining, as in this case, one should be very cautious in interpreting these as natural fractures, because they could be (and in this case, most likely are) a result of coring, core storage and sample desiccation or thin section preparation.

PPL, RDI, Scale bar = 0.51 mm

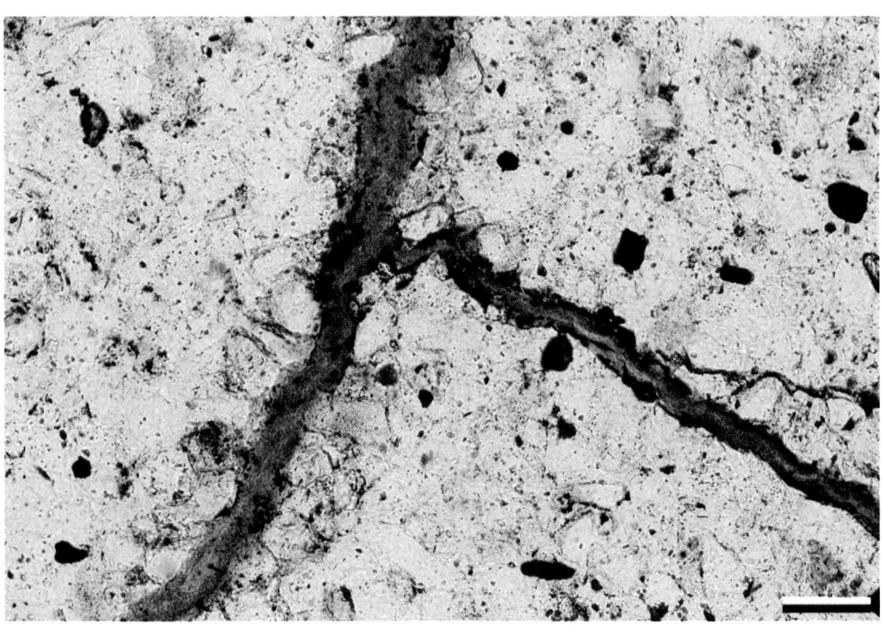

Mississippian Paradise Fm., Hidalgo Co., New Mexico

An example of intersecting fractures that clearly are natural and not induced by sampling or section preparation because they contain some diagenetic precipitates. Although they still have porosity, they show irregular formation of cement linings, consisting of clays as well as aphanocrystalline calcite. So this is an example of cement-reduced fracture porosity.

PPL, BDI, Scale bar = 0.10 mm

Up. Cretaceous Rio Ancho Fm., La Guajira Department, Colombia

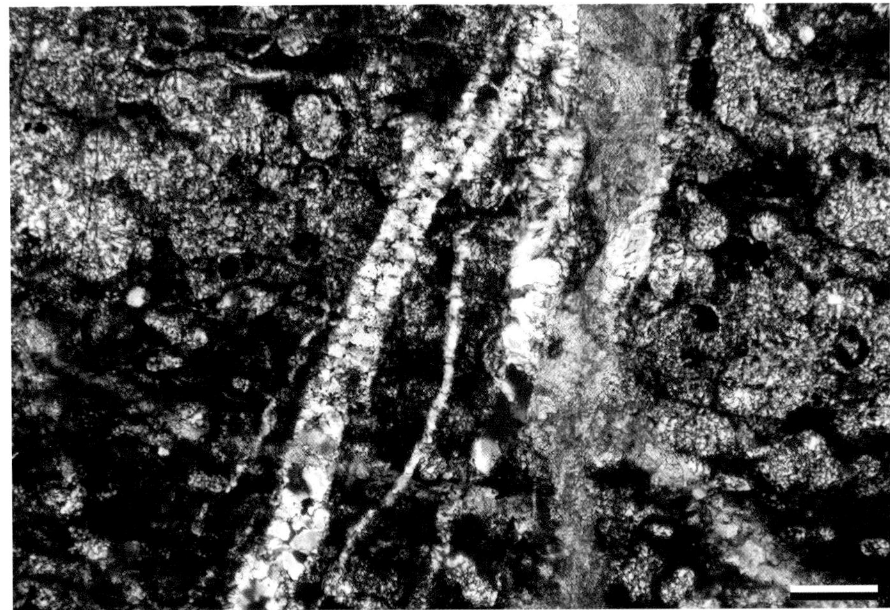

Multiple generations of healed, cross-cutting fractures are seen in this radiolarian chert The earlier set of almost exclusively vertical fractures are quartz-filled (not surprising in a radiolarian chert in which opaline silica would be available early in the diagenetic history). The second set of fractures have horizontal and vertical orientations, clearly cut the earlier fractures and are completely filled with calcite that has undergone continued shearing. This is an example of quartz- and calcite-occluded fracture porosity.

XPL, Scale bar = 0.26 mm

Mid. Jurassic Entrada Fm., Moffat Co., Colorado

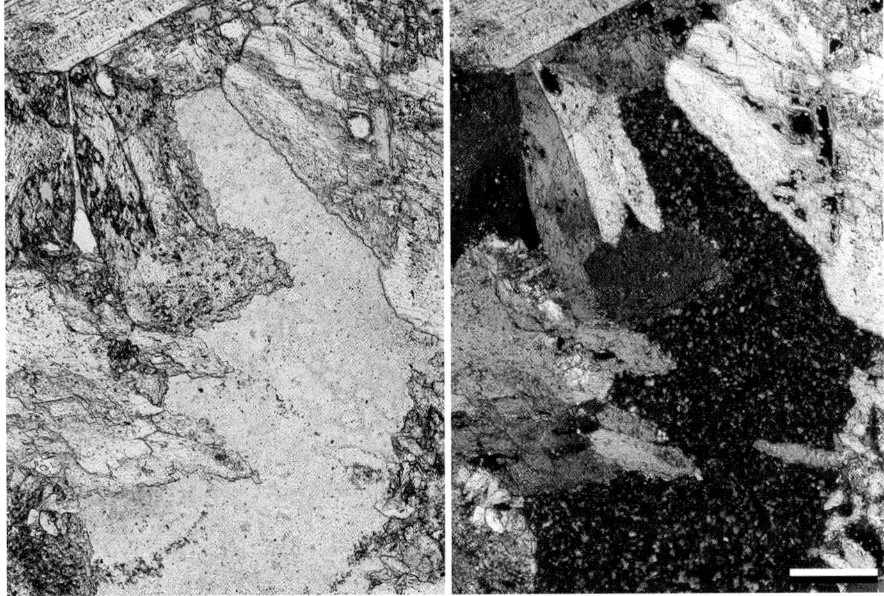

This image shows a small portion of a large fracture that was occluded by a combination of calcite and kaolinite cements. In this example, the fracture retains none of its original porosity and even the kaolinite has virtually no microporosity (as shown by lack of blue dye in the central area of the image). The calcite-filled areas were initially reduced to essentially no porosity. However, subsequent dissolution of small, irregular areas in the calcite created pores (now marked by blue dye). These are best described as secondary cement-dissolution pores rather than fracture or crystal-moldic pores.

PPL | XPL, BDI, Scale bar = 0.10 mm

Cretaceous Mesaverde Gp., Piceance Basin, Colorado

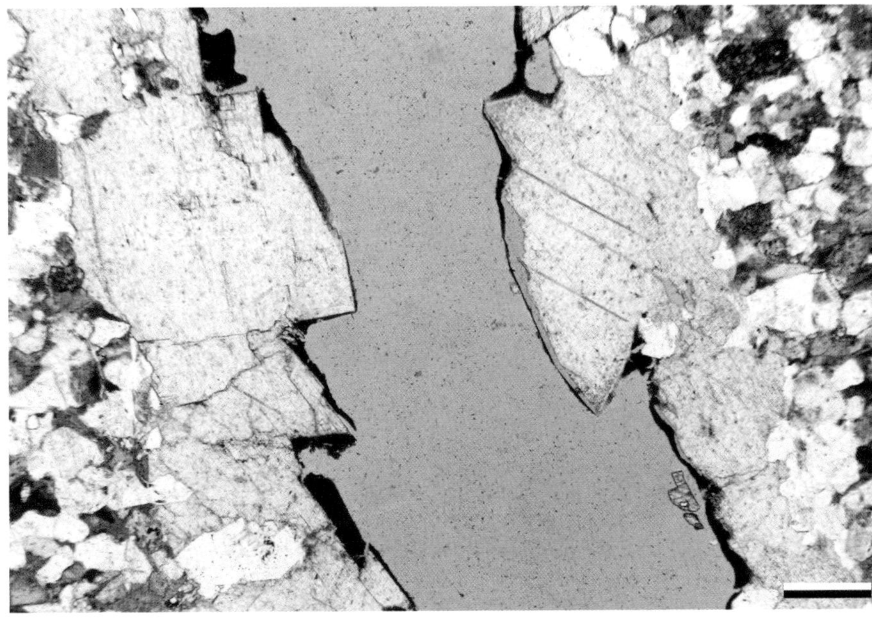

This large fracture retains substantial porosity despite thick linings of calcite (and possibly some dolomite) on both walls. Equally significant from a hydrocarbon exploration viewpoint are the thin hydrocarbon residues on the pore walls indicating that these fractures, in a core from about 1,750 m (5,740 ft) depth, have acted as conduits for oil migration. Some leached porosity also is visible in the sandstone matrix as well as in the fracture cements.

PPL, BDI, Scale bar = 0.51 mm

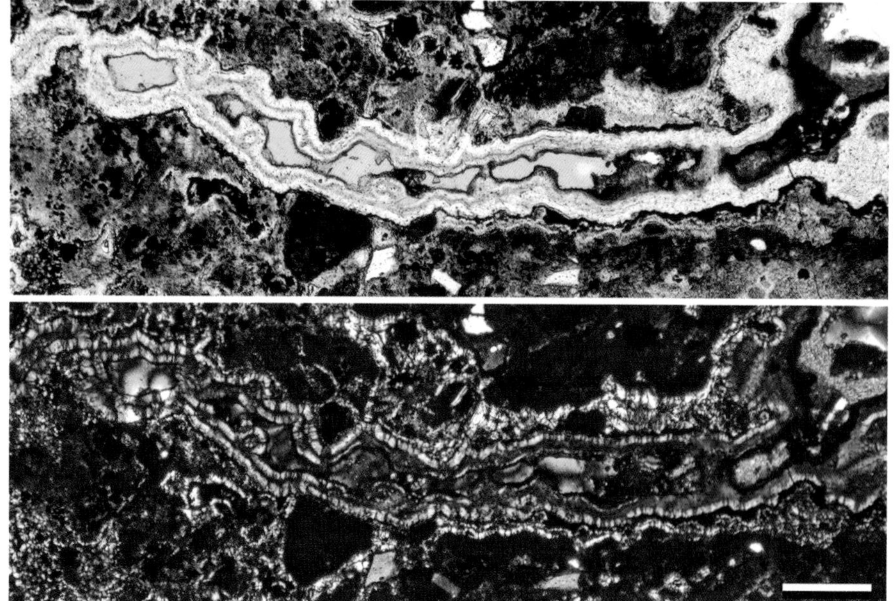

Pliocene – Pleistocene silcrete, east coast, Aruba

Not all fractures, fracture fills and fracture pores are mesogenetic or telogenetic. This example shows a silcrete developed on a reef terrace in rocks that have never left their surficial setting and thus have undergone only eogenetic diagenesis. Nonetheless, fractures and vugs were generated by shrinkage, root bioturbation and other soil-related processes and the fracture pores were substantially filled with multiple crusts of microquartz and iron oxide/hydroxide. The result is a strongly lithified duricrust (silcrete).

PPL / XPL, BDI, Scale bar = 0.26 mm

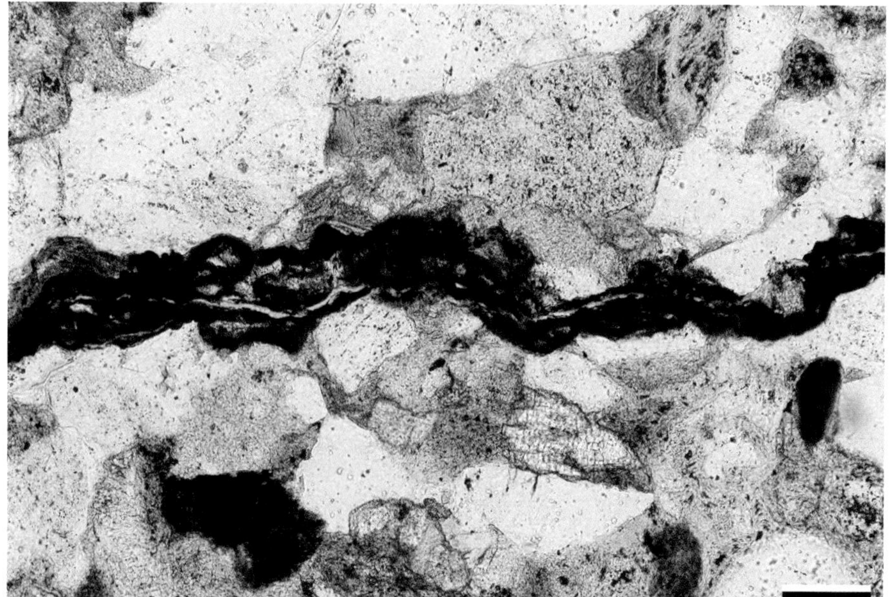

Up. Cretaceous Mesaverde Gp., Piceance Basin, Colorado

A pressure solution surface in a sandstone with small, elongate stylolite-associated pores. It has been shown that load release and/or focused fluid flow along impermeable clay-rich stylolites can produce economically significant porosity, especially in carbonate-rich rocks (Carozzi and Von Bergen, 1987; Dawson, 1988) as can stylolite-associated fractures (Nelson, 1981). However, such porosity also can result from coring, return of unpressurized core to the surface or desiccation of stored samples prior to sectioning (and that may be the case in this example).

PPL, BDI, Scale bar = 0.10 mm

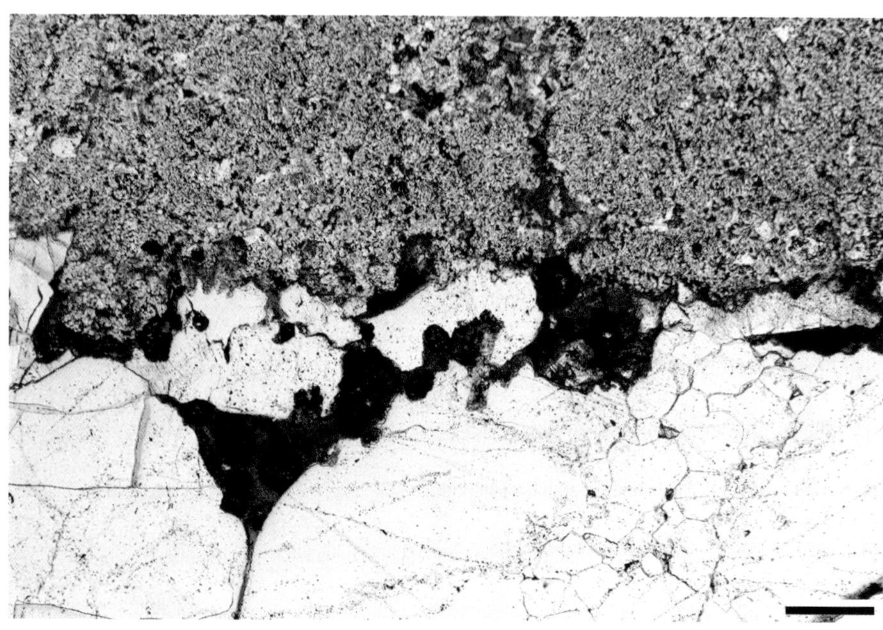

Lo. Pennsylvanian Morrow B Sandstone, Ochiltree Co., Texas

Stylolite-associated porosity along a thin solution seam (and some smaller intergranular stylolitic surfaces). The porosity seen here may be related to load release during Tertiary uplift of the area and/or with siderite dissolution along the solution surface (some siderite remnants are still visible). Additional pores are associated with small fractures in adjacent crystals that may have resulted from stress along this zone (Nelson, 1981).

PPL, RDI, Scale bar = 0.10 mm

Ordovician Utica Shale, New York

SEM image of an ion-milled sample showing porous organic matter (kerogen) in a carbonaceous shale. The organic matter contains abundant very tiny pores (foam pores) and large pores (bubble pores) that show 3D pore throats connecting to pores deeper in the sample; all the visible pores are likely interconnected. This shale has been heated into the dry-gas window. For more information on such pore varieties in shales see Loucks et al. (2012), Schieber (2013) and Loucks and Reed, (2014).

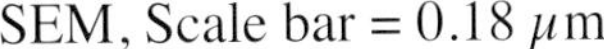

SEM, Scale bar = 0.18 μm

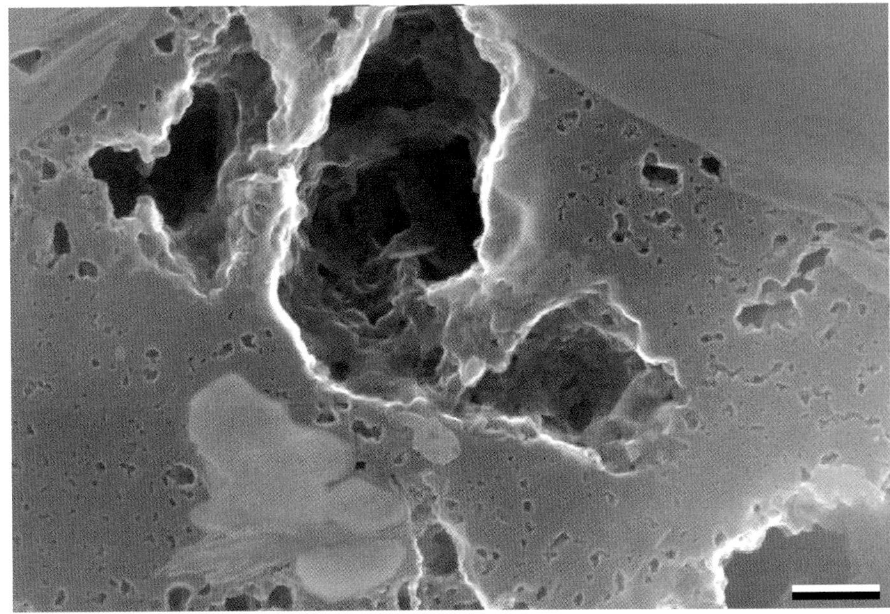

Ordovician Utica Shale, New York

SEM image of an ion-milled shale sample showing another example of organic matter porosity in the Utica Shale. Organic matter (OM) is interstitial between mineral grains (Qtz=quartz; cc=calcite) and has well developed pores (arrow). Note the bimodal distribution of "foam" and "bubble" pores.

SEM, Scale bar = 0.25 μm

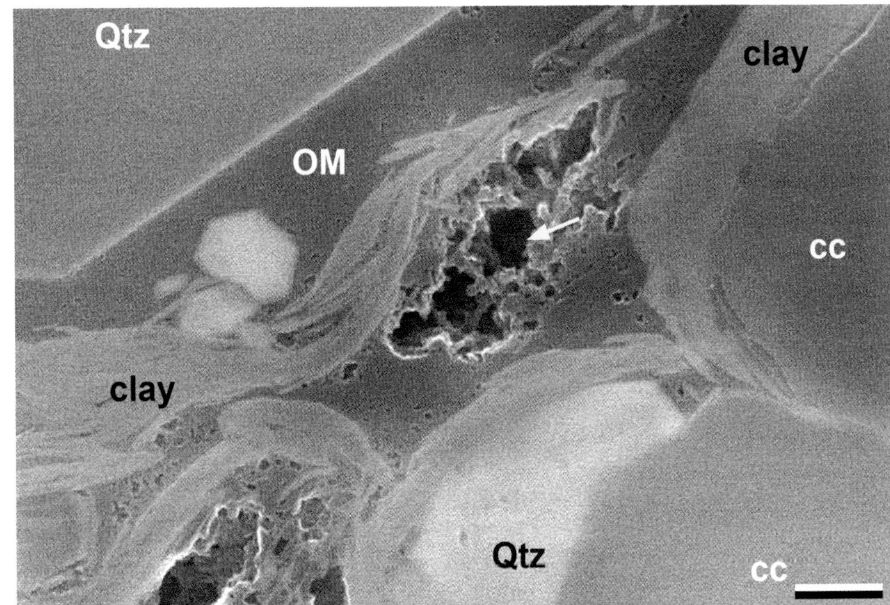

Ordovician Utica Shale, New York

SEM image of an ion-milled shale sample. The conspicuous pore southwest of the image center is framed by mineral grains (calcite). It is an intergranular pore, and this type also has been described as carbonate framework pores (Schieber, 2013).

SEM, Scale bar = 0.93 μm

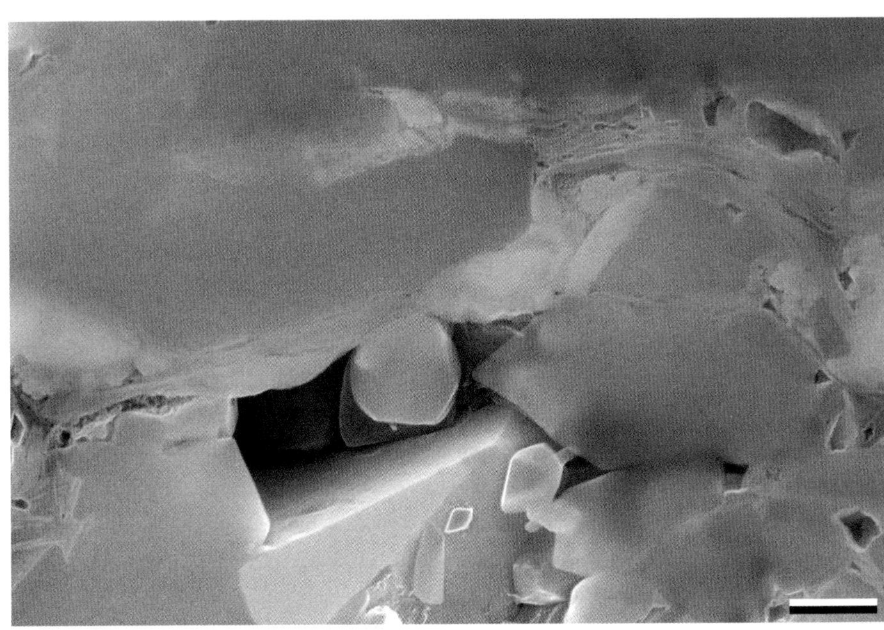

Devonian Marcellus Shale, New York

SEM image of an ion-milled shale sample showing another pore category. These pores are framed by clay minerals that were cemented between diagenetic quartz grains. Such intergranular pores have been described as phyllosilicate framework pores (Schieber, 2013) and as clay framework pores in this book. They have characteristic triangular cross-sections.

SEM, Scale bar = 0.54 μm

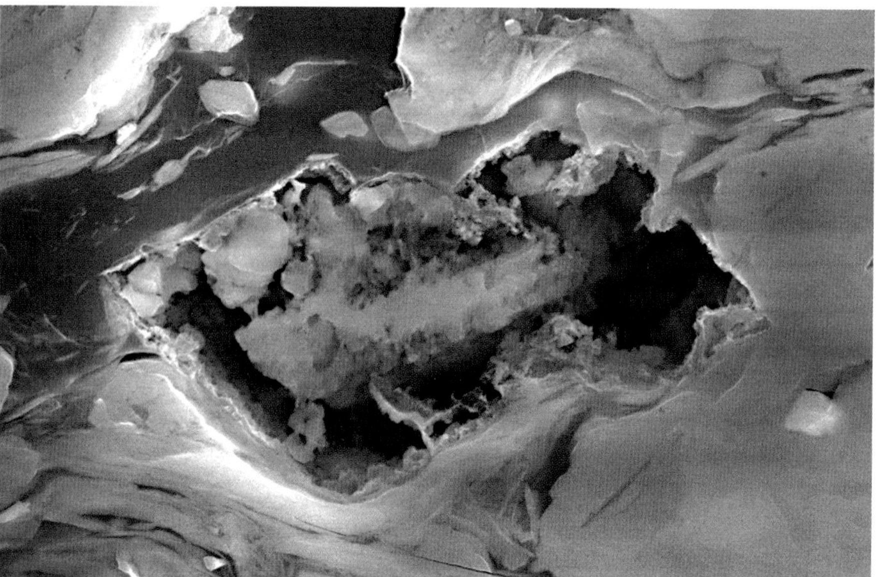

Devonian New Albany Shale, Indiana

SEM image of an ion-milled shale sample showing secondary porosity in a carbonaceous shale. The pore in the center of the image was once occupied by dolomite, which largely was dissolved due to contact with organic acids that formed during diagenesis (Schieber, 2013). Such secondary pores can contribute significantly to overall porosity.

SEM, Scale bar = 1.83 μm

Cited References and Additional Information Sources

Sandstone porosity (primary and secondary):

Appoloni, C. R., C. P. Fernandes, and C. R. O. Rodrigues, 2007, X-ray microtomography study of a sandstone reservoir rock: Nuclear Instruments and Methods in Physics Research Section A: Accelerators, Spectrometers, Detectors and Associated Equipment, v. 580, p. 629-632, doi: 10.1016/j.nima.2007.05.027.

Atwater, G. I., and E. E. Miller, 1965, The effect of decrease in porosity with depth on future development of oil and gas reserves in south Louisiana [abs.]: AAPG Bulletin, v. 49, p. 334.

Barwis, J. H., J. G. McPherson, and J. R. J. Studlick, eds., 1990, Sandstone Petroleum Reservoirs: New York, Springer-Verlag, 583 p.

Bjørlykke, K., 1984, Formation of secondary porosity: How important is it?: Part 2. Aspects of porosity modification, *in* D. A. McDonald, and R. C. Surdam, eds., Clastic Diagenesis: Tulsa, OK, AAPG Memoir 37, p. 277-286.

Bjørlykke, K., 1988, Sandstone diagenesis in relation to preservation, destruction and creation of porosity, *in* G. V. Chilingarian, and K. H. Wolf, eds., Diagenesis: New York, Elsevier Science Publ., p. 555-588, doi: 10.1016/S0070-4571(08)70180-8.

Blatt, H., G. V. Middleton, and R. C. Murray, 1980, Origin of Sedimentary Rocks: Englewood Cliffs, NJ, Prentice Hall Inc., 782 p.

Bloch, S., 1994, Secondary porosity in sandstones: Significance, origin, relationship to subaerial unconformities, and effect on predrill reservoir quality prediction, *in* M. D. Wilson, ed., Reservoir Quality Assessment and Prediction in Clastic Rocks: Tulsa, OK, SEPM Short Course Notes 30, p. 137-159, doi: 10.2110/scn.94.30.0137.

Burley, S. D., and J. D. Kantorowicz, 1986, Thin section and SEM textural criteria for the recognition of cement-dissolution porosity in sandstones: Sedimentology, v. 33, p. 587-604, doi: 10.1111/j.1365-3091.1986.tb00764.x.

Carozzi, A. V., and D. Von Bergen, 1987, Stylolitic porosity in carbonates: A critical factor for deep hydrocarbon production: Journal of Petroleum Geology, v. 10, p. 267-282, doi: 10.1111/j.1747-5457.1987.tb00946.x.

Choquette, P. W., and L. C. Pray, 1970, Geologic nomenclature and classification of porosity in sedimentary carbonates: AAPG Bulletin, v. 54, p. 207-250.

Coskun, S. B., and N. C. Wardlaw, 1993, Estimation of permeability from image analysis of reservoir sandstones: Journal of Petroleum Science and Engineering, v. 10, p. 1-16, doi: 10.1016/0920-4105(93)90046-H.

Coskun, S. B., N. C. Wardlaw, and B. Haverslew, 1993, Effects of composition, texture and diagenesis on porosity, permeability and oil recovery in a sandstone reservoir: Journal of Petroleum Science and Engineering, v. 8, p. 279-292, doi: 10.1016/0920-4105(93)90005-Y.

Dawson, W. C., 1988, Stylolite porosity in carbonate reservoirs [abs.]: AAPG Bulletin, v. 72, p. 176.

Dutton, S. P., and R. G. Loucks, 2010, Diagenetic controls on evolution of porosity and permeability in lower Tertiary Wilcox sandstones from shallow to ultradeep (200–6700m) burial, Gulf of Mexico Basin, U.S.A: Marine and Petroleum Geology, v. 27, p. 69-81, doi: 10.1016/j.marpetgeo.2009.08.008.

Ehrlich, R., S. J. Crabtree, K. O. Horkowitz, and J. P. Horkowitz, 1991, Petrography and reservoir physics, I: Objective classification of reservoir porosity: AAPG Bulletin, v. 75, p. 1547-1562.

Ehrlich, R., S. K. Kennedy, S. J. Crabtree, and R. L. Cannon, 1984, Petrographic image analysis: I. Analysis of reservoir pore complexes: Journal of Sedimentary Research, v. 54, p. 1365-1378, doi: 10.1306/212F85DF-2B24-11D7-8648000102C1865D.

Fassi-Fihri, O., M. Robin, and E. Rosenberg, 1995, Wettability studies at the pore level: A new approach by the use of cryo-scanning electron microscopy: SPE Formation Evaluation, v. 10, no. 1, p. 11-19, http://dx.doi.org/10.2118/22596-PA.

Gerard, R. E., C. A. Philipson, F. M. Ballentine, and D. M. Marschall, 1992, Petrographic image analysis: An alternate method for determining petrophysical properties, *in* I. Palaz, and S. K. Sengupta, eds., Automated Pattern Analysis in Petroleum Exploration: New York, Springer-Verlag, p. 249-263, doi: 10.1007/978-1-4612-4388-5_13.

Giles, M. R., and R. B. De Boer, 1990, Origin and significance of redistributional secondary porosity: Marine and Petroleum Geology, v. 7, p. 378-397, doi: 10.1016/0264-8172(90)90016-A.

Halley, R. B., 1978, Estimating pore and cement volumes in thin section: Journal of Sedimentary Research, v. 48, p. 641-650.

Hayes, J. B., 1979, Sandstone diagenesis—the hole truth, *in* P. A. Scholle, and P. R. Schluger, eds., Aspects of Diagenesis: Tulsa, OK, SEPM Special Publication 26, p. 127-139, doi: 10.2110/pec.79.26.0127.

Hurst, A., and P. H. Nadeau, 1995, Clay microporosity in reservoir sandstones: An application of quantitative electron microscopy in petrophysical evaluation: AAPG Bulletin, v. 79, p. 563-573.

Lindquist, W. B., and A. Venkatarangan, 1999, Investigating 3D geometry of porous media from high resolution images: Physics and Chemistry of the Earth, Part A: Solid Earth and Geodesy, v. 24, p. 593-599, doi: 10.1016/S1464-1895(99)00085-X.

Loucks, R. G., 2005, Revisiting the importance of secondary dissolution pores in Tertiary sandstones along the Texas Gulf Coast: Gulf Coast Association of Geological Societies Transactions, v. 55, p. 447-455.

Loucks, R. G., M. M. Dodge, and W. E. Galloway, 1979, Importance of secondary leached porosity in Lower Tertiary sandstone reservoirs along the Texas Gulf Coast: Gulf Coast Association of Geological Societies Transactions, v. 29, p. 164-171

Lundegard, P. D., 1992, Sandstone porosity loss—a "big picture" view of the importance of compaction: Journal of Sedimentary Research, v. 62, p. 250-260, doi: 10.1306/D42678D4-2B26-11D7-8648000102C1865D.

Lundegard, P. D., L. S. Land, and W. E. Galloway, 1984, Problem of secondary porosity: Frio Formation (Oligocene), Texas Gulf Coast: Geology, v. 12, p. 399-402, doi: 10.1130/0091-7613(1984)12<399:POSPFF>2.0.CO;2.

Morad, S., K. Al-Ramadan, J. M. Ketzer, and L. F. De Ros, 2010, The impact of diagenesis on the heterogeneity of sandstone reservoirs: A review of the role of depositional facies and sequence stratigraphy: AAPG Bulletin, v. 94, p. 1267-1309, doi: 10.1306/04211009178, doi: 10.1306/04211009178.

Nelson, P. H., 2009, Pore-throat sizes in sandstones, tight sandstones, and shales: AAPG Bulletin, v. 93, p. 329-340, doi: 10.1306/10240808059.

Nelson, R. A., 1981, Significance of fracture sets associated with stylolite zones: AAPG Bulletin, v. 65, p. 2417-2425.

Pittman, E. D., 1979, Porosity, diagenesis and productive capability of sandstone reservoirs, *in* P. A. Scholle, and P. Schluger, eds., Aspects of Diagenesis: Tulsa, OK, SEPM Special Publication 26, p. 159-173, doi: 10.2110/pec.79.26.0159.

Pittman, E. D., 1984, The pore geometries of reservoir rocks, *in* D. L. Johnson, and P. N. Sen, eds., Physics and Chemistry of Porous Media (AIP Conference Proceedings 107): Melville, NY, AIP Publishing, p. 1-20, http://dx.doi.org/10.1063/1.34299.

Pittman, E. D., 1992a, Artifact porosity in thin sections of sandstones: Journal of Sedimentary Research, v. 62, p. 734-737.

Pittman, E. D., 1992b, Relationship of porosity to permeability to various parameters derived from mercury injection–capillary pressure curves for sandstone: AAPG Bulletin, v. 76, p. 191-198.

Pittman, E. D., and R. W. Duschatko, 1970, Use of pore casts and scanning electron microscopy to study pore geometry: Journal of Sedimentary Research, v. 40, p. 1153-1157, doi: 10.1306/74D72158-2B21-11D7-8648000102C1865D.

Rouquerol, J., D. Avnir, C. W. Fairbridge, D. H. Everett, J. M. Haynes, N. Pernicone, J. D. F. Ramsay, K. S. W. Sing, and K. K. Unger, 1994, Recommendations for the characterization of porous solids: Pure and Applied Chemistry, v. 66, p. 1739-1758, doi: 10.1351/pac199466081739.

Ruzyla, K., and D. I. Jezek, 1987, Staining method for recognition of pore space in thin and polished sections: Journal of Sedimentary Research, v. 57, p. 777-778.

Schmidt, V., and D. A. McDonald, 1979, Texture and recognition of secondary porosity in sandstones, *in* P. A. Scholle, and P. Schluger, eds., Aspects of Diagenesis: Tulsa, OK, SEPM Special Publication 26, p. 209-225, doi: 10.2110/pec.79.26.0209.

Schmidt, V., and D. A. McDonald, 1980, Secondary Reservoir Porosity in the Course of Sandstone Diagenesis: Tulsa, OK, AAPG Continuing Education Course Note Series No. 12, 125 p.

Schmidt, V., D. A. McDonald, and R. L. Platt, 1977, Pore geometry and reservoir aspects of secondary porosity in sandstones: Bulletin of Canadian Petroleum Geology, v. 25, p. 271-290.

Shanmugam, G., 1985, Types of porosity in sandstones and their significance in interpreting provenance, *in* G. G. Zuffa, ed., Provenance of Arenites (NATO Science Series C): Dordrecht, Netherlands, D. Reidel Publishing, p. 115-137, doi: 10.1007/978-94-017-2809-6_6.

Shanmugam, G., and J. B. Higgins, 1988, Porosity enhancement from chert dissolution beneath Neocomian unconformity, Ivishak Formation, North Slope, Alaska: AAPG Bulletin, v. 72, p. 523-535.

Tandon, S. K., and P. F. Friend, 1989, Near-surface shrinkage and carbonate replacement processes, Arran Cornstone Formation, Scotland: Sedimentology, v. 36, p. 1113-1126, doi: 10.1111/j.1365-3091.1989.tb01545.x.

Waldschmidt, W. A., P. E. Fitzgerald, and C. L. Lunsford, 1956, Classification of porosity and fractures in reservoir rocks: AAPG Bulletin, v. 40, p. 953-974.

Wardlaw, N. C., 1976, Pore geometry of carbonate rocks as revealed by pore casts and capillary pressure: AAPG Bulletin, v. 60, p. 245-257.

Wardlaw, N. C., and J. P. Cassan, 1978, Estimation of recovery efficiency by visual observation of pore systems in reservoir rocks: Bulletin of Canadian Petroleum Geology, v. 26, p. 572-585.

Wilkinson, M., D. Darby, R. S. Haszeldine, and G. D. Couples, 1997, Secondary porosity generation during deep burial associated with overpressure leak-off: Fulmar Formation, UK Central Graben: AAPG Bulletin, v. 81, p. 803-813.

Wilkinson, M., and R. S. Haszeldine, 2011, Oil charge preserves exceptional porosity in deeply buried, overpressured, sandstones: Central North Sea, UK: Journal of the Geological Society, v. 168, p. 1285-1295, doi: 10.1144/0016-76492011-007.

Wilson, M. D., 1992, Inherited grain-rimming clays in sandstones from eolian and shelf environments: Their origin and control on reservoir properties, *in* D. W. Houseknecht, and E. D. Pittman, eds., Origin, Diagenesis, and Petrophysics of Clay Minerals in Sandstones: Tulsa, OK, SEPM Special Publication 47, p. 209-225, doi: 10.2110/pec.92.47.0209.

Mudstone/Shale Porosity:

Blatt, H., 1985, Provenance studies and mudrocks: Journal of Sedimentary Research, v. 55, p. 69-75, doi: 10.1306/212F8611-2B24-11D7-8648000102C1865D.

Blatt, H., and B. Sutherland, 1969, Intrastratal solution and non-opaque heavy minerals in shales: Journal of Sedimentary Research, v. 39, p. 591-600, doi: 10.1306/74D71CDA-2B21-11D7-8648000102C1865D.

Driskill, B., J. Walls, S. W. Sinclair, and J. DeVito, 2013, Applications of SEM imaging to reservoir characterization in the Eagle Ford Shale, South Texas, U.S.A., *in* W. Camp, E. Diaz, and B. Wawak, eds., Electron Microscopy of Shale Hydrocarbon Reservoirs: Tulsa, OK, AAPG Memoir 102, p. 115-136, doi: 10.1306/13391709M1023587.

Jennings, D. S., and J. Antia, 2013, Petrographic characterization of the Eagle Ford Shale, south Texas: Mineralogy, common constituents, and distribution of nanometer-scale pores, *in* W. K. Camp, E. Diaz, and B. Wawak, eds., Electron Microscopy of Shale Hydrocarbon Reservoirs: Tulsa, OK, AAPG

Memoir 102, p. 101-113, doi: 10.1306/13391708M1023586.

Loucks, R. G., and R. M. Reed, 2014, Scanning-electron-microscope petrographic evidence for distinguishing organic-matter pores associated with depositional organic matter versus migrated organic matter in mudrocks: GCAGS Journal, v. 3, p. 51-60.

Loucks, R. G., R. M. Reed, S. C. Ruppel, and U. Hammes, 2012, Spectrum of pore types and networks in mudrocks and a descriptive classification for matrix-related mudrock pores: AAPG Bulletin, v. 96, p. 1071-1098, doi: 10.1306/08171111061.

Loucks, R. G., R. M. Reed, S. C. Ruppel, and D. M. Jarvie, 2009, Morphology, genesis, and distribution of nanometer-scale pores in siliceous mudstones of the Mississippian Barnett Shale: Journal of Sedimentary Research, v. 79, p. 848-861, doi: 10.2110/jsr.2009.092.

Milliken, K. L., 1992, Chemical behavior of detrital feldspars in mudrocks versus sandstones, Frio Formation (Oligocene), South Texas: Journal of Sedimentary Research, v. 62, p. 790-801, doi: 10.1306/D42679DD-2B26-11D7-8648000102C1865D.

Milliken, K. L., M. Rudnicki, D. N. Awwiller, and T. Zhang, 2013, Organic matter-hosted pore system, Marcellus Formation (Devonian), Pennsylvania: AAPG Bulletin, v. 97, p. 177-200, doi: 10.1306/07231212048.

Milner, M., R. McLin, and J. Petriello, 2010, Imaging and porosity in mudstones and shales: Comparison of secondary and ion-milled backscattered SEM methods, *in* SPE Canadian Unconventional Resources and International Petroleum Conference, October 19–21, Calgary, Alberta, SPE Paper 138975, 10 p.

Ruppert, L. F., R. Sakurovs, T. P. Blach, L. He, Y. B. Melnichenko, D. F. R. Mildner, and L. Alcantar-Lopez, 2013, A USANS/SANS study of the accessibility of pores in the Barnett Shale to methane and water: Energy & Fuels, v. 27, p. 772-779, doi: 10.1021/ef301859s.

Schieber, J., 2013, SEM observations on ion-milled samples of Devonian black shales from Indiana and New York: The petrographic context of multiple pore types, *in* W. K. Camp, E. Diaz, and B. Wawak, eds., Electron Microscopy of Shale Hydrocarbon Reservoirs: Tulsa, OK, AAPG Memoir 102, p. 153-172, doi: 10.1306/13391711M1023589.

Slatt, R. M., and N. R. O'Brien, 2011, Pore types in the Barnett and Woodford gas shales: Contribution to understanding gas storage and migration pathways in fine-grained rocks: AAPG Bulletin, v. 95, p. 2017-2030, doi: 10.1306/03301110145.

Soeder, D. J. J., 1990, Applications of fluorescence microscopy to study of pores in tight rocks: AAPG Bulletin, v. 74, p. 30-40.

Yang, Yunlai, and A. C. Aplin, 2010, A permeability-porosity relationship for mudstones: Marine and Petroleum Geology, v. 27, p. 1692-1697, doi: 10.1016/j.marpetgeo.2009.07.001.

Porosity models:

Ajdukiewicz, J. M., and R. H. Lander, 2010, Sandstone reservoir quality prediction: The state of the art: AAPG Bulletin, v. 94, p. 1083-1091, doi: 10.1306/intro060110.

Ajdukiewicz, J. M., and R. E. Larese, 2012, How clay grain coats inhibit quartz cement and preserve porosity in deeply buried sandstones: Observations and experiments: AAPG Bulletin, v. 96, p. 2091-2119, doi: 10.1306/02211211075.

Beard, D. C., and P. K. Weyl, 1973, Influence of texture on porosity and permeability of unconsolidated sand: AAPG Bulletin, v. 57, p. 359-369.

Bjørkum, P. A., E. H. Oelkers, P. H. Nadeau, O. Walderhaug, and W. M. Murphy, 1998, Porosity prediction in quartzose sandstones as a function of time, temperature, depth, stylolite frequency, and hydrocarbon saturation: AAPG Bulletin, v. 82, p. 637-648.

Bjørlykke, K., M. Ramm, and G. C. Saigal, 1989, Sandstone diagenesis and porosity modification during basin evolution: Geologische Rundschau, v. 78, p. 243-268, doi: 10.1007/BF01988363.

Bloch, S., 1991, Empirical prediction of porosity and permeability in sandstones: AAPG Bulletin, v. 75, p. 1145-1160.

Bloch, S., and S. G. Franks, 1993, Preservation of shallow plagioclase dissolution porosity during burial: Implications for porosity prediction and aluminum mass balance: AAPG Bulletin, v. 77, p. 1488-1501.

Bloch, S., R. H. Lander, and L. Bonnell, 2002, Anomalously high porosity and permeability in deeply buried sandstone reservoirs: Origin and predictability: AAPG Bulletin, v. 86, p. 301-328, doi: 10.1306/61EEDABC-173E-11D7-8645000102C1865D.

Coskun, S. B., N. C. Wardlaw, and B. Haverslew, 1993, Effects of composition, texture and diagenesis on porosity, permeability and oil recovery in a sandstone reservoir: Journal of Petroleum Science and Engineering, v. 8, p. 279-292, doi: 10.1016/0920-4105(93)90005-Y.

Ehrenberg, S. N., 1990, Relationship between diagenesis and reservoir quality in sandstones of the Garn Formation, Haltenbanken, Mid-Norwegian Continental Shelf: AAPG Bulletin, v. 74, p. 1538-1558, doi: 10.1306/44B4B03B-170A-11D7-8645000102C1865D.

Ehrenberg, S. N., and P. H. Nadeau, 2005, Sandstone vs. carbonate petroleum reservoirs: A global perspective on porosity-depth and porosity-permeability relationships: AAPG Bulletin, v. 89, p. 435-445, doi: 10.1306/11230404071.

Lander, R. H., and O. Walderhaug, 1999, Predicting porosity through simulating sandstone compaction and quartz cementation: AAPG Bulletin, v. 83, p. 433-449.

Maxwell, J. C., 1964, Influence of depth, temperature, and geologic age on porosity of quartzose sandstones: AAPG Bulletin, v. 48, p. 697-709.

Nagtegaal, P. J. C., 1978, Sandstone-framework instability as a function of burial diagenesis: Journal of the Geological Society, v. 135, p. 101-105, doi: 10.1144/gsjgs.135.1.0101.

Primmer, T. J., C. A. Cade, J. Evans, J. G. Gluyas, M. S. Hopkins, N. H. Oxtoby, P. C. Smalley, E. A. Warren, and R. H. Worden, 1997, Global patterns in sandstone diagenesis: Their application to reservoir quality prediction for petroleum exploration, *in* J. A. Kupecz, J. Gluyas, and S. Bloch, eds., Reservoir Quality Prediction in Sandstones and Carbonates: Tulsa, OK, AAPG Memoir 69, p. 61-77.

Ramm, M., 1992, Porosity-depth trends in reservoir sandstones: Theoretical models related to Jurassic sandstones offshore Norway: Marine and Petroleum Geology, v. 9, p. 553-567, doi:10.1016/0264-8172(92)90066-N.

Ramm, M., and K. Bjørlykke, 1994, Porosity/depth trends in reservoir sandstones: Assessing the quantitative effects of varying pore-pressure, temperature history and mineralogy, Norwegian Shelf data: Clay Minerals, v. 29, p. 475-490, doi: 10.1180/claymin.1994.029.4.07.

Scherer, M., 1987, Parameters influencing porosity in sandstones: A model for sandstone porosity prediction: AAPG Bulletin, v. 71, p. 485-491.

Schmoker, J. W., 1997, Porosity prediction in deeply buried sandstones, with examples from Cretaceous formations of the Rocky Mountain region: Washington, DC, USGS Bulletin 2146-H, p. 89-104.

Schmoker, J. W., and D. L. Gautier, 1988, Sandstone porosity as a function of thermal maturity: Geology, v. 16, p. 1007-1010, doi: 10.1130/0091-7613(1988)016<1007:SPAAFO>2.3.CO;2.

Taylor, T. R., M. R. Giles, L. A. Hathon, T. N. Diggs, N. R. Braunsdorf, G. V. Birbiglia, M. G. Kittridge, C. I. Macaulay, and I. S. Espejo, 2010, Sandstone diagenesis and reservoir quality prediction: Models, myths, and reality: AAPG Bulletin, v. 94, p. 1093-1132, doi: 10.1306/04211009123.

Walderhaug, O., 1996, Kinetic modeling of quartz cementation and porosity loss in deeply buried sandstone reservoirs: AAPG Bulletin, v. 80, p. 731-745.

Walderhaug, O., R. H. Lander, P. A. Bjørkum, E. H. Oelkers, K. Bjørlykke, and P. H. Nadeau, 2009, Modelling quartz cementation and porosity in reservoir sandstones: Examples from the Norwegian Continental Shelf, *in* R. H. Worden, and S. Morad, eds., Quartz Cementation in Sandstones (IAS Special Publication 29): Oxford, UK, Blackwell Publishing Ltd., p. 39-49, doi: 10.1002/9781444304237.ch3.

Wangen, M., 1998, Modeling porosity evolution and cementation of sandstones: Marine and Petroleum Geology, v. 15, p. 453-465, doi: 10.1016/S0264-8172(98)00026-9.

Worden, R. H., M. J. Mayall, and I. J. Evans, 1997, Predicting reservoir quality during exploration: Lithic grains, porosity and permeability in Tertiary clastic rocks of the South China Sea basin, *in* A. J. Fraser, and S. J. Matthews, eds., Petroleum Geology of Southeast Asia: . London, GSL Special Publication 126, p. 107-115, doi: 10.1144/GSL.SP.1997.126.01.08.

Facing Page: Top — A detailed paragenetic sequence, Pennsylvanian Strawn Gp.., Texas (redrawn from Newell et al., 2003). Bottom — Calcite replacing plagioclase grain and opal cement, Cretaceous Mowry Fm., Moffat Co., Colorado (PPL | XPL).

Paragenesis

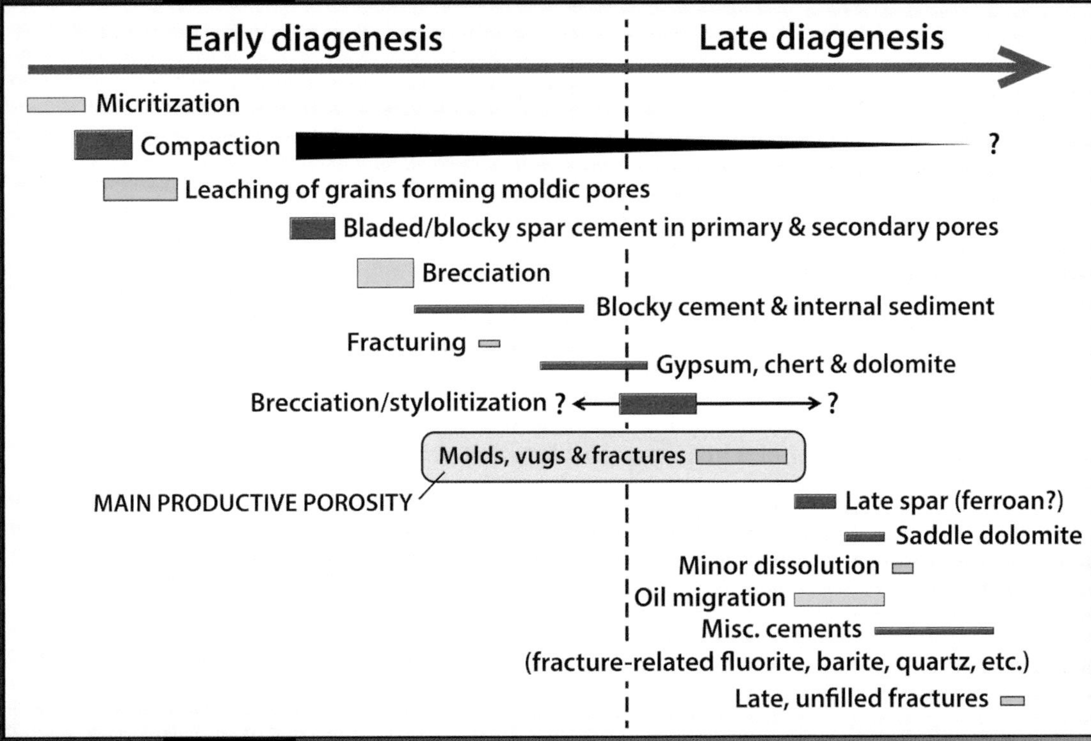

CHAPTER 23

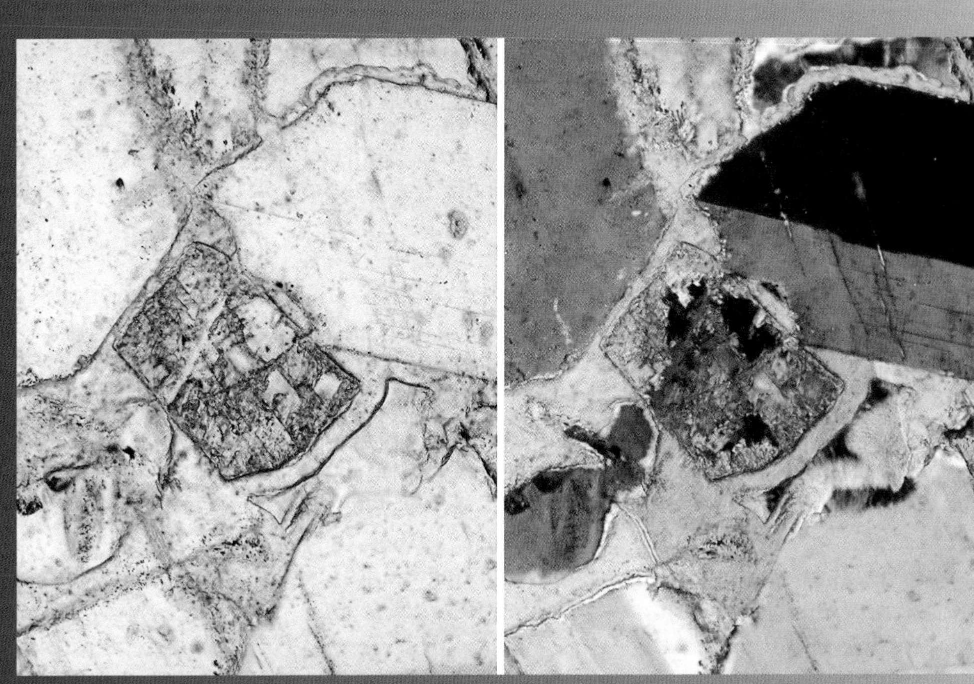

Introduction to Paragenesis

Paragenesis describes the relative sequence or timing of "events" in the diagenetic history of a rock. "Events" refers to any and all of the diagenetic processes discussed in this book: compaction, cementation by one or more minerals, dissolution, replacement, structural deformation, hydrocarbon emplacement and others. As with any geological dating, paragenesis can include relative age determinations (Event B happened before Event C but after Event A), or radiometric dating (Event A happened 426.8 m.y. ago; B occurred 316.5 m.y. ago) or a combination of the two (Event A happened 426.8 m.y. ago and B occurred at some later time). Paragenesis is not just a petrographic exercise; it almost requires integration with other geologic information such as regional depositional models and sequence stratigraphy, burial history, thermal history, timing of tectonic events, and the like. From an economic perspective, that integration may lead to valuable insights on the presence or absence of porosity in units of interest relative to the timing of hydrocarbon generation and migration—the essence of petrographic value in petroleum systems analysis.

A geoscientist working on a paragenetic study can have a variety of purposes, much like a writer of a history book. The historian sorts through masses of information related to personal histories and social interactions; the petrographer deals with observations on grains, cements and pores in a geologic unit. A good historian has a clearly defined goal, as does a good petrographer. The history book may concern the life of an individual or the story of mankind; the paragenetic sequence may be that of a single unit at a single locality, or a synthesis of diagenetic events in rocks within a basin, a broader region or the world. Clearly, the strategies and the time and effort needed to make such different products will vary greatly, but the basic research techniques remain much the same. So step one in any paragenetic study is establishing clearly defined and realistically achievable goals.

It is worth recognizing that no two petrographers are likely to produce precisely the same paragenetic account, just as no two historians are likely to agree on all details in their writings. It is especially true for petrographic studies, however, because diagenetic processes can be long-lived, depend on many factors, and vary spatially, even from pore to pore and from substrate mineral to substrate mineral. Geologic events, just like historical events, do not always happen sequentially; rather they may overlap temporally, with compaction, dissolution and cementation, for example, all potentially occurring more or less simultaneously.

In addition to temporal variations, geologic processes also can vary spatially. The history of diagenetic events in adjacent sandstones and mudstones, for example, may be very different. Likewise, a sequence of events in one area of a thin section may not apply to the entire unit, much less to an entire region, especially where fluid movements are localized or where cementation is patchy.

Constraints on observations add to uncertainty in paragenetic studies. For example, the observation of 3-D relationships of minerals (and the determination of which one came first and which ones formed later) is greatly affected by observation in 2-D thin sections. Additionally, "operator bias" can be an issue, especially with random scans of thin sections. The recognition of a pattern in one place can easily lead to the exclusion of observations that contradict the initial pattern. Point counting and accurate recording of observations (as opposed to preliminary conclusions) are essential in achieving valid paragenetic results. In other words, reliable paragenetic sequences require extensive observation on multiple samples and should be a final, long-considered conclusion of a substantial project, not a rapid judgment based on a small amount of work. Even with such care, paragenetic determinations, relative or absolute, are seldom easy or unequivocal.

Given these ever-present uncertainties, paragenetic information needs to be conveyed in a way that fairly expresses both what is known and the limits of certainty of that knowledge. To do that, paragenetic data generally are presented in unscaled bar charts that depict the fact that many diagenetic events are long-lived, can overlap with each other and commonly have uncertain starting and ending points. A useful example of an effective paragenetic data presentation, based on regional petrographic, geochemical and fluid inclusion studies, is shown in Figure 23.1 (and another is shown on the title page of this chapter).

The fact that there are uncertainties and disagreements in paragenetic conclusions, however, should not invalidate the importance of such studies. Quite the contrary, a well done paragenetic investigation can provide insights into the diagenetic history of rocks that cannot be obtained in any other way and may play a vital role in synthesis of a wide variety of geological data.

As a guide for paragenetic studies, here are six examples of specific kinds of petrographic observations that are useful for such investigations, along with potential integrations with allied data.

1. Cement stratigraphy – Essentially the diagenetic equivalent of the stratigraphic law of superposition in which cements that formed directly on grain substrates are older than cements that overlie those initial precipitates. EXAMPLE: A clay cement is

found as nearly complete coatings on detrital grains; other cements in the rock are virtually never in direct contact with the grains, but instead are found atop the clays. That indicates that the clays formed prior to other cements, although it would not give an absolute time of formation for any of the cements. In some cases, clay coatings can be radiometrically dated, potentially adding temporal constraints for one stage in the chain of events. However, diagenetic recrystallization of early-formed clays can complicate such age determinations.

2. Packing density and grain deformation – This criterion relates compaction and cementation histories. Rocks that have loosely packed grains and little or no grain deformation, despite substantial burial, are likely to have had early cementation (prior to significant burial and mechanical compaction). A rock with closely packed grains could either have escaped early cementation or could have had early cements that were later removed by dissolution. EXAMPLE: Quartz cement was precipitated within a framework of exceptionally closely packed grains—that leads to the conclusion that such cement formed after significant burial and compaction (otherwise compaction would have been prevented by the rigid cements). Using generalized porosity/depth plots (similar to those shown in Figure 10.1 in this book), in conjunction with specific burial history data from the area of interest, could provide an estimate of the time of cementation.

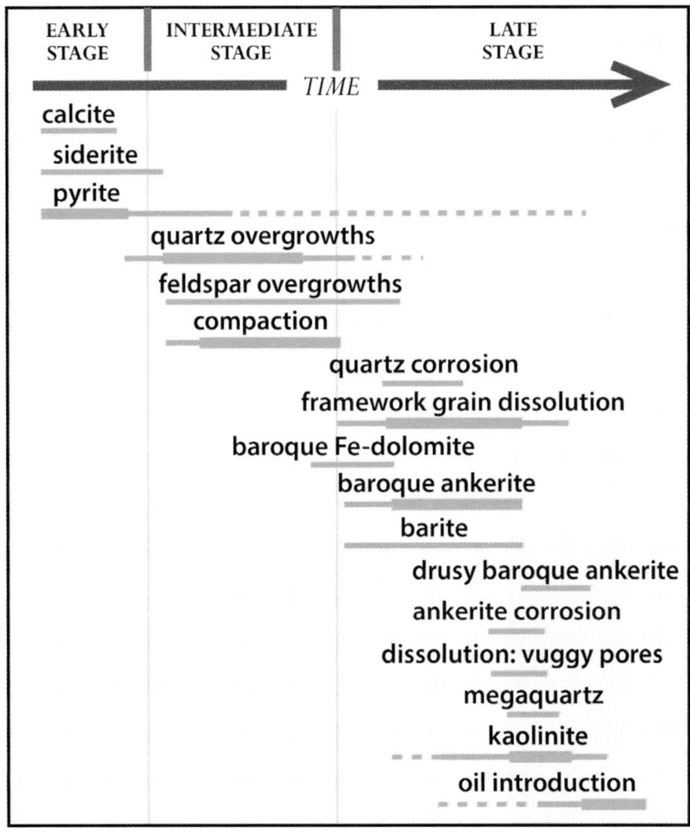

Figure 23.1 A summary paragenetic sequence for Pennsylvanian sandstones of the Cherokee Basin in Kansas showing diagenetic events and their relative (and extensively overlapping) timing. The thicknesses of the lines reflect the relative intensity of processes through time and dashed lines indicate the possible range over which a given alteration may have occurred. The data are based on petrographic, geochemical and fluid-inclusion studies. Adapted from Wojcik et al. (1997, p. 237).

3. Integration of dissolution and cementation patterns – Cements can form in primary pores at any time, but cements (not replacements) within secondary pores can only precipitate after formation of those pores by dissolution of a precursor mineral. EXAMPLE: A clay (such as kaolinite) is present mainly or exclusively within or near pores recognizably formed by feldspar dissolution. That would reliably indicate that kaolinite formation was contemporaneous with or postdated feldspar removal.

4. Cross-cutting relationships – This refers primarily to structural features (primarily fractures, faults deformation bands and stylolites) that cut each other or precursor cements within a rock. EXAMPLE: Quartz-cemented fractures with consistent orientations cross-cut dolomite crystal cements in a sandstone and those fractures, in turn, are cut by calcite-filled fractures with a different orientation. That would logically lead to the conclusion that dolomite cementation occurred prior to fracturing and that two stages of fracturing took place with quartz cementation in the first-stage fractures likely predating later fracturing and calcite precipitation. Integration of this data with an understanding of the tectonic history of the area might provide a semi-quantitative time line for these events.

5. Timing relative to hydrocarbon entry – Cements, fractures or other features that do or do not contain inclusions of hydrocarbons can help to date materials as being pre- or post-hydrocarbon migration. EXAMPLE: Hydrocarbon residues are consistently found only in contact with saddle dolomite cements that occur exclusively in fractures that crosscut all other cements in the rock. This could be interpreted to indicate late-stage influx of warm to hot brines, associated with hydrocarbon migration, through fractures in an otherwise impermeable rock. Studies of fluid inclusion chemistry and geothermometry, coupled with knowledge of the thermal history and hydrocarbon maturation patterns in this area again, could allow refinement of those conclusions.

6. **Diagnostic fabrics and morphologies – In some cases, diagenetic fabrics can be so distinctive that they can be directly tied to a time or place of formation. EXAMPLES: This is especially common in near-surface settings where soils or travertine deposits with microbial precipitates, laminated crusts, desiccation or illuviation fabrics and other exposure-related features can be definitely tied to eogenetic or telogenetic diagenesis. In mesogenetic settings there are fewer diagnostic fabrics, although saddle dolomite, especially when associated with fluorite, barite and/or metallic sulfides (e.g., galena or sphalerite), is a good indicator of alteration by high salinity and/or hydrothermal fluids. However, it still requires cross-cutting or superimposed mineral relationships to establish the relative time of such mineralization.**

As noted above, relative paragenetic sequences are strengthened by having dating information. This can come from radiometric or fission track dating of cements (glauconite, illite, zircon or apatite overgrowths and others) and, less precisely, from fluid-inclusion or isotopic geochemical studies on cements (e.g., Burley et al., 1989; Vagle et al., 1994; Goldstein, 2001; Polito et al., 2006; Blamey et al., 2014; but see also more extensive citations in Chapter 24). Unfortunately, these techniques are fraught with uncertainties (especially in terms of accurate sampling and analysis of minute grains or grain overgrowths or potential effects of later diagenesis on the integrity of earlier generations of fluid inclusions or isotopic ratios). Nonetheless, they generally can provide some useful constraints on relative paragenetic models.

Mid. Pennsylvanian Ricker Sandstone Mbr., Mineral Wells Fm., Brown Co., Texas

A simple paragenesis with two substantive stages of cementation. Extensive syntaxial quartz overgrowths are the first cements (arrows show patchy and very thin "dust rims" that help to distinguish the original grain outlines from the euhedral overgrowths and may reflect minor surficial iron oxide precipitation). The relatively loose grain packing indicates that quartz cementation began prior to substantial burial, compaction or grain-to-grain dissolution. Subsequently, hematite cement (now partly altered to goethite) filled all remaining pores and may be of telogenetic origin.

PPL, Scale bar = 0.10 mm

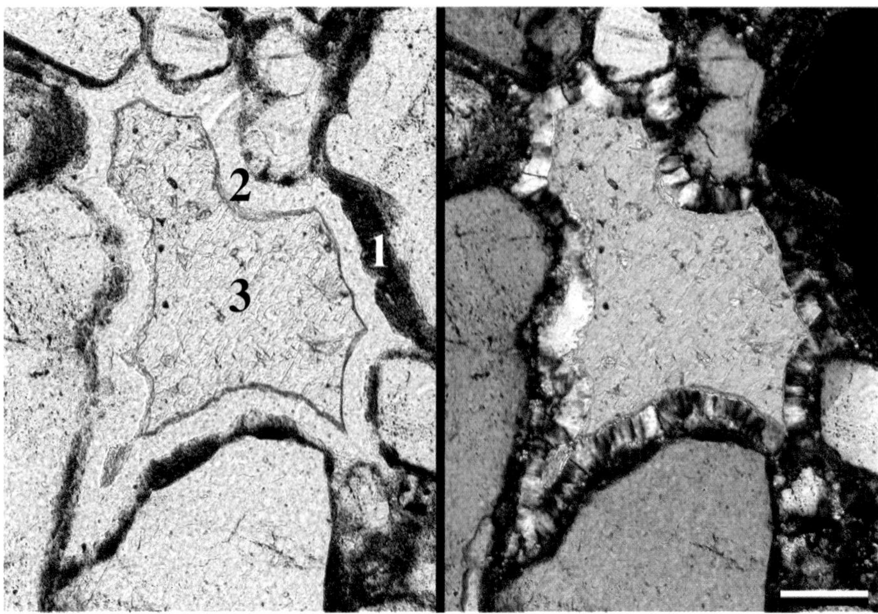

Mid. Eocene Baca Fm., Socorro Co., New Mexico

A clear example of a large pore with direct superposition of cements. Grain-coating clays (#1, dark brown) are followed by an isopachous zone of bladed to equant megaquartz and chalcedony (#2) and a final pore filling of single-crystal calcite (#3). However, that paragenetic sequence is only fully represented in the central area of the image. The smaller peripheral pores were substantially filled after clay and silica precipitation and so calcite is largely, but not completely, absent in most areas.

PPL | XPL, Scale bar = 0.10 mm

Up. Cretaceous Prince Creek Fm., North Slope, Alaska

An example of a cement superposition history that is consistent throughout the entire sample. Large overgrowth cements with euhedral terminations (#1) formed on an orthoclase feldspar. That was followed by precipitation of illite/smectite clay coatings (#2) that surround all grains and may have helped to prevent further cementation. The grain contact (#3) between the feldspar overgrowth and an adjacent quartz grain may indicate minor pressure dissolution although in most places these contacts appear to reflect compromise crystal growth boundaries.

PPL | XPL, BDI, Scale bar = 0.15 mm

Lo. Permian (Wolfcamp) Powwow Conglomerate, Hudspeth Co., Texas

Where carbonate minerals are the main cements in a sandstone, staining commonly can be of considerable help in differentiating generations. This chert arenite shows at least three carbonate cements accentuated by staining. The pink-stained nonferroan calcite probably formed as overgrowths on calcitic shell fragments. The rhombic dolomite crystals (unstained) formed next and underwent some dissolution during compaction (arrows). The ferroan calcite (blue-stained) postdates compaction and completely filled remaining primary and secondary (fracture) porosity.

PPL, AFeS, Scale bar = 0.10 mm

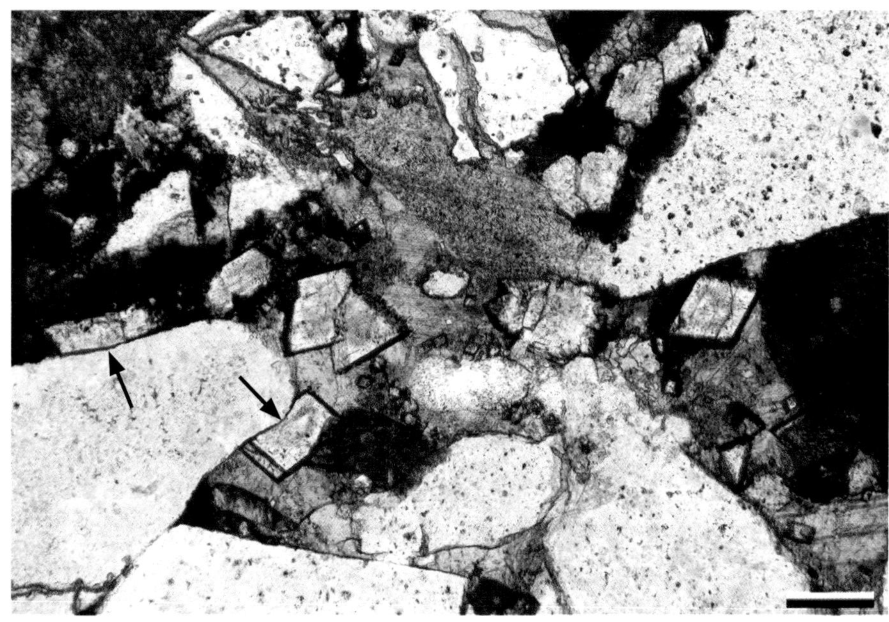

Mid. Eocene – Lo. Oligocene Spears Gp., Catron Co., New Mexico

A somewhat more complicated example of cement superposition. Thin, dark, clay rims surround most grains in this volcaniclastic sandstone. That is followed by zeolite (clinoptilolite) cements (white zones of irregular thickness). However, a thin zone of blue-impregnated porosity separates those two cements in places (arrows). This could indicate the presence of a former, now-dissolved, intermediate-stage cement. A less likely explanation is that it could reflect fracturing and spalling of the clinoptilolite crusts, either during uplift and deformation or during sampling and section preparation.

PPL, BDI, Scale bar = 0.09 mm

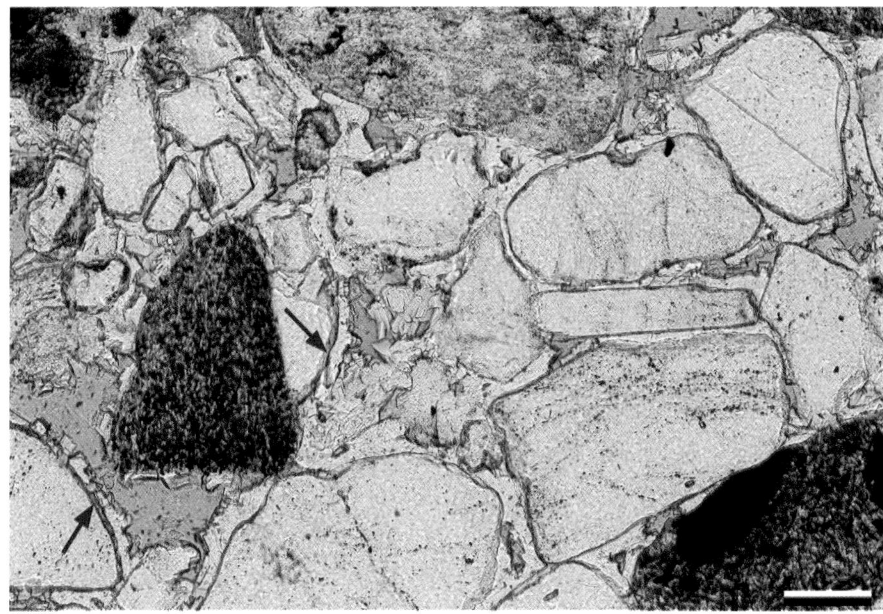

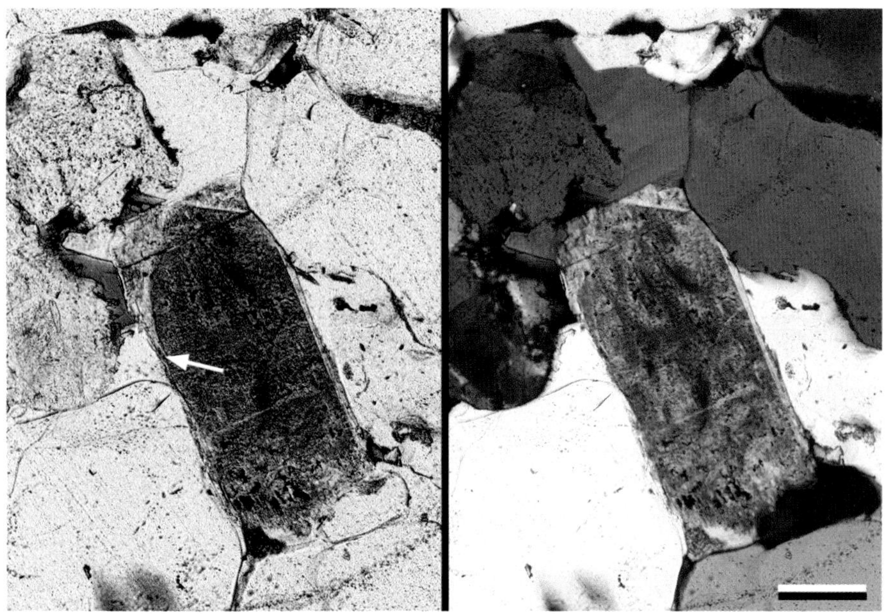

Cretaceous Dakota Gp., Moffat Co., Colorado

This vacuole-rich K-feldspar grain is the substrate for K-feldspar overgrowths (containing fewer vacuoles). Adjacent quartz grains have overgrowths and where the quartz and feldspar overgrowths adjoin their temporal relationships can be determined. The well-developed feldspar overgrowths may have started forming first, but in most areas (as at arrow) quartz and feldspar overgrowths appear to have grown simultaneously, forming compromise boundaries.

PPL | XPL, RDI, Scale bar = 0.12 mm

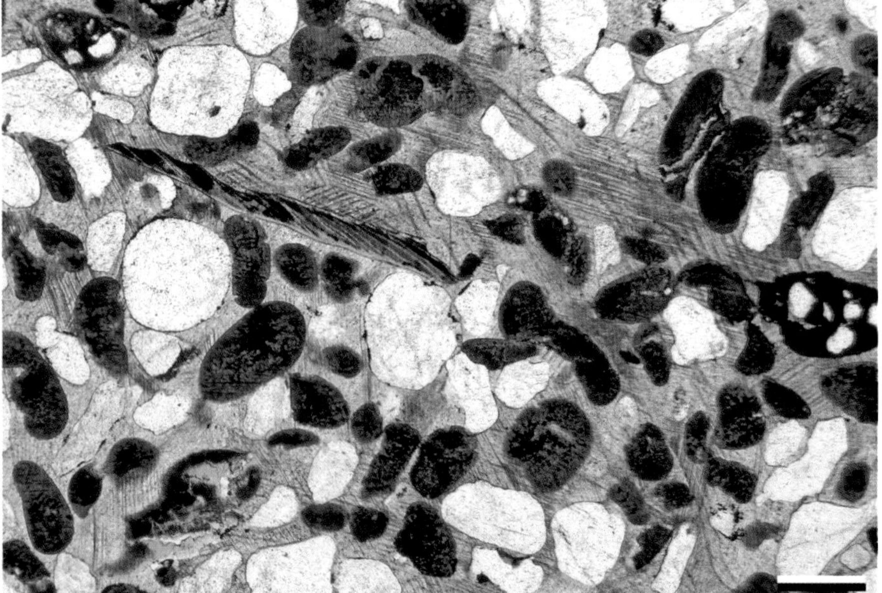

Mid. Cambrian Riley Fm., Lion Mountain Mbr., Llano Co., Texas

Units with easily deformable grains can be especially useful in determining the timing of cementation. In this case, the original sediment was a mix of soft glauconite peloids, thin and brittle brachiopod shells and hard and durable quartz grains. Calcite cement completely occludes porosity in this sample while preserving, with little to no overall fabric compaction or grain deformation, the ductile and brittle components of the rock. This indicates that cementation was very early, probably prior to any substantial burial.

PPL, Scale bar = 0.51 mm

Mid. Permian Karstryggen Fm., Jameson Land, East Greenland

This submature lithic arenite shows some compaction that probably reflects minimal early cementation. The unit is completely cemented by ferroan calcite cements ranging from moderately ferroan (purple-stained) to more strongly ferroan (blue-stained). Ferroan calcites rarely form in near-surface or shallow-burial settings because iron is incorporated in the calcite lattice only under reducing conditions. An additional indication that these are burial-stage cements comes from the extensive (dark-brown to black) asphaltic hydrocarbon residues that predate the ferroan calcite cements.

PPL, AFeS, BDI, Scale bar = 0.51 mm

Lo.–Mid. Eocene Cub Mountain Fm., Lincoln Co., New Mexico

This rock has quartz overgrowths, but they appear to largely postdate compactional crushing of feldspars and rock fragments and the formation of a very closely-packed grain fabric. Selective leaching of rock fragments and K-feldspars largely postdated the last stages of compaction because those weakened grains have not been deformed significantly.

PPL, KFS, BDI, Scale bar = 0.26 mm

Lo. Cambrian Unicoi Fm., Shenandoah Co., Virginia

Although there are some relatively early quartz overgrowths in this rock, they were too insubstantial to preclude compaction. The numerous argillaceous SRFs deformed readily, flowing into adjacent pores where they formed a pseudomatrix that probably inhibited further cementation. As a result, this rock underwent extensive compactional porosity reduction and strain deformation that is evidenced by undulose extinction in the quartz grains. So this paragenetic sequence is dominated by compaction and deformation.

XPL, Scale bar = 0.26 mm

Lo. Triassic Ivishak Fm., North Slope, Alaska

In this chert arenite, siderite of eogenetic meteoric origin (based on chemical data) formed grain-coating cements on chert clasts and quartz grains mainly in the coarse-grained central zone in this image. In that area, these grain coatings precluded later cementation—therefor compaction and grain interpenetration dominated. In the finer-grained zones at top and bottom, sparse siderite cements did not prevent later quartz overgrowth cementation. There is a lower packing density and far less compactional grain interpenetration in those zones. Secondary leached porosity probably postdates compaction in both areas.

PPL, BDI, Scale bar = 0.26 mm

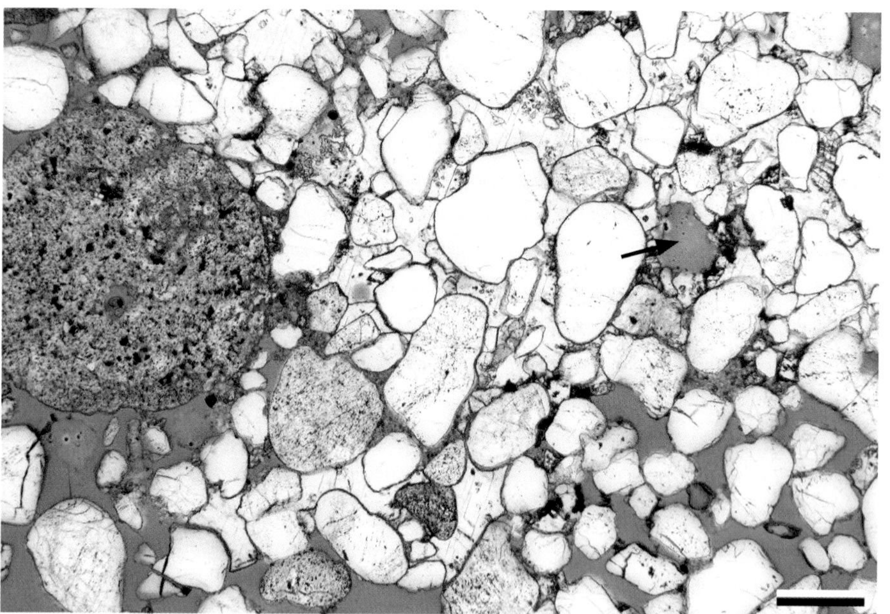

Mid. Jurassic Entrada Fm., Slick Rock Mbr., Emery Co., Utah

Dissolved grains provide an opportunity to establish temporal relations between dissolution and cementation. Here, several grains have undergone partial dissolution (best seen in the large chert clast along the left margin) or complete dissolution (an example shown at black arrow). A large, poikilotopic crystal of calcite cement (white) fills primary porosity near the leached grain but no calcite occurs within the moldic pore. This could indicate that calcite cementation predated the dissolution of at least that grain. Because similar relationships are seen in other calcite-cemented areas of this rock, it is likely to be a valid general conclusion for this rock.

PPL, RDI, Scale bar = 0.26 mm

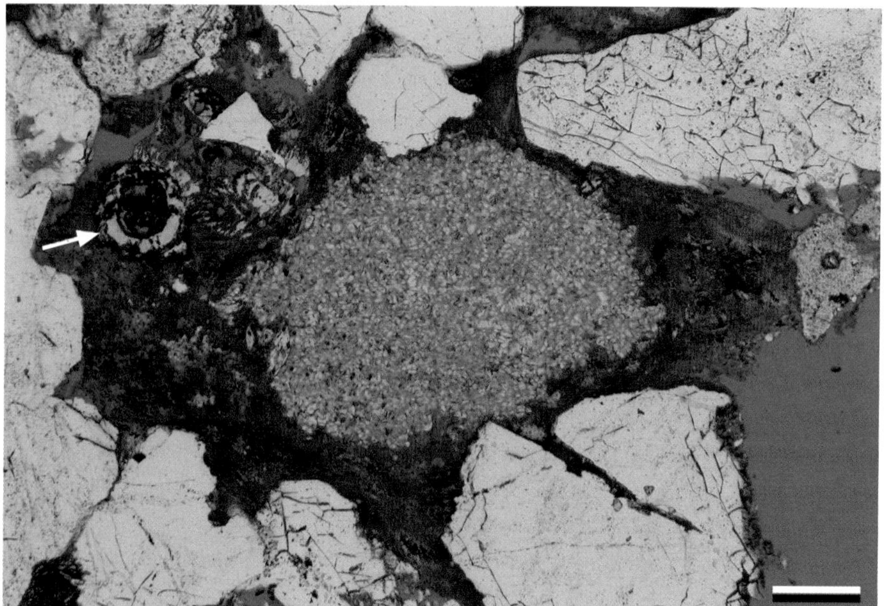

Lo. Pennsylvanian Morrow B Sandstone, Ochiltree Co., Texas

This is a more complex example of cementation-dissolution paragenesis. Sphaerosiderite (arrow) and clays were the earliest pore filling materials, and probably were of eogenetic (pedogenic) origin. Those materials outline a dissolved grain (center), probably a former feldspar, and the mold was preferentially filled with kaolinite cement. Thus the kaolinite postdates the earlier cements and is contemporaneous with or postdates feldspar dissolution. Another moldic pore (lower right) shows no cement filling and may represent a stage of dissolution that postdates kaolinite formation.

PPL, RDI, Scale bar = 0.10 mm

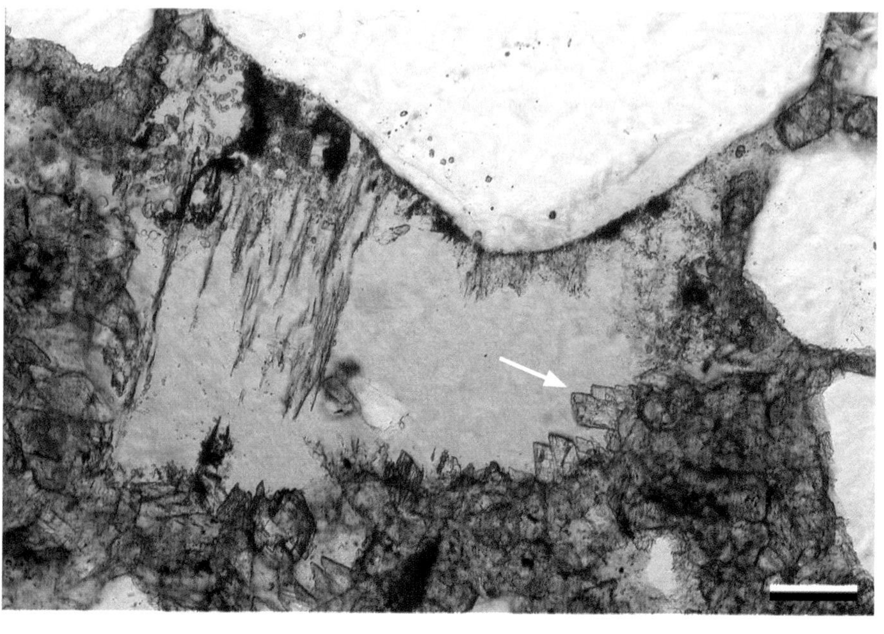

Jurassic Tilje Fm., Norwegian sector, North Sea †

This secondary pore has resulted from the nearly complete dissolution of a detrital feldspar (the traces of yellow stain suggests that it was a potassium feldspar). The quartz grain next to it has been compacted into the feldspar grain, but that occurred prior to dissolution as indicated by preservation of the delicate structure in the K-spar. The quartz also appears to show some dissolution. Siderite fills the primary pores and has infilled parts of the secondary pore (example at arrow) indicating that siderite precipitation started before K-spar dissolution (preserving the grain outline) and continued during or after K-spar dissolution.

PPL, AFeS, KFS, BDI, Scale bar = 0.05 mm

Miocene Lower Santa Fe Gp., Taos Co., New Mexico

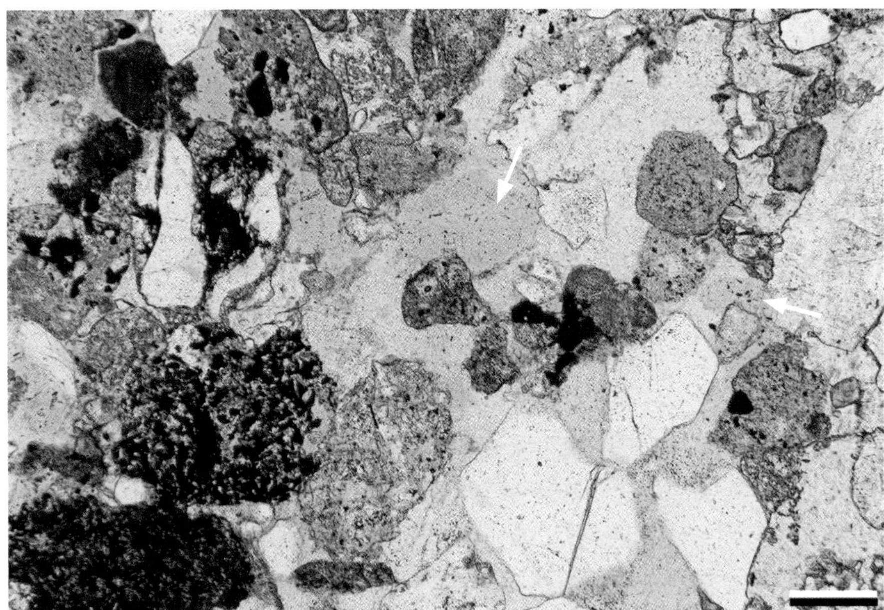

A volcaniclastic sandstone with both partially dissolved volcanic rock fragments and completely dissolved grains (probably mainly detrital feldspars). Kaolinite (cloudy pale blue) has substantially filled many primary and some possible secondary pores throughout the sample, while retaining its usual intercrystalline porosity. Additional grains were dissolved after kaolinite precipitation ended and those pores remain unfilled by cement (examples at white arrows). Thus, kaolinite cementation spanned only part of the interval during which dissolution occurred.

PPL, BDI, Scale bar = 0.26 mm

Pliocene – Pleistocene Gila Gp., New Mexico

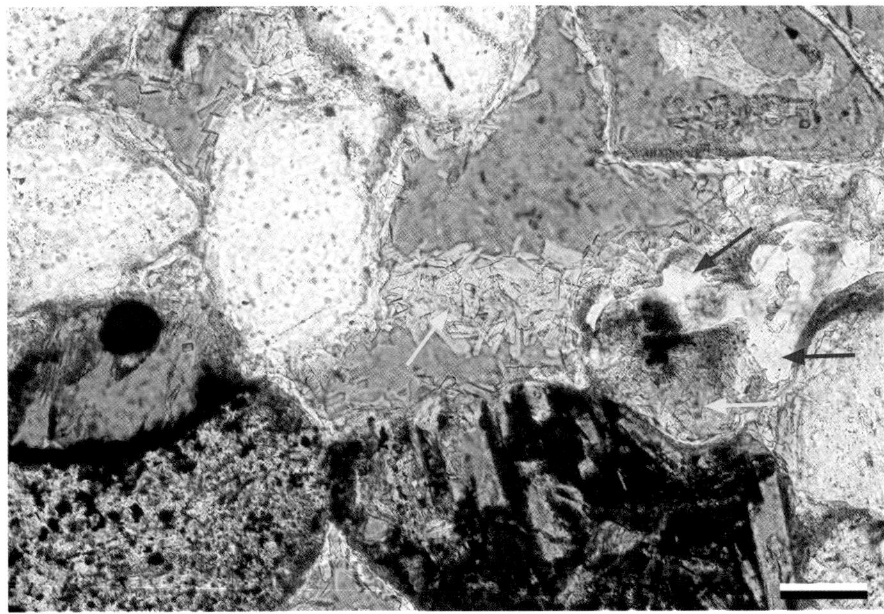

Complex paragenesis in a volcaniclastic sandstone. Clay coats (brownish illite/smectite) were the earliest formed cements or infiltrates and were followed by thin rims of clinoptilolite (white). The clays and zeolite cements provided a sufficiently rigid framework to retain the outlines of dissolved grains without compaction of the secondary pores. Several of those leached grains were partially filled with or replaced by albite (red arrows), additional zeolites (yellow arrows) and iron oxide/hydroxide cements (center bottom). Complex paragenetic sequences are common, even in young strata, in semiarid alluvial settings with fluctuating water tables.

PPL, BDI, Scale bar = 0.07 mm

Mid. Eocene – Lo. Oligocene Spears Gp., Socorro Co., New Mexico

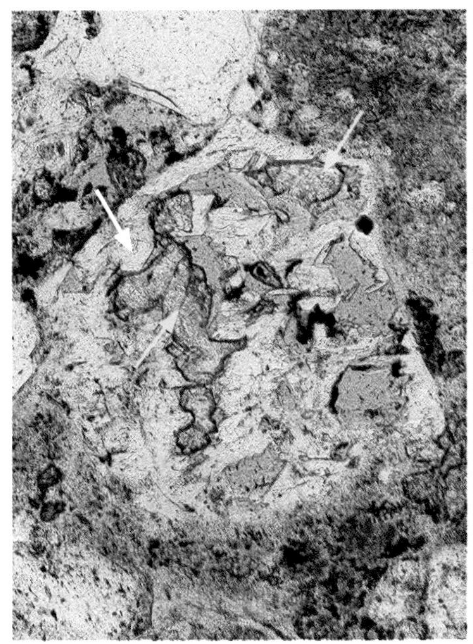

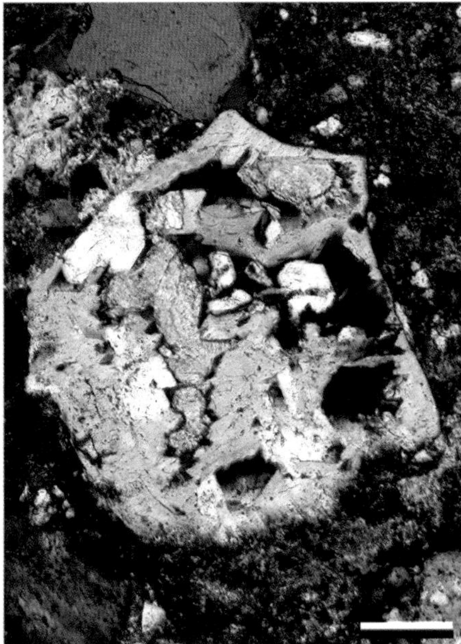

Four stages of paragenesis recorded in a single feldspar grain. The detrital grain underwent dissolution of several core areas (stage 1). Remnants of the grain then received albitic overgrowths (stage 2), some showing crystal terminations (example at white arrow). Remaining intragranular pores were then filled with calcite cement (stage 3; yellow arrows). The rounded and jagged margins of the calcite crystals, their separation from the walls of the pores and their absence in many pores all indicate substantial later corrosion of the calcite fills (stage 4).

PPL | XPL, BDI, Scale bar = 0.10 mm

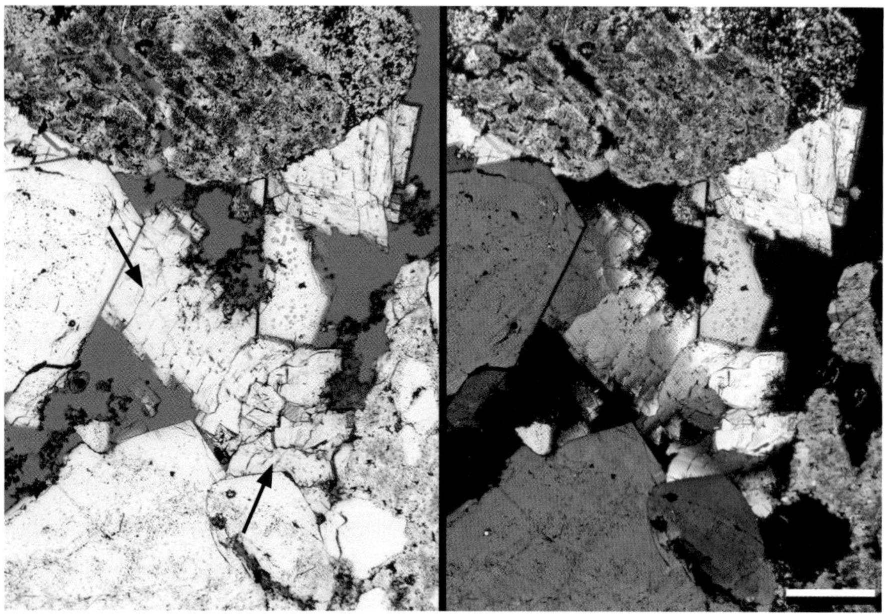

Lo. Pennsylvanian Morrow B Sandstone, Ochiltree Co., Texas

This area of a feldspathic sandstone was partially cemented by saddle dolomite cement (black arrows in PPL; note undulose extinction in XPL). The dolomite precipitated in both primary and secondary pores and postdated sphaerosiderite, illite-smecitite, quartz, kaolinite, and calcite cements as well as extensive dissolution of feldspars. As such it was clearly a relatively late-stage, mesogenetic precipitate. The corroded margins of the saddle dolomite crystals, however, indicate a still later stage of dissolution, most likely associated with hydrocarbon entry.

PPL | XPL, RDI, Scale bar = 0.26 mm

Up. Cretaceous Rio Ancho Fm., Guajira Province, Colombia

Cross-cutting paragenetic relationships generally involve fractures/veins, stylolites or other structural features cutting each other or cutting earlier cements. In this example, at least two generations of veins are present. The initial set is quartz filled and cuts previously well lithified radiolarian chert matrix. Later, calcite-filled veins transect the earlier quartz-filled veins (example near white arrow) providing a relative temporal succession.

XPL, Scale bar = 0.26 mm

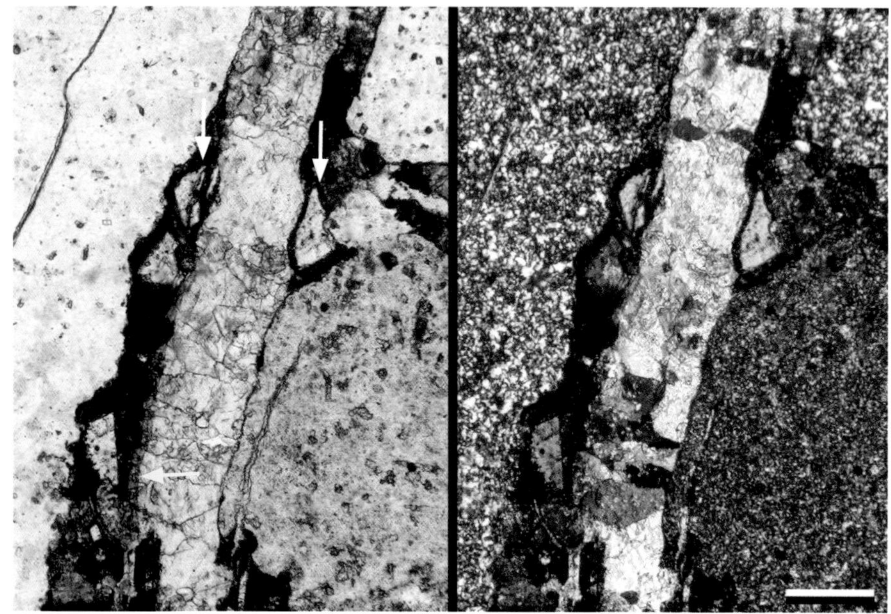

Lo. Permian Powwow Conglomerate, Hudspeth Co., Texas

Nonferroan calcite has filled fractures in this chert arenite. In cutting through the intergranular material, the fracture transected dolomite crystals (example at white arrows) that had precipitated earlier as pore-filling cements and as replacements of interstitial matrix. Later displacement along this fracture led to localized precipitation of ferroan calcite along one margin (yellow arrow).

PPL | XPL, AFeS, Scale bar = 0.20 mm

Mid. Permian Karstryggen Fm., Jameson Land, East Greenland

An example of fractures cutting more pervasive cements in a sandstone. A nonferroan calcite vein (stained pink), with irregular walls, cuts the moderately ferroan cement (stained purple) that had completely filled porosity at some earlier stage. The nonferroan calcite vein, in turn, was cut by a later fracture that was subsequently filled with highly ferroan calcite cement (stained deep blue). The changes in calcite compositions during these events presumably reflect input of waters of varied compositions from different sources. Staining is clearly essential to recognition of such fabrics.

PPL, AFeS, BDI, Scale bar = 0.10 mm

Mid. Jurassic Entrada Fm., Moffat Co., Colorado

Structural features other than fractures can be used to provide relative dating of diagenetic events. For example, these interlocked quartz grains show deformation lamellae (the dark sub-parallel planes) that mark traces of microshearing within the quartz crystals. The quartz grain at the lower center of the image has authigenic quartz overgrowths that are in optical continuity with core grain (arrows show boundary between grain and overgrowth). Because the deformation lamellae crosscut both the grain and its overgrowths, the deformation event must postdate overgrowth formation.

XPL, BDI, Scale bar = 0.04 mm

Mid. Jurassic Curtis Fm., Moffat Co., Colorado

Euhedral dolomite cements grew displacively in this glauconitic siltstone during both early and late diagenesis. The early dolomites are nonfluorescent and their presence was insufficient to prevent compactional deformation of the shale SRFs. A second generation of dolomites formed during or after hydrocarbon migration—they are strongly fluorescent (the pale green colors) because they contain hydrocarbon microinclusions. A later dissolution event created porosity within selected zones of the dolomites (marked by yellow-orange fluorescence of the colored dye injected into the porosity).

PPL | FL470, RDI, Scale bar = 0.10 mm

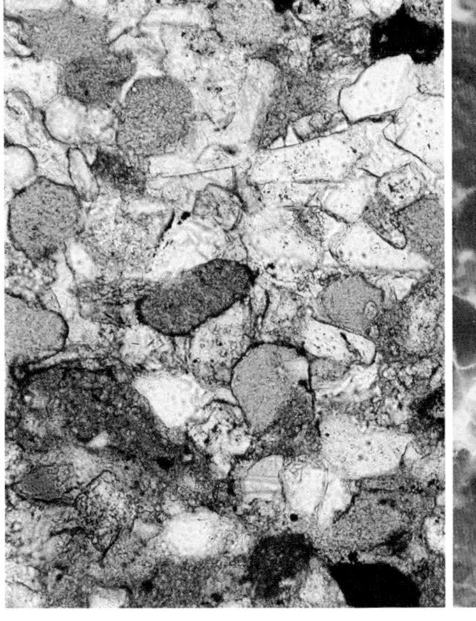

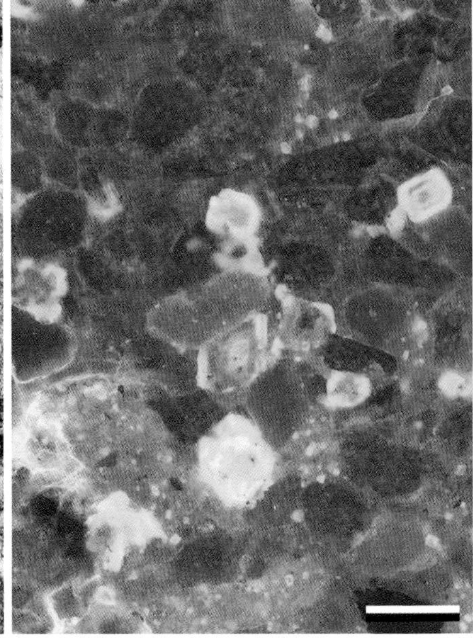

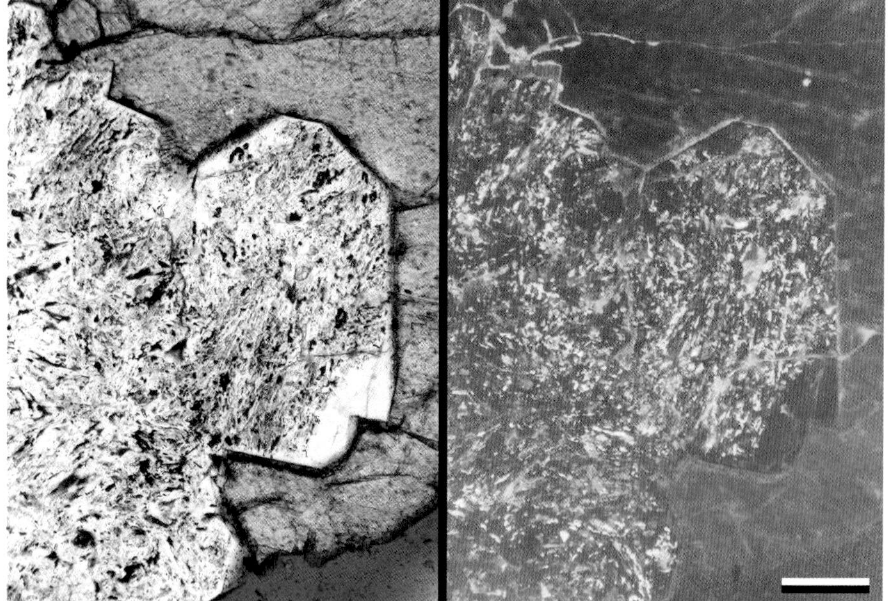

Permian Park City Fm., Tosi Chert Mbr., Park Co., Wyoming

Anhydrite nodules, formed through brine reflux from an evaporative shelf, are the earliest diagenetic phase in this deposit. The anhydrite nodules were later replaced by various forms of silica, including chert, chalcedony and megaquartz. Silicification occurred in the presence of hydrocarbons (either syn- or post migration) as shown by the presence of abundant oil inclusions—dead oil (reddish fluorescence) and live oil (yellow to pale green). After silicification, the remaining anhydrite was dissolved, and the resultant porosity was partially filled by clear quartz overgrowths and nonferroan to slightly ferroan calcite.

PPL | FL470, AFeS, BDI, Scale bar = 0.26 mm

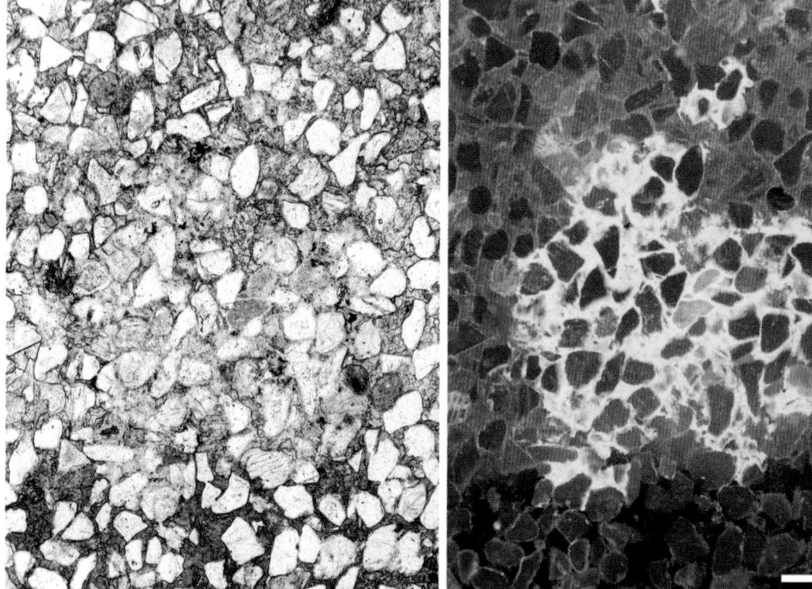

Mid. Permian Rush Springs Fm., Caddo Co., Oklahoma

A loosely packed sandstone cemented by calcite and impregnated, in places, by live oil. Petrographic and geochemical information may not be enough to establish an unequivocal paragenesis. Two very different interpretations have been published for the origin of the carbonate cements in this rock (here calcite but, in other areas, also dolomite). Donovan (1974) inferred an uplift-related (telogenetic) origin for these calcites that formed from the interactions of hydrocarbons (leaking from the underlying Cement oilfield) and meteoric waters. Kirkland et al. (1995), however, believed that they formed eogenetically. Calcite stain applied only along bottom edge.

PPL | FL470, AS, Scale bar = 0.26 mm

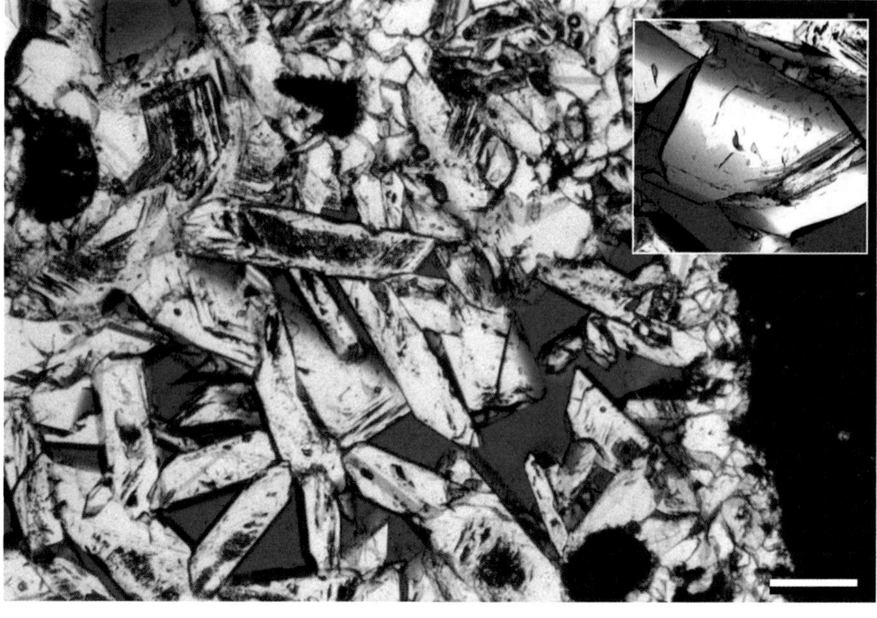

Lo. Cretaceous (Barremian) Kharaib Fm., offshore Qatar

This large open fracture in a carbonate rock was partially filled by lath-like calcite crystals. The calcites contain abundant hydrocarbon inclusions (the brownish lines and patches) that parallel calcite crystal growth zones. The inset (an image at the same scale from a different area of the same sample) shows several, large, isolated, brown, oil-filled inclusions. The unusual lath-like morphology of these calcite crystals may be related to lattice defects produced by hydrocarbons.

PPL, BDI, Scale bar = 0.48 mm

Pleistocene – Holocene coastal alluvium, Canterbury, New Zealand

Circumgranular shrinkage and porosity is produced by repeated desiccation and wetting of matrix around rigid and durable grains. It therefore is an exclusively eogenetic feature, commonly found associated with other soil-related fabrics (clay cutans, rhizoliths, laminated crusts, and others). Considerable lithification may occur in near-surface settings and recognition of these distinctive features helps to constrain the timing of such alterations. The red stain shows that the matrix material here has a high calcium carbonate content.

PPL, AFeS, BDI, Scale bar = 0.10 mm

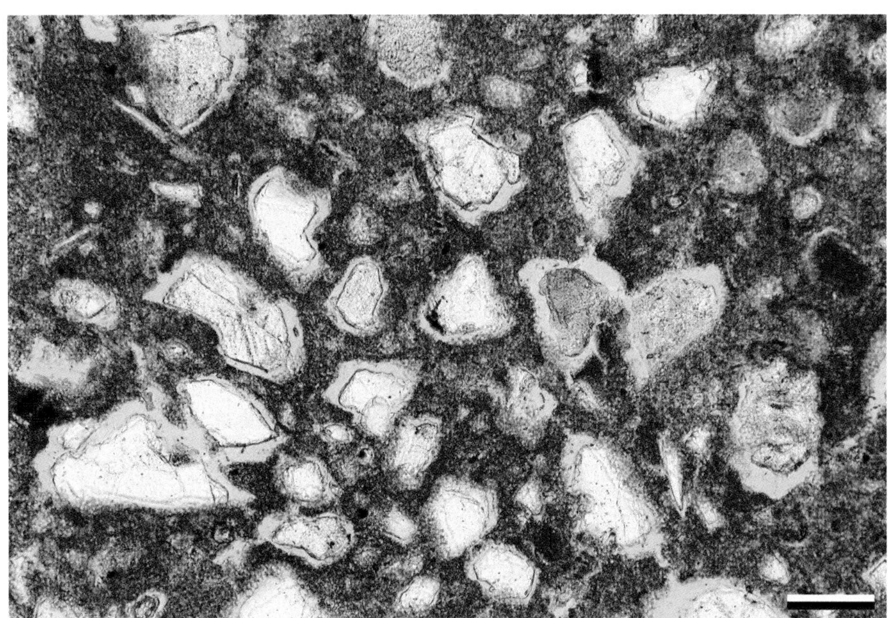

Up. Pennsylvanian Holder Fm., Otero Co., New Mexico

Most temporally definitive fabrics relate to early diagenesis when sediments are still soft and are subjected to processes that can only be found at the Earth's surface. In this example, root traces of higher plants (rhizoliths) have penetrated some tens of centimeters (about 5-10 inches) below a weathering surface underlain by marine limestones and overlain by nonmarine shales. This paleosol has laminated crusts as well as carbonate-lined root traces and the dark color reflects entrained insoluble materials from weathering (clays, organic matter, and iron- and manganese-bearing minerals).

PPL, Scale bar = 0.51 mm

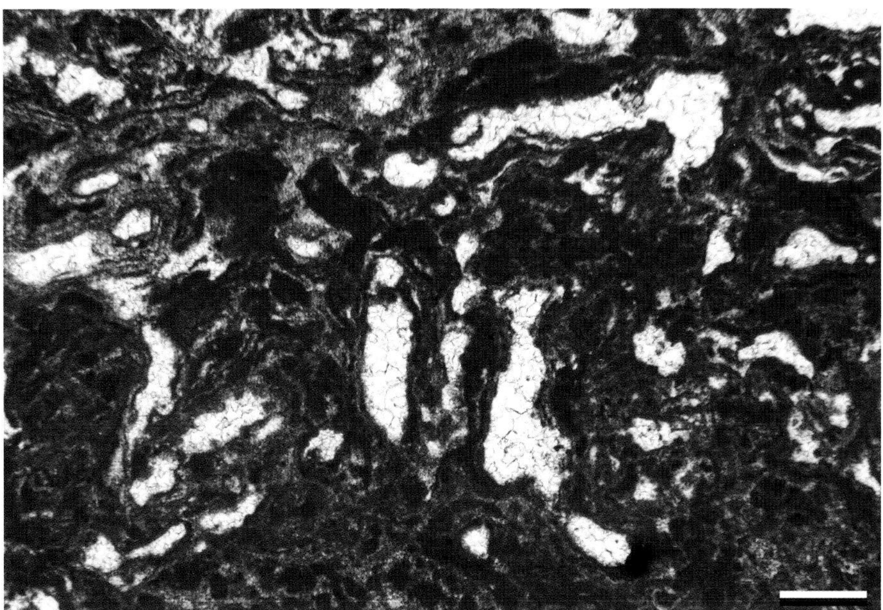

Quaternary? travertine, Bonneville Co., Idaho

These microbial (bacterial) shrubs are from a warm-spring travertine and represent another fabric indicative of surficial processes (eogenetic in this case, but potentially also telogenetic in other cases). Microbially-induced precipitates can be found in subsurface settings, but microbial shrubs, which grow into large open pores are unlikely to occur in any settings other than surficial ones. Warm to hot waters foster microbial growth so areas with emerging hydrothermal waters (travertine springs, interstitial spaces within pillow basalts, and others) are especially favored for such deposits.

PPL, AFeS, BDI, Scale bar = 0.26 mm

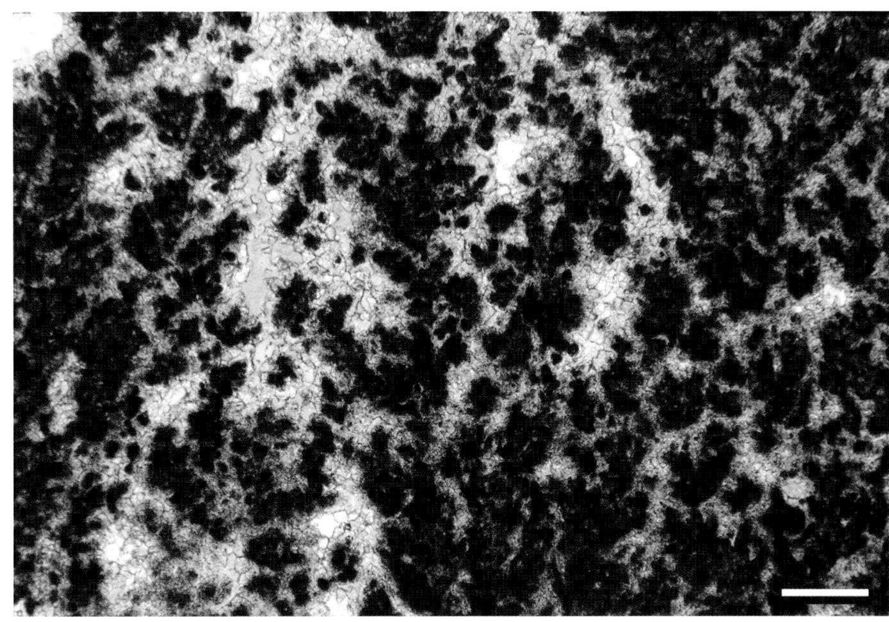

Cited References and Additional Information Sources

Baker, J. C., 1991, Diagenesis and reservoir quality of the Aldebaran Sandstone, Denison Trough, east-central Queensland, Australia: Sedimentology, v. 38, p. 819-838, doi: 10.1111/j.1365-3091.1991.tb01874.x.

Beitler, B., W. T. Parry, and M. A. Chan, 2005, Fingerprints of fluid flow: chemical diagenetic history of the Jurassic Navajo Sandstone, southern Utah, USA: Journal of Sedimentary Research, v. 75, p. 547-561, doi: 10.2110/jsr.2005.045.

Blamey, N. J. F., K. Azmy, and U. Brand, 2014, Provenance and burial history of cement in sandstones of the Northbrook Formation (Carboniferous), western Newfoundland, Canada: A geochemical investigation: Sedimentary Geology, v. 299, p. 30-41, doi: 10.1016/j.sedgeo.2013.10.005.

Burley, S. D., 1984, Patterns of diagenesis in the Sherwood Sandstone Group (Triassic), United Kingdom: Clay Minerals, v. 19, p. 403-440, doi: 10.1180/claymin.1984.019.3.11.

Burley, S. D., J. Mullis, and A. Matter, 1989, Timing diagenesis in the Tartan Reservoir (UK North Sea): Constraints from combined cathodoluminescence microscopy and fluid inclusion studies: Marine and Petroleum Geology, v. 6, p. 98-120, doi: 10.1016/0264-8172(89)90014-7.

Crossey, L. J., and D. Larsen, 1992, Authigenic mineralogy of sandstones intercalated with organic-rich mudstones: Integrating diagenesis and burial history of the Mesaverde Group, Piceance Basin, NW Colorado, *in* D. W. Houseknecht, and E. D. Pittman, eds., Origin, Diagenesis, and Petrophysics of Clay Minerals in Sandstones: Tulsa, OK, SEPM Special Publication 47, p. 125-144, doi: 10.2110/pec.92.47.0125.

Donovan, T. J., 1974, Petroleum microseepage at Cement, Oklahoma: evidence and mechanism: AAPG Bulletin, v. 58, p. 429-446.

Dutton, S. P., 1997, Timing of compaction and quartz cementation from integrated petrographic and burial-history analyses, Lower Cretaceous Fall River Formation, Wyoming and South Dakota: Journal of Sedimentary Research, v. 67, p. 186-196, doi: 10.1306/D426852C-2B26-11D7-8648000102C1865D.

Edman, J. D., and R. C. Surdam, 1984, Diagenetic history of the Phosphoria, Tensleep and Madison Formations, Tip Top field, Wyoming, *in* D. A. McDonald, and R. C. Surdam, eds., Clastic Diagenesis: Tulsa, OK, AAPG Memoir 37, p. 317-345.

Goldstein, R. H., 2001, Fluid inclusions in sedimentary and diagenetic systems: Lithos, v. 55, p. 159-193, doi: 10.1016/S0024-4937(00)00044-X.

Goodchild, M. W., and J. H. M. Whitaker, 1986, A petrographic study of the Rotliegendes Sandstone reservoir (Lower Permian) in the Rough Gas Field: Clay Minerals, v. 21, p. 459-477, doi: 10.1180/claymin.1986.021.4.04.

Hansley, P. L., and V. F. Nuccio, 1992, Upper Cretaceous Shannon Sandstone reservoirs, Powder River Basin, Wyoming: Evidence for organic acid diagenesis?: AAPG Bulletin, v. 76, p. 781-791.

Hawkins, P. J., 1978, Relationship between diagenesis, porosity reduction, and oil emplacement in late Carboniferous sandstone reservoirs, Bothamsall Oilfield, E Midlands: Journal of the Geological Society, v. 135, p. 7-24, doi: 10.1144/gsjgs.135.1.0007.

Higley, D. K., 1992, Petrology and reservoir paragenesis in the Sussex "B" sandstone of the Upper Cretaceous Cody Shale, House Creek and Porcupine fields, Powder River basin, Wyoming: Washington, DC, USGS Bulletin 1917-G, p. G1-G16.

Houseknecht, D. W., 1987, Assessing the relative importance of compaction processes and cementation to reduction of porosity in sandstones: AAPG Bulletin, v. 71, p. 633-642.

Houseknecht, D. W., 1991, Use of cathodoluminescence petrography for understanding compaction, quartz cementation, and porosity in sandstones, *in* C. E. Barker, and O. C. Kopp, eds., Luminescence Microscopy and Spectroscopy: Qualitative and Quantitative Applications: Tulsa, OK, SEPM Short Course 25, p. 59-66, doi: 10.2110/scn.91.25.0059.

Kirkland, D. W., R. E. Denison, and M. A. Rooney, 1995, Diagenetic alteration of Permian strata at oil fields of south central Oklahoma, USA: Marine and Petroleum Geology, v. 12, p. 629-644, doi: 10.1016/0264-8172(95)98089-N.

Larsen, D., J. L. Loomis, and L. J. Crossey, 2000, Diagenesis in the Point Lookout Sandstone, San Juan Basin, New Mexico and Colorado: Influence of depositional conditions, cyclic stratigraphy, and changing hydrologic regime: The Mountain Geologist, v. 37, p. 109-133.

Milliken, K. L., L. S. Land, and R. G. Loucks, 1981, History of burial diagenesis determined from isotopic geochemistry, Frio Formation, Brazoria County, Texas: AAPG Bulletin, v. 65, p. 1397-1413.

Morad, S., N. E. Serrhini, H. N. Ben Ismail, L. F. De Ros, and I. S. Al-Aasm, 1994, Diagenesis and formation water chemistry of Triassic reservoir sandstones from southern Tunisia: Sedimentology, v. 41, p. 1253-1272, doi: 10.1111/j.1365-3091.1994.tb01452.x.

Naeser, N. D., and T. H. McCulloh, eds., 1988, Thermal History of Sedimentary Basins: New York, Springer-Verlag, 500 p.

Newell, K. D., R. H. Goldstein, and C. J. Burdick, 2003, Diagenesis and late-stage porosity development in the Pennsylvanian Strawn Formation, Val Verde basin, Texas, U.S.A., *in* W. M. Ahr, P. M. Harris, W. A. Morgan, and I. D. Somerville, eds., Permo-Carboniferous Carbonate Platforms and Reefs: Tulsa, OK, SEPM Special Publication 78, p. 333-350, doi: 10.2110/pec.03.78.0333.

Odom, I. E., T. N. Willand, and R. J. Lassin, 1979, Paragenesis of diagenetic minerals in the St. Peter Sandstone (Ordovician), Wisconsin and Illinois, *in* P. A. Scholle, and P. R. Schluger, eds., Aspects of Diagenesis: Tulsa, OK, SEPM Special Publication 26, p. 425-443, doi: 10.2110/pec.79.26.0425.

Parnell, J., 1994, Hydrocarbons and other fluids: Paragenesis, interactions and exploration potential inferred from petrographic studies, *in* J. Parnell, ed., Geofluids: Origin, Migration and Evolution of Fluids in Sedimentary Basins: London, GSL Special Publication 78, p. 275-291, doi: 10.1144/GSL.SP.1994.078.01.19.

Pitman, J. K., M. B. Goldhaber, and C. Spöetl, 1996, Regional diagenetic patterns in the St. Peter Sandstone: Implications for brine migration in the Illinois Basin: Washington, DC, USGS Professional Paper 2094-A, 17 p.

Polito, P. A., T. K. Kyser, P. N. Southgate, and M. J. Jackson, 2006, Sandstone diagenesis in the Mount Isa basin: An isotopic and fluid inclusion perspective in relationship to district-wide Zn, Pb, and Cu mineralization: Economic Geology, v. 101, p. 1159-1188, doi: 10.2113/gsecongeo.101.6.1159.

Purvis, K., 1992, Lower Permian Rotliegend sandstones, southern North Sea: A case study of sandstone diagenesis in evaporite-associated sequences: Sedimentary Geology, v. 77, p. 155-171, doi: 10.1016/0037-0738(92)90123-9.

Rossi, C., R. H Goldstein, A. Ceriani, and R. Marfil, 2002, Fluid inclusions record thermal and fluid evolution in reservoir sandstones, Khatatba Formation, Western Desert, Egypt: A case for fluid injection: AAPG Bulletin, v. 86, p. 1773-1799, doi: 10.1306/61EEDD78-173E-11D7-8645000102C1865D.

Vagle, G. B., A. Hurst, and H. Dypvik, 1994, Origin of quartz cements in some sandstones from the Jurassic of the Inner Moray Firth (UK): Sedimentology, v. 41, p. 363-377, doi: 10.1111/j.1365-3091.1995.tb02107.x.

Weedman, S. D., S. L. Brantley, R. Shiraki, and S. R. Poulson, 1996, Diagenesis, compaction, and fluid chemistry modeling of a sandstone near a pressure seal: Lower Tuscaloosa Formation, Gulf Coast: AAPG Bulletin, v. 80, p. 1045-1063.

Wojcik, K. M., R. H. Goldstein, and A. W. Walton, 1997, Regional and local controls of diagenesis driven by basin-wide flow system: Pennsylvanian sandstones and limestones, Cherokee basin, southeastern Kansas, *in* I. P. Montañez, J. M. Gregg, and K. L. Shelton, eds., Basin-Wide Diagenetic Patterns: Tulsa, OK, SEPM Special Publication 57, p. 235-252, doi: 10.2110/pec.97.57.0235.

Facing Page: A detailed mineral map, based on SEM, BSE and EDS analysis, of a thin-section from the Upper Pennsylvanian Bursum Fm., Socorro Co., New Mexico. From FEI, Inc.

Emerging Techniques

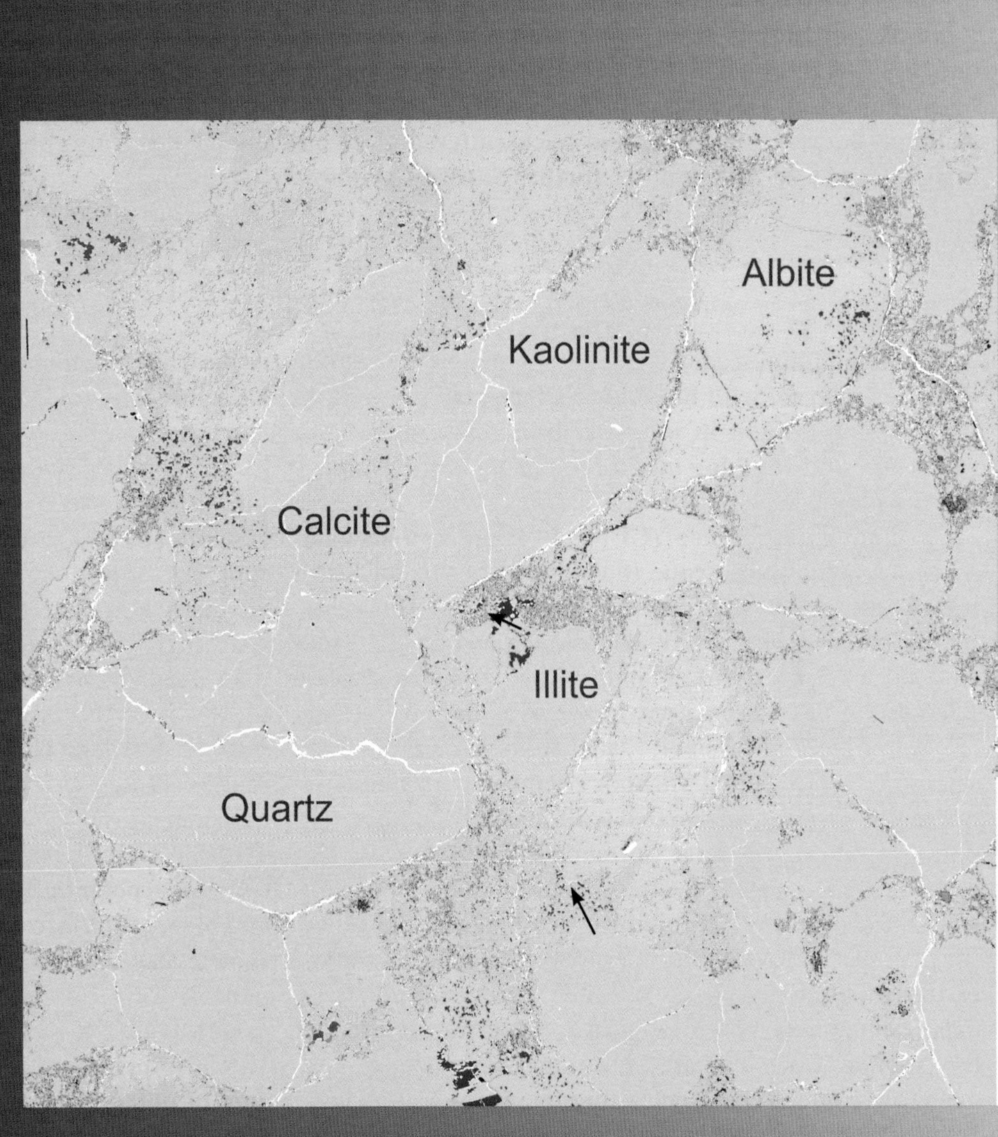

CHAPTER 24

INTRODUCTION TO EMERGING TECHNIQUES

There are many observational technologies that can be used to supplement or quantify optical petrographic observations. Some apply directly to light microscopy (staining, for example), some are done with light microscopes but require specialized additional equipment (fluid inclusion geothermometry, some forms of cathodoluminescence and fluorescence microscopy), and others require completely different and typically more expensive instrumentation (SEM/TEM, microprobe/ion probe, X-ray computed tomography and many others). Examples of output from some of the most common and best established of these techniques have already been shown and referenced throughout this book (e.g., staining, fluorescence, SEM and backscattered electron imaging) and so will not be further illustrated here, although useful summary and case-study references are provided at the close of this chapter. One well established technique, cathodoluminescence, has undergone new instrumentation developments, including diversification from light microscopes to scanning electron microscopes, and so it will be featured in this chapter along with two other relatively new and promising imaging and analysis technologies, including X-ray computed tomography, grouped here under the term "digital rock physics". These technologies are evolving rapidly and the descriptions in this chapter will soon be superceded by new developments. Nonetheless, they should serve to show readers to the potential of such technologies to enhance petrographic investigations.

CATHODOLUMINESCENCE

Cathodoluminescence is a phenomenon where electrons interact with luminescent material and induce the emission of light in wavelengths of the visible light spectrum (and reaching into the UV range). In essence, it is the photoelectric effect in reverse.

Cathodoluminescence (CL) can be observed with a petrographic microscope (floodgun) or with an SEM where other detectors can aid mineral identification. In the petrographic study of sedimentary rocks, "cold cathode" CL systems mounted on a petrographic microscope have been used for several decades (Long and Agrell, 1965; Marshall, 1988). In such a setup, a thin section is placed into a vacuum chamber with windows and a built-in electron gun that is attached to the stage of a petrographic microscope. Coronal discharge from a cold cathode at high negative voltage "floods" the sample with energetic electrons and induces cathodoluminescence that can be viewed with the microscope through a window in the vacuum chamber. A widely used commercially produced system is the Luminoscope that was developed by Nuclide Corporation, but other designs also are available commercially, including one from Cambridge Image Technology Ltd. (see also Boggs and Krinsley, 2006).

Several rock-forming minerals (e.g., quartz, feldspar, apatite and carbonates) can be identified with reasonable ease on the basis of their CL colors. Early examples in the geologic literature are studies by Long and Agrell (1965), Smith and Stenstrom (1965), Sippel (1968) and Zinkernagel (1978). A limitation for application of the "floodgun" approach is that flooding the specimen with high densities of energetic electrons can induce undesirable high specimen temperatures, cause reduced CL emissions and even permanent damage to the CL emission capability of the specimen.

In contrast, scanning for CL with an electron microscope (SEM-CL) gives much higher spatial resolutions, and because the beam "rests" on a given spot for a limited number of microseconds, heating effects are greatly reduced. The light emitted by the specimen is collected by a parabolic mirror and, via a light guide, conveyed to a photomultiplier tube. Early SEM-CL detectors were able to produce only gray-scale images, and for color imaging, it was necessary to take multiple images with color filters and then combine those component images with graphics software. The problem was that CL response between imaging passes was liable to change, potentially resulting in unreproducible images and petrographic features. More recent detector designs, such as the GATAN ChromaCL (the equipment used for most of the CL images in this book), enable live digital color imaging at a range of pixel resolutions. In this way, all CL information is acquired in a single imaging pass and the potential damage to CL-related features is greatly reduced. The fact that CL-related petrographic features can be observed without delay greatly accelerates the investigation of petrographic characteristics.

CL emissions are dependent on a number of factors including, but not limited to:

- crystal structure,
- lattice defects, and
- distribution of trace or minor elements.

As a tool for detecting chemical differences, CL, in most cases, is much more sensitive than, for example, energy dispersive X-ray spectroscopy (EDS) and can allow visualization of differences in the ppm range. By showing the distribution of trace and minor elements, as well as crystal growth patterns, SEM-CL may show features that are undetectable by other methods. Such features can be valuable for determining the provenance of sandstones, siltstones and mudstones (e.g., Matter and Ramseyer, 1985; Seyedolali et al., 1997).

The CL color (or spectrum) of detrital quartz grains reflects the source rocks of the sedimentary material. This was first comprehensively documented in a study by Zinkernagel (1978) and has been confirmed by subsequent studies. For example, blue colors (450, 505 nm) are typical for high-temperature metamorphic and magmatic quartz grains (Figures 24.1 and 24.2; Götte and Ramseyer, 2012), and brown to reddish colors (637-650 nm) are common for hydrothermal and low-temperature metamorphic quartz (Figure 24.3; Götte and Ramseyer, 2012). Violet-blue colors (~630 nm) are thought to be typical for volcanic rocks, and greenish blue tones (500 nm) are associated with pegmatites (Götze et al., 2005; Götte and Ramseyer, 2012). Yet, although there are general trends that have been recognized, when looked at closely, there appear to be multiple signatures for given provenances (Ramseyer and Mullis, 1990; Götze, 2009) and thus a need for additional investigations. In recent work (Leeman et al., 2012), a correlation between titanium concentration (down to ppm levels) and the blue CL emission band (450 nm) in volcanic quartz has been noted and allows CL to be used as a thermobarometer.

Traditionally, provenance assessments of shales and mudstones utilized bulk rock geochemical parameters that were prone to alteration by weathering and diagenesis. Thus, for a long time these rocks have been underutilized for provenance studies, even though they constitute the bulk of most basin fills. The new generation of color CL

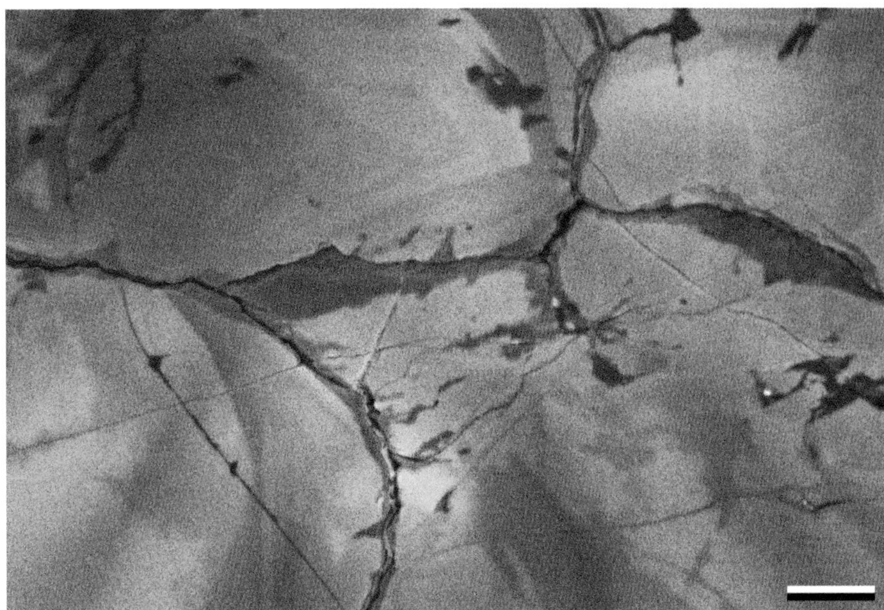

Figure 24.1: Scanned color CL of quartz from a granite pluton (Proterozoic Town Mountain Granite) near Llano, Texas. The plutonic quartz has bluish CL colors, and internal textures include mottling and growth zonation. There are red colors along the grain margins that relate to hydrothermal fluids while the pluton was cooling, and thin black features and black blotches that probably represent late and relatively low-temperature quartz formation. Scale bar = 70 μm.

Figure 24.2: Large areas of a sandstone can be viewed with color CL on an SEM and the multiple images can be stitched together with software. Scale bar = 367 μm

detectors (like ChromaCL) makes it possible to get good quality provenance information not only from large quartz grains but also from silt-size grains by combined use of CL color and internal grain textures (Figure 24.4; Schieber and Wintsch, 2005; Krinsley et al., 2009). Applying SEM-CL to quartz grains, the chemically and mechanically most robust major component of clastic terrigenous rocks, will enable us to greatly improve understanding of provenance and uplift history of the source areas of sedimentary basin fills.

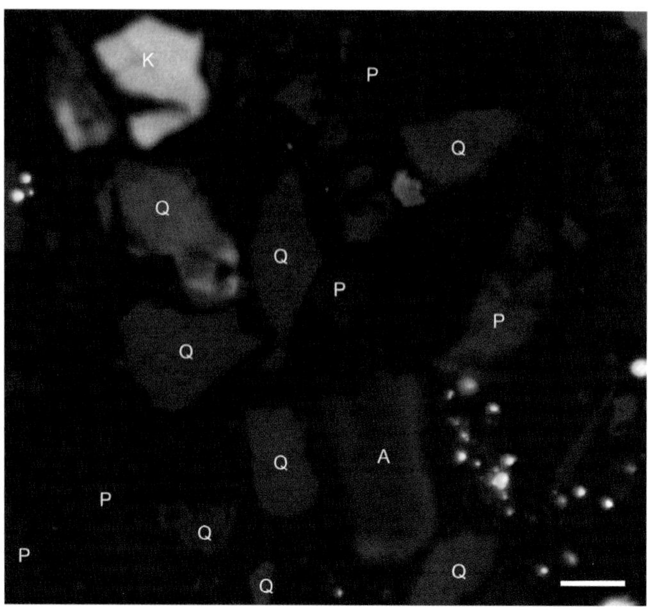

Figure 24.3: Low- and medium-grade metamorphic grains in the Cretaceous Mancos Shale (Utah) that were derived from the Sevier Orogenic belt (Q = quartz, P = plagioclase, K = K-feldspar, A = albite). Scale bar = 10.5 μm

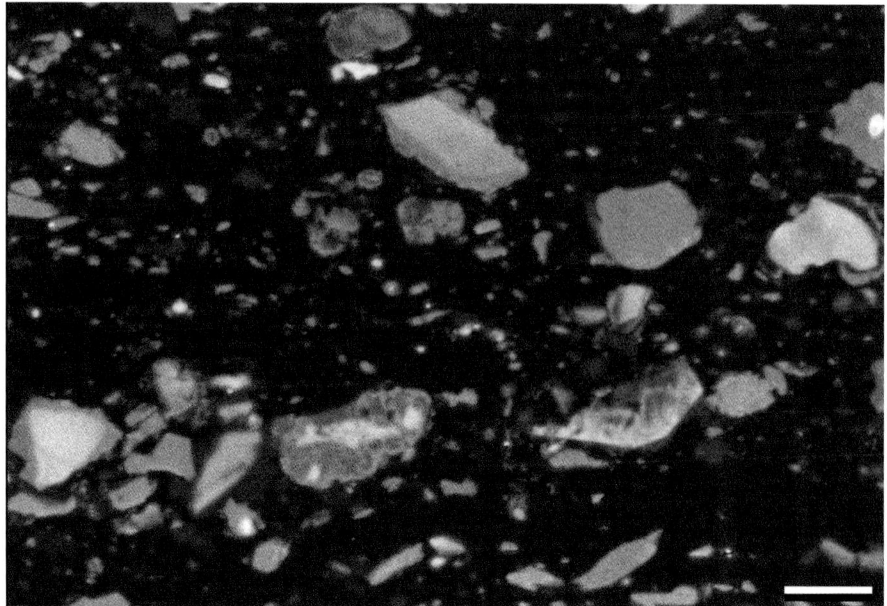

Figure 24.4: Silt-size quartz grains from the Up. Devonian New Albany Shale, Indiana. In scanned color CL a wide range of different types of quartz grains is apparent. Many have interior cores that indicate previous episodes of sediment recycling, from sources that are plutonic as well as metamorphic. Some grains show nonluminescent diagenetic overgrowths, followed by later diagenetic (higher temperature, reddish) overgrowths. Scale bar = 14 μm

Automated Mineralogy and Petrographic Analyses

Technologies that allow mapping of quantitative compositional data at thin-section scale can aid petrographic analysis. One such technique is FEI's QEMSCAN Quanta FEG with iDiscover analysis software. Other, not necessarily exactly equivalent systems, including Robertson USA/Carl Zeiss RoqSCAN, Oxford Instrument's INCAMineral system and TIMA by TESCAN. The hardware utilizes integrated SEM, BSE, energy-dispersive X-ray spectrometry (EDS) and digital image analysis to identify the different mineral phases within a sample.

The types of quantitative results include:

- Mineral maps,
- Elemental maps,
- Modal mineralogy (Table 24.1),
- Textural analysis,
- Porosity maps,
- Calculated matrix density (Table 24.2),
- Calculated porosity (Table 24.2), and
- Lithotyping,

The sample used here is a thin section of the Upper Pennsylvanian Bursum Formation from central New Mexico. It was analyzed by FEI to provide an elemental and mineralogical characterization that could be directly integrated with petrographic observations. The entire thin section (Figure 24.5) was analyzed at a 5-μm spacing with selected areas at a 1-μm spacing. Both the FEI analysis and the petrographic study identified the sample as an arkose (Folk classification) or feldspathic arenite (Dott classification), and the mineralogical data provided by the analytical work provided valuable compositional information that may have been overlooked by petrographic characterization (Figures 24.6 and 24.7). These types of analyses can be important aids to traditional petrographic techniques (Figures 24.8 to 24.11). However, they are expensive and should be integrated with petrographic studies and geologic framework information for maximum benefit and cost effectiveness.

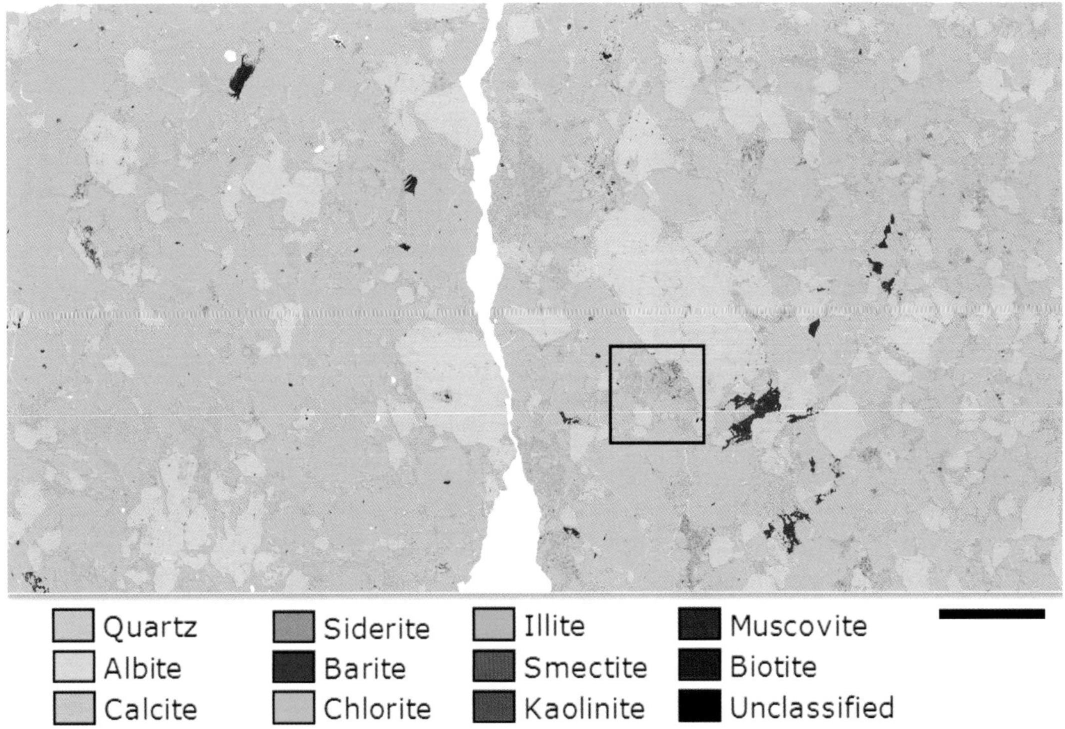

Figure 24.5: A mineral map of the entire Up. Pennsylvanian Bursum Fm. thin section showing the distribution of mineral phases. Based on the outcrop samples, the sample was thought to comprise mostly quartz and potassium feldspars. Based on analysis, however, it was shown that quartz (pink) and albite (pale blue) dominate. Scale bar = 4.8 mm.

Table 24.1: Mineral volumes within the full thin section and the 1-μm sampled areas. Area 2 is highlighted because it is the area covered by the black box in Figure 24.5 and Figures 24.6 - 24.7.

Minerals	Volume (%) Full Thin Section	Area 1	Area 2	Area 3
Quartz	63.1	78.0	55.4	62.4
K-Feldspar	0.0	0.0	0.0	0.0
Albite	21.7	13.0	30.7	20.3
Calcite	11.2	3.7	4.4	13.5
Dolomite	0.0	0.0	0.0	0.0
Siderite	0.1	0.2	0.3	0.1
Illite	0.2	0.1	0.5	0.1
Smectite	0.4	0.3	0.9	0.4
Kaolinite	0.2	0.1	0.1	0.3
Chlorite	2.0	2.7	4.9	2.0
Muscovite	0.2	0.0	0.4	0.0
Biotite	0.2	0.1	0.5	0.1
Pyrite	0.0	0.0	0.0	0.0
Rutile	0.0	0.0	0.1	0.0
Apatite	0.0	0.0	0.1	0.0
Zircon	0.0	0.0	0.0	0.0
Gypsum/Anhydrite	0.0	0.0	0.1	0.0
Barite	0.6	1.1	0.8	0.3
Other	0.0	0.0	0.0	0.1
Unclassified	0.1	0.5	0.9	0.3

Table 24.2: Porosity within the slide and selected areas was estimated by applying a background value to the BSE intensities. Porosity includes inter- and intragranular pores, fractures and cracks (natural or generated) and any organic matter. Porosity estimates cannot include micropores (~<2 μm) because of EDS resolution limits. Applying the porosity values and the data in Table 24.1, grain densities can be calculated for the thin section and 1-μm areas.

Volume (%)	Full Thin Section	Area 1	Area 2	Area 3
Porosity	2.1	2.3	4.0	2.6

gr/cc	Full Thin Section	Area 1	Area 2	Area 3
Grain Density	2.66	2.66	2.66	2.67

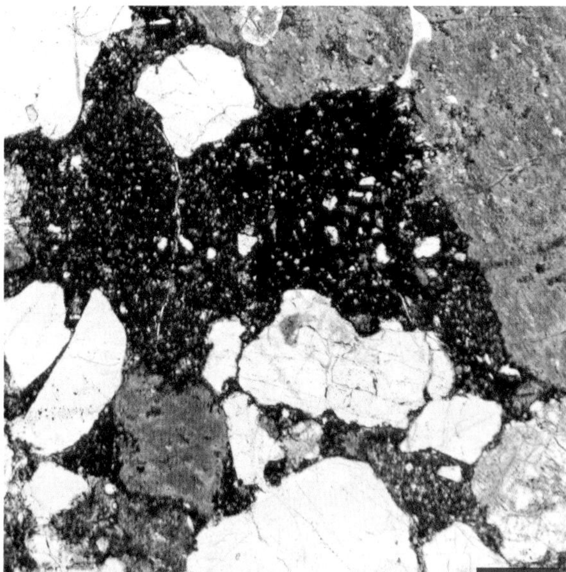

Figure 24.6: A plane-polarized light view of the thin section from the box area in Figure 24.5. Scale bar = 0.3 mm.

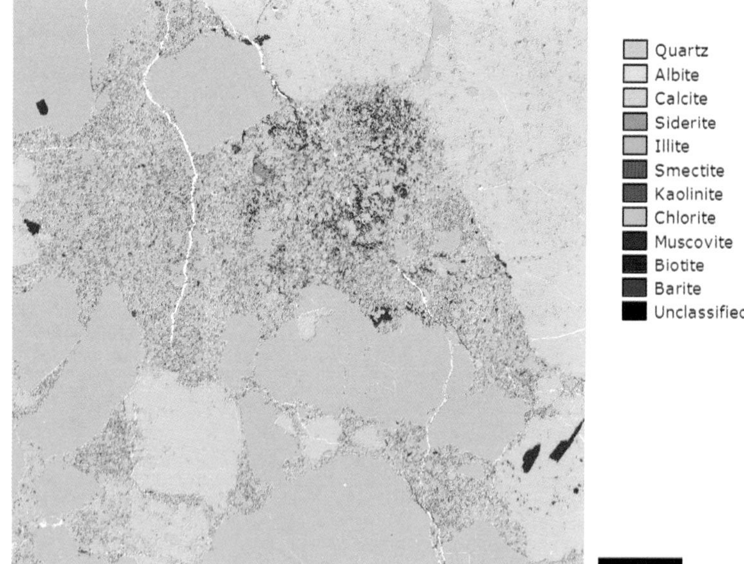

Figure 24.7: A mineral map of the of the area shown in Fig. 24.6. The reddish brown areas in Fig. 24.6 that might easily be mistaken for K-feldspar are actually albite-replaced K-feldspar (blue areas in this figure). Scale bar = 0.3 mm.

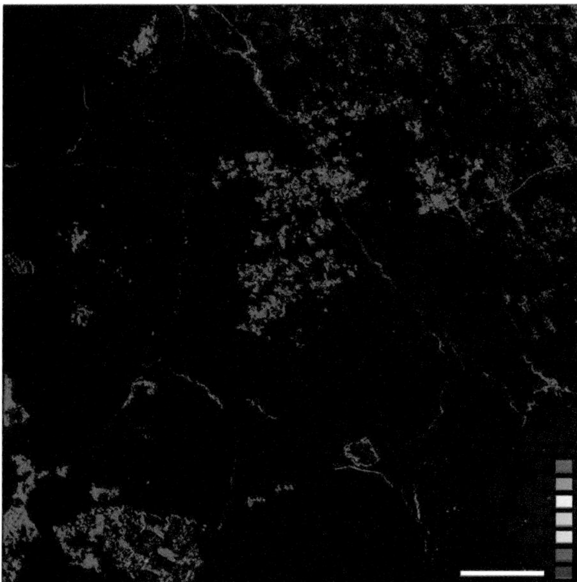

Figure 24.8: A Ca^{2+} BSE elemental map from the area in Figure 24.7. Most of the red areas represent calcite cements. Scale bar = 0.3 mm. Highest concentrations are red and lowest are violet.

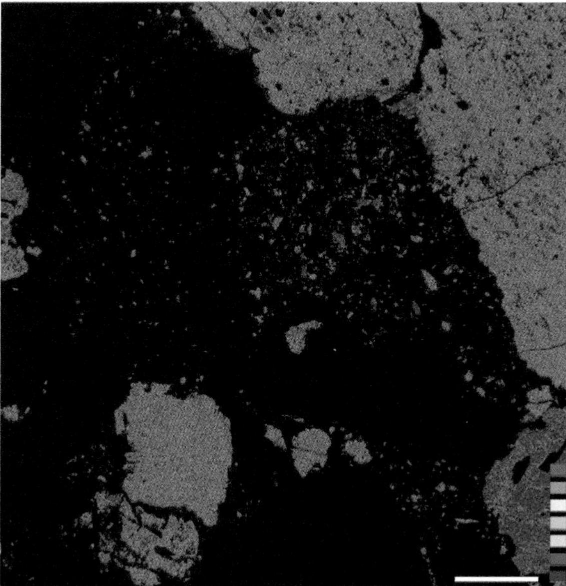

Figure 24.9: A Na^{+} BSE elemental map of the area in Figure 24.7. The sodium distribution marks the location of albite replacements (orange color) and an original albite grain (purple color). Scale bar = 0.3 mm. Highest concentrations are red and lowest are violet.

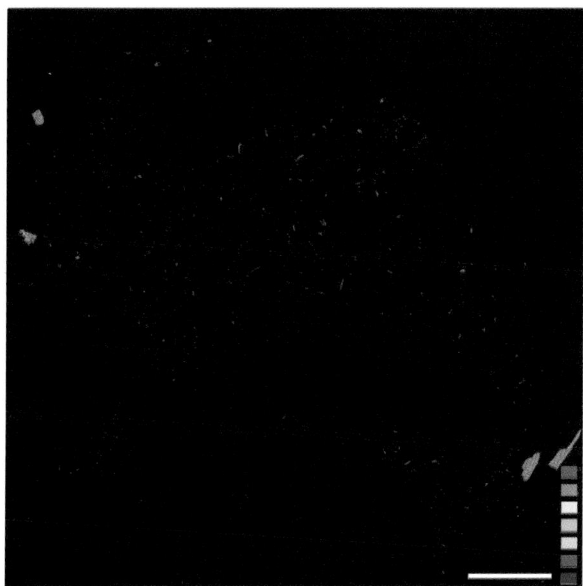

Figure 24.10: A K^{+} BSE elemental map shows little potassium in the sample area, therefore no orthoclase or K-feldspars. The green areas are muscovite grains. Scale bar = 0.3 mm. Highest concentrations are red and lowest are violet.

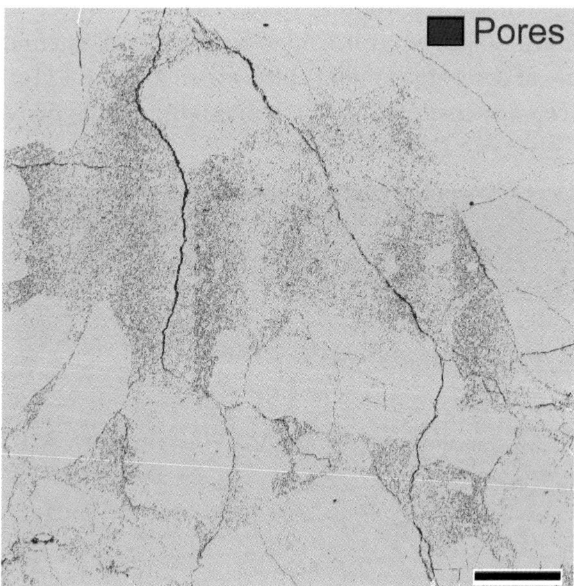

Figure 24.11: A map of pore distribution in the area from Figure 24.7. Scale bar = 0.3 mm.

Multiscale Analysis Using Digital Rock Physics

Juliana Anderson, Dava DeBauche, and Joel Walls, Ingrain, Houston, USA

Digital Rock Physics (DRP) encompasses many techniques used for both quantitative and qualitative rock analysis. DRP merges three key technologies that have evolved rapidly over the last decade. First, high-resolution diagnostic imaging methods permit detailed examination of the internal structure of rock samples on a wide range of scales. The second is advanced numerical methods for simulating complex physical phenomena, and finally, high-speed parallel computation uses powerful graphical processor units (GPUs) for data visualization. DRP methods are mostly noninvasive, nondestructive and provide relevant data rapidly.

The equipment used for image analysis in DRP includes X-ray Computed Tomography (CT) scans for whole core, MicroCT scans for plug size and micro samples (Figures 24.12 and 24.13), NanoCT for rocks containing micrite, chalk and other nano-level porous material, and a focused ion beam scanning electron microscope (FIB-SEM) for mudstone and shale.

X-ray computed tomography scans for whole core studies are acquired at 0.5 mm/voxel (a voxel is each of an array of discrete elements into which a representation of a three-dimensional object is divided). Scans provide detailed 3-D images, obtained while the core is sealed and protected inside an aluminum core barrel. Data are both visual and quantitative, as the X-ray CT scanning is run with a dual-energy process which allows for computation of bulk density (RHOB) and photoelectric factor (PEF).

Industrial MicroCT scanners are utilized for samples ranging from core-plug size (2.5-3.3 cm [1 to 1.5 inch] diameter) to small pillars (0.5 mm diameter). MicroCT resolution ranges from 40 μm/voxel to 500 nm/voxel. The Nano CT scanner is used to image rocks at a resolution of 65 nm/voxel. The FIB-SEM uses mm-sized samples to image at resolutions ranging from 750 μm to 2.5 nm/pixel. For all rock types, scanning begins at the lowest resolution (largest sample size) incrementally increasing resolution and decreasing sample size until porosity and connectivity can be computed.

Once scanning is complete at the MicroCT and Nano scale (traditionally sandstones and carbonates), images are constructed. Total porosity, connectivity, absolute permeability in the x, y and z directions, formation factor, elasticity and cementation exponent "m" can be computed on representative volumes. Grain-size and pore size distributions also can be computed at this resolution.

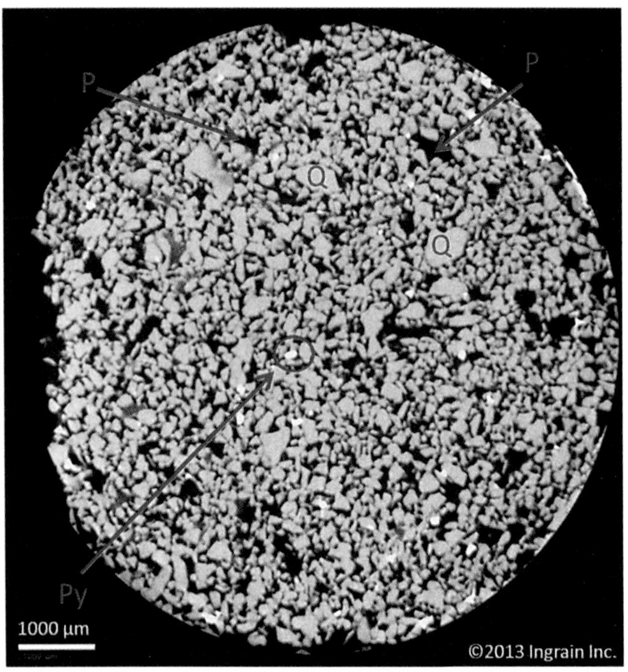

Figure 24.12: A relatively pure quartzarenite sample imaged using the MicroCT scanner. Single axial image above has a resolution of approximately 8 μm/voxel. Scan is based on density, therefore pores (P) appear dark (near black) and grains light gray. In this scan almost all grains are quartz (Q). Small, high-density pyrite (Py) clusters are present.

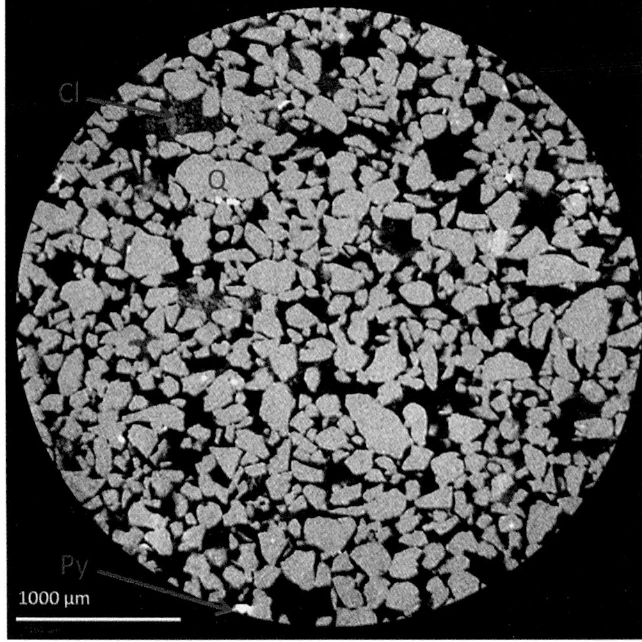

Figure 24.13: Sandstone from Figure 24.12, imaged at a resolution of 4 μm/voxel using a MicroCT scanner. Note as resolution increases, field of view decreases. At this resolution, sparse clay (Cl), which appears dark gray but lighter than pores, can be seen between quartz (Q) grains. High-density pyrite (Py) clusters are present.

FIB-SEM is used in analysis of shale samples (Figures 24.14 to 24.17). In order to view pore systems, shale samples are imaged at a resolution of 5-15 nm. FIB-SEM images provide a clear distinction between porosity, organic matter, mineral matrix and high-density materials (typically pyrite) within segmented images. Volume percentages of each phase can be calculated in each direction (Figures 24.14 to 24.16). FIB-SEM data is also used to quantify intergranular porosity and porosity associated with organic matter.

Pore-size distribution, absolute permeability, relative permeability, and capillary pressure can be computed on the 3-D FIB-SEM volume using a Lattice-Boltzmann numerical method (Figures 24.16 and 24.17). As a final step, the fine-scale micro-CT and FIB-SEM data can be upscaled back to whole core and well-log scale. Following segmentation, pore-size distribution, absolute permeability, relative permeability, and capillary pressure can be computed on the segmented 3-D FIB-SEM volume using a Lattice-Boltzmann numerical method (Figures 24.16 and 24.17). As a final step, the fine scale micro-CT and FIB-SEM data can be upscaled to whole core and well-log scale.

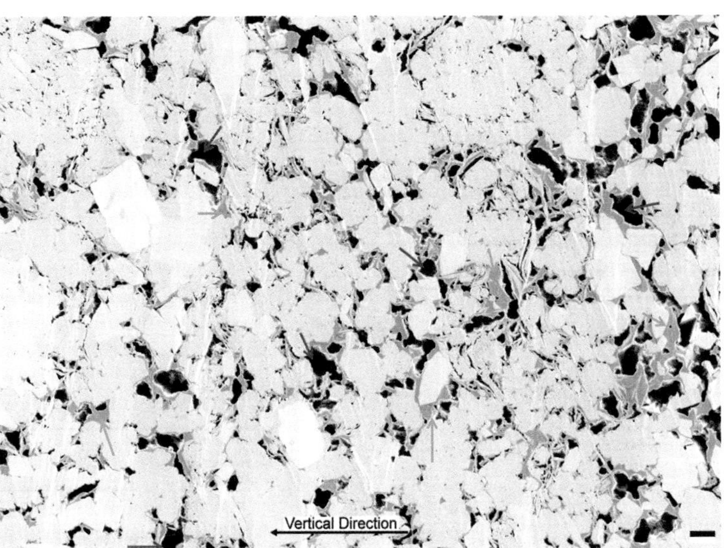

Figure 24.14: A shale sample imaged with an SEM at a resolution of 10 nm/pixel. Porosity (red arrows) is associated with organic matter (blue arrows) in a predominantly quartz matrix. The image aids in understanding the distribution of the porosity and organics in shale, and allows computation of pore-size distribution. Scale bar = 1 μm.

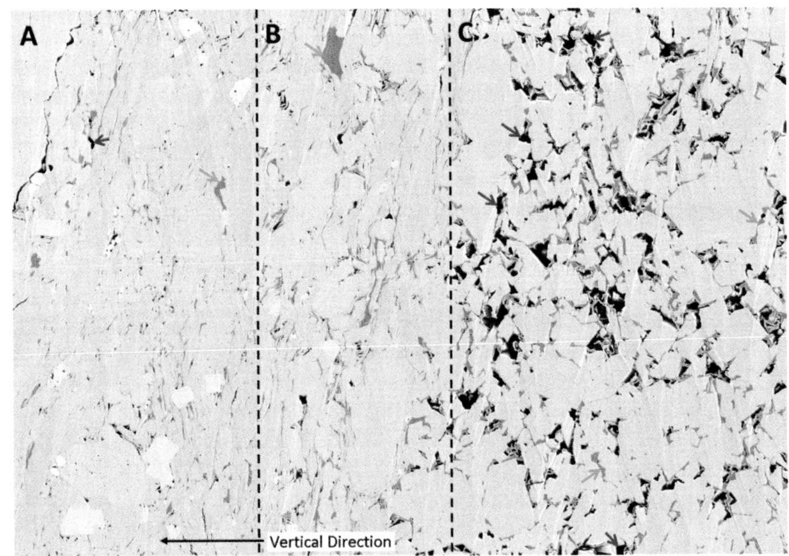

Figure 24.15: A shale sample imaged with an SEM at a resolution of 10 nm/pixel. This is a slice cut perpendicular to bedding (thus bedding is vertical in this view, with sample top to the left). Pores (small red arrows) are both intergranular and associated with organic matter (small blue arrows). The image highlights heterogeneities even at the nanoscale. Porosity and organic matter increase from section A to C. Section A contains little to no organic matter, and pores areu both intergranular and associated with organics. Pores in section C are all associated with organic matter, and the pore size is greater. In section B, organic matter is laminated parallel to the bedding plane. Scale bar = 1 μm.

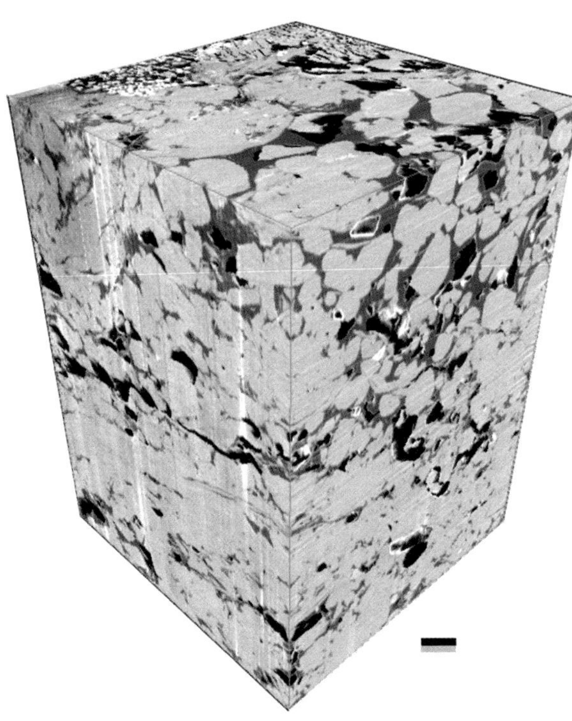

Figure 24.16: A high-resolution 3-D FIB-SEM volume. Pores (black) are associated with organic matter (dark gray) and the mineral matrix (light gray). Absolute permeability is calculated in the x, y, and z directions. Scale bar = 1 μm.

Figure 24.17: A high-resolution, segmented 3-D FIB-SEM cube where green represents organic matter, blue is connected porosity, and red is isolated porosity.

Cited References and Additional Information Sources

General

Carver, R. E., 1971, Procedures in Sedimentary Petrology: New York, Wiley-Interscience, 672 p.

Milner, H. B., 1962, Sedimentary Petrography. Volume I, Methods in Sedimentary Petrography (4th Edition): London, George Allen & Unwin, 643 p.

Tucker, M. E., ed., 1988, Techniques in Sedimentology: Palo Alto, CA, Blackwell Scientific Publications, 394 p.

Cathodoluminescence

Augustsson, C., and H. Bahlburg, 2003, Cathodoluminescence spectra of detrital quartz as provenance indicators for Paleozoic metasediments in southern Andean Patagonia: Journal of South American Earth Sciences, v. 16, p. 15-26, doi: 10.1016/S0895-9811(03)00016-6.

Augustsson, C., and A. Reker, 2012, Cathodoluminescence spectra of quartz as provenance indicators revisited: Journal of Sedimentary Research, v. 82, p. 559-570, doi: 10.2110/jsr.2012.51.

Barker, C. E., and T. Wood, 1986, Notes on cathodoluminescence microscopy using the Technosyn stage, and a bibliography of applied cathodoluminescence: Washington, DC, USGS Open-file Report 86-85, 35 p.

Bernet, M., and K. Bassett, 2005, Provenance analysis by single-quartz-grain SEM-CL/optical microscopy: Journal of Sedimentary Research, v. 75, p. 492-500, doi: 10.2110/jsr.2005.038.

Boggs, S., Jr., and D. Krinsley, 2006, Application of Cathodoluminescence Imaging to the Study of Sedimentary Rocks: Cambridge, UK, Cambridge University Press, 176 p.

Boggs, S., Jr., Y. I. Kwon, G. G. Goles, B. G. Rusk, D. Krinsley, and A. Seyedolali, 2002, Is quartz cathodoluminescence color a reliable provenance tool? A quantitative examination: Journal of Sedimentary Research, v. 72, p. 408-415, doi: 10.1306/102501720408.

Bruhn, F., P. Bruckschen, J. Meijer, A. Stephan, D. K. Richter, and J. Veizer, 1996, Cathodoluminescence investigations and trace-element analysis of quartz by micro-PIXE: Implications for diagenetic and provenance studies in sandstone: The Canadian Mineralogist, v. 34, p. 1223-1232.

Demars, C., M. Pagel, E. Deloule, and P. Blanc, 1996, Cathodoluminescence of quartz from sandstones: Interpretation of the UV range by determination of trace element distributions and fluid-inclusion PTX properties in authigenic quartz: American Mineralogist, v. 81, p. 891-901.

Evans, J., A. J. C. Hogg, M. S. Hopkins, and R. J. Howarth, 1994, Quantification of quartz cements using combined SEM, CL, and image analysis: Journal of Sedimentary Research, v. 64, p. 334-338, doi: 10.1306/D4267D93-2B26-11D7-8648000102C1865D.

Girard, J.-P., I. A. Munz, H. Johansen, S. Hill, and A. Canham, 2001, Conditions and timing of quartz cementation in Brent reservoirs, Hild Field, North Sea: Constraints from fluid inclusions and SIMS oxygen isotope microanalysis: Chemical Geology, v. 176, p. 73-92, doi: 10.1016/S0009-2541(00)00350-8.

Götte, T., T. Pettke, K. Ramseyer, M. Koch-Müller, and J. Mullis, 2011, Cathodoluminescence properties and trace element signature of hydrothermal quartz: A fingerprint of growth dynamics: American Mineralogist, v. 96, p. 802-813, doi: 10.2138/am.2011.3639.

Götte, T., and K. Ramseyer, 2012, Trace element characteristics, luminescence properties and real structure of quartz, *in* J. Götze, and R. Möckel, eds., Quartz: Deposits, Mineralogy and Analytics: New York, Springer-Verlag, p. 265-285, doi: 10.1007/978-3-642-22161-3_12.

Götte, T., and D. K. Richter, 2006, Cathodoluminescence characterization of quartz particles in mature arenites: Sedimentology, v. 53, p. 1347-1359, doi: 10.1111/j.1365-3091.2006.00818.x.

Götze, J., 2012a, Mineralogy, geochemistry and cathodoluminescence of authigenic quartz from different sedimentary rocks, *in* J. Götze, and R. Möckel, eds., Quartz: Deposits, Mineralogy and Analytics: New York, Springer, p. 287-306, doi: 10.1007/978-3-642-22161-3_13.

Götze, J., 2012b, Application of cathodoluminescence microscopy and spectroscopy in geosciences: Microscopy and Microanalysis, v. 18, p. 1270-1284, doi: 10.1017/S1431927612001122.

Götze, J., and R. Möckel, eds., 2012, Quartz: Deposits, Mineralogy and Analytics: New York, Springer, 360 p.

Götze, J., M. Plötze, and T. Trautmann, 2005, Structure and luminescence characteristics of quartz from pegmatites: American Mineralogist, v.

90, p. 13-21, doi: 10.2138/am.2005.1582.

Götze, J., H.-P. Schertl, R. D. Neuser, U. Kempe, and J. M. Hanchar, 2013, Optical microscope-cathodoluminescence (OM–CL) imaging as a powerful tool to reveal internal textures of minerals: Mineralogy and Petrology, v. 107, p. 373-392, doi: 10.1007/s00710-012-0256-0.

Gucsik, A., ed., 2009, Cathodoluminescence and its Application in the Planetary Sciences: Berlin, Springer-Verlag, 160 p., doi: 10.1007/978-3-540-87529-1.

Hamers, M. F., and M. R. Drury, 2011, Scanning electron microscope-cathodoluminescence (SEM-CL) imaging of planar deformation features and tectonic deformation lamellae in quartz: Meteoritics & Planetary Science, v. 46, p. 1814-1831, doi: 10.1111/j.1945-5100.2011.01295.x.

Hanchar, J. M., and C. F. Miller, 1993, Zircon zonation patterns as revealed by cathodoluminescence and backscattered electron images: Implications for interpretation of complex crustal histories: Chemical Geology, v. 110, p. 1-13, doi: 10.1016/0009-2541(93)90244-D.

Hogg, A. J. C., E. Sellier, and A. J. Jourdan, 1992, Cathodoluminescence of quartz cements in Brent Group sandstones, Alwyn South, UK North Sea, *in* A. C. Morton, R. S. Haszeldine, M. R. Giles, and S. Brown, eds., Geology of the Brent Group: London, GSL Special Publication 61, p. 421-440, doi: 10.1144/GSL.SP.1992.061.01.21.

Houseknecht, D. W., 1991, Use of cathodoluminescence petrography for understanding compaction, quartz cementation, and porosity in sandstones, *in* C. E. Barker, and O. C. Kopp, eds., Luminescence Microscopy and Spectroscopy: Qualitative and Quantitative Applications: Tulsa, OK, SEPM Short Course 25, p. 59-66, doi: 10.2110/scn.91.25.0059.

Kalceff, M. A. S., and M. R. Phillips, 1995, Cathodoluminescence microcharacterization of the defect structure of quartz: Physical review, B, Condensed matter, v. 52, p. 3122-3134, doi: 10.1103/PhysRevB.52.3122.

Krinsley, D., M. R. Stokes, and J. Schieber, 2009, Application of scanned color cathodoluminescence of quartz to provenance investigations of shales and mudstones [abs.]: GSA Abstracts with Programs, v. 41, p. 392.

Leeman, W. P., C. M. Macrae, N. C. Wilson, A. Torpy, C. T. A. Lee, J. J. Student, J. B. Thomas, and E. P. Vicenzi, 2012, A Study of cathodoluminescence and trace element compositional zoning in natural quartz from volcanic rocks: Mapping titanium content in quartz: Microscopy and Microanalysis, v. 18, p. 1322-1341, doi: 10.1017/S1431927612013426.

Long, J. V. P., and S. O. Agrell, 1965, The cathodo-luminescence of minerals in thin section: Mineralogical Magazine, v. 34, p. 318-326.

Marshall, D. J., 1988, Cathodoluminescence of Geological Materials: Winchester, MA, Allen & Unwin, 128 p.

Matter, A., and K. Ramseyer, 1985, Cathodoluminescence microscopy as a tool for provenance studies of sandstones, *in* G. G. Zuffa, ed., Provenance of Arenites (NATO Science Series C): Dordrecht, Netherlands, D. Reidel Publishing, p. 191-211, doi: 10.1007/978-94-017-2809-6_9.

Milliken, K. L., and S. E. Laubach, 2000, Brittle deformation in sandstone diagenesis as revealed by scanned cathodoluminescence imaging with application to characterization of fractured reservoirs, *in* M. Pagel, V. Barbin, P. Blanc, and D. Ohnenstetter, eds., Cathodoluminescence in Geosciences: Berlin, Springer, p. 225-243, doi: 10.1007/978-3-662-04086-7_9.

Owen, M. R., 1991, Application of cathodoluminescence to sandstone provenance, *in* C. E. Barker, and O. C. Kopp, eds., Luminescence Microscopy and Spectroscopy: Qualitative and Quantitative Applications: Tulsa, OK, SEPM Short Course 25, p. 67-74, doi: 10.2110/scn.91.25.

Pagel, M., V. Barbin, P. Blanc, and D. Ohnenstetter, eds., 2000, Cathodoluminescence in Geosciences: Berlin, Springer, 514 p.

Ramseyer, K. and J. Mullis, 1990, Factors influencing short-lived blue cathodoluminescence of alpha-quartz: American Mineralogist, v. 75, p. 791-800.

Richter, D. K., T. Götte, J. Götze, and R. D. Neuser, 2003, Progress in application of cathodoluminescence (CL) in sedimentary petrology: Mineralogy and Petrology, v. 79, p. 127-166, doi: 10.1007/s00710-003-0237-4.

Richter, D. K., T. Götte, and D. Habermann, 2002, Cathodoluminescence of authigenic albite: Sedimentary Geology, v. 150, p. 367-374, doi: 10.1016/S0037-0738(01)00227-5.

Rusk, B., 2012, Cathodoluminescent textures and trace elements in hydrothermal quartz, *in* J. Götze, and R. Möckel, eds., Quartz: Deposits, Mineralogy and Analytics: New York, Springer, p. 307-329, doi: 10.1007/978-3-642-22161-3_14.

Sawatzky, C. C., and G. Pe-Piper, 2013, Detrital quartz sources in the Scotian Basin, eastern Canada, using hot-cathode cathodoluminescence: Availability of coarse-grained sand for reservoirs: AAPG Bulletin, v. 97, p. 1503-1520, doi: 10.1306/03251312185.

Schieber, J., and R. P. Wintsch, 2005, Scanned colour cathodoluminescence establishes a slate belt provenance for detrital quartz in Devonian black shales of the Appalachian Basin [abs.]: Geochemica et Cosmochemica Acta, v. 69, no. 10 supl. 1, p. A592.

Seyedolali, A., D. H. Krinsley, S. Boggs, Jr., P. F. O'Hara, H. Dypvik, and G. G. Goles, 1997, Provenance interpretation of quartz by scanning electron microscope–cathodoluminescence fabric analysis: Geology, v. 25, p. 787-790, doi: 10.1130/0091-7613(1997)025<0787:PIOQBS>2.3.CO;2.

Sippel, R. F., 1968, Sandstone petrology, evidence from luminescence petrography: Journal of Sedimentary Research, v. 38, p. 530-554, doi: 10.1306/74D719DD-2B21-11D7-8648000102C1865D.

Smith, J. V., and R. C. Stenstrom, 1965, Electron-excited luminescence as a petrologic tool: Journal of Geology, v. 73, p. 627-635, doi: 10.1086/627098.

Walderhaug, O., and J. Rykkje, 2000, Some examples of the effect of crystallographic orientation on the cathodoluminescence colors of quartz: Journal of Sedimentary Research, v. 70, p. 545-548, doi: 10.1306/2DC40926-0E47-11D7-8643000102C1865D.

Wycherley, H. L., J. Parnell, G. R. Watt, H. Chen, and A. J. Boyce, 2003, Indicators of hot fluid migration in sedimentary basins: Evidence from the UK Atlantic Margin: Petroleum Geoscience, v. 9, p. 357-374, doi: 10.1144/1354-079303-569.

Zinkernagel, U., 1978, Cathodoluminescence of Quartz and its Application to Sandstone Petrology: Stuttgart, E. Schweizerbart'sche Verlagsbuchandlung, 69 p.

Fluid inclusion analysis

Burruss, R. C., 1981, Hydrocarbon fluid inclusions in studies of sedimentary diagenesis, *in* L. S. Hollister, and M. L. Crawford, eds., Fluid Inclusions. Applications to Petrology: Toronto, Mineralogical Association of Canada Short Course Handbook Vol. 6, p. 138-156.

Goldstein, R.H., 2001, Fluid inclusions in sedimentary and diagenetic systems: Lithos, v. 55, p. 159-193, doi: 10.1016/S0024-4937(00)00044-X.

Goldstein, R.H., and T. J. Reynolds, 1994, Systematics of Fluid Inclusions in Diagenetic Minerals: Tulsa, OK, SEPM Short Course 31, 198 p.

Kelly, J., J. Parnell, and H. H. Chen, 2000, Application of fluid inclusions to studies of fractured sandstone reservoirs: Journal of Geochemical Exploration, v. 69, p. 705-709, doi: 10.1016/S0375-6742(00)00096-0.

McLimans, R.K., 1987, The application of fluid inclusions to migration of oil and diagenesis of petroleum reservoirs: Applied Geochemistry, v. 2, p. 585-604, doi: 10.1016/0883-2927(87)90011-4.

McLimans, R. K., 1991, Studies of reservoir diagenesis, burial history, and petroleum migration using luminescence microscopy, *in* C. E. Barker, and O. C. Kopp, eds., Luminescence Microscopy and Spectroscopy: Qualitative and Quantitative Applications: Tulsa, OK, SEPM Short Course 25, p. 97-106, doi: 10.2110/scn.91.25.0097.

Munz, I.A., 2001, Petroleum inclusions in sedimentary basins: Systematics, analytical methods and applications: Lithos, v. 55, p. 195-212, doi: 10.1016/S0024-4937(00)00045-1.

Rossi, C., R. H. Goldstein, A. Ceriani, and R. Marfil, 2002, Fluid inclusions record thermal and fluid evolution in reservoir sandstones, Khatatba Formation, Western Desert, Egypt: A case for fluid injection: AAPG Bulletin, v. 86, p. 1773-1799, doi: 10.1306/61EEDD78-173E-11D7-8645000102C1865D.

Shepherd, T. J., A. H. Rankin, and D. H. M. Alderton, 1985, A Practical Guide to Fluid Inclusion Studies: Glasgow, Blackie & Sons, 239 p.

Van den Kerkhof, A.M., and U. F. Hein, 2001, Fluid inclusion petrography: Lithos, v. 55, p. 27-47, doi: 10.1016/S0024-4937(00)00037-2.

Walderhaug, O., 1990, A fluid inclusion study of quartz-cemented sandstones from offshore mid-Norway—possible evidence for continued quartz cementation during oil emplacement: Journal of Sedimentary Research, v. 60, p. 203-210, doi: 10.1306/212F9151-2B24-11D7-8648000102C1865D.

Fluorescence and infrared microscopy

Astin, T. R., 1987, Petrology (including fluorescence microscopy) of cherts from the Portlandian of Wiltshire, UK—evidence of an episode of meteoric water circulation, *in* J. D. Marshall, ed., Diagenesis of Sedimentary Sequences: London, GSL Special Publication 36, p. 73-85, doi: 10.1144/GSL.SP.1987.036.01.07.

Bertrand, P., J.-L. Pittion, and C. Bernaud, 1986, Fluorescence of sedimentary organic matter in relation to its chemical composition: Organic Geochemistry, v. 10, p. 641-647, doi: 10.1016/0146-6380(86)90061-6.

Bourdet, J., and P. Eadington, 2012, Fluorescence and Infrared Spectroscopy of Inclusion Oil: CSIRO Earth Science and Resource Engineering Report EP129625, 60 p.

Burruss, R.C., 1991, Practical aspects of fluorescence microscopy of petroleum fluid inclusions: *in* C. E. Barker, and O. C. Kopp, eds., Luminescence Microscopy and Spectroscopy: Qualitative and Quantitative Applications: Tulsa, OK, SEPM Short Course 25, p. 1-8, doi: 10.2110/scn.91.25.0001.

Campbell, A. R., 1991, Geologic applications of infrared microscopy, *in* C. E. Barker, and O. C. Kopp, eds., Luminescence Microscopy and Spectroscopy: Qualitative and Quantitative Applications: Tulsa, OK, SEPM Short Course 25, p. 161-171, doi: 10.2110/scn.91.25.0161.

Guilhaumou, N., N. Szydlowskii, and B. Padier, 1990, Characterization of hydrocarbon fluid inclusions by infra-red and fluorescence microspectrometry: Mineralogical Magazine, v. 54, p. 311-324, doi: 10.1180/minmag.1990.054.375.17.

Lemon, N. M., and C. J. Cubitt, 2003, Illite fluorescence microscopy: A new technique in the study of illite in the Merrimelia Formation, Cooper Basin, Australia, *in* R. H. Worden, and S. Morad, eds., Clay Mineral Cements in Sandstones (IAS Special Publication 34): Oxford, Wiley-Blackwell, p. 409-424, doi: 10.1002/9781444304336.ch18.

Liu, K., and P. Eadington, 2005, Quantitative fluorescence techniques for detecting residual oils and reconstructing hydrocarbon charge history: Organic Geochemistry, v. 36, p. 1023-1036, doi: 10.1016/j.orggeochem.2005.02.008.

Rost, F. W. D., 1992, Fluorescence Microscopy: New York, Cambridge University Press, 253 p.

Soeder, D. J., 1990, Applications of fluorescence microscopy to study of pores in tight rocks: AAPG Bulletin, v. 74, p. 30-40.

Stasiuk, L. D., and L. R. Snowdon, 1997, Fluorescence micro-spectrometry of synthetic and natural hydrocarbon fluid inclusions: Crude oil chemistry, density and application to petroleum migration: Applied Geochemistry, v. 12, p. 229-241, doi: 10.1016/S0883-2927(96)00047-9.

Wajima, T., S. Okuda, Y. Chen, M. Bessho, and T. Nishiyama, 2000, Observation of pore spaces and microcracks using a fluorescent technique in some reservoir rocks of oil, gas and geothermal fields in the Green Tuff region, Japan: Resource Geology, v. 50, p. 191-200, doi: 10.1111/j.1751-3928.2000.tb00069.x.

Zhang, L., G. Bai, X. Luo, X. Ma, M. Chen, M. Wu, and W. Yang, 2009, Diagenetic history of tight sandstones and gas entrapment in the Yulin Gas Field in the central area of the Ordos Basin, China: Marine and Petroleum Geology, v. 26, p. 974-989, doi: 10.1016/j.marpetgeo.2008.05.003.

Microprobe/EDAX/Backscattered Electron imaging/Ion probe

De Andrade, V., O. Vidal, E. Lewin, P. O'Brien, and P. Agard, 2006, Quantification of electron microprobe compositional maps of rock thin sections: An optimized method and examples: Journal of Metamorphic Geology, v. 24, p. 655-668, doi: 10.1111/j.1525-1314.2006.00660.x.

Dilks, A., and S. C. Graham, 1985, Quantitative mineralogical characterization of sandstones by back-scattered electron image analysis: Journal of Sedimentary Research, v. 55, p. 347-355, doi: 10.1306/212F86C5-2B24-11D7-8648000102C1865D.

Hay, D. C., and T. J. Dempster, 2009, Zircon alteration, formation and preservation in sandstones: Sedimentology, v. 56, p. 2175-2191, doi: 10.1111/j.1365-3091.2009.01075.x.

Hillier, S., 1994, Pore-lining chlorites in siliciclastic reservoir sandstones: electron microprobe, SEM and XRD data, and implications for their origin: Clay Minerals, v. 29, p. 665-680, doi: 10.1180/claymin.1994.029.4.20.

Kelly, J. L., B. Fu, N. T. Kita, and J. W. Valley, 2007, Optically continuous silcrete quartz cements of the St. Peter Sandstone: High precision oxygen isotope analysis by ion microprobe: Geochimica et Cosmochimica Acta, v. 71, p. 3812-3832, doi: 10.1016/j.gca.2007.05.014.

Krinsley, D. H., K. Pye, S. Boggs, Jr., and N. K. Tovey, 1998, Backscattered Scanning Electron Microscopy and Image Analysis of Sediments and Sedimentary Rocks: Cambridge, UK, Cambridge University Press, 203 p.

Laubach, S. E., R. H. Lander, L. M. Bonnell, J. E. Olson, and R. M. Reed, 2004, Opening histories of fractures in sandstone, *in* J. W. Cosgrove, and T. Engelder, eds., The Initiation, Propagation, and Arrest of Joints and Other Fractures: London, GSL Special Publication 231, p. 1-9, doi: 10.1144/GSL.SP.2004.231.01.01.

Lyon, I. C., S. D. Burley, P. J. McKeever, J. M. Saxton, and C. Macaulay, 2000, Oxygen isotope analysis of authigenic quartz in sandstones: A comparison of ion microprobe and conventional analytical techniques, *in* R. H. Worden, and S. Morad, eds., Quartz Cementation in Sandstones (IAS Special Publication 29): Oxford, Wiley-Blackwell, p. 299-316, doi: 10.1002/9781444304237.ch20.

Mørk, M.B.E., and K. Moen, 2007, Compaction microstructures in quartz grains and quartz cement in deeply buried reservoir sandstones using combined petrography and EBSD analysis: Journal of Structural Geology, v. 29, p. 1843-1854, doi: 10.1016/j.jsg.2007.08.004.

Morton, A. C., 1985, A new approach to provenance studies: Electron microprobe analysis of detrital garnets from Middle Jurassic sandstones of the northern North Sea: Sedimentology, v. 32, p. 553-566, doi: 10.1111/j.1365-3091.1985.tb00470.x.

Potts, P. J., 1987, A Handbook of Silicate Rock Analysis: Glasgow, Blackie, 622 p.

Potts, P. J., J. F. Bowles, S. J. Reed, and R. Cave, eds., 1995, Microprobe Techniques in the Earth Sciences (Mineralogical Society Series #6): New York, Chapman & Hall, 419 p.

Prêt, D., S. Sammartino, D. Beaufort, M. Fialin, P. Sardini, P. Cosenza, and A. Meunier, 2010, A new method for quantitative petrography based on image processing of chemical element maps: Part II. Semi-quantitative porosity maps superimposed on mineral maps: American Mineralogist, v. 95, p. 1389-1398, doi: 10.2138/am.2010.3433.

Pye, K., 1984, Rapid estimation of porosity and mineral abundance in backscattered electron images using a simple SEM image analyser: Geological Magazine, v. 121, p. 81, doi: 10.1017/S0016756800028041.

Pye, K., and D. H. Krinsley, 1984, Petrographic examination of sedimentary rocks in the SEM using backscattered electron detectors: Journal of Sedimentary Research, v. 54, p. 877-888, doi: 10.1017/S0016756800028041.

Reed, S. J. B., 1989, Ion microprobe analysis—a review of geological applications: Mineralogical Magazine, v. 53, p. 3-24, doi: 10.1180/minmag.1989.053.369.02.

Reed, S. J. B., 1993, Electron Microprobe Analysis (2nd Edition): Cambridge, UK, Cambridge University Press, 350 p.

Reed, S. J. B., 2005, Electron Microprobe Analysis and Scanning Electron Microscopy in Geology (2nd Edition), Cambridge, UK, Cambridge University Press, p. 232, doi: 10.1017/CBO9780511610561.

Schieber, J., D. Krinsley, and L. Riciputi, 2000, Diagenetic origin of quartz silt in mudstones and implications for silica cycling: Nature, v. 406, p. 981-985, doi: 10.1038/35023143.

White, S. H., H. F. Shaw, and J. M. Huggett, 1984, The use of back-scattered electron imaging for the petrographic study of sandstones and shales: Journal of Sedimentary Research, v. 54, p. 487-494, doi: 10.1306/212F844F-2B24-11D7-8648000102C1865.

SEM/TEM

Baker, J. C., P. J. R. Uwins, and I. D. R. Mackinnon, 1993, ESEM study of authigenic chlorite acid sensitivity in sandstone reservoirs: Journal of Petroleum Science and Engineering, v. 8, p. 269-277, doi: 10.1016/0920-4105(93)90004-X.

Bernet, M., and K. Bassett, 2005, Provenance analysis by single-quartz-grain SEM-CL/optical microscopy: Journal of Sedimentary Research, v. 75, p. 492-500, doi: 10.2110/jsr.2005.038.

Burley, S. D., and J. D. Kantorowicz, 1986, Thin section and S.E.M. textural criteria for the recognition of cement-dissolution porosity in sandstones: Sedimentology, v. 33, p. 587-604, doi: 10.1111/j.1365-3091.1986.tb00764.x.

Camp, W. K., E. Diaz, and B. Wawak, 2013, Electron Microscopy of Shale Hydrocarbon Reservoirs: Tulsa, OK, AAPG Memoir 102, 260 p.

Curtis, M. E., C. H. Sondergeld, R. J. Ambrose, and C. S. Rai, 2012, Microstructural investigation of gas shales in two and three dimensions using nanometer-scale resolution imaging: AAPG Bulletin, v. 96, p. 665-677, doi: 10.1306/08151110188.

Driskill, B., J. Walls, S. W. Sinclair, and J. DeVito, 2013, Applications of SEM imaging to reservoir characterization in the Eagle Ford Shale, South Texas, U.S.A., *in* W. K. Camp, E. Diaz, and B. Wawak, eds., Electron Microscopy of Shale Hydrocarbon Reservoirs: Tulsa, OK, AAPG Memoir 102, p. 115-136, doi: 10.1306/13391709M1023587.

Evans, J., A. J. C. Hogg, M. S. Hopkins, and R. J. Howarth, 1994, Quantification of quartz cements using combined SEM, CL, and image analysis: Journal of Sedimentary Research, v. 64, p. 334-338, doi: 10.1306/D4267D93-2B26-11D7-8648000102C1865D.

Hirt, W. G., H.-R. Wenk, and J. R. Boles, 1993, Albitization of plagioclase crystals in the Stevens sandstone (Miocene), San Joaquin Basin, California, and the Frio Formation (Oligocene), Gulf Coast, Texas: A TEM/AEM study: GSA Bulletin, v. 105, p. 708-714, doi: 10.1130/0016-7606(1993)105<0708:AOPCIT>2.3.CO;2.

Krinsley, D. H., and J. C. Doornkamp, 1973, Atlas of Quartz Sand Surface Textures: Cambridge, UK, Cambridge University Press, 91 p.

Slatt, R. M., P. R. Philp, Y. Abousleiman, P. Singh, R. Perez, R. Portas, K. J. Marfurt, S. Madrid-Arroyo, N. O'Brien, and E. Eslinger, 2012, Pore-to-regional-scale integrated characterization workflow for unconventional gas shales, *in* J. A. Breyer, ed., Shale Reservoirs—Giant Resources for the 21st Century: Tulsa, OK, AAPG Memoir 97, p. 127-150, doi: 10.1306/13321461M97441.

Welton, J. E., 1984, SEM Petrology Atlas: Tulsa, OK, AAPG Methods in Exploration Series No. 4, 237 p.

Staining/dye impregnation

Cather, M. E., N. R. Morrow, and I. Klich, 1991, Characterization of porosity and pore quality in sedimentary rocks, *in* F. Rodriguez-Reinoso, J. Rouquerol, K. S. W. Sing, and K. K. Unger, eds., Characterization of Porous Solids II (Studies in Surface Science and Catalysis 62): Amsterdam, Elsevier, p. 727-736, doi: 10.1016/S0167-2991(08)61381-6.

Dickson, J. A. D., 1966, Carbonate identification and genesis as revealed by staining: Journal of Sedimentary Research, v. 36, p. 491-505, doi: 10.1306/74D714F6-2B21-11D7-8648000102C1865D.

Friedman, G. M., 1959, Identification of carbonate minerals by staining methods: Journal of Sedimentary Research, v. 29, p. 87-97, doi: 10.1306/74D70894-2B21-11D7-8648000102C1865D.

Houghton, H. F., 1980, Refined technique for staining plagioclase and alkali feldspars in thin section: Journal of Sedimentary Research, v. 50, p. 629-631.

Laniz, R. V., R. E. Stevens, and M. B. Norman, 1964, Staining of plagioclase feldspar and other minerals with F.D. and C. Red No. 2: Washington, DC, USGS Professional Paper 501-B, p. 152-153.

Ruzyla, K., and D. I. Jezek, 1987, Staining method for recognition of pore space in thin and polished sections: Journal of Sedimentary Research, v. 57, p. 777-778.

Sardini, P., A. El Albani, D. Pret, S. Gaboreau, M. Siitari-Kauppi, and D. Beaufort, 2009, Mapping and quantifying the clay aggregate microporosity in medium- to coarse-grained sandstones using the ^{14}C-PMMA method: Journal of Sedimentary Research, v. 79, p. 584-592, doi: 10.2110/jsr.2009.063.

Widmark, T., 1980, Staining of albite: Geologiska Föreningen i Stockholm Förhandlingar, v. 101, p. 357-358.

Wilson, M. D., and S. S. Sedora, 1979, An improved thin section stain for potash feldspar: Journal of Sedimentary Research, v. 49, p. 637-638.

Whitlatch, R. B., and R. G. Johnson, 1974, Methods for staining organic matter in marine sediments: Journal of Sedimentary Research, v. 44, p. 1310-1312.

Yanguas, J. E., and J. J. Dravis, 1985, Blue fluorescent dye technique for recognition of microporosity in sedimentary rocks: Journal of Sedimentary Research, v. 55, p. 600-602.

X-ray tomography

Appoloni, C. R., C. P. Fernandes, and C. R. O. Rodrigues, 2007, X-ray microtomography study of a sandstone reservoir rock: Nuclear Instruments and Methods in Physics Research, Section A: Accelerators, Spectrometers, Detectors and Associated Equipment, v. 580, p. 629-632, doi: 10.1016/j.nima.2007.05.027.

Gualda, G. A. R., and M. Rivers, 2006, Quantitative 3D petrography using X-ray tomography: Application to Bishop Tuff pumice clasts: Journal of Volcanology and Geothermal Research, v. 154, p. 48-62, doi: 10.1016/j.jvolgeores.2005.09.019.

Josh, M., L. Esteban, C. Delle Piane, J. Sarout, D. N. Dewhurst, and M. B. Clennell, 2012, Laboratory characterisation of shale properties: Journal of Petroleum Science and Engineering, v. 88-89, p. 107-124, doi: 10.1016/j.petrol.2012.01.023.

Knight, R., R. Klassen, and P. Hunt, 2002, Mineralogy of fine-grained sediment by energy-dispersive spectrometry (EDS) image analysis–a methodology: Environmental Geology, v. 42, p. 32-40, doi: 10.1007/s00254-002-0538-7.

Mees, F., R. Swennen, M. Van Geet, and P. Jacobs, 2003, Applications of X-ray computed tomography in the geosciences, *in* F. Mees, R. Swennen, M. Van Geet, and P. Jacobs, eds., Applications of X-ray Computed Tomography in the Geosciences: London, GSL Special Publication 215, p. 1-6, doi: 10.1144/GSL.SP.2003.215.01.01.

Van Geet, M., D. Lagrou, R. Swennen, and P. Jacobs, 2003, Porosity measurements of sedimentary rocks by means of microfocus X-ray computed tomography (µCT), *in* F. Mees, R. Swennen, M. Van Geet, and P. Jacobs, eds., Applications of X-ray Computed Tomography in the Geosciences: London, GSL Special Publication 215, p. 51-60, doi: 10.1144/GSL.SP.2003.215.01.05.

To the Lab Assistants that provided inspiration during the writing of this book . . .

Kelly checked out the outcrops and . . .

Mojo checked out everything else.

GLOSSARY

Accessory heavy mineral – A mineral whose presence in a rock is not essential to the proper classification of the rock and that has a specific gravity above 2.85.

Accessory light mineral – A mineral whose presence in a rock is not essential to the proper classification of the rock and that has a specific gravity of 2.85 or less.

Acicular – used to describe needle-like shape. Commonly refers to crystals, especially mineral inclusions or authigenic minerals.

Agglutinated – Refers to a shell, test or other structure produced by an organism by gluing together (agglutinating) sedimentary particles. Especially common in foraminifera, tintinnids and tube-building worms (serpulids and others). Some organisms selectively choose specific particles to form such structures (such as the terrigenous sand grains used to produce the arenaceous tests of some benthic foraminifera).

Albite twin (Albite law twinning) – Twinning in triclinic feldspars in which the twin plane and compositional plane are (010). Such twinning is generally lamellar and multiple and shows fine striations on the (010) cleavage plane.

Albitization – Replacement of a grain, typically a more calcic plagioclase feldspar, by albite.

Alkali feldspars – The group of feldspars falling along the solid-solution line between potassium (orthoclase) and sodium (albite) feldspars. The alkali feldspars include anorthoclase, microcline, orthoclase and sanidine; albite is considered the end–member in the alkali feldspar group or the plagioclase feldspar group depending on its calcium content.

Allochthonous – Refers to material formed or produced at a site other than its present location; material of foreign origin. In the context of carbonate strata, the term normally refers to grains produced in one environment that are later reworked to another setting through the action of storms, debris flows or other transport processes.

Amygdule (alt. **amygdale)** – A vesicle in a volcanic rock formed by a gas bubble in the original lava; generally a cavity filed with a cement such as quartz, calcite, clays or zeolites.

Anaerobic – Refers to an environment characterized by an absence of oxygen or a process that proceeds in such an environment. An anaerobic organism or anaerobe is any organism that can or must live without oxygen.

Anhedral – Descriptive of a single crystal or crystal fabric that does not show well-defined or typical crystallographic forms (i.e., crystal faces are absent). Coined by Pettijohn (1957); see also Friedman (1965).

Ankerite – A white, red or grayish iron–rich mineral related to dolomite: $Ca(Fe,Mg,Mn)(CO_3)_2$. Some use the term as synonymous with ferroan dolomite; others restrict it to crystals in which the Mg:Fe ratio is less than 4. It is associated with iron ores and commonly forms as thin veins associated with coal seams.

Antiperthite – An alkali feldspar in which exsolution has created parallel or subparallel intergrowths of potassium–rich feldspar (typically orthoclase) within a sodium–rich feldspar host (albite, oligoclase or andesine).

Apatite – A group of natural, variously colored, hexagonal minerals with the general composition of calcium fluoride phosphate, $Ca_5F(PO_4)_3$ and with chlorine, hydroxyl or carbonate sometimes replacing the fluoride. Includes specific minerals such fluorapatite, hydroxylapatite, chlorapatite, carbonate–apatite and francolite. Most commonly, when no more specific composition is indicated, the mineral referred to is fluorapatite.

Arenite – A consolidated sedimentary rock consisting primarily of sand–sized particles, regardless of composition. For finer grained rocks the terms lutite or wacke are used; for coarser grained ones, rudite is applied.

Argillaceous – Pertaining to a sediment or sedimentary rock containing clay–size particles; clayey or shaly.

Arkose – A sandstone rich in feldspars (although generally the feldspars are subordinate to quartz). Arkoses are commonly coarse grained and can resemble a granite in texture and color, mainly because many arkoses are the product of weathering and disintegration of igneous and metamorphic (and especially granitic) rocks. The exact percentage of feldspar required to call a rock an arkose varies in different compositional classification schemes (see chapter on classification), but in most 25% is the lower limit. Feldspathic arenite is a synonymous term used in some recent classifications.

Authigenic – Rock constituents and minerals that crystallized or precipitated locally at the spot where they are now found (Holmes, 1928). Contrast with allogenic.

Axiolitic - A microscopic fabric composed of elongate fibers of alkali feldspar intergrown with cristobalite; commonly these crystals have nucleated and grown from the sides of a linear fracture within rhyolitic glass.

Baroque dolomite – *see Saddle dolomite*

Bauxite (lateritic bauxite) – A rock formed by tropical weathering of other rocks. It is the main ore for aluminum production. It consists mainly of aluminum hydroxide minerals (gibbsite, boehmite, and diaspore) along with iron oxides and commonly displays a pisolitic texture.

Bentonite – An absorbant, impure clay consisting mostly of montmorillonite and generally formed though the weathering of volcanic ash, typically in the presence of water. Completely devitrified ash-fall deposits are commonly referred to as K-bentonites when the dominant clay species is illite.

Biaxial – A biaxial mineral has two optic axes. This means that there are two directions along which light shows no birefringence and vibrates in a circular section with a unique constant refractive index. Biaxial minerals crystallize in the orthorhombic, monoclinic and triclinic systems.

Bimodal sediment – A sediment whose particle size distribution shows two maxima (with lesser abundances of material in other size grades). Common in some river gravels and eolian sand deposits.

Bioturbation – The mixing or stirring of sediment by plants or burrowing organisms.

Birefringence – The property of a crystal to split a beam of light into two beams of unequal velocities based on the difference between the greatest and the least indices of refraction of that crystal. Under a polarizing microscope (in cross–polarized light), the degree of birefringence is shown as "interference colors" which are a function of the mineral type, orientation and thickness of the sample, as well as the character of the light. Crystals in the cubic class typically do not exhibit birefringence and are known as isotropic minerals. Crystals in the hexagonal, tetragonal, and trigonal crystal classes exhibit birefringence and are termed uniaxial minerals. Orthorhombic, monoclinic, triclinic classes exhibit three indices of refraction and are therefore trirefringent (generally termed biaxial). Birefringence is also known as double refraction.

Bitumen – Generally refers to the spectrum of natural flammable hydrocarbons (petroleum, asphalt, mineral wax, etc.), including semisolid and solid admixtures with mineral matter.

Boehm lamellae – Multiple, subparallel, planar structural lines or bands in deformed mineral grains, especially quartz, consisting of minute (micron–sized or smaller) fluid–filled inclusions. They can be found in detrital grains or can be produced *in situ* through deformation of a sedimentary rock.

Botryoid (botryoidal) – texture or mineral habit is one in which the mineral has a globular, spherical or hemispherical external form resembling a bunch of grapes. Generally produced by layered precipitation with numerous small crystals oriented either parallel or perpendicular to the botryoid surface. A common growth form for hematite, aragonite, chalcedony, malachite and many other minerals.

Boudinage – A lenticular structure common in deformed sedimentary and metamorphic rocks, resulting from the stretching, thinning and breaking of a competent bed within less competent strata resembling boudins (a string of sausages) in cross–section.

Breccia – A rock structure marked by an accumulation of angular fragments, or of an ore texture showing mineral fragments without notable rounding. Major types of breccias in carbonate strata include fracture breccias associated with structural features or solution–collapse breccias typically associated with removal of associated evaporites or cavern formation and collapse in limestones.

Breccia porosity – The type of interparticle porosity in a breccia. Breccias are rather common in many carbonate facies, but breccia porosity is only locally of quantitative importance, especially along fracture zones (fracture breccias), dissolution features (solution breccias), or, less commonly, in debris flows (depositional breccias). See Choquette and Pray (1970, p. 244).

Burrow / burrow porosity – A feature created by organic burrowing in relatively unconsolidated sediment, in contrast to borings, which form in lithified particles or sediments. Most burrows collapse, become filled with sediment or are back–filled by the burrow–forming organism itself. Thus, burrows rarely form discrete macroporosity although they may affect interparticle pore space distribution (Choquette and Pray, 1970, p. 244).

Calclithite – A rock formed chiefly of carbonate clasts (extraclasts) derived from older, lithified limestone, generally external to the contemporaneous depositional system. Commonly located along downthrown sides of fault scarps. Term coined by Folk (1959).

Calcrete – A hard, erosion-resistant layer of surficial sand and/or gravel that has been cemented by calcium carbonate (typically low-Mg calcite) to form a duricrust.

Caliche – Surficial material such as sand-, gravel-, or cobble-sized materials that are cemented by calcium carbonate in arid climates as a result of evaporative concentration of $CaCO_3$ in surface pore waters. Commonly characterized by crusts, pisoids, reverse grading, autofracturing, and microstalactitic textures.

Carlsbad twin (Carlsbad law twinning) – Penetration twinning, particularly in orthoclase, in which the twin axis parallels the crystallographic c–axis and the twin surface is irregular.

Cathodoluminescence – The emission of characteristic visible luminescence by a substance when bombarded by an electron stream or ionized gas beam.

Cement – Mineral material, generally precipitated, that occurs in the spaces between or within individual grains (or grain molds) in a consolidated or partially consolidated sedimentary rock. The process of cement formation is termed cementation.

Chalcedony – A cryptocrystalline variety of silica, commonly microscopically fibrous, with lower indices of refraction and mineral density than quartz. Chalcedony was formerly considered to be fibrous quartz with crystallites oriented either parallel or perpendicular to long axes of the fibers. It is now considered to be an intergrowth of triclinic cryptocrystalline quartz and a quartz polymorph, monoclinic moganite.

Chert – A hard, dense, dull to semi–vitreous, cryptocrystalline sedimentary rock, composed of variable amounts of silica mainly in the form of microcrystalline quartz; may contain minor carbonate, iron oxide or other impurities.

Circumgranular cracks – Irregular to globular masses of sediment separated by nontectonic fractures and produced by alternate shrinkage and expansion are termed circumgranular cracks (Swineford, et al., 1958; Esteban and Klappa, 1983). A common feature in soils in general and caliche in particular.

Clastic – As used by most sedimentary petrologists the term refers a rock composed of particles that have been mechanically transported, at least locally. (Note that most field mappers use clastic for terrigenous rocks as opposed to limestones).

Clay – A dual use term that to most sedimentologists denotes particles smaller than 4-5 microns in size. Soil scientists prefer to use the term for materials composed of particles smaller than 2 microns in size. Sometimes the term "clay" is also used to refer to clay minerals. Because of that, considerable confusion arises from the fact that the term is also used to describe sediments and soil residuals that are strongly dominated by clay minerals (these have been termed claystones in various studies).

Clay mineral – Hydrous phyllosilicate minerals (e.g. kaolin, chlorite, smectite or illite group) that contain aluminum, magnesium, iron and potassium and whose basic structure comprises layers of silica tetrahedrons that are arranged in a hexagonal mesh pattern, similar to what is seen in micas. Clay minerals are common as alteration products of silicate minerals and may also form authigenic cements. Clay minerals are commonly, but not always, fine grained and require XRD and/or SEM for accurate identification.

Claystone – A clastic sedimentary rock composed primarily of clay-sized particles (the minimum percentage of clay varies in specific classifications). Where fissile or laminated it may also be termed a shale, although shale, claystone and mudstone are used as virtual synonyms by many workers (see discussion in Chapter 7).

Cleavage – The fracturing or breakage of a mineral along its crystallographic planes; cleavage is, therefore, a reflection of crystal structure.

Coccolith / coccosphere – Coccoliths are individual, typically micron-scale plates or shields of calcite formed by coccolithophores (single-celled algae). In life, the coccoliths overlap and interlock to form a coccosphere which surrounds the protoplasm of the coccolithophore. Coccoliths and their constituent calcite crystals are the dominant grains in chalks (principally Cretaceous-Recent), and are relatively common in other marine sediments.

Collophane – A carbonate–hydroxyl–fluorapatite. This colorless, gray or yellowish brown to dark brown, amorphous, calcium phosphate mineral is a common constituent of some skeletal materials and phosphatic marine sediments. It has extremely low birefringence (virtually isotropic) when viewed in thin section.

Comb structure – A fabric of crystals (commonly quartz) growing perpendicular to the wall of a vein and indicating infilling of an open fracture.

Conchoidal fracture – A term denoting a type of fracture in a brittle materials (such as flint or quartz) that does not follow natural planes of separation but rather results in a smooth rounded or scalloped surface.

Concretion – A hard, compact mass within a sedimentary rock formed by the localized postdepositional precipitation of mineral cement within the spaces between the sediment grains. Most, but not all, concretions are spherical or oblate and form around a nucleation site (commonly a fossil shell or particulate organic matter). Concretions are distinct in composition from surrounding rock and can consist of a variety of materials including calcite or dolomite, calcium phosphate, silica, pyrite or iron oxides.

Cutan – A pedological feature (crust or grain coating) which can be used as diagnostic indicator of paleosol formation when composed of clay minerals. Defined by Brewer (1964) as "a modification of the texture, structure or fabric at natural surfaces in soil materials due to the concentration of soil constituents or in–place modifications of the plasma (relatively unstable soil matrix)." See Esteban and Klappa (1983).

Dedolomite – an informal term used to describe the diagenetic replacement of dolomite by calcite, a process that occurs most commonly in eogenetic or telogenetic settings where calcium sulfates are undergoing dissolution. "Calcitized dolomite" is generally a better term to use as it more clearly specifies the end product.

Deformation lamellae – Narrow, planar crystallographically oriented zones with refractive index slightly different from that of the adjacent grain. Typical of low-temperature deformation.

Deformation twins – Lamellar (not simple) twins formed by deformation. They are commonly tapering or lenticular, with pointed terminations.

Desiccation – The state or process of being desiccated; a state of extreme water removal and dryness.

Detrital – Used in different ways by different authors and hence largely undefinable out of context. Sometimes synonymous with clastic, sometimes with terrigenous, and sometimes restricted to rocks composed of broken fragments of older rocks.

Devitrified – A formerly glassy volcanic deposit that has been replaced by crystalline material; devitrification refers to the process of such replacement.

Diagenesis – Any physical or chemical changes in sediments or sedimentary rocks that occur after deposition, excluding processes involving high enough temperature and pressure to be called metamorphism.

Diatoms – A siliceous group of algae (phytoplankton) that is the main constituent of the sediment diatomite. Diatom shells are called frustules and are made of opal-A. With time and temperature this changes to opal-CT and eventually quartz. The resulting rock from this change within diatomite is called porcellanite (a type of chert). Diatoms are most commonly preserved in rocks younger than Cretaceous age.

Diffusion – The movement of matter by the mixing or transport of molecules and ions from regions of higher concentration to lower concentration. Diffusion of elements within pore fluids is a slow, complex and often poorly understood process.

Dissolution – A surface or subsurface diagenetic process by which minerals are removed in aqueous solutions. The solubility of individual minerals, temperature, pH, fluid flow rates and other conditions govern the removal rates of material.

Dolomite – A term used for both a mineral and a rock. Dolomite is a widespread, rock–forming, rhombohedral mineral consisting of $CaMg(CO_3)_2$. Part of the magnesium may be replaced by iron or manganese. Dolomite is typically colorless or white but may be tinted reddish, brown, yellow, etc. It has perfect cleavage and effervesces feebly in cold, dilute HCl. Dolomite occurs most commonly as a replacement of calcium carbonate minerals. The term is also used (following Kay, 1951) for a rock composed predominantly of the mineral dolomite although the term "dolostone" is preferable.

Dolomitization – The formation of dolomite via the alteration and replacement of calcium carbonate. Replacement involves substituting Mg for roughly half the Ca in the calcium carbonate and in true (ordered) dolomite, that substitution produces alternating $CaCO_3$ and $MgCO_3$ layers.

Ductile deformation – Behavior in which rocks, at a critical stress, do not rupture but instead become permanently deformed by flowing (as opposed to brittle deformation in which materials rupture by fracturing and faulting).

Dysaerobic – Broadly refers to an environment or biozone with lower than normal oxygen levels; specific dissolved oxygen levels differ amongst users but are generally between 0.1 and 2.0 ml/l (e.g., Tyson and Pearson., 1991).

Energy-Dispersive X-ray Spectrometry (EDS or EDX) – A microanalytical technique used in conjunction with scanning electron microscopy to identify the elemental composition of a sample by measuring the energy spectrum of X-rays emitted from the specimen as a result of electron bombardment.

Eogenetic stage or eogenesis – Broadly equivalent to early diagenesis. More formally "the time interval between final deposition and burial of the newly deposited sediment or rock below the depth of significant influence by processes that either operate from the surface or depend for their effectiveness on proximity to the surface" (Choquette and Pray, 1970, p. 219). Depositional and meteoric pore waters dominate and processes such as soil formation, weathering and bacterially mediated reactions may take place.

Eolian (alt. **aeolian)** – Refers to processes and products related to the activity of winds. Winds may erode, transport, and deposit materials, and are most effective in arid regions with sparse vegetation and a large supply of unconsolidated materials.

Epigenetic – Pertaining to sedimentary structures, minerals, and mineral deposits formed after deposition, at low temperatures and pressures (a definable stage of diagenesis).

Equant – Referring to equal dimensions in a crystal, either within a clast or an authigenic mineral.

Equigranular – Even grained. This is a term mainly used for igneous and metamorphic rocks, but is also useful in sandstones and when describing textures in rock fragment.

Euhedral – A term used in reference to crystals that are well–formed with sharp, clearly recognizable faces. The antonym is "anhedral".

Evaporite – A nonclastic sedimentary rock composed primarily of minerals produced from a saline solution as a result of extensive or total evaporation of the parent solution. Gypsum, anhydrite and halite are the most common evaporite minerals in carbonate strata, but celestine, sylvite, and many other less common minerals also fall in this category.

Exsolution (exsolution lamellae) – Exsolution is a process in which a solid solution phase unmixes into two separate phases in the solid state. That process typically proceeds during cooling through the growth of lamellae within the original host crystal. This process is especially common in alkali feldspars and pyroxenes and alkali feldspars that show exsolution lamellae are termed perthites.

Extinction – The complete dimming of cross-polarized light when viewed through a crystal in thin section. Isotropic minerals, opaque minerals, or amorphous materials (glass) display constant extinction. Anisotropic minerals will show one extinction for each 90 degrees of stage rotation. The extinction angle is the measure between the cleavage direction or habit of a mineral and the extinction.

Extraclast – A detrital grain of lithified carbonate sediment (lithoclast) derived from outside the depositional area of current sedimentation. The rock composed of these grains would be a calclithite. See also intraclast (Folk, 1959).

Fabric – The arrangement of grains and/or crystals in a rock in terms of their sizes, shapes and orientations.

Fecal pellet – organic excrement, mainly from invertebrate organisms generally found as ovoid or rod-shaped grains in modern sediment or ancient rocks.

Feldspar – Any of a number of common rock–forming minerals with the general formula $XAl(AlSi)_3O_8$ where X most commonly is K, Ca or Na but can rarely also be Ba, Rb, Sr or Fe. Feldspars are the most common group of minerals, constituting nearly 60% of the Earth's crust. Their susceptibility to physical and chemical weathering and abrasion, however, make them less abundant in terrigenous sediments than quartz.

Feldspathic – Refers to a rock containing feldspars. Used in some rock classifications to connote a feldspar–rich rock (as in feldspathic arenite, subfeldspathic arenite, feldspathic wacke, etc.; other classifications use the terms arkose or subarkose for such rocks).

Felsic – Having the composition of granite or rhyolite.

Fenestrae (fenestral fabric or fenestral pores) – Primary or penecontemporaneous gaps in rock framework larger than grain-supported interstices. Such features may be open pores or may have been partially or completely filled with internal sediment and/or cement. Fenestrae occur as somewhat rounded features of spherical, lenticular, or more irregular shapes; their large size in comparison to normal intergranular openings and their multigranular roofs, floors, and other margins are key characteristics. Fenestrae are commonly somewhat flattened parallel with the laminae or stratigraphic planes of the rock. They may, however, be round or very irregular, and some are elongate in a vertical dimension. Fenestrae are commonly associated with carbonate (and also siliciclastic) microbial mats and can result from shrinkage, gas formation, organic decay, or other synsedimentary processes.

Ferricrete – A hard, erosion-resistant layer of surficial sand and/or gravel that has been cemented by iron oxides (primarily hematite) to form a duricrust.

Ferroan dolomite – A mineral that is intermediate in composition between dolomite and ferrodolomite. That is, it has some degree of substitution of Fe for Mg in the dolomite lattice (typically from 1% to slightly more than 50%). Some workers use it as a synonym of ankerite; others restrict it to crystals in which the Mg:Fe ratio is greater than 4.

Flame structure – A sedimentary structure consisting of wave- or flame-shaped plumes of mudrock that have been squeezed into overlying layers by a combination of sediment load casting and horizontal hydraulic shear or drag. Term introduced by Walton (1956).

Flocculation – The separation of solid particles, especially clays, from a liquid to form loose, fluffy aggregations or soft flakes (flocs). Flocculation typically occurs as a result of interactions between the clay particles and another substance, most commonly salt water.

Fluorescence microscopy – The use of light of specific wave lengths to stimulate a sample and excite the natural emission of light by certain minerals or compounds in the sample. To make the emissions visible, the emitted light is separated from the brighter excitation light through the use of a wavelength-specific filter.

Foliation – A repeated planar fabric within a rock, either as a compositional layering or pervasive fracture. In metamorphic usage it is usually caused by elongation of minerals and growth of clay minerals under pressure. Foliation can include slaty cleavage, schistosity, or gneissic layering.

Fracture porosity – Porosity formed by fracturing. The term is generally used for porosity occurring along breaks in a sediment or rock body where there has been little shear displacement of the opposing blocks. Fracture porosity grades into breccia porosity with increasing dislocation.

Framboid – Microscopic spheroidal clusters of pyrite grains commonly associated with bits of organic material. The term is derived from the French word for raspberry.

Frustule – The siliceous skeleton of a diatom.

Gazzi–Dickinson method – A point–counting technique used to statistically measure the components of a sedimentary rock, typically sandstone. The main distinguishing feature of the technique is counting all sand–sized components as separate grains, regardless of what they are incorporated within (e.g. a sand–sized feldspar within a large volcanic rock fragment would be counted as a feldspar). Gazzi–Dickinson point counting is used in the construction of ternary diagrams, such as QFL diagrams (see Ingersoll et al., 1984).

Geopetal structure – Any internal structure or organization of a rock indicating original orientation such as top and bottom of strata. Common examples are internal sediment accumulating on the floor of a cavity which it partly fills or solution–collapse residue that has fallen to the bottom of a vug or cave.

Glaebule – A soil feature, usually equant, prolate or irregular in shape, and generally a nodule or concretion that has not precipitated in a preexisting void space (see Esteban and Klappa, 1983, p. 28–29). In caliche profiles, glaebules consist of discrete, powdery to indurated concentrations of calcite, commonly with some degree of concentric structure (see also pisoids).

Grains – A general term used to describe silt– and larger–sized sediment particles. Grains can be monocrystalline (e.g. a feldspar grain) or can be can be polycrystalline (e.g., rock fragment).

Granophyric texture – A micrographic intergrowth of quartz and alkali feldspar, where the minerals have crystallized together. The term is not synonymous with micrographic since this can be used for intergrowth of other minerals.

Grid twinning – *see Microcline twinning*

Hardground – A zone at the sea floor, generally a few centimeters thick, in which the sediment was lithified to form a hardened surface; commonly encrusted, discolored, hardened by calcium carbonate, phosphate and/or glauconite impregnation and encrusted or bored by organisms. Hardgrounds generally reflect a hiatus in sedimentation, and may be preserved stratigraphically as a disconformity. A softer, incipient hardground is sometimes termed a "firmground."

Heavy minerals – A mineral that has a specific gravity above 2.85. Although this definition is derived from the gravitational separation of minerals in unconsolidated or friable materials, it is also applied to grains in fully lithified rocks. Heavy minerals can be naturally segregated in placer deposits.

Hiatus surface – A stratigraphic surface that represents a break in the continuity of the geologic record; a surface where sediments were either not deposited or where erosion removed material prior to deposition of overlying strata.

Hydrophyllic – Having a tendency to mix with, dissolve in, or be wetted by water.

Hydrothermal – Relating to hot water (specifically "aqueous solutions that are warm or hot relative to [their] surrounding environment" White, 1985). Normally used in sandstone petrography to describe mineral precipitates such as sphalerite, barite, saddle dolomite, epidote, fluorite, galena and others, which occur as grain replacements, cements or vein fills.

Impregnation medium – Epoxy or plastic material that is pressure/vacuum injected into sediments and rocks prior to cutting of thin sections. The material is commonly, but not always, colored with blue, green or red dyes to enhance the recognition of porosity (and distinguish it from holes created during section preparation). In some cases, fluorescent dyes are incorporated to allow study of pores (especially micropores) using fluorescence microscopy. Petrographic recognition of strained, discolored, or bubble–filled impregnation materials is important so as not to misinterpret these.

Inclusion – A fragment of older rock incorporated within a younger rock to which it may or may not be genetically related. Equally commonly, inclusions refer to voids and imperfections in a crystal in which one or more gases, liquids or solids are entrained. Analysis of those inclusions can give clues to the provenance of grains or cements as well as their formation conditions (temperatures, pressures and fluid chemistries).

Intergranular or **intergrain porosity** – Referring to pore spaces existing between individual grains or particles of a sedimentary rock. Intergranular porosity is the most commonly used term for between–grain porosity in sandstones and carbonates (cf. "interparticle porosity"), but it is not synonymous with primary porosity — it is a nongenetic term denoting only the relative position, not the time of formation, of the pores. See Choquette and Pray (1970, p. 247).

Intermediate igneous rock – Rock containing 52–63 weight percent SiO_2, and thus a rock intermediate between felsic and mafic compositions. Andesite, dacite and trachyandesite are common intermediate volcanic rocks; diorite and granodiorite are common intermediate plutonic rocks.

Interparticle porosity – Porosity between any types of sedimentary particles. Can even be used for pores between particles of silt and clay sizes where it isn't obvious that the particles are sedimentary or have been diagenetically modified (and thus this term differs somewhat from "intergrain porosity" or "intergranular porosity"). Interparticle porosity denotes position and not genesis and the term is most commonly used for carbonate rocks.

Intraclast – A grain of penecontemporaneous, weakly–consolidated, carbonate sediment that has been eroded and redeposited, generally close to its site of original formation and within the same depositional sequence (see Folk, 1959 and 1962).

Intragranular (intragrain) or **intraparticle porosity** – The porosity existing within individual grains or particles of a rock, especially within skeletal material of a sedimentary carbonate rock. Despite wide usage of "intragrain" or "intragranular," Choquette and Pray (1970) prefer to use "intraparticle" as the general term for this type of porosity, primarily for carbonate rocks.

Isopachous – A term that typically refers to cements where the diagenetically precipitated minerals forms rims of nearly equal thickness around grains.

Isotropic – In petrography, the term refers to a crystal whose optical properties do not vary according to crystallographic direction. Thus, light travels with the same speed in any direction through the crystal and the crystal shows no birefringence under crossed polarizers. Common to cubic and amorphous substances.

Kerogen – Naturally occurring, solid, insoluble organic matter found in source rocks or other sediments that can yield oil and gas upon heating. Typical organic constituents of kerogen are algae and woody plant material. Coals and oil shales should therefore be viewed as sedimentary rocks containing special types of kerogens in very high concentrations.

Labile – Refers to grains that are prone to deformation, alteration or dissolution.

Lamellae – Layered intergrowth of one crystal with another. They are sometimes called exsolution lamellae and are observed in the minerals perthite, antiperthite and pigeonite.

Laterite (lateritic) – Laterites are soils rich in iron and aluminum that are formed in hot and wet tropical areas. Nearly all such deposits are reddish-brown because of incorporated iron oxides. Laterites develop by intensive and prolonged weathering (lateritization) of underlying parent rock. Some laterites are aluminum ores and are compositionally similar to bauxites.

Litharenite – A term, a contraction of lithic arenite, was introduced by McBride (1963) for a sandstone that contains more than 25% detrital rock fragments, and less than 10% feldspars. Depending on the sandstone classification used, litharenite is roughly equivalent to lithic sandstone or greywacke). In the Folk (1980) classification, litharenites are further subdivided into sedarenites (litharenites with predominantly sedimentary rock fragments), volcanic arenites (predominantly volcanic rock fragments) or phyllarenites (predominantly metamorphic rock fragments).

Lithification – The process of transformation of unconsolidated sediment into more solid (although potentially still porous) rock through processes including compaction, cementation and recrystallization.

Lithoclast – A mechanically formed and deposited fragment of a carbonate rock, normally > 2 mm in diameter, derived from an older limestone, dolostone, or other sedimentary rock stratum. Also termed an extraclast.

Load structure – A penecontemporaneous bulbous deformation feature produced at the base of a bed that was deposited on soft sediment. They are created by differential sinking of denser sediment into underlying, less dense and typically watery material.

Mafic – An igneous rock composed largely of dark–colored ferromagnesian minerals — i.e., having a composition similar to gabbro or basalt.

Marl – A calcium carbonate-rich mudstone or shale.

Matrix – Descriptive of the sedimentary, mechanically deposited material between grains. Includes terrigenous mud, carbonate mud (micrite) or other fine–grained (typically less than 4μm) interstitial material.

Matrix porosity – The porosity of the matrix or finer portion of a sediment or rock, in contrast to porosity associated with the coarser particles or constituents; or the porosity of "blocks" of the rock in contrast to the porosity of the fractures (Choquette and Pray, 1970).

Mesogenetic stage or mesogenesis – Broadly equivalent to burial diagenesis. More formally "the time interval or stage in which the sediments or rocks are buried at depth below the major influence of processes directly operating from or closely related to the surface. It constitutes the entire time between the geologically brief early stage of burial [eogenetic stage] and a final phase of imminent erosion [telogenetic stage]" (Choquette and Pray, 1970, p. 220).

Meteoric diagenesis – Near-surface alteration in surface-derived waters. Meteoric diagenesis mostly involves fresh water and the major processes are carbonate dissolution, localized cementation and the formation of soil. In the vadose (or unsaturated) zone, pores periodically contain water, air or both. In the phreatic (or saturated) zone, pores are always fluid filled. In coastal regions, phreatic meteoric ground water passes laterally into a mixing zone with seawater.

Methanogenesis – The conversion of organic matter into methane and carbon dioxide by anaerobic microbes (termed methanogens, these are microbes from the kingdom or domain Archaea). Because they thrive without oxygen, methanogenic organisms play an important role in the subsurface, where oxygen is commonly absent. Microbially-formed (biogenic) methane can constitute a significant fraction of natural gas deposits.

Mica – A group of sheet silicates (phyllosilicates), related to clay minerals. The common forms seen in sediments are muscovite and biotite. Because chlorite is also a layered silicate with mica-like properties it commonly termed a mica. If a rock contains abundant mica it is said to be micaceous.

Microcline twinning (also tartan, tweed or grid twinning) – Informal terms for a combination of albite and pericline law twinning found mainly in microcline feldspars. The right–angle intersection of those twins yields a distinctive reticulate pattern.

Microcrystalline – Composed of crystals that average less than 10 microns (μm) in diameter.

Micrographic – A microscopic version of graphic texture. Intergrowth of two minerals, which are both in optical continuity. It most commonly occurs with quartz and alkali feldspar. It is named after early forms of writing, such as cuneiform, which it resembles.

Microlite – A very small crystal in the groundmass of a rapidly cooled lava.

Micropores (microporosity) – Porosity that is marked by the presence of pore throats smaller than 0.5 microns (μm). Such pores are commonly located within the clay cemented and matrix-filled areas of sandstones and between clay particles in mudrocks. Dyed resin can sometimes highlight micropores.

Microstylolite – A low-relief (less than 1 mm) surface produced by tectonic or burial–related pressure dissolution of soluble carbonate, commonly marked by the presence of clays, organic matter or other insoluble material. Microstylolites generally form in groups or clusters, sometimes referred to as horsetail seams. Syn.: solution seam; also see stylolite.

Mold – A mold is a pore formed by the selective removal of a former individual constituent of the sediment or rock. Most molds in sedimentary carbonates are created by the selective dissolution of feldspars, volcanic rock fragments, carbonate grains and evaporite nodules or crystals. Brit. usage – "mould."

Moldic porosity – Descriptive of pores formed by the selective removal of a former individual constituent of the sediment or rock (see definition of mold). Moldic is often used with modifying prefixes, including oomoldic or dolomoldic, feldspar-moldic. Brit. usage – "mouldic porosity."

Mudrock – This is the general term used to describe all non-metamorphic argillaceous deposits and thus is a term at an organizational level equivalent to "sandstone". Mudrocks are divided into "siltstone", "mudstone" and "claystone" although the relative percentages of clay and silt in those subdivisions vary in different classifications (see glossary entries for those terms). Used by some workers as a synonym for shale.

Mudstone – A fine-grained, dark-colored rock consisting primarily of compacted and hardened silt and clay, similar to shale but without the lamination and fissility of shale (in its restricted meaning). The proportions of silt and clay in mudstone are approximately equal.

Myrmekite – An intergrowth of plagioclase feldspar (generally orthoclase) and quartz (generally replacing potassium feldspar during later stages of igneous consolidation or during a subsequent igneous event). The quartz intergrowths occur as blobs, droplets or vermicular fabrics.

Negative crystal – A cavity in a mineral that has the crystallographic form of a crystal. In sandstones, these are most commonly seen in quartz crystals of originally volcanic origin.

Neoformation (of clays) – (1) Authigenesis of clay minerals as a result of direct precipitation from a pore fluid. (2) The neomorphism of one clay mineral polymorph to another. (3) The replacement of a precursor mineral by a new clay mineral.

Occlusion – The reduction or replacement of porosity by mineral growth or internal sediment infilling.

Oil wet – Preference of a solid to be in contact with an oil phase rather than a water or gas phase. Generally, polar compounds or asphaltenes deposited from the crude oil onto mineral surfaces cause the oil-wet condition. Some minerals, such as kaolinite, are particularly susceptible to oil wetting.

Oncoid – In North American usage, an oncoid is a coated grain of microbial or algal (but not red algal) origin that is coarser than 2 mm in diameter; a spheroidal form of a microbial/algal stromatolite showing a series of concentric (typically irregular or scalloped) laminations produced by mechanical turning or rolling. Common European usage is less genetic as a microbial/algal origin is not a prerequisite or using the term.

Ooid – A spherical to ellipsoidal grain, 0.25 to 2.00 mm in diameter, with a nucleus covered by one or more precipitated concentric coatings that have radial and/or concentric orientation of constituent crystals. Ooids can have calcareous, ferruginous (especially hematite or chamosite), siliceous, bauxitic, phosphatic, evaporitic (gypsum, halite) or other concentric coatings.

Opal-A – An amorphous form of silica (SiO_2 nH_2O), probably colloidal in origin, that composes the precipitated skeletal material of diatoms, radiolarians and siliceous sponges. The mineral is colorless to gray or brown, has high negative relief, and is isotropic in thin section. It can contain up to 20% water, but usually has 3–9%. Typically converts to opal–CT or quartz at higher temperatures during sediment burial.

Overgrowth – Secondary material deposited in optical and crystallographic continuity around a crystal grain of the same mineral composition.

Overpressure – Subsurface pressure that is abnormally high, exceeding hydrostatic pressure at a given depth. Overpressures can be produced by rapid overburden loading of low-permeability strata (i.e., loading that exceeds rates of fluid escape).In some cases, overpressuring may be produced through exceptional thermal heating of waters, organic maturation and hydrocarbon generation, gypsum dehydration and other mechanisms (e.g. Swarbrick and Osborne, 1998).

Palagonite – A typically yellow, orange or brown alteration product (a microinclusion-rich glass) created through the interaction of water with volcanic glass in rocks with a chemical composition similar to basalt. Commonly found in interstices between basalt pillows or in amygdules.

Paleosol – A fossil soil. In petrographic terms soil textures are recognised by rhizoliths, admixing of silt and clay, cutans, circum-granular cracks, pisoids and glaebules, and cements (sphaerosiderite or calcrete).

Paragenesis – A sequential order of physical changes and mineral formation or transformation. Generally applied to the sequence of diagenetic events a sediment undergoes throughout it history from deposition through progressive burial and potentially during later episodes of uplift and even re-exposure.

Paramorphism – The transformation of internal structure of a mineral without change of external form or of chemical composition.

Pedogenic – Pertaining to soil formation.

Pegmatite – An exceptionally coarse-grained intrusive igneous rock, generally with granitic composition and found in veins and pods. A major source of coarsely crystalline quartz in gravels and often characterized by a milky white color due to incorporation of abundant water–filled inclusions.

Pelite – A sediment or sedimentary rock consisting of the finest clay–sized particles regardless of composition (but typically clay, calcium carbonate and/or minute quartz particles). Includes mudstones, calcareous mudstones and shales. Generally equivalent to the term "lutite".

Peloid – A carbonate grain (allochem) formed of cryptocrystalline or microcrystalline carbonate regardless of size or origin. This term (coined by McKee and Gutschick, 1969) allows reference to grains composed of micrite or microspar without the need to imply any particular mode of origin (can thus include pellets, some vague intraclasts, micritized fossils, degraded ooids, and other grains of problematic origin including some probably microbial precipitates (e.g., Chafetz, 1986).

Penecontemporaneous – Generally referring to cements or replacement textures indicating that, in the opinion of the user, the feature or mineral formed at almost the same time as the original sediment was deposited, that is, close to the sediment–air or sediment–water interface. Syn: syndepositional.

Pericline twin (pericline law twinning) – Twinning in triclinic feldspars in which the twin axis parallels the crystallographic b–axis and the composition surface is a rhombic section. It can occur alone or in combination with albite–law twinning and in the latter case can yield complex cross–hatched twinning (variously and informally termed gridiron, tartan, tweed or microcline twinning).

Perlite / perlitic texture – A volcanic glass with rhyolitic composition. Perlite forms when an extrusive body of rhyolite or obsidian has a high water content. The perlitic texture is characterized by concentric fractures formed around closely spaced centers.

Permeability – The property or capacity of a porous rock or sediment to transmit a fluid. It is a quantifiable measure of the relative ease of fluid flow under unequal pressure.

Perthite (microperthite) – A variety of alkali feldspar consisting of parallel to sub–parallel intergrowths of a potassium–rich feldspar (generally microcline) from which a sodium–rich phase (normally albite) has exsolved, producing a fabric with blobs, strings or veinlets of one feldspar incorporated within another. Perthite refers to such structures visible to the naked eye; microperthite refers to the same structure when visible only with a microscope.

Phenocryst – A relatively large and generally conspicuous crystal distinctly larger than the grains of the groundmass in an igneous rock.

Phreatic zone – The area in an aquifer, below the water table, in which essentially all pores and fractures are saturated with water. Sometimes termed the "saturated zone". Contrast with "vadose" or "unsaturated zone".

Phosphorite – A sedimentary rock with a sufficient content of phosphate minerals to be considered of economic interest. Primary phosphorite typically consists of bedded rocks containing shell and bone fragments, pellets, peloids, coated grains and nodules composed of one of several microcrystalline varieties of apatite (primarily calcium fluorapatite). Phosphorites also can form as guano accumulations or through weathering and concentration of phosphatic material in secondary deposits.

Phyllarenite – A term coined by Folk (1968; shown as later 1980 edition in reference list). It refers to a litharenite that contains primarily foliated, schistose, phyllosilicate-rich rock fragments interpreted to be of metamorphic origin (slate, phyllite or schist).

Phyllosilicate – The phyllosilicates, or sheet silicates, are an important group of minerals that includes the micas, chlorite, serpentine, talc and the clay minerals.

Placer – a surficial mineral deposit formed by mechanical concentration of heavy mineral grains, most commonly in beach or fluvial deposits.

Pleochroic (pleochroism) – a mineral's ability to absorb or reflect different wavelengths of light depending on the crystallographic orientation. The mineral changes color during rotation of the stage (under plane-polarized light). Some strongly pleochroic minerals are biotite, chlorite and hornblende.

Plutonic – Intrusive igneous rock believed to have solidified deep within the earth. A granite is an example of a plutonic rock.

Poikilotopic (poikilitic) – Textural term denoting a condition in which small granular crystals or grains are irregularly scattered without common orientation in a larger crystal of another mineral (generally sand or silt grains in a single, coarse cement crystal). See Friedman (1965).

Polycrystalline quartz – A single quartz grain composed of two or more optically different quartz crystal units (Conolly, 1965). Polycrystalline quartz includes subvarieties that have been termed composite quartz, schistose quartz and pressure quartz.

Polymict – Referring to a sedimentary rock containing clasts of multiple rock types, for example an intraclastic limestone or calclithite with varied clast lithologies or fabric types. (Syn.: polymictic; contrast with monomict).

Polymorph – One of two or more crystalline forms of the same chemical substance. For example, calcite (rhombohedral), aragonite (orthorhombic) and vaterite (hexagonal) are polymorphs of calcium carbonate.

Polysynthetic twinning – Three or more successive twins following the same twin law in a single crystal producing fine parallel bands or alternating lamellae of reversed crystal structure (common with albite twinning in plagioclase feldspars).

Porcellanite (alt. **porcelanite)** – A hard, dense rock that resembles unglazed porcelain. Typically it is an impure variety of chert that contains considerable clay and/or carbonate material. Generally less dense and crystalline than pure chert, it generally has blocky fracture rather than the conchoidal fracture of chert. It is an opal-CT-rich intermediate stage between a siliceous ooze (opal-A) and a diagenetically mature quartz chert.

Pore – A small opening or void in a rock. In sedimentary petrographic terminology one also can refer to pores that have been "reduced" by partial filling with cement or "occluded" or "filled" through complete cementation and thus are no longer open spaces. Pores can range in size from large caverns to minute openings in the nanometer to picometer range (see Fig. 22.1).

Pore throat – In a granular rock, the small pore space at the point where two grains meet and which connects two larger pore volumes.

Porosity – The percentage of voids (empty space/pores), whether isolated or connected, in Earth material such as soil or rock. For a classification of pore types, mainly in carbonate rocks but also applicable to clastic terrigenous rocks, see Choquette and Pray (1970).

Pressure solution / pressure dissolution – Dissolution occurring preferentially at the grain contacts in a sedimentary rock where external overburden pressures exceed confining pore-fluid pressures. The dissolution broadens grain contacts and liberates solutes that can contribute to overall cementation and porosity loss.

Primary pore – A pore formed during final sedimentation or present in the rock or sediment at the time of deposition. The term "primary porosity" includes all predepositional and depositional porosity of a particle, sediment, or rock. It also refers to any postdiagenetic remnant of primary pore space. (See Choquette and Pray, 1970, p. 249).

Provenance – The source area from which the constituent particles of a sedimentary rock were derived.

Pseudomatrix – Clay- and silt-sized material formed by ductile deformation of clay- or mica-rich lithic clasts interspersed between sand-sized detrital particles.

Pseudomorph – A mineral whose outward crystal form is that of another mineral species; it has developed by alteration, substitution, encrustation or paramorphism.

Pseudouniaxial cross – an extinction pattern under cross-polarized illumination in which domains of radially-oriented crystals (or ones tangentially oriented to spherical cores, such as in some ooids) exhibit darkened extinction in a cross-shaped pattern similar in appearance to the completely unrelated optic axis figure of a uniaxial mineral.

Ptygmatic structure – a sedimentary feature deformed by dishamonic folding.

Quartzarenite – A term coined by McBride (1963) as a contraction for the earlier and looser term "quartz arenite". As such it is a sandstone that is composed primarily of quartz (typically >95% quartz) although McBride included chert and quartzite fragments in his quartz total, something not done in some other classifications (e.g., Folk, 1980).

Radiolaria – Planktic marine protozoa that live at depths from a few meters to hundreds of meters and produce delicate and complex mineral skeletons composed of opaline silica. Individual radiolarians normally are hundredths to tenths of millimeters, and they rarely exceed a millimeter. Radiolaria range from Cambrian to Recent and are major contributors to deep–water cherts and siliceous shales, especially during the Paleozoic and early to mid–Mesozoic.

Recrystallization – The formation, essentially in the solid state, of new crystalline mineral grains in a rock. The new grains are generally larger than the original grains, and may have the same or a different mineralogical composition.

Refractive index (RI) – An important optical property used in mineral identification. The refractive index is the ratio of the velocity of light in a vacuum to the velocity in the mineral.

Replacement – In its general sense, the term refers to the transformation of one mineral to another — either a polymorph or a mineral of a different composition. In the stricter, and more widely used, definition proposed by Folk (1965), the term refers to the replacement of a mineral by one of a different composition (e.g., silica or dolomite replacement of calcite). Replacement can occur through rapid transformations occurring at micron or smaller scales or it can occur by dissolution followed, at a substantially later time, by filling of the void by another mineral (termed solution-cavity fill by Folk, 1965).

Residual oil – Staining and other remnants of largely insoluble hydrocarbons. Residual oil commonly coats quartz grains and their overgrowths or stains interstitial clays.

Resorption – Melting or partial remelting of a crystal within a magma.

Rhizoliths – Organosedimentary structures produced by accumulation and/or cementation around, cementation within, or replacement of, higher plant roots by mineral matter (Klappa, 1980). Includes root casts, tubules, and molds as well as rhizocretions and root petrifications. Rhizoliths are commonly millimeters to centimeters in diameter and centimeters to meters in length, occur just below hiatus surfaces, and may taper slightly. See Esteban and Klappa (1983).

Saddle dolomite – A variety of dolomite that has a warped crystal lattice; it is characterized by curved crystal faces, curved cleavage, and sweeping extinction. Saddle dolomite is slightly enriched in calcium (typically 50–60 mol%) and, in many cases, iron (1–33 mol%). It occurs as either a cement or a replacement, is commonly associated with hydrothermal ore mineralization, sulfate–rich carbonates, and the presence of hydrocarbons. It has been interpreted to indicate formation through sulfate reduction at elevated temperatures (60–150°C). See Radke and Mathis (1980) and Folk and Assereto (1974). Syn.: baroque dolomite.

Salina – A salt pan or other land area encrusted by salts. Generally divided into intertidal to supratidal coastal salinas and inland salinas (including saline lake shores and interdune flats). See Warren (2003).

Scanning electron microscopy (SEM) – Sample examination using a scanning electron microscope that produces images by raster-scanning a focused electron beam across the sample surface and recording the produced secondary electrons.

Secondary pore – A pore created through alteration of sediment or rock after deposition. Some secondary pores can form in near-surface settings, especially in carbonate rocks (e.g. caves). In sandstones, most secondary pores are created during burial diagenesis and are related to dissolution of unstable grains and/or fracturing.

Septarian nodule – A hard, roughly spherical, diagenetic nodule or concretion formed of calcite, siderite, iron oxides, or other materials. The most distinctive characteristic is a complex network of intersecting radial and/or concentric fractures similar to shrinkage cracks. The fractures are commonly filled with calcite cements, in many cases consisting of multiple generations of yellowish–brown to white, fibrous to bladed calcite.

Sericitzation – A process in which plagioclase feldspars and other minerals are replaced by sericite (a fine-grained, white potassium mica, essentially muscovite in composition). The process commonly is associated with relatively low-temperature hydrothermal solutions.

Shard – A vitric fragment tuffaceous (pyroclastic) rocks, commonly with characteristically curved fracture surfaces due to the disintegration of bubbles in pumice.

Silcrete – "An indurated product of surficial and penesurficial (near surface) silicification, formed by the cementation and/or replacement of bedrock, weathering deposits, unconsolidated sediments, soil or other materials and produced by low-temperature physio-chemical processes..." (Summerfield, 1983).

Siliciclastic – Pertaining to clastic noncarbonate rocks, or to sedimentary fragments of previous rocks, comprised dominantly of silicon–rich minerals such as quartz or feldspar.

Silicification – The replacement or pore filling of a rock by one or more forms of silica (opal, chert, chalcedony or megaquartz).

Siltstone – A mudrock containing particles predominantly in the silt grain-size range (i.e., finer than sandstone and coarser than claystone). The exact percentage of silt-sized material required for a rock to be termed a siltstone varies in different classifications. For example, Blatt et al. (1972) specify >2/3rd silt; Hawkins and Pinches, (1992) specify at least 75% silt.

Skeletal – Carbonate components (or the rocks they form) derived from hard material secreted directly by organisms. A substitute for the confusing term "organic" of some older literature.

Solution-cavity fill – A phrase used by Folk (1965) to describe the multi-stage process in which an unstable mineral is dissolved, leaving a void space; then, after the passage of an indeterminate period of time, the void space is filled with a newly precipitated mineral that contains essentially no relict inclusions of the original material that once occupied the area. Some workers include this under the term "replacement"; others do not.

Solution seam – A low–relief internal surface produced by subsurface pressure dissolution of soluble carbonate, generally marked by the presence of clays, organic matter or other insoluble material. Similar to stylolites but marked by lower relief and less obvious insoluble residue and generally occurring as closely spaced swarms of such dissolution surfaces. Syn. Horse–tail seam or microstylolite.

Sorting – A measure of the diversity of particle grain sizes within a sediment. This is generally expressed as the standard deviation of the grain size distribution.

Spastolith – An ooid or other coated grain that has been deformed, generally by compaction of a partially leached grain or by shearing the concentric laminations away from each other or from the nucleus. Especially common in chamositic ooids where some cortex layers were dissolved and others may have been replaced by siderite or phosphate.

Spherulitic texture – A spherulite is a cluster of radiating acicular or lath-like crystals found in igneous rocks such as rhyolitic glass. Spherulites form by devitrification during rapid cooling where nucleation is impeded.

Spicule / spiculitic / spiculite – In sediments these are grains formed of silica or calcium carbonate, which originally supported the soft tissue of sponges. The simplest (monaxon) forms are shaped like an elongated, sharply-pointed cigar; more complex (triaxon, multi-axon) forms are also common. Siliceous spicules typically have a hollow central canal. A spiculite is a rock composed largely of sponge spicules.

Staining – The process of using a dye to selectively color various minerals in thin sections or rock slabs, facilitating their identification. Unstained thin sections are best for identification of mafic minerals, determination of optical properties, and recognition of textural characteristics. In many cases, however, it is difficult to distinguish between colorless feldspars and quartz, particularly if the feldspars are untwinned or between calcite and dolomite in grains and cements. The most common stains are for plagioclase, K-feldspar, and carbonate minerals and they are especially useful where quantitative mineral proportions are needed (e.g., point counts).

Stylolite – A jagged, columnar surface in rocks which may be at any orientation relative to bedding; produced by pressure–induced dissolution and grain interpenetration and often associated with large amounts of insoluble material accumulated as a result of such dissolution.

Subaerial – A term descriptive of location, of processes, or conditions operating in open air or immediately beneath land surfaces.

Subhedral – Exhibiting some traces of crystal shape.

Sulfate reduction – The reduction of sulfate ions in the presence of iron to produce pyrite. This is normally a bacterially mediated process during early diagenesis, but it may also occur later through inorganic processes (thermochemical sulfate reduction). (alt. sulphate reduction).

Sutured boundary – Seen within heavily compacted sandstones, quartzites or metamorphic quartz-rich rocks to varying degrees. The constituent grains have complexly curved, interlocking grain boundaries kink-band boundaries or deformation twin boundaries. The sutured boundaries result from strain-induced boundary migration.

Symplectite – Worm-like (vermicular) intergrowth of minerals. Myrmekite is the most common example.

Syndepositional – See "penecontemporaneous."

Syneresis cracks (also written as **synaeresis** or **synæresis**) – Irregular, radiating, elongate to lenticular cracks, extending from bedding surfaces downward (resembling desiccation cracks). However, syneresis cracks form by subaqueous shrinkage rather than desiccation, and thus are not an indicator of subaerial exposure. The process of syneresis refers to the expulsion of fluids from a gel and in claystones results from changes in clay volume related to local variations in salinity, temperature or other factors. Some syneresis cracks have been attributed to earthquake-induced dewatering (Pratt, 1998).

Syntaxial – Refers to overgrowths that are in optical continuity with their underlying grains such that the original crystal and the overgrowth form a single larger crystal, sharing the same crystallographic axes.

Tabular – Crystal or grain shape which displays straight, parallel sides.

Tasmanitids – A group of organic microfossils (including the genus *Tasmanites*) interpreted as unicellular prasinophyte algal cysts. Their known range is Cambrian to Miocene and they are a major organic constituent of many mid- to late Paleozoic shales.

Telogenetic stage or telogenesis – Occurring in the time interval during which long-buried sediments or rocks are located near the surface again, as a result of crustal movement and erosion, and are influenced significantly by processes (e.g., karst development) associated with the formation of an unconformity (Choquette and Pray, 1970). Contrast with eogenetic and mesogenetic.

Tempestite – An academic's way of saying a storm deposit.

Terrigenous – Derived from a land area and transported mechanically to a basin of deposition; commonly, essentially synonymous with "noncarbonate" (e.g. terrigenous sand vs. carbonate sand).

Trachytic texture – A subparallel, thin-section-scale arrangement of lath-shaped feldspars in the groundmass of an igneous rocks.

Travertine – A relatively dense, banded deposit of $CaCO_3$ or silica especially common in caverns or in cold-springs where it is formed by microbial growth and evaporation of spring or river water (Pettijohn, 1957). Also common at hot springs where it is formed by degassing, cooling of waters and biogenic and/or inorganic precipitation of dissolved materials.

Tuff – A general term for consolidated pyroclastic volcanic rocks. Formerly a term for a volcaniclastic sandstone.

Twinning – The development of a twin crystal (one with reversed or reflected crystal symmetry) by growth, transformation or gliding. A common feature in strained, unrecrystallized calcite crystals.

Uniaxial – Uniaxial minerals are a class of anisotropic minerals that include all minerals that crystallize in the tetragonal and hexagonal crystal systems. They are called uniaxial because they have a single optic axis.

Unit extinction – A type of extinction behavior under crossed polarizers in which an entire grain reaches extinction at the same time as the microscope stage is turned. (Syn.: straight extinction)

Undulatory (undulose) extinction – A type of extinction of crystals under cross-polarized illumination that occurs successively in adjacent areas as the microscope stage is turned (Syn.: wavy extinction or sweeping extinction).

Vacuolization – The formation of numerous small, generally water-filled, inclusions (vacuoles) as part of the diagenetic degradation of grains such as feldspars.

Vadose – Pertaining to that zone of partial or complete groundwater saturation subject to aeration and lying between the land surface and the phreatic zone (above the groundwater table). That is, it is a zone in which both water and air may be present in pores.

Variolitic texture – A variolitic texture consists of a fan-like arrangement of elongate, locally branching, crystals. The term is used primarily to describe such textures in igneous rocks, especially diabase (dolerite).

Vermicular – A crystal or cluster of crystals having a worm–like shape. Especially common in phyllosilicate minerals such as chlorite, vermiculite, kaolinite/dickite or glauconite as well as in myrmekitic intergrowths of quartz and feldspar.

Vesicle – A cavity of variable shape in a lava, caused by the entrapment of a gas bubble in the once molten rock.

Volcaniclastic – A epiclastic terrigenous rock composed chiefly of rock fragments of volcanic origin.

Vug – A pore that is somewhat equant, is larger than 1/16 mm in diameter, and generally does not specifically conform in position, shape or boundary to particular fabric elements to the host rock (i.e., is not fabric selective). Typically formed by dissolution, but the term is descriptive, not genetic (Choquette and Pray, 1970).

Wacke – An impure sandstone containing at least 10% detrital muddy matrix in the Williams, Turner and Gilbert (1953) sandstone classification as well as in the Dott (1964) classification. Prefixes like "quartzose", "feldspathic" or "lithic" are used to express the mineralogical composition of the sand fraction in wackes.

Water wet – Preference of a solid to be in contact with a water phase rather than an oil or gas phase. Oil or water, if present, would be found as bubbles or droplets effectively isolated from contact with the mineral phases in the rock.

Weathering – A term that encompasses the entire spectrum of processes acting at or near the surface of the Earth to alter, decompose or degrade rocks Such processes can selectively alter soluble or susceptible minerals and thus affect the balance of materials derived from a source area.

Xenomorphic – Referring to a mineral grain that does not have a characteristic crystal outline because of deforming pressure from surrounding grains; anhedral.

Zebraic – A term applied to banded, coarsely-fibrous chalcedony in which, when viewed along the fiber elongation direction under cross-polarized light, the fiber bands appear as light and dark stripes as a result of twisting of the crystal axes. Can occur in cements and in silica replacements of evaporite minerals.

Zeolite – Any of a large group of minerals consisting of hydrated aluminosilicates whose lattice can enclose sodium, potassium, calcium, strontium and barium. They can be readily dehydrated and rehydrated, and commonly occur as cements in sandstones, especially in volcaniclastic deposits. About 40 natural zeolites have been identified and some of the more common ones in sandstones and mudstones are analcime, chabazite, clinoptilolite, erionite, heulandite, laumontite, mordenite and phillipsite.

Zoning (mineral zoning) – A texture found in solid-solution minerals characterized optically by changes in the color or extinction angle of the mineral from the core to the rim. This optical zoning is a reflection of chemical zoning in the mineral. For example, a plagioclase feldspar can be zoned from a core that is calcium-rich to a sodium-rich rim. Zoning also can reflect variations in precipitation rate and/or inclusion content.

REFERENCES CITED IN DEFINITIONS

Blatt, H., G. V. Middleton, and R. C. Murray, 1972, Origin of Sedimentary Rocks: Englewood Cliffs, NJ, Prentice Hall Inc., 634 p.

Brewer, R., 1964, Fabric and Mineral Analysis of Soils: New York, Wiley, 470 p.

Chafetz, H. S., 1986, Marine peloids: A product of bacterially induced precipitation of calcite: Journal of Sedimentary Research, v. 56, p. 812-817, doi: 10.1306/212F8A58-2B24-11D7-8648000102C1865D.

Choquette, P. W., and L. C. Pray, 1970, Geologic nomenclature and classification of porosity in sedimentary carbonates: AAPG Bulletin, v. 54, p. 207–250.

Conolly, J. R., 1965, The occurrence of polycrystallinity and undulatory extinction in quartz in sandstones: Journal of Sedimentary Research, v. 35, p. 116-135, doi: 10.1306/74D71208-2B21-11D7-8648000102C1865D.

Dott, R. H., 1964, Wacke, graywacke and matrix; what approach to immature sandstone classification?: Journal of Sedimentary Research, v. 34, p. 625-632, doi: 10.1306/74D71109-2B21-11D7-8648000102C1865D.

Esteban, M., and C. F. Klappa, 1983, Subaerial exposure environment, *in* P. A. Scholle, D. G. Bebout, and C. H. Moore, eds., Carbonate Depositional Environments: Tulsa, OK, AAPG Memoir 33, p. 1–54.

Folk, R. L., 1959, Practical petrographic classification of limestones: AAPG Bulletin, v. 43, p. 1–38.

Folk, R. L., 1962, Spectral subdivision of limestone types, *in* W. E. Ham, ed., Classification of Carbonate Rocks: Tulsa, OK, AAPG Memoir 1, p. 62–84.

Folk, R. L., 1965, Some aspects of recrystallization in ancient limestones, *in* L. C. Pray, and R. S. Murray, eds., Dolomitization and Limestone Diagenesis: Tulsa, OK, SEPM Special Publication No. 13, p. 14-48, doi: 10.2110/pec.65.07.0014.

Folk, R. L., 1980, Petrology of Sedimentary Rocks: Austin, TX, Hemphill's Book Store, 184 p.

Folk, R. L., and R. Assereto, 1974, Giant aragonite rays and baroque white dolomite in tepee–fillings, Triassic of Lombardy, Italy: AAPG and SEPM, Annual Meeting Abstracts, v. 1, p. 34–35.

Friedman, G. M., 1965, Terminology of crystallization textures and fabrics in sedimentary rocks: Journal of Sedimentary Research, v. 35, p. 643-655, doi: 10.1306/74D7131B-2B21-11D7-8648000102C1865D.

Hawkins, A. B., and G. M. Pinches, 1992, Engineering description of mudrocks: Quarterly Journal of Engineering Geology and Hydrogeology, v. 25, p. 17-30, doi: 10.1144/GSL.QJEG.1992.025.01.02.

Holmes, A., 1928, The Nomenclature of Petrology [2nd Edition]: London, Thomas Murby, 284 p.

Ingersoll, R. V., T. F. Fullard, R. L. Ford, J. P. Grimm, J. D. Pickle, and S. W. Sares, 1984, The effect of grain size on detrital modes: A test of the Gazzi-Dickinson point-counting method: Journal of Sedimentary Research, v. 54, p. 103-116, doi: 10.1306/212F83B9-2B24-11D7-8648000102C1865D.

Klappa, C. F., 1980, Rhizoliths in terrestrial carbonates: Classification, recognition, genesis and significance: Sedimentology, v. 27, p. 613-629, doi: 10.1111/j.1365-3091.1980.tb01651.x.

McBride, E. F., 1963, A classification of common sandstones: Journal of Sedimentary Research, v. 33, p. 664-669, doi: 10.1306/74D70EE8-2B21-11D7-8648000102C1865D.

Muskat, M., 1949, Physical Principles of Oil Production: New York, McGraw–Hill, 922 p.

McKee, E. D., and R. C. Gutschick, 1969, History of the Redwall Limestone of northern Arizona: Boulder, CO, GSA Memoir 114, 726 p., doi: 10.1130/MEM114-p1.

Pettijohn, F. J., 1957, Sedimentary Rocks (2nd Edition): New York, Harper Brothers, 718 p.

Pratt, B. R., 1998, Syneresis cracks: Subaqueous shrinkage in argillaceous sediments caused by earthquake-induced dewatering: Sedimentary Geology, v. 117, p. 1-10, doi: 10.1016/S0037-0738(98)00023-2.

Radke, B. M., and R. L. Mathis, 1980, On the formation and occurrence of saddle dolomite: Journal of Sedimentary Petrology, v. 50, p. 1149–1168, doi: 10.1306/212F7B9E-2B24-11D7-8648000102C1865D.

Summerfield, M. A., 1983, Silcrete, *in* A. S. Goudie, and K. Pye, eds., Chemical Sediments and Geomorphology: Precipitates and Residua in the Near-surface Environment: London, Academic Press, p. 59-92.

Swarbrick, R. E., and M. J. Osborne, 1998, Mechanisms that generate abnormal pressures: An overview, *in* B. E. Law, G. F. Ulmishek, and V. I. Slavin, eds., Abnormal Pressures in Hydrocarbon Environments: Tulsa, OK, AAPG Memoir 70, p. 13-34.

Swineford, A., A. B. Leonard, and J. C. Frye, 1958, Petrology of the Pliocene pisolitic limestone in the Great Plains: Lawrence, KS, Kansas Geological Survey Bulletin 130, Part 2, p. 97–116

Tyson, R. V., and T. H. Pearson, 1991, Modern and ancient continental shelf anoxia: An overview, *in* R. V. Tyson, and T. H. Pearson, eds., Modern and Ancient Continental Shelf Anoxia: London, Geological Society of London Special Publication 58, p. 1-24, doi: 10.1144/GSL.SP.1991.058.01.01.

Walton, E. K., 1956, Limitations of graded bedding: and alternative criteria of upward sequence in the rocks of the Southern Uplands: Transactions of the Edinburgh Geological Society, v. 16, p. 262-271, doi: 10.1144/transed.16.3.262.

Warren, J. K. 2006. Sabkhas, saline mudflats and pans, *in* J. K. Warren, Evaporites: Sediments, Resources and Hydrocarbons: Berlin, Springer, p. 139-220, doi: 10.1007/3-540-32344-9.

White, D. E., 1957, Thermal waters of volcanic origin: GSA Bulletin, v. 68, p. 1637-1658, doi: 10.1130/0016-7606(1957)68[1637:TWOVO]2.0.CO;2.

Williams, H., F. J. Turner, and C. M. Gilbert, 1954, Petrography: An Introduction to the Study of Rocks in Thin Section (1st Edition): San Francisco, CA, W. H. Freeman & Co., 406 p.

INDEX

Information shown in parentheses is supplementary material designed to aid users in identifying terms; stratigraphic units and their ages are shown in bold-faced type; location designations for units and features are for samples shown in this book only; county-scale and more local localities along with glossary terms and bibliographic citations are not indexed

C

D

E

I

J

N

O

P

T

U

V

W

X

Y

Z